AF251508

Advances in
Cryogenic Engineering

VOLUME 27

A Cryogenic Engineering Conference Publication

Advances in Cryogenic Engineering

VOLUME 27

Edited by

R. W. Fast

Fermi National Accelerator Laboratory
Batavia, Illinois

PLENUM PRESS · NEW YORK and LONDON

The Library of Congress cataloged the first volume of this title as follows:

Advances in cryogenic engineering. v. 1—
New York, Cryogenic Engineering Conference; distributed
by Plenum Press, 1960—
v. illus., diagrs. 26 cm.
Vols. 1—   are reprints of the Proceedings of the Cryogenic Engineering
Conference, 1954—
Editor: 1960-   K. D. Timmerhaus.

1. Low temperature engineering—Congresses.   I. Timmerhaus, K. D.,
ed. II. Cryogenic Engineering Conference.

TP490.A3   660.29368   57-35598

Proceedings of the 1981 Cryogenic Engineering Conference held in
San Diego, California, August 11 – 14, 1981

Library of Congress Catalog Card Number 57-35598
ISBN 0-306-41103-2

CONTENTS

History

THE CRYOGENIC ENGINEERING CONFERENCE--A RECORD OF
TWENTY-FIVE YEARS OF LOW TEMPERATURE PROGRESS...................1
K. D. Timmerhaus

Applications of Superconductivity – MHD and Fusion

THE FORCED FLOW COOLED COILS FOR THE INTERNATIONAL ENERGY
AGENCY LARGE COIL TASK..11
J. L. Young, C. J. Heyne, P. Komarek, H. Krauth,
G. Vecsey, and C. Marinucci

10-T, 60-cm BORE Nb_3Sn TEST MODULE COIL DESIGN..................21
T. Ando, S. Shimamoto, H. Tsuji, Y. Takahashi,
M. Nishi, K. Okuno, T. Hiyama, K. Yoshida, E. Tada,
K. Koizumi, T. Kato, H. Nakajima, K. Oka, M. Shimada,
Y. Sanada, and K. Yasukochi

OPERATION OF THE MHD COMPONENT TEST FACILITY AND FIRST
EXPERIMENTAL RESULTS...29
Y. Iwasa, J. F. Maguire, M. W. Sinclair, O. Tsukamoto,
D. J. Sliski, R. J. Thome, and J. B. Thompson

A CASK MAGNET PROTOTYPE SYSTEM (CMPS) FOR STANFORD
UNIVERSITY...37
H. G. Arrendale, R. W. Baldi, J. I. Jurek, R. C. McCool,
G. D. Magnuson, J. F. Parmer, R. A. Sutton,
W. D. Taylor, E. Bobrov, and J. E. C. Williams

Applications of Superconductivity – Cryogenic Techniques

Heat Transfer in He I

Heat Transfer in He II

Refrigeration of Superconducting Systems

Refrigeration and Liquefaction

Dilution and Magnetic Refrigerators, Refrigerators for Space Applications

Mass Transfer – Flow Phenomena

Properties of Fluids

Cryogenic Applications – Bulk Storage and Transfer of Cryogenic Fluids

Cryogenic Applications – LNG Operations

Cryogenic Applications – Space Science
and Technology

Cryogenic Applications – Cryopumping

Cryogenic Instrumentation and Controls

Indexes

FOREWORD

The 1981 Cryogenic Engineering Conference, joined by the International Cryogenic Materials Conference, was held at the Town and Country Hotel, San Diego, California. The total number of oral and poster presentations in both conferences was approximately 50% greater than at the 1979 conferences. The field of cryogenics appears to be growing at least linearly and some exponential tendencies may in fact be at work.

As your fledgling editor I would to acknowledge my gratitude to the authors, 100% of whom cooperated so willingly with the new page-limit and camera-ready requirements. The "Advances in Cryogenic Engineering" has achieved, over the past 25 years, a reputation for technical excellence. Our rigorous peer-review system is the foundation of this reputation and I wish to thank all those who assisted so thoughtfully in this review process; Volume 27 is better because of your help.

R.W. Fast, Editor

DEDICATION

I have known Dr. Klaus D. Timmerhaus for 26 years — essentially the same 26 years that span the birth and development of cryogenic engineering in the United States.

I first met Tim, as his legion of friends call him, in the fall of 1955. I had come to the University of Colorado in Boulder to do my graduate work in chemical engineering. I had been advised, and wisely so, to select a thesis advisor with care. The advice given me was not to put to much emphasis on the research topic, but rather to pick a good man to work with. There were many good men as professors at C.U. at the time, as there are now, but it was clear that "Tim" was someone I would like to hang out with, and hang-out together for 26 years we have. We began by separating the isotypes of hydrogen by fractional distillation, and have continued collaborating ever since.

Although Tim is at the University of Colorado, and I am at the National Bureau of Standards, we continue to work together, teach together, and collaborate on papers.

Tim has spent his entire professional lifetime in cryogenics. I cannot possibly relate all of his accomplishments. Tim is Associate Dean of Engineering at the University of Colorado, is past-president of the A.I.Ch.E., and is a recent recipient of the Stearns Award, the highest recognition that the University of Colorado can offer one of its citizens.

Tim's association with the Cryogenic Engineering Conference began in the mid-1950's, which was a crucial period in the development of cryogenic engineering in this country. That was the period after the original AEC interest in cryogenics, and before that of NASA. Even in these doldrums, Tim was convinced that cryogenics had a great future and in fact was on the threshold of an explosion into a great number of different disciplines. He was right.

In 1954, Russell Scott conceived in first CEC, and asked Tim to take a leadership role in guiding and developing our Conferences - a role that has never stopped.

I worked with Tim on the second CEC in 1956. We were a very small, low-overhead operation. In fact, we had only one typewriter for the whole Conference. · It was my job to carry the typewriter from place to place.

Tim's interest was not so much in typewriters, as in the cryogenic information and data that were presented at the Conferences. He has been Editor of our proceedings, "Advances in Cryogenic Engineering," for 25 years and is responsible for their world recognized excellence. He is the only man alive who has read every paper presented at every CEC. Indeed, I suspect, he has re-written most of them. He has about a dozen papers of his own in the "Advances."

The Cryogenic Engineering Conferences and the "Advances" have flourished and grown in stature because of Tim, until both are internationally recognized as the best in their field.

After 25 years of steering our meetings and editing our proceedings, Tim has chosen to do something else for a change. I can't imagine why. But I do know that wherever he directs his energies, that field will be immeasurably the richer for it.

Thomas M. Flynn, Chairman
1981 Cryogenic Engineering Conference

Klaus D. Timmerhaus
University of Colorado

RUSSELL B. SCOTT MEMORIAL AWARDS

The Russell B. Scott Memorial Awards honor the first head of the Cryogenic Engineering Laboratory of the National Bureau of Standards in Boulder. Mr. Scott was the founder of the Cryogenic Engineering Conference, the first of which was held in 1954. He is the author of the best-selling classic, "Cryogenic Engineering". He retired from the National Bureau of Standards in 1965 after a 37 year career with the Bureau. Mr. Scott died in 1967.

The objectives of the Scott Memorial Awards are to provide an incentive for high quality papers at the Cryogenic Engineering Conference and to provide recognition to those authors, who in the judgment of the CEC Board, presented the best paper in each of two categories. The award consists of a certificate and a $500 honorarium.

Papers presented at the 1979 conference in Madison and published in the "Advances in Cryogenic Engineering" were eligible for consideration. Session chairmen and conference attendees, through the session feedback cards, nominated papers which were reviewed by the Awards Committee. The final decisions were made by the entire Conference Board.

In the category of cryogenic engineering applications, three papers were found to be superior. They are:

A Novel Thermometer Sensor for the mK Region Using the Proximity Effect by H. Nagano, Y. Oda and G. Fujii, Tokyo University.

Purification and Cryogenic Separation of SNG Produced from Coal by A.A. Cassano, T.C. Li, J.C. Tao and T.R. Tsao, Air Products & Chemicals, Inc.

Progress on the Development of a 3- to 5-Year Lifetime Stirling Cycle Refrigerator for Space by A. Sherman and M. Gasser, Goddard Space Flight Center; M. Goldowsky, North American Philips Corp.; G. Benson, Energy Research & Generation, Inc.; and J. McCormick, Mechanical Technology, Inc.

The Conference is pleased to announce that the winner of the Scott Memorial Award in applications is

> Progress on the Development of a 3- to 5-Year Lifetime Stirling Cycle Refrigerator for Space by Sherman, Gasser, Goldowsky, Benson and McCormick.

The three best papers in the cryogenic engineering research category were:

> Transient Pool Boiling of Liquid Helium Using a Temperature-Controlled Heater Surface by P.J. Giarratano and N.V. Frederick, NBS Thermo-physical Properties Division.

> Techniques for Reducing Radiation Heat Transfer Between 77 and 4.2 K by E.M.W. Leung, R.W. Fast, H.L. Hart and J.R. Heim, Fermi National Accelerator Laboratory.

> Densities and Di-electric Constants of LPG Components and Mixtures at Cryogenic Storage Conditions by R.T. Thompson, Jr. and R.C. Miller, University of Wyoming.

The winning paper in the research category is

> Techniques for Reducing Radiation Heat Transfer Between 77 and 4.2 K by Leung, Fast, Hart and Heim.

1981 CRYOGENIC ENGINEERING CONFERENCE BOARD

ACKNOWLEDGMENTS

The Cryogenic Engineering Conference Board is deeply grateful for the support that the following organizations have given the 1981 conference:

AIRCO, Inc.
Air Products and Chemicals, Inc.
The Air Products Foundation
Ball Aerospace Systems Division
Beech Aircraft Corporation
Brookhaven National Laboratory
Electric Power Research Institute
Fermi National Accelerator Laboratory
Gas Research Institute
General Atomic Company
General Electric Research and Development Center
Hughes Aircraft Company
International Association for Hydrogen Energy
Lawrence Livermore National Laboratory
Louisiana Tech University
Messerschmitt – Boelkow – Blohm ,GmbH
M.W. Kellogg Company
National Aeronautics and Space Administration
National Bureau of Standards
National Science Foundation
Oak Ridge National Laboratory
Union Carbide Corporation, Linde Division
University of Colorado
University of Texas – Austin
University of Wisconsin – Madison
U.S. Department of Energy, Office of Energy Research
Westinghouse Research and Development Center

THE CRYOGENIC ENGINEERING CONFERENCE—A RECORD OF TWENTY-FIVE YEARS OF LOW TEMPERATURE PROGRESS

K. D. Timmerhaus

University of Colorado
Boulder, Colorado

INTRODUCTION

Cryogenics, a term commonly used to refer to very low temperatures, had its beginning in the latter half of the last century when man learned, for the first time, how to cool objects to a temperature lower than had ever existed naturally on the face of the earth. The air we breathe was first liquefied in 1883 by a Polish scientist named Olszewski. Ten years later he and a British scientist, Sir James Dewar liquefied hydrogen. Helium, the last of the so-called permanent gases, was finally liquefied by the Dutch physicist Kamerlingh Onnes in 1908. Thus, by the beginning of the twentieth century the door had been opened to a strange new world of experimentation where all substances, except liquid helium, are solids and where the absolute temperature is only a few microdegrees away.

The point on the temperature scale, however, at which refrigeration in the ordinary sense of the term ends and cryogenics begins has never been well defined. Most workers in the field have chosen to restrict cryogenics to a temperature range below -150°C (123 K). This is a reasonable dividing line since the normal boiling points of the more permanent gases, such as helium, hydrogen, neon, nitrogen, oxygen, and air lie below this temperature, while the more common refrigerants have boiling points that are above this temperature.

FORMATION AND EARLY HISTORY OF THE CRYOGENIC ENGINEERING CONFERENCE

Even though the liquefaction of air at 50 L/h by Linde in 1898 signalled the birth of industrial cryogenics, it was not until after World War II that cryogenic engineering emerged as a recog-

nized field and not just a branch of refrigeration. In fact, the
appointment of Professor Samuel C. Collins of MIT as Professor of
Cryogenic Engineering in 1949 was the first formal recognition of
cryogenic engineering on a university campus. The year 1950 saw
the Cambridge Corporation, though not organized for the cryogenic
field, become a large cryogenic engineering concern. Shortly
thereafter in 1952, Dr. Howard McMahon was asked to promote
cryogenic engineering research and development at Arthur D.
Little, Inc. Also in 1952, Herrick L. Johnston, Inc. entered this
field and became another important cryogenic engineering concern.

The concept of the NBS-AEC Cryogenic Laboratory is generally
credited to Dr. Edward F. Hammel who, as head of the Cryogenic
Laboratory at the Los Alamos Scientific Laboratory envisioned the
need for a government supported central cryogenic engineering
facility to meet the expanding use of very low temperatures and
low temperature refrigerants in various branches and agencies of
the Federal government for rocket propulsion, separation of gases
by distillation, production of large quantities of liquid hydrogen, and sophisticated weapons development. This central
laboratory concept with the endorsement of numerous well-known
researchers in the field received the support of the Atomic Energy
Commission in 1950 and construction was initiated in Boulder,
Colorado in the spring of 1951. Construction of the first building, to house the liquid hydrogen and liquid nitrogen plants, was
completed in the spring of 1952, while the research laboratory was
completed later that same year. The NBS-AEC Cryogenic Engineering
Laboratory was officially made a section of the NBS Heat and Power
Division that same year with Mr. Russell B. Scott as its chief.

Even though this was a new venture for NBS with but a limited
number of trained personnel in the field, research moved ahead
very rapidly. The principal work of the laboratory was divided
into two complementary areas, namely, research on the fundamental
and mechanical properties of materials used in cryogenic construction, and the development of cryogenic equipment, techniques and
processes. (A complete description of the research programs
undertaken during the first two years of the NBS-AEC Cryogenic
Laboratory is given in Volume I of the Advances in Cryogenic
Engineering.) The withdrawal of AEC support in 1954, however,
introduced considerable concern as to the future of some of the
research programs that had only recently attained a critical mass.
In an effort to more widely publicize the NBS facilities and
expertise, Mr. Scott, along with other members of the Laboratory,
(B. W. Birmingham, W. B. Hanson, R. B. Jacobs, V. J. Johnson, and
M. M. Reynolds), proposed the holding of a scientific meeting in
conjunction with the dedication ceremonies of the NBS Boulder
Laboratories. Thus, on September 3-10, 1954, the Central Radio
Propagation Laboratory and the Cryogenic Laboratory sponsored two
parallel technical meetings and invited participation from scien-

tists and engineers throughout the world. The emphasis on this first highly successful Cryogenic Engineering Conference which attracted over 200 participants was on the basic tools of cryogenic engineering, namely, equipment, instrumentation, insulation, processes and properties of materials.

No conference was held in 1955, but Russell Scott asked me, as a part-time NBS employee, to determine whether there was sufficient interest in another conference by the cryogenic engineering community. With the aid of a questionnaire, the response was determined to be positive, particularly if the conference was again held in Boulder. As a consequence, I was requested by Scott to assist Bascom Birmingham in organizing a meeting for September 1956. This meeting attracted more than 400 scientists and engineers from all parts of the world to discuss new developments covering essentially the same basic areas of cryogenic engineering as were covered at the 1954 meeting, with the exception of a special session on liquid hydrogen and liquid helium bubble chambers.

The delegates attending the 1956 conference overwhelmingly voted to have another conference in 1957 - provided it was again held in Boulder. It is really not clear whether this vote was a favorable response to the $2.00 registration fee for the conference that included a soft-cover copy of the Proceedings, the desire to again combine a vacation with business, the industrially-funded social hour at the rustic Estes Park Chalet the first night of the conference, or a general feeling that the expanded domain of cryogenic engineering really called for a separate meeting of its own since it was not finding a home in any of the other professional societies. The history of the Cryogenic Engineering Conference suggests that it was the latter argument that swayed the participants, but at the time one could very easily recognize that each one of these considerations entered into the favorable vote. All doubts relative to the establishment of an annual Cryogenic Engineering Conference were dispelled in 1957, when over forty papers were accepted for the Conference at Boulder. Once again the main emphasis of the meeting was on the further development of the basic tools associated with cryogenic engineering including presentations on cycle analysis of hydrogen liquefaction, ortho-parahydrogen conversion schemes, separation of hydrogen isotopes by distillation and operational experiences with unattended oxygen plants. To make the Proceedings of the conference as informative as possible an attempt was made to include a majority of the general discussion which followed each presentation. This concept was to go through several revisions in succeeding conferences before procedures were finally evolved that have been found to be quite effective to date.

Regardless of how careful the planning is for a conference, there always seems to be something that still can go wrong. For example, to minimize the logistics of the 1957 conference, the social events were scheduled close to Boulder. One of these events was a chuckwagon dinner on the top of Flagstaff Mountain directly overlooking Boulder. Everything seemed to be operating smoothly until the last bus broke down halfway up the mountain. Because of the large attendance at the chuckwagon dinner, the non-arrival of this bus was not detected for the better part of an hour. Fortunately, or unfortunately (depending upon your viewpoint) the missing bus was also carrying a large supply of chilled refreshments to meet the needs of the overflow crowd on top of the mountain. Since assistance for the stalled vehicle was not forthcoming, the guide associated with the bus began dispensing the liquid refreshment to the thirsty attendees. When help did arrive they found a rollicking group of delegates who really didn't care whether they got to the top of the mountain for the chuckwagon dinner or not. Nevertheless, after a transfer of buses they did arrive at their destination only to find that the Junior Chamber of Commerce who served as the hosts for the chuckwagon dinner had just packed everything away and were ready to leave. However, the J.C.'s, true to their service motto, once more unpacked everything and took care of the late arrivals.

Even with such unexpected mishaps, the attendees once again expressed a strong interest in having the next conference in Boulder. However, the work load of the NBS Cryogenic Engineering Laboratory had by this time picked up considerably and it was clear that the laboratory could not host another conference in the following year. Since Sam Collins of MIT was a member of the Conference Board, the other board members prevailed upon him to get MIT to host the conference in 1958. This he agreed to do, provided that support staff would be provided to handle the entire mechanics of the conference. Scott volunteered my services, and thus I added the duties of Program Chairman and Local Arrangements Chairman in addition to the duties of Secretary-Treasurer of the Cryogenic Engineering Conference and Editor of the Proceedings. This latter assignment was to prove during the following decade to be a busy as well as an interesting one since the Russian space shot in 1957 rapidly escalated the cryogenic engineering effort in the U.S. and made it one of the "hottest areas going".

By 1961 the U.S. space program was moving ahead very rapidly and cryogenic problems needing solutions seemed to be increasing logarithmically. With the rapidly increasing work force in the space program, it became evident to the Conference Board that the Cryogenic Engineering Conference could be of real service by instituting a carefully designed seminar program at the conference which covered many of the problem areas being encountered in the day-to-day operation of the cryogenic systems associated with the space program. Thus, in 1962 the Cryogenic Engineering Conference at UCLA presented some twenty specialized seminars led by experts in the field. These proved to be so well-received that the concept was retained for the 1963 conference held in Boulder. These seminars were augmented with several general review sessions to provide additional background information for those who felt a need for such a review.

Setting up any type of conference can be rather involved and time-consuming, particularly when it involves not only the technical program, but developing and taking care of all the local arrangements at a location many miles from your home base. The incident with the social hour at MIT in 1958 is a good example. This event was scheduled to be held at the faculty club located on the fifth floor of one of the buildings adjacent to the campus. What was not checked out was that there was only one five-passenger elevator going to the fifth floor. You can imagine the traffic jam that occurred on the first floor near the elevator when over 400 attendees and guests tried to get to the faculty club at approximately the same time! To alleviate the long wait we finally resorted to bringing the waiters to the lobby, taking orders and then bringing the drinks down with the next empty elevator. Anybody taking the elevator on the next day must have wondered what bar he or she had stumbled into because the carpeted elevator was saturated with the aroma of mixed drinks that had accidentally been spilled the night before.

Other conferences held during the early years also seem to have had their unforgettable moments. To illustrate this, consider the 1965 Cryogenic Engineering Conference hosted by Rice University in Houston where one of the special events scheduled was a visit to the San Jacinto Monument, followed by a seafood dinner at the San Jacinto Inn. Since parking was limited at the latter facility, buses were rented to accommodate the majority of the attendees and their guests. Once east of downtown Houston, the five hundred and seventy foot monument, commemorating Texas' independence from Mexico, is visible for miles from the coastal lowlands. So, rather than taking the normal route along the river to the monument, the lead driver decided to take a shortcut utilizing the country roads which paralleled many of the canals which crisscrossed the lowlands. Unfortunately, only a few of the bridges over these canals were adequately constructed to carry the weight of the buses. Thus, the parade of buses carrying the attendees kept shuttling back and forth alongside one canal and then another for more than half an hour trying to get closer to the monument clearly visible in the distance before the lead driver finally admitted that he was lost. Upon this disclosure, all of the other bus drivers volunteered that they knew the way and took turns leading the procession for the next half hour. When this did not bring any better results, they finally hailed a

local farmer who graciously led them back to the main route and
headed them in the right direction. (It goes without saying that
the bus company was rather embarrassed about the incident and with
tongue-in-cheek asked the conference not to publicize the company's recent scenic tour of the Texas coastal lowlands.) What
makes this incident even more humorous now is that essentially the
same situation had occurred at the 1964 Cryogenic Engineering
Conference when conference bus drivers lost their way between Case
Western Reserve University and downtown Cleveland.

With the attendance at the 1966 Cryogenic Engineering Conference approaching the thousand mark, it became obvious that the
task of organizing the conference and serving as secretariat
needed a more permanent office than could be provided for by the
University of Colorado. With the 1967 conference, this office was
established in the National Academy of Sciences in Washington,
D.C. It remained there until 1971 when it was returned to Boulder
and was assumed through the gracious assistance of Bascom
Birmingham by the National Bureau of Standards. Annual meetings
of the Cryogenic Engineering Conference were held through 1973,
with the exception of 1971 when the conference assisted the
International Institute of Refrigeration in co-hosting the 13th
IIR Meeting in Washington, D.C. At the 1972 conference in
Atlanta, the Conference Board decided on a biennial schedule
alternating with the Applied Superconductivity Conference. This
schedule has been maintained to date with meetings in Kingston,
Ontario in 1975, Boulder in 1977, Madison in 1979, and San Diego
in 1981.

EARLY DEVELOPMENTS IN THE PUBLICATON
OF THE CONFERENCE PROCEEDINGS

A review of the first few volumes of the Advances in Cryogenic
Engineering shows that the Proceedings of the 1954 conference was
originally assembled by W. B. Hanson and published as NBS Report
3517. (It was later reedited when it was included as part of the
Advances in Cryogenic Engineering.) The Proceedings of the next
three conferences (1956,1957,1958) under the editorship of K.D.
Timmerhaus were originally published locally in soft-cover form
and supported entirely through generous industrial contributions. To promote quality in the publication, a peer review
system with graded reviews was initiated and papers were not
accepted for publication until every comment by the three or more
reviewers had been satisfactorily taken care of. Often this meant
extensive revisions or rewrites of papers. It also meant rejection of between 5-15% of the papers presented at the conference.

With the advent of the 1959 conference, hosted by the
University of California at Berkeley and attended by nearly 700
delegates, it became evident that the Proceedings of the Cryogenic

Engineering Conference had to be made more permanent and professional in appearance. After considerable investigation, the Conference Board selected a relatively new publishing firm in New York that offered not only to reprint the first four Proceedings under hard cover, but to set up succeeding Proceedings in hot type all under the title of Advances in Cryogenic Engineering. This action permitted the editor to shift some of the publication and promotional problems to the publisher, Plenum Press. However, the move also had its disadvantages, since it added an additional delay of three or four months to the publication of the proceedings. This delay prompted the Conference Board to investigate the use of preprints at the 1960 meeting, held once more in Boulder.

The idea of the preprints was to request each author to supply 100 copies of his or her paper and these would be placed in a preprint room and attendees could pick up copies of papers that were of immediate interest to them. It was expected that only a fraction of the attendees would be interested in the preprints and then would be selective in their choice. No assumption could have been more inaccurate! The first hundred delegates who registered for the conference made such a stampede to the preprint room that student assistants hurriedly left their posts for safer duty. It was difficult to believe, but the sight was like Macy's bargain basement during a half-price sale. Everything that was on the tables including preprints, program signs, directions, conference supplies, and even newspapers, was hurriedly carted away. There was hardly a scrap of paper remaining in the room! Obviously, only a fraction of the delegates had had an opportunity to look over the preprints. By utilizing all of the secretarial and student help that was available, we managed within twenty-four hours to make spirit masters from the one set of preprints that had been retrieved earlier for editing purposes and from these another 100 copies of each preprint were reproduced. When the availability of these additional preprints was announced, there was a large exodus of delegates to the preprint room and the same bargain basement scene was reenacted for the second time. It was difficult to imagine that some two hundred delegates were going to carry back with them as many as fifty to sixty preprints. This doubt was substantiated soon after the conference adjourned, when custodians returned hundreds of discarded preprints that they had found in rest rooms, empty classrooms, closets, dormitory rooms and trash bins all over the campus. Obviously, this experience provided us with many of the preprint guidelines that are still being followed at the present time.

In an effort to upgrade the quality of the papers to be presented at the conference and published in the Advances in Cryogenic Engineering, the Conference Board in 1961 established one award for the best research paper and another one for the best applied paper. The award originally consisted of a certificate for each

author of the award-winning papers and a $100 check. This amount
was raised to $200 some years later, and has recently been in-
creased to $500.   (In the case of multi-authored papers, the
financial rewards per author can be rather minimal as evidenced a
few years ago when eleven authors were involved with the award-
winning paper.) Since the untimely death of Russell Scott, both
awards have been designated as the Russell B. Scott Memorial
Award.

Like the conference, publication of the Advances in Cryogenic
Engineering has not only had its brighter moments, but also its
darker ones. For example, in 1973 we were involved in editing
Volume 19 of the Advances in Cryogenic Engineering and coediting a
four-volume Proceedings of the 13th Low Temperature Physics
Conference. In an effort to speed up the publication of these two
proceedings, the publisher decided to subcontract the publication
of the page proofs to a firm in Israel rather than overloading the
firm in England that had performed this typesetting for earlier
volumes. The plan seemed to be working as anticipated until one
shipment containing the marked-up page proofs of the last fifteen
papers of the Advances in Cryogenic Engineering, and the first
twenty papers from the Low Temperature Physics Conference was on
the jet plane that was hijacked by terrorists and eventually blown
up in the Sinai Desert. This loss was not recognized for some
time, since the publisher assumed that the subcontractor had run
into some delays and the subcontractor assumed that we were delin-
quent in returning the marked-up page proofs. When it finally was
established that the missing page proofs had been on the destroyed
jet plane, it meant that the page proofs had to be reprinted and
the entire proofing process repeated. This incident delayed
publication of Volume 19 by approximately four months. The other
proceedings fared even worse by having their publication delayed
nearly six months, since additional scheduling conflicts were
encountered with other texts being delivered by the publisher.
Volume 25 of the Advances in Cryogenic Engineering, on the other
hand, experienced nearly a three-months delay because of an unexpected bookbinders' strike.

Obviously, not all the delays in publication were the fault of
the publisher. In general, the editor received excellent cooperation from the many authors desiring to have their papers published
in the Advances in Cryogenic Engineering. However, a number of
the volumes had one or two authors who either were always late in
their reply to requests made by the editor or who, after reading
the reviewers' comments, would return another copy of their paper
without making a single correction and then insist that they had
responded to all of the reviewers' comments. One such case probably stands out among the rest. The author of this specific paper,
after being requested to make major revisions to his paper by four
out of five reviewers of his paper, apparently decided to ignore

those suggestions and returned another copy of his original paper
and insisted that it be published in that form. In view of the
various constructive comments made by the reviewers, and the
editor's concurrence with those comments, the paper was entirely
rewritten with the assistance of the reviewers and sent back to
the author for approval. The revised version of the paper brought
back a host of objections from the author. With the assistance of
the reviewers, all of the author's objections were met and the
paper again went back to the author. Once again, negative comments were put forth by the author as to why he did not like the
revised version of his paper. However, after further minor improvements and the blessings of the reviewers, the author was
informed that because of the delay that it was causing to the
publication, he had three choices. Either he could revise the
original paper himself, taking into consideration the reviewers
comments, he could withdraw the paper, or he could accept the
revised version and avoid the task of rewriting the paper if
acceptance for publication was to be forthcoming. The author
rather reluctantly chose the latter route and the paper was published. Fortunately, this rather strong stance by the editor was
later justified when the Conference Awards Committee nominated
this paper among several for the Russell B. Scott Memorial
Award. It was, of course, not selected as the award winner when
the prior history of the paper became available to the committee.

As noted earlier, such incidents were exceptions. In most
cases relationships with the authors could not have been better.
Each exhibited an attitude that was not only highly cooperative,
but also very professional. It was a pleasure working with such
dedicated individuals and this interaction will be missed as the
editorship of the Advances in Cryogenic Engineering is assumed by
Dr. R. W. Fast in 1981. There is no doubt, however, that it is in
capable hands and that he will carry on the past traditions of the
series. We wish him well in this new responsibility.

A review of past volumes of the Advances in Cryogenic
Engineering shows that the industry has matured considerably since
the first Cryogenic Engineering Conference in 1954. At that time,
the emphasis was essentially on air separation, liquid nitrogen,
and oxygen technology, and the developmental problems associated
with small scale liquid hydrogen and liquid helium production.
The space age in the 60's greatly accelerated the need for material and thermophysical property data, new insulation concepts,
large-scale handling, transfer and storage of liquid hydrogen and
oxygen, operational safety, and other space-related problems.

The 70's, in turn, emphasized more of the applications utilizing cryogenics. Foremost among these were applied superconductivity and LNG. There are presently 26 volumes and 27 books in the
series (Volume 10 was published in two parts). This constitutes a

total of approximately 15,500 pages with over 1700 presentations. It should be pointed out, however, that Volumes 22, 24,
and 26 are devoted strictly to the Proceedings of the
International Cryogenic Materials Conference, which has been held
jointly with the Cryogenic Engineering Conference since the
Kingston, Ontario meeting in 1975.

ACKNOWLEDGMENTS

Let me once again express my appreciation to all the many
individuals who have so ably contributed to the advancement of
cryogenic engineering during the past twenty-five years that I
have been associated with the field. They are the ones to be
congratulated since, without their dedication and hard work, there
wouldn't have been any Cryogenic Engineering Conference and accompanying series, the <u>Advances in Cryogenic Engineering</u>. The more
than fifteen thousand pages covering essentially every aspect of
cryogenic engineering fundamentals and applications are a lasting
tribute to their many fine accomplishments.

THE FORCED FLOW COOLED COILS FOR THE INTERNATIONAL ENERGY AGENCY LARGE COIL TASK

J. L. Young and C. J. Heyne

Westinghouse Electric Corporation
Pittsburgh, Pennsylvania

P. Komarek and H. Krauth

Kernforschungszentrum Karlsruhe
Karlsruhe, Federal Republic of Germany

and

G. Vecsey and C. Marinucci

Schweizerisches Institut fur Nuklearforschung
Villigen, Switzerland

INTRODUCTION

The purpose of the IEA "Large Coil Task" (LCT) is to demonstrate the viability of superconducting toroidal field (TF) coils for use in the next generation of Tokamak fusion experiments. In the "Large Coil Test Facility" (LCTF) at Oak Ridge National Laboratory, six D-shaped coils will be configured in a compact torus for testing coil performance. Each of the six coils were independently designed to permit performance evaluation of various concepts and configurations. One of the American coils and the two European coils will be internally cooled with supercritical helium flowing through cooling channels in the conductor. Emphasis in the design and development phase of these three coils was placed on mechanical performance, cryogenic stability and ac losses.

Although the design details are different, several common advantages of forced flow conductors can be seen in the final coil designs. Thermal behavior, including stability, is completely predictable in all parts of the winding and no vapor blocking can occur. The electrical turn-to-turn insulation is capable of reliably withstanding high voltage, since it is a closed insulation system. High mechanical integrity is provided, and the conductor cooling concept is adaptable to a wide variety of structural designs. Last, but not least, these concepts promise scalability of performance to full size reactor magnets.

GENERAL DESIGN PHILOSOPHY

The design philosophy was to develop coil designs which would meet the performance requirements of the LCT while demonstrating the applicability of these designs for future fusion requirements. Hence, the design principles were significantly influenced by our belief that future fusion activities will ultimately require very large magnets with high magnetic field (>9T) capability.

Westinghouse decided to use a Nb_3Sn, superconductor and to develop a conductor and winding concept which minimizes strain sensitivity. The Swiss and Euratom design groups selected proven NbTi conductors, accepted a lower stability margin compared to Nb_3Sn, and concentrated their effort on the demonstration of the general advantages of forced flow cooling for TF coils. The coil designs[1-3] which were adopted embody the characteristics which are needed for very large, high field superconducting magnets: (1) a strain insensitive conductor, (2) low ac losses, (3) stable thermal operation, (4) high conductor current density, (5) efficient force transfer and distribution system and (6) high discharge voltage capability.

CONDUCTORS

Figure 1 shows the three conductors to be used in the coils. The Westinghouse conductor is a fully transposed, compacted cable which provides sufficient mechanical support for internal radial loads while still permitting axial slippage between strands during winding. This results in greatly reduced bending strain in the conductor and permits the use of Nb_3Sn in a practical, high current conductor while providing small voids in the conductor for coolant flow.

In the Swiss coil, low ac losses are provided by CuNi interleaving of the subcables. Simple, proven cable fabrication methods are paired with simple manifolding techniques for joints and helium connections.

The Euratom coil design is based on a cabled conductor with individually cryostable strands (direct cooling, all stabilizing Cu in the strands themselves) and on mechanical reinforcement by stainless steel, so that the conductor can transmit and support substantial forces and stresses. Cooling channels for the super-

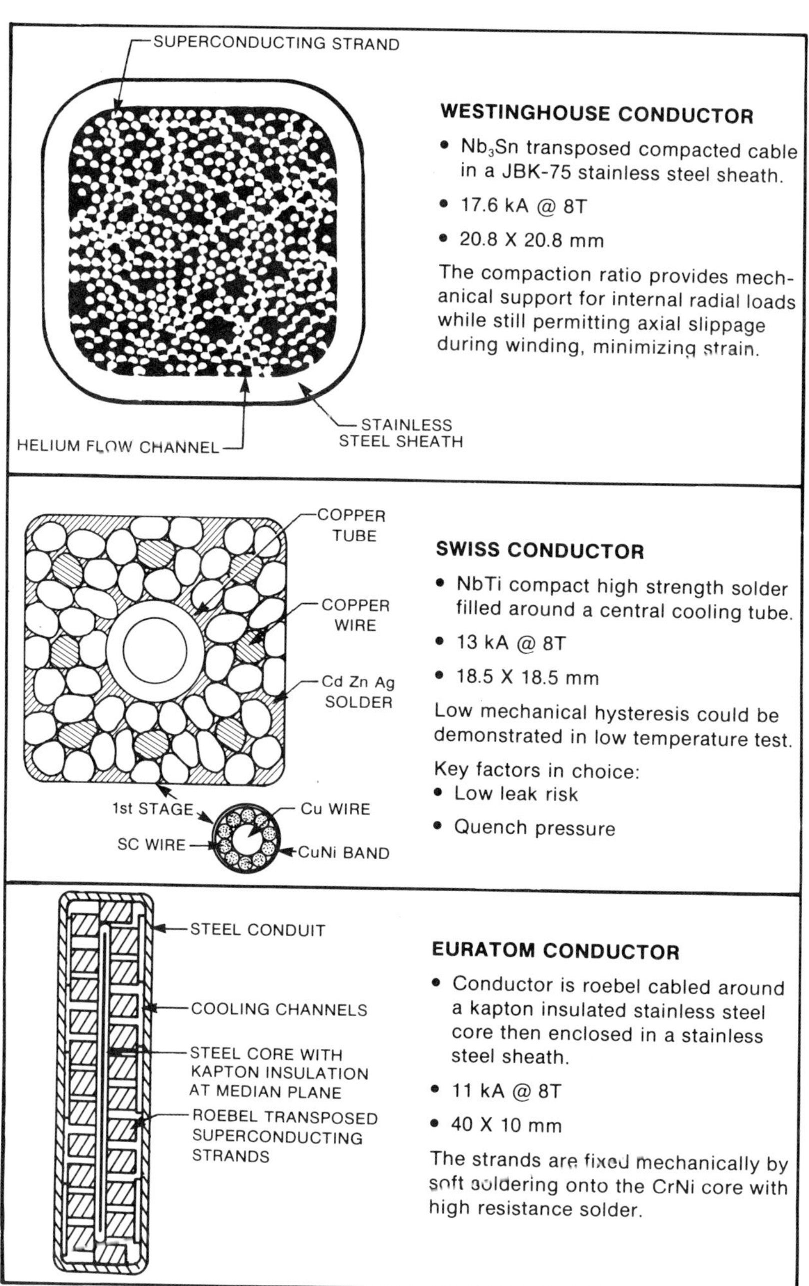

Fig. 1. Major Elements of the Three Conductors.

critical helium are provided by spacings between the individual strands and between the cable and the helium-tight jacket. The copper in the strands is work hardened to increase the force carrying capability of the conductor.

STRUCTURE AND FORCE DISTRIBUTION

Forced flow cooled conductors characteristically include self-contained conductor and helium coolant systems. This offers the advantage of decoupling the structural support from the helium containment function which, in turn, allows significantly more flexibility in the choice of structure design. The structural concepts shown in Fig. 2, vary from a box-like support structure to a completely distributed structure. While box-type coil structures for LCT magnets use steel of modest thickness (2.0 inches), a similar structural concept will require thicknesses in the range of 7-9 inches for power-plant size coils. Fabricating coil cases with plates of this thickness will be a difficult problem. In addition, some type of internal support will be required to reduce the cumulative effect of the crushing forces developed in the conductor pack. Forced flow cooling allows the adoption of a distributed structural concept where structural support members are of a modest thickness and the conductor forces are transferred to structural elements before the entire winding pack force is accumulated.

LOW AC LOSSES

Forced flow cooling is readily adaptable to conductors which have been divided into subelements which results in conductors with very low ac losses[4]. The ability to force supercritical helium through the small interstices of a cabled conductor results in decreased conductor losses relative to the losses in a similarly rated pool boiling conductor. This characteristic of forced cooled conductors will allow the use of rapidly pulsed ohmic heating coils (if needed) in future reactors without incurring unacceptably high refrigeration loads or thermal instabilities in the conductor. Table I gives the ac losses in the three coil designs.

VOLTAGE CAPABILITY

Reactor sized TF coils store large amounts of energy (4 GJ/coil) which must be removed rapidly during coil quenching to prevent damage from overheating or thermal stresses. While high current density conductors are advantageous because they can reduce coil size, they also have less volume to store heat during a magnet quench. Therefore, high current density magnets must be more rapidly discharged, but can be recooled quicker than magnets with low conductor current densities. Forced flow cooling offers

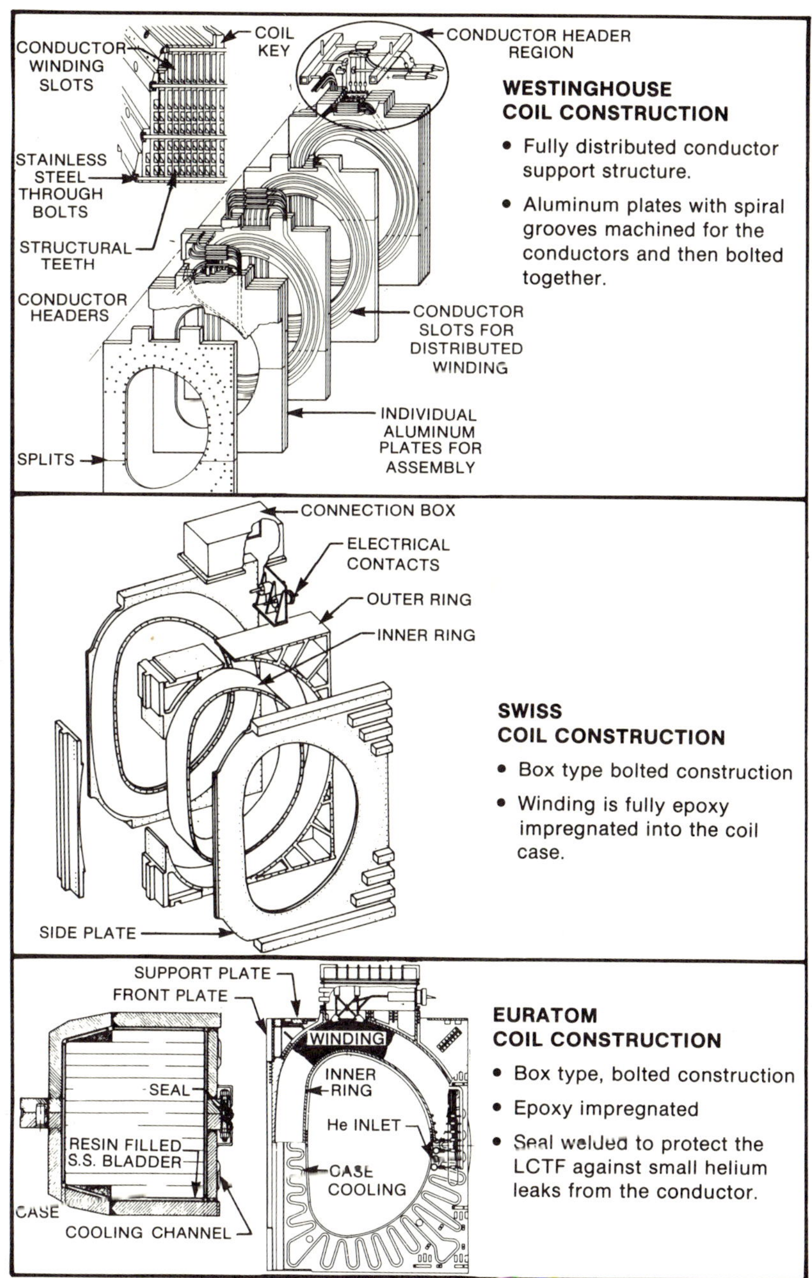

Fig. 2 Major Elements of the Three Coil Designs.

Table I. ac Losses for the Forced Flow LCT Coils at Standard
          LCTF Pulse Field Conditions

	WESTINGHOUSE DESIGN	SWISS DESIGN	EURATOM DESIGN
WINDING (W)	4	10	7
COIL CASE (PLATES) (W)	5.5	8	9
TOTAL (W)	9.5	18	16

a distinct advantage in this area since a completely closed insulation system can be provided (i.e., helium is not used as an
insulator in the magnet). Since the helium is inside the insulation system, breakdown voltages in excess of 20 kV are easily
achievable.

The helium fittings are of critical concern in the implementation of force cooling. But if designed appropriately, both
tightness and voltage capability can be achieved without problems,
as demonstrated by tests of prototypes for all three coils.

THERMAL STABILITY

One of the great potentials of the force cooled conductor
concept lies in the fact that a very high cooled surface area to
volume ratio can be obtained by subdividing the conductor in many
individual strands. This provides the ability to rapidly remove
heat input by sudden energy release and/or joule heating from the
conductor. Also, the exceptional transient heat transfer rates
can enhance the stability margin in case of rapid heat input.

This potential is utilized in an optimum manner in the
Westinghouse conductor. Here a very high current density can be
reached by the combination of a large cooled perimeter and a Nb_3Sn
superconductor. Analysis[5] programs have predicted a high level of
stability and test[6] have confirmed it. Testing by Oak Ridge
National Laboratory has shown that the Nb_3Sn conductor will
recover from a heat input of 1000 mJ/cm^3 at 80% of its quench
current with stagnant helium in the conductor.

The situation is quite different with the two NbTi conductors
where the energy absorption capability is inherently smaller due
to the much lower temperature margin. Here the design philosophy
was guided by the intention to reduce the possible heat release as
much as possible by using a rigid conductor and a monolithic
winding pack.

In case of the Euratom conductor a rather large cooled perimeter can be maintained by the adoption of the flat cable concept with separated and rather large strands. The heat absorption capabilities were calculated to be in the order of 100 mJ/cm^3 for short zones[7]. If a very long zone goes normal, e.g. a normal conducting zone extending over the whole conductor cross section, the heat absorption capabilities is 10 mJ/cm^3. However, this is not likely to occur in the chosen conductor and coil design. In addition, it was shown by analysis that the recovery capability of the conductor is much greater in cases where individual strands go normal due to the fact that the strands are thermally insulated[8].

For the Swiss coil, in normal operation for long disturbance zones, a stability factor of 1 to 1.3 is calculated based on pure cable enthalpy margin compared to maximum measured hysteretic energy release at the critical points, the sudden release of total plastic deformation energy being a pessimistic limit.

CRYOGENIC REQUIREMENTS

The LCTF helium system capability sets the envelope for the helium supply parameters for forced flow coils. The operational constraints are shown in Table II.

The individual coils will have different helium supply needs as summarized in Table III. It is important to note that all three forced flow coil designs, which are quite different, can operate within the same modest cooling supply envelope. This demonstrates the flexibility and practicality of forced flow systems for large coils. Also the properly designed refrigeration systems for forced flow cooled magnets are no more complicated

Table II. Helium Supply Characteristics

	COOLDOWN	OPERATION	WARMUP
PRESSURE (bar)	15	10-15	15
PRESSURE DROP (bar)	10	5	10
FLOW (g/s)	22-36 per coil	≈ 275	22-36 per coil
TEMPERATURE (K)	200-3.8 (variable)	>3.8	200, 300
SPECIFIED COOLING CAPACITY FOR THE COIL IN TEST POSITION (INCL. PUMPING POWER) (W)	394		

Table III. Cooling Parameters For The Three LCT-Forced
Flow Coils

	WESTINGHOUSE DESIGN	SWISS DESIGN	EURATOM DESIGN
NUMBER OF PARALLEL FLOW CHANNELS	48	22	28
FLOW RATE/CHANNEL (g/s)	2	5.5	$\leq$10
TOTAL FLOW RATE (g/s) FOR THE COIL WINDING	109	121	280
PRESSURE DROP AT RATED FLOW (bar)	~1	~3	~3
SEPARATE CASE COOLING	NONE	YES	YES
MAXIMUM QUENCH PRESSURE (bar)	100-150	150	60

than those for pool boiling magnets[9]. Any perceived complications
of conductor design and helium distribution are far outweighed by
the advantages.

POTENTIAL FOR FUTURE COILS

The foregoing discussion has highlighted the following advantages of forced flow cooled magnet systems: 1) Flexibility of
structural design and structural integrity, 2) stable and predictable thermal operation, 3) good stability at high current
density, 4) low ac loss, 5) high voltage capability.

These advantages, we believe, will become more important and
more obvious as the fusion program progresses. The forced cooled
coil designers are convinced that this cooling mode offers advantages that far outweigh any perceived additional cooling system
complexities, and in fact, offers the most realistic path to
reliable coils that will prove the viability of Tokamak fusion
reactors.

REFERENCES

1. C.J. Heyne, et. al., Westinghouse Design of a Forced Flow
      Nb$_3$Sn Test Coil for the Large Coil Program, in "Proc. 8th
      Symp. on Engr. Problems of Fusion Research", IEEE 79CH1441-
      5 NPS (1979).
2. G. Vecsey, Status of the Swiss LCT coil, IEEE Trans. on
      Magnetics, MAG-17:1738 (1981).

3. H. Krauth, et. al., Status of the European LCT coil, IEEE
    Trans. on Magnetics, MAG-17:1726 (1981).
4. G.R. Wagner, Transverse Field AC Losses in the Westinghouse
    Nb_3Sn Forced-Cooled LCP Conductor, in "Proc. 8th Symp. on
    Engr. Problems of Fusion Research, IEEE 79CH1441-5 NPS
    (1979).
5. P.W. Eckels, et. al., Designing for Low Temperature Stability
    of the Large Coil Program Nb_3Sn Coil, in "Proc. 8th Symp.
    on Engr. Problems of Fusion Research," IEEE 79CH1441-5 NPS
    (1979).
6. J.R. Miller, J.W. Lue, S.S. Shen, Nb_3Sn Cable-in-Conduit
    Conductor Tests, in "Proc. 8th Symp. on Engr. Problems of
    Fusion Research," IEEE 79CH1441-5 NPS (1979).
7. G. Krafft, et. al., Design Aspects of Forced Cooled Super-
    conductors for Large Fusion Magnets, in "Proc. 8th Symp. on
    Engr. Problems of Fusion Research," IEEE 79CH1441-5 NPS
    (1979).
8. G. Ries, Stability in superconducting multistrand cables,
    Cryogenics, 20:
9. C. Marinucci, P. Weymuth and G. Vecsey, Cooling aspects of the
    Swiss LCT coil, IEEE Trans. on Magnetics, MAG-17:1741
    (1981).

10-T, 60-cm BORE Nb₃Sn TEST MODULE COIL DESIGN

T. Ando, S. Shimamoto, H. Tsuji, Y. Takahashi, M. Nishi,
K. Okuno, T. Hiyama, K. Yoshida, E. Tada, K. Koizumi,
T. Kato, H. Nakajima, K. Oka, M. Shimada
Y. Sanada, and K. Yasukochi

Japan Atomic Energy Research Institute
Tokai-mura, Ibaraki, Japan

INTRODUCTION

At the Japan Atomic Energy Research Institute (JAERI), the cluster test program is one of the principal development programs for superconducting toroidal magnets.[1] The cluster test facility (CTF) was successfully tested in August, 1980.[2] It has two coils, called cluster test coils (CTC), which provide background field to a test module coil (TMC), as shown in Fig. 1. The first TMC is designed to generate a 10 T field within a 60 cm bore winding. Fields of 10-12 T are needed for toroidal field coils.

To achieve a magnetic field of 10-12 T, one must either use Nb_3Sn or use Nb-Ti or Nb-Ti based alloys operating below 4.2 K. JAERI chose the Nb_3Sn option because high stability is attainable from the high critical temperature and because Nb_3Sn is applicable to much higher field magnets. The role of the first TMC is to clarify the reliability of Nb_3Sn for high current conductors at fields around 10 T. This paper describes the coil concept, conductor configuration, and verification test results, particularly those results concerning mechanical aspects of Nb_3Sn conductors.

COIL CONCEPT

The main parameters of the TMC are shown in Table I. The 10 T magnetic field is produced at a operating current of 6050 A with a background field of 3 T from the CTC. The total stored energy, including CTC, is around 30 MJ. The coil is bath-cooled at 4.2 K. Two grades of conductor are used, Nb_3Sn based for fields higher than 6.2 T and Nb-Ti for the lower field region. Between

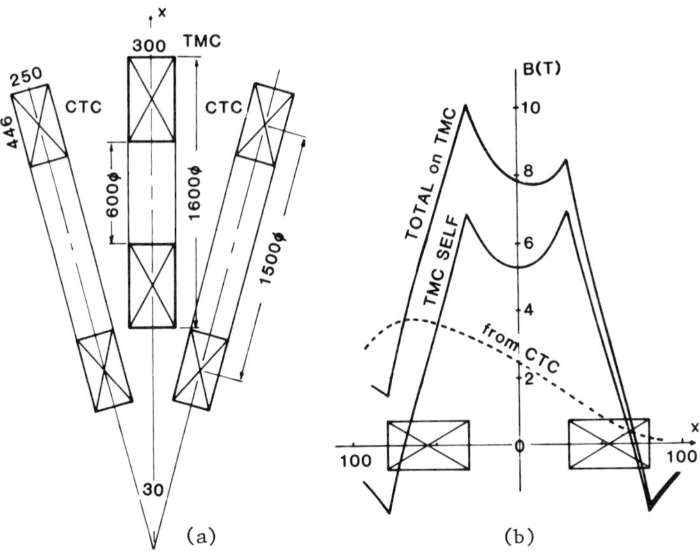

Fig. 1. Location of the TMC in the Cluster Test Facility (a) and field distribution on the TMC (b).

pancakes there is a glass epoxy, flat bar spacer whose thickness is 2.6 mm. This spacer is stuck onto the pancakes and covers ~ 25% of the surface of the conductor flat side. The insulation between turns is epoxy impregnated glass tape of 0.3 mm thickness which fully covers the surface of the narrow side of the conductor. The allowable gap between turns is less than 0.1 mm so as to avoid conductor movement and to obtain high rigidity. A

Table I. Parameters of Test Module Coil

Coil Shape	Circular
Coil Size	
Winding Inner Diameter	600 mm
Winding Outer Diameter	1600 mm
Width	300 mm
Winding Concept	Double Pancakes
Nominal Magnetic Field	10 T
Number of Turns	725
Magnetmotive Force	4.39 MA-turns
Average Current Density	30 A/mm^2
Self Inductance	0.46 H

winding tension of 350 kg is used. AISI 316L stainless steel is selected as the structural material for the outer ring of the helium vessel, where the stress is higher than in the other parts, for which 304 L stainless steel is used. Final closure of the helium vessel is performed by the electron beam welding.

CONDUCTOR DESIGN

The characteristics of the Nb_3Sn and Nb-Ti conductors are shown in Table II. Both conductors are composed of six stranded cables soldered in the center of a copper stabilizer. The Nb_3Sn strand is fabricated by the external bronze method. Each of the 129 bundles is wrapped in a niobium layer which serves as a barrier to prevent the diffusion into the copper of tin from the bronze matrix. The bronze matrix occupies 13-weight-% of the conductor. The heat treatment condition for the Nb_3Sn is 650° C for 200 hours. Figure 2(a) shows the overall conductor, and Fig. 2(b) shows the bundles of Nb_3Sn filaments in the copper matrix. The number of joints between Nb_3Sn and Nb-Ti conductors is two per double pancake; each strand is soldered with a 75 cm overlap within the continuous copper stabilizer.

The two conductor surfaces exposed to liquid helium are roughened and oxidized to obtain a high heat flux. This method was developed at JAERI and was used in the Japanese LCT conductor.[3] Figure 3 shows a stability map consisting of the heat flux curve and the joule heat generation curve of the TMC Nb_3Sn conductor at 10 T. The current at which the propagation velocity is zero, (6147 A) is obtained by:

$$\int [Q(T) - G(T)] \, K(T) \, dT = 0$$

where   Q = Heat flux to coolant per unit area of cooled surface
           G = Heat generation expressed per unit area of cooled
               surface
           K = Thermal conductivity of the conductor (2.4 W/cm
               K at 4.2 K, 10 W/cm K at 15 K)

From the above expression, it is concluded that the TMC conductor will be "cold end" stable at the operating current of 6050 A. The heat flux corresponding to 6050 A exceeds the peak nucleate boiling heat flux. This is clearly due to the high critical temperature of Nb_3Sn which leads to a large area under the cooling curve even at 10 T.

CONDUCTOR DEVELOPMENT AND VERIFICATION TESTS

For the development of large superconducting coils using Nb_3Sn conductor, the mechanical characteristics of the conductor is one of the most important problems. In the case of high

Table II. Characteristics of the TMC Nb_3Sn conductor

	Grade I (10T)	Grade II (6.2T)
Superconductor	Nb_3Sn	Nb-Ti
Conductor Size (mm x mm)	12.6 x 13.0	12.6 x 13.0
Critical Current (kA)	10	10
Copper Stabilizer		
Cold Work Reduction (%)	21	21
Resistivity (Ω-cm)	5.7×10^{-8}	$4/2 \times 10^{-8}$
0.2% Yield Stress (MPa)	333	333
Strand Cable		
Cable Size (mm x mm)	7.9 x 4.2	7.9 x 4.2
Diameter of Strand (mm)	2.48	2.48
Number of Strands	6	6
Twist Pitch of Cable (mm)	75	75
Copper/Non Copper	0.65	2.1
Diameter of Filament (μm)	4	38
Number of Filaments	331 x 234 = 77454	1,350
Twist Pitch of Strand (mm)	35	25
0.2% Yield Stress (MPa)	235	235
Effective Heat Flux (W/cm^2)	1.25	0.86
Cooling Surface Structure	Roughened and oxidized copper	Roughened and oxidized copper

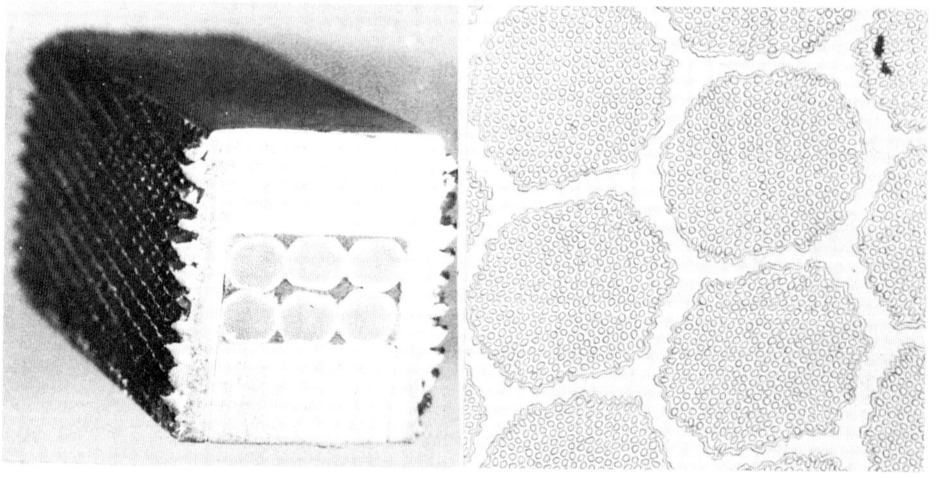

Fig. 2. TMC Nb_3Sn conductor.
(a) whole conductor   (b) a strand

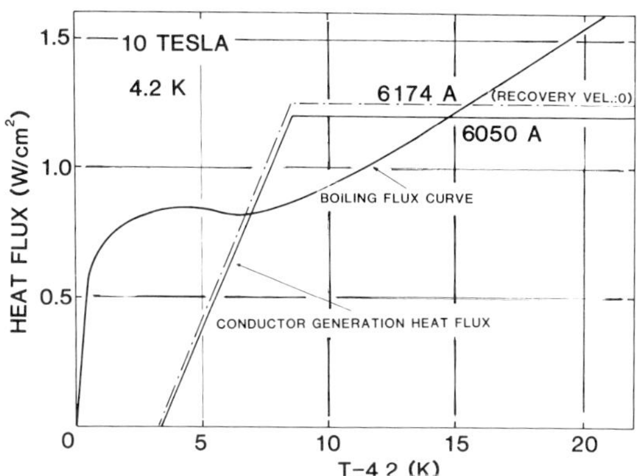

Fig. 3. Boiling heat flux curve and heat generation curve for the TMC Nb$_3$Sn conductor.

current conductors for fusion magnets, the major part of the conductor strain arises from bending.

At JAERI, it has been shown that increased tolerance to bending of the Nb$_3$Sn conductor can be obtained by heat treatment in a prebent shape. In this method, Nb$_3$Sn is reacted in a prebent

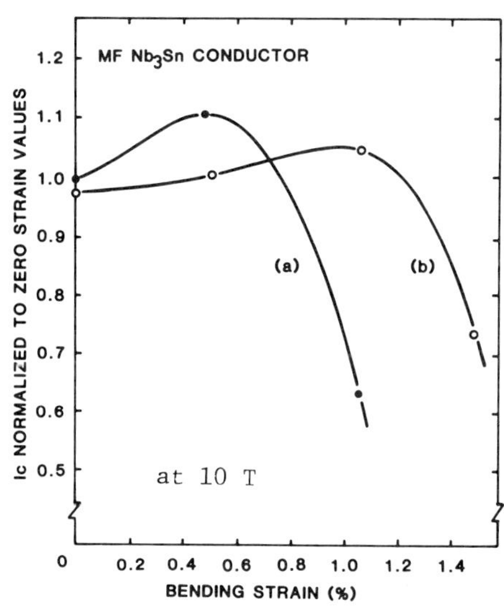

Fig. 4. Relative critical currents as a function of bending strain for the TMC strand, (a) heat treated in the straight condition (b) heat treated in a prebent (0.5% strain) condition.

shape which corresponds to 0.3 − 0.4% strain when compared to
straight conductor. After the reaction, the conductor is wound to
its final shape and the final bending strain is effectively reduced to a value which corresponds to the difference between the
prebent and final shape. Figure 4 shows the measured critical
current as a function of bending strain. In case (b), the heat
treated strand was straightened before bending to the final
shape. Strain values in this figure are calculated relative to
the straight condition. From these results it is indicated that
this new method allows bending of the conductor up to 1% strain
without a decrease in the critical current.

Figure 5 shows the stress-strain characteristics of the TMC
Nb_3Sn conductor at 4.2 K. As the maximum principal stress on the
conductor is estimated by a finite element stress analysis to be
132 MPa at 10 T operating strain should be 0.13%. The final
winding diameter is 60 cm, which corresponds to 0.7% strain.
However, the effective winding strain is reduced to 0.36% by using
a prebent diameter of 125 cm. Furthermore, about 0.1% strain will
be added due to the conductor bending from one pancake to the next
at the innermost layer. The strains are summarized in Table
III. Total effective strain on the TMC Nb_3Sn conductor is about
0.6%, although the apparent strain is about 1%.

Finally, I_c − B short sample tests on the full sized TMC
conductor were carried out. To simulate the coil winding process,
the TMC Nb_3Sn strand was heat treated in a prebent shape on a 125
cm diameter reaction spool. After the reaction, the strand was
bent from the straight condition to a 140 cm diameter three
times. Finally it was wound on a sample holder whose diameter was
54 cm. As shown in Fig. 6, the test result meets the requirements, 10 kA at 10 T, without degradation. In this case, the
definition of the critical current is based on a resistivity of 1
x 10^{-11} Ω−cm.

Table III. Summary of strains on the TMC Nb_3Sn conductor

Winding strain	Bending	0.46 %
	Tension	0.03 %
Cooldown strain		−0.01 %
Operating strain		0.13 %

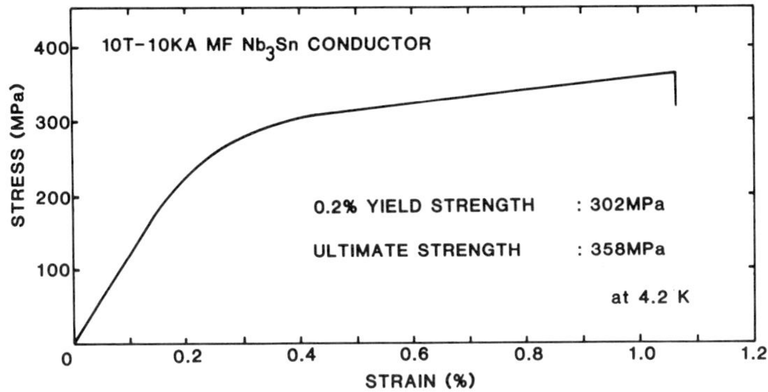

Fig. 5. Measured stress-strain characteristics of the TMC Nb₃Sn conductor at 4.2 K.

CONCLUSION

The verification tests reported here strongly indicate that the 10 T, 60 cm bore, Nb₃Sn TMC will operate reliably. A unique feature of the design that allows increased bending of the conductor up to ~ 1% strain is heat treatment in a prebent shape. It is furthermore interesting that the equal-area heat flux is higher than the maximum nucleate boiling limit. The maximum stress on the TMC conductor is calculated to be 132 MPa, which is less than 2/3 of the yield strength was measured for a full-sized conductor at 4.2 K. This TMC is the first large coil using Nb₃Sn in a high

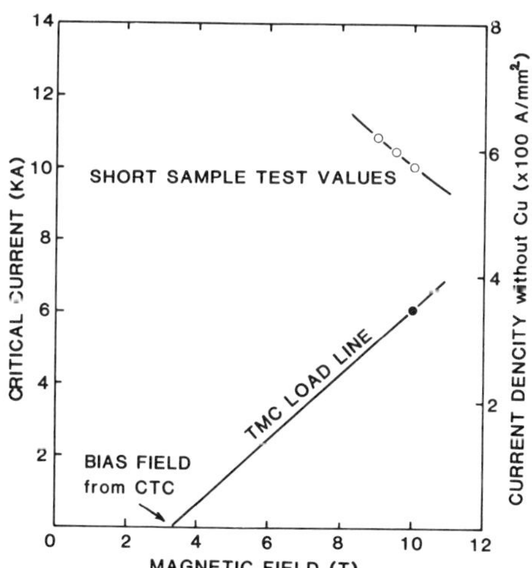

Fig. 6 Measured I_c-B characteristics of the TMC Nb₃Sn conductor.

current conductor. By the successful operation of the coil, one of the major goals in the development of high-field, large superconducting magnets will be achieved.

ACKNOWLEDGEMENTS

The authors would like to thank Drs. S. Mori, Y. Iso, and Y. Obata for their continuing encouragement. The manufacturing contribution by Hitachi Ltd. and Hitachi Cable Ltd. is gratefully acknowledged.

REFERENCES

1. T. Ando, et. al., Cluster test facility for superconducting Tokamak toroidal magnet system development, IEEE Trans on Magnetics, MAG-15:63 (1979).
2. S. Shimamoto, et. al., Construction and operation of the cluster test facility, IEEE Trans on Magnetics, MAG-17:494 (1981).
3. M. Nishi, et. al., Roughened surface study on Japanese test coil for the large coil task, IEEE Trans on Magnetics, MAG-17:904 (1981).

OPERATION OF THE MHD COMPONENT TEST FACILITY AND FIRST EXPERIMENTAL RESULTS*

Y. Iwasa,[†] J. F. Maguire, M. W. Sinclair, O. Tsukamoto,[‡]
D. J. Sliski, R. J. Thome, and J. B. Thompson

Francis Bitter National Magnet Laboratory[§]
Massachusetts Institute of Technology
Cambridge, Massachusetts

INTRODUCTION

This paper describes the first series of tests run in the MHD Component Test Facility at the Francis Bitter National Magnet Laboratory and the results obtained from those tests. Particularly interesting results include: (1) stability data on a cable conductor in a grooved subplate configuration showing non-uniform recovery across the cable cross section and (2) turn-to-turn normal zone propagation induced by streams of warm helium gas rushing through the cooling channels. The split-race-track test Facility Magnet (TFM) achieved a maximum gap field of 6 T at an operating current of 4100 A. At this field the normal state heat generation rate per unit cooled area corresponds to ~ 4 W $\cdot$ cm^{-2}, about an order of magnitude higher than the conventional, cryostable heat flux of ~ 0.3 W $\cdot$ cm^{-2}.

FACILITY DESCRIPTION AND CAPABILITIES

The facility provides the three components necessary for

---

*Work supported by the Fossil Energy Branch/MHD Division of the U.S. Department of Energy

[†]Also Department of Mechanical Engineering, MIT

[‡]Visiting Scientist, Yokohama National University, Tokiwadai, Yokohama, Japan

[§]Supported by the National Science Foundation.

testing large-scale superconducting magnet components: cryogenic support, high field, and high current. The cryogenic environment is provided by a CTI-1400 helium liquefier and various experimental Dewars, the largest of which as an o.d. of 1.2 m and a depth of 3.3 m. The high field is provided by four separate magnets: two water-cooled Bitter solenoids and two superconducting magnets. One superconducting magnet is a split-pair solenoid with a 150-mm bore diameter and 6 T field with a 30 mm gap, and the other is a large split-pair racetrack magnet (TFM) producing a field of 6 T in a 0.1 m gap of 0.97 m width and 1.8 m height. The operating characteristics are given in Table I. It is in the gap of this split-pair racetrack that the measurements described in this paper were taken. Finally, the high current for the facility is provided by three separate sources: a set of three 10 kA, 5 V DC power supplies which were used in this first series of tests, motor-generator pairs which provide up to 40 kA at 250 V, and a 25 kA, 2 V homopolar generator on loan from Argonne National Laboratory.

EXPERIMENTAL OBJECTIVES

The objectives for the first series of tests were twofold. The four-ton TFM magnet was to be charged up to its operating current and then, for given TFM field settings, an experiment which was placed in its gap was subjected to a test sequence. Since the entire facility was new, many system components were tested at successively higher field levels. These included a quench detection/dump resistor protection circuit and a computer data acquisition and storage system.

Table I. TFM Magnet Operating Characteristics

Operating Current	4100 A
Maximum Gap	100 mm
Peak Field at Maximum Gap	6.0 T
Current Density	1600 A $\cdot$ cm^{-2}
Normal State Heat Generation at 4100 A	4 W $\cdot$ cm^{-2}
Inductance W 100 mm Gap	1.3 H
Stored Energy	11 MJ

zones, could be generated. At a heat flux of 0.025 W $\cdot$ cm^{-2}, an unstable normal zone, about one-turn long (2.5 m), could be generated which would shrink once the heating pulse was removed (Fig. 2). Here and in the following data, all strand surface, regardless of strand location within the cable, was used to compute heat flux, even though all this surface was clearly not available for cooling. The traces of Fig. 2 show that the normal zone does not propagate to the second turn. At a heat flux of 0.03 W $\cdot$ cm^{-2}, partial recovery was observed following the heat-source removal. As seen from the voltage traces of Fig. 3, the heat pulse produced a normal zone that propagated only over a distance of about one turn from the heated section, as in the above case. Note that V_B taps show no resistive voltage. After a while, the normal zone shrank, but only partially; current had to be reduced for full recovery. This partial recovery was observed for initial current settings of 6500 A, 6750 A, and 7050 A, most likely because a non-uniform cooling distribution within the cable resulted from the inability of the helium, due to vapor blockage, to penetrate freely into the cable interior. At a heat flux of 0.034 W $\cdot$ cm^{-2}, the normal zone propagated, and were the current held constant, the normal zone would have spread to the entire module. In the case shown in Fig. 4, the current was reduced to keep the conductor in the inner turns from overheating. Normal-zone propagation was found to have two components. Initially, the normal zone propagates axially in the conductor at a velocity of $\sim$ 0.3 m $\cdot$s^{-1}, a computed from V_A and V_B traces of Fig. 4. This velocity seems to hold also between the 2nd turn (V_B) and the 3rd turn (V_C); but between the 3rd turn (V_C) and the 5th turn (V_E), the apparent velocity increases to about 2.5 m $\cdot$ s^{-1}. We believe that this is due to a radially outward propagation induced by streams of the warm helium gas rushing radially outward through the cooling channel.

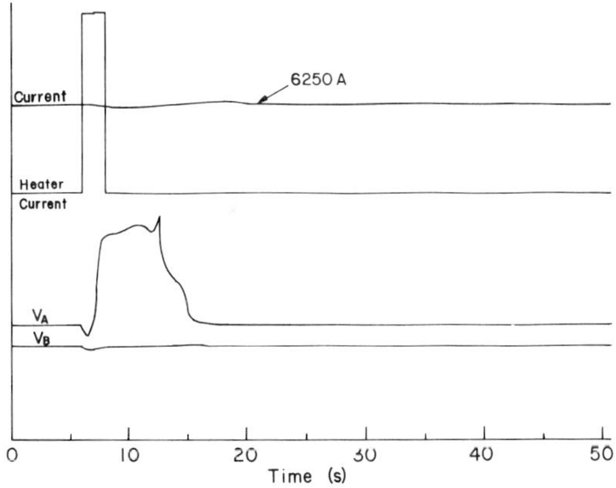

Fig. 2. Recovery traces for the CC module at 6250 A.

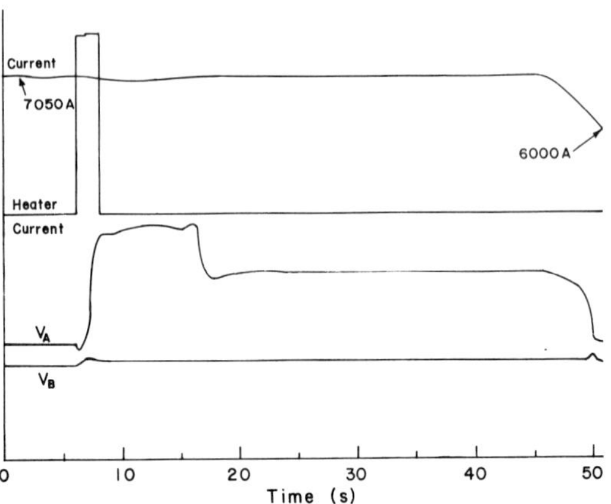

Fig. 3. Partial recovery traces for the CC module at 7050 A.

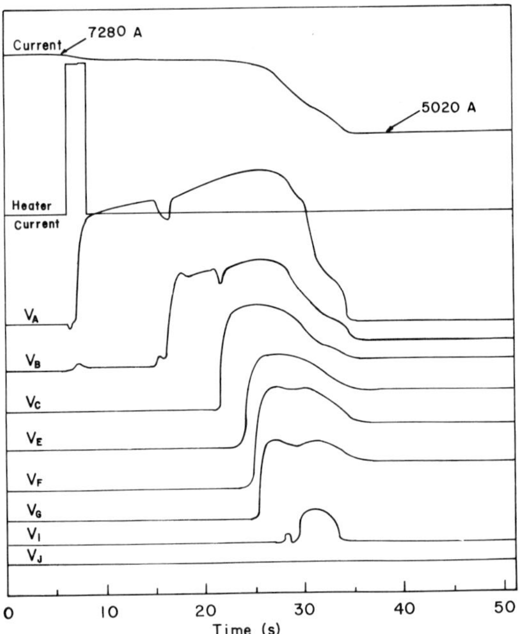

Fig. 4. Quench traces for the CC module at 7280 A.

Also measured on the cable conductor was the normal state recovery current. This is the current at which the super conductor goes from the normal state to superconducting and was measured as a function of field in the background coils. Values from this measurement were found to be 70% of critical current at field levels around 5.5 T.

Data taken on the CDIF monolithic conductor yielded similar results; persistent normal regions were again visible. At 5.5 T, propagation velocities were measured to be about $0.5 \text{ m} \cdot \text{s}^{-1}$ and the recovery current was ~ 6000 A. This value is lower than the~ 7000 A previously measured on a short sample of this conductor.[3] One possible explanation for the degraded performance may be poor cooling conditions in the present module due to the long (~ 1 m) cooling channels.

An acoustic emission (AE) sensor was placed on each coil of the TFM pair and AE signals were recorded each time TFM was energized. Many incidents of conductor slip were observed[4]; none of the slips, however, led to a TFM quench. Details of the AE observation will be presented elsewhere.

CONCLUDING REMARKS

The results presented here demonstrate the usefulness of large-scale, quality of design and manufacture-assurance tests. More specifically, the tests show that it is more difficult, as has been predicted, to achieve recovery in cables than in monoliths. Equally important, is the unverified indication that it may be more difficult to induce a non-recovering normal zone in cables than it is in monoliths. The detrimental effect of long cooling channels on recovery was also shown. This series of tests for the MHD program also has demonstrated the versatility of our facility for performing large-scale experiments. Such tests are essential to both MHD and fusion magnet system development.

REFERENCES

1. Y. Iwasa, et. al., IEEE Trans. on Magnetics MAG-17: 498 (1981).
2. R.L. Rhodenizer, The Superconducting CDIF Magnet, in "Proc. 1978 Superconducting MHD Magnet Design Conf.," A.M. Dawson, editor, (1979), p. 167.
3. J.F. Maguire, R.J. Thome, and Y. Iwasa, IEEE Trans. on Magnetics MAG-17: 474 (1981).
4. O. Tsukamoto, et. al., Identification of quench origins in a superconductor with acoustic emission and voltage measurements, Appl. Phys. Lett. 39:2 (1981).

A CASK MAGNET PROTOTYPE SYSTEM (CMPS)
FOR STANFORD UNIVERSITY

H. G. Arrendale, R. W. Baldi, J. I. Jurek, R. C. McCool, G. D. Magnuson
J. F. Parmer, R. A. Sutton, and W. D. Taylor

General Dynamics Convair Division
San Diego, California

and

E. Bobrov and J. E. C. Williams

*Francis Bitter National Magnet Laboratory**
Massachusetts Institute of Technology
Cambridge, Massachusetts

INTRODUCTION

A magnetohydrodynamics (MHD) magnet development program has been underway for approximately three years with General Dynamics Convair Division as prime contractor for the MIT/DOE project office. The objective of the program is to develop high field magnet design concepts by building test magnets to be used by Stanford University for MHD research. The first magnet design[1] used an aluminum shell with machined channels to locate and support the conductor bundles. In April 1980, Convair was directed to incorporate "Cask" structural concepts into the design.[2,3] The new design concept uses stainless-steel staves and corner blocks assembled much like a series of concentric barrels. The corner blocks divide the spaces between staves into rectangular sections that contain the conductor bundles. Trade studies have shown Cask to be an economical structure, especially when scaled-up to base-load MHD size.

---

*Work supported by the National Science Foundation.

Table I. CMPS general data and weight breakdown

General Data:

Magnet type	Dipole–MHD
Overall size	3.1 m (dia) × 4.4 m (high) × 4.8 m (long)
Stored energy	86.4 MJ
Ampere turns	12.7×10^6
Inductance	3.3 H
Charging time	50 min
Peak field	8.1 T
Central field	7.35 T
Uniformity requirements	+2.5 − 2.5%
Uniformity region	1.5 m
Discharge time constant	153 s
Operating temperature	4.5 K
Cooldown/Warm up time	480 h
Helium volume	1,500 L
Conductor cooling mode	Pool boiling

Weights:

Total	9.98×10^4 kg
Cold mass	7.85×10^4 kg
Conductor	2.40×10^4 kg
Winding structure	2.67×10^4 kg
Tension bands	N/A
N2 radiation shield	1130 kg
Vacuum vessel	1.72×10^4 kg
Magnetic shield	4.69×10^5 kg

Table I lists the design characteristics and weight breakdown of CMPS, while Fig. 1 illustrates the concept.

STRUCTURAL DESIGN

A unique feature of CMPS is its method of supporting Lorentz forces. The circular saddle winding is laid into sectors formed by a series of axial staves and corner blocks. This type of construction facilitates transmission of radial and circumferential Lorentz forces to the superstructure via the staves and corner blocks while the axial Lorentz force is reacted by end blocks attached to the staves. Figure 2 shows how the circumferential forces, tangent to the staves, kick out in the radial direction through the corner blocks. Radial loads within the conductor bundles are transmitted directly through to the superstructure because the staves are free to move in the radial direction.

A typical superstructure beam is shown in Fig. 3. The box construction is very efficient in that the double web makes the flanges fully effective and adds torsional stiffness. The maximum beam section is located at the point of maximum bending providing economy of material around the winding circumference.

Helium containment is accomplished by interposing the helium vessel wall between the superstructure and the winding substructure. This eliminates liquid helium inventory from the superstructure, thereby reducing the quantity of helium to a minimum, with an added benefit of reduced boiloff pressures during a loss of vacuum event.

Cold mass support is accomplished by seven fiberglass-epoxy tension and compression members arranged as shown in Fig. 4. The planned magnet installation near the San Andreas fault requires design for major earthquakes. A typical strut is shown in Fig. 5. The results of the dynamic analysis show the structure to be relatively stiff by virtue of its structural constraints. Table II lists the natural frequencies of the magnet. The largest acceleration load factor is 1.38 in the axial direction.

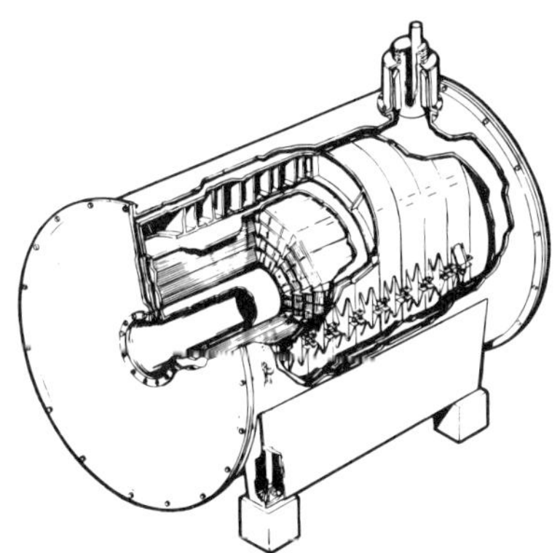

Fig. 1. The Cask Magnet Prototype System is a stave and corner-block-type of structure.

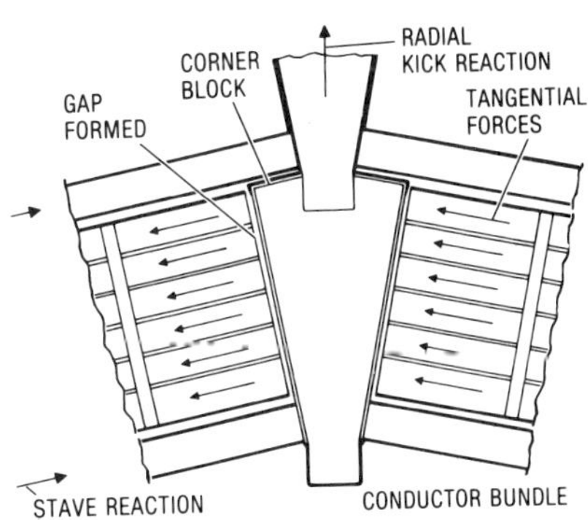

Fig. 2. Lorentz loads are reacted by the corner blocks and kicked radially to the superstructure.

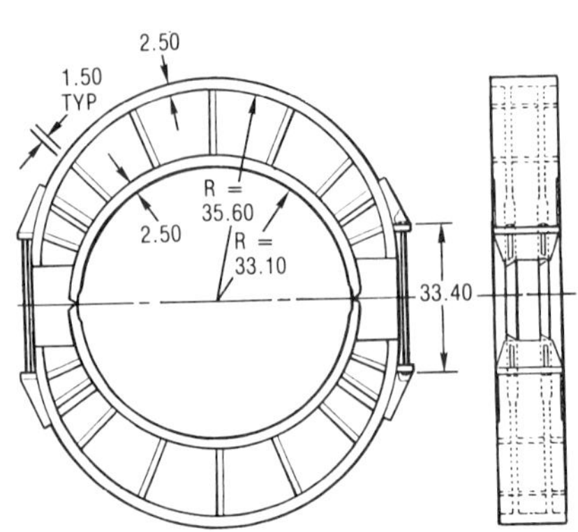

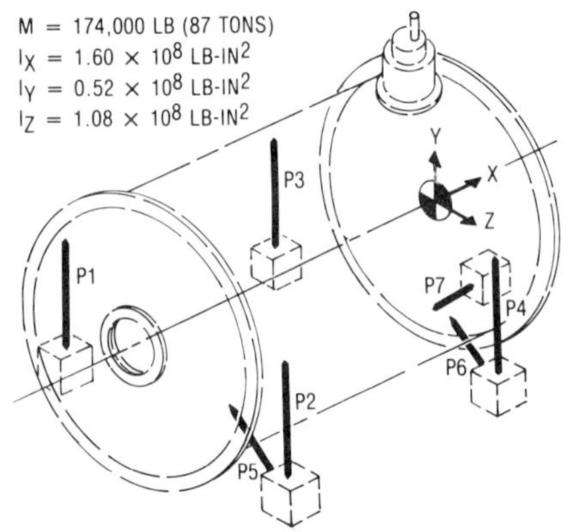

Fig. 3. Welded box beams are designed to have maximum section depth at the location of maximum bending moment.

Fig. 4. The cold mass support is designed to constrain the cold mass during major earthquake events.

CONDUCTOR DESIGN

The design goal for the CMPS conductor was to achieve cold-end stability as defined by Maddox and James at 4.5 K. To help realize this goal, the turn-to-turn insulation was designed so that all the insulation carried bearing loads with no carrier strips or excess landings blocking helium flow passages. This sort of a design, shown in Fig. 6, maximizes the amount of helium volume available to the conductor and the conductor's wetted perimeter, while minimizing the amount of excess insulation material in windings. As shown by the sample conductor in Fig. 6, the pads of the insulation mate with the protrusion of the superconducting insert, restricting vertical motion of the pads. Lateral pad motion is restricted by thin carrier strips alternately placed

Table II. Dynamic analysis determined the reasonant frequencies for CMPS

Mode Number	Resonant frequency	Vibration mode description
1	12.8 Hz	Axial
2	14.1 Hz	Lateral
3	25.3 Hz	Vertical
4	41.5 Hz	Roll rotation
5	56.7 Hz	Pitch rotation
6	74.2 Hz	Yaw rotation

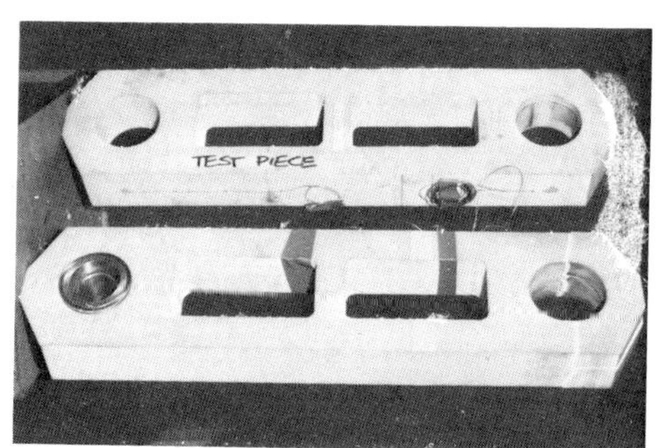

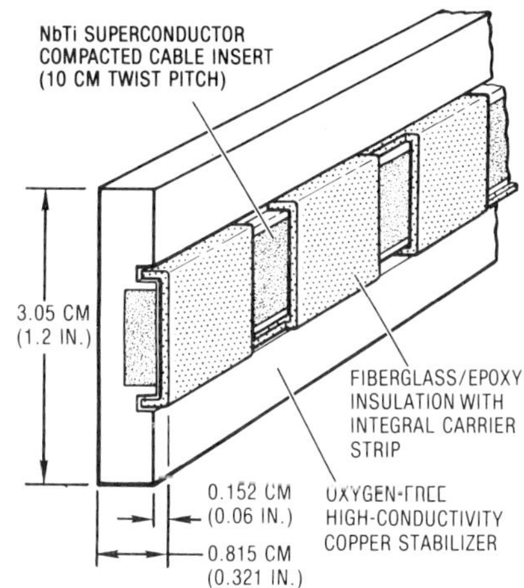

Fig. 5. Support struts are fiber-glass epoxy composites.

Fig. 6. A raised step on the conductor provides for 1.5 mm flow channels and turn to turn insulation.

on the top and bottom of the protrusion land, which also serve to make the insulation strip flexible during winding. In addition to providing the necessary mating element for the insulation pads, the superconductor protrusion creates large (approximately 1.5 mm gaps) unrestricted flow passages across the large vertical faces of the conductor.

Three grades of conductor proposed for the CMPS magnet, each having a design heat flux of 0.255 W/cm^2 and the same general shape of the grade I conductor in Fig. 6. Only the thickness changes from grade to grade. Pertinent design parameters of each of three grades are given in Table III.

To prove that a heat flux of 0.255 W/cm^2 would ensure steady-state cryostability, a detailed heat transfer test was conducted on a representative winding tier of the magnet.[5] Results of this test indicated that the conductor needed to be coated with a cupric oxide coating, made by the Enthone Corporation, and known by its trade name, Ebanol C, to sufficiently enhance the boiling characteristics. Figure 7 gives the boiling curve for the Ebanol C grade 1 conductor and shows the stability margin.

Table III. CMPS graded conductor characteristics.

Conductor Data:

Grade		I	II	III
Conductor type	–	Compacted cable		
Operating current	kA	7.235	7.235	7.235
S/C current density	kA/cm^2	33.910	44.919	60.758
Pack current density	kA/cm^2	2.761	2.953	3.146
Critical current	kA	9.798	9.660	9.044
Length	km	2.7	3.88	4.76
Superconducting matl	–	Nb Ti	Nb Ti	Nb Ti
Cu: non–Cu ratio	–	8.85	11.2	14.2
Cross sectional area	cm^2	2.1	1.965	1.81
Cond packing factor	–	0.802	0.802	0.787
He vol to cond vol	–	0.195	0.199	0.211
Stabilizer heat flux	W/cm^2	0.255	0.255	0.255
Stabilizer material	–	Copper	Copper	Copper
R.R.R. stabilizer	–	150	150	150
Solder material	–	TBD	TBD	TBD

REFRIGERATION

The refrigeration system chosen for the CMPS magnet system was the CTI 2800, capable of delivering 200 W or 100 L/h of refrigeration capacity. Heat loads to the liquid helium are tabulated in Table IV. Helium heat loads are limited by the use of liquid-nitrogen-cooled intercepts placed between the vacuum vessel and the helium vessel. Liquid nitrogen loads are given in Table V along with the location of the intercepts.

Magnet cool down could be accomplished in about 14 days, utilizing the full capacity of the refrigeration system. Helium is uniformly distributed across the bottom of the magnet windings by a distribution manifold, branches of which enter the helium vessel between superstructure flanges. Cooling of the superstructure is accomplished by conduction from the helium vessel. Good thermal contact between the helium vessel and the superstructure is assured by placing a thermal grease or metal filled epoxy between the two surfaces. Cool down gradients for the major components of the magnet system are shown in Fig. 8.

Table IV. Helium heat loads have been minimized.

Heat Source	Heat Load (W)
Conduction through helium vessel supports	5.0
Radiation from LN_2 shields through 30 layers of MLI	1.2
Joule heating in conductor splices	1.4
Conduction through MLI surrounding warm bore	5.3
Transfer line losses	7.2
Stack region	
Bayonets	
Supply	1.0
Return	1.0
Bulk transfer	0.5
Electrical instrumentation wires	1.5
Conduction and radiation from helium vessel neck, emergency vent line, and LN_2 reservoir	4.6
Total	28.7

Vapor cooled current leads at 7,400 amps — 25.5 L/hr

Table V. LN_2 is used to reduce helium boil off by intercepting room environment heat transfer.

Heat Source	LN_2 Load (W)
Radiation from vacuum vessel through 30 layers of MLI	126
Conduction through radiation shield supports	16
Bayonets	
Steady-state supply	1
Cooldown/warmup supply	1
Return	2
Reservoir level snesor	1
Thermal intercepts	
Helium vessel supports	42
Helium vessel neck	26
Total	215

LN_2 supply rate — 4.8 L/hr

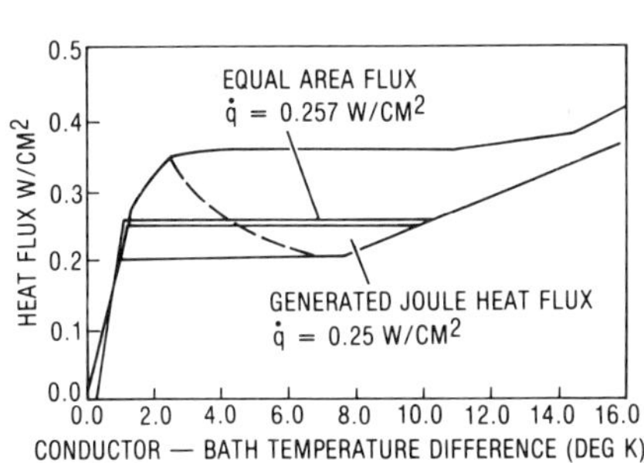

Fig. 7. The baseline conductor is cold end stable at 4.5 K.

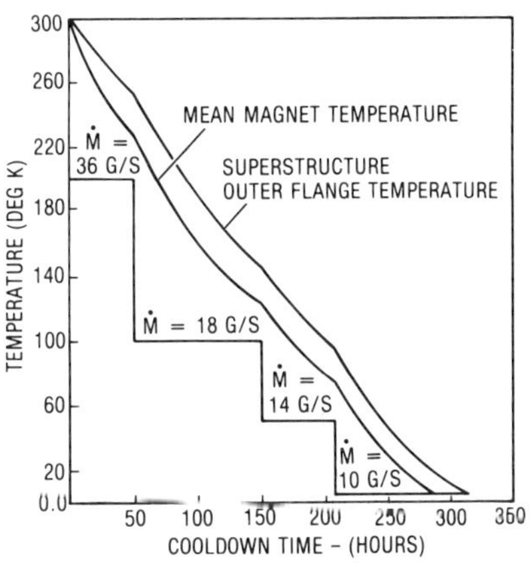

Fig. 8. Cooldown is accomplished in reasonable time with minimum gradient in the superstructure.

SAFETY, POWER, AND PROTECTION

The charge time for the CMPS magnet is 50 minutes to 7235 A. A 0.00098-ohm stainless steel resistor provides for the normal discharge of stored energy. Should a propagating normal zone or other emergency occur, a 0.022-ohm dump resistor (also stainless) will remove the stored energy with a time constant of 153 seconds. This time constant and the copper-to-superconductor ratios (Fig. 9) have been selected to limit the maximum conductor hot spot to 250 K. Figure 10 shows the temperature-time profile for a propagating normal zone.

Five quench detectors have been coupled to the coil through seven voltage taps. Any sector of the coil is observed by at least two detectors. The quench detectors differentiate between transient signals and normal zones by an integration process. A decision to dump is made when the energy lost within the magnet exceeds 27 J after a power of 32.5 W and splice joint losses are subtracted.

Figure 11 schematically represents the entire power and protection circuit. The 0.022-ohm dump resistor is center-tapped, limiting terminal-to-ground voltage to 78 V during an emergency dump. The bias current pickup provides assurance that the dump resistor is active. Additionally, circuit breaker two (CB2) will open, should a voltage breakdown to ground occur. The 12 kA main circuit breaker is driven directly by the quench detectors or protection controller.

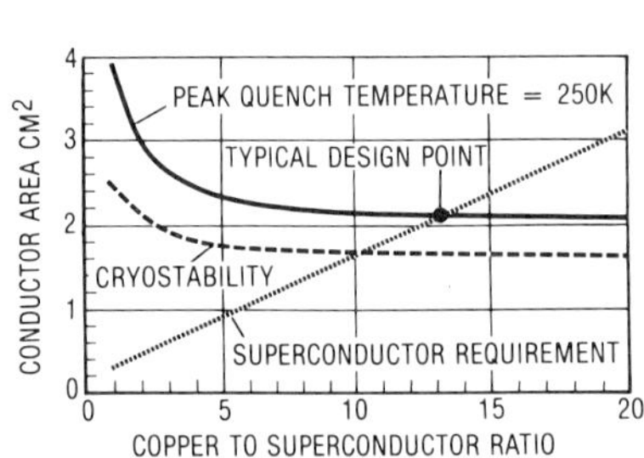

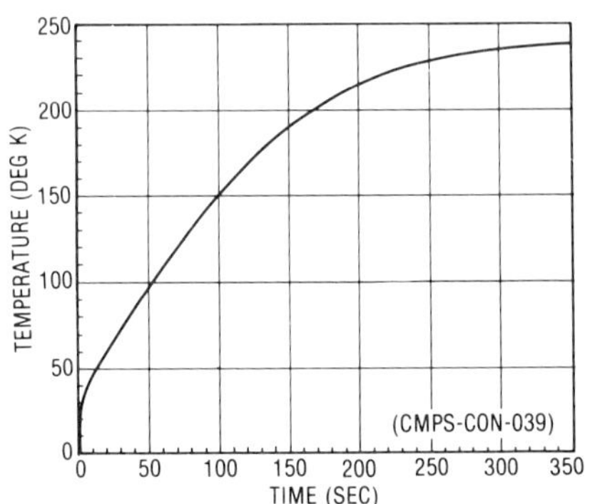

Fig. 9. The maximum allowable conductor hot spot temperature determines the CU:SC ratio and therefore the discharge time constant.

Fig. 10. Hot spot temperature reaches 240 K at the end of an adiobatic propagating quench.

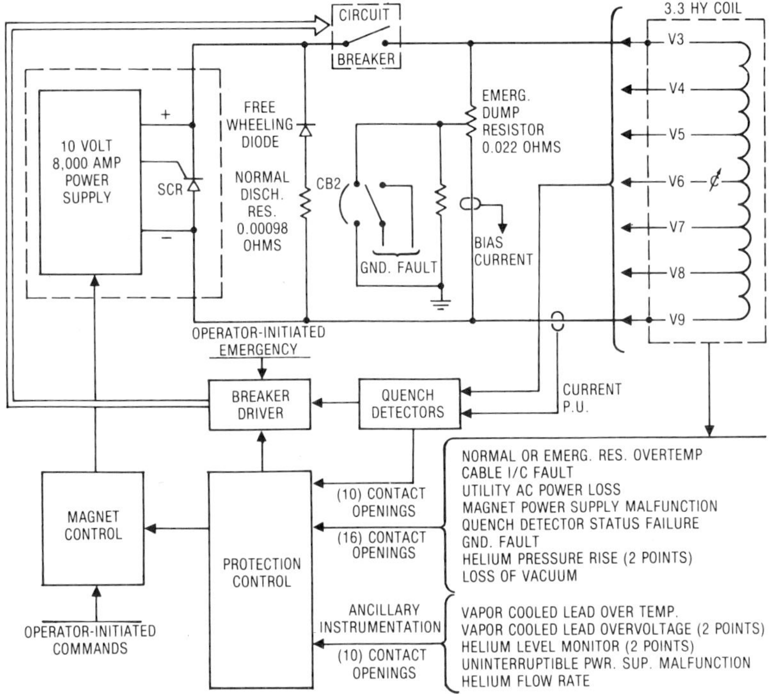

Fig. 11. CMPS power and protection circuit schematic.

The protection controller provides an automatic monitoring of 16 contact openings from switches and 10 contacts from analog instrumentation with limit sets. The controller continuously monitors the system through these contacts and, should an alarm or emergency condition develop, makes a decision to display alarm, normal discharge, or energy dump. A study and survey were performed resulting in the selection of two industrial process computers operating in a redundant mode. Criteria for selection were simple, easily learned software programming and the ability to remove the terminal device for security of the protection program.

The entire electronic subsystem, including an uninterruptible power supply, is contained in three standard rack cabinets. With the exception of the quench detectors, which are of General Dynamics design, the instrumentation is adapted from industrial process control instruments. An example of this can be seen in the lead voltage monitors where two digital panel meters with adjustable trip points are used. The first meter is used to provide a warning alarm and display, and the second will provide appropriate action. Both are mounted on a rack panel with a chassis in the rear for shielding. This approach provides a redundant fail-safe design with a minimum of detailed circuit design.

The power supply selected is a 10V 8000 A unit with SCR protection on the output. An input power of 208 VAC 3 phase was selected. Output voltage and current stability requirements are ± 0.25% over 2 hours after a 1 hour warm-up.

The circuit breaker used to initiate emergency dump is a 12 kA, 1 kV unit commonly used in industrial plants. A cabinet for mounting is provided, along with bussing and cycle counters to monitor contact life.

A detailed hazard analysis was performed to estimate possible failure modes and hazard to the magnet and personnel. All hazards anticipated may be controlled by either the automatic protection system or interlocks within the system. Should additional inter-locks from the facility be desired, the protection controller inputs are easily expanded.

REFERENCES

1. J.F. Parmer, et al., A superconducting magnet for Stanford University IEEE Trans on Magnetics, MAG-17 (1):344 (1981).
2. S.L. Ackerman, et al., Practical aspects of designing and manufacturing MHD superconducting base-load magnets in 1988 time frame, IEEE Trans on Magnetics, MAG-15 (1):310 (1979).
3. R.W. Baldi, et al., Cook commercial demo plant MHD superconducting magnet system, General Dynamics Report No. CASK-GDC-031 (December 1979).
4. B.J. Maddock, G.B. James and W.T. Norris, Superconducting composites: heat transfer and steady-state stabilization, Cryogenics, 9: (1969).
5. M. Sinclair, W.D. Taylor and J.E.C. Williams, Pool Boiling LHe Heat Transfer in an MHD Conductor Pack, in "Advances in Cryogenic Engineering, Vol. 27," Plenum Press, New York (1982).

THEORETICAL AND ENGINEERING ASPECTS OF
MOMENTLESS STRUCTURES AND COIL END TURNS
IN SUPERCONDUCTING MHD MAGNETS

E. S. Bobrov and P. G. Marston

Francis Bitter National Magnet Laboratory
Massachusetts Institute of Technology
Cambridge, Massachusetts

and

E. N. Kuznetsov

Battelle Columbus Laboratories
Columbus, Ohio

INTRODUCTION

Reliable containment of mechanical, thermal, and electromagnetic loads at minimum weight, volume and cost is the major requirement of superstructures in magnets for magnetohydrodynamic (MHD) power generation.

The most favorable stress state of the superstructure is uniform tension along the length; this is called the isotensoidal stress state. It provides maximal structural material use. An isotension structure is a theoretical ideal which cannot be surpassed by any other structural system.

Support of electromagnetic forces is not the only requirement of the isotension superstructure. Its configuration is also dictated by the configuration of the winding-substructure outer boundary, which together with the winding inner boundary defines a

---

*Work supported by the National Science Foundation.

winding volume of produce the required field intensity and homogeneity in the magnet bore. The resulting stress state of the
winding must be within certain limits depending on the mechanical
properties of the substructure and conductor.

Determining the configuration of a momentless superstructure
with bending stiffness leads to a set of integrodifferential
equations , but there is also another approach in which the momentless effect is achieved by making the superstructure of flexible structural elements, such as high-strength wires or bands.
In this concept the constraints imposed on the superstructure
configuration are less rigorous and are defined only by the profile of the winding envelope.

In this paper we analyze the possibility of supporting the
electromagnetic forces acting on the winding by means of isotension superstructures resting directly on the winding-substructure
composite. We also consider the problem momentless configurations
of saddle coil and turns.

THEORETICAL DEVELOPMENT

Description of Method

A structural element is momentless if its bending stress is
zero or negligible compared to the axial stress. If the axial
tension stress does not vary along the element length it is also
isotensoidal. A structural element can be momentless for two
different reasons. First, it can be flexible, i.e., incapable of
sustaining any appreciable bending moment. Such an element generally changes its configuration depending on the external load and
acquires the form of a funicular curve corresponding to the given
load. Second, a structural element possessing a certain flexural
rigidity can be momentless under special loads for which the
element axis is the funicular curve.

In order to be momentless a structure with flexural rigidity
should have the shape of a funicular curve corresponding to the
system of magnetic forces. This is true for the case of force
loading. It is also possible to force the system to acquire a
given displacement pattern, which can be called the displacement
loading. In reality, the loading conditions for a magnet superstructure are somewhere in between force and displacement loading.

Consider a typical cross section of the winding with the
enveloping superstructure for a square bore (Fig. 1a). Forces
acting at the winding/superstructure interface (Fig. 1b) generally
have both normal and tangential components. If the superstructure

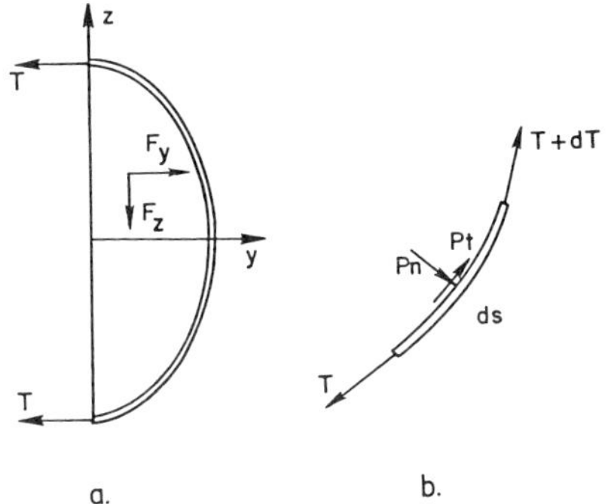

Fig. 1. Simplified scheme of substructure-superstructure interface.

is momentless, the equilibrium conditions for its infinitesimally small element of length ds in the tangential and normal directions are, respectively,

$$dT = p_t ds \qquad (1)$$

$$T = p_n \, r(s) \qquad (2)$$

where T is the axial force, r is the radius of curvature, p_t and p_n are, respectively, the tangential and normal components of the distributed load acting on the superstructure.

It follows from Eq. 1 that for the tensile force T to be constant along the superstructure length, the tangential component p_t must be zero. In the absence of a tangential load, the superstructure will be in isotension but not necessarily momentless. Only isotensile superstructures are considered here, and the effort is directed towards reducing the bending stress.

If the winding material is infinitely rigid and the superstructure inextensible no displacement will occur, the superstructure will be momentless and in isotension, and the contact pressure, according to Eq. 2 would be

$$p_n = \frac{T}{r} \qquad (3)$$

Now assume that the superstructure is made of a real material. Because of its elongation, some longitudinal (tangential) displacements will occur with superstructure elements sliding along the rigid boundary. The resulting change in element curvature is[2]

$$\Delta\kappa \, \frac{d}{ds}(u\kappa) = \frac{du}{ds}\kappa + \frac{dk}{ds}u \tag{4}$$

where $u = u(s)$ is the tangential displacement as a function of the axial coordinate s, $\kappa(s)$ is the initial curvature of the superstructure, and $\Delta\kappa$ is the change in curvature at a given point due to displacement u. This change in curvature gives rise to a bending moment

$$M = EI\Delta\kappa \tag{5}$$

where E and I are the modulus of elasticity and the cross-sectional moment of inertia, respectively.

Finally, let the winding material be real, and its boundary line displacement normal to the contour (positive if directed outwards) known and given as $v = v(s)$. The relative elongation of the superstructure can be evaluated as

$$\varepsilon = \frac{1}{S}\int_0^S \frac{v}{r}\,ds \tag{6}$$

where S is the superstructure length.

The elongation does not vary along the length of an isotensile element and can be expressed in terms of tangential and normal displacements of the superstructure as

$$\varepsilon = \frac{du}{ds} + \frac{v}{r} \tag{7}$$

The tangential displacement u can be evaluated from Eq. 7.

$$u(s) = \int_0^S \left(\varepsilon - \frac{v}{r}\right)ds \tag{8}$$

The change in curvature expressed in terms of the displacement u and v is given by

$$\Delta\kappa = \frac{d}{ds}\left(u\kappa - \frac{dv}{ds}\right) \quad \frac{du}{ds}\kappa + \frac{dk}{ds}u - \frac{d^2v}{ds^2} \tag{9}$$

and the bending moment can be determined by means of Eq. 5.

It follows from the above that the bending moment in the superstructure depends solely on boundary line displacements and its evaluation is straightforward as soon as the latter are determined.

Problem Formulation and Solution Technique

Analysis and development of an isotensile superstructure consists of two stages:

1. Development of techniques for a detailed and sufficiently accurate analysis of the winding/superstructure assembly, given all its relevant geometric parameters and material properties.

2. Identification and manipulation of parameters governing the bending moments in the superstructure so as to diminish the latter while not sacrificing other important characteristics (such as the intensity and uniformity of the induced magnetic field, etc.).

We found the finite element technique to be the most appropriate approach to the stress analysis of the multi-component structure. The structural model to be analyzed (Fig. 2 and 3) involves the winding, the bore, and the superstructure.

The winding-substructure composite is treated as an equivalent homogeneous orthotropic material, whose properties can be determined on the basis of a geometric approach.

The process of superstructure optimization consists of the gradual reduction of the maximum bending moment. This is done in a succession of trials or iterations by changing the values of the governing parameters, mainly those determining the superstructure profile. Each iteration requires solution of the entire problem involving bore/winding-substructure/superstructure interaction.

For each iteration the superstructure profile is chosen from a generic class of generalized ellipses whose shape is given by the equation

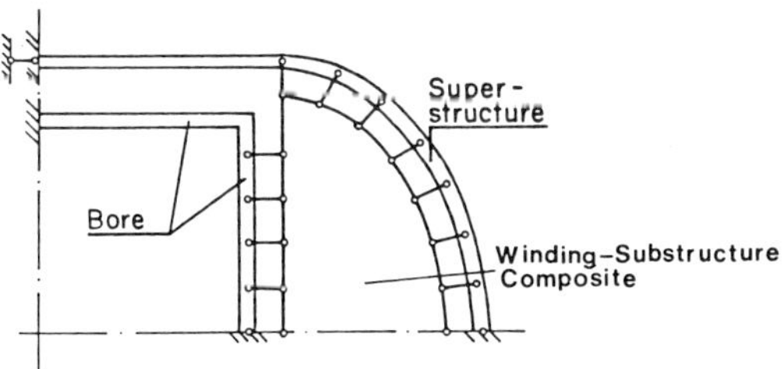

Fig. 2. Mechanical model of a magnet structure with rectangular bore.

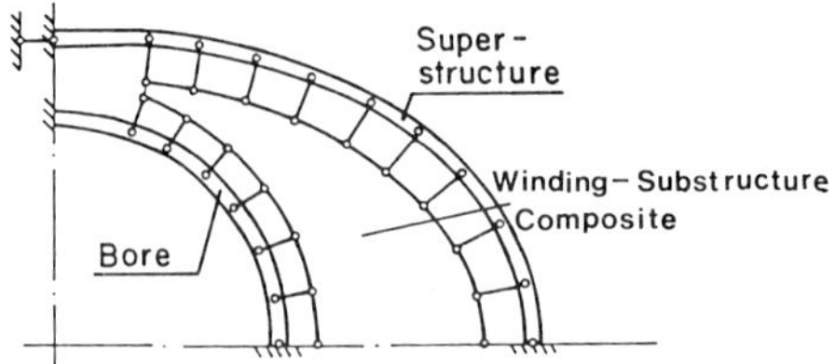

Fig. 3. Mechanical model of a magnet structure with circular bore.

$$\frac{y^n}{Y^n} + \frac{z^n}{Z^n} = 1 \tag{10}$$

Given its principal semiaxes, the generalized ellipse is determined by one parameter – the exponent n.

The approach developed permitted performance of a parametric study revealing the basic regularities of the problem in hand.

Momentless Winding End-Turn Configuration

The end-turn segments of the winding are subjected to a complex system of magnetic forces and interactions with the bore. The magnetic forces depend, in particular, on the end-turn configuration. Conceivably, the end turn can be shaped in such a way that its centroid becomes the funicular curve for the corresponding system of magnetic forces. In this case, the winding as a whole, considered as a solid bar, would be in pure tension. It could be self-supporting in the longitudinal direction while interacting with the bore in the normal direction. We considered a circular saddle end-turn configuration (Fig. 4), however the discussion is also valid for a rectangular saddle with a smooth transition between the straight and the saddle section of a coil.

To investigate the possibility of a momentless end-turn configuration, the end-turn section of the winding is considered as a spatial bar of a given cross section. Such a model can also represent a single winding element. The subsequent steps implement an approach similar to that used in the isotension superstructure analysis: reducing the maximum bending and torque moments in the bar by changing the geometry of its centroid.

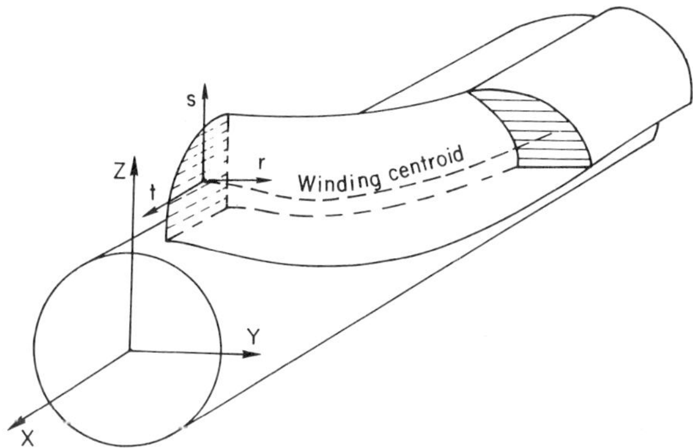

Fig. 4. Winding configuration in the end-turn region.

Conclusions on the Theoretical Development

Because of its flexibility and high tension, the momentless superstructure follows the deformation of the winding, thus creating a condition of displacement loading, rather than force loading, in the superstructure. Under displacement loading, the bending moments in the superstructure are determined by the changes in the curvature of the winding-superstructure interface.

The superstructure profile has only an implicit effect on its bending moments. Namely, the profile determines the initial outline of the winding cross section thereby affecting the deformed configuration of the latter. The single most important geometric parameter is the aspect ratio (build-up to width) of the winding.

The transverse and shear elastic moduli of the winding are crucial factors for the bending moments in the superstructure. The stiffer the winding, the smaller are the bending moments. The properties of the superstructure are also important: other things being equal, the bending moments are proportional to the superstructure flexural rigidity.

The approach developed for the design of a momentless superstructure under the action of the winding-superstructure interface pressure is fully applicable to the development of momentless or quasi-momentless conductor end-turn configurations under electromagnetic body forces, for both rectangular and circular saddle windings.

IMPLEMENTAION OF MOMENTLESS STRUCTURAL CONCEPTS IN THE CDP MHD MAGNET CONCEPTUAL DESIGN

Based on the theoretical development discussed above a conceptual design of the commercial demonstration plant (CDP) superconducting MHD magnet has been developed at FBNML. This magnet will generate the required field profile along an active length of 14.5 m, with a 6 T peak field within a square warm bore with dimensions of the inlet and outlet of 2.2 x 2.2 m and 4.0 x 4.0 m respectively. The design employs a flexible momentless superstructure made of stainless steel (304 LN) bands. The magnet is wound of internally-cooled, cabled superconductor (ICCS). The ICCS operating current is 20 kA. Each coil consists of 830 conductor turns, so that the total number of ampere-turns in this design is 33.2 x 10^6.

An isometric view of the CDP magnet conceptual design is shown in Fig. 5. The magnet assembly includes a winding-substructure

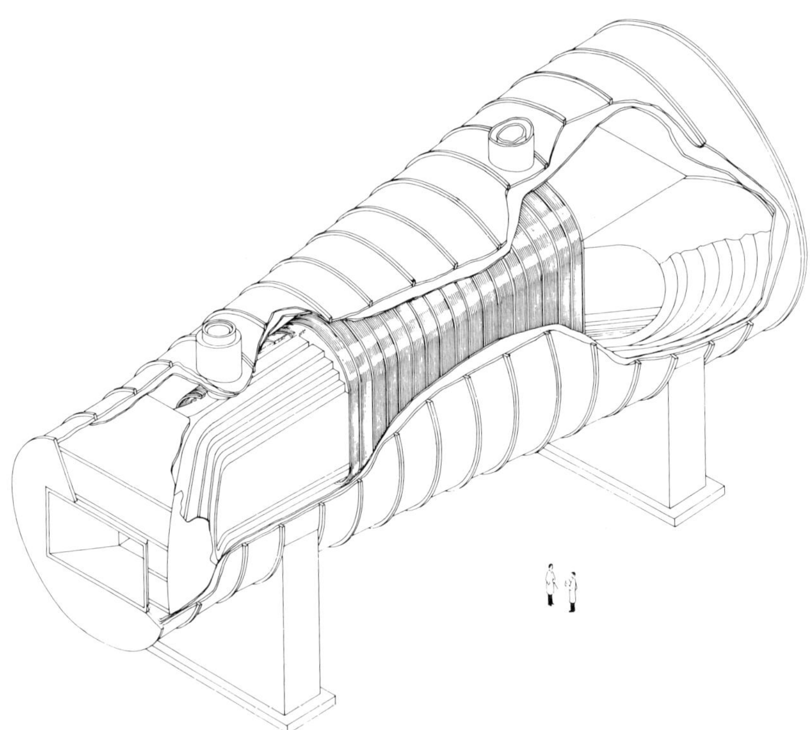

Fig. 5. General view of the CDP magnet with momentless superstructure and end turns.

subassembly, superstructure (tension bands in the straight coil-
sections and plate weldments for end-turn containment), two thermal shields, vacuum vessel, and supports.

The winding has two saddle-type coils each of which is assembled of four rectangular-saddle subcoils sized to allow a close
approximation of the curvilinear outer boundary of the winding-
substructure envelope. Cast aluminum blocks fill the space between the subcoils and the tension-band superstructure and transmit the transverse loads acting on the windings.

The cross-section of the substructure-superstructure assembly
at the inlet end is shown in Fig. 6.

The subcoils are assembled of double pancakes wound with
ICCS. G-10 pultrusions positioned between conductor turns provide
electrical insulation and compressive structural support between
the turns in both directions.

The axial force containment concept proposed in this design is
based on the approach discussed earlier in this paper. A system
of stainless-steel plates and gussets supports the components of
electromagnetic forces in the saddle region, normal to the plane
of the saddle while the tangential components of these forces are
resisted by conductor jackets and transmitted to the other end of
the coil. The end turns are shaped so that the bending stress

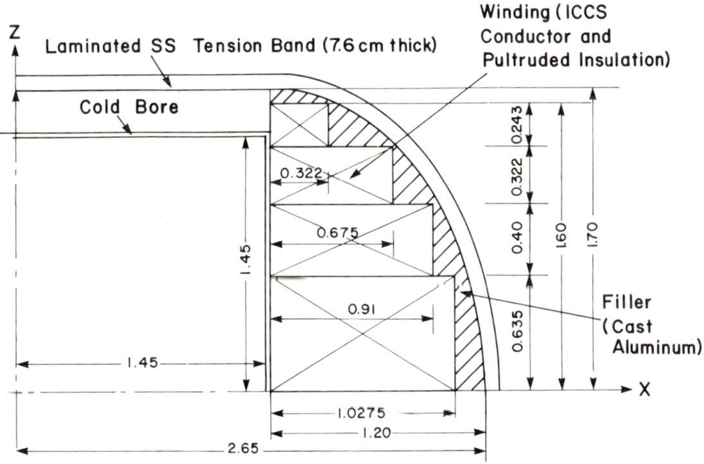

Fig. 6. Winding cross-section at the inlet end.
Dimensions are in meters.

which develops in the ICCS jacket is small compared with stress due to the tangential force.

REFERENCES

1. E.S. Bobrov, Force Containment Structure Design Considerations, in "Proc. of the 1978 Superconducting MHD Magnet Design Conf.", MIT, Cambridge, Mass., (1979).

2. A.E.H. Love, "A Treatise on the Mathematical Theory of Elasticity", Dover Publications, New York (1944).

DEVELOPMENT OF THE STATOR FOR A 20-MVA SUPERCONDUCTING GENERATOR

T. E. Laskaris, T. A. Keim, K. F. Schoch, and P. A. Rios

General Electric Company
Schenectady, New York

INTRODUCTION

The work described herein follows a detailed component development program[1] in which the critical components necessary to achieve superconducting generators for central power plant applications were addressed.

The initial effort has been focused primarily on the completion of the rotor development and test. The rotor components were fabricated and tested. Testing of the rotor has been conducted at design speed with the field winding at rated current in and environmental eddy current shield. This test has demonstrated the capability of the selected concepts for the rotor to perform within the values predicted by analysis.

Concurrent with the rotor development, an air-gap stator winding concept was selected that features a two-layer winding with two stages of transposition. Detailed design and key component development of the stator was followed by stator construction, presently under way.

Electromagnetic design of the generator determines the gross dimensions of the machine so as to achieve the desired rating and reactances with minimum size and weight, and maximum efficiency.

The overall design of the 20-MVA superconducting generator is summarized in Table I and Table II. Figure 1 is the layout of the 20-MVA generator based on the concepts that are applicable to large superconducting generators.[2] The concept and testing of the rotor is described elsewhere.[3,4]

Table I.   Principal Dimensions

Rotor Winding		
Inside Radius	, cm	5.08
Outside Radius	, cm	12.7
Active Length	, cm	132
Pole Angle	, degree	60
Electromagnetic Shield		
Outside Radius	, cm	21.8
Thickness	, cm	5.08
Resistivity	, $\mu\,\Omega\,$cm	4.35
Stator Winding		
Inside Radius	, cm	24.1
Outside Radius	, cm	35.6
Turns/Phase		14
Stator Yoke		
Inside Radius	, cm	37.8
Thickness	, cm	23.1

Table II.   Computed Parameters

Inductance		
Field Winding	, H	5.2
Armature/Phase	, H	$4.1\ 10^{-4}$
EM Shield	, H	$1.0\ 10^{-6}$
Mutual Inductance		
Field/Armature	, H	$1.9\ 10^{-2}$
Shield/Armature	, H	$1.5\ 10^{-5}$
Field/Shield	, H	$1.2\ 10^{-3}$
Resistance		
Armature/Phase	, Ω	$5.5\ 10^{-3}$
EM Shield	, Ω	$6.0\ 10^{-6}$
Reactance		
Synchronous	, p.u.	0.389
Transient	, p.u.	0.324
Subtransient	, p.u.	0.164
Rating at 0.9 PF	, MVA	21.9
Rated Voltage	, V	3600
Field Current	, A	500

STATOR CONCEPT

The stator winding contains no iron teeth and is a self contained structure fabricated from fiber-reinforced composites

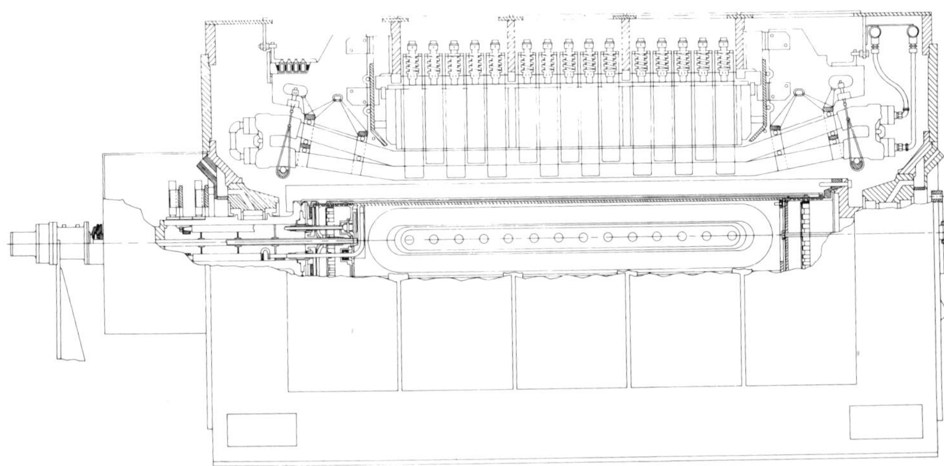

Fig. 1.   20 MVA Generator Layout.

into which the stator bars are wedged. The stator winding is
surrounded by an iron yoke to shield the environment from alternating fields. The iron yoke also enhances the flux density in
the stator winding region.

The armature winding is a water-cooled, two-layer, single-
circuit winding with 42 slots. The winding pitch is 18 slots.
Since the bars are exposed to a rotating magnetic field of relatively high intensity, it is necessary to utilize stator bars
comprised of small insulated and transposed strands to reduce the
eddy current and circulating current losses to acceptable levels.

The armature winding support system employs spring-loaded
wedges to maintain mechanical security despite the mechanical
instability of the materials employed. Plastic teeth are keyed to
the laminated iron stator core. There are half as many teeth as
there are bars in a layer (21 in this case). One bar from each
layer is placed adjacent to each side of each tooth, as shown in
Fig. 2. A fixed wedge is placed adjacent to each bar, and movable
wedges are placed between each pair of fixed wedges. As shown in
Fig. 1, the movable wedges alternate between top and bottom bars
along the length of the machine.

Each movable wedge is drawn securely into place by an epoxy-
impregnated glass tie that extends radially to the back of the
core. The tie passes radially in through a ventilation duct
axially adjacent to the first. The two ends of the tie are thus
at the same circumferential location and one duct spacing apart

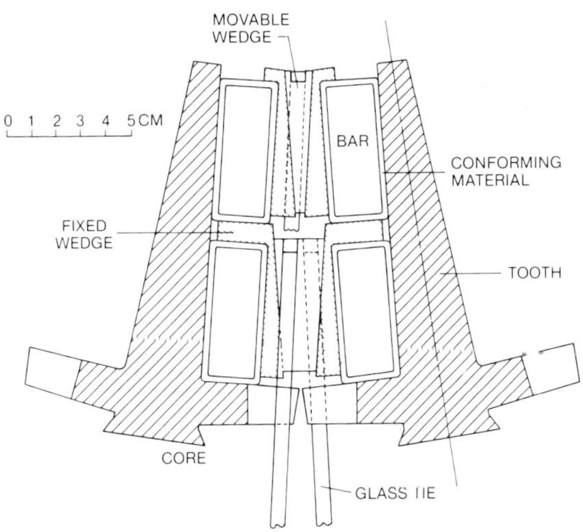

Fig. 2. Stator Winding Support Structure.

axially. The ends of the tie are looped around a steel yoke which is driven radially outward by a compression spring as illustrated in Fig. 1.

STATOR DEVELOPMENT

Each half-turn stator bar is built up of two hundred forty 0.13 cm diameter insulated wires and ten 0.84 cm × 0.42 rectangular copper nickel cooling tubes. The wires are formed into twenty 12-wire strands of rectangular section which are compacted to 0.87 cm by 0.22 cm. The strands are transposed two at a time with cooling tubes as shown in Fig. 3. The transposition is the well known 540° type with one half-cycle of crossovers occurring in each of the first and last quarters of the bar length and one half-cycle occurring between the quarter points. No transposition is performed at the ends. The conductors are molded into shape with polyester resin, and micapolyester resin insulation is applied. Water distributing boxes are brazed to the bar ends.

The stator core is stacked of punchings of 26-gauge, 2% silicon steel. Each lamination comprises a 60° sector. The stack is interrupted axially by ventilation ducts formed by insertion of radially directed I-section spacers between laminations. The ducts are typically spaced 7.5 cm apart.

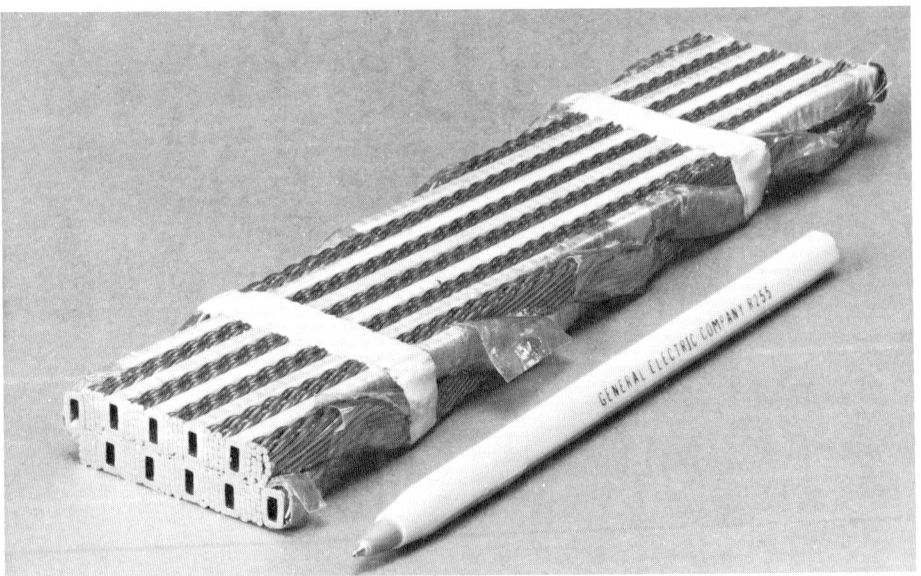

Fig. 3. Armature Bar.

The iron core is water-cooled. Aluminum extrusions extend axially along the radially outer surface of the core at locations corresponding to each row of wedge-loading springs. The spring assemblies rest on the extrusions, holding the extrusions tight to the core back. A heat transfer compound is applied to the joint between extrusions and core to ensure good thermal contact. Water passes through a hole extruded in the aluminum pads. Manufacture of the frame, core and stator bars is complete, and assembly of the stator is nearing completion. Figure 4 shows the complete armature winding assembly.

FUTURE WORK

It is anticipated that the generator will be ready for tests in 1981. At that time, no-load tests will be conducted under short-circuit and open-circuit conditions. Sudden short-circuit tests from reduced voltage are also planned. After concluding no-load tests, it is planned to proceed with fully loading this superconducting generator in 1982.

Fig. 4. Armature assembly.

REFERENCES

1. P.A. Rios, et. al., Component Development for a 20-MVA
    Superconducting Generator in "Proc. of the Electrotechnical
    Conference," Moscow, U.S.S.R., (June 1977).
2. S.H. Minnich, et. al., Design studies of superconducting
    generators, IEEE Trans. on Magnetics, Mag-15, (1):703
    (1981).
3. T.E. Laskaris, and K.F. Schoch, Superconducting rotor
    development for a 20-MVA generator, IEEE Trans. on Power
    Apparatus and Systems, PAS-99, (6):(1980).
4. T.E. Laskaris, High performance superconducting windings for
    AC generators, IEEE Trans. on Magnetics, Mag-17, (1):884
    (1981).

APPLICATION OF SUPERCONDUCTIVITY TO VAR GENERATORS*

H. Riemersma

Westinghouse Electric Corporation
Pittsburgh, Pennsylvania

INTRODUCTION

Static VAR compensators have been developed for electric power systems to provide system voltage support and reactive power control. These devices provide facilities that are similar to those traditionally provided by rotating synchronous condensers and fixed, or mechanically switched, capacitor and inductor banks. The static systems usually comprise fixed shunt capacitors and parallel inductors. The inductive VAR is varied by means of a series-connected thyristor switch which enables the inductor voltage to be readily controlled. Static VAR generators have been well received by industry because of their acceptable cost and desirable performance features such as fast response, ease of control, and low maintenance. Figure 1 (a) illustrates a basic schematic of a fixed-capacitor, thyristor-controlled inductor type VAR generator.

The losses associated with the VAR control scheme of Fig. 1(a) include reactor losses which are typically 0.6%. The high cost of losses makes it desirable to reduce them. This reduction is addressed by H. J. Boenig and W. V. Hassenzahl,[1] who propose a superconducting dc coil in which a nominally steady direct current is maintained, and to which the appropriate phases of the ac supply are periodically connected. Figure 1(b) depicts the most direct form of circuit to achieve this performance. The supercon-

---

*Work supported in part by the Los Alamos National Laboratory Contract No. 4-L20-6965P-1.

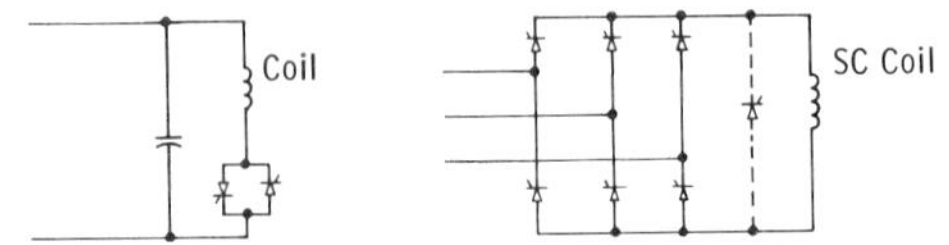

Fig. 1.　Schematic of basic VAR control.

ducting coil has zero average voltage and the alternating line currents provide lagging VAR when the thyristors are appropriately gated. In this scheme the coil would have essentially zero losses and the semiconductor losses would be about constant. The advantage of using a superconducting coil would be the economic gain of reducing the load losses.

OBJECTIVE

The objective of this work was to evaluate the technical and economic feasibility of using superconducting reactors in static VAR generation equipment and to determine if superconducting technology promises sufficient benefits to warrant additional effort. To achieve this objective a number of different conceptual superconducting VAR generator schemes was examined and the life-cycle cost of each compared with that of the conventional ac system. The conceptual designs were chosen in a rating range between 40 and 100 MVAR at a system rated voltage of 13.8 kV. Appropriate spacings, insulations system concepts, and bushings were identified for this voltage, based upon previous work.[2]

CONCEPTUAL DESIGN DESCRIPTION

Major Components

Shielded windings. In any application to power systems, reliability requirements demand that stabilization must be incorporated in superconducting windings. Stabilization usually takes the form of low resistivity "normal" material electrically parallel to and in intimate contact with the superconductor. In the presence of time-varying magnetic fields this material is a source of eddy current losses. The superconducting material itself is subject to hysteresis and eddy current losses. The key to loss reduction is to place most of the superconducting and all of the stabilizing material in regions of the coil volume where the magnetic field is very low. This goal can be achieved with shielded winding arrangements consisting of two sets of windings in which one set prevents most of the time-varying magnetic fields from reaching the other. In the case of equal length sections of infinitely long, concentric solenoids it can be shown that no alternating flux can exist in the space between the sections if the number of turns and the voltages are the same. The inner coil section thus shields the outer section and no ac current flows in the outer section. This shielding is not complete in finite length coil systems because of the effect of the fringing fields near the ends. This effect can be minimized by providing a low reluctance flux path surrounding the coil. The leakage flux that transversely cuts the windings is thereby reduced, coupling is enhanced, and the shielding is improved.

For the analytical parametric evaluation of shielded coil systems, we have used a configuration of concentric, equal length, equal turn simple solenoids with or without iron core, with the shielding capabilities represented by a variable parameter. For this initial feasibility evaluation this simple approach is adequate.

Conductor concepts. The superconducting inductor derives its viability from the reduction in load losses. The copper losses of the conventional inductor are replaced by the heat load on the superconducting system multiplied by the coefficient of performance of the refrigerator. AC losses are minimized by proper selection of filament and strand diameters, proper selection of matrix materials, and the separation of the windings into shielding and shielded windings. The analysis has been limited to multifilamentary, stranded cable conductors in single or shielded winding configurations.

Ferromagnetic core. The iron core was assumed to be of conventional construction at ambient temperature. Because of the requirements for low losses, the core uses transformer grade materials. The core losses are too large to permit removal at helium temperature with the accompanying refrigeration cost penalty. Iron core designs, therefore, require a warm bore or annular cryostat.

Cryostats. Because the ac and dc systems both have time-varying magnetic fields, it will be necessary to use non-conducting cryostats. The wall material must be strong enough to contain the 6 atmospheres pressure needed to obtain the insulating properties of supercritical helium. It must also be void-free to prevent deterioration due to corona discharges under ac voltage stresses. It must have very low permeability for helium to maintain the integrity of the vacuum jacket. Similar requirements apply for the materials that constitute the insulation system. Cryostat design and construction were explored with a number of vendors* to arrive at realistic material and manufacturing costs.

Insulation system. Insulation system concepts for the superconducting VAR inductor coil are based upon the work described in Ref. 2.

System Concepts

AC System. A circuit diagram for one phase of an ac system is shown in Fig. 1(a). The application of superconductivity involves replacing the conventional, resistive reactor with a superconducting reactor. All other components, such as capacitors, filters, and power electronics are the same for both systems. To achieve the required efficiency, a shielded coil must provide the overload capability and stabilization. Two configurations were evaluated. One uses an iron core with airgaps to control the induction. This configuration is designated FAC2. The other is an air core configuration with designation AAC2. Figure 2 shows cross section schematics.

DC systems. In the dc system only one coil system will be required to serve a three-phase power system. An electrical schematic is shown in Fig. 1(b). The coil system current consists of a dc component and an alternating current ripple with a peak value of about 10% of the dc current value. Four coil configurations have been analyzed. Two of these are air core, dc-shielded coil systems with the ac-shielding coil either resistive (ADC1-1) or superconducting (ADC2). The other two are single winding coils in which a copper stabilized superconducting coil, carrying both the dc and ac currents, is exposed to the alternating field of the ac ripple. One of these is an air core coil (ADC1) and the other uses an iron core (FDC1). These configurations are shown in Fig. 2.

METHOD OF EVALUATION

To decide if the application of superconductivity technology to static VAR generators is feasible, the benefits must be quantified in terms of some economic measure. Benefits exist if equivalent technical performance can be obtained at a lower cost. The cost of the "new" system must, therefore, be compared with that of the "conventional" system to determine whether there is a difference sufficiently large to be of interest. The term "cost" is used here to denote the approximate total expenditures that are

*Wayne Vogen, Inc., Owens-Corning, and Allegheny Ballistics Laboratory.

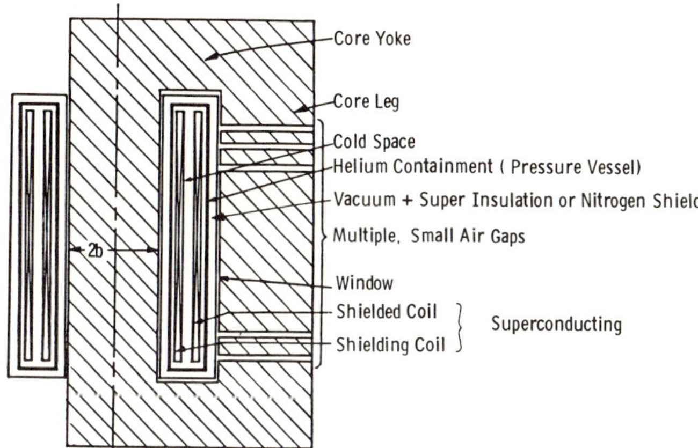

(a) Conceptual configuration for iron core systems showing coil system, warm bore, cryostat and iron core with air gaps (applicable to systems FAC and FDC)

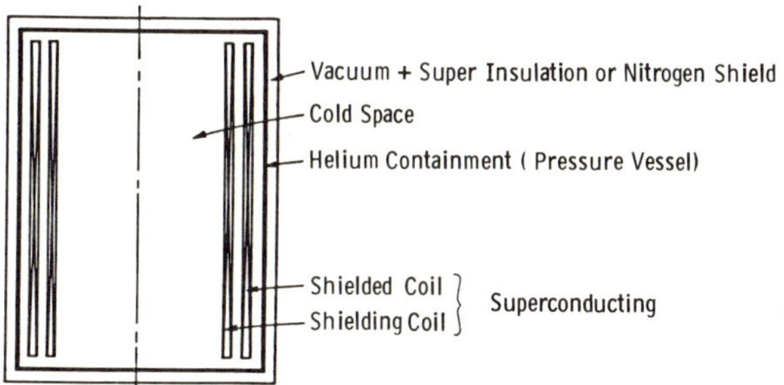

(b) Conceptual configuration for air core systems showing coil system in a pot type cryostat. (Applicable to systems AAC2, ADC1 and ADC2)

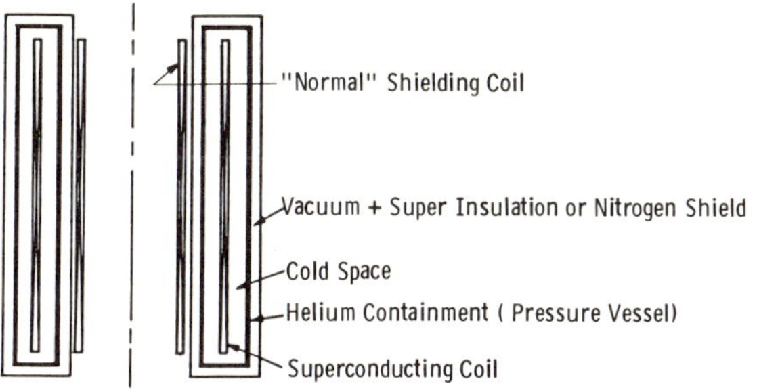

(c) Conceptual configuration for DC SAVAR system showing "Normal" resistive shielding coil in warm bore of annular cryostat (Applicable to ADC1-1)

Fig. 2. Cross section schematics of conceptual designs.

incurred in owning and operating the equipment over its lifetime and is thus a "life-cycle" cost. Generally the initial cost of superconducting equipment exceeds that of the conventional equipment because of the cost of the refrigerator. If one adds the capitalized cost of the losses over the expected life of the equipment, to obtain a life-cycle cost, then in certain applications superconducting technology can, because of improved efficiency, offer a cost benefit.

To make these comparisons, the life-cycle cost of the conventional system is needed. This cost was obtained from systems which are in existence. In the superconducting ac system, the auxiliary equipment will be the same as for the conventional system. The cost of the superconducting coils can, therefore, be compared directly with the cost of the conventional coils, to determine any cost difference between the two systems. The dc system auxiliary equipment requirements were developed under a separate contract.[3] The cost of the coil for the dc system must be compared with the available margin for dc superconducting coils.

Life-Cycle Cost of Superconducting VAR Generator

The economic life-cycle cost comparison has five steps: (1) to develop technically feasible concepts, (2) to build inductor models in parametric form, (3) to calculate cost as a function of configuration, (4) to determine the minimum cost configuration and (5) to compare with conventional system. Completion of these steps yields a minimum economic life-cycle cost for a superconducting VAR-inductor consistent with restraints imposed on the design.

Table I. Cost Elements in Economic Evaluation

I. Initial Costs ($/kVA)
    Materials – Copper, iron, superconductor,
             insulation
    Components – Cryostat, containment and bushing
    Refrigeration – Remove cold end ac losses, dewar
             losses, dielectric losses and
             lead losses
    Generation – Supply iron losses, stray losses,
             copper losses and refrigerator
             power needs

II. Operating Costs ($/kVA)
    Present worth of lifetime annual generation
      plant operating cost

Equations were derived for the superconducting inductor for all the costs listed in Table I. These were submitted to a "Geometric Programing Optimization" routine, which calculates the values of the variables that will minimize the total summation of these costs. The comparison is thus made between the life-cycle cost of actual conventional designs and the life-cycle cost of optimized conceptual superconducting models.

RESULTS

Table II shows typical results of the economic feasibility studies for coil configuration AAC2. The significant item on this is the "net gain." A positive value indicates that the life-cycle cost of the superconducting coil configuration is less than the life-cycle cost of the conventional system. The net gains are summarized on Table III where they are also expressed as a percentage of the total conventional system cost.

Sensitivity Studies

The impact on the economic feasibility of a variation in a number of parameters was examined. The most significant parameters are the energy related costs. Table IV shows the impact of these on the net gain. It is not uncommon for utilities to evaluate losses at $4 000 per kilowatt, and such evaluations may increase in the future. Systems which show a positive net gain will have this increased as loss evaluations increase. This trend is shown for coil configurations AAC2 and ADC2 which show net gain increases from 6 to 16% and 8 to 11%, respectively, when loss evaluations are increased from $2 000 to $4 000 per kilowatt.

Table II. Typical Results of Economic Study – System AAC2

	40	90	100	MVA	
Rating	40	90	100	MVA	
Coil Winding	1.8	1.6	1.0	$/kVA	
Containment + Bushing		0.5	0.42	0.21	$/kVA
S. C. Material		0.57	0.52	0.37	$/kVA
Dewar + Insulation		1.92	1.54	0.77	$/kVA
Cooling Equipment		4.6	4.2	3.2	$/kVA
Iron Core	---	---	---	$/kVA	
Losses	3.18	3.06	2.78	$/kVA	
Total	12.57	11.34	8.33	$/kVA	
Margin	15.0	15.0	15.0	$/kVA	
Net Gain	+2.43	+3.66	+6.67	$/kVA	

H. Riemersma

Table III. Summary of Results

Rating (MVA)	40		50		100	
Savings Coil	$/kVA	%	$/kVA	%	$/kVA	%
AAC-2	2.43	5.40	3.66	8.13	6.67	14.80
FAC-2	-1.85		-0.74		+0.18	0.40
ADC-1	-1.34		-0.85		+0.33	0.73
ADC-2	+4.16	9.24	+4.61	10.24	+5.67	12.60
ADC-1-1	+1.91	4.24	+2.42	5.38	+3.61	8.02
FDC-1	-1.28		-0.44		+1.30	2.90

CONCLUSIONS

1.   The technology exists in regard to structures, conductors, insulation, and cryostats, i.e., the elements unique to the superconducting aspects of the VAR generator.   However, these technologies are state of the art.   Work will be required to extend this state of the art to commercial practice particularly in regard to the conductor and insulation system.

2.   Economic benefits appear to exist in terms of a savings in life-cycle costs, particularly in coil systems AAC2 and ADC2. This benefit is strongly linked to the achievement of conductors that are cabled or braided from insulated strands and which have filaments with diameters of 80 m and 0.5 m, respectively.

Table IV. Impact of Energy Related
Cost--Coils AAC2 and ADC2

Generation Plant Cost ($/kW)	Energy Cost ($/kWHr)	Cost of Losses ($/kW)	Total Cost ($/kVA)		Net Gain ($/kVA)		% Conv.Syst.	
			AAC2	ADC2	AAC2	ADC2	AAC2	ADC2
800	0.02	2025	12.63	4.14	2.52	3.57	5.54	7.85
1000	0.03	2828	13.91	4.57	5.78	5.08	10.46	9.20
1000	0.04	3450	14.91	4.90	8.79	6.63	13.91	10.50
1000	0.05	4066	15.90	5.22	11.50	7.91	16.25	11.17

3. Using projected future parameter values of \$4 000/kW loss evaluation, and 50 μm and 0.2 μm filament diameters, the savings approaches 30%. At this level of savings a continuing effort would be warranted.

ACKNOWLEDGEMENTS

My thanks go to Mrs. Marilyn Warren for preparing this manuscript.

REFERENCES

1. J.H. Boenig and W.V. Hassenzahl, Applications of superconducting coils to reactive power control in electric power systems, <u>IEEE Trans. on Magnetics</u>, MAG 17, (1): 517 (1981).
2. H. Riemersma, et al., Application of Low Temperature Technology to Power Transformers, <u>Final Report DOE Contract No. ET-78-C-01-3290</u>, Westinghouse Electric Corp., Pittsburgh.
3. F. Cibulka, Converter Design for a VAR Generator Using a Superconducting Coil, <u>Contract No. 4-L20-6696P-1</u>, Los Alamos National Laboratory.

EXPERIMENTAL INVESTIGATION OF THE CURRENT DISTRIBUTION CHARACTERISTICS OF A SUPERCONDUCTING TRANSFORMER

J. H. Murphy

Westinghouse Electric Corporation
Pittsburgh, Pennsylvania

INTRODUCTION

Since the discovery of superconductivity, scientists have sought to apply this technology to power transformers. Studies in the 1960's using the state-of-the-art superconducting tape technology showed that the superconducting transformer was either too large or too inefficient to warrant further development.[1-7] In the early 1970's, the multifilamentary superconductor technology became state-of-the-art, and a 5 kVA experimental superconducting transformer was designed and tested based upon this technology.[8] Again, we found that the superconducting transformer was too inefficient to be considered viable. In 1977, a new winding approach, using parallel windings for both the primary and secondary windings, was developed which promised the potential of having low ac losses.[9,10] Several winding configurations have been investigated based upon the principle of a parallel winding transformer and our present analysis indicates that for generator stepup units superconducting power transformers can be made more efficient than conventional power transformers.

Figure 1 illustrates the conceptual configuration of the 1000 MVA superconducting transformer presently under consideration.

The iron core of the transformer is assumed to be of conventional construction and operating near ambient temperatures. The iron has an induction of approximately 1.75 T resulting in a core loss of approximately 2.2 W/kg for conventional materials. These losses are removed by an oil cooling system.

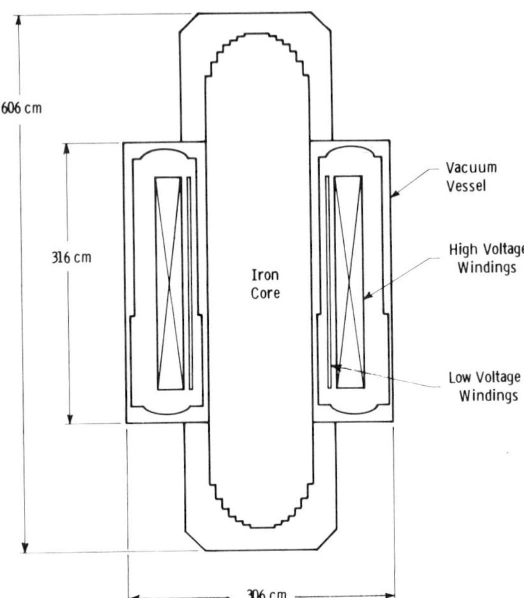

Fig. 1. Schematic diagram of a 1000 MVA
superconducting transformer.

The cryostat containment vessels are double-walled, fiber-reinforced, electrically-insulating structures. These vessels are made annular to allow the ambient temperature iron core to pass through the windings. To insure a low electric stress in these structures, the cold walls of the cryostats are covered with segmented conductive coatings that are maintained at ground potential.

To segregate the steady-state and fault currents, the superconducting transformer has four windings per phase. Nearest to the iron core is the low-voltage auxiliary winding which is connected in parallel to the low-voltage main winding (the next winding moving out in radius). Beyond this point are two high-voltage windings electrically connected in parallel: first, the main high-voltage winding then the auxiliary high-voltage winding. In either case, the main windings are designed to carry the major portion of the steady-state current and the auxiliary windings are designed to carry the fault current. The auxiliary windings are designed to remain superconducting under both steady-state and fault conditions whereas the main windings are designed to remain superconducting only for the steady-state operating conditions.

The superconducting transformer is to be wound with four different conductors: two conductors having high resistivity matrices to lower steady-state ac losses and two conductors having high conductivity matrices to stabilize the windings for operation in the supercritical helium environment. The main windings use the high resistivity matrix conductors and the auxiliary windings use the high conductivity matrix conductors. All conductors are Nb_3Sn based conductors with strand sizes of 50 μm and filament sizes of 0.3 μm.

The windings are connected to bushings which also provide a thermal and pressure transition between the ambient environment of the electrical grid and the supercritical helium cryogenic environment of the superconducting windings. The bushings proposed for this device are generically similar to the 138 kV terminations developed for the superconducting cable program.[11]

Analyses of the 1000 MVA superconducting transformer show that the efficiency should be 99.86%. Since a comparable conventional transformer has an efficiency of 99.71%, the life-cycle costs should be reduced. Based upon a cost of power of $2 000 per kilowatt, our analyses show a 30% improvement in the transformer life-cycle costs when superconductivity is used. This small economic improvement is sufficient to warrant further investigation.

The key feature of the superconducting transformer essential to its viability is the natural division of currents between the parallel sets of windings under steady-state and fault conditions. An experimental program was therefore initiated to investigate the characteristics of this type of transformer. This paper describes the design, construction and testing of a 4 kVA model superconducting transformer which demonstrates the natural current distribution necessary for superconducting transformer practicality.

SUPERCONDUCTING TRANSFORMER EXPERIMENT

There were many features of the superconducting transformer technology which needed experimental verification. In the first experimental transformer, we chose to examine only those features which were crucial to its viability. The central issues associated with the efficiency of a superconducting transformer were the divisions of currents between the main and auxiliary windings under steady-state operating conditions, and the natural switching of the current distribution during an electrical fault.

The approach formulated to demonstrate these current distribution characteristics was to operate the iron core of the experi-

mental transformer in a liquid helium bath and to make the
windings from state-of-the-art superconductors. This approach to
the experimental transformer construction and operation was based
upon the low operating voltage and the unimportance of efficiency.

Figure 2 illustrates a cutaway view of the experimental
superconducting transformer. The key feature, a transformer with
a multiplicity of superconducting windings arranged to provide
natural division of currents under steady-state and fault conditions, was retained in this experimental device.

The iron circuit used in this experiment was a C-type Hiper-
sil core with a window of approximately 5.1 cm by 14 cm. The
approximate weight of this magnetic core was 10.3 kg. Depending
upon the experimental conditions, this core carried a 60 Hz field
of up to 2 T which implied that it had a hysteresis loss of up to
5.5 W/kg. Therefore, at full voltage the magnetic core should
have had a thermal load of approximately 57 W. This load was
acceptable where the core was directly exposed to the helium bath,
which was true over approximately 75% of the surface area. However, in the region where the windings were located, the core was
not in direct contact with the helium bath and therefore vertical
cooling channels were designed of sufficient cross section to
prevent vapor binding next to the core.

The windings of this transformer have to be made of multifilamentary superconductor with matrix resistivities suitable for

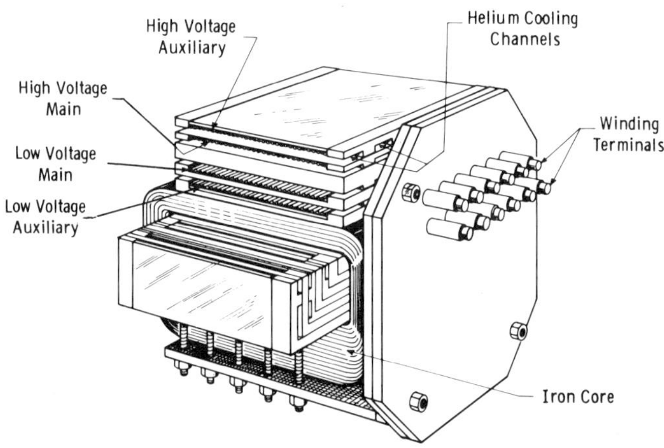

Fig. 2. Cutaway diagram of the 4 kVA
experimental superconducting transformer.

current switching during simulated electrical faults. The design approach called for using state-of-the-art superconductors therefore the conductor strand diameter and filament size were made much larger than proposed for a full-scale superconducting transformer. For the main windings, a 211 μm diameter strand with a high-resistance matrix was selected and for the auxiliary windings, a 287 μm diameter strand with a low-resistance matrix was used. NbTi filament sizes of approximately 7 and 10 μm was utilized for the main and auxiliary windings, respectively. The main winding matrix was CuNi bronze and the auxiliary winding matrix was copper. The parameters of the windings and the conductor are given in Table I.

EXPERIMENTAL

The first experiments performed on this apparatus were the traditional open circuit and short-circuit tests to determine the nominal rating of this experimental transformer. From the short-circuit tests (see Table II), the rms critical current of the low-

Table I. Winding and Conductor Parameters

Winding:[*]	LVA	LVM	HVM[+]	HVA
Active Length, cm	12.2	12.2	12.7	12.2
Volume of Conductor, cc	4.59	2.92	3.36	7.66
Number of Turns	123	123	218	218
Average Length Per Turn, cm	28.8	34.0	44.1	54.3
Conductor:				
Westinghouse Reference No.	H-60	H-61	H-61	H-60
Manufacturer	Supercon	Supercon	Supercon	Supercon
Strand Size, μm	287	210.8	210.8	287
Filament Size, μm	10	7	7	10
Number of Filaments	400	360	360	400
Twist Length, mm	2.79	2.54	2.54	2.79
Superconductor	Nb-48Ti	Nb-48Ti	Nb-48Ti	Nb-48Ti
Matrix	Cu	CuNi 30	CuNi 30	Cu
Matrix/Superconductor Ratio	1	1.5	1.5	1

[*] LVA = Low-Voltage Auxiliary Winding
LVM = Low-Voltage Main Winding
HVM = High-Voltage Main Winding
HVA = High-Voltage Auxiliary Winding.

[+] The high voltage main winding was discontinuous and therefore was not used in these experiments.

voltage superconducting windings was found to be approximately 28 A. From the open circuit tests (also see Table II) the iron core exciting current in the low-voltage windings was found to be 2.4 A (rms) for 186 V (rms). Based upon this information, the nominal rating of this experimental transformer would be approximately 4 kVA with an electrical efficiency of approximately 90% ignoring the energy requirements necessary to provide the cryogenic cooling.

The next experiment performed was a loaded secondary test and Fig. 3 illustrates the experimental arrangement. A 440 V autotransformer and a 4.5 Ω resistor bank formed the variable current supply for this experiment. Because of a manufacturing defect, only the high-voltage auxiliary winding of the superconducting transformer was connected to the secondary of the autotransformer. However, both low-voltage windings of the superconducting transformer were connected to a resistive load of approximately 0.7 Ω.

The current distribution in the experimental transformer was determined by two techniques. The first technique involved the measurement of the ac fields in the gaps between the windings using three American Aerospace Control mistors (magneto-resistive sensors). Accordingly, when the main windings were superconducting, the field in the gaps between the main and auxiliary windings should have been nearly zero, whereas, when the main windings were normal, the field in the gaps between the main and auxiliary windings should have been the same magnitude as measured on the center mistor. Figure 4 (a) illustrates mister readings before and after the low-voltage main windings normalized, confirming the projected current distribution.

Table II. Open and Short Circuit Test Results

	Short Circuit			Open Circuit			
V_1 (mV)	I_1 (A)	V_2 (V)	I_2 (A)	V_1 (V)	I_1 (A)	V_2 (V)	I_2 (A)
22	10.6	12.7	5.4	156	0.5	268	0
37	17.0	21.0	10.0	176	1.4	301	0
44	20.2	24.9	11.5	186	2.4	320	0
53	24.5	29.5	14.0	195	4.7	335	0
61	28.0	34.0	15.5	200	6.0	345	0

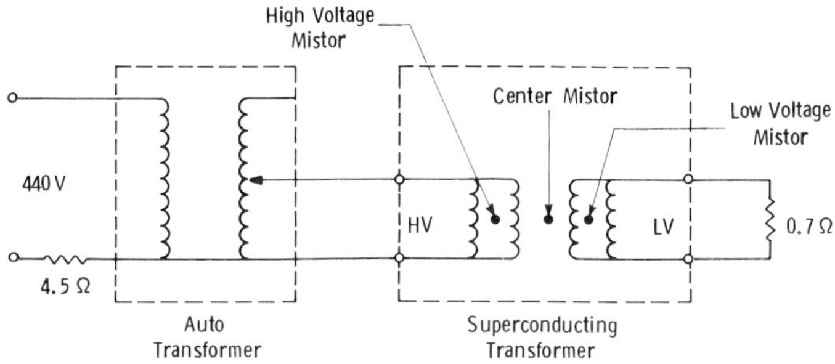

Fig. 3. Loaded secondary test arrangement used in demonstrating the superconducting transformer current distribution.

The second technique involved the measurement of the voltages and currents to and from the superconducting transformer windings and subsequent calculation of the transformer impedance. This technique should show a low transformer impedance before going normal and a high transformer impedance after normalization. Figure 4(b) summarizes these results obtained during the loaded secondary test. The solid line is the calculated performance based upon an infinite solenoidal model of the experimental super- conducting transformer. The difference between the results andthe model predictions are due to the assumptions made in this oversimplified model of the superconducting transformer. These results also confirmed the projected current distribution in the superconducting transformer.

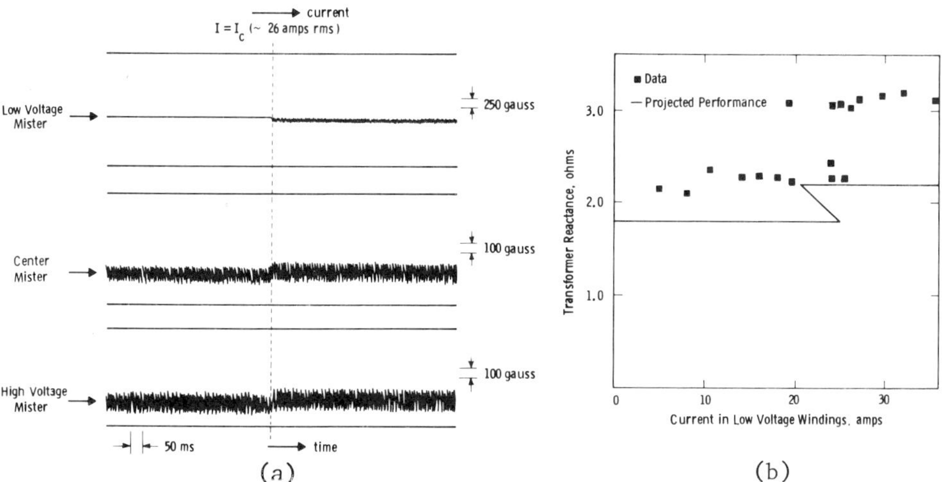

Fig. 4(a). Recorded mistor readings and (b) transformer reactances before and after low-voltage main normalization.

During the current distribution experiments, this superconducting transformer powered a resistive load for periods of up to
fifteen minutes. A steady-state loading of 0.5 kVA was sustained
for approximately ten minutes and transient loading to 0.9 kVA was
also achieved. The superconducting transformer operated reliably
as a three-winding transformer throughout these tests.

CONCLUSIONS

This experiment shows that the current distribution in a
superconducting transformer will be concentrated in the main
windings under steady-state conditions, and when a fault causes
the current to exceed the critical current in the main windings,
the current shifts to auxiliary windings. This design approach
leads to extremely low transformer losses provided that advanced
superconductors are utilized. The next step in the evolution of
superconducting transformers is the development of a prototype
conductor and verification of its ac loss characteristics.

REFERENCES

1. R. McFee, Superconducting power transformers a feasibility
       study, Elec. Eng., 80:754 (1961).
2. K.J.R. Wilkinson, Superconductive windings in power
       transformers, Proc. IEE, 110:2271 (1963).
3. K.J.R. Wilkinson, Prospects for the employment of
       superconductors in alternating magnetic fields, IEEE Trans.
       on Magnetics, 2:369 (1966).
4. P.A. Klaudy, Some Experiments Relating to the Layout of
       Superconducting Transformers, in "Advances in Cryogenic
       Engineering, Vol. 9," Plenum Press, New York (1964), p.
       349.
5. P.H. Borcherds, Physical limitations on the size of
       superconducting power transformers, Cryogenics, 113:1953
       (1966).
6. H.O. Lorch, The feasibility of superconducting power
       transformers, Proc. IEE, 9:354 (1969).
7. R.V. Harrowell, Feasibility of a power transformer with
       superconducting windings, Proc. IEE, 117:131 (1970).
8. D.W. Deis, Westinghouse, private communications
9. H. Riemersma et al., Application of superconducting technology
       to power transformers, IEEE Trans. PAS, 100:3398 (1981).
10. H. Riemersma et al., Application of low-temperature technology
       to power transformers, Final Report DOE Contract No. ET-78-
       C-01-3290 (1980).
11. S.F. Masua et al., Development of a 138 kV superconducting
       cable termination, IEEE PAS Paper F76081-0 (1976).

IMPROVED TECHNIQUE TO MEASURE ELECTRONICALLY AC LOSSES IN SUPERCONDUCTING CABLES*

F. Schauer[†] and M. Meth[‡]

Brookhaven National Laboratory[§]
Upton, New York

INTRODUCTION

The AC losses of superconductors and superconducting cables can be measured either calorimetrically or electronically. Calorimetric measurements can be done with reasonable accuracy only if background losses (such as cryostat losses, losses of instrumentation leads, etc.,) are either well-known so that they can be separated from the losses of interest or are much less than the AC losses. This method is the only one applicable to long superconducting cables, where the termination losses are negligible and dielectric losses of the high voltage insulation and losses caused by heat conduction and radiation from the environment can be separated out. In most cases, for losses of small samples up to prototypes of superconducting power transmission lines, the more accurate electronic measurement method is used and has been proven to be reliable. This method is well-known in principle; to multiply the loss voltage of a pickup coil or a probe wire attached to the superconductor by a signal proportional to the magnetic field at the superconductor surface and time-averaging the product. The field (or current) signal is the integrated voltage derived from a

*Work supported by the U.S. Department of Energy.
†Visiting Scientist, Permanent Address: Anstalt fuer Tieftemperaturforschung, Steyrergasse 19, 8010, Graz, Austria.
‡Visiting Scientist, Permanent Address: City University of New York.
§Operated by Associated Universities, Inc., under contract with the U.S. Department of Energy.

pickup coil. In most cases, the loss voltages have inductive components which may exceed the real components by several orders of magnitude. To allow reasonably accurate multiplication, the inductive components have to be compensated with voltages derived from the field pickup coils. The accuracy of this voltage compensated wattmeter technique is dependent on the inductive voltage component and is critically dependent on the phase accuracy of the signals to be processed.

This paper is concerned only with loss measurements of superconducting cables of lengths from 1 m to 100 m. A method to measure short cables of lengths to 10 m has been developed and is being steadily improved in the course of the construction of longer cables.[1,2] The method needed further improvement in order to make loss measurements of the BNL 138 kV, 4 kA, 100 meter long superconducting cable now under construction.[3] The cable system consists of two coaxial cables connected in a circular loop so that voltage and induced current can be applied simultaneously. During the ac loss measurements it will be run without high voltage.

MEASUREMENT PRINCIPLES

Current and Voltage Probes

Figure 1 shows the schematic of the existing electronic loss measurement system for a coaxial superconducting cable with a shorted end. The losses of the different cable components can be

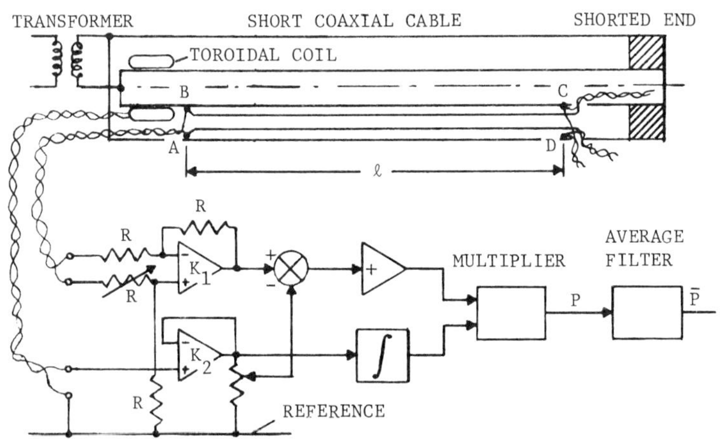

Fig. 1. Short Cable Loss Measurement Circuit.

measured separately: the inner conductor with the probe BC, the outer conductor with the probe AD, and the losses of the shorted end with the probe CD. The overall losses are measured with an AB probe. The overall current (and therefore the azimuthal magnetic field at the surface of the inner conductor) is measured with a toroidal pickup coil around the inner conductor. All probe leads are twisted together tightly to reduce inductive pickup.

The voltage probe is connected to a difference amplifier (K_1) of the wattmeter, and the current probe to the buffer amplifier (K_2) with a high input resistance. The signal derived from the toroidal coil is in phase with the inductive voltage component and is used for compensation. After integration, this signal is proportional to the current.

Phasing Problems

This whole method is very sensitive to the phase accuracy in the circuit elements before the compensation point. Phase shifts are especially critical for measurements with the AB probe. In this mode the whole inductive voltage of the cable must be compensated. For a 10 m, 4 kA cable the phase accuracy requirement is on the order of tens of microradians. The critical parts are the current and voltage pickup circuits. The input amplifiers K_1 and K_2 of the wattmeter are identical. The wiring in the wattmeter can be kept short with very low values of capacitance and inductance.

An equivalent circuit of the current pickup can be drawn (Fig. 2). From this the phase shift, δ, is:

$$\delta \simeq \omega L/R_e + \omega C R_{dc} \tag{1}$$

with ω being the power frequency, L the inductance of the toroidal coil, R_e a resistance

$$R_e = V^2/P_e, \tag{2}$$

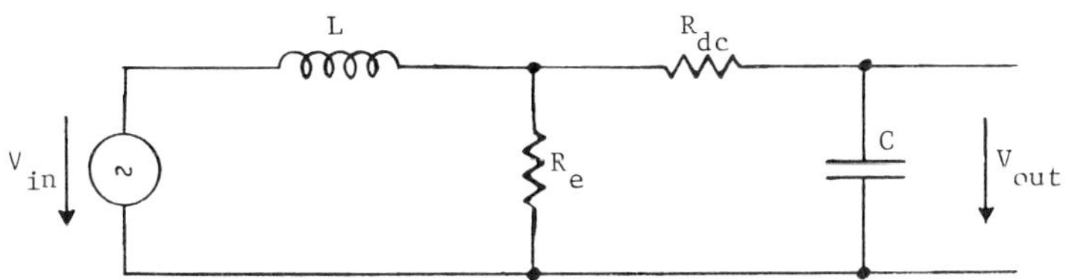

Fig. 2. Equivalent Diagram for the Current Pickup Circuit.

derived from the toroidal voltage V and the eddy current losses P_e. C is the capacity of the leads between the toroid and the wattmeter plus the input capacity of the wattmeter. R_{dc} is the dc resistance of the coil plus the leads. Since the eddy current losses of a wire in a perpendicular field are

$$P_e = \pi \, a^4 \, B^2 \, \omega^2 \, \ell/(4\rho) \tag{3}$$

(a = diameter, ℓ = length, ρ = resistivity of the wire, and B = magnetic flux density), the wire diameter a and the length ℓ should be kept small and ρ should be large. However, according to the second term of Eq. 1, ρ should be small. A good compromise is a thin copper wire with a cross section of about 0.003 to 0.01 mm^2 resulting in acceptable phase shifts in the order of 20 μrad for typical toroidal coils (for 4 kA cables) with inner diameters of 2.5 to 3 cm and 500 to 1000 turns and for lead capacitances $< \sim$ 1000 pF and lead resistances of several ohms. Another consideration regarding phase shifts in the current pickup circuit are small axial fields in the cable due to imbalances of the currents in the two oppositely wound helical tape layers of each concentric cable conductor. These axial fields are not necessarily in phase with the cable current and may induce error voltages in the toroid. To minimize these terms, the toroids have to be wound very carefully with the plane of each turn parallel to the center line of the toroidal coil. The requirement of parallelism of the toroidal axis with the cable axis is not very stringent. Errors can also be caused by axial fields inducing a voltage in the one turn loop around the cable conductor formed by the toroid itself. This one turn loop can be avoided by running either one of the adjacent wires of the beginning and the end of the toroid back around the toroid (for best results, on a mean toroid diameter). With these techniques, reasonable current pickup circuits were made and used successfully.

The phase shifts in the voltage circuits can be neglected in short cables as long as the probe wires are reasonably thin ($\leqslant$ 0.015 mm^2) for negligible eddy current losses. But for longer cables, wire resistance and the capacitance between the probe wire and the cable conductor have to be considered. In such a case, the input resistance of the amplifier K_1 also becomes important.

The loss voltage is measured between two points that are spaced ℓ meters apart. The voltage probes running from the far points (C and D) will introduce a differential phase shift with respect to the voltage at the points closer (A and B) to the cable exit port. This differential phase shift will cause a component of the reactive voltage to become a loss voltage. For a continuous and uniform injection of a voltage (A + jB) per unit of length into an RC transmission line of length ℓ , the output voltage is given by[4]

$$V_{out} = (A + B\ \theta\ell^2/6)\ell + j(B - A\ \theta\ell^2)\ell, \qquad (4)$$

where R and C are the line parameters per unit of length and θ = ωRC. The effective phase-shift of the probe is $\theta\ell^2/6$. The fractional error of the resulting loss measurement is given by

$$\Delta P/P = (B/A)\ \theta\ell^2/6. \qquad (5)$$

For the BNL 100 m cable a probe wire will be used which was constructed by slicing commercial multi-conductor flat ribbon wire. The wire size is 0.08 x 0.66 mm^2 (equivalent to # 30 AWG copper conductor) embedded in a 0.25 mm polyester film. The parameters of the conductor are: R = 0.004 Ω/meter at 8 K, C = 170 pF/meter, with θ = 0.26 x 10^{-9} rad/m^2. The probe introduces an error of less than 1% for a ratio of B/A (from previous short cable experiments) that is less than 10^4:1.

Due to the difference in wire length of the two probes running from the near and far electrical ports, the source resistances differ, introducing a differential attenuation between the two signals. A common mode input of V_c ($\approx$B/2) introduces a differential input of value $(\Delta R/R)V_c$, where ΔR is the differential resistance of the two probe wires and R is the input impedance of the operational amplifier (OPAMP). If the differential signal is V_d ($\approx$B), the effective input signal is $V_d + (\Delta R/R)V_c$, and for an error of less than 1% and a ratio of V_c/V_d of 10^4:1 $\Delta R/R$ must be less than 10^{-6}. Typically ΔR is less than 10Ω (recalling that a portion of the probe is at room temperature). Therefore, OPAMPs are required with an input resistance greater than 10 MΩ at 60 Hz.

The overall eddy current losses of the probe wires are about 10 mW each and can be neglected compared to the expected cable loss of several watts.

Balancing Problems

One must also worry about the balancing of the difference amplifier depending on the common mode voltage (CMV). For short superconducting cables the tuning of the variable resistor R of Fig. 1 can easily be done by changing R until the wattmeter shows the same result after exchanging the voltage leads to the inverting and non-inverting inputs of K_1. Of course, when doing this the current leads to the reference and to amplifier K_2 have to be exchanged. The CMV arises because usually the whole superconducting cable together with the secondary of the current transformer is electrically floating to avoid potentially dangerous ground

loops. Therefore, the potential of the cable is determined by the
ground capacities and the capacitive coupling of the transformer. The adjustment of R becomes the more sensitive as the
cable losses are reduced.

For the large cable system to be tested at BNL the CMV problem becomes crucial.[5] The loss measurement circuit for the 100 m
cable is shown in Fig. 3. The losses of the inner and outer
conductors of one cable are measured simultaneously ($V_{AD} + V_{CB}$).
The OPAMP K_1 of the wattmeter now is a combined adding and subtracting amplifier. The points A and B of the outer and inner
cable conductors are the only ones connected to the reference.
Thus the CMVs become practically zero. The probe leads and the
ground leads of the inner conductor are closely twisted together
and run from point B through the core of the potheads to the top
of the cable termination. From there, both probe wires continue
inside a conduit on the reference potential to the wattmeter. The
wires from point A of the outer conductor are brought in a similar
manner to the wattmeter.

The advantage of closely twisting the probe wires to the
reference wire or even running them inside a reference conduit is
that the loops formed by each of the probes and their returns
through the instrument ground enclose a minimal number of flux
lines that are generated by the cable and its exciting power

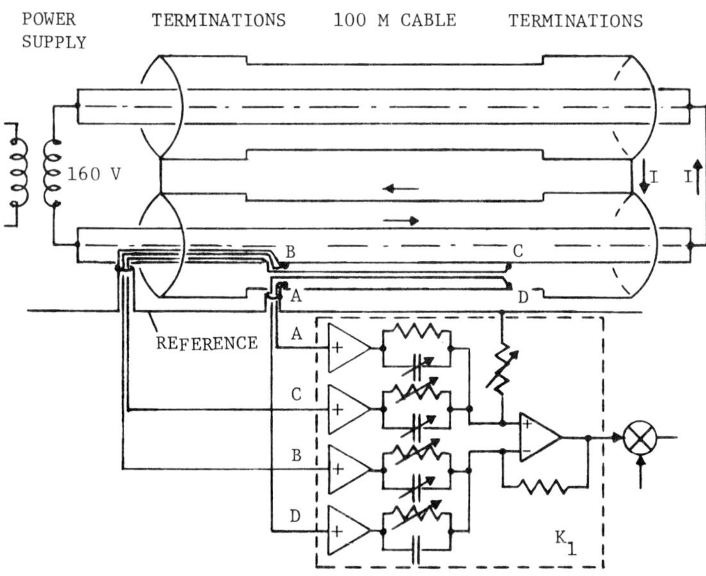

Fig. 3. 100 m Cable System Loss Measurement Circuit.

supply. Thus the wattmeter in the improved probe configuration is
not subject to the large value[5] ($\simeq$ 44V) of reactive voltage induced in the loop A–B–C–D.

EVALUATION AND CONCLUSION

To check the new instrumentation, a previously tested 0.72 m
cable (cable No. 9)[1] with helically wound superconducting tapes
(similar to the 100 m cable) was wired according to Fig. 1 (ℓ =
0.65 m). In addition to the probe wires shown in Fig. 1 "ground
wires" were soldered to each point A through D and twisted closely
to the corresponding voltage probe. A current pickup circuit was
installed with a phase shift of < 25 μrad.

The loss measurements of all cable components were performed
in three different ways: (1) they were done using the original
circuit, shown in Fig. 1, (2) the ground wire corresponding to
each probe was connected to the wattmeter reference, and (3) the
amplifier set shown in Fig. 3 was used.

With all three methods, no significant difference in the
results could be found. The dependence of the sensitivity of the
setting of the variable resistor R in Fig. 1 on the grounding was
clearly seen in the experiments. In the old configuration of Fig.
1 the wattmeter readout was very sensitive to a very small change
of R whereas in the case of grounding the corresponding probe
points, this sensitivity was much less. This effect was rapidly
increasing with decreasing losses to be measured. The effect is
consistent with the idea of the CMV stated above.

With the third method, using the new amplifier cluster, no
significant difference in the results could be found between using
the ground wires or not. Sometimes the wattmeter readout without
grounding was unstable and fluctuated around the mean value.

The results of the loss measurements for a field of 500 A/cm
(corresponding to 3550 A) at the inner conductor are listed in
Table I. The high losses are difficult to explain since the
influence of the end effects is not fully understood. The sum of
the losses of AD, BC and DC is larger by about 26% than the
overall AB losses. This suggests an inacurracy of the measurement
method of about ± 13% of a single measurement, which is
reasonable.

The new measurement method, therefore, appears to be suitable
for the BNL superconducting 100 m cable. It is believed that in
the case of a long cable the accuracy will be higher because of
larger absolute losses, corresponding to a higher signal-to-noise
ratio.

Table I. Losses of the Short Cable Components

Voltage Probes	Losses (mW)	Comments
AB	770	Overall losses
CD	590	Shorted end losses
BC	100 (154 mW/m)	Inner conductor losses
AD	280 (431 mW/m)	Outer conductor losses

REFERENCES

1. M. Garber, Resume of cable loss measurements to January 1, 1979, BNL Power Transmission Project, Technical Note, No. 86 (Jan. 1979).

2. G. H. Morgan, F. Schauer and R. A. Thomas, An improved 60 Hz superconducting power transmission cable, IEEE Trans on Magnetics, MAG 17(1):157 (1981).

3. E. B. Forsyth, Test results of AC superconducting cables, to be presented at the 1981 IEEE/PES Conference, Minneapolis, Minnesota, Sept. 20-25, 1981.

4. M. Meth, Design of voltage probes for conductor loss measurement at Fifth Avenue test site, BNL Power Transmission Project, Technical Note No. 109 (Oct. 1980).

5. F. Schauer, M. Meth and G. Morgan, Measurement of conductor loss at the Fifth Avenue test site, BNL Power Transmission Project, Technical Note No. 108 (October, 1980).

CRYOGENIC TESTING OF 100-m SUPERCONDUCTING POWER TRANSMISSION TEST FACILITY

R. J. Gibbs, J. E. Jensen, and R. A. Thomas

Brookhaven National Laboratory[*]
Upton, New York

INTRODUCTION

The final tests of the cryogenic system designed to cool the facility for testing 100 m superconducting power transmission cables have been completed. The results of tests of a three-expander configuration were reported in 1980. Following the completion of those tests, the system was modified to incorporate a fourth turbo expander remote from the refrigerator at the far end of the load. In this configuration the load, consisting of the superconducting cables and their containment vessel, becomes a long (> 100 m) counterflow heat exchanger with internal heat generation[2].

SYSTEM DESCRIPTION

The portion of the system under test is shown in Fig. 1. It consists of the transfer lines to and from the load, the load itself, and the turbines T3 and T4. The transfer lines to and from the load are each 65.6 m in length and have static vacuum, multilayer insulation. The supply line is 25.4 mm ID and the return line 50.8 mm ID, both rated for 15 atm operating pressure. The cable containment vessel is a horizontal vacuum and multilayer insulated vessel 100.7 m long made up of five ~ 20 m sections evacuated and sealed by the manufacturer, with field welded and evacuated joints between sections. The cold inner tube of the vessel is 213.5 mm ID, rated for 15 atm service. At each end of the containment vessel are large vacuum-insulated termina-

---

[*]Work supported by Associated Universities, Inc., under contract with the U.S. Department of Energy.

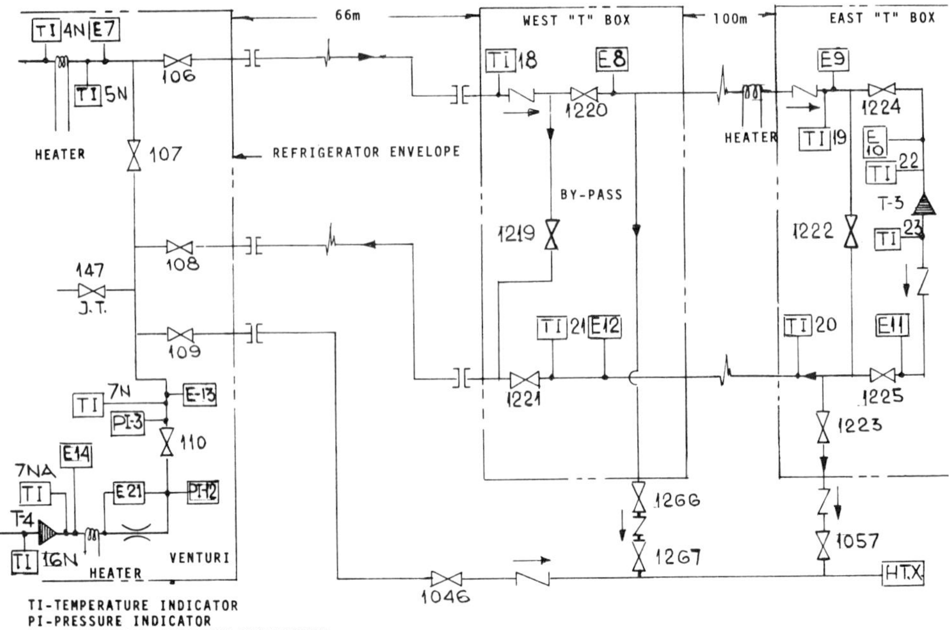

Fig. 1.   Flow Schematic.

tion boxes. The transfer lines are connected at the "west" box,
and the T3 turbine is installed in the far-end "east" box. The
turbine T4 is located in the refrigerator cold box between the
exit of the return transfer line and the cold end of the final
heat exchanger. The load flow passes through both of these turbines as in a Brayton cycle refrigerator. The higher temperature
turbines T1 and T2 are also located in the refrigerator cold
box. The rest of the system has been described earlier.[1,3]

During these and preceding tests, a corrugated transfer line,
simulating a cable, was installed in the bore of the cryogenic
enclosure. A heater ~ 4 m long was placed about 50 m from the
west end in the inner 31.8 mm corrugated tube to simulate an
electrical load on the cable. The outside corrugated tube, 60.7
mm, enclosing the inner corrugation forms an annular space which
may be evacuated or contain a small amount of helium to simulate a
cable and allow heat transfer between inner and outer flows.
During the first stages of cooldown, above 80 K, the valves are
positioned so that the flow passes through both the inner
corrugation and the outer annulus, formed by the cryogenic
enclosure and the simulated cable, in a west to east direction at
~ 15 atm. At the east end, this flow is channeled into a warm
return line back to the compressor. Upon reaching an average
temperature of 80 K in the test section, the valves are

repositioned to allow the supply stream to pass through the bore of the simulated cable to the east end and the return stream to return through the inner tube of the cryogenic enclosure in a counterflow mode. Turbine T4 is started at this time and the inlet pressure to it controlled to 7 atm. After the return stream pressure in the test section has been reduced to 8-10 atm, the turbine T3 is started, completing the flow circuit for final cooldown and normal operation.

TESTS PERFORMED AND RESULTS

Test Requirements

Since these tests were to be the last tests of the system prior to installation of the superconducting cables, data were required on the various modes of operation, including system stability and component performance. The specific data desired were: (1) total refrigeration, with and without lead flow, (2) the heat loss from the transfer line and the cable containment vessel, and (3) the performance of the remote turboexpander (T3).

Experimental Procedure

In order to simplify the loss measurements, the system was started and cooled down with the high pressure flow, through the center of the simulated cable, thermally isolated from the lower pressure return stream. This was to allow the system to be analyzed in sections without the complication of heat transfer between the two flows. After 100 hours of operation in this mode, during which time heat was added either at the inlet to turbine T4 or in the center of the load (see Fig. 1 for heater placement) to balance the refrigerator at full load, it became apparent that the desired heat leak information could not be obtained.

Temperature Measurements

Temperature readings from sensors TI 18, 19, 20 and 21 were not consistent with other temperature readings, so a component by component analysis was impossible. On reviewing the data following the run it became obvious that even with good temperature measurement at each point, the calculated losses would have a scatter of at least ± 10%. This accuracy limit is due to the physical properties of cold, dense supercritical helium and to the difficulty of accurately measuring flows at the various points in the system.

The germanium temperature sensors we used had been calibrated to an accuracy of ± 5 mK and, in most cases, it is believed that ± 10 mK was achieved in the field installation with the exception of those sensors listed above. For many cryogenic systems an error of ± 10 mK would be considered quite acceptable; however, a calculation of the total refrigeration using sensors 5 N and 7 N indicated that for this facility even this small inaccuracy could result in an uncertainty of ± 100 W depending on the sign of the errors for each sensor.

Flow Measurements

Measurement of helium flow at low temperature is difficult at best. In an effort to provide accurate flows measurement for this system, all orifices were ASME calibrated for use at 80 K or ambient. The flows through the load were found by subtraction from the data taken at the higher temperatures. The calculated accuracy of the flow meters is ± 2% which resulted in errors of ± 2.6 g/s at the warm inlet and outlet of the refrigerator and ± 1 g/s at the inlet to turbines T1 and T2. Therefore, the calculated error in flow through the load could be as great as 3.6 g/s. Using the same temperature points as before, this error in flow translates into an error ± 20–25 W in refrigeration capacity.

Thermal Inertia Effects

The final obstacle to good heat balance calculations is the large volume of cold helium gas contained in the return side of the cable enclosure. At the nominal operating pressure of 7 atm, the enclosure contains 280 kg of helium at 7.5 K, and 237 kg at 8.0 K, with specific heats of 10.15 and 9.441 kJ/kg-K respectively. During the eight days of testing, this volume was held within these temperatures under all conditions. The heat capacity represented by this mass of helium makes the response time to changes in load very long and at no time during the tests was equilibrium operation achieved. The load was always either heating or cooling even though the rate may have been as little as 0.1 K in 15 hours. Because of the number of different tests to be run it was not possible to remain at a given condition beyond 12–15 hours. During the periods of cooling, part of the mass flow into the load was being accumulated in the return volume. When the temperature was increasing the flow out of the load was greater than that entering. In general, the rate of heating or cooling was low so that the amount of helium accumulated or rejected by this volume was within the flow measurement errors and so was not a readily measurable quantity but only contributed to the scatter of the data.

Turbine Measurements

All of the planned tests were performed with the inlet and outlet pressures at the refrigerator controlled to ± 0.05 and ± 0.02 atm respectively. The inlet throttle valves to turbines T1 and T3 were wide open so that the only flow control to the refrigerator precooling circuit and to the load was by virtue of the specific volume of the helium entering the inlet nozzles of these two turbines. The temperature variation at the inlet to T1 was ± 1.5 K at 105.5 K for the eight day period; therefore, the variation in flow through the precooling circuit was within the measurement accuracy of the flow meter. The average mass flow through the T1 and T2 was 55.6 g/s. The specific volume of the helium entering the turbine T3 varies as a rather steep function of temperature in the 8 K region. Even though the entering temperature varied by only 0.75 K total during the tests, the specific volume varied by ± 5% which is close to the calculated mass flow errors expected. The calculated average mass flow rate through T3 was 70.6 g/s + 4% − 3%.

Results

Because of the unreliability of some of the temperature sensors, it was necessary to make the heat balance calculations using sensors known to give correct values. The results without lead flow were: total refrigeration 773 W and total losses 440 W. With lead flow the performance was: total refrigeration 654 W, total losses 440 W and lead flow 0.92 g/s. The refrigeration without lead flow is a 70 W increase in capacity over that measured in the previous configuration without the additional turboexpander. This was the first test where reliable data with lead flow were obtained. These results were discouraging when compared to the calculated capacities of 819 W of total refrigeration plus a lead flow of 1.5 g/s. A review of the performance of the various components has shown some possible reasons for the lower capacity.

A comparison of actual turbine performance against calculated performance is shown in Table I. This comparison clearly shows several problem areas. The two most obvious are the turbines T1 and T3. The problem indicated by the high inlet temperature and low mass flow entering T1 has been known for quite some time. The first or warm exchanger has been known to be too small for several years. Indicative of the problem is the warm end Δ T which is consistently 5 K and which should be 3 K to meet the design requirements. The result of this is a high inlet temperature to T1 which in turn reduces the mass flow through T1 and T2 and eventually shows up as a 0.25 K higher discharge temperature from T2 and a similarly higher temperature of the flow to the load. The low efficiency of T3 results in a loss of refrigeration. The calculated turbine work is 261 W while the actual turbine work is 160 W. The loss of 100 W of cooling power in the middle of the load is obviously a most critical loss.

Table I.   Turbine Operating Conditions

| Turbine | Temperature, K | | Pressure, atm | | Flow Rate | Efficiency |
	Inlet	Outlet	Inlet	Outlet	g/s	%
Design						
T1	90.0	77.55	15.1	8.5	59.0	67.0
T2	22.2	16.74	8.4	3.0	59.0	70.2
T3	8.0	7.06	14.0	7.0	70.3	55.1
T4	8.0	6.26	6.5	3.0	69.5	60.8
Actual						
T1	105.5	90.6	14.2	8.0	55.6	65.5
T2	21.9	17.0	7.8	3.1	55.6	71.0
T3	8.8	7.7	14.0	7.1	70.0	32.0
T4	8.0	6.3	6.7	3.2	70.0	63.0

The high heat leak of the system is also of concern. The 440
W loss is at least 300 W higher than that calculated based on
reasonable performance of the transfer lines and the enclosure
insulation system. Previously run tests of large segments of the
transfer lines indicated losses of ~ 0.4 W/m for a total of 53
W. Tests of sections similar to the enclosure have shown that 0.5
W/m should be achieved or 50 W. Allowing another 24 W for each
terminal box brings the total expected to ~ 150 W. During the
tests, the system was surveyed several times looking for possible
sources of the large losses. Infrared surveys of the transfer
lines and enclosure, both by day and night, did not show any
obvious heat leaks in the enclosure, but a number of cold areas
were found on the transfer lines. These were all localized cold
spots which showed condensation and/or frost at various times.
There were no sections that were cold the full length, indicating
that there were no vacuum failures. This was verified by vacuum
measurement wherever possible.

SUMMARY AND RECOMMENDATIONS

The actual operation of the system was quite smooth and required a minimum of manpower. The newly installed automatic
control systems on the turbines performed perfectly although,
except for speed control, they had little to do. The computers
used to monitor the temperatures, pressures, and flows worked very
well. The ability to make real time plots of temperature at

various places in the load was extremely useful in adjusting the trim heaters and lead flows. Of course the amount of data at the end of the run was tremendous, but a complete plot of some of the load temperatures against time was very useful in determining the most stable operating periods to use for calculating purposes. Without these on-line temperature plots and the later complete run profiles it is quite possible that the slight upward or downward drift in temperature would not have been noted and been of use to explain the data scatter. The only operations requiring manual operation were the adjustment of the trim heaters and the lead-flow gas.

Since four of the critical temperature sensors were not trusted, any differences in operation with and without heat exchange between the two counterflow streams in the load could not be observed. It is quite possible that the magnitude of the transfer line losses masks all such small effects.

Again it is pointed out that this system with its large mass of supercritical helium was extremely stable against short-term large perturbations, all of which are damped and disappear into this heat sink. No system or thermo-acoustic oscillations were observed and those temperature oscillations which were observed due to a change in operating mode, etc., were obviously quickly damped.

In order to be sure that the system will work after installation of the superconducting cables with their necessary electrical connections, the present heat leak must be significantly reduced and the available refrigeration capacity increased as much as possible. The transfer lines appear to be the largest offenders as far as heat leak is concerned. One line has been opened, found to have a thermal short at an elbow, and is being rebuilt. Increasing the refrigeration capacity may or may not be more difficult. Perhaps the quickest and easiest 100 W could be found by modifying the inlet nozzles and turbine wheel of the T3 turbine. Since the T4 turbine performs so well in the same temperature regime, it is reasonable to assume that a large improvement in T3 is also possible. It is somewhat more difficult to improve the precooling efficiency of the refrigerator, in particular the first heat exchanger since there is no room in the cold box for a larger heat exchanger. However, it is possible to place another heat exchanger in series, connected at the warm end and placed in an insulated enclosure external to the present cold box. This would have the same effect as increasing the size of the present exchanger. This would have the same effect as increasing the size of the present exchanger. Data exist which should allow the design of such an exchanger.

ACKNOWLEDGEMENTS

The authors gratefully acknowledge the contributions of the whole Power Transmission Group. This is a group effort, from design to operation it required the efforts of everyone at one time or another, and all should be justly proud of their accomplishments to date.

REFERENCES

1. E.B. Forsyth, et. al., The Cooling and Electrical Excitation of 100 m Superconducting Power Transmission Cables, in "Proc. 8th Intl. Cryo. Engr. Conf.," C. Rizzuto, ed., IPC Sci. and Tech. Press, Guildford (1980).
2. G.H. Morgan and J.E. Jensen, Counter-flow cooling of a transmission line by supercritical helium, Cryogenics, 17(5): 259 (1977).
3. R.J. Gibbs, Supercritical Helium Refrigerator to Cool Flexible AC Superconducting Power Transmission Cables, in "Adv. in Refrig. at the Lowest Temp.", Intl. Inst. of Refrig., Comm. A1-2, Paris (1978), p. 113.

DISCUSSION

Question by R.G. Scurlock, University of Southampton, England: What thermometry is being used on the cable? Can you comment on the $\pm$ 10 mK accuracy you are quoting?

Answer by author (RAT): The temperature sensors are germanium resistance temperature sensing elements. A large batch was obtained commercially and then cycled repeatedly between room and liquid helium temperatures. Only those sensors which gave stable and repeatable readings at liquid helium temperature and which had no excessive interlead contact resistances were chosen for calibration. All sensors were calibrated against the same standards at Dr. C.A. Swenson's laboratory at Iowa State University to an accuracy of 2 mK below 12 K. The quoted accuracy of $\pm$ 10 mK allows for deficiencies in the digitization of the measurements by the computerized data acquisition system and inherent errors arising from installation difficulties. (Note that it is important to have all the sensors calibrated against the same temperature scale. The helium vapor pressure scale differs from the newer EPT-76 by 7.1 mK at 4.2 K).

DESIGN AND CONSTRUCTION OF A SUPERCONDUCTING MAGNET SYSTEM FOR THE ABSOLUTE AMPERE EXPERIMENT*

W. Y. Chen and J. R. Purcell

General Atomic Company
San Diego, California

and

P. T. Olsen, W. D. Phillips, and E. R. Williams

National Bureau of Standards
Washington, D. C.

INTRODUCTION

The Electrical Measurements and Standards Division of the National Bureau of Standards will undertake an absolute ampere experiment,[1,2] which will involve measuring the force exerted on a current-carrying, normal conductor coil by a set of superconducting coils and also measuring the voltage induced in the normal coil as it is moved in the field of the superconducting coils. This direct measurement, if performed with precision in the neighborhood of a part per million, holds the promise of resolving small but important discrepancies between indirect methods of determining the absolute ampere from measurements of fundamental physical constants. To achieve the desired accuracy and resolution, the superconducting coils are required to generate precise radial fields of about 0.2 T at a radius of 35 cm, over a region of $\Delta R = \pm 0.8$ cm and $\Delta Z = \pm 2.5$ cm. The quality of the field is represented by the product $r \cdot B_r$ which must be held uniform within 20 ppm over the region of $\Delta R = \pm 0.8$ cm.

General Atomic Company has been awarded a contract to design and construct the complete superconducting magnet system to be

---

*Work supported by National Bureau of Standards Contract NB80SBCA0496.

utilized in the absolute ampere experiment. This includes the superconducting coils, the dewar, and other accessories. The superconducting coils consist of two main solenoids and two compensation coils used for fine trimming. Figure 1 is a sketch of the overall magnet system required for conducting the experiment.

SYSTEM REQUIREMENTS

The magnet system was designed to meet the following major design requirements:

1. The radial field must be approximately 0.22 T at a nominal diameter of 70 cm in the midplane of the coils.

2. There should be less than 200 parts per million (ppm) change in the radial field at 70 cm diameter for vertical displacements from the midplane of ±2.5 cm.

3. There should be less than 20 ppm change in the product of the radial field times the diameter for changes in the diameter of ±0.8 cm about 70 cm.

4. There should be a rigid mechanical connection between the superconducting coils and some reference plane outside the dewar, designed to minimize the drift of the coils with respect to the reference plane.

5. It is desirable to keep the LHe boil-off rate to 1 L/h or less. Furthermore, the hold time for keeping coils immersed without He transfer should be a minimum of 16 h, with 48 h hold time preferable.

6. The outside diameter of the dewar should be less than 65 cm. The suspended coil, nominally 70 cm mean diameter with 60 cm inside diameter must be able to be passed over the dewar and located at the midplane after the dewar is assembled. Furthermore, the outside of the dewar near the midplane should be concentric with the superconducting coils within 3 mm, or the coils should be adjustable to that tolerance.

7. The dewar and support structure should contain no magnetic materials. Use of metals should be minimized and construction should minimize eddy currents. The dewar must not be metallic.

8. The current required by the superconducting coils should be minimized and must not exceed 60 A.

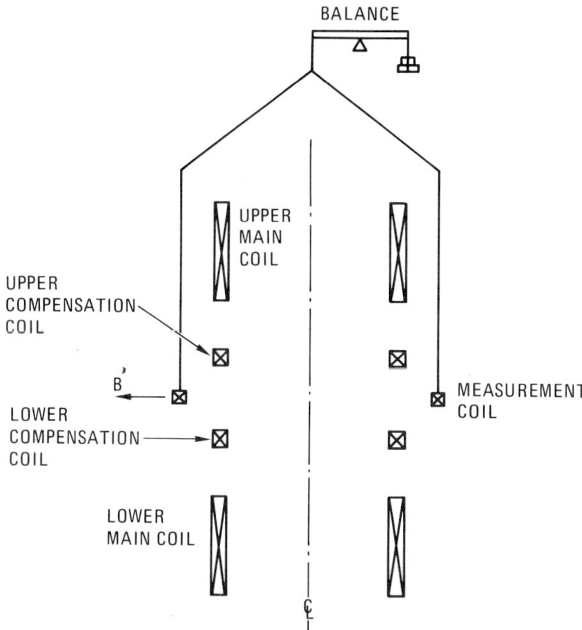

Fig. 1. Overall set-up for performing the absolute-ampere experiment. The currents in the upper and lower coils are in opposite directions, producing a radial field distribution.

9.   The main field coils and the compensating coils should be provided with enough auxiliary current taps so that the area ampere-turns of the two main coils and the two compensating coils can be equalized. The electrical midplanes of the compensating and main coils must be adjustable to coincide within 20 pm, probably by the use of auxiliary end coils with separate current leads.

DESIGN DESCRIPTION

Superconducting Coils

To meet the above requirements, it was decided to configure the superconducting coils as shown in Fig. 2. The magnet system consists of two main solenoid stacks and two compensation coils placed symmetrically about the midplane. Each main coil stack consists of ten identical modules. All coil modules including the compensation coils contain 0.254 mm NbTi-Cu superconducting wires wound on precision machined G-10 bobbins. The entire coil stack

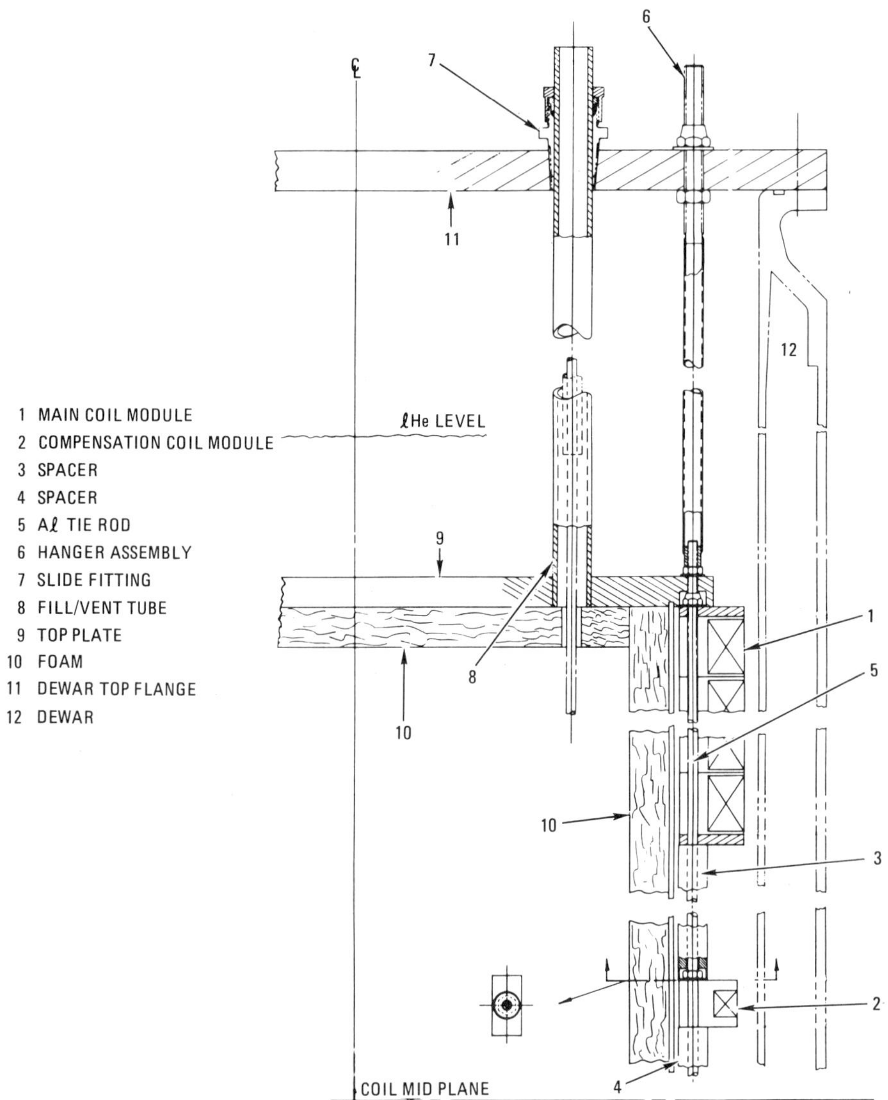

Fig. 2. Design of the absolute ampere coil system.

is precisely aligned and clamped together by 12 aluminum alloy clamping bolts.

Table I is a summary of the major design parameters. It was decided to select the operating current to be 10 A so that the heat leak due to the leads can be minimized. Furthermore, all 22 coil modules are connected in series, so that only two leads are

TABLE I

ABSOLUTE AMPERE COIL DESIGN PARAMETERS

```
OVERALL PARAMETERS
     OVERALL HEIGHT . . . . . . . . . . . . . . . . . . . . . . . . . . . . . . . 151 cm
     AVERAGE RADIUS . . . . . . . . . . . . . . . . . . . . . . . . . . . . . . . 24 cm
     NUMBER OF MODULES . . . . . . . . . . . . . . . . . . . . . . . . . . . . 22
     NUMBER OF SOLENOID STACKS . . . . . . . . . . . . . . . . . . . . . . . .4
                              (2 MAIN COILS, 2 COMPENSATION COILS)
     TOTAL NUMBER OF TURNS . . . . . . . . . . . . . . . . . . . . . . . . .220,472
     OPERATING CURRENT . . . . . . . . . . . . . . . . . . . . . . . . . 10 AMPERES
     TOTAL INDUCTANCE . . . . . . . . . . . . . . . . . . . . . . . . . . . . 8380 H
     CONDUCTOR. . . . . . . . . . . . . . . . . . . . . .10 MIL NbTi-Cu COMPOSITE
     PEAK FIELD . . . . . . . . . . . . . . . . . . . . . . . . . . . . . . . . 2.7 T
     BOBBIN MATERIAL . . . . . . . . . . . . . . . . . . . . . . . . . . . . . G–10

MAIN COIL MODULES
     NUMBER OF MODULES . . . . . . . . . . . . . . . . . . . . . . 20 (10/STACK)
     CROSS SECTION DIMENSION . . . . . . . . . . . . . . . . .3.48 cm x 2.34 cm
     NUMBER OF TURNS . . . . . . . . . . . . . . . . . . . . . . . . . . . . 10,710
     AVERAGE CURRENT DENSITY . . . . . . . . . . . . . . . . .1.32 x 10⁴ A/cm²

COMPENSATION COIL MODULES
     NUMBER OF MODULES . . . . . . . . . . . . . . . . . . . . . . . . .2 (1/STACK)
     CROSS SECTION DIMENSIONS. . . . . . . . . . . . . . . . .1.62 cm x 1.50 cm
     NUMBER OF TURNS . . . . . . . . . . . . . . . . . . . . . . . . . . . . .3136
     AVERAGE CURRENT DENSITY . . . . . . . . . . . . . . . . .1.29 x 10⁴ A/cm²
```

required. However, taps are placed at each module interconnection so that the current in any coil module can be precisely trimmed with a separate power supply. Low current (~1A) leads will be provided for the trimming operation.

To increase the packing fraction in the windings, the superconducting wires will be flattened by a set of rollers to approximately 0.20 mm x 0.36 mm just prior to being wound onto the bobbin. Test results with smaller test coils showed that the flattening process does not affect the superconducting characteristics significantly. The coil winding will be fabricated by a wet layup process. Tests indicated that the tight winding packing obtained through wire-flattening will allow full impregnation.

All coil modules will be layer-wound. An electronic wire guide positioner is used for precision wire placement. Two modes of winding are possible: (1) a helical winding mode in which the wide guide moves linearly with the spindle rotation, or (2) a step winding mode in which the wire guide is advanced only during a small angular span ($\sim 10^0$) of each turn of the spindle. Either mode can allow the production of tightly packed windings.

After all the winding modules are completed, they will be stacked together, aligned and clamped tightly with the aluminum tie bolts. G-10 spacers will be used to hold the various solenoid

stacks in position. The entire assembly will be attached to a G-
10 top plate. The assembly will then be attached via three hollow
stainless steel struts to the top flange which will be attached to
the dewar, the entire system will be held rigidly by the top
flange.

Each coil module is protected during a quench by two diodes
connected in forward and reversed direction. During the normalization of any module, the diodes will act as a low resistance
shunt so that the other modules will not dump excessive energy
into the normal module.

COIL DEWAR

To minimize eddy current effects, the dewar will be constructed mostly with epoxy fiberglass. The dewar essentially
consists of the concentric cylindrical LHe vessel and the outer
vacuum wall. The space between the two cylinders is evacuated.
Two thermal radiation shields and layers of aluminized Mylar
sheets are installed in the vacuum jacket. The radiation shields
are constructed by attaching Cu wires in the vertical direction to
Mylar sheets wrapped into a thin cylinder. The shields are
contact cooled by the boil-off He gas, thus the dewar will not
contain a separate LN_2 reservoir. Baffles will be installed above
the LHe surface to reduce direct thermal radiation and also provide contact cooling to the radiation shields by conduction
through the LHe vessel.

The LHe vessel and the outer vacuum vessel are fabricated by
wet lay-up with fiberglass filament winding. Molecular sieve
coating will be applied to the inner cylinder wall for adsorbing
the diffused helium to reduce the frequency of pump-outs.

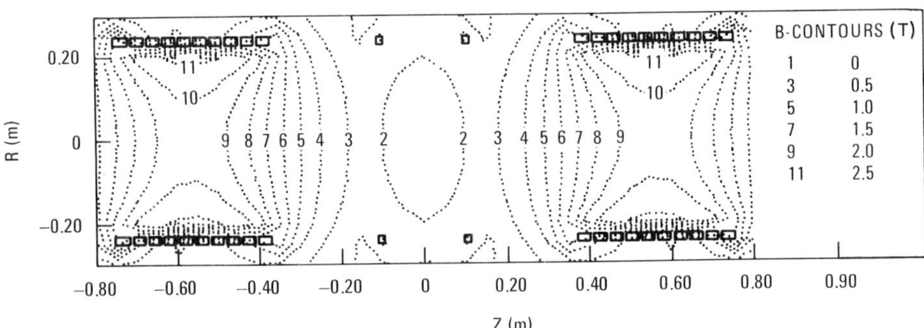

Fig. 3 Typical magnetic field contour plot of the
coil system.

To reduce eddy current effects, the metallic layer of the aluminized Mylar "superinsulation" sheets is segmented in the vertical direction. An external thermal shield consisting of parallel Cu wires running in the vertical direction and attached to a thin Mylar cylinder will be attached on the dewar outer wall to ensure that the dewar outer wall will be maintained at a uniform temperature. The heat leak of the dewar is expected to be about 1 L/h.

FIELD ANALYSIS

The magnetic field distribution of the magnet system has been analyzed. Figure 3 is a field contour plot over the coil vertical cross section. The winding cross sections are also shown. Figure 4 is a field contour plot near the region where the suspended measurement coil will be placed. Figure 5 is a computed flux line trace in the same region. It can be seen that the field lines are almost purely radial in the regions of interest.

FABRICATION

The various components of the coil system are presently (August 1981) in different stages of fabrication:

1. Coil Winding. The various coil modules are being wound. The coil bobbins have already been fabricated by Spaulding Fibre Company. The superconducting wire was provided by Magnetic Corporation of America. A helically

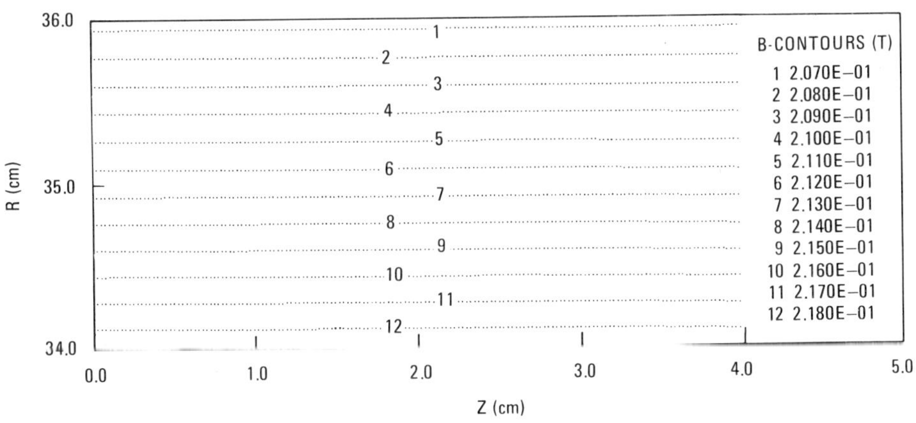

Fig. 4. Field contour plot at the region where the measurement coil will be placed (R = 35 cm, Z = 0 cm). Note: r•B is constant

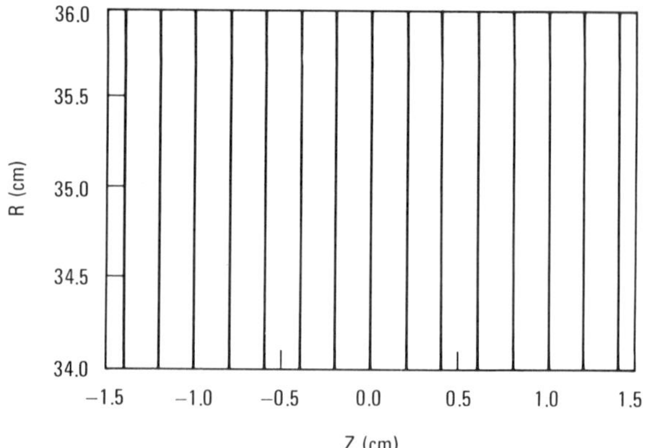

Fig. 5. Magnetic field line trace at the region where the measurement coil will be placed.

wound, wet lay-up winding process is being utilized. About three to five splices are required per module, depending on the wire length in each shipped spool. It is expected that it will require about three months to complete the winding and testing of the individual modules.

2. The Dewar. The dewar is in the final stages at Berkeley Cryogenic Technology Development Corporation. Both the helium vessel and the outer vacuum vessel are completed. The complete dewar is assembled, and liquid helium tests will be conducted shortly.

3. Complete Magnet System. It is expected that the complete magnet system will be completed and assembled in October 1981. The system will then be tested and delivered to NBS where the actual measurements will be conducted.

SUMMARY AND CONCLUSIONS

General Atomic Company has designed a complete superconducting magnet system for NBS to be utilized in the absolute ampere experiment. Key features of the magnet system are high precision, low LHe consumption, low eddy current effects, and modular construction. The complete system is expected to be assembled and tested by October 1981.

REFERENCES

1. P.T. Olsen, et al., _IEEE Trans. Inst. and Meas.,_ IM-29 (4): (1980).

2. P.T. Olsen, W. D. Phillips and E. R. Williams, _NBS Journal of Research,_ 85(4): (1980).

A 50-T/m SUPERCONDUCTING QUADRUPOLE MAGNET FOR A POLARIZED PROTON BEAM FACILITY*

R. P. Smith, J. A. Bywater, K. P. Coover, J. D. Gonczy,
J. A. Hoffman, S. H. Kim, R. C. Niemann,
D. A. Underwood, and R. R. Rezmer

*Argonne National Laboratory
Argonne, Illinois*

INTRODUCTION

High-gradient quadrupole magnets are required for a polarized
proton beam facility at Fermilab. The gradient must be 50 T/m
over a 13 cm diameter clear bore, and the field purity must be of
the order of a few parts in one thousand at a radius of 5 cm. For
cryogenic efficiency the magnets must operate at 1000 amperes.
Passive quench protection is desired in the event beam particles
are accidentally directed into the magnet windings. The design
parameters are given in Table I.

COIL WINDING DEVELOPMENT

The details of the development program and the design features of the magnet have been presented earlier.[1] The winding
shape of the magnet poles was modified during the winding development program to facilitate the machining of the ends. The two
faces of each pole are now perpendicular to one another. The pole
ends are numerically machined to be approximately constant perimeter, without the terraced shape required by the earlier approach. The revised winding shape is seen in Fig. 1. The central
angle subtended by the winding blocks was varied to optimize the
calculated field purity. This single-parameter design was seen to
provide essentially as good a field quality as the earlier design,
and well within the requirements.

---

*Work supported by the U.S. Department of Energy.

Table I. Design Characteristics of the Quadrupole

Gradient	50 T/m
Winding Bore Diameter	15 cm
Clear Bore	13 cm
Operating Current	1000 A
Effective Length	2.80 m
Stored Energy	3.4×10^5 J
Winding OD	20 cm
Sextupole at 5 cm Radius	$\leqslant 1 \times 10^{-2}$ quadrupole
Sum of Higher Multipoles at 5 cm Radius	$\leqslant 1 \times 10^{-2}$ quadrupole
Overall Cryostat Length	3.35 m
Overall Cryostat Diameter	56 cm
Iron Yoke ID	30 cm
Iron Yoke OD	45 cm
λ J Operating	25 kA/cm^2
Conductor Monolith	NbTi
Cu:SC Area Ratio	1.8 : 1.0
Conductor Insulation	Formvar with Kapton film overwrap
Conductor Support	Epoxy impregnation
Coil Support	Aluminum rings and aluminum bore tube

Fig. 1. Cross Section of Quadrupole Magnet.

The four poles are assembled on an aluminum bore tube and clamped against the Lorentz forces by aluminum rings that are closely fitted at an elevated temperature and allowed to shrink tightly onto the assembled coils. The final practice pole was wound with superconductor instead of copper wire and was sectioned into four short pieces. These sections of coil were clamped to the bore tube and banded with a wet-layup ring of epoxy and fiberglass. After curing, the ring was precisely machined. This mock-up was used to study the technique of shrink-fitting the aluminum rings.

Care was taken not to heat the ring above the glass-transition temperature of epoxy; at a preheat of $100^{\circ}C$, a 0.010 cm diameter clearance between the aluminum ring and the epoxy-glass ring was found to permit assembly. The resulting radial interference at room temperature is 0.013 cm, which increases to 0.023 cm at 4 K.

The azimuthal compression generated in the coil windings by this preload was shown by finite element stress analysis to exceed the Lorentz forces so that the coil does not separate from the winding post. It was found possible to reheat an aluminum ring applied in this manner so that it was easy to remove it to inspect the epoxy-fiberglass ring for damage.

PROTOTYPE MAGNET RESULTS

Four poles having a straight-section length of 40.64 cm were wound, vacuum impregnated, and assembled following the procedure discussed above. Steel bands were used to clamp the poles to the bore tube while the fiberglass rings were applied. These clamps were removed and the heated aluminum rings all installed from one end to demonstrate the feasibility of using this technique for a full-length magnet. The prototype is shown in Fig. 2. It has an effective length of 53 cm.

The completed coil was tested without iron in a vertical dewar. An iron shield would contribute 11% of the total field so that the aircore design gradient is 45 T/m. Potential taps on the four poles were connected to potentiometers and recording instrumentation. Small inductive pulses were observed during the first charging. The frequency of occurrence and their magnitude were much less on successive chargings.

The magnet exceeded the design current, 924 amperes, on the fourth quench, and surpassed 95% of short sample on the fifth. The first three quenches originated in the same pole (the first one wound). The fourth and fifth quenches originated in other poles. Because the poles are quite thermally isolated, only one

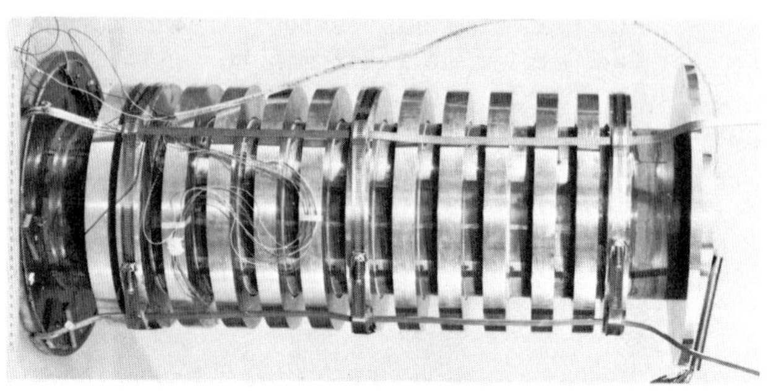

Fig. 2. Completed Quadrupole Magnet.

pole is typically driven normal during the quench. The magnet
tests support the calculated results that the magnet is completely
safe during such a quench.

FIELD QUALITY MEASUREMENT

Field quality was measured under operating conditions in the
test cryostat with a set of 24 inductively coupled stationary
coils. The coils were arranged on the surface of a cylinder 5 cm
in radius so as to couple to the radial field component of the
magnet. Outputs from the individual voltage integrators were
processed with CAMAC based microcomputer system. The phase and
magnitude of multipoles up to the twenty-second were immediately
available with an accuracy of 2×10^{-3} of the primary quadrupole
field at 5 cm whenever the magnet was charged or discharged. By
making measurements which included various proportions of end and
central field, the central field quality and multipole contributions from the winding ends were obtained. The quadrupole field
strength was found to be 49.7 ± 0.8 T/m-kA and the maximum
strengths of high multipoles were found to be 4×10^{-3} of the
quadrupole at 5 cm for the 12-pole and 20-pole in the central
region and 5×10^{-2} for the 12-pole in the ends.

REFERENCE

1. R.P. Smith, et al., A high gradient quadrupole magnet for a
    polarized beam facility, IEEE Trans on Magnetics, MAG-17
    (1):176 (1981).

MANUFACTURE OF A 6-m SUPERCONDUCTING SOLENOID INDIRECTLY COOLED BY SUPERCRITICAL HELIUM

T. Satow, T. Kawaguchi, O. Ogino, and T. Kawamura

Mitsubishi Electric Corporation
Kobe, Japan

INTRODUCTION

The technical development of a superconducting solenoid for a pulsed muon channel[1] is described. Important features of the solenoid are the epoxy-impregnated coil with sharp-cornered mono-lithic superconductor and rectangular supercritical helium (SHE) cooling tube, and the cryostat with warm-iron vacuum wall and improved suspension columns.

Prior to the construction of the solenoid, a design study was performed and background research was carried out. The main part of the development was the construction and testing of epoxy-impregnated model coils and supercritical helium experiments.

The solenoid system has been operated since the summer of 1980 by the Meson Science Laboratory of the University of Tokyo, located at the Japanese National Laboratory for High Energy Physics[2]. A photograph of the solenoid is shown as Fig. 1.

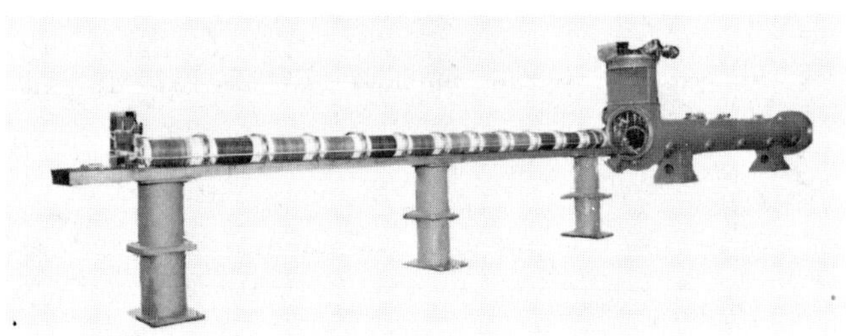

Fig. 1. Superconducting Solenoid drawn out of Vessel.

RESULTS OF DEVELOPMENT PROGRAM

Superconducting Coil

Development work was carried out in order to determine the epoxy impregnant and the impregnation method. The research consisted of: (a) tests on superconducting coil characteristics for 2 small model coils with two different impregnants, (b) tests on methods of impregnation vs cooling characteristics using model coils of copper wire, (c) 4.2 K and pressurized pool cooling test with model coils and (d) SHE test of one model unit.

The model coils were of 78 mm inner diameter, 140 mm outer diameter and 30 mm length. One epoxy formula was chosen because we had used it earlier on other superconducting coils. It was a solventless type without filler. The impregnation method was a vacuum-pressure process and the coils were cured on a rotisserie in order that the resin could not run out of the coil.

Quench tests were carried out with the model. It was found that there was inductive coupling between coil and the non-insulated cooling tube as shown in Fig. 2. The loss in the tube was estimated to be about 17% of the stored energy. Our conclusion was that it was necessary to insulate the cooling tube[3].

No degradation was found in the model unit coil; an exciting current of 930 A was achieved at 4.7 K and 970 A was reached at 4.4 K without quenching. The quench current was 943 A at about 4.6 K in the case of indirect cooling by liquid helium flowing through the cooling tube. These materials and methods were therefore used for the full-size solenoid coils.

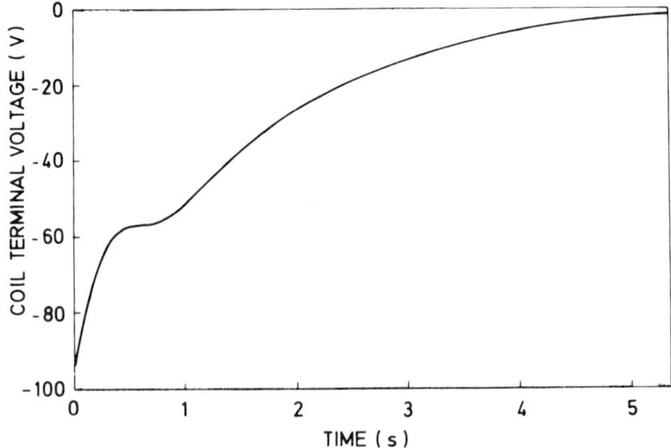

Fig. 2. Quench Test of Unit Coil with Non-insulated Cooling Tube.

Cooling Characteristics

Experiments were made on supercritical helium cooling, using a superconducting test coil. Current lead cooling-excitation tests, pressure drop measurements of the cooling tube, etc. were performed.

The cryostat made for the tests has basically the same structure as the 6 m solenoid. The refrigerator used has a capacity of 15 W at an SHE flow mass of 3.5 g/s and a pressure of about 6 atm. The refrigeration test results are shown in Fig. 3. An SHE inlet temperature of 4.9 K and an outlet temperature of 5.0 K were obtained during the test. From these values the heat losses of the SHE transfer lines and the cryostat were calculated to be 2.5 W and 1.5 W, respectively. From this test it was recognized that in order to obtain a coil inlet temperature below 4.5 K, the heat loss through the transfer tube should be below 0.7 W and a bath temperature of the SHE cooler should be kept at 4.4 K.

The friction factor of the SHE cooling tube was investigated; the test results showed that it was approximately double of that of a smooth pipe calculated with flow in the transition region.

A cooling test of the current leads was also carried out. The required liquid helium mass flow was observed to be 0.3 g/s for a pair of leads at a current of 750 A. The test circuit is shown in Fig. 4.

DESCRIPTION OF MUON SOLENOID

The superconducting solenoid has the structure shown in Fig. 5. It is composed of 12 superconducting coils, a 6 K shield, an

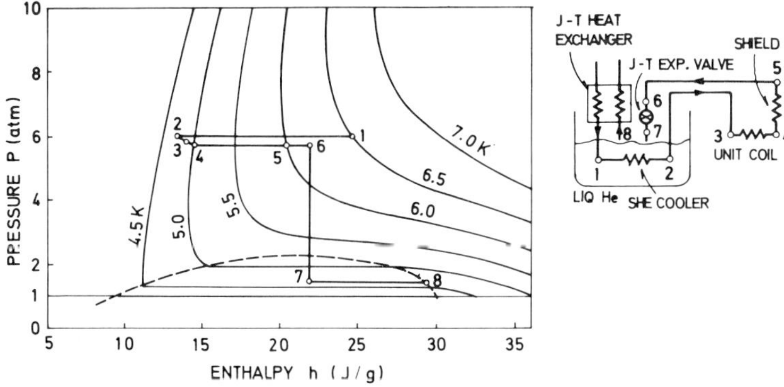

Fig. 3. Pressure-enthalpy Diagram for Test Model.

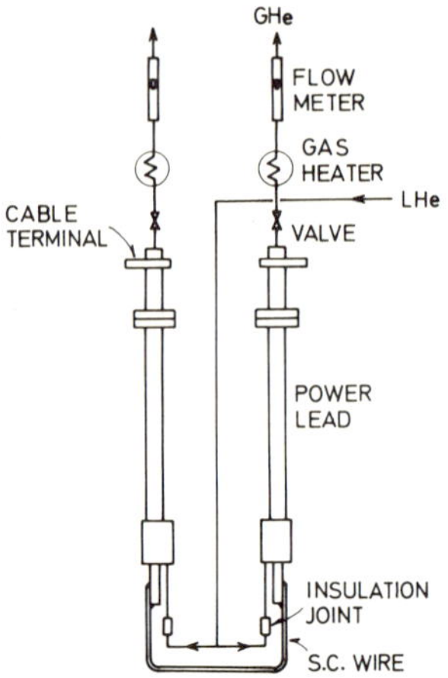

Fig. 4. Test Circuit for Current Leads.

80 K shield, 15 suspension columns, a vacuum vessel, and a port with current leads and pipes for cooling. The solenoid has a total length of 6.5 m, a vacuum bore of 12 cm and a total weight of 6.5 tons.

Superconducting Coil

The dimensions of an individual coil, shown in Fig. 6, are given in Table I. The coil is wound from a monolithic superconductor and is vacuum impregnated with the resin used in the test coils. The specification of the superconductor is given in Table II and the short sample characteristics in Fig. 7. Since the superconductor is Formvar-insulated and sharp-cornered, the

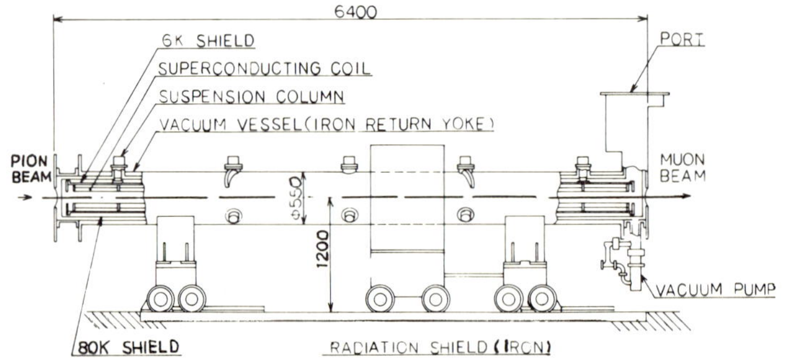

Fig. 5. Structure of Solenoid.

Table I. Dimensions of Individual Superconducting Coil

Effective Bore Diameter	12 cm
Coil Outer Diameter (including Cooling Tube)	21 cm
Coil Length	50 cm
Winding Inner Diameter	13 cm
Winding Outer Diameter	19.6 cm
Winding Length	48 cm
Number of Turns	2750

amount of resin required is reduced which enhances coil stability and thermal conduction.

The coils are wound as double pancakes so that all leads are taken out on the outside of the coil, making it easy to connect or disconnect them. The calculated maximum hoop stress due to electromagnetic forces at a current of 730 A is 3.7 kg/mm^2. The epoxy-impregnated coil is self supporting and does not require outside support members.

A 7 mm x 10 mm rectangular hollow tube for SHE cooling was wound outside the coil. The tube is insulated with glass tape in order to avoid inductive coupling to the coil.

The 12 coils were assembled on a special base plate, and were inserted into the 6 K shield. Good coil alignment (0.3 mm in 6 m) was obtained from the precisely machined inner surface of the 6 K shield.

Table II. Specification of the Superconductor

Dimension	1.6 mm by 3.2 mm
Radius of the Corner	0.2 mm
Diameter of Nb-Ti Filament (average)	3.4 μm
Number of Filaments	1259
Cu-SC Ratio	3.8
Twist Pitch	27 mm
Resistivity of Copper Matrix (at zero Field)	
	$9 \times 10^{-9}\,\Omega\text{-cm}$

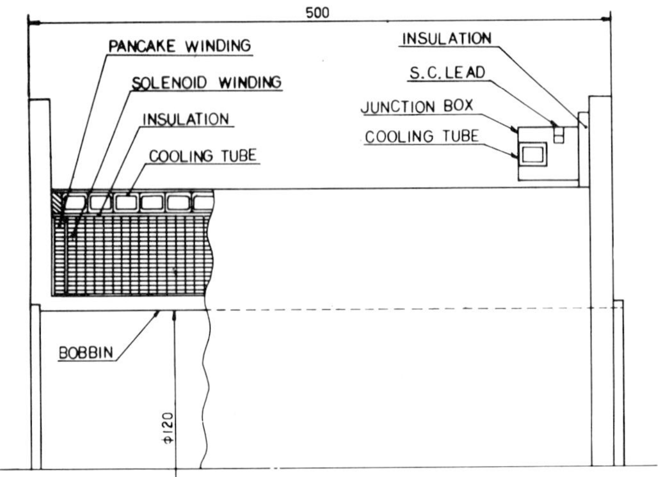

Fig. 6.   Unit Coil.

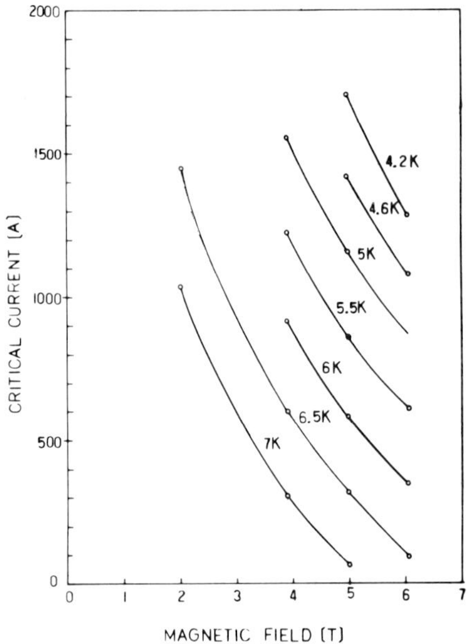

Fig. 7.   Short Sample Characteristics.

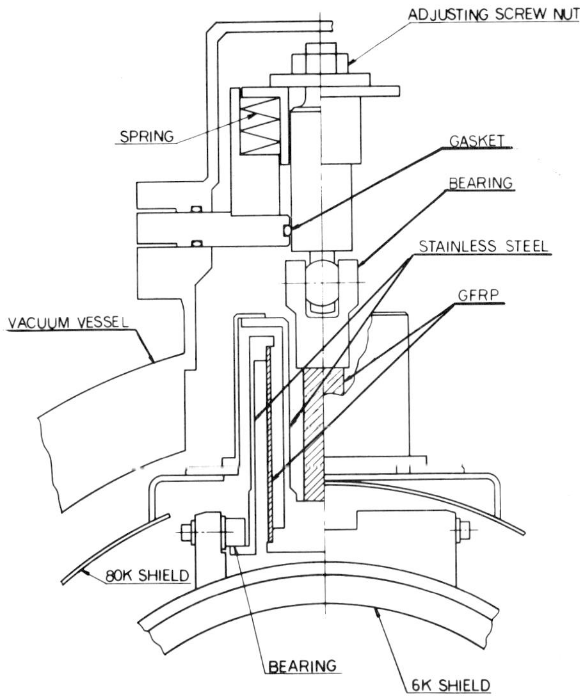

Fig. 8.   Structure of Column.

6 K Shield

The 6 K shield is a 14 mm thick stainless steel cylinder of 29 cm inner diameter and 6.1 m length.   A 10 mm diameter copper tube is soldered on the cylinder.   Supercritical helium after cooling the coil passes through the copper tube to cool the 6 K shield. In addition the shield supports the long solenoid.

Suspension Column

A suspension column consisting of GFRP and stainless steel pipe was developed to support the long solenoid and reduce heat leak to the 6 K shield.   The columns are placed at 15 points on the vacuum vessel to support the superconducting coils, 6 K and 80 K shields (total weight 2 tons) as well as any unbalanced magnetic forces.   The column, shown as Fig. 8 can tolerate thermal contractions of 1 mm in the radial direction and 20 mm in the axial direction due to the spring and bearing.

Vacuum Vessel

The vacuum vessel consists of a 4 cm thick iron cylinder with an outer diameter of 55 cm and a length of 5.9 m, two stainless steel cylinders, and two iron flanges. The iron cylinder is also a warm iron yoke which acts as a return path for magnetic flux. The stainless steel cylinder is to decrease the unbalanced magnetic forces. The iron flanges reduce axial leakage of the magnetic flux shown in Fig. 9. For the case when the coils are displaced by 10 mm in the axial direction a force of about 100 kg is estimated to be between the coils and the vacuum vessel.

Another feature is a pair of counter-flow current leads to be cooled by liquid helium separately supplied from the cold box.

Performance

The quench current of each double pancake was measured in a pressurized LHe bath at 4.8 K. The measured current exceeded 850 A, at which the overall current density is 150 A/mm^2. The coil characteristics are shown in Fig. 9.

As reported in Ref. 1, satisfactory performance data were obtained for both the 6-m magnet and the refrigeration system: (1) for the 4.5 K SHE inlet temperature, the coil temperature was 4.5–4.6 K, the 6 K shield was 6–8 K, the heat leak to the coil was 1 W and to 6 K shield was 6 W (2) the rated current of 730 A was achieved after two training quenches. The reason for this training is supposedly the change in the flux distribution at the end.

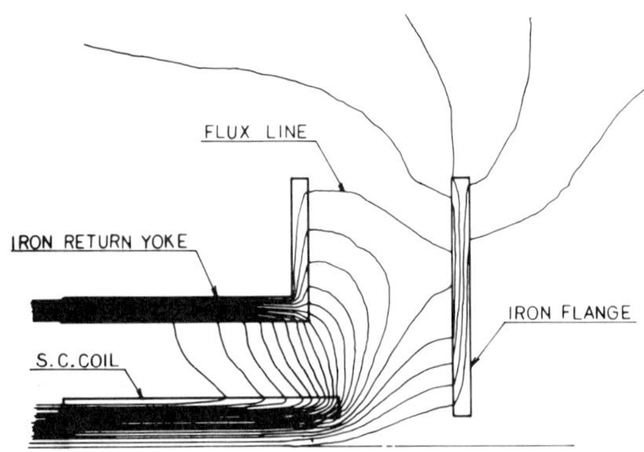

Fig. 9. Flux Distribution.

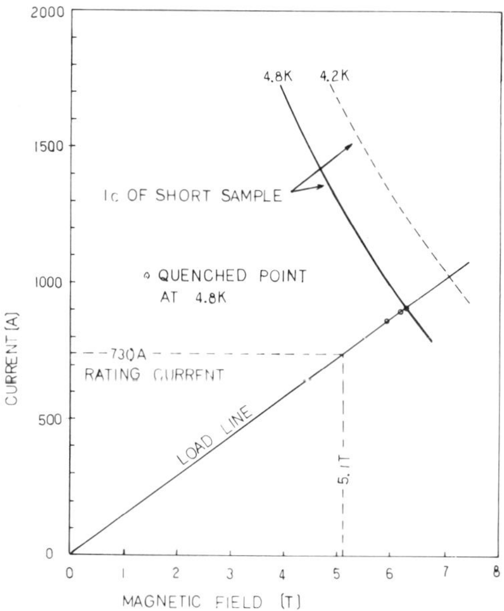

Fig. 10. Unit Coil Characteristics.

CONCLUSION

The solenoid operates satisfactorily and has been used on a high-flux pulsed muon beam. Continuous operation for 2500 hours has been achieved. This is the first long-term operation of a large scale superconducting magnet system in Japan[3].

ACKNOWLEDGEMENT

Special thanks should be devoted to Asso. Prof. K. Nagamine, Dr. H. Nakayama, Dr. J. Imazato and Prof. T. Yamazaki of University of Tokyo for valuable suggestions and enthusiastic discussions with us.

REFERENCES

1. K. Nagamine et al, Superconducting solenoid and its cooling system for pulsed muon channel, IEE Proc. In Magnetics, MAG-17:1882 (1981).
2. K. Nagamine, Pulsed SR facility at KEK booster, Hyperfine Interactions, 8:787 (1981).
3. K. Nagamine, Private communication.

COIL CONVERSION OF THE FERMILAB 30-INCH BUBBLE CHAMBER MAGNET*

W. Craddock, R. Kephart, M. Kobayashi,[†] and M. Mruzek

Fermi National Accelerator Laboratory
Batavia, Illinois

and

I. Pless

Massachusetts Institute of Technology
Cambridge, Massachusetts

INTRODUCTION

A superconducting split solenoid for the Fermilab 30-inch Bubble Chamber is presently under construction. The gravity feed pool boiling coils have independent cryogenic/vacuum systems which can move with the iron halves without disconnection for easier maintenance of the chamber. Magnet parameters are listed in Table I. This magnet will replace the existing 8 MW copper coils as part of Fermilab's energy conservation program.

DESIGN AND CONSTRUCTION

A solder filled 14 strand cable with a central copper core, identical to that used in the new Chicago Cyclotron Magnet at Fermilab[1], is layer wound on to 0.050 in. thick G-10CR periodic spacers as shown in Fig. 1. These spacers form both the cooling passages and the layer-to-layer insulation. The conductor span varies from 1.4 in. on the inside to 2.3 in. on the outside. Gas

---

*Work sponsored by U.S. Department of Energy.

†Permanent and present address: National Laboratory for High Energy Physics, Ibaraki, Japan.

Table I.   Magnet Parameters

Configuration: Split solenoid with
                                          horizontal axis
Central field: 2.75 T
Clear Bore: 1.02 m (40 in.) presently
                                          1.09 m (43 in.) future
Stored Energy: 9.9 MJ
Amp turns: 1.85×10^6 per coil
Current: 675 A
Inductance: 43 H
Winding I.D.:   1.28 m (50.3 in.)
Winding O.D.:   1.71 m (67.3)
Maximum Field on Conductor: 4.94 T
Conductor Dimensions: 0.226 cm × 0.455 cm
                                          (0.089 in. × 0.179 in.)
Conductor Current Density: 6670 A/cm^2
Coil Current Density: 4175 A/cm^2
Numbers of Layers: 61
Copper/NbTi Area ratio: 9.8
Axial Force per coil: 1.2×10^6 (2.7×10^5 lbf)
Axial Magnetic Spring Constant: 3.5×10^7 N/m
                                          (2.0×10^5 lbf/in.)
Vertical Radial Decentering 3.3×10^5 N
   force: (7.5×10^4 lbf)
Horizontal Radial Decentering 1.8×10^5 N
   force: (4.0×10^4 lbf)
Estimated Helium boil-off: 15 liters/hr
Weight of Magnet Iron: 150 metric tons

Fig. 1.   Coil winding at a layer transition showing spacers
and radial cooling grooves in cryostat insulation.

generated between the layers escapes through radial cooling grooves in the ground plane insulation. G-11 and G-30 were selected over G-10 for this insulation because of their superior heat resistance during the closure welding of the outer ring. Perforated Mylar and dimpled Mylar is placed in the cooling channels between layers as added electrical insulation. Perforated Mylar which does not significantly affect conductor stability is used in the high field region with the short conductor spans. The solid film dimpled Mylar is used in the outer layers. A 0.002" thick Mylar pressure sensitive tape is applied to both narrow edges of the conductor and forms the turn-to-turn insulation which is both cheap and surprisingly easy to apply.

The coil is clamped at each of the sixty spacer locations and baked at 125°C five different times during coil winding. Each of the spacers is coated with a B-staged epoxy which fully cures during heating. The elevated temperature and a bearing pressure of 2300 psi between conductor and G-10 imbeds the high spots of the cable into the spacers. This procedure increases the modulus of the G-10 conductor combination from 4×10^5 to 1.2×10^6 psi. Increasing the radial stiffness of the coil reduces its peak hoop stress by 25% and limits the radial strain differences between coil and cryostat. Bonding conductor to G-10 with epoxy restrains the coil when clamping fixtures are removed, and it insures axial force transfer in shear to the outer pushbars.

At the completion of winding an outer ring is slipped on and welded to the inner winding spool which forms a complete 304 stainless steel cryostat. All major cryostat welds are made with the SMAW process using special low ferrite 316L-15 electrode from Teledyne McKay. Final coil preload at each G-10 spacer column from both the I.D. and O.D. is easily accomplished with screws pressing against heavy metal bars as shown in Fig. 2.

FIELD, FORCES AND MECHANICAL CONSIDERATIONS

Figure 3 is one-half of the complex iron yoke. A magnetic field calculation with GFUN-3D was attempted, but there were an insufficient number of iron elements for convergence. Subsequently the problem was modeled as cylindrically symmetric using TRIM, Fig. 4. More iron exists at the bottom of the iron yoke than the top, and there is a slightly greater amount of iron at the beam entrance than exit. This results in a net radially decentering force downward and towards the beam entrance which is relatively insensitive to coil position. Field measurements with existing coils provide the basis for our most accurate estimate of these forces. Uncertainties in both radial and axial forces have led to a support structure designed to carry at least twice the load estimate in Table I. All supports are a minimum ten times stiffer

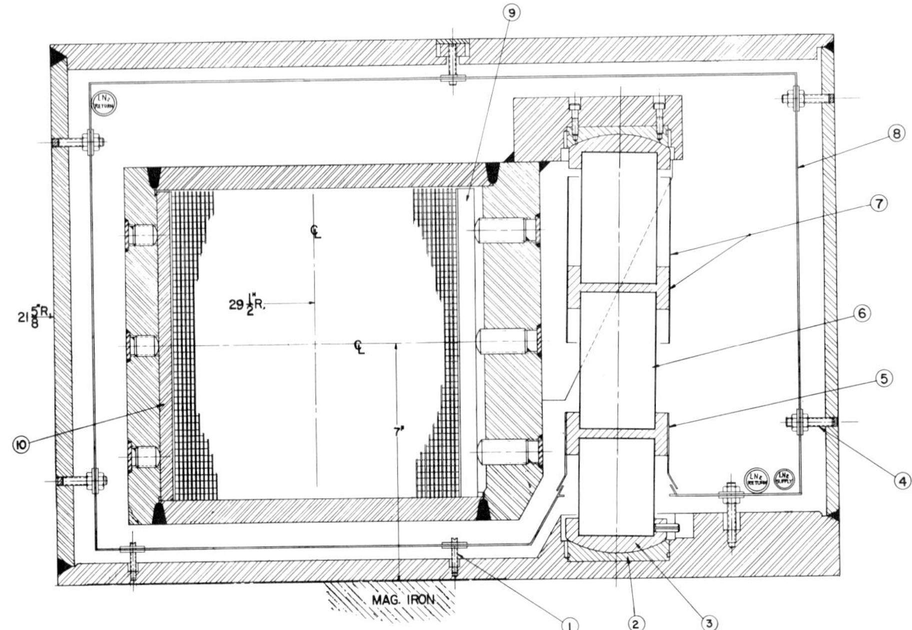

Fig. 2. Typical cross section of cryostat and vacuum shell.
Legend:(1) Ti6A14V standoff; (2,3) 7075-T6 aluminum spherical
bearing surface; (2) Microseal 100-1 impinged graphite coating;
(3) Anodized surface; (4) G-10 standoff; (5) LN_2 intercept;
(6) Randolite fiberglass axial support rod; (7) helium gas
cooled intercept and radiation shield; (8) LN_2 shield; (9,10)
Outer and inner preload bars.

than the magnetic spring constants. Coil vibration induced by the
bubble chamber pistons was studied by Fermilab and the Battelle
Memorial Institute. No problems are expected.

Figures 2 and 5 show the supports. Every support has two
spherical ends which eliminate any thermal stress in these members
during cooldown. Axial and horizontal forces are carried by 2-1/4
inch diameter Randolite* rods in compression. This material was
selected over G-10 and other laminates on the basis of strength/
conductivity ratio, thermal shock resistance, and availability.
Hexcel 1543 satin weave with 80% of the glass oriented in the
axial direction is the reinforcement in Randolite. Epon 828 (CL)
from Shell is the epoxy/curing agent matrix. Table II lists our

---

*Randolite is produced by the Randolph Company, Houston, Texas.

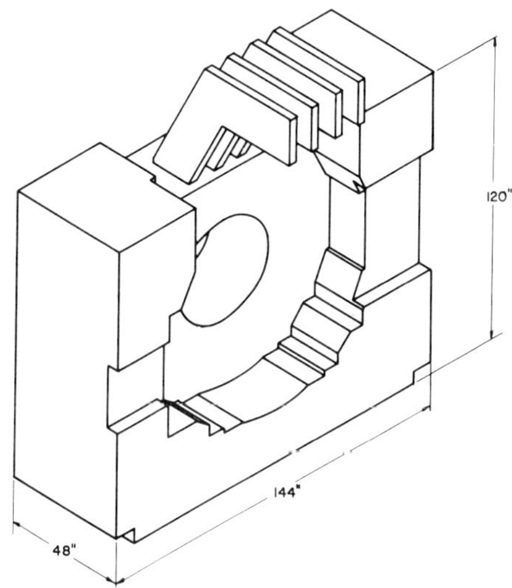

Fig. 3.   Iron yoke half.

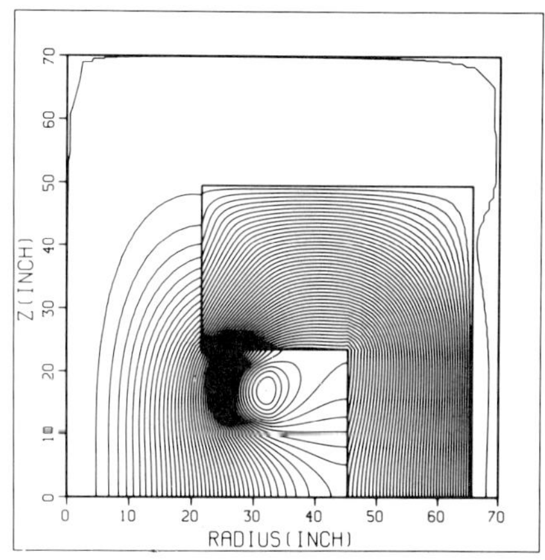

Fig. 4.   TRIM model.

measured properties of Randolite rod with properly constrained ends in the two inch diameter range.

Vertical forces are carried directly to the iron through massive stainless steel tension links and aluminum strain gage bolts. Horizontal forces are transmitted through the vacuum shell

Table II. Axial Properties of Randolite Rod

Compressive Strength	(295 K)	524 MPa $(7.6 \times 10^4 \text{ psi})$	
Compressive Modulus	(295 K)	3.1×10^4 MN/m^2 $(4.5 \times 10^6 \text{ psi})$	
Compressive Modulus	(77 K)	3.4×10^4 MN/m^2 $(4.9 \times 10^6 \text{ psi})$	
$\int_{103}^{300} k(T)dT$		0.95 W/cm $\pm$ 15%	(Ref. 2)
$\int_{20.4}^{76.7} k(T)dt$		0.089 W/cm $\pm$ 15%	
$\Delta L/L_o$ (295 K to 77 K)		0.00127	(Ref. 3)

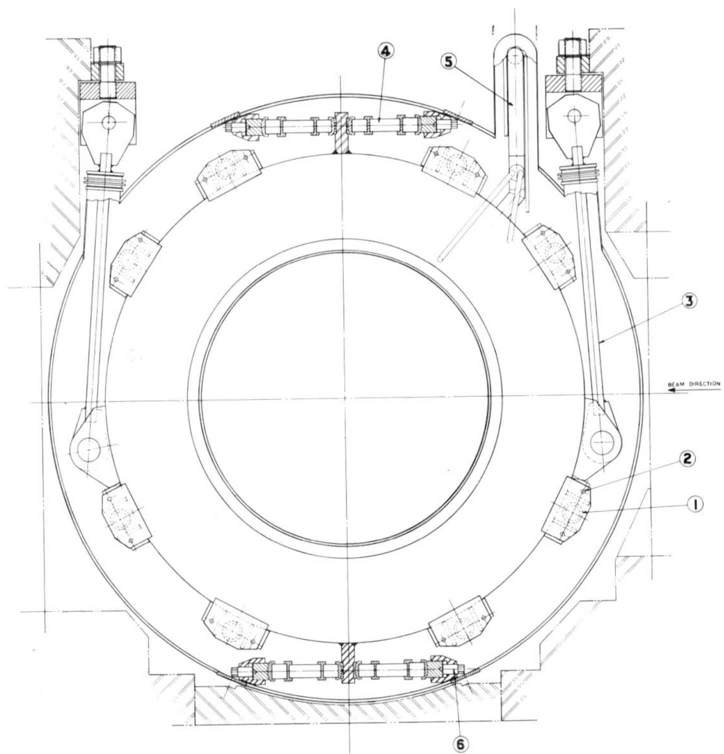

Fig. 5. Cryostat supports. Legend: (1) Randolite axial support columns; (2) Inconel 718 axial pretension rods; (3) 316 stainless steel vertical support links, 1.75" × 3.75"; (4) Randolite horizontal support columns; (5) helium vent; (6) Inconel 600 C.D. preload screws.

to the iron. Sixteen 0.3 inch diameter Inconel 718 rods hold the axial posts in compression when the coils are not energized.

Coil preload and winding tension are selected on the basis of stress results from a computer program [4]. With a 500 psi average radial preload and a winding tension that varies from 110 to 140 lbs, the coil is in radial compression at all times. With full field a maximum hoop stress of 18,000 psi exists at the innermost layer.

The original cryostat stress analysis has been verified with a finite element analysis using ANSYS performed by Fermilab and Computerized Engineering, Inc. Maximum combined stress is less than 25,000 psi anywhere in the cryostat. However, local stresses can approach 70,000 psi at the outer wall screw holes due to stress concentration.

SPLICE AND CABLE PROPERTIES

Approximately twenty lap joint splices joined on the narrow edge of the conductor are wound into both coils. Splicing technique and material characterization of three lead-tin solders (40/60, 50/50, and 60/40) were carefully studied to insure mechanical and electrical integrity[5]. All three solders had approximately the same 4.2 K ultimate strength of 25,000 psi. Elongation data was considered only sufficiently accurate to conclude that 60Pb40Sn solder has six times greater elongation than 40Pb60Sn at 4.2 K. Based on these solid wire tests, 60Pb40Sn was finally selected. Splices of differing lengths were tested in LN_2. Banding both ends with fine copper wire was found to greatly improve splice integrity by preventing a peeling failure mechanism. At liquid nitrogen temperature a properly made splice 4 inches long is stronger than the cable itself.

Electrical resistivity of all three PbSn solders at 4.2 K can be fitted within 20% by the correlation $\rho = 5.8 \times 10^{-7}(1 + 0.016B)$ Ωcm where $B < 5.5$ T is the transverse magnetic field in tesla. At 5.5 T the average resistance per unit length of our splices was measured to be 0.68 $\mu\Omega$/inch. The joule heating of our actual 28-inch splices is therefore in the low milliwatt range.

Additional tests at 4.2 K measured cable mechanical properties as well as changes in critical current and RRR with cyclic loading. Table 3 lists mechanical properties based on an overall conductor cross section of 0.0159 in^2. A room temperature strain of 1.3% did not change the critical current. At 4.2 K a sample of cable was cycled 640 times to 645 lbs (40.5 ksi) with $\Delta\varepsilon = 0.31\%$ and $\varepsilon_{max} = 0.91\%$. The critical current was reduced by 8% and the initial RRR of 107 was lowered by 12%. Any solder filled cable with the same Cu/SC ratio should have similar properties.

CRYOGENICS

Liquid helium is gravity fed from the bottom of the storage dewar to the bottom of the cryostat. Boil-off gas not needed for the current leads is routed by means of internally finned tubing to ~ 10 K intercepts on each of the Inconel rods and Randolite columns. The gas stream is split and flows in a serpentine fashion up each stainless steel support through a series of holes, finally exiting near 77K.

Optimum intercept location for the Inconel and Randolite supports has a broad minimum. An improvement factor of four in boil-off is obtained using this technique. The gas cooling scheme for the stainless steel supports reduces their non-intercepted

Table III. Cable Properties MN/m^2 (psi)

	295 K	4.2 K
Dynamic tensile modulus:	6.2×10^4 (9×10^6)	9.0×10^4 (13×10^6)
Proportional limit:	130 (1.9×10^4)	140 (2.0×10^4)
Ultimate strength:	280 (4.1×10^4)	330 (4.8×10^4)
Max. stress of stable hystersis loop	170 (2.5×10^4)	280 (4.0×10^4)

heat leak by five times. All supports have an additional LN$_2$ intercept.

CONDUCTOR STABILITY AND MAGNET SAFETY

The average surface heat flux from the inner layers of conductor is 0.47 W/cm^2. Since narrow horizontal channels provide very poor cooling for this magnet due to vapor locking, it cannot be considered fully cryostable. Two experiments measured conductor stability. Results indicate that the cold end recovery current is 710 A. Recovery also occurs at 635 A even when a heater inducing a normal zone at 700 A remained on continuously. The normal operating current is between 675 and 700 A. Based on a computer adiabatic heating model of the conductor, all the energy can be safely dumped in an air-cooled cold drawn steel or pure nickel resistor at 500 V. These two materials have a large thermal coefficient of resistivity which results in a nearly constant voltage discharge. This reduces our maximum peak voltage by a factor of two.

REFERENCES

1.   E.M.W. Leung, et. al., IEEE Trans.on Magnetics., MAG 15 (1):199 (1981).
2.   M. Kuchnir, Measurement of heat conduction between 300K and 80K, Fermilab Report TM-744 (September 1977).
3.   M. Mruzek, Experimental investigation of thermal contraction for fiber-reinforced epoxy resins, common plastics and other materials, Fermilab Report TM-974 (June 1980).
4.   W. Young, Wisconsin Superconducting Energy Storage Project, Vol. 1, Appendix VI-A (July 1974).
5.   M.T. Mruzek, Properties and methods of lead/tin splices for superconductors, Fermilab Report TM-994 (September 1980).

DESIGN STUDY FOR SUPERCONDUCTING MAIN FIELD COILS FOR THE OAK RIDGE ISOCHRONOUS CYCLOTRON*

S. W. Schwenterly, P. S. Litherland, J. K. Ballou, R. L. Brown,
W. A. Fietz, W. H. Gray, E. D. Hudson, R. S. Lord, M. S. Lubell,
J. A. Martin, J. R. Moore, R. E. Stamps, and P. L. Walstrom

Oak Ridge National Laboratory
Oak Ridge, Tennessee

INTRODUCTION

The Oak Ridge Isochronous Cyclotron (ORIC) has recently been modified for operation as a booster accelerator for the new tandem electrostatic accelerator in the Holifield Heavy Ion Research Facility. The existing ORIC main field coils produce 1.9 T, and the resulting low bending power limits available beam energies, especially for high masses. This paper summarizes a design study to replace the ORIC main field coils with 3.2 T, superconducting NbTi coils, in order to allow acceleration of ions up to a mass of 238 into the nuclear reaction regime. An important benefit is the lower electric power requirement and operating cost of such coils.

A number of superconducting cyclotrons are in various stages of planning or completion, both in the United States and abroad.[1-4] The conversion of a conventional cyclotron raises problems not encountered in these superconducting machines, but many aspects of their designs have been useful in the ORIC study. Table I shows the major parameters of the superconducting coils proposed for ORIC. The winding pack dimensions, ampere-turn requirements, and forces were determined from computer calculations.

---

*Work supported by the Office of Fusion Energy, U.S. Department of Energy under contract W-7405-eng-26 with the Union Carbide Corporation.

Table I. Major Coil Parameters

Field: 3.26 T in gap, 3.62 T in windings
Winding dimensions: 2.52 m ID × 3.12 m OD × 0.25 m width
Overall current density: 3161 A/cm^2
Stored energy: 53 MJ
Mass per coil: 7730 kg (17,000 lb)
Axial Force: 3.9 MN (880,000 lb) toward midplane at full current
             0.3 MN (76,000 lb) away from midplane at 17% current
Allowable deflections – Coil axial suspension: 0.5 mm (0.02 in.)
Coil alignment tolerances: 1.0 mm (0.04 in.) in all degrees of
                            freedom

MECHANICAL DESIGN

The overall magnet configuration is shown in Fig. 1. The
coil case serves as the winding bobbin, helium containment vessel,
and winding support structure. Axial magnetic loads are reacted
through 24 epoxy-fiberglass support links equally spaced around
the outer and inner diameter of the coil case (Fig. 2). Each link
is pinned between a pair of gussets on the coil case and a clevis
on the end of a 9 cm schedule 40 stainless steel support tube.
The 77 K radiation shield is also supported and heatsunk at this
point. The other end of each tube is welded to a 5 cm thick backplate that is bolted directly to the yoke side pieces, using the
existing coil mounting holes. The mounting holes in the backplate

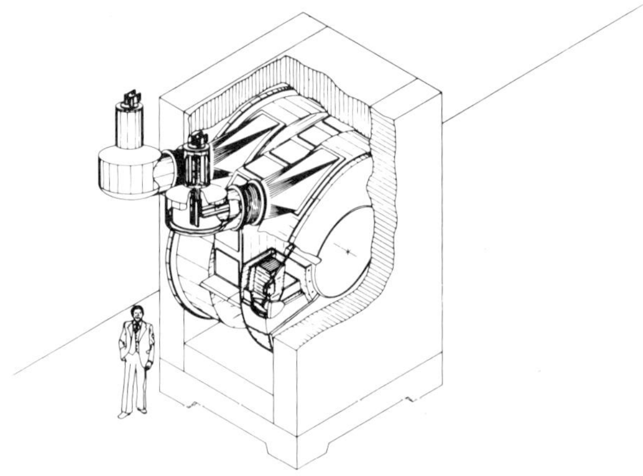

Fig. 1. Isometric view of overall assembly.

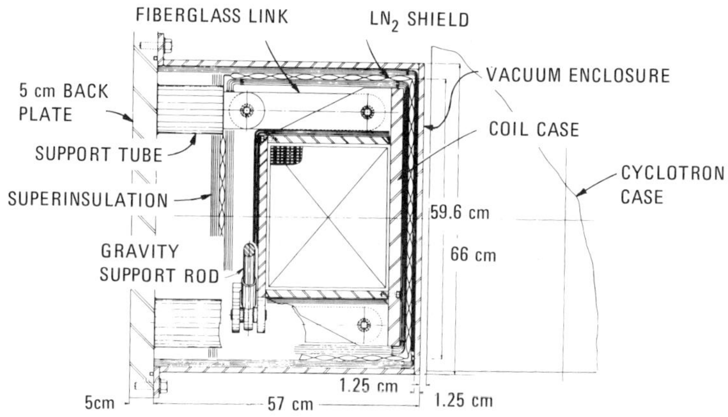

Fig. 2. Detailed assembly cross section.

are drilled oversize to allow final adjustment of the coil radial position. Extra stiffening structure will be added to the sides of the cyclotron yoke to support the large axial forces.

The total heat leak of the epoxy-fiberglass links is limited to 6.3 W, as explained later, and the full-load deflection must be much less than 0.50 mm. When a support link must operate between two temperatures with limited deflection and heat leak, one well-known figure of merit for choice of the material is E/k the ratio of the Young's modulus to the average thermal conductivity over the temperature interval. Uniaxial fiberglass in a NASA Resin 2 matrix[5] has been chosen for the suspension links because it has a particularly high value of this parameter. The design configuration is shown in Fig. 3. The link is fabricated by winding uni-axial fiberglass tape to form a band. After curing, the bands are potted into stainless steel end caps which transfer compressive loads from the bearings to the fiberglass.

Assuming an active length of 20.3 cm and an E value of 69 GPa[6] for uniaxial fiberglass, the 6.3 W heat leak criterion leads to a combined cross-sectional area of 17.2 cm^2, or a band cross section 2.86 cm square. The calculated tensile yield and compres-sive buckling safety factors for the link are 7 and 45 respective-ly. Reported test results[7] indicate excellent long-term dimen-sional stability for a member similar to the proposed design. The deflection of the links is calculated to be 0.28 mm at full field. The stainless steel support tubes deflect 0.22 mm.

The radial forces, including the gravity load are supported by the octagonal array of 122 cm stainless steel rods, 1.9 cm in

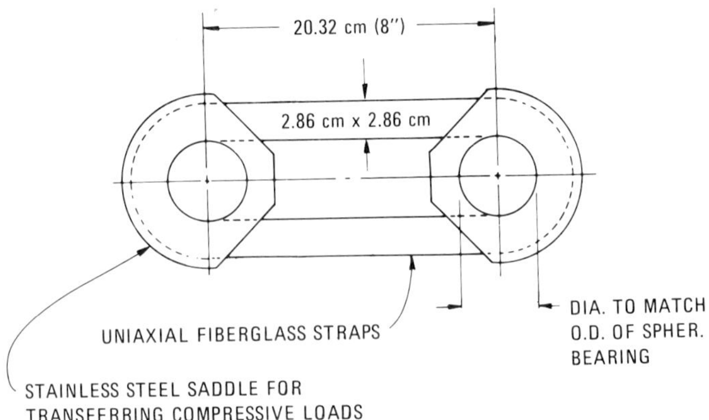

Fig. 3. Fiberglass support link.

diameter shown in Fig. 4. The rods are pinned to the coil at the four 90° positions and to the cold ends of four of the stainless steel axial support posts near the 45° positions. The location of the cyclotron in a Zone 2 seismic area requires a 7% increase in the static design loads. The chosen diameter permits a tensile safety factor of at least three under full gravity and thermal contraction loads. All radial and axial supports are attached with spherical bearings to allow for small misalignments and adjustments and to minimize bending stresses.

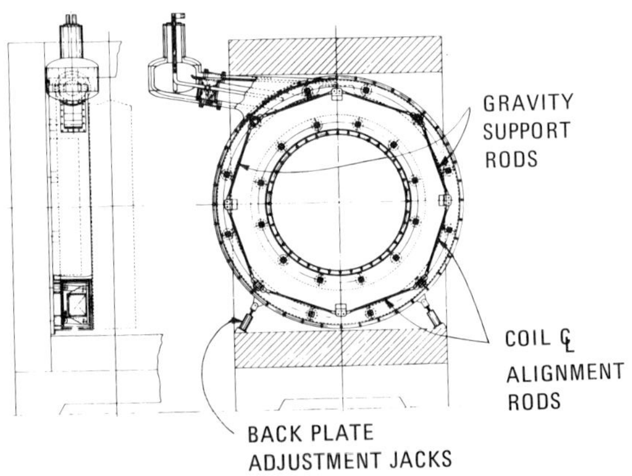

Fig. 4. Assembly side view.

A 77 K liquid nitrogen radiation shield surrounds the coil case. The cold shield extends into the bus duct and terminates around the vapor-cooled lead dewar. Both sides of the LN_2 shield are wrapped with 10 layers of super insulation. A common vacuum envelope extends around the coil case, bus duct, and vapor-cooled lead dewar.

CONDUCTOR AND WINDING SCHEME

For maximum magnet stability, a fully cryostable pool-boiling design has been chosen. The Joule heat flux is about 0.26 W/cm^2 for a normal zone in the conductor. The cooling passages occupy 15% of the winding-pack volume and one-half of the conductor surface is exposed to helium. The conductor area and current are maximized, subject to 5000 A limit on the existing bus work to the cyclotron. This lowers the number of turns and inductance, which in turn reduces the coil dump voltage and allows faster routine charging and discharging of the coils. Coil fabrication is also simplified with the large conductor.

With the above concerns and a consideration of the winding pack dimensions, the available standard insulation thicknesses, and the ampere-turn requirement, a set of conductor and coil parameters has been chosen as shown in Table II. Basically, the conductor will consist of a solid copper stabilizer bar, with a superconducting insert soldered into a 3 × 7 mm^2 groove in one face. The insert will contain twisted filaments of niobium 46.5 wt % titanium alloy, no greater than 60 μm in diameter. The critical current will be between 7000 and 7500 A at 4.6 K and 4 T, and the overall Cu:SC ratio will be about 40:1. The insulation will be composed of perforated sheets of G-10 epoxy fiberglass composite between adjacent turns and pancakes. Figure 5 shows a detail of the proposed conductor and insulation scheme. The coils are wound in continuous double pancakes with no internal joints, because of several coil design factors. Among these are: better transmission of radial forces, reduced conductor motion, better

Table II. Conductor Parameters

Conductor dimensions	1.97 × 0.52 cm^2 (0.78 × 0.20 in^2.)
Pancake insulation	0.16 cm (0.063 in.) thick
Turn insulation	0.12 cm (0.047 in.) thick
Coil width × height	12 × 47 turns
Turns per coil	564
Current	4255 A
Heat Flux	0.26 W/cm^2

cooling, and more efficient packing of the available winding space.

To protect the coils in case of an unanticipated quench, an external air-cooled dump resistor will be connected in parallel with the coil. The dump resistor is sized to limit conductor temperatures to about 100 K in a quench. Assuming all the current transfers to the copper matrix, a computer calculation is made of the temperature rise versus time, using the temperature-dependent resistivities and heat capacities of the conductor materials. The program also considers the effects on the inductive coil time constant due to the iron in the cyclotron yoke. No cooling by the surrounding helium is assumed. The dump resistor is varied until the desired temperature is reached. With the conductor dimensions and current given previously, the computer solution of the heat-up equations leads to resistance and peak voltage of 35 mμ and 150 V, which is below the Paschen minimum voltage for helium breakdown. A protection circuit time delay of 0.1 s is assumed. The current and stored energy are reduced to safe levels within about 10 min and the peak temperature is 104 K.

Radial and axial stresses must be considered in the design of the winding pack, to prevent excessive deflection of the coil cases and to maintain compressive stresses between all layers of conductor and insulation. These two concerns may be satisfied by the design of a sufficiently rigid bobbin and coil case and by the use of appropriate pretension on the conductor during winding. A computer program called STANSOL[8] is used to predict the mechanical stresses in the bobbin and windings during preload, cooldown, and energization. The program considers all loads resulting from pretension internal pressure, thermal contraction, and Lorentz forces

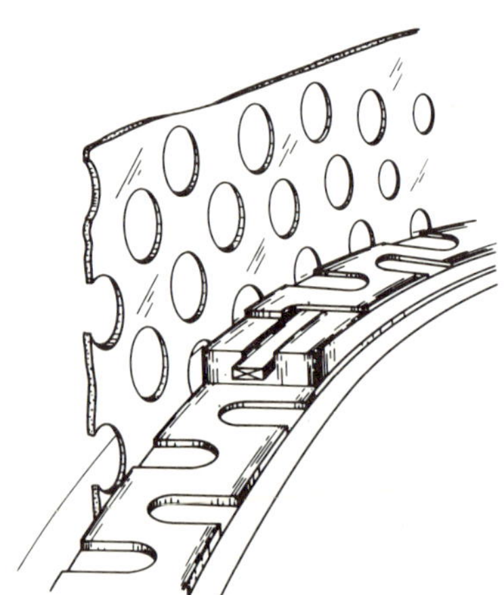

Fig. 5. Conductor and insulation scheme.

and incorporates all the desired material properties and magnetic field distributions. A winding pretension is first assumed; then, STANSOL is used to determine the radial and tangential stresses in the various layers of the coil. The preload and bobbin thickness are adjusted until it is verified that a compressive radial stress of over 0.7 MPa exists at the inner bobbin-conductor insulation interface over the whole range of coil excitation. For the assumed design, a 1.6 cm bobbin thickness and a 3780 N winding tension have been found to be adequate. No external banding is required. The calculated hoop stresses versus radius under full coil excitation are well within safe operating limits. The radial pressure at the bobbin is 0.74 MPa and is compressive as desired.

THERMAL ANALYSIS

The principal heat loads are thermal radiation, conduction and Joule heating in the current leads, conduction through the mechanical suspension, and parasitic losses in the coil lead dewars, main liquid storage dewar, and transfer lines. Minimum heat load is obtained when the total radiation plus conduction heat leak to the coil just balances the "self-sufficient" heat load of the vapor-cooled leads, in this case 7.2 W per coil at 4.6 K and 4255 A. Radiation is calculated to supply 0.3 W per coil, and the radial suspension rods contribute another 0.6 W. The dimensions of the fiberglass links are chosen to pass the remaining 6.3 W.

The total radiation and conduction heat loads amount to 14.4 W at 4.6 K and 1176 W at 77 K. The equivalent liquid requirements are 25 L/h of liquid helium and 26 L/h of liquid nitrogen. In order to handle the parasitic and other unforeseen losses it is good practice to size the helium refrigerator for roughly twice the coil heat load. With a standard commercial refrigerator-liquefier of 100 W or 40 L/h capacity a cooldown time of 160 h is estimated. The refrigerator will draw a maximum of 200 kW of electrical power, as compared to the 1.75 MW consumption of the existing coils.

CONCLUSION

The design study described here demonstrates the feasibility of replacing the existing ORIC coils with superconducting ones. The design is quite conservative, requires no unusual technology, and should result in a coil system with good reliability and durability. The operating regime of ORIC will be considerably extended, and running costs should be reduced. A proposal to continue with detailed design and coil fabrication is currently under review and has been submitted to the Nuclear Science Advisory Committee.

REFERENCES

1. H.G. Blosser, IEEE Trans. Nucl. Sci. NS-26:2040 (1979).
2. H.G. Blosser and F. Resmini, IEEE Trans. Nucl. Sci. NS-26: 3653 (1979).
3. J.H. Ormrod et al., IEEE Trans. Nucl. Sci. NS-26:2034 (1979).
4. E. Acerbi et al., IEEE Trans. Nucl. Sci. NS-26:2048 (1979).
5. L.M. Soffer and R. Molho, NASA CR-72114 (1967).
6. M.B. Kasen, Cryogenics 15:331 (1975).
7. V. Buchanan, D. Dassner, and I. Polk, IEEE Trans Nucl. Sci. NS-26:4039 (1979).
8. N.E. Johnson, W. H. Gray, and R. A. Weed,in "Proc. 6th Symp. Eng. Problems of Fusion Research", IEEE Science Center, Piscataway, New Jersey (1976), p. 243.

THE SUPERCONDUCTING CHICAGO CYCLOTRON MAGNET

E. M. W. Leung, R. D. Kephart, A. S. Ito, and R. W. Fast

*Fermi National Accelerator Laboratory**
Batavia, Illinois

INTRODUCTION

During the evening of February 21, 1981, the superconducting
Chicago Cyclotron Magnet (CCM) reached a full field of 14.52 kG at
a current of 900 A. This magnet, whose design, construction and
cryogenic testing without iron have been reported earlier[1], became
the world's second largest superconducting solenoid (in terms of
radial dimensions) after the BEBC bubble chamber magnet at CERN.
This paper describes the completed magnet test with iron (Fig. 1)
and compares measurements to calculations. Performance of the 24
slider type four tube (three G-10 and one AISI 304 stainless
steel) composite support columns, each capable of a collapse load
of 1.33×10^6 N (3×10^5 lbs), is given in a separate paper[2].
Essential magnet parameters are included in Table I.

STABILITY CONSIDERATIONS

To insure cryogenic stability in a pool boiling magnet, one
might choose the Stekly parameter[3], α, to be $\leqslant 1$. Another typical
criterion is the maximum surface heat flux Y, from the conductor
upon a complete transfer of current to the stabilizer. Keeping Y
$\leqslant 0.3$ Wcm^{-2} has been a conservative magnet design guideline for
years. The authors feel that defining a parameter $\beta = I_{op}I_R$, is
useful for the analysis of coil stability.

---

*Work supported by Universities Research Association, Inc. under
contract with the U.S. Department of Energy.

Fig. 1. A fish-eye view of the superconducting
Chicago Cyclotron Magnet.

Table I. Magnet Parameters

Configuration:	Split solenoid (2 coils)
Winding I.D.:	5.19 m (204.4 in.)
Winding O.D.:	5.48 m (215.6 in.)
Cross section of each 100 turn coil:	142 mm x 117 mm (5.6 in. x 4.6 in.)
Spacing between coils:	1.85 m (73 in.)
Conductor specifics:	See Ref 1
Maximum test current:	900 A
Coil current density:	5415 A cm^{-2}
Conductor current density:	8611 A cm^{-2}
Stored energy at 900 A:	~ 26 MJ
Dump resistor:	0.2 Ω, center tap grounded
LHe refrigeration:	pool boiling, gravity fed, 2000 liter storage in magnet cryostat, intermittent transfer
Steady state LHe boil-off:	given in details later
Steady state LN$_2$ boil-off:	200 L/day (calculated) 288 L/day (measured)
Magnetic field & forces:	Calculated using TRIM & GFUN [1,4,5,6]

Following Stekly,

$$I_R = I_c \sqrt{\alpha} \qquad\qquad \alpha = \rho I_c^2 / hAf'p(T_c - T_b)$$

hence,

$$\beta = \sqrt{\alpha}(I_{op}/I_c)$$

> = square root of the Stekly parameter × fraction of the conductor short sample current at which the magnet is operated.

where

ρ = electrical resistivity of the substrate of a conductor (including magnetic effect) at 4.2 K

T_b = bath temperature

T_c = critical temperature at I_c and B_m

B_m = maximum magnetic field in the coil

p = perimeter of conductor directly cooled by liquid helium

$f'A$ = cross section of portion of conductor carrying current when in normal mode

h = heat transfer coefficient to helium, depends on the construction details of the coil (e.g., width and orientation of cooling channels) [6]

α = Stekly parameter

I_c = critical current of conductor at B_m and T_b

I_R = full recovery current

I_{op} = magnet operating current

The availability of cooling to the conductor is indicated by $\sqrt{\alpha}$ while I_{op} can be adjusted, such that $\beta \leqslant 1$. This approach is most useful when a compact coil has to be built. A choice of $0.9 \leqslant \beta \leqslant 0.95$ should be a good design criterion.

For the CCM coil, using a vertical channel correlation developed by M. Wilson [7], h is calculated to be 0.251 Wcm^{-2}K^{-1}, T_c = 6.5 K, B_m = 2.8 T, I_c = 2500 A (measured), T_b = 4.2 K, ρ = 2.68 × 10^{-8} Ωcm, f' = 0.74, p = 0.548 cm, A = 0.1045 cm^2. Substituting into expressions given above, we get α = 6.85 and I_R = 955 A. Experimentally we have built a small test coil, to simulate the CCM coil cooling situation. It was found that the conductor operating at 900 A with a background field of 2.85 T will recover fully from an R-C heat pulse (τ= 30 ms) of 7.2 J over ~ 1 cm of conductor length. This experiment plus the detection of conductor motion during the initial charge up of CCM convinced us that the $\beta \leqslant 1$ criterion is a useful design aid. For CCM at a I_{op} of 900 A, Y is 0.512 Wcm^{-2}.

An interesting plot of β vs. E, the magnetic store energy, for a number of pool boiling type super conducting solenoids, is shown in Fig. 2. For a given E, a high β represents a more aggressive design.

MAGNET TESTING AND COMPARISON TO CALCULATIONS

Cooldown, Charge Up and Field Measurements

We started the second LN_2 cooldown on February 2, 1981, at approximately the same rate as reported before.[1] By Feb. 8, both cryostats were filled with LN_2. After allowing the whole system to sit in liquid nitrogen for 2 days and removing the remaining LN_2 by pressurizing the cryostats, we then blew gaseous helium through the system at low pressure ($\leqslant$ 3 psig), to displace the residual and LN_2 and GN_2 and to subcool the coils to ~ 72 K. By adopting such a procedure, we required 1500 liters of LHe for further cooldown to 4.2 K instead of the 7000 liters estimated to be required for starting LHe cooldown at 90 K. This represents substantial monetary and time savings and hence a worthwhile procedure when one does not have a liquefier for cooldown. The steady state LHe boil-off without current dropped from 22.4 L/h 19 hours after first fill with LHe to 12.2 L/h 14 days later. It took a long time to reach equilibrium because of the extreme difficulty for the heat trapped in the intermediate AISI 304 column to escape, through long lengths of G-10, into the LHe temperature environment. We started the electrical testing by ramping up and down between 0 A and 200 A to make sure that all the interlocks and dump systems were working. On Feb. 21, we brought CCM up to 900 A in approximately 100 A steps.

The imbalanced voltage between the upper and the lower coils was continuously monitored as part of the quench detection

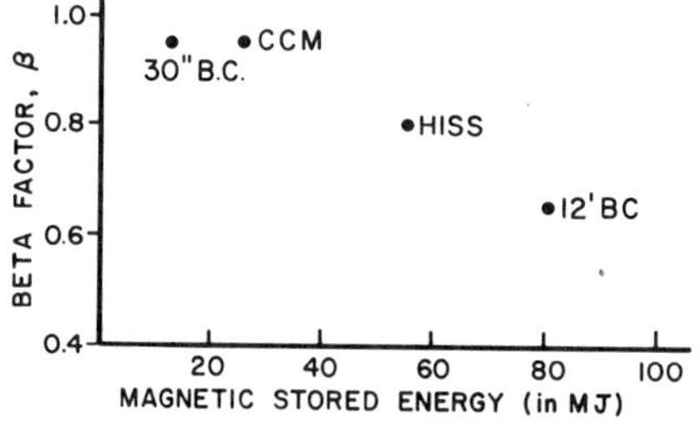

Fig. 2. β vs E for a few superconducting magnets of the same type Ref 8, 9, 10.

system. Prominent spikes of the order of a few volts which disappeared (or were present at smaller magnitude) on subsequent runs were observed. These observations during the first two charges of the magnet up to 700 A are presented in Fig. 3.

It can be seen that there were many more voltage excursions in the first charge. Simple calculations indicate that they are consistent with conductor motion. Such disturbances can grow into quench if the magnet is operating above the full recovery current. Hence, it is important to observe the $\beta \leqslant 1$ criterion.

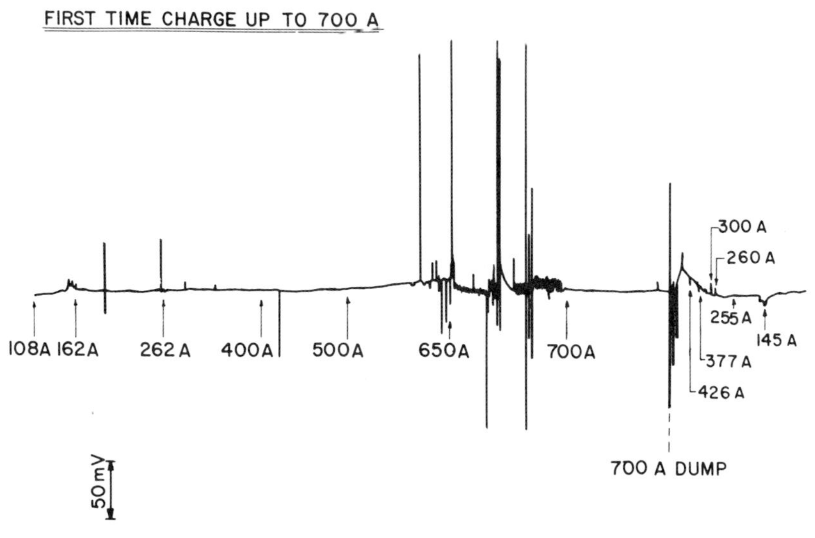

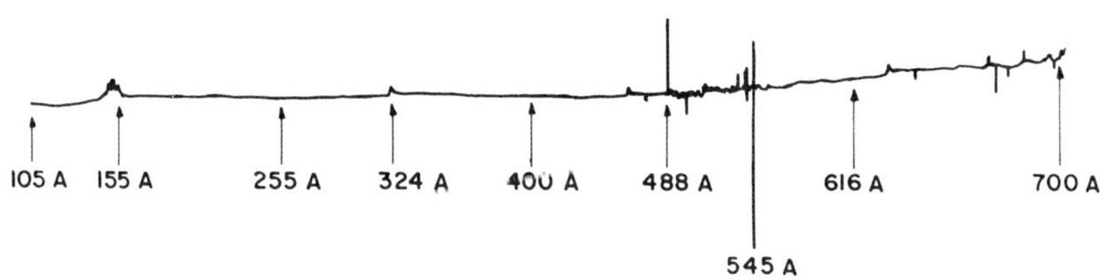

Fig. 3. Imbalanced voltages during charge.

Thermometry, LHe Consumption Rate and Cryogenic System Performance

The electrical resistance of the superconducting coils gives us two useful thermometers during cooldown, helping to generate graphs indicating the average cooldown rate of the magnet. Nine chromel-constantan thermocouples are located on the radiation shields, LN_2 intercepts of selected support columns and the power chimney. Cryogenic strain gages (MicroMeasurements WK-09-250 BP-120) mounted on the AISI 304 tubes of 3 support columns provide the temperature by virtue of the apparent strain vs. temperature curves provided by the manufacturer. During the run, the average temperature of the LN_2 intercept on the support columns was $\sim$ 97 K and that of the radiation shield $\sim$ 87 K. Improvements to the LN_2 system to increase LN_2 flow will probably lower these temperatures and the LHe usage. With the present system, it is estimated that the equilibrium LHe consumption rate will be $\sim$ 11 L/h (Fig. 4) with a T_{ss} = 62 K. Table II gives a breakdown of the heat leak in three different cases.

Table II. Heat Leak Into LHe System of CCM

	Best Measurements in Test without current	Anticipated steady state without current	Ideal case with Improved LN_2 System
Temperatures defined in Fig. 4.	T_{RS} = 87.0 K T_{CI} = 97.0 K T_{SS} = 70.5 K	T_{RS} = 87 K T_{CI} = 97 K T_{SS} = 62 K	T_{RS} = 78 K T_{CI} = 80 K T_{SS} = 48 K
Breakdown			
Columns	4.7 W	3.9 W	2.7 W
Thermal Radiation	2.4 W	2.4 W	$\sim$ 1.5 W
Strain Gages	0.1 W	0.1 W	0.1 W
Chimney	0.3 W	0.3 W	0.3 W
Current Lead & Others	1.1 W	1.1 W	1.1 W
TOTAL	8.6 W (12.2 L/h)	7.8 W (11.1 L/h)	5.7 W (8.1 L/h)
With Current	10.0 W	implying that current leads do contribute additional heat load when in operation.	

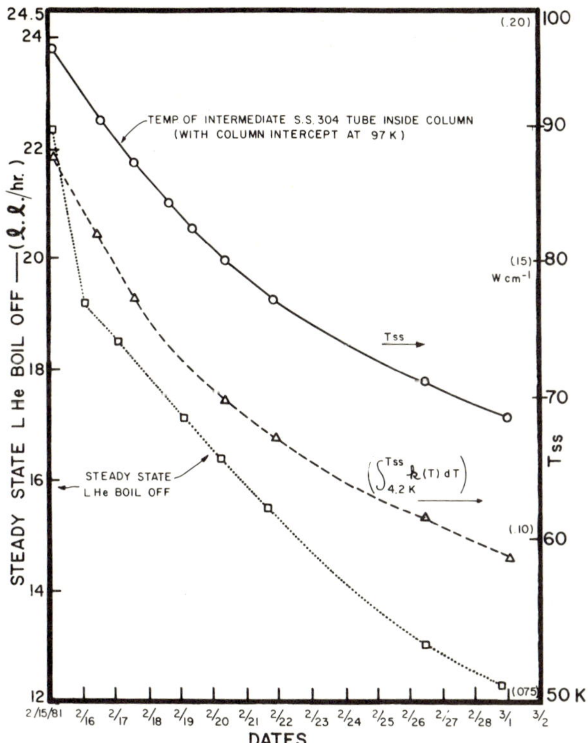

Fig. 4. Magnet LHe boil-off rate (without current), the temperature (T_{SS}) of the intermediate AISI 304 tube of the support column and the thermal conductivity integral of G-10 from T_{SS} to

$$4.2 \text{ K, } \int_{4.2}^{T_{SS}} k(T)dT, \text{ , against time.}$$

where T_{RS} = temperature of radiation shield
      T_{CI} = temperature of column liquid nitrogen intercept
      T_{SS} = temperature of intermediate stainless tube

CONCLUSION AND REMARKS

1. A very low heat leak pool boiling gravity fed superconducting magnet has been built. A new insulating method between 78 K and 4.2 K[1,11], which involves using an aluminum tape and 12 layers of NRC-2 superinsulation, is one of the contributing factors for low heat leak.

2. A beta factor has been introduced which should be useful in designing pool boiling type magnet, especially in the case

where the coil has to be very compact due to a limitation in space.

3. For a small additional heat leak, it is worthwhile to have ample instrumentation, to provide feedback for future if the magnet is a success and diagnostics if there are problems. For example, by having sufficient thermometry, we know that a lower LHe consumption rate can be achieved if the LN_2 is improved.

4. Substantial savings can be realized by conversion projects of this type. CCM will save 99% of the electrical power required to run the conventional magnet and $\sim$ \$200,000 for each year of continuous operation.

REFERENCES

1. E.M.W. Leung, et.al., The superconducting Chicago cyclotron magnet—an old magnet with a new pair of energy efficient coils, <u>IEEE Trans. on Magnetics</u>, MAG-17 (1):199 (1981).

2. E.M.W. Leung, R.D. Kephart and C.P. Grozis, A Low Heat Leak Support Structural Member for the Superconducting Chicago Cyclotron Magnet, in "Advances in Cryogenic Engineering, Vol.27", Plenum Press, New York (1982).

3. Z.J.J. Stekly and J.L. Zar, Stable superconducting coils, <u>IEEE Trans. on Nucl. Sci.</u> NS-12:367 (1965).

4. R.J. Lari, Graphics, Time-Sharing Magnet Design Computer Programs at ANL, in "Proc. 5th Int. Conf. on Magnet Technology," Rome (1975), p. 244.

5. E.M.W. Leung, Magnetic Field Calculation of the Superconducting Version of the Chicago Cyclotron Magnet using GFUN-3D, Fermilab Technical Memorandum TM-759 (1978).

6. C.W. Trowbridge, Progress in Magnet Design by Computer, in "Proc. 4th Int. Conf. On Magnet Technology," Brookhaven (1972) p.555.

7. M.N. Wilson, Heat Transfer to Boiling Liquid Helium in Narrow Vertical Channels, in "Pure and Applied Cryogenics, Vol 6. Liquid Helium Technology," Pergamon Press, Oxford, England (1966).

8. Private communication, W.W. Craddock, Fermilab.

9. Private communication, R.C. Wolgast, Lawrence Berkeley Laboratory.

10. J.R. Purcell, The Superconducting Magnet System for the 12-Foot Bubble Chamber, Argonne National Laboratory Report, HEP-6813 (1968).

11. E.M.W. Leung, R.W. Fast, H.L. Hart, J.R. Heim, Technique for Reducing Radiation Heat Transfer Between 77 K and 4.2 K, in "Advances in Cryogenic Engineering, Vol. 25", Plenum Press, New York (1980) p. 489.

A SUPERCONDUCTING SOLENOID FOR COLLIDING BEAM EXPERIMENTS*

D. Andrews and the CLEO Collaboration

Cornell University
Ithaca, New York

H. Coffey and K. Efferson

American Magnetics, Inc.
Oak Ridge, Tennessee

and

P. C. Vander Arend

Cryogenics Consultants
Allentown, Pennsylvania

INTRODUCTION

The "CLEO" detector (Fig. 1) at the Cornell Electron Storage Ring (CESR) is a large (1000 ton) general purpose instrument for studying electron-positron collisions at high energies. Central to the detector is a 2 meter diameter, 3.2 meter long solenoid coil in an iron yoke. The momenta of ionizing particles produced in high energy collisions at the center of the coil are determined by the curvature of their trajectories in the magnetic field. Detectors outside the coil discriminate between particle species and measure the energy of gamma rays originating in the collisions. For these detectors to function usefully, the number of particles which interact while passing through the coil must be minimized. The coil must therefore be designed for a high current density and/or must utilize low atomic number materials. A water cooled aluminum coil, which produces a 0.42 tesla field when excited with 2 megawatts of electric power, has been in use for

---

*Work supported by the National Science Foundation.

the past two years. A superconducting coil, designed for 1.5
tesla and tested outside the iron yoke to 1.05 tesla, has recently
been installed and is currently undergoing its first tests in the
iron yoke. The new coil is somewhat more transparent to high
energy particles, will improve CLEO's momentum resolution, and of
course will allow substantial savings in electrical power costs.
The savings in operating costs will equal the capital cost of the
coil and associated cryogenic system after approximately one year.

   The high current density requirement precluded the use of a
fully stabilized superconducting coil, the "conventional" design
for coils of comparable size. Protection of the coil against
overheating in a quench was therefore a critical design consideration. The design chosen utilizes a dump resistor across a fast
switch and an inductively coupled soft aluminum bobbin to dissipate the magnetic energy gracefully in a quench. The performance
of this system will be described in a later section. A bath cool-

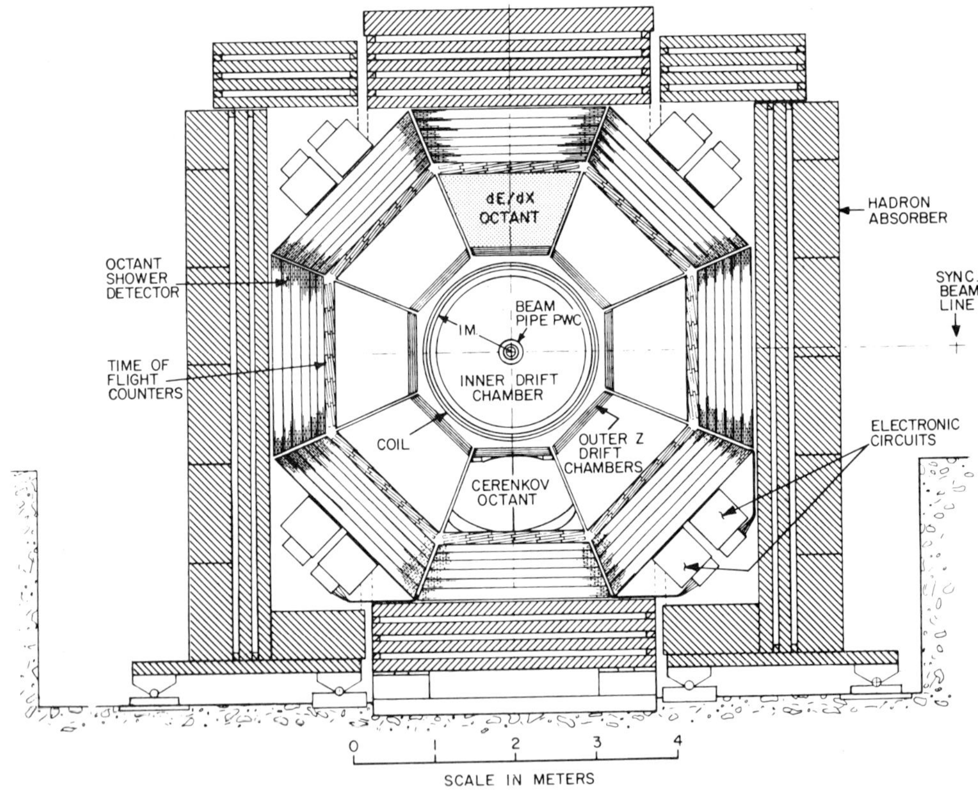

Fig. 1.   End view of the CLEO detector.

ing system would have necessitated a helium pressure vessel, undesirable if the total thickness of the coil and cryostat was to be kept low. Consequently the coil was designed to be cooled by conduction to liquid helium flowing in tubes.

Coils of different design details but similar design goals are in use at the CERN Intersecting Storage Rings,[1] and in the "CELLO" detector at PETRA.[2] A coil of similar design is under construction for the TPC detector at PEP.[3] We relied heavily on the development program at Berkeley for large thin solenoids in the initial phases of our project.

THE COIL

Figure 2 shows a section through the coil. Two layers of niobium titanium wire in a monolithic copper matrix are wound on an 1100-H14 aluminum cylinder (the bobbin). Three layers of 6061 aluminum wire band the coil. The coil and banding were wound with a wet layup technique using a filled epoxy. A continuous spiral of finned aluminum tubing carries the helium coolant. Table I lists some of the coil parameters.

The superconductor was manufactured by Magnetic Corporation of America. Table II lists some of its properties. It consisted of seven pieces with a total length of 36000 feet. Splices between pieces were made by soft soldering the edges of the conductor over one full turn of the winding.

The superconductor was insulated with Formvar 0.004 to 0.005 inches thick. An additional thickness of 0.01 inch of epoxy, on

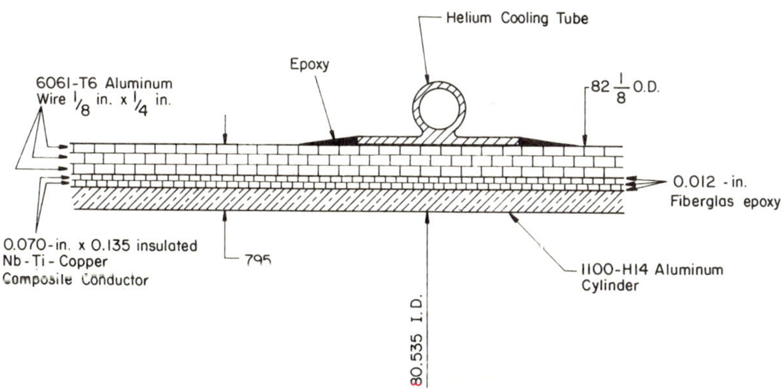

Fig. 2. A section of the coil package. (All dimensions in inches)

Table I. Coil Parameters

Length: 124 inches
Diameter: 81 inches
Number of turns: 1654
Number of layers: 2
Inductance (no iron): 2.8 henries
Inductance (with iron): 3.7 henries
Current required for 1.5 tesla: 2260 amperes
Stored energy at 1.5 tesla: 9.4×10^6 joules
Thickness (including cryostat): 0.213 lb (m)/in^2
                                 .7 radiation lengths
Current density at 1.5 tesla: 2.6×10^5 A/in^2
Cold mass: 3600 lbs

the average, separated turns. Insulation between layers consisted of 3 layers of 0.004 inch glass tape, impregnated with epoxy. Kapton tape was added in the region of the splices. The coil to bobin resistance was measured to be of order 100 megohms, and the insulation withstood 1500 volts in hipot tests.

Both the differential thermal contraction between aluminum and the coil composite during cooldown, and the magnetic hoop forces tend to separate the coil from the bobbin. An important design goal was to avoid tensile stress in the epoxy joint between the coil and bobbin, which could lead to "training." This was accomplished by winding the coil and banding under tension, which puts a compressive prestress on the bobbin. The conductor was wound with 120 pounds of tension (1400 psi) and the first layer of banding with 200 pounds (6500 psi). Strain gauges mounted on the bobbin indicated a compressive prestress of about 8000 psi. The outer two layers of banding were wound with reduced tension, after the strain gauges indicated the bobbin was beginning to creep. We calculate that the bobbin will remain in compression up to magnetic fields of about 1.2 tesla.

THE CRYOGENIC SYSTEM

A section of the all aluminum cryostat is shown in Fig. 3. The radiation shields, 1/8 inch thick, are cooled on one edge only; the temperature at the uncooled end is 110 K. The shields are slit in the median plane and the separate halves are held apart with epoxy-fiberglass (G-10) spacers. The slits interrupt eddy currents and prevent magnetic forces which could distort the shields during a rapid magnet discharge. Eighty layers of NRC-2 superinsulation are wrapped between the shields and the vacuum walls. An additional ten layers are between the shields and the coil.

Table II. Superconductor Parameters

Material: Nb-45% Ti in a copper matrix
Dimensions: .065 x .130 inches
Number of filaments: 1250
Filament diameter: 33.7 microns
Cu/SC: 3.89/1
Insulation: Heavy Formvar (0.004 to 0.005 inch)
Critical current at 1.5 tesla, 4.6 kelvin: 3525 amperes
RRR of copper: 130

The spacing of the coil and shields from each other and from
the vacuum walls is set by a system of G-10 posts. Additional
posts at one end only (the "fixed" end) hold the coil against the
axial magnetic decentering force between the coil and iron
poles. At 1.5 tesla, this force is calculated to be 51 tons for
each inch of misalignment. The free end of the coil slides
towards the fixed end on Teflon pads during cooldown.

The cooling tubes spiral around the outside of the coil on a
one foot pitch. The eleven turns have a total length of 230 feet,
and were manufactured as a single extension. The inside diameter
of the tube is 0.375 inches, and the fin is 2 inches in width.

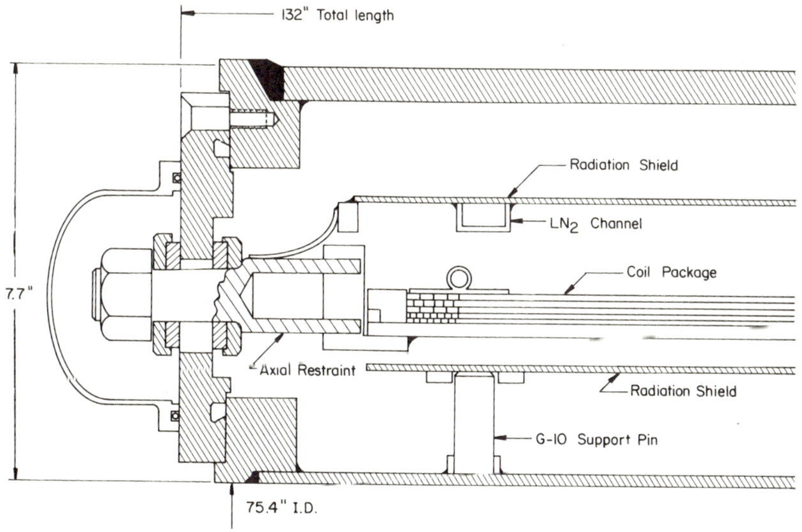

Fig. 3. A section of the fixed end of the cryostat.

All connections to the coil (helium, liquid nitrogen, power leads, and instrumentation leads) come through a four inch square pipe to the fixed end of the coil. The pipe penetrates the 34 inch steel yoke and leads to the "control dewar." The control dewar is a shielded cold box containing control valves, relief valves, vacuum ports, and vapor cooled power leads. The magnet leads are superconducting through the four inch pipe, being soldered to 3/4 inch square hollow copper bus bars carrying helium to and from the coil. The helium plumbing is electrically isolated at each end of the pipe by ceramic tubes on the bus bars.

The control dewar is fed cryogen by a 40 foot long shielded transfer line from a Helix Model 1430 helium refrigerator. During cooldown, cold helium gas is taken directly from the refrigerator. Cooldown from room temperature is accomplished in 48 hours. Until the coil reaches 100 K, the temperature difference between supply and return gas is not permitted to exceed 50 K. For steady state operation, liquid helium is drawn from a 1000 liter storage dewar which is fed from the refrigerator. The total pressure drop through the cooling tube is typically 3 psi. The total heat leak to the coil package, exclusive of the current leads and transfer lines, has been measured to be 10 watts. The refrigerator can keep the entire system cold at the operating current, and liquify an additional 15-20 liters of helium per hour.

QUENCH PROTECTION

A key design problem for a superconducting coil with high stored energy and current density is limiting the temperatures and internal voltage reached if the coil quenches. The solution chosen here was to use an inductively coupled secondary (the coil bobbin) to remove current from the primary (the coil) as the primary resistance grows. Switching a resistor in series with the coil when a quench is detected shifts the ampere-turns to the secondary more rapidly. The performance of this system is illustrated by data from a quench at 1600 amperes, shown in Fig. 4.

The quench detector consists of a bipolar comparator looking for a difference between the voltages across the two coil halves. Inductive voltage during charging are nearly the same for the two halves, making the comparator sensitive only to a resistance difference between the two layers. For a reasonable threshold setting, the comparator can detect a quench in about 10 milliseconds; consequently this part of the quench is too rapid to be seen in Fig. 4.

When the comparator trips, an SCR in series with the coil is opened, switching a 200 milliohm dump resistor in series with the

coil. The ratio of ampere-turns in two closely coupled inductors
discharging together is given by the ratio of their constants.
After the dump resistor is switched in, the time constant of the
primary circuit is 2.8 henries/0.2 ohms = 14 seconds, compared
with 3 seconds for the secondary. Consequently the current in the
coil falls abruptly to 14/17 of its initial value, or 1320
amperes. An eddy current of 0.5 megamperes is now flowing in the
bobbin. In 200 milliseconds the bobbin heats to about 20 K, and
the thermal contact between the coil and bobbin is sufficient to
drive the entire coil normal. The resistance of the normal (but
still cold) coil is 300 milliohms. The primary circuit time
constant has decreased further to 5.6 seconds, and the coil current drops again to 1040 amperes. By now I^2 has decreased to 40%
of its initial value, even though only about 2% of the magnetic
field has decayed. Since the whole coil is normal, the remainder
of the discharge is smooth and safe. The peak temperature in the
winding after the quench, estimated from the time integral of I^2,
is 75 K.

OPERATING TESTS

During June, 1981, the coil was run continuously and largely
unattended in order to gain experience with the reliability of the

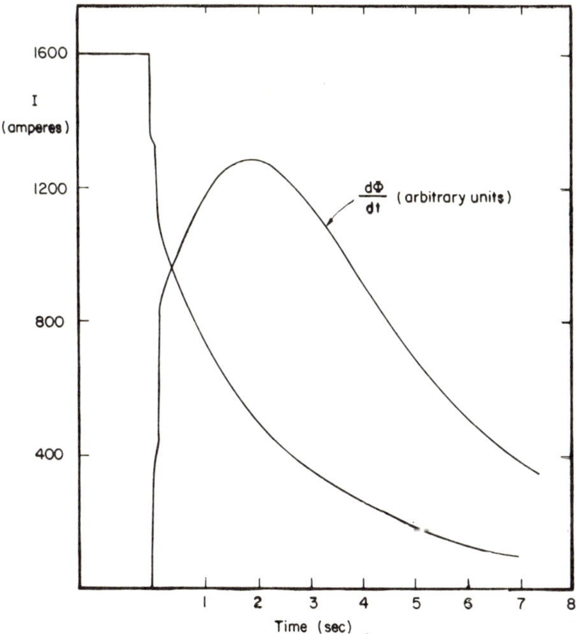

Fig. 4. The coil current (I) and voltage induced on a
pickup coil (dΦ/dt) for a 1600 ampere quench.

entire system. The operating current was 1600 amperes, which would have produced a uniform 1.05 tesla field had the iron yoke been present. The coil was on for a total of 600 hours.

During this period five unintentional quenches were recorded. In each case the waveform of $d\Phi/dt$ and the current versus time indicated that the protection system described above had functioned as intended, and no excessive voltages or temperatures were experienced. The first two quenches were caused by a momentary loss of pressure driving helium through the cooling tubes. In one case a manual adjustment was being made to the compressor suction pressure, and in the other the compressor circuit breaker tripped. A third quench occurred when the voltage drop across one of the vapor cooled leads exceeded the set point of an interlock and initiated a trip. We believe that all three of these failure modes can be avoided in the future. The final two quenches were apparently due to electrical noise in the protection circuits. One occurred during an electrical storm and the other was associated with a step in the line voltage. Further study of these failure modes has been deferred until the coil is operated in the iron yoke.

ACKNOWLEDGEMENTS

Many conversations with Mike Green, Phillipe Eberhard, and John Taylor at LBL are gratefully acknowledged.

REFERENCES

1. M. Morpurgo, Cryogenics, 17(2):89 (1977).
2. H. Desportes, J. LeBars, and G. Mayaux, Construction and Test of the CELLO Thin-wall Solenoid, in "Advances in Cryogenic Engineering, Vol. 25," Plenum Press, New York (1980).
3. M. A. Green, et. al., Construction and Testing of the Two-meter diameter TPC Thin Superconducting Solenoid, in "Advances in Cryogenic Engineering, Vol. 25," Plenum Press, New York (1980).

DISCUSSION

Question by G. Morgan, Brookhaven National Laboratory: What are the temperatures of the liquid helium and the superconductor during steady-state operation?

Answer by author: The helium temperature was 4.6 K; the temperature of the coil may have been as high as 5.5 K at one location, but this value is quite uncertain due to questions of sensor calibration.

CONSTRUCTION AND TESTING OF SUPERCONDUCTING SOLENOID MAGNET MODEL FOR COLLIDING BEAM DETECTOR*

S. Mori, R. Yoshizaki, H. Kawakami and K. Kondo

*University of Tsukuba
Ibaraki, Japan*

H. Hirabayashi, K. Morimoto, M. Wake, and A. Yamamoto

*National Laboratory for High Energy Physics
Ibaraki, Japan*

K. Aihara, Y. Kazawa, H. Kimura, Y. Miyake,
H. Ogata, R. Saito, and S. Suzuki

*Hitachi Ltd., and Hitachi Cable Ltd.
Ibaraki, Japan*

and

R. Kephart and R. Yamada

*Fermi National Accelerator Laboratory[†]
Batavia, Illinois*

INTRODUCTION

Several thin walled aluminum stabilized superconducting solenoid magnets have been constructed and are being used for high energy colliding beam experiments (e.g., the CELLO and CLEO detectors). Construction of a similar, but larger solenoid magnet, 3 meters in diameter and 5 meters long with a central field

---

*Work sponsored by Japan/U.S. Collaboration fund and the U.S. Department of Energy.
†Operated by Universities Research Association, Inc. under contract with the U.S. Department of Energy.

of 1.5 T is being contemplated for the Fermilab Collider Detector
Facility. To begin the study of the problems associated with the
design of this magnet, a small R&D solenoid was constructed and
tested. It has a diameter of one meter and is one meter long.

DESIGN AND FABRICATION OF R&D SOLENOID MAGNET

The main parameters of the solenoid are listed in Table I and
the geometry of the solenoid is shown in Fig. 1.

Mode of Operation

Refrigeration for the coil was provided by flowing LHe at 30
L/h through a cooling pipe wound over the coil. The coil is
stabile as a result of the large thermal diffusivity of the high
purity aluminum stabilizer. The coil is protected from damage
during a quench by insuring that the stored energy is distributed
over the entire coil, thus limiting the maximum temperature rise
in the conductor.

Table I. Parameters of R&D Solenoid Magnet

DIMENSIONS OF COIL	Diameter:	1 m
	Length:	1 m
	Turn number:	269 (excluding 2 joint layers)
	Number of joints:	2
	Joint:	1 turn overlay, 3/4 turn weld
ELECTRICAL	Inductance:	47 mH
	Stored Energy	600 kJ at 5 kA
	Resistance:	315 mΩ at 290 K (0.24 mΩ at 10K)
MAGNETIC	Central field:	1.4 T at 6 kA
	Conductor field:	1.6 T at 6 kA
CONDUCTOR	Filament:	50 μm ϕ × 1400
	Material ratio:	Al:Cu:NbTi = 24:1:1
	Purity of aluminum:	99.99% (RRR $\geqslant$ 1000)
	Short sample current:	7.7 kA at 4.2 K and 2 T

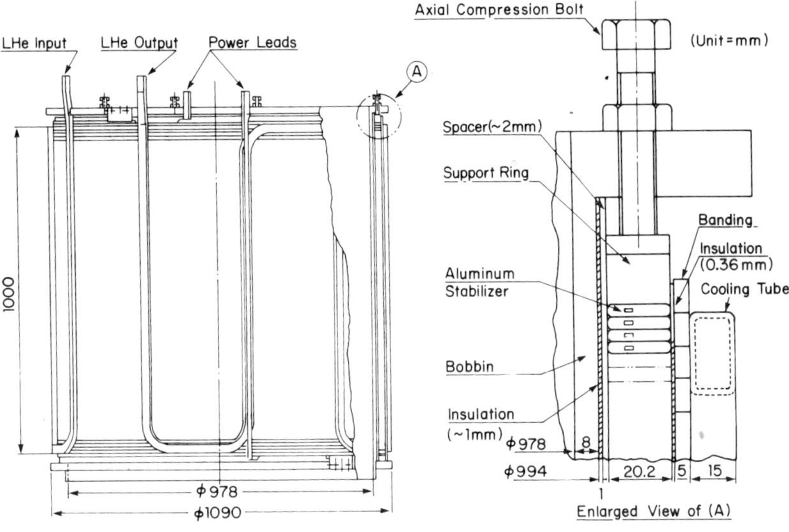

Fig. 1. Geometry of R&D solenoid magnet.

Conductor

The parameters of the conductor used for this magnet are given in Table I and a cross section shown in Fig. 2. A monolithic multifilamentary Cu/NbT: superconductor with a copper to superconductor area ratio of 1 to 1, was stabilized with pure aluminum. The metallurgical bonding between the Cu/NbT: and the aluminum is done by the method developed at Hitachi[6]. The method uses extrusion with front tension, (EFT) in which aluminum is extruded over the superconductor wire while the wire is pulled with forward tension. During this process, the aluminum diffuses into the copper to a thickness of about 2 μm. The cross-section area of the Cu/NbT: monolith is not reduced in the process. The characteristics of the EFT boundary is listed in Table II compared with those obtained by soldering.

The measured resistivity of the high purity aluminum stabilizer after extruding was 1.9×10^{-9} Ωcm at 10 K, while the value at 290 K was 2.5×10^{-6} Ωcm. The residual resistivity ratio of the aluminum before extruding was about 1250. From tests which will be described later, we estimate that the resistivity was 3.5×10^{-9} and 4.0×10^{-9} Ωcm at ~0.5 and ~1.0 T respectively with the increase due to magneto-resistivity. The conductor joints are made by welding the adjacent turns as shown in Fig. 3. In Table III the resistances of welded joints measured with 1 kA at 4.2 K

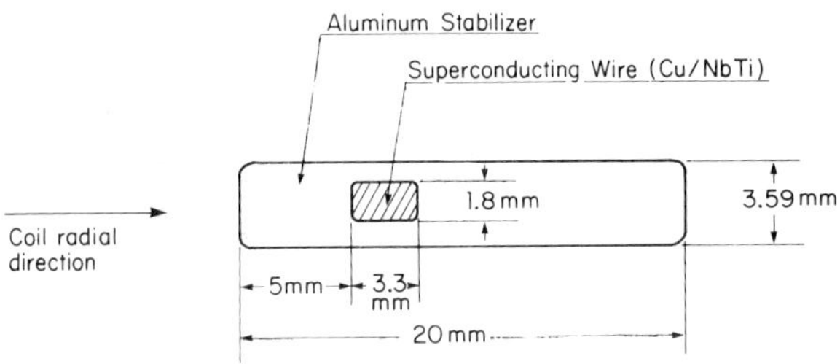

Fig. 2. Cross section of the EFT conductor.

and 1 T are compared with soldered joints. The resistance of the
welded joint is inversely proportional to the total length of the
joint as expected. The total resistance of two welded joints of
400 mm each was measured to be 6.5 × 10^{-10} Ω.

The expected short sample curves at 4.2 and 4.7 K and the
measured curve at 4.2 K are shown in Fig. 4. The load line of the
conductor at the maximum field, which is at the center turn of the
coil, is also shown in Fig. 4. The magnet was designed to operate
at 4.5 kA with the maximum field of 1.21 T. The superconductor is
in a magnetic field, which varies from the central field at the
inner edge to almost zero field at the outer edge. The average
field over the conductor is about one half of the central field.
This should give an additional safety factor for the stability of
the conductor.

Construction Method

The flow chart of the R&D coil construction is shown in Fig.
5. The coil bobbin was made of 5083 aluminum, with a one millimeter thick insulation layer, consisting of mica sheets and glass
tape impregnated with polyester resin. We made the insulation
layer rather thick to prevent ground faults from the coil winding
to the bobbin. The conductor itself was wrapped with 0.1 mm thick
polyamide-imide tape impregnated with B-staged epoxy.

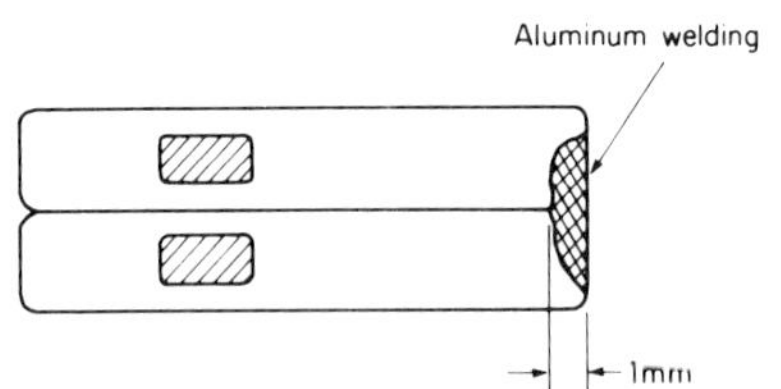

Fig. 3. Joint for EFT conductor. The welding is made keeping
superconductor part cool with watercooling heat sinks.

Table II. Comparison between EFT and Soldering

	EFT	Soldering
Thickness (mm)	$< 2 \times 10^{-3}$	~ 0.1
Resistance (cm)	2.4×10^{-9}	2.1×10^{-8}
Shear strength (kg/mm^2)	> 1	–

Table III. Conductor Joint Methods and Their Resistance

	Length (mm)	Al Welding	Al Soldering
Measured resistances	50	9.3×10^{-9}	$2.4 \quad 10^{-6}$
(Ω)	100	5.0×10^{-9}	$1.6 \quad 10^{-7}$
(at 1 T, 4.2 K, 1 kA)	200	2.5×10^{-9}	$2.5 \quad 10^{-8}$

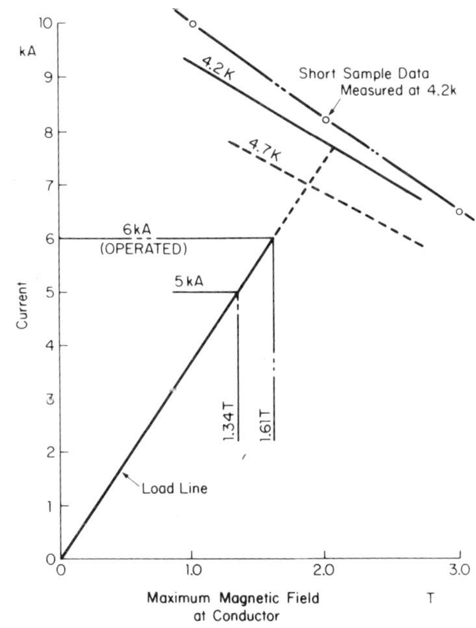

Fig. 4. Load line of R&D solenoid and short sample data.

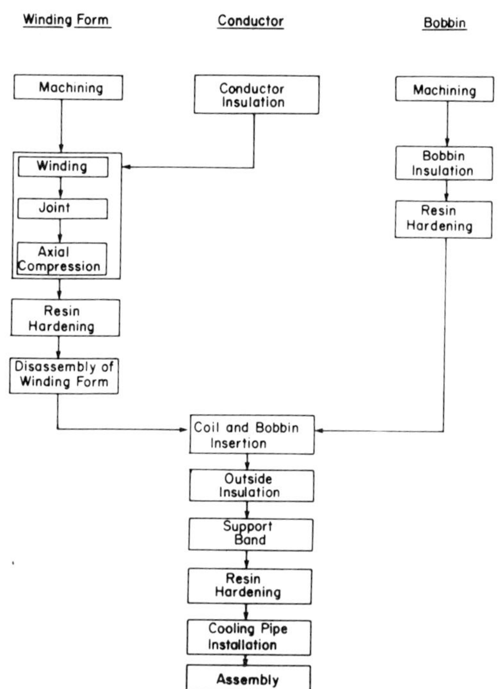

Fig. 5. Flow chart for
construction of R&D coil.

Instead of winding the conductor directly on the insulated
bobbin, the conductor was wound and cured with polyester resin on
a separate metal winding form. With this procedure we could
easily apply an axial compression of 82 kg/cm^2 to the conductor
during winding to prevent conductor movement in the axial direction when excited[7]. After curing the resin, the winding form was
disassembled and removed. The cured conductor winding was then
placed over the insulated bobbin, and the ~ 2 mm gap between was
filled with glass tape. An insulation layer of mica tape and
polyimide tape impregnated with polyester resin was wound over the
conductor winding. Then 5083 aluminum banding, 10 mm wide and 5
mm high, was wound over this insulation with a tension of 100
kg. After curing of the polyester resin at room temperature, the
cooling pipe was glued to the surface of the banding with Stycast
resin and banded again with aluminum wire. The coil was then
placed in a LN$_2$ shielded vacuum can and instrumentation and power
leads were installed. The conductors between the power leads and
the ends of winding were carefully cooled with separate liquid
helium pipes.

The hoop stresses in these component layers were calculated for two cases: a bonded insulation layer and an unbonded insulation layer. The stress after banding, cooldown and excitation to 1.5 T are shown in Fig. 6. In both cases the stress in the pure aluminum conductor is below the elastic limit of 2 kg/mm^2. The stress in the banding is below 4 kg/mm^2, which is lower than the design stress of 9.4 kg/mm^2 for 5083 aluminum. The quality of the insulation layer is not clear but the whole system seems sound.

A heater was embedded into the coil at the mid-plane and about a dozen voltage taps were installed along the whole conductor length. Signals from these taps were used to determine when the normal zone arrived at known points. The voltage increases yielded information about the change of resistance of the conductor in that region.

TEST RESULTS

Two air-core test runs were done on the R&D magnet, in April, 1981 and in July, 1981. In the first test, the magnet was cooled down from room temperature to 4.5 K in 36 hours, using liquid nitrogen and liquid helium from a storage dewar. During the test liquid helium flow from a dewar was used. In the first test run, the magnet reached the design excitation current of 4.5 kA, which would give 1.5 T, if there had been an iron yoke. An attempt was made to raise the excitation current to 5 kA, but it was unsuccessful due to an unexpectedly high local temperature in the coil. It was estimated that the temperature of the conductor was about 6.5 K at some point. After the test a bundle of instrumentation wires was found thermally shorting between the nitrogen shield and the coil structure, and there was some indication that the power leads were not cooled adequately. During the first test

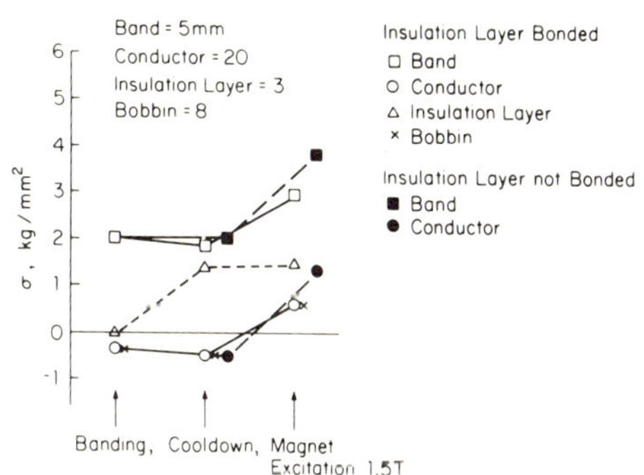

Fig. 6. Stress in coil components.

the propagation velocity of the normal zone was measured at 2, 3 and 3.5 kA during heater induced quenches. The heater was powered with a pulse current of 0.12 s duration, and the energy needed to start the normal region was 22.5 J at 3.5 kA. The measured propagation velocity is shown in Fig. 7.

During the first test run, an estimated energy of about 1.3 MJ was unintentionally dumped into the magnet, while the stored energy was about 400 kJ. The current remained at ~ 4 kA for 30 s after the quench. The temperature of the coil went to 65 K. The interesting fact is that the coil was very stable and can absorb quite a lot of energy without causing any serious damage.

The thermal defects noticed in the first heat run were corrected and several other modifications made to cool the coil more efficiently. In the second test run the magnet was cooled down to 4.5 ± 0.1 K and ran successfully at currents of 5, 5.5 and 6 kA without a quench. In the 6 kA excitation run the current was raised at the rate of 400 A/min from 0 to 4 kA and at 100 A/min from 4 to 6 kA. No unusual behavior such as temperature rise was observed. The heater quench tests were performed at 3.5, 4, 4.5, and 5 kA to measure the propagation velocity of the normal zone. Approximately 13 J was needed to initiate a quench at 5 kA.

The velocities are shown in Fig. 7 with solid lines for the first and second test runs. The estimated velocity of the normal zone along the conductor length (if we assume the normal zone spreads only in that way) is shown on the left ordinate. The effective velocity along the axis of the solenoid is shown on the right. The propagation speeds of the normal zone during the spontaneous quenches of the first test run are shown with dashed lines. One velocity was measured near the quench origin and the other near the center of the coil. The velocity near the quench origin was about twice as much as near the center of the coil. These quench origins were at 20 to 30 turns from the top of the coil. Other data indicated a local abnormally high operating temperature in this region. On the other hand, the velocity data near the center of the coil in the first test run indicated the temperature around there were not abnormally high.

In the second test run, the power supply was kept on intentionally for 3.6 s at 5 kA, after inducing a quench with the heater. No damage was noted. The quench properties were investigated by using computer programs[8]. The agreement between the data and calculated numbers on the temperature rise is fairly good.

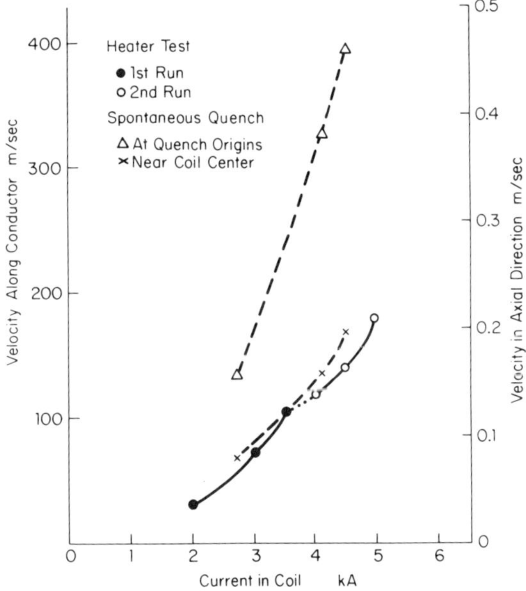

Fig. 7. Propagation speed of normal zones.

CONCLUSION

This R&D coil was excited beyond the designed values without spontaneous quenching. The effective value of I × H was 40% more than the design value. The coil was subjected to quite strong axial stress because the test was done without iron yokes. With this successful test, it seems the whole construction scheme, especially the EFT aluminum stabilized conductor together with its weld joints, is satisfactory for the construction of a large solenoid.

REFERENCES

1. H. Desportes, J. LeBars and G. Mayaux. Construction and Test
    of the CELLO Thin-Walled Solenoid, in "Advances in
    Cryogenic Engineering, Vol. 25," Plenum Press, New York
    (1979), p. 175.
2. D. Andrews, et. al., A Superconducting Solenoid for Colliding
    Beam Experiments, CEC 81 paper FA-4 in "Advances in
    Cryogenic Engineering, Vol. 27," Plenum Press, New York
    (1982).
3. S. Mori, Status Report on the R&D Superconducting Solenoid
    Magnet for the pp Colliding Beam Detector (1) to (7),
    unpublished.
4. Hitachi Ltd., Engineering Reports HES-1 (Nov. 1979) to HES-49
    (July 1980).
5. R. Fast (editor), Conceptual Design of a Large, Thin Coil
    Superconducting Solenoid Magnet for Colliding Beam
    Experiments at Fermilab, Technical Memorandum, TM-826
    Fermilab (Oct. 25, 1978).
6. K. Yamaji, Light Metals 23:87 (1973) in Japanese.
7. R. Yamada, Can we test large solenoid coils safely without
    yoke, CDF Design Note-86, Fermilab (1981).
8. H. Hirabayashi, et. al., Quench properties of superconducting
    solenoid magnet at Fermilab colliding beam experiment,
    Tsukuba Report HEAP-1, Univ. of Tsukuba, Japan.

A HIGH VOLTAGE, HIGH CURRENT CRYOGENIC BUSHING FOR THE WESTINGHOUSE LCP MAGNET MAIN LEADS*

M. A. Janocko, J. F. Quirk, and E. J. Shestak

*Westinghouse Electric Corporation
Pittsburgh, Pennsylvania*

INTRODUCTION

The Westinghouse magnet for the Large Coil Project (LCP) employs a forced flow supercritical helium conductor cooling scheme which precludes the use of the usual gas cooled, current leads, in which the cooling helium gas exits at room temperature. Instead, the Westinghouse design will incorporate a lead with heat leak loads transmitted by conduction in the copper to the supercritical helium stream. The main lead seal which incorporates this copper bus must have an electrically insulating, vacuum tight bushing which can repeatedly withstand the thermal stresses of cooling to cryogenic temperatures. The vacuum integrity of the bushing is especially important, because leakage from either the helium system at the conductor end, or from the environment external to the coil container into the insulating vacuum will cause the magnet to warmup and quench. Additionally, the bushing must support mechanical loading applied by the external leads and thermal stresses accumulating within the coil and transmitted via the attached superconducting conductor.

MAIN LEAD DESIGN

Figure 1 shows the location of the main leads on the header box at the top of the coil. Fig. 2 shows a cross-section of the main lead assembly. The copper lead has a customer-specified taper on the current supply end for a bolted solder joint to the incoming supply bus. A customer-specified flange is provided for

---

*Work supported by Union Carbide Corporation under Contract No. 22X-31747C.

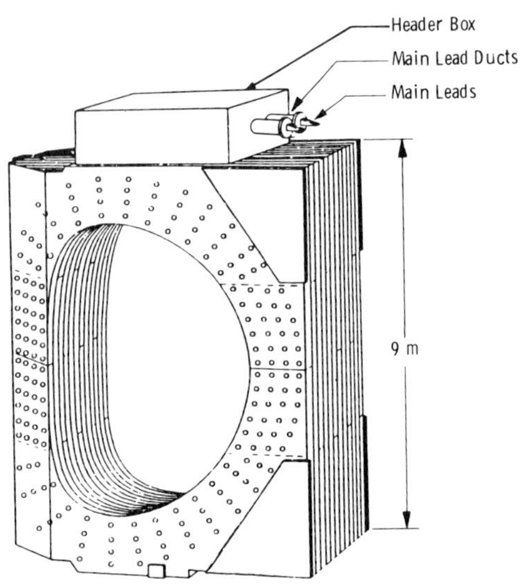

Fig. 1. LCP coil main lead location.

attachment and sealing to the lead duct. A length of superconductor swaged down inside a copper sleeve is brazed into the magnet
end of the copper through-lead and is resistance welded to the
magnet conductor (not shown) in the same manner as all joints in
the magnet winding.[1] Supercritical helium coolant at ~ 4 K is
introduced via a connecting chimney which is electrically isolated
from the rest of the stainless steel helium plumbing system by a
braided fiberglass reinforced epoxy tube (not shown) which itself
presents a challenging materials selection problem which has been
described elsewhere.[2] The chimney is welded to a Nitronic 40
stainless steel adapter collar, which is in turn welded, on the
magnet side, to the header enclosing the superconducting joint and
on the lead side to a lead collar which is brazed into the copper
through-lead. The helium coolant is thus hermetically contained
within the through-lead and header assembly, and exits into the
conductor header after passing through and cooling the copper
through-lead.

The problem of interest here is the electrical and thermal
isolation of the copper through-lead from the lead duct flange
with a vacuum tight, loading bearing seal.

The design is an adaptation of a bushing used for sealing the
lead penetrations of large transformers and of power feedthroughs
in nuclear plant containment walls. The insulation, sealing, and

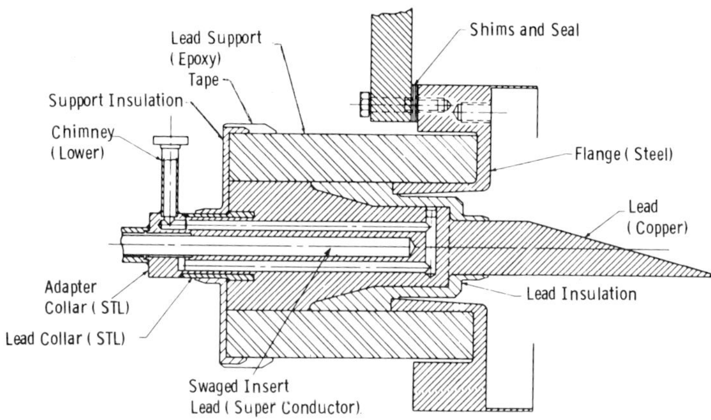

Fig. 2. Main lead assembly cross section.

structural support are all provided by a massive cast epoxy bushing which seals by compressive loading against the roughened
copper bus and the reentrant stainless collar. This compressive
loading is the result of two additive contributions. The first is
the purposeful mismatch of the thermal contraction of the epoxy
bushing to that of the stainless and copper parts. This contraction results from the cooling of the bushing from its 125°C fabrication temperature to the 4 K operating temperature. The second
contribution is obtained by adjusting the bushing geometry so that
the epoxy contraction during curing is directed inward onto the
embedded metal parts. It is also necessary, because of the re-
entrant design of the steel flange, to incorporate an electrical
insulator piece between the copper bus and the reentrant part of
the flange. This insulating piece is made of Micarta and is held
in place by the compression loading of the epoxy compound.

The main lead is subject to stresses, applied via the copper
bus and transmitted through the epoxy bushing to the steel
flange. For design purposes, the stress sharing contribution of
the Micarta insulating piece is ignored. The stresses have gravitational and thermal contraction components but are primarily due
to the Lorentz forces caused by the current in the bus and its
attached conductor interacting with the high magnetic fringe field
of the LCP coil and adjacent coils. This magnetic loading in the
most severe test configuration averages over the main leads to be
20.7 kN/m of conductor vertically, and 6.3 kN/m of conductor
horizontally.

The maximum stress in the copper lead was calculated to be
35.3 MPa. The annealed OFHC copper of the bus has a yield
strength at 4 K of 44.1 MPa giving a safety factor of 1.25.

The maximum shear stress at the interface of the epoxy bushing and copper bus is 2.21 MPa as calculated on the assumption that the loads on the bus are resisted entirely by shear at the bushing interface. Since some redistribution of the normal stress on the interface can be expected, the calculated shear stress is conservative. Although there is no data available for cryogenic temperatures, a bushing of similar wall thickness had an interface shear strength of 10.6 MPa at room temperature, or 4.7 times the calculated stress. Because the bushing properties depend on the overall geometry and the curing dynamics of the epoxy, and because the design permits it, the main leads are to be manufactured and mechanically proof tested as subassemblies prior to assembly on the complete coil.

BUSHING GEOMETRY

A technique has been developed[4] to produce high shear strength, leak tight sealing at the resin-metal interface of large conductors embedded in a cured, filled resin. This is accomplished by designing the mold shape and choosing resin composition to achieve a cumulative curing contraction of the resin mass onto the surface of the embedded conductor. The crucial feature employed in the design is the temperature accelerated curing rate of the epoxy, with curing beginning close to the conductor surface and progressing outward in a traveling wave of curing and contracting resin. The resin shrinkage during cure results in an effect similar to the cumulative compression obtained in filament winding of cylinders, as subsequent surfaces of cured resin convert their tensile shrinkage into compression at the mandrel surface. This produces compressive loads at the interface which are much higher than could be produced by uniform shrinkage of a monolithic mass. The heat necessary to accelerate the curing and produce the mass of cured resin is produced in part by the heat of the exothermic curing reaction. If the bushing mold is of the proper geometry and the conductor preheated, the comparatively poor heat conduction of the resin mass will result in a heat buildup near the center of the resin mass. It has been found experimentally that a mold shaped as a right circular cylinder with a ratio of total surface area to total volume of 1.75 cm^2/cm^3 or less will produce these preferred conditions. The formation of an exothermic curing core in the resin mass can be adversely affected, yielding an unsatisfactory seal, if the effects of the thermal conductivity and heat capacity of the massive conductor are not compensated by supplying heat to it. When preheated, the conductor is in thermal equilibrium with or is supplying heat to the exotherming epoxy mass during the formation of a seal.

A side benefit of this sealing technique is the elimination of conductor surface treatment, such as mechanical or acid cleaning, surface passivation, or antioxidation precautions, to prepare the metal surface for adhesion. The only conductor surface preparation employed is vapor degreasing and shot or sand blasting.

CRYOGENIC TEST OF BUSHING MOCKUP

The bushing fabrication technique was developed originally for ambient temperature use. The suitability of the design in terms of leak tightness and structural integrity at cryogenic temperatures had to be verified. A full-scale facsimile main lead assembly, as shown in Fig. 3, was fabricated. The epoxy cylinder had an O.D. of 19.1 cm, and I.D. of 10.2 cm and a length of 30.5 cm. The filled epoxy composition used consisted of: 72.0 wt% (9.07 kg) Silica, 15.6 wt% (1.96 kg) Epoxy Resin, and 12.4 wt% (1.57 kg) Curing Agent. These materials were mixed, degassed, cast and cured by conventional techniques. The completed bushing is shown in Fig. 4. The stainless steel pipe extending from the mockup bushing was welded to a copper gasketed ultra-high vacuum flange, which was mated to the test vessel, thus providing a sealed space around the bushing. This space was filled with helium gas at 1 atm to serve as cooling exchange gas and to provide the leak indication on a mass spectrometer leak detector. The test arrangement is shown schematically in Fig. 5. The temperatures of the stainless pipe (TC No. 1), the copper conductor (TC No. 2) and the surface of the bushing (TC No. 3) were monitored during cooldown.

The test assembly was cooled over a period of 30 hours to 77 K (Fig. 6) and then in three hours to 4 K (Fig. 7). The copper conductor is cooled only via conduction through the poorly conducting epoxy or by heat exchange with the helium gas. Thus it

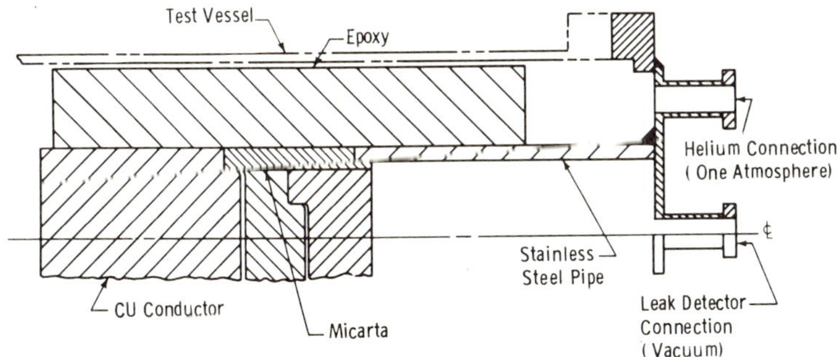

Fig. 3. Main lead bushing test piece cross section.

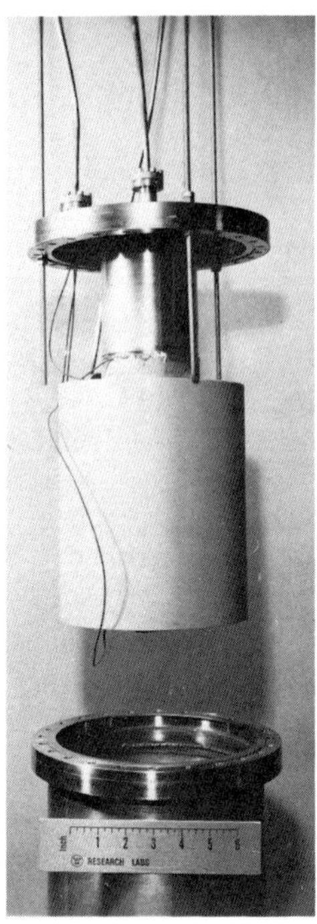

Fig. 4. Main lead bushing
test assembly.

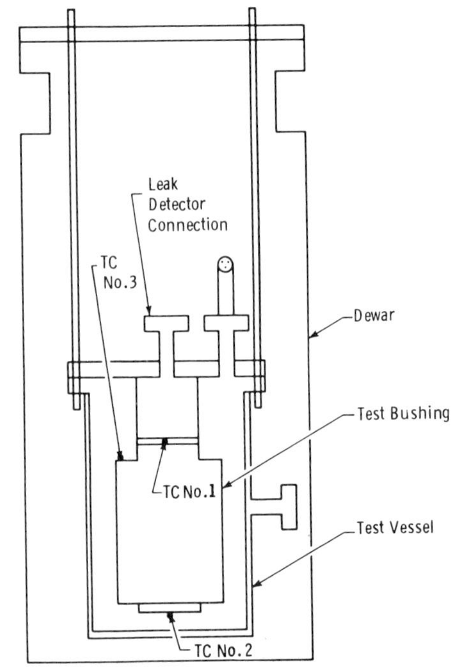

Fig. 5. Main lead
cryogenic test arrangement.

lags behind the bushing parts both on cooling and warming. During
the 58 hour warmup, the cooper conductor temperature lagged as
much as 40 K behind the temperature of the epoxy. Despite the
large temperature gradients created during this thermal cycling,
there was no failure of the epoxy or leaks greater than the
sensitivity threshold of the helium leak detector (6×10^{-11} std.
cc/sec, air equivalent).

CONCLUSION

A satisfactory method has been devised and tested for
producing large helium leak tight insulating bushings for use at
4 K.

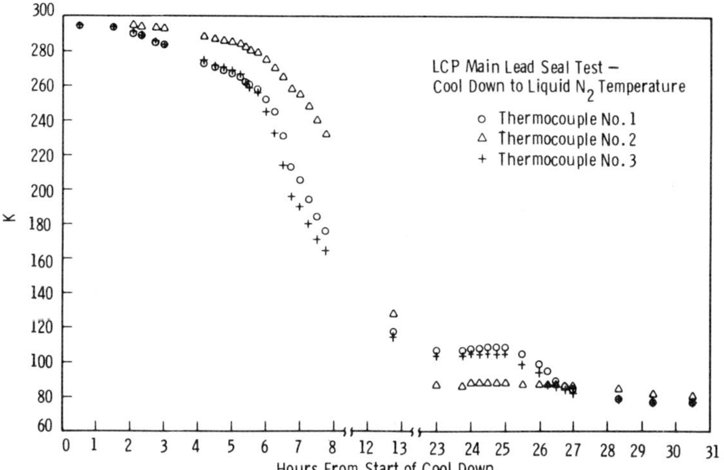

Fig. 6. Main lead test, cooldown to liquid
nitrogen temperature.

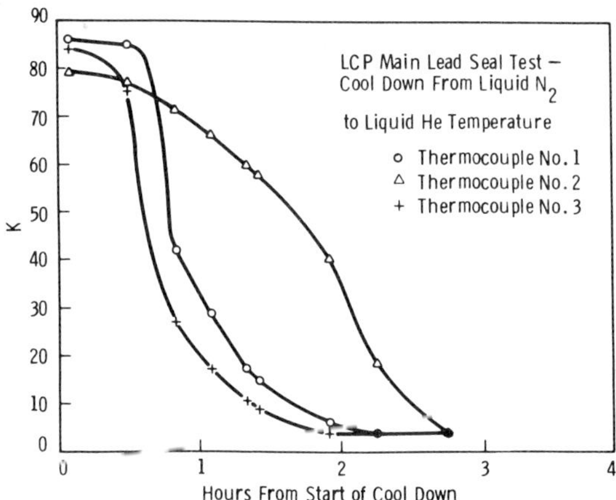

Fig. 7. Main lead test, cooldown from liquid
nitrogen temperature to liquid helium temperature.

REFERENCES

1. R.D. Blaugher, et.al., Experimental Test and Evaluation of the
   Nb_3Sn Joint and Header Region for the Westinghouse LCP
   Coil, IEEE Trans. on Magnetics MAG-17 (1):467 (1981).
2. R. D. Blaugher, Insulating Helium Flow Tubes for Force-Cooled
   Magnets, "Advances in Cryogenic Engineering, Vol 27",
   Plenum Press, New York (1982).
3. "Handbook on Materials for Superconducting Machinery,"MCIC-HB-
   04, Battelle Columbus Laboratories, Columbus (1974),
   p.5.1.3-9.
4. J. F. Quirk, "Compression Seals for Nuclear Penetrations,"
   U.S. Patent No. 4,017,568, April 12, 1977.

GAS-COOLED ELECTRICAL LEADS FOR USE ON FORCED-COOLED SUPERCONDUCTING MAGNETS*

R. G. Smits, P. L. Andrews, W. A. Burns, C. T. Day, J. W. Gary,
G. H. Gibson, M. A. Green, M. Pripstein, R. R. Ross, and J. D. Taylor

*Lawrence Berkeley Laboratory
University of California
Berkeley, California*

INTRODUCTION

The two meter diameter thin superconducting solenoid for the
Time Projection Chamber (TPC) experiment at the Stanford Linear
Accelerator Center is cooled with two phase helium in forced flow
through a pipe.[1] This magnet, which has a design current of 2300
A (the maximum current is 3000 A), requires gas cooled electrical
leads which operate directly off the two phase flow circuit.[2] The
primary requirements of the leads were: 1) reasonable efficiency,
2) the ability to operate in the cryostat vacuum, yet withstand
internal pressures up to 100 atm, 3) low pressure drop at full gas
flow, 4) the ability to operate in a horizontal position with the
cold end slightly higher than the warm end, 5) reliability so that
the lead can run without gas flow for at least 20 minutes, and 6)
restricted length (less than 600 mm).

A simple copper pipe lead permits one to operate directly off
a forced flow helium circuit. A properly dimensioned copper pipe
can be pressurized to 100 atm while the outside of the pipe is in
vacuum. This type of lead alone is not very efficient.[3] However,
if the copper pipe lead has an insert to enhance heat transfer,

---

*Work suppported by the Director, Office of Energy Research,
Office of High Energy and Nuclear Physics, Division of High Energy
Physics of the U.S. Department of Energy under Contract No. W-
7405-Eng-48.

the efficiency of the lead can be increased to acceptable
levels. The insert also permits the lead to be operated in the
horizontal position. Such leads can be built curved and the pipes
can be bundled so that the lead can carry currents up to 10,000 A
without increasing the basic copper pipe lead length.

Each TPC magnet lead consists of a bundle of five type L
copper pipes which have an outside diameter of 15.9 mm and a wall
thickness of 1 mm. The leads are 533 mm long and they are curved
with a radius of curvature of 1.18 m. All joints in the TPC
magnet leads are either hard soldered or welded. The leads are
connected directly into the helium flow circuit at the cold end.
The helium flow and current flow are separated at the room temperature end. The leads are electrically and thermally insulated
from the cryostat and the magnet and they are in the cryostat
vacuum space. The current bus bar and the helium gas pipe are
separately fed out through the cryostat vacuum wall.

This paper describes the fabrication and testing of a pair of
single tube leads which carry one fifth of the TPC magnet
current. Since one of the TPC magnet leads is horizontal within
the vacuum chamber, the test leads were tested in a near horizontal position. From the tests, one can predict the performance of
the composite TPC magnet lead or the performance of other leads
built with multiple tube bundles.

THE BASIC THEORY OF ENHANCED HEAT TRANSFER IN COPPER PIPE LEADS

The current carrying part of the lead is a commercial copper
pipe. Electrically the copper in commercial pipe is not very good
(RRR ≈5). The thermal conductivity of the lead is nearly constant
as a function of temperature (around 300 $Wm^{-1}K^{-1}$). The leads are
operated at relatively low design current densities (less than 1.5
$\times 10^7 Am^{-2}$). Therefore the leads are resistant to burnout.

Since all of the current is in the copper tube, one must
increase the heat transfer coefficient between the helium gas and
the copper in order to operate efficiently. This is done by
reducing the boundary layer thickness on the tube wall surface,
which we do with a coaxial insert. Flow through the resulting
annulus can have a Reynolds number below 2000; therefore, it is
laminar. In tubes which are long compared to the annular dimension, the Nusselt number has a minimum value of around 4.[4] The
heat transfer coefficient at the tube wall becomes two times the
helium thermal conductivity divided by the thickness of the annulus.

The annular thickness is dictated by several factors. It should be less than 0.6 mm in order to prevent thermal acoustic oscillations for a lead of this length. Practical considerations such as clearances and clogging with dirt suggest that the annulus thickness should be greater than about 0.2 mm. Using the thermal conductivity of helium one calculates heat transfer coefficients at the tube wall of 100 · $Wm^{-2}K^{-1}$ at 20 K and 620 $Wm^{-2}K^{-1}$ at 300 K.[5] The gas temperature in the lead should therefore be within 10 K of the copper tube temperature.

THE TEST LEADS AND THE TEST SET-UP

The test leads represented one-fifth of a TPC magnet lead and were the same length as the TPC magnet lead (533 mm). The test lead was straight rather than curved. The test leads were made from a single 15.9 mm OD (5/8 inch OD) copper pipe with 1.0 mm (0.040 inch) walls. The insert was made from 304 stainless steel tube which is 12.7 mm OD (0.50 inch OD). The annular space between the OD of the stainless steel tube and the ID of copper tube was about 0.58 mm (0.023 inches). The stainless steel tube had metal pieces welded into the ends, and was filled with glass wool, with a small hole in the upper end. It was spiral wrapped with a strip of stainless steel which served as a spacer to center the insert within the copper tube. Figure 1 shows the construction of the straight tube test leads.

A pair of test leads were installed in a vacuum tight test box which was connected to the TPC magnet pumped two phase cooling system. At the upper end of each lead, the warm helium gas was separated from the current. The current bus and the tube carrying helium gas passed through the vacuum wall separately. The cold ends of the leads were connected into a common pipe which carries two phase helium from the TPC magnet cryogenic system. Figure 2

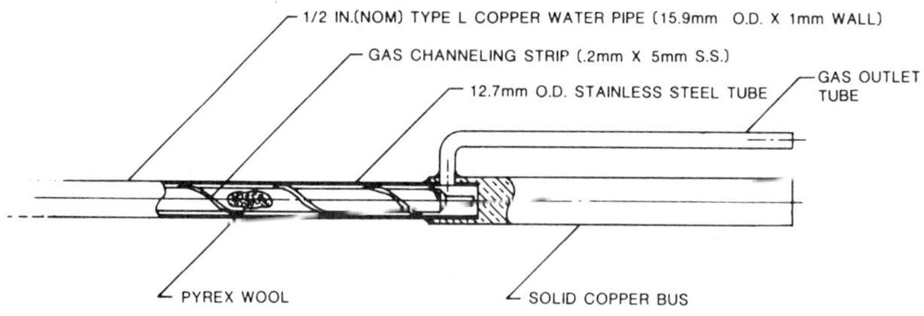

Fig. 1. The construction of a single tube enhanced heat transfer lead.

is a schematic diagram of the electrical lead experiment in a two
phase helium circuit supplied by a helium pump. A "U" shaped
piece of superconductor was extended from three quarters of the
way up one lead to the common helium carrier then three quarters
of the way up one lead to the common helium carrier then three
quarters of the way up the other lead. This super-conductor
allows the lower portion of the lead to become superconducting and
eliminates Joule heating. Figure 3 shows the experimental set up.

Five silicon diode thermometers were used in the experiment. One diode was attached to the downstream leg of the helium
circuit, with the other four evenly spaced along one of the
leads. Figure 3 shows the location of diode thermometers. In
addition to the silicon diodes, there was a pressure tap on the
manifold between the leads so that the pressure drop across either
lead could be measured while helium flowed through the leads. The
rate of gas flow through the leads was measured using rotometer
type flow meters at room temperature.

The lead experiment was powered by the 3000 A, 10 V power
supply to be used with the TPC magnet. The voltage drop across
the leads could be measured from the gas pipe at the room temperature end of each lead to the pressure tap in the manifold between
the leads.

RESULTS OF THE LEAD EXPERIMENT

During the course of testing, the total heat load into the
system was measured using gas boil off. (Note: The flow through

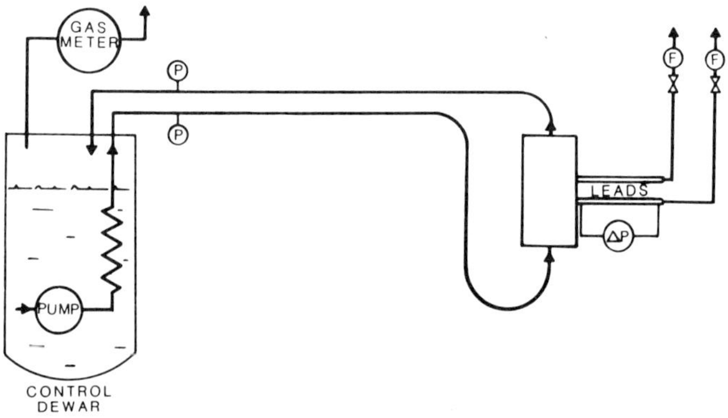

Fig. 2. Helium flow circuit for the gas cooled lead experiment.

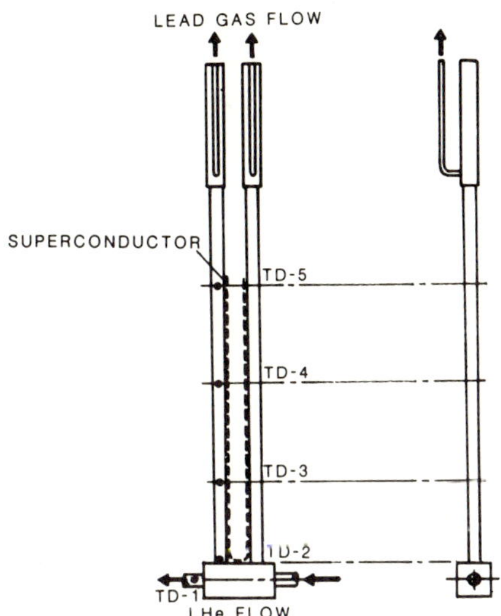

Fig. 3. The location of temperature diodes
on the electrical lead experiment.

the leads was included in this measurement.) In addition to total
heat load into the system, the temperature profile along the lead,
the pressure drop across the lead, and the voltage drop across the
lead pair were measured.

The total heat load into the experiment as measured by helium
boil off comes from several sources. They are: 1) the heat leak
into the control dewar, 2) the heat leak into the transfer line
and lead box, 3) helium pump work, and 4) the heat leak down the
leads themselves. The heat leak into the control dewar varied
from 2 to 5 W depending on the liquid level. The transfer line
heat leak was estimated to be between 14 and 16 W. The helium
pump work was a function of pump speed, pump stroke and pressure
rise across the pump (see Ref. 6). When the pump speed was 24 RPM
with a pressure rise of 0.1 to 0.2 atm and a stroke of 2.54 cm,
the mass flow through the circuit was around 18 gs^{-1}. Most of the
experimental work was done under these conditions, with the pump
work estimated as 15 W.

When no gas or current flowed through the leads, there was a
15 W heat leak down the leads. Without current, gas flows as
little as 0.02 gs^{-1} shut the heat flow off into the helium circuit. With sufficient lead gas flow, powering the leads from 0 to
800 A did not add heat to the helium system (see Fig. 4). In-

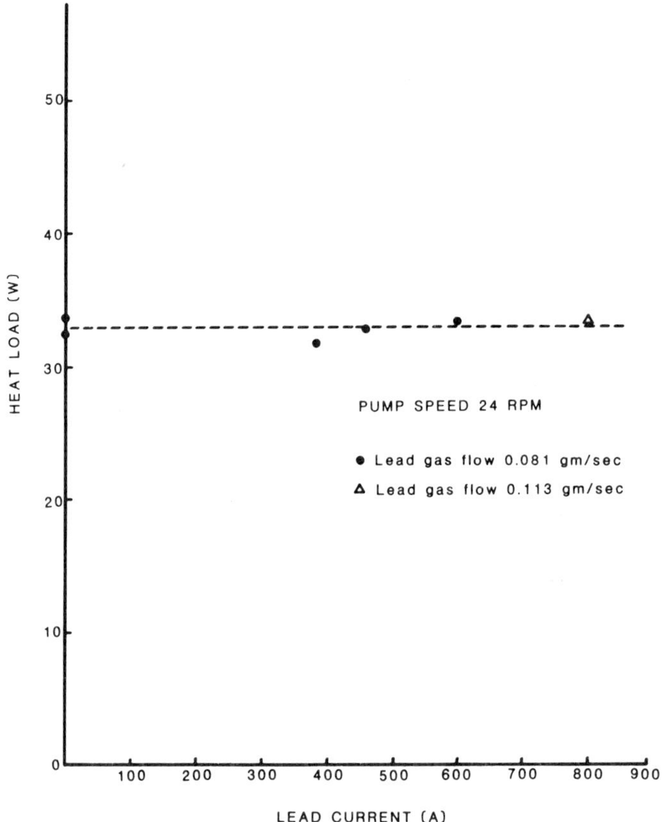

Fig. 4. Total experiment heat lead as a function
of lead current at lead angle of zero.

creasing the current in the leads did change the temperature
profile in the lead (see Fig. 5), but there was no change in the
heat flow into the helium circuit.

Most of the lead tests were done with the leads in the horizontal orientation (the cold end and warm end are at the same
level). The lead angle could be adjusted from +22 degrees (the
cold end down) to -10 degrees (the cold end about 10 cm above the
warm end. There was little or no system heat load dependence with
angle when the current in the leads was zero or 383 A. (See Fig.
6). From the results of this experiment, we found that the enhanced heat transfer lead can be operated directly from a two
phase flow circuit in a variety of orientations.

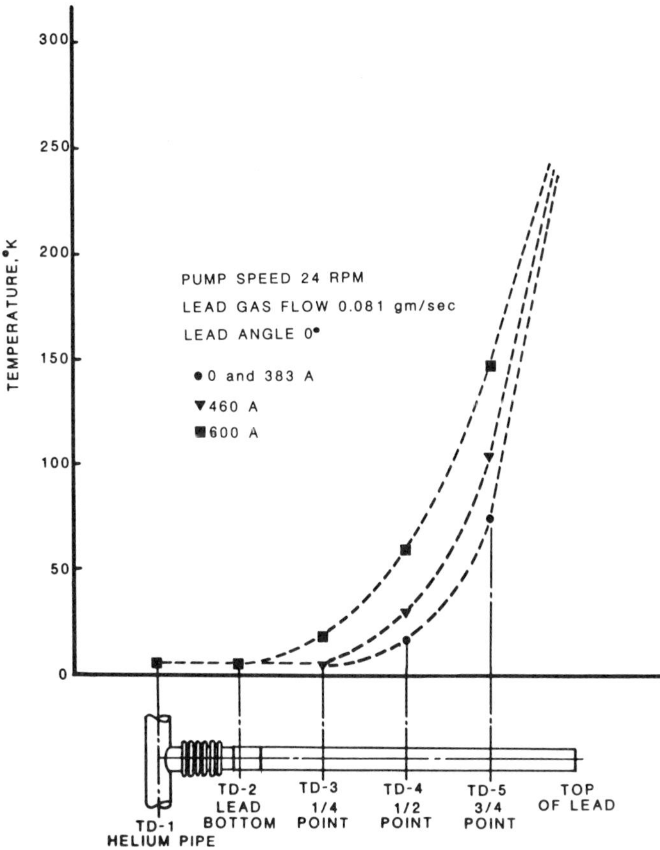

Fig. 5. Measured temperature profile versus
position along the lead for various currents.

The experiments showed that the test leads operated well at
460 A with gas flows as low as 0.044 gs^{-1}. The leads ran stabily
at currents up to 800 A with increased gas flow. Experiments with
leads of this type in a liquid bath suggest these leads can be
operated at currents above 1000 A.

At 460 A (equivalent to the design current of the TPC magnet), the lead current density in the copper was 10^7 Am^{-2}. We
calculated that these leads should operate at least 10 minutes
without burnout at this current density using an adiabatic
theory. The test leads were tested with the gas flow shut off for
27 minutes at a current of 460 A. The maximum temperature measured in the lead was 360 K with the temperature profile shown in
Fig. 7. During the 27-minute test without gas flow, the heat load

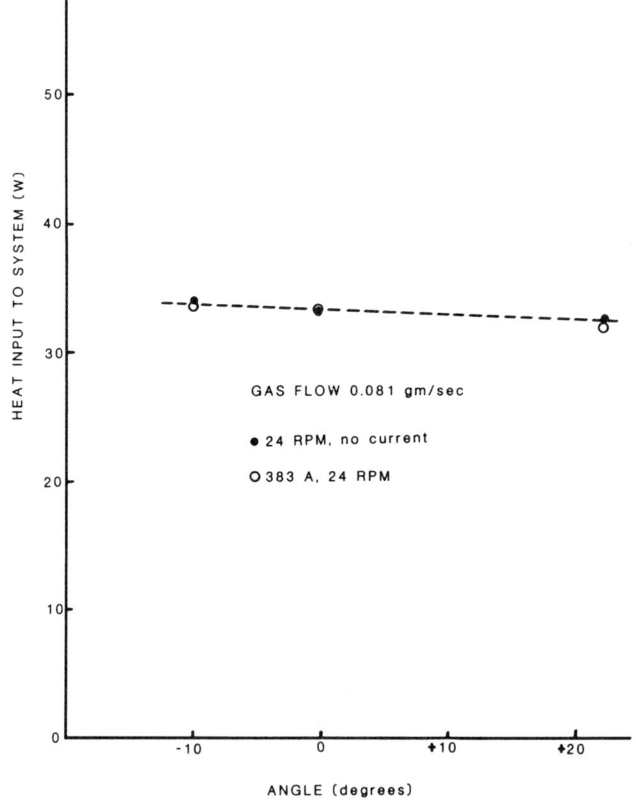

Fig. 6. Total experiment heat load versus lead angle.

into the helium circuit increased from 33 W to 63 W. The resistance across the lead pair increased from 140 $\mu\Omega$ to 450 $\mu\Omega$. The voltage drop across the lead was found to be a good indication of changes in lead gas flow.

SUMMARY

The experiment showed that stable performance of the test lead could be obtained at 460 A when the gas flow was maintained at 0.044 gs^{-1}. The pressure drop across the leads was less than 0.03 bar. The measured resistance across the lead pair was around 140 $\mu\Omega$. There is no apparent need for an active control system for the leads at this current, but if a control system is desired, the voltage drop across the lead is a good control parameter.

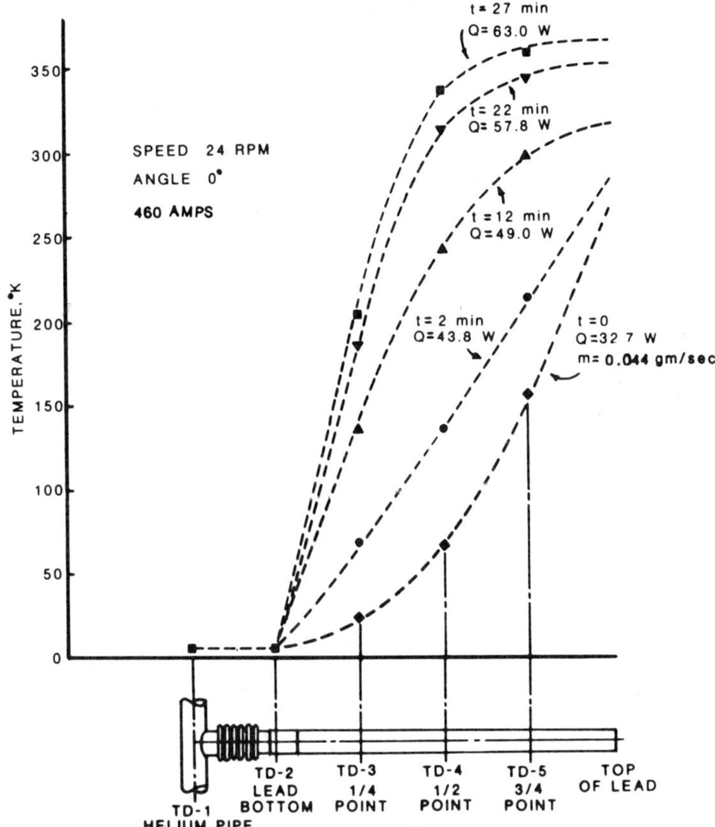

Fig. 7. Temperature profile in the lead with
no gas flow versus time at a current of 460 A.

ACKNOWLEDGMENTS

The idea of enhanced heat transfer leads is an old one. The
first leads of this type were built at LBL by R. Hintz or W.
Chamberlain in the late 60's. To our knowledge this is the first
application of this types of lead to a position other than vertical. The authors thank H. Van Slyke, C. Covey, E. Lee, and P.
Harding for their efforts. P. Eberhard is thanked for his advice
and encouragement.

REFERENCES

1. M.A. Green, W.A. Burns, and J.D. Taylor, "Forced Two Phase
     Helium Cooling of Large Superconducting Magnets, " Advances
     in Cryogenic Engineering Vol. 25, Plenum Press, New York
     (1979), p. 420.

2. M.A. Green, et al.,"The TPC Magnet Cryogenic System," Lawrence
    Berkley Laboratory Report LBL-10552, March 1980.
3. C.D. Henning, "Cryogenic Electrical Leads," Proceedings of the
    1968 Summer Study on Superconducting Devices and
    Accelerators, BNL-50155 (c55), 1968, p. 304.
4. "Thermophysical Properties of Helium 4 from 2 to 1500k with
    Pressures to 1000 Atmospheres," NBS Technical Note 631,
    November 1972.
5. F. Kreith, Principles of Heat Transfer, International Textbook
    Co., Scranton, Pennsylvania, 1961.
6. W.A. Burns, et al.,"The Construction and Testing of a Double-
    Acting Bellows Liquid Helium Pump," Proceedings of the 8th
    International Cryogenic Engineering Conference, Genova,
    Italy, June 1980, p. 383.

DISCUSSION

Comment by G. H. Morgan, Brookhaven National Laboratory: It
should be mentioned that horizontal current leads are the rule
rather than the exception for superconducting power transmission. References to this work are found in:

K. G. Lewis, et al., Current tests on a flexible superconducting core for a 2 GVA AC cable, Cryogenics, 18:143 (1978).

Yo L. Blinkov, et al., Experimental measurements of the temperature dependence of the critical current in a prototype
section of a superconducting power transmission cable, IEEE
Trans. on Magnetics, MAG-15 (1):169 (1979).

G. N. Morgan, F. Schauer, R. A. Thomas, A improved 60 Hz
superconducting power transmission cable, IEEE Trans. on
Magnetics, MAG-17 (1):157 (1981).

The Russian lead is noteworthy because it is just a section of
rigid coaxial cable with a non-metallic plug in the annulus to
enhance heat transfer in a fashion similar to that presented here.

INSULATING HELIUM FLOW TUBES FOR FORCE-COOLED MAGNETS

R. D. Blaugher

Westinghouse Electric Corporation
Pittsburgh, Pennsylvania

INTRODUCTION

The cooling schemes for large superconducting magnets can be divided into two main categories:

pool cooling, where the magnet is immersed in two-phase boiling, single-phase supercritical, or superfluid helium.

force-cooled, two-phase or single-phase helium circulated through conduits which normally enclose the superconductor.

The force-cooled technique is utilized in a number of major superconducting magnet programs. The Department of Energy Large Coil Program (LCP) is basically a feasibility demonstration of superconducting toroidal-field coils, as considered for a Tokamak fusion power reactor. Three of the six coils under construction for the LCP use forced-flow designs.[1,2] In addition the fusion related forced-flow magnet designs, Massachusetts Institute of Technology and others have evaluated forced-flow cooling for MHD magnets.[3] Forced-flow designs have also been constructed and tested for high-energy physics applications.[4]

The Westinghouse LCP design is a 3.5 m × 4.5 m, D-shaped coil, which is forced cooled.

As in virtually all forced-flow designs, it is necessary to electrically isolate sections of the Westinghouse LCP magnet and still maintain continuous uninterrupted flow of helium. This paper describes an insulating helium flow tube which has been designed, built and tested for the Westinghouse LCP magnet.

DESIGN AND TESTING OF FLOW TUBE

The Westinghouse LCP conductor utilizes a supercritical helium forced flow cooled Nb_3Sn cable enclosed by a hollow stainless steel sheath. The magnet design incorporates a modular concept which, during winding, confines the conductor to slots machined into aluminum plate supports. The plate modules are then bolted together to form the complete magnet. Helium flow is supplied to the individual conductors by a header system mounted on the top of the magnet. Helium supply and return is provided by a manifold arrangement which uniformly distributes helium to different sections of the magnet. Each manifold supplies flow to as many as five conductors, the manifold being connected to the individual conductor headers by means of an "insulating" flow tube.

The insulating tube must satisfy two main functions: (a) to pass supercritical helium with no leakage at approximately 15 atm pressure between the manifold and header and (b) to electrically isolate the conductor header from the manifold to eliminate possible shorts between the turns and ground. The plate and conductor ground wall insulation are separately applied to eliminate turn-to-turn and plate-to-plate shorting. This overall insulation system eliminates the large circulating currents that could occur in the structure. The flow tube must also accommodate differential thermal contraction between its stainless feed tubes and the aluminum support. In addition, the tube must be rigidly attached to the stainless header and manifold to be helium leak tight but also facilitate possible demounting if repairs are necessary.

An insulating flow tube has been designed and tested under conditions similar to that expected for the actual magnet. The design criteria were:

      a. Pressure capability to better than 15 atm
      b. No observed diffusion of helium through the
          insulation at 15 atm
      c. No mechanical or electrical degradation on
          thermal cycling
      d. Thermal contraction compatibility
      e. High electrical dielectric strength
      f. Simple assembly and repair

The insulating flow tube was designed around a commercially available epoxy fiberglass tube with a 0.250" OD and 0.160" ID. The tubes obtained from the Polygon Company, utilized a braided fiberglass filament reinforcement. The Polygon process actually combines filament weaving with conventional filament winding to provide superior mechanical properties over standard filament wound tubing. Typical mechanical and electrical properties for Polygon fiberglass tube are as follows:

Tensile Strength	34,000 psi
Compressive Strength	24,000 psi
Burst Strength	49,000 psi
Shear Strength	7,500 psi
Insulation Resistance	2.38×10^{12} ohm/8" length
Dielectric Strength	100 V/mil

The insulating tubing for the Westinghouse design provides a transition from a stainless manifold to a stainless conductor header. It is thus essential that the tube match the thermal contraction of stainless to minimize thermal stresses. The axial thermal contraction of the fiberglass tubes are controlled by the pitch of the woven braid or adjustment of the winding angle for a filament wound reinforcement.

Two different tubing samples were evaluated for thermal contraction down to nitrogen temperature. The observed thermal contraction was slightly nonlinear and showed an average thermal expansion coefficient of 7.13 to 7.80×10^{-6} in/in $^{\circ}$F or 12.8 to 14×10^{-6} in/in $^{\circ}$C. The handbook value for the average thermal expansion of 304 stainless steel is 7.4×10^{-6} in/in $^{\circ}$F or 13.3×10^{-6} in/in $^{\circ}$C. The observed thermal contraction, which is already quite close to that of stainless, can be adjusted in either direction by varying the pitch angle of the braid.

The as-tested tube was then fired at 580°C to drive off the epoxy binder. The structure of the fiberglass reinforcement could then be readily observed as shown in Fig. 1. This structure is clearly a woven fiberglass hollow-braid which was apparently fabricated in multiple layers. The original as-supplied tubing is shown for comparison.

The termination and sealing of the insulated tube was accomplished by a specially designed external copper ferrule and internal stainless mandrel which supports the tube during compression. The special ferrule and mandrel can be housed in most commercial high pressure fittings such as a Swagelock connection. A cross section of the complete insulated tube and ferrule design is shown in Fig. 2.

The fiberglass tubing fitting configuration as described was fabricated and assembled for helium leak testing at various pressures and temperatures down to 77 K by submersion in liquid nitro-

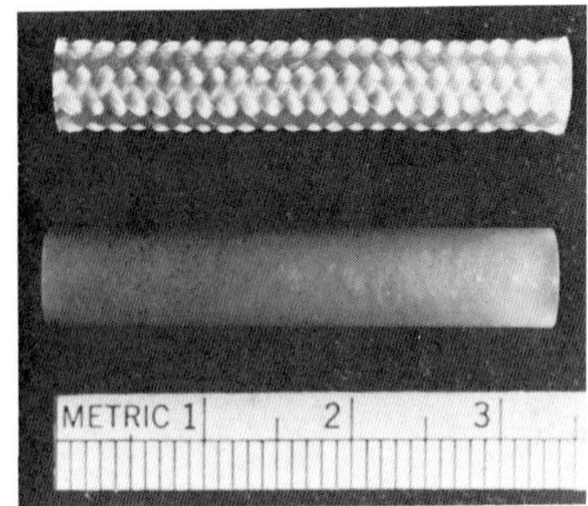

Fig. 1. Structure of Insulating Tube.

gen. Helium gas pressure was applied through a supply line to
the fitting and tubing assembly. The assembly was then placed in
a vacuum containment which was monitored for helium leakage by a
helium mass spectrometer leak detector.

The helium mass spectrometer leak detector was calibrated
with a standard leak ($2.6 \times 10^{-8} \pm 10\%$ Std. cc/s) and helium
pressure to the tube increased, at nitrogen temperature, to 100
psi. The indicated leak rate was observed to be 0.35×10^{-8} cc/s.
The assembly was allowed to warm up to room temperature and then
cooled back down to nitrogen temperature while under 100 psi
helium pressure. The pressure was then increased to 150 psi with
no change observed in the observed leak rate for the complete
cycle. No evidence of tubing failure or leakage was observed
after repeated pressure and temperature cycling.

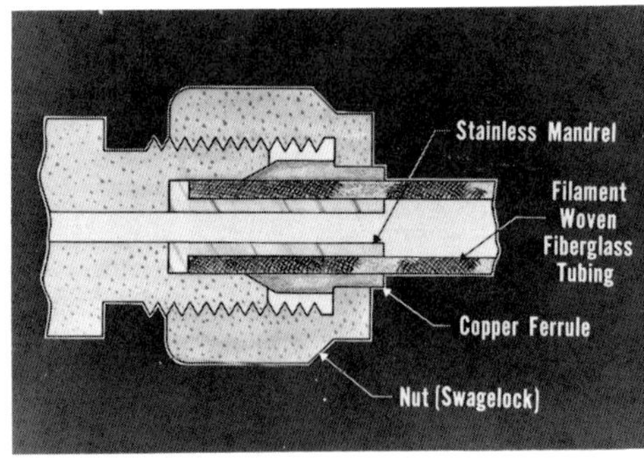

Fig. 2. Cross Section of Helium Tube Fitting.

The applied hoop stress on the tube can be calculated from

$$S = P_i \frac{r_i^{\,2}}{r_o^{\,2} - r_i^{\,2}} \left[1 + \frac{r_o^{\,2}}{r_i^{\,2}} \right] = P_i \frac{(r_i^{\,2} + r_o^{\,2})}{(r_o^{\,2} - r_i^{\,2})}$$

$$= P_i \times 2.394 \text{ (for tested tube)}$$

where P_i is the internal applied pressure and r_o and r_i are the outer and inner radii. The hoop stress for 100 psi applied pressure is approximately 239 psi which goes to 359 psi for 150 psi applied pressure. The above cold tests were run at applied pressures up to 150 psi. A room temperature test was run up to 500 psi without evidence of tubing failure. The leak detector was not used for the 500 psi test. After the room temperature test at 500 psi the tubing was cooled in liquid nitrogen while maintaining a 500 psi applied pressure. No evidence of leakage or tube failure could be observed. The insulated tubes have thus been tested to hoop stress levels in excess of approximately 1200 psi which is well in excess of the 15 atm operating pressure without any evidence of tube failure.

These results indicate that the woven fiberglass reinforced tube and special ferrule construction apear to satisfy all of the design requirements for the Westinghouse LCP coil and offer an attractive insulating flow tube for other forced-flow application.

REFERENCES

1. P.N. Haubenreich, J.N. Luton and P.B. Thompson, The role of the large coil program in the development of superconducting magnets for fusion reactors, IEEE Trans. on Magnetics, MAG-15 (1):520, (1979).
2. P. Komarek, Superconducting Magnets for Tokamaks in "Proc. 8th Int'l Cryo. Engr. Conf.", IPC Press, Genoa, Italy (1980), pp. 303.
3. M.O. Hoenig, Internally-colled cabled superconductors, Cryogenics, 20:427 (1980).
4. M.A. Green et al., Operation of a Forced Two-Phase Cooling Syotem on a Large Superconducting Magnet, in "Proc. 8th Int'l Cryo Engr. Conf.", IPC Press, Genoa, Italy (1980), p.72.

A HORIZONTAL CRYOGENIC BUSHING FOR THE TERMINATION OF AN AC SUPERCONDUCTING POWER TRANSMISSION LINE

G. H. Morgan, J. E. Jensen, A. J. McNerney,
K. F. Minati, and R. D. Schmidt

*Brookhaven National Laboratory**
Upton, New York

INTRODUCTION

High current vapor-cooled leads for superconducting magnets are common, but very few attempts have been made to combine them with high-voltage capability. Two superconducting transmission line projects, one at the University of Graz in Austria, and the other at Linde in New York have earlier designs.

The termination for the BNL superconducting ac power transmission line has three main components: (1) a commercial, air-entrance SF_6-insulated, vertical bushing with a 90 degree elbow at the bottom, (2) a horizontal cryogenic bushing linking the 6 K, pressurized-helium cable environment to the ambient temperature bushing, and (3) a stress cone which terminates the cable outer shield and transforms the large radial voltage gradient in the cable dielectric into a much lower radial voltage gradient in the high density helium at the cold end of the horizontal cryogenic bushing. Each phase of the 138 kV 3 phase superconducting power transmission line consists of a coaxial cable, and each cable has its own, fully coaxial termination. The rated current of the cables to be tested in the BNL 100 meter cable test facility will be 4.0 kA continuous and 6.0 kA intermittent, and the line to neutral voltage is 80 kV (corresponding to 138 kV phase-to-phase). The termination must be free of partial discharge to 160 kV rms, and withstand 240 kV rms and impulses in excess of 650 kV.

---

*Work supported by Associated Universities, Inc., under contract with the U.S. Department of Energy.

DESIGN CONSIDERATIONS

Apart from the cable itself, the most difficult component of the transmission line to design has been the horizontal cryogenic bushing. It consists of two coaxial, helium-cooled aluminum current leads, shown in Fig. 1. The inner lead is 63.5 mm in diameter and is 2.0 m long. The outer lead is 196.9 mm in diameter and is 0.93 m long. The space between the leads is filled with a loaded-epoxy casting which is cooled by contact with the conductors.

The temperature transition of the bushing takes place over the length of the outer conductor, so that the conical ends of the epoxy are relatively uniform in temperature The inner conductor has three independent heat exchangers: the warm end is cooled by a water glycol to air loop entirely at high voltage, the central temperature transition region is helium cooled, and the cold end is also helium-cooled, but with a simpler heat exchanger than the center section. All of the cold end of the inner, and about 1/3 of the transition region of both inner and outer conductors is covered with a layer of thin Nb_3Sn superconducting tape.

The decision to use a solid dielectric rather than vacuum was dictated by the impulse rating and by a desire for compactness. The cone taper at each end follows from the allowable surface stress being lower than the allowable stress in the solid. The allowable surface impulse stress in the warm, compressed SF_6 (5 MV/m) is higher than in the cold, compressed helium (3 MV/m) and therefore the warm end has a shorter cone. The radial stress in the solid temperature transition region is 20 MV/m. Epoxy config-urations arrived at from elementary principles were confirmed by a computer solution of the electrostatic problem. The dielectric loss (about 5 W at 80 kV rms) is small compared to other thermal loads and is not a factor in the choice of dielectric or in the thermal design. The alumina filler in the epoxy (about 70% by volume) is necessary to match the contraction of the epoxy (L/L_o = 0.00429, 5 K to 293 K) to that of aluminum (0.00415). If there is a gap between the metal conductor surfaces and the epoxy casting in the temperature transition region, several pro-blems may arise. Such gaps must be electrically shielded to prevent voltage gradients in the low density, warm helium gas. They also impede heat transfer from the epoxy and may cause ther-mal oscillations in the helium, resulting in excessive heat leak. In the present design, the epoxy is cast with the inner and outer conductors in place; during curing a void averaging 0.9 mm develops between the epoxy and outer conductor. As shown in Fig. 1, the helium in the gap is shielded by a screen embedded in the epoxy and is not stressed electrically.

The design of vapor-cooled current leads has been the subject of many publications; a 1975 review paper[1] lists 64 references and new papers continue to appear. The present leads were designed for minimum gas flow using a numerical solution to the steady-state equations with experimental values of materials properties and published helium property correlations. The program also computes the pressure drop in the leads, which is low because of the high gas pressure (12 atm). In the early design stages, data for the thermal conductivity of the filled epoxy was not available and it was estimated by comparison with epoxy-fiberglass to have a thermal conductivity integral (6 K to 300 K) of 3.1 W/cm. This would lead to a heat leak (without helium cooling) of 14 W. Each aluminum lead, by comparison, would have a conductance of about 30 W if uncooled and would dissipate about 280 W at 4 kA. Recent measurements of the epoxy thermal conductivity indicate that the thermal conductivity integral is 11.7 W/cm giving a heat leak in the present bushing (which is longer) of 38 W.

CONSTRUCTION

The outer conductor consists of a thin-wall (1.2 mm) tube, two external flanges and an extended surface coolant channel (see Fig. 1) spiral wrapped on the outside of the tube. The tube alloy is 6061-T6 for strength and low resistivity ratio which increases wall thickness. The rectangular 6061 extrusion used for the coolant channel provides some reinforcement for the tube, which may be exposed to a 15 atm internal pressure. The bronze wire cloth field control screen consists of a smooth inner cylinder reinforced with corrugated wire cloth. The two flared wire-cloth ends are dieformed and soldered to the cylinder.

The inner conductor temperature-transition section is fabricated from a unique "T" shaped extrusion (see Fig. 1) that is spirally wound over a mandrel, like a close-wound spring. The spiral winding of the extrusion is done on a lathe, after the extrusion had been annealed. The flanged portion of the "T" section faces out and forms the outside surface of the conducting tube after the adjacent edges are dip brazed to form a continuous spiral joint. The "T" shaped extrusion is made of 1100 aluminum because the higher resistivity ratio (RRR) is required for the smaller cross sectional area of the inner conductor. The total wetted surface area of the three channels in the extrusion is approximately five times the conducting tube outer surface area. The ID of the spiral is 43 mm. A short section of 43 mm OD × 32 mm ID tubing is attached to each end of the spiral heat exchanger section during the dip braze operation. After brazing, all joints must be mass spectrometer leak tight. The ends of the spiral extrusions are machined circular exposing the three flow channels, which are then pressure tested and flow checked.

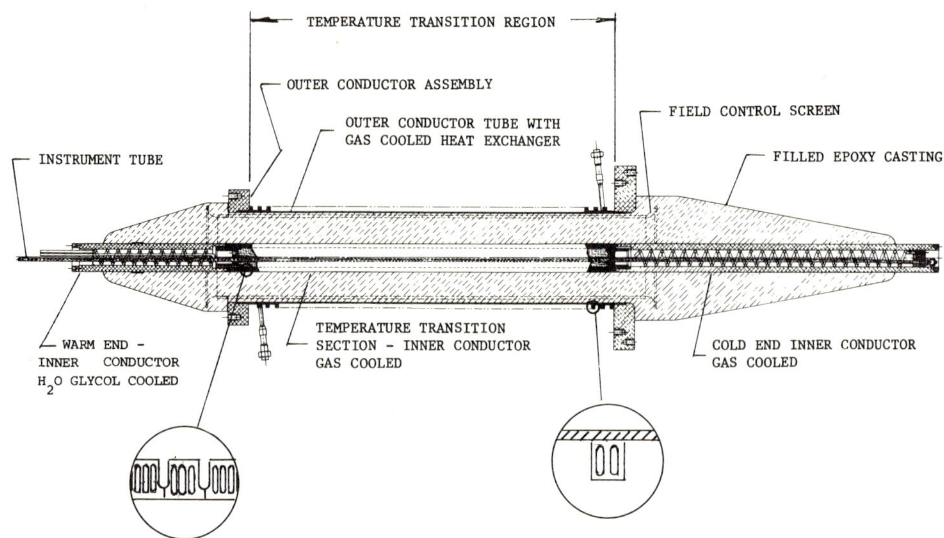

Fig. 1.   Axial Section of the Horizontal Cryogenic Bushing.

Both the warm and cold inner conductor end sections consist
of a heavy-walled tube, 63.5 mm OD × 12.7 mm wall 1100 alloy.
Circular, 6.4 mm OD aluminum tubing is spirally wound and dip
brazed to the inside surface for the coolant, which is water-
glycol at the warm end and helium at the cold end (the same helium
used in the central temperature-transition section). The two ends
are welded to the center section; the joint is reinforced by the
short 43 mm OD tubing mentioned above. The entire cold section of
the inner conductor, and one-third of the temperature transition
sections of both inner and outer conductor are gold plated over a
nickel flash to provide a base for soldering. Strips of superconductor are placed axially along the plated portion of the inner
conductor and held in place with Teflon shrink tubing.   An
aluminum mandrel covered with Teflon tubing is inserted into the
outer conductor tube. The assembly is placed into an oven for
soldering. By pressurizing the Teflon tubing, the superconductor
strips are pressed uniformly against the inside of the outer
conductor during soldering. After soldering, the Teflon is
removed and the conductor assembly is cleaned and leak checked.

The single most persistent problem has been cracks in the
epoxy due to residual curing stress. The highly filled epoxy is
brittle and has relatively low tensile strength. Vacuum leaks in
the welds occur but are usually correctable by rewelding. The
mismatch in contraction between Nb_3Sn and aluminum sometimes
causes bond failure or tape buckling, and thermal stress probably

degrades the Nb_3Sn performance. Leaks in the many mechanical seals are especially troublesome. Many of these seals are a consequence of the temporary nature of the laboratory facility, and will be absent in the outdoor test facility.

The total cost for each bushing was approximately $20 000, with the plating, brazing and casting done by outside vendors.

PERFORMANCE

Test Facilities

The laboratory test facility consists of a CTI Model 1430 refrigerator with a nominal rating of 100 W at 4.6 K and 1.2 MPa when operated with J-T expansion. The power of the rotary screw compressor is nominally 140 kW and it can deliver 20 g/s at 1.5 MPa with 0.1 MPa input. The maximum pressure attained in the test setup is about 1.0 MPa, and the maximum bleed gas attained is 1.25 g/s at 9.7 K. Three cryostats are used; two are about 12 m in length and are used when testing cables and the third is just long enough to accommodate the bushing and either a ball on the cold end for high voltage tests or, for current tests an 0.7 m length of superconducting cable with a cold, coaxial copper can for current return. For current tests, six, 2 kVA transformers are connected in parallel through 12 pairs of 500 MCM cable. The transformers are force-cooled and run at twice their rating. For voltage tests, the laboratory supplies can deliver 400 kV impulse and 275 kV rms. Impulses to 850 kV are available at the outdoor facility.[2] Temperatures are recorded by an HP 9825 computer in the laboratory facility.

Voltage Test Bushings

There are two results of interest; impulse breakdown voltage and partial discharge inception (PDI) voltage. Impulse breakdown can occur either in the epoxy or along the surface; in the latter case it is dependent on insulating gas characteristics, especially density. The PDI voltage is also gas density dependent. Tests are done first with both ends in SF_6 at ambient temperature and 0.5 MPa pressure. There have been no surface flashovers in SF_6 up to 800 kV impulse.

Two bushings with a configuration similar to Fig. 1 but without an outer conductor have been voltage tested cold, while connected to a 10 meter long cable. They differed from the present design in having epoxy instead of metal flanges. Curing cracks, wavyness and epoxy flow during bolting of seals resulted in the metal flange design of Fig. 1.

High Current Leads

Three 4 kA nominal coaxial current leads have been tested.
The first of these was 1.2 m long and made of copper tubing having
a resistivity ratio of 40, with perforated disc heat exchangers,
suitable for vertical operation only. This lead was intended for
fault current tests of short cables; it was impulse tested at 56
kA for one cycle as well as with steady rated current. The second
lead was also tested in a vertical cryostat with liquid helium, as
a quick check to verify the new conductor and heat exchanger
design. The outer conductor apparently had excessive cross-section of the 1100 aluminum alloy used, so the third coaxial lead
was made using the more resistive 3003 alloy. This third lead was
used to power a 12 m long superconducting cable at currents to 8
kA. All three leads required coolant flows somewhat higher than
calculated. Table I gives helium flow rates at 0 kA and 4 kA.
These numbers should be compared with optimum values of about 0.11
g/sec at zero current and 0.20 g/sec at 4 kA (the latter includes
a 10% allowance because with optimum gas flow the conductor bor-
ders on instability). The flows at 0 kA of lead no. 1 and 2 include cryostat heat leak and are not given in Table I. The third
lead was refrigerated, so the flow was set by a control valve to
give a temperature profile which matched computer predictions as
nearly as possible. The flow required was about 50% high at zero
current and 10 to 20% high at 4 kA. The reason for this was
realized later during the testing of the first combined voltage
and current bushing. Table I also lists coolant flows determined
for two of these.

High Voltage and Current

Three combined voltage and current bushings have been tested
cold. The first (No. 103) was voltage tested warm then cooled for
voltage and current tests. It was first connected to the same
cable used with earlier voltage bushings. It was cooled to 9 K
without cracking and had no leaks, but helium pressure dependent
leaks in the test equipment prevented attaining sufficient helium
density for partial discharge or full voltage measurements. It
was impulsed to 100 kV at a helium density of 40 kg/m^3. After
further tests with a corona ball replacing the cable, the corona
ball was removed and the bushing reconnected for current tests.
It was cooled to 10 K at the inner conductor and data taken with
currents to 6 kA, but with insufficient coolant flow for stable
operation above 2 kA. The excessive heat load was attributed in
part to high purity copper used to stabilize the Nb_3Sn tapes.
This copper was also present in the earlier current leads. The
next bushing tested (No. 102) had this copper removed, and a much
thinner outer conductor. This lead was tested to 7 kA, but refrigerator capacity limited the maximum stable current to about 3.5

Table I. Coolant Flow in Current Leads, Gram/Second

| Serial | Zero Current | | 4 kA | |
No.	Inner	Outer	Inner	Outer
1			0.25	0.20
2			0.27	0.32
3	0.15	0.17	0.22	0.24
102	0.20	0.17	0.54	0.40
112	0.16	0.15	0.33	0.31

kA. At this time higher temperature measurements of the thermal conductivity of the filled epoxy became available. The bushing was redesigned with the temperature transition region almost doubled in length and the outer conductor thickness increased in proportion. The first of these new, longer bushings (No. 112) has been tested cold with currents to 6 kA. Near the end of the planned tests, the warm cone cracked, probably because of combined residual curing, thermal and tensile stress from the massive cables attached there. This bushing operated stably at more than 4 kA, but was not stable at 6 kA with coolant flow available. As before, coolant flows were determined by temperature profile.

SUMMARY

Three combined voltage and current bushings have been tested. Vacuum leaks and refrigerator capacity have limited the helium density attained and prevented complete voltage tests while cold. The bushings have attained satisfactory PDI and excellent impulse breakdown when tested warm in SF_6. The excessive coolant flow required in the second of the three led to lengthening the temperature transition region in the third. This bushing was run stably with currents in excess of 4 kA, and was tested at 6 kA, but without sufficient coolant flow to prevent a very slow temperature rise. The coolant flows required are 40% greater than expected. A possible cause is a radial curing void under the outer conductor which will be backfilled in the next bushing. The epoxy filler presently used results in a rather high thermal conductance, and a new formulation may be tried.

ACKNOWLEDGEMENTS

The authors are indebted to E. B. Forsyth, the Project Manager, for continued advice and encouragement, to R. Thomas who managed the computerized data recording, and to the craftsmanship and dedication of W. Kristiansen and N. Houvener who helped assemble and test the bushings.

REFERENCES

1. Yu. L. Buyanov, A.B. Fradkov and I. You Shebalin, A Review of current leads for cryogenic devices, Cryogenics 15:193 (1975).
2. E.B. Forsyth, A.J. McNerney and M. Meth, Load Excitation at the Superconducting Cable Test Facility, BNL Report 26641, Brookhaven National Laboratory (1979).

A LOW-HEAT-LEAK SUPPORT STRUCTURAL MEMBER FOR THE SUPERCONDUCTING CHICAGO CYCLOTRON MAGNET

E. M. W. Leung, R. D. Kephart, and C. P. Grozis

*Fermi National Accelerator Laboratory**
Batavia, Illinois

INTRODUCTION

The superconducting Chicago Cyclotron Magnet (CCM) at Fermilab has a pair of 5.33 m diameter split solenoid coils; the design, construction and testing of which had been reported earlier.[1,2] The magnetic field and forces on these coils were calculated using the magnetic codes TRIM and GFUN-3D.[3-5] Each coil is subjected to a very high axial attractive force (4.7×10^6 N) towards the iron yoke and this force increases at a rate of 6.34×10^6 N/m as the coil is displaced towards the yoke. In addition, the radial decentering force acting on each coil amounts to 7.88×10^5 N/m. The break-even point (the point at which LHe plus operation cost = electrical power cost) for the CCM conversion project is a liquid helium boil-off rate of $\sim$ 40 L/h or $\sim$ 28 W. As a result, to make the project cost effective requires the development of a support system that can react reliably and safely the aforementioned forces while at the same time achieving minimal heat leak into the 4.2 K environment.

We decided to react the vertical and de-centering force components with 12 slider-type composite support columns, equally spaced at 30° around the circumference of each coil. The major design goals for each of the 24 columns required were a collapse load of 1.33×10^6 N (300 KIPS) at operating condition, a heat load of $\leqslant$ 150 mW into the LHe temperature environment, and the

---

*Operated by Universities Research Association, Inc. under contract with the U.S. Department of Energy.

ability to withstand the estimated decentering forces. In addition, the support system had to allow for the ~ 8 mm (on the radius) of radial differential thermal contraction that occurs during cooldown of the magnet.

DESIGN OF A SINGLE COLUMN

The primary structural support unit is a four-tube composite column. A short epoxy fiberglass (G-10) tube connects a slider (~ 300 K) to a LN_2 temperature heat sink. Between this heat intercept and the cryostat, there are three tubes, two G-10 and one AISI 304 stainless steel, connected together as shown in Fig. 1. A finished 4-tube assembly is shown in Fig. 2.

The stainless steel/G-10 transition joints are glued (Epon 815) and pinned such that they can take both tension and compression. It is also important for the flanges, (with the grooves accepting the ends of the G-10 tubes), at the ends of the intermediate stainless tube to be exactly parallel to each other. This facilitates assembly and reduces the possibility of the G-10 tubes breaking under compressive local end stresses. This was achieved by using electron beam welding which provided extremely small warpage (the flanges are parallel to each other to within 0.1 mm (0.004 inch) and high welding efficiency (measured to be over 98%). The large radial differential thermal contraction between the coil and the vacuum shell is taken care of by a slider mechanism designed into the column (Fig. 3). The sliding material is made of bronze impregnated Teflon, which has an extremely low coefficient of sliding friction (< 0.05). A thin-walled (0.25 mm) bellows closes the high vacuum circuit while permitting motion of the columns.

We considered using G-10CR instead of G-10 for the composite tubes. Although G-10CR has a higher strength than G-10, the strength/thermal conductivity ratio is about the same. We chose G-10 because of quicker delivery. We measured variations in the material properties of G-10 from different vendors and thus recommend careful testing procedures if G-10 is chosen for support structures. AISI 304 was chosen for the metallic tube because it lent itself better to welding when compared to either 6061-T6 aluminum or titanium.

PROTOTYPE TESTING

A minimum safety factor of three at normal operating condition was our goal. Room temperature mechanical testing was carried out and results extrapolated to lower operating temperatures since the low temperature strength characteristics of both G-10 and AISI 304 are fairly well known.[6] The tube members are sized so that buckling is not possible.

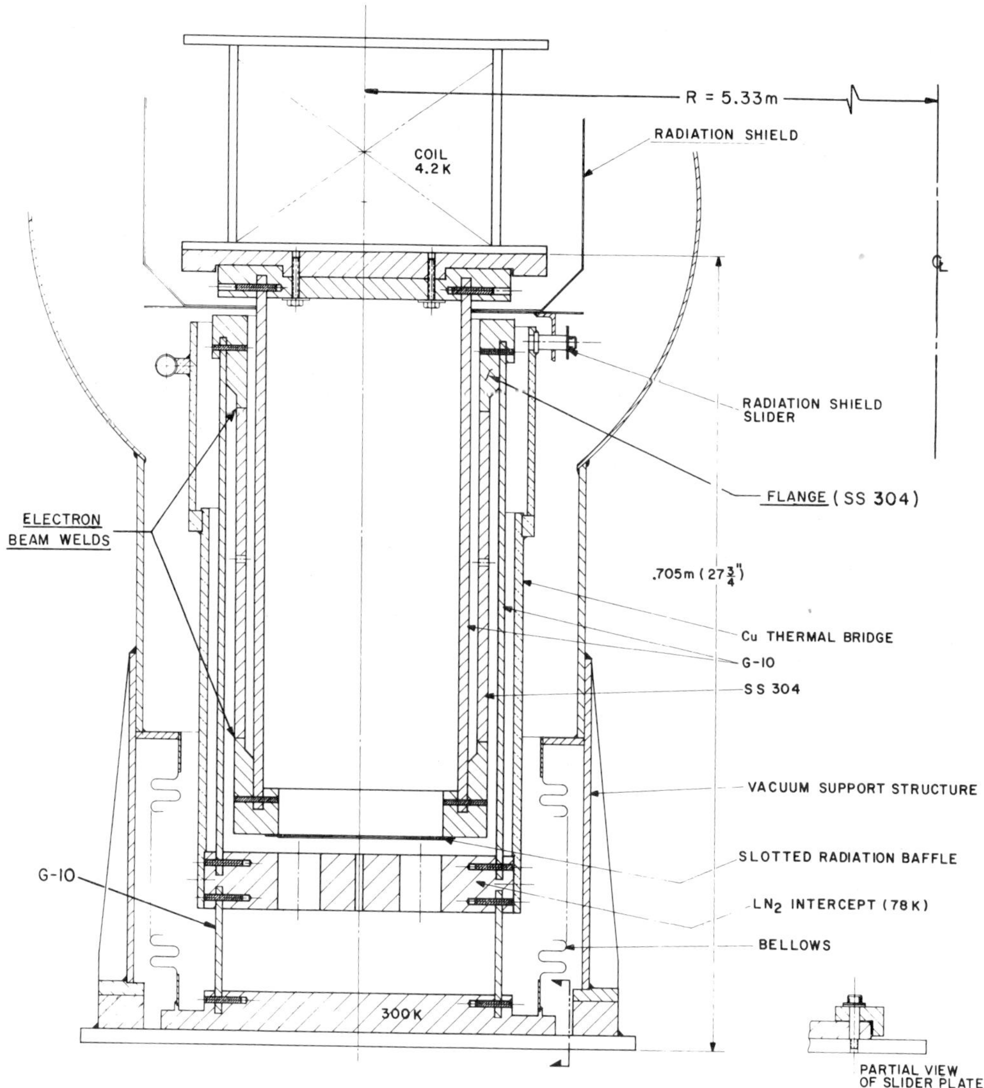

Fig. 1.   Schematic of a Column.

Compression and Tension

The prototype was loaded to 8.9×10^5N (200,000 lbs) corresponding to a load capacity of 1.78×10^6 N at operating condition.

Fig. 2. A single column without   Fig. 3. An installed column
slider.   on the lower coil.

Before the magnet is energized, the columns on the upper coil must support its weight. Each column has to hold 5.4×10^3 N in tension. The prototype was measured to have a yield strength of 1.82×10^4 N and calculated to have an ultimate tensile strength of 5.18×10^4 N.

Side Loading and Loading at an Angle

The lateral stiffness constant was calculated to be 2.6×10^{-6} m/N (4.5×10^{-5} in/lb) and measured to be 3.4×10^{-6} m/N (6.0×10^{-5} in/lb). Actual testing performed on the prototype showed yielding at a side load of 4504 N (1012.5 lbs). The radial de-centering force for CCM is 7.9×10^5 N/m (4500 lb/in). If the coil were 2.54 cm from the magnetic center, the maximum side load that a column has to hold was calculated to be 2682 N (603 lbs). By careful surveying during the installation of the coils, we were able to locate the coils to within 6.4 mm (0.25 in) of the geometric center. Therefore, if the geometric and magnetic centers are the same, we can expect a safety factor of 6.7. The prototype was deliberately subjected to an excessive side load of 1.33×10^4 N (3000 lbs) and retested in compression. No significant degradation in performance (Fig. 4) was observed. We also loaded the prototype at an angle (gradient = 1/64) to simulate a situation

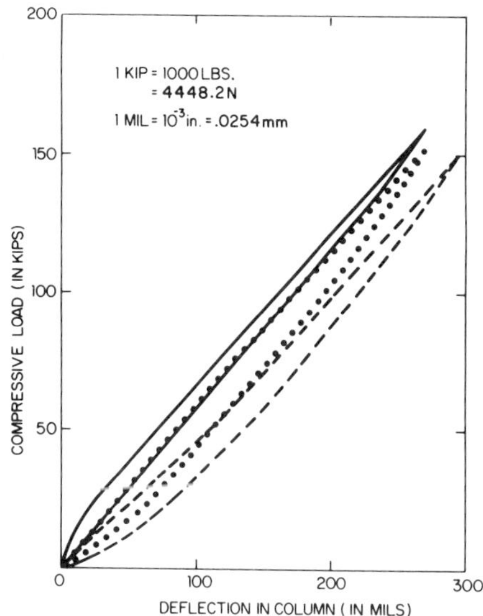

Fig. 4. A typical loading and unloading curve
. . . . .before the prototype was yielded sideways
——————after the prototype was yielded sideways
– – – –after the prototype was yielded sideways and
now loaded at an angle.

where a column is not well shimmed. The loading curve is presented in Fig. 4 also.

Slider

The frictional force from the slider and a retarding force
from the bellows both contribute to a slider side loading effect
on a column during cooldown. The system was optimized during a
series of small tests. The final performance is presented in Fig.
5.

Proof Testing of All Columns Built

Twenty-six more support columns were built after the prototype testings. Each of them were subjected to a compression test
up to 6.7×10^5 N (150,000 lbs) at room temperature and a creep
test. They were all cycled up and down in the compression mode
three to four times to make sure that the loading and unloading
curves were repeatable. The creep test helps to make sure that
all the tubes are correctly mated together.

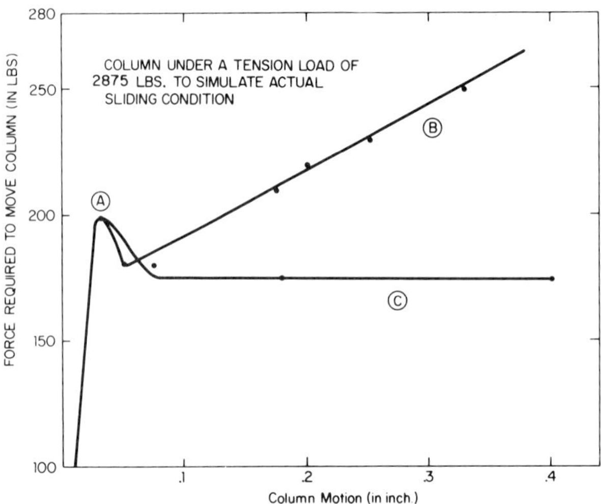

Fig. 5. Slider mechanism test
(A) indicates that the coefficient of static
     friction is 0.070
(B) indicates that the bellows lateral stiffness
     constant is $\Delta.82 \times 10^4$ N/m (275 lb/in)
(C) indicates that the coefficient of sliding
     friction is 0.060.

System Performance During Magnet Testing

The three cryogenic strain gages on three separate intermediate stainless tubes and seven other gages calibrated to be used
on G-10[7], distributed over the various parts of different columns
give an indication of the stress levels induced in the supports
while the magnet is running (Table I).

A maximum stress of 142.8 MPa (20.7 Ksi) is found on the
stainless intermediate tube close to the flange area. This is
close to the calculated 129 MPa (18.7 Ksi). AISI 304 stainless
steel in annealed form (assumed annealed close to weld) has a
yield strength of 427.6 MPa (62 Ksi) at a temperature of 62 K,
hence again a safety factor of three.

During cooldown and warmup, a linear potentiometer system
with digital readout monitors the movement of the support columns
relative to the coil cryostat, and a double-acting hydraulic
system can be activated to push or pull on the base of the individual column to facilitate their motion.

Table I. Column Stresses

	Measured Value	Calculated Value
Average stress on G-10 tube of 3 upper columns	58.5 MPa (8,487 psi)	56.2 MPa (8,150 psi)
Average stress on G-10 tube of 3 lower columns	48.3 MPa (7,000 psi)	56.2 MPa (8,150 psi)
Average stress on AISI 304 stainless steel tubes (maximum stress location)	137.9 MPa (20.0 Ksi)	129.0 MPa (18.7 Ksi)

Measurements during the actual operation indicated that the CCM thermal radiation shield was actually at a temperature of 87 K instead of the desired 78 K and that the LN_2 intercept of the columns was 97 K instead of 80 K. Estimated heat load into cryostat via each column amounts to 162.5 mW, which is within 10% of our design goal of 150 mW per column.

CONCLUSION

We have successfully built a support structure capable of high mechanical load (1.33×10^6 N in compression) and low heat leak ($\sim$ 150 mW). It contributed much to the achievement of the low LHe usage rate of CCM and therefore to the success of the whole project.

REFERENCES

1. E.M.W. Leung, et al., The Superconducting Chicago Cyclotron Magnet -- An Old Magnet with a New Pair of Energy Efficient Coils, IEEE Trans. on Magnetics 17:199 (1981).
2. E.M.W. Leung, et al., "The Superconducting Chicago Cyclotron Magnet," "Advances in Cryogenic Engineering, Vol 27," Plenum Press, New York (1982).
3. R.J. Lari, "Graphic, Time-Sharing Magnet Design Computer Programs at ANL," Proc. 5th Int.Conf. on Magnet Technology, Rome (1975), p. 244.
4. E.M.W. Leung, "Magnetic Field Calculation of the Superconducting Version of the Chicago Cyclotron Magnet using CFUN-3D," Fermi National Accelerator Laboratory Internal Report TM-759, January 1978.
5. C.W. Trowbridge, "Progress in Magnet Design by Computer," Proc. 4th Int. Conf. on Magnet Technology, Brookhaven (1972), p. 555.
6. F.R. Schwartzberg, et al., "Cryogenic Materials Data Handbook, Vol. 2," Air Force Materials Laboratory Report AFML-TDR-64-280 (1970).

7. S. Bonifas and E.M.W. Leung, "Measurement of Apparent Strain
    Curves of Two MicroMeasurements Strain Gauges Mounted on G-
    10," Fermi National Accelerator Laboratory Internal Report
    TM-987, August 1980.

HIGH-FREQUENCY MULTI-STRIPLINE CRYOINSERT FOR SUPERCONDUCTING ELECTRONICS

P. A. Moskowitz, R. W. Guernsey, and J. W. Stasiak

IBM Thomas J. Watson Research Center
Yorktown Heights, New York

INTRODUCTION

High speed Josephson switching circuits operate in a 4.2 K liquid helium environment.[1,2] This and other unique conditions under which Josephson circuits operate place severe restrictions on the type of contact system for chip testing. The length of each individual contact must be minimized, so as to provide a low inductance to be consistent with fast switching times. Also, the entire sample holder and cryoinsert assembly must be made of non-magnetic materials, in order to maintain a low ambient magnetic field for the proper operation of circuitry.

Any contact array for Josephson chip testing must make a reliable and non-destructive contact at room temperature, and maintain that contact through a temperature change of approximately 300 K. Differential contraction experienced upon cooling the contact array to 4.2 K is a serious problem. A 6 mm silicon chip will shrink only 1 µm when cooled to 4.2 K. However, for some materials the contraction across the width of the contact array is on the order of the contact pad size. Thus, materials having small coefficients of contraction must be used.

The complete testing package, known as the cryoinsert assembly, consists of a metal flange which mates with an LHe vessel, a 130 cm long Pyrex support structure and envelope, a high frequency multiple-stripline input/output (I/O) cable and a unique chip connector.

FLANGE AND HOUSING

The upper flange shown in Fig. 1 mates with the top flange of

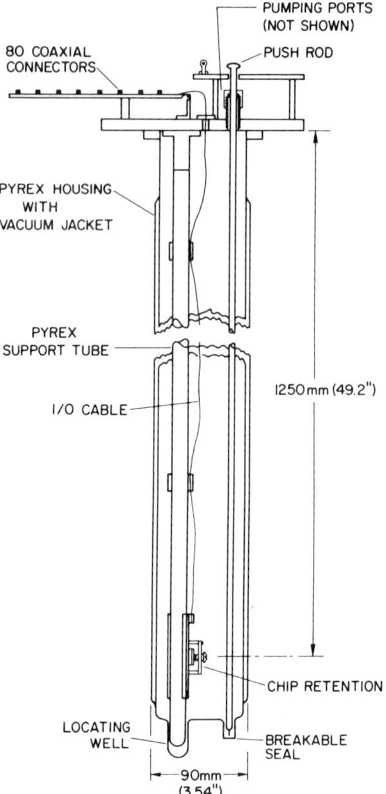

Fig. 1. The Cryoinsert.

a helium dewar and holds the Pyrex center support tube and the
Pyrex housing. Pumping ports, a Pyrex push-rod, the cable feed-
through, and bulkhead support for the coaxial cable connectors are
mounted on the flange. The housing core and vacuum jacket are
kept at a low pressure during insertion so that the chip stays
above its critical temperature (Tc) until it is placed within the
low field region of the dewar. In order to cool the chip, additional helium gas is let into the core and vacuum jacket. Then
the liquid is admitted into the core by using a push rod to break
a thin polyimide seal at the bottom.[3]

It is sometimes necessary to warm the system above Tc during
testing to eliminate magnetic flux trapped in the superconducting
structures. This is accomplished by evacuating the housing jacket, slightly pressurizing the core so as to drive the liquid out,
and then energizing heaters as necessary. Once the temperature of

the system has exceeded Tc, the heaters are turned off and the liquid is readmitted.

CONTACT ARRAY

The Josephson chips to be tested have a double peripheral row of 228 solder pads. Each pad is approximately 100 μm in diameter and the pads are placed on 200 μm centers. An array of Cobra springs, shown in Fig. 2, developed by IBM/East Fishkill, contacts these pads.[4-6] The Cobras are formed from 1.6 mm lengths of 63.5 μm diameter Neyoro-G, a wire manufactured by the J.M. Ney Company. This wire is an alloy of gold, copper, platinum, silver, and zinc. It is non-magnetic, does not oxidize, and maintains its resilience at 4.2 K. The Cobras are held in polyimide dies in such a way as to allow free movement at both ends. While one end of each Cobra presses into a pad on the chip, the other end makes contact with a similar pad on a 25 mm silicon wafer, the adapter, that has thin film wiring from the contact pads to the I/O cable connection (Fig. 3).

I/O CABLE

The flexible polyimide I/O cable consists of eighty 0.036 mm thick copper lines over a continuous ground plane. Each line is 0.18 mm wide, spaced on a 0.51 mm pitch with a fan-in to 0.36 mm at the adapter end. The room temperature end of the cable is terminated by eighty SMA coaxial connectors, while at the low temperature end there is a spring-loaded connection to the adapter (Fig. 4). Initial plans to solder the adapter connection were abandoned because of difficulty in achieving a joint that would withstand the differential contraction on multiple cycles to 4.2 K.

The nominal cable impedance of 50 ohms has been met to within five percent, by the manufacturer, the Rogers Corporation. When

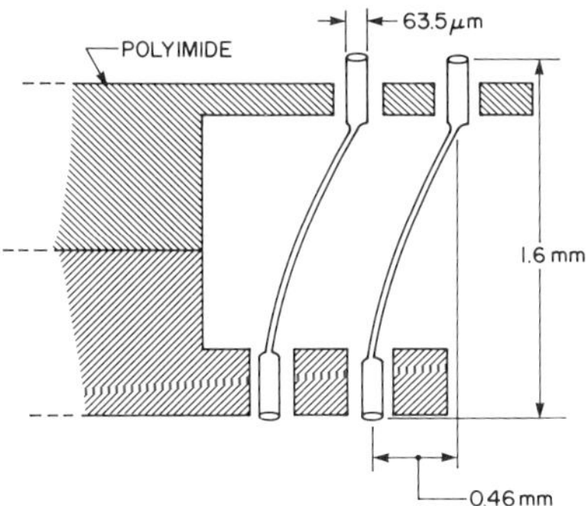

Fig. 2. Cobra contacts.

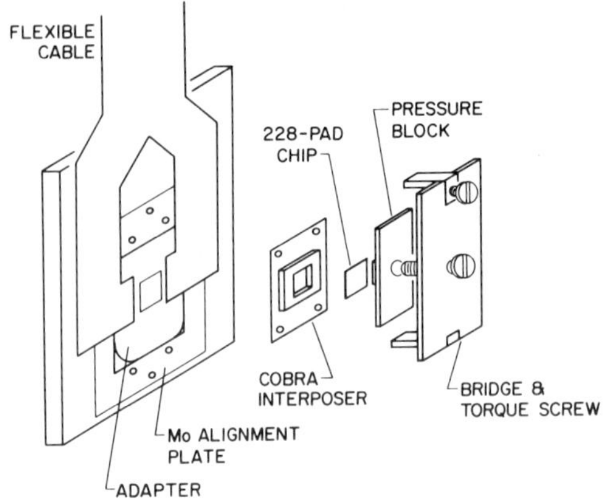

Fig. 3. Chip retention system.

in use, one end of the cable is at 4.2 K and the other at 300 K. A signal that is introduced with a 200 ps risetime at one end of the 1.4 m cable arrives with a 400 ps risetime at the other end, and is attenuated about 15% in amplitude. Under these conditions, the worst-case forward crosstalk is about 5.0 to 7.0%, while the backward crosstalk is 3.5%. These results are in substantial agreement with theoretical expectations.[8] The propagation velocity is 0.6 c. Heat loss, ameliorated by vapor cooling, is on the order of 1.7 mW per lead. Cables have been cycled over thirty times between room temperature and 4.2 K without apparent degradation of their characteristics.

CONCLUSION

To date, four cryoinsert systems have been constructed. Two

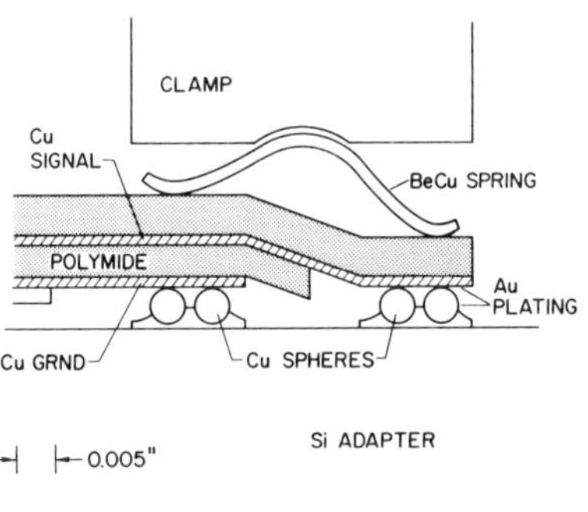

Fig. 4. Cable to adapter connector.

have been devoted to chip testing, while two others have been used to provide the I/O system for Josephson system test vehicles. The chip testing application provides the most rigorous exercise of the system. The greatest number of cycles between room temperature and liquid helium temperature that any of the cryoinserts has experienced has been about thirty. If faults appear, they are usually at the connections to the adapter. Methods to improve cyclability are under investigation. The cryoinserts have, in general, proven to be reliable laboratory instruments.

Work is now under way to define the next generation of cryoinserts. By enlarging the adapter from 25 mm to 57 mm, we may be able to provide as many as 250 I/Os. Since, it would be inconvenient to make individual connections for each of 250 lines at the room temperature interface, a multiple stripline connector should replace the coaxial connectors now in use. More data will be accumulated on the performance of the present instruments, so that we may be able to evaluate the long-term reliability of the systems.

REFERENCES

1. Josephson computer technology, IBM J. Res. Develop., 24:105 (1980).
2. J. Matisoo, The superconducting computer, Scientific American, 242:50, (1980).
3. J.W. Stasiak and R.W. Guernsey, Slip-on vacuum connection for low temperature use, IBM Technical Disclosure Bulletin, 23:3384, (1980).
4. H.P. Byrnes and R. Wahl, Contact for an Electrical Contactor Assembly, U.S. Patent No. 4,027,935, June 7, 1977.
5. H.P. Byrnes, P.A. Moskowitz, and R. Wahl, Cryogenic testing of Josephson chips, IBM Technical Disclosure Bulletin, 23:4363 (1981).
6. P.A. Moskowitz, Electrical performance of high density probe array for testing Josephson circuit chips, IEEE Trans on Magnetics, Mag-17:761 (1981).
7. H.R. Bickford, et. al., Flexible I/O cable to thin film connector, IBM Technical Disclosure Bulletin (to be published).
8. C.J. Anderson, private communication, (to be published).

THE DEVELOPMENT OF FORCE-COOLED
SUPERCONDUCTORS FOR USE IN LARGE MAGNETS*

J. R. Miller

*Oak Ridge National Laboratory
Oak Ridge, Tennessee*

INTRODUCTION

A high degree of stability is required of the large superconducting magnets being developed for fusion research and other
applications. In recent years considerable effort has been expended toward achieving the required stability with conductor
designs incorporating forced convection of the helium coolant. A
variety of force-cooled conductors is currently being applied to
the construction of large magnets. Although there are some similarities, there are many differences both in appearance and performance even when the designers have worked to the same set of
specifications as in the Large Coil Program (LCP) at Oak Ridge
National Laboratory (ORNL). In the case of the LCP, we are afforded the opportunity of examining three different force-cooled
designs in one application.

Force-cooled conductors offer some advantages over pool-boiling types; however, they also suffer some disadvantages. For
example, since cooling is internal they allow continuous insulation between windings, but the finite inventory of helium available to a section of conductor puts finite limits on the
stability. The Large Coil Program will offer a useful comparison
of force-cooled to pool-boiling types.

Despite finite limits to the stability, force-cooled conduc-

---

*Research sponsored by the Office of Fusion Energy, U.S. Depart-
ment of Energy, under contract W-7405-eng-26 with the Union
Carbide Corporation.

tors can be extremely stable, even with low or zero net helium
flow.[1,2]   Consequently, we often drop the force-cooled nomen-
clature and refer to them as Internally Cooled Superconductors
(ICSs) to emphasize that high net flow is not always necessary for
stability. In recent years we have gained an understanding of the
low flow stability as well as of the surprising multiple stability.[3,5] These results suggest that a large ICS magnet can possess
about the same stability as one cooled by He II[6] without the
increased refrigeration costs and complexity. Formulas have been
derived for extrapolation of the experimental data that allow
avoidance of multiple stability regimes.[4,5]

There is safety to be had with force-cooled designs because
the helium inventory is usually small and contained in a small,
strong tube. The designer must, however, pay attention to the
maximum pressure rise in the event of a quench. Theory and recent
experimental results that provide a simple analytic formula give a
powerful tool to the designer considering this aspect.

GENERAL CONSIDERATIONS

For the designer who will use an ICS for a magnet, it is
important from the outset to be keenly aware of the features
peculiar to such conductors. If his design is to be successful,
he must make full use of the good features and eliminate or minimize the others. Possibly the clearest distinction from a design
standpoint between the ICS and a conductor cooled by pool-boiling
helium is: the stability of the latter is limited by heat transfer whereas the stability of the ICS, if properly designed, is
limited by the enthalpy change of the internal helium between its
initial bulk value and that at the current-sharing temperature of
the conductor.

Heat transfer in boiling helium, being extremely geometry and
orientation dependent, presents uncertainties to the designer that
are usually overcome only by extensive and expensive modeling. We
shall show, however, that it is possible to ensure that heat
transfer in the ICS is good enough that all the helium in the
vicinity of a thermally perturbed section participates in its
recovery to the superconducting state. As a result, the physical
properties of helium, which are much better known, can be used to
calculate stability.

Pressurized He II stabilization also eliminates the need for
precise knowledge of heat transfer and takes advantage of the
enthalpy of the bath. However, the stability margins afforded in
a large coil[6] are not clearly superior to those obtainable in an
ICS, and the refrigeration penalty imposed by the lower operating

temperature and unusual cryogenic constraints must be weighed carefully.

Insulation integrity can be better achieved in an ICS because the helium does not permeate the space between turns. Also, the windings can be impregnated to provide a solid, electrically insulating barrier. Possibly just as important, the potting can increase the strength of the windings and inhibit conductor motion. Additional strength may also be obtained from the sheath or jacket that must be used for containment of the helium. If a distributed structure is required, it is a natural part of the ICS concept. Also, the conductor jacket, the impregnant, and the outer casing of the coil can act as triple insurance against leaks and as a sturdy container for the helium in the event of a quench.

STABILITY

Not long ago it was supposed[7] that stability in an ICS was achieved by flowing the helium, with resultant penalties in pumping power, to obtain a high heat transfer coefficient. Then experimental data began to show that, in the cable-in-conduit conductors at least, surprisingly high stability was achieved with low or zero net flow.[1,2] More careful study of stability margins (the sudden energy deposition required to just quench the conductor) versus current showed that the stability under certain conditions is multivalued.[3] Some fairly recent data,[5] shown in Fig. 1, dramatically illustrate this phenomenon. In the range of currents shown, a thermal perturbation of around 50 mJ/cm^3 of conductor might cause a quench whereas a larger perturbation at the same current level would not. Further increases in the heat pulse would not bring about a quench until the level was increased almost an order of magnitude above that necessary initially.

In recent years we have studied this region of multiple stability both experimentally and theoretically.[4,5] Others have examined the transient phenomena, both experimentally and with computer calculations, that lead to this behavior.[8,9] Generally, the lower region is explained in terms of transient conductive heat transfer, which is insufficient for higher heat inputs. The upper region of stability, on the other hand, results from high induced flow caused by the transient pressure rise following a heat pulse. We have found that the occurrence of these multivalued stability regions can be predicted and, by proper choice of conductor parameters and operating conditions, can be avoided. The limiting current density, J_{lim}, below which stability is single-valued and has only the upper value scales according to the formula[4,5,10]

$$J_{lim} \propto \left[\frac{f_{Cu}(1 - f_{cond})}{f_{cond}}\right]^{1/2} (T_c - T_B)^{1/2} \rho_{Cu}^{-1/2} \tau_H^{-1/5} \ell_H^{2/5} D_h^{-1}, \qquad (1)$$

where f_{cond} is the fraction of the cable space occupied by conductor, f_{Cu} is the fraction of conductor made up of copper, ρ_{Cu} is the resistivity of the copper at operating conditions, T_c is the critical temperature of the conductor, T_B is the initial bulk fluid temperature, τ_H is the duration of the heat pulse being stabilized against, ℓ_H is the heated length, and D_h is the hydraulic diameter of the flow path. We have thus far found scaling with this formula to be quite good, with the possible exception of the parameters τ_H and ℓ_H.[5] However, the dependence on these is fairly weak, which is also fortunate, because in a real coil they are not so well known. Experimental results from a test coil recently built and operated at ORNL seem to indicate that the scaling holds as well in a real coil as in the small-scale conductor tests.[11]

Figure 2 shows that the addition of a steady helium flow can also remove the multivalued region,[5] as should be expected. However, another possibly more interesting point is illustrated in

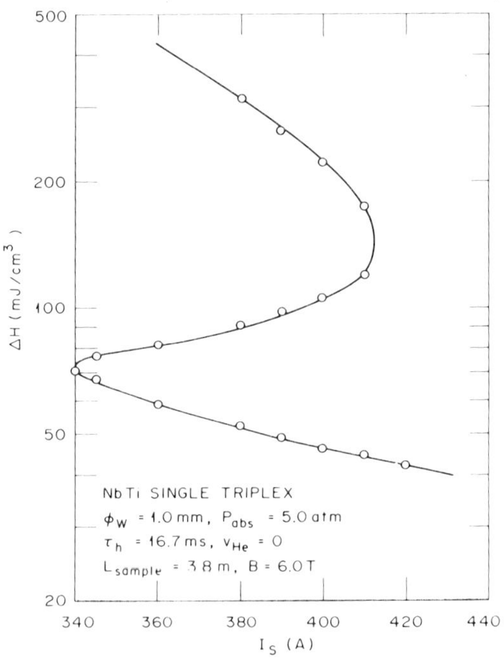

Fig. 1. Stability margin vs current in a single triplex ICS.

the figure. As shown by the dashed curve, the value of the stability margin ΔH in the upper region has the lower bound

$$\Delta H \gtrsim \frac{A_{He}}{A_{cond}} \int_{T_B}^{T_{cs}} \rho c_p \, dT, \tag{2}$$

where A_{He}/A_{cond} is the ratio of conductor cross section to the
helium cross section and ρc_p is the product of helium density and
specific heat. The integral is evaluated from the inital bulk
fluid temperature T_B to the current-sharing temperature T_{cs} of the
conductor. The pressure is taken as the initial value, which of
course is not true over the entire recovery event but which gives
a simple formula that has proven to be extremely useful in estimating available stability margins.

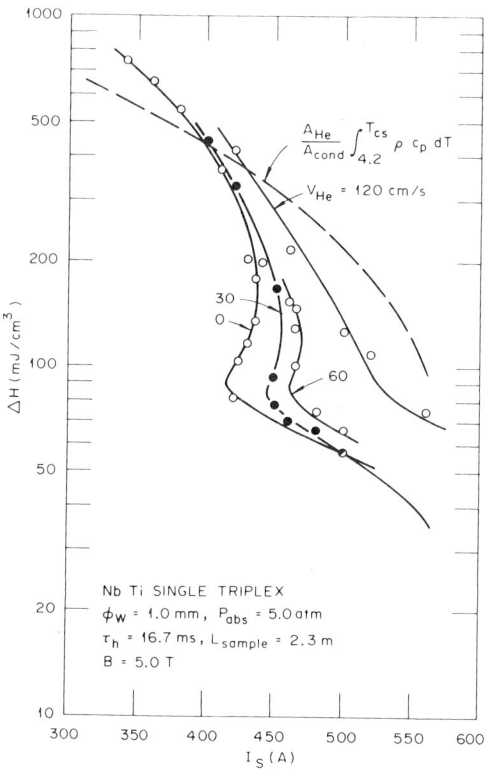

Fig. 2. Stability margin vs current in an ICS showing
variation of the multivalued nature with net helium flow
velocity.

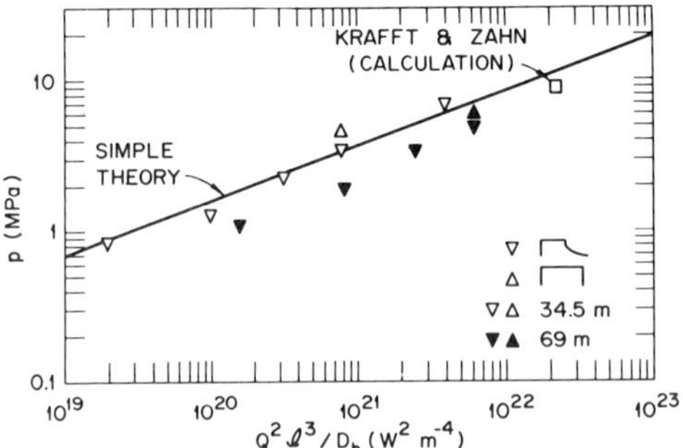

Fig. 3. Experimental simulation of the maximum pressure in a quenching ICS magnet compared with a simple formula involving the scaling perimeter $Q^2\ell^3/D_h$ and with a more detailed computer calculation by Krafft and Zahn (see Ref. 13).

PROTECTION

As mentioned, containment of the helium in a small high strength channel might appear to be an advantage in the event of a quench in a large coil. However, the long and restricted flow paths that must accompany the use of an ICS will give rise to high pressures. This problem has also been the subject of both theoretical and experimental studies.[12-14] For scoping studies and initial design a simple formula has been developed for the maximum pressure p_{max} expected in the event of a quench of an entire coil:[12]

$$p_{max} = 0.10 \left(\frac{Q^2 \ell^3}{D_h} \right)^{0.36} \tag{3}$$

Here, Q is the initial heating rate per unit volume of helium, ℓ is the half-length of a flow path, and D_h is the hydraulic diameter. Figure 3 gives a comparison of this formula with experimental results and a detailed calculation from an independent source.

Equation 3 was developed from a more detailed theory which also makes predictions of the maximum temperature rise during a quench.[12] Shown in Fig. 4 is a comparison of the calculated and measured temperature rises in the same quench experiment.[12] The temperature rises in this case were modest because this type of conductor keeps the helium near and in good thermal contact, allowing advantage to be taken of its comparatively high heat capacity. The theory overestimates the temperature rise probably

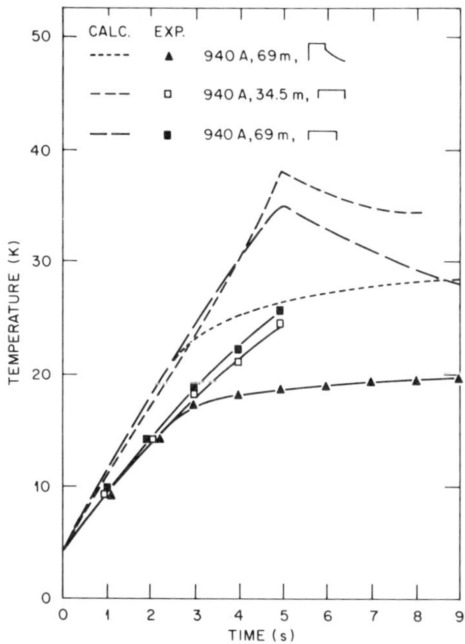

Fig. 4. Calculation and measurement of the conductor temperature rise during simulated quenches of an ICS magnet.

because it does not take into account other heat absorbing materials such as the jacket and insulation. More detailed calculations give temperature histories in close agreement with the experimental ones.[15]

DESIGN PROCEDURE

From the above discussion a straightforward design procedure can be outlined for selecting ICS parameters for a particular application. Once the operating field B, overall current density J, and operating current i_{op} have been chosen, the required stability margin must be determined. Margins of $\sim$ 100 mJ/cm^3 of conductor may be desired for very large magnets. By assuming that the conductor can be designed to operate in the range below the region of multiple stability, it is assured that the stability margin has the lower bound given in Eq. 2. This bound (referred in this case to the helium fraction), is plotted in Fig. 5 for different values of T_B. A stability margin of 300 mJ/cm^3 of helium is indicated on the figure for purposes of illustration. If the choice of superconductor has already been made, then a choice of T_B determines the corresponding critical current $i_c(T_B,B)$ required since

$$T_{cs} - T_B = \left(T_c - T_B\right)\left[1 - \frac{i_{op}}{i_c(T_B,B)}\right]. \tag{4}$$

For example, $T_c \simeq 9.1 - 0.44B$ and $i_c(T_B,B) \propto (0.55 - 0.027B)$ $(T_c - T_B)$ for commercial NbTi conductors. The ratio of critical current to operating current puts constraints on the choice of the relative fractions of copper, superconductor, and helium. Other considerations such as the required amounts of insulation or structure also must be taken into account, and a preliminary strand diameter must be chosen.

With these choices made, the resultant conductor current density J_{cond} must be compared with the limiting value for multiple stability J_{lim} by using Eq. 1 to extrapolate empirical results like those in Fig. 1. Adjustments can be made to the strand diameter to bring these values into line, or another iteration can be made in which T_B or the relative fractions of materials in the winding are varied.

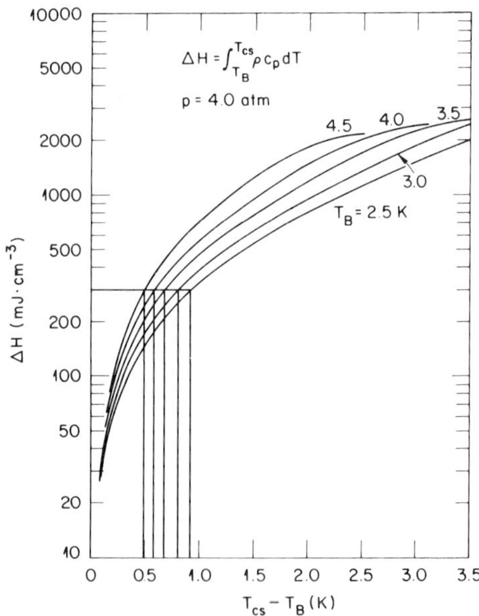

Fig. 5. Available stability margin in an ICS (referred to the helium fraction) if operated below the region of multiple stability (helium properties from Ref. 16).

When adequate stability has been assured by arranging for $J_{cond} < J_{lim}$, the maximum possible quench pressure should be checked according to Eq. 3, using

$$Q = \frac{J^2_{cond} \, \rho_{Cu}}{f_{Cu}} \, \frac{A_{cond}}{A_{He}} \ .$$

If p_{max} cannot be made acceptable by adopting a winding scheme with short enough flow paths, another iteration will be necessary. More detailed calculations can be made of quench pressures and maximum temperature rise after the conductor design is more firm.

CONCLUSION

Force-cooled conductor or ICSs offer an attractive option for designers of large magnets. Knowledge of the performance of this type of conductor has advanced sufficiently in recent years for reliable designs to be made. Preliminary designs can be made with the help of relatively simple formulae and a rather extensive data base from small conductor tests.

ACKNOWLEDGMENTS

The contributions, aid, advice, and criticisms of my co-workers, J. W. Lue and L. Dresner, appear throughout this work and are gratefully acknowledged. I also wish to thank M.S. Lubell for his encouragement and his suggestion of this paper.

REFERENCES

1. M.O. Hoenig and A.G. Montgomery, in "Proc. 7th Symp. Engineering Problems of Fusion Research", Vol. I, IEEE Science Center, Piscataway, New Jersey (1977), p. 780.
2. J.R. Miller, et. al., IEEE Trans. Magn., MAG-15:351 (1979).
3. J.W. Lue, J.R. Miller, and L. Dresner, J. Appl. Phys. 51:772 (1980).
4. L. Dresner, IEEE Trans. Magn., MAG-17:753 (1981).
5. J.W. Lue and J.R. Miller, IEEE Trans. Magn., MAG-17:757 (1981).
6. S.W. Van Sciver, IEEE Trans. Magn., MAG-17:747 (1981).
7. L. Dresner and J.W. Lue, in "Proc. 7th Symp. on Engineering Problems of Fusion Research", Vol. I, IEEE Science Center, Piscataway, New Jersey (1977), p. 703.
8. S.R. Shanfield, et. al, IEEE Trans. Magn., MAG-17:2019 (1981).
9. G. Reis, "Quench Induced Helium Flow and Recovery in the EURATOM LCT Coil," IEEE Trans. Magn., MAG-17:2097 (1981).

10. J.R. Miller,et. al., in "Advances in Cryogenic Engineering, Vol. 26", A.F. Clark and R.P. Reed, eds., Plenum Press, New York (1980), p. 654.
11. J.W. Lue and J.R. Miller, Performance of an Internally Cooled Superconducting Solenoid, in "Advances in Cryogenic Engineering, Vol. 27", Plenum Press, New York (1982).
12. J.R. Miller, et. al., in "Proc. 8th Intern. Cryogenic Engineering Conference", IPC Science and Technology Press Ltd., Guildford, England (1980), p. 321.
13. G. Krafft and G. Zahn, in "Proc. 8th Symp. on Engineering Problems of Fusion Research", Vol. IV, IEEE Science Center, Piscataway, New Jersey (1979), p. 1724.
14. G. Krafft and G. Reis, in "Proc. of the 8th Intern. Cryogenic Engineering Conference", IPC Science and Technology Press Ltd., Guildford, England (1980), p. 330.
15. G. Reis, private communication (July 1980).
16. R. D. McCarty, Thermophysical Properties of Helium, <u>Normal Bureau of Standards Report NBS-631</u>, Boulder, Colorado, (1972).

DISCUSSION

Question S. Wipf, Los Alamos Scientific Laboratory: You give the stability margin in terms of so many joules per cm^3 from which recovery is possible. Do you have a feeling what actual disturbances are likely to occur in your conductor when employed in a large device; one expects conductor movement within the conduit.

Answer by Author: In design studies for coils using an ICS we usually use as a goal for stability margin a few tenths of a Joule per cm^3 of conductor, because this is usually the <u>maximum</u> heat energy that might conceivably be deposited by conductor motion. However, we have tested ICSs with demonstrated stability margins much less than this level without observing spontaneous quenchs. It could be that the strands in an ICS don't move nearly as much as is generally thought, or all motions occur at low enough field or current level to make them harmless.

CRYOGENIC ASPECTS OF THE INTERNALLY COOLED, CABLED SUPERCONDUCTOR (ICCS) FOR THE 12-TESLA PROGRAM*

M. O. Hoenig, A. G. Montgomery, M. M. Steeves, and M. M. Olmstead

Plasma Fusion Center
Massachusetts Institute of Technology
Cambridge, Massachusetts

INTRODUCTION

The idea of an ICCS was first proposed by Montgomery and Hoenig in 1974.[1] Much work has been done since by Oak Ridge National Laboratory, MIT and others and has been reviewed by Hoenig.[2] The ICCS for the US-DOE 12 T coil program has been developed by a joint effort involving MIT, Westinghouse, Airco, and Supercon. This ICCS, a prototypical conductor for the Fusion Energy Device (FED), will operate in a 11-12 T field while carrying not less than 10 kA at 4.2 K. It will be tested in the coil shown in Fig. 1, called the ICCS/HFTF Test Coil, which consists of three series-connected double pancake subcoils (57 turns total). It is scheduled for testing at the Lawrence Livermore Laboratory High Field Test Facility (HFTF) in 1982. This paper summarizes progress made in the past year in support of the 12 T program.

CRYOGENIC STABILITY OF THE ICCS

The need for an ICCS conductor for fusion and MHD coils arose when it was found that more conventional superconducting hollow conductors[3] required unacceptably high steady-state flow velocities in order to provide adequate heat transfer. The increased surface area (A_H) of an ICCS reduces the need for a large heat transfer coefficient (h). This can be readily seen from

$$\dot{q} = h \, A_H \, \Delta T \, , \tag{1}$$

---

*Supported by the US-DOE Division of Magnetic Fusion Energy

where $\dot{q}$ is the heat flux and ΔT the temperature differential between wire or tube wall and coolant.

In subsequent MIT tests of low current (555 A at 12 T) ICCS conductors it was found that the recovery from transient heating pulses was nearly independent of steady-state flow.[4] The stagnant supercritical helium recovery rates observed were appreciably in excess of those predicted by the theoretical thermal diffusivity heat transfer coefficient given by

$$h = K \ (\pi \alpha t)^{-1/2} \ , \qquad (2)$$

where K is thermal conductivity, α is thermal diffusivity and t is elapsed time.   It was thus concluded that heat transfer to the helium within the tightly packed ICCS is enhanced by locally induced turbulence resulting from the transient heat pulse.

In order to better understand the mechanism of the self in-duced heat transfer enhancement, an experiment was performed using stagnant helium in capillary tubes.[5]   In this experiment the heat transfer characteristics of stagnant supercritical helium were studied by resistively heating the capillaries with pulse cur-rents.   The results are relevant to the helium space in an ICCS since each tube diameter was matched to the hydraulic diameter of an ICCS.   Typical results are shown in Fig. 2 for a 2 m long heated zone, where X = ± 1 m either end of the heated zone.   Time

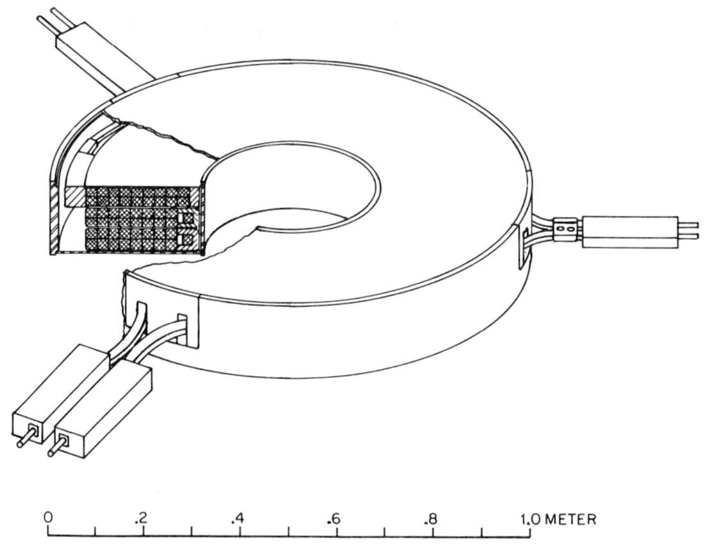

Fig. 1.   ICCS/HFTF Test Coil.

and position - dependent heat transfer rates were observed which were in excess of the theoretical diffusion heat transfer rate of Eq. 2. The values of h indicate a rapid buildup of heat transfer for specific locations along the length of the heated capillary. The closer the location is to the center of the capillary, the greater the delay time between heating pulse and start of the buildup. It can be shown[5] that this delay time (Δt) can be expressed by

$$\Delta t = S/C, \tag{3}$$

where S is the distance from the end of heated tube length ($X = \pm$ 1 m) to the specific location and C is the speed of sound in helium prior to heating. It can also be shown that the peak value of h at a given location decreases $X \to 0$. Furthermore, the peak h is seen to decrease as the time delay increases.

By extrapolating the data of Fig. 2 to a 20 m heated length, two features stand out: (1) the rapid buildup of h near the center of the heated zone is not initiated until 50 ms have elapsed; and (2) the magnitude of the peak value of h, at the center has fallen down to or even below the value predicted from the thermal diffusivity (Eq. 2).

A possible conclusion based on data obtained in the capillary tube experiments is that there is a limit to the length of an ICCS

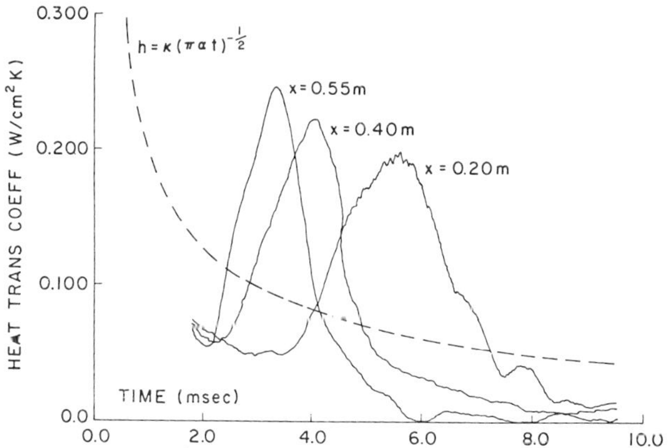

Fig. 2. Inferred heat transfer coefficient at three locations (X) along a 2.0 m heated section following a 1.75 ms pulse. Also plotted is an approximate theoretical prediction assuming no induced flow (broken line).

which can recover from a large transient heat pulse by self induced heat transfer. This limit would appear to be around 20 m.
In addition to pulse heating, the ICCS is subject to joule heating
during the delay time. If the heated length is not continuous,
such as in a coil with a high field gradient, the delay time will
be short. Since a toroidal field coil of an FED has a relatively
high field gradient in its high field region, its cryostability
may be enhanced by pulsed energy induced turbulence.

SUPERCRITICAL HELIUM PUMP OPERATION

A bellows pump was built at MIT for the ICCS/HFTF Test
Coil. The pump consists of two enclosed and opposing expansion
bellows and four rectifying check valves. The pump has been
tested with a flow impedance consisting of a simulated load and
orifice flow meter. A schematic of the test setup is shown in
Fig. 3. Figure 4 shows a plot of measured performance with a test
load and orifice meter as well as the estimated pump performances
with the ICCS/HFTF Test Coil subcoils connected in parallel or in
series but without the orifice meter. The pump should readily be
able to operate at 3.8 seconds per stroke for the parallel operation, delivering a total flow of 4.2 g/s with a pressure drop
across the parallel subcoils of 2 psi. For the series operation

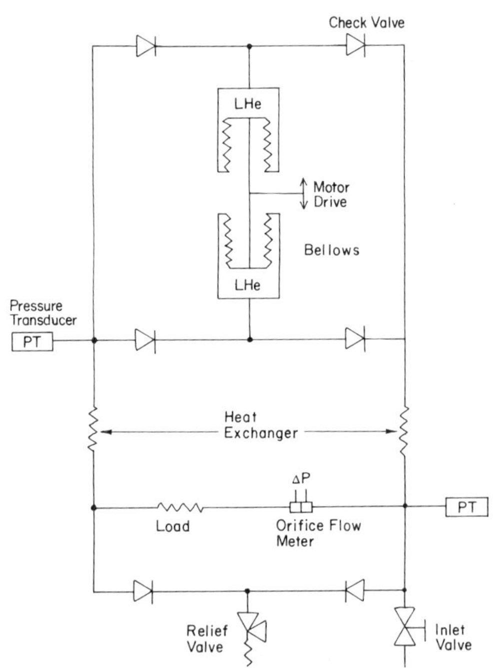

Fig. 3. Schematic for Pump
Test.

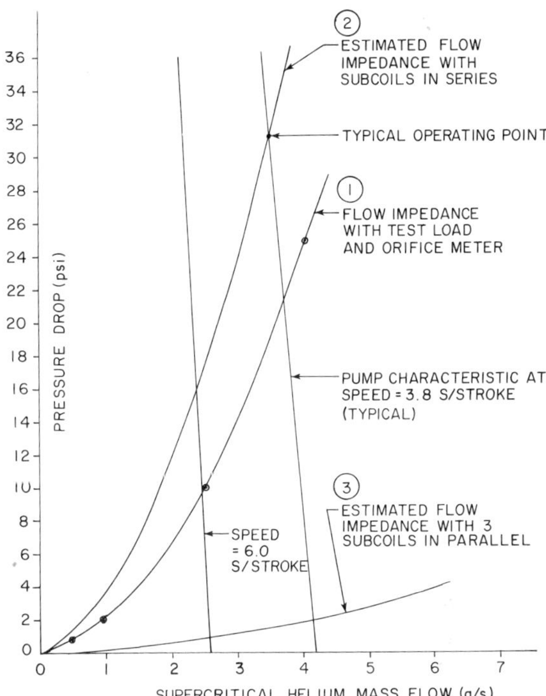

Fig. 4. Bellows Pump
Performance

the pump speed will be about 9 seconds per stroke to produce a
mass flow of 1.5 g/s at a pressure drop of 7 psi. To maintain the
supercritical helium state a minimum pressure of 19.1 psig must be
maintained. Allowing a 10.9 psi safety margin, maximum system
pressure will be 32 and 37 psig for parallel and series operations
respectively.

SUBSIZE CABLE TESTS

A subsize cable is one with 27 strands and a nominal current
rating of 555 A at 12 T. The full scale cable has 486 strands.
Airco's Nb_3Sn multifilamentary wire has been selected, following
tests at MIT of the critical current density (J_c) in subsize
cables sheathed in stainless steel and copper. Figure 5 is a plot
of the MIT test data at 4.2 K with Airco's single strand test
results[6] shown for comparison. Critical currents were evaluated
by sample voltage drop criterion of 1.5 μv/cm in all cases. J_c is
calculated as the measured critical current divided by the total
noncopper area. Conductor parameters are listed in Table I, and
while all variables have not yet been tested, the observed critical current densities (Fig. 5) vary by less than 9% over the
parameters tested at each field examined.

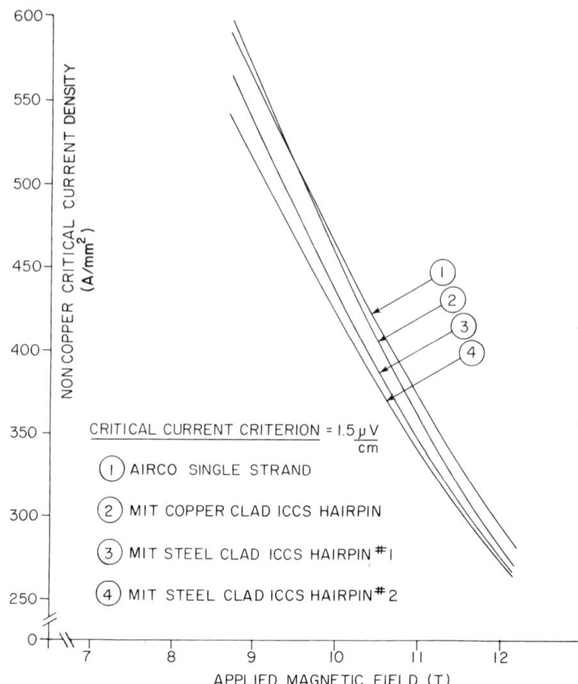

Fig. 5. Dependence of critical current density on applied field for subsize ICCS and single strands.

Figure 6 shows an updated ICCS/HFTF Test Coil load line and three possible ICCS wire options (dashed curves are estimates of critical current based on single wire data, solid curves on 27 strand ICCS data). The original design called for the use of a developmental Supercon wire (with the current density optimized by a 750°C-30h heat treatment) in two out of three of the sub-coils.[7] Due to unexpectedly low critical currents of 27 strand Supercon cables sheathed in stainless steel, this wire was rejected. However, testing has continued with Airco samples activated under the original (Supercon optimized) conditions. By holding firing conditions constant, the number of experimental variables is minimized and other variables such as ICCS compaction, twist pitch and sheath material can be studied. Fully optimized Airco wire (735°C for 240h) would give the potential of 13 T operation, while 12 T operation can be achieved even with nonoptimized wire (750°C-30h), provided that all 486 strands of the fullscale cable contain Nb_3Sn. A hybrid ICCS (2 multifilamentary strands to each pure copper strand) appears to have adequate margin to meet the 10 kA current requirement.

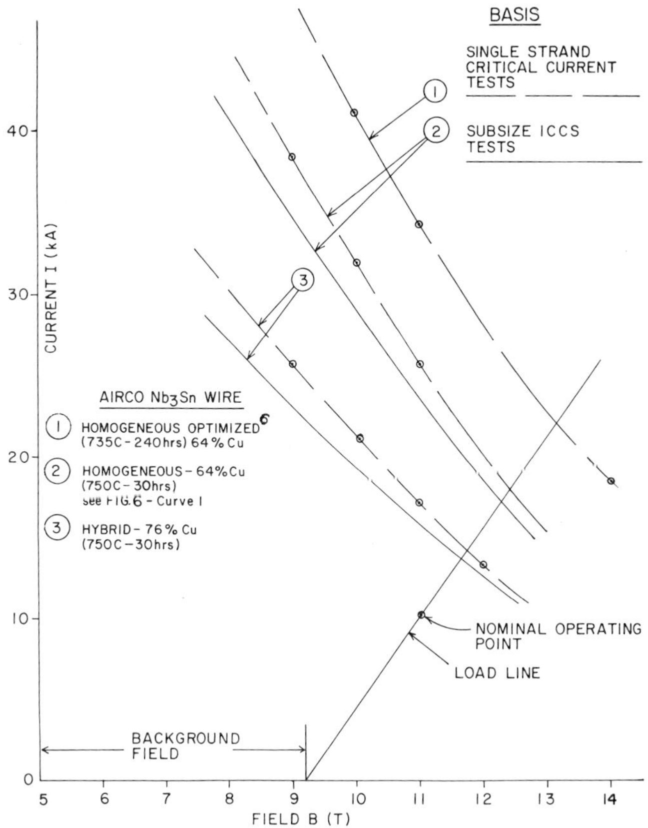

Fig. 6. Operating Characteristics for the
ICCS/HFTF Test Coil.

PULSE COIL

The ICCS/HFTF Test Coil will be equipped with an electromagnetic pulse coil wound around the ICCS at the crossover of the
central double pancake as shown in Fig. 7.

In order to evaluate the efficiency of this pulse coil a
reduced length version of the coil was tested using the subsize
ICCS, previously tested under steady-state conditions (See Fig. 5,
sample 4). A section view of the test pulse coil on a sample of
square A-286 stainless steel tube is shown in Fig. 8. Measurements of the stability of the 27 strand ICCS were made under the
following conditions: (1) steady-state current equal to 75% of
the critical current, (2) pulsed energy deposits in an 11 T

Table I. Subsize ICCS Parameter Study

1. Parameters which vary from hairpin to hairpin.

Hairpin Reference No.	2*	3*	4*	5	6	7	8
Twist Pitch	1 in	1 in	1 cm	1 in	1 in	1 in	1 in
Percent Void	23	44	37	37	23	5	15
Sheath Material	OFHC Cu	321 SS	321 SS	321 SS	321SS	321 SS	321 SS
Sheath Wall	0.031 in	0.010 in	0.010 in	0.010 in	0.010 in	0.010 in	0.010 in
Sheath OD	0.250 in	0.218 in	0.208 in	0.208 in	0.208 in	0.208 in	0.208 in
Strand Insulation	Oil	Oil	Cu S	Oil	Oil	Oil	Oil

* Test Completed

2. Parameters held fixed.

Wire manufacturer	Airco
Matrix	Bronze
Number of filaments per strand	2869
Filament diameter	3.5μm
Weight percent tin	13
Wire copper to noncopper ratio	1.8/1
Wire diameter	0.71mm
Cable configuration	3 x 3 x 3
Cable makeup	2:1 hybrid
Reaction parameters	750C–30 hrs.

field. A marginal normalcy was generated with an imposed energy density (q) of approximately 135 mJ/cm^3 using stagnant helium at 1 atm and 4.2 K. When the conductor was filled with pressurized supercritical helium at 3 atm, the conductor was stable up to q $\simeq$ 580 mJ/cm^3. Further experiments are planned to extend the stability measurements and to measure the flux penetration of the heavy ICCS sheath.

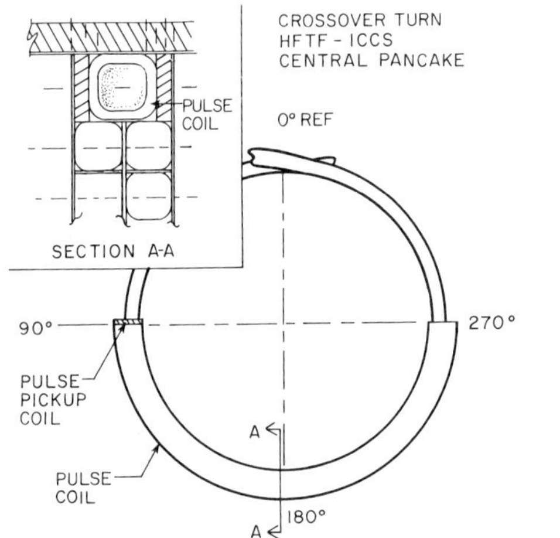

Fig. 7. Placement of the pulse coil in the ICCS/HFTF.

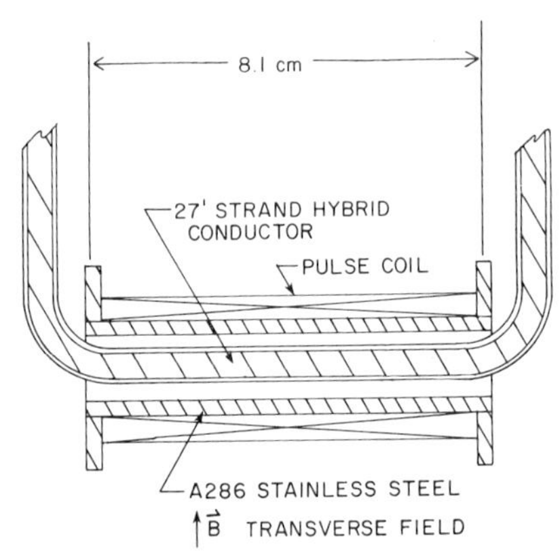

Fig. 8. Section view of Test Pulse Coil.

ICCS TERMINATION DEVELOPMENT

A 2 m long full size ICCS hybrid cable has been terminated on both ends by Airco. The cable termination procedure[7] basically consists of the following: end compaction and sheath removal, application of a copper tube and insertion of a monel cooling tube, application of a monel coupling and final squaring of the copper end. The test cable, when operated at 10 kA in a hairpin configuration with its center in a 12 T cross field, will provide data concerning the performance of the superconductor, the terminations and their joints. The resistive voltages developed across both the high field segment and the terminations will be measured.

ICCS ENCAPSULATION AND COIL WINDING

Several 45 m lengths of the full scale cable (10 kA at 12 T) were encapsulated by Airco in various metal sheaths. The initial sheath material was Nitronic 40. A 45 m length of the resultant ICCS, using a model (copper) cable was used to wind a double pancake subcoil model which was then subjected to a typical activation procedure (30 h at 700°C). The subcoil was subsequently released from its restraints in order to install a copper pulse coil around its crossover region. Insulated with epoxy-fiberglass (G10) strips and a fiberglass overwrap, the model coil has been rewound and is ready for vacuum epoxy potting. This exercise successfully demonstrated the feasibility of the manufacturing process.

For the ICCS/HFTF Test Coil the original Nitronic 40 sheath has been replaced with JBK-75, a variant of the more common A-286 alloy, both of which are considered metallurgically stable at cryogenic temperatures following the superconductor activation cycle.[8] Metallurgical compositions are given in Table II. A 45 m length of the JBK-75 alloy-encapsulated cable has been furnished to Westinghouse who have started the winding process.

CONCLUSION

The ICCS/HFTF Test Coil program has contributed significantly to the understanding of supercritical helium thermodynamics in relation to ICCS stability. The helium pump module, recently tested, appears satisfactory for use in conjunction with the ICCS/

Table II. Weight Compositions of ICCS Sheath Materials

	Ni	Cr	Mn	Mo	Fe
Nitronic 40	7.19	20.48	8.54	0.17	Bal
A-286	24.55	14.06	0.12	–	Bal
JBK-75	30.6	14.47	0.23	1.21	Bal

HFTF Test Coil. High field tests of Airco Nb_3Sn superconducting cables, encapsulated in stainless steel sheaths, show only slight reductions in critical current density from those of single wires. Magnetic pulse penetration of the sheath to produce detectable normal zones has also been demonstrated. Cryogenically compatible sheaths have been fabricated on a production basis. Prototypical terminations for the ICCS/HFTF Test Coil have been fabricated and should be tested shortly.

REFERENCES

1. M.O. Hoenig and D.B. Montgomery, Dense supercritical-helium cooled superconductors for large high field stabilized magnets, IEEE Trans on Magnetics, MAG-11:569 (1975).
2. M.O. Hoenig, Internally cooled cabled superconductors – Part I and Part II, Cryogenics, 20:373 (1980), and 20:427 (1980).
3. M. Morpurgo, The design of the superconducting magnet for the 'Omega' project, Particle Accelerators, 1:(1970).
4. M.O. Hoenig, J.W. Lue and D.B. Montgomery, in "Proc. of the Sixth Intl. Conf. on Magnet Technology," ALFA, Bratislava, (1978) p. 1021.
5. S.R. Shanfield, et.al., Transient cooling in internally cooled, cabled superconductors, IEEE Trans on Magnetics, MAG-17:2019 (1981).
6. P.A. Sanger, et.al., Critical properties of multifilamentary Nb_3Sn between 8 and 14 tesla. IEEE Trans on Magnetics, MAG-17:666 (1981).
7. M.O. Hoenig, et.al., Progress in the ICCS-HFTF 12 tesla coil program, IEEE Trans on Magnetics, MAG-17:638 (1981).
8. E.A. Erez and H. Becker, Evaluation of Materials for Internally Cooled, Cabled Superconductor Jackets, in "Advances in Cryogenic Engineering, Vol. 28," Plenum Press, New York (1982).

PERFORMANCE OF AN INTERNALLY COOLED
SUPERCONDUCTING SOLENOID*

J. W. Lue and J. R. Miller

*Oak Ridge National Laboratory
Oak Ridge, Tennessee*

INTRODUCTION

Not long ago, the use of cable-in conduit type of conductors cooled by forced flow supercritical helium was proposed as an alternative to bath cooling large superconducting magnets.[1] The rationale behind this proposal was to use the large surface area of the cable strands and turbulent flow to ensure good heat transfer to the coolant. The price one expected to pay for forcing super critical helium through the conduit in the turbulent region was the pumping power loss. Subsequent experiments on samples up to several meters long however showed that because of the existence of high conductive transient heat transfer a reasonably high stability margin could be obtained with little or no flow.[2,3] Furthermore, under proper conditions, heat transferred to helium from the initially quenched superconductor could induce a transient fluid flow which produces very high local heat transfer.[4] Thus, very high stability margins, limited only by the available helium enthalpy, could be observed with no net flow. If it could be shown that these phenomena hold in a magnet, then high pumping power loss may not be a necessary characteristic of Internally Cooled Superconductors (ICS).

Another concern regarding ICS is the vulnerability to strand motion in a magnetic field. This could cause degradation of the magnet performance. Experiments on samples with short, straight sections in high fields[5,6] alleviated this worry.

---

*Research sponsored by the Office of Fusion Energy, U.S. Department of Energy, under contract W-7405-eng-26 with the Union Carbide Corporation.

A superconducting solenoid using an ICS has been completed and tested. Details of the design, construction, and preliminary testing have been reported elsewhere.[7] The performance of this solenoid and, in particular, its stability margin as a function of operating field is reported here.

SOLENOID CONSTRUCTION

We used an ICS with Cu/NbTi composite strands for the construction of the solenoid. The cable, enclosed in a stainless steel jacket, consists of 12 triplex units of composite wires cabled around a core of 7 triplex units of copper wire. The superconducting strands are thus fully transposed. A cross section of the conductor is shown in Fig. 1b. The round stainless steel conduit provides a cable space of 7.36 mm in diameter, where the void fraction for helium flow is about 0.43.

The solenoid was designed to produce the maximum field possible with the available length of conductor (about 360 m). The desire not to overbend the conductor led to a winding diameter of 113 mm. The coil was layer wound in 19 layers with an average of 21 turns per layer. A wrap of Nomex* braid was applied as a standoff. Vacuum impregnation with epoxy resin after winding, completed the insulation and provided insurance against leaks. A "saddle tee" was soldered to the end of each layer to make possible various series-parallel combinations of flow paths. Figure 1a shows the finished coil with flow manifolds and electric current junctions. A brief list of the conductor and coil parameters is given in Table I.

INSTRUMENTATION AND TEST SETUP

One of the primary objectives of building this test solenoid was to investigate the stability of an ICS in a magnet. Since the innermost layer of the solenoid sees the highest field, we instrumented it for stability testing. Four induction heating coils, each extending about 26 cm, were wound on the conductor at the middle of this layer. They were driven by a capacitive discharge circuit similar to that described elsewhere.[8] Cowound voltage leads were placed over this heated zone to observe the response of the conductor following a pulse discharge.

---

*Reference to a company or product name does not imply approval or recommendation of the product by Union Carbide Corporation or the U.S. Department of Energy to the exclusion of others that may be suitable.

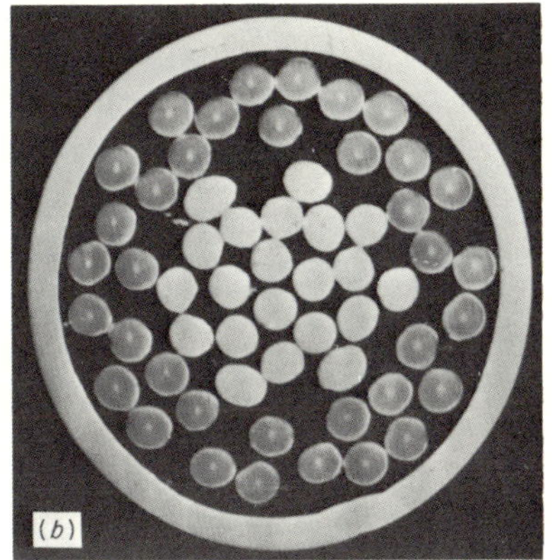

Fig. 1. (a) Finished solenoid and (b) Conductor
cross section. The solenoid was mounted in the
position shown (upside down).

Quenching of the magnet was detected by the unbalanced voltage signal between the inner 12 layers and the outer 7 layers,
which have approximately the same self-inductance. The quench
protection circuit had a room temperature dump resistor of 0.1 Ω
which gave a discharge time constant of about 0.23 s.

A helium pump[9] was available for flowing helium through the
magnet and was housed in the same dewar with the magnet. Hydraulic connections to the magnet in this series of tests were made in
such a way such that the inner two layers had the shortest paths
and thus received the most flow. The outer layers, having several
layers in series, received very little flow. This arrangement
forced helium to flow through the parts of the coil where it was
most needed.

Table I. Conductor and Coil Parameters.

Cable pattern	12 × 3 NbTi composite around 7 × 3 Cu core
Strand size	0.72 mm
Cu:SC ratio	1.8:1 in composite strands; 3.4:1 overall
Conduit ID × OD	7.36 mm × 8.56 mm
Void fraction in cable space	0.43
Winding ID × OD × height	113 mm × 414 mm × 203 mm
Total number of turns	399
Total conductor length	340 m
Coil inductance	23 mH

The solenoid was mounted (upside down) with the current junctions facing the bottom of the dewar. With the flow circuit heat exchanger located at the bottom of the dewar, the magnet could be operated with its main body suspended above the helium bath.

MAGNET PERFORMANCE

Quench Field

At a bath temperature of 4.2 K, quench currents of about 4800 A were observed on the first and several subsequent charges. This corresponds to a maximum field at the winding of about 7.65 T. During these tests the liquid helium level in the dewar was either high enough to immerse the whole magnet or had fallen below the bottom of the magnet. No difference in performance was observed. The helium flow inside the conductor conduit was maintained at pressures from 1.1 to 2.5 atm. It was either stagnant or given a flow velocity of up to about 0.6 m/s through the innermost layer. Comparison of the magnet load line with the short sample critical current as shown in Fig. 2 indicated that the critical current limit was reached. When the dewar helium temperature was lowered to 3.9 K, a quench current of 5100 A or a maximum field of about 8.13 T was achieved. Scaling the critical current measured at 4.2 K to 3.9 K, shown as the broken curve in Fig. 2, also indicates that the achieved quench current is very close to the critical current limit.

Stability Margins

Stability margins of the magnet were measured by discharging a capacitive voltage into the induction heating coils. All four coils with a total heating length of about 1.1 m were used in the present experiment. Pulse currents decayed in about three cycles, indicating that the initial capacitive energy was dissipated in the conductor and the pulse coils themselves in about 3.5 ms. In the field range of T studied (5 to 7 T) there was no measurable difference in the decay time constant. Thus, the total effective resistance of the coupling circuit remained constant, but the fraction of energy dissipated in the primary (pulse coil) circuit varied as the pulse coil resistance changed with the field. Considering the magnetoresistance of the pulse coil windings, it was determined that the fraction of energy deposited directly into the conductor ranges from 46% at 5 T to 32% at 7 T. Furthermore, since the magnetic diffusion time constant of the stainless steel sheath is much shorter than the time constant of the cable strands, essentially all of this energy went to the strands.

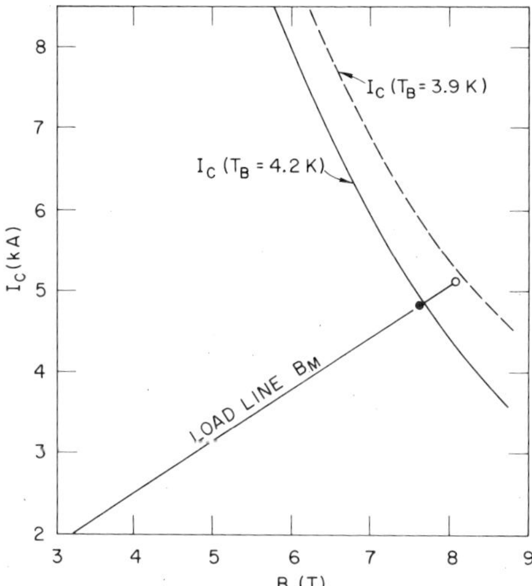

Fig. 2. Short sample critical currents and magnet
load line at the winding. The data points are
measured quench fields (currents) at 4.2 K and 3.9 K.

The measured stability margins as a function of the magnetic
field or magnet current are shown in Fig. 3. Similar to the
results of short sample stability experiments,[4,10] it exhibits
regions of high stability margin and regions of low stability
margin. For fields below 5.8 T, stability margins of 300–600
mJ/cm^3 were observed, while margins of 120 mJ/cm^3 or less were
found for fields of 6.1 T and above. There is a clear discontinuity in the stability margin between 5.8 and 6.1 T. Because of the
similarity of this stability curve to that observed in short
samples, we infer that multiple stabilities could exist in this
region. Based on the results of short samples and the scaling
relationships developed,[11] we have calculated that the low stability margin should disappear at fields below 5.7 T for 5.0 atm
helium and no flow operation. Although we do not know precisely
how this limiting current (field) scales with helium pressure or
flow rate, earlier results[4,10] showed that both the lower pressure
and the flow used in the present experiment should push this limit
higher. Therefore, we view the present results as excellent
agreement between the performance of an ICS in a magnet and the
scaling relationship developed on the basis of earlier short
sample results.

Two data points with no helium flow are also shown in Fig.
3. The slightly lower stability margin at 5.1 T indicates that
the coil temperature was probably a little bit high when there was
no flow, since this high stability margin is limited by the

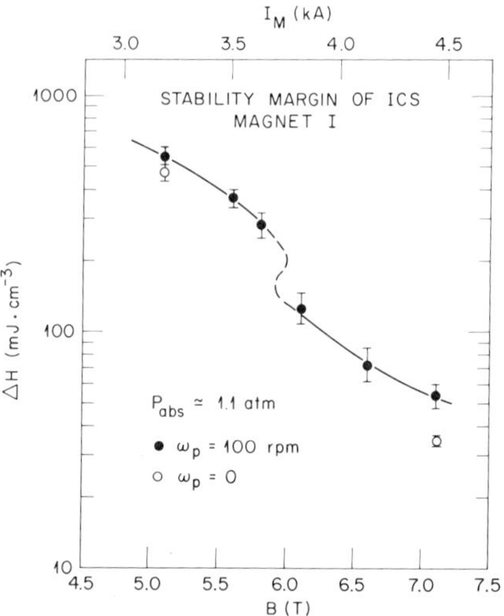

Fig. 3. Stability margin of the solenoid as a
function of the magnetic field or magnet current.
Data point limits are measured quench and
recovery levels. ω_p is the pump velocity.
At ω_p = 100 RPM, an estimated maximum flow
velocity of 0.6 m/s is applied to the first layer.

available helium enthalpy. The larger difference at 7.1 T is
probably due to both the elevated temperature and the degraded
heat transfer at no flow.

Pulse Fields

During the pulse discharge for stability testing, the peak
pulse field on the conductor generated by the pulse coils ranged
from 0.34 to 1.14 T. Within 0.6 ms, pulse coil current was swept
from positive peak to negative peak and the negative peak had
decayed to about half of the positive peak value. Thus, pulse
field rates of 900-3000 T/s were experienced by the conductor in
the high field region of the magnet, and recovery was observed for
fields of 7.1-5.1 T. Table II summarizes the coil performance
characteristics described above.

CONCLUSION

A superconducting magnet constructed of ICS has been built
and tested. Particular emphasis was placed on measurement of the

Table II. Coil performance characteristics.

Maximum current/field attained @ 4.2 K	4815 A/7.68 T
Maximum current/field attained @ 3.9 K	5100 A/8.13 T
Magnetic stored energy @ 7.68 T	267 kJ
Current density over the cable space @ 7.68 T	11.3 kA/cm^2
Maximum charge rate (limited by power supply)	45 s to 7.1 T
Stability margins @ 1.1 atm 4.2 K	56 mJ/cm^3 @ 7.0 T
	590 mJ/cm^3 @ 5.0 T.

stability margins of the magnet. Test results showed that this
magnet has demonstrated the following merits:

1. It reached short sample critical current limit with no
    training.
2. No degradation due to strand motion in fields was appar-
    ent.
3. It can be charged and discharged to high current densi-
    ties and high fields at a fast rate.
4. It has reasonable stability margin up to 90% of the
    critical current.
5. Stability measurements verify the scaling relationship
    for a limiting current below which there is single valued
    high stability margin.
6. It is not really necessary to have bulk helium flow to
    ensure high stability.
7. Very high pulse field changes can be accommodated as long
    as the energy deposition is within its stability margin.

ACKNOWLEDGEMENTS

   The authors would like to express their appreciation to J. P.
Rudd, R. C. Dixon, C. W. Proctor, D. L. James, and H. C. DeArmond
for their help in the construction and the setup of the experiment
and to S. S. Shen for his help in solving the energy deposition of
the pulse energy.

REFERENCES

1. M.O. Hoenig, D.B. Montgomery, *IEEE Trans. on Magnetics*, MAG-11
    (2):569 (1975).
2. Y. Iwasa, M.O. Hoenig, D.B. Montgomery, *IEEE Trans. on
    Magnetics*, MAG-13 (1):678 (1977).
3. J.R. Miller, et. al., *IEEE Trans. on Magnetics*, MAG-15 (1):351
    (1979).
4. J.W. Lue, J.R. Miller, and L. Dresner, *J. Appl. Phys.* 51
    (1):772 (1980).

5. M.O. Hoenig, A.G. Montgomery, and S.J. Waldman, in "Advances
     in Cryogenic Engineering, Vol. 25," Plenum Press, New York
     (1980), p. 251.
6. J.R. Miller, et. al., in "Advances in Cryogenic Engineering,
     Vol. 26," Plenum Press, New York (1980), p. 654.
7. J.R. Miller, et. al., Design construction and test of a 113-mm
     bore solenoid with NbTi cable-in-conduit superconducting
     windings, IEEE Trans. on Magnetics, MAG-17 (5):2250 (1981).
8. R.B. Easter, J.R. Miller, and S.S. Shen, in "Proc. 8th Symp.
     on Engineering Problems of Fusion Research", IEEE
     Publishing Services (1979), p. 1718.
9. J.W. Lue, et. al., Test of a Cryogenic Helium Pump, in
     "Advances in Cryogenic Engineering, Vol. 27," Plenum Press,
     New York (1982).
10. J.W. Lue, and J.R. Miller, IEEE Trans. on Magnetics, MAG-17
     (1):757 (1981).
11. L. Dresner, IEEE Trans. on Magnetics, MAG-17 (1):753 (1981).

DISCUSSION

Question by J. Alcorn, General Atomic Co.: Since high stability
margins in ICS can be achieved at no flow, would you expect comparable performance for a similar bath cooled cable?

Answer by author: Yes, if the heat transfer is sufficient. This
depends on how the cooling channels are arranged , and how good
the helium replenishment is. Note that in a bath there would be
no induced-flow enhanced heat transfer.

Question by M. Hoenig, Massachusetts Institute of Technology:
What is J/J_{crit} for $\Delta H = 500$ mJ/cm^3 at corresponding field?

Answer by author: For a stability margin of 500 mJ/cm^3, the
magnet is operating at 3.3 kA and 5.2 T. The operating current to
critical current ratio is about 30%.

THERMOHYDRAULIC ANALYSIS OF INTERNALLY COOLED SUPERCONDUCTORS

E. A. Ibrahim

Westinghouse Electric Corporation
Pittsburgh, Pennsylvania

INTRODUCTION

Recent experimental tests of Internally Cooled Superconductors (ICS) have shown surprisingly high stability margins even at zero flow rates.[1-3] The tests have also shown that the helium flow rate through a conductor has very little effect on its stability margin. The early analysis of ICS systems, which assumed a steady-state heat transfer coefficient and ignored the induced helium flow after a quench, is certainly inadequate to explain the existence of the high stability margin, particularly at zero flow. Some recent publications studied the ICS systems with varying degrees of approximation.[4-6]

The work described in this paper is a thermohydraulic analysis of ICS systems that takes into account the high heat transfer coefficient due to transient effects and predicts an induced flow for an originally stagnant helium. The model used in this work takes into account the effect of the heat transfer from the warm helium to the steel jacket, the possible decay of the current in the superconductor during a quench, and the corresponding decay of the magnetic field. Properties of helium and copper are based on NBS data.

The Westinghouse Thermohydraulic Analysis Code (WESTAC) is based on this approach. WESTAC has been tested successfully in special cases where analytical solutions exist. In a future publication, we will show that WESTAC predicts the results of the quench pressure rise experiment at Oak Ridge National Laboratory.[7]

MODEL AND GOVERNING EQUATIONS

The conductor model to be analyzed by WESTAC code consists of strands of Nb_3Sn or NbTi superconductor and copper stabilizer encased in a stainless steel jacket. Helium fills the spaces around the strands for cooling.

The fluid flow in this one-dimensional model is governed by the conservation laws for mass, momentum and energy.[8]

$$\partial\rho/\partial t = - (\rho v)', \tag{1}$$

$$\partial(\rho v)/\partial t = - (\rho v^2)' - P' - \tau\, p/A, \tag{2}$$

and

$$\partial/\partial t \left[\rho(u + 1/2\ v^2)\right] = - \left[\rho v(u + 1/2\ v^2)\right]' - (Pv)' + s, \tag{3}$$

Where the source, s, is the heat transferred to the helium from contacting surfaces. The primes denote differentiation with respect to x.

We found it advantageous to use ρ, v and T as the dependent variables. The use of ρ rather than P gives more stable solutions. It is also easier to construct helium property tables in terms of ρ and T since in all stability and quench problems the maximum value density is determined by the initial condition. The lower limit of the tables could be taken very close to zero kelvin. Conversely, the upper limit of the pressure is not known. The internal energy and the pressure can be expressed in terms of ρ and T using thermodynamic equations.[9] Equations 1, 2 and 3 then become

$$\partial\rho/\partial t = - pv' - vp', \tag{4}$$

$$\partial v/\partial t = - vv' - c_1\rho' - c_2\,T' - c_3, \tag{5}$$

and

$$\partial T/\partial t = - vT' - c_4 v' + c_5, \tag{6}$$

where

$$\alpha = (\partial P/\partial\rho)_T, \qquad\qquad \beta = (\partial P/\partial T)_\rho,$$

$$c_1 = \alpha/\rho, \qquad\qquad c_2 - \beta/\rho$$

$$c_3 - \tau p/\rho A = 1/2\ fpv^2/A,$$

$$c_4 = \beta T/\rho c_v, \qquad\qquad c_5 = \left[s + (\tau pv/A)\right]/\rho c_v,$$

Heat conduction in the model is governed by two separate conduction equations, one for each region. For the copper and superconductor region

$$\partial T_c/\partial t = \left[A_c \partial/\partial x \ (k_c \cdot \partial T_c/\partial x) + q - hp_c \ (T_c - T)\right]/\rho_c A_c c_{pc} \tag{7}$$

and for the jacket region:

$$\partial T_s/\partial t = \left[A_s \partial/\partial x \ (k_s \cdot \partial T_s/\partial x) + hp_s \ (T - T_s)\right]/\rho_s A_s c_{ps}, \tag{8}$$

with no heat generation term.

The term $\rho_c A_c c_{pc}$ in Eq. 7 is averaged over the copper and the superconductor in the conductor material. The heat generation term is due to joule heating in the conductor. It is equal to zero below the current sharing temperature, assumed to be linear between the current sharing and the critical temperatures and proportional to the square of the current and the resistivity above the critical temperature. The current in the conductor is assumed to decay during a quench based on the inductance of the coil, the resistance of the normal part of the conductor, the delay time of the dump circuit and the value of the dump resistor. The details of this analysis can be found in Ref. 9.

The transient conduction heat transfer coefficient which is proportional to $t^{-1/2}$ is used in Eq. 7 and 8. As time increases, the value of the heat transfer coefficient goes down. When the value of the heat transfer coefficient drops below the steady state value as calculated by Dittus-Boelter correlation, the steady state value is used.

INITIAL AND BOUNDARY CONDITIONS

To solve Eq. 4-8 we have to know the initial and boundary conditions. The initial conditions for the fluid, namely ρ, v and T, are determined by solving Eq. 4, 5 and 6 with the left hand sides set equal to zero. The resulting differential equations in x are

$$v' = - (c_3 v + c_2 c_5)/D, \tag{9}$$

$$T' = \left[c_5 v \ (1 - \alpha/v^2) + c_3 c_4\right]/D, \tag{10}$$

where

$$D = v^2 - \alpha - c_2 c_4.$$

Numerical solution of Eq. 9, 10 gives the initial conditions of v and T. The density is then found by dividing the constant steady state mass flux by the velocity. The DGEAR of the IMSL[10] library which uses Adam's variable order predictor corrector method is used for the numerical solution. The initial temperature distribution of the solids is taken as identical to that of the fluid.

The boundaries of the solids are assumed to be insulated. One of the two boundary conditions is assumed for the helium at the conductor ends: either closed end or open end. The closed end represents a symmetry plane, as the center of a conductor with initial zero velocity and a uniform or symmetric disturbance, ρ', for which v and T' are all equal to zero. For the open end the helium pressure is assumed constant giving:

$$\partial\rho/\partial t = - (\beta/\alpha) \, \partial T/\partial t \tag{11}$$

which is the only boundary condition for an open end.

NUMERICAL SOLUTION OF THE GOVERNING EQUATIONS

To solve the transient Eq. 4-8 we express the derivatives with respect to x in the right hand side as differences using second order polynomial approximation and solve the resulting differential equations with DGEAR. The derivatives with respect to time need not be expressed in difference form as this is done internally in DGEAR. For example Eq. 4 becomes:

$$\partial\rho_i/\partial t = - \rho_i v_i' - v_i \rho_i', \qquad i = 1, 2, \ldots n. \tag{12}$$

with similar expressions for the other equations. In all, we have five equations for each node or 5n equations. With another differential equation for the current we have 5n +1 differential equations to solve by DGEAR.

ANALYSIS OF CONDUCTOR STABILITY

To show of WESTAC's capabilities we analyzed a 40 m conductor filled with stagnant helium at 1.5 MPa and 4.2 K. The Nb_3Sn superconductor area is 0.69×10^{-4} m^2, the copper area is 1.18×10^{-4} m^2 and each of the helium and the jacket has an area of 1.00×10^{-4} m^2. The wetted perimeters of the strands and the jacket are 1.00 and 0.08m respectively. The copper has a residual resistivity ratio of 100 and the friction factor is assumed to be 0.0146. The conductor is carrying a current of 17625A at a magnetic field of 8.0 T. At time t = 0, the copper and superconductor temperature is assumed to be 20 K. Because of symmetry, only half the conductor length is analyzed with the left end closed and the right end open.

Figure 1 gives the helium temperature as a function of distance and time. It shows that the temperature in the middle of the conductor is higher than at the ends because the fluid gets compressed more in the middle. Before the relief pressure wave reaches the middle section of the conductor, the helium in that section acts like a constant volume with uniform temperature and pressure. The helium temperature keeps going up with time because of heat transfer from the copper. The sharp temperature increase of the helium nodes close to the end is due to numerical instabilities.

The helium pressure is shown in Fig. 2. As expected, the pressure in the middle part of the conductor stays uniform (zero slope in pressure) until the relief wave arrives from the conductor end. After the relief wave arrives at the center of the conductor, a pressure gradient exists everywhere in the conductor except at the center where the gradient is always equal to zero.

The induced velocity in the conductor is shown in Fig. 3. The velocity is highest at the conductor ends and zero at the center. Before the arrival of the relief wave the velocity in the middle section of the conductor is zero. The induced flow is created by the arrival of the relief wave from the conductor end. Before the wave arrives to the middle section of the

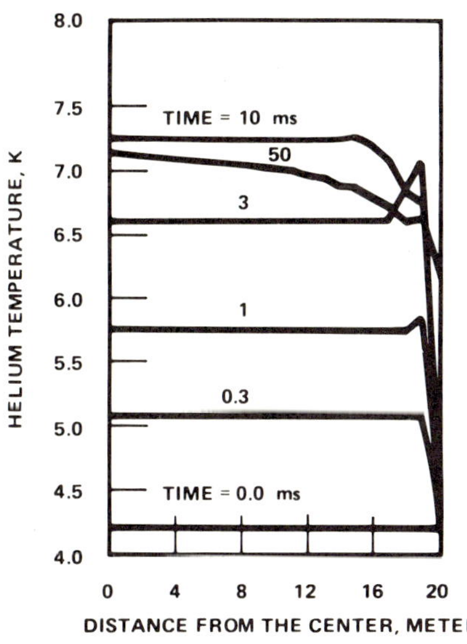

Fig. 1.   Helium Temperature vs. Time and Distance

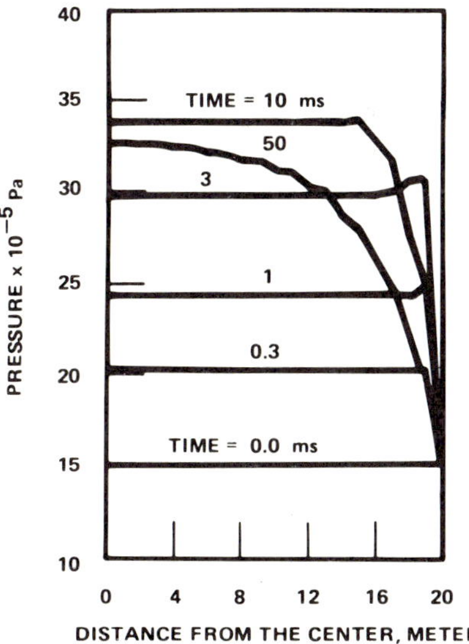

Fig. 2.   Helium Pressure vs. Time and distance

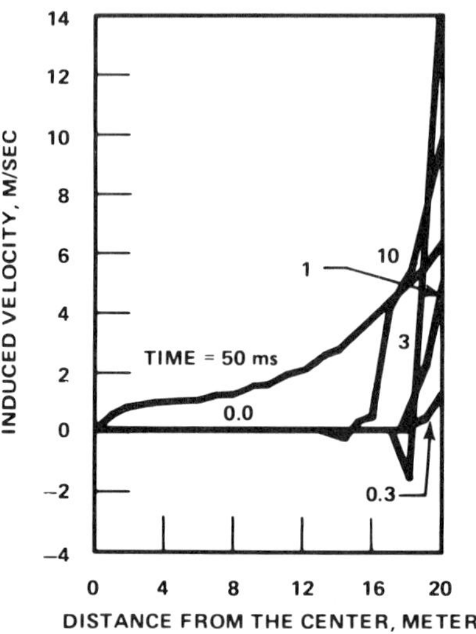

Fig. 3. Induced Velocity
vs. Time and Distance

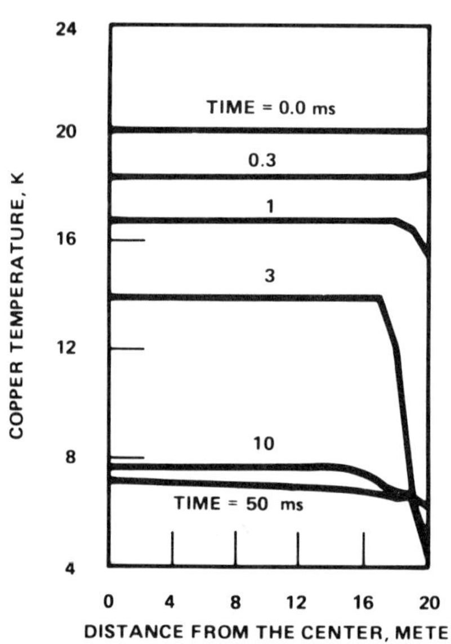

Fig. 4. Copper Temperature
vs. Time and Distance

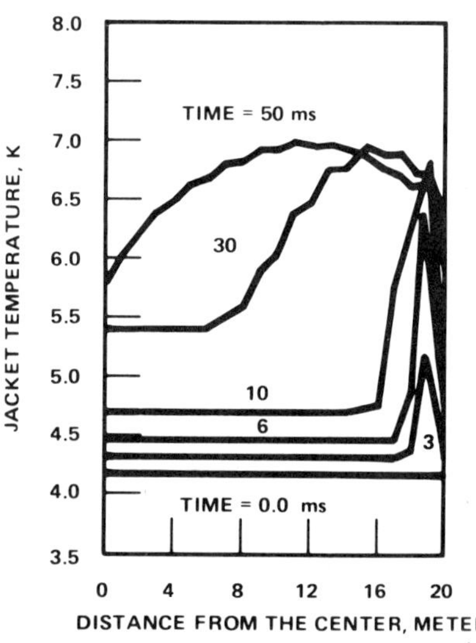

Fig. 5. Jacket Temperature
vs. Time and Distance

conductor, the induced velocity is zero. After the wave arrives at the center a finite velocity exists everywhere in the conductor except in the center. The negative velocities shown are due to numerical instabilities.

The temperature distribution in the copper is plotted in Fig. 4 as a function of distance and time. Initially, the heat transfer from the copper to the helium is by conduction with a heat transfer coefficient proportional to $t^{-1/2}$ and the temperature all over the conductor goes down quickly. Later the heat transfer due to the induced flow increases and dominates especially at the conductor ends resulting in a quicker temperature drop at the ends than the drop at the center where there is no induced flow. Eventually, in this case, the conductor recovers.

The temperature of the steel jacket is plotted in Fig. 5. Initially, the jacket is in thermal equilibrium with the helium at 4.2 K. As the helium warms up, it transfers heat to the jacket. At the end of the conductor, the helium temperature is relatively low as shown in Fig. 1, causing the jacket temperature to increase only slightly. Close to the center of the conductor the helium temperature is high, but the heat transfer coefficient is low because of the low induced flow rate. Elsewhere in the conductor, the combined effect of the temperature levels and the heat transfer coefficient causes the jacket temperature to increase and the conductor temperature takes the shape shown in Fig. 5 with the maximum temperature away from the center. The low thermal conductivity of steel precludes a more uniform temperature distribution in the jacket.

CONCLUSIONS

The computer code WESTAC appears to correctly analyze the thermal characteristics of internally cooled superconductors. In the special case of steady state flow, the resulting pressure drop or temperature increase calculated by WESTAC agrees well with closed form solutions. In the case of constant volume heat addition WESTAC also checks successfully. The results of the stability analysis of an internally cooled superconductor as calculated by WESTAC are as predicted from the physics of the problem. We believe that WESTAC is a useful tool for analyzing internally cooled superconductors for such design aspects as stability, quench pressure and temperature rise, joint analysis and current lead design.

NOTATION

A = area, m^2
c_p = specific heat at constant pressure, J/kg-K

c_v = specific heat at constant volume, J/kg-K
f = fanning friction coefficient
h = heat transfer coefficient, W/m^2-K
k = thermal conductivity, W/m-K
ℓ = conductor length, m
n = number of nodes
P = pressure, Pa
p = total wetted perimeter = $p_c + p_s$, m
q = heat generation per meter, W/m
s = volumetric heat source in the fluid, W/m^3
T = temperature, K
t = time, s
u = internal energy, J/kg
v = helium velocity, m/s
x = distance in the flow direction, m
ρ = density, kg/m^3
τ = shear stress, Pa

Subscripts

none = helium
c = conductor
s = steel jacket

REFERENCES

1. A. G. Montgomery and M. O. Hoenig, Multifilament Nb_3Sn multistrand cable performance using supercritical helium, IEEE Trans. on Magnetics, MAG-15(1):794 (1978).

2. J. R. Miller, et. al., Measurements of Stability of Cabled Superconductors Cooled by Flowing Supercritical Helium, IEEE Trans. on Magnetics, MAG-15(1):351 (1978).

3. J. W. Lue, J. R. Miller and L. Dresner, Stability of cable-in-conduit superconductors, Appl. Physics, 51, (1):772 (1980).

4. V. Arp, Computer Analysis of Quench Transients in Force-Flow-Cooled Superconductors for Large MHD Magnets, in "Proc. Superconducting MHD Magnet Design Conference," MIT, Cambridge (1978).

5. G. Krafft and G. Zahn, Experimental and Theoretical Investigations of Heat Induced Transients in Forced Flow Helium Cooling Systems, in "Proc. Eighth Symp. on Engr. Prob of Fusion Research," IEEE, 79 CH1441-5, NPS:1724 (1979).

6. M. A. Hilal, et. al., Transient Stability of Forced Flow Cooled Conductors, in "Proc. Eighth Symp. on Engr. Prob. of Fusion Research," IEEE, 79CH1441-5, NPS:1774 (1979).

7. J. R. Miller, et. al., Pressure Rise During the Quench of a Superconducting Magnet Using Internally Cooled Conductors, in "Proc. 8th Intl. Cryo. Engr. Conf.," IPC Science and Technology Press, Guildford (1980).

8. R. B. Bird, W. E. Stewart and E. N. Lightfoot, "Transport Phenomena," John Wiley & Sons, Inc., New York (1960).

9. E. A. Ibrahim, Thermohydraulic analysis of internally cooled superconductors with WESTAC code, Westinghouse Advanced Programs, Report: WAPD-TN-81-32 (1981).

10. "The IMSL Library Volume 1," International Mathematical & Statistical Libraries, Inc., Houston, Texas (1980).

ANALYSIS OF SOME RESISTIVE TRANSITIONS IN THE ISR SUPERCONDUCTING QUADRUPOLE MAGNETS

K. N. Henrichsen, H. Laeger, Ph. Lebrun, and L. Walckiers

CERN
Geneva, Switzerland

INTRODUCTION

CERN has installed and is operating in its Intersection Storage Rings (ISR) a superconducting high-luminosity insertion,[1] the main elements of which are eight superconducting quadrupole magnets,[2] housed in individual liquid helium bath cryostats.[3] Prior to site installation, all magnets were thoroughly tested in the laboratory with respect to maximum performance, magnetic field quality,[4] and cryogenic behavior; those tests involved a number of provoked resistive transitions ("quenches") which were analyzed in detail, both from their electrical and cryogenic aspects.

MAGNETS AND TEST EQUIPMENT

The quadrupole magnets have a nominal field gradient of 43 T m^{-1} at 1600 A excitation current. The diameter of the warm bore is 173 mm. Figure 1 shows two cross sections of a magnet in its cryostat with windings and steel yoke. Each coil has 290 turns of solid composite conductor of 1.8×3.6 mm^2 cross section. The magnet coils are protected by a parallel resistor of 0.215 Ω mounted inside the cryostat. Two types of magnets are installed, short magnets of 0.65 m magnetic length and stored energy of 295 kJ and long magnets of 1.15 m magnetic length and a stored energy of 535 kJ at nominal excitation currents. Each magnet is equipped with auxiliary windings, a sextupole for beam chromaticity correction, and a dodecapole to correct steel yoke saturation effects at different field values.

Cryogenics

Each magnet operates in a bath of saturated liquid helium at a temperature of 4.3 to 4.4 K, contained in a cylindrical helium

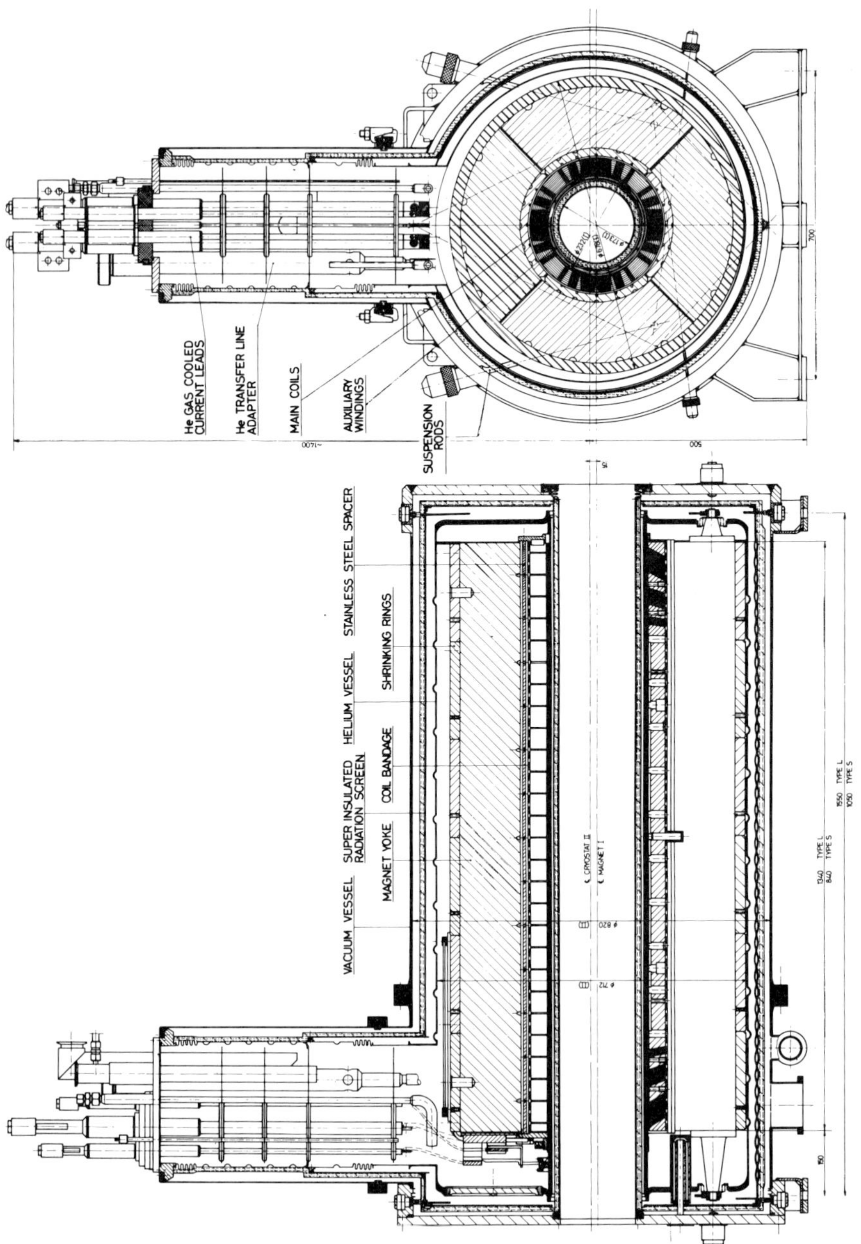

Fig. 1. Cross-sections of a long magnet in its cryostat.

vessel made of all-welded austenitic stainless steel. Normally, the boil-off vapor cools the radiation screen, service funnel and current leads of the cryostat, and is then returned warm to the compressor of the cryogenic plant via small diameter piping which is inadequate for high vaporization rates. In order to limit the helium inventory in the system, the helium vessel closely fits around the magnet with an eccentricity of 15 mm yielding extra ullage volume in the upper half. Moreover, a low design pressure (2 bar) and a corrugated shell construction based on finite-element structural calculations make it possible to keep to a minimum the transverse dimensions and weight of the helium vessel. Consequently, helium vaporized by the energy release at a magnet quench must be efficiently discharged from the cryostat. The relief device is a standard weight-loaded safety relief valve with full opening, directly actuated by the pressure in the cryostat. For the tests discussed here, its cracking pressure has been adjusted

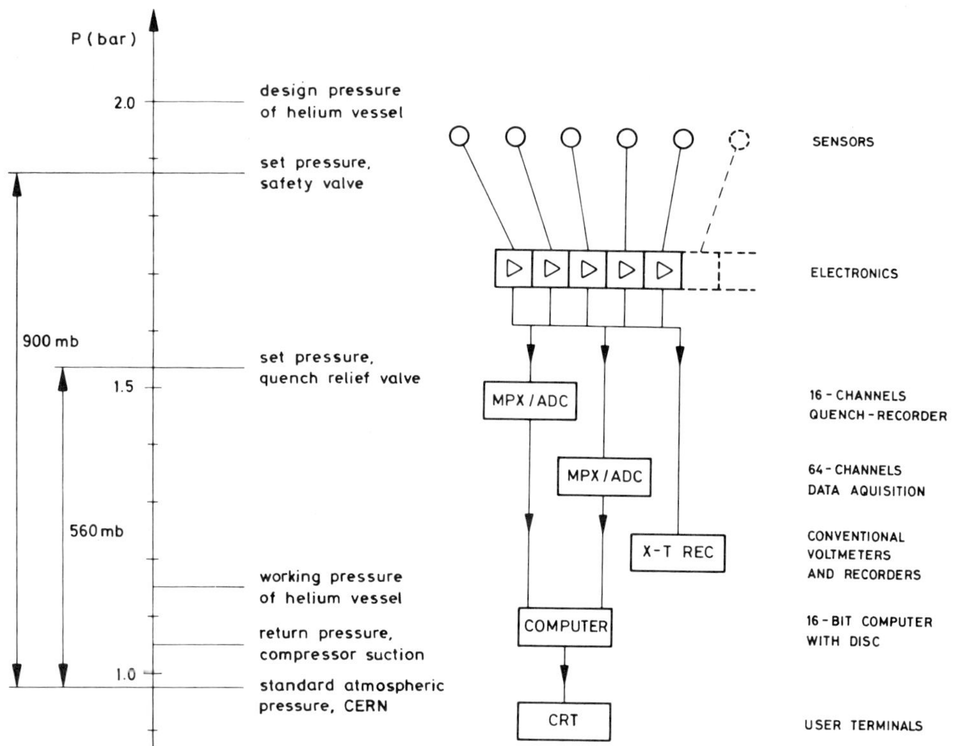

Fig. 2. Pressure levels relevant to cryostat design and operation.

Fig. 3. Data acquisition and processing system.

at 560 mbar, above which it pops open and discharges through its 198 mm^2 nozzle and 150 mm diameter, 40 m long recovery line into a rubber balloon at atmospheric pressure.[5] In parallel, a similar valve, set at 900 mbar, and opening directly to the atmosphere, acts as a safety device for ultimate protection of the cryostat. Figure 2 shows the relevant pressure levels.

Measurement Equipment

The magnets are equipped with comprehensive instrumentation for control, monitoring and diagnoses. The measured parameters are converted into analog voltage signals. A computer-controlled data acquisition system[6] provides access to all relevant parameters for alarm programs, logging programs and real-time display of measured or computed values. This system is shown in Fig. 3. It includes a quench recorder, which proved very useful for the detailed analysis of transient phenomena related to the magnet quenches. With a time resolution of 10 ms, it permits the recording of events including those occurring in a time interval prior to the detection of the quench. Internal coil voltages and cryostat pressure rise can be recorded. The precise measurement of the decay of the excitation current permits the calculation of some important parameters, such as the instantaneous power dissipation, the total induced voltage and the total resistance of the excitation winding. The average temperature of the winding can in turn be deduced from the calculated value of the coil resistance. The maximum possible temperature which could have occurred in the winding can be calculated as follows (adiabatic assumption):

$$T_1 = T_0 + \int_{t_0}^{t_1} J^2 \rho(T) / C(T) \, dt \, ,$$

where J = current density, ρ = resistivity, C = specific heat per unit volume and t_0 = quench start time.

RESULTS OF MEASUREMENTS

Typical recordings of magnet current, cryostat pressure and discharged flow rate during the first ten seconds following detection of a quench are shown in Fig. 4 for a long quadrupole magnet initially excited at 1600 A, without and with energy extraction. Immediate observations can be made from these curves:

1. Although starting almost at the same time as the quench, the cryogenic effects extend over a much longer time scale than

the electrical aspects, e.g., the current decay. In fact, helium is vaporized at a high rate for several minutes after a quench, whereas the current decays to zero within a few seconds.

2. The pressure rise in the cryostat is faster when energy is extracted from the magnet: this energy is in fact dissipated in the protection resistor which is in much better thermal contact with the helium than the magnet coils.

3. The relief valve shows a lag in opening pressure of about 200 mbar, which is only slightly dependent on the rate of pressure rise.

Energy Dissipation

From the cryogenic point of view, two distinct processes occur after a magnet quench: first, a quasi-isochoric pressure rise until the relief valve opens and secondly, a high flow-rate discharge of cold helium. During the first phase, the rate of energy transfer to the helium can be inferred from the rate of pressure rise in the cryostat, using a simple isochoric model. Assuming uniform energy release to the helium and thermal equilibrium in the cryostat, the amount of heat received by the cryogen equals the increase in its internal energy: this transformation can be illustrated by a segment of isostere on the pressure-internal energy diagram for helium,[7] shown in Fig. 5. For the nominal helium fill of the cryostat, i.e., 63 L liquid and 84 L vapor at saturation, the segment of isostere lies entirely within the two-phase region, which results in a moderate rate of pressure rise of 0.018 bar/kJ. The corresponding rate of energy release to the helium, dQ/dt, appears in Fig. 6, together with the rate of dissipation of magnetic energy dE/dt, as calculated from the relation between current and stored energy. The calculations show that:

1. All the energy stored in the magnet, i.e., 535 kJ, is dissipated within a few seconds, at a peak power of 600 to 700 kW.

2. This energy is released to the helium at a much slower rate, yielding a thermal power of the order of 10 kW.

3. The difference between those two dissipation rates results in thermal energy accumulation in the coils, as can be seen from the variation of their electrical resistance. Assuming a homogeneous temperature distribution in the coil, one can express electrical resistance measurements in terms of average coil temperature (except below 30 K due to magnetoresistance effects), and hence assess the thermal energy stored in it.

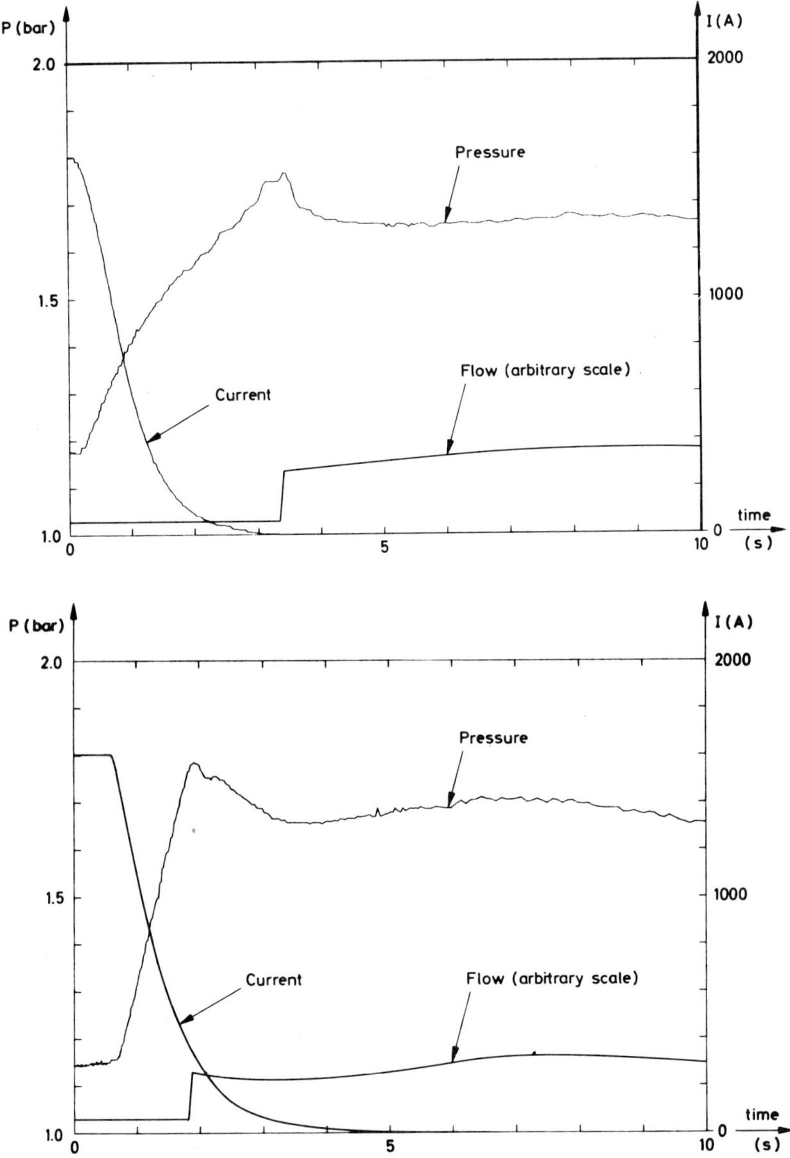

Fig. 4. A quench at nominal level without (a) and with (b) energy extraction.

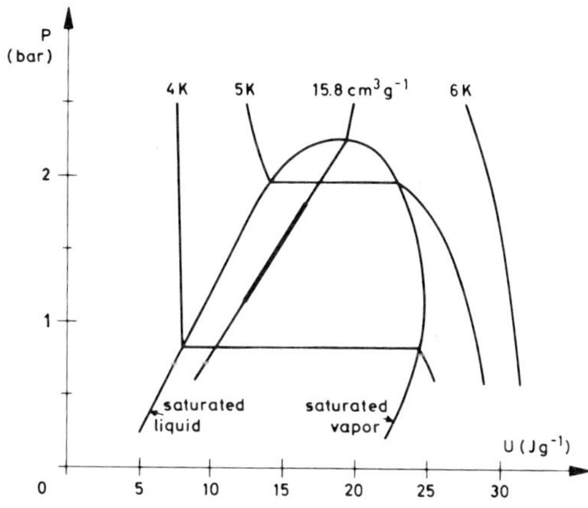

Fig. 5.   Isochoric pressure rise after a quench.

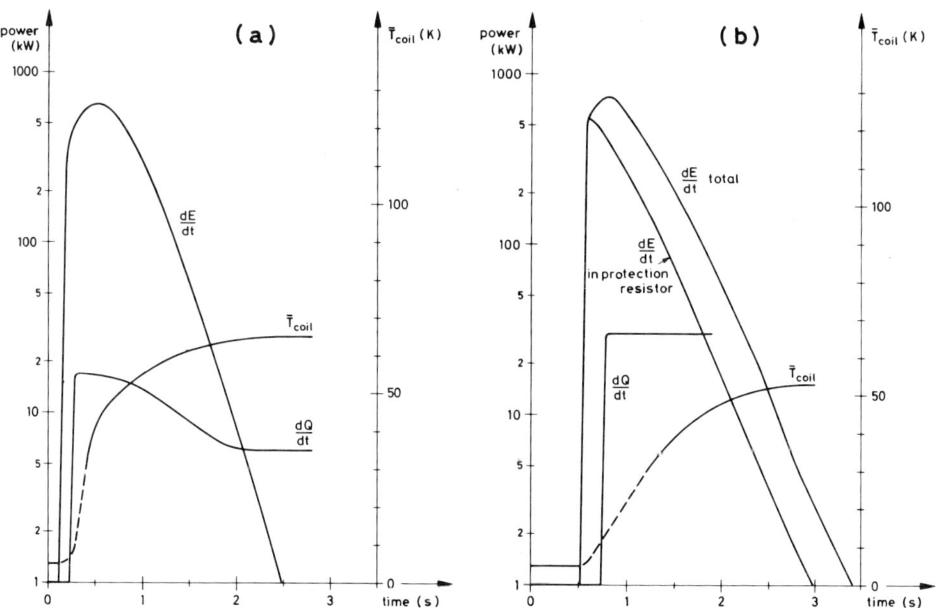

Fig. 6.   Power dissipation and coil temperature after a quench
at nominal level without (a) and with (b) energy extraction.

At the end of current decay, one finds 65 K and 460 kJ for a quench without energy extraction, and 53 K and 300 kJ for a quench with energy extraction: almost all the energy dissipated in the coil is still there, in the form of heat.

4. Energy starts reaching the helium about 0.1 s after it begins to be dissipated in the coil; this time lag can be accounted for by the thermal propagation across the fiberglass-epoxy insulation.

5. In the case of energy extraction, 50% of the stored energy is dissipated in the protection resistor, so that the average coil temperature shows a smaller increase than without energy extraction. The difference between average coil temperature and maximum temperature calculated with the adiabatic model is only about 10 K. The rapid variation of magnetic field due to the energy extraction spreads the resistive transition in the whole coil within a few tens of milliseconds. However, for quenches without energy extraction, maximum temperatures up to 120 K were calculated.

Helium Discharge

After the relief valve has opened, cold helium is dicharged through the valve nozzle into the recovery line, along which it progressively warms up. Typical flow rates, calculated from level-rise observations in the recovery balloon, are 150 to 200 g s^{-1} for quenches performed under nominal conditions. Recovery

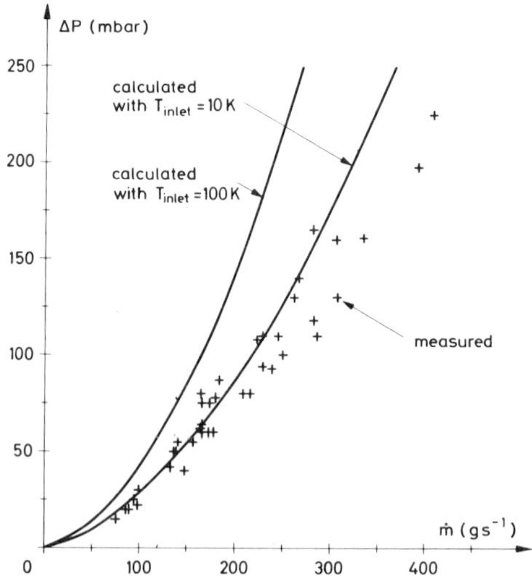

Fig. 7. Hydrodynamics of helium recovery line.

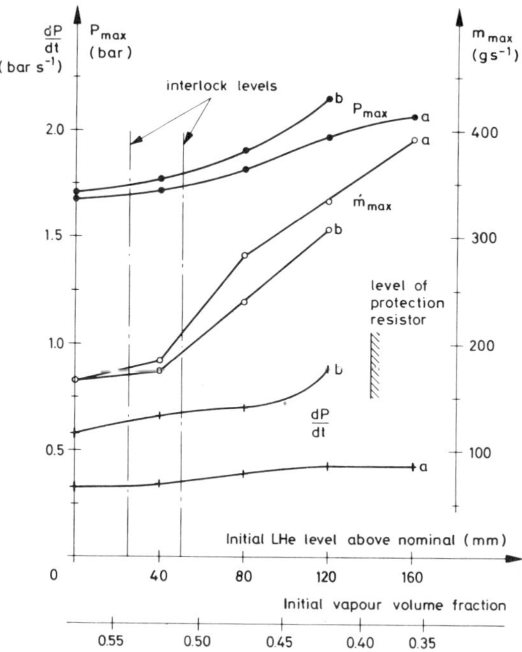

Fig. 8. Influence of initial liquid level

line pressure drop versus flow rate has been plotted for a number
of quenches in Fig. 7, together with theoretical curves calculated
by a hydraulic flow model taking into account convective heat
exchange between flowing fluid and line wall, as well as variation
of helium properties with temperature and pressure.[7] The predictive power of the model seems to be adequate for sizing such lines
with confidence; the design of the quench discharge piping for the
final set-up in the ISR relies on such calculations.

Influence of Initial Liquid Level

The effect of varying the initial liquid level was investigated (Fig. 8). The rate of pressure rise after a quench is
little dependent on the initial filling of the cryostat, in accordance with the predictions of the isochoric model: a marked increase in dP/dt can only be seen when energy is dissipated in the
protection resistor close to a high liquid level. After opening
of the relief valve, the discharge rate and the maximum pressure
reached in the cryostat increase significantly with higher initial
liquid level, up to values beyond the design pressure of the
helium vessel. Although the vessel can withstand pressures well

above this design value, interlocks[8] triggered by excessive liquid level were set on the helium transfer and the magnet current, in order to prevent the occurrence of abnormal conditions during magnet operation.

All eight cryomagnets successfully underwent their testing program; they have been installed and are presently operating in the ISR machine.

REFERENCES

1. J. Billan et al., IEEE Trans. Nucl. Sci., NS-26, 3:3179 (1979).
2. J. Billan et al., in "Proc. XIth Int. Conf. on High-Energy Accelerators", CERN, Geneva, (1980), P. 848.
3. H. Laeger, Ph. Lebrun, and P. Rohmig, in Proc. 8th International Cryogenic Engineering Conference," Genoa (1980), p. 124.
4. L. Walckiers, IEEE Trans. on Magnetics, MAG-17 (5):1872 (1981).
5. H. Laeger and Ph. Lebrun, in Proc. 8th International Cryogenic Engineering Conference, London (1978) p. 186.
6. K.N. Henrichsen and L. Walckiers, in "Proc. 6th Int. Conf. on Magnet Technology," Bratislava, (1977), p. 955.
7. R.D. McCarty, N.B.S. Technical Note 631 (1972).
8. K.N. Henrichsen, et al., IEEE Trans. on Magnetics, MAG-17 (5): 2059 (1981).

THERMAL CONTACT RESISTANCE AND CRYOGENIC STABILITY OF LARGE CONDUCTORS

M. A. Hilal

Michigan Technological University
Houghton, Michigan

INTRODUCTION

It is anticipated that commercial size superconducting magnets for MHD, fusion and energy storage applications will require
high current conductors as a result of cost, safety, fabrication
and reliability considerations. The two types of conductors
usually considered for these magnets are: a) the built-up conductors with composite or monolith regions soldered to the stabilizer
and b) the cable type conductors with twisted and transposed
wires. The performance of built-up type conductors is significantly affected by the thermal contact resistance of the solder
bond. A simple model is considered to determine the maximum
allowable thermal contact resistance to ensure conductor recovery. The analysis is extended to consider the two regions of
the conductor, the composite and the stabilizer, with the temperature distributions simultaneously determined. In this case the
minimum propagation zone (MPZ) energy is calculated as a measure
of the conductor stability. It is found that the MPZ energy
abruptly decreases if the thermal contact resistance exceeds
certain critical value. Such phenomenon can be explained by
comparing the heat generation and the heat removal in the conductor.

ZERO DIMENSION MODEL

Consider the conductor shown in Fig. 1. Region 1 represents
the stabilizer and region 2 represents the monolith or composite. The following assumptions are made:

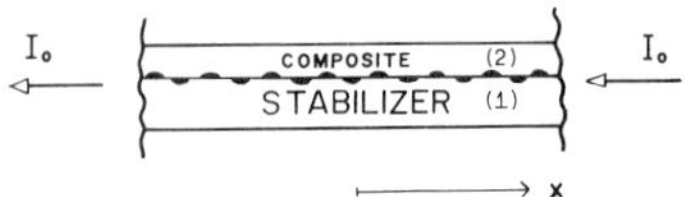

Fig. 1.   A conductor with a monolith soldered to the stabilizer.

1.   Two region conductors
2.   Uniform temperature over the cross sectional area of the
     stabilizer and the composite
3.   Uniform thermal contact resistance along the length of
     the conductor
4.   No longitudinal heat conduction in the bond
5.   Negligible axial heat conduction in the composite
6.   No exposure of the composite surface to liquid helium
7.   The heat generation due to current transfer from the
     composite region to the stabilizer region, is negligible

The heat conduction from the composite to the stabilizer is given by

$$q_{g2} = (T_2 - T_1)/R_t \tag{1}$$

where q_{g2} = heat generated in the composite,
T_1, T_2 = stabilizer and monolith temperature respectively
R_t = thermal contact resistance.

In the current-sharing region, the heat generated in the composite is given by

$$= 0 \quad \text{for} \quad T_2 < T_s \tag{2}$$

$$q_{g2} = Q_0\{(T_t - T_2) + \gamma_A(T_2 - T_s)\}(T_2 - T_s)/(T_t - T_s)^2 \quad \text{for} \quad T_s < T_2 < T_t$$

$$= \gamma_A Q_0 \quad \text{for} \quad T_2 > T_t$$

where $Q_0 = I_0^2 \rho/A_0$, I_0 = total current, ρ = resistivity, A_0 = total stabilizer cross-sectional area, T_s, T_t = saturation and transition temperature[1], respectively, γ_A = ratio of the stabilizer cross-sectional area in the composite to the total stabilizer cross-sectional area.

Combining Eq. 1 and 2 gives

$$(\gamma_A - 1)T_2^2 + (T_t + T_s - 2\gamma_A T_s - \alpha)T_2 + \gamma_A T_s^2 - T_t T_s + \alpha T_1 = 0 \tag{3}$$

where $\alpha = (T_t - T_s)^2 / (Q_o R_t)$

The minimum value of T_2 is given by:

$$T_2 = 0.5 \left[T_t - \alpha + T_s (1 - 2\gamma_A) \right] / (1 - \gamma_A) \qquad (4)$$

The minimum composite temperature occurs if the following relation is satisfied:

$$\alpha^2 + 2\alpha(2\gamma_A T_s - T_t - T_s - 2\gamma_A T_1 + 2T_1) + T_s^2 - 2T_t T_s + T_t^2 = 0 \qquad (5)$$

If we take $T_1 = T_b$, the helium bath temperature, α is given by

$$\alpha = 0.5 \left[-B + \sqrt{B^2 - 4C} \right] \qquad (6)$$

where $B = 2(2\gamma_A T_s - T_t - T_s - 2\gamma_A T_b + T_b)$

$C = T_s^2 - 2T_t T_s + T_t^2$

Figure 2 shows $1/\alpha \left[= Q_o R_t / (T_t - T_s)^2 \right]$ for different values of T_t and T_s for $\gamma A = 0.1$. For the CDIF conductor[2] parameters given in Table I, the critical fraction of the monolith-stabilizer contact area is 0.54.

CONDUCTORS WITH NEGLIGIBLE AXIAL HEAT
CONDUCTION IN THE COMPOSITE REGION

The thermal conductivity of superconductors is small compared to stabilizing metals such as copper or aluminum. Also the cross

Table I. Design Parameters for the CDIF Conductor

Field	7 T
Current	6130 A
ρ	4×10^{-10} Ωm
$\gamma A (A_2/A_o)$	0.1
T_b	4.5 K
T_s	4.9 K
T_t	6.2 K
Conductance of cu-solder interface[3]	1000.0 W/m^{2}K
Critical Fraction of contact area	0.54

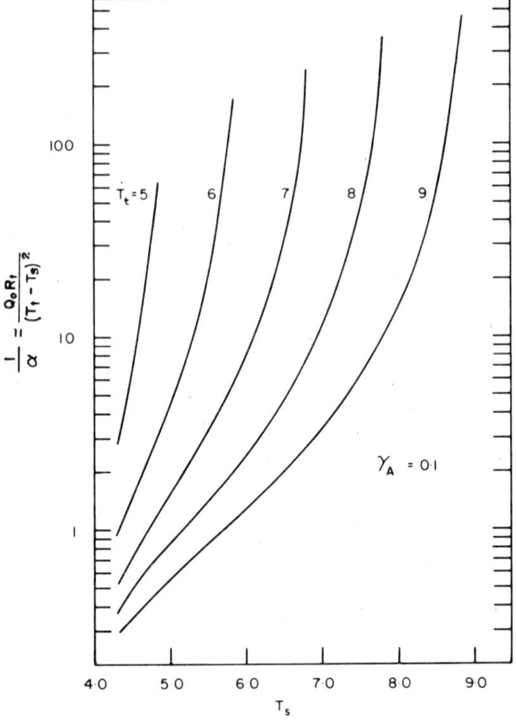

Fig. 2. $1/\alpha$ at different values of T_t and T_s.

sectional area of the superconductor is relatively small so heat conduction along the composite can be neglected. Assuming that the composite has a small fraction of the total stabilizer cross sectional area, the axial heat conduction in the composite region can also be neglected. This is valid as long as

$$R_t \overline{k} \, \gamma_A A_o \ll 1$$

where $\overline{k}$ is the average thermal conductivity of the stabilizer.

The heat conduction equation in the stabilizer region is:

$$\frac{Q_1 dQ_1}{dt} = k_1 A_1 \left[s_1 q_b(T_1) - q_g(T_1) \right] \tag{7}$$

where s = effective cooling surface area, q_b = boiling heat flux, q_g = heat generated per unit length, $Q = KA \, dt/dx$. The subscript 1 refers to the stabilizer.

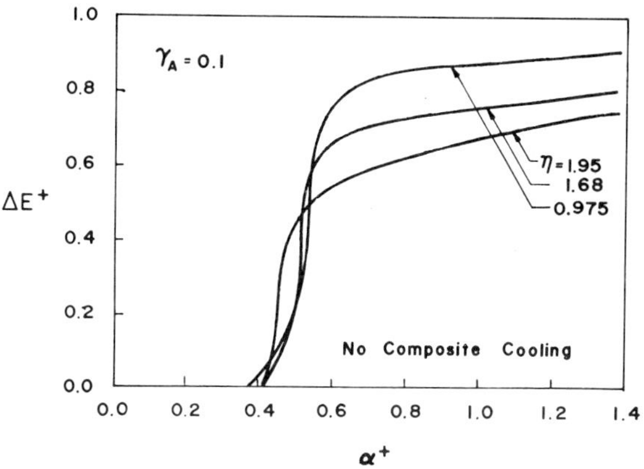

Fig. 3. ΔE_+ versus α^+ - no heat
conduction in the composite.

Equation 3 is used to determine T_2 given T_1. Equation 7 can
be rearranged

$$0.5 \frac{dQ_1'^2}{dT_1} = \eta\gamma_{s1}q_b^*(T_1) - f(T_2) \tag{8}$$

where
$$Q_1'^2 = Q_1^2/(I_o^2 L), \quad \eta = (s_o q_{b,max})/Q_o, \quad \gamma_{s1} = s_1/s_o, \quad q_b^* = q_b/q_{b,max},$$

L = Lorenz ratio, $q_{b,max}$ = nucleate heat flux, and

$$f(T_2) \begin{cases} = 0 \text{ for } T_s > T_2 \\ = (T_c - T_2)/(T_c - T_s) \text{ for } T_s < T_2 < T_c, \ T_s > T_2, \text{ or } T_c > T_2 \\ = 1 \text{ for } T_2 > T_t \end{cases} \tag{9}$$

The temperature distributions both in the stabilizer region and
the composite region can be determined using Eq. 2, 3 and 8. A
normalized MPZ energy, ΔE^+, is defined as

$$\Delta E^+ = \Delta E / \Delta E_o \tag{10}$$

Where ΔE_o is the MPZ energy assuming perfect thermal contact. ΔE^+
is plotted versus $\alpha^+ [= \alpha/(T_t - T_s)]$ for different values of η in
Fig. 3 for $\gamma_A = 0.1$. The abrupt decrease in ΔE^+ can be clearly
seen in this figure. Based on the above analysis, the critical
fraction of the monolith-stabilizer contact area for the CDIF conductor is 0.6 compared to 0.54 obtained using the zero dimension
model.

COMPLETE ANALYSIS OF THE EFFECTS OF THERMAL CONTACT
RESISTANCE ON CONDUCTOR STABILITY

This analysis includes the axial heat conduction both in the composite and the stabilizer regions. The composite surface can be partially exposed to liquid helium and heat generation due to current transfer is also included. The other assumptions previously listed are still valid. Considering both heat conduction and current transfer we have

$$\frac{d}{dx}(k_1 A_1 \frac{dT_1}{dx}) = S_1 q_b - q_{g1} - f_g q_{gs} - \frac{T_2 - T_1}{R_t} \tag{11}$$

$$\frac{d}{dx}(k A_2 \frac{dT_2}{dx}) = S_2 q_b - q_{g2} - (1-f_g) q_{gs} + \frac{T_2 - T_1}{R_t} \tag{12}$$

$$\frac{dV_1}{dx} = I_1 \frac{p}{A_1} \tag{13}$$

$$\frac{dV_2}{dx} = (I_o - I_{sc} - I_1) \frac{\rho}{A_2} \tag{14}$$

$$\frac{dI_1}{dx} = \frac{V_2 - V_1}{R_c} \tag{15}$$

where q_{gs} = heat generated in the bond region, f_g = fraction of the heat generated in the bond and transferred to the stabilizer region, V = voltage, I = current, I_{sc} = current in the superconductor, R_c = electric contact resistance.

The fraction f_g is difficult to determine since it depends on the nature of the bond. Assuming identical bond surface and uniform heat generation in the bond, f_g is equal to 0.5. To simplify the analysis we assume that the current transfer length is small compared to the length of the current sharing region and Eq. 13, 14 and 15 can be eliminated. The quantities q_{g1}, q_g and q_{gs} are

For $T_2 < T_t$

$$q_{g1} = (1-\gamma A)Q^o, \tag{16}$$

$$q_{g2} = \gamma A \, Q_o, \text{ and} \tag{17}$$

$$q_{gs} = 0 \tag{18}$$

For $T_s < T_2 < T_c$

$$q_{g1} = (1-\gamma_A)\ Q_o\left[(T_2-T_s)^2/(T_t-T_s)^2\right] \tag{19}$$

$$q_{g2} = \gamma_A Q_o(T_2-T_s)\left[(T_t-T_2)+\gamma_A(T_2-T_s)\right]/(T_t-T_s)^2] \tag{20}$$

$$q_{gs} = \left[I_o^2/(T_t-T_s)^2\right]\ \left[Q_2^2/(k_2^2 A_2^2)\right]R_c(1-\gamma_A)^2 \tag{21}$$

For $T_2 < T_s$

$$q_{g1} = q_{g2} = q_{gs} = 0 \tag{22}$$

The boundary value non-linear differential Eq. 11 and 12 are solved numerically. Figure 4 shows ΔE^+ versus α^+ for $\gamma_A = 0.1$. The value of the critical contact area for the CDIF conductor in this case is 0.42 compared to 0.6 using the previous analysis.

DISCUSSION OF THE RESULTS

Figure 5 shows ΔE^+ versus α^+ for 1) a conductor with electric contact resistance of 5×10^{-9} $\Omega \cdot m^2$, a hundred times greater than measured values,[4] 2) zero electric contact resistance assuming no conduction in the composite region, 3) zero electric contact resistance considering conduction both in the composite and the stabilizer regions and 4) same as 3) but assuming partial cooling $\gamma_{s1} = 0.9$. As can be seen from the figure, the electric contact resistance does not have strong influence on the MPZ energy. In most of the cases the heat generated in the bond is a small fraction of the heat generated in the conductor regions. The partial cooling, strongly affects the critical value for α^+ Fig. 4, and

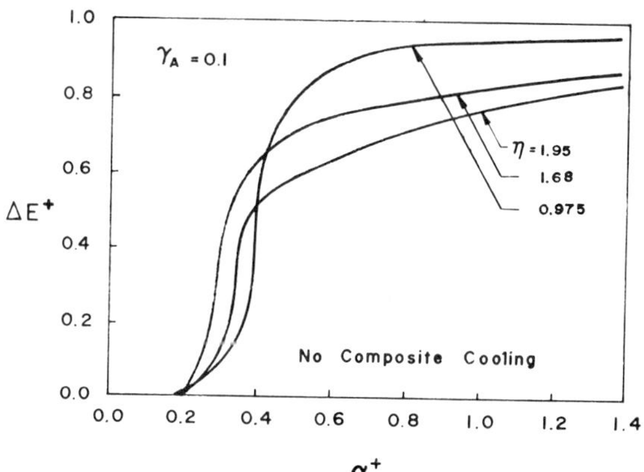

Fig. 4. ΔE^+ versus α^+ at different values of η considering heat conduction in the composite and the stabilizer regions.

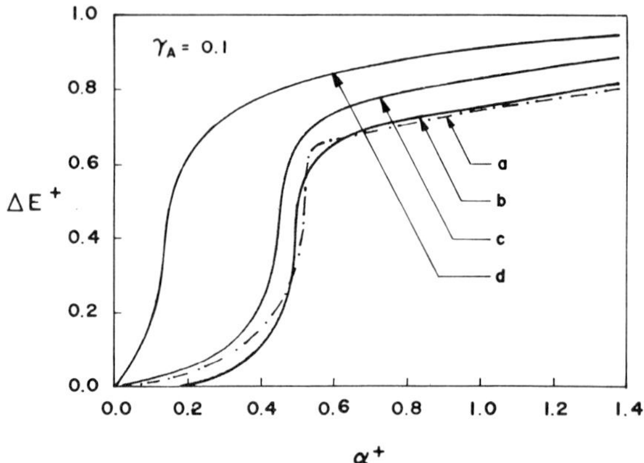

Fig. 5. ΔE^+ versus α^{+1} for the different cases mentioned in the test.

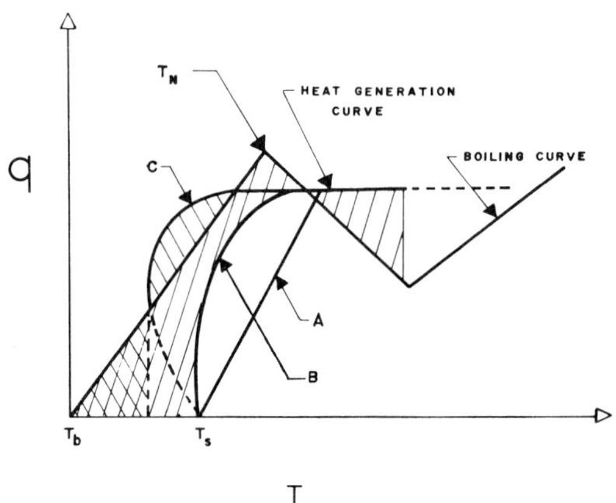

Fig. 6. q_b and q_g for built-up conduction.

the sudden decrease in ΔE^+ can be eliminated having sufficient cooling area for the composite region.

The abrupt decrease in ΔE^+ can be explained by referring to Fig. 6, where A, B, and C represent the heat generation curves with different thermal contact resistances. Curve A ($R_t = 0$) and curve B ($R_t > 0$) do not intersect the nucleate part of the boiling curve and satisfy the equal area criteria having maximum temperature higher than T_N, Fig. 5. Curve C intersects the nucleate region of the boiling curve and maximum temperatures higher than T_N cannot be obtained. If the thermal contact resistance of C is smaller, the maximum temperature can be higher than T_N. The abrupt change in the conductor maximum temperature corresponds to abrupt change in ΔE^+.

The abrupt decrease of ΔE^+ was not predicted by Willig.[5] His analysis is more conservative since it requires having the slope of the heat generation curve always less than the slope of the boiling curve.

REFERENCES

1. R.L. Powell and A. F. Clark, Definition of Terms for Practial Superconductors: 2 Critical Parameters, _Cryogenics_, 18:137 (1978).
2. R.L. Rhodenizer, et al., Status of the CDIF Superconductor Magnets, in "Proc. of the 1980 Superconducting MHD Magnet Design Conference, "FBNML/MIT, Cambridge, Massachusetts (in press).
3. L.S. Challis and J.D.N. Cheeke, Thermal conduction across copper-lead-copper sandwiches at helium temperatures, _Proc. Phys. Soc._ 83:109 (1964).
4. E.L. Foster, Some Properties of Materials at Very Low Temperatures, _D. Phil Thesis_, Oxford University (1955).
5. R.L. Willig, A Solder Bond Requirement for Large, Built-up, High Performance Conductors, _IEEE Trans on Magnetics_, MAG-17:1036 (1981).

THE SUPERCONDUCTING-TO-NORMAL TRANSITION
IN A FULLY STABILIZED WINDING*

Z. J. J. Stekly and W. F. B. Punchard

*Magnetic Corporation of America
Waltham, Massachusetts*

INTRODUCTION

One of the major design considerations in large fully stable superconducting coils is that of the initiation and propagation of a normally resistive region within the windings. In high current density coils with little or no liquid helium (LHE), once started the propagation of the normal region proceeds as a result of heat conduction from the resistive region into the superconducting region. The usual procedure in estimating the propagation or quench consists of computing the velocity of propagation along, and transverse to, the conductor and treating the normal region as essentially adiabatic.[1] When small amounts of LHE are present in the windings the procedure is modified by taking into account heat transfer to the liquid in calculating the propagation velocity and its sensible heat in the normal region once the front has passed.

As long as LHE is present in coils that are designed for full stability the normal region does not propagate; it recedes. In estimating the voltages and temperatures that would exist during propagation of the normal region, the assumption is generally made that the internal winding resistance is small so that the current decay as a function of time is determined primarily by the external resistance. Once the current time is known as a function of the maximum temperature in the winding is estimated by treating the conductor as adiabatic. Use of an effective specific heat can take into account the presence of other materials such as struc-

---

*Supported by the U.S. Department of Energy, MHD Division, through the Magnet Technology Field Office at the Massachusetts Institute of Technology, Francis Bitter National Magnet Laboratory.

ture, insulation and helium gas. This sort of calculation results in a worst case estimate of the voltages and temperatures, but because these are major design variables it is desirable that they not be unrealistically conservative.

The purpose of this paper is to present some of the results of a more comprehensive study that specifically deals with the transitional stage of the quench of fully stabilized superconducting coils that bridges the time between when the normal region is large enough to affect the LHE level and the time the helium has been vaporized.

DISCUSSION OF THE MODEL

The initiation of a normal region within a fully stabilized winding is a result of either steady or transient heat input to the windings. When helium liquid is present these heat fluxes will generate bubbles. Because of the stability of the winding a transient heat input creates a normal region that is transient; it recedes as soon as the heat input is over. A steady heat input creates a steady normal region.

For small amounts of heat input the bubbles generated will rise to the liquid surface. However, if the heat input is large enough the vaporized helium cannot escape and remains in the vicinity of the conductor. This paper will limit itself to the case where the bubbles can get to the surface (the case where they cannot will be the subject of a subsequent paper).

If the heat inputs are the result of the operating conditions such as changing magnetic fields, neutron heating, local heat leaks, etc., the refrigeration system should keep the helium liquid at the appropriate operating level. However, should unexpected heat inputs due to shorts in the winding, poor joints, mechanical movement, etc., exist the liquid level will drop. So long as the heat inputs to the winding are such that bubbles can get to the surface, and so long as the winding is immersed in the liquid, the stability will prevent local normal regions from propagating. As the helium level drops however the uppermost turns become exposed and are no longer fully stable. If these turns become normal, the normal region can propagate only to the liquid helium level. Propagation below the liquid level is possible only for windings that are not fully stabilized in LHe.

The model to be analyzed is summarized in Fig. 1. It consists of a single turn with a total length 2ℓ above the LHe. The exposed length of conductor is assumed to be normal (or in the current sharing mode). It is assumed that the heat conducted into the helium bath by the conductor vaporizes the liquid helium. The analysis is carried out for two limiting cases as far as the

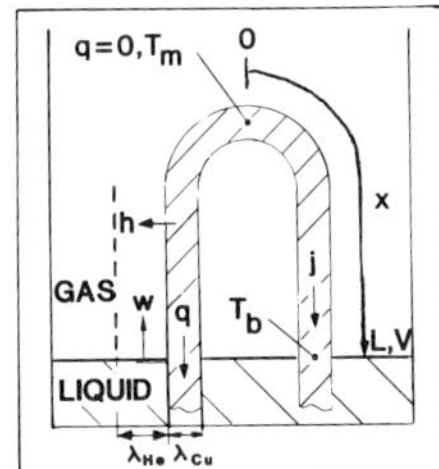

Fig. 1. Model of exposed conductor.

interaction of the conductor with the vaporized helium. In the
first case it is assumed that the helium vapor flows away or does
not exchange heat with the conductor. One-dimensional solutions
using either constant properties or a constant Lorentz number have
been treated earlier[2,3]. In the second case it is assumed that
the helium is constrained to flow along the conductor and this is
the calculation normally done for vapor cooled power leads[4]. For
this model the maximum temperatures result if the conductor is
allowed to reach a steady state temperature distribution. Transient conductor temperatures would be lower than the steady
state. As a consequence the analysis will be restricted to the
steady or quasi-steady temperature distributions in the conductor.

ANALYSIS

The one-dimensional equation for the steady state temperature
distribution T as a function of the distance x along the conductor
with the above assumptions is:

$$d(k\,dT/dx)/dx + \rho j^2 f + (w/A)c(dT/dx) = 0 \qquad (1)$$

The first term is a result of conduction with a thermal conductivity $k(T)$. The second term is the joule heating in a conductor of
resistivity $\rho(T)$ carrying a current density j. The superconducting to normal transition is taken into account in the temperature
dependent parameter f which is equal to the fraction of the total
current that flows in the normal portion of the conductor. The
third term is a result of heat transfer with a mass flow rate of
helium w associated with the conductor of surface area A. The
helium gas has a temperature dependent specific heat c.

As written, Eq. 1 is valid for the case where the vaporized
helium flows along the conductor and is in good thermal contact
with it. If the helium flows away or is not in good thermal
contact the last term in the equation can be eliminated.

It is convenient to introduce the conduction heat flux per
unit area along the conductor $q = k\ dT/dx$ and eliminate x by

setting $dx = kdT/q$. Performing this and rearranging variables to normalized the equation results in:

$$(q/j)d(q/j) + f\rho kdT + (wc/Aj)(q/j)dT = 0 \qquad (2)$$

Because all of the parameters are functions of temperature the normalized variable q/j is a function of temperature alone. The required boundary conditions are: the temperature T of the conductor at $x = 0$ is T_m, and at this location by symmetry the heat flux $q = 0$.

The helium flow rate w is a result of the heat conduction at the conductor helium liquid interface:

$$w\, h_{fg} = q_{T_b} A = (q/j)_{T_b} (jA) \qquad (3)$$

where h_{fg} is the latent heat of vaporization. The parameter α has been introduced for simplicity; $\alpha = 1$ represents the case where all the vaporized helium gas flows along the conductor with perfect heat exchange assumed ($h = \infty$), and $\alpha = 0$ represents no helium flow or no heat exchange with the conductor ($h = 0$).

With this substitution we have:

$$(q/j)d(q/j) + f\rho kdT + (\alpha c/h_{fg})(q/j)_{T_b} (q/j)dT = 0 \qquad (4)$$

which must be solved for $(q/j)_T$, T_m as a function of T for each value of T_m. Once this has been done the following can also be obtained:

1. The conduction heat flux at the conductor helium interface $(q/j)_{T_b, T_m}$

2. Normalized length ℓj of the exposed conductor as a function of maximum temperature T_m:

$$\ell j = \int_o^\ell d(xj) = \int_{T_b}^{T_m} \left[\kappa/(q/j)\right]\, dT \qquad (5)$$

3. Voltage along the conductor:

$$V = \int_o^\ell \rho jf dx = \int_{T_b}^{T_m} \left[f\rho k/(q/j)\right]\, dT \qquad (6)$$

4. The rate of change of helium liquid level v is obtained directly from the heat conducted into the helium. In normalized form this is simply:

$$(v\Lambda/j) = (q/j)_{T_b, T_m} / (\gamma h_{fg}) \tag{7}$$

where γ is the density of liquid helium and Λ is the ratio of helium area to conductor area.

5. Since both the length of the exposed region and the rate of change of the liquid helium level are functions of the maximum conductor temperature T_m, the time t to reach a given value of T_m can be obtained. In normalized form this becomes:

$$(j^2 t/\Lambda)_{T_b, T_m} = \int_o^{\ell j} d(\ell j)/(v\Lambda/j) \tag{8}$$

which must be integrated numerically.

Some of the results for $\alpha = 0$ ($h = 0$) can be obtained from the differential equation in integral form:

$$(q/j) = (2\int_T^{T_m} f\rho k \, dT)^{1/2} \tag{9}$$

$$\ell j = \int_{T_b}^{T_m} \left[k/(2\int_T^{T_m} f\rho k dT)^{1/2} \right] dT \tag{10}$$

$$V = \int_{T_b}^{T_m} \left[f\rho k/(2\int_T^{T_m} f\rho k dT)^{1/2} \right] dT \tag{11}$$

The results for $\alpha = 1$ are obtained by numerical integration starting with a chosen value of (q/j) at $T = T_b$ and integrating until $q/j = 0$ is reached which determines T_m.

DISCUSSION OF COMPUTED RESULTS

The equations were numerically integrated using the properties of copper, with the results shown in Fig. 2-6. Each figure contains two curves; one for the zero heat exchange case ($\alpha = 0$) and one for the perfect heat exchange case ($\alpha = 1$).

Figure 2 shows the length of the normal region (T_m to T_b) as a function of T_m. As T_m increases from T_b the length of the normal region increases initially, peaks, and then starts to decrease. This peak is associated with a peak in the thermal conductivity of copper versus temperature curve which occurs at around 20 K. The peaks in the length versus maximum temperature curve occur in a region where the thermal conductivity is decreasing. It means that in order to satisfy the boundary conditions, the length of the normal region must decrease to compensate for the lower average thermal conductivity of the material over the region T_b to T_m. Since this is physically unrealizable in a real situation (the length of the normal region having always to increase), when this temperature is reached an unstable situation arises whereby there are no quasi steady-state solutions and the maximum temperature will increase at a rate determined by the heating rate and the thermal capacities of the materials. The behavior of the hotspot becomes decoupled from the bath.

Figure 2 also shows that at a given T_m the scaled length ℓj is always greater for the perfect heat exchange case than for the zero heat exchange case and that the temperature at which the instability occurs is greater for the perfect heat exchange case.

The normalized heat flux into the helium bath as a function of hotspot temperature is shown in Fig. 3. The heat flux for the zero heat exchange case is higher than that for the perfect heat exchange case. For no heat transfer all the joule heat in the conductor must be conducted along the conductor to the helium liquid. As a result higher maximum temperatures result in higher heat conduction at the helium liquid interface. Alternately, for good heat exchange the helium gas intercepts the heat conducted from higher temperatures and as a result the heat conducted at the low temperature end is essentially independent of the maximum temperature above 20 K to 30 K.

The voltage (from $x = 0$ to ℓ) as a function of T_m is given in Fig. 4. Note that the voltage itself is a normalized variable and does not vary with current density. As such it is a basic variable which can be used to scale from one coil to another or to correlate the data from several experiments. The figure shows relatively monotonic increases in voltage up to the instability point for both limits of heat transfer considered. For no heat transfer the voltage at the instability point is about 5 mv. This would correspond to 10 mv for both halves of the conductor ($-\ell < x < \ell$). The corresponding values for the good heat transfer case are 14.6 mv for half and 29.2 mv for the full conductor. These values are of importance in setting upper limits on the detection voltage in a quench detection and discharge system. Since the instability conditions represent the upper limits for quasi-steady behavior of the exposed turns the voltages can be used to estimate the maximum power being generated within the windings. For a 10 000 A conductor the power generated for each turn would range from 100 W ($h = 0$) to 292 W ($h = \infty$). These powers should be readily detectable either in terms of helium flow rate, pressure rise in the dewar, or helium liquid level drop.

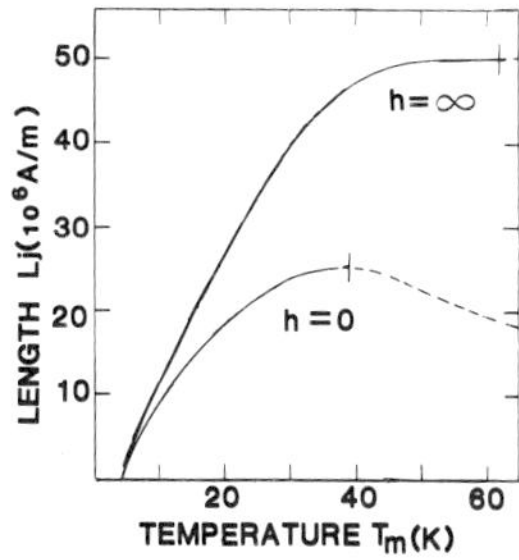

Fig. 2. Normalized exposed
conductor length vs max
conductor temp.

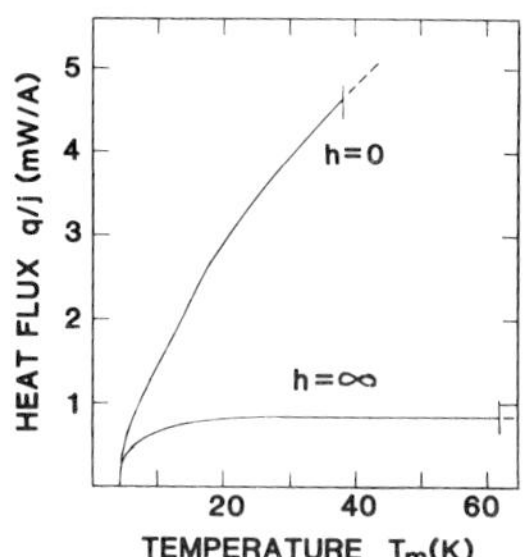

Fig. 3. Normalized heat
flux to liquid helium
vs max conductor temp.

Figure 5 shows the normalized time tj^2/Λ to reach a given maximum temperature T_m.

The normalized length is plotted in Fig. 6 as a function of normalized time tj^2/Λ (a cross plot of Figs. 4 and 5). The length is plotted with zero at the top left increasing in the downward direction. Plotted this way it is simply a pictorial representation of the position (normalized) of the liquid helium level as a function of time (normalized). The h = 0 curve shows an increasingly rapid drop of the helium level. This would be expected from the variation of heat flux to the helium for the h = 0 case in Fig. 3. The h = ∞ curve shows a liquid level which after an initial time drops linearly with time. This is to be expected because the heat flux to the liquid helium becomes independent of maximum temperature (or length) of the exposed turn.

Figures 7 and 8 show the length of the exposed turn and the time at the point where instability occurs as a function of current density. At around 2×10^7 A/m^2 the half length of exposed turns

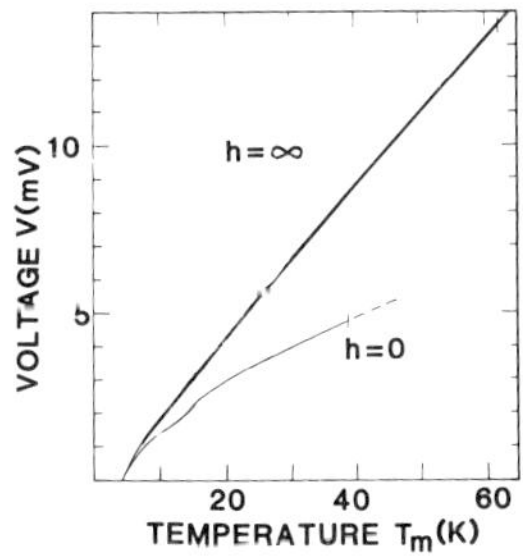

Fig. 4. Voltages vs max
conductor temp.

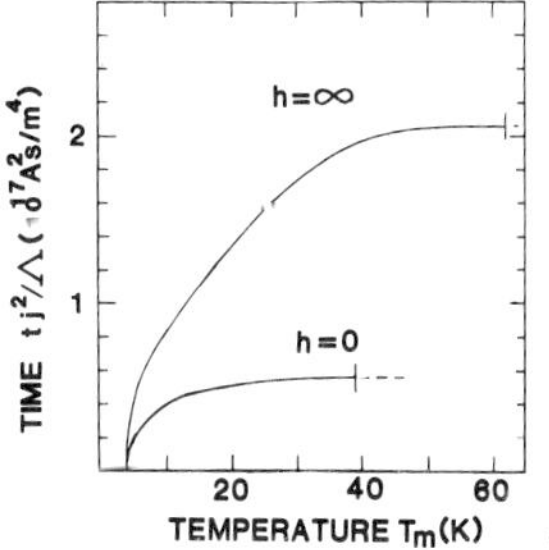

Fig. 5 Normalized time
to reach given max temp
vs maximum temp.

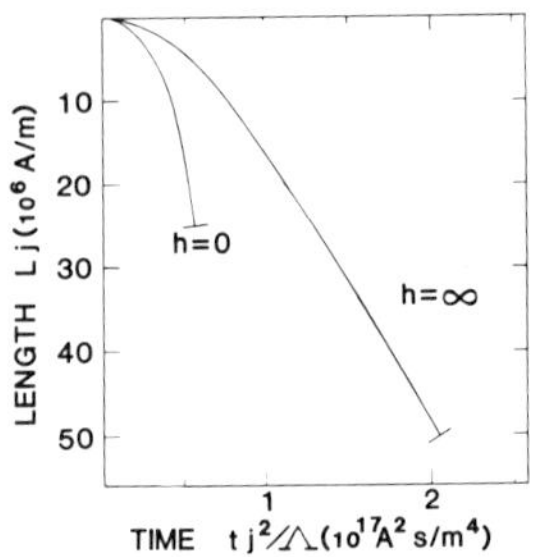

Fig. 6. Normalized length vs normalized time.

is of the order of 1.5 m to 3 m (full length 3-6 m). These are
not generally lengths that are small compared with the coil
dimensions even in some of the largest coils. At higher current
densities these lengths decrease inversely with j.

The time it takes to expose the critical length of conductor
is given by Fig. 8. Depending on the choice of parameters, at 2 ×
10^7 A/m^2 this time can vary from 16 to 100 seconds.

CONCLUSIONS

The initial stage of the quench process of a fully stabilized
coil has been modeled and analyzed. The results have shown that
the conductor temperatures and turn voltages remain relatively low
while the conductor is in a quasi-steady state condition where it
conducts heat to a slowly dropping liquid helium level. Once the
limits of this quasi-steady condition are exceeded the conductor
temperature and voltage have a rapid rate of rise with time.

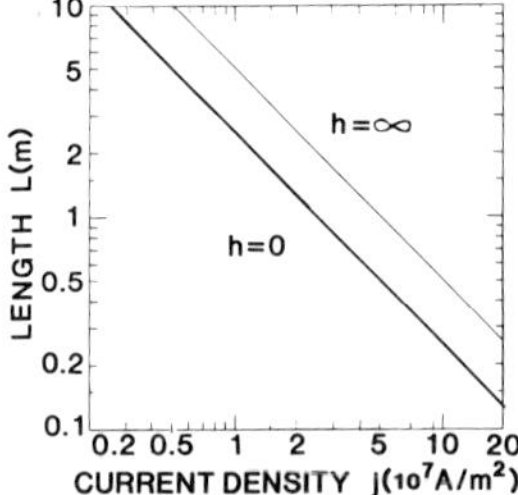

Fig. 7. Maximum stable
length exposed conductor
vs current density.

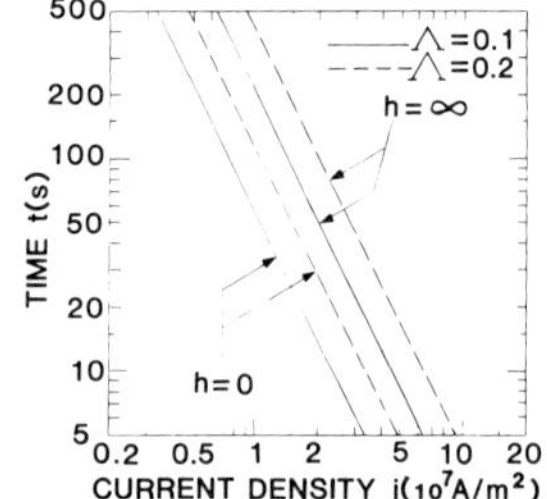

Fig. 8. Time to reach
max stable length
vs current density.

REFERENCES

1. D. Eckert, et al., Numerical treatment of the quenching process in superconducting magnet systems, Cryogenics 21 (6):307.
2. Z.J.J. Stekly, et al., Advanced superconducting magnets investigation, Avco Everett Research Laboratory, Final Report No. NS 8-21037 (Feb 1968).
3. R.L. Willig and M.A. Hilal, Reliability of pool cooling for tokamak fusion reactor magnets, Proc. 6th Symp Engr. Problems of Fusion Research, (1975) p 1928.
4. G. Aharonian, et al., Behavior of power leads for superconducting magnets, Argonne National Laboratory Report ANL-HEP-PR-80-55 (1980).

THERMAL CHARACTERISTICS AND THERMAL STABILITY
OF THE AC SUPERCONDUCTING COMPOSITES

V. A. Altov, E. V. Blagov, and N. A. Kulysov

*All-Union Institute of Metrological Service
Moscow, USSR*

and

V. V. Sytchev

*USSR State Committee for Science and Technology
Moscow, USSR*

INTRODUCTION

At the present time a number of methods stabilizing composite superconductors has been developed for use in dc electrotechnical devices. The development of superconductors having small losses in ac operation enables one to begin planning for ac superconducting electrotechnical devices. Meanwhile, the problem of ac stabilization and reliability remains one of the important tasks.

An analysis of thermal stability of an ac composite superconductor is based on the one-dimensional thermal conduction equation:

$$AC(T) \frac{\partial T}{\partial t} = \frac{\partial}{\partial x}\left(KA \frac{\partial T}{\partial x}\right) + \frac{\rho(T)}{A}I_o^2 \sin^2 2\pi\nu t - hP\ (T-T_o) + W \qquad (1)$$

where $C(T)$ = the conductor heat capacity per unit volume, $K(T)$ = the conductor thermal conductivity, $\rho(T)$ = the conductor specific resistance, I_o = the ac amplitude, T_q = the helium bath temperature, h = the heat transfer coefficient, x = the coordinate along the conductor, ν = frequency, and t = time.

The first term in the right-hand part of Eq. 1 characterizes heat transfer along the conductor due to a thermal conductivity.

Since a copper which is usually used as a stabilizer has high
thermal conductivity one can neglect a change in the temperature
over cross-section. The second term determines an electric power
dissipation due to ac flow along the conductor in the normal and
resistive states. The third term describes the heat flux from the
conductor surface into a helium bath. The fourth term characterizes the matrix and hysteretic losses in a composite conductor due
to a change in the magnetic field (self and external). As shown
by Carr[1,2] the total losses are mainly hysteretic losses value at
low frequencies of the field change. At intermediate frequencies
the matrix losses dominate, and at high frequencies the skin-
effect predominate.

For the most practical cases, one can neglect a contribution
of losses to the total heat exchange in the conductor, although
they can somewhat affect the critical current I_c.

The behavior of an ac composite superconductor at any moment
with a homogeneous temperature distribution along its length (KA
$\partial T/\partial x$ = 0– an isothermal case) and the thermal stability were
considered earlier[3,4,5,6].

ANALYSIS USING DIMENSIONLESS QUANTITIES

For the isothermal case Eq. 1 can be transformed into a more
convenient dimensionless form:

$$\Omega C (\tau) \frac{d\tau}{d\theta} = \alpha r i_o^2 \sin^2 - \gamma(\tau) \tau \qquad (2)$$

where

$$\Omega = \frac{2\pi\nu C_o A}{h_o P} = \text{dimensionless frequency}^6$$

$$\alpha = \frac{\rho_o I_c^2}{h_o PA(T_c-T_o)} = \text{Stekly stabilization parameter}^7$$

$$\tau = (T-T_o)/(T_c-T_o) = \text{dimensionless temperature}$$

$$i_o = I_o/I_c = \text{dimensionless current amplitude}$$

$$\theta = 2\pi n t = \text{dimensionless time}$$

$$r = \begin{bmatrix} 0; & \tau < 1-i \\ (\tau+i-1)/i; & 1-i<\tau<1 \\ \rho/\rho_o; & \rho >1 \end{bmatrix} \begin{array}{l} \text{dimensionless resistance} \\ \text{of the composite} \\ \text{superconductor} \end{array}$$

$$\gamma(\tau) = h(\tau)/h_o = \text{dimensionless heat transfer coefficient}$$

C_o, ρ_o, h_o are the conductor heat capacity, specific resistance and heat transfer coefficient at the critical temperature T_c.

A typical volt-ampere characteristics of the ac composite superconductor for constant heat transfer coefficient (γ = const.) are represented in dimensionless form in Fig. 1. A dotted line shows the dc equilibrium states.

One of the main parameters of a dc conductor is the maximum equilibrium current $\bar{i}_*$ above which a transition to the normal state is characterized by a non-controlled growth of the conductor temperature. It is shown[3] that the ac conductor thermal states at $\tau=\tau_{min}$ and $\tau=\tau_{max}$ coincide with the equilibrium states of this dc conductor at the corresponding instantaneous values of the alternating current. Moreover, as for the dc case, an ac threshold amplitude $\tilde{i}_*$ can be marked above which the conductor temperature grows in a non-controlled way from a cycle to cycle.

Another typical parameter is the ac amplitude i_o^n at which the minimum conductor temperature $\tau_{min}=1$. It is clear that at $i<1$ the condition $\tau_{min}<1$ corresponds to a conductor transition into a superconducting state.

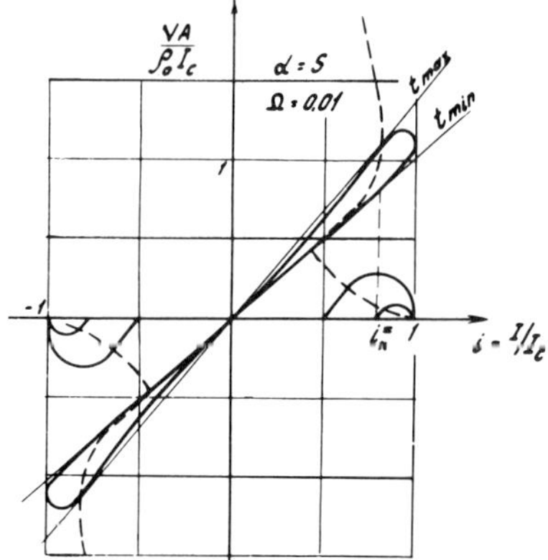

Fig. 1. Calculated I-V characteristics of the ac composite superconductor for constant heat transfer coefficient at Ω=0.01, α=5.

Figure 2 illustrates a qualitative appearance of the typical parameters $\tilde{i}_*$ and i_o^n as a function of the frequency[5]. It is easy to see that for the ac amplitudes lying within the range $i_o^n - \tilde{i}_*$, for instance, point 'a' (coordinates i_1 and Ω_1) a conductor can be found in two equilibrium thermal stable cycles – fully superconducting and fully normal. This means that in a sufficiently long ac conductor there can coexist extended sections being both normal and superconducting with a boundary character- ized by an inhomogeneous temperature distribution in a transient region. The future behavior of these sections, i.e. how they grow or shrink can be found from Eq. 1 for the superconducting and normal sections by using the boundary conditions. An analytical solution of this task appears difficult since it is nonstationary and Eq. 1 is nonlinear. At the same time earlier representations concerning ac composite superconductor behavior with a homogeneous temperature distribution along its length enable one to imagine a possible behavior of the conductor in the limiting cases. As was shown[3] with an increase in frequency due to a thermal conductor inertia, the temperature oscillations decrease. The threshold amplitude $\tilde{i}_*$ tends to some limiting value; for thin conductors where the skin-effect can be neglected this value is equal to $\sqrt{2}\ i_*^=$.

Analogously, the amplitude i_o^n corresponding to the condition $i=1$ with an increase in frequency tends to a limiting value $\sqrt{2}\ i_m^=$ where $i_m^=$ is the minimum current for the existence of the

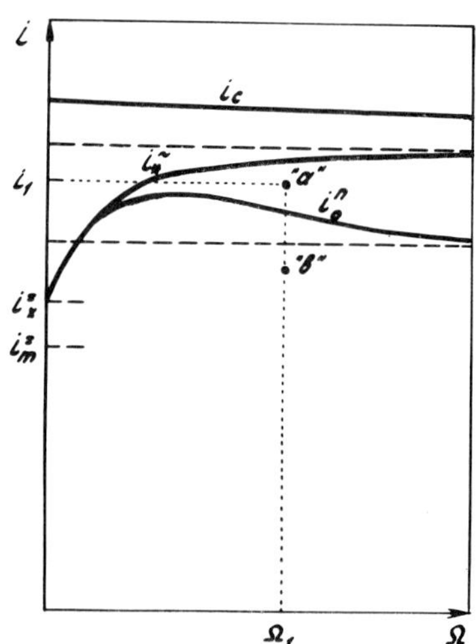

Fig. 2. A diagram of the frequency dependence of typical parameters of an ac composite superconductor.

normal zone for the dc case characterized by the normal zone shrinking with infinite velocity ($v \to -\infty$). Thus, at sufficiently high frequencies for the description of the process of the normal zone propagation along the conductor carrying alternating current one can use models developed for the dc case. In fact, the estimates for $\Omega \gg 1$ have shown[8] that the mean velocity $\bar{v}$ of the ac normal zone propagation coincides with that for the dc with an error of the order of $1/\Omega$.

It is evident that at low frequencies ($\Omega \to 0$) the process of normal zone propagation can be considered as quasistationary and for its description one can use the relations obtained for the dc case. In this case the error[8] when determining the instantaneous velocity of the ac normal zone propagation is of the order Ω.

The region of intermediate frequencies is the most difficult to analyze, although even here it is possible to find some regularities. We can affirm that in this region there exists the amplitude $i_o^n < \tilde{i}_o < \tilde{i}_*$ at which the mean velocity of the normal zone propagation is equal to zero ($\bar{v} = 0$). It is evident that at some frequency Ω_1 and $i_o > i_p$ the normal and superconducting sections can coexist in the conductor pulsing in the dimensions; however, a constant growth of normal sections is observed. This means for a conductor of finite length, that it will gradually pass to a normal state. At $i_o^n < \tilde{i}_o < \tilde{i}$ the normal sections will shrink until they completely vanish. At $i_o < i_o^n$ (see, point 'b' in Fig. 2) the normal sections can exist only when there is an additional source of heat (for instance, a heater). In this case with a growth in amplitude the normal zone pulsing in the dimensions will increase up to $i_o = \tilde{i}_p$ in a controlled way. Schematically this situation is illustrated in Fig. 3. The dotted line 1 characterizes a volt-ampere characteristics of the dc conductor under power from a point microheater[9]. It is seen from Fig. 3 that this curve asymptotically approaches some threshold current value corresponding to the minimum current of the normal zone propagation. By analogy with an isothermal case for sufficiently high frequencies ($\Omega \gg 1$) when the conductor temperature remains practically constant within the cycle there can be constructed a volt-ampere characteristic for the given power (curve 2 in Fig. 3). For sufficiently thin conductors, where the skin-effect thickness essentially exceeds the conductor diameter, $\tilde{i}_p = \sqrt{2}\, \bar{i}_p$. In Fig. 3 the curve 3 characterizes the thermal cycle when the ac amplitude essentially exceeds the minimum current of the dc normal zone propagation $\bar{i}_p$ and the normal zone increases beginning with some moment within the cycle in a non-controlled way. With a reduction within the cycle of the instantaneous current i the velocity of the normal zone propagation falls, and after the normal zone achieves the maximum size it begins to shrink. At the instantaneous current values close to the minimum current for the existence of the

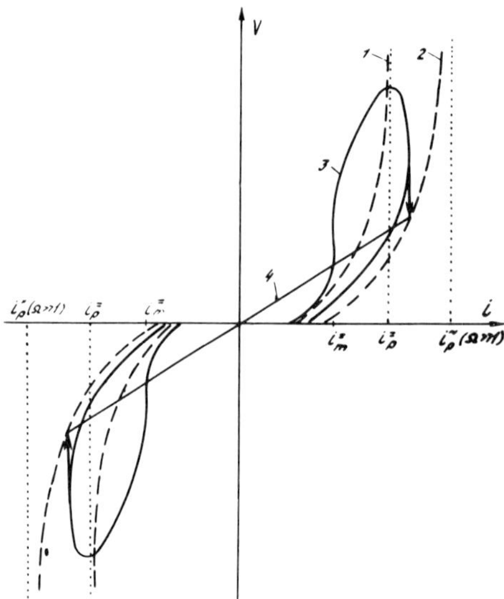

Fig. 3. A diagram of the composite superconductor states with a point power source for the dc (curve 1) and ac (curves 2,3,4) cases.

normal zone $i_m^=$ the normal zone shrinks with infinite velocity. It is easy to show that the points of intersection of the volt-ampere characteristics with the dotted line 1 characterizing the dc conductor equilibrium states correspond to the maximum values of the length (resistance) of the normal zone for each cycle. It is clear that at the fixed amplitude with an increase in frequency an amplitude of the length oscillation of the normal section will reduce and at sufficiently large frequencies ($\Omega \gg 1$) the volt-ampere characteristics (I-V) of the conductor (for instance, curve 3 in Fig. 3) will transform into a section of a straight line passing through the origin (line 4 in Fig. 3).

These regularities are well confirmed by experiments to study the process of normal zone propagation in an ac composite conductor. The sample under study was a bifiller one-layer coil wound with multifilament Nb-Ti superconductor 0.3 mm in diameter a fixed distance between the turns to provide liquid helium access to the conductor surface. Along the length of the conductor there are potential leads to measure a voltage drop both for the total sample and for separate sections. There is a microheater in the central section to initiate a resistive region. The sample is placed in an external homogeneous magnetic field created by a solenoid.

In Fig. 4 are illustrated typical volt-ampere characteristics of the conductor in an external field B=6 T for various values of the ac amplitude. In the upper part of Fig. 4 are depicted volt-ampere characteristics taken from the central section of the conductor (1 cm long) where the microheater is placed. Volt-ampere characteristics depicted in the lower part of Fig. 4 are taken from the whole length of the sample. The heater power is 0.1 W and is chosen so that the conductor in the central section will pass into the normal state at the instantaneous current values close to the minimum current of the dc normal zone propagation $I_p^=$. The analysis of the experimental characteristics given in Fig. 4 shows that the total resistance of the normal zone within the cycle reaches its maximum value at the instantaneous current value $I \approx I_p^=$.

Curve 4 of Fig. 4 corresponds to the case when the sample under study is fully filled by a normal zone during a part of the

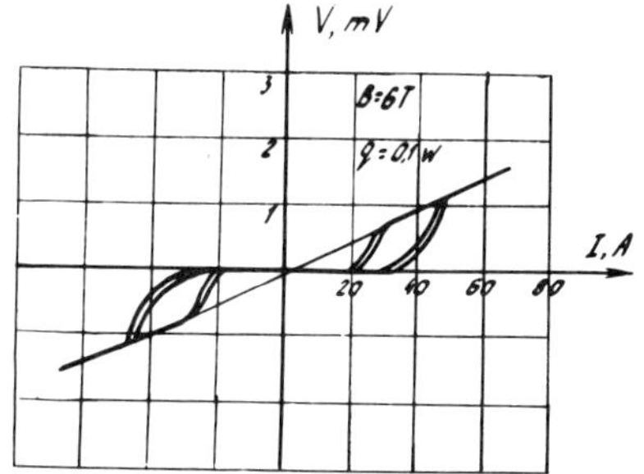

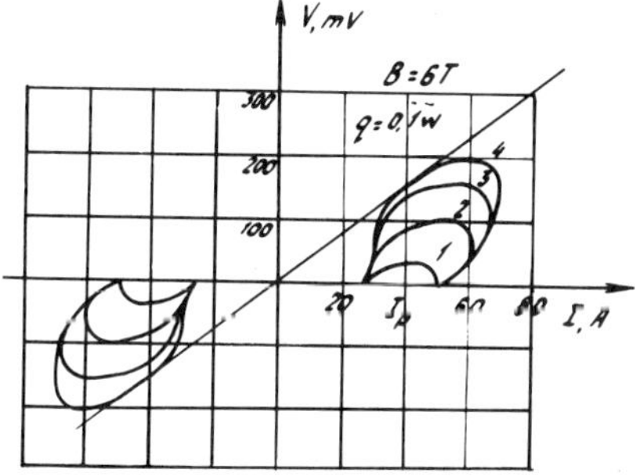

Fig. 4. The typical I-V characteristics of the composite superconductor with a point heater at various ac amplitudes.

period. In this part the volt-ampere characteristic of the conductor is a section of the straight line passing through the origin. Under a reduction of the instantaneous current value I within the cycle the size of the normal zone sharply decreases near the current I_m. In this case the volt-ampere characteristic on the central section sharply deviates from the straight line corresponding to the normal state and that speaks about the drop of the conductor temperature below T_c. It is interesting to notice that inspite of a wide range of the ac amplitudes (curves 1-4) I-V characteristics on the central section change insignificantly.

The results obtained and the model representations developed enable one to forecast the behavior of ac stabilized conductors at their transition into a normal state.

REFERENCES

1. W.J. Carr, Jr., Conductivity, permeability, and dielectric constant in a multifilament superconductor, J. Appl. Phys. 46:4043 (1976).
2. W.J. Carr, Jr., M.S. Walker, and J.H. Murphy, Alternating field loss in a multifilament wire for weak ac fields superposed on a constant bias, J. Appl. Phys., 46:4048 (1975).
3. V.A. Altov, N.A. Kulysov, and V.V. Sytchev, Stability of Composite Superconductors under AC Conditions, in "Advances in Cryogenic Engineering, Vol. 22," Plenum Press, New York (1977) p. 408.
4. V.A. Altov, N.A. Kulysov, and V.V. Sytchev, Volt-ampere characteristics of composite superconductors under A.C. conditions, IEEE Trans. on Magnetics, 15:383 (1979).
5. Yu. M. Lvovsky, The model of the heat exchange of the cooled ac conductor, Inzhenerno-fizicheskij zhurnal (in Russian), 35:75 (1978).
6. V.A. Altov, et. al., Thermal characteristics of an ac composite conductor, Izv.AN SSSR Energetika i transport (in Russian) 6:39 (1980).
7. A.R. Kantrowitz and Z.J.J. Stekly, A new principle for the construction of stabilized superconducting coils. Appl. Phys. Lett, 6:56 (1965).
8. S.S. Kutateladze, M.O. Lutset, and Yu. M. Lvovsky, Normal zone propagation in a superconductor carrying time-dependent current, Cryogenics 18:310 (1978).
9. V.V. Sytchev, et. al., Terminal characteristics and thermal stability of composite superconducting materials, Cryogenics, 13:644 (1973).

HEAT TRANSPORT TO He I FROM A POLISHED SILICON SURFACE

E. Flint, J. Van Cleve, L. Jenkins, and R. Guernsey

IBM Thomas J. Watson Research Center
Yorktown Heights, New York

INTRODUCTION

Superconducting tunnel junctions (Josephson junctions) may replace transistors in future high performance computers. Circuit chips, fabricated on silicon using thin film techniques, might be 0.64×0.64 cm^2 and dissipate 2-50 mW/cm^2. The operating temperature would be maintained by immersing the computer in liquid He. It is important to know the upper bound on the chip temperature at a given bath temperature for the expected power inputs. We have studied the thermal transport from a polished silicon surface to He I (2.5 K $\leqslant T_o \leqslant$ 4.8 K) in the non-boiling and pool boiling regimes. We find that the liquid superheat at the onset of widespread boiling, as indicated by a sharp increase in the differential heat transfer coefficient, approaches that predicted by the theory of homogeneous nucleation if the helium vapor and liquid properties are accurately included. Hence, we are able to set a fundamental upper limit on the chip temperature and this is within the operating range of the circuits. Such a limit is not realized with room temperature coolants.

Published data[1-4] show a significant spread ($\sim$ a factor of 5) in the temperature difference (ΔT) between the heated surface and the bath for $\dot{Q}(\Delta T)<300$mW/cm^2. The differences can be attributed to geometry, surface orientation[5] and finish[2], and experimental procedures. The effect of the surface finish is of particular interest when considering heat transport from polished silicon. The data in the literature indicate that rough surfaces exhibit substantially similar heat transfer characteristics but that polished surfaces (e.g., 400 grit) are hotter for the same heat flux[2,4]. Polished silicon surfaces (SiO$_2$) are very smooth when compared to most metal surfaces and so it is necessary to examine the onset of pool boiling and heat transfer for such a surface.

The mechanisms governing the nucleation of bubbles at a heated surface in helium are different from those in room temperature coolants. In the latter, surface cavities in which gas or vapor is trapped before heating constitute the nucleation centers[6]. The superheat for activation then depends on the fluid properties and the cavity geometry in at least two different ways. For liquids with a non-zero contact angle the vapor and liquid phase can coexist in re-entrant cavities over a range of temperature (including the saturation temperature) which depends on the surface tension and cavity geometry. In addition, the gas phase of a different substance (which is only partially soluble in the coolant) may become trapped in surface cavities. Neither of these mechanisms apply to liquid helium[7]. Liquid helium has a zero contact angle and wets all solid surfaces below the critical temperature, 5.2 K. It is not possible for the vapor and liquid phases to coexist below the surface of the liquid before heating, although a heated surface may trap vapor in re-entrant cavities during the filling of the bath. Furthermore at these temperatures all other substances (except ^{3}He) solidify and hence can not lower the activation energy of cavities. It is necessary, therefore, to consider nucleation in the bulk fluid.

APPARATUS

Several surfaces were studied in these experiments. In each experiment a 6.4 mm x 6.4 mm polished Si chip was glued to a vertical phenolic mount and was immersed in a temperature regulated ($\pm$ 2 mK) He I bath. Heat was applied to the surface of the chip facing the phenolic (using a deposited $Pt_{30}Si_{70}$ heater) and the temperature of the chip was measured on the opposite surface using implanted thermometers[8]. Calculations indicate that roughly 1% of the heat flowed through the phenolic to the bath.

The surfaces of the chips were examined with a scanning electron microscope and an optical microscope. The highest quality surfaces are smooth to 100Å over most of the surface. The chips actually studied had 1000Å–20,000Å protrusions and 100Å–2000Å pits or channels over some of the surface. For comparison, a copper surface polished with 400 grit paper is covered with grooves 50,000Å–100,000Å across. The surfaces studied in this work are relatively smooth but it is difficult to anticipate the resulting effect on $\dot{Q}$ (ΔT).

The data were taken by regulating the bath temperature (2.4 K $\leq T_0 \leq$ 4.8 K) and measuring the chip temperature at the four thermometer sites. Power was applied to the heater and the ΔT recorded after waiting for a stable reading. By increasing the power in steps to a level near the peak nucleate boiling flux (PNBF) the transfer characteristic for convective cooling and pool boiling to the liquid helium was determined. Hysteresis effects were also studied by measuring the superheat while decreasing the heater power (Fig.1).

RESULTS: MEASUREMENTS OF THE HEAT TRANSFER CHARACTERISTIC

In this section the results from four runs including three samples are presented. Figure 1 shows the results for the heat transfer characteristic as obtained in a run for a bath temperature (T_0) of 4.2 K. The sudden increase in the differential transport coefficient occurs at $\Delta T]$ 0.33 K and is believed to mark the onset of widespread bubble nucleation. The ΔT required for onset was greatest during this run. The PNBF (measured separately) is also shown. Figure 2 gives $\dot{Q}$ (ΔT) for increasing heat flux at the temperatures studied. The onset ΔT increases with decreasing bath temperature and becomes more difficult to identify. These results are qualitatively similar to other results found in the literature.[2-6]

The inset in Fig. 1 shows the heater voltage $(\sim \dot{Q}^{1/2})$ vs. ΔT as the power was swept continuously. We believe the temperature backstep results from the sudden activation of sites across the chip. Numerous other examples were observed in other runs, but rarely were the backsteps as large. An important feature of the temperture backsteps, which is discussed below, is that the ΔT at which they were observed, ΔT_{bs}, never exceeded $\Delta T \sim$ 0.33 K.

ANALYSIS: THE UPPER BOUND TO BOILING ONSET IN HE I

Theories of homogeneous nucleation[9-11] describe bubble formation in a bulk fluid in the absence of dissolved gas. In liquids such as water, the superheat required for homogeneous nucleation is much larger than the observed ΔT required to initiate bubbling at surfaces. This is because gas trapped in cavities on the surface or bubbles of the critical radius formed in the liquid from dissolved gas serve as nucleation sites at much lower superheats[6]. In He I, as mentioned above, none of these conditions are present if precautions are taken against vapor trapping in re-entrant cavities during the filling of the experimental chamber. Furthermore, the system operates near the critical point; nucleation rates increase drastically as the surface tension and latent heat of vaporization go to zero.

These experiments measure the heat flux from the surface to liquid helium as a function of ΔT. Relating this data quantitatively to

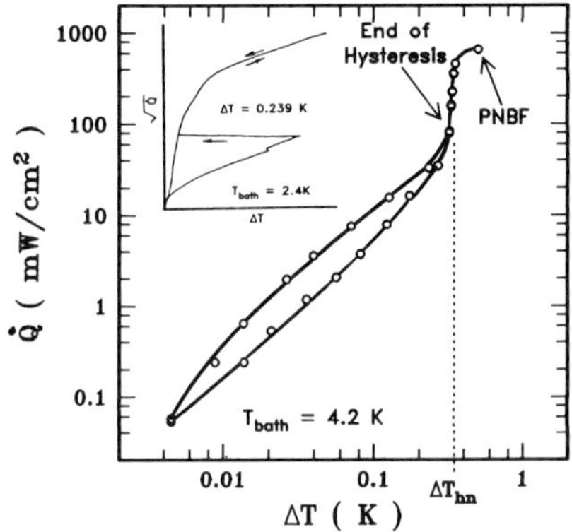

Fig. 1. $\dot{Q}$ (ΔT) at 4.2 K showing hysteresis and (inset) temperature backstep.

the rate of bubble nucleation is a difficult hydrodynamic and thermodynamic problem which is not within the scope of this work. However, a sudden rise in the differential heat transfer coefficient indicates that a transition has occurred between the relatively low heat transfer of predominantly single phase convective flow to the high heat transport capability of two-phase flow with vapor formation near the surface. Our operational definition of the onset of homogeneous nucleation near the surface will be discussed below.

Our model is a modification of Frenkel's thermodynamic theory of homogeneous nucleation[12]. That is, we consider such quantities

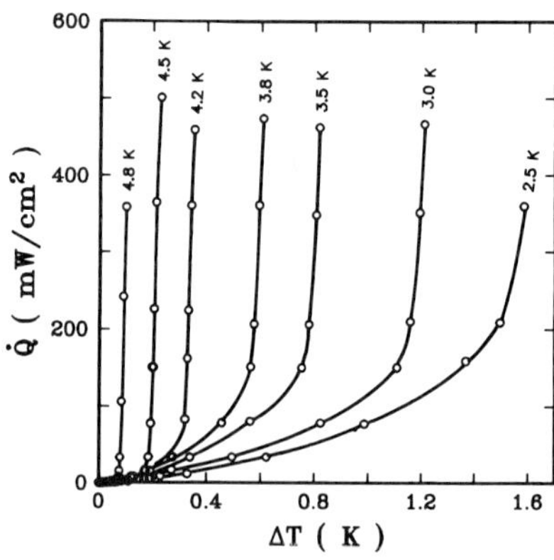

Fig. 2. $\dot{Q}$ (ΔT) for several bath temperatures.

as surface tension and pressure to be well-defined in bubbles near the critical radius, r_c. For He, r_c is on the order of 50–100Å near the predicted onset of nucleation except near the critical point. Such a bubble would contain > 1000 atoms at p = 1 atm.; so a bulk calculation seems to be a reasonable approach.

Frenkel's model[12] of homogeneous nucleation, when applied to the nucleation of bubbles in superheated liquid rather than drops in a supercooled vapor, gives the following expression for the rate of formation of bubbles of the critical radius per unit volume per unit time:

$$R = n_\ell(T) \sqrt{\frac{\sigma(T)}{m_{He}}} \exp\left\{-\frac{4\pi}{3}\sigma(T)r_c^2(T_o,\Delta T_\ell)/kT\right\}.* \qquad (1)$$

The critical radius is found by setting the chemical potentials of the gas and liquid phases equal, and requiring that the pressure inside the bubble be greater than that outside the bubble by $2\sigma/r_c$. If $v_\ell \ll v_g$ and r_c is presumed to be large enough to consider the equilibrium vapor pressure inside the bubble to be that of a plane interface at the same T, then one obtains $r_c = 2\sigma(T)/p_s(T)-p_s(T_o)$. Using the Clausius-Clapeyron equation $r_c = 2\sigma(T_o+\Delta T_\ell)kT_o^2/q_{g\ell}p_s(T_o)\Delta T_\ell$.

The first modification we make is to compute $p_s(T_o + \Delta T_\ell)$ directly from the vapor pressure curve rather than from the Clausius-Clapeyron equation. This eliminates substantial errors resulting from the non-linearity of $p_s(T)$ between T_o and $T = T_o + \Delta T_\ell$.

We also account for the fact that the pressure inside the bubble is not $p_s(T_o + \Delta T_\ell)$ but somewhat less – call it $p_g(T,r)$ – as the equilibrium vapor pressure over a concave surface is less than that over a plane at the same temperature. Our derivation proceeds as follows, with all temperature-dependent quantities evaluated at T unless otherwise noted:

$$\mu_g(p_g,T) = \mu_\ell(p_\ell,T) \qquad (2)$$

---

*This is similar to the form of Blander and Katz [AICHE J 21:833 (1975)]. The theory of absolute reaction rates [S. Glasstone, K, J. Laidler and H. Eyring, "The Theory of Rate Processes". McGraw-Hill, N.Y. (1941), p. 184] give a factor $n_\ell kT/h$ instead of $n_\ell\sqrt{\sigma/m_{He}}$ before the exponential, while the energy of formation is $(4/3)\pi\sigma r_c^2$ in both cases. kT/h and $\sqrt{\sigma/m_{He}}$ are within an order of magnitude of each other for 2.2 K $<T<$ 5.0 K; the rapid variation of the exponential insures that the choice of factors, in this case, changes ΔT_ℓ very slightly for a given R.

which implies

$$\int_{p_g}^{p_s} v_g dp = v_\ell (p_s - p_\ell) \tag{3}$$

assuming the liquid to be incompressible.

In most calculations $v_g = kT/p$ however, helium departs significantly from ideality near saturation conditions. A numerical comparison of the experimental values of v_g and p at various temperatures indicates that, for $(p_s-p_g)/p_s < 0.3$ a linear approximation is a much more accurate representation of the gas equation of state. This condition is satisfied in the range of interest for bubbles of the critical radius. Then we have

$$v_g(T,p) = v_{go} + M(T)(p_s(T)-p), \tag{4}$$

where $v_{go} = v_g(T,p_s(T))$ and $M(T) = \dfrac{v_{go}^2}{m_{He}}\left(\dfrac{\partial p}{\partial \rho_g}\Big|p_s,T\right)^{-1}$. Both v_{go}

and $\dfrac{\partial p}{\partial \rho_g}\Big|p_s,T$ may be found in the standard NBS tables[13]. Then, from Eq. 4, we find:

$$p_g = \frac{v_{go} + Mp_s - [v_{go}^2 + 2Mv_\ell(p_s-p_\ell)]^{1/2}}{M} \tag{5}$$

and

$$r_c = \frac{2\sigma}{p_g-p_\ell} = \frac{2\sigma}{p_g(T)-p_s(T_o)} \tag{6}$$

since the liquid pressure is the vapor pressure at the liquid surface, $p_s(T_o)$, if the gravitational head is ignored.

By writing p_g explicitly in terms of r, allowing v_ℓ/v_g to become negligibly small, and considering the gas to be ideal, one gets the familiar result $p_g(T) = p_s(T) \exp\{-2\sigma v_\ell/rkT\}$. Thus our equation accounts for the curvature of the surface and the non-ideality of ^{4}He near saturation.

Knowing $r_c(T_o,\Delta T_\ell)$ we can calculate the rate of bubble nucleation. The adjustable parameter is the number of nucleations per cm^3 per sec which we wish to define as the onset of nucleation. As a minimum we would like to have one bubble occur near the chip in the time lapse between the increase of power supplied to the chip and the measurement of ΔT, on the order of 1 min. With the length of a chip side being L, a volume L^3 might conceivably be superheated. Setting L = 4 mm gives us $R \sim 10^{-1}/cm^3$ sec as a lower bound. One bubble/cm^3 sec is an appropriate value for the onset of homogeneous nucleation; in any event ΔT_{hn}, the predicted onset superheat, is not affected greatly by the choice of the onset rate. At T_o = 4.22 K, for example, R = 1 cm^{-3} s^{-1} at ΔT_{hn} = 0.345 K, R = 10^{-3} cm^{-3} s^{-1} at ΔT_{hn} = 0.340 K, and R = 10^3 $cm^{-3}s^{-1}$ at ΔT_{hn} = 0.355 K.

Thus, for a given T_o, we know that the bulk liquid will start to boil at a superheat near ΔT_{hn}. Even for a perfectly smooth surface in a pure liquid, the onset of pool boiling must occur at or below this superheat. As the liquid near the chip reaches $T_o + \Delta T_{hn}$, it begins to boil spontaneously, and $\dot{Q}$ rises drastically. No surface structure or impurity in the liquid is required for this phenomenon, though these factors may reduce the superheat required for pool boiling, and ΔT_{hn} is the upper bound for the onset of pool boiling.

In analyzing our data we need to identify the liquid superheat at which nucleation in the liquid begins. We do this by taking the chip ΔT, above which there is no hysteresis, and subtracting the Kapitza drop, $\Delta T_k = 2\dot{Q}/c_v c_s$ where c_v is the specific heat and c_s is the sound velocity of silicon[14]. The resulting quantity, ΔT_h, is an experimental lower bound on the liquid superheat for homogeneous nucleation since the observed hysteresis at lower ΔT implies that activated nucleation sites at the surface are required to sustain boiling. ΔT_{hn} must be greater than or equal to ΔT_h, since for all superheats less than ΔT_h bubble nucleation is not a bulk phenomenon. In Fig. 3 we compare our modified theory of Frenkel with the ΔT_h values from our experiments. ΔT_h and ΔT_{hn} are in good agreement, and we note that there are no $\Delta T_h > \Delta T_{hn}$, as predicted. It is also noted that there is a dramatic increase in the differential heat transport coefficient at this ΔT (Fig. 1).

The temperature backsteps noted in several experimental runs, as in the inset of Fig. 1, should also be constrained by homogeneous nucleation. These backsteps are conceivably the result of the activation of a small number of previously vapor-filled sites; site activation increased $\dot{Q}$ for a given ΔT. Since nucleation begins in the bulk liquid at a superheat ΔT_{hn}, the sites must be activated for $\Delta T_\ell \lesssim \Delta T_{hn}$. We observe that the superheat at which the backsteps occurred is less than ΔT_{hn}, a result consistent with the predictions of homogeneous nucleation.

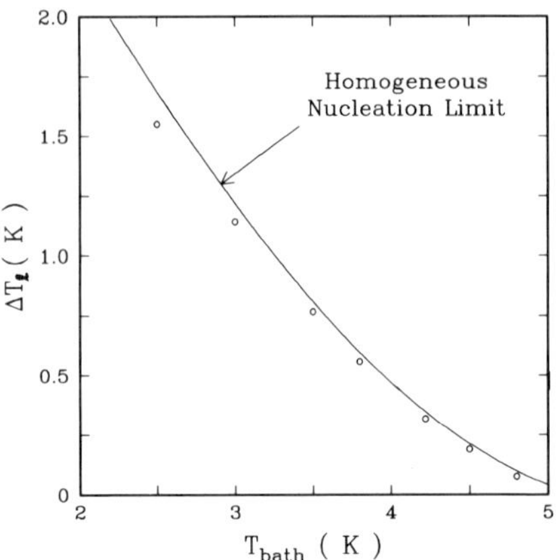

Fig. 3. After Frenkel, using modified formula for r_c. Experimental lower bound for the onset of homogeneous nucleation.

CONCLUSION

In this work we have found evidence that the superheat at boiling onset near a heated surface in He I is limited by the physics of homogeneous nucleation. This implies that there is a fundamentally derived superheat, independent of surface finish, above which one can anticipate the high thermal transport of pool boiling. For example, in a bath at 4.2 K homogeneous nucleation occurs at a superheat of 0.345 K. Thus, one can expect that with a heat flux of 200 mW/cm^2 to a 4.2 K bath that the superheat will be less than 0.345 K.

NOTATION

k	Boltzman's constant (erg/K)
m_{He}	mass of a helium atom (gms)
n	number density (atoms/cm^3)
p	pressure (dynes/cm^2)
$q_{g\ell}$	latent heat of vaporization (ergs/atom)
Q	heat flux (mW/cm^2)
R	rate of bubble formation (sec^{-1} cm^{-3})
r_c	minimum radius of bubble stable against collapse (cm)
T	temperature of liquid or vapor in superheated region = T_0 + ΔT_ℓ.
T_0	bath temperature (K)
ΔT	temperature difference between chip and bath (K)
ΔT_h	experimentally determined, lower bound for the homogeneous nucleation superheat (K).

ΔT_{hn} liquid superheat required for the homogeneous nucleation of bubbles (K).

ΔT_{ℓ} temperature difference between the superheated liquid and T_0 (K).

v   volume per atom (cm^3/atom)

μ   chemical potential (erg/atom)

ρ   density (gms/cm^3)

σ   surface tension (dynes/cm).

Subscripts

g   gas parameter

ℓ liquid parameter

s   evaluated at saturated conditions

o   ambient bath conditions.

REFERENCES

1. R.V. Smith, Cryogenics 9:11 (1969).
2. J.C. Boissin, et. al., in "Advances in Cryogenic Engineering, Vol. B", Plenum Press, New York, (1967), p. 607.
3. V. Purdy, C. Linnet and T.H.K. Frederking, in "Advances in Cryogenic Engineering Vol. 16", Plenum Press, New York, (1970), p. 359.
4. K.R. Efferson, J. Appl. Phys., 40:1995 (1969).
5. D.N. Lyon in "Advances in Cryogenic Engineering, Vol. 10B", Plenum Press, New York, (1964), p. 371.
6. P. Griffith and J.D. Wallis, Chem. Eng. Prog. Symp. Series, 56:49, (1960).
7. R.V. Smith, J. Eng. Ind. 91B:1217 (1969).
8. S. Early and T.H. Geballe, "Silicon on Sapphire Bolometer for Low Temperature Small Sample Calorimetry", preprint, unpublished.
9. D. Turnbull and J.C. Fisher, J. Chem. Phys., 17:71 (1949).
10. S. Takagi, J. Appl. Phys., 24:1453.
11. M. Volmer, "Kinetic der Phasenbildung", Theodor Steinkopff, Dresden and Leipzig: (1939).
12. J. Frenkel, "Kinetic Theory of Liquids" Dover, New York: 1955 Chapter VII.
13. R.D. McCarty, "Thermodynamic Properties of Helium 4 from 2 to 1500 K at Pressures to 10^8 Pa.", Reprint No. 40 from J. Phys. and Chem. Ref. Data 2:4 (1973), p. 923.
14. Private communication, Humphrey Maris, Brown University.

DISCUSSION

Question by L. J. Challis, Nottingham University, England:

Is it known what determines the points in the liquid at which the bubbles form?

Answer by author: According to the nucleation model, bubbles appear at random points in a homogeneous system.   In this experiment it is likely that there are points at the heated surface with a slightly higher probability for nucleation, but we were unable to look for them.

Question by S. Caspi, Lawrence Berkeley Laboratory:

What is the effect of convection on the model due to the vertical orientation of the sample?

Answer by author: Fluid convection near the heated surface will reduce the volume of the fluid superheated to ΔT_{hn}.   In the calculation of ΔT_{hn}, choosing a smaller volume is equivalent to assuming a larger critical rate of bubble production, but this does not affect the result significantly.

Question by D. Petrac, Jet Propulsion Laboratory:

Did you investigate the effect of hydrostatic pressure on the ΔT at the same power?

Answer by author: No, we did not.   The effect of the hydrostatic pressure head on ΔT_{hn} is less than 1.3 mK/cm which for our system is a very small effect.

EMISSIVITY MEASUREMENTS OF METALLIC SURFACES USED IN CRYOGENIC APPLICATIONS

W. Obert

JET Joint Undertaking
Abingdon, Oxon, England

J. R. Coupland and D. P. Hammond

Euratom-UKAEA Association for Fusion Research
Abingdon, Oxon, England

and

T. Cook and K. Harwood

Oxford Instrument Co.
Osney Mead, Oxford, England

INTRODUCTION

The use of present and future cryopump systems for fusion devices, having typical values of 10^6-10^7 liters sec^{-1} for pumping speed and pumping areas in the range of 100 m^2, requires optimization for both performance and economics. A common factor in any design is the type of material and surface finish used for the black radiation shields at 77 K and the polished pumping areas at liquid helium temperature. The 77 K surfaces require high values of emissivity whereas the values for the 4.2 K surfaces need to be as low as possible. Vacuum and thermal considerations in general restrict the choice of material to stainless steel, copper, and aluminium substrates with various surface finishes. The final choice of material, manufacturing procedure, and surface treatment for both surfaces determines the cost and overall economics of the system.

The total emissivity of a number of samples of such materials with different surface finishes has been measured, under conditions that exist in practical vacuum systems, by the evaporation loss method. This method consists of metering the gas boil-off from cylindrical samples filled with liquid helium or liquid nitrogen immersed in a vacuum cryostat, and this can be directly related to thermal input and hence emissivity. The criterion used in preparing the samples was that only economic standard industrial techniques should be used, ensuring that there would be no limitations on scaling to larger areas.

APPARATUS

A schematic layout of the apparatus is shown in Fig. 1. It consists of a standard dewar with an annular nitrogen jacket, the inside of which is used as the emitter surface (emissivity E_2) and which is painted with Nextel Velvet paint.* The sample under test (emissivity E_1) is suspended in vacuum by means of a thin wall stainless steel tube attached to the neck. These samples were usually 610 mm long, and either 102 mm or 73 mm, in diameter which is a fair representation of a practical system and is also of sufficient size to enable the boil-off to be measured accurately. Two radiation baffles are connected to the neck, the lower one being anchored at 77 K. In the case of stainless steel samples, a copper sleeve was inserted in the sample to minimize any axial temperature gradient as the liquid level dropped. The gas boil-off was monitored for the LN_2 case with a Gapmeter[+] and for the helium case with either a Gapmeter or Mass Flowmeter[‡], depending on the flow rate. The temperature of the emitter during the tests from ambient temperature to 77 K was monitored with a digital thermometer. As a check on the experimental accuracy of the evaporation technique, a heater was installed inside one of the samples and the linearity between boil-off and power was confirmed.

ANALYSIS AND ERRORS

Radiation Exchange

The evaporation loss method yields a measurement of the effective emissivity Em given by

$$Q = \sigma \, Em \, A_{\perp} \left[T_2^{\,4} - T_1^{\,4} \right] \tag{1}$$

---

*Nextel Velvet Coating 101-C-10, 3M Co., Decorative Products Division.

+Fisher Controls Series 1100, Croydon, England.

‡Hastings Flowmeter EALL 500P, Teledyne Hastings-Raydist, Hampton, Virginia.

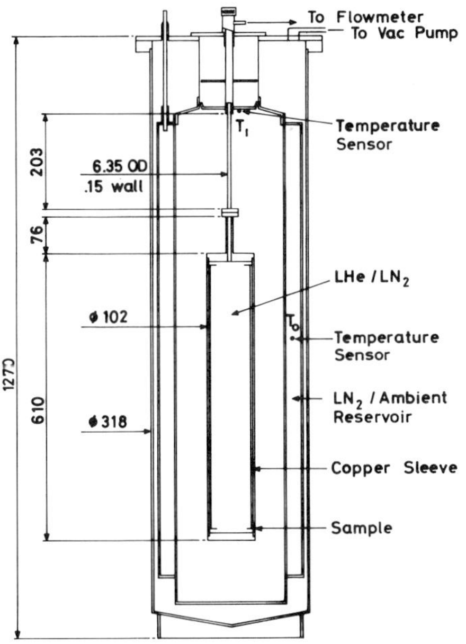

Fig. 1.   Schematic Layout of apparatus.

and the true emissivity of the sample surface (E_1) can be obtained
from Em using the expression

$$E_1 = \frac{Em}{1 - \dfrac{A_1}{A_2}\left[\dfrac{Em}{E_2} - Em\right]} , \qquad (2)$$

where Q is the heat transfer determined from the boil-off rate, E_1
E_2 the emissivities, A_1 A_2 the areas, T_2 T_1 the temperatures of
the sample and emitter respectively, and σ the Stefan Boltzmann
constant.   When E_2 is unity E_1 = Em, but for values less than
unity depends on both Em and E_2, the magnitude of which is shown
in graphical form in Fig. 2. For values of Em $\geqslant$ 0.1 the error is
$\leqslant$ 1% for E_2 = 0.8 which was taken for the calculation for the 77 K
to 4.2 K experimental results.   For the ambient to 77 K case E_2
was taken as 0.98.   It is only at the higher emissivity values
that the error needs to be taken into account.

End Effects

Equation 1 applies strictly only to long cylinders.   For our

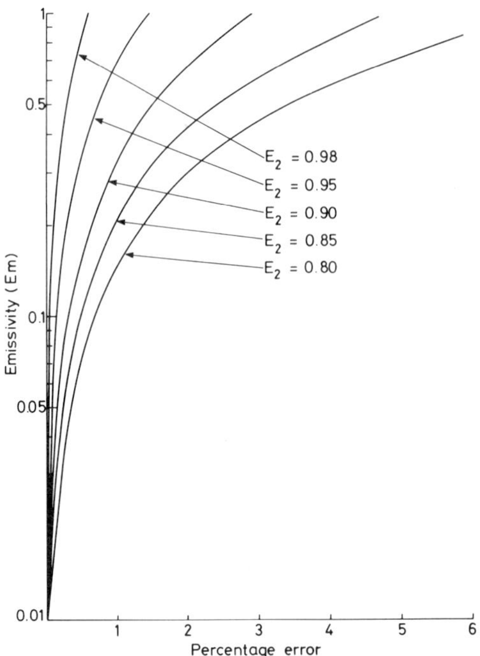

Fig. 2. Relationship between emissivity and error due to E_2.

case, deviation from ideal geometry was calculated to give a correction of less than 1% and was ignored in our calculation of E_1. As a check on this we measured a sample of the same diameter but half the length of standard samples. The results were within 5% of those obtained with a standard sample having the same surface preparation.

Neck Losses

The calculated heat input from conduction is of the order of 8% and $\leqslant 0.1\%$ at $E_1 = 0.01$ and 0.9 respectively and only becomes an appreciable fraction of the total heat input at low emissivity values. The anchoring of the neck at 77 K was confirmed by a temperature sensor (see Fig. 1).

Flow Measurement

The errors caused by the Gapmeters are not a fixed percentage but ± 3% of indicated flow ± 0.3% of full scale deflection. The Flowmeter has a quoted accuracy of ± 1%. In practice it was found that the flow was never perfectly stable and an extra percentage error has been included for this effect.

MATERIALS AND SURFACE FINISH

The choice of materials was dictated by thermal, vacuum, and manufacturing considerations and was limited to Stainless Steel (St. St.), Copper (Cu), and Aluminium (Al) substrates with various surface finishes. For the surface finishes only those which are suitable for large scale application by standard economical industrial techniques have been studied. For most of the materials the following surface finishes were investigated: material as found, mechanically polished (Mech. Pol.) and electropolished (Elect. Pol.).

Additional surfaces studied include: shot blasted St. St., silver plated St. St., aluminized Mylar and aluminium foil. The effect of a thin oxide layer ($\sim$ 2 μm) on polished aluminium surfaces, a standard industrial technique for protection of metallic mirror surfaces, was also studied.

For the high emissivity surfaces (required for the LN_2 radiation shields) standard paints and chemical oxidization processes have been studied with particular interest in a specially developed black anodizing procedure, which produces comparatively thick oxide layers ($\leqslant$ 40 μm). The effect of variations in the porous structure of the layer and the initial basic surface roughness of the substrate were investigated.

For all measurements precautions were taken to ensure that grease and oil contaminants were not present at a level to affect the results.

RESULTS AND DISCUSSION

Measurements at Ambient Temperature

To obtain comparison with infrared reflectometer measurements and information on the effect of sample temperature a preliminary experiment was carried out at ambient temperature where both the annular nitrogen container and sample were filled with water with a temperature difference of $\sim$ 20-30° C. By measuring the ΔT of the sample water as a function of time it is then possible to calculate a value for the emissivity. This technique is only applicable to the high emissivity samples, in that a measurable ΔT can be obtained within a reasonable time. Some of the results are shown in Fig. 3. These results are in good agreement with data[1] for infra-red reflectometer measurements made at ambient temperature.

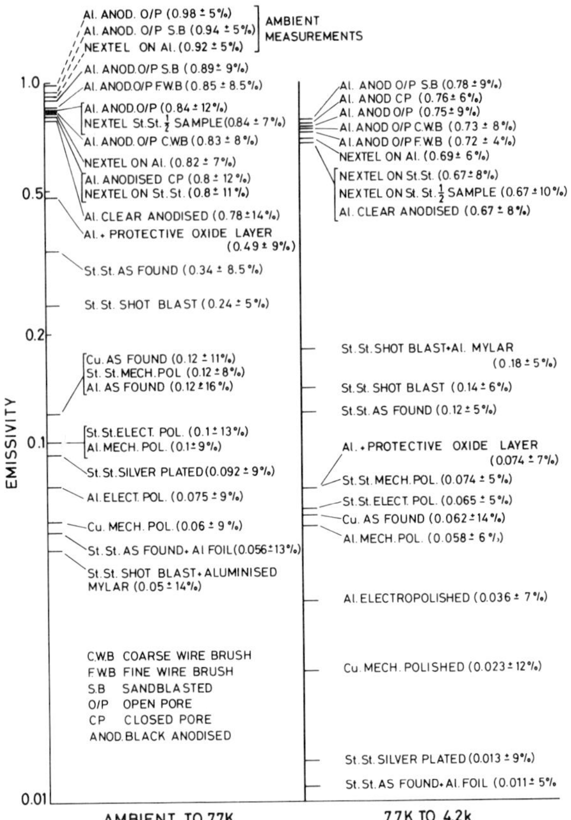

Fig. 3. Graphical representation of results.

Measurements at Low Temperature

The results are shown in a graphical representation form, for ease of comparison, in Fig. 3.

For the following discussion we will confine the remarks to results relevant to large-scale cryopump design.

High emissivity (black) surfaces. The results, show the highest emissivity for the black anodized, open pore, sandblasted aluminium surface. The value is even higher than measured for black paint. The measurements also indicate the importance of the preparation of the black anodized samples. It should be noted that standard black anodizing is normally not applicable for the absorption of 300 K (10 μm) radiation because it becomes reflective at a wavelength greater than several μm.[2] The measurements clearly show that this type of black anodizing is suitable for cryopump applications and that it is possible to take advantage of the various benefits of this technique with regard to low cost, low residual gas constituents, bakeability, tritium compatibility, and mechanical stability.

It is important to note also the temperture dependence of the sample surface emissivity. Both black paints and black anodizing show, for the same radiation source, a clear decrease in emissivity when the sample temperature is lowered from 300 K to 77 K. This effect must be taken into account in any design where black surfaces such as we have studied are operated at cryogenic temperatures. M. M. Fulk and M. M Reynolds[3] observed a similar effect for low emissivity polished surfaces but such an effect does not seem to have been mentioned in publications for black surfaces.

<u>Low emissivity (polished) surfaces.</u> In some cases the measured emissivities of the polished surfaces were found to be up to an order of magnitude higher than published data. This may be caused by the fact that, rather than small highly sophisticated laboratory surfaces, quite large standard industrially produced surface samples have been studied. Silver plated surfaces give, as is known,[4] the lowest emissivity values for a nude surface which can be used for pumping. The results for the much cheaper polished aluminium surfaces show an emissivity which is ~ 3.0 times higher but this value is still acceptable with the general design requirements for large scale cryopump systems.

It is worth noting that, for a surface not used for pumping, i.e., no main gas condensate, the cheaper combination of stainless steel covered with two layers of aluminium foil gives even lower values than the silver coated surface. The use of aluminized Mylar foil as a cover was found to improve the reflectivity for the 300 K to 77 K case by a factor of five; however for 77 K radiation to a 4.2 K surface the emissivity was found to deteriorate drastically and to be even higher than for St. St. as found.

The protective oxide layer showed an unacceptable emissivity and we therefore do not recommend this for cryopump design.

CONCLUSION

The lowest value of emissivity was found for the silver plated steel sample; however the more economical polished aluminium was found to be acceptable with the design requirements for the large scale JET cryopump system[5]. The emissivity measurements for aluminium, with various surface finishes, show that it is suitable and most ecomonical for both the low emissivity (polished) and high emissivity (black anodized) surfaces. This, together with its manufacturing advantages (e.g., the ease of producing extrusions with integrated cooling channels, thermal properties and high vacuum compatibility) makes aluminium an excellent choice of material for large scale cryopumps.

The black surfaces showed a clear drop in emissivity with decreasing sample temperature and a specially developed black anodizing procedure was found to be superior to the standard organic paint usually used. In the final results absolute limits of error have been assigned, but it should be noted that, in general, relative values could be reproduced within much closer limits, of the order of 3%.

REFERENCES

1. W. Obert and J. Schmid, Production of Black Surfaces and Determination of their Emissivity for Infrared Radiation, JET Report CSN /C (80) 20.
2. S. Isobe, et al., in "Proc. of the 8th Symposium on Engineering Problems of Fusion Research", Vol. I, San Francisco, California, (1979).
3. M.M. Fulk and M.M Reynolds, J. Appl. Phys. 28:1464 (1957).
4. C. Benvenuti, J. Vac. Sci. Technol. 11 (1974).
5. W. Obert, P.H. Rebut, and G. Duesing, Cryopumps for JET – Design of a Largescale System of Open Configuration for High Specific Pumping Speed, in "Proc. of the 9th Symposium on Engineering Problems of Fusion Research", Chicago, Illinois, (1981).

A COMPUTER MODEL FOR TRANSIENT HEAT TRANSFER TO LIQUID HELIUM

D. S. Holmes

Aero Propulsion Laboratory
Wright Patterson Air Force Base, Ohio

and

A. R. Menard

Saginaw Valley State College
University Center, Michigan

INTRODUCTION

Transient boiling heat transfer rates can greatly exceed steady state nucleate boiling critical heat fluxes for short periods of time. This enhanced heat transfer rate might be useful for cooling superconducting devices subject to intense heat inputs over a short time span. Efficient cryostability analysis relies on an ability to model accurately the transient heat transfer characteristics of liquid helium since the use of steady state values could result in an overly conservative design.

The approach used here was to develop a computer model for one-dimensional heat transfer from a flat surface in contact with a pool of liquid helium at saturated conditions. The intended application was to study distributed heat inputs to a superconductor winding such as might be experienced during the fast discharge of an energy storage coil. This allowed heat transfer along the conductor to be neglected. The most critical aspect of transient heat transfer analysis is the prediction of the transition between the nucleate and film boiling regimes. The large temperature differences which characterize film boiling will raise the superconductor above its critical temperature, quenching the magnet and causing further joule heating.

MODELING

The analogy between electric circuits and heat transfer is used so that the sophisticated computer codes that have been developed to simulate electric circuit transient response can be used to simulate transient heat transfer. In this analogy (shown in Fig. 1) electric resistance corresponds to the inverse of thermal conductivity, capacitance is the analogy of specific heat, current is equivalent to heat flux and voltage corresponds to temperature. SCEPTRE, a powerful computer code capable of handling properties which vary with temperature, was used in this study.

A lumped parameter, thermal transmission line model is used to describe the surface. The elements making up the surface include thermal resistances and capacitances whose values are temperature dependent, and heat sources to model eddy current or joule heating. A Kapitza resistance is added at the surface to liquid helium interface. The Kapitza boundary resistance is assumed to follow the form

$$R_K = 1/h_K A = c_K T_s^{-3} A^{-1} \left[1 - \frac{3}{2}(\Delta T/T_s) + (\Delta T/T_s)^2 - \frac{1}{4}(\Delta T/T_s)^3 \right]^{-1} \quad (1)$$

in which h_k refers to a specific surface.

The liquid helium is modeled by dividing it into two parts: at thin boundary layer and the bulk bath. Two dependent heat sources are used, one to control heat transfer from the surface to the boundary layer and the other to control the heat transfer from the boundary layer to the bath. The heat flux which passes through these two heat sources is dependent upon the instantaneous energy in the boundary layer (E_{BL}), the instantaneous temperature of the helium directly adjacent to the surface (T_ℓ), and the steady state heat transfer characteristics of the surface. Time and heat flux from the surface are avoided as independent variables. The temperature of the helium at the surface is given by a very small thermal capacitance (C_{BL}) which acts as an ideal thermometer. Since the heat transfer was defined in terms of the boundary layer energy and not the average temperature of the boundary layer, an arbitrary value of specific heat could be used and the energy defined as

$$E_{BL} = T_{BL} C_{BL} \quad (2)$$

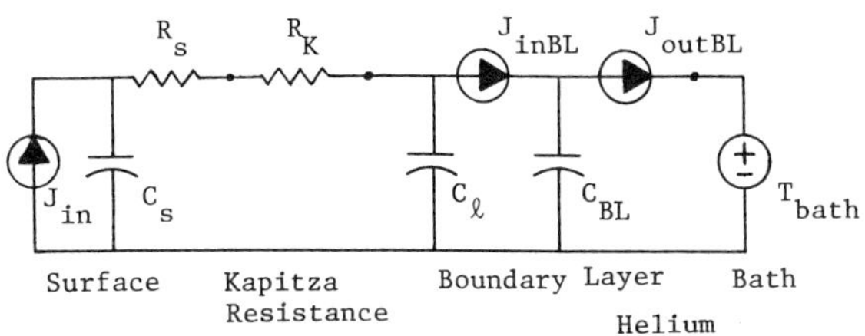

Fig. 1. Electrical circuit model of liquid helium
in contact with a heated surface.

thus eliminating the need to integrate the specific heat over a
complex function of temperature to determine the energy in the
boundary layer.

The heat transfer out of the boundary layer is assumed to be
a function of only the instantaneous energy in the boundary layer. A typical plot of heat flux out of the boundary layer vs.
boundary layer energy is presented in Fig. 2. The energy in the
boundary layer as a function of steady state heat flux was difficult to estimate since little data exists which can be put in this
form. The energy in nucleate and transition boiling boundary
layers was estimated to vary linearly with $T_\ell - T_{bath}$. The energy
in a film boiling boundary layer was calculated by using a slightly modified form of the equation given by Iwasa and Apgar[1] for
vapor layer thickness,

$$\delta_{FB} = 5\theta_\ell/L + .53\theta_\ell^3/3L \tag{3}$$

and assuming a linear temperature profile between T_ℓ at the surface and T_{sat} at $x=\delta_{FB}$.

DETERMINATION OF HEAT TRANSFER MODES AND HEAT FLUXES

Conduction was assumed to occur for small values of θ_ℓ and
E_{BL}. Heat transfer during conduction was calculated by assuming a
temperature profile between T_ℓ and T_{bath} of the form

$$T = (T_\ell - T_{bath}) (1 - x/\delta)^n + T_{bath} \tag{4}$$

where δ is the thermal boundary layer thickness for convective
heat transfer, which was estimated to be about 2×10^{-6} m. The
value of the exponent n in Eq. 3 is calculated by matching E_{BL}
with the energy in a boundary layer with a linear temperature
profile (n=1), assuming constant liquid properties, as follows

$$E = \int_{o}^{\delta} \int_{T_{bath}}^{T_\ell} \rho \, c_p \, dT \, dx = \rho_\ell c_p \, \delta(T_\ell - T_{bath})/(n + 1) \qquad (5)$$

When Eq. 5 is equated with E_{BL}, the exponent becomes

$$n = \left[E_{BL}/(\rho_\ell c_p \, \delta \, \theta_\ell) \right] - 1 \qquad (6)$$

Once the temperature profile is thus specified, the conduction heat transfer is easily calculated using

$$q = - k_\ell dT/dx \qquad (7)$$

Heat transfer out of the boundary layer is calculated using a value of $x/\delta = 0.9$ since the slope of the temperature profile at $x=\delta$ is either zero or infinite unless n=1.

A criterion for the beginning of nucleate boiling was derived from an equation by Han and Griffith[7] for the critical radii of nucleation sites

$$r_c = \frac{\delta \, (T_w - T_{sat})}{3(T_w - T_{bath})} \left[1 \pm \sqrt{1 - \frac{12 \, (T_w - T_{bath}) \, T_{sat} \, \sigma}{(T_w - T_{sat})^2 \, \delta \, \rho_v \, \lambda}} \right] \qquad (8)$$

The critical radii become real when the term under the square root equals or exceeds zero. For the case where $T_{bath}=T_{sat}$, the incipience criterion becomes

$$(T - T_{bath}) = 12 \, T_{bath} \sigma / (\rho_v \lambda) \qquad (9)$$

This can be related to an energy by multiplying it by the volumetric specific heat of the liquid to get

$$E_{T1} = 12 \, \rho_\ell \, c_p \, T_{bath} \, \sigma/(\rho_v \, \lambda) \qquad (10)$$

For saturated liquid at 4 K, $E_{T1} = 1.04 \times 10^{-2}$ J/m^2. A value of 2×10^{-2} predicts the incipience of nucleate boiling surprisingly well. Ordinary fluids usually require a wider range of activated cavities before nucleate boiling is observed.

For nucleate boiling heat transfer, the equation

$$q = c \, (T_\ell - T_{bath})^n \qquad (11)$$

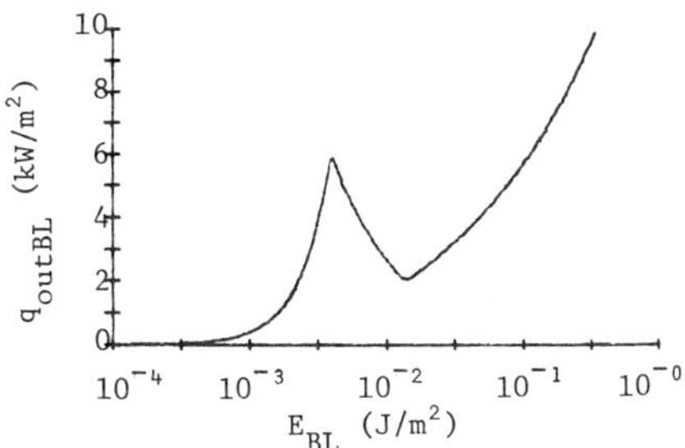

Fig. 2. Heat flux out of the liquid helium boundary
layer as a function of the energy in the boundary layer.

is used, where the values of the constants c and n are evaluated
from the steady state heat flux vs. temperature difference curve
for the surface to be modeled. The heat flux appears to be independent of the energy in the boundary layer in the nucleate boiling heat transfer regime. Evidence to this effect is given by the
very weak influence of heating or cooling rates on the position of
the boiling curve observed by other workers[10,11].

Transition to film boiling is assumed to occur when the
energy in the heated boundary layer is equal to the latent heat of
the helium within the boundary layer. For short times, the boundary layer grows at a rate given by the equation for thermal diffusion into a solid. The resulting criterion for transition can
be expressed as

$$E_{BL}(T_\ell - T_{bath})^3{}_T = \text{const} \qquad (12)$$

when the exponent n for nucleate boiling equals 2. This correlates well with the data of Steward except at very short times
when the finite rise time of the heater current makes the total
heat dissipated difficult to determine. The data of Steward was
used to determine values of $c_{T1} = 1$ and $c_{T2} = 15 \times 10^4$ J-K^3/m for
the constant at incipience and completion of the transition be-
tween nucleate and film boiling.

Transition boiling was modeled as a combination of nucleate
and film boiling heat transfer. A good discussion of the appearance of dry spots on the surface prior to reaching the peak heat
flux is given by Yu and Mesler[12]. The dry patches decrease the
heat transfer possible with pure nucleate boiling until dryout
occurs. The fraction of nucleate boiling contributing to the heat
transfer was calculated from

$$f = (1 - \Delta T^*)^{10} \tag{13}$$

where

$$\Delta T^* = \left((E_{BL}\theta_\ell^3)^{.25} - (c_{T1})^{.25} \right) \bigg/ \left((c_{T2})^{.25} - (c_{T1})^{.25} \right) \tag{14}$$

which is a modified version of the relation given by Kalinin, et al[13]. Transition boiling is assumed to occur between departure from nucleate boiling and departure from film boiling. The peak nucleate and minimum film boiling points occur within the transition boiling regime. The dimensionless temperature difference was stated in terms of the $1/4$ power of $E_{BL}\theta_\ell^3$ and the transition conditions since the critical temperature differences are not constants in transient heating. If E_{BL} is proportional to θ_ℓ, the $1/4$ power of Eq. 12 should be proportional to θ_ℓ.

The equation used for heat transfer in film boiling was an empirical fit to the correlation of Breen and Westwater[14], evaluated at P=1 atm and modified by a function of the steady state and instantaneous boundary layer energies as follows

$$q_{FB} = \left(200\ \theta_\ell^{1.2} \right) \left[(E_{FBss} + E_{BL})/2E_{BL} \right] \tag{15}$$

to model transient film boiling.

MODEL RESULTS AND PREDICTIONS

A lumped parameter model was developed for a thin copper surface in order to predict the transient response of this technically important surface. Peak nucleate and minimum film boiling heat fluxes of 6 and 2 kW/m^2 were chosen as representative for pool boiling in saturated liquid helium at 4 K. The value of the Kapitza constant was taken as 12×10^{-4} m^2-K^4/W, which corresponds to a Kapitza conductance of 5.7 kW/m^2-K at 1.9 K (see Snyder[15]). Square wave heat fluxes of varying magnitudes were input to the surface and the surface temperature was plotted as a function of time (see Fig. 3). Attempts to model the graphite/quartz surfaces of Steward[6] have yielded only qualitative agreement due to difficulties in defining a minimum film boiling point[10]. Attempts to model recovery failed. This failure was traced to the assumption implicit within this model that the temperature profile in the boundary layer can always be modeled as a simple curve of the form presented in Eq. 4. A simple curve is not an accurate representation of the temperature profile during periods of rapidly decreasing heat flux.

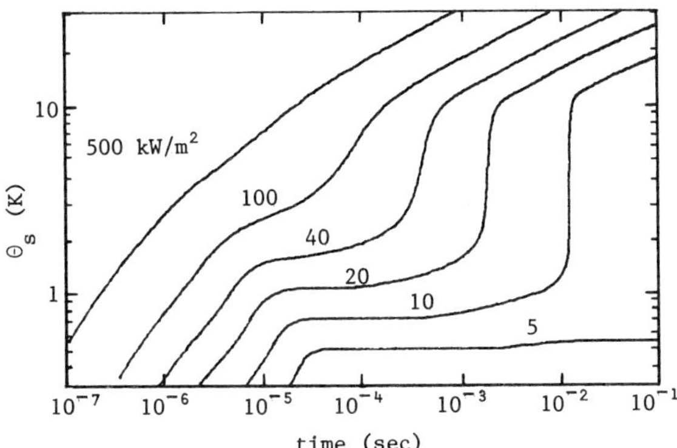

Fig. 3. Transient response of a copper surface subject to square wave heat inputs in a bath of saturated liquid helium at 4 K.

DISCUSSION OF THE TRANSITION CRITERION

The criterion developed to predict transition from nucleate boiling deserves further investigation. Equation 12 can be restated by substituting qt for E_{BL} and Eq. 11 for $(T_L - T_{bath})$ to obtain

$$t_T \, q^{(3+n)/n} = \text{const} \tag{16}$$

The value of the exponent n is usually given to be in the range of 2-3, which correlates very well with the data of other researchers[2,4] where the exponent is seen to vary between 2.3 and 2.8.

SUMMARY

The one-dimensional model for transient heat transfer to liquid is adequate for modeling distributed heat inputs to a superconductor. A criterion for the onset of transition to film boiling has been developed. An analysis of the effect of coatings on the time to reach film boiling will be reported at a later date.

NOTATION

A = surface area, m^2 **Subscripts**
C = capacitance, J/K
c = constant   bath = bulk liquid property
c_p= specific heat, J/kg-K   BL = boundary layer

E = boundary layer energy per unit area, J/m^2

h = heat transfer coefficient, $W/m^2\text{-}K$

k = thermal conductivity, $W/m\text{-}K$

L = volumetric specific heat, $J/m^2\text{-}K$

n = exponent

q = heat flux, W/m^2

R = resistance, K/W

t = time, s

T = Temperature, K

x = distance from surface, m

FB = film boiling

K = Kapitza conductance

ℓ = liquid property evaluated adjacent to the surface

sat = saturation

s = surface

ss = steady state

T = transition

ν = vapor

1 = at nucleate to transition boiling point

2 = at film to transition boiling point

δ = boundary layer thickness, m

λ = latent heat of vaporization, J/kg

ρ = density, kg/m^3

σ = surface tension, N/m

θ = related to bath temperature $(T\text{-}T_{bath})$, K

REFERENCES

1. Y. Iwasa and B.A. Apgar, Cryogenics 18:267 (1978).
2. O. Tsukamoto and S. Kobayashi, J. Appl. Phys. 46:1359 (1975).
3. D.E. Baynham, V.W. Edwards, and M.N. Wilson, IEEE Trans. on Magnetics, MAG-17(1):732 (1981).
4. C. Schmidt, IEEE Trans. on Magnetics, MAG-17(1):736 (1981).
5. Y. Iwasa, Cryogenics (19):705 (1979).
6. W.G. Steward, Int. J. Heat. Mass Trans. 21:863 (1978).
7. J. Jackson, Cryogenics 9:103 (1969).
8. D. Gentile, W. Hassenzahl, and M. Polak, J. Appl. Phys. 51(5):2758 (1980).
9. C.Y. Han and P. Griffith, Int. J. Heat Mass Trans. 8(6):887 (1965).
10. P.J. Giarratano and N.V. Frederick, in "Advances in Cryogenic Engineering, Vol. 25," Plenum Press, New York (1980), p. 455.
11. V.A. Grigoriev, et. al., in "Proc. 6th Intl. Cryo. Engr. Conf.," IPC Science and Technology Press, Guildford (1976).
12. C.L. Yu, R.B. Mesler, Int. J. Heat Mass Trans., 20:827 (1977).
13. E.K. Kalinin, I.I. Berlin, V.V. Kostyuk, and E.M. Nosova, in "Advances in Cryogenic Engineering, Vol. 21," Plenum Press, New York (1976), p. 273.
14. B.P. Breen, and J.W. Westwater, Chem. Eng. Progress, 58:67 (1962).
15. N.S. Snyder, Thermal conductance at the interface of a solid and Helium II (Kapitza conductance), NBS Technical Note 385 (1969).

HEAT TRANSFER TO BOILING HELIUM FROM MACHINED AND CHEMICALLY TREATED COPPER SURFACES

H. Ogata and W. Nakayama

Hitachi, Ltd.
Tsuchiura, Japan

INTRODUCTION

Heat transfer to boiling liquid helium plays an important role in stabilizing superconducting magnets which are cooled by a saturated liquid helium-I bath. Improvement in the heat transfer from the conductor to the liquid helium reduces the amount of stabilizing material, such as copper, and increases overall current density in the magnet.

In 1966, Cummings and Smith[1] found a marked enhancement of boiling heat transfer using a thickly-frosted surface. Butler et al.[2] reported the increase of critical heat flux by coating or partly coating the surface with a thin layer of insulating material. Krause et al.[3] increased the cooling rate of the superconductor for the LCT (Large Coil Task) by machining deep triangular grooves (~ 2 mm deep) in the surface. In JAERI (Japan Atomic Energy Research Institute), various heat transfer surfaces were tested[4], and the finely grooved surface of the special structure (Thermoexcel-C*), which was chemically oxidized, was adopted as a cooling surface of the Japanese LCT conductor[5].

However, the mechanism of heat transfer enhancement for these surfaces has not been well clarified. The purpose of this report is to investigate more closely the effect of surface conditions on boiling of liquid helium.

---

*Commercial name of the surface originally developed as a condensing surface by Hitachi, Ltd. and Hitachi Cable, Ltd.

EXPERIMENTAL METHOD

Liquid helium at atmospheric pressure was boiled from a rectangular test surface suspended in a glass dewar. The test specimen, as is shown in Fig. 1, consisted of a copper block, fitted with a carbon resistance thermometer and an electric heater and encased within thermal insulation on the sides and bottom. The projected area of the heat transfer surface was 1.5 cm × 1.5 cm.

Specimen A, a smooth surface, produced a basic boiling curve which is used to assess the performance of the structured surfaces shown in Fig. 2. The structured specimens were provided with machined grooves of about 1 mm depth on the surfaces. They were rectangular grooves (B), triangular grooves (C), crossed triangular grooves (D), trapezoidal grooves (E) and Thermoexcel-C (F).

The surfaces were rinsed with solvent, acid, and water, and dried in air just before each experiment, yielding good reproducibility of the data. After the data were taken, the surfaces of some specimens (A, B1, C, D, F) were chemically oxidized with alkali. This treatment produced an oxidized copper layer of a few microns in thickness on the surfaces.

Most specimens were tested twice by setting the orientation of grooves vertical and horizontal, since the orientation of conductors in Tokamak field coils varies from vertical to horizontal.

The heater current was increased stepwise from zero to the point where the temperature of the block reached about 30 K, and again decreased stepwise to zero.

The heat flux through the boiling surface, based on the projected area, was determined by subtracting the heat leak from

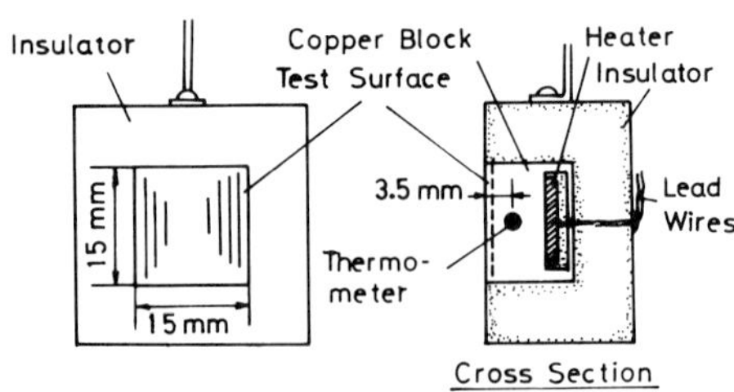

Fig. 1.   Arrangement of a test specimen.

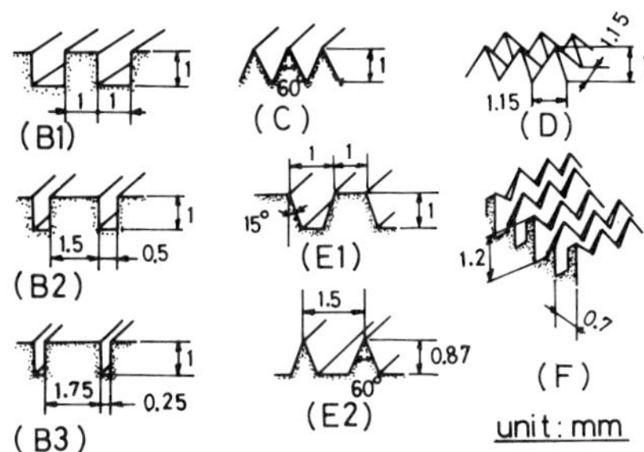

Fig. 2.   Sketch of heat transfer surface.

the electrical input to the heater. The relationship between heat leak and block temperature was found in advance by a calibration experiment, in which a thermal insulator was cemented to the boiling surface of specimen A. The estimated error in the heat flux measurement ranged from ± 30% (at the lowest heat flux of 0.03 W/cm^2) to ± 1% in the nucleate boiling regime and from ± 10% to ± 20% in the film boiling regime. The error in the temperature difference between the surface and the saturated liquid, ΔT, was within ± 10%.

Figure 3 shows the boiling on the specimen E2 which has horizontal grooves. They were taken through the transparent window of the dewar.

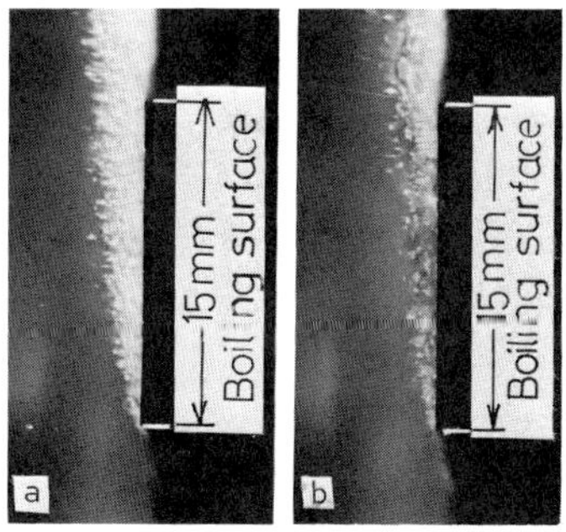

Fig. 3.   Photographs of typical boiling. (a) nucleate boiling, q = 0.25 W/cm^2 (b) film boiling, q = 0.35 W/cm^2.

EXPERIMENTAL RESULTS

Smooth Surface

The boiling curves are shown in Fig. 4 together with data reported in the literature[1,7-9], most of which were obtained with horizontal surfaces facing upward.

Machined Surfaces

The experimental results obtained with the machined surfaces are shown in Fig. 5. Only the curves from nucleate boiling data, obtained with increasing heat flux are shown. Hysteresis of nucleate boiling was observed, which was of a similar amount to that observed with the smooth surface. The effect of the orientation of the grooves was significant to the maximum nucleate boiling heat flux of and to the film boiling data. It can be seen that the surfaces B2 and B3, which have horizontal grooves, produced the curves closest to that of the smooth surface in the film boiling regime. The data from specimen F showed the highest heat flux among the surfaces tested and a weak dependence on the orientation.

Oxidized Surfaces

The results obtained with the oxidized surfaces of A, B1 and F are shown in Fig. 6, compared with the data of the original surfaces shown in Fig. 5. Significant differences were found in the nucleate boiling regime between the original and oxidized surfaces. With increasing heat flux, the boiling curves of the oxidized surfaces are located at higher ΔT by a factor of 2 ~ 4 in the nucleate boiling regime. But during decreasing heat flux ΔT

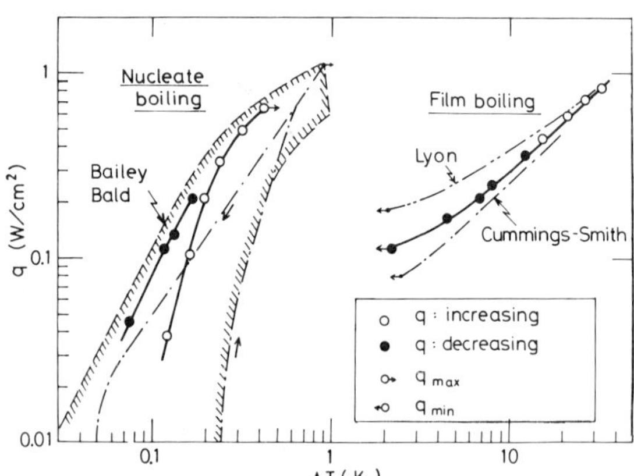

Fig. 4. Pool boiling curve for the smooth surface (specimen A).

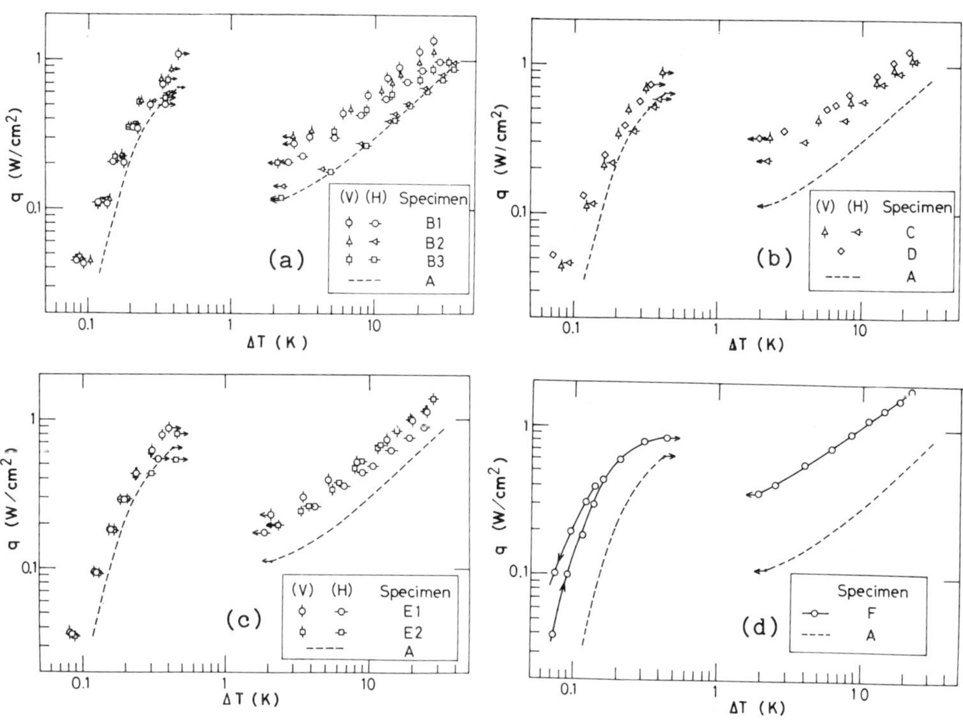

Fig. 5. Pool boiling curves for machined surfaces.
(V): vertical grooves, (H): horizontal grooves

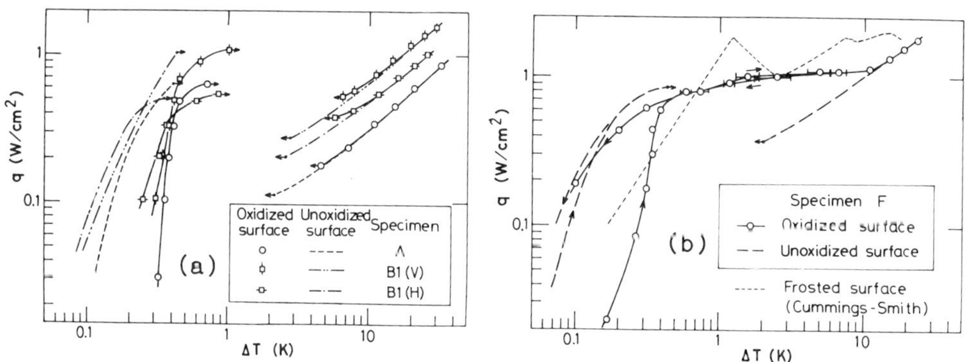

Fig. 6. Pool boiling curves for oxidized surfaces.

of the oxidized surface deviates only about 20% from that of the original surface at an equal level of heat flux. The minimum heat flux for film boiling, is also affected by oxidization. Aside from the minimum heat flux, heat transfer in the film boiling regime was not influenced by oxidization.

The oxidized specimen F showed remarkable characteristics different from other specimens; q_{min} is raised and the difference from q_{max} is reduced, the transition from nucleate boiling to film boiling was not so abrupt as was seen with other specimens. For comparison shown are the data of the thickly ($\sim$ 1 mm) frosted surface by Cummings and Smith[1].

DISCUSSION

Film Boiling

Frederking derived the following correlation of experimental data for the horizontal surface facing upward,

$$Nu_B = 0.2 \ (Ra_B)^{1/3} \tag{1}$$

The authors attempted to see whether Eq. 1 and several other correlations could be applied to the data from vertical surfaces. A good fit of the data was found if the modified Frederking equation is used;

$$Nu_L = 0.2 \ (Ra_L{*})^{1/3} \ (A/A_o) \tag{2}$$

Where, as the length scale, the height of the surface, L, is used. Properties of the fluid are those at the mean film temperature. The calculated A/A_o was 1.67 for the specimen E2, 1.77 for E1, 2.0 for B1, B2, B3, C, D and 4.0 for F. Most of the experimental data were well correlated as shown in Fig. 7; however, the data for the specimens with horizontal grooves fall below the prediction of Eq. 2. From these results, it can be concluded that the heat flux in the film boiling regime is increased in proportion to A/A_o, except for the case of the surface with narrow, horizontal rectangular grooves.

Deterioration of heat transfer in the horizontal grooves could be attributed to the stagnant vapor film covering the downward facing wall of the grooves. The crossed grooves are considered an effective means to solve the problem. Inclined faces of the triangular and trapezoidal grooves are also good for preventing the stagnation of vapor.

Deterioration of heat transfer on the specimen B3 with vertical grooves of 0.25 mm width seems to imply the overlapping

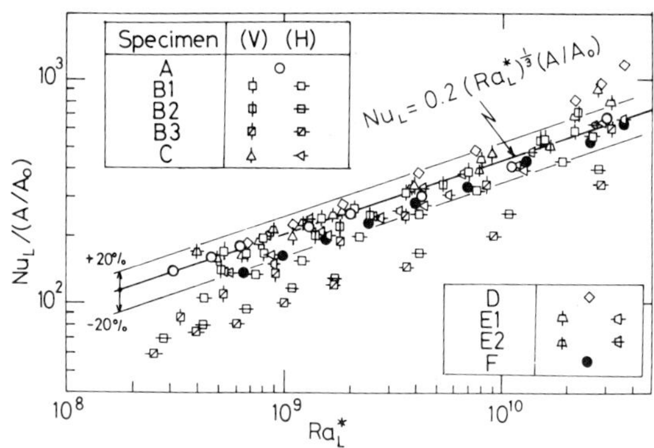

Fig. 7. Correlation of film boiling.

of vapor films developed on the groove walls. The calculated thickness of vapor film in film boiling is of the order of 0.1 mm for helium.

Nucleate boiling

Effect of roughness. The heat flux in the nucleate boiling regime for the machined surfaces increases in proportion to A/A_o at low heat fluxes, indicating that the local boiling phenomena on the surface are unchanged from that of the smooth surface. As the heat flux increases, however, the boiling curves for the machined surfaces approach that for the smooth surface. Since a layer of bubbles in nucleate boiling grows to 2-3 mm in thickness as shown in Fig. 4, the width and the depth of the grooves on the tested surfaces, (from 0.25 to 1.2 mm), are too narrow to let fresh liquid replace the bubbles.

Effect of oxidization. An oxidized layer on the surface causes a temperature drop within itself due to its low thermal conductivity. According to an approximate estimate, this temperature drop is 0.15 K at 0.5 W/cm^2. Except for a few data at heat fluxes around 0.5 W/cm^2 and those with decreasing heat flux, however, most data show the shift of the nucleate boiling curves toward larger ΔT than the sum of the above temperature drop and the superheat on the original surface. This implies that the non-porous thermally insulating layer suppresses nucleation of bubbles.

Critical heat flux. Figure 8 summarizes the characteristics of various tested surfaces. The following are concluded:

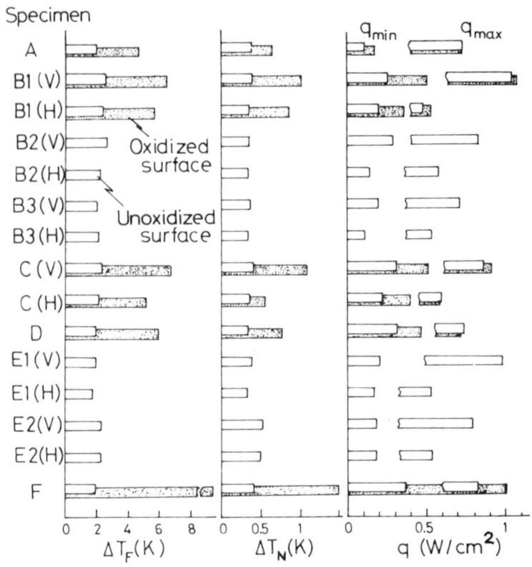

Fig. 8. Summary of ΔT_F, ΔT_N, q_{min} and q_{max}.

(1) ΔT_N and ΔT_F are found to be independent of the surface configuration and orientation.

(2) q_{max} is not much influenced by surface oxidization, but q_{min}
for the oxidized surface is raised 1.5 to 2 times compared
with the unoxidized surface.

(3) q_{max} is increased on the surfaces with vertical grooves,
crossed grooves, and Thermoexcel-C; however, it slightly
decreases on the surface with horizontal grooves.

(4) In general, an approximate linear correlation between q_{min} and
A/A_o is found for the unoxidized surfaces. q_{min} for the
oxidized surfaces increases to a far greater extent than
expected from the surface area increase.

CONCLUSION

Heat transfer, especially in film and transition boiling
regimes, is improved by methods which
(1) Increase the real surface area by means of grooves and fins.
(2) Employ surface structures that facilitate removal of vapor
bubbles. Recommended minimum width of the grooves is about
0.3 mm to avoid blanketing the grooves by vapor.
(3) Oxidation of the surface helps to improve transition boiling
heat transfer.

The oxidized Thermoexcel-C surface satisfies the above conditions giving a surface area increase of a factor 4 and the minimum
heat flux of factor 10 compared with the untreated smooth surface.

ACKNOWLEDGEMENT

The authors wish to thank Mr. H. Yoshida, Hitachi Cable, Ltd., for his cooperation in preparing the test surfaces.

NOTATION

A_o = heat transfer area
A = real surface area
B = subscript indicating that the length scale is the Laplace length
C_p = specific heat
g = acceleration of gravity
k = thermal conductivity
Nu – Nusselt number
Nu_L = $qL/k\Delta T$
q = surface heat flux
q_{max} = maximum nucleate boiling heat flux
q_{min} = minimum film boiling heat flux
Ra = Rayleigh number
Ra_L = $[L^3 g \rho_v C_p (\rho_L - \rho_v)/k\mu] [0.5 + \lambda/C_p T]$
T = temperature
ΔT = temperature difference between surface and saturated liquid
ΔT_N = ΔT at q_{max}
ΔT_F = ΔT at q_{min}
λ = latent heat
μ = viscosity
ρ_L = density of liquid
ρ_V = density of vapor

REFERENCES

1. R.D. Cummings and J.L. Smith, "Liquid helium Technology", Pergamon Press, Oxford (1966), p. 85.
2. A.P. Butler, et. al., Int. J. Heat Mass Transfer, 13:105 (1970).
3. R.P. Krause, et. al., IEEE Trans. on Magnetics, MAG-15:748 (1979).
4. M. Nishi, T. Ando and S. Shimamoto, Japan Atomic Energy Research Institute Report, JAERI-M 8771 (1980).
5. S. Shimamoto, et. al., in "Proc. 8th Symposium on Engineering Problems of Fusion Research, Vol. 3," IEEE, New York, (1979), p. 1174.
6. W. Nakayama, et. al., Hitachi Review, 24(8):329 (1975).
7. R.L. Bailey, Rutherford Laboratory Report, RL-73-089 (1973).
8. W.B. Bald and T.Y.Wang, Cryogenics, 16:314 (1976).
9. D.N. Lyon, in "International Advances in Cryogenic Engineering, Vol. 10," Plenum Press, New York, (1965), p. 371.
10. T.H.K. Frederking, A.I.Ch.E.Journal, 12(2):238 (1966).

STEADY-STATE HEAT TRANSFER TO BOILING LIQUID HELIUM IN SIMULATED COIL WINDINGS*

P. L. Walstrom

*Oak Ridge National Laboratory
Oak Ridge, Tennessee*

INTRODUCTION

Three of the six superconducting toroidal field magnets being built for testing in the Large Coil Test (LCT) Facility at Oak Ridge National Laboratory are to be cooled by pool boiling in liquid helium at 4.2 K; these include two U.S. coils, made by General Electric (GE) and General Dynamics (GD), and the Japanese coil. All three coils are designed to be cryostable. In the present investigation, a test bundle simulating the GE/LCT winding pack, instrumented with heaters and thermocouples, was tested to determine its steady-state heat transfer behavior. The results indicate that the GE/LCT magnet will recover only cold-end conduction in small regions where the conductor is nearly horizontal but elsewhere is unconditionally cryostable.

TEST BUNDLE CONFIGURATION AND TEST PROCEDURES

The GE/LCT conducting element is formed by cabling sixteen 3.4 mm square subelements around a rounded 3.5 by 35 mm copper core, soldering in a bath, and squaring in a Turk's head die to final dimensions of 42 by 10 mm. The subelements are made by inserting a squared NbTi multifilamentary composite wire into a U-shaped copper carrier. The subelements, separated by 1.8 mm wide gaps and angled at 20° from the conductor axis, form a pattern of parallel channels approximately 12 cm long.

---

*Research sponsored by the Office of Fusion Energy, U.S. Department of Energy, under contract W-7405-eng-26 with the Union Carbide Corporation.

The conductor is pancake wound with the wide face down. Turn-to-turn insulation is provided by Nomex* paper. Pancake-to-pancake insulation is provided by grooved, unperforated epoxy-fiberglass sheets. The three winding pack constituents are shown in Fig. 1. The test bundle is a stack 12 pieces high of 30.5 cm lengths of GE/LCT 8 T grade conductors separated by the Nomex paper strips. Heaters were installed in three of the conductor pieces by pulling the superconducting multifilamentary composite wire from the surrounding copper carriers and soldering stainless-steel-sheathed, MgO-insulated Nichrome wires into the resultant grooves. The stack was divided into three groups of four conductor pieces, the heated piece being the second one from the bottom in each group of four pieces. The conductors in the lower two groups were modified by drilling 1 mm diameter holes (with an average of two per groove) between subelements and through the core, where grooves on the top side coincided with grooves on the bottom side. Eight Au-0.07 at % Fe vs copper thermocouples with reference junctions in the bath were installed in each of the three heated conductor pieces (four in the subelements and four in the core between subelements). The unheated pieces immediately above and below the heated pieces were each instrumented with two thermocouples in the core.

In assembling the stack, Nomex strips, modified by punching them with 4 mm diameter holes on 13 mm centers in a rectangular pattern with 2.4 holes per centimeter of conductor, were placed between the conductor pieces in the lowermost group of four pieces. Unperforated Nomex strips were placed between pieces of the other two groups. The stack of 12 instrumented pieces was placed between two grooved pieces of G-10 epoxy-fiberglass insulating sheets with the same groove and land dimensions as the GE/LCP interpancake spacers. The stack with surrounding insulation was placed in a clamping mechanism and the aluminum bolts tightened for an estimated pressure of about 14 MPa (2000 psi) after cooldown, a typical value for the GE/LCT coil windings when energized. The clamping mechanism and the stack were placed in a pivot mechanism with an activating rod extending to room temperature which allowed continuous variation of angle from -3° to 90° from horizontal when the sample was in a liquid helium bath.

---

*Reference to a company or product name does not imply approval or recommendation of the product by Union Carbide Corporation or the U. S. Department of Energy to the exclusion of others that may be suitable

TEST RESULTS AND DISCUSSION

The heater current was supplied by an operational power supply programmed to produce 39 five second steps at equal power intervals in a staircase-up, staircase-down fashion. The heater voltage and current signals, together with the 12 thermocouple signals, were digitized and recorded by an automatic data acquisition system. Figure 2 is a computer-generated plot of data from the sample with unmodified conductor and insulation showing the temperature traces derived from four thermocouple signals by use of the NBS tables. The data were smoothed by averaging ten points at a time. The traces for the end of the sample are consistently lower than the traces for the middle, as would be expected as a result of the fact that the channels are shorter at the ends. The subelement temperature at the center of the sample is consistently greater than the core temperature. An elementary calculation shows that even if all of the heater power flux were conducted through the subelement to the core without heat transfer to the liquid from the subelement, the temperature drop through the height of the subelement could only be a few hundredths of a kelvin; the much larger observed temperature difference must be due to incomplete solder bonding of the subelement to the core. The latter conclusion is borne out by destructive examination of production specimens. The constituents of the sample, however, are sufficiently well coupled thermally to exhibit breakaway (defined here as an abrupt change in slope at the lower end of the film boiling portion of the heat transfer curve) at the same power level; therefore, the breakaway flux data can be applied without large errors to the magnet itself, in which heat is generated bothin the core and in the subelements in the event a normal zone is formed in the conductor.

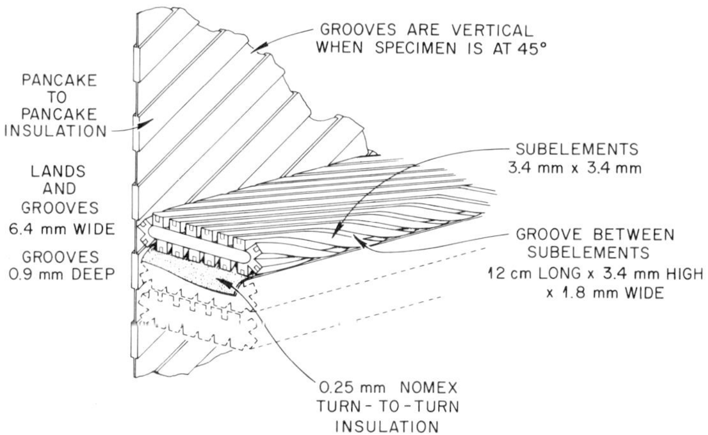

Fig. 1.   GE/LCT coil winding pack elements.

The temperature traces in Fig. 2 exhibit, in addition to hysteresis (i.e., different temperatures for increasing and decreasing power levels at the same flux), transient behavior on a long time scale. At nucleate boiling flux levels the sample reaches fairly stable temperatures (fluctuations of less than $\pm$ 0.05 K) in less than a second; at the higher power levels, stable temperatures are not reached in 5 s for some cases. The time scale for this behavior must be related to the time required to set up stable, thermally driven two-phase flow patterns because relevant conduction time constants in solid constituents are an order of magnitude less. For the unmodified specimen, at 5 and $10°$, an oscillation with a period of about 1 s was observed at a power level near the breakaway value.

Figures 3 and 4 are plots of the heat flux vs the stabilized temperature rise measured for subelements at the middle of the heated samples at various angles of the conductor axis from horizontal. Two scales on the ordinate are shown: (1) the usual power/wetted area scale, with an assumed wetted perimeter of 13.8 cm, and (2) the power per unit cell volume, where the unit cell, a rectangular parallelepiped, here has a height equal to the conduc-

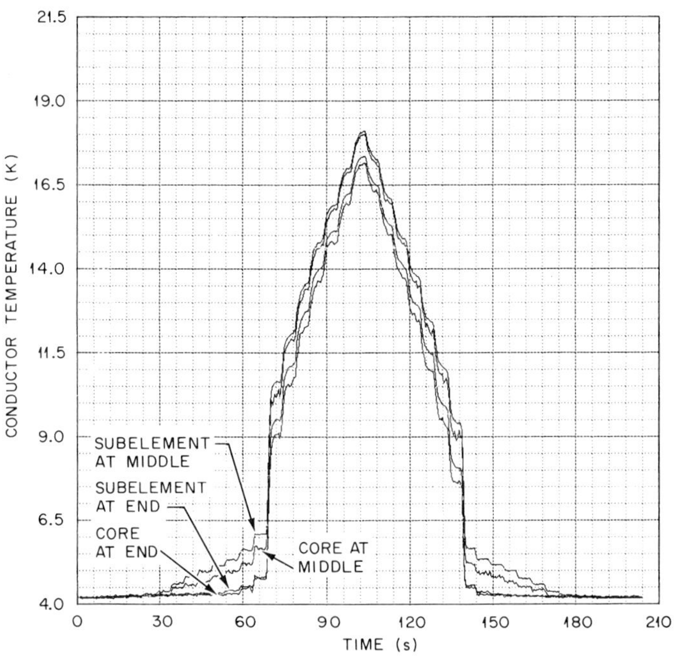

Fig. 2. Computer-generated plot of temperature vs time for four locations on the unmodified specimen at $0°$.

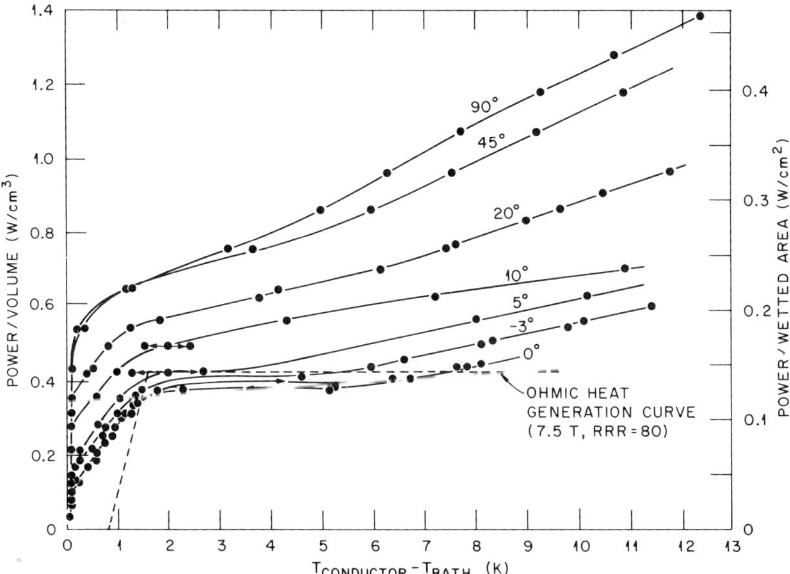

Fig. 3. Curves of steady-state heating power density
vs temperature rise above the bath temperature for the
unmodified conductor sample at various angles from
horizontal. The points at the ends of the double-
headed arrows represent the temperature extremes of
the oscillations observed at the angles and power
levels indicated.

tor height plus the thickness of the turn-to-turn insulation and a
width equal to the conductor width plus the thickness of the
interpancake insulation. The latter scale is useful in comparing
the effectiveness of different conductor/cooling channel designs
based on steady-state heat transfer because it does not depend on
inherently arbitrary definitions of wetted perimeter and because
it exhibits the tradeoff between wetted perimeter and heat transfer degradation due to channel restrictions. Significant hysteresis is observed only for the unmodified (top) heated sample at
$0°$ and for the middle (holes in core but not in Nomex) sample at
$0°$ and $-3°$. (Data were taken at $-3°$ as a check on the angular
accuracy of the pivot mechanism near $0°$.) Comparison of the data
demonstrates that addition of the holes to the core alone improves
the $0°$ heat transfer only slightly (curves not shown) while the
addition of holes to the turn-to-turn insulation, which allows
vertical flow, increases the breakaway flux significantly - about
35% (see Fig. 4). The latter modification to the GE/LCT coil
design was not made because of concern about turn-to-turn shorts;

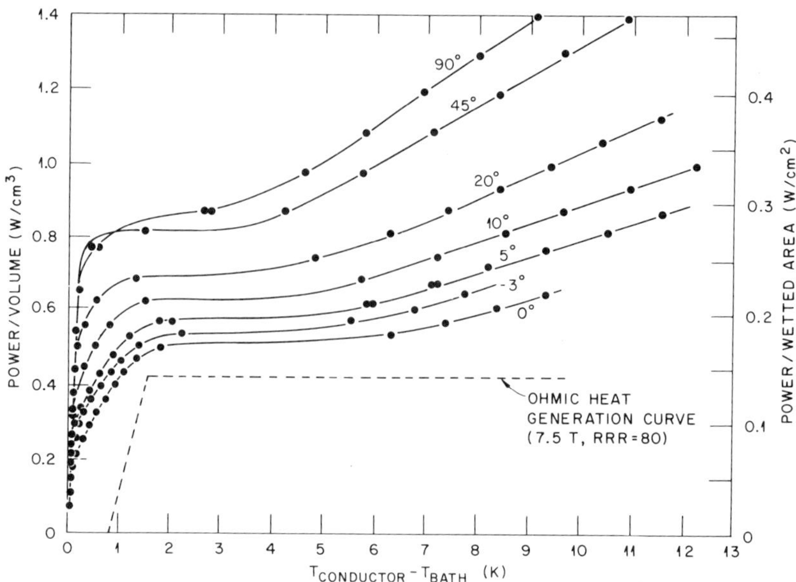

Fig. 4. Curves of steady-state heating power density
vs temperature rise above the bath temperature for a
conductor sample with holes drilled in the core and
with perforated turn-to-turn insulation.

such a modification would be feasible if a thicker turn-to-turn
insulation were to be used.

In an attempt to determine the effect of the grooved G-10
interpancake insulation on the steady-state heat transfer, an
additional set of curves was obtained for which the interpancake
insulation and side clamping plates were removed, allowing unrestricted flow from the channels between the subelements to the
bath. For comparison, data for two angles (0° and 20°) of the top
(no holes) sample are shown in Fig. 5 along with the data for the
same sample with the interpancake insulation in place. Only a
very slight improvement in heat transfer is observed for the 0°
data after removal of the interpancake insulation, but the 20°
curve with the interpancake insulation removed lies considerably
above the curve from the data for which the insulation was in
place. Behavior similar to the 20° curve was observed for the
other angles away from horizontal. The data demonstrate the
limiting effect on the heat transfer of the long horizontal heated
channels.

Table I is a comparison of the data from the present test
with data from similar tests performed on test bundles simulating

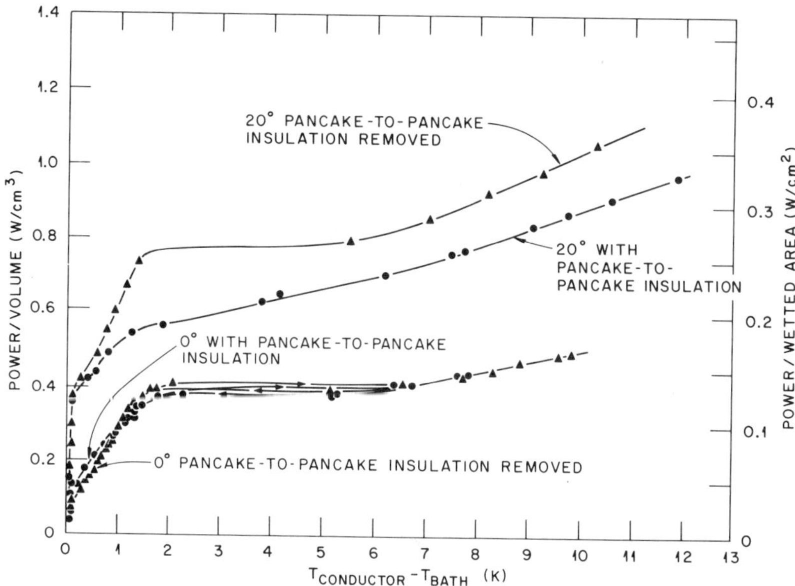

Fig. 5. Comparison of steady-state heat transfer data for an unmodified sample at 0° and 20° from horizontal with and without interpancake insulation.

the windings of the other pool-boiling LCT coils.[1-3] Also shown for comparison are data from a previously reported steady-state heat transfer experiment for a similar conductor[4] with channels of approximately the same length (13 cm) but with more area than the GE/LCT coil (3.1×3.1 mm vs 3.4×1.8 mm). The Japanese coil data are considerably higher than the two U.S. coils, even when power/unit cell volume, a fairer basis of comparison, rather than power/unit area is used.

The high breakaway flux obtained for the Japenese winding design can be attributed to the coated, highly developed surface and the large (8 cm × 2.9 mm) cooling channels between the pancakes. The GD/LCT coil designers also tested the effects of modifying their conductor with a chemical treatment (Ebanol C) and observed a 40% increase in heat transfer, but they did not use it for reasons of cost.

CONCLUSIONS

The present data show that the worst case steady state stability in the GE/LCT magnet windings is at a horizontal conductor orientation. The heat transfer improves with inclination of the conductor from horizontal. Calculations[5] show that for these

**Table I. Comparison of Steady-State Cryostability
Parameters for the Pool-Boiling LCT Coils**

Coil	I_{op} (kA)	Unit cell dims. (cm)	j_{av} over unit cell (kA/cm^2)	Wetted perimeter (cm)	Cu cross section (cm^2)	Cu resistivity at field (Ω-cm)	Meas. min heat flux (W/cm^2)	Ohmic heating flux (W/cm^2)	Meas. min power/unit cell vol. (W/cm^3)
GE	10.9	1.06 high 4.38 wide	2.4	13.8	3.0	5.5×10^{-8} at 7.5 T	0.13 at 0°	0.14	0.38 at 0°
GD	10.3	3.42 high 1.15 wide	2.6	12.0[a]	2.4	4.0×10^{-8} at 6.8 T	0.17 at 90°	0.15	0.52 at 90°
Japan	10.2	2.73 high 1.55 wide	2.4	5.4[b]	2.3	5.1×10^{-8} at 8 T	0.86 at 90°	0.42	1.09 at 90°
LCS		1.30 high 3.20 wide		10.0			0.18 at 0°		0.40 at 0°

[a] Includes area of fins.

[b] Derived from projected area of developed surface.

small regions normal zones will recover by cold-end conduction
from the inclined conductor on either end.

ACKNOWLEDGEMENTS

The author is indebted to P. F. Michaelson and C. L.
Linkenhoker of the General Electric Co. Energy Systems Department
for assistance in design of the experiment and for providing the
test samples, to J. P. Rudd for installing the instrumentation on
the samples, and to J. S. Goddard and L. R. Layman for programming
the data acquisition system and preparing the data reduction
software.

REFERENCES

1. E.H. Christianson, Large coil program conductor LHe heat
   transfer and heater verification test analysis, General
   Dynamics Report No. GDC 91Z0122 (1980).
2. E.H. Christianson and S. D. Peck, Pool Boiling LHe Heat
   Transfer in Vertical Conductor Packs, in "Advances in
   Cryogenic Engineering, Vol 27," Plenum Press, New York
   (1982).
3. S.Shimamoto, et. al., Japanese Design of a Test Coil for the
   Large Coil Task, in "Proc. of the 8th Symposium on
   Engineering Problems of Fusion Research," IEEE, New York
   (1979).
4. P.F. Michaelson, et. al., Heat Transfer and Helium
   Replenishment in Cabled Conductor Cooling Channels, in
   "Advances in Cryogenic Engineering, Vol. 25," Plenum Press,
   New York (1980), p. 398.
5. C.L. Linkenhoker, private communication.

POOL BOILING LIQUID HELIUM HEAT TRANSFER IN VERTICAL CONDUCTOR PACKS*

E. H. Christensen and S. D. Peck

*General Dynamics Convair Division
San Diego, California*

INTRODUCTION

In 1979 a series of pool boiling heat transfer tests were made on one-foot (0.305m) long sections of a bundle of 25 General Dynamics Convair Division Large Coil program (LCP) conductors with intervening insulation and helium passages at varying orientations.[1] Three different insulation systems were designed and tested for this vertically standing, D-shaped Tokamak fusion field coil. Of these, the only acceptable insulation to provide sufficient ventilation and steady-state heat transfer for unconditional cryostability in this layer-wound configuration consisted of slat-like "snowfence" insulation between layers, with slats orthogonal to the conductors and "paperdoll" insulation between turns consisting of bearing pads on each side of a thin, web-like carrier. This insulation provides for helium flow in the three orthogonal directions around and along the conductor.

Early in the program, suggested by the above tests, concern arose that heat transfer might be degraded upward in vertical channels parallel to the conductor due to accumulation of energy in the helium as it rises along the conductor and also near walls where at least one transverse direction of helium ventilation is obstructed. To further confirm and quantify these effects, a 1.07m high, 3 x 5 bundle of full size LCP Grade I conductors with intervening insulation was constructed. The bundle was bounded on two adjacent sides by solid insulation surfaces simulating the

---

*Work supported by the U.S. Department of Energy under Contract No. 22X-31746C with the Union Carbide Corporation.

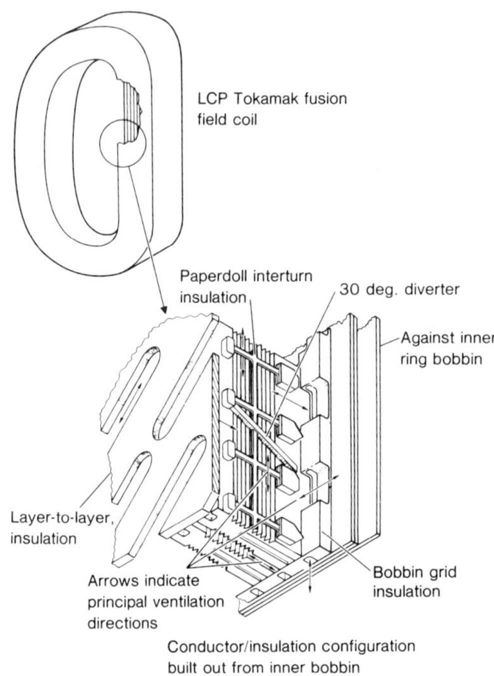

Fig. 1. "Snowfence" interlayer and "paperdoll" interturn insulations combine to provide liberal three dimensional ventilation.

boundaries of the inboard corner of the magnet in the vertical leg. Figure 1 shows the conductor and winding pack insulations along with the unique grid insulation against the inner bobbin ring which provides added helium flow and volume to the highest field inner layer of the magnet. The first insulation tested had layer-to-layer insulation with the slots normal to the conductor and only "paperdoll" insulation between turns without the angled flow diverters. The 45° angled slotted layer insulation and the 30° angled diverters were added later to channel flow upward and away from the inner ring and sidewall to improve the heat transfer.

TEST PROCEDURE

The 1.07 m high test package shown in Fig. 2 was the maximum height that could be run in the General Dynamics LHe heat transfer test dewar. It was anticipated that with the original 0.305 m specimen data and these data the results on heat transfer could be extrapolated at least to the 2.5 m high, near-vertical conductors

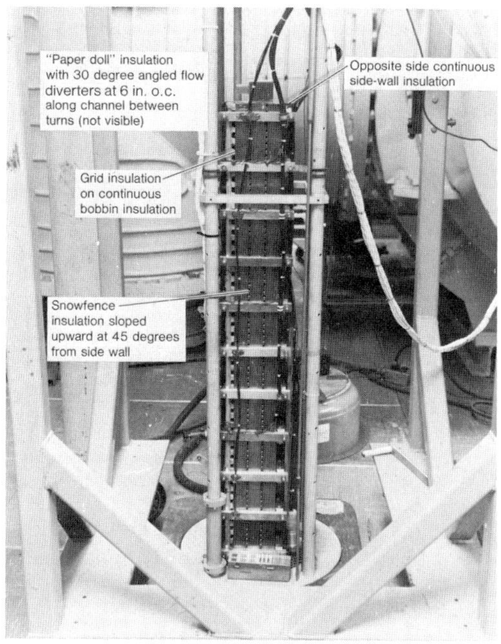

Fig. 2. 1.07 m high, 3 layer x 5 turn conductor package installed in a dewar insert.

of the LCP magnet. The 3-layer by 5-turn experimental conductor array (Fig.3) contained four heated and temperature-instrumented test conductors and was assembled in a fixture simulating the high-field corner of our LCP magnet design. Two conductors (A and B in Fig. 3) contained heaters fabricated of nichrome foil in printed-circuit fashion laminated between 0.076 mm Kapton films and then installed by bonding into the center plane of the conductors (Fig. 4). Each heater (Heater A was made from three, longitudinally equal sections) had a total maximum power capability of 3 000 W. The other two heated conductors (C and D in Fig. 3) contained superconducting cable and LCP prototype half turn heaters fabricated from a hairpin loop of 23-gage 0.56 mm nichrome wire insulated with 0.051 mm Kapton. The heater loop was contained in a small recess slot at the edge of the superconducting cable as shown in Fig. 4. The heater loop in Conductor D was attached by a soldering method with Woods metal. The heater was installed in Conductor C by first forming indium foil into the bottom of the slot and then laying the hairpin heater into the slot formed by the indium foil. The foil was then folded over the wire and

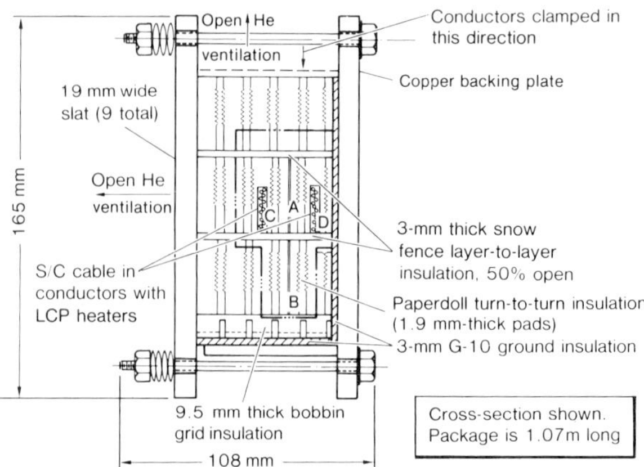

A, Three-segment heat transfer heater conductor;

B, Single, full-length heat transfer heater; C&D conductors have soldered
cable & LCP ½-turn heater; 11 remaining conductors solid copper.

Fig. 3. Test package simulates coil pack
and ventilation paths near inboard corner
of LCP magnet.

pressed into place, ultimately cold-welding the indium foil to the
solder-covered superconductor cable.

All heaters could be run at steady-state levels from zero to
full power or could be pulsed at constant power levels up to full
power for pulse durations down to 5 ms. The power and control
circuitry, instrumentation, data handling and reduction, and
analysis and results of transient performance of the prototype
half-turn heaters have been described in an earlier paper.[2] The
four heated conductors were instrumented with gold plus 0.07% iron
versus copper thermocouples. Conductor A, with the three-segment
heater, had 10 thermocouples arrayed at 6 stations along the
length of the conductor with the end stations 7.6 cm (3 in.) from
each end. Conductors B, C, and D had thermocouples at 23 cm (9
in.) from the lower end and 7.6 cm (3 in.) from the upper end.
The distribution of sensors permitted assessment of the degree of
uniformity of heat flux to the helium along the length of the
vertical conductors.

Figure 5 shows the package and conductor bars before assembly. The final version of insulation for the vertical high-field
conductors is shown with turn-to-turn diverters and angled slot
layer insulation to channel flow upward and away from the inner
ring grid insulation and sidewall and to induce cross-flow on the
conductors.

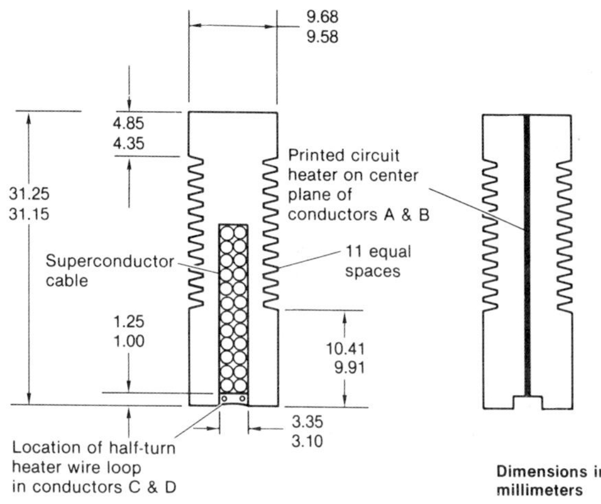

Fig. 4. Grade I conductor with superconducting
cable and modifications for heaters.

Steady-state heat transfer was measured by stepping the power
to the heaters, measured by voltage times current, waiting 3 to 5
s for the conductor temperature to stabilize, and recording all
temperatures and heater power. Heat flux at each step is the
total power of the heater divided by total wetted area of the
conductor.

Transient heat transfer was measured by setting the heater
power supply at power levels up to full power, then pulsing the
heater with a square wave for a preset duration. Durations of 5
ms to 250 ms were used. The temperature response of the conductor
was recorded as a function of time. The transient temperature
response of the conductors to the various heat pulses provides the
fundamental data to determine LHe transient heat fluxes during the
heating and cooling.

A computer program called TRNRDC was developed to extract the
transient boiling heat transfer data from high-power pulse transient tests, using a calorimetric technique.[3] The program computes boiling heat flux as a function of temperature differential
between conductor surface and liquid helium from the transient
energy balance on a length of conductor:

$$q_{He}A_{He} = Q_H - Q_I \Sigma m_i C_i \frac{dT_i}{d\tau} \qquad (1)$$

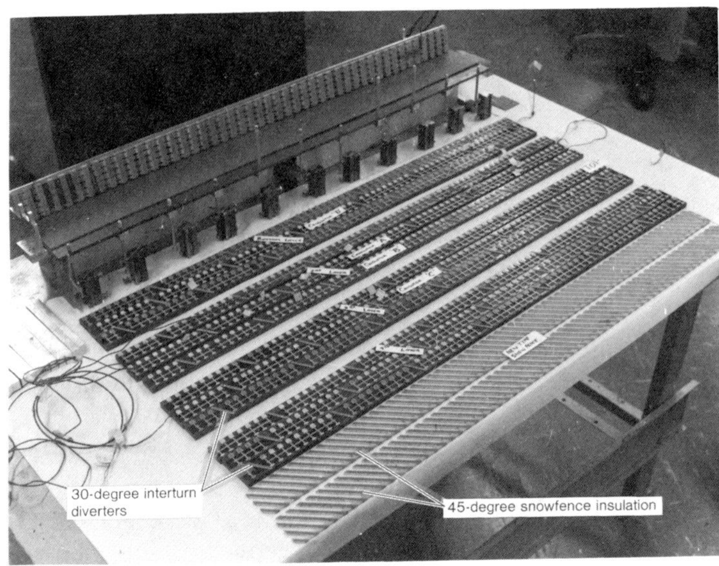

Fig. 5. The insulations were modified to induce helium flow across the vertical conductors and away from walls.

where q_{He} is boiling heat flux, A_{He} is measured wetted surface area of test conductor, Q_H is measured heater power input to the test conductor, and Q_I is a correction term accounting for the heat flow rate from the conductor into the insulation bearing pads. The mass of the ith component of the test conductor, which consists of a heater, heater plastic encapsulation material, and the metals of the conductor is m_i, C_i is the specific heat of the ith component of the test conductor and $dT_i/d\tau$ is the time rate of change of temperature of the ith component.

The quantities Q_I and $dT_i/d\tau$ cannot be measured directly. These terms are computed using information obtained from the solution of a detailed thermal model by Convair program P4560[4] which is an integral part of the data reduction process. The model is driven to the measured conductor surface temperature response and heater power input to obtain the heat flows and temperature responses of the other components of the test conductor during the transient. TRNRDC computes Q_I by summing the appropriate heat flows and numerically differentiating the component temperature responses; then, using input values for A_{He}, Q_H, m_i, and C_i, it computes q_{He} as a function of temperature difference.

TEST RESULTS AND DISCUSSION

Steady-State Heat Flux Data

The steady-state heat transfer performance of the system is directly presented in terms of LHe boiling heat flux (W /cm^2 wetted area) as a function of measured temperature differential between the conductor copper stabilizer and helium bath. Figure 6 shows the local heat flux along the length of Conductor A with the orthogonal "snowfence" insulation and clear vertical-channeled "paperdoll" insulation. The heat flux is seen to degrade with height along the bar. When the fluxes are compared on the basis of the Maddock "equal area" criterion to the joule heat flux at selected comparative locations in the higher-field regions of the vertical leg of the magnet, the conductor is not even cold-end-conduction stable over the majority of its height.

A further surprise was that unheated conductors adjacent to the heated conductor rise in temperature, particularly near the top, in response to increased heating on the energized conductor. Since this cannot be explained by conductive heating through the insulation and gets worse toward the top, it is attributed to the progressively increasing accumulation of energy in the helium as it rises parallel to the conductors of the pack. The boundary of this energy laden helium would be expected to be an inverted conical shape, probably skewed away from the side wall of the pack. The temperature increases at the top of unheated bars are as much as one-third the rise of the corresponding location on the heated bars. The authors believe the upper reaches of the channels were largely surrounded by vapor with a small amount of liquid, that the gas was considerably warmer than the small amount of liquid, and that the gas was a conduction and convection heat-transfer path from the heated conductor to the cooler adjacent unheated bars. The cooler adjacent bars tend to be cooled from their opposite side by boiling liquid helium.

Conductor/insulation configurations that channel helium laterally across and away from the heated conductor were then expected to preclude the vertical degradation. Further benefit would accrue for channeling the helium transversely away from the high field region (away from bobbin and away from the side-walls). Other data[5] suggested that cross-flow in a direction 45 to 60 degrees from the conductor axis yielded highest fluxes for this conductor. This was the reasoning which led to choice of the 45-degree snowfence and 30-degree diverters. The 30° was a compromise to minimize the elimination of "paperdoll" bearing pads. An interval of 15.2 cm for the diverters were based on the fact that Ref. 5 had shown 30.5 cm vertical channels to be acceptable but that some bypass of the diverter could occur in the fin

grooves and by expected aspiration or "blow back." This modified
insulation system was installed (Fig. 5) and tested.

The modified insulation in transversely diverting the vertical flow yielded completely uniform heat flux over the full length
of Conductor A with fluxes somewhat higher than the highest fluxes
of Fig. 6. Figure 7 shows the degree of vertical uniformity of
heat transfer on Conductor B and also the considerably higher
fluxes attributable to the greater availability and ventilation of
helium from the adjacent deep-channel bobbin grid insulation. The
conductor is unconditionally cryostable in this highest field
location of the magnet. The total data vindicates the lateral
diversion concept of the insulations in the high-field regions and
against the walls in eliminating the vertical degradation of heat
transfer.

Transient Heat Flux Data

Typical transient heat flux data are shown in Fig. 8 for a
25-ms duration, 3 000 W/1.07 m pulse which deposits an energy

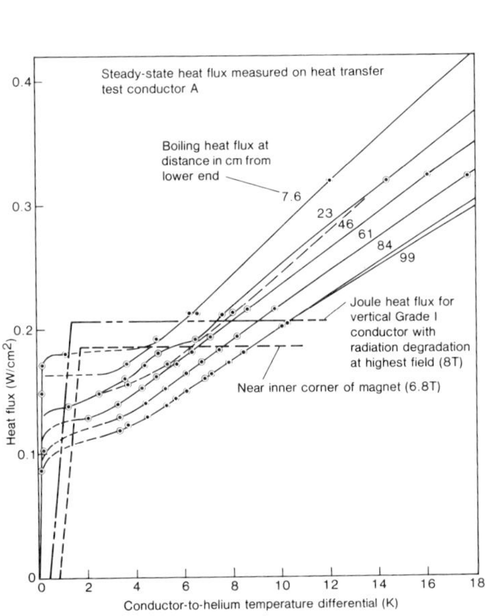

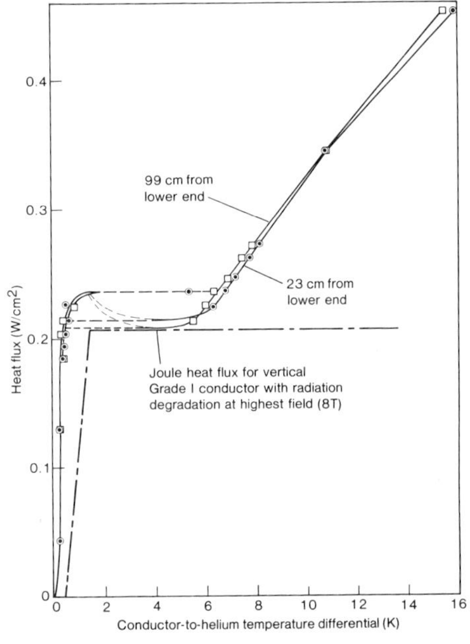

Fig. 6. Vertical Grade I conductor without angled flow
diverters experiences unaccept-
bly degraded heat flux in upper
reaches of channel.

Fig. 7. Steady-state heat flux
with angled insulation is uniform along conductor and yields
unconditional cryostability for
Conductor B on the bobbin grid
insulation.

density of about 250mJ/cm^3. The high-peak nucleate boiling flux
is evident and was found to vary directly with the power intensity
of the pulse. From a magnet cryostability standpoint, the higher
the power intensity of the fault energy pulse, the more energy
that must be deposited to drive the conductor above the nucleate
boiling regime and into normalization temperature ranges. For
short, high-power pulse intensities the heating transient flux
becomes asymptotic to the steady-state data in the film boiling
regime. For longer-duration pulses yielding energy deposition
densities exceeding 1,400 mJ/cm^3, the channels vapor lock and dry
out and heat flux drops to near zero at conductor temperature
rises of 23 to 25 K, but when the heat pulse stops and joule heat
flux levels are simulated, the channels rewet and cooling returns. During cooling or recovery the transient flux was generally not hysteretic like the steady-state flux and was generally
above the steady-state flux until the temperature dropped to 1 to
3 K differential where it then fell below the steady-state flux
(Fig. 8).

Wall and Insulation Helium Gap Effects

Freedom of ventilation of vapor laden helium away from the
conductor and replenishment of liquid to the conductor is known to
affect the total boiling heat flux and, of greatest importance to
unconditional cryostability, the minimum film boiling or unconditional recovery flux. Proximity of the wall was anticipated to
have a degrading influence on ventilation and heat flux but it was
expected that increased thickness of helium gaps in the insulation
next to walls would compensate for this effect. Figure 9 depicts
the local measured unconditional recovery flux at 8 T for conductors progressively removed from the wall boundaries. The heat
flux at the sidewall with the intervening relatively thin paperdoll turn-to-turn insulation is degraded 30% relative to the flux
in the pack at least four conductors away from the wall. The
steady-state heat flux variation with location in the other plane
is similar from the second conductor and out into the pack.
However, the much deeper bobbin insulation helium channels have
compensated the wall effect and improved the heat flux for the
conductor at the wall by 26%. A similar benefit of thicker wall
channels is shown in the transient recovery flux data.

CONCLUSION

Insulations which allow helium to flow upward along the near-
vertical heated conductor result in progressively reduced heat
flux upward in the channel and non-cryostable conditions on the
conductors in the higher field zones. Angled insulations were
configured to induce the helium to flow laterally across the
vertical conductor and away from the heated conductor which pro-

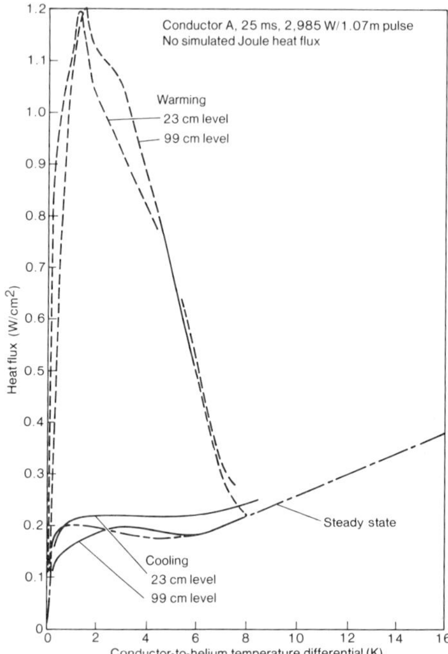

Fig. 8. Highest power pulse yields maximum transient breakaway flux during heating.

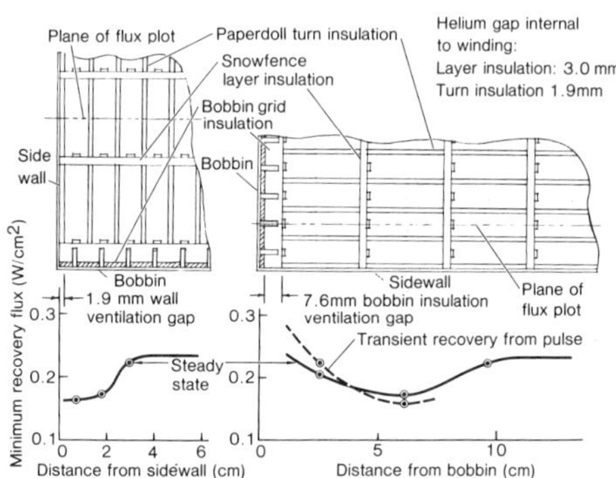

Fig. 9. Recovery flux at 8 T varies with distance from wall and helium gap in boundary insulation.

duced uniform and cryostable steady-state and transient heat flux along the high-field conductors in the vertical leg.

Solid walls adjacent to vertical conductors and their standard insulations attenuate the ventilation and flow of helium and reduce heat flux up to 30% from that which prevails on conductors well removed from the walls. The heat flux near the wall can be improved by deeper helium channels and better ventilation in the standoff insulations between the wall and conductor and around the conductors. Increasing the bobbin wall channel depths three-fold over the basic layer insulation results in flux degraded only 10% from that in the pack well away from the wall.

High-power pulse transient heat flux data is characterized by a high peak nucleate boiling flux which varies directly with the intensity of the power of the pulse and which tends to stabilize the conductor against moderate fault energy depositions. Transient cooling fluxes are generally higher and lower than corresponding steady-state flux above and below 1-3 K conductor

temperature differentials respectively. The General Dynamics LCP
conductor and laterally diverted insulation system can accommodate
pulse-energy deposition densities exceeding 1 J/cm^3 before the
channels around the vertical conductor show evidence of vapor-lock
and dry-out. The channels rewet after the intense fault-pulse
energy stops and recovery occurs at simulated joule heat flux
levels.

REFERENCES

1. E.H. Christensen, Pool Boiling Helium Heat Transfer in Typical
   Conductor Packs, in "Proc. of the Eighth Symp. on Eng.
   Probs. of Fusion Res.", IV:1769 (1979).
2. R.E. Bailey and E.H. Christensen, LCP heater thermal
   performance test results and unique test techniques, IEEE
   Trans. Magn. Mag - 17:486 (1981).
3. E.H. Christensen, et. al., Superconducting magnet
   thermodynamic design analysis programs and large
   conductor/insulation systems heat transfer test, General
   Dymanics Convair (GDC) Report No. GDC-ERR-80-075, (Dec.
   1980).
4. R.F. O'Neill and E.R. Neuharth, Convair thermal analyzer
   computer program P4560, version C, GDC Report No. GDC-
   BTD69-005A, (May 1969).
5. E.H Christensen and R.P Krause, Configuration dependent heat
   transfer data and quench pressure computer program for
   superconducting magnets, GDC Report No. GDC-ERR-79-059,
   (Dec. 1979).

DISCUSSION

Question by P. W. Eckels, Westinghouse R & D Center:　In the
transient case, do you normalize the added heat in terms of the
heat capacity of the helium in the channel to determine the vapor
fraction?

Answer by author:　The heat delivered to the helium before the
vapor lock was detected far exceeded that needed to vaporize the
liquid in the channel. Therefore, there was considerable outflow
of vapor and inflow of LHe before the channel vapor locked and
dried out.

Question by P. Seyfert, CEN-Grenoble, France:　Did you try to
compare your transient heat transfer results to published results
of other authors?

Answer by author: Yes, we got a good comparison of peak nucleate
boiling correlation with intensity of heating a rate of
temperature rise. The correlation was not possible in the film

boiling minimum regime since our transients were constrained to the calorimetric response of a real, heavy conductor whereas most other transient data has been for light, rapid response systems such as heated foils or ribbons.

BOILING HEAT TRANSFER FROM BUNDLED CONDUCTORS IN NORMAL HELIUM*

A. Khalil

University of Wisconsin
Madison, Wisconsin

INTRODUCTION

The knowledge of the boiling heat transfer characteristics on bundled conductors in liquid helium is essential for defining the cryogenic stability limits for superconducting energy storage magnets build with such conductors. Bundled conductors provide more surface area in contact with liquid helium than solid conductors of the same total cross-sectional area, and hence have higher stability limits set by

$$I^2 R = q''_{cr} S \tag{1}$$

While pool boiling heat transfer to liquid helium had been thoroughly investigated[1-4] there had been only few studies of boiling on bundles in liquid helium. Whetstone and Boom[5] investigated the nucleate cooling stability for superconductor-copper composite bundled conductors in liquid helium by heating the central wire in the bundle. They also investigated the effect of the gap between wires on the stability limits. Onishi et al[6] investigated the transient stability of a bundled superconductor cable against thermal disturbances and defined the transient stability limits for the pulsed energy and duration.

Although the total steady state nucleate stability is a conservative criteria, it provides a main guideline for magnet design. Therefore, the present study has been carried out to experimentally investigate the nucleate and film boiling charac-

---

*Supported by the U.S. Department of Energy.

teristics of horizontal conductor bundles cooled with normal helium. The effect of wire size, gap between wires and number of wires in the conductor bundle on the critical and recovery heat fluxes is also investigated.

EXPERIMENT

Samples of conductor bundles simulating a cryogenic cable cooled with pool boiling heat transfer are tested in a saturated liquid helium bath at atmospheric pressure. Conductor bundles of 1, 3, 7 and 19 wires with gaps from 0 to 0.041 cm, and with wires 0.125 or 0.317 cm in diameter and 3 cm long are mounted horizontally on an epoxy fiber glass 8 cm diameter disc. Uniform heat generation is applied to all wires by passing dc current through the current lead soldered to the ends of each bundle. Voltage taps are soldered to the ends of each bundle to measure the voltage drop across the sample.

Thin wall 304 stainless steel hypodermic tubes are used to simulate the individual cable wires because of ease of installing thin thermocouple wires through them in order to measure the inside surface temperature. The average outside surface temperature is obtained through a simple temperature correction. The thermocouples used are Chromel vs Au 0.07 at/o Fe with the hot junction electrically insulated with a thin layer of GE varnish and glued in the center of the tube. The cold junction is immersed in the saturated liquid helium bath (4.2 K).

Because of the high electrical resistivity of stainless steel and the small wall thickness of the tubes (0.015 cm) the need for high currents can be avoided. The stainless steel tubes have almost flat resistivity characteristics in the considered temperature range in contrast to copper or aluminum. Thus, overheating of the bundles can be avoided at burnout. The gaps between the individual wires (tubes) are adjusted by placing a spacer wire of specified diameter around the tubes. This spacer wire is removed after soldering the tubes to thin current lead plates. The tolerance in the size of the gap between wires is 0.002 cm.

The boiling curve of each sample is obtained by passing a steady dc current through the tube and recording the voltage drop and tube surface temperature superheat above bath temperature. The same procedure is repeated at different power levels by slowly increasing the current. Power and temperature are recorded simultaneously up to 72 K in the film boiling regime then the current is slowly reduced until recovery to the nucleate boiling regime occurs. Surface heat flux is plotted as a function of surface temperature rise in Fig. 1, 2 and 3.

The experimental error in measuring the wall temperature in the bundle does not exceed 4% below 6 K and 2% above 16 K. The end losses of the wires are calculated using the solution of the heat diffusion equation for long fins with internal heat generation[7], and it is found to be less than 1% of the total power dissipated in each wire.

RESULTS AND DISCUSSION

Table I shows a summary of the critical and recovery heat fluxes for the cable samples considered. The case of a single wire cooled with infinite helium bath represents the optimum conditions compared to different bundles. A sudden transition from nucleate to film boiling occurs at a heat flux of 0.73 W/cm^2 and is associated with a surface temperature rise from 5.3 K to 29 K, as shown in Fig. 3. As the heat flux is reduced, recovery to nucleate boiling occurs at 0.46 W/cm^2 and a surface superheat of a 0.6 K. Both the nucleate and film boiling curves are smooth with no oscillations before transitions. Figure 3 also shows the boiling curves for bundles of 3 and 7 wires with zero gap. The calculated surface heat fluxes are based on the total surface area and this might be a conservative approach because liquid helium cools the exposed surfaces only. However, it is useful for comparison purposes with other bundles. Curves 3 and 4 represent the boiling curves of the central and outer wires in the seven wire bundle and they show much lower heat transfer coefficients in both nucleate and film boiling regimes compared to the single wire. A gradual to sharp transition to film boiling is observed with some temperature oscillations before the sharp transition. The boiling characteristics of each wire in the bundle is expected to be different because of its orientation. Therefore, the outer wire

Table I. Critical and Recovery Heat Fluxes
for Different Conductor Bundles

d cm	N	$\frac{\delta}{d}$	outer wire		center wire	
			q''_{cr} W/cm^2	q''_{rec} W/cm^2	q''_{cr} W/cm^2	q''_{rec} W/cm^2
0.125	1	∞	0.731	0.465	–	–
0.317	1	∞	0.494	0.401	–	–
0.125	3	0	0.362	0.300	–	–
0.125	7	0	0.263	0.230	0.185	0.160
0.125	7	0.13	0.501	0.443	0.385	0.361
0.125	7	0.27	0.625	0.560	0.413	0.380
0.125	19	0.33	0.665	0.500	0.492	0.443

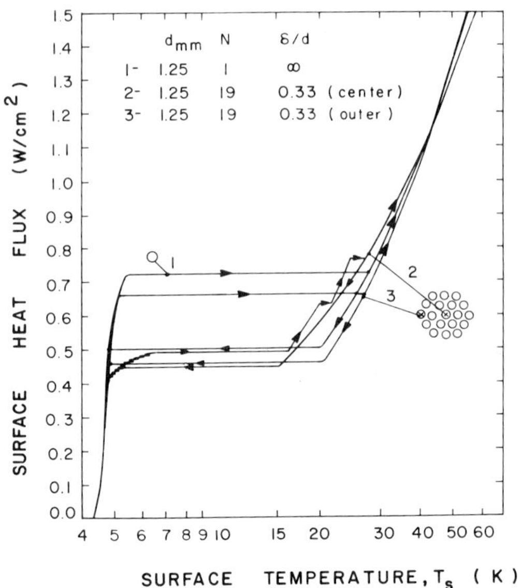

Fig. 1. Heat transfer characteristics for a 19 wire bundle compared with a single wire.

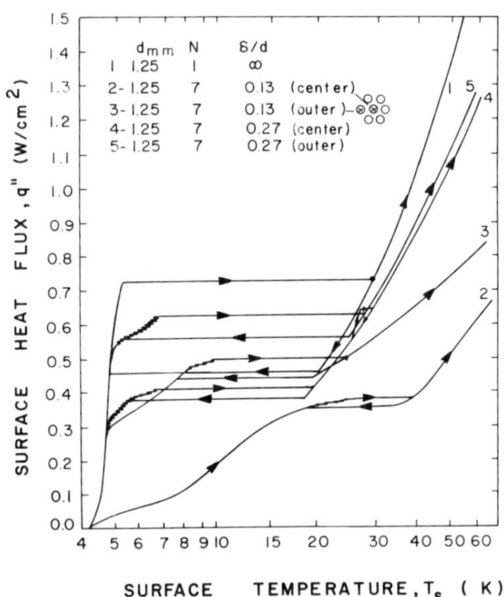

Fig. 2. Heat transfer characteristics for a 7 wire bundle compared with a single wire.

with zero angle from the mid-horizontal plane represents an average for the other outer wires.

As the gap between the wires increases, the critical heat flux improves and the film boiling curve moves closer to that of the single wire. Figure 1 shows that when increasing the gap to 0.041 cm, the critical heat flux of the outer wire approaches 90% of that of a single wire. However, the central wire of this bundle shows some temperature oscillations before the sudden transition to film boiling at 0.49 W/cm^2. Also film boiling secondary hysteresis is observed, and these two phenomena can be attributed to transitions to film boiling (burnouts) of wires below and above the central wire.

The same phenomenon is observed for the seven wire bundle with gaps of 0.033 and 0.016 cm as illustrated in Fig. 2. It is also observed that the difference between the critical and recovery heat fluxes is larger for surface wires than for central wires.

Figure 4 shows the effect of the gap on the critical heat flux for both central and outer wires. As the gap is increased the critical heat flux improves and becomes closer to the single wire value, because of improved bubble circulation and venting. This can be explained on the basis of the bubble departure diameter[8] (D_b) defined by

$$d_b = C_d \ \beta \ [2 \ g_c \ \sigma/g(\rho_1 - \rho_v)]^{\frac{1}{2}} \tag{2}$$

where $C_d\beta = 1.332$ for a contact angle (β) of 90°, which yields a bubble departure diameter of about 0.57 mm. The critical heat flux for the outer wires can also be correlated to the hydraulic diameter for the area between wires using a least square fit as follows

$$q''_{cr} = 1.67 \ (D_e)^{0.42} \tag{3}$$

where

$$D_e = \left[\frac{2}{\pi} \ \sqrt{3} \ (1 + \delta/d) - 1\right] d. \tag{4}$$

The extrapolation of this correlation to 0.73 W/cm^2 yields a critical gap of about 0.48 mm which is close to the D_b predicted by Eq. 2.

The critical heat flux for the 0.125 cm wire agrees within 13% with that predicted by the Kutateladze correlation[9] which yields a critical heat flux of 0.63 W/cm^2. However for the 0.317

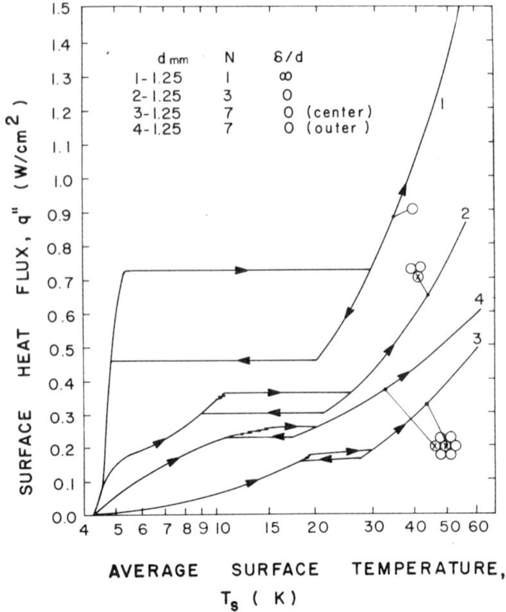

Fig. 3. Heat transfer characteristics of
bundles with zero gap.

cm wire the critical and recovery heat fluxes are reduced to 0.494
and 0.401 W/cm^2 respectively (as shown in Fig. 5) which is expected because the wire size is no longer of the same order as the
bubble diameter. The size effect is also clear in the film boiling region. The behavior of the 0.317 cm wire in the bundle form
is expected to be qualitatively similar to that of the 0.125 cm
wire bundle.

Since the present experiments are carried out on bundles of
commercial surface roughness, no attempt is made to introduce
roughness as a new parameter because it is beyond the scope of the
present work. However, previous work[10,11] shows the enhancement
of boiling heat transfer by increased surface roughness or using
certain surface coatings.

CONCLUSIONS

The present work illustrates the enhancement of boiling heat
transfer when using bundled conductors instead of solid cables
provided that the wires in the bundles are separated by a sufficient gap. For example, comparing a single 0.317 cm wire with a
bundle of seven 0.125 cm wires we find that they have the same

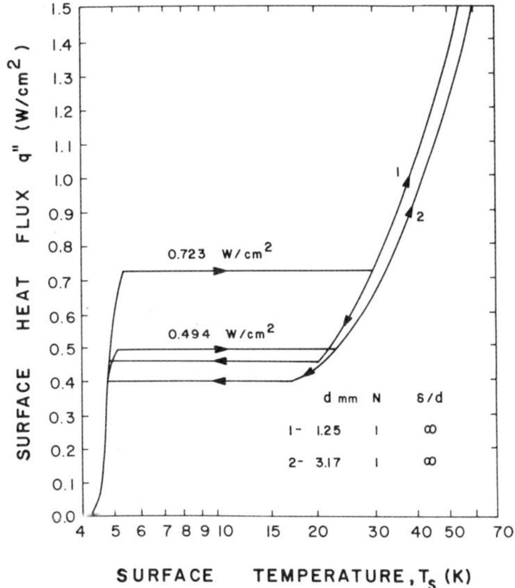

Fig. 4. Critical heat flux as a function of wire gap.

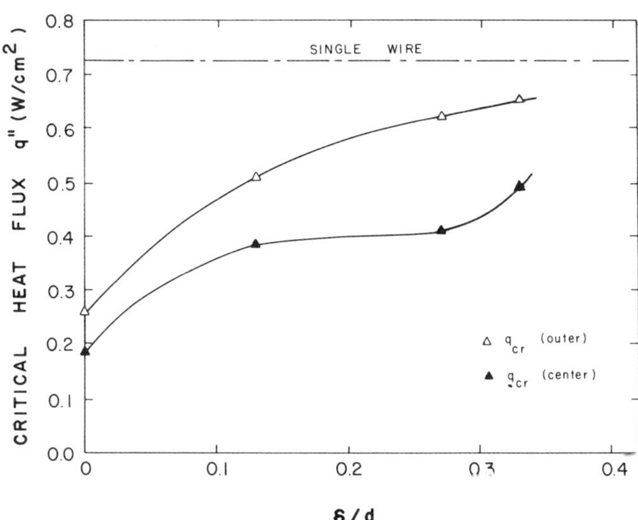

Fig. 5. Effect of wire size on the heat
transfer characteristics.

total cross sectional area. However, a seven wire bundle with 0.041 cm gap can dissipate 1.75 watts per unit length at burnout while a single 0.317 cm wire dissipates only 0.49 watts per unit length.

ACKNOWLEDGEMENTS

The author would like to thank G. McIntosh, S. Van Sciver and Y. Eyssa for their valuable discussions and comments.

NOTATION

C_d = empirical correlation constant
d = wire diameter, cm
D_b = bubble departure diameter
D_e = hydraulic diameter
g = local gravitational acceleration, cm/s^2
g_c = standard gravitational constant ($980\ cm/s^2$)
h = heat transfer coefficient W/cm^2 K
I = electrical current, A
N = number of wires in the bundle
q'' = surface heat flux = $h\ \Delta T_o$, Wcm^2
q''_{cr} = critical heat flux, W/cm^2
q''_{rec} = recovery heat flux, W/cm^2
R = electrical resistance per unit length, Ω/cm
S = surface area per unit length, cm
T_b = bath temperature, K
T_s = surface temperature, K
ΔT_o = $T_s - T_b$, K
B = contact angle
δ = gap between wires, cm
σ = surface tension, dyne/cm
ρ_1 and ρ_v = liquid and vapor densities, gm/cm^2

REFERENCES

1. R.D. Cummings and J.L. Smith. Boiling Heat Transfer to Liquid Helium in "Liquid Helium Technology", Pergamon Press, Oxford (1966).
2. J. Jackson and A.S. Erwin, Heat Transfer to Liquid Helium from Bare and Enamelled Wires, in "Proc. Second Intl. Conf. on Magnet Techn.", Oxford, (1967), p. 494.
3. D.N. Lyon, Boiling Heat Transfer and Peak Nucleate Boiling Fluxes in Saturated Liquid Helium Between the λ and Critical Temperatures, in "Advances in Cryogenic Engineering, Vol. 10", Plenum Press, New York (1964), p. 371.
4. R.M. Holdredge and P.W. McFadden, Heat Transfer from Horizontal Cylinders to Saturated Helium I Bath in

"Advances in Cryogenic Engineering, Vol. 16", Plenum Press, New York (1970) p. 352.

5. C.N. Whetstone and R.W. Boom, Nucleate Cooling Stability for Superconductor-Normal Metal Composite Conductors in Liquid Helium, in "Advances in Cryogenic Engineering, Vol. 13", New York (1968), p. 68.

6. T. Onishi, et. al., Transient stability of superconducting cables against thermal disturbances, Cryogenics, 21:431 (1981).

7. D.Q. Kern and A.D. Kraus, "Extended Surface Heat Transfer", McGraw-Hill, New York, (1972), p. 194.

8. W. Fritz, Maximum volume of vapor bubbles, Phys. Z, 36:379 (1935).

9. S.S. Kutateladze, Heat transfer in condensation and boiling, USAEC Report AEC-tr-3770 (1952).

10. M. Nishi, et. al, Roughened surface study on Japanese test coil for the large test coil task, IEEE Trans on Magnetics MAG-17:904 (1981).

11. B.J. Maddock, G.B. James and W.T. Norris, Superconductive composites, heat transfer and steady state stabilization, Cryogenics, 9:261 (1969).

POOL BOILING LHe HEAT TRANSFER IN AN MHD CONDUCTOR PACK*

W. D. Taylor

General Dynamics Convair Division
San Diego, California

and

J. E. C. Williams and M. Sinclair

Francis Bitter National Magnet Laboratory[†]
Massachusetts Institute of Technology
Cambridge, Massachusetts

INTRODUCTION

As part of the Cask Magnet Prototype System (CMPS) design effort[1], a boiling heat transfer test was conducted to determine if the baseline conductor (Fig. 1), or some variation thereof, would be cold-end cryostable[2,3] in all possible conductor orientations within the magnet. Other tests[4,5] have shown variation of boiling fluxes with angular orientation in a vertical plane. However, no data could be found on the effects of angular orientation about a horizontal axis, which characterizes CMPS. To obtain this data, seven variations of the CMPS baseline conductor (Fig. 2) were tested in up to 12 different angular orientations (Fig. 3). Three surface conditions and four types of turn-to-turn insulation were studied in this test. The three surface conditions were machined copper, machined copper coated with 0.05 to 0.075 mm of lead-tin solder, and machined copper coated with Ebanol C, (a commercial cupric oxide coating). Turn-to-turn insulation was also varied, using the four types shown in Fig. 4.

---

*Sponsored by the U.S. Department of Energy.

†Supported by the National Science Foundation.

Fig. 1.   CMPS baseline conductor.

Layer-to-layer and conductor-to-ground insulations (Fig. 2) were used to simulate the helium cooling passages that would exist in the final magnet.

TEST PROCEDURE

The test package, cross-sectioned in Fig. 2, consisted of twenty-nine 30.5-cm long dummy (non-superconducting) conductors. These were arranged in two rows of 14 and 15 conductors, with simulated staves and corner blocks[1] (the substructure for cask type MHD magnets).   The dummy conductors were built at General Dynamics Convair Division and sent to the Francis Bitter National Magnet Laboratory where they were instrumented and incorporated into the total test package. Seven of the 29 conductors contained heaters.   The heaters were made of two stainless steel ribbons slightly narrower and shorter than the dummy conductors. Kraft paper insulated the steel ribbons from each other and from the copper conductors as shown in Fig. 5. The heaters were noninductive, the two strips being joined to each other at one end of the test specimen and to copper terminal strips at the other.

Gold plus 0.07% iron vs chromel differential thermocouples were embedded in the seven samples with heaters, and also in five adjacent samples without heaters. Twenty-one thermocouples were installed. The thermocouple wires were brought out of the package

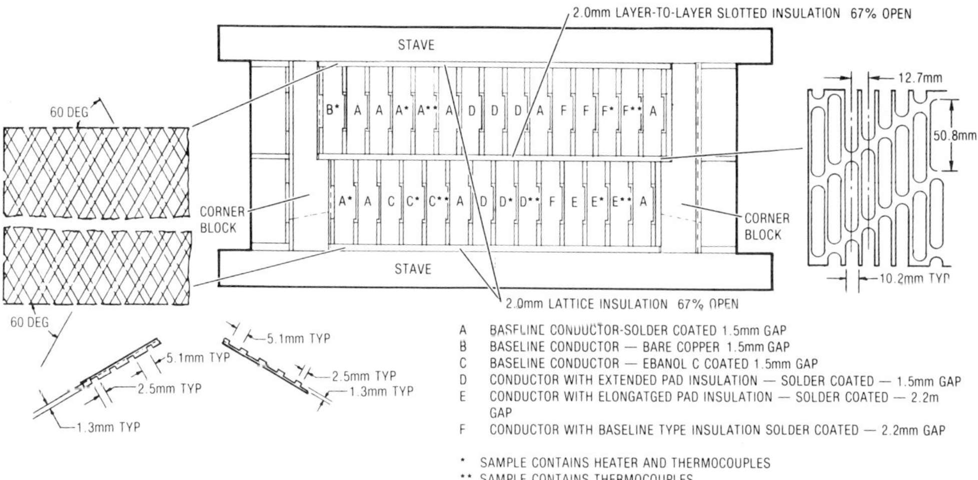

Fig. 2.   Test Package simulates CMPS winding conditions.

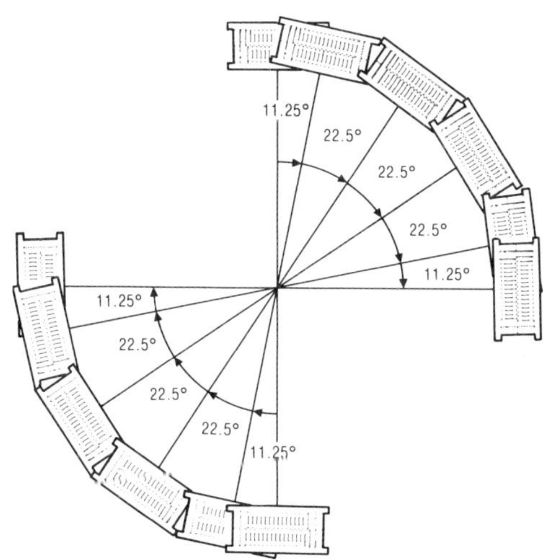

Fig. 3.   Test package was tested in up to twelve angular positions.

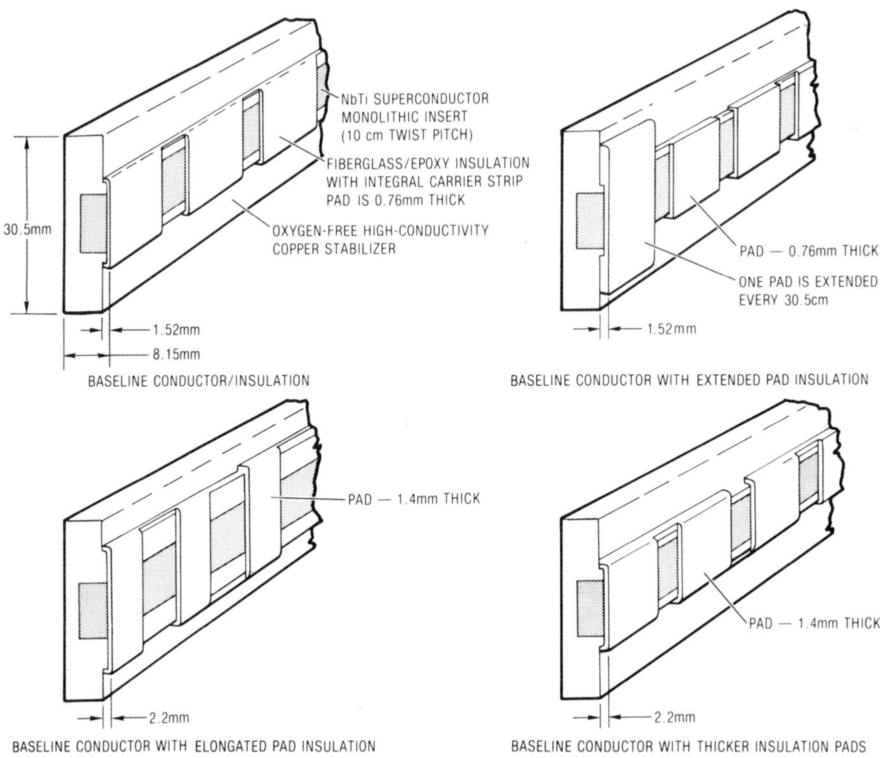

Fig. 4. The baseline conductor and three variations of its insulation were tested.

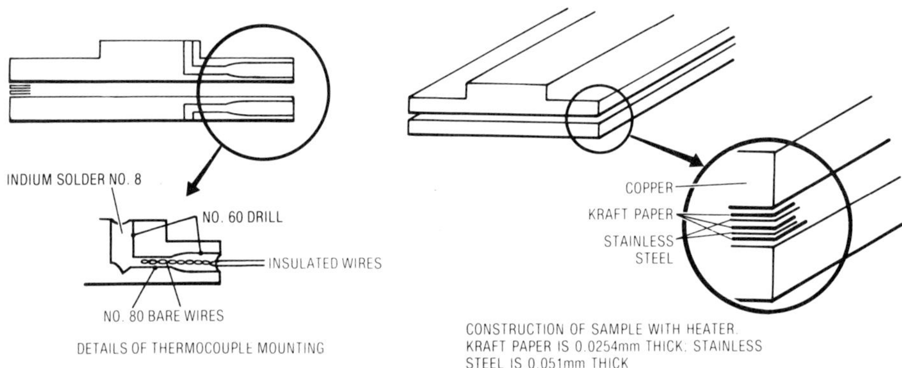

Fig. 5. Instrumentation installation provided reliable data measurements.

by way of the layer insulation and through slots cut into the
dummy corner blocks. Reference junctions were formed by soldering
the wires to terminals located on one of the dummy staves.

Test conductors were clamped between simulated staves and
corner blocks with sufficient pressure to simulate expected magnet
conditions. This assembly was then suspended on trunnions from
the support system of a 60-cm cryostat. A worm gear and wheel
provided a means for rotating the assembly about a horizontal axis
to an accuracy of one degree. The apparatus was precooled with
nitrogen gas and then immersed in liquid helium.

The measurement of quasi steady-state heat transfer was made
by sweeping the current in the heaters slowly and measuring the
signal generated by the thermocouples. This technique generated
continuous heat flux versus temperature difference data for each
of the test samples. Heater current was increased linearly until
film boiling occurred and then decreased at the same rate. Preliminary tests indicated that for current ramping rates equal to
or less than 0.5 A/s (about 2 to 3% of maximum heater current),
peak nucleate boiling and minimum film boiling fluxes would not
change. All test runs were then made at a current sweep rate of
0.5 A/s.

Data was recorded directly on an x-y recorder plot, and
digitally on magnetic tape. The x-y recorder plotted heater
current and thermocouple voltage in real time. A typical x-y plot
is shown in Fig. 6. Digital data was collected on a Biomation
transient data recorder collecting one data point every 100 ms.
This data was subsequently written on magnetic tape for permanent
storage.

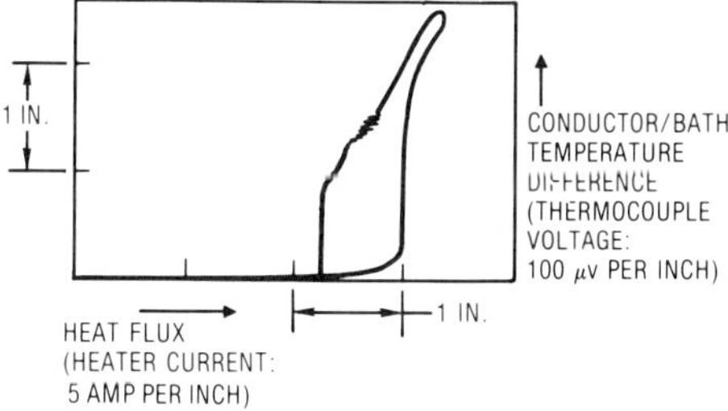

Fig. 6. Typical x-y trace of steady-state boiling curve.

RESULTS

Test results are presented in Fig. 7 and 8 as breakaway and recovery heat fluxes plotted as a function of angle-to-vertical, and proximity to the lower stave. General trends from these figures indicate a maximum heat transfer to the helium when the conductor is rolled 30 to 40 degrees from the vertical (zero-degree orientation in the figures). Breakaway fluxes vary more widely with angle than do the recovery fluxes. However, both heat fluxes shows a dramatic decrease when the conductor's broad cooling faces are horizontal (90 degrees orientation in the figures).

Restriction of helium replenishment, caused by the relatively thin insulation between the conductors and the unventilated stave beneath, lowers both the breakaway and recovery fluxes. Another study[4] indicated that unless helium cooling channels next to an unventilated wall are approximately three times larger than those between turns, heat transfer characteristics of those conductors next to the wall are degraded 20 to 25 percent from those well-ventilated wall areas. This effect exists in the present arrangement because the insulation next to the staves is almost the same thickness as the turn-to-turn insulation. This restricted ventilation may also be masking any variation in recovery flux for different turn-to-turn insulation thickness, because increasing the inter-turn insulation thickness does not improve helium purging if the conductor-to-stave insulation thickness is not increased. Again, this phenomenon is more important for recovery fluxes than for breakaway heat fluxes.

The hysteresis, shown in Fig. 6, which existed in all the data, indicated that the sweep rate used was fast enough to introduce significant transient effects. Correction for these effects consists of accurately accounting for the transient energy storage. Figure 9 shows both the raw data and the data corrected for the energy storage transient. Further testing is required to verify this correction technique; Fig. 7 and 8 are uncorrected data.

Figures 7 and 8 show that conductors coated with Ebanol C perform significantly better than those coated with solder or plain copper. Solder-coated conductors generally performed better than the uncoated. However, only the conductor coated with Ebanol C met the cold-end stability requirements for CMPS. Figure 10 shows that the heat flux generated by a normalized conductor is slightly less than the equal area heat flux, thus ensuring cold-end stability. This stability calculation is conservative because future data is expected to show an increase in the equal-area flux once the film-boiling hysteresis is removed. The resulting stability should offer an ample safety margin.

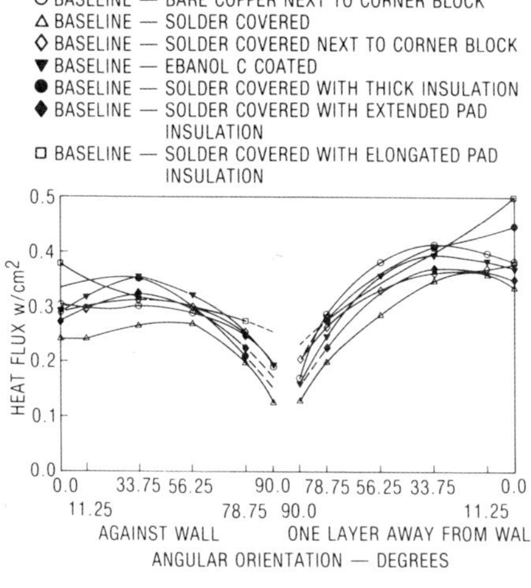

Fig. 7. Peak nucleate boiling fluxes are dependent upon angular orientation.

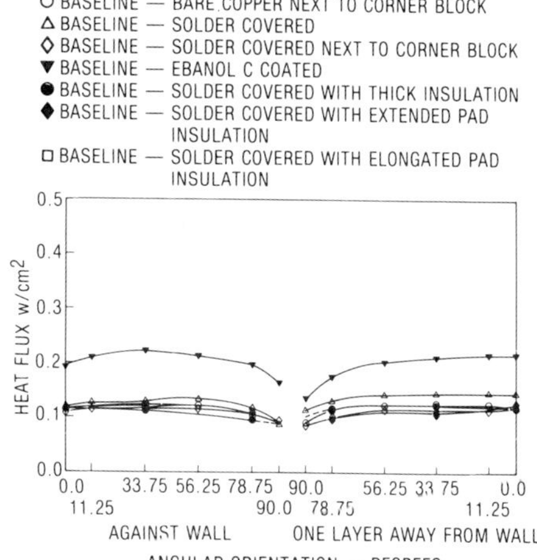

Fig. 8. Recovery fluxes are relatively sensitive to angular orientations.

CONCLUSIONS

The CMPS baseline conductor must be coated with Ebanol C to meet cryostability requirements of the design. Degradation of heat flux, due to angular orientation and wall effects, contributed to the need for Ebanol C coating. In the testing procedure,

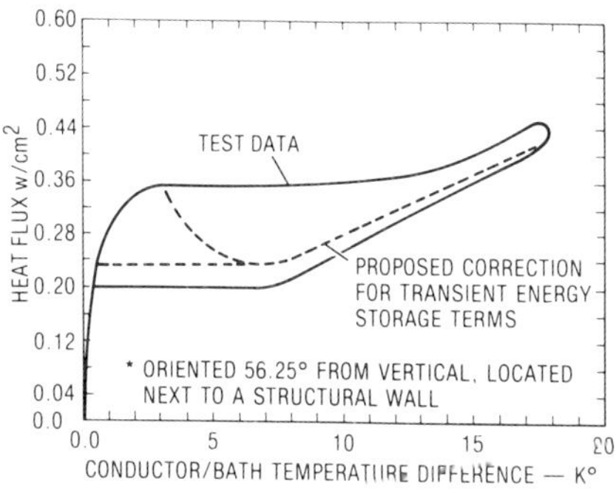

Fig. 9. Steady-state boiling curve for an Ebanol C-coated conductor.*

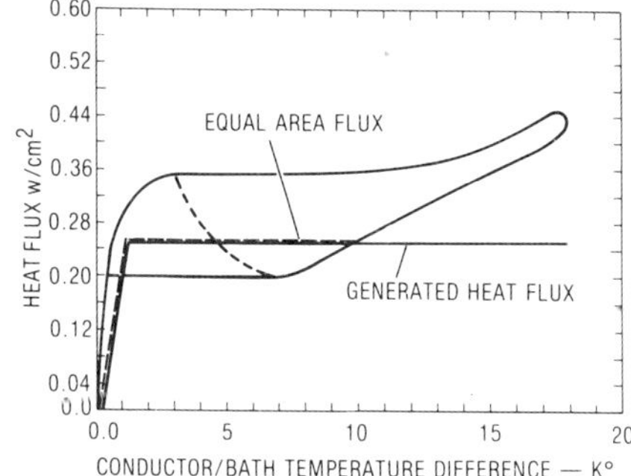

Fig. 10. Cryostability criteria is met using a conservative boiling curve

hysteresis in the film boiling regime could have been eliminated either by laborious data processing or by stepping the heater current instead of sweeping it.

ACKNOWLEDGEMENTS

MIT-FBNML personnel who made significant contributions to this program are: E.S. Bobrov, and P.G. Marston, Project Advisors.

The General Dynamics Convair Division personnel who made significant technical contributions to this program are: R.W. Baldi, Chief Engineer, E.H. Christensen, Thermodynamics, R.G. Jones, Structural Design, R.C. McCool, Structural Analysis, and Dr. J.F. Parmer, Program Manager.

REFERENCES

1. J.F. Parmer, et. al., Cask Magnet Prototype System (CMPS) for Stanford University, in "Advances in Cryogenic Engineering, Vol. 27", Plenum Press, New York (1982).
2. B.J.Maddock, G.B.James and W.T. Norris, Superconducting composites: heat transfer and steady-state stabilization, Cryogenics, 9: (1969).
3. Cask Magnet Prototype System Specification, Request for Proposal, Superconducting MHD Magnet for Stanford University, Francis Bitter National Magnet Laboratory, MIT (February 1978).
4. E.H. Christensen, and S.D. Peck, Large coil program conductor LHe heat transfer and heater verification test data analysis, General Dynamics Convair Division report number 91Z0122, (10 July 1980).
5. E.H. Christensen, and R.P. Krause, Configuration dependent heat transfer data and quench pressure computer program for superconducting magnets, General Dynamics Convair Division GDC-ERR-79-059, (December 1979).

HEAT TRANSFER CORRELATIONS FOR A CRYOSTABLE ALTERNATOR FIELD WINDING*

P. W. Eckels, J. H. Parker, Jr., A. Patterson, and J. H. Murphy

Westinghouse Electric Corporation
Pittsburgh, Pennsylvania

INTRODUCTION

The EPRI-Westinghouse 300 MVA superconducting generator features a cryostabilized field winding. The conductor is wound in slots with radial spacers between layers to form channels that are supplied helium from reservoirs external to the winding region. In the event of a resistive transition, helium is circulated to the winding by natural convection. Early theoretical considerations based on the work of Eckert[1] predicted heat transfer in the thermosyphon well in excess of that of turbulent natural convection. Coriolis induced stabilization of the boundary layers and spiral form secondary flows were identified as potential problems among the secondary effects of fluids flowing in rotating channels.

As a step in the 300 MVA development, an effort was initiated to measure the heat transfer in rotating winding thermosyphons. An exact replica of the winding 9.1 cm long containing heaters and temperature sensors was spun in a rotating dewar equipped with a data transmission and processing system.[2] The copper bars simulating conductors have bare surfaces and were handled as we would expect the conductor to be handled. Kapitza conductance effects are included in the convection coefficient as is the effect of helium expansion in the radial channels -- the latter being negligible. Data for the straight section of the winding are reported here.

---

*Cofunded by the Electric Power Research Institute, Contract No. RP 1473-1 and the Westinghouse Electric Corporation.

EXPERIMENTAL APPARATUS

Figure 1 shows several views of the conductor module with the various major components identified. Shown are three stacks of conductors radially insulated from each other by 0.025 mm of Kapton. Each stack is insulated from the adjacent stack by 0.381 mm strips of G-10CR which form the cooling channels. Two-thirds of the conductor surface is wetted and the channel aspect ratio is 20. The unheated stack of conductors on each side of the heated test stack models the spreading of heat from the heated channel to adjacent channels.

Silicon diode temperature sensors are soldered to the copper conductors in recessed slots between the conductors. Leads run in the slots for 3.8 cm and are bonded to the copper with GE laquer over laquer impregnated cigarette paper. Soldering the cathode of the diode to the copper results in undesirably strong capacitive coupling between the heater and sensor, and which in the future lead to exploration of epoxy bonding of diodes despite their reduced response time. Figure 1 shows the location of the sensors, indicating the temperature of the first, middle and last conductor in the stack. The data show that the middle diode temperature is always the average of the other two so it is used as wall temperature for computing heat transfer coefficients. Inlet and outlet helium temperatures are indicated by diodes supported by stainless steel rods in the flow stream. Their leads are wrapped around the rod to reduce conduction temperature measurement errors. Lake Shore Cryotronics produced a special epoxy potted carbon glass thermometer for these tests to calibrate the Si diodes which has successfully performed in acceleration fields of 2000 g's.

The heaters are fabricated from type K (Chromel-Alumel) sheathed thermocouple wire 1.01 mm in diameter. The 304 stainless steel sheath serpentines back and forth through the conductor stack and is soldered into slots in the bars. The return bends overhanging the conductors are soldered into copper cooling strips that are axially insulated against heat flow to and from the conductor stack. Radial subdivisions prevent radial heat flow along the end strip which is in effect a guard heater.

The transient heat transfer data collection required a small-scale data acquisition and control system. For the microcomputer, a S-100 (IEEE Standard 696) bus and a 4 MHz Zilog Z80 8-bit microcomputer are used as the central processor unit and the system includes a dot-matrix printer, dual double-density mini-floppy disk drive, CRT display, ASCII keyboard, 50 K of RAM, 12 K of ROM, and several serial and parallel I/O boards. The system, using a Digital Research CP/M operating system and a Microsoft BASIC compiler, is programmed to collect data from the rotating telemetry

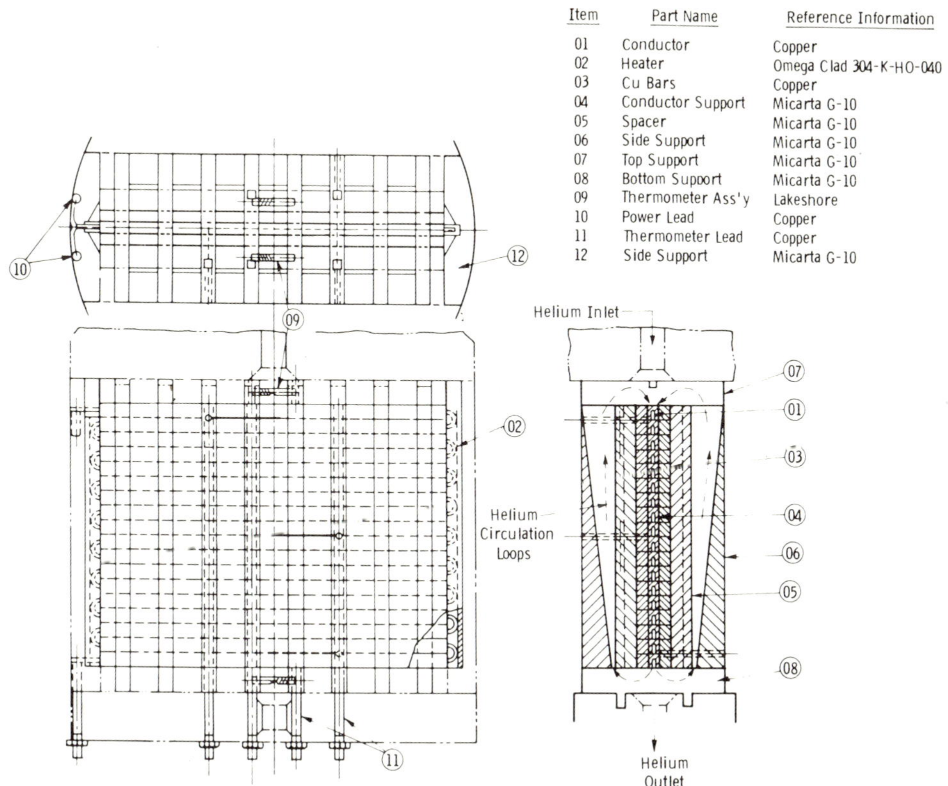

Fig. 1.   Slot simulation test module.

system at a rate of 1416 bits per second, to pulse the experimental heater circuitry up to 88 times per second, and to convert the voltage readings into temperature for real-time display or print-out after an experimental run. The data is transmitted as 16 bits per word -- 10 data and 6 channel identifiers. Available RAM limited experimental runs to one-minute duration.

EXPERIMENTAL TECHNIQUE

Following cooldown, the rotating dewar is filled with liquid helium and spun to establish an isothermal condition in the test chamber. The number of heat pulses in a one-minute run, their time interval and voltage are entered. Typically six one-half second pulses are initiated with various voltages (50, 75, 100, 150, 200, and 250 in ascending or descending sequence) applied· to the heater. Conductor, dewar, and fluid temperatures, voltage and current are logged during the run. Helium temperature is measured at the inlet and outlet. Mean pressure in the test section is estimated by assuming a linear decay of pool depth and calculating the hydrostatic pressure. Uncertainty in the pressure is the largest source of error in the correlated data and work is continuing to more accurately define the pool depth vs time relationship.

Reduction of the data to heat flux, heat transfer coefficient and temperature difference is accomplished by first computing the heat flux from $\dot{q}'' = V^2/(R \cdot A_s)$, with the conductor spacers presumed adiabatic which introduces an error of less than 1%. Heat transfer coefficient is then calculated from $h = \dot{q}''/(\Delta T)$, with ΔT taken between the mean copper and inlet helium.

The data correlation variables used are derived from an early analysis by Eckert[1] in which turbulent thermosyphon flows were found to be described by the non-dimensional parameters Nu and Ra. Closed-cell thermosyphon data and some free-convection experiments have been correlated at high Gr numbers by $(GrPr)^2$ to eliminate the viscosity dependence. Because this data spans the transposed critical and because of the associated strong Pr variations, we were able to evaluate the correlation form $[GrPr$ vs $(GrPr)^2]$ for convection loops. Correlation using $(GrPr)^2$ decreased the data correlation coefficient in this case.

The thermodynamic condition at which properties should be evaluated is not clearly defined by existing experiments. Constant wall temperature natural draft cooling channels have been found[3] to have the highest correlation coefficient using wall temperature for all properties except the coefficient of thermal expansion, which is evaluated at the inlet fluid temperature. That method is used in this presentation of the data.

Another consideration in compressible fluid thermosyphon analysis in the rotating frame is that adiabatic expansion and compression of the coolant produce fluid temperature changes. Compression temperature rises are eliminated by the copper heat exchanger but radial inflow adiabatic expansion decreases the temperature between channel inlet and outlet by 0.2 K. No attempt has been made to isolate that effect.

EXPERIMENTAL RESULTS

In the winding straight section, the Coriolis acceleration is perpendicular to the heated surface as shown in Fig. 2. Considering for a moment isothermal flow, the leading side of the channel is stabilized (tending toward laminar flow) and the trailing side is destabilized and tends to form streamwise Taylor-Gortler vor-

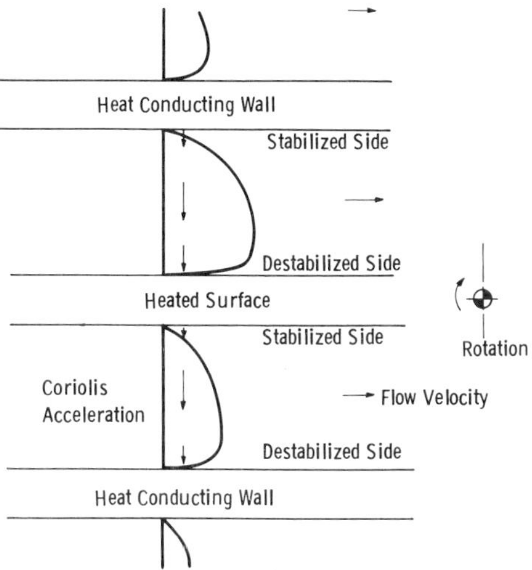

Fig. 2. Diagram of the velocity profiles in the straight section.

tices.[4] Heating the wall intensifies both of these phenomena. Although it is not possible to a priori predict the effect of streamwise vortices on heat transfer, stratification will reduce the heat transfer coefficient because heat transmitted from both sides of the heated surface encounters a stabilized boundary layer eventually. The stabilized boundary layer is analogous to flow on a flat plate cooled from below. Various investigators[5,6] find the stability criterion for that case to be Ri > 0.0417, so a laminar-turbulent transition is expected.

Figure 3 shows the heat flux from the wetted surface vs temperature difference for several values of velocity. Notable features are the repeatability of the low ΔT data, the strong non-linear variations in $\dot{q}''$ vs ΔT and the span of the transposed critical temperatures. The temperatures with the lowest signal-to-noise ratio are the low values of ΔT and even these correlate exceeding well. We see very strong non-linear variations in $\dot{q}''$ vs ΔT which can be typical of data spanning the transposed critical or data recording a transition in flow regime. Also shown is a band indicating the range of transposed critical temperatures encountered in these experiments. The occurrence of the steep rise in $\dot{q}''$ precedes the approach to the transposed critical band indicating a probable transition in the flow regime to full turbulence. Note that the high-speed data are not influenced by the transposed critical region.

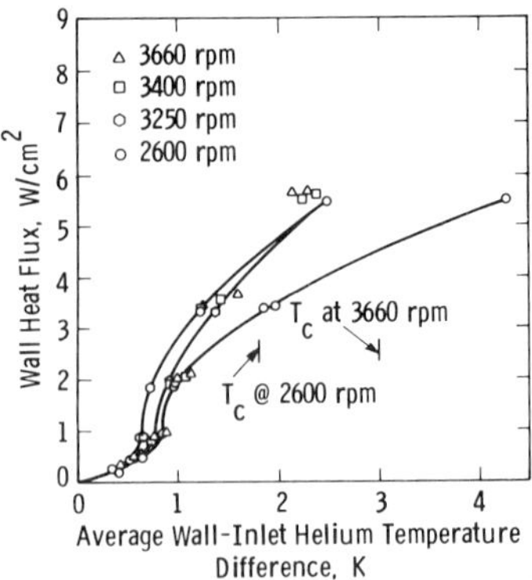

Fig. 3 Heat flux vs temperature difference for the
winding straight section for several speeds.

Figure 4 shows the data of Fig. 3 correlated by the variables
Nu and Ra. The linearized equation of performance of the thermosyphon computed by conventional buoyancy induced circulation
techniques is also shown. The circulation computation is for a
rectangular channel with three adiabatic sides and one long side
heated but it differs from the actual test channel which transfers
heat through the opposite wall to a channel beyond. If the heat
transfer to the second channel is impedance free an increase of
40% over the adiabatic wall computation might be expected, thus
explaining some of the theory-measurement difference. The most

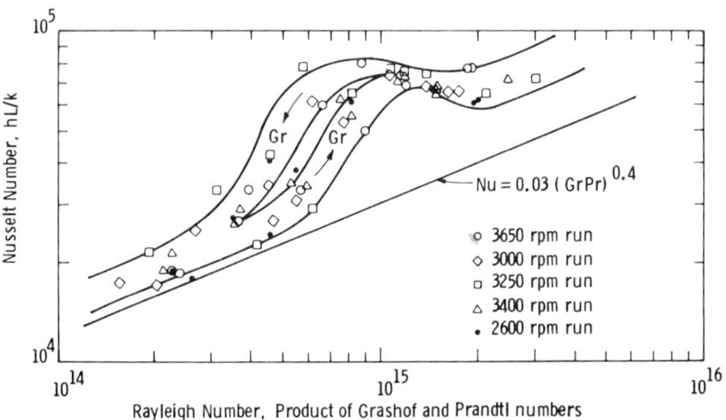

Fig. 4. Correlation of heat transfer in the winding
straight section. Coriolis vector is perpendicular
to the heated surface.

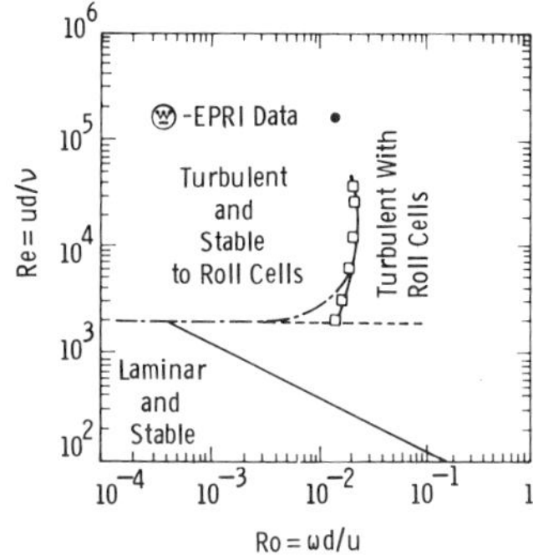

Fig. 5 Neutral stability boundary for laminar
Poiseuille and turbulent channel flow.

striking feature of the data is the hysteresis type loop generated
by increasing and decreasing heat flux to vary Gr. With increasing heat flux the lower curve is generated with a steep rise as
the flow becomes fully turbulent. Decreasing heat flux generates
the upper curves because turbulence persists in the channel for
the 10 s between pulses despite all thermometers indicating thermal equilibrium for 8.5 s. If pulses are initiated 20 s apart the
lower curve is retraced with decreasing heat flux. The steep rise
goes well above any correlation line to a possible correlation
that may be 25% higher than the lower line.

Another interesting aspect of the data is the comparison of
the transition indicated by heat transfer with existing theory and
experiment. Figure 5 is reproduced from Lezius and Johnston[4] and
indicates by a " • " our estimate of the transition point at 3650
rpm (Ra = 6 × 10^{14}). Considering the approximate computational
scaling and non-smooth inlet, the agreement is fortuitous.

Figure 6 shows the measured heat transmission capability of
the cooling surface compared to the heat generated per unit cooling surface area. That stability map shows that the conductor is
cryostable.

SUMMARY

A cryostable field winding of NbTi has been proposed for
large superconducting generators using high-performance thermosyphons. Thermosyphons can be designed to transfer heat at rates
well in excess of natural convection limits but other data[7] shows
that under certain conditions the effects of Coriolis acceleration
can reduce heat transfer in rotating channels by as much as 60%.
Conventional buoyancy induced flow analysis and conventional heat

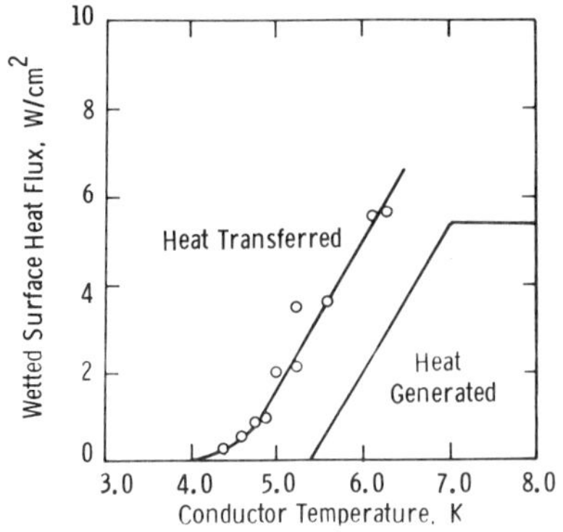

Fig. 6. Conductor stability map for the winding straight section. Conductor is 3.84 × 2.03 mm² and the speed is 60 Hz.

transfer correlations provide a good indicator of heat transfer coefficients in rotating thermosyphons if regimes of strong secondary flow are avoided. Turbulence is shown to exist in liquid helium filled channels for more than 10 s after the turbulence generating mechanism has ceased.

ACKNOWLEDGEMENT

The efforts of Prof. J. L. Smith in reviewing and improving this experiment are acknowledged with appreciation. Our thanks go to Marilyn Warren for preparation of this manuscript and Paul Steve and Joseph Buttyan for their help in setting up and running the experiments.

NOMENCLATURE

A_s	= wetted cooling surface		R	= heater resistance
d	= Channel depth = 0.381 mm		Ra	= Ragleigh number = GrPr
Gr	= Grashof number based on L		Ri	= Richardson number[6]
h	= heat transfer coefficient		T_c	= transposed critical temp
k	= thermal conductivity of He		U	= helium velocity
L	= channel length = 6.858 cm		V	= heater voltage
Nu	= Nusselt number = hL/k		ΔT	= temperature difference
Pr	= Prandtl number		ν	= kinematic viscosity
$\dot{q}''$	= surface heat flux			

REFERENCES

1. E.R.G. Eckert and T.W. Jackson, Analytic Investigation of Flow
    and Heat Transfer in Coolant Passages of Free Convection
    Liquid-Cooled Turbines, National Committee for Aeronautics
    Report RM E50D25 (July, 1950).
2. D.C. Litz, et. al., High Tip Speed Test Rig to Study Natural
    Convection in Liquid Helium, in "Advances in Cryogenic
    Engineering, Vol 27", Plenum Press, New York (1982).
3. W. Elenbaas, Dissipation of Heat by Free Convection, Parts I
    and II, Philips Research Report 3, N. V. Philips
    Gloeilampenfabrieken, Eindhoven, Netherlands (1949).
4. D.K. Lezius and J.P. Johnston, Roll-Cell Instabilities in
    Rotating Laminar and Turbulent Channel Flows, J. of Fluid
    Mechanics, Part 1 77:153, (1976).
5. S. Chandrasekhar, "Hydrodynamic and Hydromagnetic Stability,"
    Oxford University Press, London (1961).
6. H. Schlicting, "Boundary Layer Theory," McGraw-Hill, New York
    (1960).
7. 300 MVA Superconducting Generator Development, Appendix N,
    Westinghouse Report #9 to EPRI (1981).

DISCUSSION

Question by M. A. Hilal, Michigan Tech. University: Can the
pressure in the windings approach the critical pressure and if so,
what correlation is appropriate?

Answer by author: Pressure in the test cell varied between 3.5
atm and 9.0 atm from test to test. The thermodynamic state in the
channel did traverse the transposed critical trajectory of states
in some cases.   An additional complication exists because the
hydrostatic pressure varies by more than 1.5 atm from inlet to
outlet of the heated channel. The outlet, therefore, may traverse
the transposed critical while the inlet sees no such effect. More
effort could be usefully applied to determine the best property
evaluation technique.

INDIRECT CONDUCTION COOLING OF
SUPERCONDUCTING MAGNET

Y. Matsubara, K. Kanbara,* T. Munekata, and K. Yasukochi

Atomic Energy Research Institute
Nihon University
Tokyo, Japan

INTRODUCTION

Epoxy impregnated superconducting magnets with channels for
forced helium cooling have recently been reported.[1-3] The purpose
of this paper is to investigate the behavior of solid conduction
cooling in an epoxy impregnated superconducting magnet with a
forced convection helium channel. The effect of the weak thermal
coupling between the coil and the cooling helium has been clarified.

TEST COIL

The superconducting magnet used for the experiments was a
pancake coil with a rectangular helium cooling channel bonded on
the outside of the coil as shown in Fig. 1. A copper composite
conductor (Nb-48 wt.%Ti) with a circular cross section was used.
The specifications of the conductor and the coil parameters are
given in Table I. The epoxy resin used for potting the coil is
EPIKOTE 815 with EPICURE V-40 as the hardener.

Two phase liquid helium or supercritical helium gas precooled
to near 5 K flows in the cooling channel. A fraction of the flow
is diverted for the gas cooled current leads. An automatically
controlled CTI Model 1430 was used for the refrigeration system.

---

*Present Address: Department of Physics, College of Humanities
and Sciences, Nihon University.

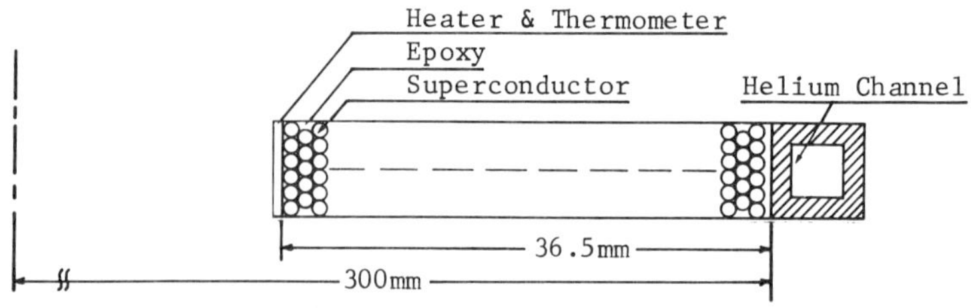

Fig. 1. Cross sectional view of the superconducting winding.

With the test coil placed in a vacuum chamber, the coil is cooled by means of the solid conduction to the cooling channel. The thermal coupling between the coil and the channel is weak because of the thin (1.5 mm) layer of epoxy. An electric heater made of thin sheet nichrome was installed inside the first layer of the coil for quench experiments.

EXPERIMENTAL RESULTS

Most of the coil performance tests have been done with the two phase liquid helium cooling. Figure 2 shows test results at a mean coil temperature of 5 K. A spontaneous quench occured at 544 A corresponding to an overall current density of 327 A/mm^2, which is approximately the short sample critical current density at 5 K and 3 T. Following the quench, the coil temperature increased to about 35 K. The temperature rise of the cooling channel however was ~1.3 K. The helium temperature at the outlet changed from 4.7 K to 4.8 K. These temperatures corresponded to saturated vapor pressures of 1.52 atm and 1.65 atm. Therefore the thermal disturbance caused by the quench does not affect the cooling helium significantly. No training of the test coil was observed.

Table I. Conductor Specifications and Coil Parameters

Wire Diameter	1.23 mm	Coil Outer Diameter	300 mm
Filament Diameter	41 µm	Cross Section	7.8 x 36.5 mm^2
No. of Filaments	144	No. of Turns	171
Twist Pitch	25.4 mm	Current	500 A
Cu/Sc Ratio	5	Max. Field at Wire	2.4 T
		Inductance	0.013 H
		Stored Energy	1.62 kJ

Quench experiments have been done by using the nichrome heater. At the operating current of 500 A, the minimum pulse width required to quench the coil was about 50 msec at a power rate of 4.7 watts (0.235 J). At 300 A, the pulse width was 160 msec at the same power rate (0.752 J). Decreasing the power increases the minimum pulse width at a given operating current. However, the minimum thermal energy required to quench the coil decreased with increasing power rate.

The propagation velocity of the normal zone in the radial direction was calculated from the terminal voltage change. Results are shown in Fig. 3 as a function of the operating current. It is found that the propagation velocity obtained from the continuous heating data is higher than that from pulsed heating. This higher velocity may be caused by decreased temperature margin, since the mean coil temperature has been increased by the continuous heating.

COMPUTER SIMULATION

In order to clarify the thermal performance of this coil system, a numerical analysis has been performed. Heat conduction equation of the segment, i, in the radial direction is given by,

$$C_i(\partial T_i/\partial t) = Q_J + Q_H - Q_c + \left[\lambda_{i-1}(T_{i-1}-T_i)-\lambda_i(T_i-T_{i+1})\right]/\delta R \quad (1)$$

Heat transfer coefficient of forced convection helium is well known[4], however we used a constant heat transfer coefficient

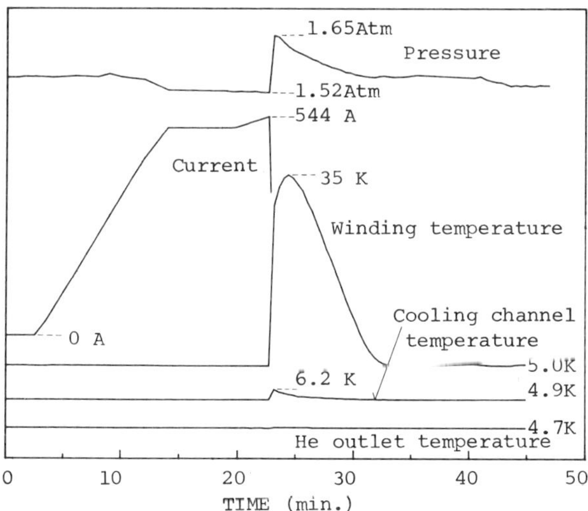

Fig. 2. Temperature and pressure response following the quench.

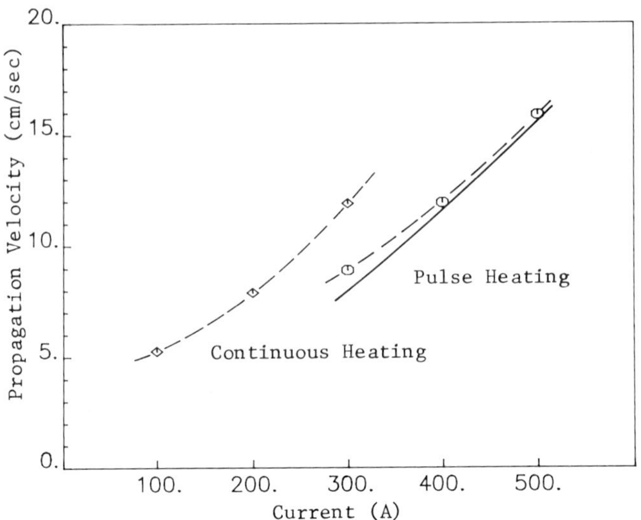

Fig. 3. Radial propagation velocity of normal zone.

(h = 1.25 × 10^3 J/sec •m^2) to avoid the complexity of hydrodynamics.

Pulse heating at the inner segment of i=1 is considered as
the disturbance for the quench simulation. A thermal diffusion
time constant was applied to the equation for heat Q_H generated by
the pulse heating[5],

$$Q_H = \begin{cases} Q_0 \left[1-\exp(-t/\tau_d) \right]/A_c & t < \tau_p \quad \text{(heater on)} \\ Q_0 (1-\exp(-\tau_p/\tau_d))(\exp\left[(\tau_p-t)/\tau_d \right])/A_c & t \geq \tau_p \quad \text{(heater off)} \end{cases} \quad (2)$$

One of the important factors of this computer simulation is
finding the mean thermal conductivity of the epoxy impregnated
coil. Figure 4 shows the winding arrangement of the coil with the
superconducting composite wire having circular cross sectional
area. The effective thermal conductivity of the coil can be
expressed as a function of the thermal conductivity of the
epoxy. The inside of the conductor is assumed to be in the isothermal condition. Thus the heat conducted in the radial direction is given by,

$$Q = \lambda_e \cdot \delta T/\delta R \quad (3)$$

Thermal flow between each conductor is expressed as a line perpendicular to the isothermal line as shown in Fig. 4.

After some calculations, we get an equation for λ_e,

$$\lambda_e/\lambda = 2\cos(\pi/6)\sum_{n=1}^{N_0}\left\{\sin\theta/\left[N_0(b/a - \cos\theta]\right]\right\}\tag{4}$$

Packing density of the test coil was 0.79 which corresponds to a/b = 0.93 and λ_e/λ = 8.

CALCULATED RESULTS AND DISCUSSION

A numerical solution of Eq. 1, employing parameters used for the experiment, is plotted in Fig. 5. A perturbation energy of 0.234 J was applied by a 50 msec pulse. The diffusion time constant τ_d = 0.2 s was used which was derived from the preliminary calculation using Eq. 1. This long time constant may depend on the heat capacity of epoxy surrounding the heater.

Figure 5 (a) shows the temperature distribution after the heater pulse lasting 0.5 sec, was applied at the operating current of 500 A, and (b) shows that of the recovery mode following current shut down at 0.38 s. The maximum coil temperature at 0.38 s agreed with the experimental result. The thermal expansion coefficient of epoxy does not change significantly below 40 K and the radial temperature gradient is always negative, hence the thermal stress in the coil winding is due to the compression force. The propagation velocity is also calculated, is plotted on Fig. 3 as a solid line, and agrees well with the experimental results.

Both $(\partial T/\partial R)$ and $(\partial T/\partial t)$ decrease with increasing temperature, caused by the increasing thermal conductivity and heat capacity. This effect is useful in keeping the coil temperature low enough from the stand point of thermal stress.

A large temperature gradient $(\partial T/\partial R)$ appeared between the outermost layer of the coil and the cooling channel because of the poor thermal conductivity of the epoxy layer. However this effect is also important in keeping the helium temperature constant. In the case of Fig. 5 the normal zone thermal wave reached the cooling channel in 0.37 s. The temperature of the first layer of the coil was 32 K at this time. Therefore the cooling effect has no influence up to this period. A significant temperature increase can be seen at the outermost layer for times greater than 0.37 s, therefore if the epoxy layer were not present an abrupt thermal disturbance might be given to the cooling helium.

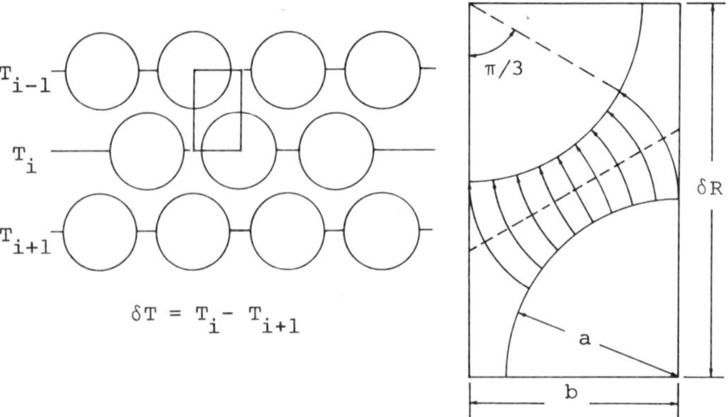

Fig. 4.   Analytical model of thermal flow.

The stability behavior of this coil system has also been simulated. The result for the operating current of 300 A is shown in Fig. 6. The minimum perturbation energy to quench the coil is obtained by increasing the pulse width with constant heater wattage Q_0. This perturbation energy is found to be 0.73 J for Q_0 = 4.7 W and 0.93 J for Q_0 = 1.0 W. These results agree with the experimental result. Therefore it was indicated that the intensity of the disturbance is important in finding the stability limit.

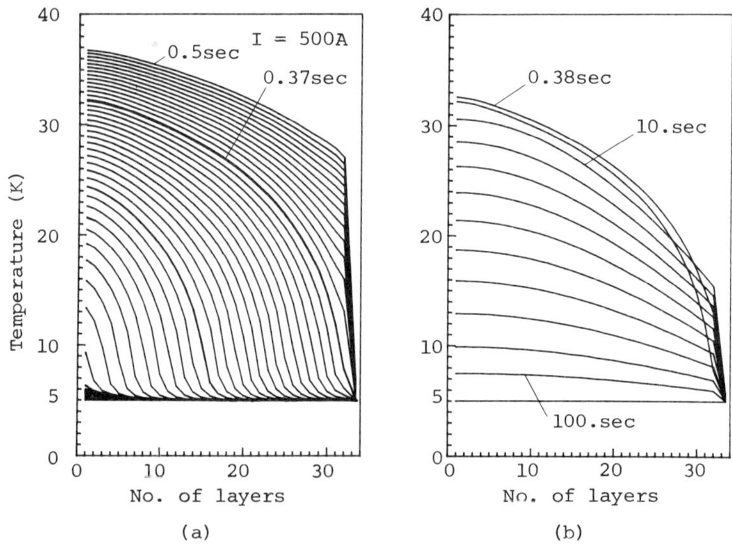

Fig. 5.   Computer plots of temperature profiles.   Layer number 1 is at the inner diameter.

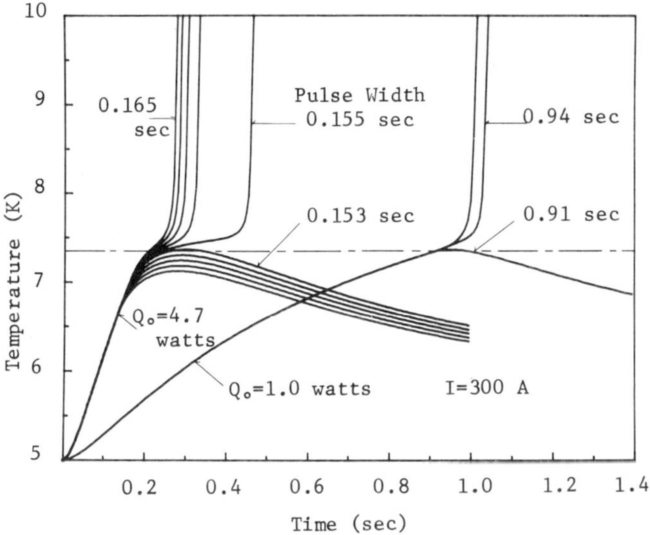

Fig. 6. Computer simulation of stability limit.

CONCLUSION

Indirect conduction cooling of a superconducting magnet has been discussed. The model coil used for the experiments has been successfully operated. A computer simulation method for the conduction cooling effect was developed to clarify the temperature distribution in the coil winding following the quench.

The experimental results and the computer simulation indicate that the thermally weak coupling between coil and cooling has following features.

1. If the coil winding and cooling channel are weakly coupled, the disturbance on the temperature and pressure balance of the refrigeration system caused by a quench can be minimized. The recovery time of the whole system may also be reduced.

2. The stability of the magnet does not depend significantly on the cooling condition of the convecting helium. Therefore the magnet could be designed without the elaborate consideration to the heat transfer rate to the cooling helium.

ACKNOWLEDGEMENTS

The authors acknowledge T. Ogasawara and Y. Hiresaki for useful discussions.

NOTATION

A_c = cross sectional area of the coil
a = radius of the conductor without coating
b = equivalent radius defined in Fig. 4.
C_i = heat capacity of the segment i
N_0 = $\pi/6/\delta\theta$, number of individual units present
Q = heat conducted in the radial direction
Q_c = heat removed from the unit area of the coil
Q_J = heat generated in the segment i
Q_H = heat added to the unit area of the coil
T_i = temperature of the segment i
δT = T_i-T_{i+1}, temperature difference of adjacent unit volume
δR = length of the segment i to the radial direction of the coil
$\delta\theta$ = equivalent angle inclement for the heat flux
θ = $n\,\delta\theta$, for n=1 to N_0
λ = thermal conductivity of the impregnant
λ_e = effective thermal conductivity of the coil
λ_i = effective thermal conductivity of the segment i
τ_d = equivalent diffusion time constant
τ_p = duration of the heater pulse

REFERENCES

1. J. Zellweger, G. Vecsey and I. Horvath, Superconducting
   Magnets of the Biomedical Facility at SIN, in "Advances in
   Cryogenic Engineering, Vol 25," Plenum Press, New York
   (1980), p. 232.
2. M.A. Green, W.A. Burns and J.D. Taylor, Forced Two-phase
   Helium Cooling of Large Superconducting Magnets, in
   "Advances in Cryogenic Engineering, Vol. 25," Plenum Press
   New York (1980), p. 420.
3. W.H. Hassenzahl, Quenches in the Superconducting Magnet CELLO,
   in "Advances in Cryogenic Engineering, Vol. 25," Plenum
   Press New York (1980), p. 185.
4. M.C. Jones and W.W. Johnson, Heat transfer and flow of helium
   in channels-practical limits for applications in
   superconductivity, NBS Technical Note 675 (1976).
5. M. Sinclair and Y. Iwasa, Transient resistive zones in NbTi
   composites, IEEE Trans, on Magnetics MAG 15 (1):347 (1979).

DEVELOPMENTS IN He II HEAT TRANSFER AND APPLICATIONS TO SUPERCONDUCTING MAGNETS

S. W. Van Sciver

University of Wisconsin
Madison, Wisconsin

INTRODUCTION

Within the past ten years there has been a growing interest in
He II and its application to superconducting magnet technology.[1-3] Numerous heat transfer experiments and magnet tests[4,5]
have been carried out showing generally improved behavior over
that of He I. It now appears that He II is gaining acceptance as
a coolant for superconducting magnets and that there exists a
reasonable understanding of its heat transfer characteristics.

This paper has two sections. First, a survey of the heat
transfer in He II describes the working principles for the cooling
capability of this fluid. In the second section, a discussion of
composite superconductor stability is presented specifically
within the context of He II cooling. Where possible direct reference is made with recent experimental investigations.

THEORY OF HEAT TRANSFER IN HE II

Understanding the heat transfer characteristics of He II
requires an appreciation for the principles which govern heat flow
in this unique medium. Since He II behaves so differently from
any other liquid, a two fluid model[6] has been devised to describe
its properties. This model is a direct analogy to that used in
the description of superconductivity.

The two fluid model characterizes He II as a mixture of two
interpenetrating fluids. The normal fluid component is thought to
behave as an ordinary liquid having finite viscosity η, entropy s
and density ρ_n. The superfluid component, with density ρ_s, represents the fraction of the fluid which is in a quantum mechanical
ground state. The unique properties of the superfluid component

are its vanishing entropy and viscosity. The model further assumes that the density of liquid He II is a sum of the two components,

$$\rho = \rho_s + \rho_n \tag{1}$$

Although ρ is approximately constant for $T < T_\lambda$, ρ_s and ρ_n are
strong functions of temperature having equal value at 1.95 K.[7]

When He II is subjected to a heat current, a process described
by the two fluid model is developed. This process, known as
internal convection, associates the heat flow with the normal
fluid carrying entropy away from the heater surface and the superfluid flowing toward the heater. Internal convection obeys momentum conservation,

$$\rho\vec{v} = \rho_s\vec{v}_s + \rho_n\vec{v}_n \tag{2}$$

where $\rho\vec{v}$ is zero when there is no net mass flow. Since the
superfluid carries no entropy, the heat flow is described by the
flow of normal fluid and is given by the relation

$$\vec{q} = \rho ST\vec{v}_n \tag{3}$$

where v_n is the normal fluid velocity. Note that for a heat flux
of 1 W/cm^2, the normal fluid velocity is about 5 cm/sec at 1.9 K.

There exist two processes associated with internal convection
which give rise to temperature and pressure gradients in the
fluid. These processes are due to different mechanisms each
dominating the behavior for different heat fluxes and channel
dimensions.

Shown schematically in Fig. 1 is a channel of diameter d
containing He II. The two mechanisms of heat flow are the thermo-
molecular fountain effect and mutual friction. The fountain
effect gives rise to a temperature gradient as a result of the
normal fluid interaction with the walls of the tube. It is described by a relation which has an explicit diameter dependence.

$$\nabla T = g(T)\ q/d^2 \tag{4}$$

where $g(T) = \dfrac{32\eta}{\rho^2 S^2 T}$ for cylindrical cross section tubes.

This quantity is well known from heat flow measurements and is
tabulated in the literature.[8]

The other mechanism, known as mutual friction, results from
the turbulent interaction between the normal and superfluid compo-

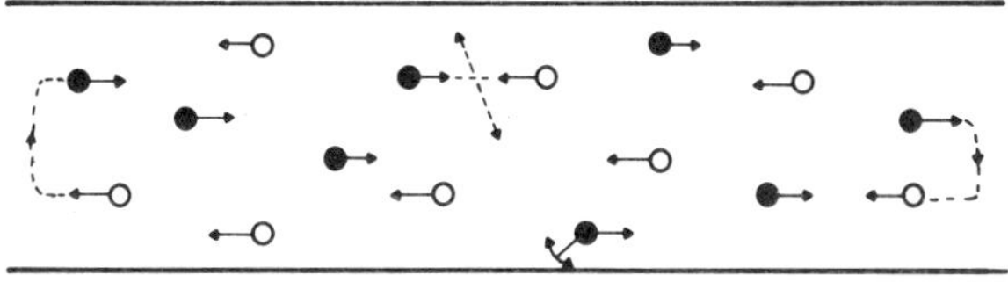

Fig. 1. Schematic representation of two-fluid
model: ●, normal fluid o, superfluid components.

nents. This is a nonlinear frictional process and is described by

$$\nabla T = f(T) \, q^m \tag{5}$$

where m, the exponent, is approximately 3 and

$f(T) = \dfrac{A \, \rho_n}{S} \, (\rho_s ST)^{-m}$, A being the empirically determined Gorter-

Mellink parameter.[9,10] One should note that the units of f(T) are consistent only when m=3 and that there is no explicit diameter dependence in Eq. 5.

Because of the different functional dependences to Eq. 4 and Eq. 5, each relation dominates the heat flow for different sets of conditions. For example a tube of diameter 10 μm is limited by the normal fluid interacting with the walls, and therefore has a linear relationship between ∇T and q. For this channel a temperature gradient of about 2.5 mK/cm is typically observed at 1.9 K and 1 W/cm^2. For channels with diameter in excess of 100 μm, mutual friction is the determining mechanism for the heat flow and the temperature gradient. At 1.9 K and 1 W/cm^2 the mutual friction interaction gives rise to a temperature gradient of about 0.5 mK/cm, independent of tube diameter.

For most applications, channels are large enough to have their heat flow controlled by the mutual friction mechanism. When using Eq. 5, it is possible to define an effective thermal conductivity, K_e, as a function of temperature and heat flux. A smoothed fit to one set of measurements[11] of K_e is plotted in Fig. 2 versus temperature. Note that K_e is a function of heat flux due to the nonlinear nature of the heat flow equation. Other data[12-15] give similar results within ±20% of these values. Also, there exists approximately a 10% decrease in K_e per bar (100 kPa) of applied pressure at 1.9 K.[16]

HEAT TRANSFER LIMITS

Heat transfer between a solid and He II is primarily controlled by the properties of the bulk fluid. It is surprising

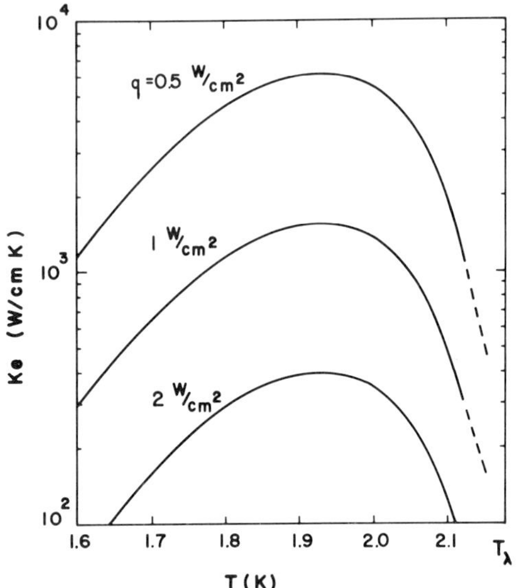

Fig. 2. Effective thermal conductivity of He II.

therefore that the heat transfer character is rather similar in appearance to that of normal helium. This fact can be seen in Fig. 3, which is a plot of surface heat transfer for He I[17] compared to saturated and subcooled He II.[16] Four aspects to these heat transfer curves are discussed below:

1)   Steady state peak heat flux, q*

2)   Transient peak heat flux, q*(t)

3)   Kapitza conductance, $q/\Delta T_s$ for small ΔT_s

4)   Film boiling, $q/\Delta T_s$ for large ΔT_s.

Steady State Peak Heat Flux

The limiting heat flux in He II is best understood by reference to the helium phase diagram, Fig. 4. Since He II is such a highly conducting medium, we have reasonable confidence in assuming its behavior is governed by equilibrium thermodynamics. Consequently, when a channel containing He II is subjected to a heat flux, a temperature difference is established due to the frictional processes described above. If the cold end is maintained at a constant temperature, the maximum temperature difference is determined by the allowable temperature excursion on equilibrium phase diagram. If the system is near saturation, this

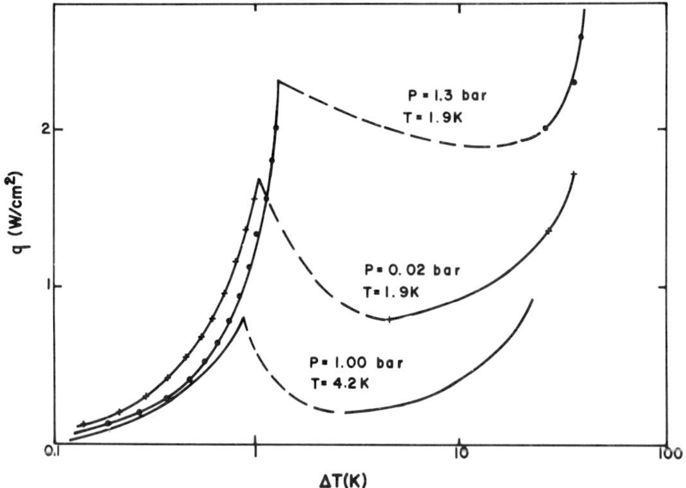

Fig. 3   Typical heat transfer and boiling curves
for He I and He II.

excursion is limited by the He II – vapor phase boundary as shown
by position 1 in Fig. 4.   The actual point at which the hot end
temperature crosses the phase boundary is determined by the hydrostatic head, $\Delta p = \rho g h$.[18]   It is worth noting that the pressure at
the λ-point corresponds to a hydrostatic heat of about 3.5 meters
of saturated He II, above which the transition occurs at the He
II-He I phase boundary.

An alternative operating condition which is becoming popular
in many systems is that of 1 bar (100 kPa) subcooled He II.   This
condition is shown on the phase diagram by position 2.   In subcooled He II the transition always occurs at the λ-temperature and
is accompanied by the creation of He I at the heater surface.
Usually the formation of He I induces the onset of film boiling
and an associated coexistence of three phases.[19]

In linear systems, the peak heat flux is determined by two
factors.   The bath temperature T_b sets the allowable maximum
temperature difference across the channel.   Because of the rapidly
decreasing effective thermal conductivity at lower temperature,
1.8 K is often chosen as a good practical operating temperature.
The other determining factor for the peak heat flux is the channel
length, L.   For very long channels, the steady state peak heat
flux is low because the allowable temperature gradient is smaller
than in short channels.

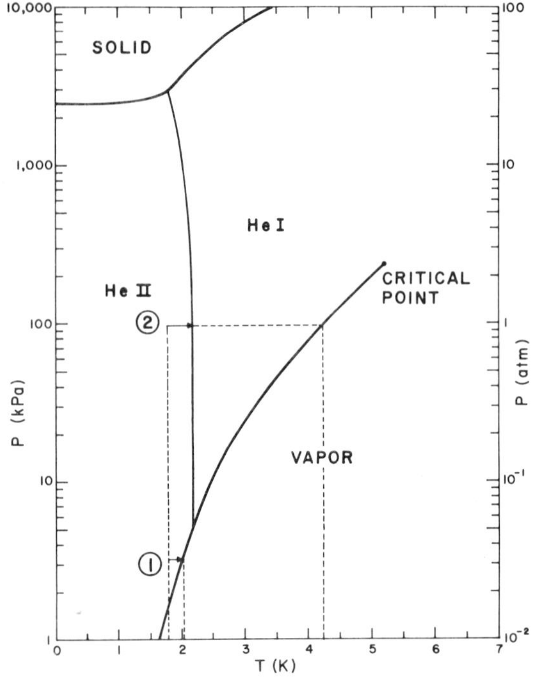

Fig. 4. Phase diagram of helium showing location of (1)
saturated and (2) subcooled operation.

For actual experiments, the peak heat flux can be interpreted
by integration of Eq. 5. This analysis gives rise to a relation

$$q* = \frac{F(T_m, T_b)}{L^{1/3}} \tag{6}$$

where the function $F(T_m, T_b) = \int_{T_b}^{T_m} \frac{dT}{f(T)}$. Thus, for experiments

performed in subcooled helium, where $T_m = T_\lambda$, the quantity $q* \, L^{1/3}$
should be a function of bath temperature only. Plotted in Fig. 5
along   with   data   for   peat   heat   fluxes   from   the
literature,[13,14,16,20,21] is the caluclated function $F(T_b, T_m)$
based on one set of measurements.[11] The fit to the data is good
considering the channel lengths vary over four orders of magnitude
and is generally within ±20% for most bath temperatures.

The fact that the steady state peak heat flux is determined
only by the channel geometry has been visually shown in a   recent

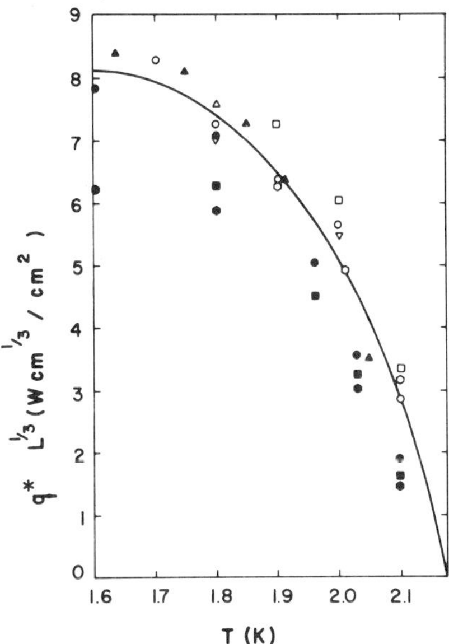

Fig. 5. Normalized steady state peak heat flux in He II.

experiment by Gentile and Francois.[22] In this study, a uniform
section channel is connected to a conical piece with a larger
helium cross sectional area. Because of the variability of heat
flux across the conical section, the workers were able to induce
the onset of vaporization in the channel rather than on the heater
surface. A schematic representation of their visual observations
are shown in Fig. 6. All measurements were performed in saturated
He II.

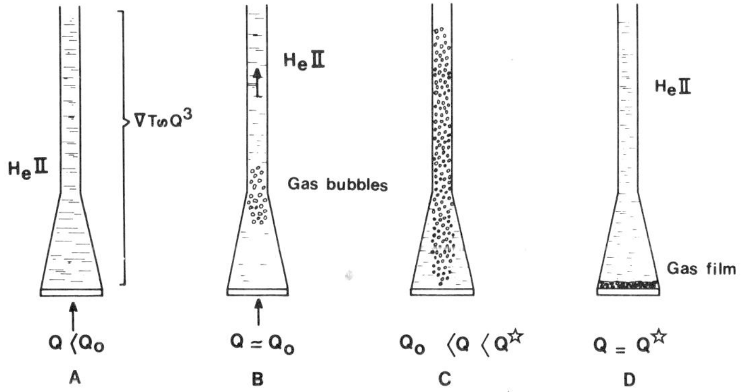

Fig. 6. Visual observations of heat transport in a non-
uniform channel.

Up to a heat Q_o, the heat flow was seen to obey the mutual friction relation. However, the Q_o vapor bubbles were observed to occur in the lower end of the straight section of the channel at a heat flux consistent with the temperature reaching the saturation condition. Once the bubble formation occured, it was possible to increase the heat flux to substantially higher values until Q* initiated film boiling at the surface. Thus, a two phase mixture of He II and vapor was seen to enhance the heat transport capability of the bulk fluid.

Although apparently correct for wide channels ($d \geqslant 1mm$), the above description of steady-state heat transfer must be modified somewhat when considering smaller dimensions. This feature has been shown in some recent experiments on fine capillaries[23]. In these studies, multibore tubing with holes 40 μm in diameter were filled with saturated He II and subjected to steady state heat transport. Two quite new observations were reported and can be seen in Fig. 7. First, independent of bath temperature the peak temperature at the heated end of the channel reached the λ-transition. Thus, the He II-vapor transition arguments raised to describe the peak heat flux in the saturated system do not appear to apply in this case. This fact may be associated with a superheating mechanism discussed in connection with some experiments in saturated He II.[24]

The other observation is the appearance of some modified heat transport behavior at large temperature differences. Plotted along with the data in Fig. 7 is the relation for temperature differences across the channel achieved by integrating the sum of Eq. 4 and Eq. 5. The expression fits the data quite well except at high temperature difference, $\Delta T \geqslant 0.1$ K, where the helium appears to become more conductive. This characteristic has the effect of increasing the peak heat flux by about a factor of four.

Transient Heat Transfer

When He II is subjected to a large heat pulse, its temperature rises and heat diffuses into the fluid by apparently the same mechanisms that govern the steady state heat transfer. Therefore, the transient peak heat flux is established by the requirement that the temperature of the helium rise to T_λ or the saturation temperature at that hydrostatic head.

The above hypothesis has been used to describe transient heat transfer measurements in channel geometries.[25,26] The approach is to assume that Eq. 5 correctly describes the local temperature gradient with a time varying heat flux. Combining this assumption with the energy balance equation, a relation for the maximum energy disturbance, $\Delta E = Q\Delta t$, versus the total allowable enthalpy rise of the helium

$$\Delta E_O = V_H \int_{T_b}^{T_\lambda} C_P dT \text{ results,}$$

$$\frac{\Delta E}{\Delta E_O} = X (q \ L^{1/3})$$

(7)

where L is the length of the channel and X is a numerically deter-mined function. The quantity $q \ L^{1/3}$ is a scaling parameter re-sulting from the non-linear heat flow equation.

Plotted in Fig. 8 is the calculated relationship $\Delta E/\Delta E_O$ vs $qL^{1/3}$. It should be noted that there are no adjustable parameters in this formulation. Also displayed in the figure are measure-ments of $\Delta E/\Delta E_O$ for various channel configurations. There is some variability in the data but generally the agreement between the calculations and analysis confirms the assumptions.

There have been several measurements[27,28] of the time depend-ent temperature gradient in He II which confirm the correctness of

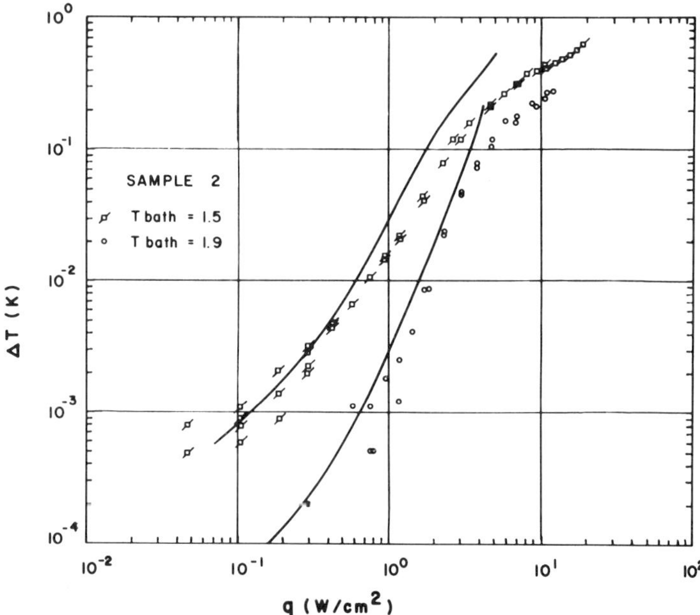

Fig. 7. Temperature difference across 40 μm diameter capillaries filled with He II.

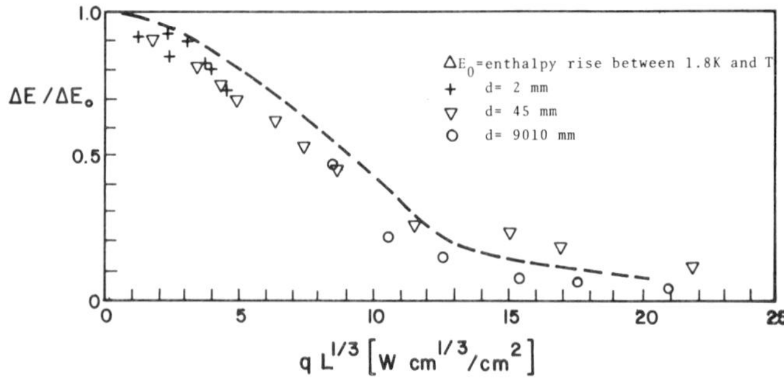

Fig. 8. Normalized transient heat transfer in He II.

the above assumption. One recent experiment[29] was performed on a
6 mm I.D. tube 2.3 m long which was open at both ends to a subcooled, 1 bar (100 kPa), bath of He II. A copper heated section
of 10 cm length located in the center of the tube was used to
apply high power short duration heat pulses.

The temperature distribution in the helium channel as a function of distance and time for one set of measurements is shown in
Fig. 9. The pulse size was 0.92 J and was applied in the first 20
ms. Two important features of these data should be noticed.
First, the sharp temperature distribution which is apparent in the
first several hundred milliseconds is the result of the heat being
transferred to the helium in the tube. Integration of the enthalpy rise contained in the temperature distribution at 200 ms is
equal to the total energy in the applied pulse, 0.92 J. Secondly,
it requires longer times, greater than 500 ms, before a substantial fraction of the heat is transferred to the bath. This fact
is evident by the establishment of a temperature gradient at the
ends of the channels. Analytic interpretation of these data
requires a time dependent solution for the temperature distribution similar to that used for the peak heat flux. Although, the
analysis has not been performed for the data presented in Fig. 9,
the general form of the results is conferred by similar calculations on different geometries.[30]

One final comment regarding the time to achieve steady-state
and its connection with transient heat transfer should be made.
Measurements for long channels[13,27] have shown a strong length
dependence on the time required to establish a steady-state
temperature gradient, sometimes as much as several hours for long
tubes. For applications this long time constant can impact the
type of analysis since the heat transfer process is either steady-
state or transient depending on the distance between the source
and the heat sink. Thus, in order to apply heat transfer analysis

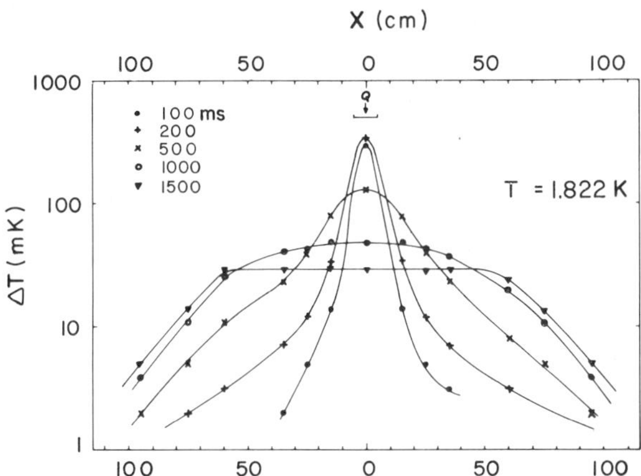

Fig. 9. Temperature distribution in a He II channel after a 0.92 J pulse.

to a He II cooled system, it is necessary to compare the time constant of the disturbance with the thermal time constant of the cooling system. This point will be further discussed in the section on stability applications.

Kapitza Conductance

At the interface between a solid and He II, a thermal boundary resistance is present due to a phenomenon called Kapitza conductance.[31,32] When the solid body is a pure metal like copper or aluminum, this boundary conductance can lead to the largest temperature difference in the system. In Fig. 3, the small ΔT region of the heat transfer for He II is caused by this effect.

Present understanding of Kapitza conductance associates heat transport across the boundary with an acoustic mismatch between two very different materials.[33] Models used to interpret this effect generally provide a relation,

$$q = \sigma(T_s^n - T_b^n) \tag{8}$$

where n is around 4. Unfortunately, experiments do not correlate well with this expression. Empirical values for n vary between 2 and 4 while the magnitude of σ is known to no more than a factor of two. This problem appears to be a consequence of the complicated effects of surface preparation, sample purity and method of measurement.

Examples of practical Kapitza conductance data are plotted in Fig. 10. It is interesting to note that there exists about a factor of two variation in T_s depending on which data is considered. As has been reported elsewhere,[34] samples with apparently identical preparation have significant variations although perhaps not as large as is shown in Fig. 10.

Another feature to Kapitza conductance is that different applications require different parts of the heat transfer curve. Therefore, care must be exercised when applying a fit made at high heat fluxes, $q > 1$ W/cm^2, to an application which operates at heat flows below 100 mW/cm^2. This problem has been demonstrated in a recent experiment[37] involving heat transfer from a thick walled tube filled with He II. In this work, the tube was heated from one end and insulated by vacuum so the heat was tranferred from the copper to the helium much like a heat transfer fin. In interpreting the data, it was not possible to provide a good fit for the entire tube using one set of parameters in Eq. 8. This illustrates the problem of modeling complex heat transfer problems involving Kapitza conductance. In order to model the system well, one needs to understand Kapitza conductance for the particular material. However, the approach defeats the purpose of the analytical effort, which should be useable as a productive tool.

Film Boiling

Under most circumstances, the peak heat flux is associated with a transition to the film boiling state. If the system is subcooled, film boiling is preceded by the formation of He I. However, if the heat flux is higher than the peak nucleate boiling heat flux for He I, film boiling is rapidly initiated at the He I – solid interface. Thus the presence of He I usually has little impact on the film boiling heat transfer.

Most experimental work on film boiling heat transfer has attempted to understand the steady-state processes. Numerous measurements on fine wires and flat plates have been reported in the literature.[38-41] For cylindrical wires, the film boiling heat transfer coefficient, h_{fb}, depends on wire diameter, surface and bath temperature as well as hydrostatic head in saturated systems. Data for flat surfaces show similar variation although the heater dimensions have less significance. Typical heat transfer coefficients for large samples, $d \sim 1$ cm, are 0.02 W/cm^2 K at saturation increasing to 0.06 W/cm^2 K at 1 bar (100 kPa) subcooling.

There appears to be some fundamental understanding of film boiling heat transfer in He II based on non-equilibrium kinetic theory.[42] This interpretation, which associates the heat transfer

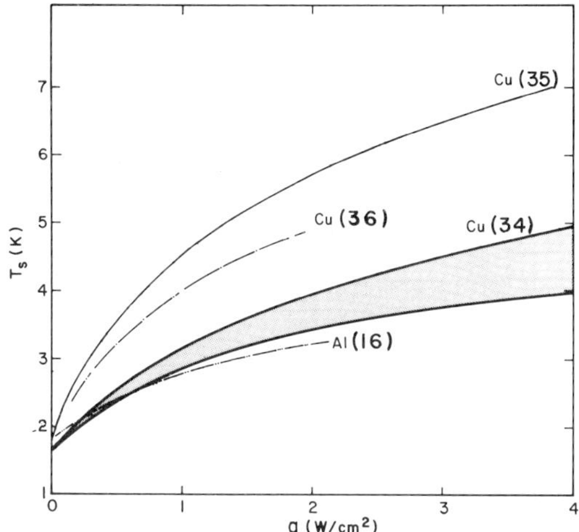

Fig. 10. Practical Kapitza conductance data for Cu and Al.
Numbers refer to references.

to pure conduction in the vapor film, predicts the film boiling
heat transfer coefficient for fine wires to within 10-20%. Similar interpretation has been applied to understand the recovery
process from film boiling to the Kapitza regime.[43] Although film
boiling is an important heat transfer process in He II, it appears
to be the least understood owing to its strong dependence on
experimental geometry, applied pressure and temperature.

APPLICATION TO MAGNET STABILITY

In order to apply the understanding of He II heat transfer to
the stability of composite superconductors, it is necessary to
address the entire system. Such an analysis requires the combination of surface heat transfer from the conductor to the helium
with the heat transported in the helium channels. Within the
context of this approach, it is possible to define various stability criteria limited by the heat transfer characteristics. Three
such criteria discussed below are; steady-state, cold-end recovery, and enthalpy stability.

Steady-State Stability

This criterion is established based on the requirement that
the heat generated by the conductor while fully normal must be
transferable by the adjacent channels.[44] Shown schematically in
Fig. 11 is a typical composite conductor with insulating spacers
and cooling channels, allowing heat flow in several directions.

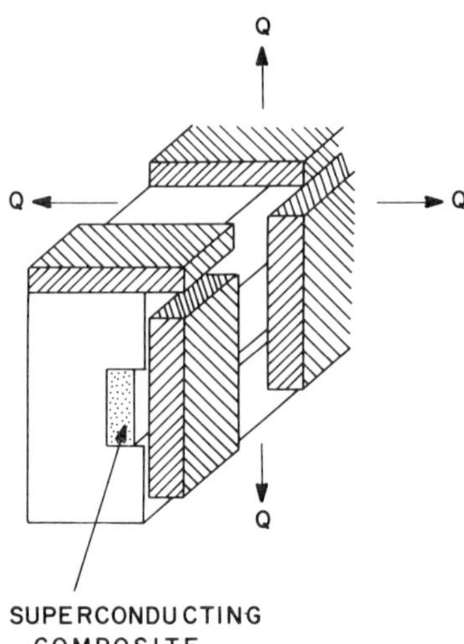

Fig. 11. Typical bath cooled composite superconductor.

The steady-state stability criterion is defined according to the relation

$$\frac{I^2 \tilde{\rho}}{A_S} = \frac{F(T_m, T_b)}{L^{1/3}} A_c = q^* A_c \tag{9}$$

where the left hand side is the heat generated in fully normal condition, A_S being the stabilizer cross section area. The right hand side is the peak channel heat flux q^* defined by Eq. 6 times A_c, the channel cross section per unit length of conductor. In this case, it is usually assumed that one can take advantage of all channels, horizontal and vertical.

Caution must be exercised when applying the above stability analysis. First, because the time constants for steady-state in long channels can be greater than the duration of the instability, it may be incorrect to assume steady-state heat transport. Second, should a substantial fraction of the conductor enter the normal state and stay there, the heat generated would be sufficient to increase the bath temperature with time, thereby reducing the peak heat flux, see Fig. 5. These two features encourage defining the above stability criterion as a limited time steady-state approach. Fortunately, the transient heat transfer in He II is always better than steady-state allowing for improved stability for short duration normal zones.

An interesting and perhaps important aspect to steady-state stability in He II, as described by Eq. 9, is that there appears to be a tradeoff between stabilizer which reduces the heat generation term and channel cross section which increases the heat transfer rate. For example, for the specific conductor considered in Fig. 11, a relation exists for the overall current density of the winding, J_o, versus physical aspects,

$$J_o = \left[\frac{F(T_m,T_b)}{L^{1/3}} \frac{N\gamma\alpha}{\tilde{\rho}}\right]^{1/2} \frac{1A_S}{[A_S+A_I]} \qquad (10)$$

where N is the number of cooling channels, γ the fraction of the conductor surface that is cooled and α the fraction of the conductor that is stabilizer. Optimization of Eq. 10 shows that J_o has a maximum value for a thickness 1 which is of the order of the conductor width. This indicates a tendency for He II stabilized magnets to increase their helium to conductor ratio over that for He I cooling.

Cold End Recovery

As in the case of He I cooled composite conductors, there is some evidence that a criterion based on end cooling[45] may be applicable to He II. To approach this problem, it is necessary to construct an entire heat transfer curve normalized on a per unit surface area basis.

Such a curve is shown in Fig. 12 by the solid line. The peak heat flux is defined according to Eq. 9. Below that value Kapitza conductance governs the temperature difference, while at higher fluxes film boiling is present. Unfortunately, the film boiling heat transfer coefficient is not well-known, which presently weakens the application of this method. The recovery heat flux is assumed to equal the peak flux for reasons of simplicity and uncertainty as to its value. Also plotted in Fig. 12 is a heat generation curve showing a current sharing temperature T_{cs} and critical temperature T_c above which the conductor is fully normal.

The cold end recovery criterion is established by the requirement that the area of region A be less than the area of B in Fig. 12, under which conditions a limited size normal zone will contract and recover. The criterion benefits by the use of high T_c materials such as Nb_3Sn, where area B can be of substantial size. This shows some incentive to consider He II as a coolant for superconducting materials with high T_c.

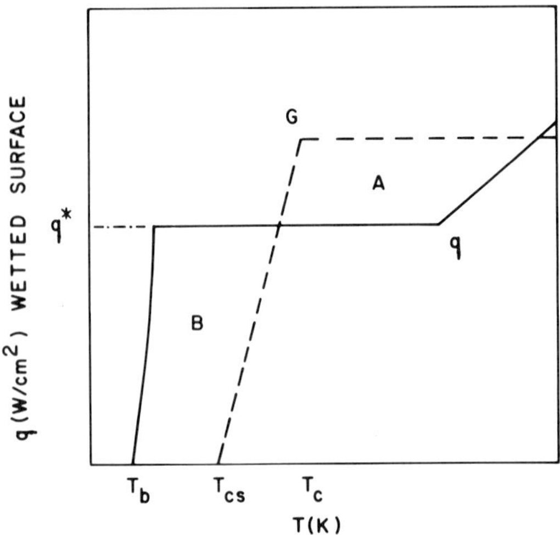

Fig. 12. Cold end recovery stability applied to He II.

Enthalpy Stability

The concept of limited time stability[46] has great importance
to He II cooling because of the long times to achieve steady-state
heat transfer compared to the duration of most conductor instabilities.

Enthalpy stability is defined as follows: an energy pulse is
deposited locally in a winding cross section, Fig. 13. The heat
diffuses into the helium channels some distance ℓ, which is governed by the processes in transient heat transfer described
above. The maximum energy deposition ΔE can therefore be defined
by the rate of heat generation as well as the channel cross section. If the heat transfer is controlled by one dimensional
processes, such as for hollow conductors, the maximum energy
disturbance is defined according to the results presented in Fig.
8. For a multi-dimensional winding such as in Fig. 11, the maxi-

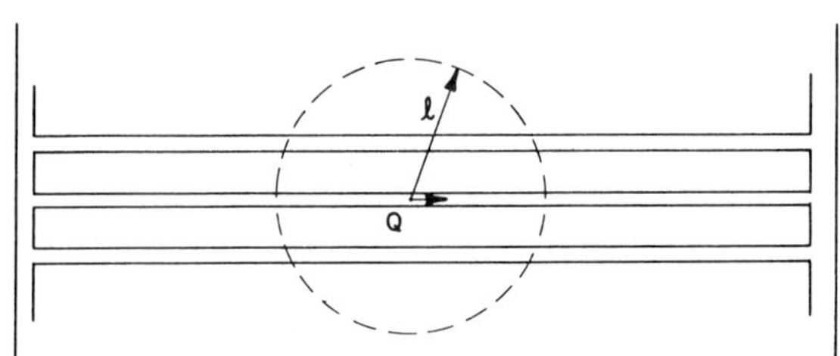

Fig. 13. Schematic representation of enthalpy stabilization.

mum energy deposition can be approximated by adding to total channel cross section per unit conductor length much as has been achieved for steady-state criteria.

In practice, calculation of the maximum disturbance ΔE in one dimensional channels is achieved by taking into account transient heat transfer,

$$\Delta E = X(qL^{1/3})\, V_{He}\, \Delta \varepsilon_o \qquad (11)$$

where the transient heat transfer function, X, is given in Fig. 8. The helium volume, V_{He}, depends on the winding geometry and represents only the liquid in the channels which are transferring the heat. It is important to note that the specific enthalpy $\Delta \varepsilon_o = 270$ mJ/cm^3 between 1.8 K and T_λ. As in other transient stability analyses, the above calculation suffers from the lack of knowledge of the actual disturbance spectrum and the description of the heat generation term.

Stability Experiments

The clearest example of an experimental test of composite conductor stability in He II has been reported by Claudet, et al.[47] A schematic of the test section used in this experiment is shown in Fig. 14. The coil was wound of a composite NbTi in copper similar to that proposed for use in the Torus Supra toroidal field coils.

The conductor with a copper to superconductor ratio of 1.6 was insulated by intermittent composite spacers which provided vertical cooling channels. Different spacer widths, 1, and thicknesses, e, were used to investigate the effect of varying the channel dimensions. The test section was instrumented with a local heater which applied pulses up to 1 J. All tests were performed in a helium bath at 1 bar (100 kPa) pressure at either 4.2 K or 1.8 K and in a background magnetic field up to 12.5 T.

A summary of the results are shown in Fig. 15. which is a plot of sample current versus applied field. A number of important observations can be gleaned from this work. First, on an overall basis it is clear that the stability of the test section is much higher in He II than in He I. For example at 8 T, the stable current that recovers from 1 J pulse is a factor of two or three higher in He II than in He I. This fact is due to the higher heat transfer. Furthermore, the channel width, e, has no apparent effect on the stability at 4.2 K but strongly impacts the performance at 1.8 K, with the wider channels having higher stability.

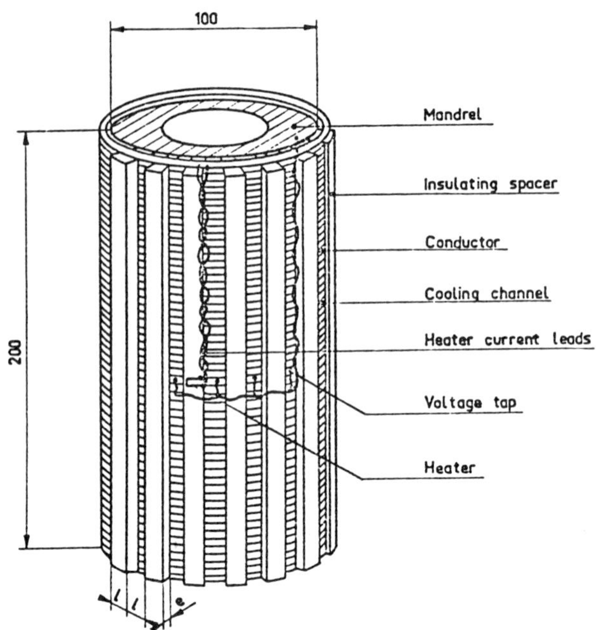

Fig. 14. Test section for stability experiment.
Channel width 1 = 7, 10 or 14 mm; e = 0.9 and 1.8 mm.

To test the experimental results, the experimenters calculated
the steady state stability limit given by Eq. 9 for the given
channel geometries. These results are shown in Fig. 15 by the
dashed line for He II and the dotted line for He I. Generally,
the analysis confirms the enhanced stability in He II and the
reason for better performance in He II with the wider cooling
channels. The difference in magnitude between the measurement and
calculation at low field, B < 8 T, appears to be due to end
cooling stabilization in both cases.

An additional artifact to the He II stability data is observable at high fields, B > 8T. At this point there occurs a clear
break in the stable current line with the spacer width, 1 being
the important parameter. This behavior has been attributed to the
existence of stable normal zones in uncooled regions of the conductor.[48,49] It is relatively more important in He II because of
the lower thermal conductivity of the stabilizer at low temperatures.

Other stability experiments[50,51] in He II have shown comparable improvements in performance although not as clearly described
in terms of heat transfer. As a rule these studies have shown

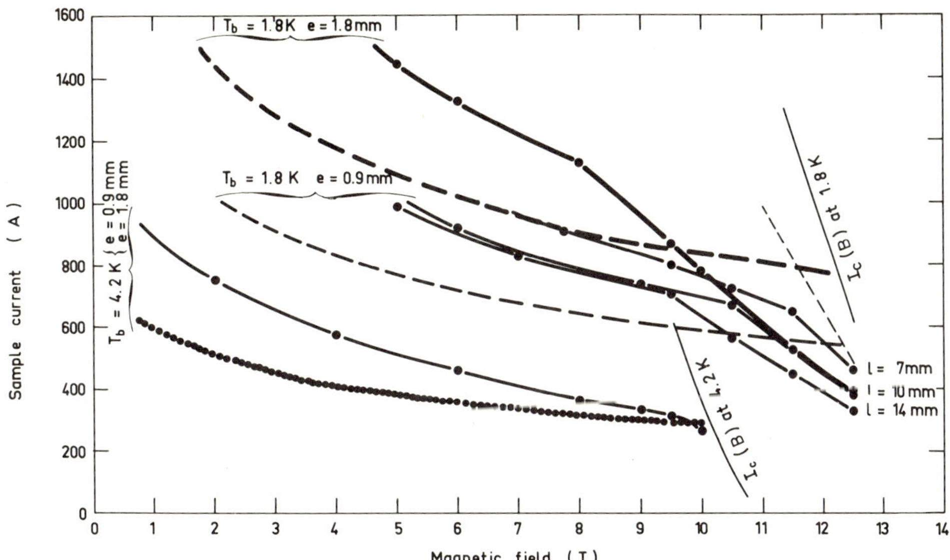

Fig. 15. Stable current for test section cooled
by He I and He II.

improved performance in He II over that in He I provided the
cooling channels are of adequate size.

CONCLUSION

Numerous contributions to the field of He II heat transfer
have established the principles governing its behavior and application to superconducting technology. There exists several problems that require further work. A few examples of these include: multidimensional heat transfer (2D or 3D), studies of
restricted geometry applicable to configurations with small channels, and further investigations of film boiling. More studies of
stability along the lines of the work surveyed in this paper are
also in order. Major engineering advances are being lead by
specific applications groups, the most notable being Toruo
Supra. In this work, magneto cooled by He II will operate as part
of large complex experiment requiring high duty cycle and reliable
performance.

ACKNOWLEDGEMENTS

The author wishes to express appreciation to the numerous
workers who have contributed to this review. Special thanks go to

J.C. Lottin, C. Meuris, P. Seyfert, G. Claudet, B. Turck, D. Gentile and M. Francois.

NOMENCLATURE

A	– Gorter Mellink parameter
A_c	– Channel cross section area
A_I	– Insulator cross sectional area
A_S	– Stablizer cross sectional area
C_p	– Constant pressure specific heat of helium
d	– Channel diameter
e	– Spacer (channel) thickness
ΔE	– Energy disturbance, $Q\Delta t$
ΔE_o	– Enthalpy change in helium
$\Delta \varepsilon_o$	– Specific enthalpy change
$F(T_m, T_b)$	– Heat conductivity integral
$f(T)$	– Turbulent heat conductivity function
$g(T)$	– Laminar heat conductivity function
g	– Acceleration of gravity
h	– Hydrostatic head
h_{fb}	– Film boiling heat transfer coefficient
I	– Transport current
J_o	– Overall current density
K_e	– Effective thermal conductivity
ℓ	– Diffusion length
1	– Spacer (channel) width
L	– Length of channel
m	– Exponent in heat flow equation
n	– Exponent in Kapitza equation
N	– Number of cooling channels
p	– Pressure
q	– Heat flux per unit area
$q*$	– Peak heat flux per unit area
Q	– Total heat flux
Q_o	– Critical total heat flux
$Q*$	– Peak total heat flux
S	– Entropy
T	– Temperature
T_λ	– He II–He I transition temperature
T_S	– Surface temperature
T_b	– Bath temperature
T_c	– Critical temperature
v	– Velocity
v_s	– Superfluid velocity
v_n	– Normal fluid velocity
V_{He}	– Helium Volume
X	– Transient heat transfer function
α	– Stabilizer fraction in conductor
γ	– Fraction of conductor surface cooled

$\tilde{\rho}$	– Electrical resistivity
ρ	– Density
ρ_s	– Superfluid density
ρ_n	– Normal fluid density
η	– Viscosity
σ	– Kapitza conductance coefficient

REFERENCES

1. R. Aymar et al., Conceptual design of a superconducting tokamak: "Torus II-Supra", IEEE Trans on Magnetics Mag 15:542 (1979).

2. Wisconsin Superconductive Energy Storage Project, Vol. 1, Engineering Experiment Station, University of Wisconsin, July 1, 1974.

3. Y.H. Hsu et al., An ETF TF-coil concept employing NbTi alloy, bath cooled with superfluid helium, IEEE Trans on Magnetics Mag-17:626 (1981).

4. G. Bon Mardion, G. Claudet and J.C. Vallier, Superfluid Helium Bath for Superconducting Magnets, in "Proceedings 6th International Cryogenic Engineering Conference" IPC Science and Technology Press, Grenoble (1976).

5. R. Aymar et al., A NbTi 10T 388 mm bore magnet working at 1.8 K in pressurized superfluid helium, Cryogenics 20:521 (1980).

6. J. Wilks, "The Properties of Liquid and Solid Helium", Clarendon Press, Oxford (1967).

7. Tabular Values for the Thermodynamic Functions of He II are available in J.S. Brooks and R.J. Donnelly "The Calculated Properties of Helium II", A Technical Report for the Institute of Theoretical Science and Department of Physics, University of Oregon, Eugene (1973).

8. V. Arp, Heat transport through Helium II, Cryogenics 10:96 (1970).

9. This coefficient has been reported to increase slightly with temperature approaching a value of 4 very near T_λ. Some workers have resolved this problem by choosing an average value for m = 3.4 over the entire range from 1.4 K to T_λ. On physical grounds, the present discussion will use a coefficient of 3 with the understanding that there may be some discrepancy introduced.

10. C.J. Gorter and J.H. Mellink, On the irreversible processes in liquid helium II, Physica XV:285 (1949).

11. S.W. Van Sciver, Kaptiza Conductance of Aluminum and Heat Transport from a Flat Surface Through a Large Diameter Tube to Saturated Helium II, in "Advances in Cryogenic Engineering, Vol. 23", Plenum Press, New York (1978), p. 340.

12. W.F. Vinen, Mutual friction in a heat current in liquid Helium
    II, I. Experiments on steady heat currents, Proc. Roy.
    Soc. London A240:114 (1957).
13. J.C. Lottin, Helium II Experimental Facilities at Saclay, in
    "Advances in Cryogenic Engineering, Vol. 27", Plenum Press,
    New York (1982).
14. G. Bon Mardion, G. Claudet and P. Seyfert, Steady State Heat
    Transport in Superfluid Helium at 1 Bar: in "Proceedings
    7th International Cryogenic Engineering Conference", IPC
    Science and Technology Press, London (1978).
15. B.W. Clement and T.H.K. Frederking, Thermal Boundary
    Resistance and Related Peak Heat Flux during Supercritical
    Heat Transport from a Horizontal Surface Through a Short
    Tube to a Saturated Bath of Liquid Helium I, in "Liquid
    Helium Technology, Proceedings International Institute Ref.
    Commission I", Pergamon Press, Oxford (1966).
16. S.W. Van Sciver, Kapitza conductance of aluminum and heat
    transport through subcooled He II, Cryogenics 18:521
    (1978).
17. D.N. Lyon, Boiling Heat Transfer and Peak Nucleate Boiling
    Heat Fluxes in Saturated Liquid Helium Between the λ and
    Critical Temperatures, in "Advances in Cryogenic
    Engineering, Vol. 10", Plenum Press, New York (1965), p.
    371.
18. There exists some disagreement on whether superheating occurs
    in He II allowing the temperature to achieve a higher value
    than saturation. This is a detail which will be returned
    to below.
19. S. Caspi and T.H.K. Frederking, Triple-phase phenomena during
    quenches of superconductors cooled by pressurized
    superfluid Helium II, Cryogenics 19:513 (1979).
20. C. Linnet and T.H.K. Frederking, Thermal conditions at the
    Gorter-Mellink counterflow limit between 0.01 and 3 bar, J.
    Low Temp. Phys., 21:447 (1975).
21. G. Krafft, Superheating and bubble formation in Helium II, J.
    Low Temp. Phys., 31:441 (1978).
22. D. Gentile and M.X. Francois, Thermal Instabilities in a
    Helium II Channel, in "Advances in Cryogenic Engineering,
    Vol. 27", Plenum Press, New York (1982).
23. B. Helvensteijn, S. Breon and S.W. Van Sciver, Heat Flux in He
    II in Microbore Tubing, in "Advances in Cryogenic
    Engineering, Vol. 27", Plenum Press, New York (1982).
24. L.J. Rybarcyk and J.T. Tough, Superheating in He II and the
    extension of the lambda line, J. of Low Temp. Phys. 43:197
    (1981)
25. "Torus II Supra", Association Euratom-CEA Eur-CEA-FC-1021,
    October 1979.

26. R. Aymar et al., Tore Supra-status report concerning the superconducting magnet after the qualifying development program, IEEE Trans. on Magnetics, MAG-17:1911 (1981).

27. S.W. Van Sciver, Transient heat transport in He II, Cryogenics 19:385 (1979).

28. S.W. Van Sciver and R.L. Lee, Heat Transfer from Circular Cylinders in He II, in "Cryogenic Processes and Equipment in Energy Systems", ASME Publication H00164 (1980).

29. J.C. Lottin, private communication.

30. E. Canavan, private communication.

31. N.S. Snyder, Heat transport through He II: Kapitza conductance, Cryogenics 10:89 (1970).

32. T.H.K. Frederking, Thermal Transport Phenomena at Liquid Helium II Temperatures, Chem. Eng. Progr. Sympos. Series, 64:21 (1968).

33. I.M. Khalatnikov, "Introduction to the Theory of Superfluidity", W.A. Benjamin Inc., New York (1966).

34. G. Claudet and P. Seyfert, Bath Cooling with Subcooled Superfluid Helium", in "Advances in Cryogenic Engineering, Vol. 27", Plenum Press, New York (1981).

35. S.W. Van Sciver, Heat Transfer in Superfluid Helium II, in "Proceedings 8th International Cryogenic Engineering Conference", IPC Technology Press, Genova (1980).

36. J.C. Lottin, private communication.

37. E. Canavan and S.W. Van Sciver, Axial Heat Transport in Helium II in a Thick Walled Copper Channel, in "Advances in Cryogenic Engineering, Vol. 27", Plenum Press, New York (1981).

38. R.C. Steed and R.K. Irey, Correlation of the Depth Effect on Film Boiling Heat Transfer in Liquid Helium II, in "Advances in Cryogenic Engineering, Vol. 15", Plenum Press, New York (1970) p. 299.

39. R.M. Holdredge and P.W. McFadden, Boiling Heat Transfer from Cylinders in Superfluid Liquid Helium II Bath, in "Advances in Cryogenic Engineering, Vol. 11", Plenum Press, New York (1966) p. 507.

40. K.R. Betts and A.C. Leonard, Free Convection Film Boiling From a Flat Horizontal Surface in Saturated He II, in "Advances in Cryogenic Engineering, Vol. 21", Plenum Press, New York (1975) p. 282.

41. H. Kobayashi and K. Yasukochi, Recovery heat flux from silent and noisy film boiling states in saturated liquid He II, Cryogenics 19:93 (1979).

42. D.A. Labuntzov and Ye.V. Ametistov, Analysis of Helium II film boiling, Cryogenics 19:401 (1979).

43. A.P. Kryukov and S.W. Van Sciver, Calculation of the recovery heat flux from film boiling in superfluid helium, Cryogenics 21:525 (1981).

44. This criterion is analogous to the cryostatic stability
    criterion for normal helium cooling, see A.R. Kantrowitz
    and Z.J.J. Stekly, A new principle for the construction of
    stabilized superconducting coils, Appl. Phys. Lett. 6:56
    (1965).

45. B.J. Maddock, G.B. James and W.T. Norris, Superconductive
    composites: heat transfer and steady-state stabilization,
    Cryogenics 9:261 (1969).

46. S.W. Van Sciver, Enthalpy stability criterion for magnets
    cooled with superfluid Helium II, IEEE Trans on Magnetics,
    Mag-17:747 (1981).

47. G. Claudet et al., Superfluid helium for stabilizing
    superconductors against local disturbances, IEEE Trans on
    Magnetics, Mag-15:340 (1979).

48. C. Meuris and A. Mailfert, Influence of the spacers on the
    stability of channel cooled superconducting coils, IEEE
    Trans on Magnetics, Mag-17:1079 (1981).

49. P. Seyfert, et al., The effect of evenly spaced channels on
    the stability behavior of superconducting windings cooled
    by superfluid helium, IEEE Trans on Magnetics, MAG-17:1757
    (1981).

50. Y.H. Hsu et al., Measurement of stability of cooled
    conductors cooled by He I at reduced temperatures or He
    II., IEEE Trans on Magnetics, Mag-17:750 (1981).

51. R. Aymar, et al., Global test of the conductor for "Tore
    Supra" under actual working conditions, IEEE Trans on
    Magnetics, MAG-17:2205 (1981).

He II THERMOHYDRODYNAMICS AND PHASE TRANSITION DYNAMICS: NEWTONIAN FLUID ASPECTS RELATED TO THE STABILITY OF SUPERFLUID-COOLED MAGNETS*

T. H. K. Frederking

University of California
Los Angeles, California

INTRODUCTION

Large NbTi superconducting magnets based on critical currents achievable by operation near 2 K are being evaluated at present by several groups[1-4]. The magnets require extensive superfluid refrigeration systems with the benefit of excellent thermostatic bath stability. Not all superfluid properties are known in the range of heat flux densities (q) needed for the dimensioning of stabilizer and coolant duct cross sections. The purpose of this report is to outline the knowledge evolved from our research of the last few years.

During operation, magnet disturbances at high q may have to be handled. Perturbations may be caused, for instance, by a local quench. The He II must remove the energy dissipated. At high q, turbulent systems of quantized vortices are established which show, in part, Newtonian behavior. These Newtonian aspects are stressed in the present survey.

Before going into details, several special characteristics of the fluid system are enumerated. Rotons[5,6] are the main contributors to the thermal energy at 2 K. Convection modes dominate transport, and special Knudsen-Casimir diffusion[7,8] is restricted to very narrow fluid spaces (of the order 10^{-6} cm) near 2 K. The two-fluid model to describe heat transfer in He II is referred to

---

*Research supported in part by the National Science Foundation and NASA Lewis Research Center, Cleveland, Ohio.

as having zero net mass flow. Superfluid flows toward the hot
spot in the magnet while the entropy carrier, normal fluid, flows
from "hot" to "cold". The possibility of laminar flow is recognized but turbulent motion dominates at high q-values[5,6].

SINGLE-PHASE OPERATION OF BULK HELIUM II

In this regime, forced flow with speeds above ~ 1 cm/s is
excluded. No important forced flow effects have been found in
recent work in this area[9,10]. The most frequently studied configurations are long ducts with insulated walls (Gorter-Mellink [GM]
ducts) and single horizontal conductors surrounded by a wide He II
space.

Long GM-ducts

This case is referred to as GM-convection[11]. The GM-duct is
capable of a transport rate which may be expressed in terms of an
effective counterflow speed of the two-fluid system, w_{eff}, and the
entropy (ρS) per unit volume of liquid (ρ = density). There are
three non-classical effects: (1) The thermal energy transported
is proportional to ($S\nabla T$), in contrast to the classical Newtonian
case which involves the difference $c_p \Delta T$, (c_p = specific heat at
constant pressure P); (2) The thermo-osmotic driving gradient,
$S\nabla T$, is dominate; (3) The thermodynamic state of order, expressed
in terms of the superfluid density ratio (ρ_s/ρ), shown in Fig. 1a,
is a unique function of temperature T for a specified P. The GM-
heat current per duct cross section may be written as[12]

$$q = \rho \, S \, T \, w_{eff}(\rho_s/\rho) \tag{1}$$

$$w_{eff} = K_{GM}\{S|\nabla T|\nu_n \, (\rho_s/\rho)\}^{1/3} \tag{2}$$

(K_{GM} =const=11.3; $\nu_n = \eta_n/\rho_n$ = kinematic viscosity of normal fluid). At sufficiently low T, for instance at about 1.2 K, only a
few rotons contribute to ($\rho S T$), and (ρ_s/ρ)$\to$1. In this "roton
depletion" limit the <u>form</u> of Eq. 1 and 2 is reminiscent of
Newtonian fluid flow in so far as there exists a constant ratio of
a transport parameter N_q to the cube root of a driving force
parameter $N_{\nabla T}$,

$$N_q = q \, D_h \, / \, (\eta_n \, S \, T) \tag{3}$$

$$N_{\nabla T} = D_h^{\,3} \, \rho^2 \, S|\nabla T| \, / \, \eta_n^{\,2} \tag{4}$$

where D_h = hydraulic diameter of G-M duct and η_n = shear viscosity. Because of the formal similarity, Table I compares GM-convection and Newtonian convection induced by gravitational buoyancy
in the turbulent regime.

Table I. Comparison of classical free convection with GM-
          convection at roton depletion (zero net mass flow).

FLUID   He II: $(\rho_s/\rho)\to 1$ Newtonian

Driving Gradient   Thermo-osmotic   Buoyancy

Thermal Energy $\Delta E_T = \rho\, S\, T$ $\Delta E_T = \rho\, c_p\, \Delta T$
  Transported   (per   unit   volume)

Flow Rate Parameter $N^q = qD_h/(\eta_n ST)$ $q\ D_h/c_p\Delta T\ \eta)\cdot Pr$

Dimensionless
Driving Force $N_{\nabla T}=D_h{}^3\,\rho^2\,S|\nabla|/\eta_n{}^2$ Rayleigh number Ra

Transport Function $N_q = K_{GM}(N_{\nabla T}\rho/\rho_n)^{1/3}$ $N_q \sim Ra^{1/3}$

Important points are elucidated further by a "third fluid"
model[13], modified to incorporate an eddy Prandtl number (Pr = η
c_p/k) of unity (hypothetical "pseudo-classical convection"
model). This model has a simple classical analog for $ST\to c_p\Delta T$,
$N_{\nabla T}\to$Grashof number, and N_q = ratio of the Nusselt number to Pr
(Ref 14). The transformation from classical to pseudo-classical
convection has to take into account the non-classical effects
mentioned earlier.

Another consequence of the GM analysis is the existence of a
finite speed (w_{eff}) associated with propagation of "hot" turbulent
domains. This speed, shown in Fig. 1b, is smaller than the speed
of second sound. At the lower boundary of the GM-regime, the

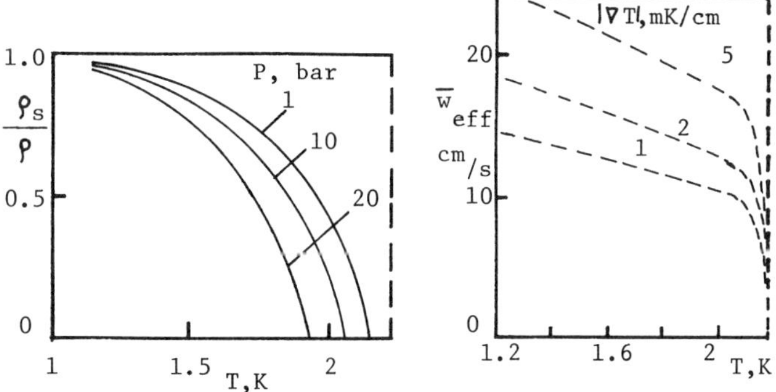

transition from laminar to turbulent flow has been described
phenomenologically by Dimotak is[15]. He proposed a critical GM-
scale useful for the formulation of analog conditions at roton
depletion (Table I). An example of the flow transition is shown
in Fig. 2 for a porous medium[16].

Single Horizontal Conductor

The He II thermohydrodynamics has been evaluated indirectly
from our experimental data, as thin entropy-rich layers around the
heated conductor are not resolved readily with solid state thermometers. Two properties of the data support Newtonian fluid
behavior: First, q (T), at constant ΔT, is consistent with GM-
convection in ducts; second, the diameter influence q (D) agrees
with Newtonian convection around single cylinders[17-19]. Therefore, it is possible to deduce convection properties of the heated
layer. For roton depletion the experimental results are consistent with the Newtonian pseudo-classical model[20-25]. There are
three different possibilities of thermohydrodynamic layers around
conductors. The layer thickness of Fig. 3a is of the order of the
cylinder diameter D, which is very small, and q varies approximately at $q \sim D^{-1}$. Figure 3b presents an intermediate diameter
with a very thin boundary layer thickness ($q \sim D^{-1/4}$) and Fig. 3c
indicates turbulent conditions with locally homogeneous turbulence
($q \sim D^{0}$). In view of this Newtonian behavior, several authors
have proposed semi-empirical equations for single conductors. For
the last case of homogeneous turbulence, three different equations
are listed in Table II based on data correlations[21-23]. Aside
from w_{eff} (Eq.1), Table II includes constants and reference
lengths. A numerical example[24], (Fig. 4), illustrates Eq. 5-7 for
near-saturated He II. For this case only a few data are available
near 2 K whose scatter does not permit a critical choice of the
best equation.

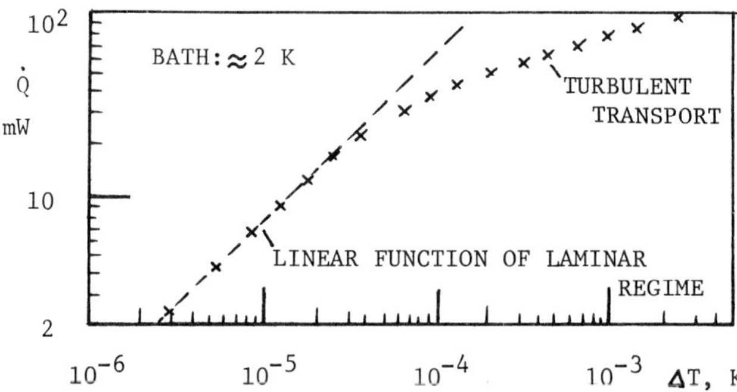

Fig. 2. GM-heat transport versus temperature difference at
zero net mass flow; (Q = $\overline{q}_o$ A_o; A_o total cross section).

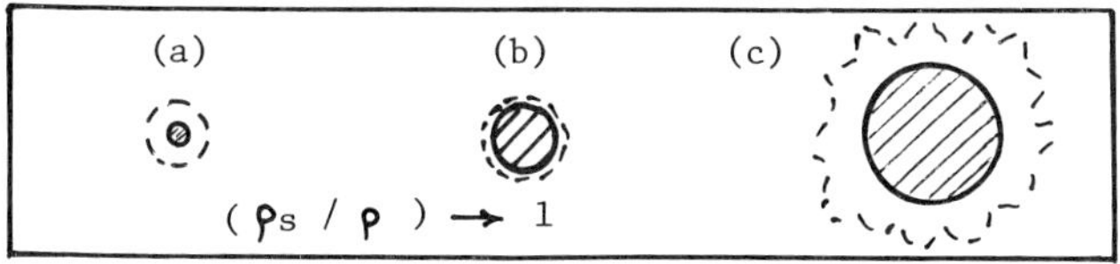

Fig. 3. Horizontal cylinder conditions (schematically).

PHASE TRANSITION LIMITS of He II

In principle many thermodynamic paths are possible which lead from He II to another fluid phase. Examples of isobaric paths (P=const) appear to have been discussed most frequently, e.g. Ref. 10. Exploratory phase transition work in the 1960's, reviewed by Irey[25], has been conducted primarily with non-superconducting specimens. Results up to 1975[26] and other He I topics have been surveyed[27-29] including an introductory discussion[30]. The overall Kapitza resistance[31] determines the temperature of the supercon- ductor. It may reach 3 K at the lambda transition. Kapitza data have been reviewed in Refs. 32 and 33.

The He^4 pressure-temperature phase diagram is depicted in Fig. 5 along with two isobaric T-excursions starting from He II. Below the pressure of the fluid triple point of saturated liquid

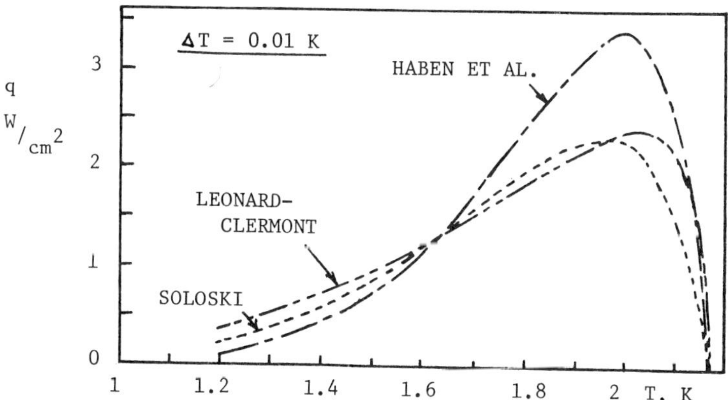

Fig. 4. Functions q(T) (Table II) for localized homogeneous turbulence (q~D^0) in He II[24].

Table II. Asymptotic equations for single horizontal cylinders in He II, (superscript $^{\circ}$ for value at T=0 K).

	SPEED w_{eff}	Eq.	K_{ij}=const Ref.length
Haben et al.[20]	$K_{HC}\left\{\nu_n \dfrac{S\Delta T}{L^{\circ}_{HC}} \cdot \dfrac{\rho_s}{\rho} \cdot \dfrac{\rho_n}{\rho}\right\}^{1/3}$	5	Ratio $K_{HC}/L^{\circ}_{HC}{}^{1/3}=16$ cm$^{-1/3}$
Amar[21]: Asymptote of Leonard et al.[22]	$K_{LC}\left\{\nu_n \cdot \dfrac{S\Delta T}{L^{\circ}_{LC}} \dfrac{\rho}{\rho_s} \dfrac{\rho}{\rho_n}\right\}^{1/3}$	6	$K_{LC}\approx 4.6$ $L^{\circ}_{LC}=1$ cm
Soloski[23]	$K_{SC}\left\{\nu_n \cdot \dfrac{S\Delta T}{L_{SC}} \cdot \dfrac{\rho}{\rho}s\right\}^{1/3}$	7	$K_{SC}=9.3$ $L_{SC}=1$ cm

(P = 50 mbar), a T-increase leads to a first order transition with a latent heat (λ) of vaporization (path 1). The other isobar (path 2) involves a lambda transition from He II to He I without latent heat. Subsequently there is a first order transition in the range $P_{\lambda}<P<P_c$, (P_c thermodynamic critical pressure). Departures from isobars are quite common however, e.g. dP < 0 when local accelations occur, or dP > 0 in long He II coolant ducts in magnets. Nevertheless, the path dP = 0 has been so popular that departures from isobars have been accounted for partially as "superheat data"[34-36]. In contrast, theoretical thermodynamic models and plausibility arguments exclude superheating of of HeII[37,9,10]. In short GM-ducts no superheating has been observed[38].

In near-saturated He II pools, the isothermal change in P from the vapor-liquid boundary to the conductor may be quite small ($\Delta P\ll P$). Further, in terrestrial dewars the isothermal distance from the phase boundary to the location of the conductor in the P-T-plane is the hydrostatic difference ΔP_g. When dissipation occurs, the isobaric temperature distance from the conductor location to the He II phase boundary often is $\Delta T^* \ll T$. Therefore, the limiting ΔT^* is expressed in terms of the derivative of the vapor pressure curve $dP/dT \approx \rho_v \lambda/T$, ρ_v = density of saturated vapor.

$$\Delta T^* = \Delta P_g\, T/(\rho_v \lambda) \tag{8}$$

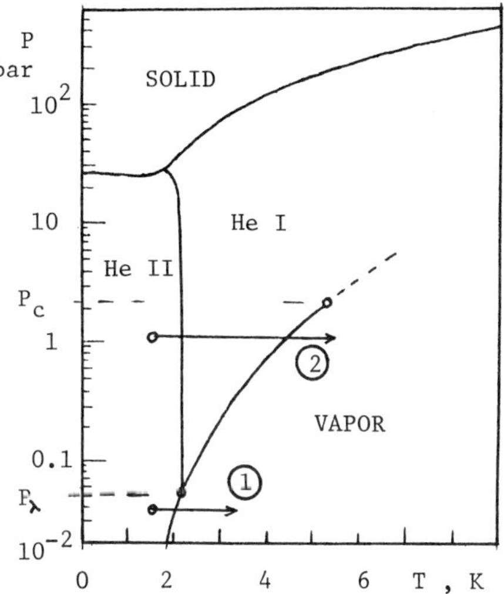

Fig. 5. Equilibrium state diagram of He^4 P(T) including
two examples of isobaric paths, where C.P. = critical point,
P_c = critical pressure, and P_λ = pressure at lambda
transition of saturated liquid.

This limit may be used directly in the appropriate equation q(ΔT)
in order to arrive at the isobaric limit q*(ΔT*). Evidence
forshort GM-ducts supports this condition when cross sections are
large enough[39].

For large isobaric ΔT-values, the strong temperature influence on some thermophysical properties has to be considered, i.e.
integration is required. In this context it is noted that both
the GM-duct, Eq. 1, and the cylinder, Eq. 7, analysis contain the
same T-dependent property combination. Therefore, the same integral has to be evaluated for dP=0. For instance, after integration over a distance L of a GM-duct from a bath value T to T_λ the
result is expressed in terms of a property integral I for a reference length L_{ref} = 1 cm. For a change from 0 K to T, I has the
value I_{max}

$$q_{Lref} = q_{o\lambda} \cdot I/I_{max} \tag{9}$$

Figure 6 shows that $q_{o\lambda}$ is a relatively weak function of P. Also
the influence of P on the ratio (I/I_{max}) appears to be quite weak
(insert in Fig. 6). For a length L, the limiting heat flux densities of isobars are obtained as [40]

$$q_L(L)/q_{Lref} = (L_{ref}/L)^{1/3} \tag{10}$$

For arbitrary ΔT limits, two terms have to be considered:

$$\int_{T_1}^{T_2} f \cdot dT = \int_{T_1}^{T_\lambda} f \cdot dT \quad - \int_{T_2}^{T_\lambda} f \cdot dT \tag{11}$$

Reference 41 treats the case of small departures from isobars.

PHASE TRANSITION DYNAMICS DURING TRANSIENTS

So far, the thermostatics along isobars and phase boundaries have been emphasized. For $dP \approx 0$, early results[38] indicate a direct transition to film boiling, not only for near-saturated liquid, but also for pressurized He II ($P<P_c$). Related studies are Ref. 42 - 52. Recent work has shown that triple-phase phenomena are possible in general[53], and this has been confirmed recently[36]. In other words, the three fluid phases, He II, He I and He^4 vapor may coexist simultaneously in various boiling and non-boiling conditions.

Triple-phase modes complicate the phase transition dynamics. In general, the pressure is not constant in the long, narrow coolant channels found in magnets. In addition, the surface area of a dissipative region wetted by fluid is not equal to the duct cross section. Further, triple-phase modes, such as nucleate and film boiling, have received little attention so far[36,53]. Figure

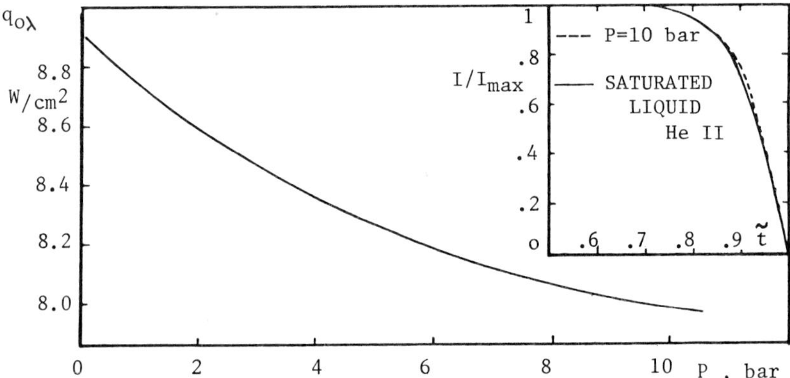

Fig. 6. Limit values of pressurized GM-duct operation at dP=0, (Insert: Integral ratio, Eq. 9).

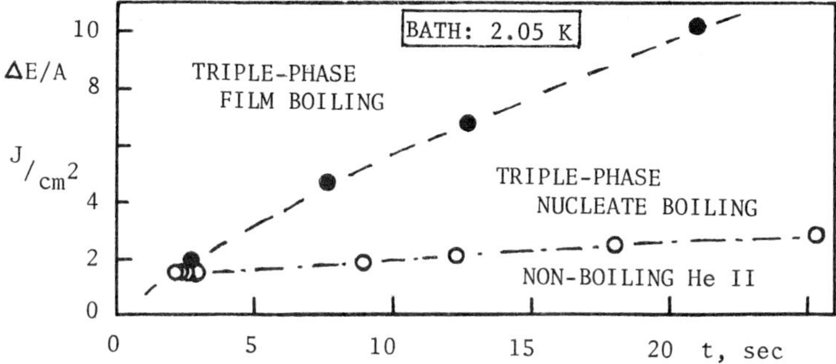

Fig. 7. Energies per fluid-wetted surface area of composite conductor NbTi/Cu at regime boundaries[54].

7 is an example of various phase boundaries in the energy – time domain evaluated by Chuang[54]. The energy per surface area of a NbTi/Cu-composite is displayed for step inputs in heating power. At low energies, non-boiling He II provides excellent cooling conditions. This regime is followed by triple-phase nucleate boiling which keeps T of the composite below the transition temperature of the superconductor. Finally triple-phase film boiling is established at high ΔE/A and short times, and a quench cannot

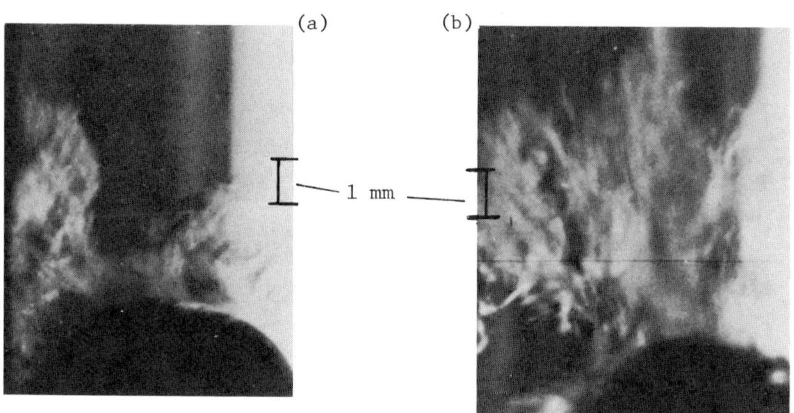

Fig. 8. Cooldown of Cu sphere involving triple-phase phenomena: Prominences reaching into He II at a spherical surface (sphere dia.0.635 cm); (a) He II bath: 1.53 K; Time ratio t/t_{tot}=0.47 of total cooldown time t_{tot}=5.9 sec; (b) Bath: 1.65 K; t/t_{tot} = 0.3 of t_{tot} = 7.7 sec.

be avoided. For details see Ref. 54. During cooldown, film
boiling is very ineffective as far as a speedy recovery is concerned[55]. Also note that despite low T, removal of He I layers
may take some time. Triple-phase film boiling has the appearance
of subcooled, classical film boiling in some cases[56]. Figure 8
however shows complex triple-phase phenomena during cooldown.
There are peculiar "structures" which appear to be prominences of
entropy-rich fluid tending to propagate into bulk He II.

CONCLUSIONS

Despite a lack of many details of triple-phase transients,
data obtained provide guidance for the prediction of He II systems
from first principles: First, on the basis of the He^4 equilibrium
state thermodynamics, phase boundaries may be located; second, on
the basis of Newtonian fluid mechanics, limits may be evaluated in
conjunction with superfluid constraints. The latter eliminate
gravity effects in the non-boiling He II regime. In general He II
has shown very promising coolant properties for use with advanced
superconducting magnet systems[57-61].

ACKNOWLEDGMENTS

During the work on superconductor-He II interaction the input
of many former students, Drs. C. Linnet, Y.W. Chang, R.C. Amar,
S.C. Soloski, S. Caspi and Messrs. Richard J. Allen, J.Y. Lee, and
of present collaborators, C. Chuang, Y. Kamioka, Y.I. Kim, S.W.K.
Yuan, Jeffrey M. Lee and M. Reardon is gratefully acknowledged.

REFERENCES

1. R.W. Boom, et al., in "Advances in Cryogenic Engineering, Vol.
    19," Plenum Press, New York (1974), p. 117.
2. G. Bon Mardion, et al., in "Advances in Cryogenic Engineering,
    Vol. 23," Plenum Press, New York (1978), p. 358.
3. R.P. Warren, et al., in "Proc. 8th Int'l Cryo. Eng. Conf.,"
    IPC Science and Technology Press, Guildford (1980), p. 373.
4. R. Aymar, et al., IEEE Trans. in Magnetics, MAG-17:38 (1981).
5. L.D. Landau and E.M. Lifshitz, "Statistical Physics" (1958),
    "Fluid Mechanics" (1959), Pergamon, London.
6. S.J. Putterman, "Superfluid Hydrodynamics", North Holland (1974)
7. D.S. Greywall, Phys. Rev. B23:2152 (1981).
8. R. Schmidt and H. Wieschert, Z. Phys. B36:1 (1979).
9. S. Caspi, Ph.D. thesis, Univ. of California, Los Angeles (1978).
10. S. Caspi and T.H.K. Frederking, AIChE 72nd Annual Meeting.
    1979, paper 94b.
11. C.J. Gorter and J.H. Mellink, Physica 15:285 (1949).
12. S.C. Soloski and T.H.K. Frederking, Int. J. Heat Mass Transf.
    23:437 (1980).

13. R.J. Donnelly, _Bull. Am. Phys. Soc._ 25:107 (1980).
14. W.G. Brown and K.R. Solvason, _Int. J. Heat Mass Trans_ 5:859
    (1962)
15. P.E. Dimotakis, _Phys. Rev._ A10:1721 (1974).
16. T.H.K. Frederking, et al., in "Proc. 16th Int'l Conf. Low
    Temp. Phys.," North Holland (1982).
17. R.B. Bird, W.E. Stewart and E.D. Lightfoot, "Transport
    Phenomena," Wiley, New York (1960), p. 412.
18. E.R.G. Eckert and R.M. Drake, "Heat and Mass Transfer," Mc-
    Graw-Hill, New York (1959), p. 412.
19. H. Senftleben, _Z. angew Phys._ 3:361 (1951); 5:267 (1953).
20. R.L. Haben, et al., in "Advances in Cryogenic Engineering, Vol.
    17," Plenum Press, New York (1972), p. 323.
21. R.C. Amar, Ph.D. thesis, Univ of California, Los Angeles (1974).
22. A.C. Leonard and M.A. Clermont, in "Proc. 4th Int'l Cryo Eng.
    Conf.," IPC Science & Technology Press, Guildford (1972),
    p. 301.
23. S.C. Soloski, Ph.D. thesis, Univ. of California, Los Angeles,
    (1977).
24. S.C. Soloski, _Rept. CR 3262_, NASA (April 1978), p. 22.
25. R.K. Irey, Heat Transport in Liquid Helium II, in "Heat Trans-
    fer at Low Temperature," W. Frost, ed., Plenum Press, New
    York (1975), p. 325.
26. T.H.K. Frederking, _Progr. Refrig. Sci. Technol IIR_, 1:71 (1978).
27. E.G. Brentari, P.G. Giarratano and R.V. Smith, NBS Technical
    Note 317, (1965).
28. J.A. Clark and R.M. Thorogood, in "Cryogenic Fundamentals,"
    G. Haselden, ed., Academic Press, New York (1971).
29. J.A. Clark, Cryogenic heat transfer, _Adv. Heat Transf.,_ 5:325
    (1968).
30. I.R. McDougall, _Cryogenics_ 11:260 (1971).
31. T.H.K. Frederking, _Chem. Eng. Progr. Sympos. Ser.,_ 64 (87):21
    (1968).
32. N.S. Snyder, _Cryogenics_ 10:89 (1970).
33. L.J. Challis, _J. Phys._ C 7:481 (1974).
34. G. Kraft, _J. Low Temp. Phys._ 31:441 (1978).
35. S.W. Van Sciver, _Cryogenics_ 19:385 (1979).
36. D. Gentile and M.X. Francois, _Cryogenics_ 21:234 (1981).
37. A.B. Pippard, "Elements of Classical Thermodynamics", Cambridge
    Univ. Press, (1961).
38. C. Linnet and T.H.K. Frederking, _J. Low Temp. Phys._ 21:447
    (1975).
39. C. Linnet and T.H.K. Frederking, _Warme u. Stoffubertragung_
    5:141 (1972).
40. Y. Kamioka, et al., _Rept. UCLA-ENG-8109_, p. 105 and C. Chuang,
    Ph.D. thesis, Univ. of California, Los Angeles (1981).
41. R.K. Childers and J.T. Tough, _Phys. Rev._ B13:1040 (1976).
42. H. Kobayashi and K. Yasukoshi, _Cryogenics_ 19:93 (1979).

43. H. Kobayaski and K. Yasukoshi, in "Advances in Cryogenic Engineering, Vol. 25," Plenum Press, New York (1980), p. 372.

44. H. Kobayashi, K. Yasukoshi, K. Tokuyama, in "Proc. 6th Int'l Cryo. Engr. Conf.," IPC Science & Technology Press. Guildford (1976), p. 307.

45. H. Kobayaski and K. Yasukoshi, in "Proc. 8th Int'l Cryo. Engr. Conf.," IPC Science & Technology Press, Guildford (1980) p. 171.

46. M.A. Hilal, in "Advances in Cryogenic Engineering, Vol. 25", Plenum Press, New York (1980), p. 358.

47. S.W. van Sciver and R.L. Lee, in "Advances in Cryogenic Engineering, Vol. 25," Plenum Press, New York (1980), p. 363.

48. S.W. van Sciver, in "Proc. 8th Int'l Cryo. Eng. Conf.," IPC Science & Technology Press, Guildford (1980), p. 228.

49. D.A. Labuntzow and Ye. V. Ametistov, Cryogenics, 19:401 (1979).

50. D.A. Labuntzow and Ye. V. Ametistov, Cryogenics, 21:51 (1981).

51. V.E. Keilin, et al., in "Proc. 8th Int'l Cryo. Engr. Conf., IPC Science & Technology Press, Guildford (1980), p. 403.

52. P. Seyfert, G. Claudet and M.J. McCall, in "Advances in Cryogenic Engineering, Vol. 25," Plenum Press, New York (1980), p. 378.

53. S. Caspi and T.H.K. Frederking, Cryogenics, 19:513 (1979).

54. C. Chuang, Y. Kamioka and T.H.K. Frederking, in "Advances in Cryogenic Engineering, Vol. 27," Plenum Press, New York (1982).

55. C. Chuang, Y. Kamioka and T.H.K. Frederking, in "Advances in Cryogenic Engineering, Vol. 26," Plenum Press, New York (1980), p. 667.

56. T.H.K. Frederking, R.C. Chapman and S. Wang, in "International Advances in Cryogenic Engineering, Vol. 10," Plenum Press, New York (1965), p. 353.

57. D. Gentile and W.V. Hassenzahl, in "Advances in Cryogenic Engineering, Vol. 25," (1980), p. 385.

58. R.P. Warren, et al., in "Proc. 8th Int'l. Cryo. Eng. Conf.," IPC Science & Technology Press, Guildford (1980), p. 373.

59. S. Caspi, in "Proc. 8th Int'l. Cryo. Engr. Conf.," IPC Science & Technology Press, Guildford (1980), p. 238.

60. W.Y. Chen, et al., IEEE Trans. in Magnetics, MAG-17:57 (1981).

61. C. Taylor, et al., IEEE Trans. in Magnetics, MAG-17:1571 (1981).

TRANSIENT HEAT TRANSFER IN SUPERFLUID HELIUM

L. Dresner*

Japan Atomic Energy Research Institute
Tokai-mura, Ibaraki, Japan

INTRODUCTION

According to the Gorter–Mellink relation,[1] the heat flux in
superfluid helium is proportional to the cube root of the temperature gradient. If we use this proportionality in place of
Fourier's linear law to derive an equation of heat transport, we
obtain a non-linear partial differential equation. Such equations
are usually difficult to solve because we cannot superpose solutions to obtain others. In spite of this, the problem of this
paper, the constant-flux problem, can be solved because its
temperature profiles are self-similar.

Self-similarity means that the temperature profile at one
time can be obtained from that at a different time by suitable
(different) stretching of the distance and temperature axes of the
latter profile. The self-similarity of the temperature profiles
is connected, as we shall see, with the invariance of the non-
linear partial differential equation to certain groups of
transformations. The advantages of self-similarity, as expressed
by Barenblatt,[2] are these: "In handling experimental data, self-
similarity has reduced what would seem to be a random cloud of
empirical points so as to lie on a single curve or surface, constructed using self-similar variables chosen in some special
way. Self-similarity of the solutions of partial differential
equations has allowed their reduction to ordinary differential
equations, which often simplifies the investigation."

---

*On assignment by the Oak Ridge National Laboratory of the U.S.
Dept. of Energy operated under contract W–7405–eng–26 by Union
Carbide Corp. Research sponsored by the Office of Fusion Energy.

This quotation describes exactly the accomplishments of this paper. We reduce the partial differential equation of heat transport to an ordinary differential equation, the appropriate solution of which we find without extensive computation. The reduction involves the similarity variables $\Delta T/\sqrt{t}$ and $z/\sqrt{t}$, where ΔT is the temperature rise at a distance z from the heated face at a time t after the (constant) heating has begun. Use of these variables should, and does, reduce all of the experimental temperature profiles reported by van Sciver[3] to a single, universal curve. We obtain this curve as well by solving the differential equation; agreement is excellent. In fact agreement with all the experimental data reported by van Sciver is excellent, so that the Gorter-Mellink law seems to be a very successful basis for describing transient heat transfer in superfluid helium.

BASIC EQUATIONS

The Gorter-Mellink relation states that in superfluid helium $q = -K(\nabla T)^{1/3}$. Combining this with the equation of energy conservation, $\nabla \cdot q + \rho c(\partial T/\partial t) = 0$, we obtain the non-linear equation of heat transport,

$$\rho c(\partial T/\partial t) = \nabla \cdot (K(\nabla T)^{1/3}) \tag{1}$$

In the problem of this paper, the heat source is a slab, located at z = 0, emitting a constant heat flux that has been turned on suddenly at t = 0. In slab geometry, and with constant thermo-physical properties*, (1) becomes

$$\partial \theta/\partial \tau = \partial[(\partial \theta/\partial z)^{1/3}]/\partial z \tag{2}$$

where $\theta = (T - T_b)/(T_\lambda - T_b)$ and $\tau = Kt/\rho c(T_\lambda - T_b)^{2/3}$. In terms of these variables, the constant-flux boundary condition at z = 0 can be written

$$(\partial \theta/\partial z)_{z=0} = -(q/K)^3/(T_\lambda - T_b) \tag{3}$$

We seek the dimensionless temperature θ as a function of z and τ.

GROUP INVARIANCE

Equation (2) is invariant to the one-parameter family of groups of stretching transformations

$$z' = \lambda z, \quad \tau' = \lambda^\beta \tau, \quad \theta' = \lambda^\alpha \theta \qquad 0 < \lambda < \infty \tag{4}$$

---

*This assumption is discussed in the last section of the paper.

where $2\alpha - 3\beta + 4 = 0$. In order to satisfy the boundary condition (3) we must choose $\alpha = 1$ and $\beta = 2$.

Solutions of (2) invariant to group (4) must be of the form

$$\theta/\tau^{\alpha/\beta} = y(z/\tau^{1/\beta}) \tag{5}$$

where y is an as yet undetermined function and $\theta/\tau^{\alpha/\beta}$ and $z/\tau^{1/\beta}$ are two independent invariants of group (4). When (5) is substituted into (2), the latter reduces to an ordinary differential equation for $y(x)$, where x is an abbreviation for the invariant $z/\tau^{1/\beta}$:

$$\beta\frac{d}{dx}\left\{\frac{dy}{dx}\right\}^{1/3} + x\frac{dy}{dx} - \alpha y = 0 \tag{6}$$

As we know from a previously proved theorem,[4] the ordinary differential equation (6) must be invariant to the group of transformations

$$x' = \mu x, \quad y' = \mu^{-2}y \qquad 0 < \mu < \infty \tag{7}$$

This can be verified directly by substitution. According to a theorem of Lie's,[5] if we use as new dependent and independent variables in (6) a first differential invariant and an invariant, respectively, (6) will reduce to a first-order equation. We take $v = \dot{x}\dot{y}^{1/3}$ as the first differential invariant ($\dot{y} \equiv dy/dx$) and $u = xy^{1/2}$ as the invariant. Then

$$\frac{dv}{du} = \frac{u(2\beta v - 2v^3 + 2\alpha u^2)}{2\beta u^2 + \beta v^3} \tag{8a}$$

or when $\alpha = 1$ and $\beta = 2$

$$\frac{dv}{du} = \frac{u(2v - v^3 + u^2)}{2u^2 + v^3} \tag{8b}$$

We now proceed by analyzing the direction field of the first-order differential equation (8b).

DIRECTION FIELD OF EQ.(8b)

Because the temperature rise is positive and falls as we move away from the heated plate, $u > 0$ and $v < 0$. Hence we shall only be interested in the fourth quadrant of the (u,v) plane (see Fig. 1). First we find the curves on which dv/du equals either zero or infinity. Only on these curves does dv/du change sign, and only where curves of $dv/du = 0$ and $dv/du = \infty$ intersect are there singular points. Now $dv/du = 0$ on the v-axis, $u = 0$, and on the curve C_1: $u^2 v^3 - 2v$. furthermore, $dv/du = \infty$ on the curve

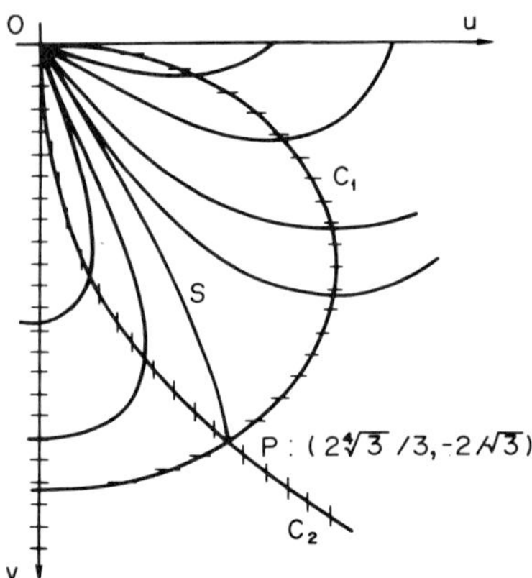

Fig. 1. The direction field of Eq. (8b) in the fourth quadrant.

C_2: $2u^2 + v^3 = 0$. The origin 0 and the point P are singular points.

When $\alpha = 1$ and $\beta = 2$, the boundary condition (3) is $\dot{y}(0) = -(q/K)^3/(T_\lambda - T_b)$, a finite constant. Since $y(0)$ is also finite, when $z = 0$, u and v also equal zero. Therefore, we are interested in integral curves that pass through the origin 0, and thus in the nature of the singularity at 0. As we approach 0, three behaviors are possible for v and u: (i) $v \ll u^2$, (ii) $v \sim u^2$, and (iii) $v \gg u^2$. In case (i), the differential equation (8b) reduces to $dv/du = u/2$ since near 0, $v^3 \ll v$, too. Thus $v = u^2/4 + $ constant, which contradicts hypothesis (i). In case (ii), $v = bu^2$. Substituting this into (8b) and keeping only lowest order terms, we find $b = 1/2$. Since this curve does not lie in the fourth quadrant, we are left only with alternative (iii). When $v \gg u^2$, (8b) becomes $dv/du = 2uv/(2u^2 + v^3)$. We cannot drop the v^3-term in the denominator because we do not know if $u^2 \gg v^3$. But the differential equation can be integrated by making use of its invariance to the group $u' = \lambda u$, $v' = \lambda^{3/2}v$. Using another theorem of Lie's,[5] we find that $[v(u^2 - v^3)]^{-1}$ is an integrating factor. A short computation then gives as the integral $\Phi(u,v) = \ln[(u^2 - v^3)/v^2]$ so that $u^2 = a^2v^2 + v^3$ where a is an arbitrary constant. When v is small enough, $a^2v^2 \gg v^3$, so close enough to the origin, $u = -av$, $a > 0$.

This linear relation between u and v enables us to reach an interesting and useful conclusion. Written in terms of x and y, it becomes $y(0) = a^2 \left| \dot{y}(0) \right|^{2/3} = a^2 (q/K)^2 (T-T_b)^{-2/3}$. Thus

$$\theta(0) = a^2 \, q^2 \, t^{1/2}/K^{3/2} \, (\rho c)^{1/2} \, (T_\lambda - T_b) \qquad (9)$$

We hypothesize that the transient phase of heat transfer comes to
an end when the temperature of the heated surface reaches the
lambda point. Thereafter, the heat transfer is much reduced
because of the phase change, and the surface temperature rises
(burnout). According to this hypothesis, at time t_{bo}, $\theta(0) = 1$.
Then (9) gives

$$q^4 t_{bo} = K^3 \rho c (T_\lambda - T_b)^2 / a^4 \tag{10}$$

Van Sciver's experiments[3] confirm the constancy of $q^4 t_{bo}$ for
various q with high accuracy. Once we find the value of the
constant a we shall be able to calculate the right-hand side of
(10) and compare it with van Sciver's measured values.

THE SEPARATRIX

The integral curves in the fourth quadrant passing through
the origin are of two kinds, those that eventually intersect the
curve C_1 and those that eventually intersect the curve C_2. These
two kinds are separated by a single, exceptional integral curve,
S, called a separatrix, that joins the singularities O and P.
Curves intersecting C_1 eventually intersect the u-axis. There v =
0 and u > 0, i.e., $\dot{y}$ = 0 and y > 0. Such curves have a minimum
and do not conform to the temperature profiles we are seeking.
Curves intersecting C_2 eventually intersect the v-axis: u = 0,
v < 0, i.e., y = 0, $\dot{y}$ < 0. Such curves reach zero temperature
with a negative slope, a possibility we also reject. All that is
left is the separatrix.

Other work from ref. 4 shows that points of the separatrix
near the singularity P correspond to the asymptotic behavior y ~
$4\sqrt{3}/9x^2$. Such behavior is what we are looking for in the
solution we seek. In fact, we can check this asymptotic behavior,
which is primarily a consequence of the group invariance, against
van Sciver's experiments. If we plot $\Delta T / \sqrt{t}$ ~ y against
$z / \sqrt{t}$ ~ x on log-log paper, then (i) all of van Sciver's points
for ΔT versus z at different t should fall on a single curve, and
(ii) this curve should asymptotically assume a slope of -2.
Figure 2 shows the results and confirms both expectations.
Furthermore, we expect the non-asymptotic portion of the experimental curve in Fig. 2 to correspond with the portion of the
separatrix between O and P.

To find the separatrix, we must perform a numerical integration. Owing to the divergence of the integral curves as we
approach P and their convergence as we approach O, we expect
integration in the direction P → O to be stable and integration in
the direction O → P to be unstable. Furthermore integration in
the direction P → O involves no trial and error, as integration in

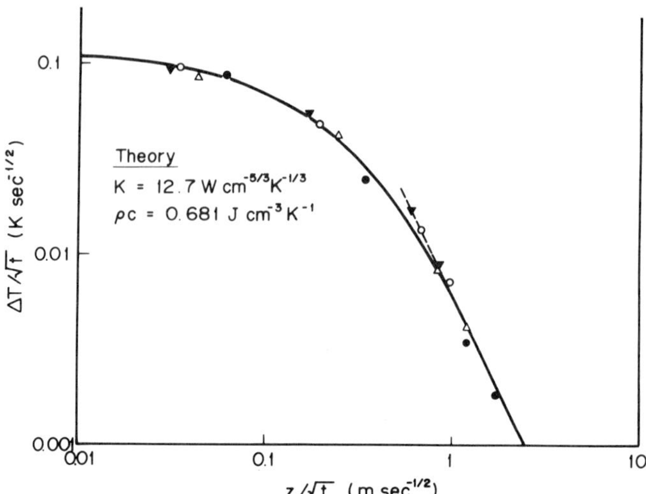

Fig. 2. Comparison of van Sciver's experimental points with the theoretical curve (T_b=1.8 K, q=2.2 W cm^{-2}).

the direction 0 $\rightarrow$ P would. Application of L'Hospital's rule shows that $(du/dv)p = - (3+\sqrt{17})\ 2 \cdot 3^{3/4} = -1.5624$. We use this slope to advance a short distance from P and then we continue by numerical integration towards 0. The resulting curve for the separatrix approaches the origin with a slope of -1.095, so that $a = (1.095)^{-1} = 0.9132$.

To find $\dot{y}$ (x), one procedure is the following. Near the origin $\dot{y}^{1/3}/y^{1/2} = v/u -1/a$ so that if we arbitrarily normalize y (0) = 1, we then find $\dot{y}$ (0) = -1.313. With this initial condition, we can integrate forwards numerically to obtain y (x). Because we are integrating in the direction 0 $\rightarrow$ P, the integration eventually becomes unstable. But we can integrate close enough to the asymptotic limit $y = 4\ \sqrt{3}/9x^2$ that there is no difficulty in graphically continuing the numerical solution valid for small x. To find integral curves corresponding to other values of y(0) we use the transformations (4) which on a log-log plot correspond to a translation. Thus the calculated curve can be compared directly by superposition with the experimental points of Fig. 2. This we do as follows.

COMPARISON WITH EXPERIMENT

Written in terms of time, temperature, and distance, (9) is

$$(\Delta T/\sqrt{t})_{z=0} = a^2\ (q/K)^2\ (K/\rho c)^{1/2} \qquad (11)$$

and the asymptotic limit $y = 4\sqrt{3}/9x^2$ is

$$z^2 \Delta T/t^{3/2} = (4\sqrt{3}/9)\,(K/\rho c)^{3/2} \qquad (12)$$

Thus by choosing $K/\rho c = 18.6$ $cm^{4/3}$ sec^{-1} $K^{2/3}$ we can make the asymptotic portion (12) of the calculated curve coincide with the straight line of slope -2 in Fig. 2. If we choose $K = 12.7$ W $cm^{-5/3}$ $K^{-1/3}$ and $\rho c = 0.681$ J $cm^{-3}K^{-1}$, corresponding to the value of $K/\rho c$ just determined, we make $\Delta T/\sqrt{t}\,|_{z=0} = 0.110$ K $sec^{-1/2}$ when calculated from (11). This brings the entire calculated curve into the excellent agreement with the experimental points shown in Fig. 2. The values of K and ρc determined by this fitting procedure are in good agreement with the point data given in Wilk's book[1] (see Table I).

When the values of K and ρc just found are substituted into (10) we find $q^4 t_{bo} = 290$ W^4 cm^{-8} sec compared with the measured value of 110 W^4 cm^{-8} sec reported by van Sciver. If we suppose the deterioration of heat transfer begins not at $T_\lambda = 2.17$ K, but, say, at 2.04 K, then the calculated value of $q^4 t_{bo}$ drops to 116 W^4 cm^{-8} sec. For $T_b = 2.0$ K, this gives 3.2 W^4 cm^{-8} sec calculated compared with 17 W^4 cm^{-8} sec measured. A compromise value for the burnout temperature of 2.08 K gives $q^4 t_{bo} = 157$ W^4 cm^{-8} sec at $T_b = 1.8$ K and 12.9 W^4 cm^{-8} sec at 2.0 K calculated versus 110 and 17 W^4 cm^{-8} sec, respectively, measured.

Table I. Properties of Superfluid Helium

T	P_n	P_s	S	A	K	c	c
(K)	(g/cm^3)	(g/cm^3)	(J/g-K)	(cm-s/g)	*	(J/g-K)	$(J/cm^{3}$-K)
1.8	0.047	0.099	0.54	88	10.4	2.81	0.410
1.9	0.063	0.083	0.72	105	11.6	3.79	0.553
2.0	0.082	0.064	0.94	130	11.6	5.18	0.756
2.1	0.108	0.038	1.24	190	8.35	7.51	1.10
2.17	0.146	0.0	–	–	0.0	–	–

$^*K = \rho_s ST(S/A\rho_n)^{1/3}$, W-$cm^{-5/3}K^{-1/3}$

Shown in Table I are the constants in the theory, K and ρc, in the temperature range 1.8 – 2.7 K covered by the extra money experiments. Only very close to T_λ does either constant vary sharply with temperature. In the range 1.8 – 2.1 K, for example, K varies by only 18% from the geometric mean of its extremes, and ρc by 64%. This may explain the success of the constant property theory in attaining the good fit of Fig. 2: the shape of the curve there depends mainly on the values of K and ρc at temperatures below T_λ. On the other hand, it seems likely that the compromise value of the burnout temperature of 2.08 K is lower than T_λ because of the sharp decrease of K in the range 2.1 – 2.17 K. But all in all, agreement between the constant-property theory and experiment seems good.

CONCLUDING REMARK

In view of how well the theory reproduces every aspect of van Sciver's measurements, the Gorter-Mellink relation, as mentioned in the introduction, seems to be a very successful basis for describing transient heat transfer in superfluid helium.

NOTATION

A = mutual friction parameter
a = reciprocal slope of the integral curves near the origin in the (u,v) plane
c = specific heat at constant pressure
K = coefficient in the Gorter-Mellink relation
q = heat flux vector; constant wall heat flux
S = entropy of fluid
t = time after heating has begun
t_{bo} = time at burnout
T = temperature
ΔT = temperature rise
T_b = bath temperature
T_λ = lambda temperature
u = $xy^{1/2}$
v = $x\dot{y}^{1/3}$
x = $z/\tau^{1/\beta}$
y = function defined in Eq. (5)
$\dot{y}$ = dy/dx
z = distance from heated plate
α, β, λ = constants in definition of transformation group (4)
μ = constant in definition of transformation group (7)
θ = $(T - T_b)/(T_\lambda - T_b)$
ρ = He II density
ρ_n = density of normal fluid component
ρ_s = density of superfluid component
τ = $Kt/\rho c(T_\lambda - T_b)^{2/3}$

REFERENCES

1. J. Wilks, "The Properties of Liquid and Solid Helium," Oxford
    University Press, Oxford, (1967).
2. G. I. Barenblatt, "Similarity, Self-Similarity, and
    Intermediate Asymptotics", N. Stein, translator, M. van
    Dyke, translation editor, Consultants Bureau, New York.
3. S. W. van Sciver, Cryogenics 19: 385 (Aug., 1979).
4. L. Dresner, On the calculation of similarity solutions of
    partial differential equations, Oak Ridge National
    Laboratory Report ORNL/TM-7407, (Aug., 1980).
5. A. Cohen, "An Introduction to the Lie Theory of One-Parameter
    Groups," G. E. Stechert and Co., New York, (1931).

FLOW STATES AND HEAT TRANSFER PROPERTIES
OF He II PHASE SEPARATORS*

U. Schotte

Freie Universität Berlin
Berlin, Federal Republic of Germany

INTRODUCTION

Phase separators, which are being developed for the use on spacecraft, are special valves intended to keep liquid cryogen separated from its vapor while allowing heat to be carried away by the vapor into the space vacuum.

Work on He II phase separators[1-3] has shown that the special properties of the superfluid, namely the fountain effect and the ideal heat transfer properties, make the construction principle straightforward. In practice, however, deviations from the ideal superfluid behavior occur so easily that the phase separators actually studied work mostly in the non-ideal regime. Therefore a close look at these deviations is necessary to make sure of the separation and heat transfer qualities of such a device.

WORKING PRINCIPLE OF A He II PHASE SEPARATOR

The most interesting phenomena of He II are observed in narrow geometries, like the He II film, capilleries and sintered materials or compressed powders. The theory is easiest for straight cylindrical tubes which we will treat extensively. The results can, in principle, be extended to more complicated structures.

The classical arrangement[4] consists of two baths connected by a narrow tube with a length of the order of centimeters and width of a few micrometers. One bath is slowly heated and the heat

---

*This work was supported by the DFVLR: 01 TQ 029 ZA/RT-WRT 2079.

transfer to the other one is observed in a stationary state with
no net mass transfer of He II. In contrast, a He II phase separator works in a non zero net mass flow stationary state. How it is
related to the classical arrangement can be explained by a simple
experiment (Fig. 1). Consider two containers each with some
amount of He II in equilibrium with its vapor, each connected to a
pump which maintains T_o and p_o in the first dewar and $T_o + \Delta T$, p_o
$+ \Delta p$ in the second. The Clausius–Clapeyron equation holds giving
to a good approximation.

$$\Delta T/\Delta p = (\rho_v L/T)_{T=T_o} \tag{1}$$

where ρ_v is the density of the He vapor and L the latent heat of
evaporation. If we now connect both dewars by a narrow tube of
length ℓ what will happen? We will observe the fountain effect
because He II obeys the London equation. There must be no chemical potential gradient in the system:

$$\nabla \mu = \frac{\nabla p}{\rho} - S \nabla T = 0 \text{ with } |\nabla T| \approx \frac{\Delta T}{\ell} \text{ and } |\nabla p| \approx \frac{\Delta p}{\ell} \tag{2}$$

ρ and S are the density and entropy per gram of He II. The fountain pressure $p^F = \Delta p$ is easily seen by comparing the factors in
Eq. 1 and 2, to be more than ten times the applied Δp. Therefore
all He II will rush into the warmer dewar. The stationary state
is reached when dewar with T_o and p_o contains only vapor (and a
film on the walls) and the He II is in the other dewar and in the
tube such that along the tube Eq. 2 holds as before. Now we can
write expressions for the pressure and temperature gradients,

$$|\nabla p| = \frac{\Delta p + \rho g h}{\ell} = \frac{P_F}{\ell}$$

$$|\nabla T| = (T_o + \Delta T - T_F)/\ell \tag{3}$$

with $\qquad T_F = T_o + \Delta T - (\Delta p + \rho g h)/\rho S.$

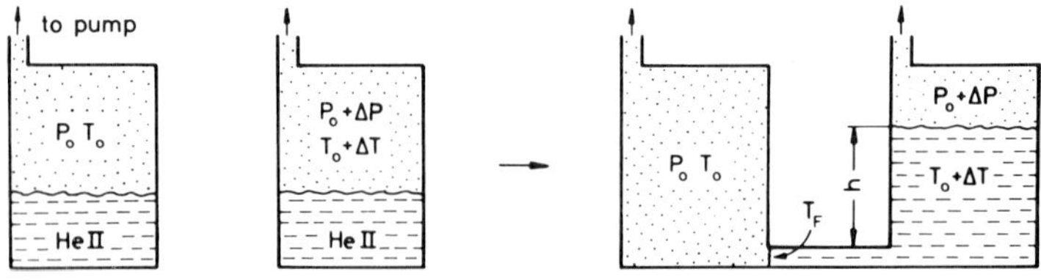

Fig. 1 Explaining the working principle of a He II phase
separator, see text.

A steep gradient in μ, proportional to $T_o - T_F$, occurs at the liquid vapor interface at the tube end which drives an evaporation process and thereby the transfer of heat between the dewars. The fountain temperature T_F, calculated from Eq. 6 is found to be very close to $T_o + \Delta T$ reflecting the tendency of He II to stay isothermal.

This arrangement is the phase separator: the liquid is held back in the warmer "cryogen storage" by a fountain pressure always opposite and equal to the applied pressure difference while heat is transferred by evaporation and the vapor can be pumped off. The system remains stable as long as

$$T_o + \Delta T - T_F > T_o$$

or

$$pgh < \Delta p(\rho ST/\rho_v L - 1). \tag{4}$$

So the device will work in a zero gravity environment, but can be endangered by high accelerations giving rise to extra mechanical pressure, for instance at a rocket launch. In what follows we will not mention the hydrostatic pressure explicitly; it is always corrected for in the experimental results discussed.

The details of He II phase separation have been published before[5]; we will repeat the basic relations we need later on which are the expressions for the mass and heat transport. The latter are obtained from the two fluid hydrodynamic equations linearized with respect to velocities and solved for the stationary state. The subscripts s and n stand for the normal and super components. The mass flow rate

$$\overset{\circ}{m} = \rho v = \rho_n v_n + \rho_s v_s \qquad \text{with} \tag{5}$$

$$\rho_n + \rho_s = \rho$$

is related to the heat flow rate:

$$L\dot{m} = \dot{q} . \tag{6}$$

The normal component transports the heat, and flows according to the Hagen-Poiseuille law like an ordinary viscous fluid

$$\dot{q} = \rho \, S \, T(v_n - v) \tag{7}$$

$$v_n = \frac{d^2}{32\eta} \frac{\Delta p}{\ell} \text{ for a cylindrical tube of diameter d.} \tag{8}$$

Combining these equations one find for the total mass flow

$$\dot{M} = Z\rho d^2 \ S \ T \ \Delta p / \eta(ST+L)\ell \qquad (9)$$

where $Z = 2\pi Rd/12$ for an annular gap of gap widths d and radius $R \gg d$ and $Z = \pi d^2/128$ for the cylindrical tube.

THE ACTIVE PHASE SEPARATOR

The active phase separator (APS) is the device closest to the one we discussed and is shown schematically in Fig. 2. "Active" refers to the fact that the length of the slit can be varied mechanically. Typical mass flow characteristics are shown in Fig. 3. The flat part, for low $\Delta p(=p_B-p_i)$, is described quantitatively by Eq. 9 up to a small correction due to the heat flow resistance contributed by the liquid vapor interface (for details see Ref. 3). As one sees from Eq. 9 the heat and mass flow can be predetermined by choosing the geometry. Since, as we will see, the tendency of He II to exhibit turbulence sets an upper limit to d, only the variation of R and ℓ are feasible in the ideal regime described.

Varying the length of the slit has a rather small effect in the flat, low flow regime, but it turns out that the change in length and therefore pressure gradient allows one to pass into the

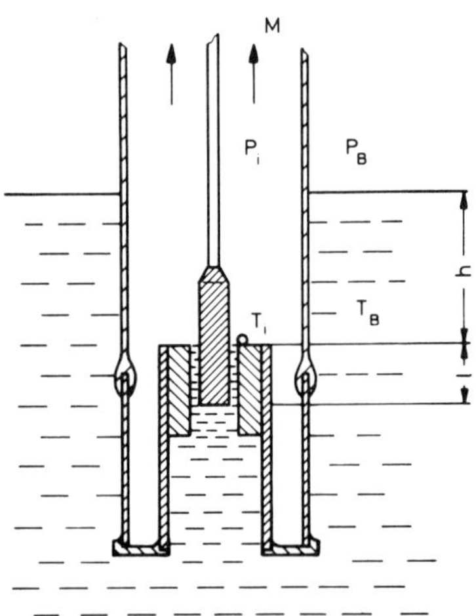

Fig. 2. Schematics of the active phase separator. The movable pin is connected to an electrodynamic driving system and a temperature control unit.

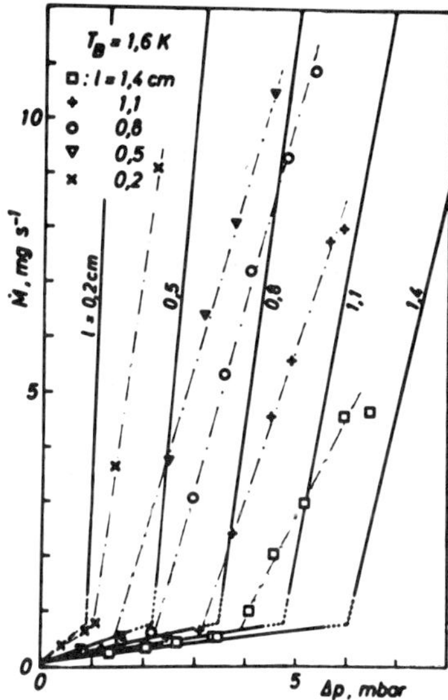

Fig. 3. Mass flow rates vs. pressure drop for a 14 µm gap at T_B = 1.60 K. The broken lines, are experimental results, full lines are calculated.

regime beyond the critical velocity of v_s, at which the friction between the two He components gives rise to a new flow state, marked by the steep high flow part of the curves in Fig. 3. If one knows the critical velocity v_s^c one can calculate from Eq. 5-9, a critical pressure difference for each length.

$$\Delta p_c \approx l\rho_s v_s^c 12\eta/\rho\ d^2 \ . \tag{10}$$

Using the empirical rule $\quad v_s^c d^{1/4} = 1 \quad$ (cgs units) $\tag{11}$

the turn-off points seen experimentally are very well described by Eq. 10 and 11. Beyond this point bulk helium flows into the insert tube and is evaporated from the walls which are in contact with the bath. In this regime phase separation works only because of the special construction which has to consist of a bath, a narrow tube and a heat exchanger. In this state, He II still behaves very special and flows out at a much smaller rate than He I would.

This Gorter Mellink dissipative flow regime must occur in any He II phase separator since for practical reasons the tube or pore widths must be in a range where sizeable heat is transferred. If the widths are chosen between d = 5 and 15 μm, the critical Δp_c falls well into the region of Δp available. Since the phase separator is a system where temperature and pressure at both ends of the tube are varied according to the needs, there will be a finite mass flow which in the Gorter Mellink regime is determined by

$$\nabla \mu = \frac{\nabla p}{\rho} - S \nabla T = A(T)\, \rho_n (v_n - v_s)^3 \text{ for } v_s > v_s^c \qquad (12)$$

A (T) is the Gorter Mellink constant, and

$$\dot{m} \equiv \rho v_n - \rho_s (v_n - v_s) = \rho v_n - \rho_s \sqrt[3]{|\nabla \mu| / A(T)\rho_n} \qquad (13)$$

where Eq. 1 is to be used for $\nabla \mu$ to obtain $\dot{m}(\Delta p)$. This means $\dot{m}$ is much higher than before, it approaches instead of $(ST/L)\rho v_n$. If this flow rate can be completely evaporated, the heat flow from the system is much higher than before, while the liquid itself carries in heat through the tube less effectively than before. One sees in Fig. 3 that the Gorter Mellink approach describes the high flow regime quite well. Important for the application is the fact that phase separation is still managable, and vapor and heat flow can be raised drastically in this regime where He II behaves nearly like a normal fluid.

THE PASSIVE PHASE SEPARATOR

A typical mass flow result[6] is shown in Fig. 4 for a sintered stainless steel plug with pores 2-5 μm wide at $T_B = 2.05$ K. The passive phase separator works without mechanical manipulation: when the bath temperature rises, so does the pressure and the fountain pressure, and the mass and heat flow along with them. If one characterizes a porous plug by its porosity and its o permeability κ, the mass flow should be given by

$$\dot{M} = F\, \phi \kappa S T \Delta p / L \eta \ell \quad (F = \text{cross section}). \qquad (15)$$

Looking at these (Fig. 3) and other porous plug results, we found the concept of permeability is problematic. The straight line in Fig. 3 is calculated with $\overline{d} = 2\mu m$, using Eq. 9 would give a line with half the slope. Since for d = 2-5 μm the Δp_c lies between 4 and 32 mbar (at $T_B = 2.05$ K) one can not be sure that the observed flattening of the mass flow is caused by Gorter Mellink dissipation, excess liquid now being sucked back by smaller pores where the fountain effect still works ideally. We suspect that

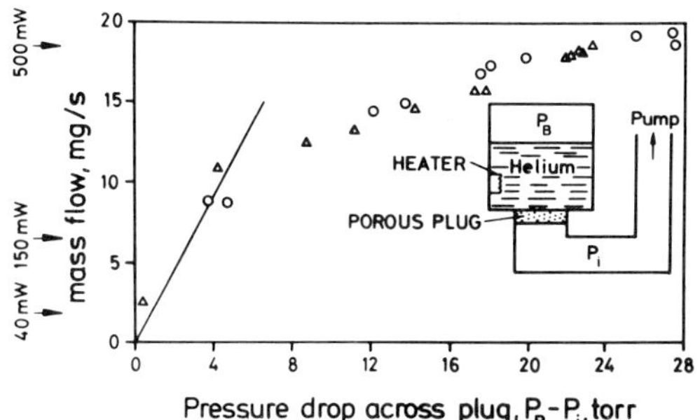

Fig. 4. Vapor mass flow from 4.7 cm^2 sintered stainless
steel plug at He II-temperature 2.05 K, porosity 18%, pore
size 2 – 5 μm, thickness 0.67 cm, permeability κ = 5.5·10^{-10}
cm^2, from Petrac and Mason. [6]

the flow state, discussed in the next section, is more effective
in reducing the mass flow at higher Δp.

INFLUENCE OF THE MATERIAL THROUGH KAPITZA CONDUCTANCE

We have shown that, in the ideal flow state, along the tube
the temperature of He II drops off from T_B to T_F which differ only
by a few mK. In contrast $T_B - T_i$ can be up to several hundred mK
distributed along the tube wall. So some heat conduction should
take place between fluid and solid. There are a few hints that
effects from this are to be expected for very narrow tubes or low
temperatures: The "bad heat exchanging qualities of sintered
copper for He II" have been observed by Hartoog[7]. He found that
heat can be drained from the fluid into the walls along a rather
short path thus lowering the apparent heat conductance of the heat
exchanger. For passive phase separators the observation is com-
municated that phase separation seems to take place somewhere
within the plug. The APS can get into a state with extremely low
mass flow at high Δp, that is the turn to Gorter Mellink friction
does not appear. We believe that the draining heat from He II
along the flow path not only spoils the evaporation process but
also the fountain effect in a surprising way. A model calculation
for the temperature distribution in the tube from which heat is
drained at a rate

$$d\dot{q} = - \pi d\lambda \ (T(x) - T_{wall})dx \qquad (16)$$

where λ is the Kapitza conductivity and the wall temperature
varies linearly between T_B and T_i can be performed similarly as in
Ref. 5, with the boundary conditions adapted to the phase separa-

tor arrangement. The result for the temperature of the liquid at the tube end is:

$$T(\ell) = T_i + (T_B - T_i)/\left((L\rho_v/\rho ST) + \frac{\ell}{\Lambda} \coth \frac{\ell}{\Lambda}\right) \tag{17}$$

with $\Lambda^2 = d^3(\rho S)^2 T/128\eta\lambda$. One can easily check that for $\lambda \to 0$, which means $\Lambda \to \infty$, one recovers the fountain temperature (compare to Eq. 3) and for λ dominating, that is $\Lambda/1 \ll 1$, we have

$$T(\ell) = T_i + (T_B - T_i) \frac{\Lambda}{\ell} \tanh \frac{\ell}{\Lambda}. \tag{18}$$

This means that $T(\ell)$ can get close to T_i. From Hartoog's paper who plotted $(\Lambda/\ell) \tanh \ell/\Lambda$ for $\lambda = 0.05\ T^3 W/Kcm^2$, effects are to be expected from about d=10 μm down. The finite heat conduction of the material itself, which is assumed infinite in the calculation, will somewhat reduce the heat drainage from liquid into walls. At first glance one expects the mass transport to decrease because the evaporation rate proportional to $T(\ell) - T_i \sim \Lambda/\ell$ tanh (ℓ/Λ) does. This will, however, not be the main effect. Due to heat exchange with the walls, a stronger temperature gradient is imposed on the superfluid than is compatible with the London equation when the slit is just filled, the liquid responds to $T(\ell) < T_F$ by an extra fountain pressure which pulls the liquid back towards the bath until the bulk fluid again has a surface temperature close to T_F. It will probably form a pronounced meniscus with a gradual transition to a film getting quite thin towards the exit. The film takes up the extra pressure $\Delta p \approx \rho S(T(\ell) - T_B)$. We recall that for the ideal superfluid the Landau equations contain the van der Waals force which can balance an extra pressure gradient. It is imaginable that due to this effect the superfluid will hardly enter a porous plug when $d \lesssim 1$ μm, except for a film, quite contrary to the behavior of He II in a "superleak" between two baths. Experimentally and theoretically it is found that if this withdrawal sets in before the Δp_c (calculated from Eq. 10) is reached one does not get into the Gorter Mellink regime at all with accessible Δp range. It is not easy to calculate the heat transfer properties of a system partly filled with liquid, vapor and film. We expect that, depending on the heat transfer parameters of the material, the porous plug will show a low heat and mass flow dominated by vapor flow in a similar way as the APS. This could be checked by recording T_i and p_i for the porous plug and watching for strong deviations from the Clausius Clapeyron relation.

CONCLUSION WITH RESPECT TO APPLICATION

We have seen that three heat and mass flow states in He II phase separators are possible: (1) the ideal flow, which uses the ideal heat conductance of He II alone, along with perfect phase

separation, (2) the Gorter Mellink flow where the liquid is not
completely held back by the fountain effect and for the phase
separation an extra heat exchanger is needed, and (3) a state of
reduced heat and mass transfer with "too good" phase separation
hampered by vapor within the phase separator. For the porous plug
it is difficult to keep these states apart, also data on porous
plugs are scarce and incomplete. The APS is planned to be used on
GIRL[8] where the detectors and bolometers are in contact by metal
leads with the He II reservoir which has to have a reliably
constant temperature T_B = 1.6 K. The IRT[9] cryostat will have a
porous plug and everything including the detectors will be cooled
by cold vapor from the vent line. While the porous plug is a
simpler device, it seems that the advantage of the APS lies in its
fast response to external heat loads.

ACKNOWLEDGEMENTS

Thanks are due to D. Schotte, H. van Beelen and the colleagues
of the FU Berlin cryogenics group, especially H.D. Denner and Z.
Szücs, for stimulating discussions.

NOTATION

A(T)	Gorter Mellink constant
d	slit or gap width
ℓ	length of narrow channel
L	latent heat of evaporation
$\dot{m}$	mass flow rate per cm^2
$\dot{M}$	total flow rate
p	pressure
$\dot{q}$	heat flow rate per cm^2
R	radius of annular gap
S	entropy of helium per gram
T	temperature
v	fluid velocity

Greek letters

μ	chemical potential
η	viscosity of helium
κ	permeability
λ	Kapitza conductivity
Λ	thermal adjustment length of He II flow along solid boundary
ρ	density
ϕ	porosity
$\Delta\phi$	porosity difference between bath and vent line

Subscripts

n	normal component
s	superfluid component
B	bath
i	for temperature and pressure in the vent line
v	vapor
F	for Fountain temperature and pressure
c	critical

REFERENCES

1. A. Elsner, G. Klipping, in "Advances in Cryogenic Engineering, Vol. 14", Plenum Press, New York (1969), p. 416.
2. A. Elsner, G. Klipping, in "Advances in Cryogenic Engineering, Vol. 18", Plenum Press, New York (1973), p. 317.
3. P.M. Selzer, W.M. Fairbank, C.W.F. Everitt, in "Advances in Cryogenic Engineering, Vol. 16", Plenum Press, New York, (1971), p. 277.
4. W.E. Keller, E.F. Hammel, Annals of Physics 10:202 (1960).
5. U. Schotte, H.D. Denner, Z. Phys. B41:139 (1981).
6. D. Petrac, P.V. Mason, in "Proc. Int'l Cryo. Engr. Conf.", IPC Science and Technology Press, Genova (1980), p. 97.
7. A. Hartoog, Cryogenics 18:435 (1978).
8. GIRL = German Infrared Laboratory, to be carried with the Space Shuttle in 1986.
9. IRT = Infrared Telescope, to be flown on Shuttle Spacelab 2, see G.R. Karr, et. al., in "Proc. 8th Int'l. Cryo. Engr. Conf.", IPC Science and Technology Press. Genova (1980) p. 38.

He II EXPERIMENTAL FACILITIES AT SACLAY

J. C. Lottin

Centre d'Études Nucléaires de Saclay
Gif-sur-Yvette, France

INTRODUCTION

In this paper we describe two laboratory test facilities constructed and operated at Saclay for general studies on thermal properties of He II at variable temperatures and pressures related to the cooling of superconducting magnets.

Two other He II facilities also in operation at Saclay, built in a joint collaboration of three CEA Laboratories (Fontenay-aux-Roses, Grenoble and Saclay) for the qualification tests of a new Tokomak "Tore Supra", under construction in France, have already been reported in detail[1,2] and will not be described here.

The areas of investigation for the basic He II studies include:

- Heat transfer in the fluid and at the interface between fluid and solid for both steady state and transient regimes.

- Properties of materials in magnetic fields for temperatures between 1.5 K and 20 K.

- Analysis of the stability of superconductors in specific configurations as a function of temperature.

The laboratory equipment consists of several cryostats, of which the two primary ones are described here, connected to a common helium supply and pumping system. The computerized data aquisition system permits rapid storage and visualization of the system physical parameters, especially the temperature, in such a way as to give on-line information.

CRYOSTAT "MARINETTE"

We have constructed and operated a cryostat, called "Marinette" which is a flexible experimental facility usable either with or without an 8.5 T magnet in He II (Fig. 1).

The overall cryostat is 30 cm in diameter and 110 cm in height. Without the magnet, the subcooled He II experimental volume has an 18 cm diameter by 22 cm height in the temperature range above 1.4 K. With the superconducting coils installed, a 10 cm diameter is available. For experimentation above the He II temperature range, a vacuum can is available to be inserted inside the coil. This cryostat is based on the conventional stratified temperature bath concept which uses an insulating fiberglass epoxy separator between the upper He I and the lower He II regions. This separator is easily inserted into the cryostat at room temperature and it is sealed at low temperature due to differential thermal contraction. Cooling power at reduced temperature is achieved by means of a boiler surrounding the experimental vessel. A heat transfer area of 1100 cm^2 permits the utilization of stainless steel for the wall between the two baths insuring good leak tightness. The two radiation shields are helium vapor cooled, thus requiring no liquid nitrogen and providing better insulation due to lower shield temperatures. Further details of the cryostat performance are reported elsewhere[3].

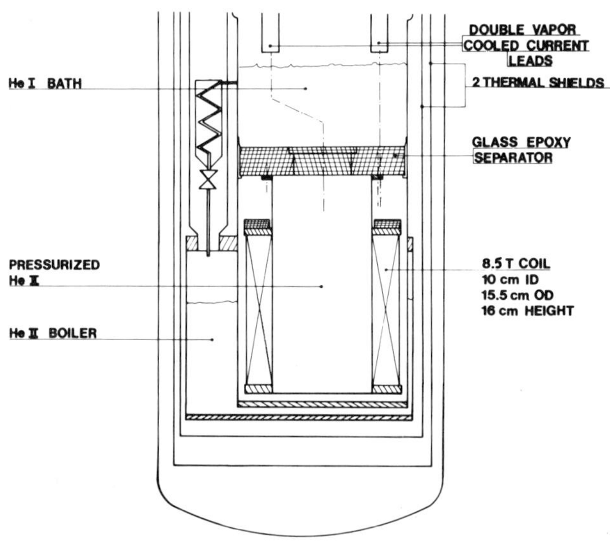

Fig. 1.   "MARINETTE" cryostat with the 8.5 T coil.

The magnet is a 20 layer solenoid, vacuum grease impregnated with a 50 μm Mylar insulator between each layer. It provides at 1.8 K a central field of 8.5 T with a current of 840 A. The peak field at 1.8 K with the critical current of 868 A is 9.1 T, while 6.9 T is the maximum achieved at 4.2 K. A summary of the performance characteristics of "MARINETTE" cryostat and magnet is given in Table I.

CRYOSTAT "SUPERDUC"

A large heat transfer cryostat, which we call "Superduc", is designed to provide a well insulated environment for studies of He II heat transfer. We have adopted this approach in order to carry out quantitative studies of heat pulses and steady state heat transport in He II. Additional flexibility has been built in by allowing for variable pressure operation.

A general schematic of the cryostat is shown in Fig. 2. A 25 cm ID vacuum can, with an available height of 65 cm for installation of heat transfer experiments, is immersed in a 40 cm

Table I.　Cryogenic and magnet performance of "MARINETTE"

Helium cooldown	
Room temperature to 4.2 K	5h, requiring 30 L of LHe
4.2 K to 1.8 K	< 1 hour
Normal operation at 840 Amps	
Cryostat boil off	0.2 m^3/h at STP
Vapor cooled lead	1.8 m^3/h at STP
He II boiler	0.5 m^3/h at STP
Conductor	
Dimension	2x1 mm^2
Copper:NbTi ratio	2:1
Coil dimensions	
OD	15.4 cm
ID	10 cm
Height	16 cm
Coil total weight	12 kg
Peak field at 1.8 K	9.1 T
Overall critical current density at 1.8 K	35 kA/cm^2
Stored energy	20 kJ

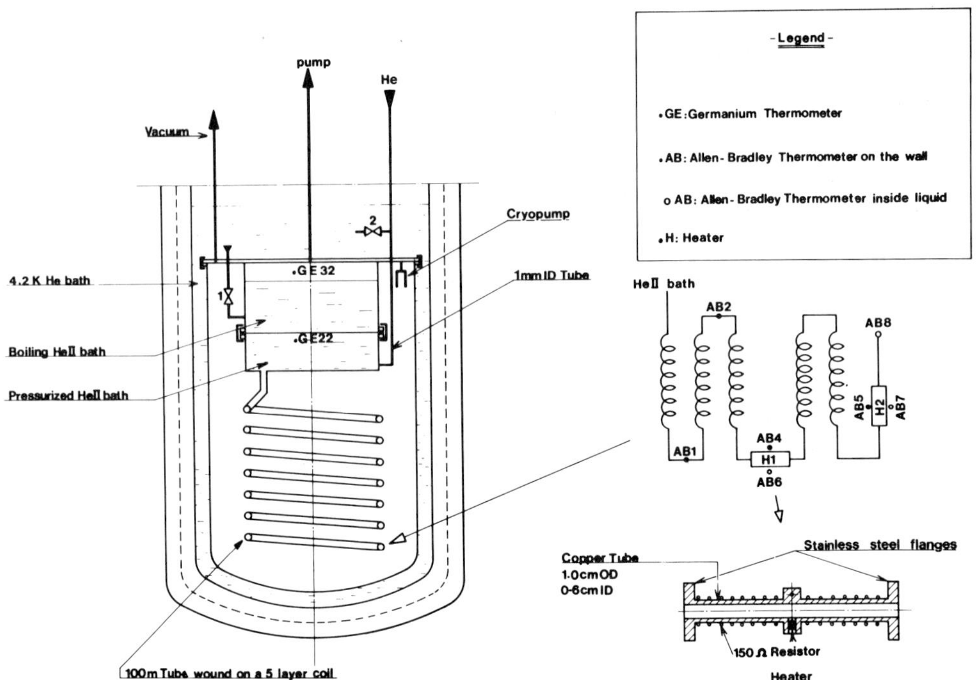

Fig. 2. "SUPERDUC" cryostat with the 100 m long tube experiment.

standard liquid He I cryostat. The top flange of the vacuum can
supports the detachable indium sealed vacuum enclosure and the
helium boiler which provides cooling power at reduced temperature
for the experimental volume attached underneath. The boiler is
supplied through the level controlling valve[1]. The experimental
volume is filled through a 1mm ID 30 cm long capillary tube in
series with valve 2 opening to the surrounding 4.2 K helium bath
and is cooled by heat transfer to the boiler through a 6mm thick
annealed copper plate 12 cm in diameter.

At 1.8 K, the temperature difference between the upper and
lower baths, due to Kapitza resistance, is 0.03 K for a 2 W experimental heat load. With valve 2 closed it is possible to pressurize the experimental volume up to 0.4 MPa. Four holes in the
bottom of the lower bath are available for connecting experimental
heat transfer systems. Insulating vacuum has an external pump and
an in situ activated charcoal cryopump for both improving the
vacuum during an experiment or for cooling down the experiment by
turning the pump heater off or on.

DATA ACQUISITION SYSTEM

The measurement and data acquisition system for the two facilities described above is designed to accept low voltage signals such as originate from temperature sensors or heaters. It then calculates the relevant physical parameters. The system contains a 64 Kbyte memory microprocessor with two input signal interfaces:

- A 16 channel multiplexer gives 12 bit definition to 10 V input signals. The rate of measurement is 20 ms per channel.

- A 4 channel signal recorder collects data in parallel within adjustable sampling times from 10 µs to 10 s per point. It works as a buffer; each channel can collect and store 1024 data points for subsequent transfer to a chart recorder or to the microprocessor for data handling.

The system, as presently configured, has a temperature measurement accuracy of 2 mK and can have a time resolution down to 10 µs, which is sufficient to measure temperature variations with time constants in the millisecond range.

EXPERIMENTS

We are presently performing two experiments on heat transfer through tubes containing He II, one a 100 m long tube and the other a 30 cm long tube.

The long tube experiment gives us the ability to measure steady state temperature profiles at very low heat flux and to perform pulsed experiments in an effectively infinitely long system. It is constructed with an 100 m long copper tube of 0.6 cm ID and 0.8 cm OD wound into a five layer coil and installed in the "Superduc" cryostat. The lower end of the tube is sealed while the upper end is connected to subcooled He II bath. Two 10 cm long test sections, located 50 m and 100 m from the heat sink, are instrumented with a heater fabricated by winding a 150 Ω manganin wire on the surface of the copper tube and two carbon temperature sensors. The test sections are connected to the tube by means of stainless steel flanges. Heater power is determined by simultaneous measurement of the current and voltage.

The temperature of subcooled helium bath is kept constant to within 1 mK with an electronic temperature regulator using a 100 Ω carbon resistor as sensor. Eight other carbon resistors are distributed along the experiment (Fig. 2). All carbon resistors are calibrated in situ against a germanium resistor located in the subcooled He bath. The four probe method is used with constant current of 0.3, 1, or 3 µA.

Figure 3 shows the temperature profile along the tube for different heat fluxes. In Fig. 4, we plot the temperature gradient versus steady-state heat flux for several different experiments. The data are plotted as log-log to demonstrate the well known relationship that ∇T is proportional to Q^m. In the 1.6 to 1.9 K range, the fits to the data from the long tube experiment give m = 3 ± 0.15.

For transient heat pulses, negligible energy is transferred to the heat sink because of the length of the tube. Shown in Fig. 5 are temperature profile aspects at different times after the application of a constant 100 mW heat flux. This result illustrates that during the first few minutes, most of the energy is contained in the helium close to the heater, while 50 m from the heater, the helium temperature increase is only 2 mK. Figure 5

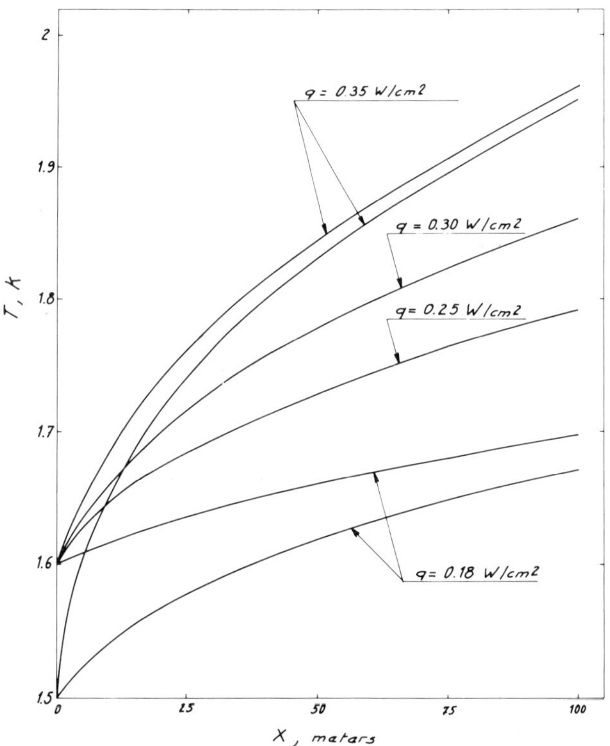

Fig. 3. Temperature profiles along the 100 m long pipe.

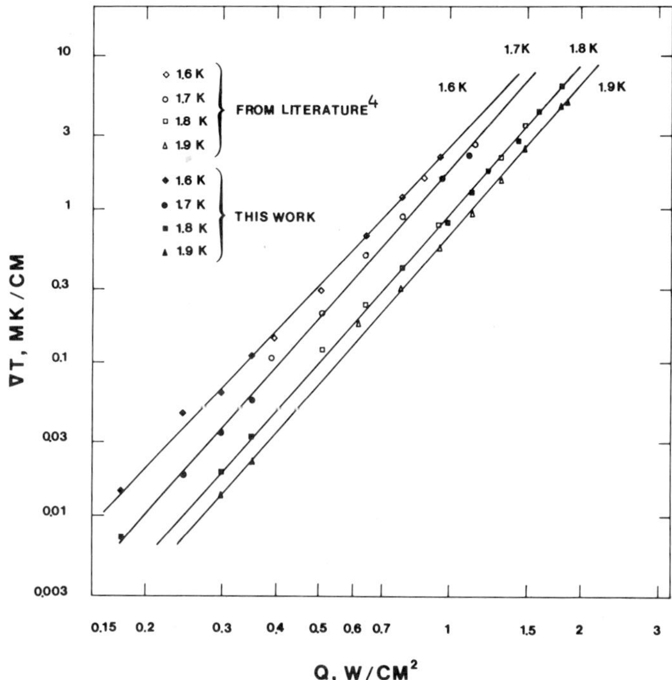

Fig. 4. Temperature gradient versus heat flux. Low and high flux are respectively from the 100 m and 0.3 m tubes.

also shows that several hours is required to reach steady state. This fact makes stabilization of the experiment and sensor calibration a difficult task.

The 30 cm long tube experiment consists of a 0.7 cm ID fiberglass epoxy vacuum insulated tube opened at the bottom end to the 6 L subcooled He II "Marinette" cryostat bath and terminated at the upper end by a copper plug instrumented with a heater and three temperature sensors. The copper plug interface with He II has been machined and degreased. A germanium resistor is used for calibration of the two carbon resistors on the copper bloc, while the bath temperature is measured with a second germanium resistor used for the calibration of the three sensors located inside the helium channel.

This configuration was selected to compare the heat transfer curve with data from surfaces facing upward, which is published in the literature.

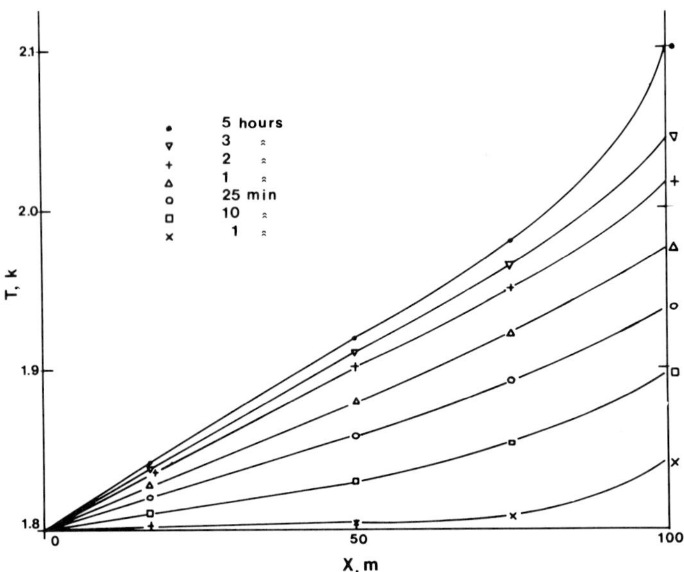

Fig. 5. Transient temperature distributions along the
100 m long pipe for a 100 mW step function heating power.

The temperature gradient versus heat flux for this experiment, plotted in Fig. 4 for high heat fluxes at 1.7, 1.8 and 1.9 K
is in good agreement with measurements from the 100 m tube as well
as the data in the literature[4]. We have also measured the temperature difference between the copper bloc and the helium bath below
and above the critical heat flux, Q*. Below Q*, temperature
differences are small being about 2 K at 1.8 K for 1 W/cm^2
(Kapitza resistance), while for a heat flux slightly larger than
Q*, ΔT can exceed 30 K. Further work will be reported in the
future.

CONCLUSION

We have performed experiments with the two facilities described above which enabled us to check the overall set up for
satisfactory operation. These results provide additional data in
good agreement with that available in the literature and illustrate the high capability of the system with regard to precision
and fine resolution. These experiments will be extended to a
series of He II heat transfer and superconductor stability studies.

ACKNOWLEDGMENT

The author gratefully acknowledge Dr. Roubeau for his active contribution especially in the "Superduc" system.

REFERENCES

1. J.L. Augueres, et al, 700 mm diameter cryostats operating at 1.8 K and atmospheric pressure, Cryogenics 20:529 (1980).
2. R. Aymar, et al, Test of a model coil of Tore Supra, IEEE Trans. on Magnetics, MAG-17:38 (1981).
3. M.X. Francois, J.C. Lottin and J. Plancoulaine, Control of Pressurized Superfluid Helium, in "Advances in Cryogenic Engineering, Vol. 25", Plenum Press, New York (1979), p. 541.
4. S.W. Van Sciver, Kapitza Conductance of Aluminum and Heat Transport from a Flat Surface through a Large-diameter Tube to Saturated Helium II, in "Advances in Cryogenic Engineering, Vol. 23", Plenum Press, New York (1977), p. 340.

BATH COOLING WITH SUBCOOLED SUPERFLUID HELIUM

G. Claudet and P. Seyfert

Centre d'Études Nucléaires de Grenoble
Grenoble, France

INTRODUCTION

Several years ago, after the successful completion of some preliminary test experiments, we began a systematic investigation of the potential of superfluid helium bath cooling. The present paper is intended to give an overview of results obtained from work done in our laboratory or to which we have made substantial contributions as a part of the Tore Supra project.

Throughout our program we have chosen to investigate subcooled superfluid helium at atmospheric pressure. Our experience of the past few years has fully validated the arguments presented in favor of this choice.[1]

We will survey the fundamental aspects of subcooled superfluid helium with emphasis on its application to superconductive magnet technology and discuss some particular problems of the Tore Supra project.

RESULTS OF GENERAL INTEREST FOR HELIUM II TECHNOLOGY

Steady State Heat Flow

In contrast to the behaviour of classical fluids, heat transport in He II is governed by a peculiar mode of conduction. Heat transport along extended channels with diameters above approximately 0.1 mm is found to obey the equation

$$q\, L^{1/n} = [f(T)\, dT/dx]^{1/n} \tag{1}$$

with q the heat flux per unit cross-sectional area of channel and x the relative space co-ordinate (x = 1 means full length L).

Values of n~3 are reported in literature. The function $f(T)$ is
sometimes referred to as the heat conductivity function of He
II. It has a maximum near 1.9 K and drops to zero at the lambda
temperature, T_λ.

We have extensively investigated one dimensional heat transport in subcooled superfluid helium over comparatively large
temperature differences[2]. The integral heat equation

$$qL^{1/3\cdot4} = W(T_c, T) \qquad\qquad (2)$$

appeared to give the best fit to our results. T_c and T represent
the cold and hot end temperature of the channel, respectively.
The function $W(T_c, T)$ exhibits a monotonic, strongly non-linear
increase with T and approaches a peak value as $T \to T_\lambda$.

Kapitza Boundary Resistance

Interfaces between solids and He II create the so-called
Kapitza boundary resistance to the flow of heat. The resulting
temperature drop is in practice by far the largest contribution to
the overall temperature difference in He II heat transfer.

The strong influence of the metallurgical state of the solid
surface and the non-linear character of the transfer law are well
known. For this reason we have carried out measurements on
technical copper with heat flux values up to 9 W/cm^2. Several
samples were given a surface treatment which was thought to
approximate the final stages of superconductor manufacture and
magnet assembly.

Figure 1 shows the results for four types of samples. Each
curve represents in fact an average of the data from three "identically prepared" samples. We generally observed differences of
the same order of magnitude between samples of different types and
samples of apparently identical condition. According to the
literature the irreproducible character of Kapitza boundary resistance is due to small variations of the surface condition which
are almost uncontrollable in practice.

We can nevertheless, make a reasonably realistic recommendation for design situations using commercial copper. This is to
use the upper bound (with regard to temperature) of the average
experimental results.

We should finally add that there is recent experimental
evidence[3] for invariable Kapitza boundary resistance with oscillating heat fluxes up to frequencies of 600 Hz.

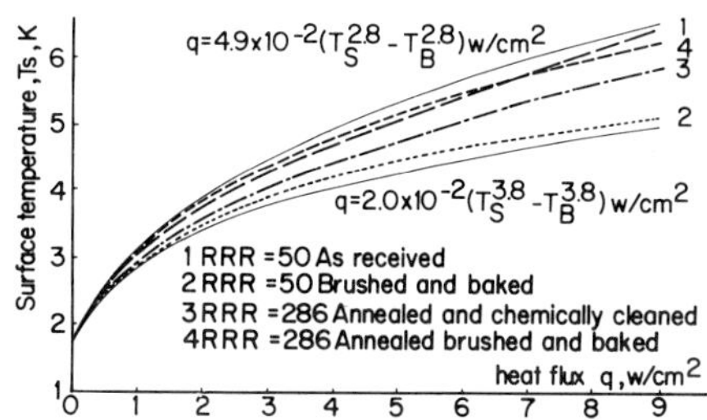

Fig. 1. Kapitza boundary resistance of technical copper (T_B is the bath temperature.)

Transient Heat Absorption

We have studied two aspects of transient heat absorption which are of particular concern for superconducting magnet technology[4].

We first focused our attention on the amount of heat a He II specimen can absorb when a step function heat pulse is applied to it. We determined by experiment the time between release of the pulse and onset of thermal runaway at the heated surface. Our test specimens ranged from closed cells containing fluid layers with thicknesses of some millimeters to open cells with lengths of some centimeters connected to a He II bath at constant temperature.

Besides the experimental studies we have worked out a theoretical model of one-dimensional heat propagation in He II. It is based on the assumption that the high heat fluxes considered here immediately establish a state of He II which inhibits heat transfer by second sound and only allows heat diffusion by local temperature gradients. The heat conduction equation (Eq. 1) and the classical energy conservation law together with appropriate boundary conditions form the mathematical basis of our model. A specially written computer program solves the system numerically.

Figure 2 shows experimental and computed results. The ratio of the heat actually transferred to the enthalpy increase of He II that would result form an isothermal rise between the initial temperature and T_λ is plotted as the ordinate. The abscissa is divided into units of qd$^{1/3}$.4 q being the heat flux per unit area of fluid section and d the length of the test specimen. Results with our extended test cells have been given a correction which roughly accounts for the two-dimensional geometry of the heat

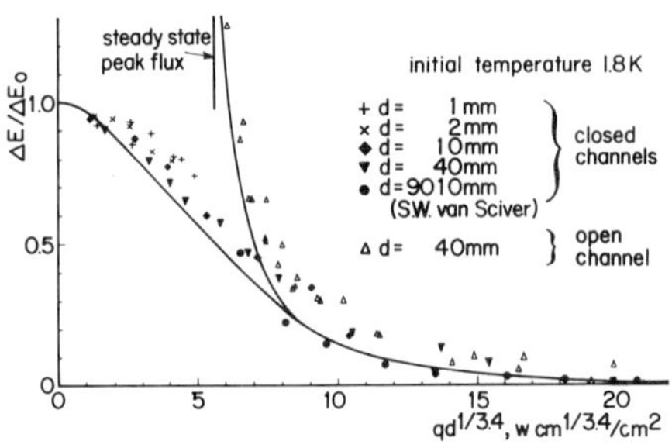

Fig. 2. Fraction of enthalpy used during transient
heat transfer.

inlet. Our model results are shown by the full lines and appear
to describe the experiments in a satisfactory manner.

We have in addition directed our interest towards recovery
from film boiling. In experiments we subjected a He II specimen
with open geometry (d = 4 cm) to an intense heat pulse which
initiated considerable thermal runaway of the heated surface (10 –
20 K typically). Immediately following the pulse a reduced heater
power was maintained, thus simulating joule heating within a
segment of superconductor that has been driven normal. The principal finding was that even if post-heating exceeded the peak
steady state flux, recovery from film boiling could take place, although only for a limited duration. Stabilization of superconductors may well benefit, however, from this temporarily restored
normal Kapitza transfer regime. Figure 3 illustrates the effect
for our specimen. It shows how the post-heating initial pulse
energy correlates with the post-heating amplitude (power ratio 10)
for the limiting situation of a nearly vanishing duration of the
restored Kapitza transfer regime. Both quantities are referred to
unit cross-sectional area of He channel.

Our theoretical model supplemented with the appropriate
sequence of boundary conditions appeared to reproduce the recovery
behaviour in a qualitatively correct manner. Quantitave agreement
could only be obtained, however, with some arbitrary assumptions. As a matter of fact our one-dimensional model cannot
adequately describe the region close to the heated surface where
transient recovery originates. We are presently trying to extend
the model to two dimensions.

PRACTICAL APPLICATIONS TO THE TORE SUPRA PROJECT

Numerous and difficult specifications are imposed on superconducting magnet systems which are to be constituent parts of

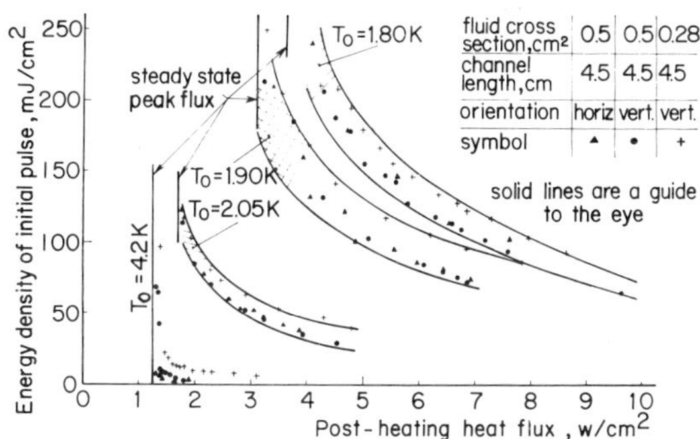

Fig. 3. Limits of transient recovery from film boiling

plasma fusion Tokamaks. High magnetic field and current density as well as fast field variations and large stored energy set high goals for the design of the conductor and the associated cooling mode.

The Tore Supra Tokamak, a project supported by the EURATOM-CEA association, is designed with superconducting toroidal field magnets[5]. The windings of monolithic NbTi conductor will be cooled by superfluid helium at atmospheric pressure. The very severe requirements can thus be met including a magnetic field of 9T with superimposed field variations in the range of 50 T/s and an average current density in the winding of 50 A/mm^2. The additional cost to be paid with respect to 4.2 K bath cooling is represented by a 300 W refrigerator operated at 1.75 K.

Before the final decision on the project could be reached, an experimental program was carried out including qualifying tests at a significant scale. This program, executed by a joint team with groups working at Fontenay-aux-Roses and Saclay, has been reported on elsewhere[5]. We are concerned here only with some experiments which will focus on implementation and cryogenic aspects.

We briefly mention the stability experiments performed on channel cooled test solenoids. They demonstrated that with He II at 1.8 K the recovery current can be more than doubled with respect to 4.2 K bath cooling. For fields up to about 11 T the critical current, of NbTi cooled with superfluid helium is thus essentially equal to that of Nb$_3$Sn at 4.2 K.

Large He II Cryostat, 70 cm I.D.[5]

With a view towards large scale applications, we have built two identical cryostats of significant size, thus testing the

feasibility of He II technology from an industrial point of
view. Both cryostats, designed as constituent parts of high field
test facilities, were successfully implemented. One, installed at
Saclay, is dedicated to the Tore Supra program and the other,
installed at the Service National des Champs Intenses in Grenoble,
will be used for high field superconductor development.

Figure 4 shows the three possible configurations in which the
cryostats can be operated: a) operation in the 338 mm I.D. bore
of NbTi coil giving a peak field of 10 T at center; with a sample
is immersed in the main He II bath, b) inside the same coil field,
using an additional metallic insert-cryostat, 281 mm I.D., to
provide temperatures from 1.8 K up to 300 K for experiments in a
magnetostatic field, c) by use of an insert-cryostat made of
insulating material, experiments with superimposed time varying
fields can be carried out between 1.8 K and 4.2 K within a 150 mm
I.D. Configurations a and b have been frequently used for more
than one year. Configuration c is in preparation. The cryogenic
test of the glass/epoxy insert has been successfully completed.

The lower part of the main cryostat which constitutes the He
II compartment has a volume of about 435 L. The insulating separator (2) made of a 50 mm thick epoxy fiberglass plate is equipped
with electrical feed-throughs, an access for a cooldown line (5)
and safety (6) or emergency (7) cold valves. The refrigeration
circuit with counterflow heat exchanger (9) and expansion valve
(10) supplies saturated liquid helium to heat exchanger (11) which
constitutes the cold source.

With a 400 m^3/h pumping unit, 2 W of refrigeration is available at 1.8 K for experiments. Total liquid helium consumption of
the cryostat without current leads is then 5.5 L/h. Both cryostats were manufactured by industrial firms with the same methods
as used in He I technology (welding and leak detection). NbTi
superconducting magnets with 330 mm I.D. bore are now operated up
to 10 T in these cryostats in fully safe and automatic conditions.

Model Coil of Tore Supra[5]

We will only discuss here the cryogenic and safety aspects of
this experiment.

The model coil consisting of 9 double pancakes contains about
120 kg of conductor and 14 L of liquid helium. With its 888 mm
O.D. it fits into the so-called BIM magnet which adds 4 T to the
self field of 2.4 T. A hydraulic jack applied along the 500 mm
I.D. allows us to introduce mechanical deformation and shear
stresses to the winding. The model coil is in horizontal orienta-

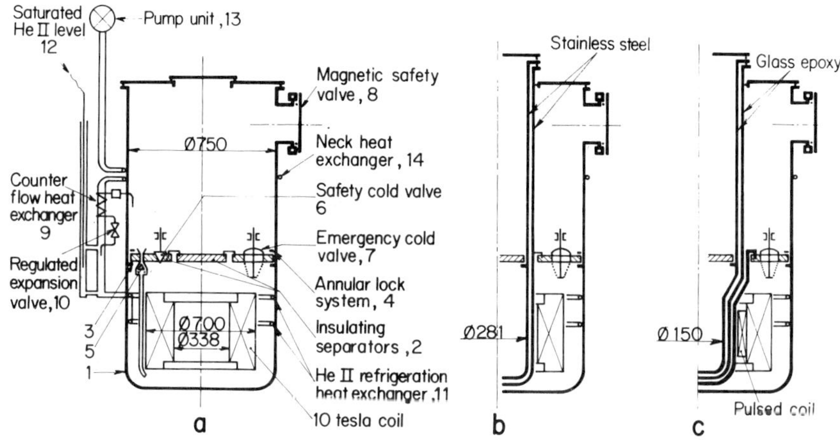

Fig. 4. The different configurations of our large He II cryostats

tion. Figure 5 represents the cryogenic circuit which is similar to that of Tore Supra.

The external casing of the coil is cooled at 4.2 K by circulation of liquid He I. The He II refrigeration circuit with counterflow heat exchanger (5) and expansion valve (6) removes heat from the heat exchanger (9). A heat pipe (2), 2.8cm I.D. and 7m long, ensures thermal connection between the winding and the cold source. It operates without any fluid motion. Filling proceeds through valve (13). Fluid removal in the case of heating is possible through the safety valve (12). We have verified that there was no particular problem with the filling of the model coil circuit and the cooling to below 4.2 K. In continuous operation

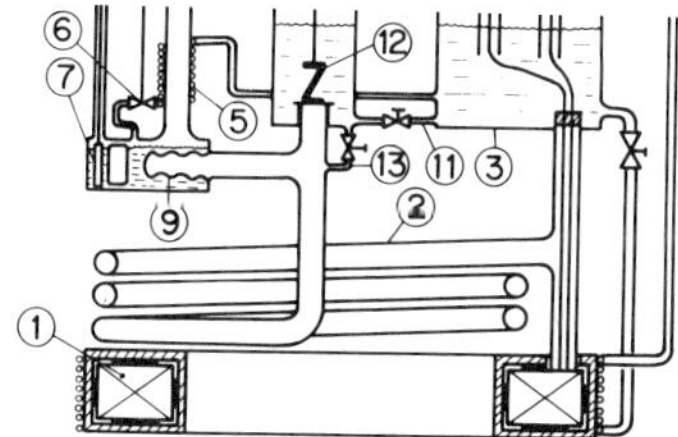

Fig. 5. Cryogenic circuit used in the Tore Supra model coil test experiment.

with a pumping unit of 400 m^3/h the available refrigeration capacity is 3 W at 1.8 K.

The quench behaviour of a full scale toroidal field coil could be approached by special experimental conditions. The BIM magnet and the model coil were series connected and supplied with 1200 A. Coil transitions were then initiated by means of an electrical heater. During the test 167 kJ were released in the coil. Pancakes driven normal reached temperatures between 45 K and 60 K. According to our estimates about half of the released energy was absorbed by the fluid.

CONCLUSION

The results of the work summarized here together with results obtained by other groups, in our opinion, have provided a conservative data base for a conservative design of He II-bath cooled superconducting magnets. The large transient heat absorption of He II should stimulate more research aimed at the actual utilization of this potential benefit, thus opening the way to even higher performance magnet systems. The development program and qualifying tests carried out for the Tore Supra project have demonstrated that the subcooled superfluid helium cooling mode constitutes at present a technology suitable to solve problems of considerable intricacy.

REFERENCES

1. G. Bon Mardion, et al., Helium II in Low Temperature and
    Superconductive Magnet Engineering, in "Advances in
    Cryogenic Engineering, Vol. 23", Plenum Press, New York
    (1978).
2. G. Bon Mardion, G. Claudet, P. Seyfert, Practical data on
    steady state heat transport in superfluid helium at
    atmospheric pressure, Cryogenics, 19:45 (1979).
3. J.A. Katerberg, A.C. Anderson, Comparison of steady state and
    second sound measurements of the Kapitza resistance, J. Low
    Temp. Phys., 42:165 (1981).
4. P. Seyfert, G. Claudet, M.J. McCall, Transient Heat Transfer
    in Boiling Helium-I and Subcooled Helium-II, in "Advances
    in Cryogenic Engineering, Vol. 25", Plenum Press, New York,
    (1980).
5. R. Aymar, et al, Tore Supra-status report concerning the
    superconducting magnet after the qualifying development
    program, IEEE Trans. on Magnetics, MAG-17:1911 (1981).

DISCUSSION

Question by Ph. Lebrun, CERN:
Did you have to implement special quality control methods for superfluid He piping and vessels?

Answer by Author:

Plumbing for superfluid helium was controlled according to standard methods of LHe technology. It is our experience that there is no need for additional quality control.

Question by J. Alcorn, General Atomic Co:

In Tore Supra is there a separate He II_s/He II_p heat exchanger for each toroidal field coil?

Answer by Author:

Saturated He II is contained in three vessels located around the torus and maintained at uniform pressure by a common pumping system. Each of the 18 toroidal field windings has its own pressurized He II circuit. In this way a saturated He II bath is provided for every six He II_p/He II_s heat exchangers.

STABILITY TEST OF SUPERCONDUCTING MAGNETS WHEN COOLED IN DIFFERENT PHASES OF LIQUID HELIUM

H. Kobayashi and K. Yasukōchi

Atomic Energy Research Institute
Nihon University
Tokyo, Japan

INTRODUCTION

We evaluated the cooling capacity of liquid helium at different temperatures, both at saturated and at atmospheric pressures, through the observation of the stability performance of superconducting Nb-Ti magnets with respect to various coolants. Of course, an overall evaluation must take into account the refrigeration system. The stability of the superconducting magnet is estimated by the behavior of the quench current I_q vs. the sweep rate of the operating current $\dot{I}_c$, as well as the behavior of I_q triggered by artificial disturbances of energy E_d. From these criteria, the pressurized helium-II (He IIp) below 2 K is confirmed as an outstanding coolant for cryostaticly stabilized magnets.

EXPERIMENTAL PROCEDURE

Stability tests were carried out with three small superconducting magnets wound with Nb-Ti wire. Basic parameters of these magnets are listed in Table I. The third magnet has voltage taps both at the center and the ends of each layer, a search coil and an insulated manganin heater (12.2Ω) for applying disturbances to the center of the innermost layer. The heater was buried inside the epoxy fiberglass bobbin of the coil in such a way that only a few turns of the conductor were in tight contact with the surface of the heater. To prevent the heater power from passing directly into the cooling channel through the very narrow interturn gaps several turns at the site of the heater were impregnated with grease. Fiberglass epoxy spacers of rectangular cross-section (thickness e = 0.05 cm x width ℓ' = 0.15 cm x length L = 6 cm) were interleaved between the first several layers except the innermost space between the first layer and the bobbin. The spacers and layers together form cooling channels with cross-sectional area A_{ch} of 0.01 cm^2 along the length of the magnet and at right angles to the windings. The disturbance energy E_d was measured in terms of the pulse current through the heater with a duration of 5 ms.

Both pressurized and saturated liquid helium cooling at temperatures from 1.6 K to 4.2 K was provided in a Claudet-type cryostat.[1] The liquid temperature T_b was measured by two standard Ge thermometers. Unlike He II, the temperature of He I below 4.2 K, especially, of He I_p is difficult to control because of the low thermal conductivity, and thus it is nearly impossible to set the temperature to a specific value.

Table I. Magnet parameters

Parameter	Coil Number		
	1	2	3
Core diameter mm	0.254	0.254	0.01[a]
	(single)	(single)	(multi.)
Composite diameter mm[b]	0.35	~0.35	0.35
Cu/SC ratio	0.9	~0.9	1.1
Number of turns	9070	8050	7478
Number of layers	70	36	48
Inner diameter cm	4.2	1.8	0.8
Outer diameter cm	14.2	4.5	7.4
Length, L, cm	5.0	8.7	6.0
Packing factor, %	30	35	40
Space for liquid helium, %c	12	30	41
Field constant, G_o, kG/A	1.47	1.10	1.30
Inductance, H	3.6	0.44	0.82

[a]600 filaments
[b]All composites are insulated with 10 μm Formvar.
[c]Volumetric percentage of the space into which the liquid helium permeates.

RESULTS

Magnet No.1, closely wound (no spacers) showed no degradation and no training effects. Figure 1 shows the dependence of I_q on T_b, for different sweep rates $\dot{I}_o$. For the low $\dot{I}_o$, $I_q(T_b)$ is determined by the T_b-dependence of the critical operating current I_{co}. For rapid sweeps, relative reduction (=degradation) of I_q increases towards T_λ but I_q recovers steadily both in He II_p and saturated helium-II(He II_s). The fact that I_q/I_{co} is almost unity

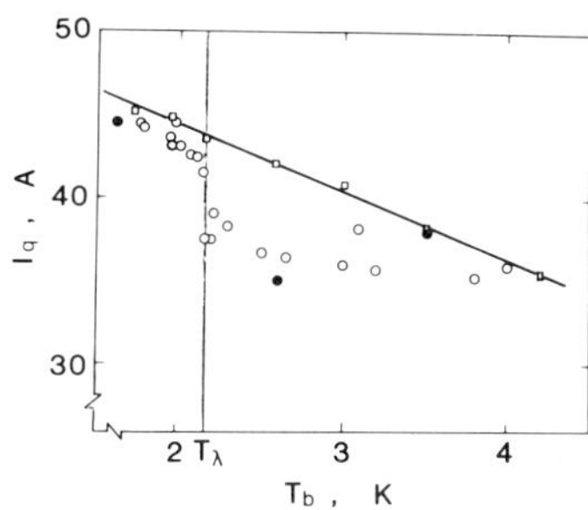

Fig. 1.   Bath temperature dependence of the quench current for magnet No. 1 in pressurized helium at $\dot{I}_0$ = 5A/min ($\square$) and 200 A/min ($\circ$) and in saturated liquid at $\dot{I}_0$ = 200   A/min ($\bullet$).

below 2 K, even though the magnet is subjected to large magnetic instabilities, reflects the 'isothermal' operation mode due to the high cooling capacity.

In Fig. 2, we give an example of the degradation of I_q which is enhanced by lowering T_b. The superconducting wire used for the second magnet has poor contact between the superconducting core and the electrolytically deposited copper matrix. Although the degraded I_q somewhat recovers in the He II$_p$ region, I_q does not become larger than that at 4.2 K. Severe training effects still remain in He II$_p$. The first or the second flux jump during increases of I_0 always trips quenching in the He I region between T_λ and about 3 K.

In the preliminary experiments up to this point, pressurized and saturated liquid helium show no essential difference in cooling capacity as long as the cooling ability is 'measured' by means of I_q as a function of $\dot{I}_0$. To refine the definition of

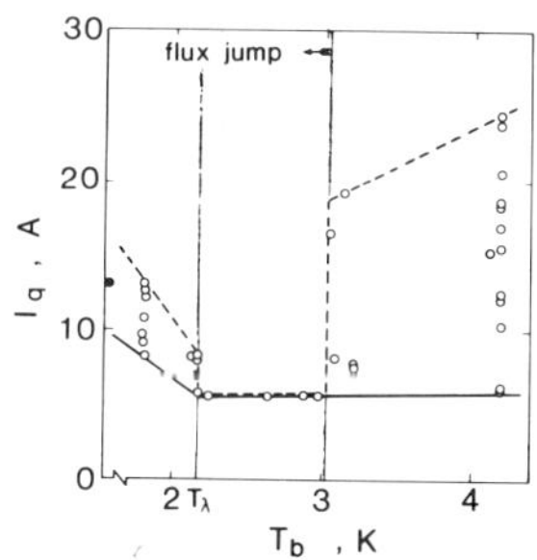

Fig. 2.   Bath temperature dependence of the quench current for magnet No. 2 in He I$_p$ and He II. $\dot{I}_0$ = 5 A/min (Solid line: estimated lower limit, dashed line: estimated upper limit.)

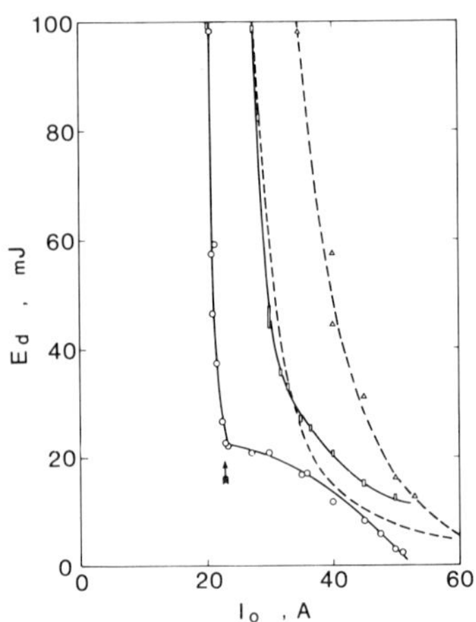

Fig. 3. Critical quenching pulse energy vs. operation current, in He II$_p$ and He I at 4.2 K. (o) 4.2 K, (□) 2.10 K, (Δ) 1.87 K. Dashed lines are calculated from Eq. 5.

cooling capacity, we measured I_q while applying an artificial disturbance of energy E_d to the third magnet. Some examples of the dependence of E_d on I_o for He II$_p$ and He I at 4.2 K are shown in Fig. 3. The same steep decrease in E_d with increasing I_o was observed as had been obtained by Keilin for noninductive samples.[2] However, with careful measurement, we find a distinct difference not only in the amount but also in the shape of the curves for He I and He II. To compare the cooling capacity in each phase, I_q with constant E_d, and E_d at a given I_o as a function of T_b are shown in Fig. 4(a) and (b). Clear differences in cooling capacity can be seen among the four phases.

DISCUSSION

The maximum heat fluxes, Q_{max}, in He I and Q_λ in He II play important roles in the behavior of the quenching mechanism. The E_d and I_q vs. T_b curves shown in Fig. 4 resemble the Q_λ and Q_{max} vs. T_b curves except for He II$_s$ in which Q_λ has a peak at about 1.9 K.[3,4] However, Q_{max} and Q_λ alone are not sufficient to determine I_q and E_d. To explain the quenching mechanism in He II, we introduce a two dimensional expansion coefficient γ which indicates over how many (fractional) turns the initial normal zone has expanded.

The quench at a relatively low field occurs when the total heat flux G_T summed up in a channel reaches the critical heat flux Q_T[5]; $Q_T = 2\ Q_\lambda\ A_{ch}$. (The factor 2 expresses the number of the

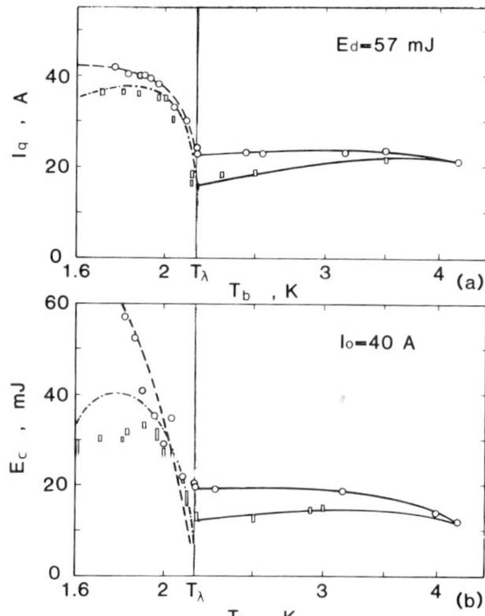

Fig. 4. Bath temperature dependence (a) of the quench current at E_d = 57 mJ, and (b) of the quenching pulse energy at I_o = 40 A. Dashed lines in the He II region are calculated from Eq. 5. ------, (o) pressurized, -•-•-•-•-, ($\square$) saturated, (for H = 20 cm) (The solid lines in He I are experimental.)

heat-flow passages per channel of our magnet.) Although we should include the transient effect for the stability criteria[6], only the steady-heat transport has been taken into account here. The joule heat g per unit surface area of the conductor at the center of the first layer is given by

$$g = I_o^2 \, \rho(H_o)/A_{cu} \, P_W \tag{1}$$

where $\rho(H_o)$ is the resistivity of the copper matrix, at the maximum generated field, A_{cu} the cross sectional area of the copper matrix and P_W is the wetted perimeter.

$$\rho = A + B \, H_o = A + B \, G_o \, I_o \tag{2}$$

where A = 1.74 x 10^{-8} Ωcm, B = 0.53 x 10^{-8} Ωcm/T for our composite. If γ turns switch into normal state, the channel needs to release the total heat $G_T = \gamma g \, A_s$, where A_s is the surface area per turn in which one channel should bear the cooling of the heat: so that $A_s = P_W \, (\ell+\ell')$ ℓ and ℓ' are the channel and the space width respectively. Thus we can define I_q as I_o which satisfies the condition of

$$I_o = I_q = \{ \, G \, Q_\lambda \, (T_b)/\gamma\rho\}^{1/2} \tag{3}$$

where $G = 2\,A_{cu}\,A_{ch}/(\ell + \ell')$. On the simple assumption that the initial normal zone due to the disturbance increases with the energy of the disturbances E_d and with the inverse ratio of the critical current $I_c(T_b, H_o)$, γ takes the form of

$$\gamma = a\,E_d^{\,m}\{I_c(T_b, H_o)\}^{-n} \tag{4}$$

where a is an adjustable parameter which includes such effects as the thermal conductivity both between turns and between the heater and not the layer in question. Thus far, the exponents m and n have not yet been independently determined, but a choice of m = 1 and n = 2 [cf. the dashed lines in Figs. 3 and 4] seems to explain most of the experimental results consistently. γ is not necessarily integer because the dissipated power is not discrete if a resistive state appears. γ expresses something like a critical current margin with respect to the stabilization i.e. how difficult it is for the normal zone to spread when a conductor with a large I_c is disturbed.

Finally, inserting γ into Eq. 3, we obtain

$$I_q = \left[G\,Q_\lambda(T_b)\{I_c(T_b, H_o)\}^2 / a\,E_d\,\rho(H_o) \right]^{1/2} \tag{5}$$

I_q does not correspond to the minimum propagating current I_{mp} below which the quench does not occur even for considerable disturbances. In He I, I_{mp} is well defined because Q_{max} is smaller than Q_λ of He II. Before the blow-off of liquid occurs in a channel, the conditions for film-boiling or the equal area criterion of Maddock[7] are met at the surface of the innermost layer at the current level of I_{mp}. I_{mp} in He II would be larger than I_{co} in the case of the magnet with relatively large cooling channels like our magnet. This can be seen rather clearly in Fig. 3 at the sharp bend (arrow mark) in He I. The dashed line represents the value derived from Eq. 5 by assuming a linear dependence of $I_c(T_b, H_o)$ on T_b. In order to obtain $Q_\lambda(T_b)$ for the calculations of Eq. 5, we used the diagram worked out by Bon Mardion[8] for He II_p and the relation of $Q_\lambda(T_b) \propto (H/L)^{1/3}$ together with the data obtained by Linnet and Frederking[9] for He II_s, where H is the immersion depth. Equation 5 explains the characteristics shown in Figs. 3 and 4 when a is fitted to the data at 1.840 K.

CONCLUSIONS

From the experimental results and the theory, we can deduce some recommendations for the stabilization in He II: (i) Use of He II_p below 2.0 K is the best way to make use of the margin I_c and Q_λ, as long as the thermal boundary resistance inside the conductor is small. (ii) He I_s, He I_p near T_λ and He II_s are not to be recommended as coolants. The inappropriateness of He II_s for superconducting magnets is supported by the fact that the temperature-rises during the quenching process[10] are considerably larger than those of He II_p. (iii) To keep γ small, the neighboring turns should be separated thermally. But then, there arises the danger of a burn out due to localized heating.[11] For this reason it would be advisable to reduce the Cu/SC ratio without varying the size of the superconducting composite, and to operate without changing I_o both for the protection and the stabilization of the magnet.

ACKNOWLEDGMENTS

We wish to thank K. Fuse and K. Yasuda for their work on the experiments. We also thank T. Ogasawara, K. Yasohama, Y. Kubota, H. Okubo and L. Boesten for their useful suggestions and technical assistance.

REFERENCES

1. G. Claudet, et.al., The Design and Operation of a Refrigerator System using Superfluid Helium, in "Proc. 5th Intl. Cryo. Engr. Conf.," IPS Science & Technology Press, Guildford (1974), p.265.
2. V. E. Keilin, et.al., Superconductor stability against heat pulses in saturated and pressurized superfluid helium, Cryogenics. 20 (12):694 (1980).
3. C. N. Whetstone and R. W. Boom, Nucleate Cooling Stability for Superconductor-normal Metal Composite Conductors in Liquid Helium, in "Advances in Cryogenic Engineering, Vol. 13",: Plenum Press, New York (1968) p. 68.
4. T. H. K. Frederking and C. Linnet, Liquid Temperature Effects on Thermally Influenced Transition Currents of Nb-alloy Superconducting Solenoids in He I and He II, in "Advances in Cryogenic Engineering, Vol. 13," Plenum Press, New York (1968) p. 80.
5. G. Claudet, et.al., Superfluid helium for stabilizing superconductors against local disturbances, IEEE Trans. on Magnetics, MAG-15:340 (1979).
6. S. W. Van Sciver, Enthalpy stability criterion for magnets cooled with superfluid helium II, IEEE Trans. on Magnetics, MAG-17:747 (1981).
7. B. J. Maddock, G. B. James and W. T. Norris, Superconductive composites: heat transfer and steady state stabilization, Cryogenics, 9 (4):261 (1969).
8. G. Bon Mardion, G. Claudet and P. Seyfert, Practical data on steady state heat transport in superfluid helium at atmospheric pressure, Cryogenics, 19 (1):45 (1979).

9. C. Linnet and T. H. K. Frederking, Thermal conditions at the
    Gorter-Mellink counterflow limit between 0.01 and 3 bar, J
    Low Temp. Phys., 21:447 (1975).
10. V. A. Al tov, et.al., "Stabilization of Superconducting
    Magnetic Systems," Plenum Press, New York and London
    (1977).
11. C. Meuris and A. Mailfert, Influence of the stability of
    channel cooled superconducting coils, IEEE Trans. on
    Magnetics, MAG-17: 1079 (1981).

MEASUREMENTS OF HEAT TRANSFER TO He II AT ATMOSPHERIC PRESSURE IN A CONFINED GEOMETRY*

R. P. Warren and S. Caspi

*Lawrence Berkeley Laboratory
University of California
Berkeley, California*

INTRODUCTION

Recently the enhanced heat removal capability of pressurized superfluid helium II has been exploited in fusion[1] and accelerator dipole[2] magnets. In superfluid the internal convection mechanism dominates the heat removal process while orientation with respect to gravity is of secondary importance. Heat transfer, however, can be influenced by the thermodynamic state of the liquid, especially with regard to possible phase transitions. The transformation from non-saturated He II must involve an He I state before the film boiling transition is experienced. Some steady state measurements of heat transfer to non-saturated He II have been previously reported.[3-10]

In typical magnet designs, cooling passages between turns are provided by gaps in the electrical insulation, and are typically a fraction of a millimeter wide. The purpose of the work reported here is to measure the attenuation of the heat transfer in such a restrictive geometry.

EXPERIMENTAL APPARATUS

The heated surface is shown in Fig. 1. It consists of a cop-

---

*This work was supported by the Director, Office of Energy Research, Office of High Energy and Nuclear Physics, Division of High Energy Physics of the U.S. Department of Energy under Contract No. W-7405-Eng-48.

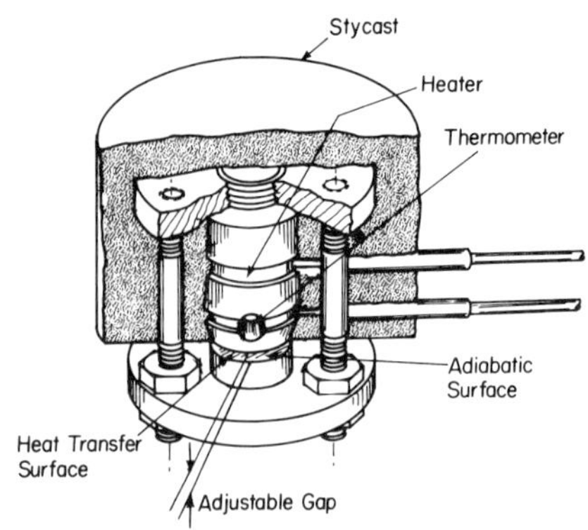

Fig. 1. Experimental assembly.

per cylinder with one end (A = 1.27 cm^2) exposed to the bath. A heater winding is wrapped in a circumferential groove and a thermometer (Lake Shore Cryotronics "carbon glass") is located in a transversely drilled hole which is positioned between the heater and the exposed end. This assembly is potted in Stycast. Following machining, the exposed copper surface was allowed to oxidize with no subsequent surface treatment. An adiabatic surface, machined from Nema G-10 epoxy-fiberglass to the same diameter as the heated surface, was located coaxial with the heated cylinder. The spacing, δ, between the two surfaces, henceforth referred to as the gap, was adjustable. The assembly was located near the bottom of a vertical magnet test vessel with a height and volume of approximately 1.3 m and 100 L (with magnet in place). The exposed surface faced <u>downward</u> for all but one test series (δ = 0.25 mm) in which case the exposed surface was oriented vertically. A magnet test facility , which has been previously described,[11] allows testing at liquid temperatures from about 4.5 K down to about 1.5 K at a pressure of approximately 1 atm. The bulk liquid temperature is measured with a thermometer (of the same type as in the heated cylinder) that is located aproximately 50 cm from the heated surface. Thermometer read out and heater control are accomplished by a Hewlett-Packard (HP) 9845B desk top computer, a programmable scanner (HP 3495A), and a programmable power supply (HP 6002A).

EXPERIMENTAL PROCEDURE

Once a selected bath temperature is reached, the refrigerant flow is reduced to a low level and manually modulated in response to bath temperature changes. At predetermined time intervals the computer commands the power supply to increase the current through the heater (steps of 1 to 4 mA are typical). After allowing 2 seconds for stability, the heater voltage and current are measured and converted into power. The scanner then switches approximately 1 μA into the heater and bath thermometers. The thermometer current and voltage are computed using calibration tables and a spline fit interpolation equation. Temperature accuracy is estimated to be better than ± 5 mK. The temperature of the heated surface is extrapolated from the measured cylinder temperature, using the heater power and the temperature dependent thermal conductivity of copper. The Kapitza resistance is included in the measured surface conductance but this effect is not accounted for in arriving at the surface temperature.

Up to 100 data points were collected and stored in each run. During the run the data are displayed as a plot of heat flux vs surface temperature on the computer CRT and are available in a tabular print out.

The test pressure for all runs was approximately 1.2 atm.

TEST RESULTS AND DISCUSSION

Some results are shown in Figs. 2 and 3 in the form of heat flux density vs the temperature difference, $T_s - T_b$ (s = surface, b = bath). When the bath temperature approaches T_λ, ($T_\lambda - T_b <$ 50 mK) the transitions between the various heat transfer regimes become very apparent (Fig. 2). As the bath temperature is lowered (Fig. 3) the transitions become less apparent but the overall characteristics remain.

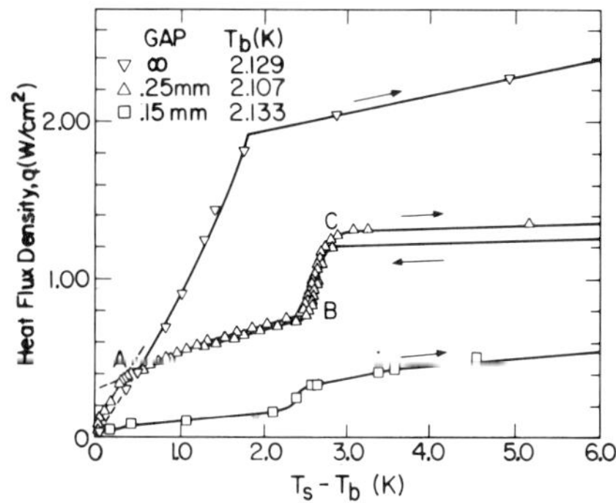

Fig. 2. Characteristic results near T_λ.

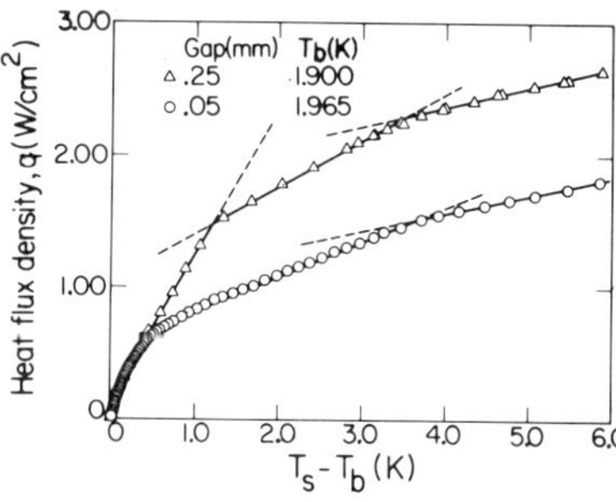

Fig. 3. Characteristic results near 1.9 K.

With the unrestricted geometry ($\delta = \infty$) there is a monotonic rise in ΔT with increasing heat flux (Fig. 2). When the liquid temperature at the heated surface reaches T_λ a transition to He I occurs followed by an almost simultaneous transition to film boiling. We refer to the heat flux density at the transition, q_{max}, as the limiting heat flux. Heat transfer to a confined geometry reveals some new details (shown in Fig. 2) for the 0.25 mm vertical gap. Above a critical heat flux, on the order of 0.1 W/cm^2, mutual friction dominates the heat removal rate through the gap. This Gorter–Mellink region terminates at point A where the heat flux, q_λ, is sufficient to create a normal He I layer adjacent to the surface. The coexistence in the steady–state of an He I layer in a He II bath prior to the inception of film boiling has previously been observed.[4] Increasing the heat flux up to point B is accompanied by an increase of the temperature of the subcooled He I layer up to the corresponding saturated value. At this temperature nucleate boiling commences and subsequent increases in the heat flux result in markedly enhanced heat transfer. When point C is reached the surface dries out and a transition to film boiling occurs. Except for a small hysteresis around the film boiling transition the original curve is traced with a decrease in power. A similar but less pronounced characteristic is seen for the horizontal channel ($\delta = 0.15$ mm, Fig. 2).

As the bath temperature is lowered from T_λ, the nucleate boiling transition (point B) becomes increasingly less obvious, until it completely disappears (Fig. 3), however, the transition points (A,C) corresponding to q_λ and q_{max} can always be seen with a restricted geometry. The equivalence betwen q_λ and q_{max} for the unrestricted case is typical.

One case where the bath temperature dropped thru the lambda point during the course of the run is shown in Fig. 4. On the increasing portion of the heat flux cycle the bath temperature was on the He I side of T_λ. With decreasing heat flux, the He II/He I interface propagated[12] toward and reached the heated surface at point B. The difference in the heat transfer characteristics on either side of the lambda point are apparent.

CRITICAL HEAT FLUX

The critical heat flux values are plotted in Fig. 5 as a function of T_b. For the unrestricted geometry, q_{max} ($= q_\lambda$) increases from about 0.4 W/cm^2 near the lambda point ($T_\lambda - T_b \lesssim 5$ mK), to 10.5 W/cm^2 at $T_b = 1.837$ K. In the He I region at $T_b - T_\lambda \approx 6$ mK we have found the transition to film boiling of this subcooled liquid to occur at about 0.4 W/cm^2. Lyon[13] has found that the film boiling transition of saturated He I at 1 atm for a

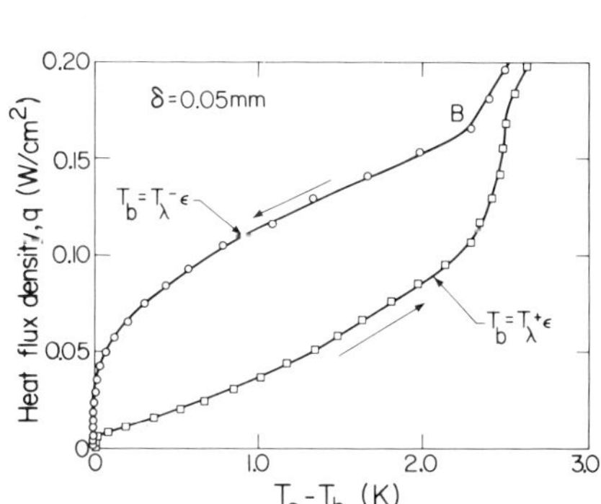

Fig. 4. Heat transfer at the lambda point.

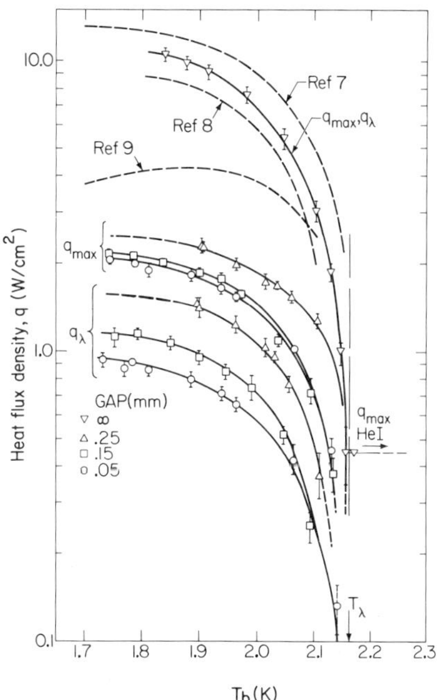

Fig. 5. Critical heat flux.

downward facing surface occurs at 0.2 W/cm^2. Therefore, we expect that in the region of $T_\lambda - T_b \leqslant 5$ mK, for a downward surface, the curves for q_λ and q_{max} should diverge. Also shown for comparison are two sets of measurements by Kobayashi for upward facing Nichrome[6] and Nichrome and copper[7] surfaces and measurements by Van Sciver[8] from a cylindrical copper surface. The present data falls very close to the center of the data by Kobayashi and significantly above the data by Van Sciver. Kobayashi comments on the difficulty of maintaining steady temperatures in his 5 L test vessel at the higher heat rates.[6] Our 100 L liquid volume very much attenuates this problem. The temperature drift during a run was typically 5 mK or less and never exceeded about 20 mK.

Both q_λ and q_{max} are shown for the restricted geometries. The dominant role of mutual friction in these geometries is clearly evident. The limiting heat flux is significantly reduced relative to the unrestricted case and the formation of an He-I layer typically occurs at $q_\lambda/q_{max} \approx 0.5$. As a result of the presence of the He-I layer, heat transfer is influenced by channel orientation when the heat flux is increased above q_λ.

SURFACE CONDUCTANCE

The surface conductance ($\equiv q/\Delta T$) for the unrestricted case is plotted vs $T_s - T_b$ for various T_b (Fig. 6). No significant dependence on bath temperature is found. In the range of $T_s - T_b$ of 1 to 4.5 K, the conductance is seen to vary from about 1 to 2.3 W/cm^{2}K, the conductance h_{max}, at the limiting heat flux (upper line Fig. 7) falls from about 2.6 at $T_b = 1.8$ K to about .7 W/cm^{2}K as T_λ is approached.

The Kapitza conductance, which is defined as

$$h_0 = \left[\frac{q}{\Delta T} \right]_{\Delta T \to 0} \quad ,$$

was computed from the above values of q_{max} using the expanded form of the T^3 dependence as originally suggested by Frederking[14]:

$$\frac{h_{max}}{h_0} = 1 + \frac{3}{2}\left[\frac{\Delta T_{max}}{T}\right] + \left[\frac{\Delta T_{max}}{T}\right]^2 + \frac{1}{4}\left[\frac{\Delta T_{max}}{T}\right]^3 \quad .$$

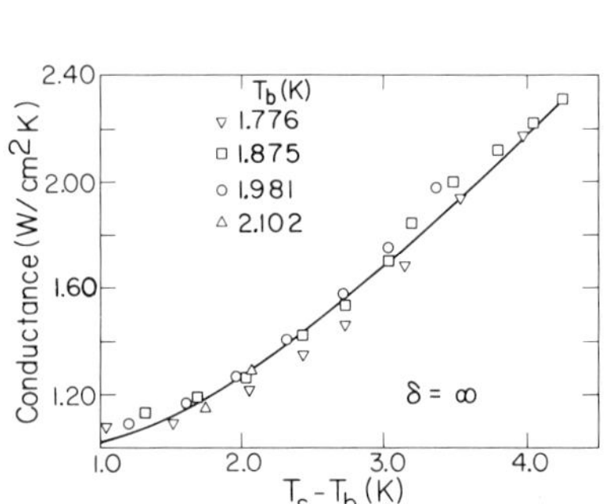

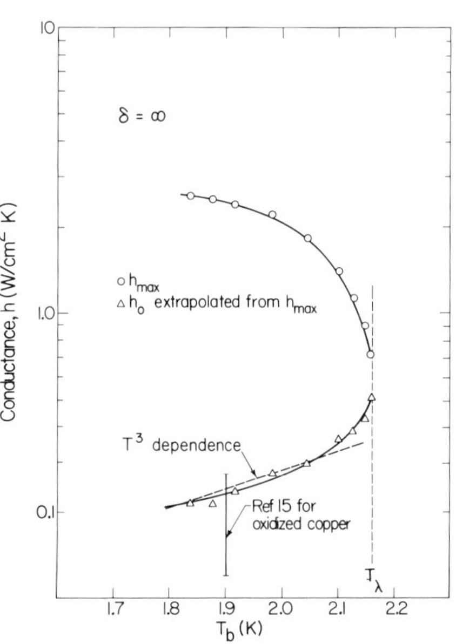

Fig. 6. Conductance as function of temperature difference.

Fig. 7. Conductance at the limiting heat flux and Kapitza conductance.

The computed h_0 values show the anticipated T^3 dependence at low temperatures, and near the lambda point, h_0 approaches the conductance at the limiting heat flux. For comparison, the range in h_0 for oxidized copper in saturated He II as reported by Snyder[15] is shown.

CONCLUSIONS

Appreciable heat flux densities are possible in pressurized (1 atm) He II. For an unrestricted geometry the transition to film boiling reaches over 10 W/cm^2 at 1.8 K. This transition is a simultaneous transition to He I and vapor. For downward facing surface at $T_\lambda - T_b < 5$ mK a steady-state existence of a He I layer at the heated surface prior to a film boiling is possible. In a restricted geometry in which mutual friction dominates, a transition to He I film on the heated surface occurs prior to the film boiling. When an He I layer is formed normal convection in He I and nucleate boiling can play important roles. Heat flux densities of the order of 1 W/cm^2 are achievable in narrow passages as small as 0.05 mm with an aspect ratio typical of that found in magnet design.

REFERENCES

1. R. Aymar, et. al., TORE-SUPRA status report concerning the superconducting magnet after the qualifying development program, _IEEE Trans on Magnetics_, MAG-17:1911 (1981).
2. C. Taylor, et al., Design of epoxy-free superconducting dipole magnets and performance in both Helium I and pressurized Helium II, _IEEE Trans on Magnetics_, MAG-17:1571 (1981).
3. C. Linnet, "Limiting Heat Current Densities of Insulated Vertical Channels in Helium II Under Pressures Between the Saturated Vapor Pressure and 3.3 Atmospheres," Ph.D. Dissertation, University of California, Los Angeles, Calif., (1971), University Microfilms, Ann Arbor, Michigan, 72-5843.
4. S. Caspi and T.H.K. Frederking, Triple-phase phenomena during quenches of superconductors cooled by pressurized superfluid He II, _Cryogenics_, 19:513 (1979).
5. R.A. Madsen, Heat Transfer to an Unsaturated Bath of Liquid Helium, in "Advances in Cryogenic Engineering, Vol. 13," Plenum Press, New York (1968) p. 617.
6. H. Kobayashi and K. Yasukochi, Maximum and Minimum Heat Flux and Temperature Fluctuation in Film Boiling States in Superfluid Helium, in "Advances in Cryogenic Engineering, Vol. 25," Plenum Press, New York (1980) p. 372.

7. H. Kobayashi and K. Yasukochi, A Sample Configuation Effect on the Heat Transfer from Metal Surfaces to Pressurized He II, in "Proc. 8th Intl. Cryo. Engr. Conf.," IPC Science and Technology Press, Guildford (1980) p. 171.

8. S.W. Van Sciver and R.L. Lee, Heat Transfer to Helium-II in Cylindrical Geometries, in "Advances in Cryogenic Engineering, Vol. 25," Plenum Press, New York (1980) p. 383.

9. Bon Mardion, G. Claudet and P. Seyfert, Practical data and steady state heat transport in superfluid helium at atmospheric pressure, Cryogenics, 19:45 (1979).

10. D. Gentile and M.X. Francois, Heat transfer properties in a vertical channel filled with saturated and pressurized Helium II, Cryogenics, 21:234 (1981).

11. R.P. Warren, et al., A Pressurized Helium II-Cooled Magnet Test Facility, in "Proc. 8th Intl. Cryo. Engr. Conf.," IPC Science and Technology Press, Guildford (1980) p. 373.

12. S. Caspi, Gravitational Convection of Subcooled He I and the Transition into Superfluid He II at Atmospheric Pressure, in "Proc. 8th Intl. Cryo. Engr. Conf.," IPC Science and Technology Press, Guildford (1980) p. 238.

13. D.N. Lyon, Boiling Heat Transfer and Peak Nucleate Boiling Fluxes in Saturated Helium Between the λ and Critical Temperatures, in "International Advances in Cryogenic Engineering, Vol. 10," Plenum Press, New York (1965), p. 371.

14. B.W. Clement and T.H.K. Frederking, Thermal Boundary Resistance and Related Peak Flux During Heat Transport From a Horizontal Surface Through a Short Tube to a Saturated Bath of He II, in "Proc. of Intl. Inst. of Refrig., Commission I," Pergamon Press, Oxford (1966), p. 45.

15. N.S. Snyder, Thermal conductance at the interface of a solid and Helium II (Kapitza conductance), National Bureau of Standards Note 385, (Dec. 1969).

THERMAL INSTABILITIES IN AN He II CHANNEL

D. Gentile

Centre d'Études Nucléaires de Saclay
Gif-sur-Yvette, France

and

M. X. Francois

Laboratoire de Thermodynamique des Fluides
Orsay, France

INTRODUCTION

Cooling of superconducting magnets by helium II is proving to be a very useful technique which has been studied for a number of years.[1,2] The advantages of He II are the enhanced thermal stability of the conductor due to the superfluid properties and, consequently, the greater heat fluxes which can be transferred from a solid surface to the coolant. So far, most studies of the stationary and transient thermal properties of cooling channels have been performed in experimental configurations where the heat flux is uniform. In a previous paper,[3] the different stationary heat flow regimes were described for a channel filled with saturated helium II with a non-uniform heat flux. It was shown that above a certain value of the heat flux Q_o the value of Q greatly increases with a very little increase in average temperature, and in this regime the temperature oscillates by a few mK. It appears that an additional heat transfer process is present in parallel with the classical Gorter-Mellink (G.M.) process.

This paper is a more detailed experimental study of these temperature fluctuations and an interpretation is given in terms of local nucleation in the fluid.

APPARATUS AND PROCEDURE

Figure 1 shows the experimental cell consisting of a vertical axis channel of length 15 cm, inner diameter 1 cm, heated by a

"

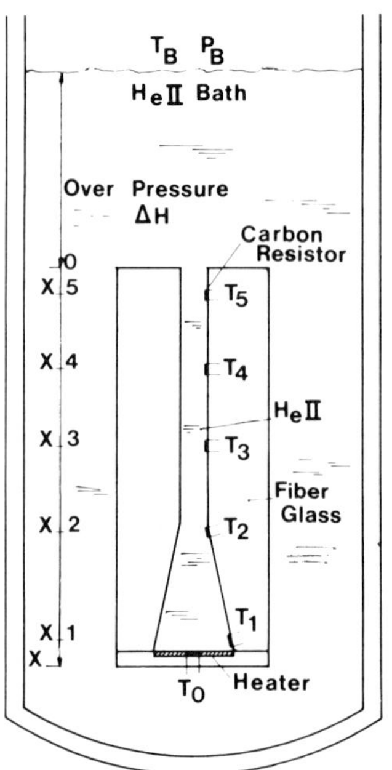

Fig. 1. Experimental set up.

copper block attached to the bottom of a conical part, height 5
cm, maximum inner diameter 3.2 cm. The cell is made of transparent fiberglass placed vertically in a glass helium dewar. This
particular geometry is simpler than that used earlier[3] and allows
a greater heat flux gradient. The experiments were made in saturated He II. The heat flux is generated in the copper by a thin
layer constantan resistor ($R = 10\Omega$) stuck on the copper. Temperatures are measured by six calibrated carbon thermometers, one
located on the copper block and five in the channel, indicated as
T_1 to T_5 in Fig. 1. Measurement of these resistors is achieved
with an automatic ac bridge which gives a resolution of a few
μK. The helium bath level, measured by a superconducting level
gauge, is measured to $\pm$ 1 mm.

EXPERIMENTAL RESULTS

Visual Observations

A sketch of the visual observations is given in Fig. 2 for
different regimes. Q_o is defined as the maximum heat flux removed
by the G.M. process; Q^* as the onset heat flux for film formation.

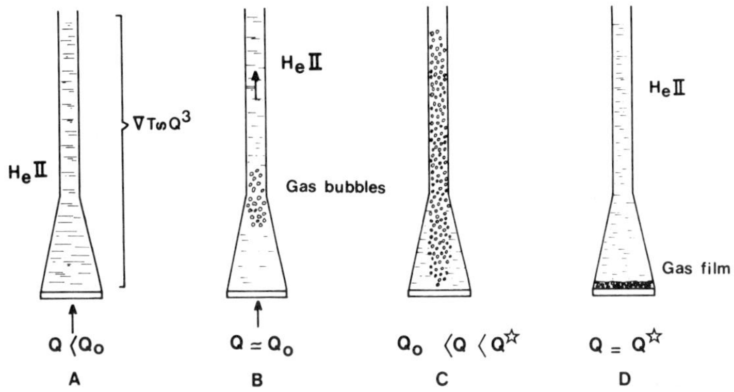

Fig. 2.   Schematic representation of the different heat transfer regimes.

For $Q \approx Q_0$ (Regime B) we observed a periodic phenomenon. For a well defined time interval τ_1, only liquid phase helium can be seen. Then a cloud of bubbles appears suddenly at the top of the cone, remains for a time $T_2 \ll T_1$ and then disappears to leave the helium in a single phase. It is not possible at present to see whether the bubbles collapse or rise rapidly out of the channel.

The frequency $(\tau_1 + \tau_2)^{-1}$, the volume in which the bubbles are nucleated and the bubble density all increase with the heat flux. The frequency increases from $\sim 10^{-2}$ Hz to several Hz. As the heat flux is increased a stage is reached when the whole channel remains full of bubbles all the time (Regime C) corresponding presumably to homogeneous nucleate boiling of the fluid, that is to say the boiling is not specifically associated with the surface.

This behavior is replaced at $Q = Q*$ by film boiling at the copper surface (Regime D). Further increases in Q increase the film thickness to several mm.

Temperature Instabilities

The behavior in these different regimes is reflected as temperature excursions. Thus, in regime B, the creation and disappearance of the bubble cloud coincide with the appearance of a negative temperature pulse on thermometers T_1 and T_2 and a positive pulse on T_3, T_4 and T_5 (This is shown in Fig. 3). The regularity of the temperature variation is shown for T_2 in Fig. 4.

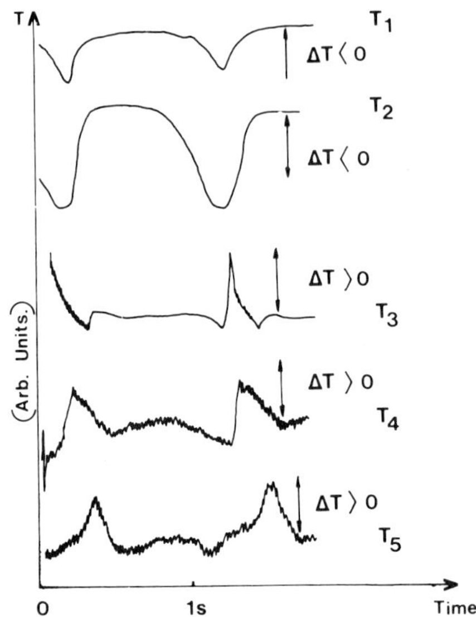

Fig. 3 Typical temperature excursions in the channel. We stress that at present we have not investigated the relative phases of the temperature excursion. They are shown more or less in phase purely for convenience.

The negative pulses on T_2 and T_1 are presumably the result of the removal of latent heat from these regions of the liquid to form bubbles. The positive pulse on the other thermometers may represent the collapse of the bubbles in these cooler regions of the fluid.

For $Q = Q^*$, the film formation on the copper surface induces a negative temperature pulse, which propagates in the fluid and could be used to analyze the kinetics of gas layer nucleation.[3,4]

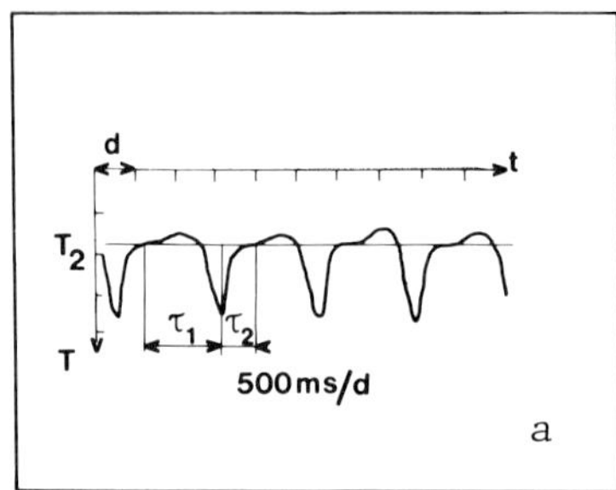
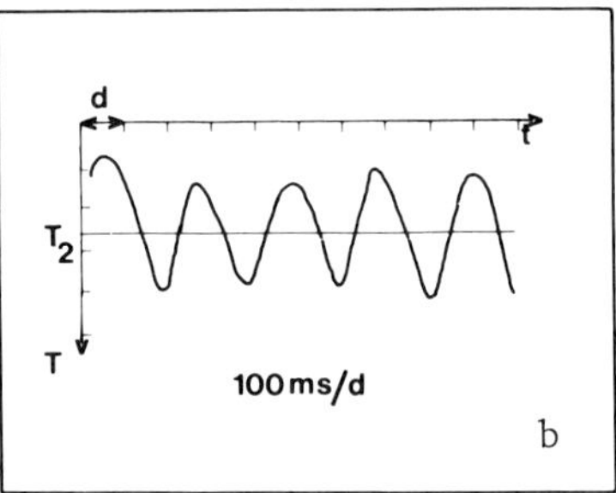

Fig. 4. (a) The temperature excursion of T_2 for several periods demonstrating the regular periodicity. (b) The temperature excursion of T_2 at heat fluxes just below Q^*, the critical value for film formation.

INTERPRETATION

We present here the main features of the interpretation. A theoretical thermodynamic analysis and detailed calculations will be published elsewhere.

Thermal Instability

It has been extensively verified in these experiments that regime B in Fig. 2 appears if $T_2 > \sim T_{sat}$ (where T_{sat} is the equilibrium temperature corresponding to the local hydrostatic pressure). The dependence of the frequency on the heat fluxes, i.e. on $\delta T = T_2 - T_{sat}$ at constant hydrostatic pressure, and the dependence of the frequency on the hydrostatic pressure at constant heat fluxes have both been determined. They are completely equivalent if δP is related to δT by the Clausius Clapeyron equation. The frequency increases with δT. Moreover to create a small ΔT and slow periodic oscillations, it is necessary to first increase ΔT and then reduce it to the final small value. Figure 5 shows the minimum temperature T_2 for which the instabilities have been observed. It is clear that T_{2min} is very close to T_{sat}.

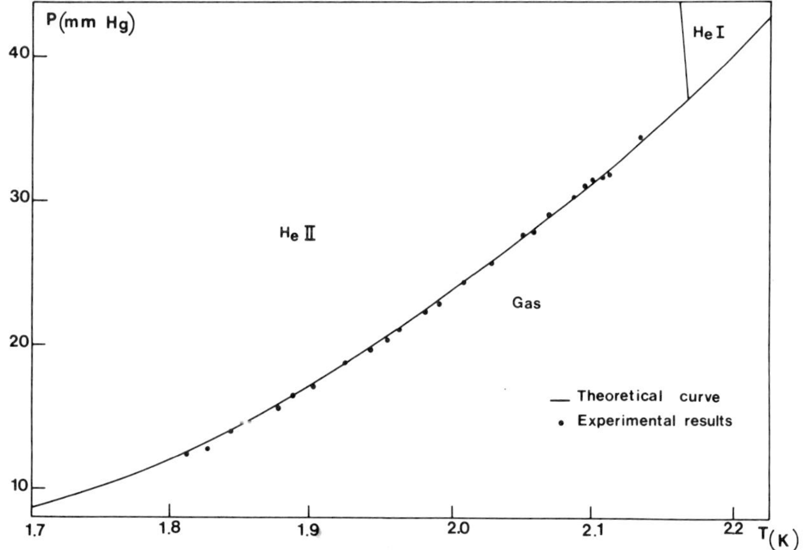

Fig. 5. The points show the smallest measured values of T_2 at which instabilities have been seen as a function of pressure P where $P = P_{bath} + \rho g \Delta H(X_2)$. The curve shows $T_{sat}(P)$.

This sudden nucleation process (negative pulse) followed by relaxation back to the initial temperature seems to be the result of a transition between two thermodynamic states of the liquid: the equilibrium state on the vaporization curve and the metastable state on superheated curve. The transition is achieved by the nucleation of a cloud of bubbles.[5]

Temperature Profile

The temperature profile in the channel can be calculated using the G.M. equation. We show in Fig. 6 the experimental measurements compared with theoretical calculations for three different heat fluxes. The helium level was effectively constant during the experiment. It can be seen, that, due to the geometry, the region around X_2 (T_2) reaches and crosses, the vaporization curve first (superheating), although other parts of the channel still remain above the vaporization curve. This is presumably the reason why the nucleation occurs first in the conical part. In the case of a constant cross section or no heat flux gradient, the monotonic curvature of the temperature profile implies that the gas appears suddenly at the copper surface.

Calculation and measurement of temperatures show that the situation shown in Fig. 6 is difficult to obtain for $T_{bath} < 1.8$ K. This is in agreement with experimental results on the instabilities.

CONCLUSION

It has been shown that in an experimental configuration in which the greatest heat flux does not occur at the heater surface, bubble nucleation can take place, greatly increasing the efficiency of heat transfer in the channel. In the present arrangement fluxes as high as 10 W/cm^2 at 2.09 K could be maintained in the channel before a film appeared at the heater surface. Furthermore it seems likely that higher values are possible with different geometries.

The creation and disappearance of the bubble cloud occurs periodically with a corresponding periodic excursion in fluid temperature. This process seems to be the result of a transition between two thermodynamic states, one superheated and the other saturated helium II. These results emphasize the need, when discussing the cooling characteristics of saturated helium, to separate the regions of the solid-liquid interface from the helium in the channel.

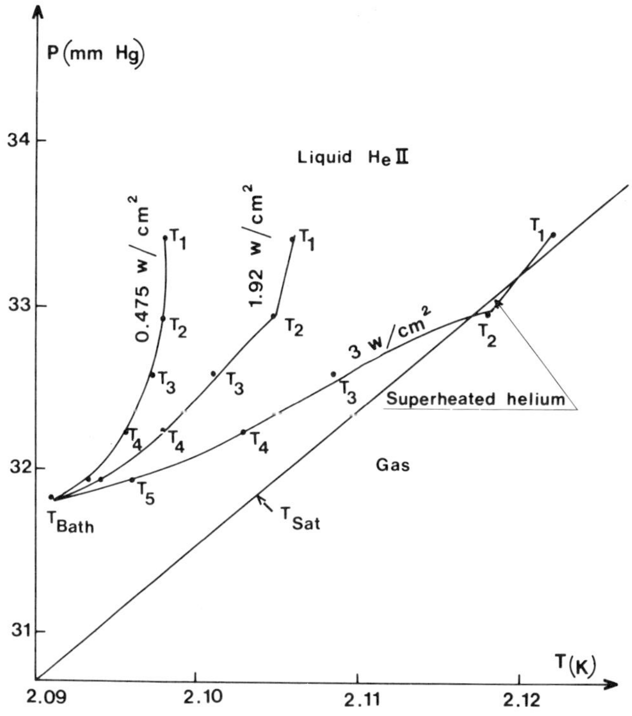

Fig. 6. Temperature profile in the channel. The points show for various heat fluxes in the cylindrical channel the values of temperature as a function of thermometer depth, expressed as a hydrostatic pressure. The lines are calculated from the G.M. equation (note the change in slope between T_2 and T_1 which reflect the change in cross section and so heat fluxes).

ACKNOWLEDGMENT

We wish to thank Professor L.J. Challis for continuous interest in the the work and very helpful discussions.

REFERENCES

1. R. Aymar, et al, Test of a model coil of Tore Supra, _IEEE Trans. on Magnetics_, MAG-17:38 (1981).
2. Y H. Hsu, et al, Measurement of stability of cabled conductors cooled by He I at reduced temperature or He II, _IEEE Trans. on Magnetics_, MAG-17:750 (1981).

3. D. Gentile and M.X. Francois, Heat transfer properties in a
    vertical channel filled with saturated and pressurized
    helium II, Cryogenics 21 (4):234 (1981).
4. D. Gentile, Contribution à l'étude des mécanismes de
    transferts thermiques à l'hélium, Thesis, University Paris
    VI (1980).
5. J.E. Broadwell and H.W. Liepmann, Phys. of Fluids, 12:1533
    (1969).

DISCUSSION

Question by S. Caspi, Lawrence Berkeley Laboratory: Were any
pressure fluctuations observed, and what is the possibility of
cavitation?

Answer by L.J. Challis (University of Nottingham): No attempt has
been made yet to detect pressure fluctuations although that would
be an interesting experiment. At this stage I would say that the
formation of bubbles after the helium crosses the saturation curve
seems a more likely process than cavitation.

Question by P. Seyfert, CEN-Grenoble, France: Do you expect these
oscillations to occur in pressurized superfluid helium?

Answer by L.J. Challis: This is an interesting question. No
measurements have yet been made with pressurized helium in the
present apparatus but with the previous apparatus (Ref. 3) the
oscillations disappeared when the helium was pressurized. The
ratio of heat area to channel area was 5 in the earlier
experiments (10 in the present) so experimentally one could argue
that oscillations might appear if larger ratios and heat fluxes
were used. Theoretically this is less clear however. The
negative slope of the λ line seems to imply that the helium next
to the heater is going to cross the phase boundary first whatever
the area ratio.

AXIAL HEAT TRANSPORT IN He II IN A THICK-WALLED COPPER CHANNEL*

E. Canavan and S. W. Van Sciver

University of Wisconsin
Madison, Wisconsin

INTRODUCTION

An understanding of the stabilization of large superconducting magnets by superfluid helium requires a knowledge of the factors influencing growth and decay of normal zones. In addition to the distribution of heat production, which is controlled by current diffusion[1], the evolution of normal zones is determined by heat diffusion in the conductor, heat transfer across its surface (Kapitza conductance) and heat transport in He II. This paper reports an experimental and computational study of a configuration similar to a short section of a cooling channel in a magnet stabilized with copper and saturated He II. By combining theories of heat diffusion in solids, Kapitza conductance, and Gorter-Mellink counterflow, a simple one dimensional model is developed which yields temperature distributions that match fairly well those measured experimentally.

EXPERIMENT

The experimental arrangement is illustrated schematically in Fig. 1. The helium channel, a tube of OFHC copper (RRR=200), is 9.6 cm long, 0.95 cm in inner diameter, and has a 0.32 cm wall thickness. The tube is supported inside a vacuum can and thermally insulated from the can by a 4 cm section of thin wall stainless steel tubing. The bottom of the channel is coated with a layer of epoxy sufficient to reduce the fraction of the heat flow entering

---

*Work partially funded by National Science Foundation Grant #CME-8011583.

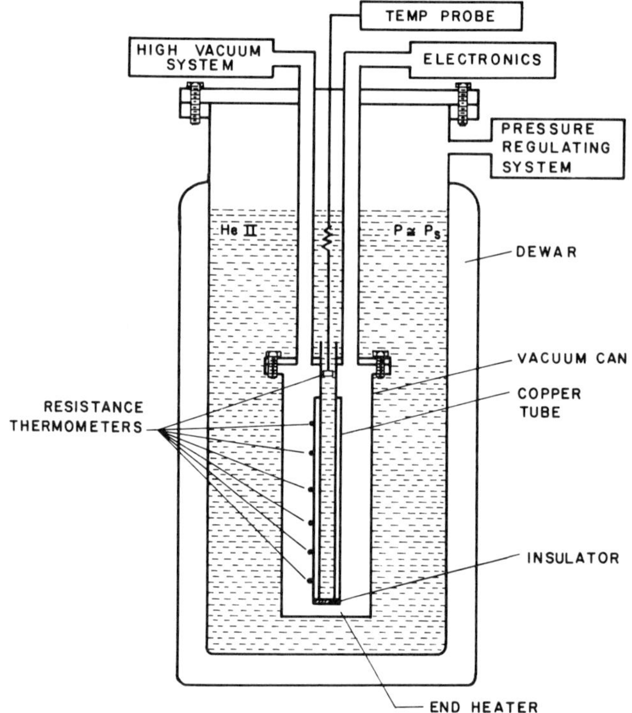

Fig. 1. Schematic of experimental apparatus.

the helium through the end plug to less than 1%. An 80 ohm heater
of manganin wire is wrapped around the outside of the tube just
below the bottom of the channel. The temperature distribution of
the copper is measured by nine 75 ohm Allen Bradley resistors
affixed to a flat face machined into the tube; they are mounted at
centimeter intervals beginning one centimeter above the heater.
The leads are wrapped around the resistors and CryCon* thermally
conducting grease is applied to improve thermal contact. The
temperature distribution of the helium is measured by a 75 ohm
resistor on a stainless steel capillary tube which can be moved
axially into the channel. The casing of the resistor is partially
ground away to improve thermal contact with the liquid. Bath
temperature is measured with a Cryo-Cal calibrated germanium
resistor suspended above the vacuum can. The conductances of all
the   resistors   are   measured   using   two   SHE   Corporation

---

*Air Products and Chemicals, Inc.

Potentiometric Conductance Bridges, one to continuously monitor the bath temperature and one to observe the temperature at various points in the system.

The resistors are recalibrated against the germanium resistor each time they are brought to liquid helium temperature. The calibration data for each resistor is fit to a polynomial which is then used to convert conductances to temperatures. As the conductance bridges are accurate to 0.1% this technique should measure temperature differences to within 2 mK. Heater power was measured to within approximately 0.5%.

For each set of data, the power is raised to the desired level, then the conductance of the nine resistors on the copper, or the conductance of the probe resistor at eleven one centimeter intervals inside the channel, is measured. Measurements are made after the system achieves steady state. In all cases, the bath temperature is held constant to within less than 2 mK during the recording of conductance values by adjusting a valve to the vacuum system. Measurements were made at bath temperatures of 1.8 and 2.0 K.

THEORY

The three processes taking place in the thick-walled channel configuration (heat transport through the copper, across the interface, and through the helium) have each been studied separately by previous workers. Heat transport in the copper is described by the familiar diffusion equation:

$$\rho_c C_c(T_c) \frac{\partial T_c}{\partial t} = \nabla (k(T_c)\nabla T_c) \tag{1}$$

The strong temperature dependence of heat capacity, and thermal conductivity, must be accounted for in this regime.

Heat is transported in He II by the counterflow of its normal and superfluid components. At the fairly high heat fluxes studied here, and in channels of diameter >0.1 cm, the steady-state can be fairly well described by adding to the Landau two-fluid equation the term $A\rho_s\rho_n|v_n-v_s|^3$ to account for the mutual friction between the components. This yields a solution for the channel problem:

$$q = \rho_s ST_h \left(\frac{S}{\rho_n A} \nabla T_h\right)^{1/3} \tag{2}$$

High heat fluxes force the establishment of mutual friction flow very quickly.[2] Previous workers have assumed the equation may be taken as accurate from the time of onset of heat input.

Heat transfer across interfaces at liquid helium temperatures, a process called Kapitza conductance, can lead to large temperature differences across the boundary.[5] Recent theoretical advances provide a good explanation of the behavior of extremely clean surfaces at low fluxes. Data from practical materials show large variations, even between identical samples, due to the extreme sensitivity of Kapitza conductance to surface imperfections and impurities. However, most data can be fit to the empirical form[6]:

$$q_s = h_o(T_c^n - T_h^n).$$

(3)

The parameter n has been found to vary from 2.8 to 3.8.

To simplify the analysis, radial temperature gradients in the helium and copper are ignored. This is justified by the very high effective thermal conductivity of both materials. For the steady state of this one-dimensional system, two nonlinear ordinary differential equations, coupled by the Kapitza term, are obtained. These can be solved numerically by converting them to a series of finite difference equations. Unfortunately, several standard techniques for solving systems of nonlinear equations fail to converge for this set. This problem can be circumvented by assuming that Eq. 2 describes heat transport in He II at all times, which leads to a time dependent description of the system of the form:

$$\rho_c C_c(T_c)\frac{\partial T_c}{\partial t} = \frac{\partial}{\partial x}\left[(k(T_c)\frac{\partial T_c}{\partial x}\right] - \frac{P}{A_c} h_o(T_c^n - T_h^n)$$

(4)

$$\rho_h C_h(T_h)\frac{\partial T_h}{\partial t} = \frac{\partial}{\partial x}\left[f(T_h)(\frac{\partial T_h}{\partial x})^{1/3}\right] + \frac{P}{A_h} h_o(T_c^n - T_h^n)$$

where

$$f = \rho_s S T_h (S/\rho_n A)^{1/3}$$

(5)

This set of equations can be easily solved by a standard finite difference method. The method, though requiring more computer time, also shows the evolution of the system through time.

In attempting to apply this model to the experimental configuration, boundary conditions were applied: for helium, zero heat flow at the warm end and constant temperature at the cold end; and for copper, specified heat flow at the warm end and zero heat flow at the cold end. For simplicity, the properties of helium were taken at 0.05 atm, but this should make little differences as these properties are only weakly pressure dependent.

RESULTS

Temperature distributions in the copper for bath temperatures of 1.8 and 2.0 K are shown in Figs. 2 and 3. The solid lines were generated by the computer model. Since h_o varies between samples and is not known for this particular sample, we have chosen h_o and n so that Eq. 3 becomes

$$q_s = 0.035 \ (T_c^{2.8} - T_h^{2.8}) \ \text{in W/cm}^2 \tag{6}$$

which gives the best fit to the data. This value is consistent with the range of Kapitza conductance coefficients reported in the literature.[6] The fit follows the general form of the data. However, there is some discrepancy between the results of computation and measurement, illustrating the difficulty of modelling heat transfer in systems involving Kapitza conductance, which varies greatly from sample to sample and perhaps even between different regions of the same sample.

Typical helium temperature results are shown in Fig. 4. The solid lines are generated by the model using h_o = .035, as for Fig. 2. While the computer model uses a constant temperature boundary condition, the copper channel in the experiment is linked to the bath through a section of thin walled stainless steel tubing. Since negligible heat enters the helium in this region, the temperature gradient will be constant. To facilitate comparison, the helium temperature profiles are set equal to the bath temperature at the 9 cm point. Again, it is seen that the theory matches the data fairly well in form. However, discrepancies in magnitude clearly exist. Using the model it is possible to fit to individual data sets very well by adjusting the Kapitza parameter h_o or the thermal conductivity. In Fig. 5 are two calculations for the same heat flux, one using the same parameters as the previous calculations, and one using a much lower thermal conductivity. By reducing thermal conductivity or increasing h_o, the effective length over which the bulk of the heat is transferred to the helium is reduced. This increases the length over which the helium must carry the heat, and thus increases the temperature rise in the helium.

As the code is based on pure Gorter–Mellink counterflow it should be expected that agreement with experimental measurements break down above the critical power for boiling. Boiling, as indicated by the onset of large temperature fluctuations and an audible noise, was first observed at a heater power of 1.75 W at bath temperatures of 1.8 and 2.0 K. No sharp rise in temperature

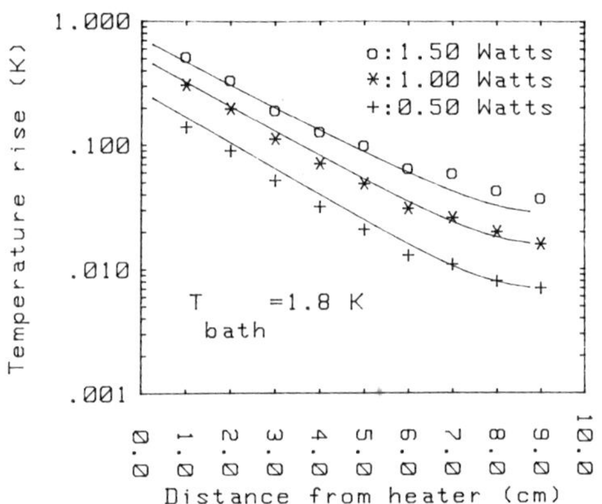

Fig. 2. Temperature distribution in copper at 1.8 K.

over the non-boiling regime was observed. The reduction in heat transfer with boiling should cause heat to penetrate further along the copper, but these powers are not great enough to cause the film boiling zone to propagate up the entire length of the tube.

In these experiments, the surface of the liquid was approximately 27 cm above the bottom of the tube, so the saturation at the warm end of the channel is approximately 1.86 K for a 1.800 K bath temperature and 2.09 K for a 2.000 K bath temperature. Note that peak helium temperature for T_b = 1.8 K, Q = 1.50 W is just below 1.86 K, indicating that this power is close to the critical

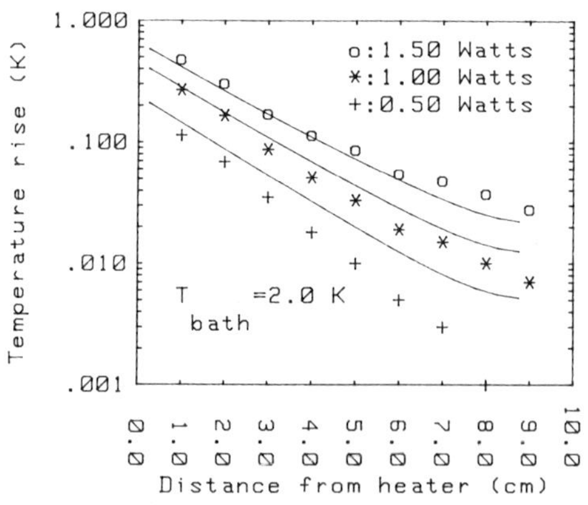

Fig. 3. Temperature distribution in copper at 2.0 K.

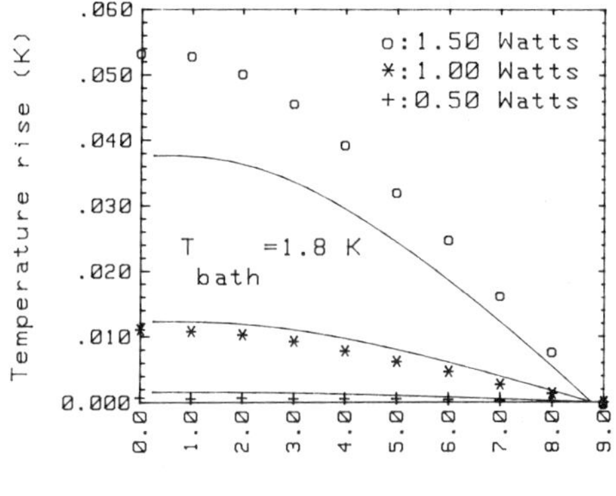

Fig. 4. Helium temperature distribution at 1.8 K.

power. Above the critical power, boiling gives rise to large
(>10mK) temperature fluctuations, but the average temperature is
observed to drop in some cases, and remain approximately the same
in others probably due to mixing caused by the bubbles. Also, in
this incipient boiling region the probe itself might have had some
effect on the measured temperature, for example, by allowing cold
helium to flow down its surface.

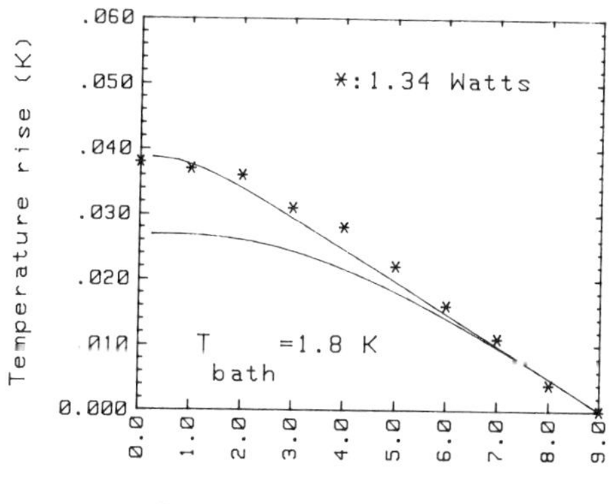

Fig. 5. Helium temperature distribution for two different
choices of copper thermal conductivity.

CONCLUSION

A computer model is developed to simulate a configuration similar to a short section of a He II cooling channel in a copper-stabilized magnet. The model, which includes Gorter-Mellink counterflow transport, Kapitza boundary resistance, and thermal diffusion, produces steady state copper and helium temperature distributions that agree in form with experimental measurements of the configuration. However, for more exact simulation it is necessary to measure independently the Kapitza conductance and thermal conductivity. Such a model, if extended somewhat and included in a larger system, could prove a valuable tool in the design of He-II stabilized magnet.

ACKNOWLEDGEMENTS

Thanks go to Mark Steinhoff for help in the initial phases of this experiment.

NOTATION

NOMENCLATURE

A = Gorter Mellink parameter
A_c = cross sectional area of copper
A_h = cross sectional area of helium
C_c = heat capacity of copper
C_h = heat capacity of helium
f = heat conductivity function for He II
h_o = Kapitza conductance coefficient
k = thermal conductivity
p = wetted perimeter of channel
q = bulk heat flux density
q_s = surface heat flux density
S = entropy
T_c = temperature of copper
T_h = temperature of helium
t = time
v_n = normal fluid velocity
v_s = superfluid velocity
x = position coordinate
ρ_c = density of copper
ρ_h = density of helium
ρ_n = normal fluid density
ρ_s = superfluid density

REFERENCES

1. M. A. Hilal and R. W. Boom, IEEE Trans. on Magnetics, MAG-11:544 (1975).
2. W. F. Vinen, Proc. Roy. Soc. A240:128 (1957).
3. G. Claudet and P. Seyfert, in "Advances in Cryogenic Engineering, Vol. 27", Plenum Press, New York (1982).
4. S. W. Van Sciver, Cryogenics 19:385 (1979).
5. N. S. Snyder, Cryogenics 4:89 (1970).
6. S. W. Van Sciver, in "Proc. 8th Intl. Cryo. Conf." IPC Science and Technology Press, London (1981), p. 228.

DISCUSSION

Question by P. Seyfert, CEN. Grenoble, France: When you put the Kapitza coefficient $h_o = 0$ in your model, a simple channel heated at one end is described. Did you use this particular version to calculate the temperature profile in the case of step function heat pulses?

Answer by author: No, we have not used the model with $h_o = 0$, $k_c = 0$. However, we have run the model using a pulsed heat input and the form of the results seem similar to the above mentioned experimental results.

HEAT TRANSFER IN He II IN MICROBORE TUBING*

B. P. M. Helvensteijn, S. R. Breon, and S. W. Van Sciver

University of Wisconsin
Madison, Wisconsin

INTRODUCTION

Helium II has long been recognized as an excellent heat transfer medium. In 1970, Arp[1] suggested that enhanced heat transport in the transition region between laminar and turbulent flow may be achievable by decreasing the cross-sectional area of the channel. The laminar region is distinguished by the functional dependence to temperature gradient, being proportional to heat flux rather than the well known cubic relationship occuring in the turbulent region. Experiments [2] show some potential for extending the laminar region beyond the onset of turbulence, thus reducing substantially the actual temperature gradient in the fluid. The present paper reports on experiments intended to investigate this possibility.

THEORY

We begin by reviewing the theory of temperature and pressure gradients.

Temperature Gradient

For a long tube of diameter d, the temperature gradient ∇T as a function of heat flux Q is given by two relationships each applying in different flow regions:

$$\nabla T_L = 32 \ \eta Q/(\rho^2 S^2 T d^2) \ \text{---} \ \text{for laminar flow} \qquad (1a);$$

---

*Work supported in part by National Science Foundation grant #CME-8011583.

$$\nabla T_T = A\rho_n \left[Q/(\rho_s ST)\right]^m /S \text{ --- for turbulent flow} \qquad (1b).$$

The subscripts, L and T refer to the laminar and turbulent regions. The symbols are: η – normal fluid viscosity; ρ, ρ_n, ρ_s – density of He II, of the normal component, of the superfluid component; S – entropy; A – Gorter-Mellink constant, η, ρ and S are given by Wilks[3], A by Vinen[4]. The exponent m ~3 for T $\leqslant$ 1.9 K, but increases to nearly 4 as T $\rightarrow$ T_λ. It should be noted that only for m=3 are the units in Eq. 1b consistent. To arrive at a formula that applies both for the laminar and turbulent region, we assume that one can add the temperature gradients in (1a and (1b):

$$\nabla T = 32 \ \eta Q/(\rho^2 S^2 T d^2) + A \ \rho_n \left[Q/(\rho_s ST)\right]^m /S \qquad (2)$$

Equation 2 predicts a gradual change from laminar to turbulent flow. Experiments [2] show an intermediate region where enchanced heat flux will occur due to the suppression of the turbulent term. This effect is strongly geometry dependent but occurs in the range of 1 W/cm^2 for tubes of diameter 100μm.

Pressure Gradient

There exists a pressure gradient ∇P for the laminar and turbulent regions. In the laminar region ∇P is due to the well known fountain effect, and is given by:

$$\nabla P_L = 32 \ \eta Q/(\rho S T d^2) \qquad (3a);$$

In the turbulent region ∇P is caused by viscous force on the normal fluid and has a corresponding correlation[5]:

$$\nabla P_T = 0.158 \ \eta^2 \left[Qd/(\eta ST)\right]^{1.75}/(\rho d^3) \qquad (3b).$$

To develop an equation that can be applied in both laminar and turbulent regions we assume the total ∇P is given by a sum of the two terms:

$$\nabla P = 32\eta Q/(\rho S T d^2) + 0.158 \ \eta^2 \left[Qd/(\eta ST)\right]^{1.75}/(\rho d^3) \qquad (4)$$

The goal of the present experiment is to investigate the applicability of Eqs. 2 and 4 to heat transport in fine capillaries.

EXPERIMENT

For the investigation of heat transport by He II in small channels an experiment is set up as shown in Fig. 1. Steady-state measurements are performed in which a constant heat is resistively generated in a small reservoir of liquid helium at the base of a stainless steel tube containing either bulk He II or glass microbore tubing. The top of the tubing opens to a bath of saturated He II. From temperature measurements of the helium at either end of the tube one obtains $\nabla T(Q)$ characteristics. The helium-filled cavity and tubing are mounted in a vacuum chamber to minimize parallel path heat flow.

Two samples of multi-or microbore glass tubing are used. Sample 1, with a length of 20 cm, has 1040 holes of 10 μm diameter; sample 2, 4.2 cm long, has 90 holes of 40 μm diameter. A single stainless steel tube (sample 3) with an I.D. of 4 mm and a length of 20 cm is used for comparison of the multibore results with the bulk properties.

To mount the glass tubing in the vacuum can a thin wall SS tube tightly fitting around the glass tubing is soldered into a brass mounting piece. The void space between the glass and stainless steel tubing is filled with low viscosity epoxy to eliminate He II in that space. This piece is sealed to the top flange of the vacuum can by means of an indium ring. At the lower end of the stainless steel tube a hollow brass plug is soldered. A heater of manganin wire is wound around the plug. For the temperature measurements an Allen Bradley carbon resistor R_c (nominal value 75Ω) is mounted inside the plug and calibrated against a germanium resistor on top of the vacuum can.

The experimental procedure is as follows. As the vacuum can is immersed in a He II bath the cavity in the brass plug is filled

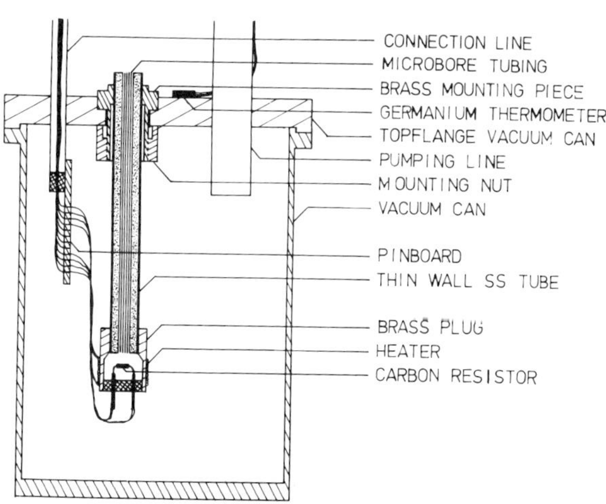

Fig. 1.   Experimental apparatus.

with the liquid helium via the tubing. Care is taken that the capillaries are open by purging with N_2 gas at room temperature and with He gas at 77 K. After the calibration of R_c, the temperature at each end of the tubing is measured at a given heat input.

RESULTS AND DISCUSSION

We have carried out measurements of $\nabla T(Q)$ for all three tubes at two bath temperatures, 1.5 K and 1.9 K.* These data are plotted in Figs. 2, 3 and 4. It is interesting to note that for both samples of microbore tubing, the peak heat flux corresponds to the helium in the heater plug achieving T_λ. This observation is quite different from that which occurs for sample 3, for which the peak temperature is lower and corresponds to the saturation condition. Taking account for the hydrostatic head of helium, it follows that $T_{max} \simeq \int pgh(dp/dT)sat\ dT + T_b$, resulting in $T_{max} =$ 1.60 K for $T_b = 1.5$ K and $T_{max} = 1.94$ K for $T_b = 1.9$ K. (T_b is the bath temperature and h = 20 cm at the top).

The measured peak temperatures and heat fluxes for all three samples are listed in Table I. A surprising result is observed in

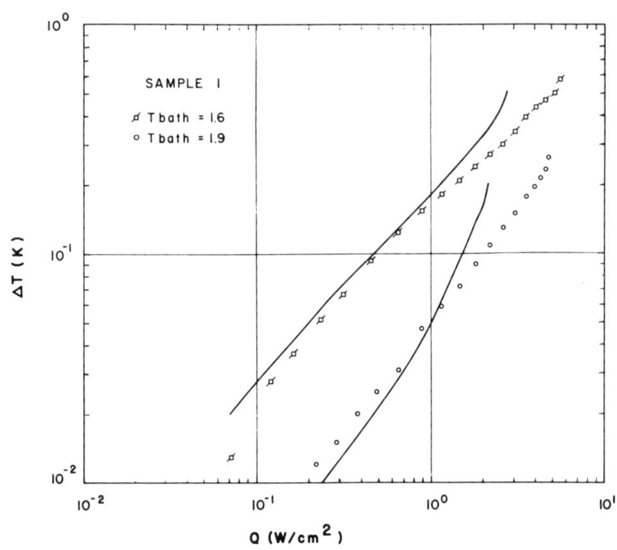

Fig. 2. Temperature difference across sample 1 as a function of thermal flux. Solid lines were calculated using Eq. 2.

---

*Bath temperatures for sample 1 are 1.6 K and 1.9 K.

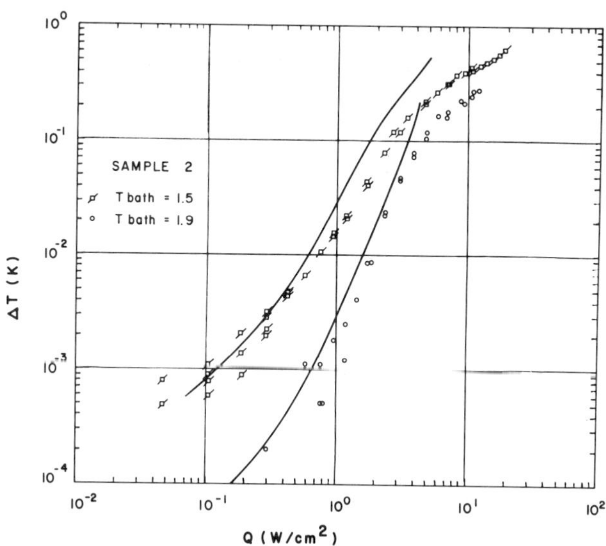

Fig. 3. Temperature differences across sample 2 as a function of thermal flux. Solid lines were calculated using Eq. 2.

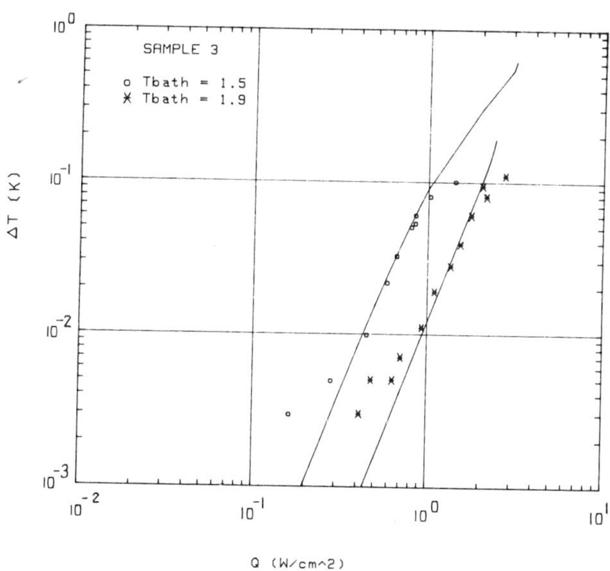

Fig. 4. Temperature difference across sample 3 as a function of thermal flux. Solid lines were calculated using Eq. 2.

sample 2 where a peak heat flux of 10.7 W/cm^2 is observed for a bath temperature of 1.9 K. Since this particular sample is in the turbulent regime at high heat fluxes, one would expect a peak heat flux corresponding to bulk and thus to be approximately 4 W/cm^2. At this time, we have no physical explanation for this phenomenon.

The measurements of $\nabla T(Q)$ in the present experiment can be used to test the form of Eq. 2. Since the results give temperature differences, it is necessary to integrate Eq. 2 over the length of the channel. This process was achieved numerically with the aid of an Hewlett-Packard 9845B computer. For simplicity, we have assumed the coefficient m=3 in the turbulent term. The results of the above calculations are also plotted in Figs. 2, 3 and 4. The agreement, although not exact, gives good correspondence with the data. Moreover, we observe no enhanced heat transport in the intermediate region between laminar and turbulence.

To attempt to understand the reason for the temperature going to T_λ in the multibore tubing we have calculated absolute pressure as a function of temperature in the heated plug for all three channels by simultaneous integration of Eqs. 2 and 4. These calculations are plotted versus temperature on a helium phase diagram in Fig. 5. One can see that for the wide channel (sample 3) no substantial pressure change occurs allowing the temperature to be limited by the vaporization curve. However, for samples 1 and 2 with multibore tubing the condition is different. For sample 1 the limit occurs when the temperature crosses T_λ at a higher pressure and thus should be a He II-He I transition. However, for sample 2, the pressure increase is not sufficient to cause a He II-He I transition. Rybarcyk and Tough[6] have made similar observations which have been interpreted as superheating. In order to test this concept, we intend to perform pressure difference measurements on the present configuration and test the performance of Eq. 4.

Table I. Maximum in Heat Flux and Temperature Measured in Different Experiments

	Sample 1		Sample 2		Sample 3	
$T_b[K]$	1.595	1.900	1.505	1.895	1.525	1.885
$T_{max}[L]$	2.165	2.15	2.145	2.165	1.625	1.995
$Q_{max}[Wcm^{-2}]$	5.6	4.8	18.7	10.7	1.43	2.73

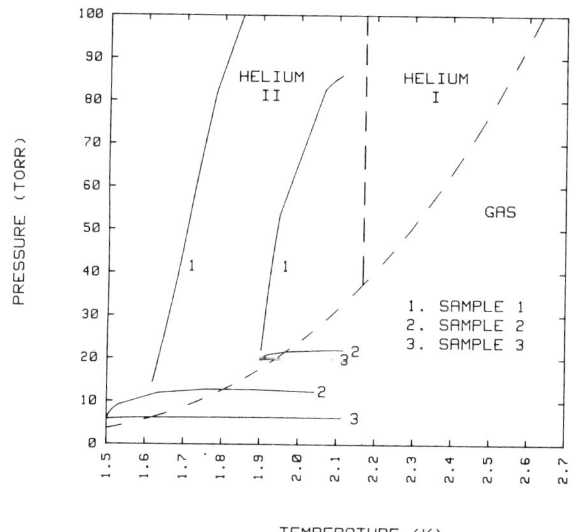

Fig. 5. Pressure as a function of temperature for various tube diameters calculated from Eq. 4.

CONCLUSIONS

We have observed an enhancement of heat transport in He II contained in multibore tubing. This result has two aspects. First, for smallest diameter tubing, pressure enhancement allows the temperature to rise to T_λ at the peak heat flux. This is different from wide channels in saturated He II which have T limited by the liquid vapor transition. However, for the intermediate size tube, apparent superheating is observed. Second, we have measured a higher effective thermal conductivity in the turbulent regime, which yields a higher peak heat flux. The latter results are in disagreement with our simple interpretation based on the sum of the two transport regimes.

ACKNOWLEDGEMENTS

The authors wish to express their gratitude to Mike Meisner for his invaluable computational assistance and to Ed Dreier for his experimental assistance.

REFERENCES

1. V. Arp, Cryogenics, 10:96 (1970).
2. R.K. Childers and J.T. Tough, Phys. Rev. B, 13:1040 (1976).

3. J. Wilks,"The Properties of Liquid and Solid Helium,"
    Clarendon Press, Oxford (1967).
4. W.F. Vinen, Proceedings Roy. Soc. A 240:114 (1957).
5. F.A. Staas, K.W. Taconis and W.M. Van Alphen, Physica 27:893
    (1961).
6. L.J. Rybarcyk and J.T. Tough J. of Low Temp. Physics, 43:197
    (1981).

TRANSIENT TRIPLE-PHASE TRANSITIONS
IN SUPERFLUID LIQUID He II*

C. Chuang, Y. Kamioka, and T. H. K. Frederking

University of California
Los Angeles, California

INTRODUCTION

Thermodynamic equilibrium of the superfluid phase (He II) of He^4 exists adjacent to two other fluid phases, the disordered vapor phase, and liquid He I which behaves in many respects as a "classical" liquid. Because of the existence of these two other phases, special triple-phase transitions are possible which are unknown in other fluids. The He II - He I phase change is a second order transition, while the He I - vapor phase change is of first order. In the past triple-phase transitions have received relatively little attention, as special configurations of liquid He II have been studied, e.g. long insulated tubes known as Gorter - Mellink (GM) ducts.[1,2] However, it is expected that in general a thermodynamic change of state, in helium such as an isobaric temperature excursion, will involve equilibration heat and mass transfer with the appearance of the three phases. Indeed, recent studies have shown for a single conductor in a bath that triple - phase phenomena do occur during quasi - steady operation.[3,4] During temperature changes associated with near - isobaric heat supply, there is an initial transition from non-boiling He II to a triple - phase mode in the vicinity of the He I boiling point. A subsequent change occurs from this mode to film boiling. Because of a nearly complete lack of data on transients involving triple phases, this paper has the following purposes: (1) to present data which show that triple-phase phenomena are not restricted to quasi-steady changes, but do extend to transients as well, (2) to present data which document special stability properties, of

---

*Work supported in part by the National Science Foundation.

superfluid He II, absent in classical fluids. The present research program aims at a better resolution of transient heat transfer to superfluid He II from a superconducting NbTi/Cu composite of an intermediate size.

EXPERIMENTS

The superconducting composite used in the present studies contains 48 filaments of the alloy Nb 48a/oTi embedded in a copper substrate. The conductor has a cross section 0.2×0.2 cm^2. The original Formvar insulation has been removed, and prior to immersion in liquid He, the Cu − surface is wiped with acetone. As indicated in Fig. 1, two pieces of the composite are connected by soft soldering along two faces of the central conductor containing a carbon thermometer (39 Ω at room temperature, 1/8 W). Heater windings are mounted on a stainless steel tube contained in each of the outer conductors. Epoxy-fiberglass (G-10) end pieces of 0.3 cm thickness are used to define the length of the heating section. The surface area in contact with fluid, $A_s = 4.44$ cm^2. Details of the composite have been described elsewhere.[5]

The low temperature apparatus is shown in Fig. 2. During the runs, this apparatus is located at the bottom of the liquid He dewar. The composite specimen is mounted on the lower chamber of the vacuum-insulated apparatus. The fluid volume of the chamber is 54.9 cm^3. A Gorter-Mellink tube (304 stainless steel, O.D. 0.635 cm, wall thickness 0.0152 cm, length 12.66 cm) connects the lower chamber with an upper reservoir. A system of heat exchangers permits thermal equilibration of the inner pressurized fluid with an outer He II bath. The top flange is sealed with an indium O-ring during the runs with cold pressurized helium. For these conditions helium gas is admitted via a heat exchanger system upstream, and the pressure is adjusted to the desired value. An outer bath temperature is selected below the lambda temperature. For runs with near-saturated He II at low pressures, the upper flange is removed.

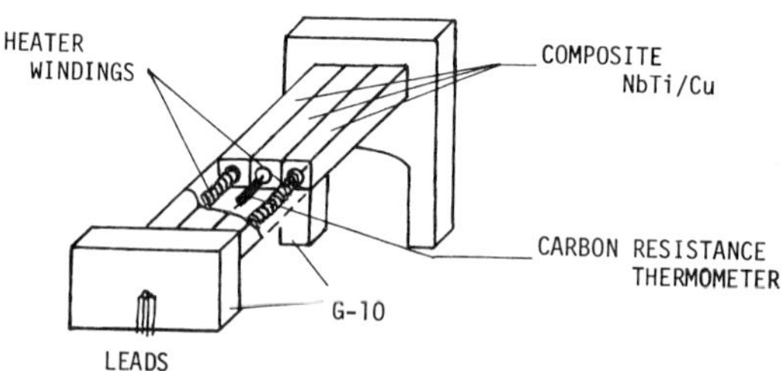

Fig. 1. Composite superconductor specimen (schematic).

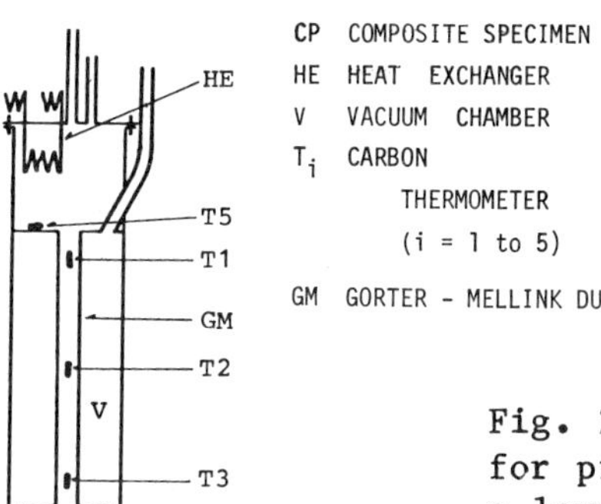

Fig. 2. Experimental apparatus for pressurized He II containing a lower chamber connected by a GM – duct to an upper pressurized reservoir.

The heater windings are energized in power steps stepping the current through the heater made of constantan wire (0.0075 cm diameter, 105 Ω). As a consequence of the heat input, the thermometer on the composite (T4 in Fig. 2) starts to rise first, followed by the other thermometers. Thermometers T1 to T3 inside the GM-tube are mounted using a tube (321 stainless steel, O.D. 0.127 cm) arranged in the center of the GM-duct. Thermometer wires are routed upward. The resulting cross section of the GM-duct, A_c = 0.274 cm^2 i.e. the ratio (A_s/A_c) is 16.2. Thus, the various q-values reported as power density per fluid-wetted composite area, become one order of magnitude larger when they are expressed as power density of the GM-tube.

RESULTS AND DISCUSSION

Figure 3 is an example of a set of three triple-phase conditions (marked as 1 to 3) evolving as a function of time. The insert in the figure shows the equilibrium state boundaries of helium in the pressure – temperature diagram. It includes an isobaric excursion in T crossing two phase boundaries.

As the power step is imposed, the temperature of the composite conductor rises relatively fast from 2.05 K toward T-values typical of the Kapitza-resistance – controlled regime of non-boiling He II. Subsequently there are two relatively fast changes (α) and (β). The first transition (α) takes the composite from point A_1 to point A_2. The second transition (β) causes T to rise from point B_1 to point B_2. Point A_1 is the last point of the non-boiling regime of He II, and point A_2 is near the boiling point T = 4.7 K at this pressure. From point A_2 to B_1 the temperature undergoes a very small change. This is interpreted as strong evidence for the existence of local nucleate boiling at the surface of the composite conductor. As nucleate boiling is absent in

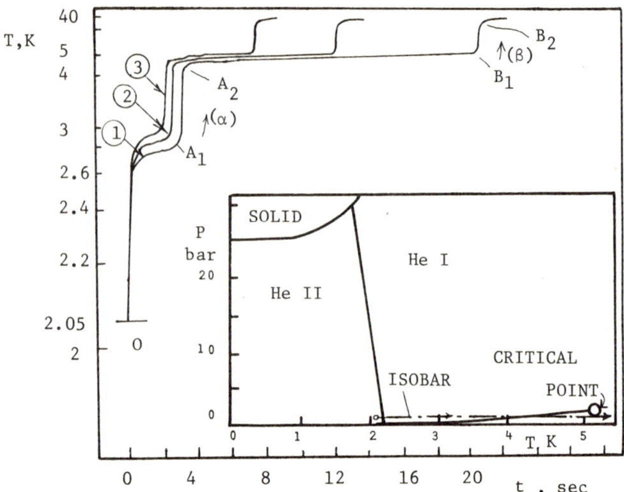

Fig. 3. Composite temperature (T4) as a function of
time for triple-phase changes at various heat flux
densities, ($q_1 = 0.470$ W/cm^2, $q_2 = 0.532$ W/cm^2,
$q_3 = 0.613$ W/cm^2); (Power density per unit area of
fluid-wetted composite surface) ; P = 1.53 bar.
(α) = Transition from non-boiling He II to triple-phase
        nucleate boiling mode;
(β) = Transition from nucleate to film boiling mode.
Insert: Phase diagram P(T) of He^4 with equilibrium state
         boundaries and isobaric T-excursion leading to
         triple − phase phenomena.

He II, and known to originate from He I, this particular equilibration mode involves the three fluid phases He II (outside the
wall-adjacent "boiling layer"), a He I layer near the composite
surface, and helium vapor in the nucleate boiling regime. Because
of thermodynamic constraints imposed on the He I − to − vapor
phase change of nucleate boiling, the temperature is kept within
limits given by the maximum metastable superheat of He I.[6] Thus,
the critical value T_c = 5.2 K is an upper bound for the wall-
adjacent fluid. The temperature of the composite is slightly
above the fluid value because of a finite Kapitza resistance. At
point B_1, the second change in T, designated as (β)-transition is
initiated toward film boiling temperatures. The helium vapor film
insulates the composite whose temperature is brought quickly
beyond the transition temperature of the superconductor, i.e. the
superconductor is quenched.

A few quasi-steady data points have been taken in order to
check consistency of the present data with other results for
quasi-steady triple-phase transitions. The present observations
appear to be consistent with results of Ref. 3.   However, in
contrast to Ref. 3, the range of observable triple-phase changes
is not restricted to a narrow T-interval near the lambda temperature ($T\lambda$). Instead, between 1 bar and P_c the present results show
that triple-phase transitions appear in the entire T-range covered
from about 2 K to $T\lambda$.

Figure 4 presents the times needed for various transitions
for externally imposed steps in heat input. The latter are expressed as power per fluid-wetted surface area of the composite.
Times smaller than $t(q_L)$ are located in the non-boiling regime of
He II. At times which exceed $t(q_p)$ film boiling takes place. In
between, nucleate boiling cools the superconductor below its
transition temperature provided there are no excessive thermal
resistances within the composite itself. At low q-values the
width of the nucleate boiling regime is quite large. At high
power inputs, however, the two times approach each other more and
more. At q-values beyond 1 W/cm^2 the two transitions occur nearly
simultaneously. The corresponding power densities per cross
section of the GM-duct are of the order 10 W/cm^2 and higher.

In contrast to the preceding results for pressurized He II,
the records for low pressure liquid may be expected to lead to a
relatively early film boiling transition during an isobaric T-
excursion. This may imply a premature transition to high temperatures in classical liquids, such as He I, liquid nitrogen and

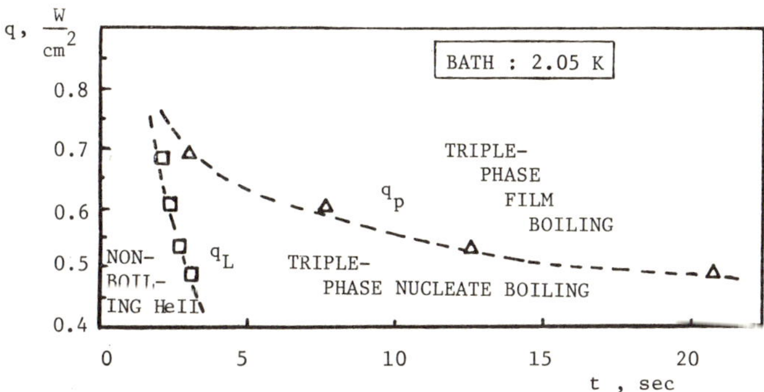

Fig. 4. Heat flux densities versus times required for the
occurence of the transitions; [$t(q_L)$=time at point A_1,
$t(q_p)$=time at point B_1].

water.[7] In contrast, the present He II data reveal a stabilizing
influence caused by the presence of the superfluid. Figure 5
shows the composite temperature vs. time. Power ratings, in q,
W/cm^2, are (1) 0.0851, (2) 0.104, (3) 0.133, (4) 0.159, (5) 0.183,
(6) 0.218, (7) 0.251, (8) 0.286, (9) 0.387, (10) 0.428. Starting
from T= 2.05 K, the composite temperature is seen to rise relatively fast initially. Subsequently the rate of change with time,
dT/dt, decreases and increases again. Finally a maximum T_{max} is
reached at t_{max}. For $t > t_{max}$, T decreases toward a minimum. A
subsequent relatively small rise of T brings the composite to a
steady state whose final temperature turns out to be smaller than
T_{max}. The maximum T_{max} is indicative of local vapor formation.
Because of the good entropy removal capability of He II, the vapor
is recondensed. Thus, the system is brought back to an equilibration mode in the non-boiling regime of He II. During run #1, T-
oscillations are observed for $t > t_{max}$. These oscillations are
reminiscent of He II – He I transitions during near-isobaric heat
supply.

Figure 6 displays t_{max} as a function of $q(t_{max})$. In a limited time domain the data may be described in terms of power law.
The exponent d log q/d log t_{max} = –1 is indicated in Fig. 6. At
shorter times the exponent exceeds unity.

CONCLUSIONS

The present results on transient heat transfer from a superconducting composite to superfluid pressurized He II show different triple-phase modes after equilibration in a non-boiling He II
mode has ceased. There are three transient regimes: Non-boiling
He II terminated at a heat flux density q_L, triple-phase nucleate
boiling terminated at q_p, and triple-phase film boiling at

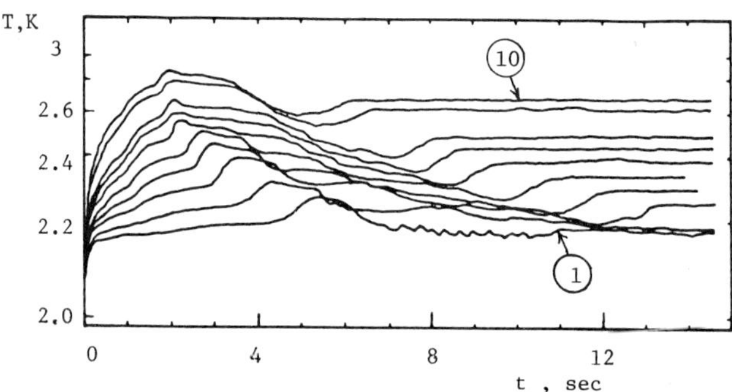

Fig. 5. Composite temperature (T4) versus time for
transient heat supply to near-saturated He II at 2.05 K.

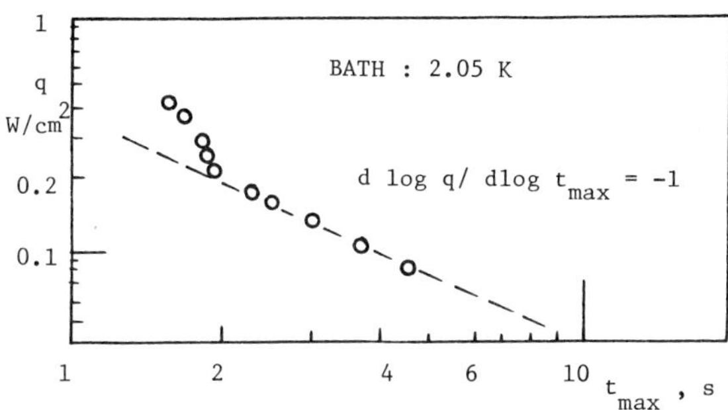

Fig. 6. Transients in near-saturated liquid He II:
$q(t_{max})$ at the maximum temperature established during
the runs; q = power density per unit area of heater
surface wetted by fluid; Values of the power per cross
section of the GM-duct are one order of magnitude larger.

$q>q_p$. There exist continuous functions q_L (t), $q_p(t)$ extending
from very long times of quasi-steady operation to short times of
the order 1 s. Below approximately 1 s, the difference between q_L
and q_p disappears.

Temperature excursions during transients in near-saturated He
II show a remarkable capability of the fluid system to recover low
temperatures in a limited q-range subsequent to the attainment of
a maximum temperature. For this type of operation in He II there
has been no indication of a premature transition to film boiling
(known to occur in classical liquids[7].)

Results for the present geometry of a GM-duct in contact with
a chamber suggest that it may be possible to utilize similar
chamber-like terminations as quench arrestors for winding sections. The chamber has the purpose of inhibiting quench propagation once a particular magnet section has quenched.

ACKNOWLEDGMENTS

The inputs of Y.I. Kim, Sidney W.K. Yuan and Jeffrey Marshall
Lee during various phases of this work are gratefully acknowledged.

REFERENCES

1. C. Linnet, Ph.D. thesis, Univ. of California, Los Angeles; and
    C. Linnet et al, Thermal conditions at the Gorter-Mellink
    counterflow limit between 0.01 and 3 bar, J. Low Temp.
    Phys. 21:447 (1975).

2. S.W. Van Sciver, Transient heat transport in He II, Cryogenics
    19:385 (1979).
3. S. Caspi and T.H.K. Frederking, Triple-phase phenomena during
    quenches of superconductors cooled by pressurized super-
    fluid He II, Cryogenics 19:513 (1979).
4. D. Gentile and M.X. Francois, Heat transfer properties in a
    vertical channel filled with saturated and pressurized
    helium II, Cryogenics 21:234 (1981).
5. S. Caspi et al, Report NASA CR 2963, (April 1978).
6. L.C. Brodie, et al, Transient heat transfer into liquid
    helium I, J. Appl. Phys. 48:2882 (1977).
7. D.N. Sinha, et al, Premature transition to stable film boiling
    initiated by power transients in liquid nitrogen,
    Cryogenics 19:225 (1979).

CYCLE DESIGN FOR THE ISABELLE HELIUM REFRIGERATOR

D. P. Brown, A. P. Schlafke, and K. C. Wu

*Brookhaven National Laboratory**
Upton, New York

and

R. W. Moore

Koch Process Systems, Inc.
Westborough, Massachusetts

INTRODUCTION

The superconducting magnets for the ISABELLE storage ring/accelerator are designed to be operated at 3.8 K using a forced-flow supercritical helium cooling system.[1] The ISABELLE refrigerator has been designed subject to these requirements. The design output is 13.65 kW of refrigeration below 4.2 K (for cooling the magnet and distribution system), 55 kW at 55 K (to cool heat shields for the whole system) and 100 g/s of liquefaction (for magnet power lead cooling). The system incorporates a subcooler section that produces liquid helium at 5.3 atm and 2.6 K and circulates it through the loads, and a Claude-type main refrigerator. The main refrigerator section has five stages of cooling, with four of them below liquid nitrogen temperature. Liquid nitrogen precooling is not used. With compressors which have isothermal efficiencies of 60% the efficiency of the refrigerator system will be about 26% of Carnot.

CYCLE OVERVIEW AND LOAD DESCRIPTION

The cycle schematic for the ISABELLE Refrigerator with a simplified load is given in Fig. 1. The load and the subcooler

---

*Operated by Associated Universities, Inc. under contract with the U.S. Department of Energy.

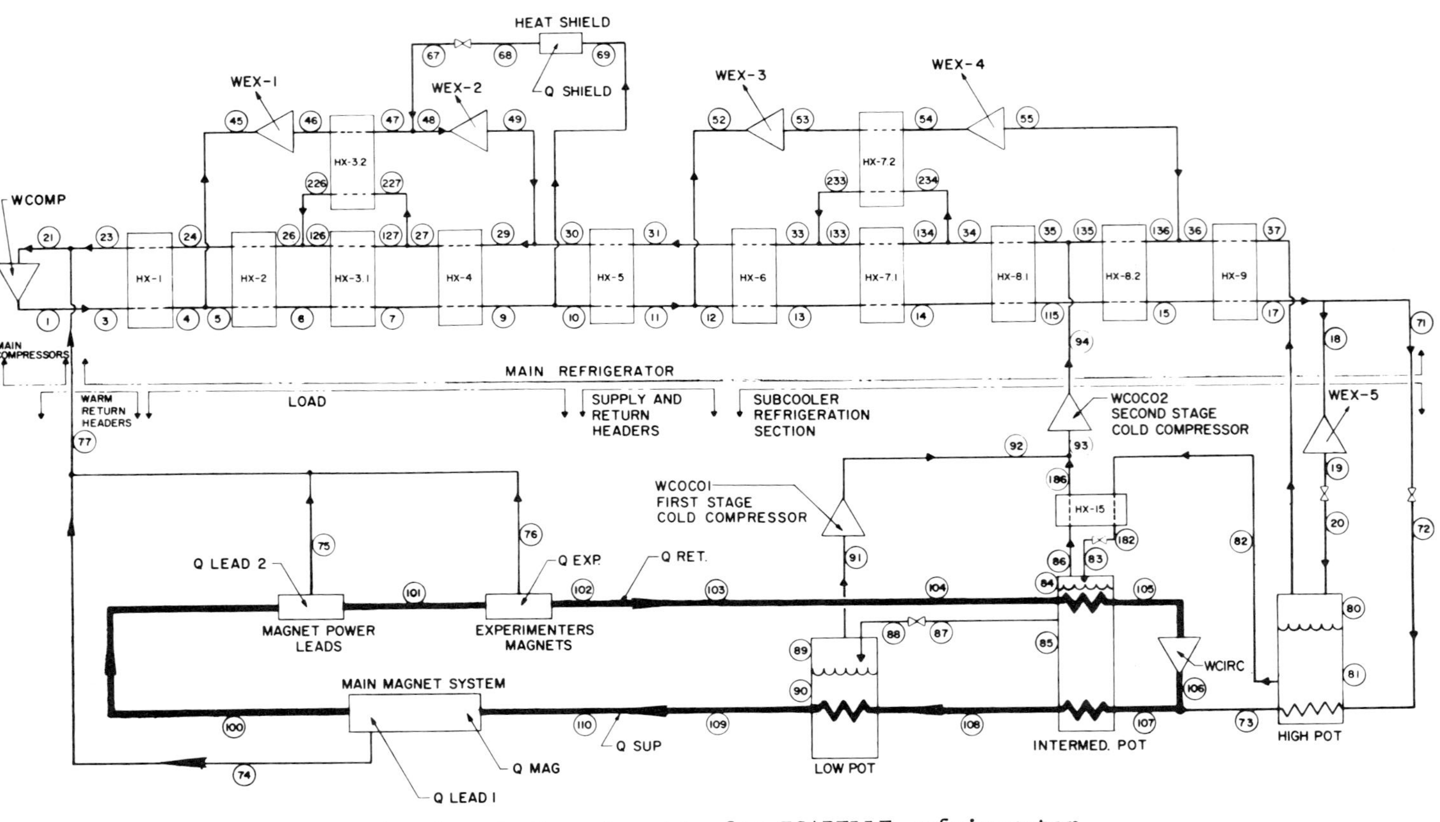

Fig. 1. Cycle schematic for ISABELLE refrigerator.

section are connected by the supply and return headers. A circulating compressor in the subcooler section is used to pump helium through the headers and deliver refrigeration to the magnets. Liquid helium is supplied for lead cooling and for superconducting magnets in physics experiments set up in the six intersection regions of ISABELLE. The over-all load is therefore a combination of refrigeration and liquefaction.

The minimum operating pressure for the ISABELLE load has been chosen[1] as 5 atm. Higher pressures will be used during cooldown and other periods of high heat load when the temperature of the system rises above the design level. The control system will be set to maintain essentially constant helium density at the inlet of the circulating compressor. Minor adjustments in the rotational speed of the circulating compressor will then be used to trim for constant flow rate to the load. This method of operation is intended to reduce the need for venting of helium and subsequent "reliquefaction."

Based on theoretical and experimental results, the loads for four possible modes of operation have been estimated. The performance requirements in terms of forced flow are summarized in Table I. For the base design mode, the refrigeration is 13.7 kW and the liquefaction rate is 100 g/s. The heat shield load is 55 kW, provided by 350 g/s of helium flowing between 40 K and 70 K, for all conditions.

SUBCOOLER SECTION

The subcooler section of the refrigerator, shown in the lower half of Fig. 1, incorporates some unique features: cold vacuum compressors, a circulating compressor and heat exchangers/pots. The vaccum compressors are designed to pump the intermediate pot to 0.35 atm (3.2 K) and the low pot to 0.1 atm (2.5 K). Cold vacuum compressors were selected rather than using a subatmospheric suction for the main warm compressor, to minimize system power input, to avoid the risk of leaking air into the system, and to reduce the cross section area of the heat exchangers. The circulating compressor pumps approximately 4 kg/s of helium through the load. This is several times the flow rate that could be provided by the cold end of the refrigerator and is comparable to the main compressor flow.

A performance summary for the cold compressors in the base design mode is given as part of Fig. 2. These three units contribute approximately 10 kW heat load at 4.5 K and 20 kW between 4.6 K and 9 K. As a result, the ISABELLE refrigerator must generate substantial amounts of refrigeration between 4.5 K and 10 K in order to satisfy the requirement to operate below 4.2 K.

```
                              PROGRAM ISA62

         CALCULATED PERFORMANCE OF A HELIUM REFRIGERATOR WHICH UTILIZES
         5 EXPANDERS AND 3 COLD COMPRESSORS.  DELIVERY OF THE REFRIGERANT
         IS IN THE FORM OF COMPRESSED LIQUID HELIUM WHICH IS CIRCULATED
         BY ONE OF THE THREE COMPRESSORS IN THE CYCLE.


                        SUMMARY OF SYSTEM PARAMETERS


                   SYSTEM DESIGN STEADY-STATE LOAD REQUIREMENTS


                  REFRIGERATION REQUIRED-WATTS                        MASS FLOW REQUIRED-GM/SEC
      QMAG    QLEAD1    QLEAD2    QSUP    QRET    QEXP    QSHLD          F74      F75      F76
      7650.    200.     2300.    1500.   2000.    0.    55000.          6.       44.      50.


                          OTHER ESTIMATED REFRIGERATION LOADS-WATTS
   QHX1     QHX2     QHX3     QHX4     QHX5     QHX6     QHX7    QHX8     QHX9    QHX10    QHX11    QHX12
  3600.0   1200.0   2400.0   800.0    600.0    500.0    500.0   200.0    150.0   100.0    100.0    100.0


                        EXPANDER/COMPRESSOR PARAMETERS

                          MAIN    COLD     COLD    CIRC.   EXPANDER EXPANDER EXPANDER EXPANDER EXPANDER
                          COMPR.  COMPR. 1 COMPR. 2 COMPR.     1        2        3        4        5

ADIABATIC EFFICIENCY              .730     .730     .620     .810     .820     .800     .780     .760
ISOTHERMAL EFFICIENCY     .60
INLET PRESSURE-ATM        1.050   .095     .340    4.150   16.227    8.860   15.579    7.922   15.494
OUTLET PRESSURE-ATM      17.250   .350    1.389    5.450    9.000    1.301    8.000    1.423    2.500
INLET TEMPERATURE-K     302.00   2.46     4.56     3.47   185.00    69.48    25.00    12.38     6.29
OUTLET TEMPERATURE-K    305.00   4.69     9.18     3.76   153.64    38.99    20.19     7.10     5.14
FLOW RATE-GM/SEC       3997.5   324.0    945.9   4054.0    648.3   1002.9   1590.4   1590.4   1304.2
WORK-WATTS         11794440.   3562.   21637.    5812.  -107172. -160487.  -38752.  -33790.   -9992.
WORK-HP              15810.       5.      29.       8.     -144.    -215.     -52.     -45.     -13.


                        HEAT EXCHANGER EFFECTIVENESS RATIO

      HX1     HX2     HX3.1    HX3.2     HX4     HX5     HX6    HX7.1    HX7.2    HX8.1     HX9
      .976    .940    .978     .978     .966    .942    .960    .980     .980     .783     .946


      HX8.2   HX15
      .950    .928


              LOAD SUMMARY

                  PRIMARY LOAD        SECONDARY LOAD
                  SUPPLY   RETURN     SUPPLY   RETURN
FLOW RATE-GM/SEC  4153.98  4053.98    354.61   354.61
TEMPERATURE-K        2.59     4.19     40.00    69.45
PRESSURE-ATM         5.35     4.20     15.67     9.67
DENSITY-GM/CC         .154     .139    .01842    .00666
ENTHALPY-J/GM        7.32    10.63    222.00   377.10
```

Fig. 2. Summary on the performances requirements of ISABELLE refrigerator.

Table I. Performance Requirements for ISABELLE Refrigerator

Mode of Operation	Supply Condition			Return Condition			Refrig. Effect	Liquid He Produced
	Flow (kg/s)	Press. (atm)	Temp. (kg/s)	Flow (kg/s)	Press. (atm)	Temp. (kg/s)	(kW)	(kg/s)
Base Design	4.154	5.35	2.59	4.054	4.20	4.19	13.7	0.100
High Refrigeration	4.462	5.35	2.59	4.379	4.20	4.16	14.4	0.086
High Liquefaction	3.022	5.35	2.59	2.878	4.20	4.31	11.1	0.144
4.3 K Magnet Temp.	3.959	6.35	2.98	3.859	5.20	4.64	16.2	0.100

The addition of heat exchanger HX-15 between the high pot and the intermediate pot has been found to increase the plant efficiency by approximately 5%.

MAIN REFRIGERATOR

The refrigerator uses a Claude-type cycle. Because of its large capacity, features have been incorporated that are not customary in smaller refrigerators.

The efficiency of a helium refrigerator increases with the number of expanders and has been reported to level off at five.[2,3] Numerical studies of cycles using three and five expanders were performed as part of the studies leading to the ISABELLE cycle. These verified the economical advantages of using the five expander cycle which has been selected.

Expansion stages 1 and 2 and stages 3 and 4 (see Fig. 1) are connected in series between the high and low pressure streams. The series arrangement allows a higher flow rate across a lower expansion ratio for each turboexpander, as compared to individual units arranged in parallel, and makes possible higher turbine efficiencies.[4]

No liquid nitrogen precooling is used because the system is primarily a refrigerator (as opposed to a liquefier). The use of the warm expansion stage to substitute for liquid nitrogen precooling can be justified on the basis of economic and operational considerations.

The heat shield circuit parallels heat exchanger 4 and expansion stage 2, with the return helium from the circuit providing about 35% of the flow into the expander.

DETAIL DESIGN

With loads and cycle configuration defined, the state varia-

FLUID PROPERTIES AND FLOW RATES

PRESSURE(ATM), TEMPERATURE(K), ENTHALPY(J/GM) AND FLOW RATE(GM/SEC)

POINT	PRESS.	TEMP.	ENTHAL.	FLOW
1	17.250	305.00	1604.33	3997.48
3	16.400	305.00	1604.04	3997.48
4	16.278	185.00	980.61	3997.48
5	16.278	185.00	980.61	3349.15
6	16.181	153.64	817.52	3349.15
7	15.999	69.48	378.39	3349.15
9	15.676	40.00	222.00	3349.15
10	15.666	40.00	222.00	2994.54
11	15.636	25.00	139.41	2994.54
12	15.636	25.00	139.41	1404.15
13	15.606	20.19	111.54	1404.15
14	15.575	12.38	62.75	1404.15
115	15.558	9.88	44.58	1404.15
15	15.543	7.28	28.14	1404.15
17	15.533	6.28	23.59	1404.15
18	15.494	6.29	23.59	1304.15
19	2.500	5.14	15.93	1304.15
20	1.434	4.61	15.93	1304.15

LIQUID FRACTION AT PT. 20 IS .778

POINT	PRESS.	TEMP.	ENTHAL.	FLOW
80	1.424	4.61	29.54	358.27
81	1.424	4.61	11.99	1011.12
82	1.424	4.61	11.99	945.88
182	1.414	3.47	7.07	945.88
83	.350	3.27	7.07	945.88

LIQUID FRACTION AT PT. 83 IS .953

POINT	PRESS.	TEMP.	ENTHAL.	FLOW
84	.350	3.27	29.03	621.86
85	.350	3.27	5.99	901.56
86	.350	3.27	29.03	621.86
186	.340	4.50	36.51	621.86
87	.350	3.27	5.99	324.02
88	.100	2.49	5.99	324.02

LIQUID FRACTION AT PT. 88 IS .913

POINT	PRESS.	TEMP.	ENTHAL.	FLOW
89	.100	2.49	26.63	324.02
90	.100	2.49	4.02	295.75
91	.095	2.46	26.53	324.02
92	.350	4.69	37.52	324.02
93	.340	4.56	36.86	945.88
94	1.389	9.18	59.73	945.88

POINT	PRESS.	TEMP.	ENTHAL.	FLOW
21	1.050	302.00	1583.34	3997.48
23	1.100	302.00	1583.34	3897.48
24	1.139	178.69	942.99	3897.48
26	1.166	151.65	802.53	3897.48
126	1.166	151.65	802.53	3264.68
127	1.243	64.86	351.68	3264.68
226	1.166	151.65	802.53	632.80
227	1.243	64.86	351.68	632.80
27	1.243	64.86	351.68	3897.48
29	1.294	38.99	217.08	3897.48
30	1.306	38.99	217.08	2894.54
31	1.318	22.60	131.43	2894.54
33	1.330	20.00	117.74	2894.54
133	1.330	20.00	117.74	1399.63
134	1.348	10.78	68.61	1399.63
233	1.330	20.00	117.74	1494.92
234	1.348	10.78	68.61	1494.92
34	1.348	10.78	68.61	2894.54
35	1.389	9.18	59.73	2894.54
135	1.389	9.18	59.73	1948.66
136	1.404	7.10	47.78	1948.66
36	1.404	7.10	47.78	358.27
37	1.411	4.60	29.54	358.27
100	4.550	3.80	9.57	4147.98
101	4.550	3.99	10.13	4103.98
102	4.550	3.99	10.13	4053.98
103	4.200	4.19	10.63	4053.98
104	4.200	4.19	10.63	4053.98
105	4.150	3.47	8.47	4053.98
106	5.450	3.76	9.91	4053.98
107	5.450	3.78	9.99	4153.98
108	5.400	3.37	8.91	4153.98
109	5.350	2.59	7.32	4153.98
110	5.000	2.90	7.68	4153.98

POINT	PRESS.	TEMP.	ENTHAL.	FLOW
45	16.227	185.00	980.61	648.33
46	9.000	153.64	815.31	648.33
47	8.904	69.48	377.10	648.33
48	8.860	69.48	377.10	1002.94
49	1.301	38.99	217.08	1002.94
52	15.579	25.00	139.41	1590.39
53	8.000	20.19	115.05	1590.39
54	7.922	12.38	69.03	1590.39
55	1.423	7.10	47.78	1590.39
67	8.904	69.48	377.10	354.61
68	9.666	69.45	377.10	354.61
69	15.666	40.00	222.00	354.61
71	15.494	6.29	23.59	100.00
72	5.550	6.34	23.59	100.00
73	5.450	4.71	13.14	100.00
74	4.975	3.35	8.63	6.00
75	4.550	3.80	9.57	44.00
76	4.550	3.99	10.13	50.00
77	1.050	302.00	1583.32	100.00

Fig. 3. State variables at each process point in the ISABELLE cycle.

bles at each process point can be determined from basic thermodynamic relationships. Iteration is required until component efficiencies, pressure drops and heat leaks are consistent with hardware design. A computer program utilizing the NBS helium properties[5] was developed by BNL to do the necessary computations. The program first calculates the circulating flow requirement for a given load, and the cooling required at the three pots in the subcooler. The interface condition between subcooler and refrigerator is then obtained from vacuum compressor performance projections. Finally the sizing of turbines and heat exchangers for the refrigerator is performed. A typical summary describing the load conditions, performance requirement of turbines, compressors and heat exchangers, and the state at each process point is in given Figs. 2 and 3.

At base design condition, the pressure in the low pot of the subcooler is chosen to be 0.1 atm corresponding to 2.49 K. This is the lowest temperature in the system. The pressure in the intermediate pot is chosen to be 0.35 atm, corresponding to 3.27 K.

The inlet temperatures for the expansion stages have been chosen to optimize the cycle, given the load and subcooler section requirements. All expansion stages but the first are below liquid nitrogen temperature. The outlet pressure of expander[5] is set at 2.5 atm to avoid liquid formation, although in practice, this may not be necessary. The total power extraction from the expansion stages is around 350 kW. The turboexpanders all have variable inlet nozzles to permit achieving near optimum efficiency over a wide flow rate range. This is particularly important during cooldown.

High effectiveness heat exchangers are used. Temperature differences across heat exchangers have been adjusted so that most of heat exchangers have an effectiveness greater than 0.94, but none exceed 0.98. The product of the heat transfer coefficient and the total heat exchanger area is about 2.10 MW/K.

The high and low pressure streams in the refrigerator are at 16.4 atm and 1.1 atm. The mid pressure between series-connected expansion stages is about 8 atm. The choice of 1.1 atm as the lowest pressure is to avoid subatmospheric suction in the main compressors. No significant cycle improvements were observed for high pressures greater than 16.4 atm. The total compressor flow required is about 4 kg/s. It is estimated that 12 MW to 15 MW of input power will be required depending on the efficiency of the compressors which are used.

REMARKS

In order to enhance the reliability and flexibility of this plant, redundant heat exchangers (1-2) and expanders, (1, 2, 3 and 4) will be provided. The redundant expanders 1 and 2 serve in a way similar to liquid nitrogen precooling to increase the cooling capacity of the system significantly during cooldown. An analysis of the cooldown process is given in Ref. 6. With redundant expanders 1 and 2 operating, the ISABELLE plant can be operated as a liquefier which will be capable of producing 11 000 L/h of liquid helium at 4.6 K.

The refrigerator is now under construction and is scheduled for completion and testing in 1983.

REFERENCES

1. D.P. Brown, IEEE Transactions Nucl. Sci. NS-28 (3):3214 (1981).
2. A. Khalil and G. McIntosh, in "Advances in Cryogenic Engineering, Vol. 23," Plenum Press, New York (1978), p. 431.
3. W. M. Toscano and F.J. Kudirka, in "Advances in Cryogenic Engineering, Vol. 23," Plenum Press, New York (1978), p. 456.
4. E. Carbonell, in "Proc. 5th Int. Cryog. Engr. Conf.," IPC Technology Press, London (1974), p. 353.
5. R.D. McCarty, NBS Technical Note 631, U.S. Dept. of Commerce, (1972).
6. K.C. Wu, D.P. Brown and R.W. Moore, Analysis of Cooldown Performance for ISABELLE Refrigerator, in "Advances in Cryogenic Engineering, Vol. 27," Plenum Press, New York (1982).

DESIGN OF 24.8-kW, 3.8-K CRYOGENIC SYSTEM FOR ISABELLE

D. P. Brown, M. Afrashteh, J. A. Bamberger, A. Fresco, A. P. Schlafke,
W. J. Schneider, J. H. Sondericker, A. Werner, and K. C. Wu

*Brookhaven National Laboratory**
Upton, New York

INTRODUCTION

ISABELLE will consist of two proton accelerator/storage rings
in a common tunnel. There will be 1084 superconducting magnets
installed in the 3.8 km circumference tunnel. Protons will be
accelerated to an energy of 400 GeV in each ring and stored for 24
hours. The two beams are counter-rotating and intersect at six
places around the ring. The particles which emerge from proton
interactions at these intersecting points are studied using various types of detectors.

PERFORMANCE REQUIREMENTS

The accelerator magnets are to be maintained at 3.8 K or
below. The magnet vessels and their interconnecting piping have
been designed for a working pressure of 20 atm. Since 1976 the
magnets have been designed[1] for single phase (as opposed to pool
boiling) cooling of the magnet coils. The magnets use "cold
iron," i.e., the iron magnetic return path is at the temperature
of the coil. The cooldown mass of the system is high (5×10^6 kg)
and therefore cooldown of the system must be carefully planned and
executed to avoid excessive cooldown time. Maximum cooldown time
for the entire system has been administratively set at 14 days.

---

*Operated by Associated Universities, Inc. under contract with the
U.S. Department of Energy.

OVERALL DESIGN

Main Refrigerator

A block diagram showing the components of the system designed to meet these requirements is shown in Fig. 1. After studies of the economics and system reliability, it was felt that our interests would be best served by using a fully centralized, i.e., single refrigerator plant, as opposed to a distributed plant. The main refrigerator, then, is the heart of this system and produces all the refrigeration required. The cycle used[2] is an adaptation of the Claude cycle, does not use liquid nitrogen as a precoolant, and has turboexpanders at different temperature levels.[2]

The refrigerator is now under construction. It is contained in four horizontal cylindrical vacuum tanks. A fifth vacuum tank contains the subcooler/circulation system which is described below. The five cold box sections each have 4.0 m diameter vacuum tanks. Their lengths vary from 10.8 to 13.8 m.

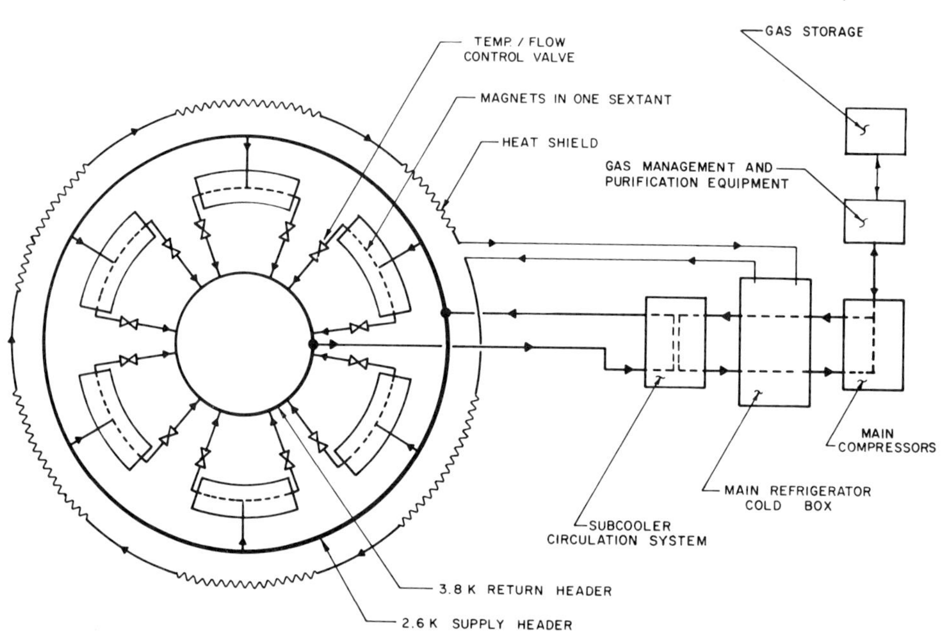

Fig. 1. ISABELLE Cryogenic System Block Diagram. Illustrates primary and shield circuits for one accelerator ring.

All heat exchangers are of the brazed aluminum, plate-fin type and have their long axes horizontal. Special care has been taken in the design of these exchangers to prevent flow maldistribution caused by the horizontal flow path. The warmest heat exchanger requires three cores (0.91 m × 1.07 m × 4.6 m) manifolded in parallel by a 0.78 m diameter header to meet the requirements of our specification.

Subcooler/Circulation System

The main refrigerator has a return pressure of 1.4 atm at its low temperature end and therefore will produce refrigeration to the level of 4.6 K. A special refrigerator section has been appended to the main refrigerator to reduce the temperature to 2.6 K on the supply side of the magnets and to provide a high flow rate to reduce the temperature rise across the magnets. To reach the low pressure (0.1 atm) required in this subcooler section, a two-stage centrifugal compressor system is used. This "vacuum pump" operates in the temperature range of 2.46 K (inlet) to 9.18 K (outlet). Another centrifugal compressor is used to circulate fluid through the load. This "circulating compressor" has an inlet temperature of 3.47 K; a high throughput, 4054 g/s, and a low pressure ratio, 1.31. The details of the subcooler may be found elsewhere.[2] The circulating helium stream heat exchanges with the low temperature liquid produced by the "vacuum pump."

Distribution and Load Matching

The magnets in the ISABELLE rings are arranged in sextants. The forty-five magnets in half of a typical sextant are cooled in series. For the two rings, then, there are 24 parallel flow paths. Vacuum-jacketed piping with a heat shield is used as supply and return headers to these 24 loops.

Each loop has a control valve to regulate the helium flow rate in that loop. The flow will be controlled so that the temperatures of the last magnet in each loop are equal. A process control computer will provide "global" control to orchestrate the mean temperature and pressure in the system as a function of heat load. Helium gas-cooled magnet power leads are used to transfer electrical power in and out of the magnet string. These leads require an adequate supply of helium gas flowing from the cold to the warm end to prevent burnout. A control system for each lead is provided and appropriate interlocks will be incorporated to assure safe operation of this sub-system.

Design Heat Load

The heat load which will be presented to the refrigerator is summarized in Table I. The total load is divided into two temperature ranges. The primary load is comprised of system elements which require refrigeration near 4 K. Liquefaction loads (100 g/s in this case) have been converted to equivalent refrigeration capacity for inclusion in the table. The secondary load is primarily the heat shield which operates at a mean temperature of 55 K.

The estimates of heat load for the magnets and the power leads have been verified by measurement on prototypes. The losses for the distribution system have not yet been verified by measurement. The loss due to pressure drop in the circulating helium stream is the work supplied to the circulating compressor and is included in the distribution system loss.

Ratio of Refrigeration Capacity to Heat Load

A crucial parameter in the design of any cryogenic system is the choice of the ratio of refrigerator capacity to the design heat load. This ratio has far reaching effects. It determines, to a great extent, the length of the cooldown. It determines the ability of the refrigerator to absorb above design heat loads due to equipment problems, e.g., insulating vacuum failure, and to be independent of operator errors. It will determine, the "availability" of the system to meet its duty schedule. Budgetary constraints, of course, tend to drive this ratio as low as possible.

For the ISABELLE system a ratio of 1.5 was initially chosen. The ratio was established and the order for the refrigerator cold box placed, i.e., the capacity was fixed, before definitive heat load measurements had been made on prototypes. The design capacity and the final ratio are given at the bottom of Table I. The design capacity is 24.8 kW and 55 kW for the primary and secondary loads, respectively.

System Design Pressure

The fluid which circulates through the ISABELLE magnets is a compressed liquid. As liquid, its density is not a strong function of its pressure. The specific heat of helium in this temperature range increases with decreasing pressure. A study considering these factors, as well as the pressure drop through the system, indicated that the optimum pressure would be the lowest pressure at which the system could be operated.

The lower limit is set by the pressure drop in the magnet power lead flow loop. With the main compressors operating at a suction pressure of 1.05 atm, a minimum pressure of 3.5 atm is required at the magnets to assure proper gas flow to the leads.

TABLE   I

ISABELLE HEAT LOAD ALLOWANCE

	Primary Load W at 4 K	Secondary Load W at 55 K	
MAGNET SYSTEM			
Insulation	2860	9300	
Supports	325	600	
Connecting Piping	1365	7600	
Instrumentation	550	—	
Subtotal	5100	17500	
MAGNET POWER LEADS			
Main Coils (4250 A)	1100	—	
Correction Coil (100 A)	2200	—	
Lead Pots	300	800	
Subtotal	3600	800	
DISTRIBUTION SYSTEM			
Piping	1075	12700	
Valves	1225	3700	
Circulating Compressor	3900	—	
Subtotal	6200	16400	
EXPERIMENTAL AREA DETECTORS			
1000 L/h Equivalent	1800	—	
TOTAL EXPECTED LOAD	16700	34700	
REFRIGERATOR CAPACITY	24800	55000	
CAPACITY/LOAD RATIO	1.485	1.585	

The nominal design pressure for the system has been chosen as 5 atm. This allows a margin for control purposes between the nominal and required minimum pressures. Control of system pressure during operation and cooldown are discussed elsewhere.[2-3]

Physical Plant

Two buildings have been provided for this equipment. One building (15.5 m × 42.9 m) houses one end of each of the five cold boxes (the remainder of each box being out-of-doors) plus the allied low temperature equipment such as turboexpanders adsorber, etc. The other building (16.4 m × 61.8 m) is located nearby and serves to house the compressors. Some equipment will also be located out-of-doors.

Reliability and Redundancy

High reliability is required for this system. Each component of the system has been scrutinized to see if it offers the highest reliability for its class. Even with the most reliable components, it was realized that high reliability could not be guaranteed because of operational problems, particularly capacity degradation due to contamination.

The only reasonable way to enhance the reliability for that type of failure is through redundancy and that was the path chosen for ISABELLE. Figure 2 shows the redundant components in the cold box. In addition, the compressor system will have redundancy in each stage of compression and oil removal. Redundant heat exchanger assemblies are provided for those which operate from 300 K to 152 K. These exchangers will freeze out and moisture and oil on their surfaces. When the pressure drop through the exchanger becomes excessive, it will be warmed and cleaned after switching to the redundant exchanger.

The dual bed adsorber designed to remove nitrogen and oxygen from the helium stream, operates at a temperature of 69 K. Special precautions have been taken to isolate the beds in case of temperature upsets to prevent desorption of contaminants which would then carry to lower temperature regions.

The expanders which operate above 7.1 K have redundancy. No redundancy was provided below this temperature since (1) essentially all contaminants should be frozen out by this level and (2) the switchover valves required would impose a rather large heat load at this low temperature.

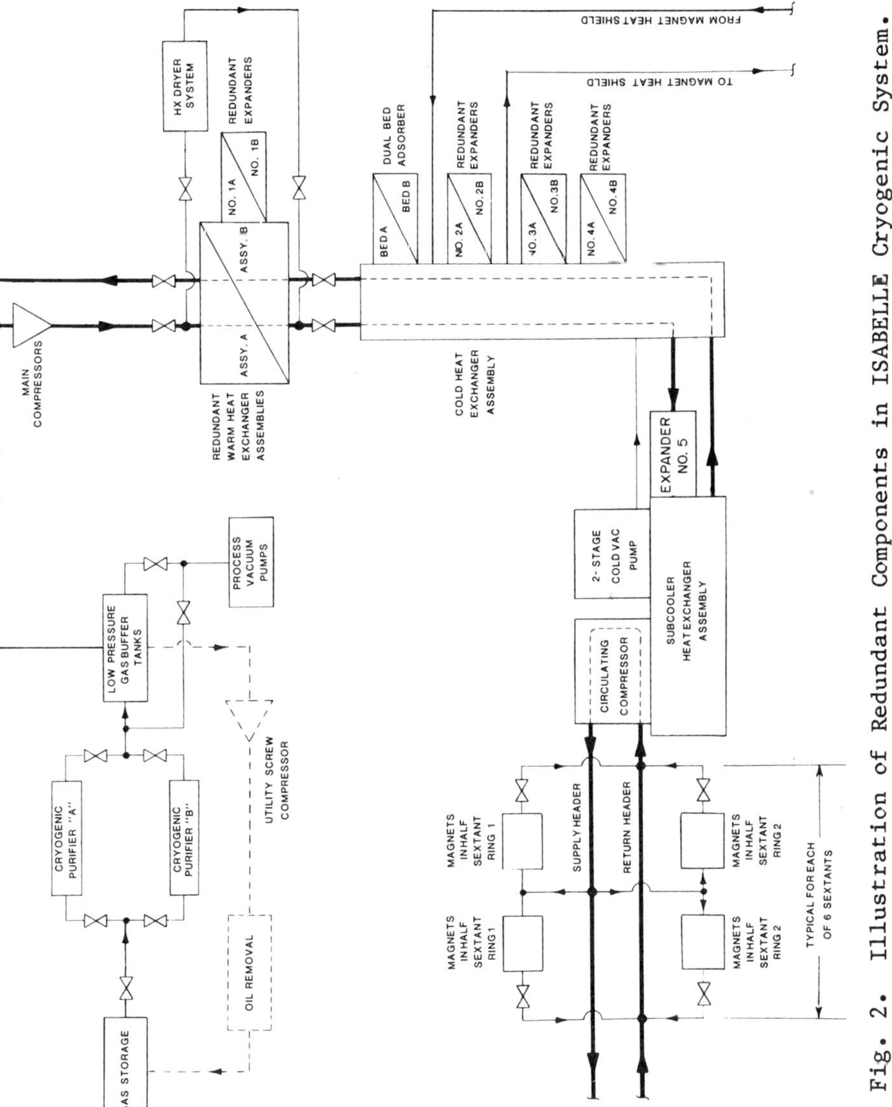

Fig. 2. Illustration of Redundant Components in ISABELLE Cryogenic System.

Process Control

The ISABELLE cryogenic system is distributed over a very
large area and demands sophisticated control techniques for plant
optimization.   Conventional analog control could be used but
initial hardware costs would be high and operations very manpower
intensive.   To reduce costs and gain extensive flexibility, a
commercial distributed process control computer system will be
utilized.   In general, cryogenic and petro-chemical control systems are so similar that "off the shelf" process control computer
systems can be easily adapted for refrigeration systems.   Temperature measurement is the only area where some additional interfacing is required.

The control center for the ISABELLE cryogenic plant will be
located in an operations area in the cryogenic wing of the main
service building.   It will consist of the host computer, color
graphics consoles for operator interface and various peripherals.   A high speed data highway will connect distributed processors, interfacing 4500 points, located around the ring and in the
compressor and refrigerator areas, to the host.   All system process variables reside in the host data base which will permit on
line process modeling, reconfigurable control tasks and remote
operation from ISABELLE main control.

REFERENCES

1. J.A. Bamberger, et al., Forced Flow Cooling of Isabelle Dipole
       Magnets, <u>IEEE Trans on Magnetics</u>, MAG-13 (1):696 (1977).
2. D.P. Brown, et al., Cycle Design for the ISABELLE
       Refrigerator, "Advances in Cryogenic Engineering, Vol. 27,"
       Plenum Press, New York (1982).
3. K.C. Wu, D.P. Brown, and R.W. Moore, Analysis of Cooldown
       Performance for ISABELLE Refrigerator, "Advances in
       Cryogenic Engineering, Vol.27," Plenum Press, New York
       (1982).

HEAT LOAD MEASUREMENT OF PROTOTYPE CRYOGENIC MAGNETS AND LEADS FOR THE ISABELLE PROJECT*

A. P. Werner, D. P. Brown, and W. J. Schneider

Brookhaven National Laboratory
Upton, New York

INTRODUCTION

A major feature in the cryogenic system development program for ISABELLE is the measurement of the heat loads of prototype hardware. In particular, superconducting magnet assemblies, magnet power leads and magnet power lead vessels have been measured. In each of these cases it has been possible to build equipment which has a heat load equal to or less than the allowance for it in the heat load budget.

There are over 1000 magnets and 3000 magnet power leads used in the ISABELLE system. When measuring a single component it is important to reduce systematic errors to a minimum because of the large multiplier in the final system. For this reason, the techniques and instrumentation used have undergone considerable refinement in order to achieve acceptable measurement accuracies.

The cryogenic elements, e.g., magnets and lead pots, are typically arranged in series. The helium flow and the insulating vacuum are continuous through the series. It is necessary in the development stages of the project to measure single elements. The ends of a single element which is removed from its mates must be carefully terminated so that the contribution of the ends to the heat load is the same during the test as it will be in service. The technique is the equivalent to that of a guarded plate calorimeter.

---

*Operated by Associated Universities, Inc. under contract with the U.S. Department of Energy.

DESCRIPTION OF TEST BED

The test bed for heat load measurement consists of two "instrumented ends," flow meters and a 100 W helium refrigerator. A schematic of the test bed is shown in Fig. 1. The heat loads for steady-state conditions are determined by:

$$q = \dot{m} \, (h_2 - h_1)$$

where,

q = rate of heat transfer $\dot{m}$ = mass flow rate
h_1 = inlet enthalpy h_2 = outlet enthalpy

Values of h_1 and h_2 are interpolated from the values given in NBS 631[1] using measured values for pressure and temperature at each point.

The "instrumented ends" are vacuum-jacketed vessels which contain high accuracy cryogenic thermometry.[2] One end is placed

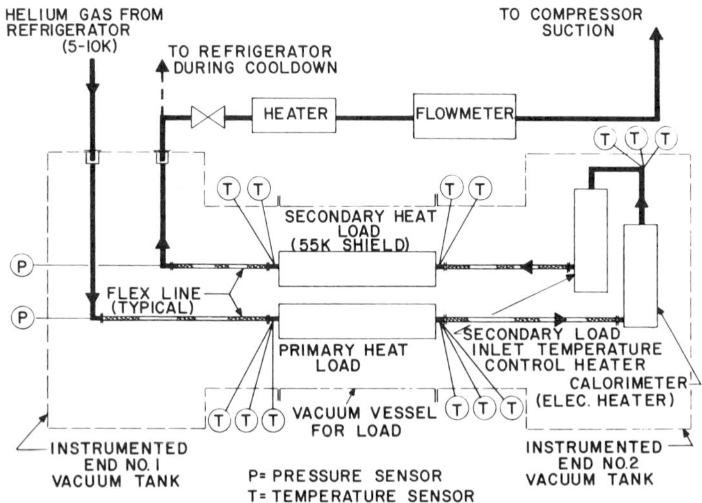

Fig. 1. Schematic of Test Bed. T represents temperature sensors, P are pressure transducers.

at the inlet to the load, the other at the outlet. Much of the ISABELLE hardware has a heat shield which operates at a nominal temperature of 55 K. In order to accommodate the dual temperature requirement, two sets of circuits and instrumentation can be used in each end. An electric heater system is located in the helium stream following the low temperature part of the load and is used to raise the gas temperature to that correct for the shield.

After the helium gas passes through the heat shield, its temperature is measured. The gas then leaves the instrumented end and passes through a heater which brings its temperature to ambient. After passing through a flow meter the gas returns to compressor suction.

The flow rate through this apparatus is controlled so that a temperature rise of about 1 K is observed across the load. Typically this results in a flow of about 1 g/s when a magnet is being measured. A large temperature difference is desirable so that temperature measurement errors are small by comparison. Operation must be such that only single phase fluid exists in the apparatus when measurements are being made. This is accomplished by operating above the critical temperature.

Wherever the measured temperatures are used to calculate the heat load, redundant temperature sensors (either two or three) are installed. This allows some estimate of the measurement error at each point and permits easy identification of sensor drift and most types of instrumentation failures. A further check of the temperature sensors was performed by flooding the apparatus with boiling liquid and comparing the offsets from the equilibrium temperature at the observed vapor pressure. Usually agreement within ± 0.005 K can be attained. The sensors with the least offset were chosen to be used in the calculations and the offsets entered as corrections to observed temperatures. A period of about seven days is required for a magnet to reach thermal equilibrium. A magnet was said to be at steady-state when no temperature change greater than 0.2 K was observed in a 24 hour period. A typical test would consist of taking readings at half hour intervals for 24 hours. A computer program was used to calculate the helium properties at each point and the magnitude of the heat load. The standard deviation from the mean of all readings taken at a point was also calculated and used as a guide to the consistency of the procedure.

TEST RESULTS

Dipole Magnets

A series of tests were run to examine the heat load charac-

teristics of a prototype ISABELLE magnet. One study was to determine the heat load as a function of insulating vacuum quality.
The primary load to approximately 4 K and the secondary load to
the heat shield at 55 K are measured. The results for two test
periods in April and June of 1978 are shown in Figs. 2 and 3. The
same dipole magnet was used in both runs.

Figure 2 shows the measurements for the primary load. The
good agreement between data from the two runs indicates adequate
repeatability for this method of heat load measurement. The error
bars indicate one standard deviation above and below the mean heat
load measured at a particular vacuum level. For groups of 20 to
50 readings, standard deviations of 0.25 to 0.50 W were observed. The goal is for each magnet to have a heat load of less than
4.6 W when the insulating vacuum is 10^{-4} torr. The measurements
indicate that this goal was achieved.

Figure 3 shows the measurements for the secondary load. The
data from the April run showed that the magnet exceeded the desired design heat load. The insulation was reworked and the magnet
retested in June. The rework was necessary to reduce effects from
the ends where the magnet would normally connect to a mate. The
June data shows that the reworked insulation system was good
enough to meet the heat load allowance (11 W) for the heat shield.

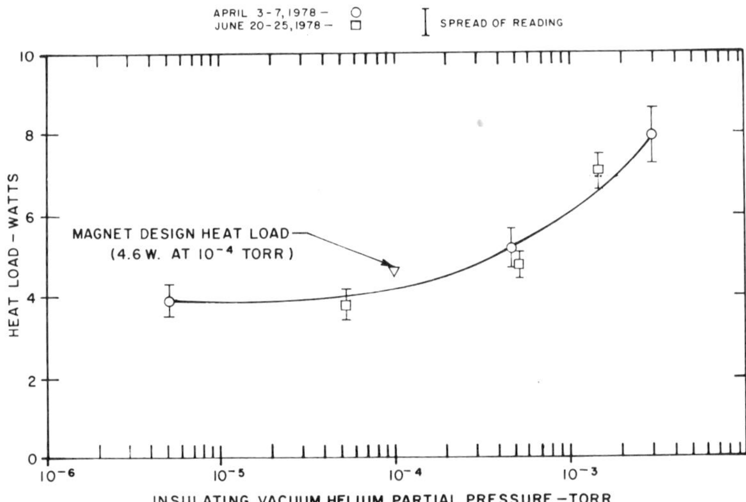

Fig. 2. Dipole Magnet 4 K Heat Load.

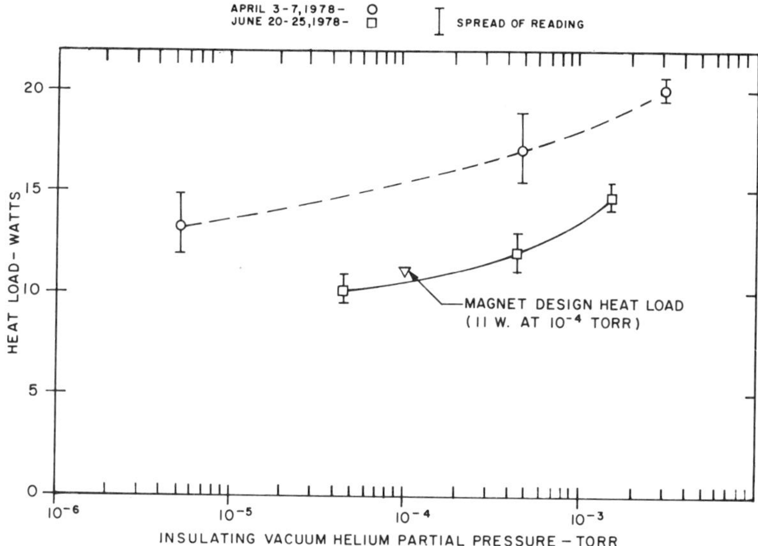

Fig. 3. Dipole Magnet 55 K Heat Load.

Table I shows our estimate of the breakdown of the design heat loads. We cannot measure each of these loads separately to verify this breakdown, so the values shown are somewhat speculative except for the total.

Heat Shield

This heat load measurement apparatus was also used to study the feasibility of inserting a heat shield between the warm and

Table I. Dipole Magnet Design Heat Load Allowance

	Allowed Dipole Load Watts	
	4 K	55 K
Insulation-radiation and conduction @ 10^{-4} torr	3.0	9.0
Supports	0.3	0.5
Connecting Piping	0.8	1.5
Instrumentation	0.5	–
Total	4.6	11.0

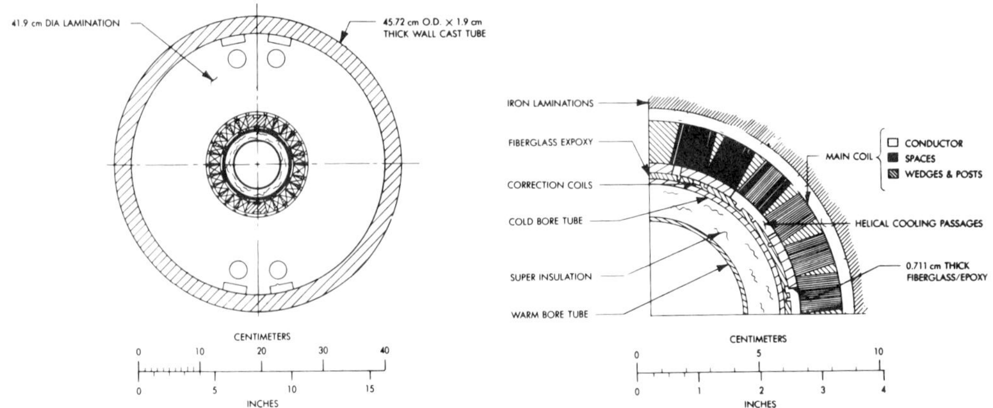

Fig. 4.   ISABELLE Dipole Magnet Cross-Section.

cold bore tubes of the magnets. In the ISABELLE design, which is shown in Fig. 4, the beam circulates in a 9.3 cm diameter room temperature tube (the "warm bore tube"). Only a 1.03 cm wide annulus separates that tube from another tube (the "cold bore tube") which is at a temperature of approximately 4 K. These tubes are about 5 m in length.

A heat shield was installed in the annulus and the magnet was retested. It was found that the primary load was essentially unchanged and that the secondary load had doubled. This result was thought to be caused by ovality and straightness limits for these tubes which make it difficult to position the tubes properly. On the basis of these tests, the heat shield in this location was dropped, from the design.

Magnet Power Leads

The same instrumented ends have been used to measure the performance of ISABELLE magnet power leads and the vacuum-insulated vessels in which they are mounted.[3] The ISABELLE design calls for 48 power leads with a rated current of 4250 A and 3312 leads with a rated current of 100 A. The 4250 A leads carry the main magnet current in and out of each end of each sextant of the two rings. The 100 A leads are used singly or in parallel groups to satisfy all other requirements.

A magnet lead vessel was first tested without any leads installed. The results are given in Table II.

Table II. Magnet Power Lead Vessel Heat Load Measurements

	Primary Load (W)	Secondary Load (W)
Calculated Heat Load	5.00	13.88
Design Allowance	6.25	16.66
Measured Heat Load	5.44 ± 0.28	10.99 ± 0.75

Pairs of leads were then installed in the previously meaured vessel. The heat load of the vessel was subtracted from the measured heat load to arrive at the figure for the leads. Table III gives the data for a typical (single) 4250 A magnet power lead. A total of 7 leads were tested: Two just met the allowance. Three were essentially right on the design values, and two were 20% over the design allowance and were rejected. Variation in performance was caused by variation in the interference fit between the copper conductor and its stainless steel tube sheath. Redesign and implementation of tight quality control during manufacture has eliminated this problem.

The design of the 100 A power leads for ISABELLE incorporates "bundles" of leads. In this space-saving innovation, 12 leads are grouped together and share a common helium gas flow. The leads are, of course, electrically insulated. These have come to be known as "12 x 100 A" bundles. Several of these bundles have been tested and all were well within the design heat load allowance for them. The results are presented in Table IV.

COMMENTS

This highly versatile facility with its accurate instrumentation has proven useful in the development of equipment which meets the heat load requirements for ISABELLE. The resolution and repeatability available allow comparisons between different designs and a realistic evaluation of the overall expected heat load for ISABELLE.

Table III. Thermal Performance of a Single 4250 A Power Lead

	Conduction (W)	Mass Flow (g/s)	Total Equivalent Load (W)
Calculated Heat Load	6.1	0.24	19.1
Design Allowance	7.4	0.29	22.9
Measured Heat Load	5.1	0.25	18.6 ± 2.01

Table IV. Thermal Performance of a 12 x 100 A Power Lead Bundle

	Conduction to 4 K (W)	Mass Flow (g/s)	Total Equivalent Load (W)
Calculated Heat Load	1.65	0.066	5.2
Design Allowance	2.5	0.1	7.9
Measured Heat Load	2.0	0.073	5.9 ± 1.0

ACKNOWLEDGEMENTS

The authors wish to acknowledge the assistance of many members of the ISABELLE staff in design, fabrication, and operation of the test facility. Special thanks to T. Cromer for gathering and processing the data, W. Kollmer, D. Zantopp and the members of the electronic group for their superb instrumentation, and the Word Processing Center for graciously typing this paper.

REFERENCES

1. R.D. McCarty, Thermophysical Properties of Helium-4 from 2 to 1500 K with Pressures to 1000 Atmosphere, NBS Tech. Note 631, (1972).
2. J. Sondericker, Production and Use of High Grade Silicon Diode Temperature Sensors, in "Advances in Cryogenic Engineering, Vol. 27", Plenum Press, New York (1982).
3. D.P. Brown and W.J. Schneider, in "Advances in Cryogenic Engineering, Vol. 25", Plenum Press, New York, (1980), p. 300.

ANALYSIS OF COOLDOWN PERFORMANCE FOR ISABELLE
HELIUM REFRIGERATOR

K. C. Wu and D. P. Brown

*Brookhaven National Laboratory**
Upton, New York

and

R. W. Moore

Koch Process Systems, Inc.
Westborough, Massachusetts

INTRODUCTION

The ISABELLE refrigerator[1] was designed to provide cooling for
1000 superconducting magnets. It has the dual duty of providing
cooling at steady-state operating conditions (3.8 K) and to cooldown massive magnets from ambient temperature. During the cooldown, tremendous quantities of heat ($\sim 4.8 \times 10^{11}$ J) must be
removed from the magnets (5.9×10^6 kg of iron) and 75 000 liters
of cold helium must be supplied to fill the system. In the planning for ISABELLE, this cooldown has been estimated to require
approximately two weeks. Because of this long cooldown time and
the fact that during cooldown the refrigerator will be operated
quite differently from its steady-state design configuration, it
is important to identify the cooldown procedure and maximize the
cooling capacity.

RELATIONSHIP BETWEEN A REFRIGERATOR AND ITS LOAD

A cooldown analysis involves matching performance between the
refrigerator and the load. The cooling process can be character-

---

*Operated by Associated Universities, Inc. under contract with the
U.S. Department of Energy.

ized by three parameters: 1) the rate at which heat is removed from the load or apparent refrigeration, 2) the rate at which helium is deposited in the load or apparent liquefaction rate and 3) the temperature at which the process is carried out.

The capacity of the refrigerator may be limited by any component in the refrigerator, but the most common limits are compressor capacity, expander throughput and heat exchanger surface area. Within operating ranges, the apparent refrigeration and apparent liquefaction rate can be traded for each other. A performance curve for the refrigerator is given in Fig. 1. The load behaves oppositely to the refrigerator during cooldown with apparent liquefaction rate directly proportional to the apparent refrigeration as shown in Fig. 1. This can be understood on the simple physical ground that the more cooling a refrigerator provides, the faster the load will be cooled down and the more rapidly helium will be accumulated in the load. The intersecting point for the two curves determines the operating point at that particular temperature.

CYCLE DESCRIPTION AND PROCEDURE FOR COOLDOWN

A flow diagram of the ISABELLE refrigerator is shown in Fig. 2. The system has five turboexpanders in its cycle, four of which have redundant mates. During steady-state operation there is no excess compressor flow to drive the redundant expanders. In cooldown operation, flow can be allocated to the redundant expanders by rearranging the flow inside the refrigerator or by using helium returned from the magnet. The latter is permitted because the ISABELLE magnet system is designed for 20 atm pressure and a small pressure drop is anticipated through the load.

During initial stage of cooldown, the magnet is warm and the

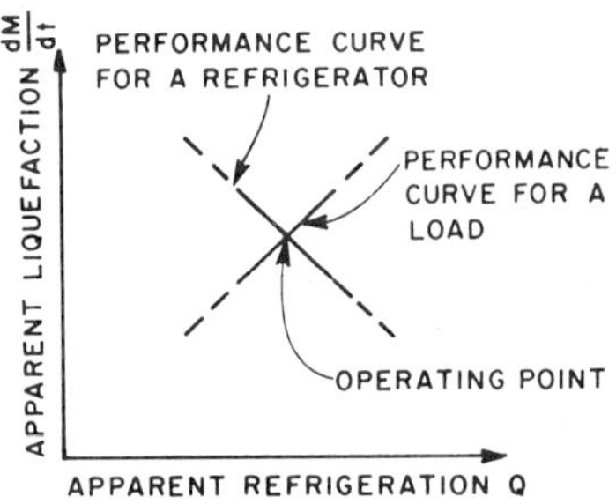

Fig. 1. Relationship between a refrigerator and its load.

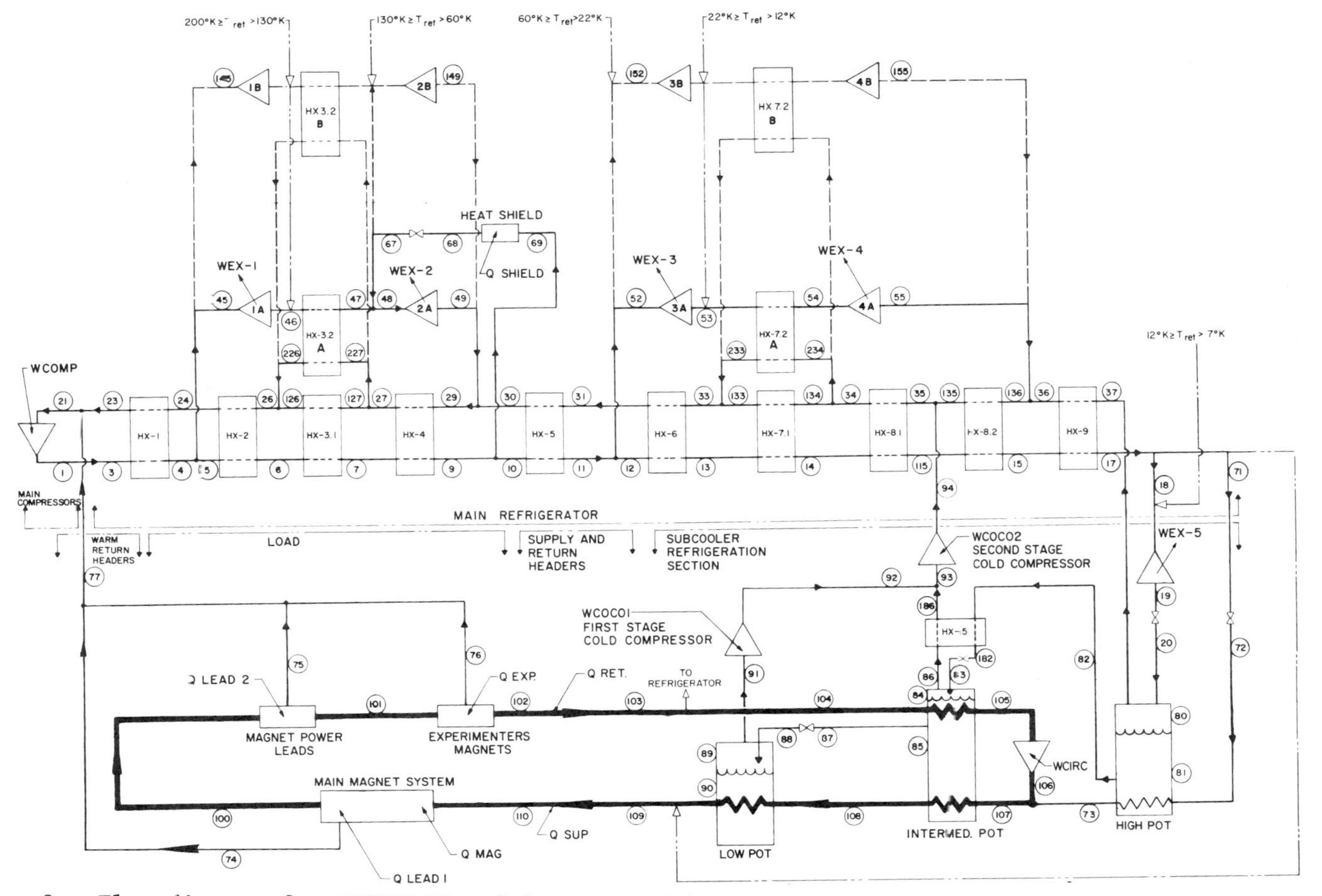

Fig. 2. Flow diagram for ISABELLE refrigerator with its redundant turbines and heat exchangers.

return helium cannot be used for cooling i.e.,·the refrigerator behaves as a liquefier. During this state, two sets of expanders 1 and 2 in parallel and one set of expanders 3 and 4 will be operated. High pressure (~ 15 atm)helium from the outlet of heat exchanger 8 will be used to cool the headers and magnets. The return gas will be returned to suction through the cryogenic purifier until the temperature reaches approximately 200 K.

As the return temperature becomes lower than 200 K, helium will be introduced to the proper redundant turbine. Five locations with their operating temperatures are shown in Fig. 2. The switch of return helium from one turbine to the other is similar to the switch of return helium from one heat exchanger to the other in the cooldown operation of a conventional helium refrigerator. Proper throttling of flow through each turbine is required to maintain reasonable temperatures throughout the refrigerator. Apparent liquefaction rate and refrigeration supplied must also be balanced by holding the magnet system pressure constant.

When liquid helium temperature is reached at the outlet of turbine 5, the refrigerator will be switched into its steady-state configuration. The circulating compressor is then started and liquid level in the pots of the subcooler must be maintained. When the average temperature in the magnet is around 5.5 K, the pressure in the magnet will be allowed to fall along a constant density line with decreasing temperature as shown in Fig. 3 until

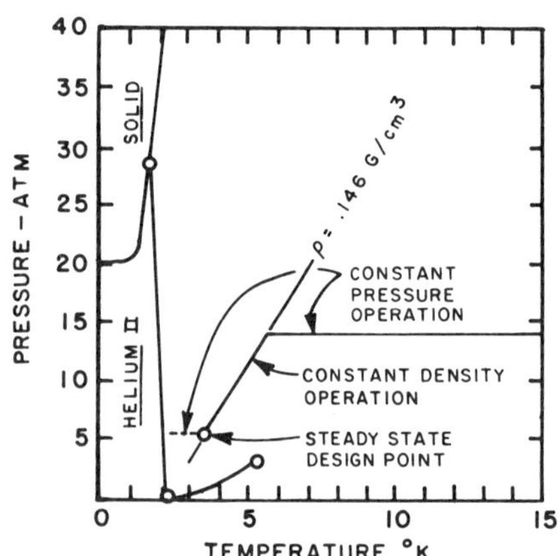

Fig. 3. Pressure vs. temperature for ISABELLE load.

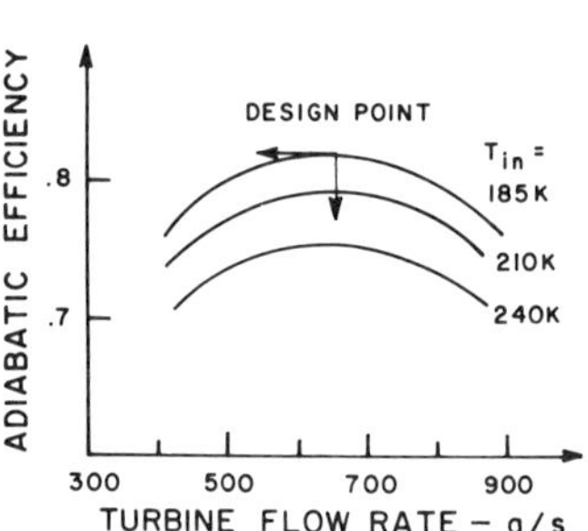

Fig. 4. Efficiency for turbine 1.

the 5 atm operating condition is reached. Finally the two vacuum compressors are turned on and steady-state operation is reached.

MODELING THE REFRIGERATOR

We assume quasi-steady-state operation, so the problem of modeling cooldown is the same as other off-design computations for the refrigerator. The method employed in the present analysis is based on solving terminal temperature relationships among the constituent heat exchangers and turbines. The terminal temperatures for heat exchangers are related by the effectiveness and energy equations.[2] The inlet and outlet conditions for a turbine are related by the adiabatic efficiency.[7] Given a flow distribution, the heat transfer area times overall heat transfer coefficient AU, or the number of heat transfer units NTU, for the heat exchangers and η for turbines, solutions can be obtained by inverting the temperature relationship matrix with iteration against specific heat.

Typical performance for a turbine, with efficiency versus flow rate at selected inlet temperature is given in Fig. 4. For simulating heat exchanger performance, we assume values of AU for heat exchangers with flow rate greater than design value and values of NTU for those flow rates less than design value. This results in a conservative prediction for performance.

CONCEPTS OF MAXIMIZING COOLDOWN CAPACITY

For proper operation of a refrigerator, both refrigeration, liquefaction and load temperature must be considered at the same time. For the initial state of cooldown, one could produce more refrigeration (or work extraction) by expanding helium across a fixed pressure drop at higher temperature than at lower temperature. For example, 1 g/s of helium produces 345 W in warm expander 1 and 2, but only 55 W in expander 3 and 4. Six times more cooling could be obtained by allocating helium from EX3 and EX4 into EX1 and EX2. Therefore, in the initial stage of cooldown, both sets of EX1 and EX2 should be operated at their full flow capacity to obtain the maximum refrigeration for that temperature.

As the temperature decreases, more helium will be allocated into EX3 and EX4 accompanied with a reduction in EX1 and EX2. At this stage, the refrigeration could only be enhanced by using redundant EX3, EX4 and EX5 in series with the magnets.

RESULTS AND DISCUSSION

Cooldown calculations have been made for load return temperatures of 300 K, 175 K, 85 K, 40 K, 25 K, 20 K, 15 K, 12 K and 10

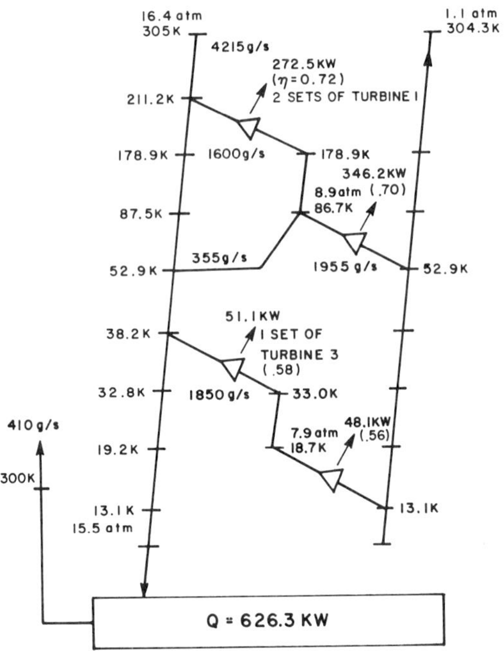

Fig. 5. T-S diagram for initial stage of cooldown.

K. Apparent liquefaction rates between 80 g/s and 400 g/s are examined. A typical T-S (temperature-entropy) diagram for cooldown operation is given in Fig. 5. At high temperature (> 85 K) a high ratio of refrigeration to liquefaction is anticipated due to the high specific heat of iron and the low density of helium in this region. As the temperature decreases, this ratio decreases correspondingly. For a load temperature below 40 K, the apparent liquefaction rate is estimated to be between 150 g/s and 200 g/s for the present analysis. The apparent liquefaction rate versus refrigeration rate for $T_{ret} < 40$ K is plotted on Fig. 6. The capacity curve for the refrigerator shifts to the left and tends to the steady-state capacity as temperature decreases.

CONCLUSION

The ISABELLE refrigerator with its redundant expanders properly used achieves cooldown capacity well beyond its steady-state capacity. During the initial stage of cooldown, by utilizing the two warmest redundant expanders the refrigerator can generate 2.3 times steady-state refrigeration above 50 K, which is comparable to a typical liquid nitrogen precooled refrigerator. As the

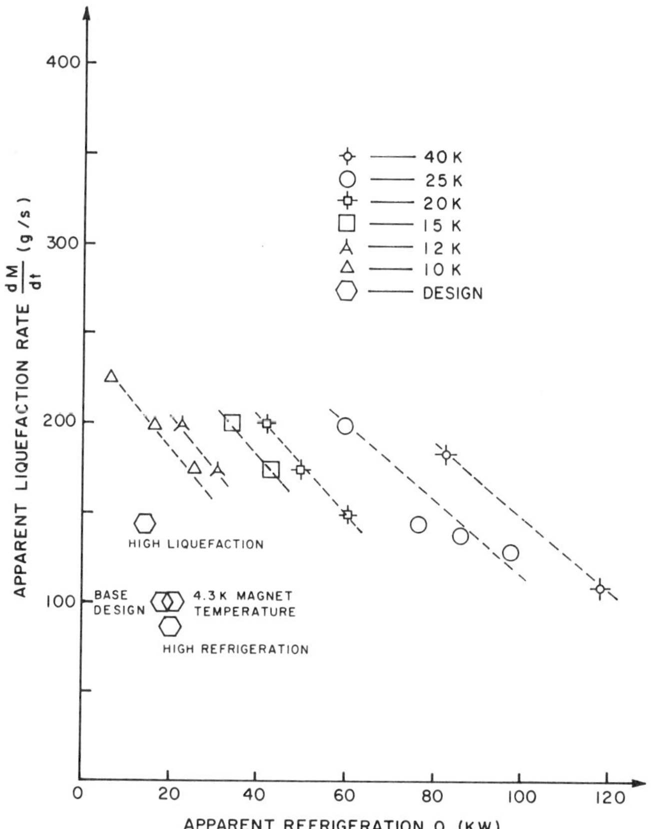

Fig. 6. Refrigerator characteristic at various return temperature.

temperature decreases, a significant portion of the refrigerator is generated through use of redundant expanders 3 and 4 and expander 5 in series with the magnets. After the refrigerator goes into its steady-state operation, the capacity of the refrigerator approaches the steady-state value. The cooldown rate at this stage relies on the design safety margin, which for the ISABELLE refrigerator is 50%.

REFERENCES

1. D.P. Brown, et al., Cycle Design for the ISABELLE Helium Refrigerator, in "Advances in Cryogenic Engineering, Vol. 27", Plenum Press, New York (1982).
2. W.M. Kays and A.L. London, "Compact Heat Exchangers,"The National Press, Palo Alto, Calif., (1955).

"MAGCOOL"—THE PRODUCTION COOLING FACILITY FOR ISABELLE MAGNETS

J. A. Bamberger, M. Afrashteh, D. P. Brown, W. J. Schneider,
J. H. Sondericker, and K. C. Wu

*Brookhaven National Laboratory**
Upton, New York

INTRODUCTION

The ISABELLE proton accelerator uses over one thousand superconducting magnets to guide the particle beams in two circular
rings, 3.8 km in circumference. Prior to their installation in
the tunnel all magnets must be tested and measured at their 3.8 K
operating temperature. This paper describes the refrigeration
system, called "MAGCOOL," to accomplish this task.

SYSTEM FUNCTIONS

The magnet test procedure involves five main tasks which must
be performed in sequence, each with distinct operating requirements for a set period of time. The MAGCOOL system provides five
test stands, each of which is capable of performing any of the
five basic tasks. The normal duty cycle is 24 hours per task;
therefore with one magnet under test at each of the stands the
production test rate is five magnets per week for a five day
operation. The system block diagram, Fig. 1, schematically shows
the arrangement of the test stands. For illustration, only one of
the five tasks is shown for each of the test stands. A system of
headers and remotely operable valves allows any of the basic
functions to be performed at any of the test stands. A process
control computer is used to monitor and control these tasks which
are described below.

---

*Operated by Associated Universities, Inc. under contract with the
U.S. Department of Energy.

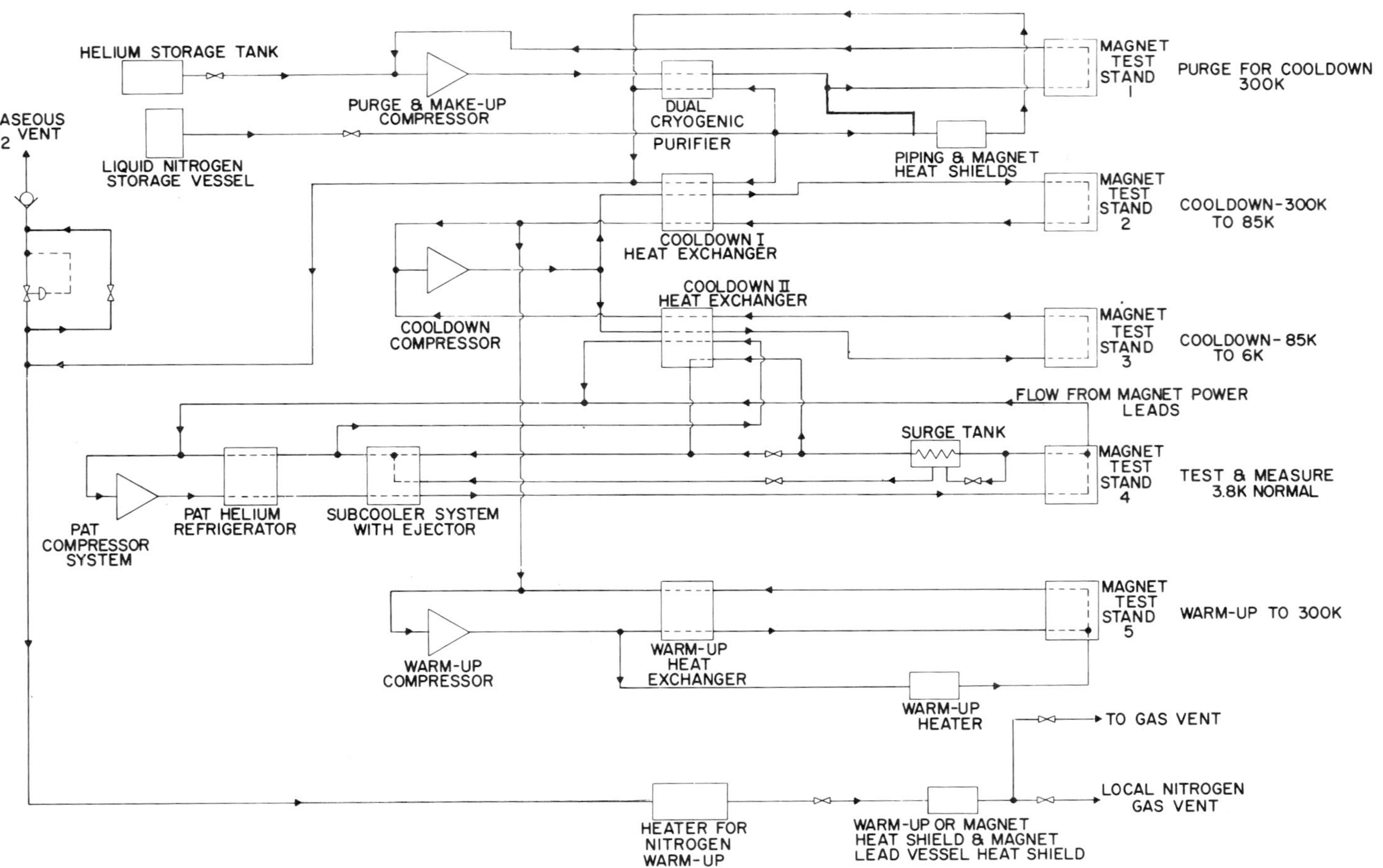

Fig. 1. MAGCOOL System Block Diagram.

PUMP AND PURGE PROCEDURE

In the ISABELLE ring 45 magnets will be cooled in series with flowing supercritical helium. All contaminants must be removed prior to the initial cooldown and test in MAGCOOL. The tested magnets must be warmed up to room temperature and delivered in clean, purged condition ready for installation in the ISABELLE ring.

After initial connection to the test stand, successive evacuations to below 50 mtorr are performed followed by backfill with pure helium. Helium gas is then circulated using forced flow from the screw compressor through one of the dual cryogenic purifiers and then through the magnet at the test stand. The purifier is designed for a flow rate of 15 g/s at a pressure of 8 atm. The capacity of the purifier is based on an assumed contamination level of 50 ppm of water and air and a duty cycle of 24 hours per dual bed, including reactivation time. An automatic fill system for the liquid nitrogen cooled charcoal bed is provided. Reactivation of the purifier is by means of nitrogen gas which is warmed by an external electrical heater.

MAGCOOL provides the following equipment for the pump and purge procedure:

- Two-stage mechanical vacuum pumps for evacuation
- Pure helium gas for purging the evacuated system
- Circulating gas system with screw compressor
- Four stage oil removal at the screw compressor outlet
- Dual cryogenic purifiers

COOLDOWN I

During Cooldown I the magnets are cooled from room temperature to 85 K. This is accomplished with a counterflow heat exchanger followed by a liquid nitrogen cooled plate-fin heat exchanger. The weight of a dipole magnet is about 6500 kg requiring removal of about 500 megajoules to bring this material, mostly iron and stainless steel, to 85 K. The cooldown rate for this process must be limited in order to avoid excessive stresses which could result from temperature gradients beyond those permitted. The refrigerant and magnet ΔT's are monitored and the helium flow rate and temperature are automatically regulated to achieve an optimum cooling rate without exceeding allowable temperature gradients. A screw compressor, similar to the one used for purging is used to circulate the helium coolant through the heat exchanger and magnet systems. The maximum available flow rate is 65 g/s with a return pressure of 8 atm and a supply pressure of 12 atm.

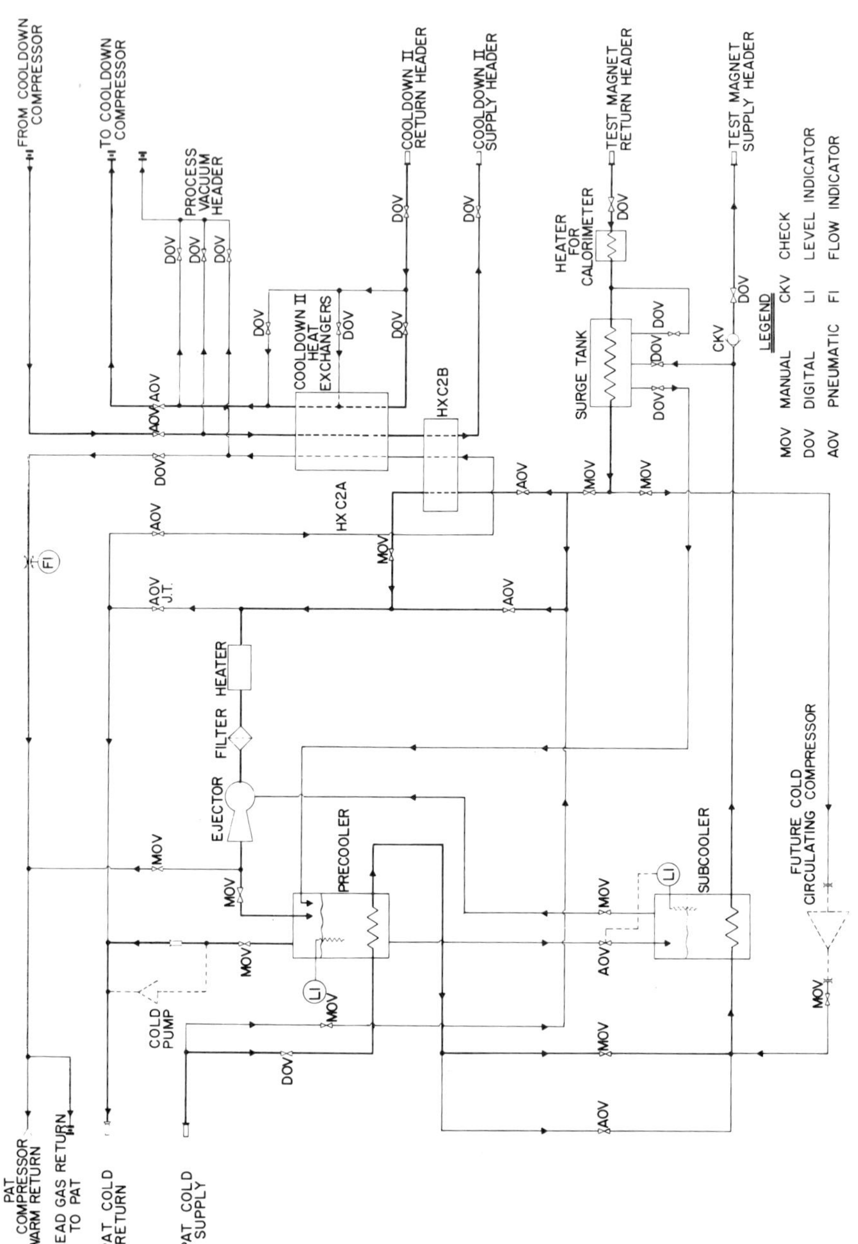

Fig. 2. MAGCOOL Low Temperature Flow Schematic.

Liquid nitrogen is circulated through the magnet heat shield during the Cooldown I operation cooling the aluminum shield to 85 K and bringing the multi-layer insulation to the equilibrium gradient.

COOLDOWN II

During this operating phase the magnets are cooled from 85 K to 6 K, requiring the removal of about 25 megajoules per magnet. An existing Claude cycle helium refrigerator ("PAT") is used for the MAGCOOL operations below 85 K. This refrigerator, with about 1 kW capacity at 4.5 K, has dual expanders, uses liquid nitrogen precooling, and is equipped with oil flooded screw compressors driven by electric motors totaling 550 kW.

Figure 2 shows the flow schematic for MAGCOOL operations below 85 K. The "PAT" refrigerator itself is not shown, but its supply and return lines are included and identified as "PAT cold supply and return."

Heat exchangers HX C2A and HX C2B are used for Cooldown II. A portion of the cold high pressure return stream from the "Test and Measure" circuit is used in cooldown heat-exchanger HX C2A. This three-pass heat exchanger cools a 15 g/s circulating side-stream from the cooldown compressor which then circulates through the Cooldown II headers to the magnet and then back through the heat exchanger. The lower heat exchanger, HX C2B, provides additional cooling capacity during magnet "training".

When the magnet has reached equilibrium conditions near the end of the Cooldown II procedure, a total heat load determination is made on each magnet by measuring the temperature rise across the magnet with germanium resistance thermometers and determining the flow rate from calibrated flow meters provided in the circulating gas system.

TEST AND MEASURE

The refrigeration circuit for the Test and Measure (T.M.) circuit is also shown in Fig. 2. During this sequence the magnet is brought to the maximum design temperature of 3.8 K and is then energized at this temperature. The magnet is first "trained" to achieve the required field strength by successive quenches.

During this training phase part of the stored magnetic energy is released to the supercritical helium refrigeration system. When the magnet has achieved design field strength, magnetic measurements are performed during steady-state conditions. These measurements are made for two reasons: 1) to determine whether the magnet performance meets the stringent required tolerances; and 2) to precisely establish the location of the magnetic field relative to the warm vacuum tank so that the magnet can later be accurately positioned and surveyed into the ISABELLE ring.

A liquid helium precooler and subcooler are used to cool the supercritical helium gas which is circulated to the magnet via the T.M. circuit headers. The subcooler is pumped by an ejector which uses the high pressure return gas stream, expanding through a nozzle.[1] Two modes of operation are indicated on Fig.2: In Mode "A" (shown bold) the J-T flow goes directly to the load at about 15 atm; in Mode "B" a cold circulating pump is provided which sends supercritical helium in a closed loop through the subcooler bath, to the magnet and back. Mode "A" is simpler, but the flow rate is limited to the refrigerator J-T flow (approximately 55 g/s). The circulating pump for Mode "B" is designed to achieve the ISABELLE design flow rate of 170 g/s, at a pressure of 5.0 atm.[2]

During the training sequence of the magnet about 1 megajoule of energy is released when the magnet quenches. A 600 liter surge tank is provided to store the cold gas evolved when the magnet is quenched. An automatic system of valves and pressure sensors allows this gas to return to the system through HX C2B making use of available sensible cooling capacity from the stored quench gas.

A small protion of the supercritical helium flow to the magnet is used to cool the current leads for the magnet; this gas flow is automatically controlled and returned warm to the compressor system.[3]

WARM-UP

During warm-up to room temperature the same temperature gradient limits as discussed previously for Cooldown I apply. For warm-up of the magnet, a third screw compressor is used to circulate helium gas through tempering heat-exchangers to the magnet and back. The heat exchangers are set up symmetrically for Cooldown I and for Warm-up: Heat exchangers previously used for a Cooldown I cycle are switched for Warm-up on the next cycle, and vice versa. Thus "cold" heat-exchangers from a previous cooldown are used for warm-up, providing tempered gas and avoiding excessive temperature gradients across a magnet during the warm-up cycle to room temperature.

The magnet heat shield is warmed by nitrogen gas which can be heated by an external electrical heater. The nitrogen gas is purged out of the shield circuit when the magnet warm-up is completed and replaced with pure helium.

INSTRUMENTATION

The complexity of MAGCOOL with its large number of process variables and control loops requires a sophisticated control system. Cryogenic systems are often complex because a large number of subsystems must operate in parallel to produce the desired result. The MAGCOOL system, in addition to subsystem control, must regulate, control, and monitor the five magnet test stands through asynchronous cryogenic cycles. If a conventional analog control system were utilized, many operators would be required making the system very costly and difficult to operate.

For simpler operation and higher reliability a commercially available process control computer system was selected to perform the control tasks. The computer control system was specified, built and purchased to interface the MAGCOOL instrumentation. The input/output gear drives 178 valves, controls 38 direct digital control loops and reads over 100 assorted analog inputs. Conventional analog stations are not needed since all display and control is accomplished through a color graphics terminal. The commercial package includes all necessary system software; "CRISP" process control language for applications programming; cursor controlled CRT display building from the operators keyboard; scrolled trending; alarm and scale routines; battery backup of main memory; FORTRAN and BASIC high level programming languages.

The one area where commercial process control capability deviates from cryogenic needs is in temperature measurement and display. Since the normally supplied 12 bit analogue to digital conversion is not sufficient for high accuracy thermometry, this task was handled separately. A high resolution low drift microprocessor was designed to linearize and convert temperature inputs to engineering units. Temperature data is then transmitted to the process computer via a serial data link.

The process control computer has been tested for the past several months. During this period the system has been completely checked out and the beginnings of MAGCOOL applications software installed. Future operators of the system have been getting some initial training at the keyboard and have enthusiastically accepted this more modern form of control.

ACKNOWLEDGEMENTS

Thanks to the ISABELLE design group, especially J. J. Agostine and R. S. Meier for their assistance, and also to the ISA Cryogenic Instrumentation group.

The magnet electrical cryogenic interconnections are provided by the ISA Magnet Division, under the direction of D. W. Gardner.

Many concepts and ideas have evolved from earlier forced flow tests performed by the ISA Cryogenic Group under H. Hildebrand's direction.

The MAGCOOL system is fabricated to BNL specifications[4] by CVI, Inc., Columbus, Ohio. The ejector design and fabrication is by T. R. Strobridge, National Bureau of Standards, Boulder, Colorado.

The process control system is built to BNL specifications[5] by Anaconda Advanced Technology, Inc., Dublin, Ohio using a Digital Equipment Company LSI-11/23 computer.

REFERENCES

1. J.A. Rietdijk, The Expansion Ejector, A New Device for
    Liquefaction and Refrigeration at 4 K and Lower, in:
    "Liquid Helium Technology, Proc. of the Int. Inst. of
    Refrig., Commission I", Pergamon Press, Oxford (1966),
    p.241.
2. D.P. Brown, et al., Design of 24.8 kW, 3.8 K Cryogenic System
    for ISABELLE, in "Advances in Cryogenic Engineering, Vol.
    27", Plenum Press, New York (1982).
3. D.P. Brown and W.J. Schneider, Magnet Leads for the First
    Cell, in "Advances in Cryogenic Engineering, Vol. 25",
    Plenum Press, New York (1980).
4. J.A. Bamberger, Specification for ISABELLE Magnet Production
    Cooling Facility MAGCOOL, Brookhaven National Laboratory
    ISA Specification No. 35 (October 1979).
5. J.H. Sondericker, Specification for the ISABELLE Magnet
    Production Cooling Facility Process Control Computer,
    Brookhaven National Laboratory ISA Specification No. 44
    (January, 1980).

SIMULATION OF THE ENERGY SAVER
REFRIGERATION SYSTEM

H. R. Barton, Jr., J. E. Nicholls, and G. T. Mulholland

*Fermi National Accelerator Laboratory**
Batavia, Illinois

INTRODUCTION

The helium refrigeration for the Fermilab Energy Saver is
supplied by a Central Helium Liquefier and 24 "Satellite"
Refrigerators installed over a 1-1/4 square mile area.[1] An inter-
active, software simulator has been developed to calculate the
refrigeration available from the cryogenic system over a wide
range of operating conditions. The refrigeration system simulator
incorporates models of the components which have been developed to
quantitatively describe changes in system performance. The simulator output is presented in a real-time display which has been
used to search for the optimal operating conditions of the Satel-
lite-Central system, to examine the effect of an extended range of
operating parameters and to identify equipment modifications which
would improve the system performance.

THE METHOD FOR SIMULATION CALCULATIONS

A cryogenic refrigerator consists of a set of counter-flow
heat exchangers, each of which must conserve energy. For a
refrigerator containing N heat exchangers, there are N simultaneous equations representing conservation of energy. Since these
equations are linear in enthalpy, it is natural to solve the
problem by calculating the enthalpy (not the temperature) at each
of the process points. This choice of variables avoids any difficulties due to the variation of specific heat with temperature.

---

*Operated by Universities Research Association, Inc., under
contract with the U.S. Department of Energy.

The number of process points is 2N, so N additional equations are required. The N additional equations are obtained by requiring that the overall heat transfer coefficient times area, UA, of each of the N heat exchangers have the correct value. This means performing N integrations to solve the first-order differential equation for heat transfer for each of the heat exchangers.

In doing the analysis of the refrigerator, the correct UA values are obtained by successively correcting the choice of enthalpies at N process points and performing trial integrations for each of the heat exchangers until the desired UA values are obtained. At each step of this iterative procedure, a matrix of partial derivatives $\partial(UA)j/\partial H_i$ is calculated by the finite difference method while satisfying the N energy equations. If $\delta(UA)j$ is the desired adjustment in the UA value for heat exchanger number j, then the first order adjustments are

$$\delta(UA)j = \sum_i \frac{\partial(UA)j}{\partial H_i} \delta H_i$$

In order to extract the enthalpy corrections δH_i, the NxN matrix of partial derivatives is inverted and multiplied times $\delta(UA)j$. Since the energy conservation equations are used at each iteration, the energy constraints are always satisfied.

A modification of this procedure is required if the capacity rate (specific heat times mass flow) of the minimum stream is so small that the counter-flowing streams approach the same temperature. In this case, the full size of the heat exchanger cannot be utilized.

The helium flow to an expander in the cycle comes from one of the process points mentioned above. The efficiency of the expander is defined as the actual enthalpy decrease produced by the expander divided by the enthalpy decrease which would result from an isentropic expansion of helium at the inlet conditions. The enthalpy at the exit of the expander is calculated by using the specified expander efficiency and pressure ratio. In cycles where the helium discharged from the expander is mixed with another helium stream, the enthalpy of the mixture is calculated using a linear mixing formula. This procedure is used at each iteration of the simulation so that the effect of the expanders is always accurately modeled.

COMPUTER SIMULATION PROGRAMS

Simulation programs have been written for the Central Liquefier and the Satellite Refrigerator. The output of the simulator is displayed on the screen of a graphics terminal. The operator of

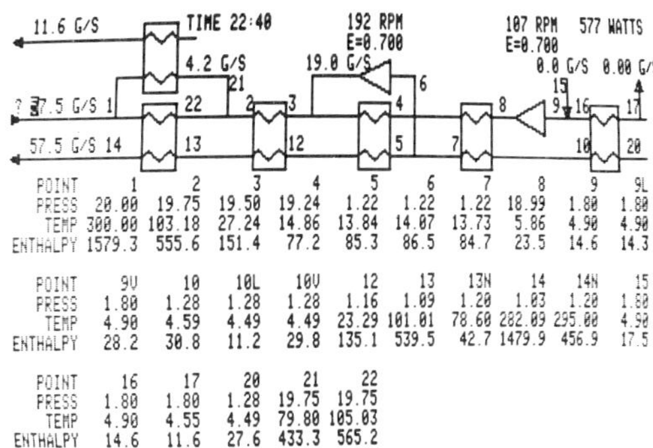


```
POINT        1       2       3       4       5       6       7       8       9      9L
PRESS     20.00   19.75   19.50   19.24    1.22    1.22    1.22   18.99    1.80    1.80
TEMP     300.00  103.18   27.24   14.86   13.84   14.07   13.73    5.86    4.90    4.90
ENTHALPY  1579.3   555.6   151.4    77.2    85.3    86.5    84.7    23.5    14.6    14.3

POINT       9U      10     10L     10U      12      13     13N      14     14N      15
PRESS      1.80    1.28    1.28    1.28    1.16    1.09    1.20    1.03    1.20    1.80
TEMP       4.90    4.59    4.49    4.49   23.29  101.01   78.60  282.09  295.00    4.90
ENTHALPY   28.2    30.8    11.2    29.8   135.1   539.5    42.7  1479.9   456.9    17.5

POINT       16      17      20      21      22
PRESS      1.80    1.80    1.28   19.75   19.75
TEMP       4.90    4.55    4.49   79.80  105.03
ENTHALPY   14.6    11.6    27.6   433.3   565.2
```

Fig. 1. Sample simulator display for the Satellite Refrigerator.

the simulator enters input at the keyboard of this terminal. A sample of this simulator display is shown in Fig. 1. This display was designed to include a flow diagram with numbered process points. Pressure, temperature and enthalpy for the process points are displayed. The flow at each process point is determined by the operator, who types numbers directly on the display of the flow diagram.

Since the UA values depend on flow, the simulator calculates the effective UA values for the heat exchangers; for finned tubing heat exchangers in the Satellite, $UA \sim f^{0.8}$. Approximately five iterations on the heat balance are required to get acceptable UA values. This takes several seconds on the Cyber 175 computer and the resulting operating conditions are displayed on the operator's terminal immediately afterward.

COMPARISON WITH CENTRAL LIQUEFIER DATA

The simulation model contains the equations which are fundamental requirements for any thermodynamic system. During testing of the Central Liquefier in April, 1981, considerable data were accumulated which could be used as an experimental check of the model. A chi-squared analysis of 25 sets of process data consisting of simultaneous measurements recorded from 16 process points in the liquefier was used to establish the accuracy of the simulation equations. The data for this analysis were taken while the plant was liquefying 1200 L/h of helium without nitrogen precooling.

The flow diagram for the Central Liquefier shown in Fig. 2 is a sample of the simulator output. The values shown are for

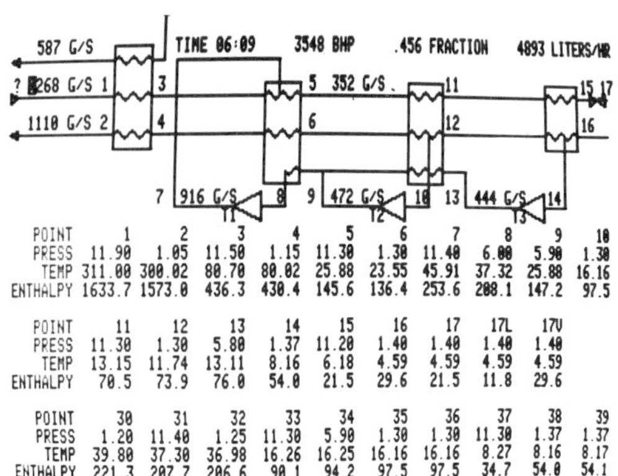

POINT	1	2	3	4	5	6	7	8	9	10
PRESS	11.90	1.05	11.50	1.15	11.30	1.30	11.40	6.00	5.90	1.30
TEMP	311.00	300.02	80.70	80.02	25.88	23.55	45.91	37.32	25.88	16.16
ENTHALPY	1633.7	1573.0	436.3	430.4	145.6	136.4	253.6	208.1	147.2	97.5

POINT	11	12	13	14	15	16	17	17L	17V
PRESS	11.30	1.30	5.80	1.37	11.20	1.40	1.40	1.40	1.40
TEMP	13.15	11.74	13.11	8.16	6.18	4.59	4.59	4.59	4.59
ENTHALPY	70.5	73.9	76.0	54.0	21.5	29.6	21.5	11.8	29.6

POINT	30	31	32	33	34	35	36	37	38	39
PRESS	1.20	11.40	1.25	11.30	5.90	1.30	1.30	11.30	1.37	1.37
TEMP	39.80	37.30	36.98	16.26	16.25	16.16	16.16	8.27	8.16	8.17
ENTHALPY	221.3	207.7	206.6	90.1	94.2	97.5	97.5	34.7	54.0	54.1

Fig. 2.　Sample simulator display for the Central Liquefier.

full-capacity operation of the Liquefier. The heat exchangers for
the Central Liquefier are grouped into four modules and are of
aluminum, finned-plate construction.[2] Because the process instrumentation is installed externally to these four modules, not all
process points are measured. The measurements taken do exceed the
minimum number required so that a seven constraint fit to the data
is possible. A fit is necessary because the raw data do not
conserve energy or helium due to small experimental errors in the
measurements. The flow measurements are made with venturis and
have an accuracy of $\pm 4\%$ of the compressor discharge flow. The
enthalpy measurements were made using strain gauge, pressure
transducers and silicon diode temperature transducers so that the
enthalpy has an accuracy of $\pm 2J/g$. With these measurement uncertainties, the data satisfy the seven constraint equations with
chi-squared values of 7.0 or lower in all but a few instances.
The enthalpy measurements at 16 process points and the flow data
at 6 points are used in the least-squares fit to the model.

This work verifies that the equations used in the simulator
are an accurate description of the liquefier. Additional data
taken while operating over a wide range of compressor flows will
be needed to verify the scaling dependence of heat exchanger UA on
flow.

The Central Liquefier is supplied with 12 atm helium by two
reciprocating compressors with 600 g/s capacity each and by the
screw compressors of the Satellites. If the flow to the plant is
increased 40% above its design value of 1268 g/s by using
additional compressors, the capacity of the plant could be
increased 37% with the present heat exchangers. The simulator has
been used to study this upgrade of the Central Liquefier. Three

replacement turboexpanders would be required to handle 1282, 660 and 622 g/s of helium flow. If these expanders have efficiencies of 82.5, 82 and 81%, they would operate with exit temperatures of 37.7, 16.4 and 8.2 K.

RESULTS FOR THE SATELLITE REFRIGERATOR

During normal accelerator running, the Satellite Refrigerator will be operated without nitrogen precooling and with the higher-temperature expander off. In this mode the heat exchangers of the Satellite Refrigerator are unbalanced by liquid helium flow obtained from the Central Liquefier. If f is the high pressure flow and F is the low pressure flow, the flow imbalance of the Satellite heat exchangers is $\beta = 1 - f/F$. At a flow of 40 g/s, the total available UA of a Satellite is 14 kW/K. The high pressure stream is expanded from 20 atm to 1.8 atm using the wet engine of the Satellite.

Figure 3 shows the calculated refrigeration available at the magnets using the simulator model for the Satellite Refrigerator. Lines labeled with values of β represent proportional changes in flow. The curvature of these lines is due to decrease in effectiveness of the heat exchangers as flow is increased.

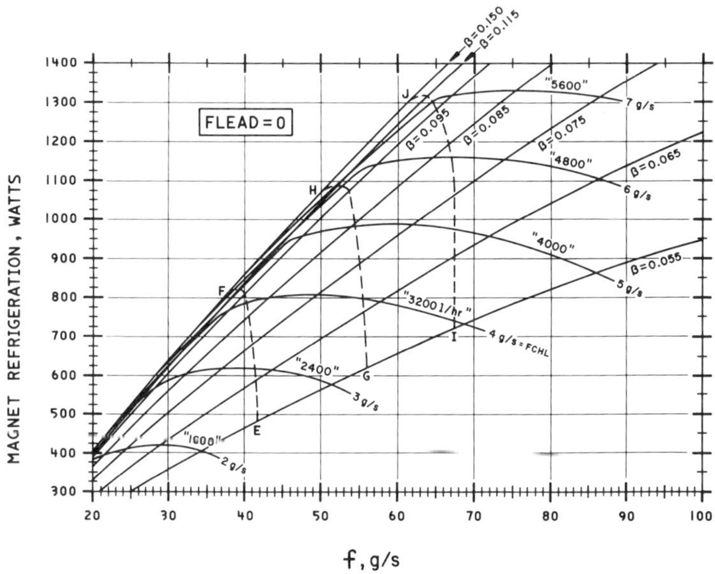

Fig. 3. Refrigeration provided by the Satellite Refrigerator as a function of compressor and Central Liquefier flows.

Operation of the Satellite Refrigerator at the lowest mass
flow results in the lowest two-phase temperature at the magnets.
The shell-side pressure drop in the model is proportional to
$F^{1.8}$. Figure 3 shows dashed isotherms; E-F is 4.4 K, G-H is 4.5 K
and I-J is 4.6 K.

There are operating modes when the Satellite Refrigerator
must function as a helium liquefier. The model has been used to
calculate the reduction in refrigeration which occurs when
liquefaction is provided. This effect is shown in Fig. 4. During
normal, steady-state operation, each Satellite Refrigerator must
supply 0.7 g/s of liquid helium on the average to cool the magnet
power leads. During cooldown of Energy Saver magnets and quench
recovery, the liquefaction requirements are greater. Non-zero
liquefaction increases the demand for helium from the Central
Liquefier.

SOME APPLICATIONS OF SIMULATION

The simulator can be used to quantitatively study abnormal
operating modes of a system of cryogenic hardware and is a powerful tool for predicting the refrigeration margin available from
the system when components are operating with reduced
performance. The simulator has been used to make design decisions

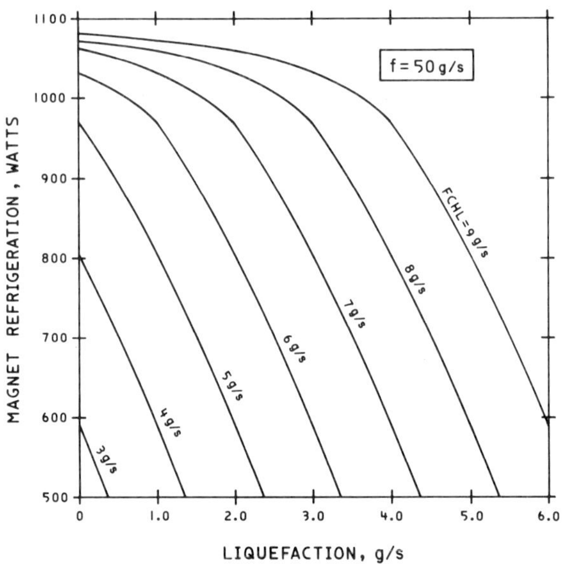

Fig. 4. Effect on magnet refrigeration
when liquefaction is required.

which will boost the cryogenic performance of the refrigeration
system. The output of the simulator serves as a guide to the
operator of the cryogenic system. Changes in the operating mode
can be tested on the simulator beforehand to determine the
magnitude of the effects the operator of the cryogenic system will
encounter.

More precise detection of out-of-limits transducer readings
is possible when the data are tested against comparable simulation
output rather than the minimum and maximum operating extremes.
Because the transducers in a cryogenic system supply information
redundant to the minimum set required by the simulator, the reading from a failed transducer can be replaced by a number which has
been calculated by the simulator.

The simulator can be incorporated in the control system of
the cryogenic system. This will allow the controls to anticipate
effects on the system before they become measurable upsets in the
process. A feedforward control system requires a model of the
process in order to predict the behavior of the cryogenic
system. Since the time response of cryogenic systems is very
slow, this type of model-driven control system would improve
stability of operation.

ACKNOWLEDGEMENTS

The simulators use subroutines written at the Cryogenics
Division of the National Bureau of Standards to calculate helium
properties, and several subroutines obtained from Brookhaven
National Laboratory.[3]

REFERENCES

1. P.C. VanderArend, Helium Refrigeration System for Fermilab
       Energy Doubler, in: "Advances in Cryogenic Engineering,
       Vol. 23", Plenum Press, New York, (1977), p.420.
2. W.M. Toscano and F.J. Kudirka, Thermodynamic and Mechanical
       Design of the FNAL Central Helium Liquefier, in: "Advances
       in Cryogenic Engineering, Vol. 23", Plenum Press, New York
       (1977), p.456.
3. R. King, Determination of the Performance and Cost of the
       Model 4000 Helium Liquefier, ISABELLE Division Technical
       Note No. 19, Brookhaven National Laboratory, July 23, 1976.

FERMILAB TEVATRON FIVE REFRIGERATOR SYSTEM TESTS

C. Rode, R. Ferry, M. Leininger, J. Makara, J. Misek,
D. Mizicko, D. Richied, and J. Theilacker

*Fermi National Accelerator Laboratory**
Batavia, Illinois

INTRODUCTION

During the latter half of 1980 the Fermilab refrigeration
test program switched from stand-alone refrigerator tests to a
subsystem test, which now consists of two compressor buildings
with four, two stage units each, five refrigerators, and the
Central Helium Liquefier (CHL). Each of the refrigerators cools
two 125 m strings of magnets.

TEVATRON REFRIGERATION SYSTEM

Description

The Fermilab superconducting accelerator is cooled with a
5000 L/hr central helium liquefier (CHL) coupled with 24 satellite
refrigerators[1] (Fig. 1). High pressure helium (20 atm) supplied
compressors is delivered to the refrigerators through a common
header. Each compressor building has four two-stage screw compressors with a 58 g/s throughput. High pressure helium is cooled
in the refrigerator through a series of four heat exchangers by
cold, low pressure return gas from the magnets. Cold, high pressure helium is then expanded to subcooled liquid with a reciprocating liquid expansion engine and sent to the magnet string.

---

*Operated by Universities Research Association, Inc. under
contract with the U.S. Department of Energy.

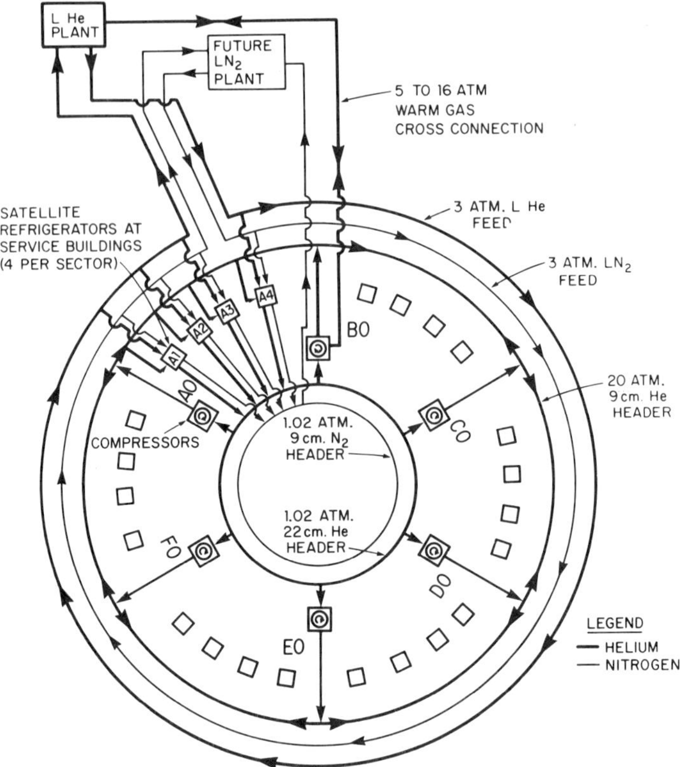

Fig. 1. Layout of the Refrigeration System.

Supplemental low pressure cooling can be supplied by a LN_2 pre-cooler, a reciprocating gas expansion engine, and/or LHe supplied by the CHL. The LHe supplied by the CHL is sent directly to the magnet string and returns through the low pressure side of the heat exchangers. This resulting flow imbalance between the shell and tube side of the heat exchanger allows the satellite refrigerator to operate as an "amplifier" with a flow gain of 12. Local microprocessors (μp) independently control the subsystems through a series of servo control loops.[2] A host computer oversees the 24 refrigerator and six compressor μp's allowing for a master control, alarm, and data acquisition station.[3]

Compressor Control Loops

The satellite refrigerator compressor subsystem consists of 24 two-stage screw compressors, located in six compressor build-ings located around the ring at A0 through F0. All compressors and refrigerators are interconnected through a common 22 cm suc-tion header and 9 cm discharge header.

Helium inventory control is achieved through a pair of control valves located at the BO compressor building, which are cross-connected to the CHL compressor discharge and a 115 m^3 helium buffer tank. One control loop regulates at 1.068 atm of suction pressure, dumping excess helium from the discharge of the satellite compressors directly into the discharge of the CHL compressor, which operates at a lower pressure. In the satellite mode at full capacity this is 10% of the satellite compressor output, but may vary from 0 to 20% as the magnet ramp is turned off or on due to the density changes from 4.5 to 4.9 K in the coil cryostats.

A second loop supplies makeup helium from the buffer when the suction falls below 1.034 atm. This would be the case if the CHL was off and also when the ramp trips off without a quench. The buffer has an active range of 1600 liquid liters equivalent and is backed up by tube trailers and (in the final system) the main CHL dewar.

Regulation of the suction pressure is crucial to the system operation. Inadequate makeup flow could lead to drawing the suction header subatmospheric and inevitably pulling in contaminates (air and water vapor). To prevent this, two independent makeup valves and control loops are used. The main loop will be controlled by the μp with an electric actuator while the backup loop will be pneumatically actuated in the event of a power or μp failure.

Control of the compressor capacity is achieved through the first stage suction slide valve. This control loop has two possible modes of operation: discharge or suction pressure regulating (Fig. 2). The normal mode for the loop is discharge pressure regulated. Suction regulation is used in compressor buildings where the suction header has become isolated (valved off) from the BO inventory control loops. Inventory control is then accomplished through the discharge header with the BO and adjacent connected buildings regulating the discharge pressure.

Each group of four compressors has a bypass control loop which is also regulated in the same two modes. Should the discharge pressure rise above the regulation point (20 atm), the loop bypasses a portion of the high pressure gas back to the compressor suction. This loop would normally only open if the compressors were fully unloaded and still had too much throughput.

We will normally be running one or two idling compressors, which can go to full load in 30 s for system stability. This additional compressor capacity is used to handle the surge of suction gas in the event of a magnet quench or ramp change.

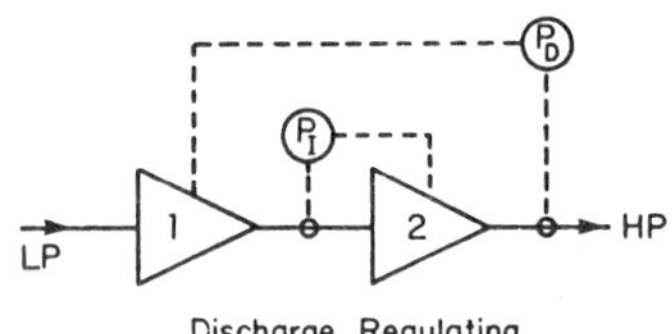

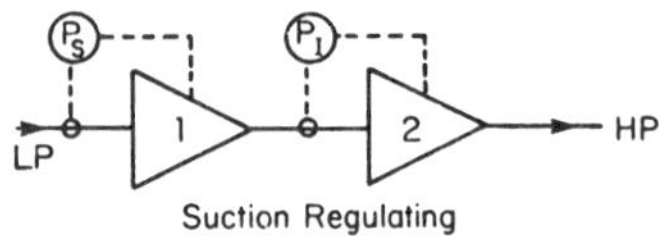

Fig. 2. Compressor Control Loops.

Compressor efficiency is maximized by regulating the second stage slider (Fig.2). Previous tests have shown that the motor horsepower required for our compressors is minimized with an interstage pressure of 3.5 atm. Regulating the second stage slider to achieve this interstage pressure maximizes the compressor efficiency as well as the thrust bearing life.

Refrigerator Control Loops

The satellite refrigerator subsystem consists of 24 refrigerators, each cooling two 125 m strings of magnets. Twelve control loops tune the refrigerator to one of its four operating modes: satellite, liquefier, refrigerator, or stand-by.[1] Under normal operating conditions, the refrigerator will be operating in the satellite mode, producing 966 W of 4.6 K refrigeration with the aid of the CHL LHe flow. In the other three modes, the refrigerator is operated independently of the CHL. LN_2 is consumed by the #1 heat exchanger (Fig. 3) to precool high pressure helium and the gas expansion engine is operated to supply supplemental shell side cooling. Design capacities for the liquefier and refrigerator modes are 126 L/h of LHe and 623 W of 4.6 K cooling, respectively. The stand-by mode is a combination of refrigeration and liquefaction used to keep the magnets cooled without the aid of the CHL. This mode is designed to produce 490 W and 26.6 L/h of LHe.

Three of the refrigerator loops control the flow of liquid nitrogen. LN_2 flow to the precooler (Fig. 3.) is controlled by valve V_1. The valve is temperature servoed by a platinum resister, T_1, located at the precooler vent. LN_2 to the upstream and downstream magnet shields (Fig. 4.) is also temperature servoed by a vapor pressure thermometer (VPT), T_7, to maintain 85 K gas at the vent.

As shown in Fig. 3. helium flow into the refrigerator is split between heat exchangers #1 and #2 by control valves V_2 and V_3. Both valves are temperature servoed by VPT, T_2. Push-pull regulation of the valves maintains the high pressure inlet to heat exchanger #3 at 81 K without wasting LN_2. For instance, when T_2 falls below the set point, V_2 will throttle closed and V_3 will open in order to take advantage of the excess shell side cooling.

The middle of the heat exchanger column is maintained at a set temperature by one of two control loops. Either the gas engine or the flow imbalance resulting from the LHe flow from the CHL (V_6) supplies the supplemental shell side cooling. This additional flow, controlled by V_6 or the gas engine speed, is servoed by a VPT, T_6, with a set point of 25 K. The LHe control loop, V_6, is only active during the satellite mode. For the other three modes the gas engine control loop is active.

Single phase helium pressure is regulated by three tightly coupled servo loops. The liquid engine speed, cold box J-T valve (V_4), and a coldbox bypass valve (V_5) are all servoed by pressure sensor P_4. The liquid engine speed control pressurizes the magnets and is servoed to maintain the pressure at 1.8 atm. As a backup to the liquid expander, the coldbox J-T valve servoes to maintain 1.7 atm. A magnet bypass valve, V_5, is used as an internal pressure clipping valve (or relief) and is set at 2.0 atm; this valve is mainly active during magnet cooldown.

The flow of LHe through the magnets is controlled by a J-T valve, V_8, located at the far end of each magnet string (Fig. 4.). This valve throttles subcooled liquid at 1.8 atm to a two phase mixture at 1.2 atm. The valve is servoed to maintain 0.1 K of superheat at the low pressure return to the heat exchanger. A

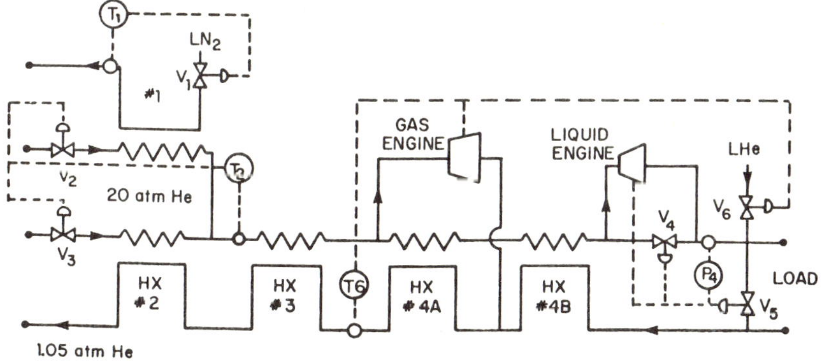

Fig. 3. Refrigerator Control Loops.

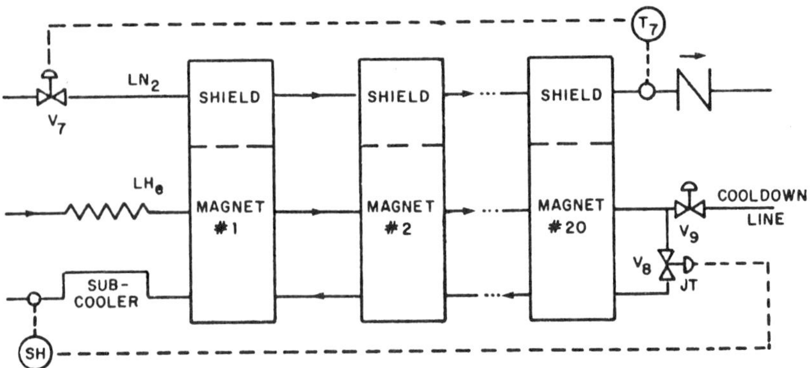

Fig. 4. Magnet Control Loops (Two Per Refrigerator).

differential pressure cell across the return pressure tap and VPT
serves as the superheat sensor.

Mobile Purifier

Purification of helium can be a costly and time consuming
task. Rather than build 24 stationary purifiers, we have built
two mobile units consisting of three trailers each. The first
trailer contains two vacuum pumps, two dehydrators, and a 77 K
charcoal bed with a heat exchanger. The second trailer contains a
small helium compressor while the third contains contamination
detection equipment, including a gas chromatograph.

Auxiliary piping on the refrigerator and compressors permits
us to clean up individual elements. A heat exchanger or expander
can be derimed and the refrigerator restarted in a few hours while
the balance of the system operates unaffected. Individual compressors, with its oil removal system, can be decontaminated in
one day.

FIVE REFRIGERATOR TEST

Test Using Two Compressor Buildings

The configuration for this test is given in Fig. 5, with the
two compressor buildings located 1000 m apart. The A1, A2, A3,
A4, and B1 refrigerators ran with independent schedules.

The first test was to see if there were any control instabilities caused by the coupling: however none showed up in the
test. (Later tests showed several 8 to 24 h instabilities caused
by magnet quenching.) A major goal was to gain operating experi-

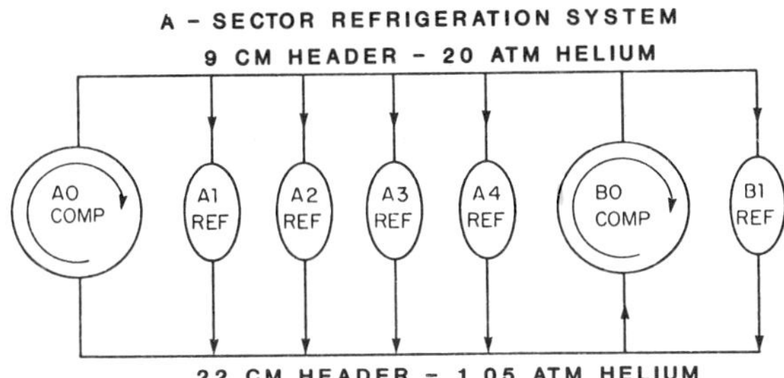

Fig. 5. Five Refrigerator Configuration.

ence in control of contaminates: initial system clean up cooldown, long term operation, individual coldbox derime, and cold box warm-up. We obtained several refrigerator-years of experience, including twice crashing the entire system because of contamination.

The final contamination detection layout which evolved from this test is:
1.   At each compressor building there is a water detector, a nitrogen detector and, in the future, an oil dectector.
2.   In the mobile purifier there is a water detector, a nitrogen detector and, a gas chromatograph.
3.   There is no monitoring of impurities in the refrigerator buildings.

Satellite Refrigerator Concept Test

The goals of this test were to obtain operating experience with the satellite mode and to measure the transfer line heat load.   For the test the A1 refrigerator simulated the CHL by operating in the liquefier mode, sending supercritical helium through the transfer line to the A2 refrigerator.   The A2 refrigerator operated in the satellite mode and cooled two 125 m strings of magnets.   Normally, A2 ran with only the 6 K expander operating, but for short periods it was turned off and the coldbox J-T was used.

We tested the 266 m A1-A2 transfer line and measured a helium heat load of 9 W and a nitrogen shield load of 140 W.[4]   We had many problems with contamination plugging venturies as well as flow instabilities and reversals.   By carefully choosing our operating pressures and temperatures in the line we were able to cool two magnets strings at A2.

Current Testing

We have completed and are operating (August, 1981) four sections of transfer line for a total length of 975 m.   Again the

Al refrigerator is simulating the CHL and we are studying flow instabilities and measuring heat leaks.[4]

In the accelerator tunnel we are assembling a 732 m string of magnets consisting of six cryoloops which are being hooked to the Al, A2, and A3 refrigerators. A normal cryoloop consists of 16 dipole magnets (for bending), four quadrupole magnets (for focusing), four spool pieces (for steering, tuning and quench protection), and two end modules (for instrumentation and cryogenic control). For this run we will operate the CHL to supply LHe as well as LN_2 to the refrigerators Al to A4.

The A4 refrigerator will continue to be used for refrigeration studies. The B1 refrigerator will be used to cool our 125 m above ground magnet testing facility. The transfer line from the CHL has not yet been installed to B1, so it will continue to run in standby mode with a temporary LN_2 supply.

REFERENCES

1. C.H. Rode, et al, "Advances in Cryogenic Engineering, Vol. 25," Plenum Press, New York (1980), p. 326.
2. M. Martin, _IEEE Trans. Nucl. Sci._, Vol. NS-28 (3): 3251 (1981).
3. J.R. Zagel, _IEEE Trans. Nucl. Sci._, Vol. NS-28 (3): 2152 (1981).
4. C. Rode, Fermilab Tevatron Transfer Line, "Advances in Cryogenic Engineering, Vol 27," Plenum Press, New York (1982).

MIRROR FUSION TEST FACILITY:
MAGNET CRYOGENICS*

J. H. VanSant

Lawrence Livermore National Laboratory
University of California
Livermore, California

INTRODUCTION

Construction of the superconducting magnet system for the Mirror Fusion Test Facility (MFTF) at Lawrence Livermore National Laboratory (LLNL) is nearing completion with engineering tests scheduled to begin later in 1981. The original magnet system, designed for physics experiments on fusion plasmas with single-cell-mirror confinement, is composed of a pair of large C-coil magnets in a yin-yang arrangement, which will be cooled by liquid helium (LHe) to 4.5 K.

Congressional approval was obtained in 1980 to expand the MFTF to a multiple-cell configuration of 22 superconducting magnets. This rescoping has resulted in the deletion of physics experiments with the single-cell MFTF. However, acceptance tests will still be performed this year to both clarify design requirements and provide data for the design of the other magnets. The existing yin-yang magnets as well as an identical pair will be used in the new system.

The primary requirement on the cryogenics system of the magnets is to assure a superconducting state for the magnet coils, a large task considering their enormous size.

---

*This work was done under the auspices of the U.S. Department of Energy under Contract NO. W-7405-Eng-48.

MAGNET DESIGN

The assembled yin-yang pair, illustrated in Figs. 1 and 2, have a combined diameter of 8.5 m and weigh over 340 t. They will produce a peak magnetic field in the winding of nearly 8 T. With over 25 km and 1392 turns of conductor in each magnet coil, the total stored magnetic energy will exceed 420 MJ.

Figure 3 is a cross section of the magnet structure. The coil and electrical insulation are contained in the 1.3 to 2.5cm thick jacket that is supported by a 7.6 to 12.7cm thick structural case. To provide good transfer of the mechanical load to this case, a high-density plastic is injected into the space between the jacket and a thin steel bladder. Steel buttons keep the bladder from contacting the case. This space is left between the bladder and the case for helium (He) flow to shorten the cool-down and warm-up cycles as well as to provide for vacuum pumping during operation.

Fig. 1. Relocation of the MFTF coil vessels to LLNL Bldg. 431 for installation of instrumentation and LN shields before placement in the vacuum vessel.

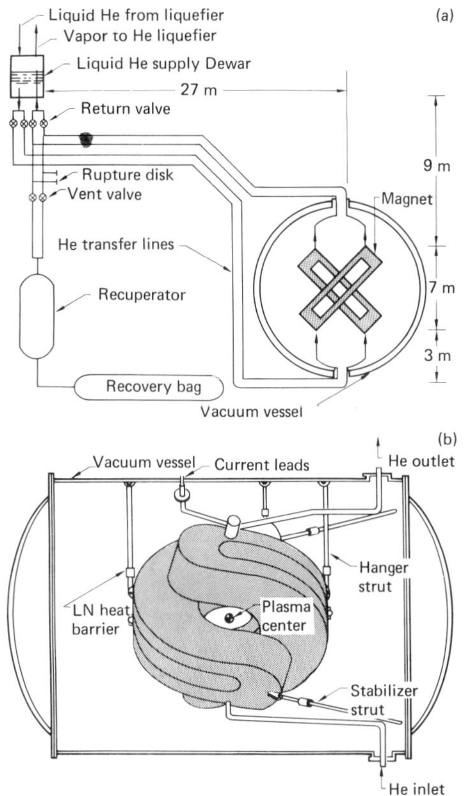

Fig. 2. MFTF Magnet, vacuum vessel, and LHe flow system.

The conductor, originally developed and tested for the MFTF magnets,[1] is copper-stabilized niobium-titanium (Nb-Ti). Internal cooling passages allow transverse and axial flow of coolant and more than double the heat transfer area. We have determined the cryostability limits of the superconductor from test coil measurements[2]: the conductor exhibits cold-end recovery, and the current-field limits provide an adequate margin above the magnet design point of 5775 A, 7.68 T, and 4.5 K. (Additional design details are given in Ref. 3.)

MAGNET COOLING

The MFTF Magnets--oriented at 45° to enhance the flow of LHe and inhibit vapor trapping--are cooled by the natural convection flow of LHe from an elevated Dewar (Fig. 2). Also, plenum spaces are included at the top and bottom of each magnet both to dis-

tribute the flow entering and leaving the coils and to provide a
space for vapor to flow outside the cold pack. To regulate flow
during cool-down and warm-up cycles, each coil has a dedicated
piping circuit; in-line valves keep the temperatures relatively
uniform and thereby limit thermal stresses.

Using finite difference analysis of a representative flow-
system model,[4] we have estimated steady-state LHe flow rates
through the magnet system: the flow is 6 L/S for our heat load of
460 W, and the average vapor quality in the coils is less than
1%. Conventional engineering methods were used to estimate pressure loss of the turbulent flow in the piping system; calculated
pressure loss through the coils was less than 10% of the overall
system pressure loss. Based on our results, we selected a 15cm
diameter for the LHe transfer lines.

The results of the cool-down/warm up analyses (performed by
General Dynamics Convair* under contract to LLNL[5]) indicate we can
cool the magnet system down in 4 to 5 days. The longitudinal
temperature differential from top to bottom will be modest, resulting in thermal stresses of less than 100 MPa (14,500 psi),
well below the allowable limit of 550 MPa (~80,000 psi). Our
anticipated 5-day warm-up schedule will also cause very low
stresses. To cool the structural cases that make up most of the
cold mass, approximately 80% of the He flow will be routed through
the guard vacuum spaces.

THERMAL PROTECTION

Good thermal protection of the magnets is essential for
maintaining a superconducting condition in the coils. We have
eliminated all possible heat sources to the extent that the total
heat load on the magnets should not exceed 460 W. These heat
sources include: radiation from 320 m^2 of liquid nitrogen (LN)
thermal shields, heat conduction from seven magnet supports,
thermal shield supports, instrumentation wires, current leads and
Joule heat from electrical joints in the coils. We prevent gas
convection and conduction by operating the insulating vacuum at
10^{-6} torr.

The LN Shields that surround the magnets are estimated to
contribute the largest share of the heat load--approximately 180 W

---

*Reference to a company or product name does not imply approval or
recommendation of the product by the University of California or
the U.S. Department of Energy to the exclusion of others that may
be suitable.

of radiation and 90 W of conduction through the shield supports. These estimates are approximately double the expected initial heat load because thermal emissivity values will probably be lower than those used for design. The shields will become less effective during periodic surface regeneration cycles (evaporation of condensed gases that could desorb and poison the fusion plasma).

More than 1000 brackets support the LN shields from the 4.5 K surface, but space limitations require that most of them be no more than 3 cm high. These LLNL-designed brackets are made of a high-density, fiberglass-epoxy composite to NEMA G-10 specifications, satisfying our requirements for heat transfer and mechanical strength. We performed room-temperature mechanical tests on sample brackets and found that they can support compressive loads up to 4500 N.

The magnets will be supported by two hanger and five stabilizer struts attached to the vacuum vessel (Fig. 2). We have minimized conduction heat transfer through the struts by the use of stainless steel as well as by LN-cooling, which serves as a thermal barrier to heat flow from the vacuum vessel. Also, we will place aluminum shields around each strut to reduce thermal radiation heating. The total heat load on the magnets from the struts is estimated to be less than 60 W.

The two magnets contain 120 conductor joints, made by a cold-welding method developed by LLNL,[6] each less than 10^{-8} ohms. The joint is soldered into a tapered copper busbar that gives added mechanical support and increased heat transfer area.[3] The heat load contributed by the joints is less than 45 W.

Vapor-cooled current leads are used to penetrate the vacuum vessel wall and make the temperature transition from 4.5 to 300 K. Each lead is constructed of 60 6.4 mm copper tubes, contained in a stainless-steel tube and assembled to copper terminal blocks with a high-temperature braze. Helium vapor flows inside the copper tubes at 0.4 g/s when the current is 6000 A. The heat load from each lead is approximately 8 W, only a few watts more than ideal leads.[7] During thermal acceptance tests, we estimated that the current leads could be operated for nearly 20 min at 6000 A without cooling before their temperature reached 400 K.

CRYOGENIC SYSTEM

To cool the magnets and the vacuum-pumping cryopanels,[8] the MFTF employs a cryogenic system with an LHe volume of 30,000 L and a refrigerator/liquefier with a capacity of 600 L/h and 3300 W. The cryopanels and magnets have separate supply Dewars that can be operated independently. Specifically, a 25,000 L storage Dewar

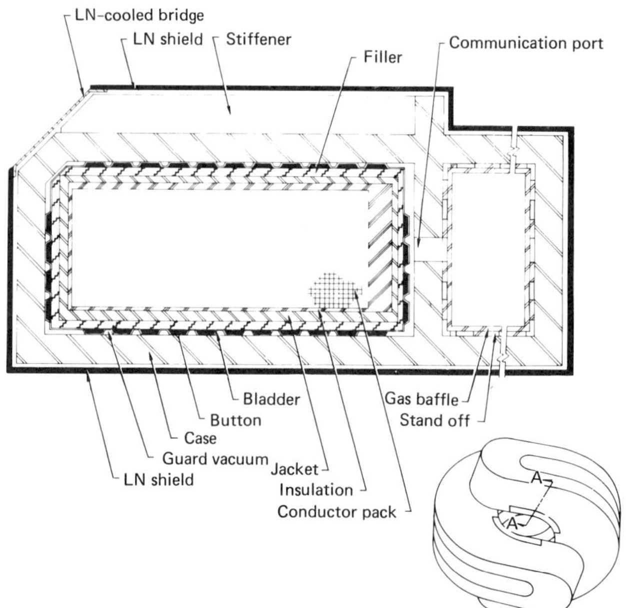

Fig. 3. Sectional ("A-A") view of the magnet coil showing its construction and thermal shields.

provides a standby source of LHe in the event that maintenance is required on the refrigerator; this Dewar also provides storage for LHe removed from the cryopanels and magnets.

If the magnets should experience a quench (i.e., cease to operate in their superconducting mode), a large amount of He vapor would be generated by Joule heating. When this occurs, the return- and vent-line valves would be actuated by an automatic detection system so that the supply Dewar would be isolated from the magnet system, and the vapor would vent to a recovery system. Should an over-pressure occur, rupture disks (shown in Fig. 3) would open at 0.55 MPa (80 psi) to protect the piping system.

The recovery system includes a 3,000 m^3 He gas bag and a 9 m^3 recuperator, which warms the vented He enough to protect the bag from thermal shock. However, the gas bag collects only one-half of the maximum volume of gaseous He vented during a quench; the remainder is vented to the atmosphere.

SUMMARY

The MFTF Magnet system has been designed for stable superconducting operations. The design current and Joule heating rate are less than the specified maxima; the expected heat load on the magnet is also less than allowed. Furthermore, cool-down and warm-up will not develop undesirable stresses. When engineering tests are performed on the magnets this year, the specific cryogenic performance will be evaluated.

REFERENCES

1. D.N. Cornish, et. al., Development Work on Superconducting Coils for a Large Mirror Fusion Test Facility, Lawrence Livermore National Laboratory, report, UCRL-78891 (1977).
2. D.N. Cornish, et. al., MFTF test coil construction and performance, IEEE Trans on Magnetics., MAG-15:530 (1979).
3. C.D. Henning, et. al., Mirror Fusion Test Facility Magnet System--Final Design Report, Lawrence Livermore National Laboratory, report, UCRL-52955 (1980).
4. J.H. VanSant and R.M. Russ, Thermal Control for the MFTF Magnet, Lawrence Livermore National Laboratory, report, UCRL-82850 (1979)
5. R.F. O'Neill and D.H. Reimer, Thermodynamic Analysis of the Magnet System for a Mirror Fusion Test Facility, General Dynamics Convair, report, CASD-LLL-78-002 (1978).
6. D.N. Cornish, D.W. Deis, and J.P. Zbasnik, Cold-Pressure-Welded Joints in Large Multifilamentary Nb-Ti Super conductors, Lawrence Livermore National Laboratory, report, UCRL-79723 (1977).
7. G. Aharonian, L.G. Hyman, and L. Roberts, Behavior of power leads for superconducting magnets, Cryogenics, 21:145 (1981).
8. J.H. VanSant, D.S. Slack, and R.L. Nelson, Cryogenic System for the Mirror Fusion Test Facility, Lawrence Livermore National Laboratory, report, UCRL-83590 (1980).

A UNIQUE HELIUM REFRIGERATION SYSTEM FOR THE LARGE COIL TEST FACILITY AT OAK RIDGE NATIONAL LABORATORY

R. L. Powell

Koch Process Systems, Inc.
Westborough, Massachusetts

INTRODUCTION

The Large Coil Program (LCP) at Oak Ridge National Laboratory (ORNL) has as its charter the testing of six superconducting Tokamak magnets manufactured with different design concepts by six different organizations. Three of these magnets will require forced flow of supercritical helium for cooling, while the other three will use the pool boiling method. The cryogenic system must therefore simultaneously supply both types of cooling with flexibility to allow any one coil to be tested to full current while the remaining five are held in a background (zero current) state.

Additional system requirements were also necessary to accomplish the goals of the test program. The forced flow magnets required a distribution system so that flow to each may be varied independently. Liquefaction to replenish the dewar was also required while holding the entire magnet system cold in a standby mode. More routine cryogenic requirements for cooldown, warm-up, and withdrawal of liquid helium for lead cooling were also necessary. A definition of the operating modes and a summary of the system refrigeration requirements are given in Tables I and II respectively.

Oak Ridge National Laboratory had purchased a helium refrigeration system for the originally scoped program, from Koch Process System, Inc. (then CTI Cryogenics) in 1976. A T-S diagram for one of the operational modes for this refrigerator is shown in Fig. 1. This unit would simultaneously supply 866 W of refrigeration at 3.56 K plus 4.5 g/s for lead cooling. However, as the refrigeration requirements for a combined pool boiling, forced

Table I. Operating Mode Definition Large Coil Test Facility

       Mode I
       -Liquid production only
        -Magnets are warm

       Mode II
       -Forced Flow Magnet Test
        -One FF magnet on test with remaining
         five magnets in background mode

       Mode III
       -Pool Boiling Magnet Test
        -One PB magnet on test with remaing five
         magnets in background mode

       Mode IV
       -Liquefaction in standby
        -Dewar is filled while holding entire
         system in cold condition

flow facility were refined, it became apparent that this unit could only supply about half of the newley defined requirements. Therefore, the final design constraint was to utilize as much of the original system as possible to minimize additional expenditures while upgrading the total capacity to meet the new requirements.

The forced flow cooling requirement could be met in several ways. One was to use a cold helium pump to circulate the required 300 g/s of supercritical helium through a heat exchanger system

Table II. Refrigeration Requirements-Large Coil Test Facility

TEST MODE	I	II	III	IV
Liquid Helium Production at 4.24 K (g/s)	12.5	12.5	12.5	12.5
Heat Load at 3.56 K (kW)	0	1.424	1.247	1.377
Heat Load at 4.24 K (W)	13	342	498	297
Supercritical Helium Flow at 15.0 atm and minimum of 3.7 K (g/s)	0	300	300	300

Note: Heat loads are vaporization heat loads only on total system and do not include heat loads on vapor returns.

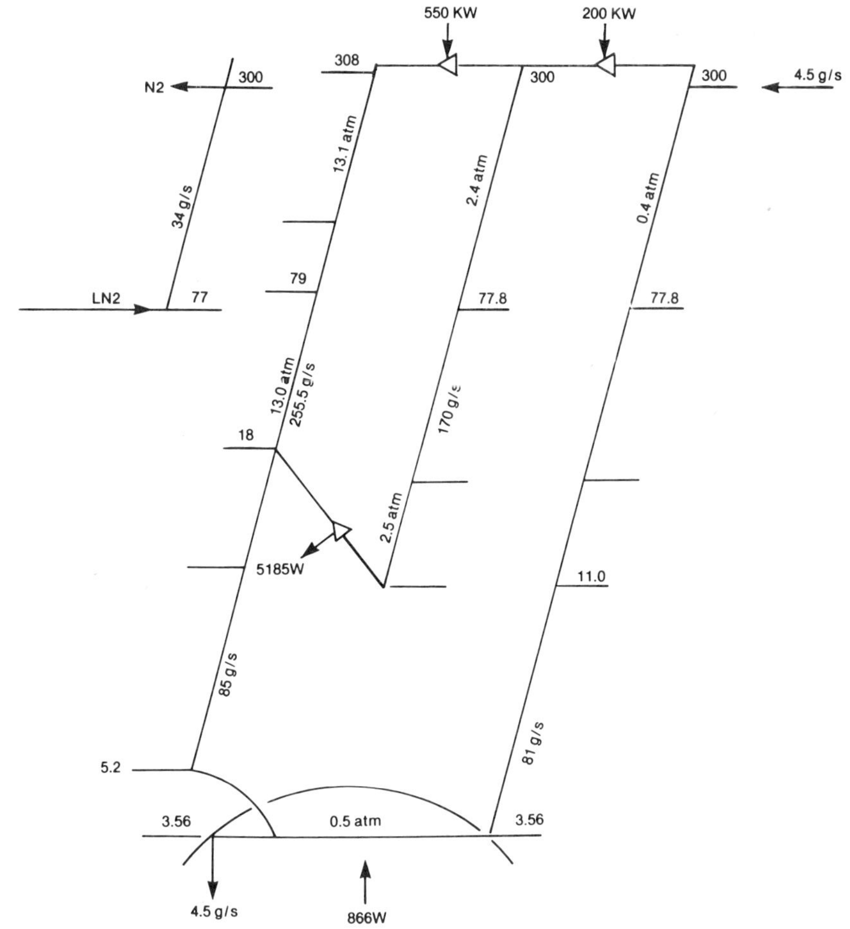

Fig. 1. T-S Diagram for Original ORNL Refrigerator.

and then through the magnet. This method would have required
development of such a pump. Another was to use an ambient
temperature compressor system to provide flow which, after cooling
in the refrigerator heat exchanger system, would pass through the
magnets and return through the heat exchanger system to ambient
compressor suction. This method, if it was to be performed
efficiently, would require additional passes on all the refrigerator heat exchangers. In both of these cases, the basic refrigerator capacity would have to be increased.

After study of the requirements and constraints, Koch Process Systems, Inc. proposed a method to combine warm temperature compression with the refrigeration cycle to meet the new capacity requirements while minimizing the number of heat exchangers which would have to be modified. The simplified flow diagram of the resulting system is shown in Fig. 2.

CYCLE DESCRIPTION

To illustrate the new system, a T-S diagram for system operation during a forced flow magnet test (Mode II) is shown in Fig. 3. A helium flow of 300 g/s is compressed in two parallel two stage compressor systems to a pressure of about 16 atm. After passing through the aftercooler and oil separation system, it is delivered to the cold box at about 15.3 atm. This flow is cooled with liquid nitrogen pre-cooling plus counter-exchange with lower pressure return flows. It reaches a temperature of about 4.5 K at a pressure of about 15 atm prior to entering the boiler heat exchanger (HX-5) immersed in a liquid helium bath above which a pressure of 0.5 atm is maintained. This supercritical helium then flows through a magnet, back through HX-5 to be recooled, through

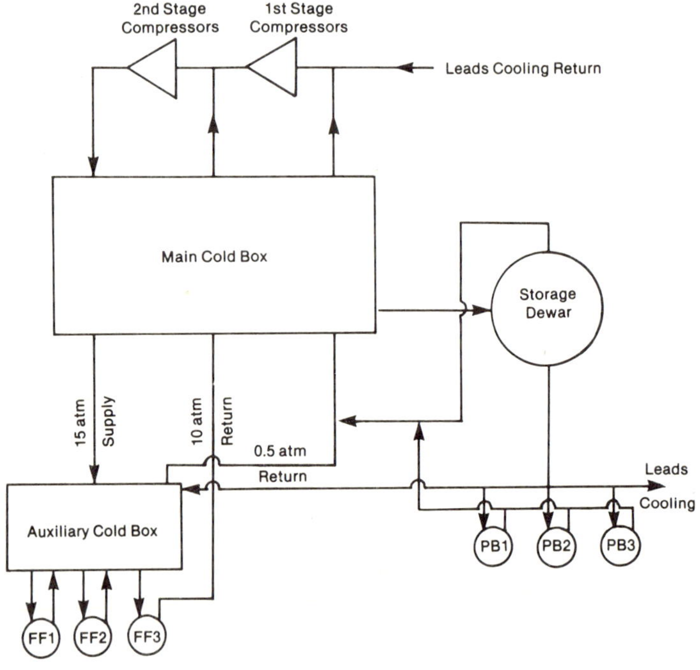

Fig. 2. Simplified Flow Diagram Modified ORNL Refrigerator.

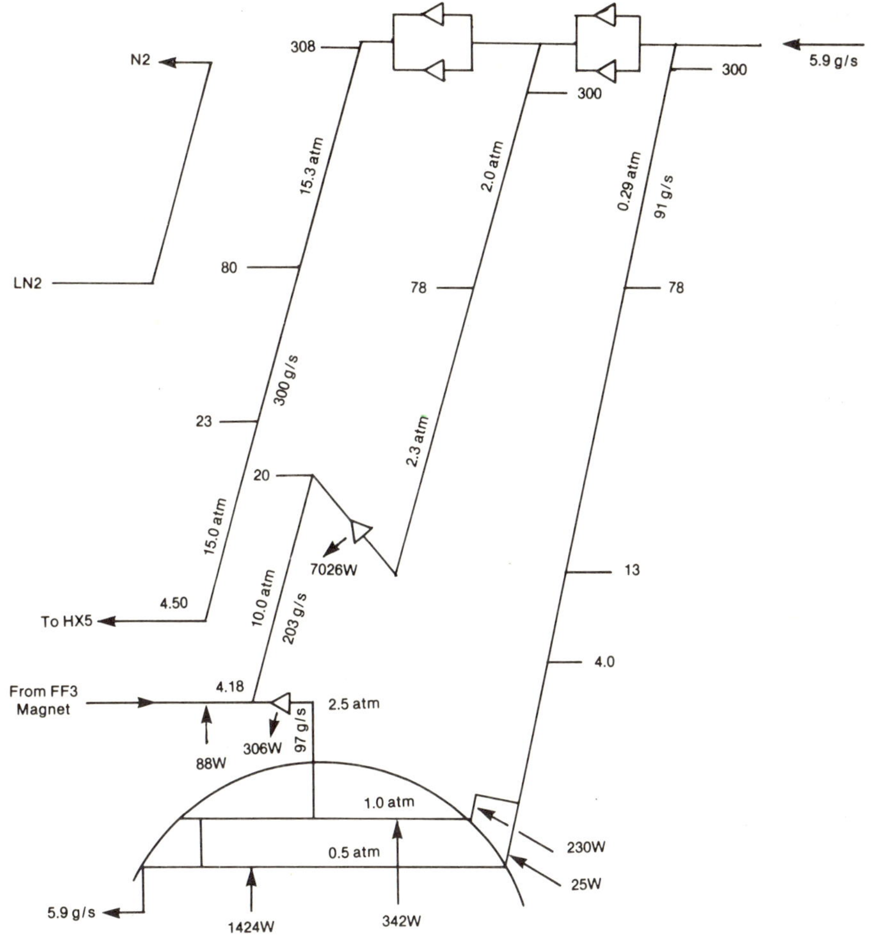

Fig. 3.   T-S Diagram for Modified ORNL Refrigerator.

the second magnet, back through HX-5 and finally through the third
forced flow cooled magnet before returning to the cold box. At
this point, the temperature is 4.18 K and the pressure has dropped
to about 10 atm. The total heat load on the 3.56 K bath is 1.424
kW in this test mode. An additional 88 W heat load is imposed on
the system due to heat losses in the transfer lines connecting the
magnet system and the main cold box.

The flow is then divided: 203 g/s is expanded through the
large turboexpander and returns through the heat exchanger system

to the suction of the second stage compressors, while the other 97 g/s is expanded through the smaller lower temperature turboexpander and then through the J-T valve to the dewar.

Liquid is made in the dewar at 4.24 K and is withdrawn for lead cooling and cooling the pool boiling magnets. Helium for lead cooling returns to the suction side of the first stage compressor at 300 K. Cold vapor from the J-T valve, dewar boil-off, and boil-off from the magnet dewars receives a total of 230 W of heat leak and returns to the first stage compressor suction after passing through the cold box heat exchangers.

Liquid helium is also withdrawn from the storage dewar and flashed to 0.5 atm to cool the forced flow supercritical helium in HX-5. Flashdown and boil-off from this bath return to the first stage compressor suction through the cold box heat exchangers after absorbing 25 W of transfer line heat leak.

EQUIPMENT DESCRIPTION

The original system consisted of two Sullair oil flooded screw compressors, one Sulzer TGL-32 gas bearing turboexpander, a single cold box with heat exchangers and a remote control console. To meet the new capacity and operational requirements, the following additions and changers were made:

1. New 400 HP first stage and 1000 HP second stage Sullair oil flooded screw compressors were added in parallel with the original units.

2. The complete compressor gas management system was modified to permit operation of any one, any series, or any combination of compressors independent of each other or cold box operation.

3. The existing Sulzer TGL-32 gas bearing turboexpander wheel and nozzle were modified to accept the higher flow requirements of the new system and a new Sulzer TGL-16 gas bearing turboexpander was added.

4. The J-T heat exchanger was replaced with a new unit to accommodate higher flows and the new flow configuration.

5. A new auxiliary cold box was added to house the 3.56 K bath, new forced flow heat exchanger and the new valves necessary for flow distribution control to the forced flow magnets.

6. The control console was completely rebuilt to accommodate the new compressors, the new turboexpander, and the new controls for the valves in the auxiliary cold box.

EXPECTED SYSTEM PERFORMANCE

The expected performance of the modified refrigeration system results from an attempt to meet the various operational requirements within the constraints of a cost effective modification of the originally supplied equipment. Table III gives the expected refrigerator performance for each mode of operation.

As can be seen, the liquefaction rate for Mode I exceeds that required. For the magnet test Modes II and III, all requirements can be met except the 12.5 g/s lead cooling flow. This simply means that the liquid level in the 5000 gallon storage dewar will fall during each test. At the end of the test, the dewar will be re-filled at a rate slightly lower than that requested while holding the entire system cold (Mode IV).

Table III. Expected Performance of Modified Refrigerator
for Large Coil Test Facility

TEST MODE	I	II	III	IV
Liquid Helium Production at 4.24 K (g/s)	16.8	5.9	7.1	10.7
Heat Load at 3.56 K (kW)	0	1.424	1.247	1.377
Heat Load at 4.24 K (W)	13	342	498	297
Supercritical Helium Flow at 15.0 atm and minimum of 3.7 K (g/s)	0	300	300	300

Note: Heat loads are vaporization heat loads only on total system and do not include heat loads on vapor returns.

THE HELIUM MANAGEMENT SYSTEM FOR THE MASSACHUSETTS INSTITUTE OF TECHNOLOGY FRANCIS BITTER NATIONAL MAGNET LABORATORY HYBRID MAGNET FACILITY*

J. R. Hale, M. J. Leupold, and J. E. C. Williams

*Francis Bitter National Magnet Laboratory
Massachusetts Institute of Technology
Cambridge, Massachusetts*

INTRODUCTION

The Francis Bitter National Magnet Laboratory has recently put into service a hybrid magnet system consisting of a 35.6 cm room temperature bore superconducting solenoid, and a family of water-cooled Bitter solenoid inserts, with bores ranging from 3 to 5 cm. The two magnets, separately energized, are concentric, with the fields additive.

While in use, the current in the superconducting magnet is constant at its maximum operating value, but the current in the insert magnet is changed to meet the field requirements of the user during the course of the experimental run. This causes the liquid helium boil-off rate to vary significantly.

THE HYBRID MAGNET SYSTEM

The design and construction of the hybrid magnets has been described elsewhere.[1,2] The characteristics can be described briefly as follows: any one of a family of high-performance water-cooled magnets is mounted in the room temperature bore of a superconducting solenoid. The fields from the two magnets add, thereby producing a more intense field than could be done in any other way with the available power. The hybrid magnet is operated with constant background field and swept insert field.

---

*Supported by the National Science Foundation.

To date, two insert magnets have been tested, and subsequently used by experimenters. With the 5 cm bore insert, the peak on-axis field is 24 T, 7.0 T contributed by the superconducting magnet operating at 1 370 A, and 17.0 T by the insert. With the 3 cm bore insert, the peak field is 28.5 T. In this case, the contributions are 7 T from the superconducting magnet, 21 T from the insert, and an additional 0.5 T from the novel iron clamping members in the insert stack. This combined field of 28.5 T is the highest steady field presently available anywhere in the world. A new insert, now under construction, will raise this peak available field strength to 30 T.

The design of the superconducting portion of the hybrid was a bold one,[2] inasmuch as it had to produce up to 7.5 T with a coil that had to fit into an existing cryostat that had originally been intended for a 6.0 T magnet. This design goal was actually exceeded — the magnet, operating with no insert — generates 7.6 T at 1 450 A, with an overall current density of nearly 6 300 A/cm^2. That corresponds to a surface heat flux of 0.5 W/cm^2.

There are three sources of helium evaporation in the superconducting magnet: (1) Lead loss; at full operating current this is approximately 5 liters per hour, (2) Radiation and Conduction; this is about 3 liters per hour, (3) Swept insert field losses; the varying insert field induces transport and eddy currents in the superconducting winding causing an additional helium evaporation which varies between zero and 27 liters per hour.

THE HELIUM MANAGEMENT SYSTEM

Source of Liquid

The primary source of liquid is the Laboratory's helium facility. Under normal circumstances, helium can be purchased as high-pressure gas, and stored in two tanks at up to 15 atm. In addition to the two tanks, the facility comprises a CTI 1400, operated only in the liquefaction mode, three reciprocating compressors, a purifier, an impurity analyzer, and a 1 000 L liquid storage vessel permanently coupled to the liquefier. There are also two portable transfer dewars, of 250 L and 500 L capacity, which are utilized to transport liquid from the 1 000 L dewar to the hybrid magnet experimental area.

Under typical conditions, two of the compressors are used to supply high-pressure gas to the liquefier, and the third is used to collect boil-off gas for storage in the high-pressure tanks, and to maintain a supply of pure 15 atm gas. However, if the need warrants it, all three compressors can be utilized for liquid production.

The liquefier and distribution system was designed to accommodate several user groups, primarily the MHD Group and the hybrid magnet.

Hybrid Magnet Refrigeration System

The hybrid facility portion of the helium management system, shown in Fig. 1, comprises the magnet cryostat itself, a reliquefier,* two transfer lines, and ancillary plumbing, valving, and pressure and flow indicators. Transfer of liquid helium to the hybrid system is by the portable transfer dewars. Evaporated helium gas is returned to the compressors. The helium management system is thus closed but with batch transfer of liquid.

The gas evaporated because of the swept insert field cannot be returned cold to the liquefier because of plumbing constraints. Because of the, at times, high eddy current dissipation the cold gas out flow therefore represents a large potential thermodynamic inefficiency in the system.

For this reason the reliquefier is incorporated into the helium management system. Under ideal conditions most of the sensible heat of the outflowing cold gas stream could be used to liquefy a high pressure in-coming gas stream. That means, for example, that an outflow of 20 L/h of liquid as vapor at 4.2 K could be reduced to a loss of only 4 L/h.

HELIUM MANAGEMENT OPERATION

There are three modes of cryogenic operation: (1) Maintenance between runs, (2) Cool-down and fill in preparation for a run; and (3) The operational run itself. Typical helium loss rates in these modes are shown in Table I.

Table I. Summary of Typical Helium Consumption Rates

Operation Mode	Typical Duration of this Mode	Average Rate of Helium Consumption	Average Overall Rate First 4 Months of Operation
Between runs	1 to 3 weeks	4 L/h	
Cool-down and fill	4 to 8 hours	60–80 L/h	6.4 L/h
Run	3 to 6 hours	18–35 L/h	

*Supplied by Cryogenic Consultants, Inc., Allentown, Pennsylvania.

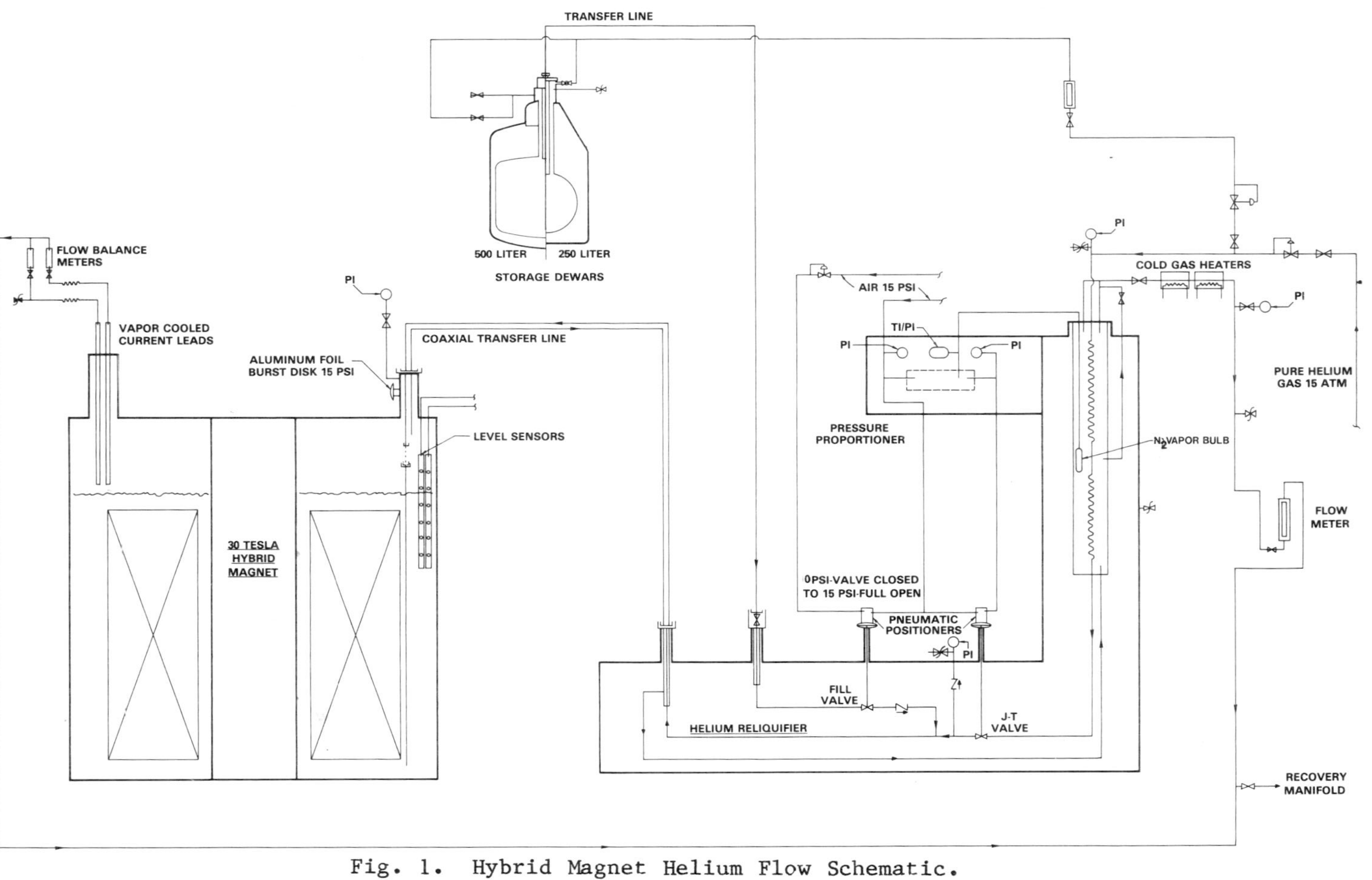

Fig. 1. Hybrid Magnet Helium Flow Schematic.

CONCLUSION

That the utilization of the reliquefier does reduce the helium consumption has been demonstrated. Furthermore, it allows the liquid level in the cryostat to be maintained constant with only occasional need to alter the delivery rate from the transfer dewar. There is an apparent inefficiency of up to 50% in the delivery of liquid to the cryostat from the transfer dewar.

Owing to heat leaks in the plumbing external to the reliquefier, full reliquefaction efficiency has not yet been achieved.

REFERENCES

1. M.J. Leupold, R.J. Weggel, and Y. Iwasa, Design and Operation of 25.4 and 30.1 T Hybrid Magnet Systems in "Proc. Sixth Intl. Conf. on Magnet Technology" Bratislava, 1977.
2. M.J. Leupold, et al., 30 tesla hybrid magnet facility at the Francis Bitter National Magnet Laboratory, IEEE Trans. on Magnetics, MAG-17:1779 (1981).

COOLING THE ARGONNE NATIONAL LABORATORY SUPERCONDUCTING HEAVY-ION LINAC WITH TWO REFRIGERATORS IN PARALLEL*

J. M. Nixon and L. M. Bollinger

Argonne National Laboratory
Argonne, Illinois

INTRODUCTION

The Argonne Superconducting Heavy Ion Linac consists of a series of niobium accelerating resonators grouped in cryostats.[1] The resonators are cooled by a forced-circulation flow of liquid helium directly from a CTI Model 1400 refrigerator through a low-loss distribution line. Helium entering each cryostat is condensed to near liquid saturation by heat exchange with the colder, lower pressure return stream.

As more resonators and cryostats have been added to the linac over the past few years, it has become necessary to increase the cooling capacity of the system without increasing the flow of helium through the outgoing primary side of the distribution line and the resonators. This has been accomplished by coupling the first operational CTI Model 2800 turbo-expander refrigerator in parallel with the Model 1400 in such a way as to avoid major modifications in the helium distribution system.

GROWTH OF THE LINAC

The modular design concept of the linac has made possible a steady increase in the number of resonators following a plan to develop and build a small superconducting linac to boost the energy of heavy ions from an existing tandem electrostatic accelerator.[2] The first prototype cryostat (A) was about 2 m long and

---

*Work supported by the U. S. Department of Energy.

contained two resonators and two superconducting focusing magnets. Later, cryostats C and D were put on line as additional resonators were completed. These cryostats are about 4 m long, and each contains six resonators and three magnets.

A new shorter resonator was developed for lower particle velocities, (low-β), and four of these have been installed in the original cryostat A.[3] Recently cryostat B, (Fig. 1) containing seven low-β resonators, one high-β resonator, and four magnets, was installed. Downstream of the four cryostats a "rebuncher/debuncher" resonator completes the current planned phase of the linac. A plan view of this linac configuration is shown in Fig. 2. A summary of the various components of the linac and their approximate heat loads are given in Table I.

Since the total heat load of the completed linac far exceeds the output of the original Model 1400 refrigerator ($\sim$ 95 W), additional refrigeration capacity was needed. This was obtained by installing a CTI-Cryogenics (now Koch Process Systems, Inc.) Model 2800 refrigerator with a rated capacity of 200 W at 4.6 K with two compressors. We now have generous capacity for the current linac, and the capability of handling future additional loads.

Table I. Linac Components and Heat Loads

Cryostats	Resonators	Magnets
A	4 low-β	2
B	7 low-β 1 high-β	4
C	6 high-β	3
D	6 high-β	3
Total 4	24	12

Approximate heat loads:

```
24 resonators × ~ 4 W each  =     96  W
            Cryostat    A    =     10
                        B    =     15
                        C    =     15
                        D    =     15
    Buncher (resonator + cryostat)=      5
  Rebuncher (resonator + cryostat)=      5
        Helium Distribution Line  =     20
                        Total          181  W
```

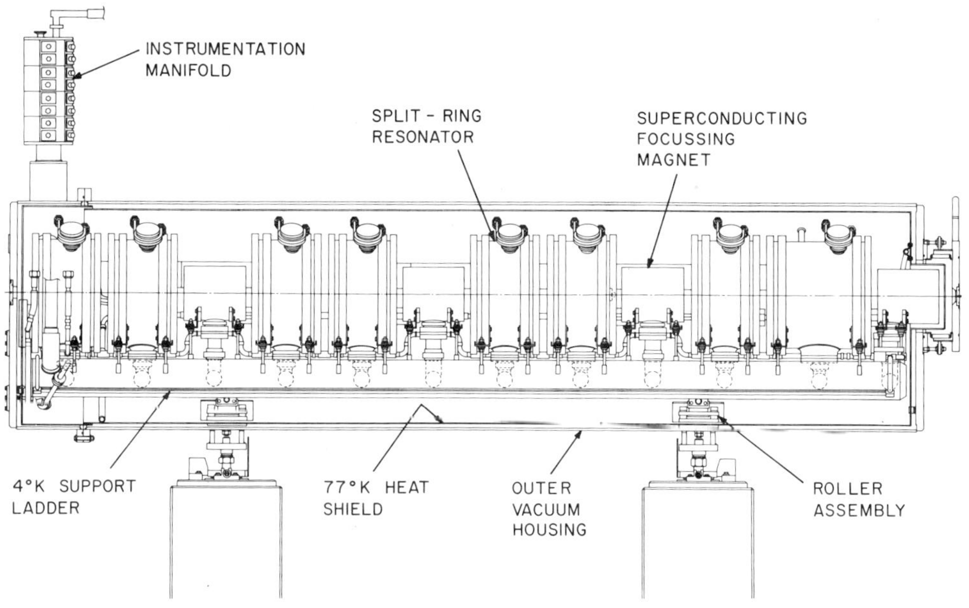

Fig. 1. Side view of Cryostat B.

HELIUM FLOW SYSTEM

In the original system the Model 1400 served as both a helium
pump and a refrigerator, circulating about 6.7 g/s of helium out
through the cryostats in series on the primary side of the circuit.[4] Two-phase flow from the refrigerator was condensed to all
liquid, first by heat exchange in a coil in the 1000 liter dewar,
and then in successive heat exchangers in distribution boxes as it
left each of the cryostats. This flow was adequate to remove the
resonator heat loads without an undue increase in gas fraction,
and the pressure drop was moderate in the piping and valves of the
distribution system.

The principal problem involved in integrating the Model 2800
refrigerator into the system was to determine how best to use its
capacity without creating an excessive pressure drop in the outgoing stream. The mass flow of the 2800 operating as a refrigerator with two compressors is about 15 g/s. This flow could not be
added to that of the 1400 without an unacceptable increase in
pressure. High pressure on the outgoing side has the undesirable
effects of raising the temperature of the superconducitng resonators, and reducing or eliminating the latent heat of vaporization of the helium as the pressure approaches or reaches the
critical point. In normal operation the maximum pressure at the
entrance of the first cryostat is usually held to about 185 kPa

absolute ($\sim$ 12 psig), making the maximum helium temperature about 4.9 K in the first resonators.

Experience with the original system demonstrated that the first indication that the heat load on the system exceeded the capacity of the refrigerator was the boiling away of liquid in the shell sides of the heat exchangers in the line adjacent to each cryostat. It was this fact that led to the scheme adopted for connecting the 2800 to the system. The solution was to add the new flow from the 2800 to the returning flow from the 1400 in the secondary side of the distribution line, where pressure drops are small.

The helium flow diagram in Fig. 3 shows the coaxial delivery tube of the 2800 connected to the return line just ahead of the shell side of HX3 in distribution box C. The supply line feeds into a small phase separator tank in the return line. The two-phase flow from upstream also enters the phase separator. A portion of the gas from the phase separator may be drawn back to the 2800, and the remaining flow replenishes the shell sides of the condensing heat exchangers HX3 in box C and HX2 in box D. Any remaining liquid accumulates in the 1000 liter dewar.

The flow from the 2800 must be adjusted so as to maintain a pressure in the phase separator which does not inhibit flow from the 1400, and at the same time provides a reasonable pressure gradient to move the increased mass flow towards the dewar. A differential pressure gage connected between the shell side of HX3 and the dewar provides a sensitive means of determining this pressure drop.

In order to complete the helium circuit of the 2800, a new transfer line was installed to return low pressure cold gas from the dewar to the 2800 just as cold gas is returned to the 1400. Thus the two machines operate in parallel as refrigerators into the same dewar, and the supply and return mass flows of each are balanced.

ADDITIONS TO THE CRYOGENIC SYSTEM

During 1980 several major additions were made to the cryogenic system to prepare for operation of the full linac as shown in Fig. 2.

Extension of the Main Helium Distribution Line

The original main helium line terminated in a capped tee near the downstream end of what is now the cryostat C position. In order to supply helium to cryostat D and the rebuncher, a 10 m

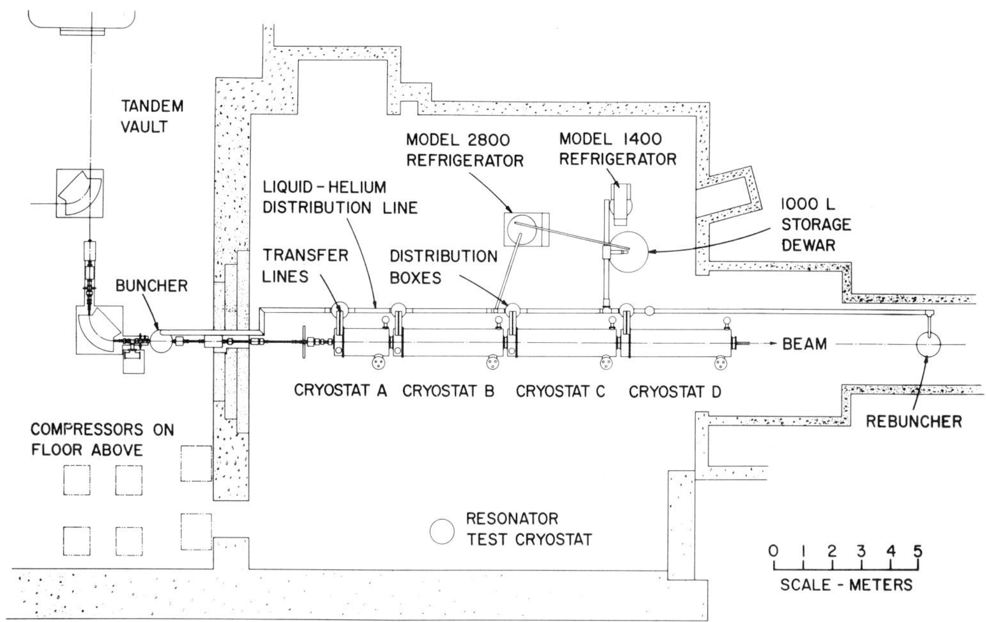

Fig. 2. Plan view of superconducting linac cyrostats
and helium distribution system.

length of line was extended from the tee to the rebuncher. Distribution box D and a rebuncher valve box each have supply, return, and bypass valves to control helium flow to cryostat D and
the rebuncher. Distribution box D also contains two transfer line
bayonet ports and condensing heat exchanger HX2. The designs of
the LN_2 shielded main line and distribution box are the same as
previously reported.[4]

Phase Separator and Transfer Lines

The output of the 2800 refrigerator connects to the main
helium line through the phase separator. A coaxial transfer line
entering near the top feeds two-phase helium from the refrigerator
so that it impinges tangentially on the vessel wall. The return
stream from the 1400, flowing in the outer annulus of the main
helium line, is interrupted and fed into the phase separator to
join the flow from the refrigerator. Separated gas can be drawn
from the vessel back to the refrigerator through the annular passage surrounding the supply line. The remaining flow continues
downstream through the shell sides of HX3 and HX2, and into the
dewar.

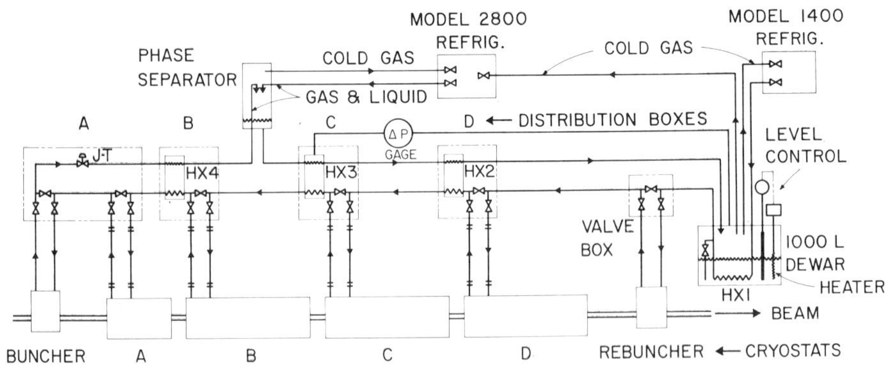

Fig. 3. Cooling system helium flow diagram.

Two transfer lines were fabricated and installed to connect the new refrigerator to the system. The coaxial line extends from the phase separator to the delivery tube port of the 2800. The insulating vacuum of this line is common with the vacuum of the entire distribution line. A separate line extends from a port on the top of the 1000 liter dewar to one of the return gas bayonet connections on the refrigerator. Both lines are insulated with multi-layer insulation.

Model 2800 Refrigerator and Compressors

The new refrigerator was installed near the Model 1400 and 1000 liter dewar as shown in Fig. 2. Its two rotary screw compressors were installed near the 1400 compressors about 50 m away on a thick concrete floor above the tandem vault. Separate supply and return lines connect the screw compressors to the 2800, but interconnections and valves at the compressors allow the screw compressors to feed the 1400, or the three reciprocating compressors to feed the 2800.

The Model 2800 refrigerator is the first of its type to become operational. It was received in June 1980, started up that October, and has been running continuously, except for several brief shutdowns, ever since. During factory tests, with LN_2 precooling, the machine produced 63 L/h, or 203 W at 4.6 K with two compressors; and 90 L/h, or 305 W at 4.6 K with three compressors. The machine has two gas bearing turbo-expanders in series, dual 80 K charcoal adsorbers with valves and heaters for regeneration while operating, a coaxial delivery tube port, and two pairs of bayonet connections for external refrigeration circuits with separate J-T and return valves for each circuit. The coaxial port and one of the return bayonet connections are now in use. The extra pair of bayonet connections will be used in the future for the ATLAS program, a proposed extension of the linac.

OPERATING EXPERIENCE

To date, the Model 2800 refrigerator has operated almost continuously for eight months with no important problems. No operator is present during nights and weekends. The only maintenance carried out has been to change the charcoal traps in the compressors.

Because the heat load imposed by the linac changes greatly from time to time, it has been necessary to operate the refrigerator under a wide range of conditions. The system performs satisfactorily under all of these conditions. However, some of the refrigerator valving and pressure regulation, which are designed primarily for steady operation with a large heat load, are not optimum for the small loads we often have.

The operation of two very different refrigerators in parallel has not been a source of trouble. Although both draw cold low-pressure gas from the same dewar, each automatically adjusts its operating pressure in such a way as to form a mass balance. It is possible for the large refrigerator to take such a large fraction of the cold gas that the small unit becomes unbalanced, but this problem can be avoided by adjusting the makeup gas regulator of the 2800.

Initially, a minor problem in the parallel operation of the two refrigerators was that the thermal time constant of the system was so long that one hesitated to leave the system in unattended operation until it had been allowed to stabilize for some hours. This inconvenience has now been removed by installing a control system that automatically varies the power to a 200 W heater in the 1000 liter dewar so as to maintain any one of the following parameters constant: (1) the net flow of warm gas between the refrigerators and the storage tank, (2) the liquid level in the 1000 liter dewar, or (3) the dewar pressure.

An example of the reliability and flexibility of the system is a recent use of the refrigerators to cool the whole linac from 100 K to 4.6 K without using the cooling power of any stored liquid helium. This cooling of a complex system weighing about 2500 kg was achieved during a period of 48 hours of largely unattended operation.

ACKNOWLEDGEMENTS

The authors gratefully acknowledge the following people for their exceptional efforts in the design, fabrication, installation and operation of the system: T. Sterling, and W. Ball, designers of the ANL Engineering Division; B. Millar, technician in charge of the cryogenic system; and personnel of the ANL Central Shops and Quality Assurance Divisions.

REFERENCES

1. J. Aron, et. al., The superconducting heavy-ion linac at Argonne, IEEE Trans. on Nucl. Science, NS-28:3458 (1981).
2. L.M. Bollinger, et. al., in "Proc. 1976 Proton Linear Accel. Conf.," AECL 5677 (1976), p. 95.
3. K.W. Shepard, Development of Superconducting Resonators for the Argonne Heavy-Ion Linac, IEEE Trans. on Nucl. Science, NS-28: (1981).
4. J.M. Nixon and L.M. Bollinger, in "Advances in Cryogenic Engineering, Vol 25," Plenum Press New York (1980), p. 317.

LIQUID NEON HEAT INTERCEPT FOR SUPERCONDUCTING ENERGY STORAGE MAGNETS*

A. Khalil

University of Wisconsin
Madison, Wisconsin

and

G. E. McIntosh

Cryogenic Technical Services, Inc.
Boulder, Colorado

INTRODUCTION

A common reaction to the proposal to cool superconducting energy storage magnets (SMES) with superfluid helium at 1.8 K is concern for the additional refrigeration power required as compared to operation at 4.2 K. Indeed, the Carnot work at 1.8 K is 2.35 times that at 4.2 K and actual efficiency at the lower temperature is likely to be lower than with normal helium. (We assume 20% of Carnot at 1.8 K and 25% at 4.2 K to account for greater heat leaks and flow losses). However, the principal consideration is the total refrigeration requirement, not just that at 1.8 K. It has been determined[1], that effective use of heat intercepts limits total refrigeration power for a 1.8 K system to only 30 to 31% more than a magnet operating at 4.2 K.

The very important function of heat intercepts in reducing overall refrigeration power to both 1.8 and 4.2 K systems has given rise to a number of analyses[1-5] aimed at minimizing heat leaks, refrigerator power, or overall cost. In the most recent of these studies, Hilal and Eyssa[5] concluded that a finite number of

---

*This work was supported by the U.S. Department of Energy and the Wisconsin Electric Utilities Research Foundation.

individually refrigerated heat intercepts could approach the absolute minimum power theoretically achievable with an infinite series of infinitesimally small refrigerators intercepting heat as it flows from ambient to 1.8 K. For energy storage magnet applications it appears that the practical maximum number of heat intercepts is three and calculations have been made to determine the optimum temperature of each heat intercept and the location of the intercepts on structural support struts. Typical intercept temperatures are 30.8 K for one intercept, 12.4 and 67.6 K respectively for two, and 7.8, 30.8, and 102.5 K for three intercepts.

One problem with the analytical heat intercept solutions typified by the above set of temperatures is that none correspond with normal boiling points of cryogenic fluids. Also, most previous calculations of heat intercepts have been based on strut cooling with insulation heat leak considered secondary. However, in the currently favored low aspect ratio ($\beta=0.01$) Wisconsin SMES concept design, radial forces are significantly lower than for previous designs and insulation heat leak is from about 30 to 50% of the total. The purpose of this work is to extend previous analyses to include both insulation and strut heat leaks and to examine the impact of using storable, boiling cryogens for heat intercept fluids specifically liquid neon and nitrogen.

SELECTION OF FLUID FOR HEAT INTERCEPTS

While the design and application of Brayton cycle refrigerators is certainly within the state of the art,[6] systems with nitrogen or helium as working fluids suffer some thermodynamic disadvantages compared to a liquid boiling at constant temperature and they cannot provide stored refrigeration. What is needed for intercepts are fluids with normal boiling points that approximate the optimized intercept temperatures and which can be stored for continuous, instant response to load variations. The fluids should also be non-flammable, non-toxic, non-corrosive, cheap, and readily available. Nitrogen is the only cryogen which truly meets these requirements. Unfortunately, its normal boiling point is too high for most effective shielding without a lower temperature heat intercept. Neon is the only other boiling cryogen for this application and it is neither cheap nor readily available in large quantities. However, at 27.1 K, neon is more effective in a single shield than nitrogen. Because nitrogen and neon are the only qualified cryogens, the present analysis is limited to one intercept cooled with either nitrogen or neon or two intercepts at ~ 28 and ~ 78 K respectively.

REFRIGERATION POWER FOR 1000 AND 5000 MWhr SMES UNITS

A computer program was set up to optimize refrigerator power input requirements by setting the location of temperature intercepts for a strut length of 1.5 m. The program utilized variable thermal conductivity values for struts based on material similar to polyester-fiberglass and incrementally variable values for the thermal conductivity of evacuated Perlite. Refrigeration power calculations assumed the following Carnot efficiencies: 20% at 1.8 K, 30% at 28 K, and 40% at 78 K.

Summarized results of the calculations for 1000 and 5000 MWhr magnets are presented in Tables I and II. Optimization curves for nitrogen, neon, and nitrogen/neon heat intercepts are shown in Figs. 1-3, for a 5000 MWhr energy storage magnet.

Figures 1 and 2 for single intercepts cooled with nitrogen and neon, respectively, show a simple relationship of power requirements to intercept position. As the intercept is moved away from the cold surface, the 1.8 K power, P_c, decreases while the intercept refrigerator power, P_1, increases under the combined influence of insulation heat leak, P_i, and strut conduction. For nitrogen the optimum intercept location is about 1.1 m from the cold end and the optimum refrigeration power is 6.65 MW. The neon

Table I. Optimum Refrigeration Power for 1000 MWhr SMES UNIT

Number of Shields	T_{s1} K	X_1 m	T_{s2} K	X_2 m	P_c MW	P_1 MW	P_2 MW	P MW
1	28	0.58	–	–	0.72	1.15	–	1.87
1	78	1.04	–	–	1.85	0.75	–	2.60
2	28	0.61	78	0.87	0.69	0.32	0.58	1.59

Table II. Optimum Refrigeration Power For a 5000 MWhr SMES UNIT

Number of Shields	T_{s1} K	X_1 m	T_{s2} K	X_2 m	P_c MW	P_1 MW	P_2 MW	P MW
1	28	0.612	–	–	1.88	2.76	–	4.64
1	78	1.102	–	–	4.84	1.81	–	6.65
2	28	0.65	78	0.91	1.8	0.8	1.3	3.9

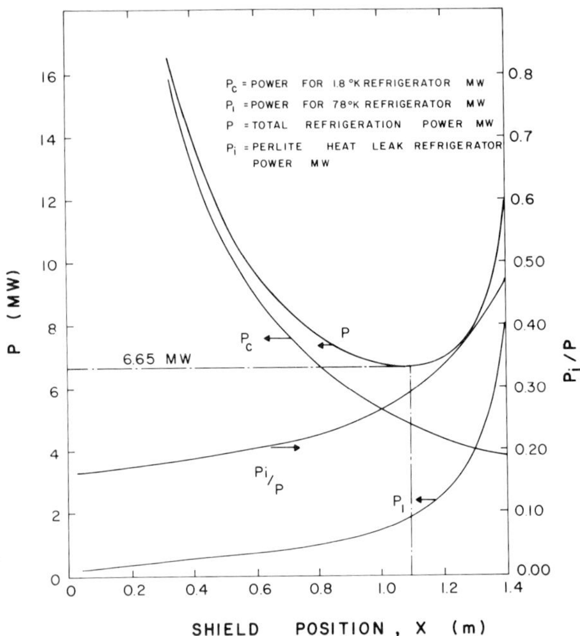

Fig. 1. Nitrogen cooled intercept location for minimum total refrigeration power for a 5000 MWhr SMES.

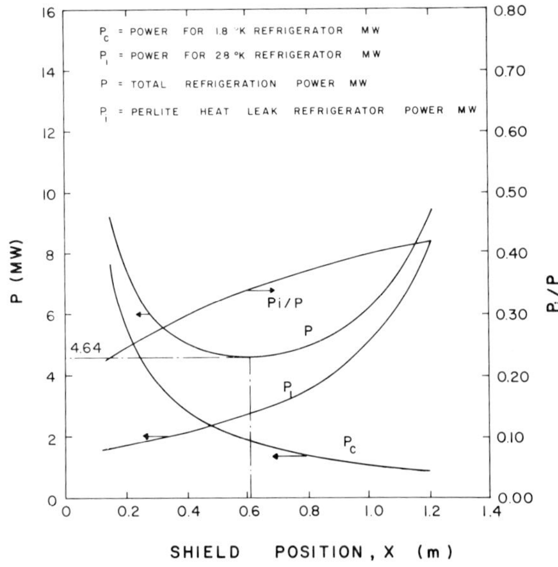

Fig. 2. Neon cooled intercept location for minimum total refrigeration power for a 5000 MWhr SMES.

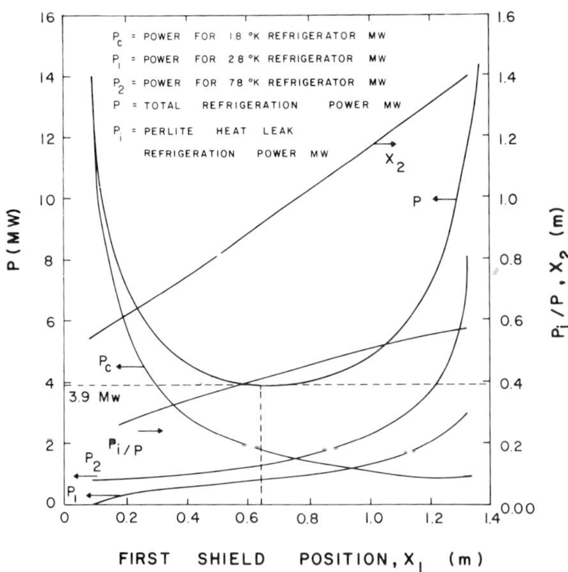

Fig. 3. Location of nitrogen and neon cooled intercepts for minimum total refrigeration power for a 5000 MWhr SMES.

intercept optimum is located about 0.61 m from the 1.8 K surface with a total refrigeration requirement of 4.64 MW, a reduction of 30% from that for nitrogen. The computation procedure for two shields is more complex because it is necessary to optimize the location of both intercepts. However, the results can be plotted to simultaneously show the interrelation of minimum total power and both intercept locations. As shown in Fig. 3, the minimum power for 78 and 28 K intercepts is 3.9 MW, a reduction of 41% from a single nitrogen intercept and 16% lower than a single neon shield. At 0.65 and 0.91 m, respectively, from the cold surface, the position of the intercepts does not vary markedly from their locations when used separately (0.65 and 1.1 m respectively).

COST COMPARISONS

Estimated capital and operating costs for cryogenic systems for 1000 and 5000 MWhr energy storage magnets are given in Tables III and IV. Capital cost estimates include the heat intercept shield or shields, refrigerators for 1.8, 28 and/or 78 K, and the heat intercept fluids. (Nitrogen and neon prices were assumed at $1 and $60 per gallon respectively.) Yearly operating costs are based on 20% of the capital investment and 4¢ per KWhr for electric power. Results are consistent for both SMES units. A single nitrogen intercept has the lowest capital cost but higher power

Table III.   Cost Estimate for 1000 MWhr SMES UNIT

Item	Nitrogen Shield		Neon Shield		Two Shields Nitrogen & Neon	
	Capital Cost 10^6	Yearly Cost 10^6	Capital Cost 10^6	Yearly Cost 10^6	Capital Cost 10^6	Yearly Cost 10^6
Shield Cost	3.500	0.7000	3.5000	0.700	7.000	1.40000
Refrigerator Cost	3.2369	0.6473	2.4867	0.497	2.184	0.4368
Ref. Power Cost	–	0.9110	–	0.6552	–	0.5571
Coolant Cost	0.0298	0.0059	1.7899	0.3579	1.8197	0.3635
Total Cost	6.7667	2.2642	7.7766	2.2104	11.0038	2.7579

Table IV.   Cost Estimate for 5000 MWhr SMES UNIT

Item	Nitrogen Shield		Neon Shield		Two Shields Nitrogen & Neon	
	Capital Cost 10^6	Yearly Cost 10^6	Capital Cost 10^6	Yearly Cost 10^6	Capital Cost 10^6	Yearly Cost 10^6
Shield Cost	6.4515	1.2923	6.4515	1.2923	12.903	2.5806
Refrigerator Cost	6.8613	1.3723	5.1458	1.0290	4.4772	0.8954
Ref. Power Cost	–	2.3302	–	1.6259	–	1.3666
Coolant Cost	0.055	0.011	3.300	0.6600	3.355	0.671
Total Cost	13.3678	5.0058	14.8963	4.6072	20.7352	5.513

requirements raise its operating expense above that of a neon shield. Use of two shields results in the highest capital cost which cannot be compensated by lower power costs so that this arrangement is most expensive to operate. The cost of neon makes the single neon heat intercept more expensive than nitrogen despite the significantly lower refrigerator cost. However, lower annual power expense holds down the operating cost of the neon system so that it is the lowest of the three. This final result is the basis for continued consideration of neon despite price and availability uncertainties.

NEON COST AND AVAILABILITY

Availability of neon is not the limiting problem to its use for intercepts. The concentration of neon in air is about 18 parts per million and, with U.S. air separation plants processing on the order of 10^9 pounds of air daily, the potential supply is in the range of 5 to 10 tons per day. Unfortunately, present market demand is such that most neon is a throw away by-product of these plants. Estimated 1978 U.S. Production of neon[7] was only about 10^6 ft^3 or the equivalent of about 5300 gallons. Clearly, considerable advance planning must be done to augment the supply in order to charge a 5000 MWhr SMES unit with the 55,000 gallons necessary for the system and its 100% reserve.

The cost of neon is the other problem restricting its potential use. Current market price of about $1/ft^3 for pure gas would be equivalent to $193/gallon for liquid. However, bulk volumes are available at lower prices and an industry source indicated that very large quantities of crude neon (75% neon, 25% helium) might be purchased at 25 to 30¢ per ft^3 of neon on a contract basis. If so, the cost of liquid would fall in the range of $50 to $60 per gallon consistent with the values used in this study.

CONCLUSIONS

For a large energy storage magnet system, liquid neon is a more effective heat intercept fluid than liquid nitrogen. Further, based on an estimated cost of $60 per gallon, 1000 and 5000 MWhr energy storage magnets project lower operating costs with a single neon heat intercept as compared to one at nitrogen temperature or two intercepts cooled by nitrogen and neon respectively. Additional work needs to be done to clarify the neon price/supply situations to validate this study. Finally, the economics and operating parameters of a hybrid system consisting of a liquid nitrogen temperature heat intercept and a low temperature shield cooled by a Brayton cycle refrigerator will be evaluated as an alternate to using liquid neon.

ACKNOWLEDGEMENTS

The authors acknowledge the technical assistance and advice of Y. M. Eyssa and M. A. Hilal.

NOTATION

β = Magnet aspect ratio, length/diameter.
P = Total refrigeration power, MW.
P_c = Cold end refrigeration power, MW.
P_1 = First shield refrigeration power, MW.
P_2 = Second shield refrigeration power, MW.
P_i = Perlite heat leak refrigeration power, MW.
T_s = Shield temperature, K
X = Distance from cold end to shield, m.
Subscripts
1   first shield.
2   second shield.

REFERENCES

1. M.A. Hilal and G. E. McIntosh, "Advances in Cryogenic Engineering, Vol. 21," Plenum Press, New York (1976), p. 69.
2. M.A. Hilal and R. W. Boom, "Advances in Cryogenic Engineering, Vol. 22," Plenum Press, New York (1977), p. 224.
3. A. Bejan and J. L. Smith, Jr., Cryogenics 3:158 (1974).
4. A. Bejan and J. L. Smith, Jr., "Advances in Cryogenic Engineering, Vol. 21," Plenum Press, New York (1976), p. 247.
5. M.A. Hilal and Y. M. Eyssa, "Advances in Cryogenic Engineering, Vol. 25," Plenum Press, New York (1980), p. 350.
6. R.C. Muhlenhaupt and T. R. Strobridge, NBS Technical Note 366, 1968.
7. B.S. Kirk and A. H. Taylor, "Kirk-Othmer: Encyclopedia of Chemical Technology, Vol. 12," Wiley & Sons, (1980), p. 249.

THERMODYNAMIC ASPECTS OF SMALL
4.2-K CRYOCOOLERS*

F. W. Pirtle, P. A. Lessard, J. M. Kaufman, and P. J. Kerney

CTI-Cryogenics
Waltham, Massachusetts

INTRODUCTION

Superconducting electronic devices are gaining much attention because of their potential benefits in size, weight, accuracy, sensitivity, and reproducibility[1]. The cryogenic system necessary to maintain the device in the superconducting state requires innovative design to guarantee success of superconducting electronics applications. Since closed cycle coolers offer potential size, weight, and longevity benefits, a study has been performed to evaluate different closed cycle system options from a thermodynamic viewpoint. Presented in this work are the thermodynamic concepts generated to satisfy the following design criteria:

- o   Operational temperature of 4.2 K
- o   Refrigeration capacity of 2.5 W
- o   Input power maximum of 2500 W (1000 W/W refrig.)
- o   Minimum size and weight

DISCUSSION

Available Technology Survey

A survey of available technology was made by expanding on the work of Strobridge[2] to determine if any cryocooler system that has been made in the past, or is available today will satisfy the design criteria. The survey revealed that:

---

*Supported by Air Force Wright Aeronautical Laboratory under Contract F33615-78-C-3413.

o   The 1000 W/W can be achieved at a capacity of approximately 10 W.

o   The goal of 1000 W/W of refrigeration exceeds the state of the art by approximately a factor of two for a capacity of 2.5 W.

o   Present system weights are 220 to 700 lb/W (100 to 317 kg/W) for a capacity of 2.5 W.

In determining which system types should be selected for detailed study, the following generic machines that provide refrigeration at liquid helium temperatures were considered:

o   Totally regenerative cycle (e.g., Stirling, Gifford-McMahon)

o   Combination regenerative cycle with Joule-Thomson (J-T) loop

o   Claude and Collins cycle

o   Cascaded Joule-Thomson

o   Exotic (e.g., electrocaloric, adiabatic demagnetization.)

The exotic approaches were eliminated from consideration because of inapplicability or because considerable development would be required to make them commercially viable in the foreseeable future. Although sections of a totally regenerative machine that will reach temperatures less than 4.2 K have been built[3], the projected specific power consumption exceeds the design goal considerably. The single cascaded J-T cycle identified in the survey exceeded the input power target by a factor of 6. Its use would also necessitate the maintenance of multiple separate cryogens.

Thermodynamic Optimization

Four thermodynamic cycles were selected for detailed study: (1) the Claude cycle with one expansion engine, (2) the Collins (multi-expander Claude) cycle with two parallel expansion engines, and (3) the Gifford-McMahon and (4) Stirling regenerative cycles with attached Joule-Thomson loops.

<u>Single engine Claude cycle.</u> A parametric analysis of this cycle revealed that the single expander Claude cycle cannot achieve the input power goal of 1000 W/W using reasonable hardware. The power requirement was greater than 3300 W.

<u>Collins cycle</u>. The Collins cycle, with precooling provided
by turbine or piston expanders, is the "standard" method of providing helium refrigeration/liquefaction in sizes greater than 10
W. The independent parameters to be optimized are expander efficiencies, heat exchanger temperature pinches, and expander outlet
temperatures, illustrated in Fig. 1 for a dual expander cycle.
Each cycle was optimized by a parametric analysis similar to the
methodology of Hubbell[4] for 2.5 W at 4.2 K plus estimated parasitic heat leaks of 0.2 W. The optimum cycle was one which
achieved a minimum total entropy generation for physically reasonable hardware. Expander efficiency was estimated to be 70% of
isentropic; compressor isentropic efficiency was estimated to be
60%. These estimates are consistent with results on larger
machines but their achievement may require innovative mechanical
design for smaller sizes. Heat exchanger effectiveness was
approximately 0.95.

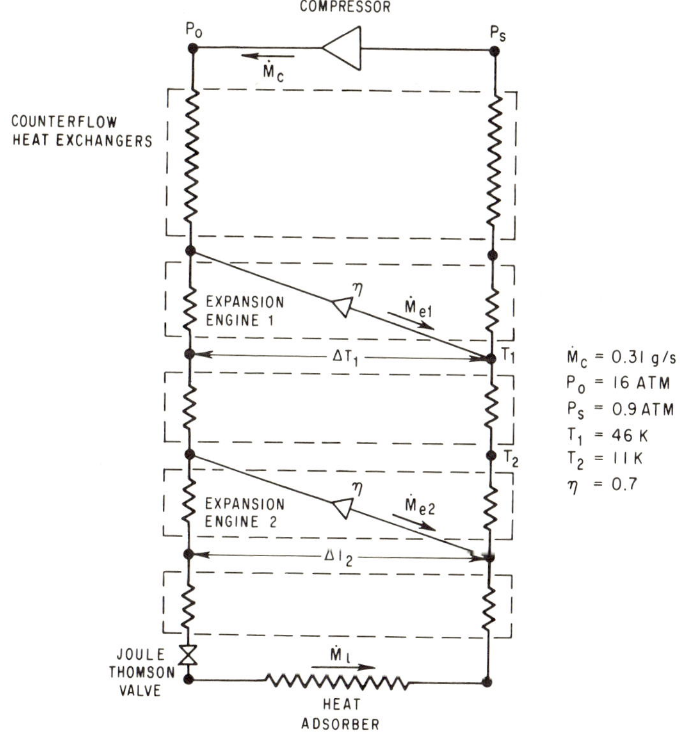

Fig. 1. Collins Cycle Schematic & Optimization Results.

The important Claude/Collins cycle results are:

o   The cycle power required for a given load is strongly
     dependent on expander isentropic efficiency as shown in
     Fig. 2 for the dual expander cycle. The input power goal
     of 1000 W/W cannot be achieved with an expander
     isentropic efficiency below approximately 50%.

o   The dual expander Collins cycle can achieve the input
     power goal with reasonable component efficiencies,
     although considerable mechanical design innovation will
     be required in the compressor, expanders, and heat
     exchangers for this service.

o   The power requirement for the optimized dual expander
     Collins cycle is 1762 W which is about 70% of the 2500 W
     goal. About 35% of the compressor mass flow is expanded
     in the two engines.

Regenerative Cycles with J-T Loops. This cycle consists of a
regenerative periodic flow cooler that is used to absorb heat from
a Joule-Thomson loop as shown in Fig. 3. The two types of regenerative systems studied were the Stirling cycle for its high
efficiency, and the Gifford-McMahon for its combination of efficiency and high reliability. The J-T loop in Fig. 3 consists of a
high pressure stream rejecting heat to the two stages of the
periodic flow machine, with final expansion through a J-T valve.
The optimization procedure for the regenerative/J-T systems was
qualitatively similar to that of the Collins cycle but differed in

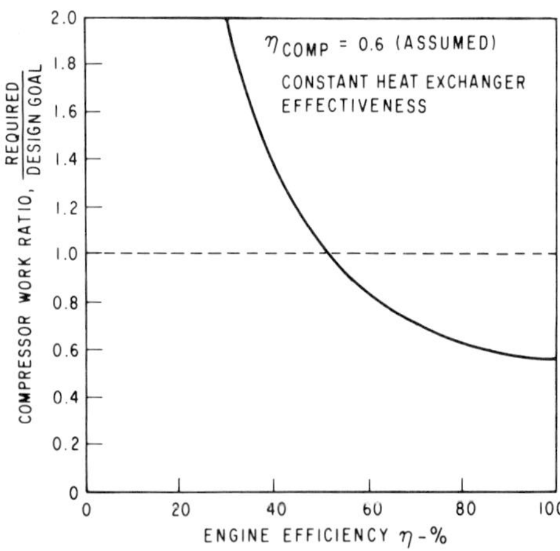

Fig. 2.   Effect of Engine Efficiency on Required Compressor Work.

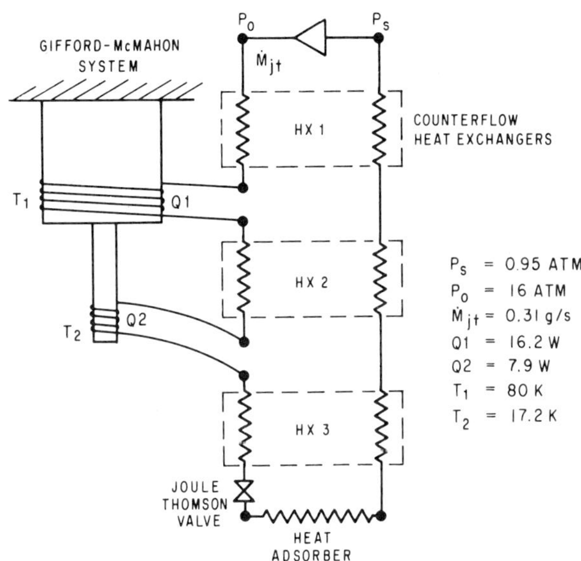

Fig. 3. Gifford-McMahon/Joule-Thomson Schematic &
Optimization Results.

the performance calculations for the periodic flow machines and
the tradeoffs evaluated in choosing a total system. For example,
the loss mechanisms evaluated for the periodic flow machines
included motional, pumping, and regenerative losses that are
unique to these machines. Optimization tradeoff evaluation included parametric studies of second stage temperature, first stage
(including radiation shield) temperature, and heat exchanger
requirements for the J-T loop.

The optimized cycle input power requirements are 1007 W for
the J-T loop and 2089 W for the Gifford-McMahon machine, a total
of 3096 W. This is about 24 percent over the 2500 W goal. The
optimized cycle using a Stirling precooler requires 1172 W for the
J-T loop and 1571 W for the Stirling machine, a total of 2743 W,
which is 10% over the input power goal.

Cycle Comparison. Comparison of the total power requirements
for the three cycles as well as an estimated system weight and
volume is shown in Table I. The weight figures for the three
systems represent an improvement of 9-20% over state of the art.

Although the Collins cycle has the lowest power requirement,
it represents an optimized thermodynamic cycle in which component

miniaturization and development may be difficult. The J-T loop components to be discussed in the next section have the potential of reducing the input power of the G-M/J-T and Stirling/J-T systems to within the program goals and to improve further the performance of the Collins cycle.

Cold End Components – Efficiency Improvements

Wet (two-phase) expanders, cold gas compressors, and ejectors were studied to determine potential improvements to the J-T sections of the various cycles.

<u>Two-Phase Expansion Engines</u>. The J-T valve is a large source of irreversibility in the various cycles because pressure is converted into fluid turbulence and lost instead of being converted into work. An attractive alternative to the isenthalpic expansion of the J-T valve is the expansion of the gas into the two-phase region by doing work on a piston or rotor in an approximately isentropic fashion. The most serious objection to two-phase or wet expansion engines, namely that heat transfer with residual liquid decreases efficiency, is less applicable to helium than to other fluids because of the relatively large ratios of heat capacity to latent heat and to cylinder/piston material heat capacity. Johnson, et al[5] describe the construction, operation, and test results for such an engine.

Figure 4 illustrates the potential effect of the wet engine on cycle input power. The improvement is significant but the achievable efficiency of miniaturized wet engines is not known at present. Larger sizes have achieved approximately 90% of isentropic[5].

<u>Cold Gas Compression</u>. The power required to compress a gas is proportional to the inlet temperature. Therefore the compres-

Table I. Cycle Comparison

Cycle	Weight Ratio Lbs/W (kg/W)	Total Component Volume Ft^3 (m^3)	Total Power W	Total Power Design Goal
Collins	180 (81.6)	8.5 (0.24)	1762	.705
Regen (Stirling/J-T)	199 (90.3)	9.5 (0.26)	2743	1.10
Regen (GM/J-T)	173 (78.5)	9.0 (0.25)	3096	1.24

sion of helium gas while it is cold offers the thermodynamic benefit of lowering input power requirements. The resultant lower pressure ratio at the warm temperature can also simplify the warm compressor by reducing the number of stages required.

Although the power required to compress the cold gas could be provided independently, one possible source is the output of a wet engine. This utilizes energy that would otherwise be thrown away and minimizes connections to ambient temperature that are sources of heat leaks. Figure 5 is a graph of overall compressor input power ratio versus inlet temperature to the wet engine for an expander and compressor with 100% isentropic efficiencies. Note that the work ratio denominator is the work with a wet engine only. The reduction in input power is approximately 20-30%.

Ejector. With an ejector, excess flow in the high temperature sections of the cycle is used to increase the pressure of the flow returning to the compressor while maintaining the low pressure at the cold head by "pumping" on the load in an ejector or jet pump. This has the same benefits as the cold gas compressor and has the operational benefit of no moving parts.

Addition of an ejector to the cycle of Fig. 1, for instance, reduces the input power from 1762 W to 1621 W for an ejector pressure ratio of 1.8:1 (state of the art) or 1265 W for a pressure ratio of 4:1 (advanced state of the art). This is a 39% increase in cycle efficiency without the addition of moving parts.

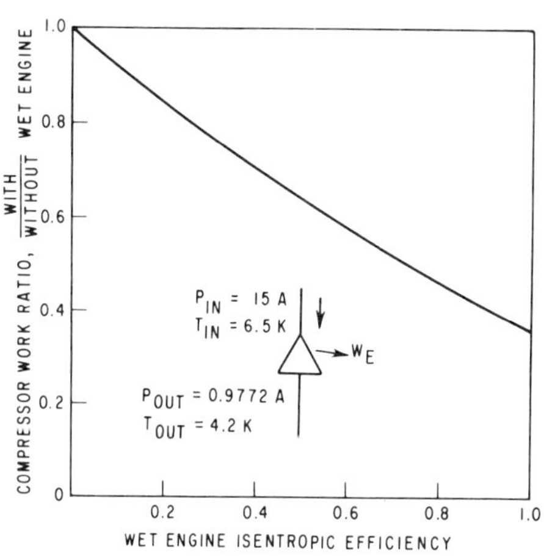

Fig. 4. Two-phase Expander Evaluation.

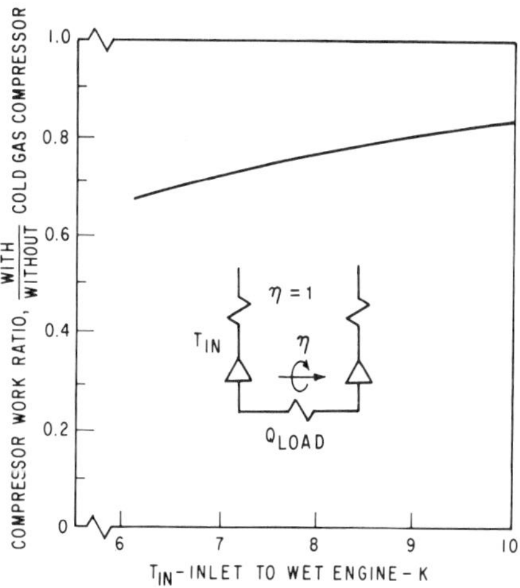

Fig. 5. Cold Gas Compressor Evaluation.

CONCLUSIONS

The Collins dual expander cycle is the only approach that thermodynamically satisfies the input power goal of 1000 W/W requiring a projected 1762 W for a 2.5 W nominal 4.2 K load (705 W/W).

Both the Stirling/J-T (1100 W/W) and the G-M/J-T (1240 W/W) exceed the input power goals but are thermodynamically competitive with each other.

Wet engines, cold compressors, and ejectors all have potential to increase overall system efficiency significantly.

REFERENCES

1. D.B. Sullivan and J.W. Vorreiter, Space applications of super-conductivity, _Cryogenics_ 19:11 (1979).
2. T.R. Strobridge and D.B.Chelton, Size and Power Requirements of 4.2 K Refrigerators, in "Advances in Cryogenic Engineering, Vol. 12," Plenum Press, New York (1966), p. 576.
3. J.E. Zimmerman, U.S. Dept. of Commerce, NBS Cryogenics Division, Boulder, Colorado, Private Communication.
4. R.H. Hubbell and W.M. Toscano, Thermodynamic Optimization of Helium Liquefaction Cycles, in "Advances in Cryogenic Engineering, Vol. 25," Plenum Press, New York (1979), p. 551.
5. R.W. Johnson, S.C. Collins, and J.L. Smith, Hydraulically Operated Two-Phase Helium Expansion Engine, in "Advances in Cryogenic Engineering, Vol. 16," Plenum Press, New York (1970), p. 171.

DISCUSSION

Question by M. Gasser, NASA/Goddard Space Flight Center: What was the origin of the 2.5 W cooling load?

Answer by author: This work studied the thermodynamic aspects of cryocoolers with capacities less then 5 W. Coolers with other capacities were considered, but only the 2.5 W analysis is presented here.

Question by R. C. Longsworth, Air Products and Chemicals Inc: Would considerations of cost and reliability favor one of the less efficient systems?

Answer by author: Yes, considerations of these factors might make one of the regenerative cycles a better choice.

HELIUM LIQUEFIER CYCLES WITH SATURATED VAPOR COMPRESSION

M. Minta and J. L. Smith, Jr.

Massachusetts Institute of Technology
Cambridge, Massachusetts

INTRODUCTION

The conventional helium liquefaction cycle employs three refrigeration stages: liquid nitrogen precooling, expansion engines, and a J–T expansion. The cycle low pressure is matched to the liquid dewar pressure. The cycle high pressure is determined by compromise between the performances of the expanders and the J–T stage. Normally, the J–T stage operates with a high pressure above the optimum level, i.e., $P_{high} > P$ for which $(\partial H/\partial P)_T = 0$. Non-liquefying helium cycles usually operate with low pressure levels above one atmosphere since this reduces the heat exchanger surface area required. Gifford McMahon cycles are typically of this design. Operation of helium refrigerators at temperatures below 4.2 K requires reduced pressure levels. This increases the compressor size and the heat exchanger surface area required. For 1.8 K refrigerators, this penalty is so large that practical cycles have required three pressure levels and the associated heat exchanger complexity.

The saturated-vapor-compression (SVC) helium cycle (Fig. 1) offers an alternative solution to these problems. The cycle high pressure is no longer limited by the performance of the J–T stage which is replaced by a supercritical wet expander. The compressor size is reduced because of the increased suction pressure, and the heat exchanger surface area required is also reduced due to the increased cycle pressure levels. The problem of an internal pinch of the heat exchange ΔT in the J–T heat exchanger is also removed and there is an overall improvement in the cycle thermodynamic performance as a result of better matching of the heat capacity rates of streams in the heat exchangers.

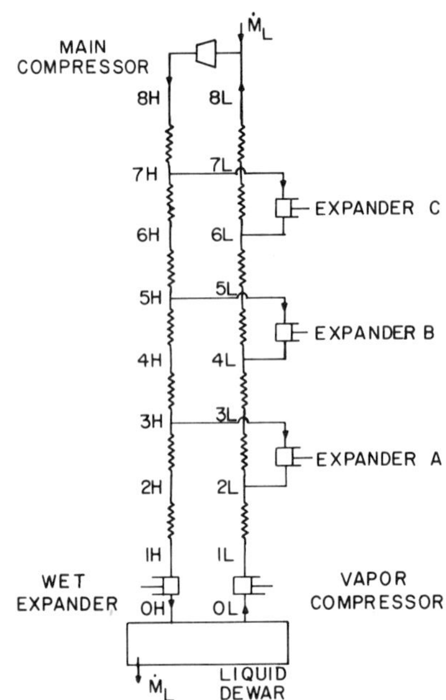

Fig. 1 Saturated-vapor-compression cycle for 4.2 K liquefier.

This paper is the design study of a SVC helium liquefier operating at elevated pressures. The study was done to show the potential of the SVC cycle by direct comparison with a conventional cycle using the same precooling expanders and a supercritical wet expander instead of a J-T valve (Fig. 2.).

DESCRIPTION OF CONVENTIONAL AND SVC HELIUM CYCLES

In the conventional helium cycle the low pressure of the cycle is fixed by the saturation pressure of helium at the required temperature. The high pressure of the cycle is selected for a high work per unit of mass flow through the expanders with good expander efficiency. For reciprocating expanders this pressure ratio is usually about 15 to 20. At these levels the high pressure to the J-T stage is above the Joule-Thomson inversion pressure corresponding to the temperature at the warm end of the J-T stage. Thus in the conventional cycle the J-T stage would actually run more efficiently at a reduced high pressure, but would require a system with three pressure levels in the heat exchangers and in the compressor.

The conventional cycle with a supercritical wet expander replacing the J-T valve, Fig. 2, is a system with a lowest temperature stage that can efficiently utilize the same high pressure

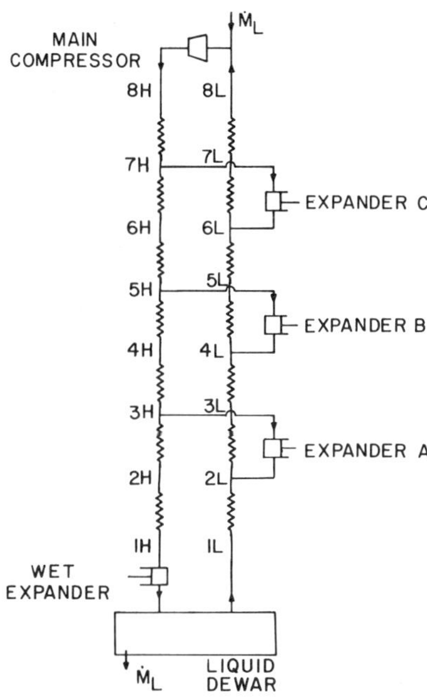

Fig. 2. Conventional 4.2. K liquefier
cycle with wet expander.

level as the higher temperature stages. A helium liquefier with a
hydraulically operated supercritical wet expander has been operating quite effectively at the MIT Cryogenic Engineering
Laboratory. The performance of the system was reported[1] several
years ago when it was first installed.

In the SVC helium cycle (Fig. 1) a cold compressor is used to
compress saturated vapor from the liquid dewar up to the low
operating pressure of the cycle. With both the wet expander and
the cold compressor, the pressures in the cycle are independent of
the liquid dewar pressure. The cycle pressures can now be selected for the best performance of the expanders and heat exchangers
in the cycle.

The SVC helium cycle has two potential applications. The
first is improved equipment for the production of atmospheric
pressure liquid helium. The second application is for refrigeration at temperatures below 4.2 K down to 1.8 K. In this application the cycle pressures are at conventional levels and the wet-
expander cold-compressor section is used to maintain sub
atmospheric pressures over the boiling liquid helium. With a two-

stage cold compressor the vapor from the liquid boiling at 1.8 K will be compressed to the normal atmospheric low side pressure of the helium cycle. The wet expander will expand supercritical helium directly to 12 mbar (1.8 K). This system will allow the conversion of a standard helium liquefier to a 1.8 K refrigerator without the addition of large low pressure (12 mbar) heat exchangers and large displacement vacuum pumps. Preliminary calculations indicate that the system will have higher efficiency than conventional 1.8 K refrigerators.

OPTIMIZATION OF SATURATED-VAPOR-COMPRESSION CYCLE

The first step in the study of the SVC cycle was the selection of the configuration for the cold end (below 80 K). The arrangement shown in Fig. 1 evolved as the best system from a family of arrangements of heat exchangers, two precooling expanders, a supercritical wet expander, a J-T valve and a cold compressor. For each configuration, the optimum expander inlet temperatures (States 3 and 5) were selected for minimum mass flow at state 6 for a given liquefaction rate, and fixed heat exchanger stream to stream temperature difference at states 2 and 4. Figure 3 shows the influence of the expander inlet temperature on cycle performance. For each configuration considered, the stream to stream ΔT in the heat exchangers were checked for internal pinch points to insure that the specified ΔT was at the location of the minimum ΔT.

The second step in the study of the SVC cycle was the evaluation of the influence of the cycle pressure level at constant cycle pressure ratio. A pressure ratio of 10 was selected for the study. The performance of the cycle clearly improves with increasing pressure level for fixed stream to stream temperature differences in the heat exchangers ($\Delta T/T$ constant at the pinch point for each stage) and for fixed expander and compressor efficiencies. For pressures of 10 atm and 1 atm, 11.6% of the high pressure stream at state 5 is liquefied. For higher pressures the values are: at 25 atm and 2.5 atm, 16.1%, at 30 atm and 3 atm, 16.8% and at 40 atm and 4 atm, 17.3%.

The last step in the study was the optimization of the conventional wet expander cycle (Fig. 2) for comparison with the optimized SVC cycle. The SVC cycle at pressures of 40 atm and 4 atm was selected because 40 atm was judged a reasonable upper limit for a main compressor of standard design. Although compressors for higher pressures are available, the higher pressures introduce additional problems. The conventional cycle was optimized with the same pressure ratio, the same expander efficiencies and the same values of $\Delta T/T$ at the minimum heat exchanger temperature difference point for each stage as were used for the SVC cycle.

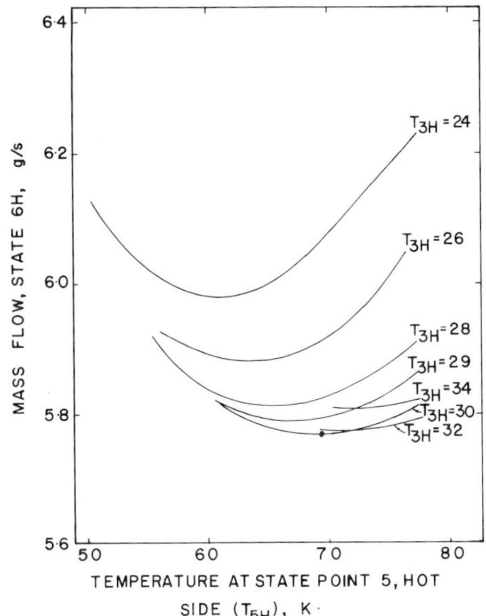

Fig. 3. Optimization of expander inlet temperatures
for $\Delta T_2/T_{2H} = \Delta T_4/T_{4H} = 1/40$, $P_H = 40$ atm, $P_L = 4$ atm.

The results shown in Table I compare the performance of the
conventional and the SVC cycles both optimized on the same basis. The SVC cycle requires 62% less heat exchanger surface while
at the same time liquefying 16% more of the main compressor
flow. Since the compressor power is determined by the mass flow
rate and the pressure ratio, not the pressure level, the SVC cycle
requires 13% less compressor power in addition to the smaller heat
exchanger. These advantages must be balanced against the additional cold compressor required for the SVC cycle.

A liquid nitrogen precooled SVC cycle was compared with a
conventional liquid nitrogen precooled cycle in the same way the
other cycles were compared. Similar results were obtained.

DISCUSSION OF RESULTS

The reduction of the heat exchanger surface required for the
SVC cycle is the result of higher heat transfer coefficients
available at the higher pressure.[2,3] Since the expander performance is effected by pressure ratio rather than the magnitude
of the pressure, the SVC cycle can have the same loss of pressure
ratio due to flow resistance as the conventional cycle. Since the
pressure has been increased by a factor of four the allowable
pressure drop can be increased. This allows a higher mass veloc-

Table I. Cycle Date

	Conventional Cycle				SVC Cycle			
state	T_{hot} (K)	T_{cold} (K)	area (m^2)	mass$_H$ (g/s)	T_{hot} (K)	T_{cold} (K)	area (m^2)	mass$_H$ (g/s)
1	8.1	4.2			11.0	7.6		
			5.57	51.9			3.69	51.1
2	11.1	10.8			16.3	15.9		
			59.76	51.9			28.97	51.1
3	22.0	18.9			32.0	28.7		
			28.65	100.2			7.03	88.5
4	29.5	28.7			37.9	37.0		
			124.2	100.2			53.34	88.5
5	60.0	52.2			77.0	68.8		
			89.72	153.0			24.54	132.7
6	94.7	92.3			105.6	102.9		
			205.1	153.0			33.18	132.7
7	180.0	162.2			200.0	182.3		
			227.5	207.1			69.2	178.5
8	300.0	295.0			300.0	295.0		

ity (mass flow rate per unit cross section) through the heat
exchanger. The higher mass velocity gives higher heat transfer
coefficients which allows a smaller surface for the same ΔT, and
thus the same cycle performance. Figure 4 is a graphic comparison
of the heat exchanger surface required for the two cycles. The
data points represent the total heat exchanger surface from the
cold end up through the stage with warm end temperatures indicated
by the data point. The straight lines are for visualization and
are not intended to represent the non linear temperature distribution in the actual heat exchangers.

The second reason that the elevated pressures of the SVC
cycle give better performance is that the heat capacity rates
(mc_p) of the two streams are better matched in the heat exchangers. As is shown in Fig. 5 the cycles have each been designed
with $\Delta T/T$ the same at each of the three low temperature pinch
points in the heat exchanger stack, and with the same ambient
temperature ΔT. It can be seen that $\Delta T/T$ is smaller at each of
the four points of maximum ΔT. Thus the SVC cycle has a lower
irreversibility from heat transfer through a finite ΔT with the
same minimums in $\Delta T/T$. This results in liquefaction of a larger
fraction of the flow in the SVC cycle.

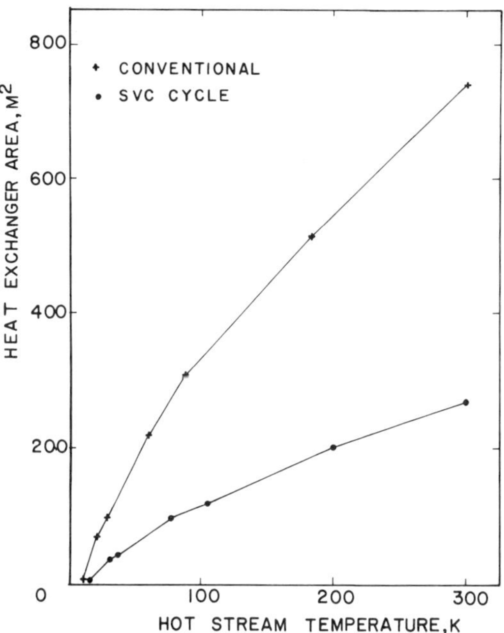

Fig. 4. Heat exchanger area in stages at temperatures below T_H.

Finally, the SVC cycle has improved performance because it takes maximum advantage of the potential of the wet expander by using the expander with a 40 to 1 pressure ratio. The wet expander liquefies 39.1% of the 11 K stream in the SVC cycle but liquefies 38.5% of the 8 K stream in the conventional cycle. Since the wet expander has a higher efficiency than the J-T valve, the SVC cycle will show a much better improvement than indicated above, when compared with the conventional cycle using a J-T valve.

SUMMARY

This optimization study has shown that the SVC cycle has the potential for significant improvements of helium liquefiers producing atmospheric pressure liquid. The SVC cycle also shows potential for refrigeration below 4.2. K down to temperatures as low as 1.8 K, even though an optimized design for a SVC 1.8 K refrigeration cycle has not been completed. The largest uncertainties in the study are the actual performances of a cold reciprocating compressor and a high-pressure-ratio wet expander. The encouraging results from this study have motivated the construction of apparatus to be used with a conventional helium liquefier

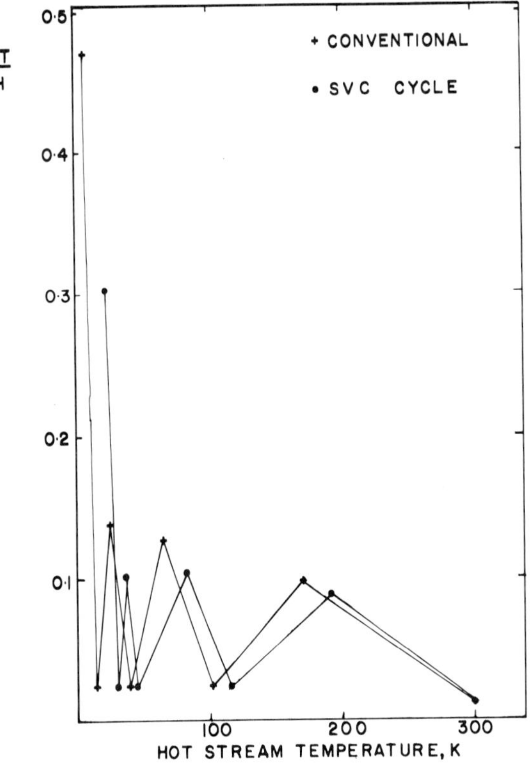

Fig. 5.   Cooling curves.

to test a wet expander and a cold compressor and to demonstrate
the potential of the SVC cycle.

REFERENCES

1. S.C. Collins, R.W. Johnson, and J.L. Smith, Jr., Hydraulically
     Operated Two-Phase Helium Expansion Engine, "Advances in
     Cryogenic Engineering, Vol. 16," Plenum Press, New York
     (1971), p. 171.
2. J.L. Smith, Jr., "The Presentation of Heat-Transfer and
     Friction-Factor Data for Heat Exchanger Design," ASME paper
     66-WA/HT-59, 1966.
3. W.M. Kays and A.L. London, "Compact Heat Exchangers," McGraw-
     Hill, New York, (1964).
4. M. Minta, M.S. Thesis, Massachusetts Institute of Technology,
     Cambridge, Massachusetts, 1981.

HEAT CAPACITY AND GEOMETRY IMPACTS ON
REGENERATOR PERFORMANCE*

B. R. Andeen

CTI-Cryogenics
Waltham, Massachusetts

INTRODUCTION

Analytic and experimental results indicate that for regenerative refrigerators providing cooling above about 70 K, the primary concern should be the matrix characteristic size. The specific matrix geometry chosen (mesh, particulate, etc.) has a direct impact on the efficiency and performance of the regenerative system and should be chosen depending upon the demands to be placed upon the system. However, for an expander with a reasonable swept volume and cycle rate, the matrix material and its volumetric heat capacity is of virtually no consequence. Also, axial variations in the characteristic size of the regenerator matrix can improve performance.

ANALYTIC RESULTS

Regenerator Materials

General considerations. In an ideal regenerator, the local variation in regenerator temperature is small in comparison to that of the gas flowing over the regenerator. This requires that the heat capacity rate of the regenerator (C_r) exceed that of the gas (C_g), or

$$(\rho c_p V/t)_r > (\rho c_p V/t)g \tag{1}$$

---

*Supported by the Air Force Wright Aeronautical Laboratory.

where V_r and V_g are the volume of regenerator active in the cyclic regenerative heat transfer, and the volume of gas processed by the regenerator (or roughly the cold end swept volume), t is the time for the thermal interaction, ρ is the density and c_p is the specific heat.

The volumetric heat capacity (ρc_p) of metals is a strong function of both temperature and the material. The wide disparity in ρc_p between different materials is shown in Fig. 1. Also shown is the volumetric heat capacity of helium, the usual working fluid in closed cycle cryogenic refrigerators. Equation 1 leads one to believe that the higher the volumetric heat capacity, the better the regenerator performance, and, therefore, a popular decision is to use nickel for the regenerator.

However, it can be shown[1] that the regenerator effectiveness is a weak function of C_r/C_g if this ratio is large. Figure 1 indicates that in the 70-300 K range a large C_r/C_g is virtually guaranteed for many materials unless:

- The cold expansion volume is large with respect to the total regenerator volume.

- The cycle time is small with respect to the time constant of the regenerator material. The resulting strong thermal gradients in the matrix elements would essentially reduce

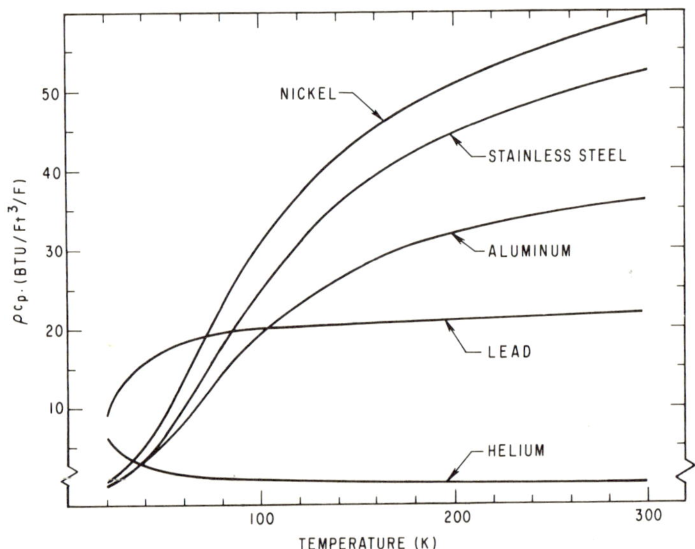

Fig. 1. Volumetric Heat Capacity of Various Materials.

the effective regenerator volume. At element sizes and cycle rates typical of current coolers (100-1500 RPM), metallics have sufficiently high thermal conductivities that they behave essentially as isothermal bodies. Materials with low conductivities (plastics, composites, etc.) will experience a reduction in the effective volume at high cycle rates and/or large matrix characteristic dimensions.

Model Results

All relevant modeling was performed using a second order[2] integral Stirling model. The model assumes that the matrix elements behave as isothermal bodies. The loss mechanisms of concern are: regenerator ineffectiveness, resistance to gas flow, and axial thermal conduction.

Table I lists the predicted losses for different wire mesh regenerator materials. The only difference between cases is the regenerator material. Several interesting observations may be made from Table I.

- The flow losses due to the regenerator pressure drop are identical: the pressure drop is a function of the matrix geometry, not material.

- The losses due to the regenerator effectiveness are all within 5% of each other.

- Those materials that exhibit high values of (ρc_p) generally are also good conductors. Therefore a smaller loss due to the regenerator ineffectiveness is usually balanced by a larger loss due to conduction down the regenerator stack. The sum of these two losses is essentially constant.

- The one nonmetallic listed has a predicted net performance superior to all others, primarily because of the low axial conduction loss. However, for the cycle rate used, a nylon element does not behave as an isothermal body, and the regenerator loss is therefore underestimated.

Particulate regenerators exhibit the similar characteristics when modelled in the same way. The result is that the net performance of typical small cryogenic coolers in the 70-300 K range is insensitive to the regenerator material, so long as the matrix element behaves as an isothermal body. The use of plastics is limited to lower cycle rates and/or smaller gas volumes than could be used with metallics.

Table I. Predicted Losses for Different Regenerator Materials (Mesh Regenerator, T_{cold} = 80 K)

MATERIAL	REGENERATOR LOSS (WATTS)		
	CONDUCTION	FLOW	EFFECTIVENESS
Stainless	0.149	0.701	0.930
Phosphor Bronze	0.187	0.701	0.927
Nickel	0.201	0.701	0.926
Lead	0.179	0.701	0.960
Nylon	0.038	0.701	0.969

Matrix Type; Mesh Vs. Balls

The two common forms of regenerators (mesh and particulate) were examined analytically. The modeled regenerators differed in characteristic size and the type of matrix. The predicted net capacity of a typical cooler for the different regenerators is plotted in Fig. 2. It can be seen that particulate regenerators have a fairly well defined optimum particle size and yield a greater cooling capacity than do meshes. Figure 3 plots the shaft power required per watt of load carrying capacity for the same cases as Fig. 2. It illustrates the relative efficiency of the cooler/regenerator combination.

These calculations indicate the following trends:

- The system performance is much more sensitive to the matrix characteristic size than to matrix material.

- For a fixed size machine, greater capacity can be achieved using a particulate regenerator.

- Meshes exhibit efficiencies superior to particulate regenerators.

Hybrid Regenerators

The total loss of the regenerator is the sum of losses due to flow resistance, ineffectiveness and axial conduction. Since the viscosity of helium decreases with decreasing temperature, it can be shown that for a homogeneous regenerator (constant hydraulic

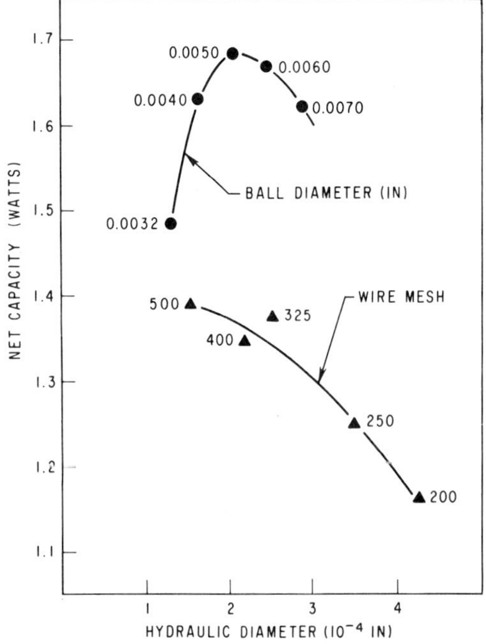

Fig. 2. Capacity at 80 K vs. Hydraulic Diameter (Mesh and Particulate Regenerators-Analysis).

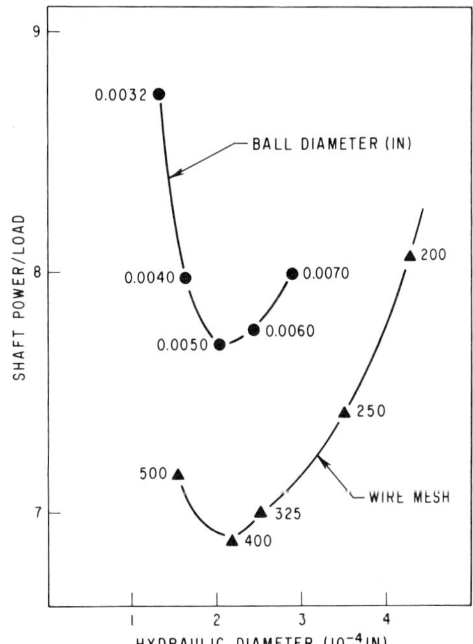

Fig. 3. Specific Power at 80 K vs. Hydraulic Diameter (Mesh and Particulate Regenerators-Analysis).

diameter) the flow resistance term is large at the warm end and the ineffectiveness term is large at the cold end. The sum of these losses can be reduced by decreasing the hydraulic diameter of the regenerator as the gas temperature reduces; i.e., employ a "smaller" matrix element at the cold end than at the warm end.

The hybrid regenerator concept was analyzed but was limited in that only two different matrix zones were considered. An optimum regenerator presumably would have a continuous variation in hydraulic diameter.

Figure 4 shows the influence which both particle size and packing fraction have on the ratio of input power to load. Several observations can be made:

- The larger diameter (and larger hydraulic diameter) matrix element should be used in the warm end of the regenerator.

- The system efficiency increases when a hybrid regenerator of proper proportions is used.

- The efficiency improves as the difference between the hydraulic diameters of the two subregenerators increases.

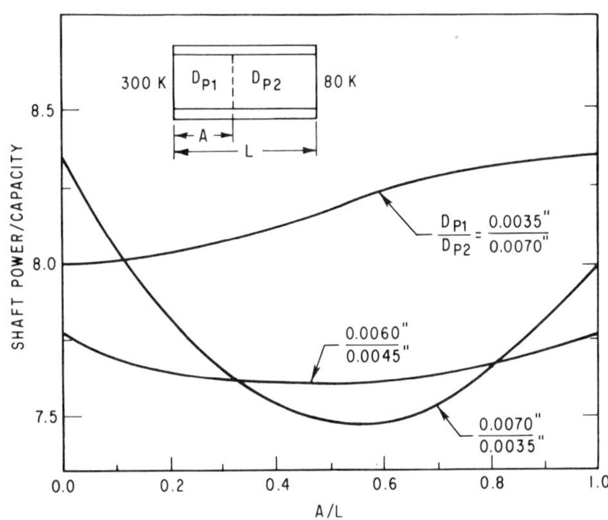

Fig. 4. Impact of Hybrid Particulate Regenerators on Specific Power. D_{p1} and D_{p2} are the diameters of the particulate material.

- A two zone hybrid regenerator can increase system efficiency by 5%. More regenerator ones probably would be more attractive.

EXPERIMENTAL RESULTS

Experimental testing was performed using a small integral Stirling machine whose swept volume was small in comparison to the regenerator volume. Figure 5 illustrates the experimental results using a phosphor bronze and a stainless steel mesh regenerator. The only experimental change in the system hardware was the material composing the matrix. Since phosphor bronze has a significantly larger volumetric heat capacity at low temperatures than stainless, traditional thinking would conclude that phosphor bronze would perform substantially better. In fact, the regenerators perform essentially the same. This experimentally demonstrates that regenerator performance can be insensitive to matrix material.

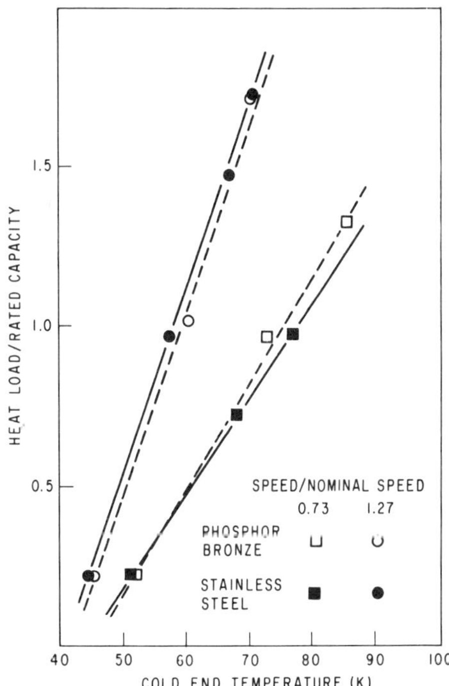

Fig. 5. Experimental Performance of Mesh Regenerators.

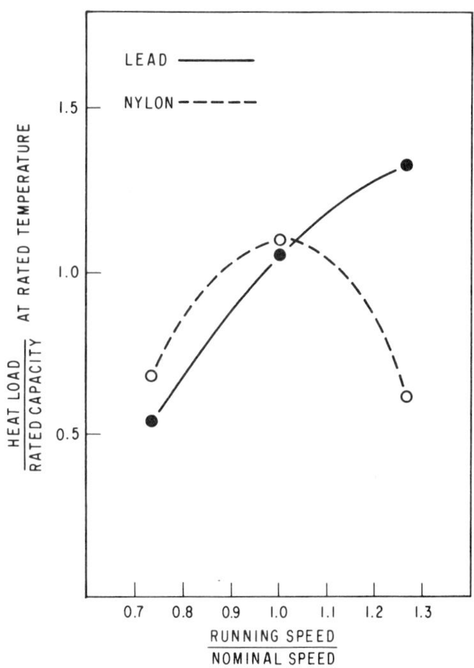

Fig. 6. Experimental Performance of Particulate Regenerators.

Experiments were also run using both lead and nylon particulate regenerators. The results are shown in Fig. 6. Specific conclusions are difficult to make because the matricies are not identical in characteristic size and shape, but trends can be identified:

- The experimental reduction in performance with increasing speed is indicative that the plastic elements are failing to act as isothermal bodies.

- Plastics can be used as regenerator material, provided the cycle rate is not too high.

No quantitative comparison can be made between the particulate and the mesh regenerators as the particulate size distribution was difficult to quantify and no attempt was made to identify and use the "optimized" particle size.

CONCLUSIONS

Several interesting and significant conclusions for small Stirling coolers result from this study:

1. Between 300 and 70 K, experimental and analytic results show that the regenerator effectiveness is a weak function of the matrix material (so long as the matrix elements behave as isothermal bodies) because the heat capacity rate ratio is large.

2. The effectiveness is primarily a function of the matrix element size. Typically, the smaller the size, the larger the effectiveness. Size is also influential in determining the flow losses through the regenerator, with smaller sizes having larger losses. The combined effects of flow loss and effectiveness are in part the reasons for the peaking of performance shown in Figs. 2 and 3.

3. The type of matrix to be used depends upon the system demand. To maximize load for a given size, a particulate regenerator is indicated. To maximize system efficiency, a mesh regenerator is indicated.

4. Materials for the regenerator for a machine reaching only to the 70–80 K level should be chosen primarily for their availability and tolerancing in the desired size. Other factors influencing the choice of a material include hardness, coefficient of thermal expansion, thermal conductivity, etc. Volumetric heat capacity has a minor influence.

5. Performance improvements may be achieved by employing a hybrid regenerator, one which varies geometrically in the axial direction. About a 5% improvement may be achieved with a two zone hybrid.

REFERENCES

1. W.M. Kays and A.L. London, in "Compact Heat Exchangers," McGraw-Hill Book Co., New York, (1964), Fig 2-34.
2. W.R. Martini, in "Stirling Engine Design Manual," Univ. of Washington, Seattle (1978), available as NTIS Document N 7823999.

DISCUSSION

Question by R.C. Longsworth, Air Products and Chemicals, Inc.:

Were composite screen regenerators studied and to what case does the 5% improvement apply?

Answer by author:

The hybrid analysis considered only two matrix zones, but each zone could have either a mesh or a particulate matrix. All four possible cases were examined (particle-particle, particle-mesh, mesh-particle, and mesh-mesh), but only the particle-particle case is presented (Fig. 4.) The other three cases exhibited the same trends as Fig. 4.

The 5% improvement refers specifically to Fig. 4, and compares the shaft power to capacity ratio of the best examined hybrid to that of the best homogeneous (A/L = 0 ~ 1.0) regenerator.

DEVELOPMENT OF AN OIL-FREE RESONANT PISTON COMPRESSOR FOR HELIUM LIQUEFACTION

P. W. Curwen and R. Hurst

Mechanical Technology Inc.
Latham, New York

INTRODUCTION

The use of superconducting electric propulsion systems for naval vessels has been under evaluation by the U.S. Navy for several years. These systems will require one or more shipboard helium liquefiers. The selected liquefaction cycle, requires a helium compressor operating at a 16:1 pressure ratio.

The major areas of concern with respect to the helium compressor, in addition to the high pressure ratio, are: oil separation (total contaminants introduced by the compressor must be less than 1.0 ppm); reliability (5000-hr MTBF); shock and vibration ruggedness; and noise level. Oil separation is important; the concern is unscheduled liquefier shutdowns due to oil carry-over and freezing in the low-temperature sections of the liquefier. Adequate performance of the oil separation equipment must be assured for maintenance intervals of not less than 2500 hours, and for all possible extremes of propulsion system operation, including environmental shock and vibration. Because of the uncertainty about the dependability of known oil separation techniques, an oil-free helium compression would be particularly desirable for this liquefier application.

DESCRIPTION OF OIL-FREE COMPRESSOR CONCEPT

The oil-free compressor to be described is the Resonant Piston Compressor (RPC). Inherent features of the RPC which match the Navy requirements are: (1) positive-displacement compression for high-pressure-ratio operation; (2) process gas lubrication (i.e., helium gas bearings) to eliminate the oil system, with attendent oil seals and oil separators; (3) simple, rugged construction; (4) potential for very long, maintenance-free life (making hermetic construction and installation a practical design approach).

The RPC is a reciprocating-piston compressor. However, unlike conventional reciprocating machines, the RPC is a spring-mass system which is driven in its resonant mode of vibration by a linear reciprocating electric motor. This mode of operation eliminates the need for converting prime-mover rotary motion into linear reciprocating motion as is normally done via oil or grease lubricated crankshaft/connecting-rod or cam/cam-follower drives. As a result, the number of bearings in the RPC is reduced to a minimum --- their primary function being guidance of the recipro-cating assembly. Since all of the dynamic forces (inertial, pressure, electrical) acting on the reciprocating assembly do so in the direction of reciprocation, the bearings are always lightly loaded, thus making gas lubrication feasible.

To achieve resonance of the RPC at the frequency of the linear motor (60 Hz), a means for dynamic energy storage must be provided. This is typically accomplished by means of one or more springs. Mechanical springs, however, are not practical for this application because of the relatively high stroke and mass of the reciprocating assembly. Instead, gas springs are used, thus making the RPC a free-piston machine (i.e., there is no mechanical connection between the reciprocating and "stationary" parts of the machine).

DESIGN FEATURES OF THE HELIUM RPC

Under contract to the U.S. Navy, Mechanical Technology Incor-porated (MTI) has designed a two-stage RPC for helium liquefac-tion.[1] Overall pressure ratio of the two stages is 16:1; design flow rate at this pressure ratio is 90 SCFM (54 lb/hr dry helium); design suction and discharge pressures are 2.0 and 250 psig re-spectively.

The compressor consists of two single-stage RPC units. The first-stage unit operates at 4.4 pressure ratio, the second-stage at 3.7. Each single-stage unit consists of two parallel-flow cylinders driven by a 17-hp linear motor. Interstage cooling is used to maintain safe operating temperatures in the second-stage cylinders. Second-stage aftercooling is used to reduce the deliv-ery temperature to 100° F. During operation helium pressure and flow for all of the hydrostatic gas bearings in the two-stage machine is provided from the second-stage discharge, with the flow returned to the second-stage inlet. The design bearing flow is about 9% of the total second-stage flow. For compressor start up, bearing flow must be supplied from a pressurized reservoir or by a small auxiliary compressor.

Figure 1 is a schematic cross section of the second-stage RPC. Except for the diameters of the compression cylinders, and the sizes of the suction and discharge valves, the designs of the first and second-stage units are identical. Immediately adjacent to each end of the linear motor are the main support bearings. Attached to the outboard end of each bearing is a 10-inch diameter double acting gas-spring piston. The combined stiffness of the four gas-spring cylinders is about 14,600 lb/in, which is about 86% of the total stiffness needed to resonate the 46-pound mass of the reciprocating assembly.

The working cylinder pistons are driven by flexure rods attached to the gas-spring pistons. Each working piston is guided by a small hydrostatic gas bearing located concentrically around the flexure rod. Conventional cantilever-reed suction and discharge valves are mounted on the cylinder head of each working cylinder.

A key feature of the helium RPC is the complete absence of any rubbing seals or sliding surfaces. Radial clearances, ranging from 0.001 to 0.002 inches, are maintained around the various pistons and in the hydrostatic bearings. Although these clearances give rise to leakage losses in both the working and gas-spring cylinders, the elimination of rubbing and sliding surfaces is key

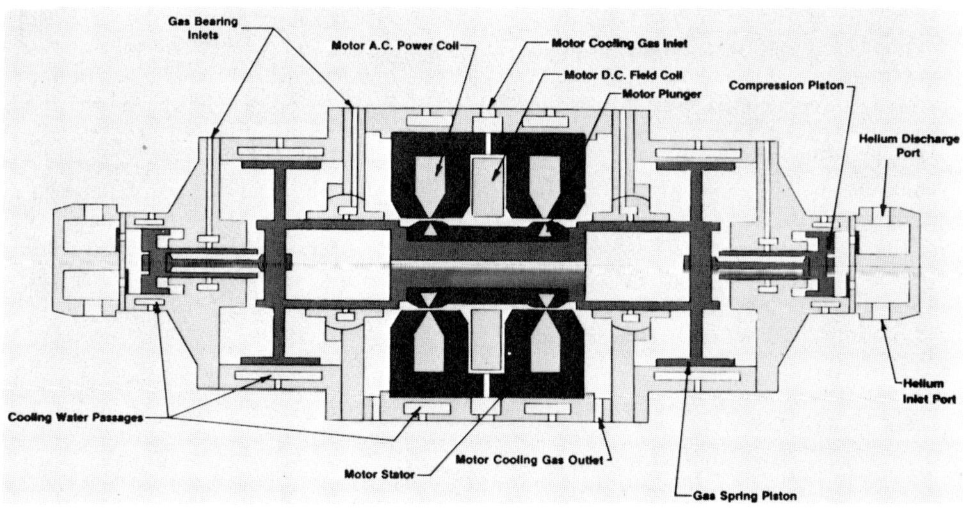

Fig. 1. Schematic Cross Section of Second-stage Helium RPC.

to providing long life and contaminant-free operation of the
RPC. The overall isentropic efficiency of the fully developed,
two-stage compressor is predicted to be a reasonable 59%, based on
net useful output flow, after deducting the bearing gas flow.

Another feature of the RPC arises from its free-piston nature. Since there is no kinematic linkage forcing the pistons to
reciprocate at fixed stroke, piston stroke can be easily varied by
changing motor voltage, or by changing gas-spring control pressure
(i.e., by changing gas-spring stiffness). The latter method is
usually simpler and requires less hardware expense. Helium flow
can thus be varied (modulated) at any pressure ratio condition to
produce the desired liquefaction rate. This modulation capability
will minimize compressor energy consumption over any mission
profile.

DEVELOPMENT OF THE HELIUM RPC

Motor Development

The major challenge for this compressor development was the
linear motor. Prior motor experience at MTI was limited to 3.5
hp, with a plunger specific power of 0.22 hp/lb. For this development, a 17-hp motor with a plunger specific power of 0.63 hp/lb
was desired. An initial motor design produced only one-half of
the design power as a consequence of insufficient pole area, thus
achieving a plunger specific power of only 0.31 hp/lb.

A second motor was developed having increased pole diameter,
pole area and plunger weight. This motor did demonstrate a 17-hp
capability at a plunger specific power of 0.45 hp/lb. Because of
its larger size, incorporation of this motor into the compressor
would have required redesign of the major compressor castings.
Furthermore, the increased weight of the motor plunger would have
required an increase in gas-spring stiffness, accompanied by an
increase in gas-spring losses. For these and also budgetary
reasons, it was decided to build and test the second-stage RPC
using the original, underpowered motor. It was recognized that
second-stage, design point conditions would not be achieved with
this motor. However, all other features of the design could still
be demonstrated and evaluated.

Compressor Test

Figure 2 shows the second-stage RPC installed in the compressor test loop. Figure 3 shows the second-stage reciprocating
assembly. All testing was done with helium.

Fig. 2. Second-stage Helium RPC in Compressor Test Loop.

Figure 4 shows the measured second-stage tare losses as a function of compressor stroke. These losses are predominately gas-spring losses. This data was obtained by removing the suction and discharge valves from the working cylinders, and shows the tare losses to be about 3.5 kW at the 1.1-inch design stroke.

Fig. 3. Reciprocating Assembly of Second-stage Helium RPC.

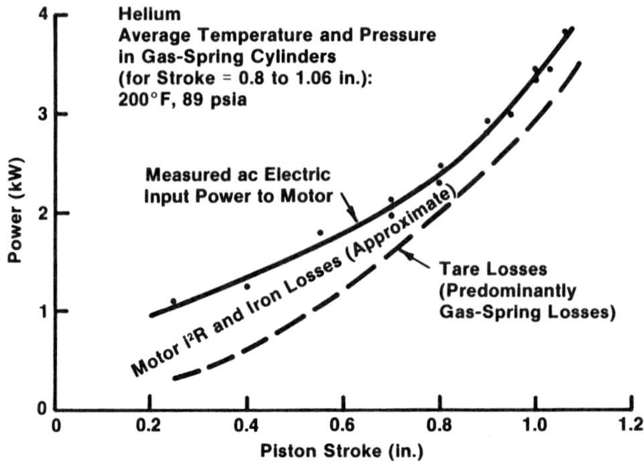

Fig. 4   Tare Losses in Second-stage Helium RPC.

This is 48% higher than predicted. Since gas-spring leakage
losses vary as the cube of piston radial clearance, it is felt
that the high tare losses were probably due to slightly excessive
piston clearance. At no time during the test program was there
any evidence of piston rubbing. Unfortunately, time and funds did
not permit any effort to reduce this clearance.

As a consequence of the power consumption of the gas springs,
only 4 to 5 horsepower of the available motor power remained to
compress helium. Consequently, to demonstrate high-pressure-ratio
operation it was necessary to greatly reduce the suction pressure
relative to second-stage design pressure. Figure 5 shows measured
P-V diagrams for the two working cylinders at a stroke of 0.88
inches. Suction and discharge pressures were 26.7 and 120.7 psia
respectively, giving a pressure ratio of 4.5 (design point ratio
being 3.7). Analysis of the P-V diagrams shows total P-V compression power (for two cylinders) to be 4.06 hp. Valve operation is
seen to be very good, with valve losses representing only 10.6% of
the total compression power.

Also plotted in Fig. 5 are the ideal (zero leakage) isentropic compression and expansion lines. Comparison of these lines
with the actual test lines clearly shows the presence of leakage
through the piston clearance seal. Without going into the details
of the alternating (oscillatory) leakage flow, it can be qualitatively shown that the relative positions of the actual and ideal
compression and expansion lines are as one would expect.

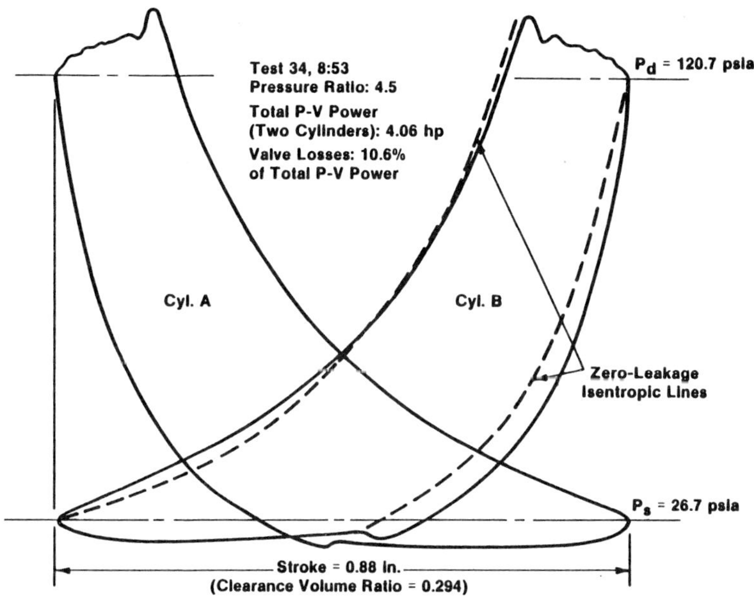

Fig. 5 Typical Pressure vs. Stroke (Volume) Diagrams
for Second-stage Helium RPC.

Because of insufficient flow instrumentation, isentropic
compression efficiency for the Fig. 5 test point could only be
determined as being within the range of 82 to 88%. Since valve
pressure-drop losses alone would yield a compression efficiency of
89.4%, it is clear that leakage losses accounted for somewhere
between 1.4 to 7.4 points of efficiency loss. As piston stroke
increases,the effect of leakage on compression efficiency is
reduced. This was clearly verified by the test data; for a 0.935-
inch stroke, the lower bound of compression efficiency increased
to 86.3%.

Efficiency of the linear motor during the (necessarily)
reduced-stroke testing was 77 to 78.5%. It is characteristic of
this type of motor that efficiency increases with increased
stroke. Efficiencies of 83% have been achieved for this type of
motor[2], and a design-stroke efficiency of 85% is a reasonable
development goal.

Fifty-four hours of compressor operation were accumulated
during the initial testing. For most of this period, the hydrostatic gas bearings were sustained by the second-stage RPC.
Operation of the bearings was very close to prediction and no
problems were encountered. Operation of the reciprocating assembly without sliding contact was fully demonstrated.

CONCLUSION

In prior development programs, small gas-lubricated RPC's
have been tested for 4000 hours and longer without failure[3]. The
shock ruggedness of properly designed gas bearings has also been
vividly demonstrated[4]. Based on the initial test results reported
here, there is every reason to believe that the U.S. Navy requirements for compressor reliability and ruggedness can be achieved by
the oil-free RPC concept.

With regard to RPC performance (flow and efficiency), projections from the test results reported here confirm that with modest
development the originally predicted performance levels should be
achieved.   However, a breakthrough in linear motor design, in
terms of plunger specific power (hp/lb), must be achieved.   Although design power was achieved with the second development
motor, a motor having less plunger weight is required to achieve
compressor design efficiency through reduction of gas-spring size
and associated losses.   To this end, a high level of in-house
activity is currently underway at MTI to evaluate two "breakthrough" motor concepts which, if successful, could be applied to
this helium compressor application.

REFERENCES

1.   J. McCormick and S. Gray, Design of a linear-motor resonant-
       piston compressor (LMRPC) for a shipboard helium cryogenic
       refrigerator, MTI Report 75TR50, prepared under U.S. Navy
       Contract N00024-73-C-0440, (June 20, 1975).
2   P.W. Curwen and A.W. Liles, Development of a Free-Piston
       Variable Stroke Compressor for Load-Following Electric Heat
       Pumps, in "Proc. of EPRI-RWE Heat Pump Conference,"
       Dusseldorf, West Germany, (June 1980).
3.   P.W. Curwen, Recent developments of oil-free linear-motor
       resonant-piston compressors, ASME Paper 69-FE-36, Applied
       Mechanics and Fluids Engineering Conference, Evanston,
       Ill., (June 1969).
4.   A. Frost, Evaluation of gas bearings for use in naval
       machinery under conditions of high impact, MTI Report
       70TR11, prepared under U.S. Navy contract N000-14-66-C-
       0282, (Feb. 1970).

DISCUSSION

Question by J.C. Riple, Garrett-Airesearch Corp.:   What are the
vibration characteristics of the machine?

Answer by Author: It is intended that the 1st and 2nd compressor
stages will be assembled together on the same axial centerline.

The piston-plunger assemblies of each stage will be operated 180°
out-of-phase resulting (ideally) in cancellation of vibration
forces and zero vibration of the compressor casing. Calculations
show that in actuality, transmitted dynamic forces due to mechanical and operational differences between the two stages should be
less than five pounds.

During testing of the second-stage unit by itself, vibration
amplitude at a 1-inch piston stroke was about 0.032 inches, which
produced a dynamic force (calculated) of about ±34 pounds transmitted to the compressor base.

Question by K. Bern: What is stroke variation when running at
resonance?

Answer by speaker: Stroke variation at true steady-state conditions if very small. "Real world" variations, e.g. changes in
voltage or frequency to the electric motor can cause changes in
stroke due to the high-Q nature of the of the resonance condition.   Under our laboratory operating conditions, we found the
stroke variation due to these effects to be about ± 0.010 inches.

Changes in compressor operating (suction or discharge) pressures, however, cause significantly larger changes in stroke. For
most applications, the stroke will have to be actively controlled,
using methods described in the paper.

ANALYSIS AND OPTIMIZATION OF A LINEAR MOTOR FOR THE COMPRESSOR OF A CRYOGENIC REFRIGERATOR

A. K. de Jonge

Philips Research Laboratories
Eindhoven, The Netherlands

and

A. Sereny*

Philips Laboratories
Briarcliff Manor, New York

INTRODUCTION

Several free-piston, free-displacer (FPFD) Stirling cryogenic refrigerators are currently under development at various Philips laboratories. While the thermodynamic design of a Stirling machine is well understood, the application of a voice-coil-type linear motor to the Stirling piston compressor is relatively recent. The FPFD technology has been applied to a 1 W, 77 K Stirling refrigerator produced by N.V. Philips, Eindhoven under the designation UA 7011. In a recent paper, de Jonge[1] gave a simplified description of the pressure fluctuations and provided a theoretical model that explains the behavior of the FPFD type machine. This cooler, with a linear motor inside a hermetically sealed case, is a concept demonstration of a unit having long life, simple construction, and negligible degradation. In this report, an analytical model for the linear electromagnetic drive system is given. This study led to the preparation of basic design curves that clearly explain the behavior of such magnetic devices under various operating conditions.

---

*Xerox Corp., Palo Alto Research Center, Palo Alto, California.

THEORETICAL MODEL

A schematic drawing of the proposed moving-coil linear electromagnetic drive system is given in Fig. 1. The major element of the system, the motor, consists of a moving coil that is directly coupled to the piston, which is free to reciprocate in a gap formed by a stationary permanent magnet and iron assembly. In a configuration where the ratio of gap length to gap width is large, a nearly uniform magnetic field is established across the gap. Within the gap, a coil can reciprocate as a result of an applied ac. The piston and linear motor are both sealed in a hermetic envelope with electric feed-throughs. The center position of the coil-piston assembly is held and controlled by a mechanical spring.

For the theoretical analysis of the linear motor, the following assumptions are introduced:

*   The magnetic field in the gap is radial, constant and uniform; the field outside the gap is zero.

*   The power source voltage, E_k, has only a first harmonic and is given by,

$$E_k = E_o \cos\beta \tag{1}$$

*   Axial coil position also has only a first harmonic, and is given by,

$$y = \hat{y}\cos(\beta - \phi) \tag{2}$$

*   The self inductance and resistance of the coil are constant and independent of coil position.

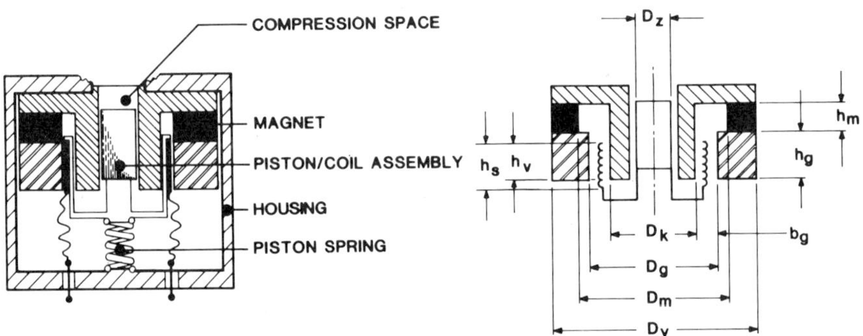

Fig. 1. Schematic of a moving-coil-type linear electromagnetic drive system showing various design parameters.

The schematic describing the electrical circuit of the linear motor is given in Fig. 2. For this electric circuit, the following elements were defined:

E_m = induced voltage $\qquad\qquad$ R_w = resistance of coil
L = self inductance of coil $\qquad$ E_k = voltage of power source
R_u = resistance of power source

The induced voltage in the motor can be calculated as,

$$E_m = B_g \ell_v \dot{y} \tag{3}$$

where, B_g = magnetic flux density in gap.

ℓ_s = length of coil wire
ℓ_v = length of coil-winding = $\ell_s h_g h_v / (h_s h_g)$.
$\dot{y}$ = velocity of coil in axial direction.

The relationship between the voltage and current, i, of the circuit described in Fig. 2 is:

$$E_k = E_m + i(R_w + R_u) + L di/dt \tag{4}$$

or,

$$R_L \frac{di}{d\beta} + i = \frac{E_o \cos\beta + B_g \ell_s \dfrac{h_v}{h_s} \dot{y}\omega \sin(\beta - \phi)}{R_w + R_u} \tag{5}$$

where, $R_L = \omega L/(R_w + R_u)$ and w = running frequency.

Note that while h_g, ℓ_s and h_s are constants, (see Fig. 1); h_v, the coil length in the field, could vary with time. One of the difficulties in solving Eq. 5 is the non-dimensional periodic function of time, h_v/h_g. For $h_v/h_g > 0$ the following two cases can be studied:

CASE 1: The coil is longer than length of the magnetic gap, i.e., $h_o/h_g > 1$.

$$\frac{h_v}{h_g} = 1 \qquad \text{for} \qquad 0 < \frac{|y|}{h_g} \leq \frac{h_s/h_g - 1}{2}$$

$$\frac{h_v}{h_g} = \frac{h_s/h_g + 1}{2} - \frac{|y|}{h_g} \qquad \text{for} \qquad \frac{h_s/h_g - 1}{2} < \frac{|y|}{h_g} \leq \frac{h_s/h_g + 1}{2} \tag{6a}$$

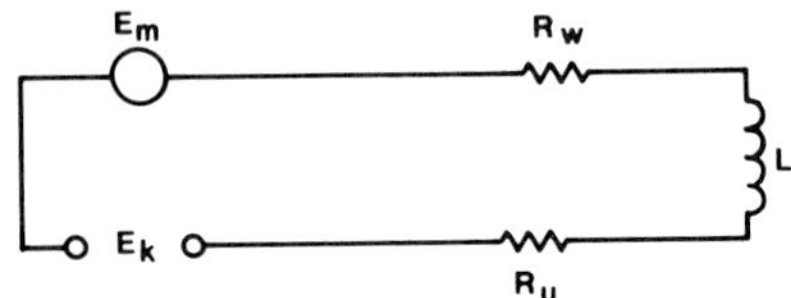

Fig. 2. Schematic describing a linear motor.

CASE 2:   The coil is shorter than length of the magnetic gap, i.e., $h_s/h_g \leqslant 1$.

$$\frac{h_v}{h_g} = \frac{h_s}{h_g} \quad \text{for} \quad 0 < \frac{|y|}{h_g} \leq \frac{1 - h_s/h_g}{2}$$

$$\frac{h_v}{h_g} = \frac{h_s/h_g + 1}{2} - \frac{|y|}{h_g} \quad \text{for} \quad \frac{1 - h_s/h_g}{2} < \frac{|y|}{h_g} \leq \frac{1 + h_s/h_g}{2} \tag{6b}$$

A schematic description of the periodic functional relation between h_v/h_g and $\beta - \phi$ is given in Fig. 3, where the angle Γ is defined as follows:

$$\cos\Gamma = \frac{|1 - h_s/h_g|}{2\hat{y}/h_g} \quad \text{for} \quad \frac{|1 - h_s/h_g|}{2\hat{y}/h_g} \leq 1 \qquad \cos\Gamma = 1 \quad \text{for} \quad \frac{|1 - h_s/h_g|}{2\hat{y}/h_g} > 1 \tag{7}$$

With the foregoing restrictions and the help of Fourier analysis Eq. 5 can be solved.

CALCULATION OF POWERS

Now, having a closed form analytical expression for the current in the coil, various powers relating to the motor operation are calculated.   The following three kinds of power are discussed.

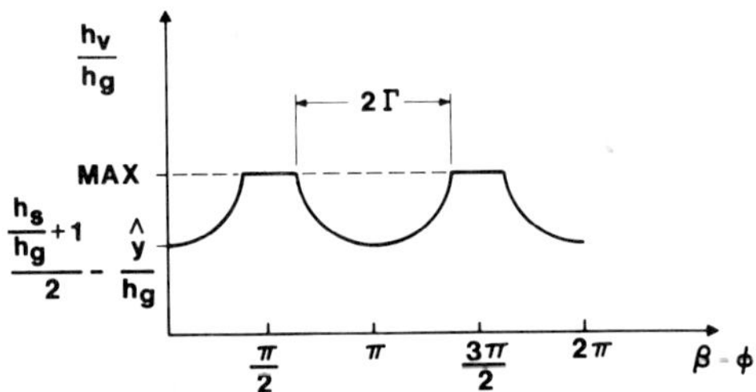

Fig. 3.   Coil length h_v/h_g as it varies through one cycle. Max = 1, where $h_s/h_g \geqslant 1$, Max = h_s/h_g where $h_s/h_g < 1$.

* Electric input power to motor, P_k.

* Output power of linear motor, P_p.

* Power dissipated in the form of heat in the course of transforming the electric to mechanical power, P_w.

To calculate the electric input power, evaluate the following integral,

$$P_k = \frac{1}{t_p} \int_o^{t_p} E_k i \, dt = \frac{1}{2\pi} \int_o^{2\pi} E_k i \, d\beta \tag{8}$$

Now, the magnetic power, P_m, is introduced. This quantity, characterizing the magnetic circuit of a linear motor at a given operating frequency is defined as:

$$P_m = (B_g \omega h_g)^2 f V_g / (2\rho_E) \tag{9}$$

where, f = fill factor = $A\ell_s / (V_g (h_s/h_g))$.

V_g = volume of gap = $h_p \pi (D_g^2 - D_k^2)/4$.

A = cross-section area of wire.

Using the definition of P_m, Eq. 8 transforms to,

$$P_k = \frac{1}{\pi} P_m \frac{1}{h_s/h_g} \int_o^{2\pi} E' i' \cos\beta \, d\beta \tag{10}$$

where $E' = E_o h_s / (B_g \omega h_g^2 \ell_s)$, $\quad i' - i\rho_E h_s / (B_g \omega h_g^2 A)$, and

ρ_E = specific resistance.

Next, an expression for the mechanical output power of the linear motor is calculated.

$$P_p = -\frac{1}{t_p} \int_o^{t_p} B_g \ell_v i \dot{y} \, dt = -\frac{1}{t_p} \int_o^{t_p} E_m i \, dt \tag{11}$$

Now, using Eqs. 1, 9 and the expression for ℓ_v the following can be shown:

$$\frac{P_p}{P_m} = - \frac{1}{\pi h_s/h_g} \int_0^{2\pi} \frac{h_v}{h_g} \frac{\hat{y}}{h_g} \sin(\beta - \phi)\, i\, d\beta \qquad (12)$$

The heat dissipated in the motor windings and in the power-source can be obtained using the principle of energy conservation:

$$P_w = P_k - P_p \qquad (13)$$

The efficiency of this linear motor is now defined as,

$$\eta = P_p/P_k = f(\phi,\ \hat{y}/h_g,\ R_L,\ h_s/h_g,\ E',\ R_u/R_w) \qquad (14)$$

A more useful expression for the efficiency can be obtained after utilizing the relation given by Eq. 12, viz.,

$$\eta = P_p/P_k = f(\phi,\ \hat{y}/h_g,\ R_L,\ h_s/h_g,\ P_p/P_m,\ R_u/R_w) \qquad (15)$$

Examine now the motor efficiency and its dependence on the phase shift, ϕ. This can be done by observing $d\eta/d\phi$ while keeping all the physical motor parameters and the motor output power P_p fixed. The phase shift for maximum motor efficiency. ϕ_m, is obtained by solving $d\eta/d\phi = 0$.

With this phase shift, the maximum efficiency can be calculated as a function of the normalized amplitude $\hat{y}/h_g$ and the motor parameters R_L, h_s/h_g, P_p/P_m and R_u/R_w. For the specific case of $R_L = 0$ the phase shift $\phi_m = \pi/2$ and the maximum efficiency is given by the simple form,

$$\eta_m = \left(1 + \frac{(P_p/P_m)(h_s/h_g)}{(\hat{y}/h_g)^2 (h_v/h_g)^2} \right)^{-1} \qquad (16)$$

for cases where R_u is negligible and $h_v/h_g = $ constant.

DISCUSSION OF RESULTS

The magnetic power, introduced in Eq. 9, characterizes the magnet-iron circuit of a linear motor at an operating frequency. The magnetic power as defined in Eq. 9 is a function of the magnetic flux density, the height of the gap, the filling factor of the gap with wire, the volume of this gap, and the specific resistance of the wire, as well as a function of the running fre-

quency. This number (in watts) can be evaluated for any given
design, and is independent of the dynamics of the system.

Figure 4 gives the maximum efficiency as a function of h_s/h_g
for a constant $\hat{y}/h_g$ and various P_p/P_m. In this curve it can be
seen that the efficiency for a fixed magnet iron assembly with a
fixed power output initially increases with coil length, h_s/h_g,
and then drops with further increase of the coil length. This
figure also shows the existence of three regions of operations
with respect to coil length in gap, h_v/h_g. For the specific case
given in Fig. 4, these regions can be specified as follows:

$$0 \leqslant h_s/h_g \leqslant 0.8 \qquad\qquad h_v/h_g = h_s/h_g$$

$$0.8 < h_s/h_g \leqslant 1.2 \qquad\qquad h_v/h_g = \text{a periodic function of time}$$

$$1.2 < h_s/h_g \qquad\qquad h_v/h_g = 1.0.$$

An important observation from the curves in Fig. 4 is that
the efficiency is maximum where the coil has the same length as
the gap of the linear motor. It can also be seen that when the
coil is longer than the gap, a portion of it is not being used and

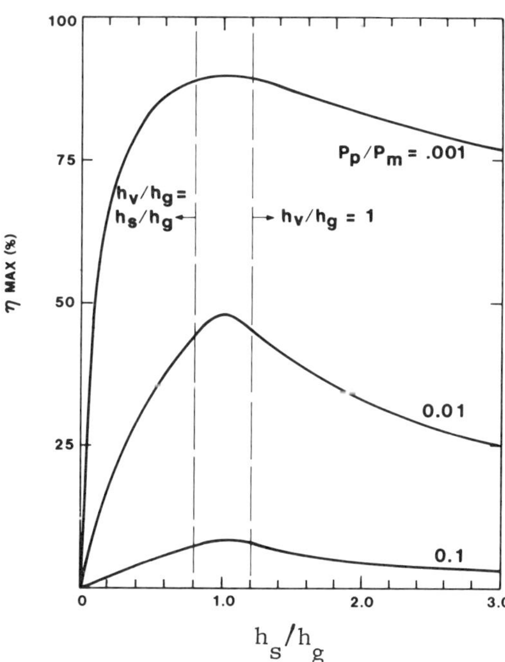

Fig. 4. Max. motor efficiency, η_{max} vs. normalized coil length, h_s/h_g, for normalized amplitude, $\hat{y}/h_g = 0.1$ and inductance $L = 0$.

the i^2R losses are larger and the efficiency drops. From the other end, when the coil is shorter than the gap, h_g, the magnetic field in the gap is wasted, and in turn the efficiency drops. From Fig. 4 it can be concluded that for small amplitudes, a good design will have a coil with a length equal to the gap length ($h_s/h_g = 1.0$).

In Fig. 5 the efficiency versus coil length for a larger amplitude is considered. Here, when a significant portion of the coil goes out of the gap, each cycle, a large dip in the efficiency can be observed in the region where h_v/h_g is a periodic function of time. The efficiency decreases because of the larger than expected currents in the coil, which is out of the magnetic field for the major portion of the cycle. It is important to note that the efficiency increases and then decreases with increasing magnetic power in the region where a portion of the coil moves out of the gap and that the efficiency increases with increasing magnetic power in the region where the coil length in the gap, h_v, is constant.

Design curves are presented in Figs. 6 and 7 for the configuration where $h_s/h_g = 1$. These curves represent maximum obtainable efficiencies, η_{max}, for given normalized power, P_p/P_m, versus normalized piston amplitude, $\hat{y}/h_g$, for specified R_L.

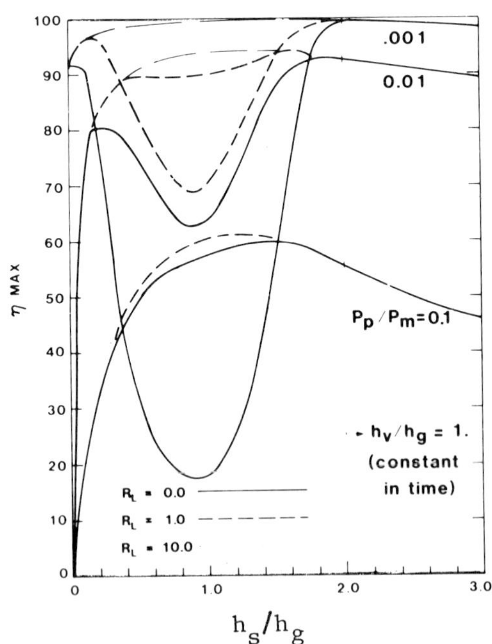

Fig. 5. Max. motor efficiency, η_{max} vs. normalized coil length, h_s/h_g, for normalized amplitude $\hat{y}/h_g = 0.5$.

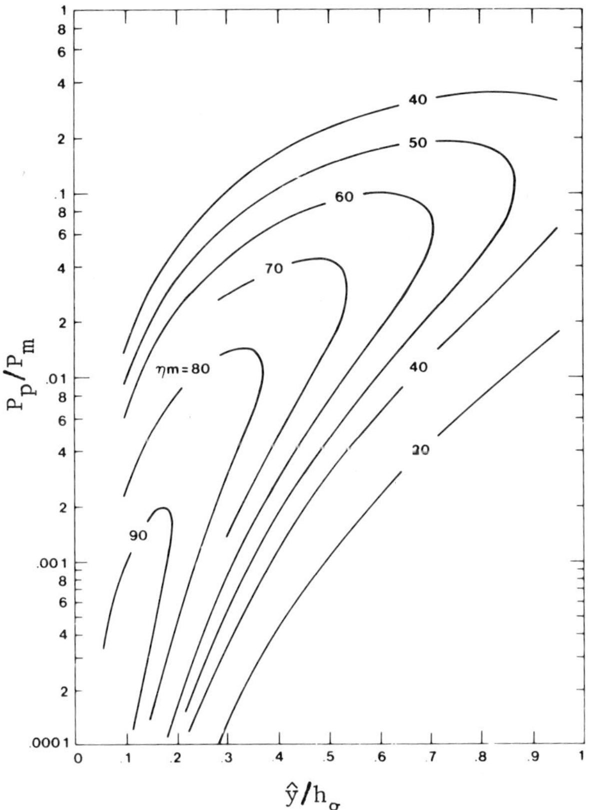

Fig. 6. Normalized motor power output, P_p/P_m vs. normalized amplitude of coil motion, $\hat{y}/h_g$, for $h_s/h_g = 1$ and $R_L = 0$.

SUMMARY AND CONCLUSIONS

1. The theory and analysis described above have been successfully applied to the design and optimization of linear motors for the compressor of cryogenic refrigerators.

2. Experimental compressors with moving coils and moving magnets have been built and tested and the results compared to this analysis.

3. Effects, such as eddy currents and hysteresis, that have not been considered in the analysis must be carefully examined for each design.

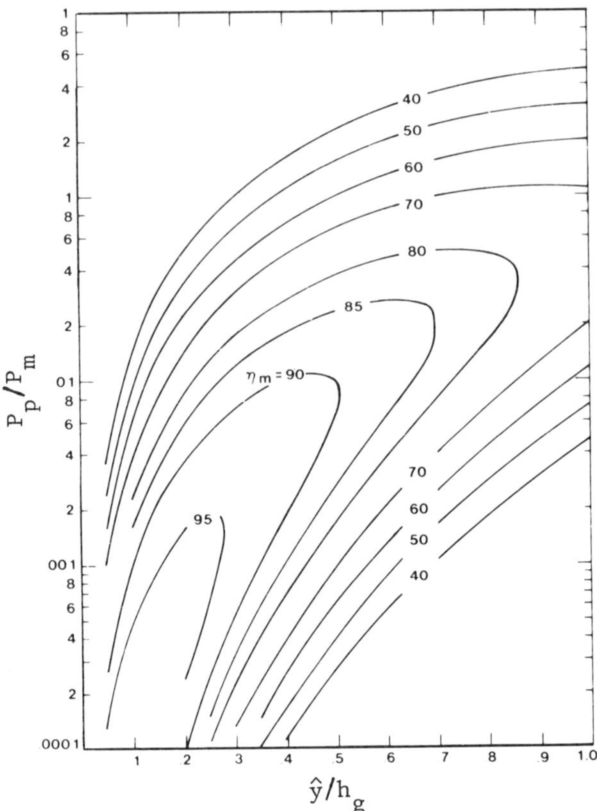

Fig. 7. Normalized motor power output, P_p/P_m vs. normalized amplitude of coil motion, $\hat{y}/h_g$ for $h_s/h_g = 1$ and $R_L = 1$.

4. Implementation of these designs to space-borne hardware is currently (1981) underway at Philips Laboratories.

REFERENCES

1. A.K. de Jonge, A Small Free-Piston Stirling Refrigerator, in "Proc. 14th Inter-Society Energy Conversion Conf.," American Chemical Society, New York, (1979).

HYDRAULICALLY CONTROLLED HELIUM EXPANSION ENGINE

G. Patton, G. Green, K. Dunn, and V. Dilling

*David Taylor Naval Ship Research and Development Center
Annapolis, Maryland*

INTRODUCTION

Helium liquefiers which use reciprocating piston expanders
have historically been designed with a crankshaft, flywheel, and
connecting rod to achieve the reciprocating motion of the piston. This mechanism is quite satisfactory for liquefiers in
stationary or laboratory service, but the stresses which arise in
this mechanism when the liquefier is exposed to a mobile environment make a reciprocating system with less mass and inertia more
attractive.

The helium liquefier, whether used as a refrigerator or
liquefier, is usually an ancillary device to cool a magnet or
electronic sensor. The personnel associated with the device may
not be experts in cryogenics or helium liquefiers. It would be
advantageous, therefore, if a liquefier could be started, cooled
down and operated automatically. The ability to adjust the refrigeration capacity and operating parameters of the device would
also lead to a system which is highly efficient and tailored to
the thermal load.

In an attempt to achieve these desirable operating features
and improve the tolerance to vibration and shock an electronically
controlled, hydraulically operated expansion engine was designed,
built and tested at David Taylor Naval Ship Research and Development Center (DTNSRDC).

HYDRAULIC ENERGY ABSORPTION

Rotary flywheel expanders store the energy extracted from the
expanded gas in rotational kinetic energy. Part of this energy is

used to drive the piston downward during the exhaust stroke with the rest of the energy being rejected into a brake mechanism.

Figure 1 shows the hydraulic expansion engine concept in its present form that has evolved from several years of development work. The present design adds a 3-way solenoid valve to the air accumulator. In previous designs[1] the hydraulic expander stores the energy of expansion as linear kinetic energy of the gas piston and the hydraulic piston. Part of this energy was used to compress air in the accumulator to drive the piston down during the exhaust stroke. The rest of the energy is rejected as heat gained in the hydraulic oil during the throttling process. This conventional type of hydraulic expander suffers from two disadvantages, however. First, it depends partially on the momentum of the reciprocating mass to compress the air in the accumulator. If the device were subjected to a shock at the appropriate time, the momentum would be cancelled and the expander would stop in mid-stroke. The second disadvantage arises when the velocity of the piston masses is insufficient to allow the gas in the expander to expand completely.

The 3-way valve in the present hydraulically operated expander permits "active control" of the expansion cycle by exhausting air from the accumulator during the expansion stroke, and pressurizing the accumulator during the exhaust stroke to drive the helium piston down. The speed of the engine is controlled by the throttle valves in the hydraulic line. The two throttle check valves allow independent control of the speed in each stroke. With the active accumulator, the expansion engine no longer depends on the mass of the reciprocating members to sustain the operation of the device. The designer is free to make these components as light as possible, with structural strength necessary for fatigue and shock resistance, and still maintain the desired thermodynamic performance.

ELECTRONIC VALVE CONTROL

The helium expansion engine follows the cycle where the cylinder is partially filled with gas at constant pressure, expanded isentropically, and exhausted at constant pressure. Figure 2 shows the pressure-volume (P-V) diagram for an expansion process. Naturally, the helium inlet and exhaust valves, shown schematically in Fig. 1, must be timed and actuated to achieve the ideal cycle. Additionally, the "active accumulator" must be vented and pressurized in time with the expansion and exhaust stroke. These timing requirements would be difficult to achieve with conventional valve actuating techniques. Expansion engines presently use various mechanical, mechanical/pneumatic, and mechanical/electric schemes to operate the helium inlet and exhaust

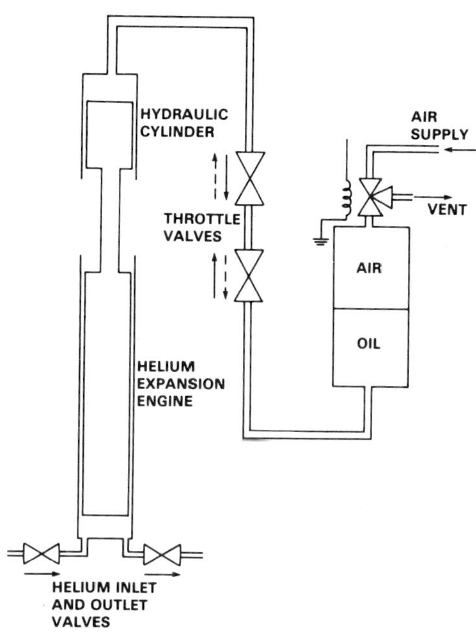

Fig. 1. Schematic of Hydraulic Expansion Engine With Active Accumulator.

valves. While these schemes work well, they do not allow the expansion cycle to be altered while the expander is in operation.

In order to achieve the desired timing, with an "active" system, the precise location of the expansion piston is required by the electronic valve control logic. The position signal is obtained from a linear variable differential transformer (LVDT) which is installed so that the core travels with the reciprocating piston assembly. One can also deduce piston velocity from this signal by differentiation. The transformer position signal is compared to a preset timing voltage by means of an electronic comparator. This comparator output controls a solenoid air valve to supply air to the helium valve operator to open the valve or vent it to close the valve. Adjusting the preset voltage levels results in a valve timing change. This control can be accomplished while the unit is in operation. Figure 3 shows a block diagram of the helium valve controls.

EXPANSION ENGINE TEST RIG

Figure 4 presents a schematic of the cryogenic portion of the

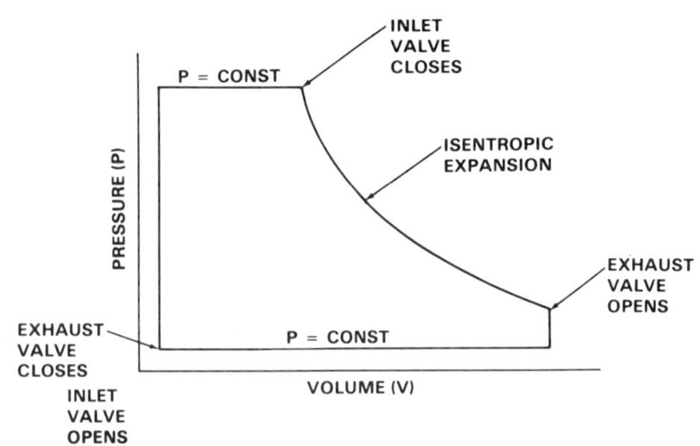

Fig. 2. Pressure-Volume (P-V) Diagram for Expansion Cycle.

expansion engine test rig. A helium compressor supplies gas to
the high pressure side of a counter flow heat exchanger. The heat
exchanger, surge volume, expansion engine and associated cold
valves are housed in a vacuum jacket to provide thermal insulation. The operation and flow path of the gas is identical to the
conventional piston expansion type refrigerator.

Figure 5 is a photograph of the expansion engine test rig.
The large cylinder in the upper portion of the photograph is the
counter flow finned tube heat exchanger and the long cylinder on
the bottom is the expansion engine. A quartz pressure transducer
is installed in the head of the expansion cylinder to monitor the
cylinder pressure during the thermodynamic cycle. The inlet and
exhaust piping of the expansion engine and the cylinder head are
instrumented with silicone diodes and thermocouple temperature
sensors. The silicone diodes are mounted in a G-10 fiberglass
holder to thermally insulate them from the inlet and exhaust
pipes. The temperature is also measured at the inlet and exhaust
of the heat exchanger by means of copper-constantan thermocouples.

The expansion engine piston is 5 cm in diameter and has a
maximum stroke of 7.5 cm. The heat exchanger, 10 cm in diameter
and 46 cm long, is installed in the system along with a 250 cc
surge volume between the heat exchanger and the inlet valve. This
entire assembly was wrapped with eight layers of superinsulation.

The integrated signal from the pressure transducer is fed to
one coordinate of an oscilloscope and the output voltage of the
LVDT, which is the piston position thus the expansion volume, is
fed to the other coordinate. The resulting P-V diagram is used in
the analysis of the thermodynamic cycle.

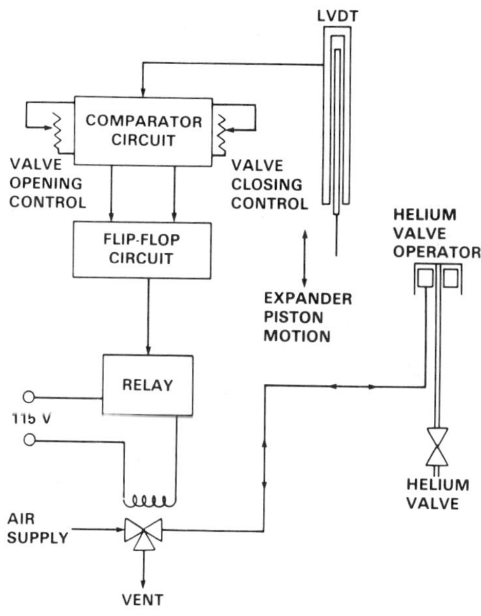

Fig. 3. Function Diagram of Electronic Valve Control.

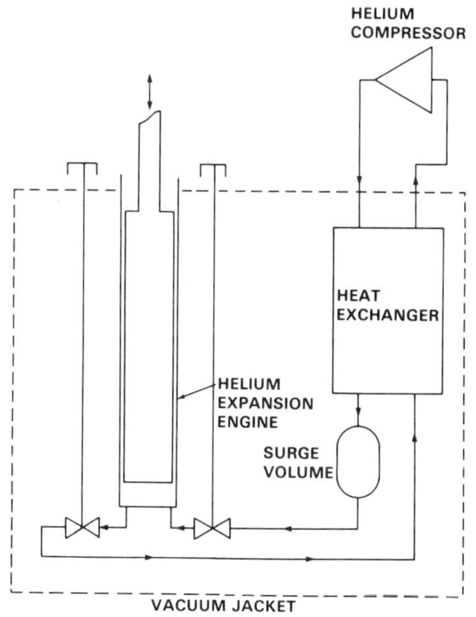

Fig. 4. Expansion Engine Test Rig.

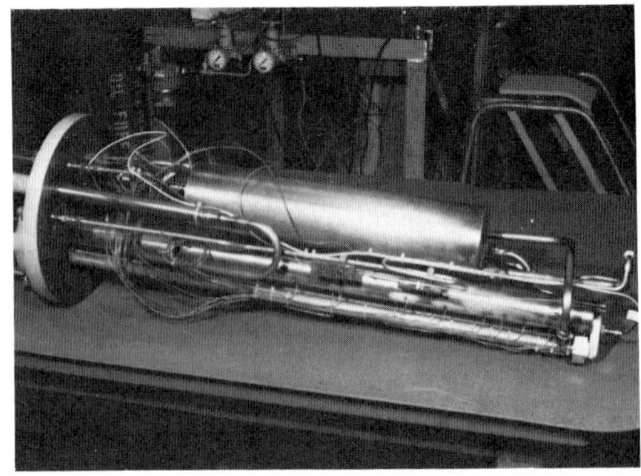

Fig. 5. Expansion Engine Test Rig.

OPERATION

High pressure helium gas is supplied by a helium compressor (18 atm discharge, 1.1 atm suction) to the hydraulically operated expansion engine. The use of a counter flow heat exchanger in the refrigeration system enabled it to achieve a temperature of 55 K.

The electronic controls enabled one to vary the point at which the inlet valve could be closed, and also to vary the stroke of the engine. Figure 6 indicates the expansion process where

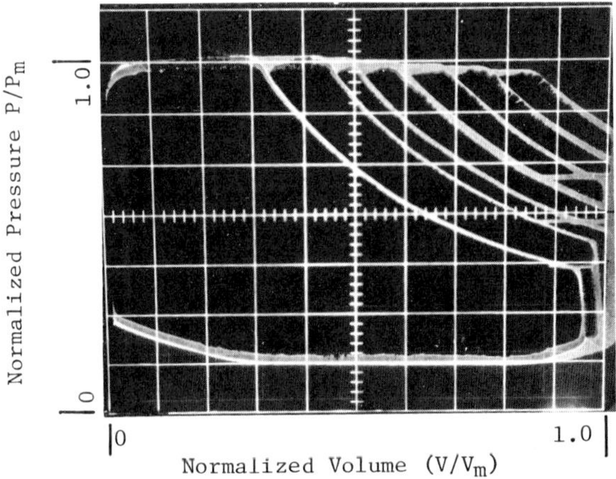

Fig. 6. P-V Diagram – Variable Cut-Off.

variable inlet valve cut-off's were accomplished electronically
while the engine was operating, which is an important factor in
the cool-down of cryogenic equipment. It should be noted that the
ordinate of this oscilloscope trace is the pressure in the expansion space and the abscissa is the stroke length or volume of the
expansion engine. These values have been normalized to the maximum pressure, P_m, and maximum volume, V_m.

Figure 7 shows an oscilloscope trace of several expansion
cycles having different stroke lengths of 96%, 90%, 84%, and 79%
of full stroke. These different stroke lengths were also changed
electronically while the expansion engine was in an operational
mode. The oscilloscope trace shown in Fig. 7 has also been normalized for the pressure and volume as indicated.

The other variable of this hydraulically operated expansion
engine is the speed. This engine has been operated at speeds from
12 strokes/minute to 160 strokes/minute. The ability to achieve
this high speed is a result of the active accumulator. Future
experiments will investigate the effects of piston speed and valve
timing on thermodynamic performance.

SUMMARY

The electrically controlled, hydraulically operated expansion
engine has undergone some initial tests at DTNSRDC. The use of
such an expansion engine has several advantages:

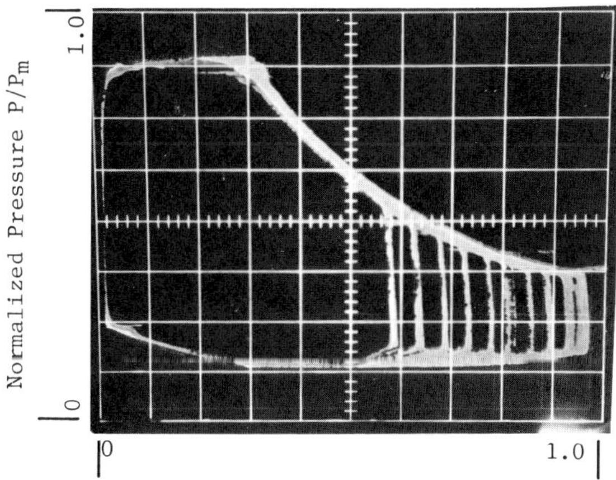

Fig. 7. P-V Diagram - Variable Stroke.

- o Elimination of the transformation from linear to rotary motion that exists on most piston expansion engines.
- o Reduction of the moving mass from the typical flywheel-crank mechanism.
- o Higher speeds than have been obtained with previous hydraulically operated expansion engines.
- o Variability of the expansion process.
- o A remote and automatically operating system.

This expansion engine and the associated instrumentation can then be used to evaluate the efficiency of the expansion process operating at various conditions. This information could then be translated to a liquefaction system that could be automatically controlled during steady state and cooldown phases by a computer.

REFERENCES

1. R.W. Johnson, S.C. Collins, and J.L. Smith, Jr., in "Advances in Cryogenic Engineering, Vol. 16," Plenum Press, New York (1971), p. 171.

A PAIR OF MINIATURE HELIUM EXPANSION TURBINES

H. Sixsmith and W. Swift

Creare Inc.
Hanover, New Hampshire

INTRODUCTION

High speed miniature turboexpanders are commonly used to provide refrigeration in medium size helium liquifiers and refrigerators. Although many of these turboexpanders operate reliably, there are occasional bearing failures. Because of the custom nature of such precision machinery, repair and replacement of bearings or rotor components can be lengthy and sometimes costly. In order to minimize system down time, spare replacement modules containing replacement shafts, rotors and bearings are commonly kept on hand in the event of a failure.

The design of a small, high speed turboexpander was pioneered by the National Bureau of Standards (NBS). The first expander operated at 600 000 rpm in externally pressurized gas bearings. A later version of this machine was scaled up by a factor of two and installed in a helium refrigerator. Although this expander has been shown to possess a high degree of reliability, it is not well adapted for the easy replacement of damaged bearings. We have designed two expanders which incorporate many of the features of the original NBS design. In addition they have been designed to broaden the range of operating conditions which can be accommodated in a given size of frame. The active components are contained in a cartridge which is held in place with a single nut. In the event of bearing failure, the damaged cartridge can be unscrewed and replaced in less than one minute. This should reduce the cost and the loss of time associated with the replacement or repair of damaged bearings.

SPECIFICATIONS AND DESCRIPTION

The turboexpanders described in this paper have been designed for the two expansion stages of a helium refrigerator. Table I shows the design operating specifications for each of the turbines. The specific speed is defined as:

$$N_s = NQ^{0.5}/(gH)^{0.75}$$

where N = Shaft speed
          Q = volume flow rate at the exit
          gH = isentropic energy across expander

The value of the specific speed in Table I is calculated with N in s^{-1}, Q in m^3/s, and (gH) in J/kg.

These figures show considerable differences in the design operating conditions between the two expanders. However, when the cross sectional areas of the flow passages were estimated, the difference was relatively small, which allowed the rotors of both turbines to have the same diameter. This in turn allowed the machines to be identical with the exception of the nozzle ring and the turbine rotor, which have different flow-channel geometries.

The turboexpander consists of an outer housing and an inner removable module. The inner module is located in the outer housing by means of a single nut. Figure 1 shows the location of most of the major components. The outer housing is fastened to the cold box flange of the refrigerator and the cold end is connected to the turbine inlet and outlet helium lines.

The warm end of the outer housing contains connectors for bearing gas lines together with a number of additional lines to provide access to the pressure at various locations along the length of the shaft. These pressures can be adjusted to optimize the performance of the bearings and the turbine. The outer housing also contains an electrical connection to a capacitance probe to monitor the speed of the shaft.

Table I. Design Operating Specifications

	Stage 1	Stage 2
Inlet Pressure (atm)	15	7
Inlet Temperature (K)	90	18.2
Outlet Pressure (atm)	7	1.2
Outlet Temperature (K)	74	12
Mass Flow Rate (g/s)	25	25
Shaft Speed (rev/sec)	8500	6000
Shaft Power (kW)	2.1	0.8
N_s	0.09	0.14

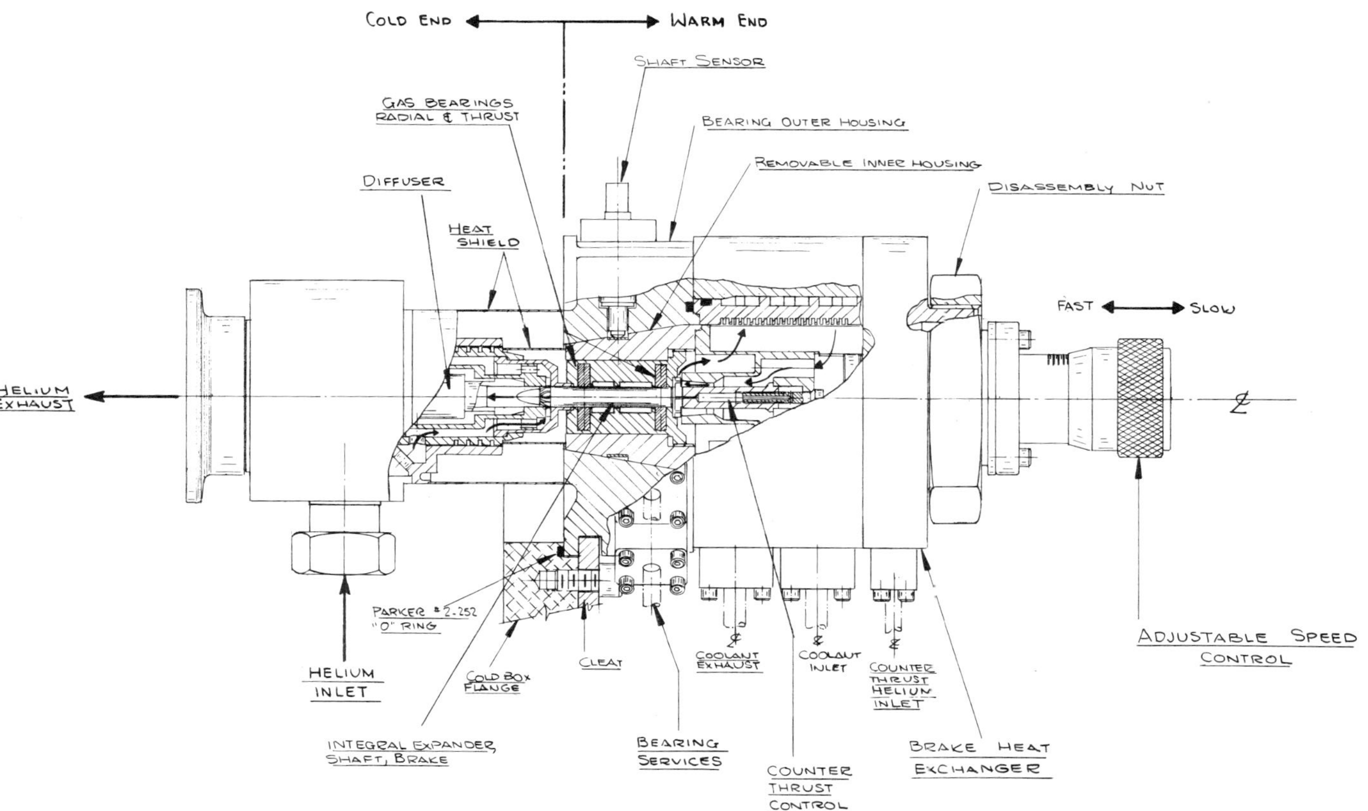

Fig. 1. Cryogenic helium turboexpander assembly.

The upper portion of the outer housing contains the brake circuit heat exchanger. This heat exchanger is designed to transfer the heat generated by the turbine work to an external heat sink.

The inner removeable module contains the vital components of the expander: the turbine rotor, the bearings and part of the brake circuit. The assembly is constructed so that all bearings services (gas supply, return, various sensing lines, and speed sensing probe) are automatically disconnected when the assembly nut is unscrewed and connected when the module is inserted and the nut is screwed on.

TURBINE DESIGN

The expander rotor is of the unshrouded, mixed flow type. Helium from the radial nozzle ring enters the rotor radially and discharges axially into a conical diffuser. The configuration is chosen to facilitate the machining of the blade profiles and to avoid the need for a precise axial location of the turbine rotor. The blade profiles are similar to those described in Ref. 3. The width and depth of the channels between the blades can be varied to accomodate a range of operating conditions.

The radial nozzle ring is designed to supply a stream of helium to the rotor inlet at a velocity close to that of the tips of the blades.

SHAFT AND BEARING DESIGN

The bearings are of the externally pressurized gas lubricated type. The bearings incorporate pneumatic stabilizers against shaft whirl in the journals and against pneumatic hammer in the thrust bearing. The stabilizers consist of volume/orifice pneumatic phase shift devices which employ a final portion of the expansion of the gas stream to generate pressures which act on the surface of the shaft and the face of the thrust bearing. These pressures have a component which is proportional to the rate of change of the clearance between the shaft and the bearing. They act in a direction to damp out any whirling in the journal bearings of any axial oscillation in the thrust bearings. Figure 2 shows the bearing assembly and the shaft. The theory of these stabilizers is given elsewhere.[1-4]

The diameters of the shaft and turbine rotor are approximately 10 mm for both expanders. The tip velocities are about 267 m/s and 188 m/s for the stage 1 and 2 rotors, respectively. The turbine rotor, the shaft and the brake impeller are manufactured as an integral piece from high tensile-strength titanium alloy.

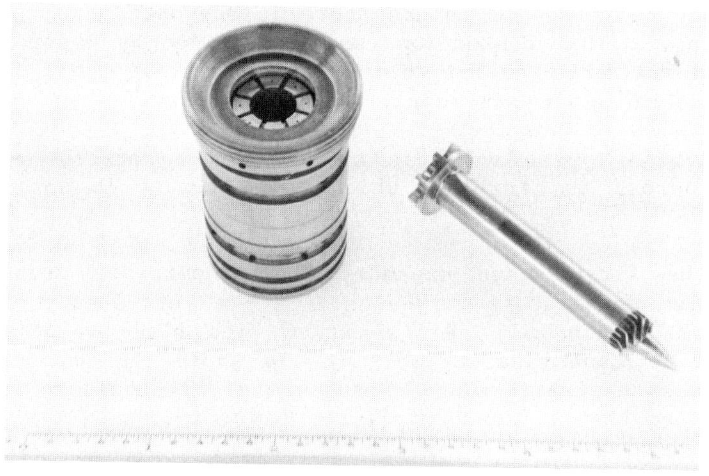

Fig. 2. Brazed bearing assembly and titanium shaft.

The journal bearing and the lower side of the thrust bearing are fabricated from a number of pieces which are silver soldered together to form an integral assembly. This unit is machined to size after soldering. The assembly incorporates the pneumatic phase shift stabilizers against shaft whirl. The upper side of the thrust bearing is built into the brake circuit assembly. Its function is to limit the axial play of the shaft and if needed, to support an upward thrust load.

The thrust load on the shaft depends on the difference between the pressure at the outlet from the turbine rotor and the pressure in the brake circuit. These pressures acting on the cross sectional area of the shaft generate thrust loads and the direction of the resultant load depends on which pressure is the greater. In most applications, the thrust is always downward and consequently a single sided thrust bearing is sufficient. In the present application, however, there is the possibility of an upward thrust in the first stage turbine so to cover this possibility a downward acting thrust bearing as shown in Fig. 1 is provided. This thrust bearing has a separate gas supply line which can be turned on if needed.

EXPANDER OPERATION

Each turbine should be provided with protective devices to prevent damage due to excessive speed or incorrect operation. It is of vital importance to ensure that the gas supply to the turbine cannot be turned on unless there is adequate gas pressure in the bearings. This protection can be provided by control valves which will not admit helium to the turbine unless there is sufficient pressure in the gas supply to the bearings.

A turbine bypass valve is also needed to prevent excessive pressure across the turbine, which would result in an excessive speed.

In the first stage expander the temperature drop at the design inlet temperature corresponding to an efficiency of 70% is about 16 K. This calls for a rather high blade velocity which limits the overspeed safety margin. To avoid the risk of excessive speed in the first stage expander the exit pressure should be controlled so that it cannot fall too low.

The machines have been constructed and tested at room temperature to obtain operating information on the bearings and to make physical adjustments to volume/orifice pneumatic phase shift network. Speeds in excess of design values have been obtained. The expanders will be installed in a refrigeration loop for performance tests and evaluation in the near future.

SUMMARY

Two miniature cryogenic turboexpanders have been developed to provide first and second stage expansion in small helium liquifiers or refrigerators. The design of each machine is virtually identical although the design operating conditions vary more than 30% in specific speed. The expanders run at high speeds in pressurized helium bearings with an integral brake heat exchanger. The vital components are contained in a cartridge, held in place by a single nut, which can be removed and replaced in less than one minute.

REFERENCES

1. D.B. Mann, et al, A Refrigeration System Incorporating A Low-Capacity, High-Speed, Gas-Bearing-Supported Expansion Turbine in "Advances in Cryogenic Engineering, Vol. 8," Plenum Press, New York (1963) p. 221.
2. B.W. Birmingham, H. Sixsmith and W.A. Wilson, The Application of Gas-Lubricated Bearings To A Miniature Helium Expansion Turbine, in "Advances in Cryogenic Engineering, Vol. 7," Plenum Press, New York (1961), p.30.
3. H. Sixsmith, Miniature Expansion Turbines, in "Advanced Cryogenics," C.A. Bailey, ed., Plenum Press, New York (1971).

4. H. Sixsmith and W.A. Wilson, The Theory Of A Stable High-Speed
    Externally Pressurized Gas-Lubricated Bearing, <u>J. of Res.
    of the National Bureau of Standards</u>, 68c:101 (1964).

DISCUSSION

Question by J.C. Riple, Garrett-Airesearch Corp: What was the
efficiency of the turbine?

Answer by the author: We did not measure the efficiency of these
turbines since we do not have a helium refrigerator at Creare.
The present design is similar to that of an NBS turbine which gave
a maximum efficiency of 79.8% (see Ref. 3).

OIL REMOVAL FROM HELIUM AT THE FERMILAB 1500-W REFRIGERATOR

W. E. Cooper, A. J. Bianchi, R. K. Barger, F. B. Johnson,
K. J. McGuire, K. D. Pinyan, and F. R. Wilson

*Fermi National Accelerator Laboratory**
Batavia, Illinois

INTRODUCTION

The Fermilab 1500 W refrigeration system for the magnet test facility uses two oil-flooded Sullair C25L screw compressors in series. We have approximately 22 000 hours of operating experience with this system. During the first 9 500 hours of operation, frequent problems were encountered with inadequate oil removal resulting in carry-over of substantial amounts of oil to the cold box. The modifications made to the system to eliminate oil problems will be described.

The refrigeration system was designed and built by Helix (now Koch) Process Systems, Inc. (HPS) (Westborough, Massachusetts) and Sulzer Brothers (Winterthur, Switzerland). The compressor skid was built by Sullair Corporation (Michigan City, Indiana). The bulk oil separators were built by Sullair to Helix specifications and the remainder of the oil removal system was provided by Helix. Nominal system specifications for combined refrigeration and liquefaction are given in Table I. A description of the original oil removal system is given in Table II.

The compressor system serves three principal functions. It provides:
1. pressurized helium for operation of the refrigerator
2. helium to purge magnets before they are cooled
3. helium to warm magnets after they have been tested.

---

*Operated by Universities Research Association, Inc. under contract with the U.S. Department of Energy.

Table I. Nominal Specifications of Refrigerator Compressor

	1st Stage	2nd Stage
Inlet pressure	1 atm	2.6 atm
Outlet pressure	2.6 atm	15.1 atm
Compressor volume ratio	2.6	3.7
Helium flow	103 g/s	211 g/s
Power required	104 kW	562 kW
Maximum oil temperature	325 K	325 K
Maximum discharge temperature	380 K	380 K
Observed helium temperature at bulk oil separator	340 K	340 K
Oil type	Ucon LB-165 (950 L)	
Refrigeration	1450 W at 4.42 K	
Liquification	3.0 g/s at 4.42 K	
LN2 pre-cooling required	27 g/s from 77 K to 300 K	

Table II. Original Oil Removal System

First Stage Compressor:

1. Bulk oil separator (vertical)
   (Exit height) - (Entrance height)   1.93 m
   Diameter   0.76 m
   Entrance flow   Centrifugal
   Lower stage: 3 reverse flow louvered canisters
   Upper stage: 3 coalescing filters

Second Stage Compressor:

1. Bulk oil separator (same as first stage)
2. After cooler (no oil drains): Patterson-Kelly 241624-25
3. HPS demister #1: 10 paralleled fiberglass packs with manifolding for 5 inlets and 6 exits
   Overall dimensions: 0.35 m diameter x 3.35 m
4a. HPS demister #2a: 8 paralleled fiberglass packs with manifolding for 4 inlets and 5 exits.
   Overall dimensions: 0.35 m diameter x 3.05 m
4b. HPS demister #2b: same as #2a with valving to run either #2a or #2b
5. Charcoal adsorber
   Bed size (approximate): 0.35 m diameter x 2.14 m
   Flow centerline: 15.2° downward from horizontal
   Charcoal: Union Carbide JXC 6x8
6. Particulate filter: Consler HHCP-4 (10 micron)

EARLY OPERATING EXPERIENCE

During early operation of the system oil punched through the charcoal bed on many occasions. This was first detected by noting an oil coating in the helium passages of magnets after they had been warmed. On at least two occasions substantial amounts (10 L) of oil reached the cold box. Because the first cold box heat exchanger is LN_2 cooled, the oil was collected primarily in this heat exchanger. On one occasion, a 0.41 MPa (60 psi) pressure differential was developed across the first heat exchanger. No evidence was found that oil had moved beyond this point. Connections were added to permit Freon flushing, the oil was drained and the heat exchanger flushed.

SYSTEM IMPROVEMENTS

At this time (9 000–9 500 system hours), a concentrated effort was made to improve the oil removal capabilities of the system. The following changes were made:

1. A vent line was added to the top of each oil cooler running to a point in the respective bulk oil separator well above the oil level. A solenoid valve in the vent lines is opened when the compressors are turned off. The routing of the original vent lines was such that helium entrapment occurred in the oil coolers (particularly the second stage). During a shutdown, helium which was mixed in the oil gradually collected in the top of the oil coolers and pushed oil from the oil coolers into the bulk oil separators. The second stage bulk oil separator level increased as much as 12.5 cm during a 24 hour shutdown. Since the solenoid-controlled vent lines were added, these variations in oil level have not been observed. Oil levels before start-up are now approximately 12 cm lower than previously and maintain this level after start-up. This change reduced carry-over from the bulk oil separator after start-up from 3 000 ppm (by weight) to 300 ppm.

2. The orifices in the bulk oil separator drain lines were increased in size. Lower drain line orifices were increased from 0.23 cm to effectively 0.32 cm; upper drain line orifices were increased from 0.08 cm to effectively 0.23 cm. These changes reduced carry-over from the bulk oil separator to approximately 30 ppm.

3. HPS demisters #2a and #2b originally had no drain lines. These demisters were designed to be used one at a time so that the unused one could be flushed with Freon to remove oil. In practice, these demisters saturated with oil in less than 24 hours because of the original

high carry over from the bulk oil separator. Drain lines were added to each of the five parallel outlet sections of each demister. Both of these demisters have been run in parallel since then, with no Freon flushing required.

4. All demister drain line orifices were removed and replaced with adjustable valves.
5. Sight glasses were installed in all drain lines.
6. Check valves were installed in drain lines to prevent oil from being pushed into the demisters during system pressurization before start-up.
7. All drain lines for oil removal elements after the second stage compressor were re-routed from the second stage oil system to a point between the first stage compressor and the first stage bulk oil separator.

At the time of the original oil problems, three coalescer filters were ordered from Balston, Inc. (Lexington, Massachusetts). These were installed in June, 1980 (12 000 system hours) between the last HPS demister and the charcoal adsorber. The specifications for these filters are given in Table III. Two oil mist detectors (PPM, Inc., Knoxville, Tennessee) were installed at the same time. The first detector is located just before the Balston filters and has a range of 0-20 mg/m^3, the second is just after the Balston filters and has a range of 0-2 mg/m^3.

PRESENT OPERATING CHARACTERISTICS

Oil mist levels have recently been sufficiently low that no measurable level is seen with either PPM detector. Therefore our only measure of oil mist level comes from collection in the Balston filters. During 10 000 hours of operation, 16 mL of oil have been collected from the first Balston filter housing blowdown. This corresponds to an oil concentration of 2 x 10^{-9} by weight. No oil has been observed at the blowdowns of either of the other two Balston filters nor has any of the three automatic drains operated. The first Balston filter housing was opened at approximately 19 000 system hours and the cartridges were found to be dry.

Based on our experience, we feel that screw compressor systems can work well in helium refrigeration systems. Oil removal is reasonably straight-forward, but we recommend a minimum of three oil removal stages after the bulk oil separator. The last stage can be used for monitoring and for protection in case of a failure upstream in the system. Careful monitoring during early system operation are important. We have used collection, the Balston Oil Check Kit, and PPM detectors for monitoring, each of which is satisfactory within its range of sensitivity.

Table III. Balston Filter Specifications

	Housing	Cartridge	Retention Eff.	No. of Cartridges
1.	KCF-480-4-500	DX	90%	4
2.	KCF-880-4-300	BX	99.99%	8
3.	KCF-880-4-300	AQ	99.9999%	8

DISCUSSION

Question by G. Gistau (L'Air Liquide, France): What is the sensitivity and reliability of the PPM laser detector?

Answer by author: Although the original instruments used lasers, the present models use light-emitting diodes. The nonimal sensitivity is $5\mu g/m^3$. We have never had sufficiently high oil aerosol levels to see a signal. However, the zero offset has been observed to drift by as much as to $40\mu g/m^3$ over a six month period, so some provision must be made to allow routine offset calibration. Aside from this, the detectors have been generally reliable, with a few mechanical malfunctions.

SEPARATION OF COMPRESSOR OIL FROM HELIUM

R. Strauss and K. A. Perrotta

Balston, Inc.
Lexington, Massachusetts

INTRODUCTION

Compression of helium by an oil-sealed rotary screw compressor entrains as much as 4000 parts per million by weight (ppm) of liquid and vapor oil impurities in the gas. For cryogenic applications, helium impurities must be reduced below ~ 0.1 ppm. In principle, the methods used for removing oil from compressed air can also be used for removing oil from helium. However, extensive experience with compressed helium has shown that oil separation equipment designed for compressed air must be modified significantly to produce the desired results with helium.

The main differences between air and helium filtration are:

1. The purity requirement for helium refrigeration systems is far more stringent than normal air filtration requirements. For most applications, filtration of oil in air to 1 ppm is satisfactory, while in cryogenic helium applications 0.1 ppm is often considered the maximum tolerable level, and even lower concentrations are desired.

2. The oil content in helium leaving a rotary screw compressor is much higher than the oil content in air from the same compressor. One reason for the high oil concentration by weight in helium is that the compressor is a volume-handling device, and a given weight of oil per unit volume of gas results in 7.25 times the weight of oil per unit weight of helium compared with weight of oil per unit weight of air. In addition, helium is commonly compressed to 200 psig or higher, while compressed air systems are usually 125 psig. The higher output pressure results in higher volumetric oil loading.

REMOVAL OF LIQUID OIL FROM HELIUM

Description of Coalescers

At least 99.99% of the oil in helium leaving the compressor is in the form of liquid droplets. Therefore, it is essential to remove virtually all the liquid oil before dealing with the vapor contaminants.

All commercial filters for high efficiency separation of liquid oil from gas are continuous coalescing filters. The typical coalescing filter element is composed of a 1/16 inch to 1/4" thick mat of borosilicate glass fibers, formed into a cylinder and structurally bonded by resin or by internal and external perforated supports. With proper design, filter elements of this construction can readily achieve initial retention efficiencies of 99.99% or higher for 0.3 to 0.6 micron particles and droplets.

If the filtered contaminant is a solid particle, it remains attached to the surface of the filter fiber against which it has impacted, held by Van der Waals forces. However, liquid droplets captured on the fibers can migrate down the length of a fiber to fiber crossover points. There the liquid droplets grow into large drops, which eventually are forced through the depth of the filter to the downstream surface by the flow of gas through the filter. By the time they reach the downstream surface, the droplets have grown so large that – at least in theory – they are too heavy to be resuspended in the gas stream. Therefore, the large drops drain down the vertical filter tube, while clean gas exits the filter housing. If gas flow direction through the filter tube is inside-to-outside (as in the case with all commercial coalescing filters), the coalesced liquid drains from the outer surface of the filter tube into the sump of the filter (Fig. 1), from which it may be removed continuously by an automatic drain. The net effect of the process is to coalesce submicron oil droplets into large drops, which are continuously drained from the system. Since the liquid is removed as rapidly as it enters the filter, the filter has an infinite life in liquid removal applications, and the efficiency of the filter remains constant indefinitely.

The theoretical picture of a coalescing filter presented above applies quite well in practice to most compressed air filtration. However, some problems arise in helium filtration. A potential difficulty with any continuous coalescing filter results from the fact that the filtered gas and coalesced liquid both exit from the downstream side of the filter element. Any re-entrainment of the coalesced liquid will contaminate the clean gas. Most manufacturers fabricate their coalescing filter elements with an

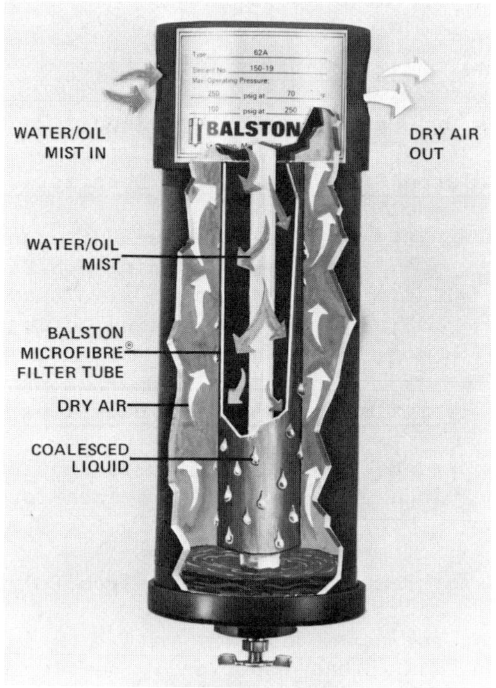

Fig. 1. Continuous Coalescing of Liquid
Mist from Air or Other Gas.

outer layer of coarse glass fiber or polyurethane foam to act as
an entrainment separator. While the entrainment separator layers
are undoubtedly beneficial, they are never 100% efficient. In a
particularly wet system, such as any helium system, there invari-
ably is significant liquid carryover from the first coalescing
filter. Therefore, more than one stage of coalescing filtration
is always needed in helium filtration.

A second practical difficulty with the picture of the theoretical coalescing filter is that coalesced oil drains more slowly
from a higher efficiency filter than from a lower efficiency
filter. If oil enters the filter more rapidly than it can drain
from the filter, the excess will be re-entrained in the exit gas,
regardless of the efficiency rating of the filter. For this
reason, a very high efficiency filter used as a first stage coalescer in a helium system is likely to give much poorer coalescing
results than a lower efficiency filter.

Table I.

1.) <u>Multiple Stage Filtration</u>

Three Stages Of Coalescing Filters Should Be Used:

<u>Stage</u>	<u>Balston Grade</u>	<u>Description</u>
1st	DX	Medium efficiency, fast draining coalescer
2nd	BX	High efficiency, slow
3rd	BX	draining coalescer

2.) <u>Maximum Allowable Linear Flow Rate</u>

Flow through second and third stages should not exceed 15 ACFM per 2 inch diameter x 18 3/4 inch long filter tube.

Recommendations for Removing Liquid Oil from Helium

Based on operating experience with a wide range of helium systems, Balston, Inc., has empirically developed recommendations for removing liquid oil from helium (Table I). Systems designed with these criteria consistently have reduced liquid oil content in helium from several thousand ppm to less than 0.1 ppm. The specific recommendations apply only to the particular construction and geometry of the Balston filter elements and filter housings, but the general principles of multiple stage and graded efficiency filtration are believed to be applicable to any manufacturer's coalescing filters.

OIL VAPOR IN HELIUM

Typical Compressor Oils

While helium producers and users have frequently installed carbon adsorption equipment to remove compressor oil vapor from helium, until now the adsorbers have been designed without benefit of quantitative information on oil vapor concentration or the effectiveness of adsorbents for removing the vapor. Balston has recently made quantitative determinations of organic volatiles in two compressor oils frequently used with helium, Union Carbide UCON LB-170X and LB-300X.

The UCON fluids are reportedly to have the composition $C_4H_9(OC_3H_6)_n$-OH. The different members of the family have different average molecular weights, and therefore different

viscosities. The numerical designation is equal to the viscosity
in SUS at 100°F. The "X" designation signifies the presence of an
aromatic amine antioxidant. Union Carbide has stated that the
product is not distilled or stripped, and therefore any low molecular weight materials produced in manufacture remain in the
fluid as delivered. In addition, the Union Carbide specification
permits up to 3000 ppm of water in the product, and a typical
water content is 1500 to 2000 ppm.

Experimental Procedure for Measuring Oil Vapor Concentration

Referring to Fig. 2, the apparatus consists of a source of
ultra zero compressed air, a heated oil reservoir, an adsorbent
cartridge housing with by-pass, a flowmeter, and a Beckman Model
400 hydrocarbon analyzer. All lines are kept at constant temperature with heating tapes. The system is first flushed with ultra
zero air, and the hydrocarbon analyzer is calibrated at the test
flow rate of 2.5 liters per minute.

About 25 grams of oil is then added to the reservoir and
heated to the test temperature. Ultra zero air is then passed
across the surface of the hot oil, through the by-pass line, to

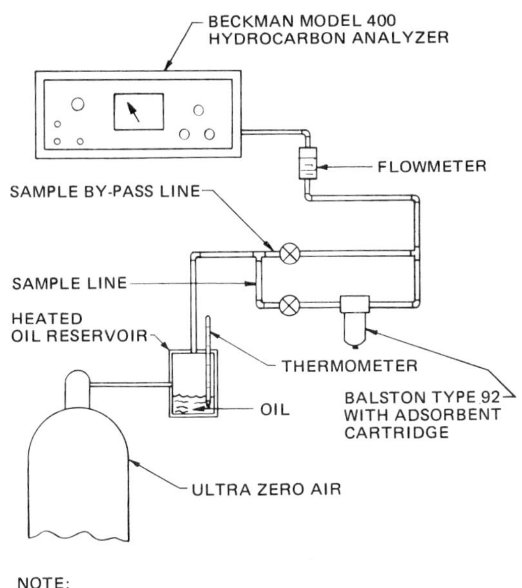

Fig. 2. Apparatus for Measuring Oil Vapor
Concentration in Gas.

the analyzer. To test the efficiency of an adsorbent, the procedure is repeated with the adsorbent cartridge installed and the air
passing through the filter holder rather than the by-pass.

Measured Volatile Hydrocarbons in the Lubricants

Figures 3 and 4 show the concentration of hydrocarbon vapors
generated by the LB oils in ultra zero air at two different
temperatures, expressed as ppm of methane by weight in air. By
summing the area under the higher temperature curve for LB 170X,
we estimate that our sample of LB 170X contained initially about
600 ppm of volatile organics by weight of oil. As noted above,
the oil as received probably will contain in addition about 1500
to 2000 ppm of water (which was not measured in our tests.)

Since the test apparatus is not intended as an efficient
stripper of volatiles, it can be assumed that any reasonably
efficient method of pretreating the oil by the user will produce
at worst, the same percentage removal of volatiles as the test
apparatus. The desorption curve for LB 170X at 200°F (Fig. 3)
shows that approximately 90% of the organic volatiles can be
removed easily, since the first 60 minutes accounts for about 90%
of the area under the desorption curve. The remaining 10% of
volatiles in the oil evolved at a much slower rate, as is evident
from the curve. If we assume that user pretreatment will remove
that fraction of the volatiles in the oil which is readily desorbed, then pretreatment will reduce the volatile hydrocarbon
contents by 90%, from 600 ppm by weight in the oil as received to
60 ppm. Union Carbide believes that it is unlikely that additional volatiles would be generated by chemical degradation in helium
compression service.

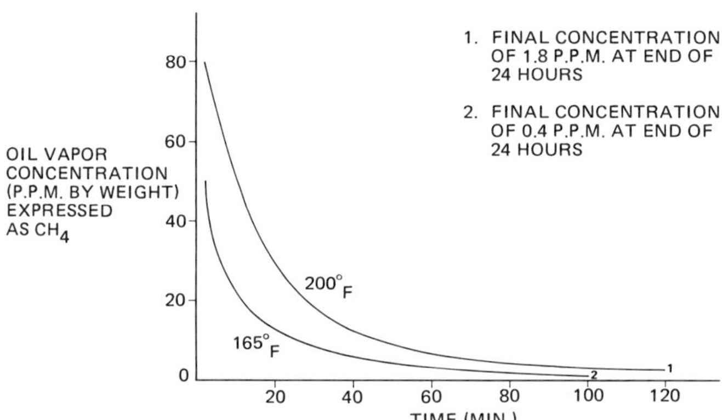

Fig. 3. Concentration of Hydrocarbon Vapor from UCON LB-170X.

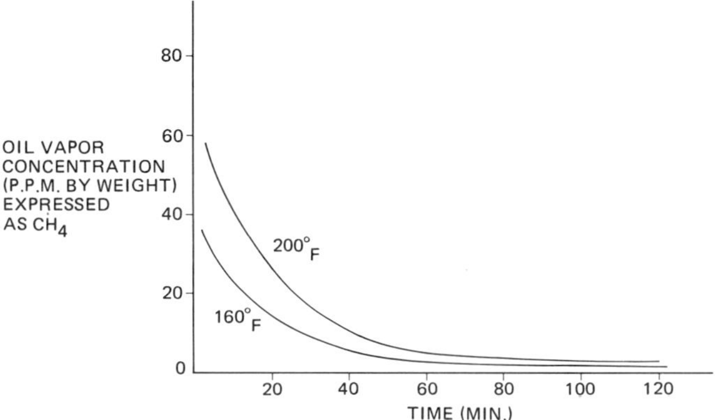

Fig. 4. Concentration of Hydrocarbon Vapor from UCON LB-300X.

Calculated Concentration of Oil Vapor in Helium

Using the derived concentration of 60 ppm organic volatiles
by weight in pretreated oil, for calculation purposes we have
taken as an example a helium compression system for 1000 SCFM,
equivalent to about 80 grams per second of helium. One manufacturer of rotary screw compressors estimates that 35 gallons
(approximately 280 lbs) of oil could be charged into a system of
this size. If the oil were pretreated to 60 ppm of volatiles, by
weight, then the total amount of volatiles in the oil charge would
be 7.8 grams. Assuming that the volatiles desorbed to produce 0.1
ppm of vapor by weight in the helium (the equilibrium concentration of volatiles in the air following the initial rapid desorption) then the 7.8 grams would be exhausted in 10 days. Thereafter, no volatiles would evolve from the oil charge.

Additional tests conducted by Balston with carbon and Type
13X molecular sieve adsorbents indicate that in the presence of
very low concentrations of vapor neither adsorbent would be more
than 50% efficient. If this were the case, an adsorbent would
reduce the downstream concentration of vapor from 0.1 ppm to 0.05
ppm in the helium, but extend the time of vapor transmission
downstream from 10 days to 20 days or longer. Over the extended
period of time, the total 7.8 grams of vapor would eventually end
up downstream of the adsorber. The adsorbent would capture a
portion of the vapor while it was being evolved and then release
the vapor gradually when there was no further evolution from the
oil, since adsorption is an equilibrium process. Considering the
cost, complexity, and added impurities which would be introduced
by a solid adsorption step, it appears to us that the potential

benefits of ·adsorption do not warrant undertaking the possible problems.

CONCLUSIONS

1)   Liquid oil contamination in helium gas can be reduced well below 0.1 ppm by a properly designed multiple state coalescing filter system containing graded efficiency filter elements.

2)   The oil vapor problem is best attacked by efficiently treating the oil to remove most of the volatiles before charging the compressor.

DISCUSSION

Comment by J.A. Bamberger (Brookhaven National Laboratory):
An additional, 4th stage, coalescer is desirable as a back-up in case of breakthrough or malfunction upstream. The possibility of eliminating the charcoal for vapor removal is quite attractive.

Response by the Author:
In most helium systems, three stages of coalescing filtration (Grades DX, BX, BX in the Balston filter group) give satisfactory oil removal and adequate short term protection against process upsets or mechanical failure. A fourth stage, while not usually recommended, will certainly give additional protection. If a fourth stage is to be used, the recommended filter is Grade AQ, essentially 100% efficient for retention of oil droplets but not a continuous coalescing filter (that is, oil remains trapped on the filter and does not drain). To obtain the desired additional system protection, it is necessary to monitor continuously the oil content of helium entering the Grade AQ filters. If any oil is detected entering the filters, action must be taken to rectify the problem upstream before the Grade AQ filters become saturated.

We agree with Mr. Bamberger that eliminating the charcoal would be quite advantageous in reducing system cost and complexity. We believe that our data justify studying this possibility on a larger scale.

Question by W. C. Chronis (CVI, Inc.):
What is the ACFM loading of the first stage coalescer?

Answer by the Author:
The flow rate through the first stage (Grade DX) can be as high as 30 ACFM per 2 inch × 18-3/4 inch cartridge. However, we usually recommend that the first stage be sized the same as the second and third stages (15 ACFM per cartridge).

AN AUTOMATIC PURIFICATION SYSTEM FOR LABORATORY HELIUM GAS RECOVERY

F. J. Kadi

Air Products & Chemicals, Inc.
Allentown, Pennsylvania

INTRODUCTION

An automatically operated liquid nitrogen cooled helium purifier has been designed and built for use with Air Products' Helifier helium liquefiers or as a stand alone purifier designed to supply repurified high pressure helium for other end uses. While the system utilizes conventional adsorption technology for the purification of helium, the present design minimizes liquid nitrogen consumption and requires a relatively short regeneration cycle time through the use of a unique single column adsorber design.*

The purifier is shown in Fig. 1 with its side panel removed. The purifier can handle up to 20% air contaminants by volume while delivering 0.8 g/s of helium with less than 50 ppm by volume of impurities. While the purifier can operate in conjunction with any impure gas recovery system capable of supplying gas at up to 15.0 MPa pressure with the requisite flow, it is usually supplied with a recovery and gas bag system which feeds the liquid equivalent of 11 or 22 L/h of gas.

DESIGN

Conventional high pressure cryogenic adsorption type purification systems used to supply pure gas to processes requiring an uninterrupted supply of gas typically rely on the use of dual switching adsorbers, i.e., one unit is processing gas while the

---

*Patent applied for.

other is being regenerated. This duplication increases both equipment costs and the complexity of the control system. The present design eliminates one of the adsorption beds while, in most applications, providing for an uninterrupted supply of pure gas. This is accomplished by accumulating a reserve supply of pure gas during the purification phase which is of sufficient volume to feed the liquefier during regeneration. This technique is practical for those cases requiring gas flow rates equivalent to 16 L/h or less.

For those applications requiring between 16 and 22 equivalent liquid L/h, make-up gas must be supplied during regeneration from an alternate source. This approach is practical if the purifier regeneration time is sufficiently short to keep pure gas make-up requirements during regeneration consistent with the quantity of gas normally needed to replenish gas losses inherent in the entire system. Typically, 80% of the liquid helium used in a research laboratory is recovered, thus the 20% required for make-up can be supplied to the liquefier during purifier regeneration. The present design meets these requirements with a minimum purification time of 8 hours and a 2 hour regeneration time. This gives a minimum on-stream duty factor of 80%. For an incoming contamination level of 1% by volume, the duty factor increases to 92%.

With reference to Fig. 1, the cold box is located in the rear of the cabinet and the controls in the front. All components housed within the cold box are indicated in the purifier flowsheet of Fig. 2. The adsorber column is the principal component in the cold box. It is designed to operate at a pressure of 9.65 MPa. Fast regeneration is achieved by containing the adsorbent in a light weight inner shell which is thermally isolated from the heavy outer pressure vessel. This design minimizes the thermal mass to be warmed and cooled thereby reducing the consumption of liquid nitrogen during the recooling of the adsorber following regeneration.

STEADY-STATE PURIFICATION

The steady-state purification phase of the cycle is monitored by a thermal conductivity gas analyzer operating in conjunction with a run timer. The gas analyzer provides continuous monitoring of the process stream sampling probe located about 85% of the way through the adsorber bed. The present analyzer can resolve the sample gas impurity to 0.3% by volume. The analyzer controls a relay which initiates regeneration. In order to protect the helium liquefier from those instances in which the total contamination level entering the purifier is less than 0.3%, a timer is placed in series with the analyzer relay so that if the analyzer

Fig. 1. Automatic Purifier Left Side View.

has not triggered a regeneration within a run time consistent with a 0.3% contamination level the timer will start regeneration.

During the purification mode of operation, impure helium enters the rear bulkhead of the purifier and is allowed to enter into the unit through the opening of SV 626. The pressure in this circuit is typically in the range of 11.0 MPa to 13.8 MPa. The incoming impure gas is brought down to the system operating pressure of 9.65 MPa by pressure regulator PCV 628. From PCV 628 the impure gas passes into the cold box. The first stage of cooling takes place in the freeze-out heat exchanger which consists of three primary heat exchanger circuits and a secondary trace line carrying boil-off nitrogen from the condenser. The impure incoming gas passes over an extended heat transfer surface where water vapor and any traces of oil are frozen out. The first pass is a vertical one with the cold end up so that any contaminants which are frozen out will collect in the bottom of the heat exchanger when the unit is warm. These contaminants can then be purged from the system through hand valve V629. The two concurrent circuits

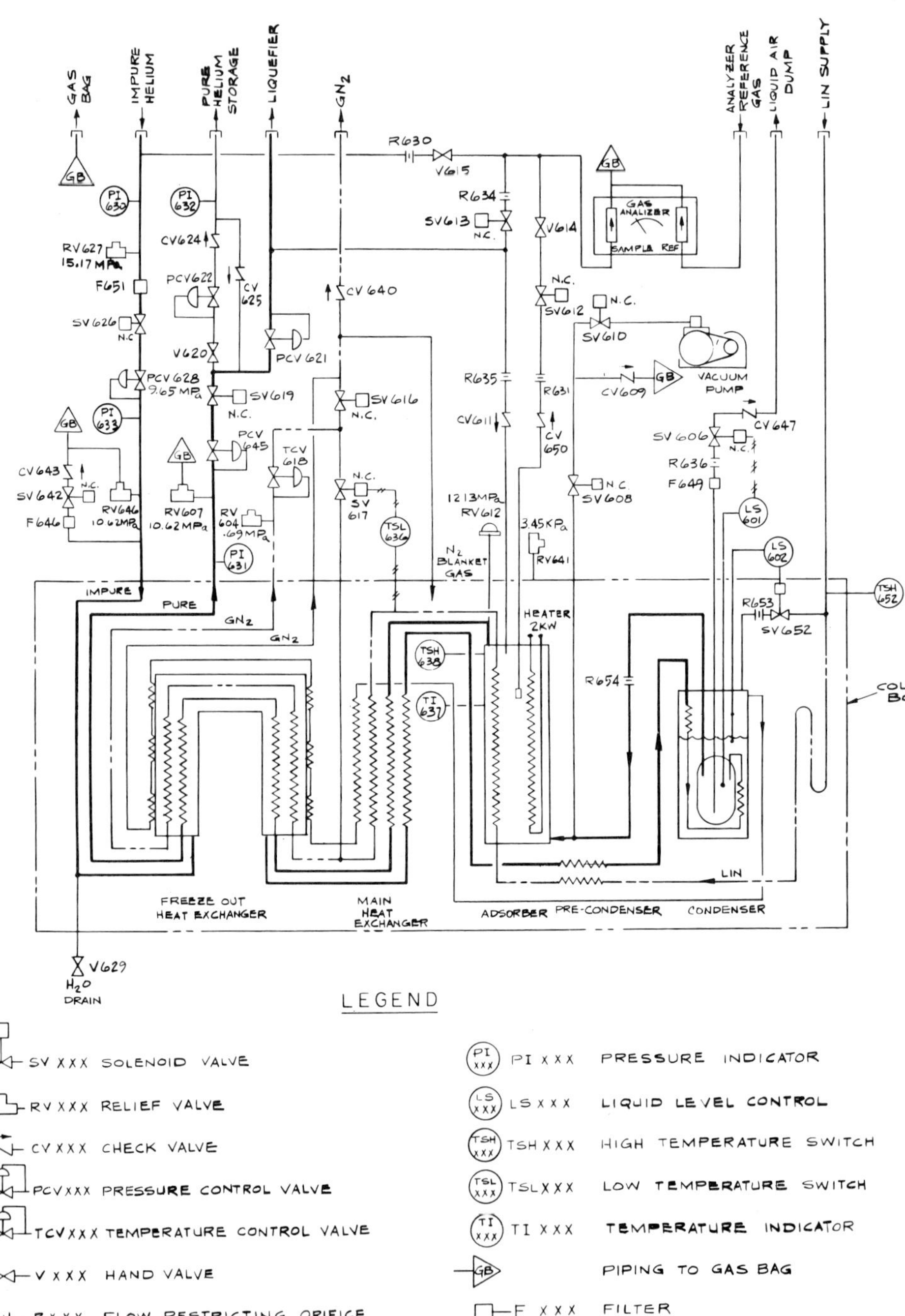

Fig. 2. Purifier Flowsheet.

in the freeze-out heat exchanger consist of the pure helium stream and the main elevated pressure nitrogen stream. Both of these circuits are inside of the extended heat transfer surface tube. The first vertical pass of the freeze-out heat exchanger generally drops the impure gas stream temperature to about 230 K.

After passing through the second vertical leg of the freeze-out heat exchanger, the impure gas is further cooled in the main heat exchanger to about 110 K. The gas mixture then enters the pre-condenser section where, for liquid nitrogen feed pressures of about 0.34 MPa absolute, any O_2 in excess of 0.6% and N_2 in excess of 4.4% is condensed and flows with the remaining vapor phase contaminants into the main condenser where any contaminants in excess of 0.15% O_2 and 1.85% N_2 are condensed. When approximately one liter of condensate collects in the pot, as measured by the liquid level sensor of controller LS601, SV 606 is opened for a period of approximately two minutes during which the condensate is purged from the condenser.

The remaining contaminants and helium pass through combination orifice and filter assembly R654 and into the adsorber. R654 is used to slow the pressurization of the adsorber with process gas following the completion of the regeneration phase; this is required in order to minimize the adsorber warm-up experienced during the gas fill. The adsorption column houses a replaceable 2 KW cartridge heater and the liquid nitrogen trace lines. The heater is used during regeneration and its maximum temperature is limited by adjustable temperature control switch TSH 638. The normal setting of this switch is 275 K to 290 K. The temperature of the adsorber can be monitored by TI 637.

The pure helium leaves the top end of the adsorption column and is used in the main heat exchanger and freeze-out heat exchanger to help cool the incoming impure gas. Upon leaving the cold box, the pure helium passes through back pressure regulator PCV 645, set to insure that the adsorber reaches at least 8.27 MPa before pure gas begins to feed to the liquefier. The pure gas passes through SV 619 and can then pass into one of two circuits. If V620 is closed and the liquefier is in operation, the gas pressure is dropped from 9.65 MPa to about 0.48 MPa by PCV 621. The pure gas leaves the purifier bulkhead at the liquefier port. A setting of 0.48 MPa on PCV 621 places the purifier feed gas about 0.07 MPa above the first stage make-up gas regulator in the liquefier (normally set at 0.41 MPa). This forces the liquefier to draw gas preferentially from the purifier rather than the pure helium make-up gas source.

If V620 is open, the purifier can begin storing any pure gas not used by the liquefier. This is accomplished by setting back

pressure regulator PCV 622 to about 9.3 MPa. Thus, if the demand
for pure helium by the liquefier is high, the pressure at the
inlet of PCV 622 will drop below 9.3 MPa and the storage of pure
gas will cease until the pressure again reaches or exceeds 9.3
MPa.

Liquid nitrogen enters the system at a pressure range of
between 0.17 MPa and 0.28 MPa and immediately splits into one of
two flow paths. One flow circuit goes directly to the pre-condenser and then through the adsorber cooling passages. The second
circuit flows, under the control of SV 652 and pressure dropping
orifice R653, into the condenser. SV 652 lets liquid nitrogen
into the condenser on an intermittent basis under the control of
liquid level detector LS 602. The pressure and, therefore, the
temperature of the liquid nitrogen in the condenser is minimized
through the use of generously sized exhaust lines.

Under steady-state conditions the flow of nitrogen through
the system is controlled solely by TCV 618 and LS 602. Cooldown
is initiated and maintained by automatically opening SV 616.
Solenoid valve SV 617 is opened only during cooldown of the cold
box. When the temperature of the nitrogen gas leaving the adsorber drops below a preset value, TSL 636 shuts off SV 617. Temperature switch TSH 652 closes the impure helium inlet valve, SV 626,
should the liquid nitrogen feed line warm up past 95 K.

REGENERATION

The two hour regeneration cycle is triggered by the gas
analyzer or the run timer and is controlled by a timer. Initially
any liquid air present in the condenser is purged by the timed
opening of SV 606. Following this, the remaining contents of the
adsorber and condenser are dumped into the gas bag by opening SV
608 and SV 642. Simultaneously, the adsorber heater is energized. If the heater is below the set point of TSH 638, the adsorber
will start to warm until the bed reaches 275 to 290 K. After
approximately 50 minutes of heating the vacuum pump and SV 610 are
activated. The evacuation of the adsorber occurs with a continual
reverse flow purge of pure helium supplied via R635 and CV 611.
At the end of the evacuation period, the vacuum pump, SV 610, SV
608, and SV 642 are de-activated and the cooldown initiated
through the opening of SV 616, SV 652, and SV 617. The adsorber
reaches liquid nitrogen temperature prior to the completion of the
two hour period. When the timer has completed the regeneration
cycle, the flow of impure gas is initiated by opening SV 626 and
SV 612.

TEST RESULTS

An important aspect of the current design is the thermal isolation achieved between the outer pressure vessel wall and the adsorber core. Figure 3 is an experimentally determined graph of the vessel wall and adsorber core temperature variations during the course of an automatic regeneration. The graph shows that the amplitude of the temperature excursion of the pressure vessel wall is only about 14% of the adsorber core temperature swing.

The effectiveness of the adsorber thermal isolation plays a major role in reducing the purifier overall liquid nitrogen consumption during normal operation. The steady state liquid nitrogen consumption, expressed in terms of liquid liters of nitrogen required per equivalent liquid liter of pure helium supplied, has been measured at various incoming contamination levels to be 0.71 at 20% contamination, 0.41 at 2% contamination and 0.31 at very low contamination levels. In addition to the steady state nitrogen consumption, one must also factor in the liquid required for cooldown following regeneration and the excess nitrogen consumption experienced before equilibrium temperatures are reached. For the case of 2% contamination, this brings the total consumption rate to about 0.6.

SUMMARY

An automatic liquid nitrogen cooled helium purifier has been developed which can handle up to 20% air contaminants by volume while delivering up to 0.8 g/s of helium with less than 50 ppm by volume of impurities. A unique fast regenerating adsorption

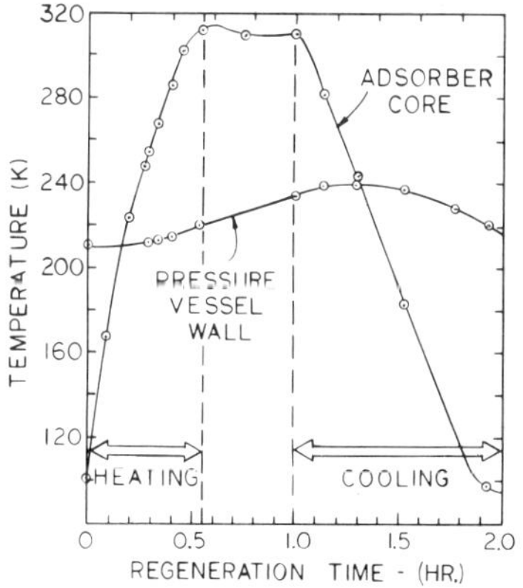

Fig. 3. Pressure vessel wall and adsorber temperature response.

column has been employed which permits an 80% on-stream time for
all contamination levels greater than 2%. Liquid nitrogen consumption for a 2% air contamination test case averaged 0.6 liquid
liters of nitrogen per equivalent liquid liter of pure helium
produced.

DISCUSSION

Question by K. D. Timmerhaus, Univ. of Colorado: Is the regeneration gas discharged to the atmosphere and considered as part of
the helium purification cost?

Answer by author: The purge gas is discharged to the atmosphere
only during the 30 min evacuation portion of the regeneration
cycle. The helium lost in regeneration is less than 0.1% of the
helium purified.

THE COOLING POWER OF ^{3}He-^{4}He DILUTION REFRIGERATORS

W. van Haeringen

Philips Research Laboratories
Eindhoven, The Netherlands

INTRODUCTION

The energy balance equation for ^{3}He transport in either the exhaust tube or the supply tube in a ^{3}He-^{4}He dilution refrigerator can be solved exactly in the low temperature range where the viscosity and heat conduction coefficients $\eta(T)$ and $\kappa(T)$ can be taken to be proportional to T^{-2} and T^{-1}, respectively. The exact solutions, $T(l)$, where l is the tube-length parameter, make it possible to give explicit estimates for the cooling power. A similar analysis applies in the case of ^{3}He-^{4}He counterflow in a superfluid injection dilution refrigerator.

ANALYSIS

In order to derive the temperature distribution along any tube in a dilution refrigeration system it is necessary to find the solutions of the appropriate enthalpy balance equation and to subject them to the corresponding boundary conditions.

The enthalpy balance for ^{3}He flow in an exhaust tube with superfluid ^{4}He background is[1]

$$\dot{n}_3 m_3 \frac{d}{dl}(\mu_3 + T(s_3 + \rho_4 s_4/\rho_3) + gz) - \frac{\pi d_1{}^2}{4} \frac{d}{dl}\left(\kappa(T)\frac{dT}{dl}\right) = 0. \tag{1}$$

Where $\dot{n}_3$ is the molar flow rate, m_3 the molar weight, μ_3 the chemical potential, s_3 the entropy, ρ_3 the density of ^{3}He; s_4 the entropy, ρ_4 the density of ^{4}He; z the vertical projection of the tube-length parameter l, d_1 the tube diameter, T the absolute temperature, $\kappa(T)$ the heat conduction coefficient, and g the gravitational acceleration. The length parameter l is in the same direction as the ^{3}He flow. It is assumed that the ^{4}He flow rate $\dot{n}_4 = 0$. If μ_4 is the ^{4}He chemical potential and p the pressure, it is straightforward to derive the Gibbs-Duhem equation:

$$\frac{d}{dl}(\mu_3 + gz) = -\frac{\rho 4}{\rho_3}\frac{d}{dl}(\mu_4 + gz) - (s_3 + \rho_4 s_4/\rho_3)\frac{dT}{dl} + \frac{1}{\rho_3}\frac{d}{dl}(p + (\rho_3 + \rho_4)gz). \tag{2}$$

It will be assumed that $d(\mu_4 + gz) = 0$ and furthermore that Poiseuille's relation applies:

$$\frac{d}{dl}(p + (\rho_3 + \rho_4)gz) = -\frac{128}{\pi d_1^4}\frac{\dot{n}_3 m_3}{\rho_3}\eta(T), \tag{3}$$

where $\eta(T)$ is the viscosity coefficient.

We concentrate on the temperature region $T < 0.01$ K where the well-known approximations $s_3 + \rho_4 s_4/\rho_3 \propto T$, $\kappa(T) \stackrel{\sim}{\propto} T^{-1}$, $\eta(T) \propto T^{-2}$ hold. Substituting Eq. 2 and 3 into Eq. 1, we get[2]:

$$\dot{n}_3 A_1 \frac{dT^2}{dl} - \frac{\dot{n}_3^2 B_1}{d_1^4}\frac{1}{T^2} - C_1 d_1^2 \frac{d}{dl}(\frac{1}{T}\frac{dT}{dl}) = 0. \tag{4}$$

The constants A_1, B_1 and C_1 turn out to be $A_1 = 12$ J/mol K^2, $B_1 = 1.1 \times 10^{-14}$ Jsm^{3}K^2/mol^2 and $C_1 = 2.6 \times 10^{-4}$ J/sm. The same equation, albeit with differing constants A_2, B_2 and C_2 and diameter d_2, results in the range $T < 0.01$ K for a ^{3}He flow in a ^{3}He supply tube. In this case the liquid consists of 100% ^{3}He, unlike in the former case where the mixture only contains 6.4% ^{3}He.

The constants are equal to $A_2 = 54$ J/mol K^2, $B_2 = 3.8 \times 10^{-13}$ Jsm^{3}K^2/mol^2 and $C_2 = 2.4 \times 10^{-4}$ J/sm.

Introducing reduced temperatures and lengths, $t_i = T/T_i$ and $\lambda 1/l_i$ with $i = 1$ or 2 and with $T_i = (B_i C_i/A_i^2)^{1/6} d_i^{-1/3}$ and $l_i = (C_i^2/A_i B_i)^{1/3} d_i^{8/3} \dot{n}_3^{-1}$, Eq. 4 reduces to:

$$t\frac{d^2 t}{d\lambda^2} - (\frac{dt}{d\lambda})^2 - 2t^3 \frac{dt}{d\lambda} + 1 = 0, \tag{5}$$

which coincides with the low temperature limit of Eq. 69 in Ref. 3. It can be verified by inspection that real-t solutions of Eq. 5 are given by:

$$t_\beta^{\,2}(x) = 2^{-2/3}\left[\left(\frac{Ai'(x) + \beta Bi'(x)}{Ai(x) + \beta Bi(x)}\right)^2 - x\right]^{-1}, \qquad (6)$$

where $x = 2^{1/3}\lambda + a$, α and β are integration constants and x is restricted to values such that the factor between the brackets is positive. The functions $Ai(x)$ and $Bi(x)$ are two linearly independent solutions of the Airy differential equation $d^2w/dx^2 - xw = 0$. A direct derivation of this result can be found in Ref. 1. The role of α is trivial: with any particular $t\beta(x)$ the function $t\beta(x + \alpha)$ is also a solution. The interesting parameter is β. In Fig. 1 the solutions of Eq. 6 have been presented for a variety of β values. Zeros in $t\beta(x)$ occur for x values fulfilling $Ai(x) + \beta Bi(x) = 0$. Except for $\beta=0$, each $t\beta(x)$ reaches ∞ for some finite $x>0$ is given by the zero of the factor between brackets in Eq. 6. No real-$t\beta$ values occur for x exceeding this value. The function $t_o(x)$ is the only solution that is real for all x values, while $t_o(x) \to \infty$ for $x \to \infty$. As a matter of course the physical interest is in those (parts of the) solutions $t\beta(x)$ which are positive.

In order to investigate which $t\beta(x)$ actually applies, one has to subject the solutions obtained to additional conditions. Let us concentrate first on the exhaust tube case. At the low-temperature side one easily derives in the case of single cycle operation as a condition in the mixing chamber M (see Ref. 1, Eq. 23).

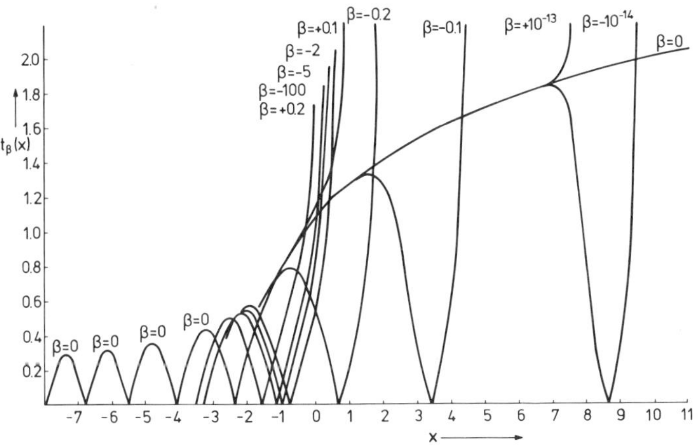

Fig. 1. Positive $t\beta(x)$ solutions, Eq. 6, for $\beta=-100, -5, -2, -0.2, -0.1, -10^{-14}, 0, 10^{-13}, 0.1, 0.2$. Solutions with $\beta>-0.2$ have a form similar to those with $\beta<-0.2$. For x values exceeding the zero of $Ai(x) + \beta Bi(x)$ no real $t\beta(x)$ exists.

However, the above described laborious procedure can be circumvented due to the circumstance that in an actual situation the average temperature gradient is at the most of the order of a few mK/cm and much less for larger d_1 and smaller $\dot{n}_3$. This implies that "practical" $t\beta(x)$ solutions should be relatively smooth. In relation to this, the interval $x_H - x_M = 2^{1/3} \, l_H/l_1$ should be sufficiently large, if appreciable temperature differences have to be bridged. As an example take $l_H = 50$ mm, $d_1 = 2$ mm, $\dot{n}_3 = 4 \times 10^{-4}$ mol/s. This leads to $x_H - x_M \approx 5$. Examination of the $t\beta(x)$ solutions in Fig. 1 in the light of these remarks leads to the conclusion that only those curves should be considered for which $|\beta| \lesssim 10^{-1}$. The other curves should be rejected because neither in the oscillating parts nor in the steep parts can they provide sufficiently long intervals $x_H - x_M$ along which $dt\beta(x)/dx$ is small enough. In Fig. 2, $t\beta(x)$ solutions with $|\beta| \lesssim 10^{-1}$ are separately displayed. It is observed that for $x \lesssim 0$ all curves practically coincide. This observation has important consequences with regard to Eq. 8. Take $\beta=0$ first. In the case of no heat load, i.e. $\dot{Q}_M = 0$, Eq. 8 leads to $x_M = -1.25$ and $t_0(-1.25) = 0.691$.

As this point is "almost" common to all $t\beta$ (x) curves with $|\beta| \lesssim 10^{-1}$, it follows that the temperature $T=0.691 \, T_1$ can be considered as the minimum temperature in a single-cycle dilution

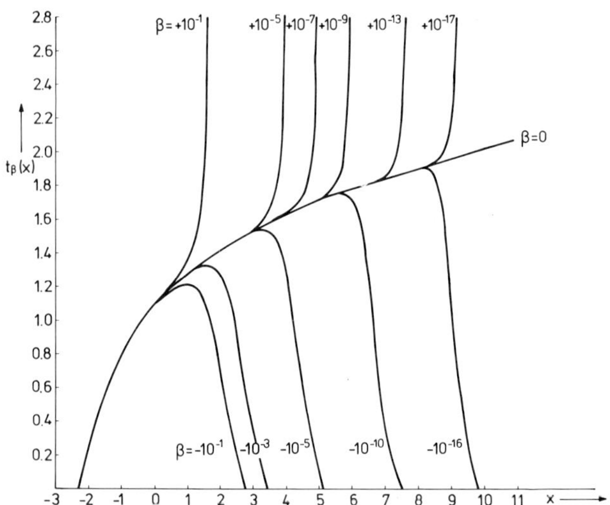

Fig. 2. Positive $t_\beta(x)$ solutions, Eq. 6, for $-10^{-1} \leqslant \beta \leqslant + 10^{+1}$. The $\beta \geqslant 0$ solutions apply in the case of an exhaust tube, while the $\beta < 0$ solutions apply in the case of a supply tube.

refrigerator, or, with d_1 expressed in m and T in K, we have

$$\dot{Q}_M = \dot{n}_3 d_1^{-2/3} \left[2.88 \times 10^{-5} t_\beta^2(x_M) - 2.13 \times 10^{-5} \left(\frac{1}{t_\beta(x)} \frac{dt_\beta(x)}{dx} \right)_M \right], \quad (7)$$

where d_1 is in m, $\dot{n}_3$ in mol/s and the cooling power $\dot{Q}_M$ in J/s Eq. 7 balances the load $\dot{Q}_M$ applied from outside, the outgoing enthalpy and the heat conduction the tube entrance in M. Substituting Eq. 6 into Eq. 7 leads to

$$\dot{Q}_M = \dot{n}_3 d_1^{-2/3} \left[1.19 \times 10^{-5} t_\beta^2(x_M) - 2.13 \times 10^{-5} \left(\frac{Ai'(x_M) + \beta Bi'(x_M)}{Ai(x_M) + \beta Bi(x_M)} \right) \right] \quad (8)$$

The straightforward procedure would now be to give a second additional condition relating the x-coordinate and the reduced temperature $t_\beta(x)$ at some other place along the tube, for instance of the simple form

$$t_\beta(x_H) = t_H, \quad (9)$$

where $t_H = T_H/T_1$ and T_H is a given (high) temperature value, while $x_H = x_M + 2^{1/3} 1_H/1_1$, 1_H being the distance between M and the heat exchanger H. Equations 8 and 9 then constitute two independent relations between x_M and β, from which x_M, β and therefore $t_\beta(x)$ can be determined at given $d_1, \dot{n}_3$ and $\dot{Q}_M$.

$$T_{min} = 0.691 \ T_1 = 3.9 \times 10^{-2} \ d_1^{-1/3}. \quad (10)$$

It follows similarly that, independent of the precise β value in the range $(-10^{-1}, 10^{-1})$, Eq. 8 can be directly used to determine the cooling power $\dot{Q}_M$, at least for not too high temperatures T_M. The procedure is to set $\beta = 0$ in Eq. 8 and to calculate $\dot{Q}_M$ for a variety of choices of points $(x_M, t_0(x_M))$ leading to

$$\frac{\dot{Q}_{M,\text{single cycle}}}{\dot{n}_3 d_1^{-2/3}} = 1.19 \times 10^{-5} t_o^2(x_M) - 2.13 \times 10^{-5} \frac{Ai'(x_M)}{Ai(x_M)}. \quad (11)$$

In Fig. 3, $\dot{Q}_M$, single cycle is given as a function of the mixing chamber temperature T_M.

The curves in Fig. 2 applying to the exhaust tube are those with $\beta \gtrless 0$, i.e. the ones that monotonously increase with x. In a ^{3}He supply tube we deal with temperatures that are decreasing with increasing 1. The descending parts of the curves in Fig. 2, i.e. the curves with $\beta < 0$, are the ones that apply in this case. If we

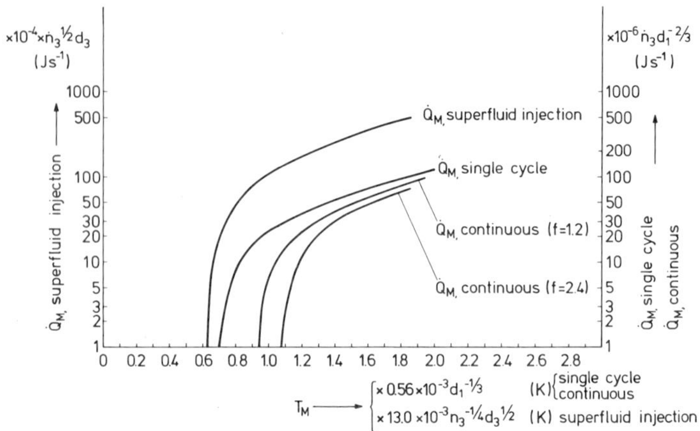

Fig. 3. Cooling power $\dot{Q}_M$ as a function of mixing chamber temperature T_M for single-cycle, continuous, and superfluid injection. The indicated scaling factors for T_M are $T_1=0.56\times10^{-3}d_1^{-1/3}$ for the single-cycle and the continuous case and $T_3=13.0\times10^{-3}\dot{n}_3^{-1/4}d_3^{1/2}$ for the superfluid injection case. The cooling power scaling factors follow from Eq. 11, 13 and 16, respectively. The flow rate $\dot{n}_3$ is to be taken in mol s^{-1} and d_1,d_3 in m. In the continuous case (see Eq. 13) two extreme f values 1.2 and 2.4 have been taken, while the ratio d_1/d_2 is chosen to be equal to 3.

now want to determine the cooling power in a dilution refrigerator with continuous circulation of ^{3}He, we have to deal with an additional condition on M in which both an incoming and outgoing flow of ^{3}He is considered. Let $t_\beta'(x')$ be the reduced temperature function applying to the supply tube with $x'=2^{1/3}1'/1_2 + \alpha'$, $1'$ being the length parameter and α' the (trivial) integration parameter, then the condition on M is easily derived from Eq. 34 of Ref 1 and is given by:

$$\dot{Q}_M=108\ \dot{n}_3 T_1^2\ \left(t_o^2(x_M) - 2^{-2/3}\left(\frac{1}{t_o(x)}\frac{dt_o(x)}{dx}\right)_M\right)$$

$$-24\ \dot{n}_3 T_2^2\left(t_\beta'^2(x_M')-2^{-2/3}\left(\frac{1}{t_\beta'(x')}\frac{dt_\beta'(x')}{dx'}\right)_M\right). \tag{12}$$

One recognizes this as a balance equation of ingoing and outgoing ^{3}He enthalpies, heat conduction terms at the entrance and exit of M and applied heat load $\dot{Q}_M$. It is observed from Fig. 2 that the descending parts of the curves with $-10^{-1}<\beta'<0$ have slopes that vary at the most by a factor of 2 in the t_β' region of

interest in Eq. 11. This observation makes it possible to estimate $\dot{Q}_M(T_M)$ in a relatively easy way, without specifying the
precise value of β'. From Fig. 2 we deduce that $f=dt_\beta'(x')/dx'$
lies roughly between the values 1.2 and 2.4. If furthermore we
make use of $t_\beta'(x'_M)=T_M/T_2=t_0(x_M)$. T_1/T_2, Eq. 12 is transformed
into

$$\frac{\dot{Q}_{M,\ continuous}}{\dot{n}_3 d_1^{-2/3}} = 0.94 \times 10^{-5} t_o^2(x_M) - 2.13 \times 10^{-5} \frac{Ai(x_M)}{Ai(x_M)}$$

$$-0.38 \times 10^{-5} f \frac{d_1}{d_2} t_o^{-1}(x_M). \tag{13}$$

The first term in the right hand side of Eq. 13 is somewhat
smaller than the corresponding enthalpy term in Eq. 11 owing to
the enthalpy of the incoming ^{3}He in M. The second term is also
present in Eq. 11. The third term is new and is due to heat
conduction in the supply tube.

A reasonable estimate of $\dot{Q}_M(T_M)$ can now be obtained from Eq.
13 by considering both the extreme cases in which $f=1.2$ and $f=2.4$
and by making a variety of choices $(x_M,t_0(x_M))$. In Fig. 3 the
function of $\dot{Q}_M$, continuous is given for $d_1/d_2=3$, which is commonly
applied value. The actual cooling power in a given dilution
refrigerator will lie somewhere between these extremes and depends
on the temperature distribution along the supply tube, which in
turn will be mainly determined by the heat exchangers used.

The above treatment can easily be extended to the case of a
counterflow tube as applied in superfluid injection methods[4,5].
In such tubes, again for $T\sim0.01$ K, the enthalpy balance equation
can be shown to be[2]

$$t\frac{d^2t}{d\lambda^2} - (\frac{dt}{d\lambda})^2 - 2t^3 \frac{dt}{d\lambda} + t^2 = 0, \tag{14}$$

with $t=T/T_3$, $\lambda=1/l_3$, $T_3=13.0\times10^{-3}d_3^{1/2}\dot{n}_3^{-1/4}$ and $l_3=0.032d_3\dot{n}_3^{-1/2}$, d_3 being the tube diameter. The solutions of Eq. 14 read:

$$t_{\beta''}(x) = \pi^{-1/4} \left[\frac{\exp(-x^2)}{\beta'' + erfc\ (x)} \right]^{1/2} \tag{15}$$

with $x=1/l_3 + \alpha''$, α'' and β'' integration constants.

The x values in Eq. 15 are restricted to such values that β''
+ erfc$(x)>0$. The analysis of Eq. 15 and the determination of the

cooling power $\dot{Q}_M$ in the injection chamber is quite similar to the one presented above for the single-cycle case and has been presented in Ref. 5. The cooling power in the superfluid injection case can be derived from Eq. 14 and 18 in Ref. 5:

$$\frac{\dot{Q}_{M,\ superfluid\ injection}}{\dot{n}_3^{1/2} d_3} = 6.13 \times 10^{-3} t_o^2(x_M) + 8.12 \times 10^{-3} x_M \quad (16)$$

In Fig. 3 $\dot{Q}_M$, superfluid injection is given as a function of T_M.

CONCLUSION

It has been shown how the exact solutions of enthalpy balance equations can be used to obtain explicit estimates, in terms of ^{3}He flow rate and tube diameters, for the cooling power in single cycle, continuous and superfluid injection dilution refrigerators. The basic cooling power equations show that apart from enthalpy terms also terms are present originating from heat conduction at the entrance and or exit of the mixing chamber. The latter terms, which can be disregarded at temperatures well above the minimum attainable temperature, play an essential role in the cooling behavior near T_{min}, and are accounted for exactly. The analysis applies in the temperature range $T < 0.01$ K. For this reason, effects due to partially diffusive ^{3}He flow, which at higher temperatures may led to an excess cooling power[6] can be disregarded.

REFERENCES

1. W. van Haeringen, F.A. Staas and J.A. Geurst, <u>Philips J. Res.</u> 34:107, (1979).
2. W. van Haeringen, <u>Cryogenics</u> 20:153, (1980).
3. J.C. Wheatley, O.E. Vilches and W.R. Abel, <u>Physics</u> 4:1, (1968).
4. F.A. Staas, A.P. Severijns and H.C.M. van der Waerden, <u>Phys. Lett.</u>, 53A:327, (1975).
5. W. van Haeringen, F.A. Staas and J.A. Geurst, <u>Philips J. Res.</u> 34:127, (1979).
6. E. Polturak, R. Rosenbaum and R.J. Soulen, Jr., <u>Cryogenics</u> 19:715, (1979).

A SIDE ACCESS TOP-LOADING DILUTION REFRIGERATOR

N. W. Kerley

Oxford Instruments
Osney Mead, Oxford, England

INTRODUCTION

The feasibility of performing nuclear orientation experiments on ions implanted at ultra-low temperatures was first demonstrated by Herzog[1]. It was confirmed that a substantial fraction of implanted ions (~80%) ended up in good sites and thus experienced the normal hyperfine field. In Herzog's experiments ions were implanted at about 200 mK and subsequently cooled to 15 mK in order to determine the nuclear orientation effects.

It was realized by Vanneste[2] that there was a requirement for ions to be implanted in a target held continuously below 25 mK. The comparatively recent advent of extremely powerful dilution refrigerators operating down to 4 mK, with several microwatts of refrigeration available outside the mixing chamber at 10 mK, opened up the possibility of performing true on-line nuclear orientation experiments for studies on short lived nucleii far from stability.

Nucleii with lifetimes down to a few minutes (limited by their spin-lattice relaxation time) can be orientated. If the ion beam is polarized outside the cryostat as described by Kaminsky[3] and Rau[4], the spin-lattice relaxation time restriction is removed thus allowing very short lived nucleii to be studied.

The dilution refrigerator described in this paper is specially adapted for this type of experiment. It is fitted with a "side access port" which allows an ion beam to pass from a mass separator onto a target cooled by the dilution refrigerator. The side access port completely eliminates window material from the beam line. In addition, a top loading facility enables the target to

be changed while the refrigerator is cold.   Three systems of this
type have been built by Oxford Instruments.

GENERAL DESIGN

Since ions cannot pass through a solid window, a method of
allowing ions to enter the dilution refrigerator directly through
a side access port was devised. The port was designed to virtually eliminate room temperature thermal radiation.   Other factors
which influenced the design were the requirement for a target/de-
tector distance of about 7 cm and the positioning of a 1.5 T
polarizing split coil superconducting magnet around the target.
The lower tail section of the cryostat (shown in Fig. 1) is rectangular, and fitted with thin aluminum windows for transmission
of γ-rays. Conduction cooled shields at 4.2 K and 77 K surround
the magnet and low temperature parts of the system. The magnet is
mounted inside a stainless steel container and immersed in liquid
helium fed from the main helium reservoir. The conduction cooled
4.2 K shield is hermetically sealed with 1 mm diameter indium wire
gaskets to contain helium exchange gas around the mixing chamber
(inner vacuum space) during the precooling stage.

For successful cooling of the target holder, good thermal
contact is required between the dilute phase inside the mixing
chamber and the target. The measured thermal impedance of 330 K/w
was somewhat better than expected indicating that screwed copper
to copper contacts are efficient, even when mated together at
about 1 K. Estimated thermal impedances of various joints are
given in Table I.

Table I.   Thermal Impedances

Kapitza boundary resistance to silver sinter in mixing chamber	100 K/W
Thermal impedance of silver sinter	100 K/W
Thermal impedance of copper plates	16 K/W
Screw connection to thick wall copper tube	< 400 K/W
Thick wall copper tube to target holder	50 K/W
M18 thread to target holder	< 400 K/W
	<1066 K/W

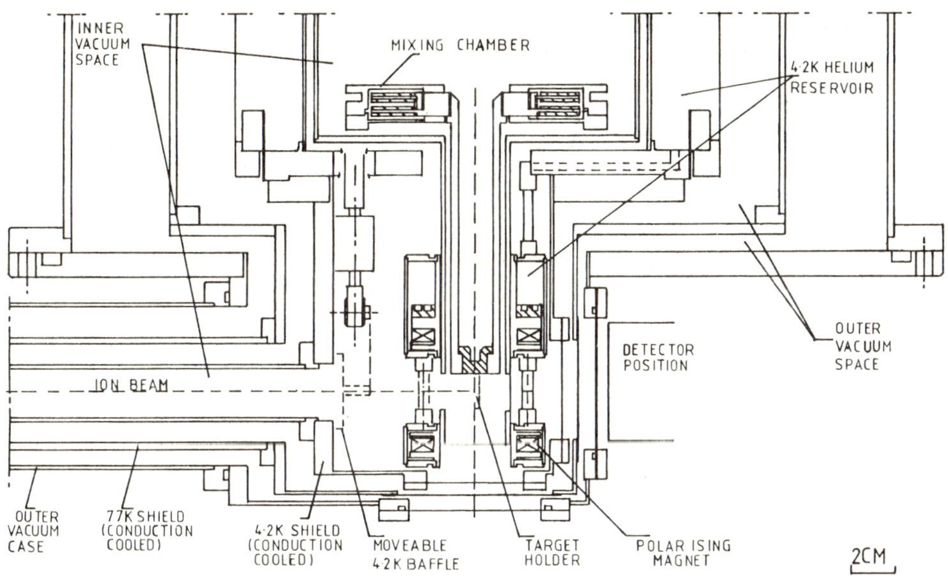

Fig. 1. Side Access Dilution Refrigerator Tail Section.

SIDE ACCESS DESIGN

To allow entry of an ion beam while simultaneously restricting
heat leaks involves attenuating thermal radiation down a long
internally blackened tube cooled to 4.2 K. The beam tube has a 20
mm bore and is 1 m long. The outer end of the beam tube is connected to the room temperature shield via two stainless steel
bellows, as shown in Fig. 2. The beam tube is surrounded by a
conduction cooled 77 K radiation shield and a room temperature
vacuum case. The stainless steel bellows have three main functions: to provide a long thermal path from room temperature to
4.2 K, to separate the main cryostat insulation vacuum from the
inner vacuum space, and to provide a contraction allowance in two
directions as the system is cooled down.

Conduction heat flux into the 4.2 K reservoir through the
bellows is only about 30 mW since the bellows are thermally grounded at 77 K. While the system is precooling with helium exchange
gas in the inner vacuum space and horizontal beam tube, the heat
leak is increased by about 2.5 W. This heat leak quickly disappears as the exchange gas is pumped away over a few hours resulting in ultra high vacuum conditions in the inner vacuum space
which minimizes surface contamination of the sample.

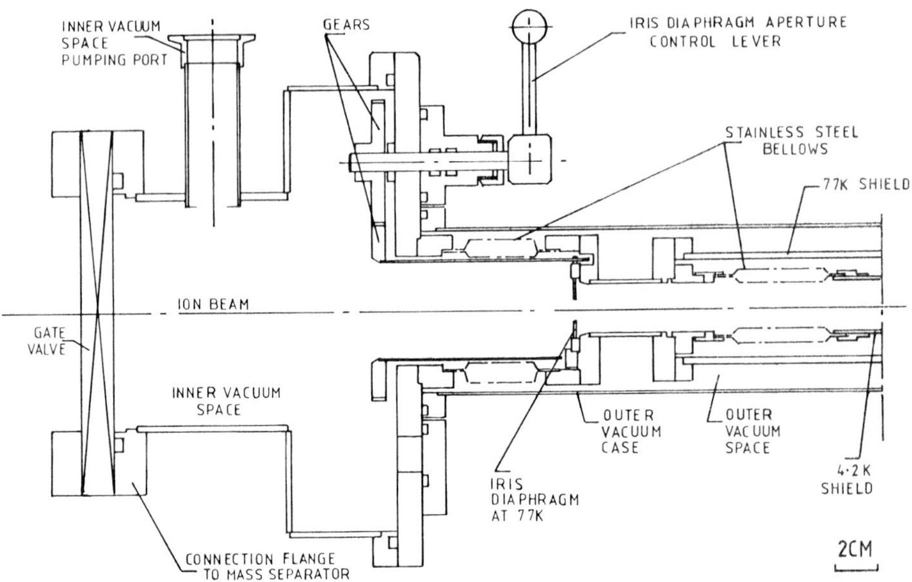

Fig. 2. Side Access Dilution Refrigerator — Inlet Connection.

The thermal radiation heat leaks down the beam tube are controlled in three ways: (1) a 4.2 K moveable radiation baffle, (2) an adjustable iris diaphragm at 77 K, and (3) coating the inside of the beam tube and the 4.2 K shield with a radiation absorbing coating.*

The 4.2 K moveable baffle allows heat leaks from the side access to be completely eliminated when not implanting ions, but it obviously also prevents entry of the ion beam. The 77 K iris diaphragm is adjustable from outside the cryostat. Its size can be varied from 1.2 mm to 20 mm. The smallest useable size of the aperture is determined by the focussing quality of the ion beam. The performance of the refrigerator as a function of aperture size with the 4.2 K baffle fully open is given in Table II.

TOP LOADING FEATURE

Changes of target are required when the original sample becomes contaminated with a build-up of daughter activity, or if a

---

*Nextel 101-C10, black velvet coating, 3M Company, St. Paul, Minnesota, emissivity 0.6 at 4.2 K.

Table II.   Refrigerator Performance

Aperture diameter mm	Sample temperature mK	Heat Leak/ cooling power µW
1.2	7.3	1.2
5.0	10.0	3.0
10.0	16.8	8.7
20.0	27.0	23.0

different target material is required. Conventional target changes can be made by dismantling the cryostat, but because of its
complexity this is a lengthy procedure. Top loading while the
system is cold is possible and is shown schematically in Fig. 3.
To load a sample, the target holder is screwed with a left handed

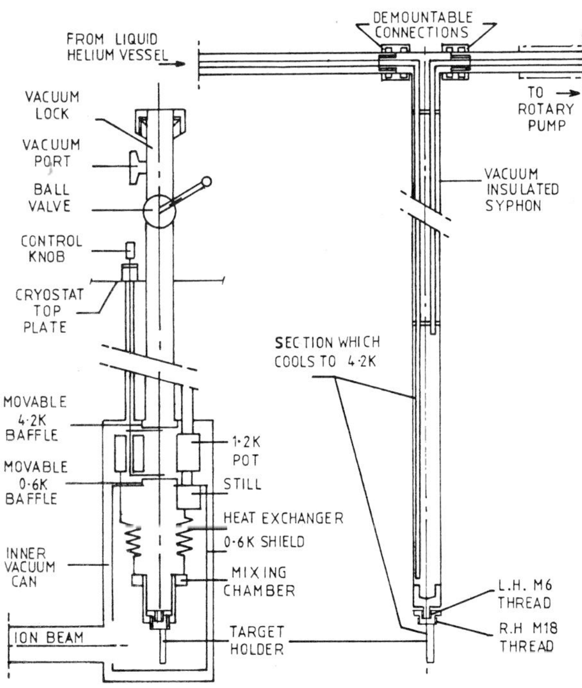

Fig. 3 Schematic Arrangement of Top Loading System.

M6 thread to the end of a special helium syphon. The target is
then introduced into the refrigerator through a vacuum lock on top
of the cryostat and precooled to 4.2 K by drawing liquid helium
through the syphon. The target is then screwed into the mixing
chamber extension tube using the righ handed M18 thread. When the
M18 thread tightens, the M6 thread unscrews, allowing the syphon
to be withdrawn. Following the reverse procedure allows withdrawal of the sample. During a top loading operation the mixing
chamber only warms to about 1 K. New samples can be loaded in a
few minutes and cooled to below 10 mK in less than 30 minutes. As
long as the M18 thread is kept clean the base temperature of the
system is not affected by top loading operations.

COMPLETED SYSTEM

The completed dilution refrigerator is shown in Fig. 4 and
Fig. 5.

Fig. 4. Overall view of dilution refrigerator
including overhead working platform.

Fig. 5. Lower end of the cryostat
showing the side access connection.

REFERENCES

1. P. Herzog, et. al., A dilution refrigerator for low
    temperature nuclear orientation experiments on line to an
    isotope separator, Nucl. Instr. & Meth. 155:421 (1978).
2. L. Vanneste, Department Natuurkunde, Katholieke Universiteit,
    Leuven, Belgium, (1979). Private communication.
3. M. Kaminsky, Phys. Rev. Lett. 28:819 (1969).
4. C. Rau, R. Sizman, Phys. Lett. 37A:113 (1971).

LARGE DILUTION REFRIGERATOR USING PRESSED METAL POWDER HEAT EXCHANGERS

Y. Oda, T. Ono and H. Nagano

University of Tokyo
Tokyo, Japan

G. Fujii

Nippon Electric Company
Tokyo, Japan

and

H. Asami

Sumitomo Heavy Industries
Hiratsuka, Kanagawa, Japan

INTRODUCTION

The technology of dilution refrigeration has progressed remarkably in recent years and temperatures of 7–8 mK have become available on a continuous basis using commercial refrigerators. A combination of the dilution refrigerator with nuclear demagnetization allows sub-mK temperatures to be attained easily, and studies of ^{3}He properties and the problems of solid state physics in the mK region are then possible. In some cases it is very effective to attain a few mK using only the dilution refrigerator. Frossati developed a continuous type heat exchanger with sintered silver powder which allows 2 mK to be reached. In this paper we describe a discrete type heat exchanger which gives a continuous low temperature of 2.85 mK at a circulation rate $\dot{n}_3$ of 270 μmol/s. Also in this paper we describe the outline of our refrigerator, its manufacturing process and the results.

DESCRIPTION OF REFRIGERATOR

This refrigerator was designed for solid state physics experiments in the temperature region from 20 K to about 3 mK, and for precooling a nuclear demagnetization stage. The cooling power of the refrigerator can be varied by the ^{3}He circulation rate $\dot{n}_3$ of 100 - 700 μmol/s. The several specimens are cooled in the same run. Since the refrigerator is run for long periods, safety precautions are taken for system operation. The temperature can be controlled easily even by untrained students.

The low temperature part consists of the 1 K pot, the still, the concentric tube heat exchangers, etc. The first time the refrigerator was tested we put a small discrete exchanger and a mixer right below the concentric tube exchanger and attained 17 mK. The second time we put five discrete heat exchangers below the concentric tube exchanger and attained 2.85 mK. The maximum cooling power at 100 mK is 600 μW. It takes one hour to cool from 300 mK to 7.5 mK.

HEAT EXCHANGER DESIGN

The heat exchanger is one of the most important components of the dilution refrigerator and much research and many types of exchangers have been reported[2].

For the high temperature region, that is the region from the still temperature to about 60 mK, the concentric tube type exchanger is generally used because of its easy construction and the properties of ^{3}He/^{4}He mixtures. Over this temperature range, the change of the concentration ratio of ^{3}He/^{4}He is large as the temperature decreases; thus, a gravitational instability may occur in the dilute phase. Also, in the concentrated phase, ^{4}He gas mixes with the main circulated ^{3}He and causes some heat leak.

In order to suppress these origins of instability and heat leaks, it is necessary to make the flow rate of ^{3}He higher than the some limiting value and reduce the cross-sectional area. The concentric tube exchanger is ideally suited for these purposes.

We construct the concentric tube exchanger as follows: A 100 cm long Cu-Ni tube (0.3mm id., 0.5mm od.) is soldered to a 200 cm long brass tube (1.4mm id., 2.0mm od.) and twisted. Then this twisted tube is inserted into the Cu-Ni outer tube (3.5mm id., 4.0mm od.) and the assembly is wound as a coil of 10cm diameter. One end of the outer Cu-Ni tube is connected to the still. The heat transfer area of the concentrated side σ_c is 97.4 cm^2. Radebaugh and Siegworth[5] have calculated the resultant mixing chamber temperature T_M when using a concentric tube exchanger. They show T_M as a function of $\sigma_c/\dot{n}_3$. The outlet temperature of the concentrated phase at the lower end of the exchanger T_{co} is given by[5]

$$T_{co} = T_M/0.36$$

In our case if we set $\dot{n}_3 = 200$ μmol/s, we estimate $T_{co} = 78$ mK.

In the temperature region lower than about 60 mK, instability does not occur, but the main problem is the Kapitza thermal resistance. As the temperature decreases, the thermal resistance increases rapidly and a large heat transfer area becomes necessary, so sintered metal powder or pressed metal powder are used for the heat exchanger. The heat exchanger which is used in this temperature region is classified into two types. One is the continuous type and the other is the discrete type. Frossati developed the former type exchanger[1]. He uses Cu-Ni foil as the partition wall between the concentrated and dilute phases, and fine silver powder is sintered on both sides of the foil. For the discrete exchanger, a copper block with two holes in it is made and metal powder is sintered or pressed into the holes.[3] In our case, we use Cu blocks with two long holes and Cu or Ag powder is pressed into them. After the powder is pressed, a small diameter hole is drilled through the center of the pressed powder sponge to act as a ^{3}He flow channel.[4]

The parameters which determine the efficiency of the discrete exchanger are the thermal conductance K_i along the flow channel and also the thermal conductance L_i perpendicular to the flow channel. The suffix i means the liquid (1), the Cu body (b) and the Kapitza conductance (k).

In the ideal discrete exchanger, the following condition is satisfied.

$$K_1 < \dot{n}_3 C < L \tag{1}$$

$$1/L = 1/L_b + 1/L_k + 1/L_1$$

where C expresses the heat capacity per mol of ^{3}He in the concentrated or in the dilute phase.

For the jth discrete exchanger, where the exchanger is counted from the high temperature side, we define $\sigma_c^{(j)}$, $T_{co}^{(j)}$, $T_{ci}^{(j)}$ as the heat transfer area, the outlet and inlet temperatures of the concentrated phase respectively. A unique relation between $\sigma_c^{(j)}/\dot{n}_3$, $T_{co}^{(j)}$ and $T_{ci}^{(j)}$ exists under the condition of Eq. 1. If we assume the still temperature (= $T_{ci}^{(1)}$) and the temperature of the mixing chamber, T_M, it is possible to calculate the number of the exchangers, $\sigma_c^{(j)}$, $T_{co}^{(j)}$ and $T_{ci}^{(j)}$.[4,5]

We assume 2.5 mK as the temperature of the mixing chamber and from the following relation we calculate the outlet temperature of the last exchanger

$$T_M = 0.36 T_{co}^{(5)}$$

Then $\sigma_c^{(5)}$, $T_{ci}^{(5)}$, $\left[=T_{co}^{(4)}\right]$ is calculated, and so on.

For the heat transfer area of dilute phase $\sigma_d^{(j)}$ use the relation

$$\sigma_d^{(j)} = 1.7\ \sigma_c^{(j)}$$

PROCEDURE FOR MAKING DISCRETE EXCHANGERS

Two holes of large diameter are drilled into a Cu block. Figure 1 shows the dimension of the first discrete exchanger (high temperature side). The diameter and the length are calculated to satisfy Eq. 1, that is, the length is fairly long compared with the diameter. The small diameter hole is used for the concentrated stream, and the large one is used for the dilute stream.

Then metal (Cu or Ag) powder is poured in the hole and is compressed by a hydraulic press. Then a small diameter hole is drilled through the center of the pressed metal powder sponge to provide the flow channel. The diameter of this flow channel is calculated to satisfy $\Delta T/T < 0.01$, for the concentrated phase and $\Delta T/T < 0.05$ for the dilute phase, where ΔT is the temperature increment due to viscous heating.

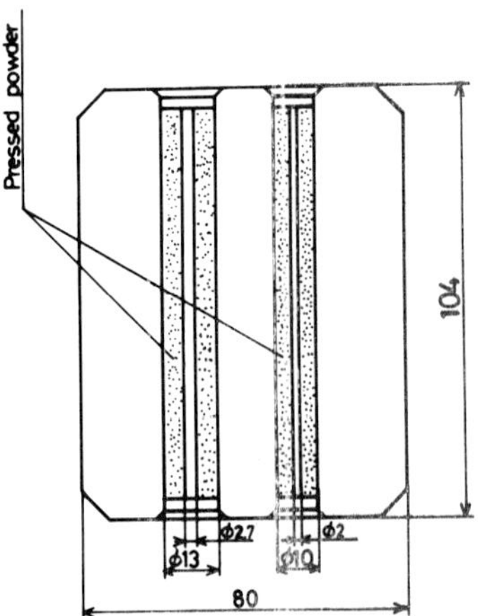

Fig. 1. Drawing of Discrete Heat Exchanger (First Stage) Dimensions are in mm.

The thermal conductance of liquid L_1 is proportional to $1/T$ and so it becomes small at the higher temperatures. From Eq. 1, it is not effective· to make σ_c so large that $L_k > L_1$ at the high temperatures. For this reason, we use Cu powder of 50 μm average diameter for the first exchanger, 30 μm Cu powder for the second exchanger, and 2.5 μm silver powder for the third exchanger. In the fourth and fifth exchangers, we use ultra fine silver powder of 700Å diameter.

In our exchangers, we use only pressed metal powder and do not sinter it, but in the case of Ag ultra fine powder, it sinters even at room temperature and the pressed powder sponge itself shrinks. Thus, we mix some larger Cu powder with the silver powder to prevent the shrinkage.

DESIGN OF STILL AND MIXING CHAMBER

In order to reduce the ^{4}He concentration in the circulating ^{3}He, we use a still with a ^{4}He film suppressor. Figure 2 shows the experimental data for the relation between the still temperature and ^{4}He concentration in the circulating ^{3}He. The total circulation rate at temperatures above 440 mK is also shown in the figure. From the experiment, the ^{4}He concentration depends only on the vapor pressure ratio above 440 mK, but not on the film flow.

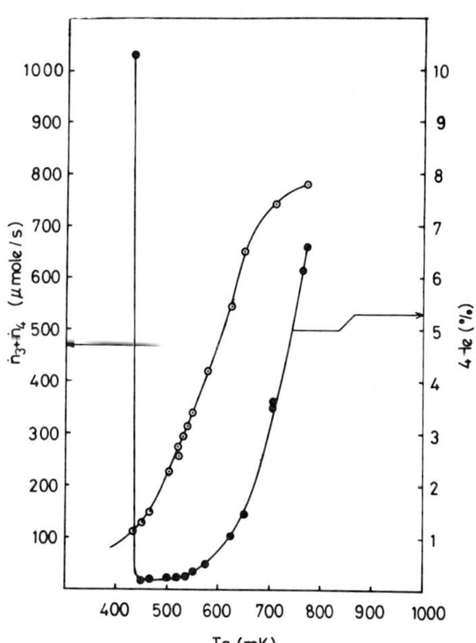

Fig. 2. Dependence of Circulation Rate $\dot{n}_3 + \dot{n}_4$ and ^{4}He Concentration on Still Temperature.

In the construction of the mixing chamber, we mix fine Ag powder with Cu powder and press them inside of it. An epoxy cell is located on the return path (the dilute phase) from the mixing chamber to the discrete exchanger, and La diluted CMN is put inside of the cell as a magnetic thermometer. The coil constant of the mutual inductance bridge is determined using NBS768 temperature standards.

Figure 3 shows the parts of the refrigerator inside of the vacuum can.

EXPERIMENTAL RESULTS

The relation between the circulation rate $\dot{n}_3$ and the minimum temperature of the mixing chamber T_M is shown in Fig. 4. From this figure, when $\dot{n}_3$ is 270 µmol/s the lowest temperature of 2.85 mK is attained. In our design of the refrigerator, we intended to attain the lowest temperature at a circulation rate of 200 µmol/s, but in practice a slightly different optimum rate occurs. This means the heat leak is larger than the designed value to some extent. There are several origins of the heat leak and the largest one might be the thermal conduction of the dilute liquid from the exchanger to the mixing chamber. If we reduce the diameter of the return tube it might be possible to reduce the temperature of

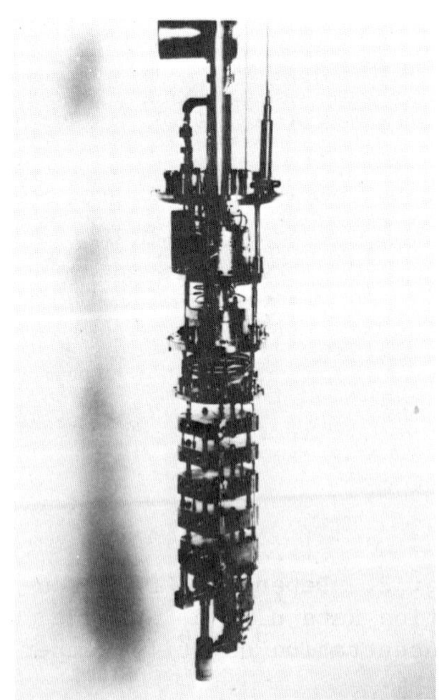

Fig. 3. Parts of Refrigerator
inside of Vacuum Can.

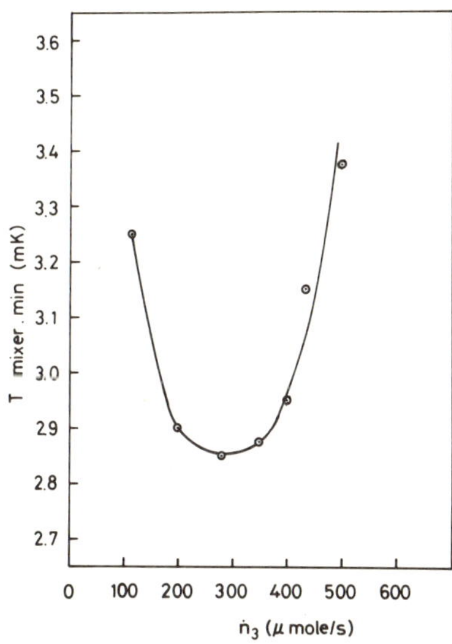

Fig. 4. Minimum Temperature of Mixing
Chamber vs. ^{3}He Circulation Rate.

the mixing chamber further. The relation between the cooling
power and the temperature of the mixing chamber is shown in Fig.
5, for two values of $\dot{n}_3$. Above 10 mK, the data fit the relation
$\dot{Q} = 84\ \dot{n}_3 T^2 M$. When the circulation rate is 730 μmol/s, the
cooling power is 5 μW at 10 mK.

From the above results, T_M is a little higher than the designed value, but this might be caused by an error in the estimation of the heat transfer area and by the fact that condition (1)
is not fully satisfied in the last exchanger. In comparison with
the continuous type exchanger of Frossati, the total volume of
liquid in the exchanger is not so large[1,6]. Our exchanger does
not require special techniques to make. This kind of discrete
heat exchanger is thought to be very useful to attain temperatures
of a few mK.

SUMMARY

A dilution refrigerator of 100 –700 μmol/s ^{3}He circulation
rate was built and the lowest temperature of 2.85 mK was
attained. In this refrigerator, the discrete heat exchangers with
pressed metal (Cu & Ag) powder are used. This type of exchanger
is made with a Cu block with two holes in it filled with pressed
metal powder through which a small diameter hole is drilled to act

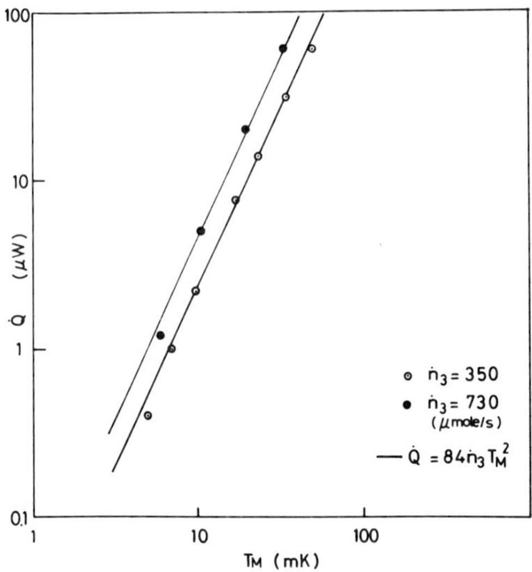

Fig. 5. Cooling vs. Temperature.

as a ^{3}He flow channel. From experiments with this type of exchanger, it is found useful for attaining a few mK continuously.

ACKNOWLEDGMENTS

The authors wish to thank Ray Radebaugh of the National
Bureau of Standards, Boulder, Colorado for his editorial assistance with this paper.

REFERENCES

1. G. Frossati, et.al., in "Proc. Hakone Int'l. Symposium," Phys.
      Soc. of Japan (1977) p. 205.
2. O.V. Lounasmaa, "Experimental Principles and Methods Below 1
      K," Academic Press, London and New York (1974) p. 17.
3. J.C. Wheatley, O.E. Vilches and W.R. Abel Physics 4:1 (1968).
4. Y. Oda, G. Fujii and H. Nagano, Cryogenics 18:73 (1978).
5. R. Radebaugh and J.D. Siegworth, Cryogenics 11:368 (1971).
6. G. Schumacher, Thesis Centre de Recherches sur les Tres Basses
      Temperature, C.N.R.S., Grenoble, France (1978).

A GADOLINIUM GALLIUM GARNET DOUBLE ACTING RECIPROCATING MAGNETIC REFRIGERATOR

R. Beranger, G. Bon Mardion, G. Claudet,
C. Delpuech, and A. F. Lacaze

Centre d'Etudes Nucléaires de Grenoble
Grenoble, France

and

A. A. Lacaze

Centre de Recherches sur les Trés Basses Températures CNRS
Grenoble, France

INTRODUCTION

Methods of adiabatic demagnetization although common in laboratory practice were limited until very recently to one-shot operations which restricted the length of the experiments. The first superconductor-switch refrigerator constructed by Heer and colleagues in 1954[1] was to provide the continuous operating conditions needed by physicists: low temperature ($<$ 1 K), low power ($<$ 1 mW).

The theoretical study of a reciprocating magnetic refrigerator with regenerator proposed by Van Geuns[2] in 1966 has not been experimentally verified to our knowledge. Steyert and coworkers at Los Alamos Scientific Laboratory devised a rotating continuously operating system in 1977[3]. To date a few machines have been proposed or tested at ambient or low temperatures.[4-8].

This paper describes the results of an experiment based on a particular type of reciprocating apparatus operating between 1.8 K and 4.2 K. Our goal was to design a test stand that would, in an initial exploratory stage, give us experience in such aspects as materials, heat transfer and mechanical guiding at low temperature, with a view to preparing the way for a rotating refrigerator.

GENERAL ANALYSIS OF MAGNETIC REFRIGERATION

The principle of a magnetic refrigerator can be described with a Carnot cycle ABCD. During isothermal magnetization of the paramagnetic material from A to B, the heat released to the warm source is $Q_W = T_W (S_B - S_A)$, where S is the entropy. During adiabatic demagnetization, the temperature is reduced to T_C and then, during isothermal demagnetization, the heat extracted from the cold source is: $Q_C = T_C (S_D - S_C)$.

Since the work of the cycle is negligible compared to the refrigeration near 5 K, a theoretical efficiency of the Carnot cycle can be defined by: $\eta_{th} = Q_C^{th} / Q_W^{th} = T_C / T_W$.

In the same way, the thermal efficiency of an actual machine can be defined by:

$$\eta_{ac} = Q_C^{ac} / Q_W^{ac} \text{ and } \eta - \eta_{ac}/\eta_{th}$$

where η_{ac} refers to the actual measured performance of the machine.

EXPERIMENTAL DEVICE

Description

The demonstration machine had to be simple and well suited for experiments to verify the calculations. The experimental unit is described in Ref. 9 and illustrated by the cross section in Fig. 1. It is essentially a double acting reciprocating magnetic

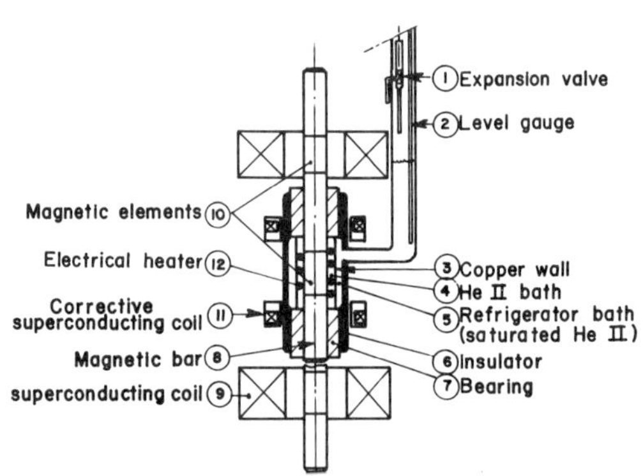

Fig. 1. Experimental device immersed in 4.2 K boiling Helium.

machine with an auxiliary liquid helium refrigerator. It operates immersed in a bath of liquid helium at 4.2 K which forms the warm reservoir of the magnetic cycle. The cold source is a superfluid bath at atmospheric pressure, that can be cooled to a temperature of about 1.4 K, either with the help of the auxiliary refrigerator or through the magnetic refrigerator, or both.

Referring to Fig. 1, the two identical magnetic elements (10) in the bar (8) are moved periodically. The upper magnetic element is magnetized in a superconducting magnet up to 5 T in the bath at 4.2 K. Then it is moved down through a guide bearing (7) and demagnetized in the middle bath (4) which constitutes the cold source. When the upper element is cooled, the lower element is magnetized in a second magnet. Then the bar is moved back up, the upper element is magnetized and the lower element is cooled. The bar slides in the guide bearings (7) that isolate the central chamber from the 4.2 K bath. The copper wall (3) surrounding the central chamber permits heat exchange with the auxiliary refrigerating bath (5).

The auxiliary refrigerator, controlled by a manual valve, can deliver a power of about 1 W at 1.8 K. Through this refrigerator it is possible to measure thermal losses in various geometrical and magnetic configurations. When auxiliary refrigeration is not required vacuum is kept in this circuit. Inside the cold reservoir (4) there is an electrical heater which allows us to determine the useful power.

The magnetic parts of the piston are alternately in contact with the 4.2 K source and the 1.8 K source. Consequently, the time taken to establish temperature equilibrium in the cylinder subjected to these variations must be short with respect to the cycle duration.

For the present experiments, we used $Gd_3Ga_5O_{12}$ (GGG) as paramagnetic substances. Each magnetic element is composed of two single crystals 2 cm long and 2.4 cm diameter. The two active components were attached to three non magnetic parts by means of screwed and glued stainless steel connecting rods. These non magnetic parts are composed of sintered alumina tube glued on a small rod of fiberglass reinforced epoxy. The resulting bar was machined with a diamond wheel.

NbTi Cryomagnetic System at 4.2 K

The magnetic induction profile illustrated in Fig. 2 could be obtained without difficulty using two main coils facing each other and two compensation coils also facing each other. In the zero field zone, the field was always less than 10^{-2} T; the maximum

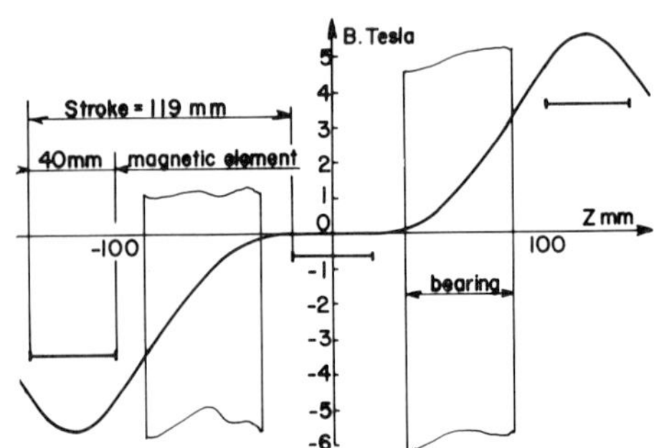

Fig. 2. Magnetic field profile.

field was about 5 T constant to ± 10% over the length of the active component. At present, the field profile is such that at the end of the adiabatic demagnetization when the sample leaves the bearing the field is nearly zero. At the end of the adiabatic magnetization coming out of the bearing, the GGG must reach 4.2 K. This is obtained when the maximum field is about 2.5 T.

Drive System

To move the bar with a maximum force of 400 N, we use a hydraulic jack having a maximum linear speed of 0.5 m/s. The movement is controlled by a function generator, which can be used to impart various movements to the bar, in particular, a sinusoidal movement, a "trapezoidal" movement, (Fig. 3) which is in fact a sequence of plateaus with sinusoidal sides and a "two slopes" movement (Fig. 3) which is a sequence of high speed movement (for isentropics) and low speed movement (for isothermals). The frequency of the movement and the duration of heat exchange at the sources in the case of the "trapezoidal" and "two slopes" movements can be varied through adjustment of the drive system.

EXPERIMENTAL RESULTS

We have tested several different paramagnetic elements and bearings from various materials. The clearance between the bar and the bearing was 0.02 mm. Preliminary experiments with $HoPO_4$ and $Gd_2(SO_4)_3$ have been reported elsewhere.[9] The experiments reported here were carried out with single crystal GGG* with the

---

*Manufactured by Laboratoire d'Electronique et de Technologie de l'Informatique, Grenoble, France.

"trapezoidal" movement and the "two slopes" movement. For these two movements, Fig. 3 compares the variation in useful power $\dot{Q}_u$ as a function of frequency for different values of magnetic field, at a temperature of 1.8 K. The total cold thermal power, $\dot{Q}_c$ removed by the garnet is due to: $\dot{Q}_u$ (simulated by the heater) and $\dot{Q}_L$ (wasted by losses), such that $\dot{Q}_c = \dot{Q}_u + \dot{Q}_L$.

As shown in Fig. 3, at low frequency $\dot{Q}_u$ is a linear function of the frequency. On the other hand $\dot{Q}_L$ is composed of static losses, $\dot{Q}_s$, independent of frequency and dynamic losses, $\dot{Q}_D$, increasing with frequency: $\dot{Q}_L = \dot{Q}_s + \dot{Q}_D$. The static losses are due to heat conduction through the connecting stainless steel tubes from 4.2 K to 1.8 K, to conduction through the liquid contained in the annular spaces between the bearings and the bar and to conduction through the bearings and the bar. The power: $\dot{Q}_s$ can be measured by linear extrapolation of $\dot{Q}_u$ toward zero frequency. Both "trapezoidal" movement and "two slopes" movement give $\dot{Q}_s$ = 0.31 W. The dynamic losses are due to friction between the bar and the bearings, to movement of liquid helium imported by the bar

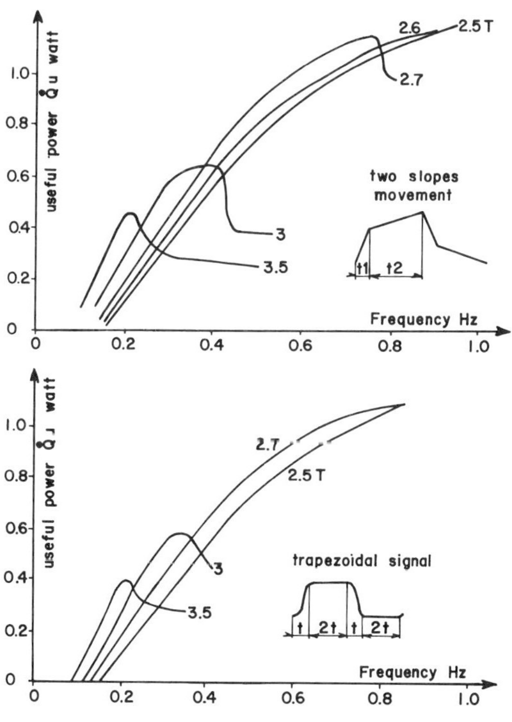

Fig. 3. Useful power at 1.8 K.

and to transfer of enthalpy due to temperature cycling of the bar itself. The quantity Q_D can be estimated by measurements in zero magnetic field with the auxiliary refrigerator and theoretical calculations. We found that the energy lost per cycle is about 0.3 J for the "trapezoidal" movement and 0.5 J for the "two slopes" movement. The higher the instantaneous velocity the higher the dynamic losses.

When we increase the frequency, the time devoted to heat transfer at sources becomes smaller and as a consequence the energy extracted per cycle decreases and the useful power P_u is no longer a linear function of frequency.

For critical values of field and frequency, a sudden decrease of useful power appears. We assume that this is due to the appearance of film boiling at the warm source.

The thermal efficiency η of the unit is computed by using: (1) the input power supplied to the cold reservoir by the electrical heater ($\dot{Q}_u$) and (2) Q_W, the power dissipated at the warm reservoir measured from the helium consumption, excluding cryostat and current lead losses. For the "two slopes" movement we have measured: (1) a limiting temperature of 1.38 K at 0.8 Hz, (2) a useful power at 1.8 K and the maximum frequency of 0.95 Hz of 1.2 W, and (3) a thermal efficiency η at 1.8 K for frequency higher than 0.4 Hz of 45%. For the "trapezoidal" movement the measured values are somewhat lower, i.e., P_u = 1.15 W at 0.85 Hz and η = 43%.

The remaining irreversibilities which arise from static losses and temperature differences which appear in the cold reservoir can be reduced, e.g., higher efficiency could be obtained by subdividing the crystal in several parts and modifying the field profile.

CONCLUSION

An experimental device has been built to study magnetic refrigeration. The choice of a double acting reciprocating machine that uses liquid helium as the warm heat reservoir gives a more simple construction and permits more versatility in operation.

The use of an auxiliary refrigerator allows the direct measure of some losses. Significant progress has been made in such technological areas as:

- the successful guide system with low friction in cryogenic medium

– improved knowledge of the thermal losses generated by the movement of a surface in a superfluid bath
– a temperature below 1.8 K in a bath of superfluid helium at atmospheric pressure was achieved from the first experiments as well as a refrigerating power close to one watt

The good understanding of the present results make it possible to design high efficiency helium refrigerators.

REFERENCES

1. C.V. Heer, C. B. Barnes, J. G. Daunt, The design and operation of magnetic refrigerator for maintaining temperatures below 1 K, Rev. Scient. Instr. 25:1088 (1954).
2. J.R. Van Geuns, A Study of a new magnetic refrigerating cycle, Philips Research Report Supplement 6, (1966).
3. W.P. Pratt, et al, A continuous demagnetization refrigerator operating near 2 K and a study of magnetic refrigerants, Cryogenics 17:689 (1977).
4. G.V. Brown, Magnetic heat pumping near room temperature, J. Appl. Phys. 47:3673 (1976).
5. W.A. Steyert, Stirling–cycle rotating magnetic refrigerators and heat engines for use near room temperature, J. Appl. Phys. 49:1216 (1978).
6. W.A. Steyert, Rotating carnot–cycle magnetic refrigerators for use near 2 K, J. Appl. Phys. 49:1227 (1978).
7. W.A. Steyert, Magnetic refrigerators for use at room temperature and below, J. de phys. Colloque C6, 39:1598 (1978).
8. J. A. Barclay, O. Moze, L. Paterson, A reciprocating magnetic refrigerator for 2–4 K operation : initial results, J. Appl. Phys. 50:5870 (1979).
9. C. Delpuech, et al, Double acting reciprocating magnetic refrigerator : first Experiments, Cryogenics 21:597 (1981).

MAGNETICALLY SUSPENDED STIRLING CRYOGENIC SPACE REFRIGERATOR: STATUS REPORT

A. Daniels

*Philips Laboratories
Briarcliff Manor, New York*

and

M. Gasser and A. Sherman

*NASA Goddard Space Flight Center
Greenbelt, Maryland*

INTRODUCTION

Conceptual designs of spaceborne cryogenic refrigeration systems capable of long term, unattended operation were presented at the 1979 Cryogenic Engineering Conference[1] and at the Conference for Cryogenic Sensors and Electronic Systems held in Boulder, Colorado in 1981[2]. Since then, efforts to translate one of those concepts into an engineering model with which to demonstrate the desired operational characteristics have continued at Philips Laboratories with the sponsorship and guidance of the NASA/Goddard Space Flight Center.

The refrigerator was designed to generate 5 W of cooling power at a temperature of 65 K, when its compression heat is dissipated at 300 K, and to maintain that output reliably for a period of 5 years or longer. The physical characteristics of the refrigerator, such as weight and size, were constrained to be consistent with spaceborne-hardware requirements. To attain the desired high efficiency, the refrigerator design is based on the Stirling cycle, which is thermodynamically reversible and, therefore, has an ideal efficiency equal to that of the Carnot cycle.

Existing Stirling cycle refrigerators have demonstrated comparatively high efficiencies, which are compatible with spaceborne applications. However, to date there is no closed cycle cryogenic cooler that could approach 5 years of maintenance-free,

stable operation in space. The cooling load requirement of the
machine presented in this paper adds further challenge to achieving this goal.

BACKGROUND: LIFE RELATED PROBLEMS

The most significant life-limiting mechanisms in a typical
cryogenic refrigerator are degradation (by contamination) of its
operating temperature, and wear of its bearing and seal surfaces. The contamination and wear mechanisms are inter-related.
The nature of that relationship is illustrated with the aid of
Fig. 1, a schematic representation of a conventional Stirling
machine.

For long, reliable operation, bearings and seals are normally
oil-lubricated. However, in Stirling machines the seal separating
the "thermodynamic" working spaces from the crankcase which contains the drive mechanism, and consequently the bearings, is not
perfect, or truly hermetic. Consequently, the lubricant is "get-
tered" (or attracted) by the low-temperature region; it therefore
migrates past the seal, and contaminates the working space.

It should be noted that substances other than oil (or hydrocarbons) are also subject to gettering; and consequently, if

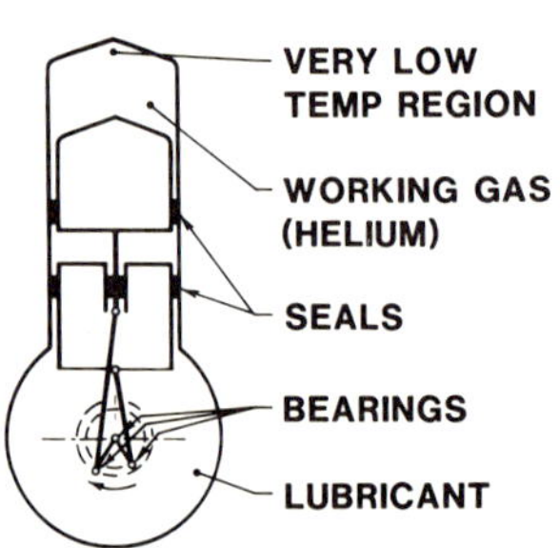

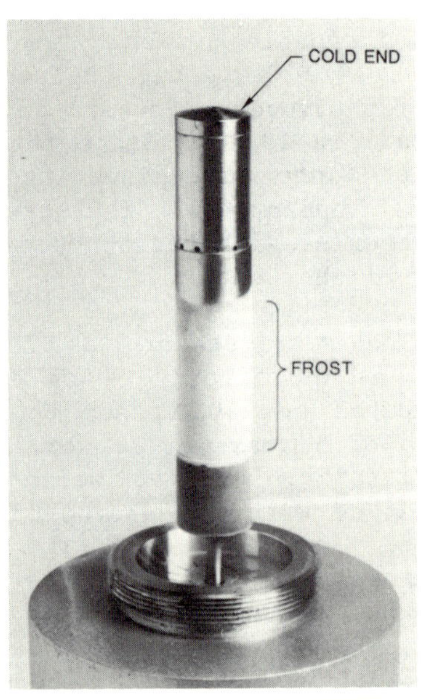

Fig. 1. Schematic representation
of a conventional Stirling refrigerator.

Fig. 2. Contaminated
displacer/regenerator.

present in the refrigerator, contribute to contamination of the
working space. The most common contaminants are air, small traces
of which are difficult to remove from the refrigerator during
purging; hydrogen, adsorbed during brazing of refrigerator components; CO and CO_2, which are outgassing products of many organic
materials; and water[2], which is adsorbed on all refrigerator
surfaces during assembly.

Figure 2 shows tangible evidence of the contamination process
just described. The crankcase of a small cryogenic refrigerator
was intentionally contaminated with very small quantities of
water, air, and CO_2, and operated for several hours. At the end
of that period, the cold head of the refrigerator was removed in a
dry helium chamber, and the displacer was thus exposed for viewing
and photographing. The contaminants, which leaked past the piston
and displacer seals (see Fig. 3), evidently froze out on the
"warm" end of the displacer/regenerator subassembly. The resulting "ice" layer increased the thermal losses which normally occur
in the annular gap between the cold finger and displaceer; the
increased losses, of course, degraded refrigerator performance by
increasing its operating temperature.

An obvious solution to the degradation problem is to eliminate the sources of contamination. This suggests either dry or
boundary lubricated seals and bearings. Indeed, many cryogenic
refrigerator designs rely on reinforced Teflon seals reciprocating
against hardened metal surfaces, and on essentially dry bearings. The problem this approach creates, however, is illustrated
in Fig. 4, a photograph of a reinforced Teflon seal and the ad-

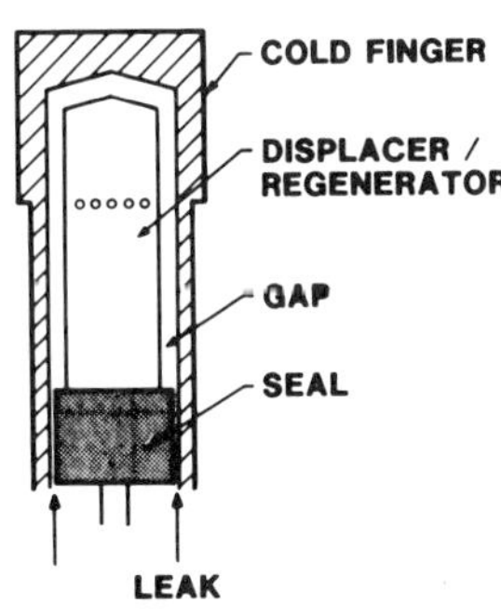

Fig. 3. Cross-sectional view
of contaminated displacer/re-
generator.

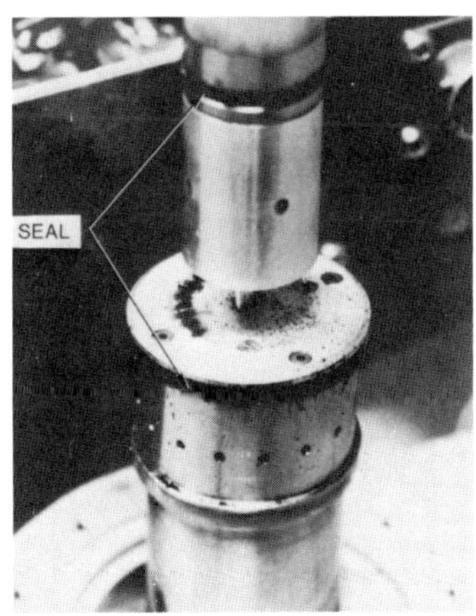

Fig. 4. Illustration of
wear associated with dry
seals.

joining surfaces after several hundred hours of refrigerator operation. As the photograph indicates, seal wear has taken place. This wear mechanism not only increases the leak past the seal, but it generates solid particles which clog critical passages in the refrigerator, a condition which also contributes to performance deterioration. Indeed, during the few hundred hours of operation, the temperature of the refrigerator degraded by several degrees.

The problems discussed and illustrated above suggest that if "years" of life are to be attained, methods must be found to either eliminate or to minimize seal and bearing wear, to prevent contaminants from reaching the refrigerator working space.

DESIGN OF REFRIGERATOR COMPONENTS

To eliminate the life-related problems just outlined, the refrigerator being developed at Philips Laboratories contains five important features: a purely rectilinear drive[3], linear electric motors, magnetic bearings, clearance seals, and "all-metal" working space surfaces.

These features have been integrated into the refrigerator design shown in cross-section in Fig. 5. The design consists of three major sections: the cold side including the cold finger, the displacer/regenerator, and the adjoining displacer-motor hous-

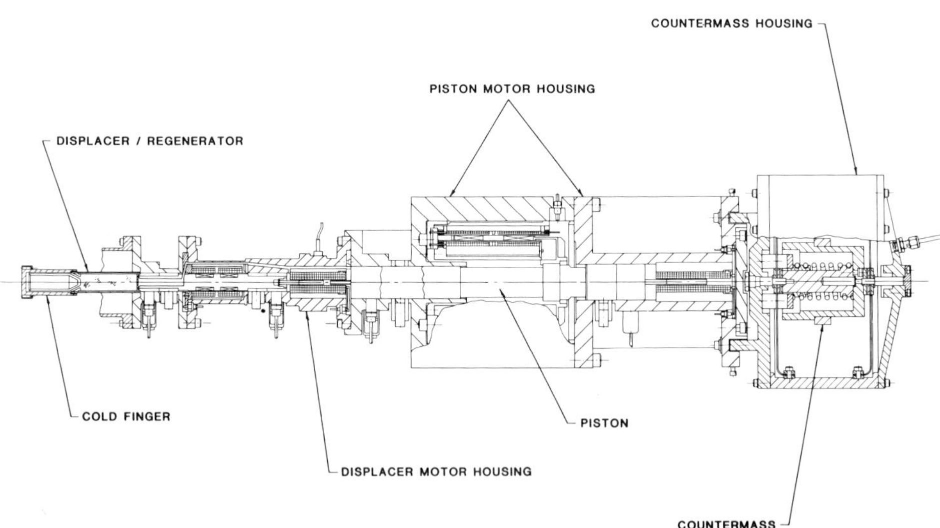

Fig. 5. Cross-sectional view of refrigerator.

ing. The piston-motor housing accommodates, as the designation
implies, the piston and the related linear motor. Finally, the
refrigerator will have provisions for accommodating either a
passive or an active balancing system, designed to minimize vibrations induced by the reciprocating masses. Figure 5 shows the
passive balancer which, in its simplest form, is a "text-book"
case of a tuned spring/mass system. The NASA/Goddard Space Flight
Center is developing an active balancing system designed to re-
ceive its input from an accelerometer with closed-loop control to
force the net vibration of the cold end of the refrigerator to
approach zero.

The major features of the moving-magnet type piston and
displacer motors are illustrated in Fig. 6. Two rows of samarium
cobalt permanent magnets are affixed to an undercarriage (armature) which, in turn, is fastened to the reciprocating member; in
this case, the piston rod. The motor coil is split into two
subassemblies which are toroidally wrapped around the permanent-
magnet structure. The coils are enveloped by ferromagnetic pole
pieces designed to provide the required magnetic flux paths, as
shown in Fig. 7. The motion of the motor armature, and therefore
that of the piston is sinusoidal, regulated by a feedback type,
closed loop control system.

To guide the motion of the motor-driven reciprocating elements, i.e., the piston and the displacer, in a closely centered
rectilinear path, without the need to use life-limiting conventional bearings, Philips Laboratories developed magnetic bearings
capable of levitating those elements. The most important features

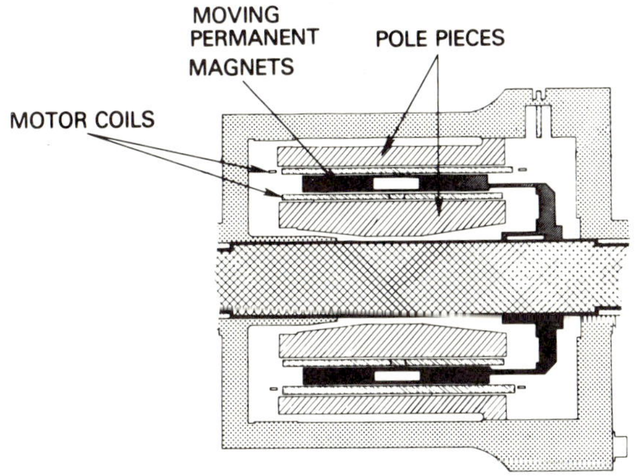

Fig. 6. Cross-sectional view of piston motor.

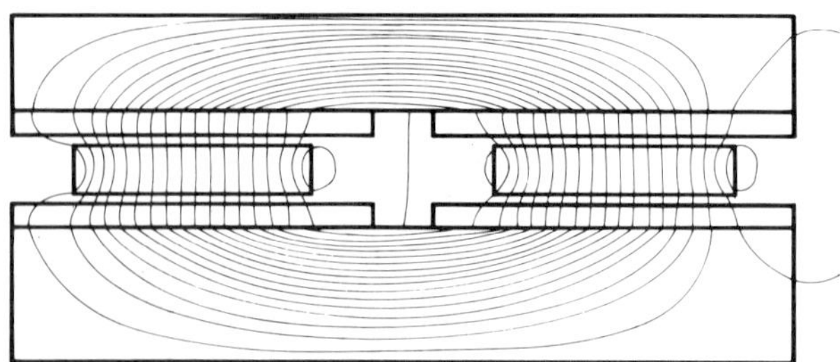

Fig. 7. Magnetic flux path in moving magnet linear motor.

of those non-contacting bearings are illustrated with the aid of Fig. 8.

Electromagnetic coils, positioned at structurally compatible places, provide the attractive forces designed to float the reciprocating ferromagnetic shaft. Radial displacements of the shaft from center position, induced by either design or stray loads, are detected with the aid of eddy current sensors which send a signal indicating the deviation in shaft position to electronic controls. The controls create a correction signal which, after amplification, causes a change in the current to the electromagnets and a corresponding radial restoring force.

The refrigerator being developed has four sets of magnetic bearings, with two each supporting the piston and displacer subassemblies. None of these bearings has to operate at low temperatures.

With the method of magnetic suspension just discussed, the radial position of a shaft can be controlled to within a fraction of a thousandth of an inch. With such close control, it is possible to attain relatively small annular gaps between a reciprocating shaft and the adjoining cylinder, or housing, with comparatively uniform geometries. Narrow annular gaps, in turn, make the use of clearance seals possible. Two such seals have been incorporated into the refrigerator design being developed: one around the piston, the other around the warm extremity of the displacer. The former ensures adequate compression; the latter prevents the working gas from short circuiting the regenerator.

Clearance seals are obviously not hermetic but, under favorable dynamic conditions and with a judicious choice of gap lengths, the leakage can be held to tolerable levels. For instance, in the design presented in this paper, the radial clearance of the seals is 0.0025 cm and the length of the displacer

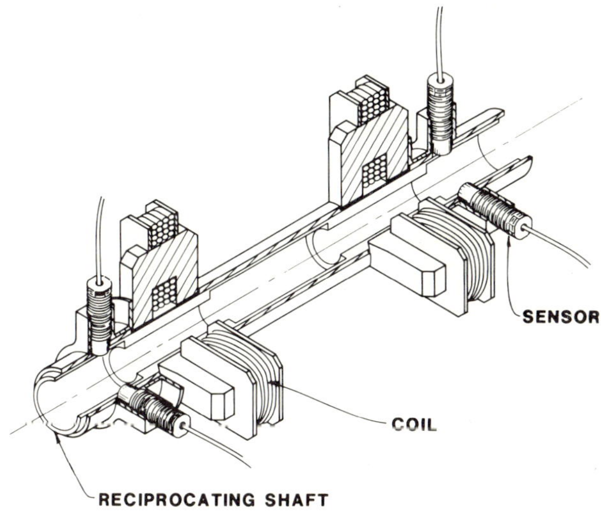

Fig. 8. Method of magnetic shaft suspension.

seal is 4 cm. At a refrigerator operating frequency of 30 Hz, and
a displacer stroke of 0.8 cm and a diameter of 2 cm, the projected
leakage through the seal is 2% of the stroke volume per cycle.
The resulting effect on efficiency is an acceptable penalty for
achieving long life.

The elimination of rotary bearings and the associated need
for lubricants, and the use of magnetic suspension and clearance
seals, make it possible to have refrigerator working spaces free
of organic materials and, consequently, of major sources of contamination. Specifically, only four materials come into contact
with the helium working gas in the refrigerator: titanium, copper, ferromagnets, and ceramics. It is therefore assumed that it
will be possible to bake out the working spaces of the refrigerator, and thus attain conditions characteristic of high vacuum
systems, i.e., freedom from contaminants.

DEVELOPMENT STATUS

Of the three features critical to the long range integrity of
the refrigerator being developed, namely the magnetic bearings,
the linear motors, and the clearance seal; the first two have been
tested at the component level and the last feature can be adequately tested only during the overall evaluation of the refrigerator.

To test the magnetic bearing performance, the prototype
bearing shown in Fig. 9 was built in a size characteristic of the

Fig. 9.   Prototype magnetic bearing.

piston shaft. An ac stiffness on the order of 10 000 kg/cm (ac) and 12 000 kg/cm (dc) was obtained with a clearance between the reciprocating parts and the adjoining walls of 0.0025 cm. This is more than adequate for meeting the requirements of the engineering model cooler.

To test the piston motor a special fixture was designed and built to simulate accurately the conditions under which the motor would operate in the actual refrigerator. At 28 Hz, the expected operating frequency of the refrigerator, the measured motor efficiency is 74%, one percentage point below that predicted by analysis (75%). Furthermore, the motor was operated in the closed-loop mode using the actual drive electronics and position control system. The side forces generated by the motor were less than 1 kg, well within the capability of the magnetic bearings. The harmonic distortion of the motion was less than 1%, which is expected to result in very satisfactory performance of the passive counterbalancing system. In general, all the static and dynamic characteristics of the motor met the refrigerator requirements.

REFERENCES

1. A. Sherman et al., Progress on the Development of a 3-5 Year
        Lifetime Stirling Cycle Refrigerator in Space, in
        "Advances in Cryogenic Engineering, Vol. 25", Plenum
        Press, New York (1980), p.791.
2. M. Gasser, A. Sherman and W. Beale, Developments Toward
        Achieving a 3-5 Year Lifetime Stirling Cycle Refrigerator

for Space Applications, in "Refrigeration for Cryogenic
Sensors and Electronic Systems", National Bureau of
Standards Special Publication 607, (May 1981).
3. A. K. de Jonge, A Small Free-Piston Stirling Refrigerator, in
    "Proc. of the 14th Intersociety Energy Conversion Engr.
    Conf."
4. P. A. Studer and M. G. Gasser, A Bi-directional Linear Motor/
    Generator with Integral Magnetic Bearings for Long
    Lifetime Stirling Cycle Refrigerators, in "Refrigeration
    for Cryogenic Sensors and Electronic Systems", National
    Bureau of Standards Special Publication 607, (May 1981).

DISCUSSION

Question by J. C. Riple, Garrett-Airesearch Corp.: How many watts
are dissipated in the magnetic bearing system?

Answer by author: We estimate about 10 W when operation in space
at zero-g and about 20 W on earth.

NOVEL TITANIUM-ALUMINUM JOINTS FOR CRYOGENIC COLD FINGER STRUCTURES

H. M. Meehan and R. C. Sweet

Philips Laboratories
Briarcliff Manor, New York

INTRODUCTION

The sensors used in airborne detection and surveillance systems often require low temperatures for optimum performance. The low temperature surface for mounting the sensors may be the containing wall of the expansion volume of a Stirling cycle cooler. It is common practice to mount infra-red detectors on copper heat exchange surfaces. Such surfaces are thermally isolated and structurally stabilized by a stainless steel member. The use of an aluminum-titanium cold finger results in a considerable weight reduction.

Previous attempts to fabricate an aluminum-titanium joint involved dipping in a molten, flux-covered aluminum bath or, diffusion bonding. Mackowiak and Shreir[1] investigated the interaction layers formed during the reaction between solid titanium and liquid aluminum. Bollenrath and Metzger[2] studied joints made by dip brazing titanium and aluminum in molten, flux-covered baths. Other investigations were carried out by Elrod and Lovell[3] and by Hocker and Metzger[4], in successful attempts to utilize aluminum as a braze material for titanium honeycomb structures. In all cases, the reaction of titanium with molten aluminum resulted in the formation of an intermetallic layer identified as $TiAl_3$. The thickness of this layer was found to increase with increasing temperature, but the length of time was not as critical. Elrod and Lovell[3] noted that contamination accelerated the conversion reaction to $TiAl_3$.

Since aluminum readily wets and bonds to clean titanium surfaces, a technique of casting molten aluminum onto an appropriately dimensioned and positioned titanium member was investigated. The casting is then machined to provide the form and structure desired.

FABRICATION AND TESTING

Procedure

Although cryogenic assemblies may require the more costly ELI* α titanium alloys and high purity aluminum, less costly grades of pure titanium and 1100 alloy aluminum make suitable materials for evaluation. After machining the parts to suitable dimensions the individual components are degreased and then degassed by heating in high vacuum (2×10^{-5} torr) for one hour at 500°C. The parts are then thoroughly cleaned. The titanium cleaning procedure is: 1) degrease ultrasonically, 2) Etch in HF/HNO$_3$, 3) Distilled water rinse, 4) Rinse in hot 10% HNO$_3$, 5) Distilled water rinse, 6) Air dry, 7) Clean ultrasonically – acetone. The aluminum procedure is: 1) Degrease ultrasonically, 2) Etch in NaOH solution, 3) Distilled water rinse, 4) De-smut in HNO$_3$, 5) Distilled water rinse, 6) Air dry, 7) Clean ultrasonically – acetone.

Elrod and Lovell[3] point out that trace contamination by fluoride salts and/or heavy metals can inhibit wetting of the titanium and cause an excessively thick and porous intermetallic layer. A careful cleaning procedure is, therefore, warranted since a thick intermetallic layer is undesirable. Figure 1 is an S.E.M. micrograph of a joint.

After cleaning, the components are assembled, jigged appropriately, and placed in a high vacuum furnace. The assembly is heated to 685°C until all the aluminum has melted. The assembly is then allowed to cool. Cooling the assembly properly is important. The solidification front begins at the coolest parts of the

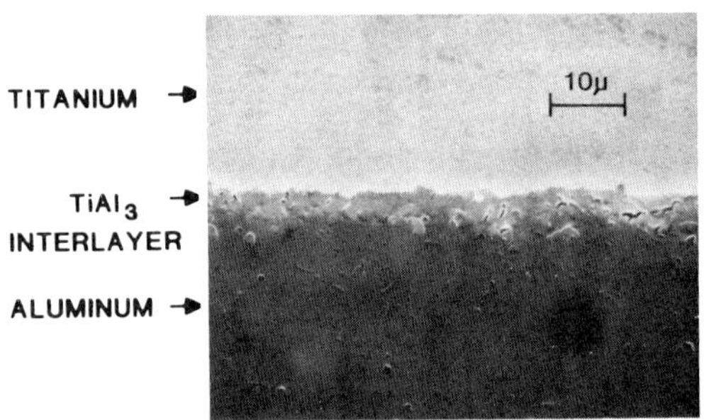

Fig. 1. Joint micrograph.

---

*ELI = Extra Low Interstitial impurities.

casting container and proceeds toward the warmest until all the liquid has solidified. The last region to solidify will frequently contain voids and cavities and should be restricted to a region of no significance to heat exchange or structural functions. Ideally, one should machine off this region, leaving a dense, cast section. In order to do this the solidification can be forced to occur in the "right" direction. Cooling the regions where solidification should first occur is a way to achieve this. One technique is to use helium gas as a quenching fluid. Helium gas can be blown into the cylindrical section of a cold finger assembly while the aluminum is molten, forcing the cooling to start at the interior surface of the cylinder. The solidification front then moves outward to the periphery of the casting which will subsequently be machined away. The interior heat exchange areas are thus more likely to be dense and well bonded.

Joint Testing, Thermal Cycling

Aluminum and titanium differ greatly in their rates of thermal expansion. Thermal cycling should be a good technique to test the durability of the cast joint.

A small test sample (Fig. 2) was prepared for cold shocking. The aluminum was contained in a ceramic crucible while the furnace was slowly cooled from an $800^{\circ}C$ casting temperature. The casting was machined into a cylinder approximately 1 $^{1}/_{4}"$ high by 3/4" diameter with a 1/4" wall thickness. The sample was moved automatically from a Dewar of liquid nitrogen into a flow of $100^{\circ}C$ heated air. This cycling from 77 K to 293 K was repeated for 1000 cycles. The sample was periodically removed and the joint leak checked with a helium mass spectrometer. No leaks were found after 1000 thermal cycles.

Fig. 2. Thermal test sample.

The same sample was then cycled five times from a liquid helium bath to room temperature. The joint remained sound and helium leak tight.

Joint Testing, Mechanical

Tensile test samples (Fig. 3) were prepared by casting pure aluminum inside a tubular titanium section at 685°C. Freedom of voids in the casting was determined by X-ray examination. After casting, the titanium pieces were machined to a test sample similar to that described by ASTM Specification E-8 with the titanium-aluminum interface approximately at the mid-point. Properly cast samples showed failure in the bulk aluminum and not at the interface.

DISCUSSION

When considering the replacement of copper-stainless steel cold fingers with aluminum-titanium cold fingers, an important question arises concerning the relative thermal conductivities of copper and aluminum. Aluminum is a lighter metal, but not as good a thermal conductor as copper. Therefore, to attain an equivalent heat transfer, more aluminum than copper contact area will be required.

Since the thermal conductivity is a function of temperature and material purity, the operating temperature of the cold finger and the purity of the aluminum used are important factors in the weight saving calculation. Table I shows a comparison of thermal conductivities and an estimate of the weight savings possible if pure aluminum is used in place of pure copper.

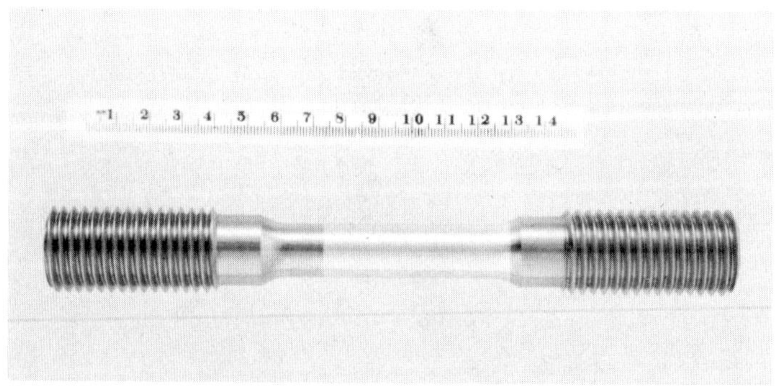

Fig. 3. Tensile test sample.

Table I. Possible Weight Savings

Temperature K	Thermal Conductivity		% Weight Savings Using Aluminum
	Al(99.996%)	Cu(99.999%)	
	W/cm – K	W/cm – K	
0	0	0	–
4	30.8	113	–110%
8	57.3	189	0
10	66.1	196	+110%
30	28.5	43	+220%
60	6.7	8.5	+260%
100	3.0	4.8	+210%
273	2.4	4.0	+190%

CONCLUSIONS

Aluminum-titanium cast structures offer good potential for use as cryogenic cold finger assemblies. Good bonding between the aluminum and titanium is exhibited if the parts are thoroughly cleaned before casting. The weight savings potential of aluminum over copper can be ascertained by examining the relative conductivity of each metal for use in a particular cryogenic temperature range. Configurations and structures other than those described are possible. Controlled cooling is necessary to produce sound, void free castings. Figure 4 is a photograph of a cast aluminum-titanium cold finger (left) next to a typical copper-stainless steel cold finger.

Fig. 4. Cast aluminum-titanium (left).
and copper-stainless steel cold fingers.

REFERENCES

1. J. Mackowiak and L. Shreir, The nature and growth of
    interaction layers formed during the reaction between solid
    titanium and liquid aluminum, J. Less-Common Metals, 1:456
    (1959).
2. F. Bollenrath and G. Metzger, The brazing of titanium to
    aluminum, Welding Journal 42 (10): 4425, (October 1963).

3. S. D. Elrod and D. T. Lovell, Aluminum brazed titanium honey-
    comb sandwich structures--a new system, Welding Journal, 52
    (10): 4255, (October 1973). Also, D. Lindh and D. T.
    Lovell, Titanium structures technical summary, F.A.A.
    Report No. FAA-55-73-27, Boeing Corporation (October 1974).
4. R. G. Hocker and G. Metzger, Weldbraze airframe components,
    Technical Report AFML-TR-77-171, Air Force Materials
    Laboratory (August 1977).

SINGLE SHOT DEMOUNTABLE
SELF-CONTAINED ^{3}He REFRIGERATOR

P. Kittel and W. F. Brooks

NASA Ames Research Center
Moffett Field, California

INTRODUCTION

We have developed a simple, modular, rapid-response ^{3}He refrigerator, which is a self-contained demountable unit. When needed, it can be bolted to the cold plate of a superfluid helium Dewar. Once installed, the refrigerator is controlled by a single heater. This approach was chosen for two reasons. First, there was a desire to extend the useful temperature range of an existing Dewar. The design allowed us to achieve 0.27 K without costly or time-consuming modifications to the Dewar. This has greatly increased the versatility of our Dewar because the ^{3}He unit can be removed when it is not needed. Second, in some space applications, a modular refrigerator will be required. For example, in the Shuttle Infrared Telescope Facility (SIRTF), a 1 to 2 K cold plate will probably be provided.[1] An experimenter requiring a lower temperature would have to provide a bolt-on refrigerator. The unit described here allowed us to experiment with a demountable design, before confronting the difficulties caused by zero gravity.

There are several uses for temperatures below 1 K in space. One is the interest in performing basic physics measurements on the critical behavior of ^{3}He/^{4}He mixtures in the absence of gravitational mixing.[2] This requires a temperature of 0.87 K. Several research groups are now considering doing tricritical point experiments in space (Private communication with L. Campbell of Los Alamos Laboratory and H. Meyer of Duke University). Another use for temperatures below 1 K is space operation of infrared telescopes. The sensitivity and response time of bolometers dramatically improves as the temperature decreases.[3,4] As an example, the noise equivalent power (NEP) of bolometers improves

"

as T^n where $3/2 < n < 5/2$. A factor of 50 decrease in NEP is possible by going from 1.5 K to 0.3 K.

SYSTEM DESCRIPTION

The refrigerator is shown in Fig. 1, and its specifications are listed in Table I. The refrigerator consists of three chambers (a pump, a condenser, and an evaporator) and interconnecting stainless steel tubing. The pump is filled with an adsorbing material; we have used Linde 5A zeolite, although activated charcoal could also be used. There is a finned copper structure inside the pump to improve the thermal conductivity of the adsorber. The control heater is attached to these fins. A copper wire, not shown, connects the pump to the cold plate. This wire cools the pump from 30 K to 20 K in 10 min when the cold plate is at 2 K. This cooling rate was selected after several trials, and represents a compromise between having a short condensation cycle (fast heating and cooling of the pump) and a small amount of heat absorbed by the cold plate (a weak thermal link while heating the pump). It would be possible to size the pump-to-condenser tube to achieve this cool-down rate, but unfortunately, such a method would put an unacceptably large heat load on the condenser during the condensation phase. The condenser is a hollow copper block that is bolted to the cold plate. A thin indium gasket is used between the condenser and the cold plate to improve the thermal conductance of this joint. Spring washers are used to compensate for differences in thermal contraction and to maintain a constant load between the aluminum cold plate, the copper condenser, and the stainless steel bolts. The evaporator is a hollow copper block hung beneath the condenser. A spiral of copper foil is

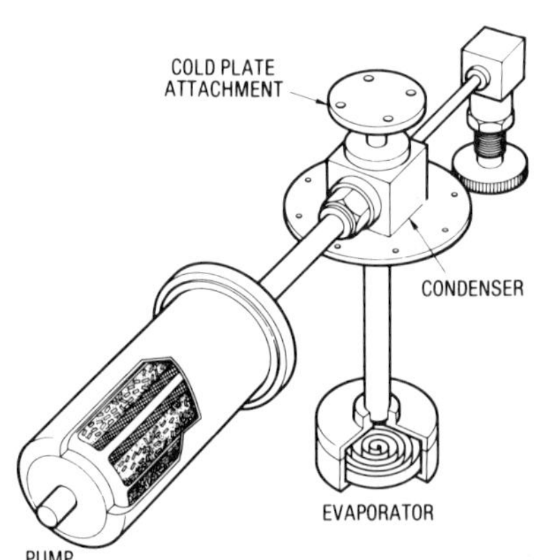

Fig. 1. Refrigerator configuration.

Table I. Refrigerator Specifications

Zeolite mass	118 gm
³He amount	0.63 moles
³He pressure at 300 K	6.2 MPa
Pump volume	0.14 liter
Total volume	0.25 liter

brazed inside the evaporator to increase the contact area between the liquid ³He and the copper block. Following thorough degassing and evacuation to 10^{-2} Pa, the unit is filled with ³He through a valve attached to the condenser; the valve is bellows-sealed and has a metal seat. After the unit is filled, the valve is shut and the open port is capped off.

While self-contained ³He refrigerators are not new,[3,5,6] this unit differs significantly from the others. The earlier refrigerators are in Dewars dedicated to them. The Dewars have usually been modified with several penetrations through the vacuum container. Several of the systems require the operation of a cold valve, the warming of the Dewar to liquid nitrogen temperature to recycle the system, and an external ³He storage chamber. The unit described here has none of those traits. The refrigerator can be removed from the Dewar by removing a few bolts, and no penetrations have been made for actuators or plumbing lines. The system can be recycled without warming the main Dewar and an external storage chamber is not required.

OPERATING PRINCIPLES

The refrigerator is simple to operate. After the system is cooled to below 2 K, the heater is turned on, heating the pump to about 30 K, by which time the ³He has been desorbed by the pump and driven out of this chamber. The ³He then condenses in the condenser (which is still cold) and drains into the evaporator. After about 15 min, equilibrium is reached and the heater may be turned off. As the pump cools, it readsorbs the gas, causing evaporative cooling in the evaporator. Refrigeration is produced until all of the liquid is evaporated or until the adsorbent becomes saturated. If further operation at 0.3 K is required, the cycle can be repeated. Since operation of the refrigerator requires only a single heater, it has been easy to automate the system.

The duration of the operating cycle is principally determined by the amount of ^{3}He in the system. A detailed method of calculating the amount needed for a given hold time has been given elsewhere,[6] and a simplified method explained here. The refrigerator has a free internal volume of V_0 and is charged with gas at density ρ_0. During condensation this free volume can be represented as the sum of three terms:

$$V_0 = V_p + V_g + V_\ell \qquad (1)$$

where V_p is the volume of the pump excluding the volume of the adsorbent, V_g is the volume of the gas in the rest of the system, and V_ℓ is the volume of the condensed liquid. During this phase of the cycle the pressure in the system is approximately uniform, but the temperature is not. The pump is being heated to a temperature T_p, while the condenser and evaporator are being held at some lower temperature T. By using the ideal gas law, the system can be treated as if the temperature were uniform at T and with an effective volume of

$$V = V_g + V_\ell + V_p(T/T_p). \qquad (2)$$

The last term on the right can be ignored if $V_p \lesssim V_g + V_\ell$ and $T \ll T_p$, which is normally the case. By reducing the effective volume ($V < V_0$), the gas density has been increased over the charging density. The effective density becomes

$$\rho = \rho_0 V_0/V \approx \rho_0\left[1 + V_p/(V_g + V_\ell)\right]. \qquad (3)$$

The density enhancement increases the amount of liquid formed during condensation. The condensation process can be easily described by considering a gas of density ρ in a fixed volume V with no adsorbent at a uniform temperature T. Starting at a temperature greater than its condensation temperature, the gas is cooled. Its density does not change until condensation begins. If cooling continues, the liquid phase (high density) and the gas phase (low density) will separate, but during this process the average density does not change.

The amount of liquid condensed is strongly dependent on ρ_0 and T. The fraction of the mass condensed is

$$f = \rho_\ell V_\ell/\rho_0 V_0 = (1 - \rho_g/\rho)\,(1 - \rho_g/\rho_\ell)^{-1} \qquad (4)$$

This relation is derived from Eq. 3 and the conservation of mass: $\rho V = V_\ell \rho_\ell + (V - V_\ell)\rho g$. The behavior of Eq. 4 is shown in

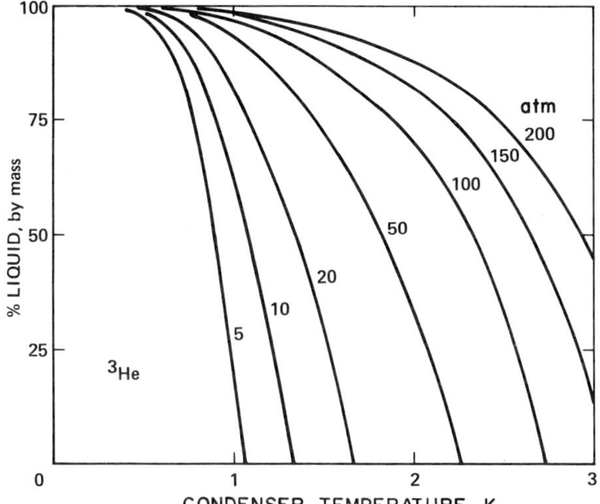

Fig. 2. Percent of ^{3}He by mass in the liquid phase in a
sealed isothermal volume. The various curves represent
different charging pressures in atmospheres at 300 K.

Fig. 2. (For ease of illustration, the curves are for the case V
= V_0, so ρ = ρ_0.) The various curves in Fig. 2 represent different charging pressures (measured in atmospheres at 300 K).
These curves show, for a given charging pressure, that the amount
condensed is very sensitive to the condenser temperature.
Furthermore, to condense a given fraction, the required charging
pressure is also strongly dependent on the condenser temperature.

Now that the amount of liquid condensed is known, there is
still the question of how much liquid remains after it is cooled
evaporatively to a lower temperature. This is easy to calculate
for the case in which the heat capacity of the evaporator is
dominated by the liquid. Consider a mass of liquid m with
specific heat C and latent heat L. Then a mass dm evaporating
will remove heat dQ = Ldm. This causes the remaining liquid to
cool: dQ = mCdt. Equating these relations gives

$$dm/m = (C/L)dt. \qquad (5)$$

This can be integrated to give the fraction remaining:

$$m_f/m_s = \exp\left[-\int_{T_f}^{T_s} (C/L)\ dt\right] \qquad (6)$$

where the subscripts s and f designate the starting and final
states.

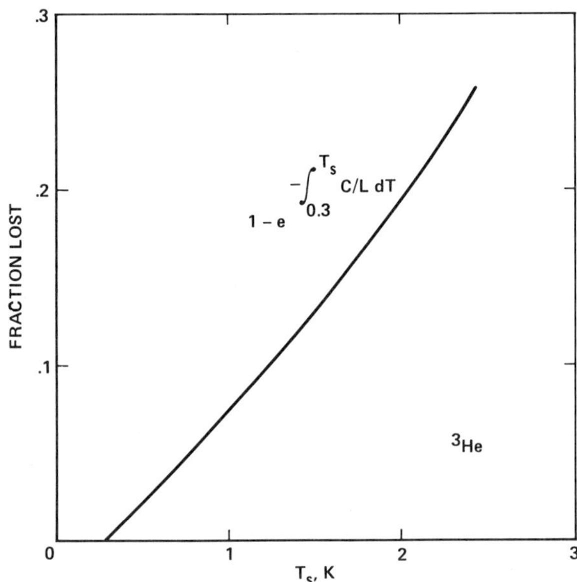

Fig. 3. Fraction of ^{3}He removed in evaporative cooling from temperature T_s to 0.3 K. The heat capacity of the evaporator and heat loads have been neglected.

Figure 3 shows the fraction evaporated ($1 - m_f/m_s$) while cooling to 0.3 K as a function of the condensing temperature. The operating temperature of 0.3 K is used here because it is typical of ^{3}He refrigerators. This limiting temperature is the result of the low vapor pressure of ^{3}He at this temperature ($\sim 10^{-1}$ Pa) that restricts the pumping speed of the interconnecting tubing. From this and from the previous discussion on condensation, it is clear that the colder the condenser, the greater the amount of liquid available at the final operating temperature. This is a result of a larger fraction being condensed and of a smaller fraction being removed during pump-down.

PERFORMANCE

A time history of the temperatures of the various parts of the refrigerator is shown in Fig. 4. The temperatures displayed are those of the pump, the condenser, and the evaporator. The pump and condenser temperatures were measured with silicon diode thermometers; that of the evaporator was measured with a germanium resistance thermometer. The lowest temperature attained by the evaporator is 0.27 K, which is lower than the 0.4 K reached by most other adsorption refrigerators.[3,5,6] Because this unit reaches a lower temperature, it probably has a larger refriger-

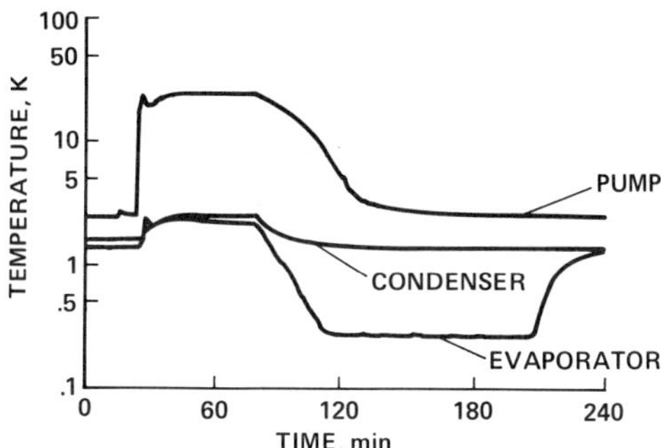

Fig. 4. Time history of the temperature of various components during a typical cycle.

ation power. However, refrigeration power of neither this unit nor the other units has been measured. The refrigeration power and ultimate temperature are limited by the vapor pressure of ³He and by the pumping speed of the system

The hold time has been a disappointing 1.5 h. Although this is adequate for our present needs, it is much less than the 12 h without load that we had planned on. The difficulty arises probably from the poor thermal conductance between the condenser and the cold plate. This is not surprising, because pressed contacts are notoriously bad. During condensation, considerable heat is conducted through the gas and connecting tubing from the pump to the condenser. This heat must pass through the condenser to the cold plate. In addition, the heat of condensation must also pass through this joint, the low conductance of which causes the temperature of the condenser to become significantly higher than that of the cold plate. Thus, the condensation occurs at a temperature greater than that of the cold plate. This greatly reduces the amount of helium condensed (see Fig. 2). In the example shown in Fig. 4, the temperature of the condenser rises from ≈ 1.6 K to ≈ 2.7 K when the pump is heated. Thus only 20% rather than 95% of the helium is condensed and 30% rather than 13% of the liquid is evaporated in cooling to 0.3 K.

CONCLUSIONS

A self-contained demountable ³He refrigerator that greatly increases the versatility of a cold-plate-style helium Dewar was developed. It is controlled by a single heater and can reach 0.27 K. Development is continuing in an effort to improve the existing unit and to eventually build a zero-gravity unit.

REFERENCES

1. F.C. Witteborn, L.S. Young, and J.H. Miller, Design
     Alternatives for the Shuttle Infrared Telescope Facility,
     in "Proceedings of SPIE", 183 (1979).
2. R.J. Donnelly and P. Kittel, eds., "Proceedings, Summary of
     the Conference on Quantum Fluids in Space", University of
     Oregon, Eugene, Oregon (1975).
3. J.V. Radostitz, et al., Portable ^{3}He cryostat for the far
     infrared, Rev. Sci. Instr. 49:86 (1978).
4. H.J. Goebel, W.F. Brooks, and P. Kittel, Ultrahigh
     responsivity ^{3}He cooled bolometers, J. Op. Soc. Am. 70:1057
     (1980); also, "Proceedings of the Workshop on Optical
     Fabrication and Testing, Optical Society of America", New
     York (1980), p. 158.
5. G. Chanin and J.P. Torre, A Portable ^{3}He Cryostat for Space
     Applications, in "Proc. Sixth Int. Cryo. Engr. Conf.", IPC
     Science and Technology Press, Guildford (1976), p. 96.
6. A. Sherman and O. Figueroa, A Portable ^{3}He Cryostat for
     Studies in Astrophysics, in "Advances in Cryogenic
     Engineering, Vol. 25", Plenum Press, New York (1980), p.
     775.

DISCUSSION

Question by N. W. Kerley, Oxford Instruments, England: Have you
considered the question of thermo-acoustic oscillations? We find
^{3}He is far more likely to oscillate than ^{4}He.

Answer by author: Thermo-acoustic oscillations are driven by
large temperature differences. In ^{4}He a difference of 300 K is
required. We have a maximum temperature difference of 30 K and
therefore think oscillations are unlikely. We have no means of
detecting oscillations.

Question by E. Troard, Jet Propulsion Laboratory: How will you
get the ^{3}He into the evaporator at zero-g?

Answer by author: The evaporator and condensor would be
combined. The condensor would no longer be connected directly to
the cold plate, but would have to be connected by a heat switch.
This switch would be closed only during the condensation phase.

MINIATURE J-T REFRIGERATORS USING ADSORPTION COMPRESSORS*

C. K. Chan, E. Tward, and D. D. Elleman

Jet Propulsion Laboratory
California Institute of Technology
Pasadena, California

INTRODUCTION

The demand for higher resolution and sensitivity of various detectors used in space missions has aroused interest in developing more sophisticated cooling systems for the detectors[1]. Because of the unique environment of space vehicles, the cooling system must be designed by taking into consideration weight, size, input power, packaging restrictions, operating life, maintenance intervals, life cycle and temperature environment. A deep space vehicle usually has an ample supply of waste heat from a radioisotope thermoelectric generator (RTG) and has a heat sink near zero Kelvin. Passive radiative coolers[2] have been used extensively for cooling temperatures above 90 K. As the cooling temperature gets lower, the weight and size of the radiator becomes a problem. Cooling below 90 K is presently handled by phase changes of solid or liquid cryogens[3] or mechanical refrigeration[4]. However, the weight of the cryogen may be a problem in a long duration space mission. The reliability and maintenance interval of mechanical refrigeration in long-term unmanned missions is a concern to many space vehicle designers.

In many spacecraft applications, refrigerators of small cooling capacity are required. For future NASA missions, the cooling power required is often only in the milliwatt range, while the temperature requirement varies from 1 K to 230 K. Furthermore, some missions may require a refrigerator having a very long lifetime. In order to meet these needs, simple refrigeration systems

---

*Work supported by the National Aeronautics and Space Administration.

incorporating few moving parts have been studied. A multistage Joule–Thomson refrigerator using adsorption compressors as the gas sources appears particularly attractive for these applications. In each refrigeration stage, gas is supplied by an adsorption-desorption compressor to a heat exchanger and Joule–Thomson valve. Heat is rejected either to an upper refrigeraton stage or by radiation to space. Waste heat, for example from an RTG, may be used as a power source[5].

The properties of four different refrigeration stages using charcoal as the adsorbent, have been studied. The adsorbed gases are nitrogen, helium, hydrogen and neon. The stages are: (1) a N_2 stage operating between 170 K and 77 K; (2) a H_2 stage operating between 77 K and 20 K; (3) a neon stage operating between 77 K and 27 K and (4) a He stage operating to 4 K. The coefficient of performance has been determined for each stage. In addition, possible combinations of the stages into an integrated multistaged system have been studied.

ADSORPTION COMPRESSOR

The adsorption compressor operation depends on the fact that certain adsorbers, such as charcoal, can adsorb very large quantities of gas[6]. The amount of gas adsorbed is a function of temperature and pressure. After adsorbing gas at some relatively low temperature and pressure, the gas may be pressurized by heating the adsorber. The gas may be readsorbed at some lower pressure and temperature by cooling the adsorber. This property of adsorbers can be used to design a compressor.

Figure 1 is a schematic of a one-stage refrigerator using two adsorption-desorption compressors. Desorption chamber A is thermally connected to the heat source. Upon heating, gas is released from the adsorbent and pressurized. The pressurized gas then flows through check valve 1 out of the compressor to the refrigerator. Adsorption chamber B is thermally connected to the upper refrigeration stage and disconnected from its heat source. The return gas from the refrigerator flows through check valve 4 and is adsorbed in B and the heat of adsorption is rejected to the upper refrigeration stage. Once the gas in A is exhausted, chamber B is connected to its heat source and disconnected from the upper stage while chamber A is disconnected from its heat source and connected to the upper refrigeration stage. Gas then flows out of chamber B through check valve 3, through the refrigerator and check valve 2 into chamber A, completing one cycle of the compressor.

COEFFICIENT OF PERFORMANCE

Single stage systems as shown in Fig. 1 have been thermo-dynamically modelled in order to determine their performance. The adsorption of the gas in charcoal as a function of pressure and temperature was first determined. The isotherm data for various gases was found to best fit an expression of the type[6]

$$C = \frac{M}{b} \, \alpha \, e^{-\gamma x} \tag{1}$$

where α and γ are correlating parameters and x is given by

$$x = \frac{RT}{\beta} \, \ell n \left[\frac{P_c}{P} \left(\frac{T}{T_c} \right)^2 \right] , \tag{2}$$

where β is a correlating parameter. The constants α, β, γ, are extracted from the experimental data outlined in Ref. 7. The values of M, b, α, γ, β, P_c and T_c for the various gases of interest are listed in Table I.

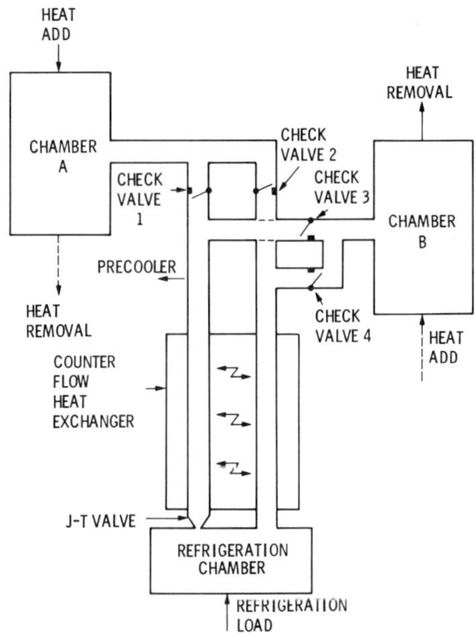

Fig. 1. Single stage J-T refrigerator using adsorption compressor.

Table I. Properties of Gases

GAS	HELIUM	HYDROGEN	NEON	NITROGEN
MOLECULAR WEIGHT (GM/GM MOLE)	4.003	2.016	20.183	28.016
INVERSION TEMPERATURE (^{0}K)	40	202	250	622
CRITICAL POINT PRESSURE, P_c (ATM)	2.26	12.8	26.9	33.5
TEMPERATURE, T_c(^{0}K)	5.3	33.3	44.5	126.2
SPECIFIC VOLUME, $V_c \frac{(CM^3)}{GM}$	14.4	32.2	2.067	3.21
VAN DER WAAL'S VOLUME b (CM3/G MOLE)	24	26.08	17	38.6
PARACHOR, β	20.4	34.2	25.0	60.0
CONSTANTS FOR ISOTHERM CORRELATION				
a	0.6722	0.40401	0.2961	0.40401
γ	0.0929	0.02870	0.0509	0.02870
	13<x<30	0<x<25	0<x<20	0<x<25

The cooling power of a Joule-Thomson refrigerator, $\dot{Q}_r$ is given by[5]

$$\dot{Q}_r = \dot{m}(h_2 - h_1) \tag{3}$$

where m is the mass flow out of the compressor and $(h_2 - h_1)$ is the enthalpy difference of the gas into and out of the heat exchanger. During the compression period, the total refrigeration load is

$$Q_r = m_a \Delta C(h_2 - h_1) \; , \tag{4}$$

where m_a is the adsorbent mass and $m_a \Delta C$ represents the gas mass liberated during the compression. The heat required for gas liberation is given by

$$Q_h = m_a C_{pa} \Delta T + m_s C_{ps} \Delta T + m_a \Delta C \Delta U \tag{5}$$

Where ΔT is the maximum temperature swing of the adsorbent and ΔU is the isoteric heat of adsorption which satisfies an equation similar to the Clausius-Clapeyron equation, i.e.,

$$\Delta U = - R \left[\frac{\partial \ell n P}{\partial \left(\frac{1}{T} \right)} \right]_C \; . \tag{6}$$

Using the expressions, one can show that the coefficient of performance (COP) of the refrigerator is given by

$$COP = \frac{Q_h}{Q_r} = \frac{(C_{pa} + m_s C_{ps}/m_a)\Delta T}{(h_2 - h_1)\Delta C} + \frac{\Delta U}{(h_2 - h_1)} \tag{7}$$

Since h_2 and h_1 are functions of P_H, P_L and T_{pre}, the COP is a
function of P_H, P_L, T_{pre}, ΔT, and ΔC.

In estimating the container mass in Eq. 6, a cylindrical
container completely filled with the adsorbent is assumed. By
imposing the restriction that the Hooke stress of the cylindrical
wall resulting from internal pressurization is limited by the
yield strength of the container material, the mass of the container to the adsorbent mass can be expressed as[5]

$$\frac{m_s}{m_a} = \frac{2\rho_s P_H}{\rho_a \sigma_y} \tag{8}$$

RESULTS

The coefficients of performance for helium, hydrogen, neon and
nitrogen stages were calculated as a function of P_H, P_L, T_{pre}, T_L
and T_H. In the calculation, the counter flow heat exchanger is
assumed to be perfect, i.e., the inlet gas temperature is equal to
the outlet gas temperature, and the precooling temperature and the
minimum temperature at the beginning of the compression cycle for
each stage is equal to the heat sink temperature of that stage.
The effects of an imperfect heat exchanger on the nitrogen stage
is reported elsewhere[5].

It can be generally concluded that, for a given precooling
temperature, a given low pressure level and a given minimum compressor temperature, there exists a minimum COP. For most efficient operation, one would like to operate the system at the
pressure and temperature values corresponding to the minimum
COP. For example, the COP for the hydrogen stage, shown in Fig.
2, has a minimum value of 40 at P_H = 60 atm and T_H = 280 K for
T_{pre} = 77 K, T_L = 77 K and P_L = 1 atm. The minimum COP decreases
as the low pressure level increases. In practice, the low pressure level is fixed by the refrigeration temperature requirement. The minimum COP for various stages and selected operating
conditions is tabulated in Table II.

MULTISTAGE DESIGN

A block diagram of a multistage refrigerator is shown in Fig.
3. For a given refrigeration load and temperature at each stage

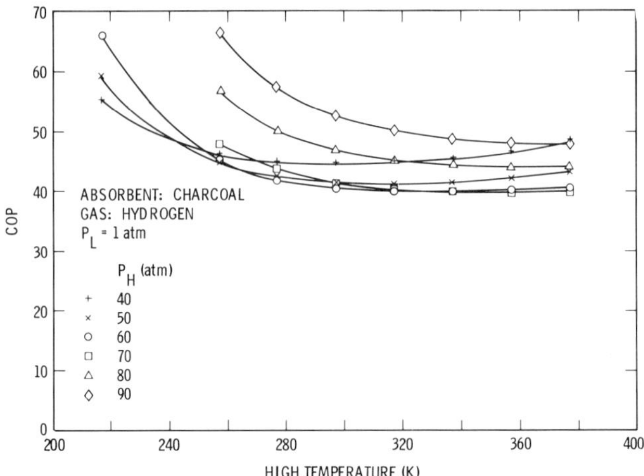

Fig. 2. Effect of maximum compressor temperature and high pressure level on coefficient of performance.

one must first determine the refrigeration system and the gas-adsorbent combination in the compressor. In the J–T process, the gas used depends on the refrigeration system. In the J–T process, the gas should have the critical point above the refrigeration temperature but a triple point below that temperature. The precooling temperature must be below the inversion temperature of that gas. The critical temperature and pressure and the inversion temperature for various gases are listed in Table I. To illustrate the design procedures, consider a helium refrigerator giving 1 mW of cooling at 4.2 K. The J–T output pressure must be helium at 1 atm. For minimum COP, the high pressure should be 25 atm. If a hydrogen stage refrigerator can provide a heat sink of 20 K for precooling the gas and cooling the helium compressor, then the maximum temperature for the thermal cycle[5] in the helium compressor is 70 K. Under these thermodynamic conditions, the COP for the helium compressor is 27 (Table II). In other words, the total heat into the hydrogen stage is 28 mW (i.e., 27 mW + 1 mW).

If a similar procedure is followed in designing the hydrogen stage, for minimum COP of 40, the adsorption compression should be operated between 1 atm and 60 atm, for a temperature swing of the compressor between 77 K and 277 K with the gas precooling temperature at 77 K. Hence, the heat dumped to the 77 K heat sink is (40 x 28 + 28) or ~ 1.2 W.

The 77 K heat sink is provided by the nitrogen stage. The COP for that stage is 22 for P_L = 1 atm, P_H = 60 atm, T_L = 150 K, T_H = 470 K and T_{pre} = 150 K. The 150 K heat sink for the nitrogen

Table II.   Minimum values of COP for specified refrigeration
                   and precooling temperatures, low pressure, and cold
                   compressor temperature.

GAS	HELIUM	HYDROGEN	NEON	NITROGEN
REFRIGERATION TEMPERATURE	4.2K	20K	25K	77K
LOW PRESSURE LEVEL	1 atm	1 atm	1 atm	1 atm
PRECOOL TEMPERATURE	20K	77K	77K	150K
COLD COMPRESSOR TEMPERATURE	20K	77K	77K	150K
HIGH PRESSURE LEVEL	25 atm	60 atm	80 atm	60 atm
HIGH COMPRESSOR TEMPERATURE	70K	277K	277K	470K
MINIMUM COP	27	42	40	22

stage can be provided by a passive radiator at 150 K. The total
heat load on the radiator will be 26 W.

The advantage of the multistage design is obvious when one
compares the radiator areas for 150 K radiator of 26 W and a 4 K
radiator of 1 mW capacity. Although the heat load of the multistage design is 26 000 times the heat load of 1 mW, the area
required at 4 K is 2 x 10^6 times the area required at 150 K for
the same heat load. Hence, staged refrigeration has an advantage
of approximately 80 in terms of radiator area. In theory, the
area required to radiate 1.2 W to deep space from a 77 K surface
is less than that required to radiate 26 W from a 150 K surface.
However, in practice, it is more difficult to construct a 77 K
radiator than a 150 K radiator.

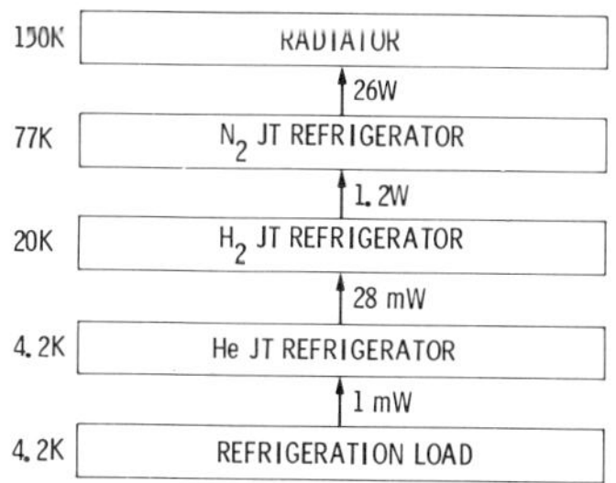

Fig. 3. Multistage Gas Adsorption
Refrigerator for 1 mW cooling at 4.2 K.

CONCLUSION

The present analysis represents a first step in examining the expected performance of an adsorption compressor used as a gas source for a J-T refrigerator providing long-term sensor cooling. A single stage refrigerator based on the design criteria has been built and is now in test.

We wish to acknowledge R. Phillips for his assistance with the calculations.

NOTATION

		SUBSCRIPTS	
b	Van der Waal's volume	a	adsorbent
C	concentration, ratio of adsorbed to absorbent mass	c	critical point properties
C_p	heat capacity	H	high pressure or maximum temperature in compressor
h	enthalpy	L	low pressure or minimum temperature in compressor
m	mass	pre	precooling
$\dot{m}$	mass flowrate	r	refrigeration
M	Molecular weight	s	container
P	pressure		
$\dot{Q}$	heat transfer rate		
Q	heat		
R	universal gas constant		
t	time		
T	temperature		
ΔU	heat of adsorption		
V	volume		
x	adsorption potential		
σ_y	yield strength		
ρ	density		
Δ	difference		

REFERENCES

1. A. Sherman, Cryogenic cooling for spacecraft sensors instruments and experiments, _Astronautics and Aeronautics_ 16:39 (1978).
2. R.V. Annable, Radiant Cooling, _Applied Optics_ 9:183 (1970).
3. R. Caren, and R. Coston, A Solid Cryogen Refrigerator for 50 K Infrared Detector Cooling, in "Advances in Cryogenic Engineering, Vol. 13", Plenum Press, New York (1976) p. 450.

4. H.L. Kroebig, Vuellemier (VM) cooler compressor/linear drive,
     AFFDL-TR1-78-760, (Nov. 1978).
5. C.K. Chan, Cryogenic refrigeration using a low temperature
     heat source, Cryogenics 21:391 (1981).
6. A.J. Kidnay, and M.J. Hiza, Physical adsorption in cryogenic
     engineering, Cryogenics 10:271 (1970).
7. A.J. Kidnay, and M.J. Hiza, High pressure adsorption isotherms
     of neon, hydrogen and helium at 76° K, in "Advances in
     Cryogenic Engineering, Vol. 12", Plenum Press, New York
     (1966) p. 730.

DISCUSSION

Question by B. Andeen, CTI-Cryogenics: How rapidly can you thermally cycle the adsorber and how does this time compare with the
desorption time of the compressor?

Answer by author: Our 7.2 g charcoal adsorber can complete the
adsorption process in a few minutes, depending on the heat input. The desorption rate was fast because we used a gas adsorption heat switch to conduct heat out of the compressor.

Question by R. C. Longsworth, Air Products and Chemicals: Has the
pulse-tube expander been studied?

Answer by author: A pulse-tube expander is being considered as a
means to eliminate the check valve system.

SUB-KELVIN TEMPERATURES IN SPACE

P. Kittel

NASA Ames Research Center
Moffett Field, California

INTRODUCTION

The use of temperatures below 1 K in space would provide a significant improvement in several areas of research. The sensitivity of bolometers improves dramatically as their temperature is lowered. The noise equivalent power (NEP) decreases as T^n, where $3/2 < n < 5/2$.[1] The sensitivity, compared with that at 1.5 K, can be therefore improved by as much as a factor of 50 at 0.3 K and up to three orders of magnitude at 0.1 K. Such detectors could be used for far-infrared (100–1000 μm) photometry and spectroscopy in the low-background conditions of space. Under such conditions, the sensitivity can be limited by the detector noise.

The study of critical phenomena in ^{3}He/^{4}He mixtures is also of interest.[2] The tricritical point (0.87 K) is of interest because it is the junction of a first-order and a second-order transition. On Earth, experiments in the region of the tricritical point are limited to a resolution of $\sim 10^{-3}$ K. This limit comes from a combination of correlation length and gravitationally induced concentration gradient effects. The present resolution is not good enough to test present theories. Reducing gravity by four orders of magnitude would increase the resolution of experiments to $\sim 10^{-5}$ K.

The temperature required for the above experiments can easily be achieved on Earth. There are several common laboratory methods available: ^{3}He refrigeration (0.3 K), dilution refrigeration (10 mK), and adiabatic demagnetization (<1 mK). These laboratory machines are usually bulky devices, however, and each presents its own problems when considered for adaptation to space. Efforts are under way at Ames Research Center and elsewhere to adapt ^{3}He refrigeration and adiabatic demagnetization for use in space. The exact configuration of the space versions of these refrigerators will, of course, depend on the application. There are, however, some general requirements. For example, the refrigerator must obviously operate in zero gravity. In some applications, the refrigerator will be inserted into a tank of superfluid helium, while in other applications it will have to operate in a vacuum bolted to a cold plate. This is the more difficult configuration to achieve, because of the poor thermal conductance of pressed contacts.[3] In either configuration, the refrigerator must operate under automatic control with good temperature stability. Finally, the mass and power consumption should be small. Preferably the mass should be <10 kg and the power consumption <100 W. The usual laboratory cryostats do not meet these requirements. However, in recent years progress has been made toward building a space-compatible refrigerator.

^{3}He REFRIGERATION

Even though ^{3}He refrigeration was developed more recently than that of adiabatic demagnetization, ^{3}He techniques are more nearly ready for use in space. The first advance in the development of a ^{3}He refrigerator for space was reported in 1976.[4] This was followed by a unit that could be bolted to a cold plate[5] and run under automatic control. The technical problem remaining is the confinement of the liquid in zero gravity. The confinement system must perform several functions: (1) it must allow for the condensation of the ^{3}He at the appropriate place; (2) it must maintain good thermal contact between the ^{3}He and the system being cooled; and (3) it must allow for effective evaporative cooling by pumping on the ^{3}He without pumping the liquid away. The most promising solution to this problem is to use a porous metal matrix.[6] In such a matrix the liquid would be confined by capillary attraction. During the condensation phase of the refrigeration cycle, the matrix would have to be the coldest part of the system. This would cause the condensation to occur preferentially in the matrix. A matrix of high-purity copper would maintain good thermal contact between the liquid and a copper cold head. Furthermore, since the thermal conductivity of copper is much higher than that of ^{3}He, evaporation should occur on the surface rather than by bubble nucleation inside the matrix, which would expel the liquid.

Although the containment of ^{3}He in a porous matrix has not yet been demonstrated, the containment of other fluids has been achieved. A series of experiments has demonstrated the feasibility of this approach.[7] These experiments were done on a variety of Ni and Cu sponges with pore sizes between 10 µm and 500 µm. The liquids used were ethanol, acetone, oxygen, argon, nitrogen, and ^{4}He. In the initial experiment, it was found that the sponges could not be completely filled because the meniscus kept the fluid out of a surface layer. Later experiments demonstrated that a gas could condense into a sponge if it was the coldest spot in the system. Furthermore, it was shown that useful refrigeration could be achieved without bubble nucleation. This was done by filling a sponge with liquid and then measuring the total heat absorbed by the evaporating liquid as a function of the power dissipated in a heater. Power inputs of up to 160 mW were used for nitrogen and 60 mW for helium without noticeable change in the total amount of heat that could be removed. The principal problem in using porous materials to confine liquid helium is the low surface tension of helium; which requires materials with very small pores.

Ames Research Center is now embarking on a program to test porous material for use in a ^{3}He refrigerator. Although the ultimate goal is to build a zero-g refrigerator, our next step will be to build one that works upside down. A schematic of a possible zero-g unit is shown in Fig. 1. It is composed of an adsorption pump, a pot filled with a porous metal matrix, and two heat switches. The pot acts as both a condenser and an evaporator. During the condensation portion of the cycle, the pot heat switch is closed, the pump heat switch is open, and the pump is heated to ~30 K. This drives the vapor out of the pump and into the pot where it condenses. The pump heater is then turned off, the pump heat switch closed, and the pot switch opened. The pump then readsorbs the ^{3}He, causing evaporative cooling in the pot. When the pot runs dry, the cycle may be repeated. This simple refrigerator should readily achieve 0.3 K in space.

ADIABATIC DEMAGNETIZATION

If temperatures below 0.3 K are required, refrigeration by adiabatic demagnetization is an ideal candidate. Because this type of refrigerator uses magnetic fields and solid refrigerants, it is inherently independent of gravitational effects. The large number of refrigerants available allows the system to be optimized

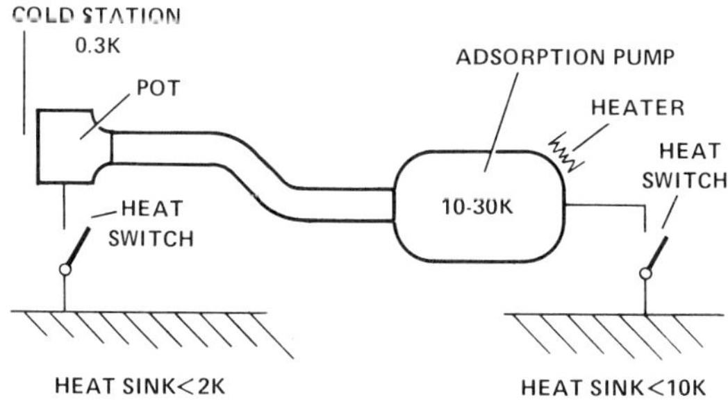

Fig. 1. Possible configuration of a ^{3}He refrigerator.

for many different operating temperatures.[8] Laboratory machines
have successfully operated as high as 300 K[9] and as low as
50 nK[10]. Recently, a refrigerator was built that operates under
automatic control and can maintain a temperature stability of
better than 1% for more than 12 h at 0.2 K.[11] Portable units have
also been built, including one that operates in a vacuum bolted to
a cold plate.[12]

Despite all of these advances, a space-compatible unit has not
yet been built. There are difficulties in adapting magnetic
refrigerators to space, because of the large power consumption and
large mass of the superconducting magnet and support equipment.
Fortunately, the cooling requirements of a bolometer are small
(the parasitic heat loads will probably be greater than the
bolometer heat load), and low-field magnets can be used.[8] Of more
concern is the torque on the magnet from the Earth's field. This
torque will have to be eliminated by compensation magnets and
magnetic shielding.

CONCLUSIONS

Either ^{3}He refrigeration or adiabatic demagnetization could be
used to produce temperatures below 1 K in space. At the present
rate of progress, one or both of these techniques will be demonstrated in space before the end of this decade.

REFERENCES

1. J.V. Radostitz, et al, Portable ^{3}He cryostat for the far
   infrared, Rev. Sci. Instr. 49:86 (1978).
2. R.J. Donnelly and P. Kittel, eds., "Proceedings of the
   Conference on Quantum Fluids in Space," University of
   Oregon, Eugene, Oregon (1975).
3. G.K. White, "Experimental Techniques in Low-Temperature
   Physics," 3rd ed., Oxford University Press, Oxford (1979),
   p. 148.
4. G. Chanin and J.P. Torre, A portable ^{3}He Cryostat for Space
   Applications, in "Proc. Sixth Intl. Cryo. Engr. Conf.," IPC
   Science and Technology Press, Guildford (1976), p. 96.
5. P. Kittel and W.F. Brooks, Demountable Self-contained ^{3}He
   Refrigerator, in "Advances in Cryogenic Engineering, Vol.
   27," Plenum Press, New York (1982).
6. R.M. Ostermeier, I.G. Nolt, and J.V. Radostitz, Capillary
   confinement of cryogens for refrigeration and liquid
   control in space. I. Theory, Cryogenics 18:83 (1978).
7. R.J. Donnelly, A study of confinement and heat transfer
   properties of cryogens, Final Report, NASA Grant NSG-2208
   (June 1979).
8. P. Kittel, Magnetic refrigeration in space - practical
   considerations, J. Energy 4:266 (1980).

9. W.A. Steyert, Stirling cycle rotating magnetic refrigerators and heat engines for use near room temperature, J. Appl. Phys. 49:1216 (1978).
10. G.J. Ehnholm, et al, Evidence of nuclear antiferromagnetism in copper, Phys. Rev. Lett. 42:1702 (1979).
11. P. Kittel, Temperature stabilized adiabatic demagnetization for space applications, Cryogenics 20:599 (1980).
12. P.L. Richards, University of California, Berkeley, private communication.

SEMIMETAL CASCADES: SOLID STATE PRECURSORS
TO SPACECRAFT SLUSH HYDROGEN REFRIGERATORS

C. A. Schalla

Lockheed Missiles & Space Company, Inc.
Sunnyvale, California

INTRODUCTION

The refrigeration of stored propellant during a spacecraft mission requires an energy conversion system unlike that of, and more challenging than, any for terrestrial use. For aerospace use other than cooling, the most advanced static energy conversion development has been in photovoltaics, thermionics, fuel cells, magnetohydrodynamics, and thermoelectrics. Thermomagnetic cooling is the facet of thermoelectric phenomena that has fostered semi-metal research. Cascaded semimetal elements are uniquely suited to cryogenic temperatures and may be developed to become solid-state propellant refrigerators for interplanetary spacecraft.

Investigations of the past two decades into the Ettingshausen effect have not improved the figures of merit of semimetals by the order of magnitude needed for bulk cryogen cooling. This deficiency, in the bismuth-antimony alloys studied, may only be due to inadequate materials development.

Some studies of magnetothermoelectric cooling have demonstrated the constraints in cooling with semimetals above 77 K.[1,2] Below 20 K, where liquid/solid hydrogen refrigeration is applicable, bismuth-antimony alloys remain promising.[3,4]

A hypothetical cooling device is described here to show the viability of a thermomagnetic cooling system and to give direction to semimetals development. The requirement for an enormous power input is shown to be an acceptable tradeoff for a spacecraft weight reduction and for long-term reliability.

SPACECRAFT APPLICATION

Nuclear electric propulsion is the ultimate system for transportation in the solar system. It will be needed to successfully complete intensive study of the outer planets. It has many advantages in a future orbital transfer system. A dual-mode electric rocket/hydrogen jet propulsion system is most promising for solar system exploratory missions. The dual-mode approach is relevant for missions requiring substantial amounts of (1) nuclear-to-thermionic electric rocket propulsion and/or secondary power and (2) hydrogen jet propulsion for both high thrust acceleration and high jet velocity trajectories. The limitations on hydrogen storage will require a refrigeration system during long-duration missions, such as a Mars sample return or a manned vehicle mission to Mars.

Propellant Storage Heat Absorption

With a composite multilayer insulation system, the solar heat leak to the propellant for a 405-day Mars mission is 264 W.[5,6] This is 0.08% of the solar radiation exposure and 67% of the latent heat capacity of the stored propellant; therefore, refrigeration is necessary to conserve approximately 2×10^4 kg of hydrogen that would be lost by vaporization and venting.

Refrigeration System Requirements

Mixtures of solid and liquid hydrogen (slush) at the triple point (13.8 K) offer increased density and heat capacity over saturated liquid hydrogen. The melting of an 80% (solids content) slush will absorb 50 kJ/kg heat input with no increase to hydrogen vapor pressure. A total of 100 kJ/kg is available when both the melting of the slush and the sensible heat increase of the subcooled liquid are controlled to a tank design limit of 1 atm of vapor pressure. Solid-state refrigeration is the most reliable and light-weight means for thermal control of the stored hydrogen.

The used of a cascaded series of thermomagnetic cooling elements could maintain a slush hydrogen temperature of 13.6 K, 0.2 K below the triple point. The gross payload savings by cooling of the propellant would be almost 2×10^4 kg. The refrigerator would require a 530 kW power supply, double the rating of a nuclear electric propulsion reactor presently being developed by NASA,[7] which is a 1.2 MW thermal heat pipe cooled reactor capable of 200 to 250 kWe from thermionic diode conversion.

The specific mass of the refrigeration system and power supply (including the reactor, thermal-to-electric conversion system, heat rejection system, and man-rated shielding) would be about 17 kg/kW. This would add 9×10^3 kg to the spacecraft to prevent the loss of 2×10^4 kg of hydrogen. Thus, 63% of the propellant storage required on a Mars mission with a vented storage system, and 18% of the spacecraft weight leaving Earth orbit, are eliminated by a refrigeration system weighing only 6%.

FIGURE OF MERIT IMPROVEMENT GOALS

The flow of heat and electricity in semimetal crystalline solids is enhanced by magnetic fields. This Ettingshausen effect results when electric current flows in such conductors at right angles to the applied field. Heat will flow in a direction mutually perpendicular to the field and the current.

A figure of merit (FOM) gives the measure of cooling capacity and embraces all the associated properties of a solid-state material. It is defined as:

$$FOM = \frac{E_{tm}^2}{\rho k}$$

where E_{tm} = thermomagnetic emf, ρ = electrical resistivity and k = thermal conductivity. In the thermomagnetic case, the resistivity is in the longitudinal direction (direction of current flow), the thermal conductivity is in the transverse direction (direction of heat flow), and the thermomagnetic emf is the longitudinal emf resulting from the mutually perpendicular magnetic field and the transverse heat flux.

A semimetal is a two-carrier system with approximately equal hole and electron mobilities. They are intermediate in electrical conductivity between metals and semiconductors. No net flow of thermal energy would be expected in semimetals, such as bismuth or antimony, which have equal numbers of electrons and holes carrying energy equally in opposite directions. In bismuth and bismuth-rich alloys, the electrons are slightly more mobile than the holes in certain crystalline directions. Discovery of bismuth alloys with higher carrier concentrations at low temperatures and greater hole mobilities at warmer temperature levels is needed to obtain higher figures of merit.

Magnetic deflections of electrons and holes causes them to accumulate and annihilate each other. This results in a release of heat at one end of the semimetal element while at the other end, the creation of electron-hole pairs absorbs heat, thereby cooling that end. The figure of merit is greater for long rather than short elements.

Improvement of the figure of merit at low temperatures is managed by the addition of neutral impurities for scattering. This reduces the heat carried by phonon vibrations of atoms in the

crystal lattice. The minimizing of phonon drag has improved since
the maturing of the dopant-gas-phase technology. With this technique the control of dopant concentrations and purity has improved
by orders of magnitude over liquid and solid dopant techniques
previously used. Ion implantation is another new technique for
administering dopant concentrations, both surface and bulk, to a
degree unobtainable with standard diffusion techniques.

For improving figures of merit, new studies may explore means
of increasing hole mobility even more. Semimetals suitable for a
refrigerator could take the form of either (1) substrated thin-
film pure bismuth elements, (2) elements either evacuated or
pressurized in staging encapsulation, or (3) elements interfacing
with hydrogen where real-time hydride formation can have a quasi-
doping effect.

Accordingly, a radical improvement of the figures of merit
(particulary for bismuth-antimony alloys) is contemplated. Projected values, as goals for solid-state refigerator development,
are plotted as functions of temperature in Fig. 1.

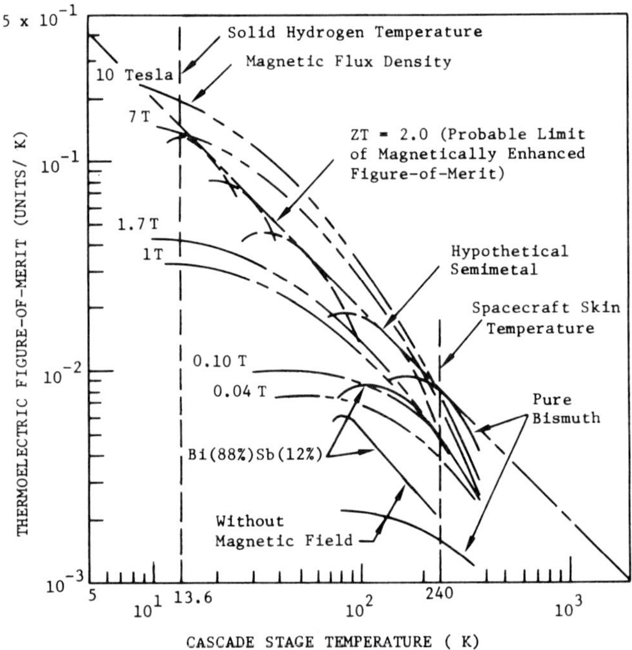

Fig. 1. Figure of merit goals for spacecraft
slush hydrogen refrigeration using semimetals.

SEMIMETAL CASCADE STAGING REQUIREMENTS

In semimetal cascading, each incremental temperature stage is comprised of a row of elements which will absorb heat from the adjacent colder stage, generate joule heat and transfer heat to the successive warmer stage. Stage size and capacity correspondingly increases toward the heat sink. The side-by-side elements in each stage are electrically and thermally in series with each other. The stage interfaces provide thermal continuity and electrical isolation. The electrical circuitry at the ends of the stages maintain a series current flow through the elements. Magnetic enhancement is provided by superconducting coils. The conceptual design of such a device, a 1 400-stage refrigerator for hydrogen aboard a spacecraft, is shown in Fig. 2.

The two-section module in Fig. 2 is arbitrarily biaxial. A number of such devices may be mounted between the cooled fluid and the standoff space of the exterior insulation. Performances contemplated with this ultrahigh staging cascade and a highly staged one of a lower coefficient of performance are compared with the ideal Carnot performance in Fig. 3. As shown in Fig. 4, the input power requirement will be enormous for the 1 400-stage

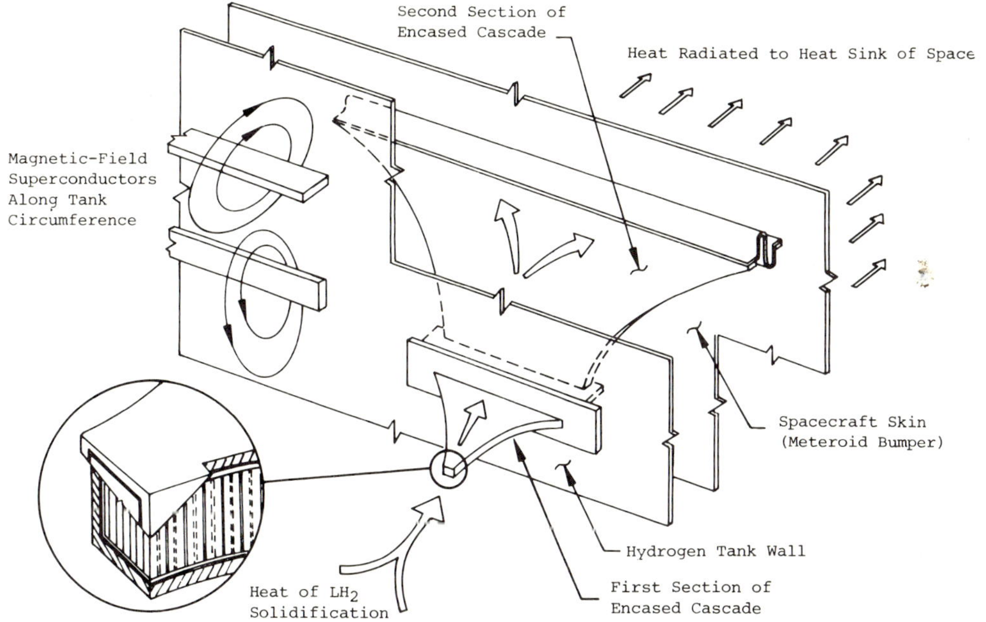

Fig. 2. Conceptual design of a cascaded semimetal refrigeration device for a spacecraft propellant storage tank.

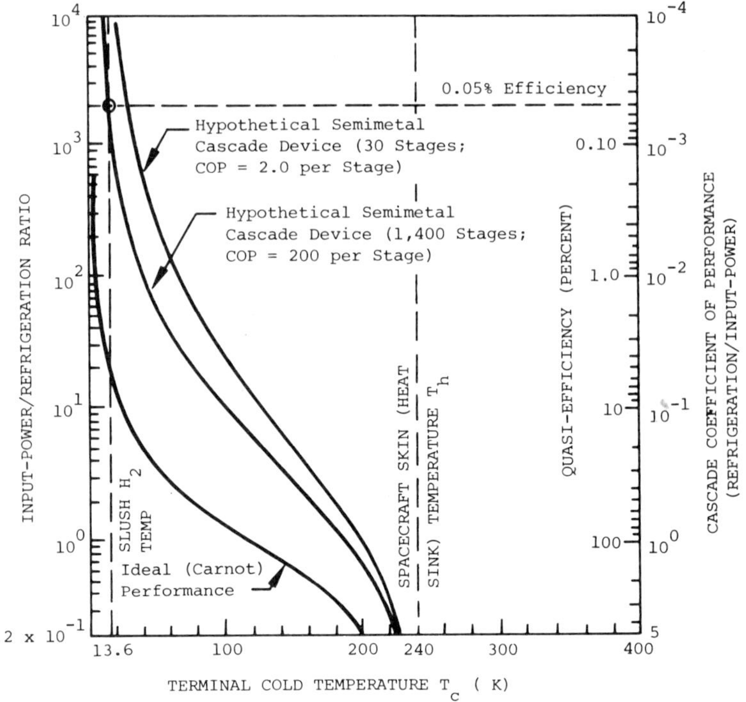

Fig. 3. Comparison of two hypothetical semimetal
cascaded cooling devices with Carnot ideal performance at spacecraft propellant storage conditions.

design and prohibitive for the 130-stage design. The design
parameters for refrigerators of different numbers of stages are
given in Table I.

Magnetic Enhancement and Other Improvements with High-Temperature Superconductors

Ordinarily the technical problems of integrating large numbers of series-connected semimetal crystals in a highly cascaded
device would seem insurmountable. The problem of maintaining a
circuitry of low resistivity is particularly difficult. A unique
advantage to this circuitry, however, is that it would exist in
the temperature environment of slush hydrogen. This is the range
where certain materials, such as niobium-aluminum-germanium could
have a useful current capacity.

The leadwire circuitry of the section of the device in the
hydrogen fluid itself would benefit from zero resistivity. The

Table I. Preliminary Calculation of Design Constants for a 130-Stage Cascaded Semimetal Refrigeration Device

Design Constants	Stage Number						
	1	20	40	80	100	120	130
Refrigerated Temperature T_c (K)	13.60	20.25	31.22	77.39	122.29	191.00	239.50
Temperature Differential ΔT (K)	0.29	0.45	0.69	1.72	2.71	4.40	5.50
Heat Sink Temperature T_h (K)	13.89	20.70	31.91	79.11	125.00	195.40	245.00
Mean Temperature T_m (K)	13.75	20.46	31.57	78.25	123.64	193.20	242.25
Figure-of-Merit Z (K^{-1})	0.135	0.096	0.063	0.025	0.0155	0.0103	0.008
Material Parameter Y $\left[(1 + ZT_m)^{0.5}\right]$	1.70	1.72	1.73	1.72	1.718	1.73	1.72
Electrical Resistivity ρ (ohm-cm)	7.00×10^{-6}	9.50×10^{-6}	1.55×10^{-5}	3.90×10^{-5}	5.00×10^{-5}	7.80×10^{-5}	9.50×10^{-5}
Thermal Conductivity k (watts/cm- K)	10.0	9.55	6.30	2.20	1.39	1.22	1.17
Seebeck Coefficient α $\left[(Z\rho k)^{0.5}\right]$	3.07×10^{-3}	2.95×10^{-3}	2.48×10^{-3}	1.41×10^{-3}	1.07×10^{-3}	9.9×10^{-4}	9.42×10^{-4}
Semimetal Element Area A (cm^2)	0.05×1.00	0.05×1.00	0.05×1.00	0.05×1.00	0.05×1.00	0.05×1.00	0.05×1.00
Semimetal Element Length L (cm)	0.25	0.25	0.125	0.032	0.018	0.012	0.010
A/L Ratio (cm)	0.20	0.20	0.40	1.56	2.76	4.24	5.00
Element Resistance R $\left(\frac{\rho}{A/L}\right)$	3.50×10^{-5}	4.75×10^{-5}	3.90×10^{-5}	2.50×10^{-5}	1.80×10^{-5}	1.84×10^{-5}	1.90×10^{-5}
Current I $[\alpha\Delta T/R(Y - 1)]$	36.4	39.0	60.7	134	225	325	379
Voltage V $[\alpha\Delta TY/(Y - 1)]$	2.16×10^{-3}	3.17×10^{-3}	4.08×10^{-3}	5.78×10^{-3}	6.93×10^{-3}	1.03×10^{-2}	1.24×10^{-2}
Element Input Power VI (watts)	7.87×10^{-2}	0.123	0.248	0.777	1.56	3.34	4.70
Stage Coefficient of Performance ϕ	11.6	11.6	11.6	11.6	11.6	11.6	11.6
Cascade Coefficient of Performance	11.6	2.38×10^{-1}	3.84×10^{-2}	1.37×10^{-3}	2.64×10^{-4}	5.08×10^{-5}	2.25×10^{-5}
Heat Removal/Element $\phi V I$ (watts)	0.910	1.42	2.88	9.00	18.10	38.7	54.0
Refrigeration Load w (watts)	264	264	264	264	264	264	264
Power Input W (watts)	287	1,096	6,850	1.92×10^5	1.00×10^6	5.19×10^6	1.18×10^7
Total Number of Elements (W/ϕVI)	315	772	2,380	2.14×10^4	5.52×10^4	1.34×10^5	2.16×10^5
Selected Number of Devices	20	20	20	20	20	20	20
Number of Elements/Device	16	39	119	1,065	2.76×10^3	6.70×10^3	1.08×10^4
Nominal Stage Width* (cm)	4.0	10.2	15.2	34.3	50.8	81.2	109
Overall Stage Thickness (cm)	2.03	2.03	2.03	2.03	2.03	2.03	2.03
Cascade Terminal Length (cm)	0.10	2.13	4.17	8.24	10.25	12.29	13.31

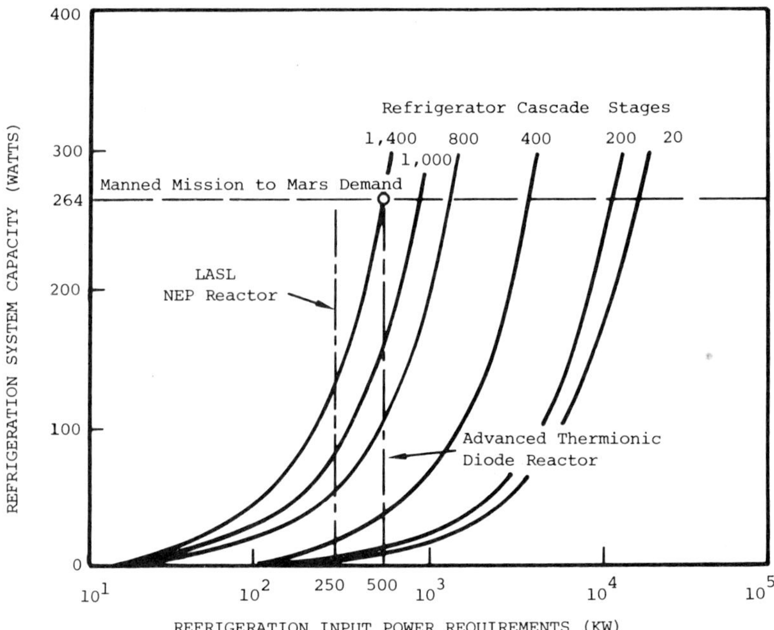

Fig. 4. Correlation of refrigerator power
requirement with advanced power supply systems.

weight reduction brought about by using superconducting coils to
generate the magnetic field required for the Ettingshausen effect
will be significant. To enhance the nonwetted part of the device,
for higher figures of merit, an arrangement of superconducting
coils along the circumferential interior tank wall will provide
suitable magnetic flux densities for both the nonwetted and wetted
sections of the type of device shown in Fig. 2. The use of superconductors at these temperatures is predicated upon a successful
materials development program and may be many years away.

The synergistic effects of slush hydrogen temperatures,
diverse pressures, and magnetic fields are yet to be investigated
fully. Therefore, future spacecraft propellant management applications could give purpose to a resurgence in semimetals development for magnetothermomagnetic refrigeration below 20 K.

REFERENCES

1. C.F. Kooi, et al, Low-Temperature Ettingshausen Coolers, in
    "Advances in Cryogenic Engineering, Vol. 9," Plenum Press,
    New York (1964) p. 367.

2. J.E. McCormick and J. Brauer, The Feasibility of Solid-State
    Cryogenic Refrigeration to 70 K, in "Advances in Cryogenic
    Engineering, Vol. 10," Plenum Press, New York (1965) p.
    493.
3. N.B. Brandt, Kh. Dittmann, and Ya.G. Ponmarev, Metal-
    semiconductor transitions in $Bi_{1-x}Sb_x$ alloys under
    pressure, Soviet Phys.-Solid State, 13:2408 (1972).
4. N.B. Brandt, N.Ya. Minina, and Yu.A. Pospelov, A new method
    for determining the dependance of the carrier density on
    pressure in semimetals, Soviet Phys.-Solid State, 10:1011
    (1968).
5. C.A. Schalla, Variable-thickness mode of insulation for
    hydrogen space storage tanks, ASME 67-HT-50 (1967).
6. C.A. Schalla, Solid-state refrigeration for interplanetary
    space storage of hydrogen, ASME 65-AV-10 (1965).
7. W.R. Hudson, NASA electric propulsion technology program, AIAA
    79-2118 (1979).

HIGH PERFORMANCE FLEXIBLE CRYOGENIC HELIUM TRANSFER LINES

H. Blessing, H. Laeger, and Ph. Lebrun

CERN
Geneva, Switzerland

and

P. Rohner and K. D. Schippl

Kabelmetal Electro GmbH
Hannover, Federal Republic of Germany

INTRODUCTION

CERN has installed and is operating in its Intersecting Storage Rings (ISR) a superconducting high-luminosity insertion[1], the main elements of which are eight superconducting quadrupole magnets[2], housed in liquid helium bath cryostats[3]. Their operation requires an efficient and reliable transfer system for liquid and gaseous helium from a refrigeration/liquefaction plant situated 50 m away. For this purpose, eight coaxial flexible transfer lines have been manufactured and installed by industry according to CERN specifications and delivery schedule. Their design, construction, manufacture and test procedures have been described elsewhere[4]. A detailed account of site installation and operational experience of the lines is given here with special emphasis on present performance, and reproducibility achieved over long-term use, as compared to results of laboratory tests performed upon delivery of each line.

DESIGN AND CONSTRUCTION

The transfer line design is based on the continuous corrugated metal tube technology developed and exploited by Kabelmetal[5,6]. Each line consists of four coaxial corrugated tubes, made of austenitic stainless steel (Fig. 1): the inner channel (14 mm I.D.) supplies liquid or gaseous helium, while the annular channel (34 mm I.D., 51 mm O.D.) between the second and third tubes takes

"

the return flow of gaseous helium and thus continuously screens the supply pipe. The other two annular channels are evacuated for thermal insulation and contain helically wound polymer spacers. An activated charcoal adsorbent fleece and thirty layers of aluminized Mylar superinsulation are wound around the return pipe. After manufacture, all tubes underwent pressure tests, thermal shocks with liquid nitrogen and helium leak-detection.

SITE ERECTION

The superconducting magnets, located inside the ISR tunnel, are not accessible during machine operation, whereas the helium refrigeration/liquefaction plant is installed in an always accessible building 50 m away. The two sites are connected by a service tunnel with several sharp bends acting as a radiation barrier against nuclear cascades from the proton beams. The eight transfer lines had to follow this existing service tunnel.

Pre-installation preparations and the actual installation of the transfer lines followed the classical scheme of electrical cable installation. In late 1979, during a normal ISR shutdown period, two standard cable trays, each 0.5 m wide, had been installed along the wall of the service tunnel. The transfer lines were then pulled during another ISR shutdown in September

Fig. 1. Cut-away view of transfer line.

Fig. 2. Transfer lines on transport drums.

1980. Kabelmetal sent two experienced cable fitters and hired five unspecialized workers locally. Two CERN technicians assisted the crew by connecting the line terminations to the necks of the cryostats and to the distribution box at the cryogenic plant.

For delivery from Kabelmetal to CERN, each line with its insulating vacuum broken to 1 bar with dry gaseous nitrogen had been wound on a wooden drum (Fig. 2), 3.6 m in diameter. The first installation day was used for preparatory work, such as placing cable rollers along the path to be followed by the lines. Pulling the lines in the tunnel required five days. Each drum was in turn suspended to the 20 kN overhead crane in the liquefier building, and the line manually unwound and pulled along the tunnel on the cable rollers. Thanks to the thin-walled corrugated tube construction, the lines have a linear mass of only 4 kg/m which eases handling. The termination on the cryostat end of each line was straightaway connected to the neck of the cryostat (Fig. 3). The line was then attached to the cable tray about every meter by standard cable straps. On the seventh and last installation day, the eight terminations on the liquefier end were connected to the distribution box (Fig. 4).

A polyethylene sheath had been extruded around the outer corrugated pipe of each line for mechanical protection of the thin-walled tubing during transport and installation. This protective sheath significantly limits the flexibility of lines so in order to achieve bending radii of about 1 m without exerting high off-axis forces on the cryostat necks and on the distribution

Fig. 3. Installed transfer lines: cryostat end.

Fig. 4. Installed transfer lines: liquefier end.

box of the liquefaction plant, two meters of the sheath at each line end were cut away after installation. The longitudinal elasticity of the corrugated tubes is sufficient to absorb thermal contraction upon line cooldown, without bellows or contraction bends. In fact, when cooled down to operating temperatures, the maximum longitudinal forces exerted on the terminations remain below 1000 N per tube, which the mechanical pieces and welds in the terminations can easily handle.

Following installation of the eight transfer lines, their insulation spaces were immediately repumped with a 100 L/s turbo-molecular pumping unit (Fig. 5); after only one day, the residual pressure had dropped to below 10^{-3} mbar, and after two days of pumping, at $\sim 10^{-3}$ bar, a thorough helium leak-check revealed that no line had suffered damage during installation.

OPERATIONAL EXPERIENCE

As part of the superconducting high-luminosity insertion, the eight flexible cryogenic transfer lines have been operating since October, 1980 to the normal ISR schedule, i.e. continuous opera-tion periods of three to ten weeks spaced by shutdown periods of up to four weeks. For each ISR shutdown period of more than one week, the cryogenic plant is stopped and the system allowed to warm up naturally. During shutdown periods, the magnet temper-

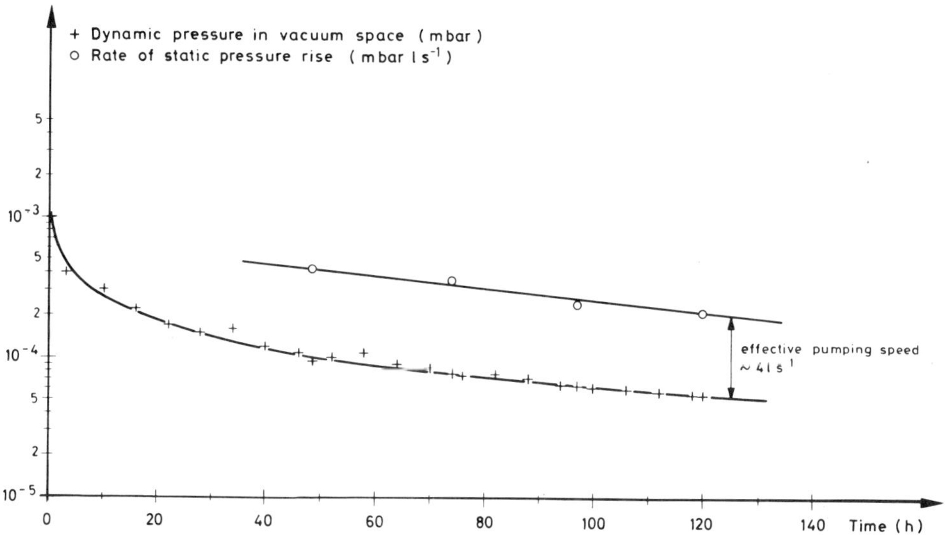

Fig. 5. Typical pumping curve of transfer line insulation space.

atures reach 100 K to 180 K, while the liquefier and transfer lines come close to room temperature.

Between October, 1980 and July, 1981, the magnets have been cooled once from room temperature and five times from the intermediate conditions mentioned above. The total cryogenic operating time has been approximately 5 000 h during which 600 000 L of liquid helium and 40 000 kg of cold helium vapor have passed through the transfer system.

Insulation Vacuum

One day before the start of a cooldown, the insulation spaces of all lines are repumped in parallel by 100 L/s turbomolecular pumping unit, which is left running until the end of the cooldown, typically for 60 h. Once liquid helium transfer starts, the vacuum spaces of the transfer lines are isolated from the pumping unit. Cryopumping maintains the residual pressure in the then sealed insulation spaces in the 10^{-6} mbar range for the entire operating period. Measurements of residual pressure after eight weeks of continuous cryogenic operation in such conditions indicate helium leak rates to the vacuum spaces below 10^{-10} mbar L/s, except for one line which shows a minor cold leak of the order of 10^{-8} mbar L/s. This line needs to have its insulation space repumped for a few hours every two or three weeks. In fact, a deterioration of its insulation vacuum to only a few 10^{-4} mbar results in a marked increase in the amplitude of the variations of the helium vaporization rate and liquid level in the corresponding cryostat, which allows fast diagnosis and swift corrective action.

Thermal Performance

Each line has to transfer a high mass flow rate (5 g/s) of gaseous helium at temperatures up to 350 K for magnet cooldown and warmup, and a low mass flow rate (0.3 g/s) of liquid helium during magnet operation. The sizing of the transfer line pipes results therefore from a technical compromise. In fact, to fulfill only the liquid transfer requirement, the lines could have been made appreciably smaller, and thus thermally more efficient. A direct consequence of the design compromise is the small fluid flow velocity during liquid transfer (a few cm/s), which could have created flow instabilities, as suggested by the Baker diagram[7] in Fig. 6. However, at no time could any sign of such instabilities be observed.

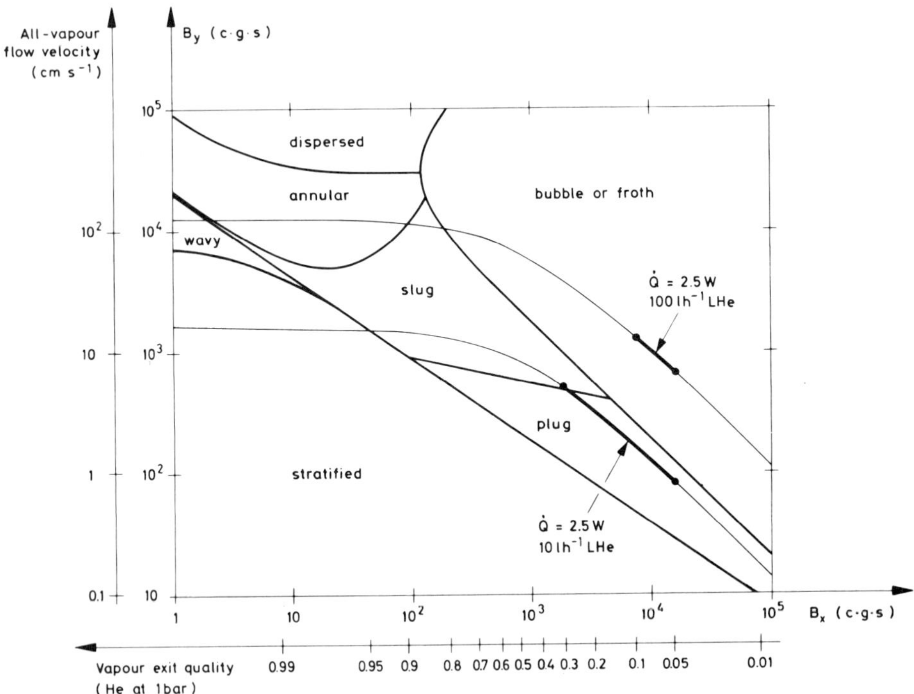

Fig. 6 Two-phase flow in transfer line according to Baker diagram.

The heat inleak to the supply line cannot be directly measured, but the global heat inleak to liquid helium for each cryostat-transfer line couple can be estimated from its helium vaporization rate. The thermal performance measured in this way shows complete agreement with heat inleak measurements performed on individual components[3,4] during laboratory tests. The heat inleak to the supply pipe of the transfer line has been deduced to be always below 3 W. The heat inleak balance into a transfer line in cryogenic operation is shown schematically in Fig. 7, which emphasizes the efficient shielding effect of the vapor-cooled screen.

ECONOMICS

The total cost of the flexible lines shows the following characteristics:

i)   Due to industrial production of the corrugated tubes on continuous automatic machines, the marginal cost per unit length is relatively low, provided enough length is manufactured in the same production run.

ii)   By comparison, the cost of terminations is relatively high, their fabrication is manpower intensive and involves precision mechanics, welding and thorough pressure and leak testing. Hence, the economics favor this type of line for relatively long transfer distances.

iii)   Installation costs are low: little field work is needed, and no highly specialized manpower is required since all critical quality control has already been done at the factory.

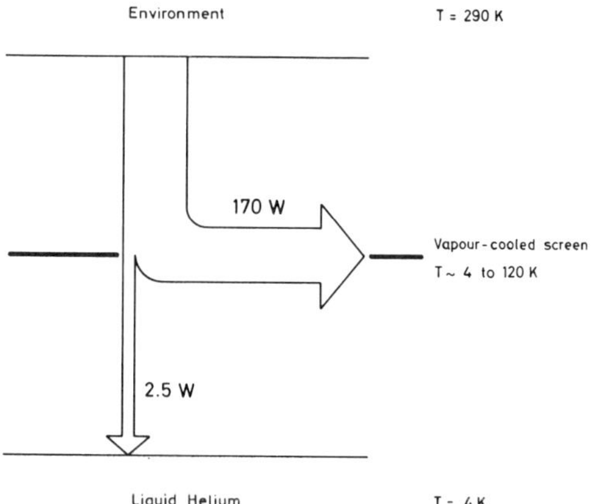

Fig. 7.   Balance of heat inleaks into 50 m-long transfer line.

On the whole, such flexible transfer lines seem to occupy a competitive position if one refers to a recently published cost survey of cryogenic components[8].

REFERENCES

1. J. Billan et al., A superconducting high-luminosity insertion in the Intersecting Storage Rings (ISR), IEEE Trans. of Nucl. Science., NS-26 (3): 3179 (1979).

2. J. Billan et al., The Eight Superconducting Quadrupoles for the ISR High-luminosity Insertion, "Proc. XIth Int'l. Conf. on High-Energy Accelerators," CERN, Geneva, (1980), p. 848.

3. H. Laeger, Ph. Lebrun, and P. Rohmig, Eight Liquid Helium Cryostats for the Superconducting Magnets of the ISR High-luminosity Insertion, in Proc. of the 8th Int'l. Cryo. Engr. Conf.," IPC Science and Technology Press, Guildford, (1980), p. 124.

4. H. Blessing, et. al., Four Hundred Meters of Flexible Cryogenic Helium Transfer Lines, in "Proc. of the 8th Int'l. Cryo. Engr. Conf.," IPC Science and Technology Press, Guildford (1980), p. 261.

5. E. Scheffler, Draht, 24:53 (1973).

6. H. Laeger, Ph. Lebrun, P. Rohner, Long flexible transfer lines for gaseous and liquid helium, Cryogenics, 18:659 (1978).

7. O. Baker, Simultaneous flow of oil and gas, Oil and Gas Journal, 185 (26 July 1954).

8. G.Y. Robinson, Jr., Economics of Cryogenic Systems for Superconducting Magnets, in "Advances in Cryogenic Engineering, Vol. 25," Plenum Press, New York (1980), p.342.

FERMILAB TEVATRON TRANSFER LINE

C. Rode, R. Ferry, M. Leininger, J. Makara, D. Richied, and J. Theilacker

Fermi National Accelerator Laboratory *
Batavia, Illinois

and

S. Stoy

Cryogenic Consultants, Inc.
Allentown, Pennsylvania

INTRODUCTION

The Tevatron refrigeration system is based on the concept of
a Central Helium Liquefier, CHL, (5000 L/h) providing liquid to 24
"satellite refrigerators" operating as amplifiers with a flow gain
of 12 producing 966 W per unit.[1,2] Both from the standpoint of
system reliability and also ease of installation, debug and start-
up, it was decided to build a complete transfer line loop starting
at the CHL and returning to it. This loop consists of 25 independent sections, each interconnected to the next with a branch
tap to the local refrigerator. This system permits the loss of
any one section and still maintains all 24 refrigerators operating
by feeding a portion of the loop in reverse. The transfer line
sections are 250 m long, interconnected with vacuum-jacketed U-
tubes. Each section contains a 4.50 cm i.d. supercritical He line
and a subcooled liquid nitrogen (LN_2) shield with an expansion
joint in the middle.

DESIGN TRANSFER LINE CROSS SECTION

Figure 1 gives the cross section of the line. The line is
made of commercial 304 stainless steel schedule 5 & 10 pipe with
epoxy fiberglass (G-10) supports. The inner pipe has an i.d. of

---

*Operated by Universities Research Association, Inc. under
contract with the U.S. Department of Energy.

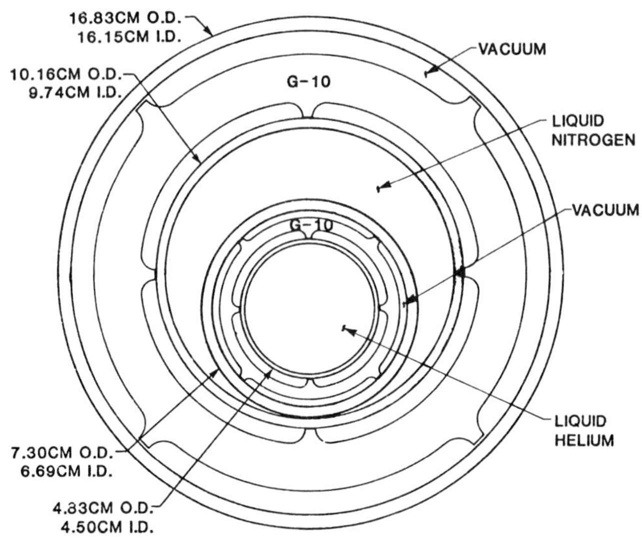

Fig. 1.   Transfer Line Cross-Section.

4.50 cm and contains supercritical He at 4.6 to 5.5 K and 3 atm. This pipe is wrapped with 15 layers of superinsulation (aluminized Mylar and Dexter paper).   A G-10 support is located every three meters.

The second and third pipes, 7.30 cm o.d. and 9.74 cm i.d. respectively, form the shield for the transfer line as well as the LN_2 supply line for the refrigerators and magnets.   The shield operates at 3 atm (subcooled liquid) with the last two sections breaking into the 2-phase region.   The flow area of the shield is 32.60 cm.^{2} The third pipe is wrapped with 60 layers of superinsulation with G-10 suspensions again located every three meters. The fourth pipe, 16.83 cm o.d., is the outer vacuum jacket.

Expansion and Bayonet Cans

The schematic of one section of the line is given in Fig. 2.   At each end is located a bayonet can, a 45.72 cm diameter vertical pipe, which contains a He and a LN_2 female bayonet for interconnecting to the next section as well as to the local refrigerator.   There is a labyrinth seal at the bottom of each bayonet which prevents thermo-acoustic oscillations in the bayonet.   A room temperature ball valve, a chevron seal, and an o-ring flange are located at the top of each bayonet.   The male bayonet inserts through both the ball valve and the chevron and makes a leak-tight seal on the o-ring.

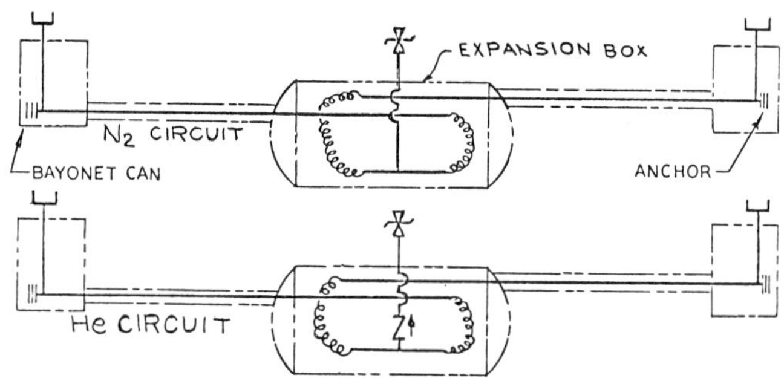

Fig. 2. One Section Tevatron Feed Line (1/25 Ring).

U–Tubes

Completing the cryogenic transfer line system are 24 helium
and nitrogen U-tubes, Fig. 3. They connect the various sections
of the transfer line and provide branch taps to the refrigerator
and magnet systems. Each U-tube has a vapor pressure thermometer
which can be read locally.

The helium U-tube consists of 4.5 cm i.d. pipe, with a 1.83 m
span, insulated with 100 layers of superinsulation, and a vacuum
jacket. A 1.22 m vertical bayonet at each end with an i.d. of 3.0
cm is inserted into the bayonet can discussed earlier. A vacuum-
jacketed flexible branch line with flow control valve joins the U-
tube to the satellite refrigerator.

The nitrogen U-tube is much more complex in that it contains
a subcooler heat exchanger which removes the shield heat leak of
the previous 250 m transfer line section. The subcooler is made
from a 3 m long, 1.09 cm i.d., 2.54 cm o.d. copper finned tube
wound on a 7.6 cm diameter mandrel in the horizontal position.
The span pipe is again 4.5 cm i.d. while the bayonets are .91 m
long with i.d.s of 3.31 cm. The inner lines are insulated with 40
layers of superinsulation and a vacuum jacket.

The main transfer line flow of subcooled LN_2 passes through
the shell side of the subcooler. Liquid (or 2-phase) nitrogen is
extracted at the top of the exchanger shell and is passed through
a Joule-Thomson valve reducing the pressure to near 1 atm, thus
decreasing its temperature to its boiling point at that pressure. This 2-phase mixture passes through the tube side of the
finned tubing, exchanging its latent heat to subcool the 3 atm
shell side nitrogen. The 2-phase nitrogen is finally transported
to the #1 heat exchanger of the cold box, providing the first
stage cooling of high pressure helium. Some LN_2 is extracted at

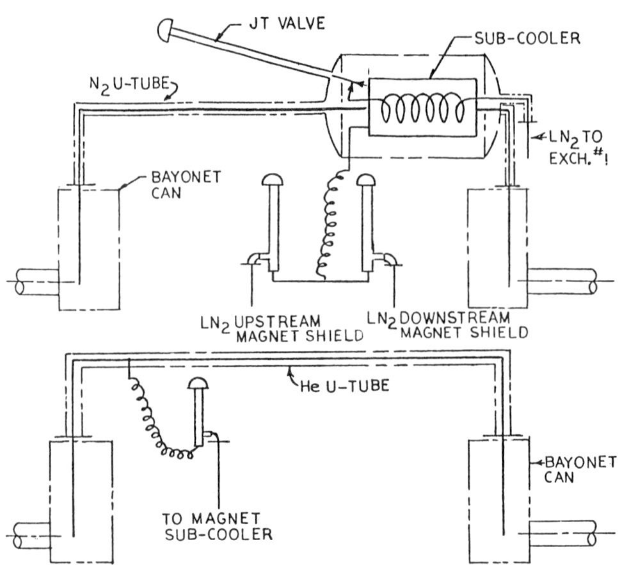

Fig. 3. One Pair Tevatron U-Tubes.

the bottom of the exchanger to provide magnet shield cooling; it flows through a vacuum-insulated flexible line with two control valves. The remaining LN_2 passes to the next bayonet can and in turn will provide shield cooling of the next transfer line section.

The transfer line itself has no isolation valves in the system. Any isolation required will be done by lifting the U-tubes until the bottom of the male bayonet clears the warm ball valve; the chevron seals will prevent a large leakage flow. The ball valves are then closed and the U-tube removed.

FABRICATION

Sections of transfer line 24.4 m long were fabricated at a specially designed facility. Fabrication was performed on four subassembly lines, each of which produced one of the four pipe sizes. Almost all of the pipe used was seam-weld pipe in 12.2 m lengths.

The first subassembly line fabricated the He pipe. Two 12.19 m lengths of the 4.5 cm inner pipe were welded, cold shocked with liquid nitrogen three times and then leak tested. Preformed cylinders with 15 layers of multilayer superinsulation, and the G-10

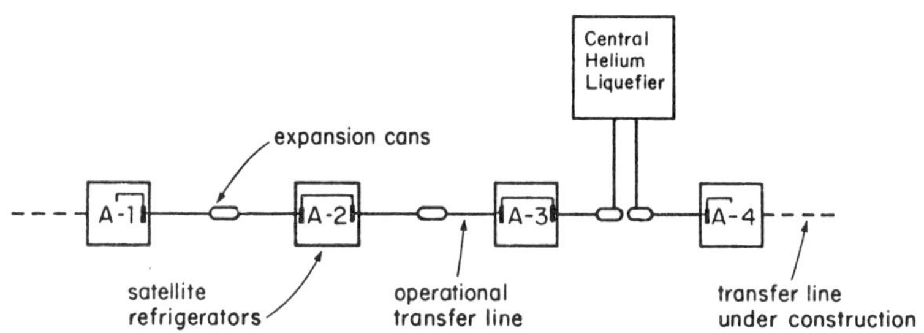

Fig. 4. Transfer Line Test Layout.

spacers were then slid over the pipe. The G-10 spacers were anchored by a 1.3 cm band on each side.

A second subassembly line was used to weld the two pieces of 7.30 cm inner liquid nitrogen shield pipe, followed by cold shocking and leak testing. The two pieces of 9.74 cm outer liquid nitrogen shield pipe were welded on the third station. The subassembly was then cold shocked and leak tested, followed by installation of preformed cylinders with 60 layers of insulation and the G-10 support rings.

The fourth station was used to weld the 16.83 cm outer vacuum shell, which was then leak tested. This assembly station was constructed so that it could be moved to a position that would permit each of the subassemblies to telescope into it on a specially designed roller system. The completed 9.74 cm subassembly was first telescoped into the vacuum jacket, followed by the 7.30 cm and 4.5 cm pipes. The completed lengths were then transported to seven predetermined locations around the accelerator ring.

Two other assembly lines are being used to fabricate the expansion boxes and the bayonet cans for the line terminations. By August, 1981 14 of the 25 expansion boxes and 11 of the 25 pairs of bayonet cans were complete. Two additional assembly lines have recently been started for He and LN_2 U-tube production.

INSTALLATION

Prior to installation of the subassemblies around the 6250 m ring, supports for the transfer line, expansion boxes and bayonet can were installed and surveyed level. Installation of the transfer line sections and the expansion boxes was carried out with a helicopter; 203 assemblies were precisely positioned around the ring from seven staging areas in 16 hours. The bayonet cans are installed in the refrigerator buildings and anchored firmly to the floor. Alignment is critical since the helium and nitrogen U-tubes are rigid without bellows.

Welding of the subassemblies in the field begins with the inner line starting at the bayonet can and ending at the expansion box. The other lines are moved as required and the bellows in the outer vacuum jacket at the bayonet cans are compressed to provide access to the inner pipe weld area. The finished line is cold shocked and leak tested. Similarly, the two liquid nitrogen shield lines are welded, cold shocked, and leak tested in sequence. Finally, the vacuum jacket line is welded by extending the bellows. The inner lines are then pressurized to 5 atm of He and a final leak test is conducted on the total system.

COOLDOWN AND WARMUP TESTS

The startup sequence of the transfer line begins with the decontamination of the flow passages. Water vapor is purged from both the helium and nitrogen passages, using a nitrogen gas purge. Our experience has shown that flow purging to less than 10 ppm water vapor is adequate for eliminating ice plugs in the refrigerator heat exchanger. The helium passage is then helium flow purged to less than 50 ppm nitrogen.

At the beginning of the transfer line cooldown sequence, the insulating vacuum is typically on the order of 0.3 torr. During cooldown, the line is cryopumped to its nominal operating vacuum of less than 10^{-7} torr. The line will cryopump as long as the residual gases are nitrogen or water vapor and not helium used for leak checking during the construction phase. The line has no permanently installed vacuum pumps or vacuum readouts.

Cooldown procedures for the transfer line were developed to minimize surging, which could result in violent pressure spikes. The circuits are first pressurized with warm gas to the nominal operating pressure of 3 atm. Liquid nitrogen and supercritical He, at 3 atm are then valved into the circuit and the flow is controlled at the outlet of the transfer line. Normal cooldown after the ring is complete will involve only one section, with both adjacent sections cold and operating. Either end is connected to the operating system and the other is tied into the low pressure He and LN_2 headers. When the new section is cold and "full" the connections to headers are removed, the normal U-tubes are installed, and full loop flow resumed.

To warm up a section the two U-tubes at both ends are removed and special L-tubes connect the refrigerator to the adjacent sections. The return leg of the transfer line is then fed backwards from the CHL, leaving the refrigerators unaffected.

The removed section is then warmed up by cross-connecting it to the warm refrigerator piping.

OPERATION AND HEAT LEAK MEASUREMENTS

There have been three transfer line runs: the first one on a 274 m section between ring service buildings A-1 and A-2 in May through July, 1980 (Fig. 4). The shield heat leak was measured, using the temperature rise in subcooled LN_2 as it traveled between buildings. The helium heat load was measured with 9 to 10 K gas; a measurement using supercritical helium was attempted but failed due to severe flow oscillations. The respective heat loads were 165 W and less than 11 W. The line was then used to cool two 125 meter strings of magnets. The A-1 refrigerator was run as a liquefier to simulate the CHL, with the transfer line shipping supercritical He, to the A-2 refrigerator, operated as a satellite, which cooled the magnets. The second run was a short heat leak run in Dec., 1980. The respective heat loads were 140 ± 20 W (0.51 W/m) and 9 ± 1 W (0.033 W/m).

The third run started in July, 1981: it tested four sections of line with a total length of 975 m. Subcooled LN_2 is shipped in both directions from the CHL. The A-1 refrigerator is again simulating the CHL, shipping supercritical He to A-4 compressor suction through A-2, A-3, and the CHL, respectively. We are currently trying to understand and control flow instabilities in both the He and LN_2 circuits so that we can proceed with a detailed set of heat leak measurements. Preliminary results indicate that heat loads are similar to previous measurements.

REFERENCES

1. C.H. Rode, et al., in "Advances in Cryogenic Engineering, Vol. 25," Plenum Press, New York, (1980), p. 326.
2. C.H. Rode, et al., Fermilab Tevatron Five Refrigerator System Tests, "Advances in Cryogenic Engineering, Vol. 27," Plenum Press, New York (1982).
3. J. Theilacker, IEEE Trans. in Nucl. Sci., NS-28 (3):3257 (1981).

A SMALL CENTRIFUGAL PUMP FOR CIRCULATING CRYOGENIC HELIUM

W. Swift and H. Sixsmith

*Creare Inc.
Hanover, New Hampshire*

and

A. Schlafke

Brookhaven National Laboratory
Upton, New York*

INTRODUCTION

A miniature centrifugal circulating pump has been developed to pump supercritical helium at 5 atm and 3.9 K from the subcooler of a refrigerator to a load which consists of magnets and gas cooled electrical leads in a prototype ISABELLE configuration. The purpose of the refrigerator/load test loop is to evaluate interactions between prototype subsystems under varying operating conditions. The loop will be used to gain operating experience in using such components as the main circulating compressor and the circulating pump to sustain sub-atmospheric baths of liquid helium.

The cooling system for ISABELLE relies on forced flow cooling of single phase helium through the magnets. The cryogenic circulating pump drives the single phase helium through a heat exchanger in which sub-atmospheric boiling removes heat from the circulating stream. The flow path of the helium is then through the magnets and finally to the cryogenic pump inlet. A single phase helium cooling loop operating at 5 atm between 2.6 K and 3.8 K

---

*Operated by Associated Universities, Inc. under contract with the U.S. Department of Energy.

through the magnet load was selected for ISABELLE over a two-phase (liquid-vapor) system for the following reasons:

1) The heat transfer characteristics of a single-phase system are more predictable.
2) Undesirable pressure/flow oscillations are less likely to occur in a single phase system.
3) Pressure drop in the return side of the distribution system of a two-phase system determines the magnet temperature. Because of the low density of the return helium vapor, the lines required would be large. The forced circulation single phase system uses smaller lines because of the higher helium density and larger allowable pressure drop.

The determination of operating pressure, mass flow rate and temperature is described in Ref. 1. These criteria were used to determine operating speed and the size of the circulating pump to meet the flow and pressure requirements. The low operating temperature controlled the choice of materials and impacted the heat leak considerations which affected the overall configuration of the pump assembly. In order to preserve cleanliness within the process stream a combination of gas bearings and magnetic thrust bearings was chosen. The drive system selected was a brushless DC motor with adjustable speed control.

PUMP DESIGN AND CONSTRUCTION

Specifications and Requirements

The specifications for the circulating pump are given in Table I. The table lists inlet and exit conditions for the pump, as well as the design mass flow rate and other information which summarize the pump configuration and operating characteristics.

The operating requirements listed are such that there is virtually no change in density of the fluid through the machine. For this reason the machine is referred to as a pump instead of a compressor. The specific speed allows the centrifugal impeller and diffuser section of the pump to be fairly conventional in design. However, the mechanical details of the machine are not conventional.

Overall Design

The pump configuration is shown in Fig. 1. The pump sub-assemblies consist of the motor drive at the top, the shaft and bearing system in the middle section, and the labyrinth seal, impeller, diffuser, inlet and outlet piping at the lower end. There is also a pressure equalizing valve (not shown) which con-

TABLE I

SPECIFICATIONS FOR PUMP

Inlet Pressure	4.67 atm
Inlet Temperature	3.925 K
Inlet Density	143.5 g/L
Exit Pressure	5.0 atm
Exit Temperature	3.989 K
Shaft Work	55.8 W
Mass Flow Rate	156.35 g/s
Volume Flow Rate	1.090 L/s
Speed	16800 rpm
Specific Speed	2600 $(rpm)\,(gpm)^{.5}/(ft)^{.75}$
Adiabatic Efficiency	0.65
Impeller O.D.	2.70 cm

nects the upper end of the housing to the inlet piping to prevent vertical shaft motion in the event of sudden pressure surges in the system.

The motor and bearings and their housing are located above the cold box flange. The interior of this upper end operates at room temperature, or slightly above, and at system pressure. The impeller, diffuser, inlet and outlet lines are located at the lower end of the assembly inside the cold box where they are in contact with fluid at 3.9 K. The warm and cold ends are separated by the rotating titanium shaft section, a labyrinth shaft seal and thin-walled structure sections which minimize heat conduction but provide for structural integrity and alignment of the entire assembly. The pump, including the motor, bearings, shaft, and impeller assembly, may be removed from the welded housing as a single unit. The overall design of the pump is the result of compromises between manufactureability, operating performance requirements, shaft speed and stability allowances and heat leak considerations.

In order to minimize heat leak by conduction to the cold end from the warm end, the titanium shaft section between the impeller and the lowest bearing is made as long and as small in diameter as

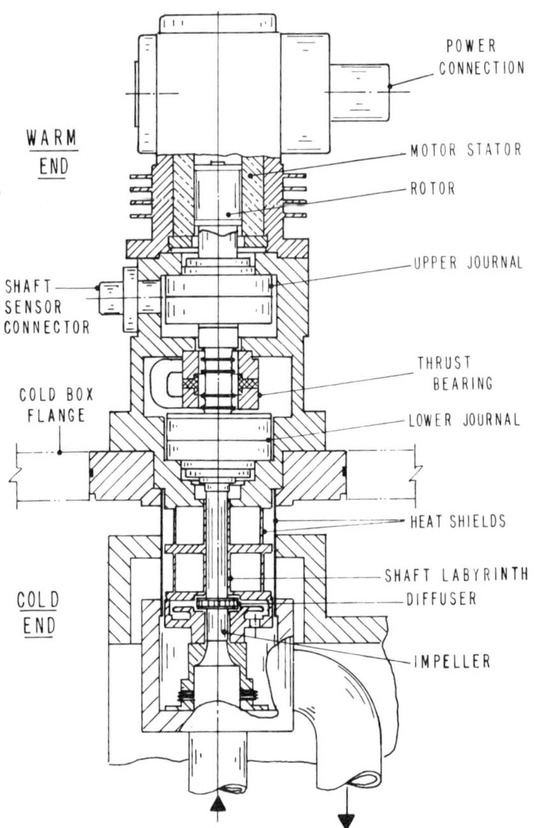

Fig. 1. Cyrogenic Centrifugal Pump Assembly.

possible. The selection of the length and diameter set constraints on the maximum operating speed which could safely be achieved without shaft instability problems. This maximum operating speed, together with the pressure and mass flow requirements set by the cycle specifications, are the remaining criteria to the impeller configuration.

Impeller and Diffuser

The mass flow rate, pressure rise, fluid density and constraints set by the shaft speed are used to determine the impeller and diffuser passageway configuration. Conventional methods used in the design of pumps were applied to develop the impeller geometry. The 2.7 cm outside diameter centrifugal pump impeller is fabricated from stainless steel. Twelve circular arc shim segments are brazed between radial discs to form the shrouded config-

uration. The blades are backswept at the exit from the impeller
with exit blade angles of about 30° from tangent. The impeller is
fastened to the cold end titanium shaft segment by means of a
threaded connection at the impeller inlet. The impeller rotates
in a labyrinth seal at the suction end to minimize leakage from
the high pressure exit end of the impeller back to the inlet.
There is also a shaft labyrinth seal between the impeller and the
lower gas bearing. The warm end of the machine is closed with
respect to the atmosphere. Therefore, convected heat leak and
fluid leakage in the shaft labyrinth is minimal.

Static pressure is recovered from the exit of the impeller
through the vaneless diffuser. The gas then proceeds from a
collector region through annular holes in the assembly to the
outlet line.

Bearing System

Temperature compensated tilting pad aerodynamic bearings are
located in the pressurized warm-end housing which operates at room
temperature. Helium which leaks up the shaft labyrinth and is
subsequentially trapped in the bearing casing at system pressure
is the lubricating fluid. In this way, contamination due to
external lubricants is eliminated. A bearing assembly is shown in
Fig. 2. The three bearing pads are each cantilevered in flexures

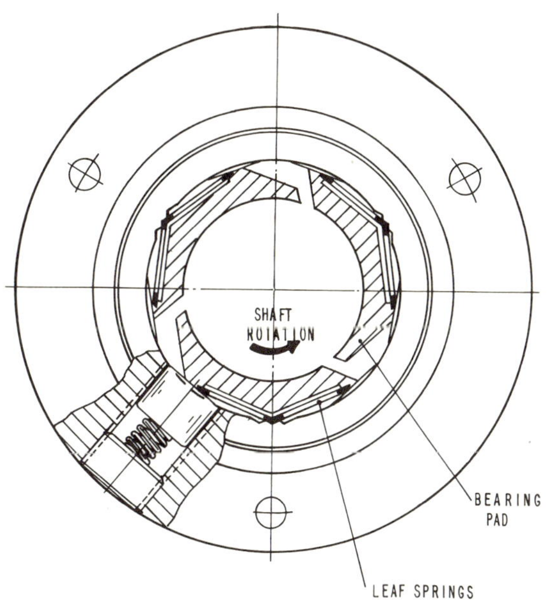

Fig. 2. Temperature-Compensated Tilting-Pad Gas Bearings.

and their camber with respect to the shaft surface is pre-set by spring-loaded manual adjustment screw for the proper running clearance. The radial clearance between the pad and the shaft is approximately 0.01 mm. The design of the bearing is such that as the gas film heats up and induces thermal distortion of the pads, the cantilevered flexures and bearing pressure in the clearance combine to reposition the bearing pads. Further description of the bearing design is given in Ref. 2.

Vertical thrust loads are carried by a magnetic thrust bearing located between the two journal bearings. The shaft segment running through the bearings is carbon steel and has magnetic field intensifying rings machined into it. Permanent magnets are located around the periphery of the shaft in an assembly which serves to provide sufficient magnetic field strength to prohibit axial motion of the shaft. The thrust capacity of the bearing is approximately 2.5 kg.

Drive System

The pump is driven by an adjustable speed DC brushless motor. The motor consists of a two-pole permanently magnetized rotor and a two-phase, two-pole wound stator. The system is capable of speeds up to 40 000 rpm, although shaft stability considerations limit the maximum practical speed to about 24 000 rpm. The motor control system consists of a DC current-limited power supply and an electronic control package which operates from 115 volt 60 Hz power. The controller consists of analog and digital microprocessor circuits that provide commutation point sensing, digital logic and synchronized application of DC power to the stator. In the event of demagnetization of the rotor due to power transients, the controller also contains circuitry which can be used to remagnetize the rotor. Speed control is obtained by adjusting the DC supply voltage to a level sufficient to maintain the desired speed.

The pump contains shaft sensors for speed monitoring and for shaft position sensing. Capacitive probes at each journal bearing provide the required signals for monitoring shaft run-out. An infra-red LED sensor provides a once-per revolution digital pulse for speed monitoring.

OPERATION AND PERFORMANCE

To date, (August 1981) the pump has been operated for several hundred hours in the test loop under a variety of conditions. Some test data have been collected on the performance of the pump, although the test data are not yet extensive enough to evaluate the complete range of operation. The test results are shown in Fig. 3 in the form of head coefficient versus flow coefficient-

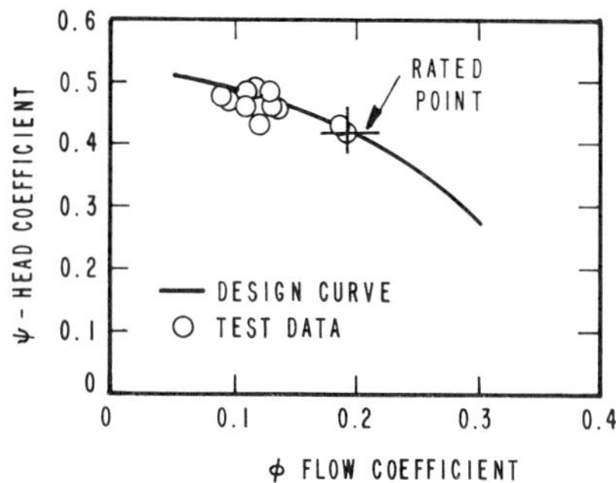

Fig. 3. Performance of Centrifugal Pump.

parameters which are conventionally used for the description of turbomachinery performance. Presentation of the data using these variables normalizes test results obtained at various speeds, temperatures and pressures and, therefore, presents the performance of the pump in a coherent way. The head coefficient is defined as:

$$\Psi = \Delta p / \rho U_2^2$$

and the flow coefficient as:

$$\Phi = \dot{m} / \rho U_2 A_2$$

where Δp is the pressure rise across the pump

$\quad\quad\dot{m}$ is the mass flow rate through the pump
$\quad\quad\rho$ is the fluid density
$\quad\quad U_2$ is the tip velocity of the impeller at its exit
$\quad\quad A_2$ is the exit area of the impeller.

Most of the data were obtained at temperatures of about 4 K. The flow and head coefficients obtained from the specifications of Table I are also shown for reference. Pump efficiency is defined as $\eta = (h_2 - h_1)'/(h_2 - h_1)$, where $(h_2 - h_1)'$ is the isentropic enthalpy rise and $h_2 - h_1$ is the actual enthalpy rise. Efficiencies as high as 62% near design point operation have been measured; however, there are not yet sufficient test data to define the peak efficiency or the flow coefficient at which it occurs.

SUMMARY

This paper describes a small centrifugal pump which has been developed to circulate supercritical helium through a test loop for superconducting magnets. The pump has a fully enclosed warm

end which contains the adjustable speed brushless DC drive motor
and self-acting bearings operating in helium gas. The drive and
bearing system is designed to minimize contamination to the circulating supercritical helium in the test loop. The performance
data which have been obtained show that the pump operates very
close to its design specifications. Additional tests are planned
to provide a more complete range of performance data for the pump.

REFERENCES

1. D.P. Brown and K.C. Wu, Choice of design pressure for helium
      cooling of ISABELLE magnet system, Brookhaven National
      Laboratory, Technical Note #88, (January 1979).
2. H. Sixsmith; A High Speed Tilting Pad Bearing; in "Proceedings
      of the Gas Bearing Symposium," Univ. of Southampton,
      England (1967).

DISCUSSION

Question by J. L. Smith, Jr., Massachusetts Institute of Technology: Is the pump shaft heat stationed and, if so, is the heat
leak to the heat station included in the efficiency quoted?

Answer by author: The rotating shaft is not stationed, but the
shaft housing was stationed for these tests. The heat leak down
the shaft, i.e., transferred to the helium, is included in the
efficiency. The heat leak down the shaft housing is not included. Calculations and test results show that an efficiency of at
least 65% is attainable with this pump, including all heat leak.

TEST OF A CRYOGENIC HELIUM PUMP*

J. W. Lue, J. R. Miller, P. L. Walstrom, and W. Herz[†]

Oak Ridge National Laboratory
Oak Ridge, Tennessee

INTRODUCTION

One way of circulating liquid or supercritical helium in a magnet built with a hollow or internally cooled superconductor (ICS) is to use a cryogenic helium pump. A developmental helium pump was acquired from Gardner Cryogenics, Bethlehem, Pennsylvania, for testing and possible use in the Large Coil Program[1] at Oak Ridge National Laboratory. It was designed to produce a helium flow from an inlet pressure of 2–3 atm and a temperature of 3.5–4.5 K to a discharge pressure of 4–7 atm. Mass flow rates of up to 20 g/s were to be expected by increasing the pump drive speed up to 333 rpm. The pump was incorporated into the helium flow loop of an ICS magnet test stand. Performance of the ICS magnet using the pump is reported elsewhere.[2] At the completion of the magnet test the pump was removed from the stand and a simplified, but better instrumented, flow loop was assembled to test it.

PUMP DESCRIPTION

The pump is a one-cylinder piston pump with the cold end 1.52 m (60 in.) below the driving unit. A schematic of the pump body is shown in Fig. 1. The cylinder (1) has a bore of 38 mm (1.5

*Research sponsored by the Office of Fusion Energy, U.S. Department of Energy, under contract W-7405-eng-26 with the Union Carbide Corporation.

† Visiting scientist from Kernforschungszentrum, Institute fur Technische Physik, Karlsruhe, Federal Republic of Germany.

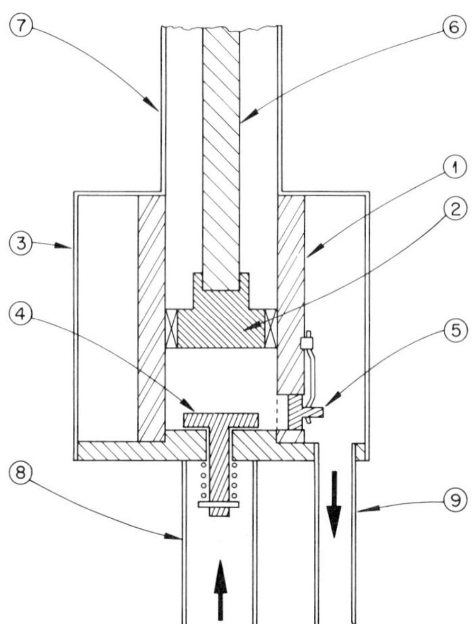

Fig. 1. Pump body cross section: (1) cylinder,
(2) piston, (3) discharge can, (4) inlet valve,
(5) discharge valves, (6) piston rod, (7) support
column, (8) inlet tube, and (9) discharge tube.

in.). Helium is sucked through the inlet "poppet" valve (4) into
the cylinder and discharged through four discharge "flapper"
valves (5) into the concentric discharge can (3). Both the inlet
and discharge valves are spring loaded. The piston (2) stroke is
about 38 mm (1.5 in.). The pump body, the piston rod (6), and the
support column (7) are housed inside a 152 mm (6 in.) diameter
sheath (not shown). Leakage through piston rings and other parts
of the pump equalizes the helium pressure in the sheath with that
inside the pump. The inlet (8) and discharge (9) tubes were
originally equipped with demountable couplings to the bottom plate
of the sheath such that the whole pump could be removed for service. Due to difficulties in sealing after cooldown, this feature
was deleted. Instead, the inlet and discharge tubes are soldered
directly into the test loop.

The driving unit is located at room temperature on top of the
dewar. It consists of a SCR-controlled electric motor, a right-
angle gear reducer, and a rocker to transmit motion of the crank
to the piston rod.

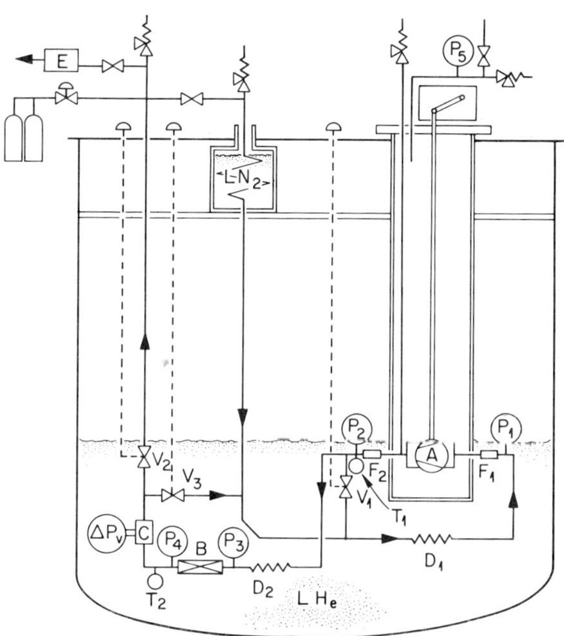

Fig. 2. Test loop schematic: (A) pump,
(B) load, (C) venturimeter, (D_1, D_2) heat
exchangers, (E) mass flow meter, and
(F_1, F_2) filters.

TEST SETUP

The helium pump and test loop were located inside a 0.76 m (30
in.) ID dewar. A schematic of the flow loop is shown in Fig.
2. The pump (A), located inside a separate sheath, hangs in the
dewar. A 5 m long section of 6.4 mm (0.25 in.) OD by 0.76 mm
(0.030 in.) wall copper tube (B) was used as a load. Flow was
measured by a venturi meter (C), which has a throat of 2.72 mm
(0.107 in.) and a throat-to-line ID ratio of 0.32. It was calibrated against a room temperature calorimetric mass flow meter (E)
made by Teledyne Hastings-Raydist.* Helium heat exchangers $(D_1$
and $D_2)$ are placed immediately before and after the pump. Each
heat exchanger has a volume of about 1500 cm^3, 34 times the pump

---

*Reference to a company or product name does not imply approval or
recommendation of the product by Union Carbide Corporation or the
U.S. Department of Energy to the exclusion of others that may be
suitable.

displacement volume, thus, serving as the surge volumes for the loop. All pressure taps are brought out to room temperature through stainless steel capillary tubes. Pressures at the pump inlet (P_1), outlet (P_2), load inlet (P_3), and outlet (P_4) are measured with Bourdon tube gages. Differential pressures across the pump (ΔP_p), the load (ΔP_{3-4}), and the venturi (ΔP_v) are measured with diaphragm-strain-gage transducers. Pump head pressure was controlled by throttling valve V_3.

RESULTS

Venturi Meter

The venturi meter was calibrated with helium gas at 77 K. This was done by filling the dewar with LN_2 and opening the bypass valve V_1. Results showed that, within experimental uncertainties, the tube-to-throat pressure drop ΔP_v can be related to the mass flow $\dot{m}$ by the simple relation:

$$\Delta P_v = \frac{8}{\pi^2 D^4} \frac{\dot{m}^2}{\rho} \left[\frac{1}{\beta^4} - 1 \right] , \qquad (1)$$

where D is the line ID, ρ is the helium density, and β is the ratio of throat to line ID. Pressure oscillations of about 0.14 atm peak to peak were observed across the venturi at helium temperatures. This was probably due to thermal acoustic oscillations in the capillary tubes. This can probably be improved by heat sink of these tubes.

Pump Characteristics

For each pump piston speed ω_p, the performance of the pump was characterized by the mass flow vs pump head.[3] Figure 3 shows results at four different pump speeds. Up to a differential pressure of 1.0 atm the mass flow rates are nearly constant, indicating a constant volumetric efficiency. As the pump head goes higher the mass flow rate starts to drop. The effect is more pronounced at lower pumping speeds. Leakage through the piston ring, other parts of the pump, or the test loop is probably responsible for lower peak differential pressures at lower pumping speeds.

This can also be illustrated by plotting the mass flow vs pump speed (Fig. 4). At high pumping speed, a linear mass flow vs pump speed relationship is observed up to a 2.5-atm pump head. Extrapolation of these results for a pump head less than 1.0 atm to the full design pumping speed of 333 rpm gives an expected mass flow of 17.3 g/s, 13% lower than the specification.

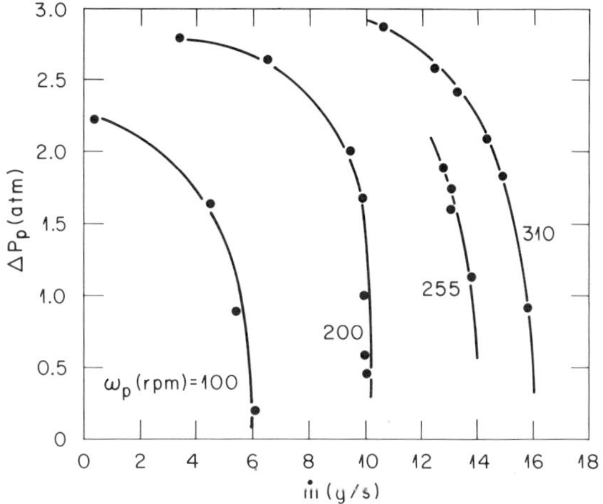

Fig. 3. Pump characteristics. Discharge
pressure = 3.5–7.5 atm; T_2 = 4.2–4.3 K.

The above results were obtained with a pump discharge pressure
in the range of 3.5–7.5 atm. Pressure variation across the pump
was about 0.95 atm peak to peak. This was damped to about 0.36
atm across the load.

Thermal Efficiency

The pump thermal efficiency η was estimated by comparing the
useful fluid power W to the measured additional cryogenic loss
when the pump was running,[4]

$$\eta = \frac{W}{Q_r - Q_s} . \tag{2}$$

where Q_r is the total heat input to the dewar with the pump running and Q_s is the total with the pump shut off. The fluid power
is given by

$$W = \frac{\dot{m} \times \Delta P_p}{\rho} . \tag{3}$$

The boiloff rates were measured using a mass flow meter at room
temperature; Q_s was ~ 14 L/h and Q_r ran from 30 to 50 L/h,
depending on pump speed and head.

The results showed that at ω_p =310 rpm, η =0.76, and at ω_p =200
rpm, η = 0.56.

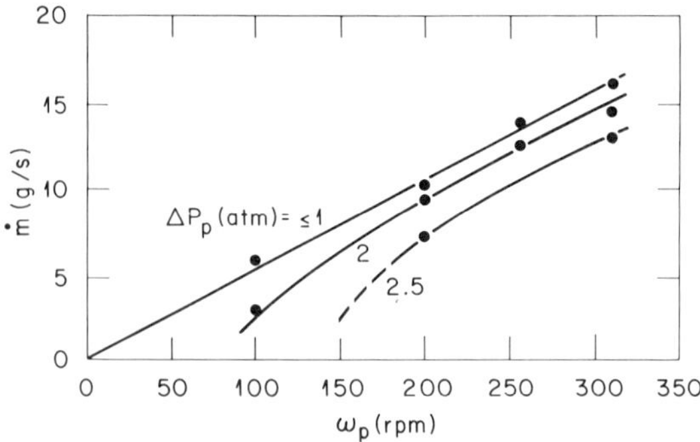

Fig. 4 Mass flow rate as a function of
pump speed. Single straight line fits
data for pump head pressure up to 1.0 atm.

CONCLUSION

A commercial cryogenic helium pump has been tested successfully. Despite flaws in the demountable connections, the piston pump
itself has performed satisfactorily. A helium pump of this type
would be suitable for the use of flowing supercritical helium
through ICS magnets. It has pumped supercritical helium up to 7.5
atm with a pump head up to 2.8 atm. The maximum mass flow rate
obtained was about 16 g/s. Performance of the pump was degraded
at lower pumping speeds. It has since been decided to use the
refrigerator compressor instead of the helium pump for circulating
supercritical helium for the force-flow coils in the Large Coil
Program.

ACKNOWLEDGMENT

The authors are indebted to J. C. Lottin for his contribution
in the initial design of the test loop.

REFERENCES

1. P.N. Haubenreich, et al., in "Proc. 8th Symp. on Engineering
      Problems of Fusion Research", IEEE Publishing Services, New
      York (1979), p. 1718.
2. J.W. Lue, and J.R. Miller, Performance of an Internally Cooled
      Superconducting Solenoid, in "Advances in Cryogenic
      Engineering , Vol. 27", Plenum Press, New York (1982).
3. G. Krafft, and G. Zahn, Cryogenics, 18(2): 112 (1978).
4. M. Morpurgo, Cryogenics, 17(2): 91 (1977).

SUPERCONDUCTING AC MOTOR FOR CENTRIFUGAL LIQUID HELIUM PUMP

A. Rivetti, R. Goria, and G. Martini

Istituto di Metrologia—"G. Colonnetti"
Torino, Italy

INTRODUCTION

For the purpose of studying the behavior of flowmeters in liquid and supercritical helium, a closed loop circuit has been designed. This circuit, located inside a liquid helium cryostat, will operate from 4.2 to 8 K, at a pressure of 0.1 MPa for liquid helium measurements and 3 MPa when supercritical helium is the working fluid. The fluid will be circulated by a centrifugal pump basically derived from a NBS study[1], with some improvements. The desired LHe flowrate of 2 to 15 g/s is expected to be obtained with a motor speed range of 3000 to 6000 RPM, requiring a power from 1 to 2 W.

Although detailed specifications of the motor features and experimental apparatus have been discussed in a previous paper[2], a brief description is reported below in order to correlate with the present results.

MOTOR DESCRIPTION

A classic three-phase asynchronous configuration was chosen for the basic design; the stator winding, made of 3 x 400 wire-turns, is similar to a traditional one, except for two differences: 1) The three 120° coils have been made of 0.15 mm diameter superconducting wire, which consists of 61 Nb-Ti filaments in a cupro-nickel matrix., and 2) The winding is supported by a slotted nylon core. These two changes are to reduce the main causes of thermal loss which, in this type of motor are principally in the stator and are due to the Joule effect in the winding, eddy-currents and magnetic hysteresis in the ferrous-core.

Three different cylindrical rotors which can be quickly replaced on the stainless-steel shaft, have been built:

<u>Rotor A</u>: a simple bird-cage rotor made of ARMCO-iron with 17 copper rods (0.4 mm diameter), located in small holes at a peripheral radius and electrically connected at the two ends

<u>Rotor B</u>: a double bird-cage rotor made of ARMCO-iron with 17 external copper rods (0.05 mm diameter) and 17 internal superconducting rods (0.4 mm diameter)

<u>Rotor C</u>: a copper rotor having the same external dimensions as the two others; i.e., diameter = 18.7 mm, length = 31.5 mm.

A special three-phase power-supply produces a current which can reach 10 A at 5 VRMS; with a frequency from 30 to 100 Hz. A three-phase set of switchable capacitors must be interposed between the power supply and the motor because the superconducting coil load is essentially inductive. At each frequency, a specific value of capacitance must be set in order to reach maximum motor power.

EXPERIMENTAL APPARATUS

The results which are discussed below have been obtained with the prototype motor running immersed in liquid helium inside a 30-liter dewar. In order to make accurate measurements of torque, speed and efficiency, a 1-m long stainless steel drive shaft was connected to the motor extended through the coverplate of the dewar through a mechanical seal. Since the first critical flexural velocity of this shaft was calculated to be around 2000 RPM, a plastic gear-box with a speed-ratio of 12:1 was interposed between the motor and the bottom of the shaft to reduce shaft speed. The rotational speed was measured by an optical sensor operated by a 12-sector wheel directly connected to the top of the shaft.

A D.C. motor, coupled outside the dewar to the top of the shaft, provided a variable counter-torque which could be precisely measured. The floating D.C. motor body was connected to the dewar through a strain-gage dynamometer allowing the value of the torque produced by the superconducting motor to be measured. The motor efficiency, which is the ratio of the resulting mechanical power (torque times rotational speed) relative to the driving power, was indirectly measured by considering the liquid helium consumption during measured time intervals. It was assumed that a helium consumption of 1 L/h corresponded to 0.7W. The efficiency, $\eta = P_m/P_m + P_{He\,l}$ where P_m = mechanical power measured from motor shaft and P_{He} = power absorbed by LHe. Helium gas flow was measured through a positive displacement flowmeter, with a heat exchanger employed to bring the gas to room-temperature. Care was taken to have negligible leakage from the shaft-dewar seal. The classical method of measuring the motor loss could not be used because the phase angle of the mostly inductive windings was ~ 90°.

Preliminary tests showed that the superconducting normal transition of the stator winding, generally occured when the current flowing in each coil exceeded 1.8 to 2.5 A, depending on the magnetic field frequency, type of rotor and counter-torque value. Therefore, the motor could be safely operated below 1.7 A. All the tests reported below have been performed with a fixed stator current of 1.5 A; however some of them have been repeated with 1 A in order to estimate the dependence of power, shift and efficiency on stator current. Results discussed in the following paragraphs have been achieved by operating the superconducting motor at frequencies between 35 and 100 Hz and increasing the counter-torque generated by the external D.C. brake-motor.

INITIAL RESULTS

Figures 1 and 2 show the initial results achieved with the three rotors. The different behavior between rotor C (massive copper) and the two others is immediately apparent.

The massive copper rotor runs with a generally small slip, never exceeding 16% even when increasing counter-torque is applied. This gives good motor stability with a prompt reaction when additional power is required. The total efficiency depends on the rotating magnetic field frequency and on the output power: it increases at highest powers and lowest frequencies,

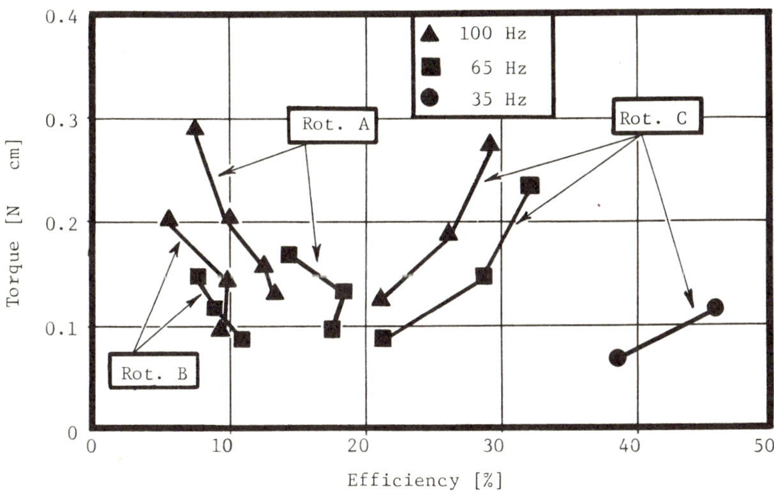

Fig. 1. Initial results - torque vs. efficiency.

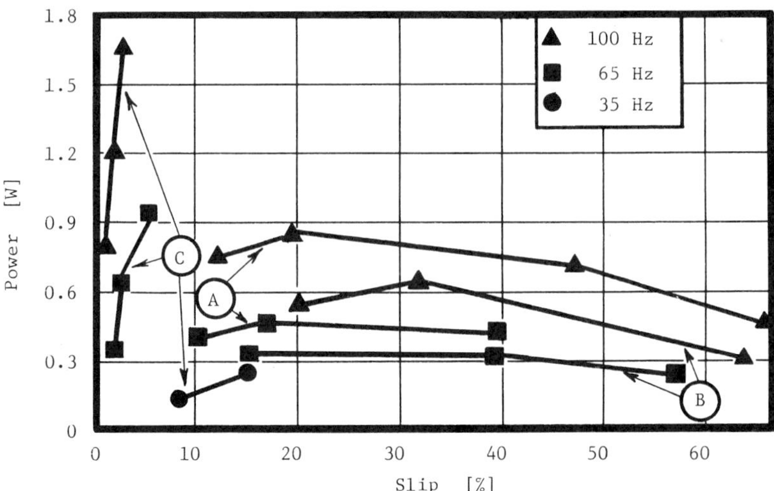

Fig. 2. Initial results - power vs. slip.

reaching the value of 46% at 35 Hz, with a power of 0.25 W. It
drops under 30% at 100 Hz when the maximum power (1.7 W) is reached.

Rotors A and B, however, show a generally low efficiency,
never exceeding 20%.   Moreover, as counter-torque increases,
efficiency decreases remarkably; at the same time large slip
values (up to 70%) occur with constant or decreasing power.
Therefore, the motor stability appears to be very limited. Although these results are very poor, some comments should be
made: rotor B (with superconducting rods) shows a behavior at
every tested frequency which is always somewhat worse than rotor A
(copper single-cage). This is probably due to the smaller size of
the external copper cage of rotor B compared with that of rotor A,
while it is apparent that no advantages come from the internal
superconducting cage where no induced current is flowing.   In
addition it is possible that the close spacing of the seventeen
superconducting rods causes the magnetic field to be rejected
outside the cage circumference thereby reducing the active cross
section of the rotor.

RESULTS OBTAINED WITH AN EXTERNAL ROTATING SHIELD

Since the results shown above were not considered to be
generally satisfactory, a new set of tests was made in order to
verify the theoretical possibility of an increase of power by
adding an external rotating shield to the rotor. Such a shield
was designed with the following aims:

1) Since the stator winding is supported by a nylon body,
the resulting high reluctance of the magnetic circuit could be
substantially reduced by surrounding the winding with a ferrous
material; by shaping it as a hollow cylinder rotating together
with the rotor, thermal losses due to eddy currents and magnetic
hysteresis in it should be as low as the slip;

2) A less dispersed magnetic field will result in lower
thermal losses in the external stainless steel body of the motor.

A 2 mm-thick shield, shaped like an overturned bell, was
built of ARMCO-iron and connected to a new shaft, specially designed for the purpose. In Fig. 3 and 4 results are reported for
the three rotors. If compared with the initial results it is
immediately apparent that the rotating shield provides the motor
with a considerable increase of power and torque, while the efficiency is generally higher and slip lower.

ANALYSIS OF RESULTS

Rotor B:

Although the shield provides remarkable advantages, the poor
characteristics of rotor design are confirmed. The torque increases 2 to 4 times depending on the frequency, while the total
efficiency reaches a maximum value of 32%. The slope of power
versus slip is higher, providing the motor with better
stability. Nevertheless one can easily see that slip, though less
than without the shield is still high, reaching values up to

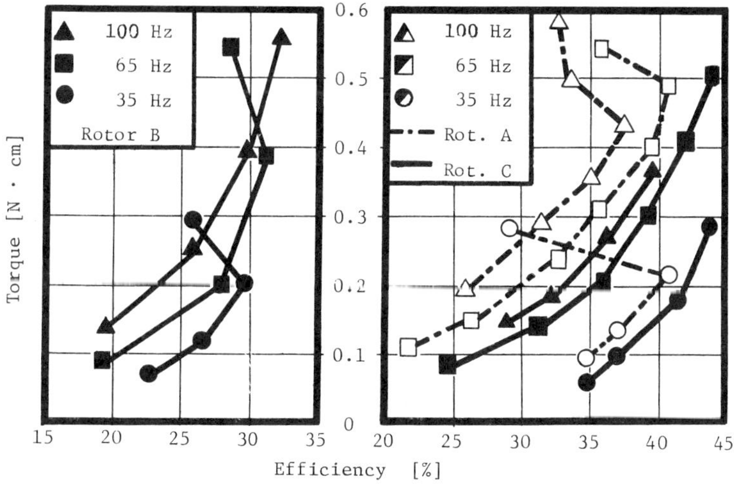

Fig. 3. Rotors with rotating shield - torque vs. efficiency.

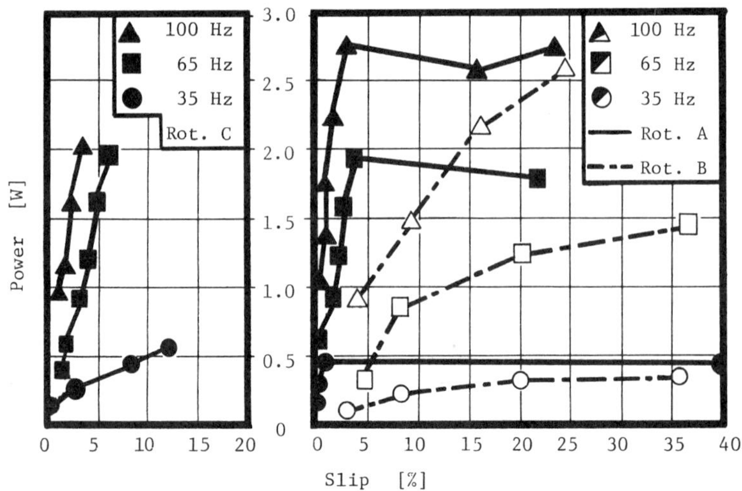

Fig. 4. Rotors with rotating shield – power vs. slip.

30%. Tests performed with 1 A stator current generally confirm
the above performance except for efficiencies which are maximum at
the lowest counter-torques values.

Rotor A:

The positive effect of the shield is apparent from results
obtained with the copper single-cage rotor. While torque and
power are similar to those of rotor B, the total efficiency is
remarkably higher and increases faster with counter-torque, reaching values over 40%.

It is interesting to observe the curves describing the behavior of the rotor when counter-torque is applied. One can easily
notice two different trends, below and above a particular counter-
torque value which is specific to each frequency. Below that
value, power, torque and efficiency increase almost vertically
versus slip which is very close to zero (in fact it is not detectable). In this region the motor is obviously very stable: an
increased power demand is well satisfied by the motor without any
unstable speed losses, indeed with a higher efficiency. Above that
value, the higher power demand causes a sudden drop in efficiency
together with a remarkable increase of slip.

It seems apparent that at this particular point the rotor
reaches magnetic saturation and there the maximum power is provided. When the critical point is exceeded, although the motor can

generate a slight increase of torque, the actual power decreases because of the considerable slackening of the rotor speed.

Data obtained from 1 A stator-current tests show, as for rotor B, that efficiency increases at low frequencies and reaches its maximum at the lowest counter-torques. A value of 56% was found for efficiency at 35 Hz with a power of 0.16 W. At higher frequencies, slip suddenly increases while efficiency drops.

Rotor C:

Results appear generally comparable to those of rotor A, even a little better; nevertheless the performance improvement due to the shield is less apparent. Although limited, the torque and power increase is notable (up to 2 times), while the efficiency is better than previously only at frequencies higher than 50 Hz. As with the other rotors, efficiency increases when higher powers are required, reaching a value of 40%. No effects of saturation were found. Slip, as in the initial results, is never zero, but is always less than 12%. Again the 1A tests showed better efficiencies (up to 56%) at the lowest frequencies and when low power is required.

CONCLUSIONS

While results with rotor B are very poor, particularly in efficiency, it is apparent that rotors A and C show acceptable performances. Generally speaking, one can say that rotor A provides a higher power particularly at the highest frequencies, provided that the critical point is not exceeded and rotor C gives a better efficiency, particularly at the lowest frequencies. The best results should then be achieved by adopting a rotor design similar to rotor A, but with a considerable increase in copper cage size, i.e., made with 3 or 4 mm diameter rods.

It has been shown that advantages expected by the use of the external rotating shield have been confirmed. They are apparent in the torque and power characteristics which increase 2 to 4 times depending on rotor configuration. Advantages in total efficiency are less conspicuous, of course, because additional losses are introduced by mechanical friction between the shield and the liquid helium. Tests to evaluate the amount of these losses were performed. By subtracting these values from the total losses measured in the previous tests, one can see that the electromagnetic efficiency, alone, is between 45 and 55%, reaching a maximum of 75% with the massive copper rotor at 100 Hz and 1.7 W.

A better rotating shield design could minimize friction losses, thereby remarkably increasing the total efficiency. Moreover, it is believed that a thin coat of niobium located on the external surface of the shield could better confine the magnetic field generated by the stator winding, making the shield itself more efficient and enhancing its advantages. Finally, from the comparison of test results performed both at 1 and 1.5 A, we believe that maximum efficiency can be reached at each speed just by adjusting the stator current to its proper value depending on the pump power requirements.

ACKNOWLEDGEMENTS

The authors are deeply indebted to Dr. G. Cignolo for his considerable contribution; the cooperation of Dr. J. R. Whetstone of NBS is also appreciated.

REFERENCES

1. P.R. Ludtke, Performance characteristics of a liquid helium pump, <u>National Bureau of Standards Internal Report</u> 75-816 (1975).
2. R. Goria, G. Martini, A. Rivetti, A Low-loss Superconducting Motor for use in Liquid Helium, in "Proc. 1° Congresso nazionale di Criogenia (A.Cr.I.)", Italy (1981).
3. S. Foner, B.B. Schwartz, eds., "Superconducting Machines and Devices", Plenum Press, New York (1974).
4. F. Pavese, et. al, Automatic Test Facility for Cryogenic Transducers in the Range 2 K to 300 K and up to 7T, in "Proc. 8th Intl. Cryo Engr. Conf." IPC Science and Technology Press, Guildford (1980), p. 812.

DISCUSSION

Question by J.L. Smith, Jr.: Was the loss due to armature currents (without the rotor) measured?

Answer by author: This loss was found to be negligible compared to the rotor and rotor/stator losses.

HIGH TIP SPEED TEST RIG TO STUDY
NATURAL CONVECTION IN LIQUID HELIUM

D. C. Litz, W. G. Moore, P. W. Eckels, and G. T. Mallick

Westinghouse Electric Corporation
Pittsburgh, Pennsylvania

INTRODUCTION

In 1978, a joint Westinghouse-EPRI Program[1] was initiated to design, fabricate and test a 300 MVA synchronous generator with a superconducting field winding. One of the major considerations in the program is the thermal and hydraulic design of the field winding cooling system. The design objective for the machine is that it have extremely high reliability on the power grid. Accomplishment of this objective necessitated a cryostable winding, i.e., if portions of the winding normalized, the cooling system would cool these areas sufficiently to re-establish superconductivity. The cooling system chosen was a natural circulation thermosyphon driven by the density differences in the liquid helium coolant and the high acceleration fields in the windings. The hydraulic performance of the cooling system and the thermal stability of the superconducting field winding are dependent upon the heat transfer coefficient between the superconductor and the coolant. Experimental measurement of heat transfer coefficients with liquid helium in a high acceleration field have been reported[2,3] but not at the "g" levels and geometry of the 300 MVA machine. In order to obtain relevant heat transfer data, a rotating test rig was designed, fabricated and tested at the Westinghouse Research Laboratories. We hope to confirm the heat transfer coefficient that was used in the thermal and hydraulic design of the liquid helium cooling system adopted for the 300 MVA generator. This paper describes the mechanical, cryogenic and data acquisition system design of the test rig and some initial results of the test program.

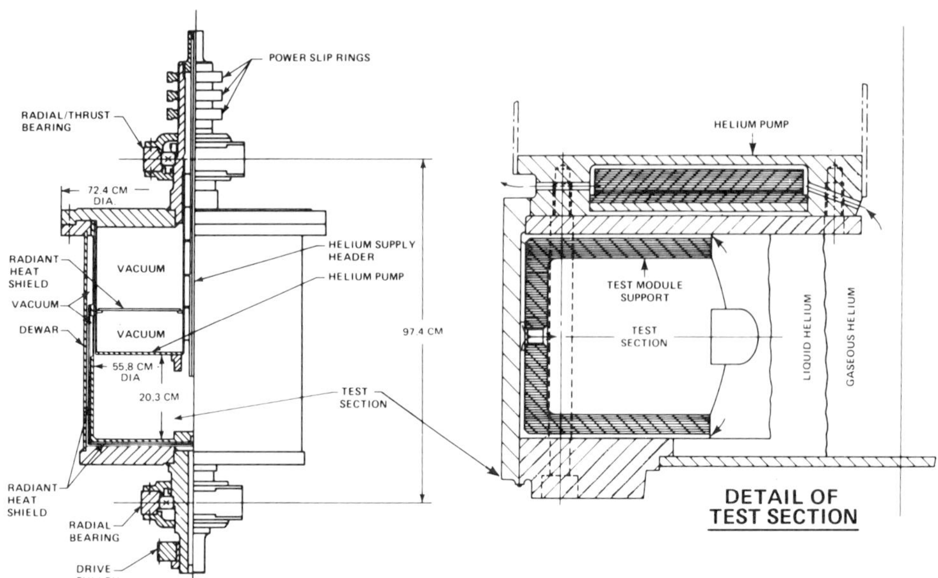

Fig. 1. Heat Transfer Test Rig.

MECHANICAL DESIGN

The high tip speed test rig (shown in Fig. 1) is a vertical
running rotor divided into top and bottom dewars. The top dewar,
connected to the upper flange, contains the test section, helium
pump and axial vacuum chambers. The bottom dewar houses the
torque tube and outer support wall. When the top dewar is inserted into the bottom dewar an "O" ring seal and bolted flanges
prevents helium leakage. The design allows the top dewar to be
pulled out of the bottom dewar giving easy access to the test
section. Disassembly, entry into the test chamber, test module
change, and reassembly of the rig can be completed in one working
day.

The test chamber was initially sized in diameter to represent
actual "g" loading that will occur in the 300 MVA generator rotor
at operating speed. Inside the outer walls of the test chamber is
the test section shown in Fig. 2a. This test section area is
readily accessible by lifting the top dewar out of the bottom one
and removing the access plate. A soft copper gasket seal is
fitted between the access plate and the end plate to prevent
helium from leaking from test section. Figure 2b shows a test
module partially inserted into one of the four radial holes in the
G-10 test module support and secured in place by a restraining
clamp. Rotational symmetry is assured by an even number

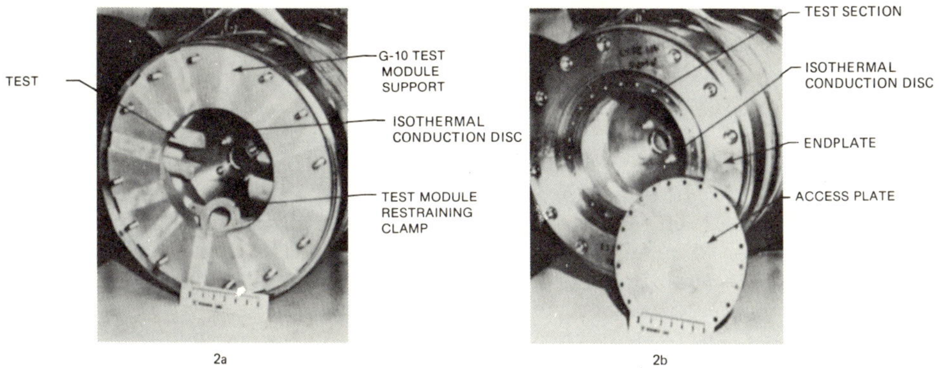

Fig. 2.   Test Section Contents.

of test modules.   One test module is designed to simulate the
rotor slot contents of the 300 MVA generator.   It can be rotated
to incremental angular positions within the G-10 support to represent the superconducting winding configuration around the end
turns.

Behind the G-10 support the thermal conduction disk made of
soft copper can be seen.   The high conductivity copper makes the
pool isothermal.   The maximum room temperature running speed was
limited to 2500 rpm due to the low yield strength of the soft
copper.

The test chamber is separated from the vacuum enclosure by
the helium pump.   The pump is shown in detail in Fig. 1.   Four
radial holes in the pump allow the helium to flow to the OD of the
pump, and from there it is forced vertically upward, cooling the
torque tube.   Helium is prevented from flowing vertically downward
and condensing on the cold test section by means of an indium "O"
ring.   The malleable indium forms an adequate seal when it is
wedged between the tight fitting helium pump and torque tube wall
and loaded by centrifugal acceleration.   Vacuum chambers with
axial and radial shields prevent excessive radiant heat losses.

The test rig, having a design operating speed of 3600 rpm, is
belt driven.   Heat input to the test modules is generated by a
slip ring assembly at the top of the rig also shown in Fig. 1.
The bayonet tube allows liquid helium to be transferred directly
into the test chamber while the rig is stationary.   Cooldown is
accomplished in 5 hours.   After filling the test section, the
bayonet tube is closed and the rig is ready for operation.

CRYOGENIC DESIGN

The rotating dewar heat transfer facility replicates the thermal environment of the 300 MVA generator. Although the rotating dewar is designed for conversion to a level regulated pool type device, it is presently a batch device which consumes a single charge of helium during several test runs. Upon spin-up, the helium forms an annular pool with a boiling interface below the vapor core. Spinup vanes are incorporated in the design, but there is no detectable difference in performance with and without the vanes. Vapor from the boiling interface moves radially inward toward the inlets to the torque tube pumps. The pumps are designed to produce more pressure rise than there is pressure drop on the torque tube so the pool interface operates at subatmospheric pressure. The assumption of isentropic compression processes in the pumps and vapor core lead to the prediction that the pool would operate at 3.7 K. In the torque tube the cold helium from the pumps cools the torque tube and radiant heat shields and is exhausted to the room near room temperature. The radiant heat shield is designed to operate at 80 K.

The dewar is supported at the top by a vapor cooled torque tube and at the bottom by spokes anchored to the radiant heat shield at a temperature of 85 K. Power and instrumentation leads parallel the central fill tube within concentric tubes whose axial chamber is subdivided with convection and radiation baffles. The significant heat leaks are conduction through the vapor cooled torque tube, conduction through the spokes from 85 K, conduction down the instrumentation and power leads and fill tubes which are anchored at 85 K, and radiation and residual gas conduction. In order to minimize heat leaks, the dewar is enclosed in a gold plated copper radiant heat shield that anchors all conduction paths at 85 K. The radiant heat shield in turn is anchored to the torque tube at the 80 K station. Radiant heat and conducted heat from the spokes and leads is transmitted to the torque tube coolant by conduction in the shield copper plating with approximately 5 K temperature differential.

The torque tube is shown in Fig. 3 and is an adaption of a design by Tepper[4]. Functionally, the tube design has components that prevent axial convection loops, vapor trap the helium flow channel, promote peripheral natural convection and transmit heat across the stratified thermal boundary layer on the radially inward tube. The G-10 thermal distance pieces have axial flow channels cut in their inner surface of such a radial width that the axial pressure drop at the design flow rate exceeds the bouyant pressure difference. The annular ring cooling fins form convection traps with the split G-10 rings and transfer heat across the stratified boundary layer on the inner cylinder. Stainless rings were soldered to the inner cylinder to form the

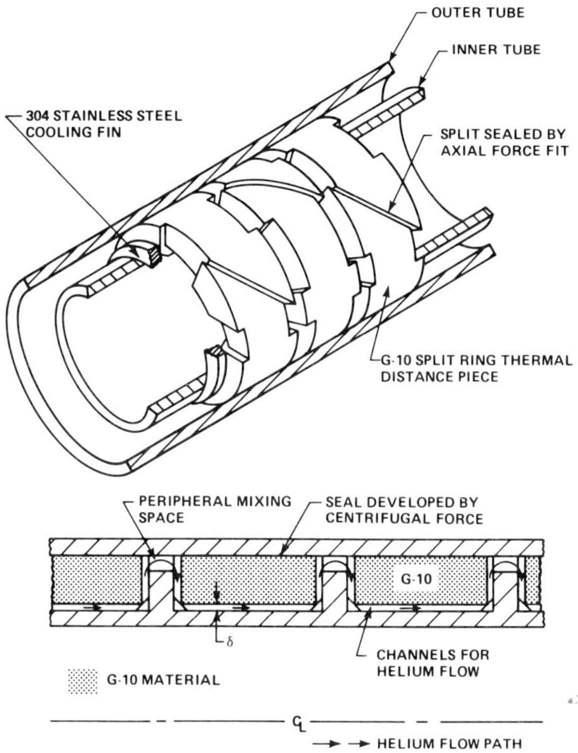

Fig. 3. Torque Tube Configuration.

fins. At the low temperature stations the rings were copper plated to increase their thermal conductivity.

Peripheral convection, which minimizes peripheral temperature gradients and hence distortion while spinning, is fostered by the small clearance volume at the tip of each fin. Figure 4 shows the predicted heat leak of the torque tube for various helium flow rates. The trajectory of self sufficient states is shown intersecting the stationary and rotating heat leak curves. Performance of the torque tube was computed by the program CRYOTL using the larger of forced convection or natural convection heat transfer coefficients. Table I is a summary of the predicted heat leaks.

RESULTS

The dewar has a measured stationary and rotating heat leak about 14% below the design value. Although the dewar has a usable helium content of 8 L and a rotating steady consumption of 7.3 L/h stratification of heat during filling and cooling to subatmos-

Table I. Heat Leaks

Radiation			Conduction		
to shield	5.5	W	spokes to shield	1.0	W
to dewar	0.04	W	spokes to dewar	0.2	W
to dewar from 300 K	0.05	W	leads	1.4	W
			torque tube	4.0	W

pheric pressures shortens the run time to between 20 and 40 minutes depending on speed. During cooldown the torque tube exhibits strong parallel channel instabilities as expected. However, upon spinup, peripheral convective mixing eliminates the peripheral temperature gradients. After several minutes of operation, balance is improved and frost patches disappear to return more uniformly distributed when again transferring helium.

Although the dewar was designed to pump itself down to 3.7 K, the lowest value obtained has been 3.9 K. Helium enters the pumps at 13 K instead of the design condition of 3.7 K. The isentropic pressure rise of the pump using the density of 13 K vapor matches the rotor performance. Thus, we have a hot centerline condition.

The rotating dewar has been used in an unusual mode. Centrifugal pressure relief valves were installed to pressurize the dewar to supercritical pressures. Pressurization occurred by warming the helium. Under these conditions a layer of warm helium was maintained over a colder outer layer for more than 600s even though the copper disk thermally shunted the hot and cold layers. For heat flow radially inward with temperatures below the neutral lapse values (adiabatic compression - expansion temperature gradient) the pool isothermalizes in less than 2s.

DATA ACQUISITION AND TELEMETRY

Because of the instrumentation requirements of this experiment, we opted to build the data collection system, rather than try to adapt a commercial unit. The data acquisition and telemetry system is built around the Intel 8748 single chip computer. The utilization of this device allows the system to be implemented with a small number of component parts, increasing its reliability. The most sensitive integrated circuit in the system can withstand 5000 g of static acceleration, considerably more than it is designed to be subject to in the present experiment.

The use of the single chip computer allows the consolidation of the control, frequency generation, and modulation functions into one package. Furthermore, it should be emphasized that the

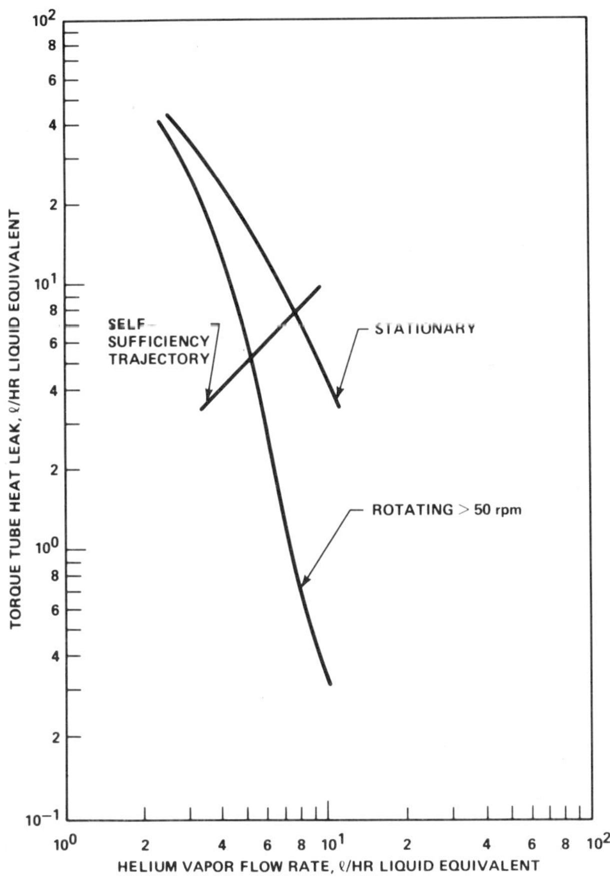

Fig. 4. Predicted Thermal Performance of the Torque Tube.

order and dwell time on any channel is determined by the "software," and hence is very easily changed. In fact, this simple expedient allows us to separate the sensors into "fast and slow" groups, having different bandwidths. (Fast channels have a bandwidth of about 12 Hz, and slow channels have a bandwidth of 1 Hz). Since the computer has a clock output of 2 MHz, this is a convenient source for RF for the telemetry. Modulation is effectively accomplished by using output ports of the chip to control a binary rate multiplier.

The most significant advantage of this approach is that it permits the conversion of the signal voltage from analog to digital form before transmission, and thereby eliminates drift as a significant source of error from the overall system. In the present system, data is transmitted in digital form using a frequency shift keying arrangement. The data is encoded as a ten bit word, giving resolution of one part per thousand. One hundred such measurements are made each second. Under these conditions, the system is capable of experimentally resolving temperature changes of about 0.05 K using silicone diode temperature sensors.

We have chosen to use a rotating transformer operating at 60 Hz for convenience. The same coil form is used for the data transmitter and receiver as is used for power transmission, which is a convenience. The two signals are separated electronically.

CONCLUSIONS

The test rig has been successfully tested at speeds to 3650 rpm. The initial test results confirmed the theoretical thermal and hydraulic performance of the 300 MVA rotor field winding cooling system. Performance of the torque tube exceeded our expectations. Complex "LSI" electronic circuits, especially in the single chip computers, lend themselves very well to use in high performance data acquisition and telemetry systems for use on rotating equipment.

REFERENCES

1. A.S. Ying, et al, Mechanical and thermal design of the EPRI/Westinghouse 300 MVA superconducting generator, IEEE Trans. on Magnetics, MAG-17:894 (1981).
2. R.G. Scurlock and G.K. Thornton, Pool heat transfer to liquid and supercritical helium in high centrifugal acceleration fields, Intl. J. Heat and Mass Transfer, 20:31 (1977).
3. C. Schnapper, The Convection and Transfer of Heat by Helium in Rotating Channels, Doctor of Science Thesis, Univ. Of Karlsruhe, (1978) Translation by A. and R. Taylor.
4. K.A. Tepper, An Experimental Liquid Helium System for a High-Tip Speed Superconducting Generator, MS Thesis Massachusetts Institute of Technology, Cambridge, (1976), p. 17.

DAMPING OF THERMOACOUSTIC OSCILLATIONS*

E. Tward and P. V. Mason

Jet Propulsion Laboratory
California Institute of Technology
Pasadena, California

INTRODUCTION

Thermoacoustic oscillations are a commonly encountered and troublesome problem in cryogenic systems[1]. Accompanying the oscillation are large heat fluxes which can cause large increases in the boiloff from dewars that seriously degrade performance. The oscillations occur in pipes which connect warm and cold sections of a cryostat or tank (e.g., fill or vent lines in a helium cryostat). The physical mechanism which causes a thermoacoustic oscillation (TAO) is understood[2-5] and criteria for the prevention of their excitation can be incorporated into cryogenic design. Since the criteria for the existence of a TAO are both temperature and geometry dependent, the modelling of TAOs becomes difficult for all but the simplest of systems, especially if there are changing temperature distributions. Under these circumstances, prudent design practice requires that parts of cryogenic systems where TAOs are possible should incorporate mechanisms for damping these oscillations.

In general, dampers that are used in the laboratory and in industry are empirical solutions. Once an oscillation is found in a cryogenic system, some modification is usually made in order to damp the oscillation. In other words, having inadvertently built an excellent organ pipe and thermoacoustic driving source, the designer then proceeds to destroy the resonant quality of the organ pipe. This is done in a number of ways, usually associated with introducing a loss mechanism into the system. These schemes

---

*Work supported by the National Aeronautics and Space Administration.

include incorporating restrictions at the cold end of the pipe, introduction of lossy materials into the pipe (e.g., pipe cleaners, glass wool, etc.) and, as is common in storage container design, addition of a capillary and surge tank at the warm end of the system.

In this paper, we will consider the design criteria for the damping mechanism required to suppress TAOs. Criteria for the design of capillary and surge tank dampers will be given.

THEORY

If a TAO exists in a pipe, the pipe may be considered to be an acoustic transmission line[6] in which standing waves are set up. The acoustic wave is excited at a point in the pipe where the temperature gradient and dimensions of the pipe are such as to make the pipe susceptible to the onset of the oscillation. The thermally excited pressure (or acoustic) wave then propagates down the pipe. If the pipe geometry is such that the wave can be reflected at some point with the proper phase relationship to the incoming wave, a resonant condition is set up and standing waves exist in the pipe. Basically, two criteria must be met for the standing waves to be set up. The reflected waves must have the proper phase relationship to the incoming wave and the acoustic losses in the pipe must be small.

The oscillation may be suppressed by the introduction of damping. As will be seen later, the damping mechanism can be localized, can be made independent of temperature distributions in the pipe, and is independent of the source strength. Since the energy transfer into the dewar is proportional to the intensity of the wave, given by the square of the pressure amplitude, reduction of the amplitude of the reflected wave will produce the desired result of reducing the heat transfer to the dewar. Figure 1a shows a pipe with a source of TAO near the cold end. This generates an acoustic incoming wave which may be reflected at the hot closed end. The reflected wave is in phase with the incoming wave and standing waves are set up if there is little damping in the pipe. Figure 1b is a general schematic of a long acoustic transmission line shown with a termination in the pipe. The source of the TAO produces an acoustic wave propagating to the right. At the termination, part of the wave is reflected back and part is transmitted into the termination. A standing wave can be set up if a sizeable amount of the power is reflected by the termination.

The incident wave pressure is given by

$$\vec{P_i} = \vec{A}\, e^{j(\omega t - kx)}, \tag{1}$$

the reflected wave is

$$\vec{p}_r = \vec{B}\, e^{j(\omega t + kx)}, \tag{2}$$

and the wave transmitted into the termination is $\vec{p}_t = \vec{C}\, e^{j(\omega t - kx)}$. The propagation of acoustic waves in a pipe is analogous to the propagation of electromagnetic waves in a waveguide and acoustic systems may be treated in the same way once the analogues of resistance, capacitance and conductance are determined.

In general, the acoustic impedance of a fluid acting through a surface area is given by $\vec{Z} = p[dV/dt]^{-1}$. The impedance can be written as $\vec{Z} = R + jX$, with a resistive and a reactive component. The resistance represents any energy loss mechanism in the system. These impedances may be either lumped components or, for example, in the case of a long pipe (length greater than $\lambda/4$) may be distributed impedances analogous to that of a transmission line. For a pipe whose radius is small compared with λ, the impedance is given[6] by $\vec{Z} = \rho c/S$.

For the case of a pipe containing a termination of impedance $\vec{Z}_t = \vec{R}_t + jX_t$ and incident and reflected waves of amplitudes $\vec{A}$ and $\vec{B}$, respectively, the ratio of the power reflected at the termination to the incident power is given by

$$P_r = \left|\frac{B}{A}\right|^2 = \frac{(R_t - \rho c/S)^2 + X_t^2}{(R_t + \rho c/S)^2 + X_t^2}. \tag{3}$$

Standing waves (TAOs) can be eliminated by introducing as a termination a damper of impedance $\vec{Z}_d = R_d + jX_d$ such that P_r, the reflected power is made small. This gives the conditions familiar from transmission line theory that

$$R_d = \frac{\rho c}{S} \qquad \text{and} \qquad X_d \ll \frac{\rho c}{S} ; \tag{4}$$

i.e., you effectively terminate the transmission line in a resistance equal to the impedance of the line at the termination. Physically, this corresponds to the situation where the energy of the acoustic wave is dissipated by the resistance R_d.

Figure 2 shows the case where the damper is placed at a branch in the line, with the impedance given by

$$\frac{1}{\vec{Z}} = \frac{1}{\vec{Z}_c} + \frac{1}{\vec{Z}_b}, \tag{5}$$

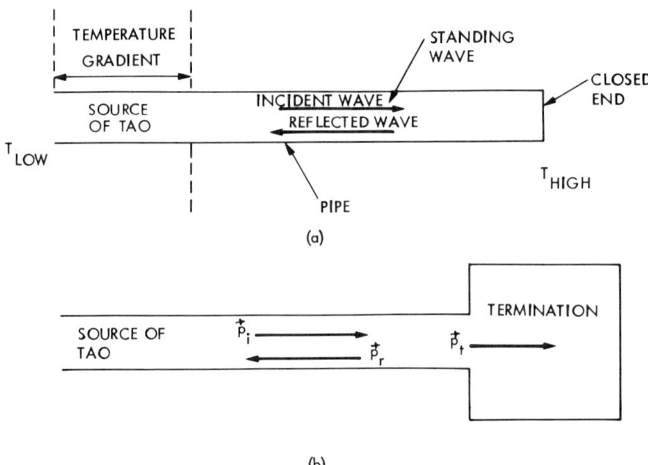

Fig. 1. (a) Incident acoustic wave generated by source of TAO is reflected by hot closed pipe end, resulting in standing waves in pipe. (b) Incident acoustic (pressure wave) is partially transmitted at the termination.

If the continuing pipe is short compared with λ and has a closed end, $\vec{Z}_c$ becomes large and one gets the same result as Eq. 3. If this is not the case, than one must take $\vec{Z}_c$ into account in order to determine the amplitude of the reflected wave. This is a more complex situation than just terminating the pipe with a damper. It is apparent that although in principle it is possible to locate the damper anywhere in a line, in practice it is usually advantageous to use the damper as the line termination. As the line termination, all the power can be routed into the damper and

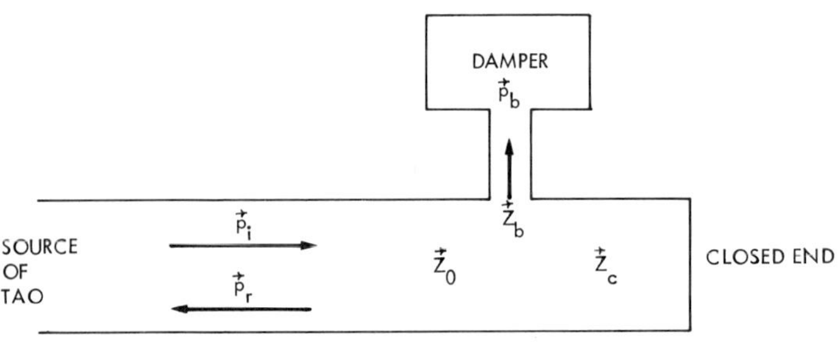

Fig. 2. Damper as branch in oscillating tube.

dissipated there. If, for example, the continuing pipe is long, its impedance $\vec{Z}_c$ = $\rho c/S$ and one finds that

$$P_r = \frac{(\rho c/2S)^2}{(\rho c/2S + R_b)^2 + x_b^2} \, . \tag{6}$$

CAPILLARY AND SURGE TANK

Figure 3 shows a damper consisting of a capillary with a surge tank. The impedance of this device can be represented as $\vec{Z}_d = \vec{Z}_c + \vec{Z}_v$, where $\vec{Z}_v$ is the impedance of the volume and the impedance of the capillary $\vec{Z}_c$ is

$$\vec{Z}_c = \frac{8\mu L}{\pi a^4} + j \, \frac{4}{3} \, \frac{\rho \omega L}{\pi a^2} \, , \tag{7}$$

Equation 7 is true to first order for small capillaries, i.e., if a $<$ $\mu/\rho\omega$. The surge tank can be shown[6] to have an impedance

$$\vec{Z}_v = -j \, \frac{\rho c^2}{\omega V} \, , \tag{8}$$

and therefore, the impedance of the orifice and damper is

$$\vec{Z}_d = \frac{8\mu L}{\pi a^4} + j \, [\frac{4}{3} \, \frac{\rho \omega L}{\pi a^2} - \frac{\rho c^2}{\omega V}] \, . \tag{9}$$

In a real installation, it may be necessary to install such a damper using a pipe to connect it to the oscillating line. Figure 4 shows such an installation in which the damper is shown as an impedance Z_d terminating the branch. Under these conditions, the impedance of the branch line must be included in the calculation. The impedance of the branch and damper is a function

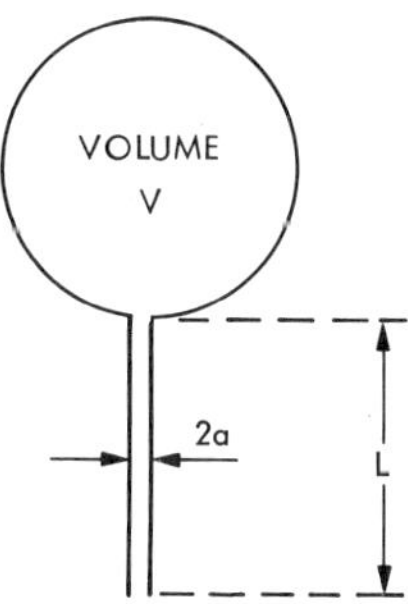

Fig. 3. Damper consisting of capillary length L, diameter 2a and surge tank of volume V.

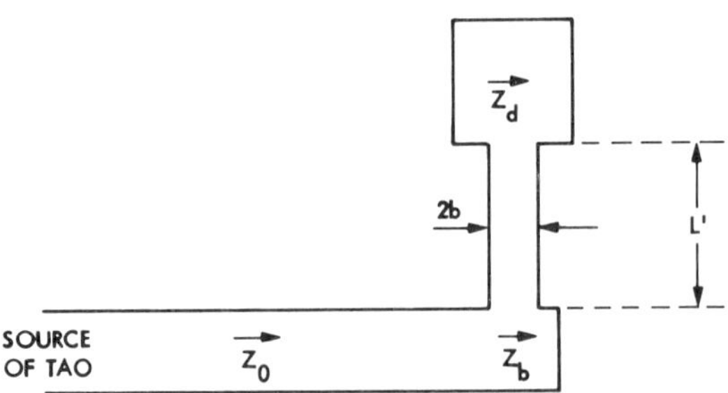

Fig. 4. Oscillating pipe of impedance $\vec{Z}_o$ with damper of impedance $\vec{Z}_d$ installed at end of branch of length L = L′ and diameter 2b.

of the dimensions of the pipe in the branch so that its length and diameter must be carefully chosen. However, as long as the length of the branch L′ $\ll$ λ, the impedance of the branch $\vec{Z}_b$ = $\vec{Z}_d$. Therefore, if the length of the pipe connecting the damper to the oscillating line is small compared with the wavelength, its impedance can be neglected. Since TAOs are usually low frequencies (long wavelength), the method of attachment of the damper to the oscillating line would usually not pose a problem. For a 10 Hz TAO in helium, for example, λ at 300 K is approximately 100 m. Therefore, connecting the damper with, for example, a 1 meter long pipe, is equivalent to locating the damper directly on the oscillating line. If an orifice and surge tank are used, the conditions for maximum damping in Eq. 4 together with Eq. 9 give the design criteria. For an oscillating pipe of cross-sectional area S the damper dimensions are given by

$$\frac{8\mu L}{\pi a^4} = \frac{\rho c}{S} \tag{10}$$

and

$$\left| \frac{4}{3}\frac{\rho \omega L}{\pi a^2} - \frac{\rho c^2}{\omega V} \right| \ll \frac{\rho c}{S} \tag{11}$$

It is worth noting that Eq. 10 which determines the orifice dimensions has a very strong fourth power dependence on the orifice radius. In addition, the orifice dimensions are both pressure and temperature dependent. The velocity of sound $c \propto T^{0.5}$ and for helium $\mu = 5.02 \times 10^{-6}\, T^{0.647}$ poise[7]. Therefore, $L/a^4 \propto p\, T^{-1.147}$, where p and T are the pressure and temperature of the helium at the orifice. Equation 11 contains two terms with differing frequency dependence. The first term, $4/3\, \rho \omega L/\pi a^2$, can in practice always be made negligible with a proper choice of L and a satisfying the condition in Eq. 10. Equation 11 then reduces to

$$V \gg \frac{cS}{\omega} \, . \tag{12}$$

If Eq. 12 is satisfied for the lowest frequency encountered, it will be even better satisfied for higher frequencies.

To illustrate the dimensions of such a device, consider a pipe containing helium gas at 1 atm which is 0.63 cm I.D. with a TAO at 10 Hz for which a damper at 300 K is required. $\rho = 1.63 \times 10^{-4}$ g.cm^{-3} C = 9.2×10^{4} cm^{-2} and V $\gg 4.6 \times 10^{2}$ cm^{3}.

Since the theory is valid only for a $< \mu/\rho\omega$, a must be less than .02 cm. A reasonable orifice size would then be one of length L = .014 cm, radius a = .02 cm with a surge tank volume V of 1 litre. Such a device will provide an attenuation in the reflected power (Eq. 3) of 95%. Use of a bigger volume V would reduce this even further.

Our own testing experience confirms the validity of these calculations[9]. Choosing an orifice and volume based on Eqs. 10 and 12 resulted in the expected dramatic reduction of TAOs. When $1/a^{4}$ did not meet the criterion of Eq. 10, no effect was seen. This is expected. If the orifice radius is too large, there is no attenuation of the sound wave. If the orifice radius is too small, the device looks like a perfectly hard wall and no sound wave can propagate down the capillary and, therefore, cannot be attenuated. When the volume was decreased so that Eq. 12 became progressively less satisfied, we could effectively generate any size TAO up to the amplitude encountered when no damper was present. Ditmars and Furukawa[8] constructed such a device; with orifice and surge tank dimensions corresponding to the dimensions which this theory predicts. In practice, if the condition a $<$ $\mu/\rho\omega$ is not satisfied, one can still get the damping expected. However, the equation for the dimension of the capillary (Eq. 10) is not strictly correct.

POROUS SOLID DAMPERS

Occasionally TAOs are damped by the introduction of porous solids (e.g., glass wool) into the oscillating pipe. The damping that occurs is due to the viscous forces in the tortuous air paths through the material. A system of this type can be modelled crudely by considering it to be a number of parallel capillaries whose impedance is given by their parallel impedance. From the previous treatment of the orifice and surge tank, it is apparent that this is effective if the capillaries are of the correct size. If the porous solid is packed tightly, it is ineffective, since it acts as a good reflector of sound waves, especially at low frequencies. If the material is too loosely packed, then an excessive length of material is required for effective damping.

In general, packing a pipe with a porous solid acts as a low pass filter. The amount of damping at the low frequencies encountered with TAOs and the cutoff frequency of the filter depend on the thickness of the material. For most TAO applications the length of material required at the frequencies encountered is too long and is not controllable or reproducible. This makes a solution of this type undesirable in most applications.

CONCLUSIONS

The problem of the damping of TAOs appears to be easily solvable in many applications through the use of an orifice and surge tank. This device can be installed either as a termination in an oscillating pipe or in a branch. The properties of the device and its effect on potential TAOs are easily calculated and show that it can be very effective in damping the oscillation. It would be prudent design practice to incorporate such a device into cryogenic systems whenever TAOs could cause a problem.

I wish to thank Dr. D. Elleman for keeping a useful book in plain view on his shelf and P. Ludtke, R. Urbach, W. Brooks and C. Marcus for help with tests.

NOTATION

$\vec{A}$	=	incident wave amplitude
a	=	radius of capillary
b	=	radius of branch
$\vec{B}$	=	reflected wave amplitude
$\vec{C}$	=	transmitted wave amplitude
c	=	speed of sound
j	=	$\sqrt{-1}$
k	=	$2\pi/\lambda$ = wavelength constant
$\vec{L}'$	=	length of capillary
p	=	pressure
P_r	=	ratio of reflected power to incident power
R	=	acoustic resistance
S	=	cross-sectional area of pipe
t	=	time
T	=	temperature
V	=	volume of surge tank
x	=	position
$\vec{X}$	=	acoustic reactance
$\vec{Z}$	=	acoustic impedance
Z_0	=	characteristic impedance of pipe
$d\vec{V}/dt$	=	volume velocity of gas
λ	=	wavelength
μ	=	viscosity
ρ	=	gas density
ω	=	angular frequency of wave

SUBSCRIPTS

b	=	branch
c	=	continuation
d	=	damper
i	=	incident
r	=	reflected
t	=	transmitted

REFERENCES

1. A. Wexler, Transfer of Liquefied Gases, in "Experimental Cryophysics", Hoare, Jackson and Kurti editors, Butterworths, (1961), p. 155.
2. N. Rott, J. App. Math. and Phys., 20:230 (1969).
3. N. Rott, J. App. Math. and Phys., 24:54 (1973).
4. N. Rott, J. App. Math. and Phys., 26:43 (1975).
5. T. Von Hoffman, U. Lienert and H. Quack, Cryogenics, 13:490.
6. L.E. Kinsler, and A.R. Frey, "Fundamentals of Acoustics", John Wiley, New York (1950).
7. R.R. Conte, "Elements de Cryogenie", Masson et Cie, Paris (1970), p. 18.
8. D.A. Ditmars, and G.T. Furukawa, Jour. Res. NBS-C 69C:35 (1965).

EXPERIMENTS ON FLOW THROUGH ONE TO FOUR INLETS OF THE ORIFICE AND BORDA TYPE*

R. C. Hendricks and T. T. Stetz[†]

NASA Lewis Research Center
Cleveland, Ohio

INTRODUCTION

Fluid machinery components and heat transfer devices contain contoured inlet configurations, most of which are sequential. Compressors, pin-finned heat exchangers, labyrinth seals and step seals, for example, consist of two or more sequential inlets. The details of heat transfer and flow dynamics in these configurations are not well understood in many cases.

In order to understand some flow phenomena in the seals of high performance turbomachines,[1] a series of choked fluid flow tests with single and multiple sharp-edge orifice and Borda type inlets were conducted,[2-8]. Borda inlets were studied so as to examine the effects of a protrusion into the "reservoir region" whereas the orifice inlets were studied to determine how the flow responded to a sharp-edge configuration; both of these basic types of geometries are found in the seal configuration.[1]

Flow jetting in single inlets[2,3,4,6] occurred over a wide range of fluid state conditions and was found to be inhibited by increasing: (i) the inlet stagnation temperature, (ii) the length to diameter ratio (L/D), and (iii) tube roughness. These tests established that the jetting phenomena could occur in a single orifice or Borda inlet passage to over 105 L/D.

---

*Presented in more detail as NASA TM-82680.

[†]Current address: Miami University, Oxford, Ohio.

Jetting was further investigated in multiple inlet configurations[5,7,8] where four sequential inlets were studied. At the nominal separation distance of 30 diameters, the flow was found to be nearly independent of the upstream reservoirs and inlets; whereas at a separation distance of nominally 0.7 diameters, jetting occurred at the lower inlet stagnation temperatures. Separation distances between these two were investigated in a water table flow visualization study, which demonstrated flow instabilities for a range of 1<L/D<10 for these sequential inlets.[5,7]

The purpose of this paper is to compare the nature of flow rates and pressure distributions in N-sequential inlets of the orifice and Borda types at nominal separation distances of 30 and 0.7 diameters over a wide range of reduced fluid state conditions. A comparison will also be drawn between applicable theoretical results and these N-sequential orifice and Borda inlet configurations.

APPARATUS AND INSTRUMENTATION

The blowdown type flow facility was basically that described in Ref. 9, but modified to accommodate the various sequential orifice and Borda inlet configurations.[5,7,10,11] A close-up photograph of the Borda inlet is given as Fig. 1(a) and the orifice inlet as Fig. 1(b). The working fluid was nitrogen.

The Borda type inlets with 1/D of 1.9 were designed to be similar to those used in Ref. 2, with spacers of 15.24 cm (6.0") and 1.03 cm (.407"). This provided two fixed spacings of 30 and 0.8 diameters respectively. A schematic of the N-sequential Borda geometry at the 0.8 separation distance is presented in Fig. 2(a), which also gives inlet geometry and pressure tap locations.

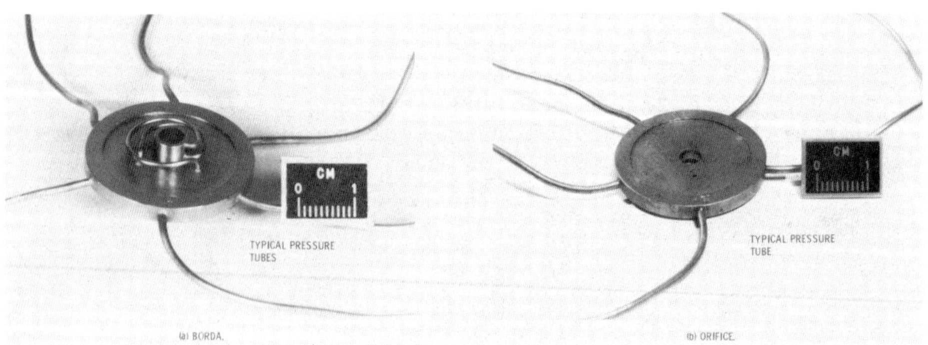

Fig. 1. Photograph of the inlet configuration.

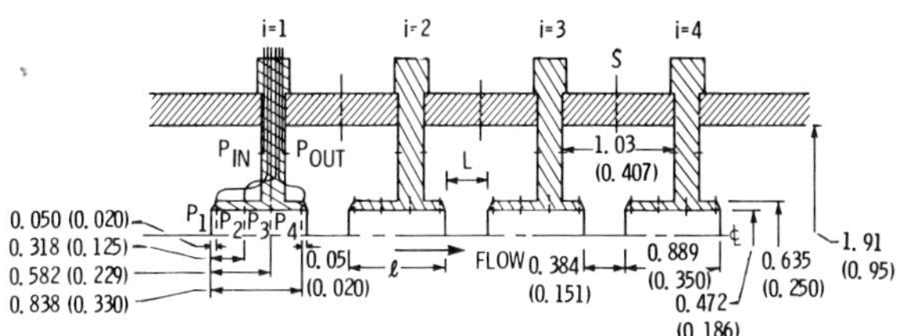

(a) FOUR BORDA WITH 1.03 cm (0.407") SPACER.

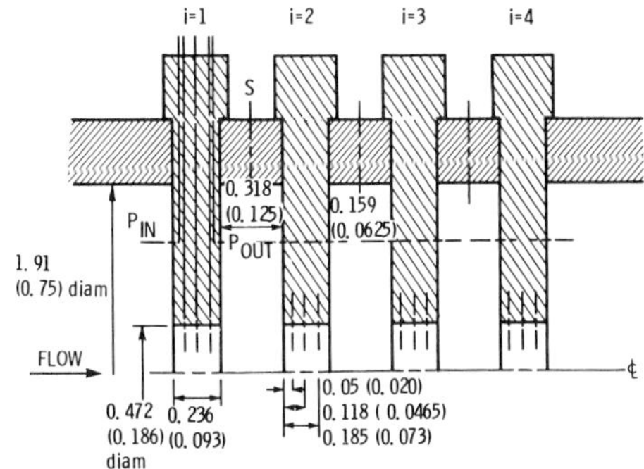

(b) FOUR ORIFICE WITH 0.32 cm (0.125") SPACER.

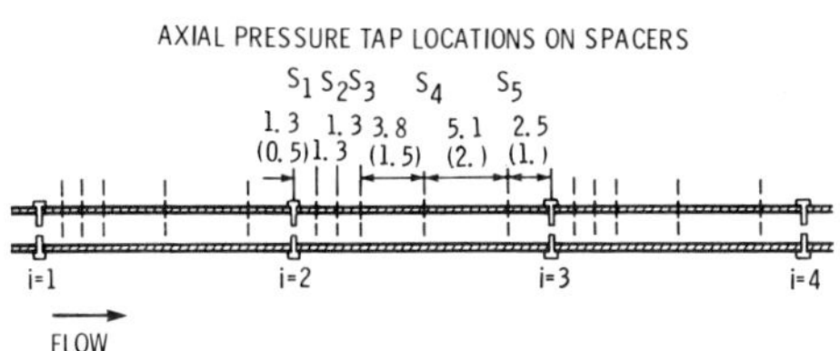

(c) FOUR ORIFICE WITH 15.24 cm (6") SPACER.

Fig. 2.   Schematic diagram of N-sequential inlet test section.

The orifice type inlets with 1/D of 0.5, similar to those of Ref. 3, were designed with spacers of 15.24 cm (6.0") and 0.32 cm (0.125"). These provided two fixed spacings of 32 and 0.66 diameters respectively. The schematic showing the pressure tap locations and inlet geometry of the N-sequential orifice at 0.66 diameters is presented as Fig. 2(b).

Figure 2(c) provides a schematic of the N-sequential orifice (or Borda) inlet configuration with the 15.24 cm (6.0") spacers. This figure also gives pressure tap locations on the 15.24 cm (6.0") spacers.

These sequential inlet configurations were fitted between inlet and outlet flange adaptors to accommodate the multiple lengths. The multiple surfaces were satisfactorily sealed by Mylar gaskets between flat faces. Pressure and flow data were recorded as described in Refs. 5 and 11.

ANALYSIS

The treatment of the simplest set of sequential inlets is quite complicated. In a thermodynamic treatment, where only the end state conditions are considered, the ith-sequential inlet may be assumed to expand isentropically, followed by an isobaric recovery in the "mixing chamber" or spacer to the adiabatic locus, with choking at the Nth inlet.[5]

The governing equations, as described in ref. 5, may be written:

$$(G/Cf)^2_i = 2\rho^2 (H_o - H_i) \tag{1}$$

where the constraints are for the isentropic case:

$$S_o\left[(P_o,T_o), i\right] = S\left[(P_e,T_e), i\right]$$

and for the isobaric case:

$$(P_e)_i = (P_o)_{i+1} \tag{2}$$

Critical Flow (choked)

$$\left[Gm^2 (dv/dp)_e\right]_{i=N} = -1$$

where

$$Gm^2 = \left[(2/v^2) \int_p^{P_o} VdP \right]_{i=N} \tag{3}$$

which upon convergence, $(G/Cf) \rightarrow Gm$, with $Cf = 0.75$. Fluid properties were calculated using the computer program GASP.[12]

At the nominal 0.7 diameter spacing, it is assumed that the flow recognizes the N-sequential inlets spaced at this separation distance as one inlet; the flow jetting condition is dominant. This assumption is based on previous experimental results and flow visualization studies, which were limited to four sequential inlets.[5,7]

The flow rates in sequential inlets can also be analyzed by the extrapolation of the flow rate through a single inlet with the appropriate constraints. Using a modified Bernoulli equation and constant area ducts, one can show[11]

$$\frac{G_{r(N)}}{G_r(N=1)} = N^{-n} \tag{4}$$

where n = 0.5. Thus knowing the mass flow rate for one inlet, the flow rate through any number of N-inlets can be determined.

RESULTS

Experimental Results

The experimental results covering a wide range of inlet stagnation conditions, where the working fluid is nitrogen and the reduced temperature ranges from $0.68 < T_r < 2.5$ (liquid to gas) with the reduced pressure to $P_r < 2.5$, are now presented in two sections, (i) flow rates and (ii) pressure profiles. Experimental data may be found in Refs. 5, 7, 10, and 11.

Flow Rates

Figures 3a,b,c,d present reduced flow rate as a function of reduced inlet stagnation pressure for various isotherms. The reduced flow rate is given by:

$$G_r = G/G* \tag{5}$$

where G* is determined by the extended corresponding states theory.[5,13,14]

As an example of the N-sequential inlets, the flow rates for the three sequential Borda and orifice inlets for the nominal 30 diameter spacing, are given in Fig. 3 (a) and (c), respectively. They are very similar, with the Borda flow rates being slightly higher.

At the nominal 0.7 diameter spacing, the N-sequential Borda and orifice, as exemplified by N = 3, are again form similar, Fig. 3(b) and (d). They are significantly higher than the flows for the nominal 30 diameter spacing, but are very similar to that of a single Borda or orifice inlet, Refs. 2, 3. The N-sequential inlets, with the data limited up to N = 4, all show these similar trends.[10,11]

Pressure Profiles

Figures 4(a), (b), (c), and (d) give a perspective of the variation of pressure profile with reduced inlet stagnation temperature. There is a significant change between the nominal spacing of 0.7 and 30 diameters.

At the nominal 30 diameter spacing for the N-sequential Borda and orifice, the pressure profiles exhibit a sharp drop at the entrance of each inlet, followed by some recovery within that inlet (Fig. 4(a) and (c)) where N = 3 is shown. At the last

　　　R. C. Hendricks and T. T. Stetz

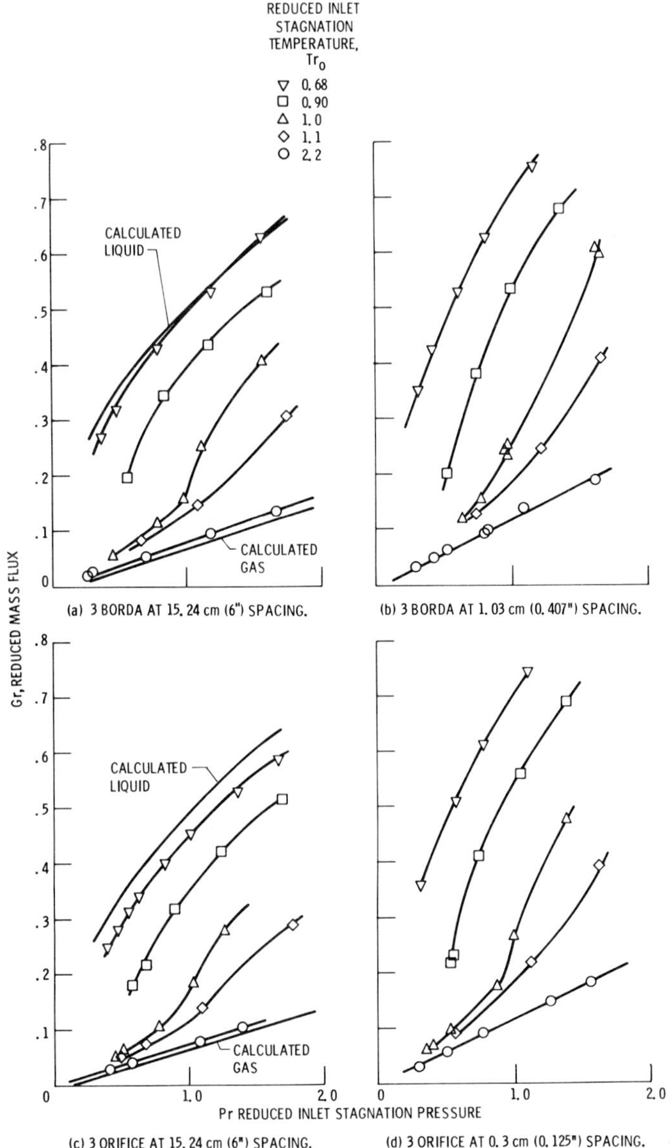

Fig. 3. Reduced mass flux as a function as reduced inlet stagnation pressure for selected reduced inlet stagnation temperatures.

inlet, i = N, jetting was found to occur, as indicated by the flat pressure profile at the lower inlet stagnation temperatures for both the orifice and Borda case.

The pressure profiles for the N-sequential orifice and Borda inlets spaced at the nominal 0.7 diameter spacing at the lower inlet stagnation temperatures resemble that of a free jet, again N = 3 is used for explanatory purposes. The fluid seems to flow unimpeded even though they are separated at a nominal distance of 0.7 diameters. At the higher inlet stagnation temperatures, the recovery is somewhere between that of a free jet and of sequential independent inlets, such as those spaced at nominally 30 diameters.

Further comparisons and experimental data can be found elsewhere,[1,2,3,5,7,8,10,11] including some of the effects of backpressure.

Analytic Comparisons

At the nominal 30 diameter spacing, we expect the sequential inlets to act independent of one another. The constant flow coefficient for each inlet used to predict the flow rates causes some problems, as well as attempting to calculate the flows near the thermodynamic critical region.

The calculated flow rates for liquid and gas are given on Figs. 3a,c for the N = 3 case. These approximate the experimental curves fairly well.

The results of the experimental flow rates and the calculated flow rates for the N-sequential inlets at the 30 diameter nominal spacing are given in Figs. 5, 6. The experimental and calculated flow rates seem to follow an empirical power law relation:

$$G_r \propto N^{-b} \tag{6}$$

where N is the number of sequential inlets and b is the undetermined function of inlet temperature and pressure.

From the experimental results and theoretical calculations, it appears that $b \sim 0.4$. The elementary relation, Eq. (4), predicts a value of $b = 0.5$, but $b \sim 0.4$ appears valid over a wide range of inlet pressure and temperatures, to the first order and excluding the thermodynamic critical region.

The reduced mass flux empirical relation can be further normalized by flow through a single inlet, thus:

$$\frac{G_r(N)}{G_r(N=1)} = N^{-b} \tag{7}$$

Hence, if $G_r(N=1)$ and b are known, the mass flux for N well separated, similar sequential inlets can be determined directly from

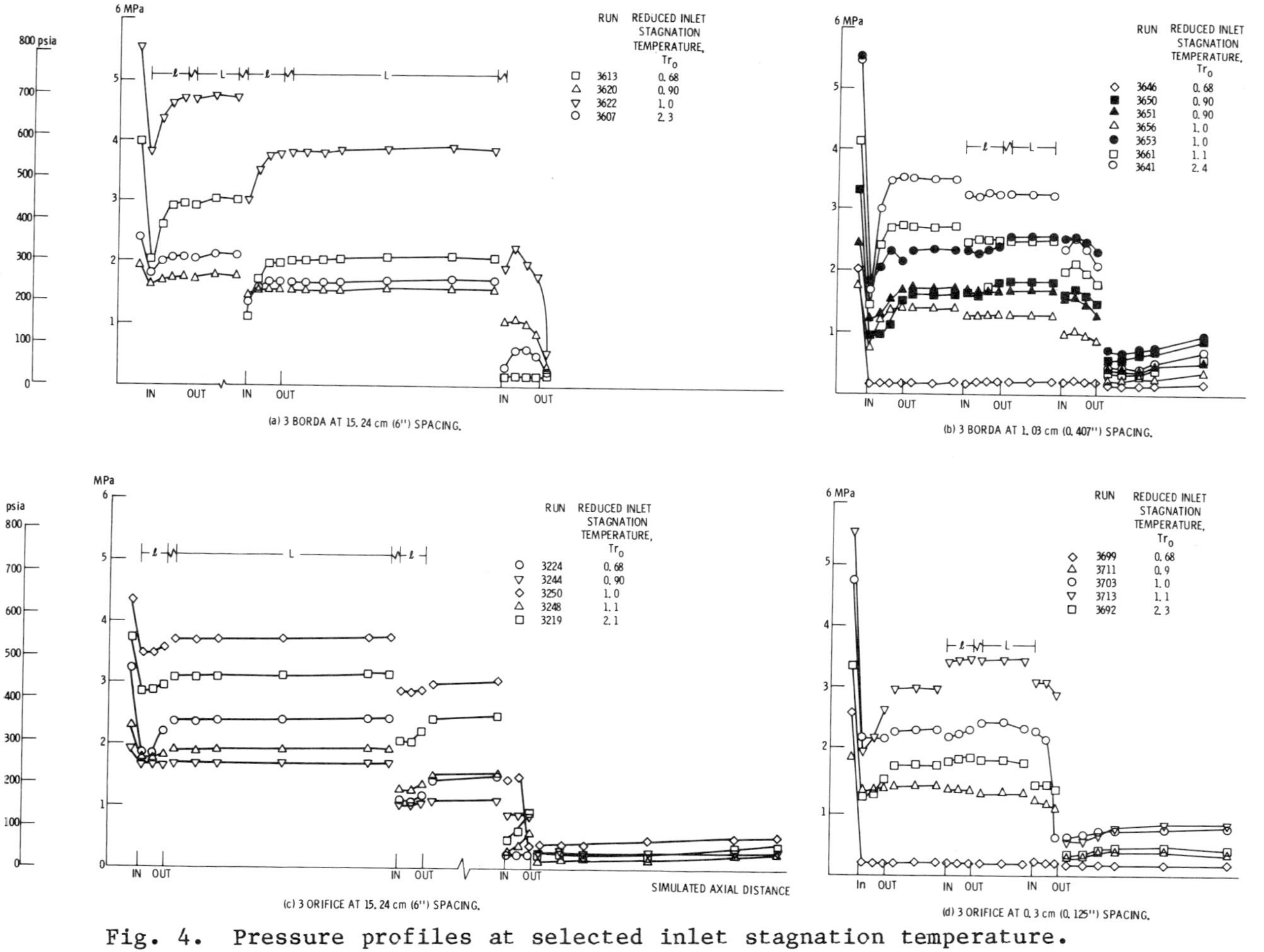

Fig. 4. Pressure profiles at selected inlet stagnation temperature.

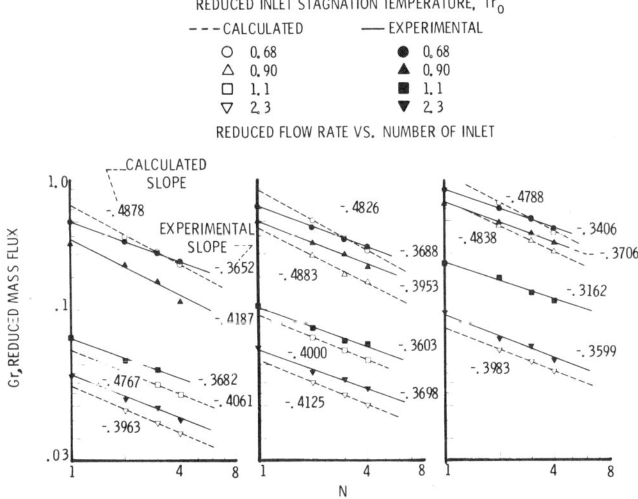

Fig. 5. Reduced mass flow rate at a function of N
for the Borda inlet at selected reduced inlet stagnation
temperatures.

knowing the mass flux through a single inlet. The flow through
this single inlet is governed by the inlet stagnation conditions.

The experimental and calculated magnitude of the exponent b
as a function of reduced inlet stagnation temperature is given as
Fig. 7 for various reduced pressures. The value of the exponent b
seems to be well behaved outside the critical region, whereas the
behavior of the exponent b in the critical region is unknown.

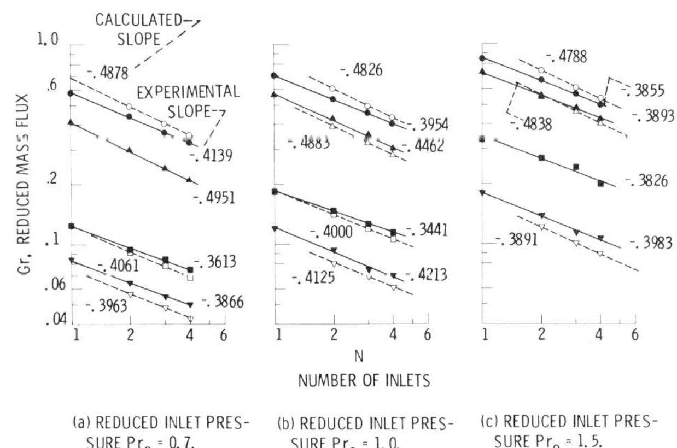

Fig. 6. Reduced mass flow rate as a function of N for the
orifice inlet at selected reduced inlet stagnation temperatures.

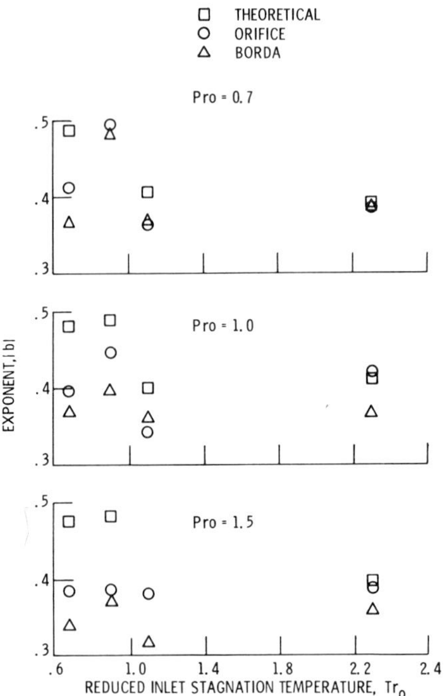

Fig. 7. The Nth inlet exponent, (b), as
a function of reduced inlet stagnation temperature
for selected reduced inlet stagnation pressures.

SUMMARY

Choked flow rate and pressure profile data were taken and
studied for a configuration consisting of up to four sequential
orifice and Borda inlets for nominal spacings of 30 and 0.7 diameters as an indicator of fluid flow through N-sequential inlets.

At a nominal spacing of 30 diameters, the pressure profiles
exhibit a sharp drop at the leading edge of each inlet followed by
some recovery within that inlet and little further recovery in the
spacer chamber. At the lower inlet stagnation temperatures, fluid
jetting can occur in the last of the sequential inlets. The Borda
and orifice inlets both showed this tendency. These sequential
inlet configurations appear to function independently with control
or choking occurring at the last of the sequential inlets, i = N.

At a nominal separation distance of 0.7 diameters at the
lower inlet stagnation temperatures, fluid jetting was prevalent
throughout all of the sequential inlets in both the Borda and
orifice case. The flow rates were the same as for a single inlet

and appeared to be controlled or choked at the first inlet, (i = 1), independent of the downstream sequential inlet configuration.

Analytic modeling is complex, but a simplistic model of the nominal 30 diameter spacing appears to give good correlation with the experimental data outside of the thermodynamic critical region. The data seem to follow a simple empirical relation:

$$\frac{G_r(N)}{G_r(N=1)} = N^{-b}$$

where b is a nonlinear function of temperature and is weakly dependent on pressure. Outside the critical region, b was found to be approximately 0.4. The principal feature of these results is that if one knows the mass flux for a single inlet, then one knows the mass flux for N-sequential well separated inlets to a first order. Further, the mass flux is governed by stagnation pressure and temperature conditions upstream of the first inlet.

NOTATION

b	slope		L	length of separation
Cf	flow coefficient		$\dot{m}$	slope
D	diameter of the inlet		N	number of inlets
f	friction factor		P	pressure
G	mass flow rate		S	entropy
G^*	flow normalizing parameter, 6010 g/cm^2-s for nitrogen, $\sqrt{P_c \rho_c / Z_c}$		T	temperature
			V	specific volume
			$\rho=1/V$	density
H	enthalpy		Z_c	compressibility
l	length of inlet			

SUBSCRIPTS

c	thermodynamic critical		m	maximum flow
e	exit		o	stagnation, or reference
I	isentropic		r	reduced by normalizing parameter
i	i-th sequential inlet		1	case for N=1, the single inlet

REFERENCES

1. R.C. Hendricks, Investigation of a 3-step cylindrical seal for high-performance turbomachines, NASA TP - 1849 (1981).
2. R.C Hendricks, Some Aspects of a Free Jet Phenomena to 105 L/D in a Constant Area Duct. in "Proc. 15th Int. Cong. of Refrig., Vol. 2." International Institute of Refrigeration, Paris (1979).

3. R.C. Hendricks, A Free Jet Phenomena in a 90-degree-Sharp Edge
    Inlet Geometry, in "Advances in Cryogenic Engineering ,
    Vol. 25," Plenum Press, New York (1980) p. 506.
4. R.C. Hendricks and N. Poolos, Critical Mass Flux Through Short
    Borda Type Inlets of Various Cross-Sections, in "Proc. 15th
    Int. Cong. of Refrig., Vol. 2," International Institute of
    Refrigeration, Paris (1979).
5. R.C. Hendricks and T.T. Stetz, Some Flow Phenomena Associated
    with Aligned, Sequential Apertures with Borda Type
    Inlets. NASA TP-1792, (1981).
6. R.C. Hendricks and R.J. Simoneau, Some Flow Phenomena in a
    Constant Area Duct with a Borda Type Inlet Including the
    Critical Region. NASA TM-78943, (1978).
7. R.C. Hendricks and T.T. Stetz, Flow Through Aligned Sequential
    Orifices. NASA TP in progress.
8. R.C. Hendricks and T.T. Stetz, Flow Through Axially Aligned
    Sequential Apertures of the Orifice and Borda Types. NASA
    TM-81681, (1981).
9. R.C. Hendricks, et al., Experimental Heat Transfer Results for
    Cryogenic Hydrogen Flowing in Tubes at Subcritical and
    Supercritical Pressures to 800 Pounds Per Square Inch
    Absolute, NASA TN D-3095, (1966).
10. R.C. Hendricks and T.T. Stetz, Flow Rate and Pressure Profiles
    for One to Four Axially Aligned Borda Type Inlets. NASA TP
    in progress.
11. R.C. Hendricks and T.T. Stetz, Flow Rate and Pressure Profiles
    for One to Four Axially Aligned Orifice Type Inlets: NASA
    TP in progress.
12. R.C. Hendricks, A.K. Baron and I.C. Peller, GASP-A Computer
    Code for Calculating the Thermodynamic and Transport
    Properties for Ten Fluids: Parahydrogen, Helium, Neon,
    Methane, Nitrogen, Carbon Monoxide, Oxygen, Fluorine,
    Argon, and Carbon Dioxide, NASA TN D-7808, (1975).
13. R.C. Hendricks, Normalizing Parameters for the Critical Flow
    Rate of Simple Fluids Through Nozzles, in "Proc. 5th Int.
    Cryo. Eng. Conf., IPC Science and Technology Press,
    England, (1974), p. 278.
14. R.C. Hendricks and J.V. Sengers, Applications of the Principle
    of Similarity to Fluid Mechanics, Water and Stream: Their
    Properties and Current Industrial Applications, in "Proc.
    9th Int. Conf. on the Properties of Steam." J. Straub and
    K. Scheffler eds., Pergamon Press, Oxford, (1979),
    p. 322. Unabriged version as NASA TM X-79258, (1979).

REVIEW OF LITERATURE DATA AND RELIABILITY OF EXPERIMENTAL METHODS FOR SOLID SOLUBILITY DETERMINATIONS IN CRYOGENIC SYSTEMS

K. D. Luks* and J. P. Kohn

University of Notre Dame
Notre Dame, Indiuna

INTRODUCTION

When one takes a natural gas stream from the field and liquefies it for the purpose of storage, transport, or purification,
one can encounter the solid solubility limits of some of the trace
species present (e.g., CO_2 and the heavier hydrocarbons). The
precipitation of these solid species in liquefaction processing
can cause problems by coating the surfaces of heat exchangers and
fouling turbo expansion devices, requiring liquefaction units to
be shut down for cleaning and/or repair.

This review will focus on experimental methods for determining the solubility of solid species that may be present in the
cryogenic liquefaction processing of natural gases (LNG), propane
gases (LPG), and ethylene. A review of the literature data on
solubility of solids in prototype binary, ternary, and higher
component systems pertinent to these processing applications is
included.

The type of equilibria that one has in liquefaction processing when solids precipitate is three-phase solid-liquid-vapor
(denoted S-L-V) equilibria; at elevated pressures, one may simply
have two-phase S-L equilibria. The S-L-V problem will be primarily addressed herein. First, the diversity of S-L-V equilibria
in thermodynamic phase space will be classified arbitrarily into

---

*Presently affiliated with the Department of Chemical Engineering,
University of Tulsa, Tulsa, Oklahoma.

four types for purposes of discussion. The nature of the S-L-V
equilibria can have a bearing on how one undertakes the experimental study of a particular system and consequently influences
the discussion of current experimental methods used for measuring
solid solubilities. Although experimental S-L-V systems can have
liquid phase solute compositions ranging from pure (at the solute
triple point) down to very dilute, the primary emphasis of the
data review will be on those systems that have been experimentally
studied at low concentrations, as it is that regime which is most
relevant to the cryogenic liquefaction processing of gases.

TYPES OF S-L-V PHASE BEHAVIOR

For the purpose of discussion, S-L-V binary system phase
behavior will be classified into four types. These are illustrated in Figs. 1 through 4, which are schematic diagrams in pressure-
temperature (P vs. T) space. In type A systems (Fig. 1), the
triple point temperature of the solute (at T_{t1} of species 1) is
usually 20 C^o or more higher than the critical temperature of the
solvent (at T_{c2} of species 2). The S-L-V locus, denoted S_1-L-V,
has an upper branch which extends from the triple point of solute
upwards in pressure to an upper critical end point, labeled K,
where the liquid and vapor phases become critical in the presence
of the solute crystals. The pressure at this "K-point" can be
quite elevated, on the order of several hundred bars. The solute
concentration is generally quite high (>10%) on this branch. The
lower branch, also denoted by S_1-L-V, extends from a second K-
point down to a four-phase S_1-S_2-L-V point (labeled Q) where S_2 is
crystals of the solvent species. The liquid phase concentration
of solute on the lower branch is often several orders or magnitude

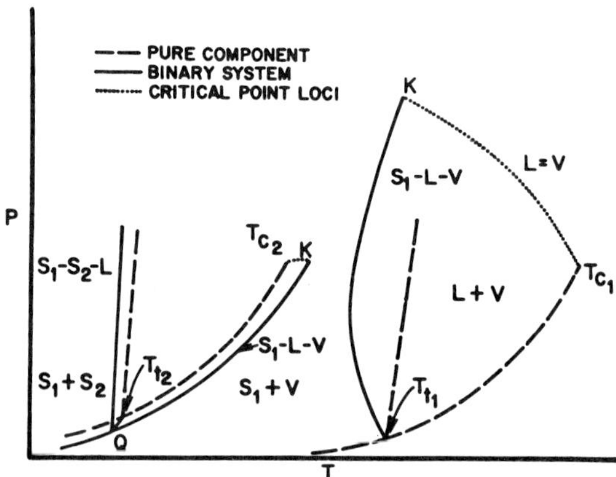

Fig. 1. Type A binary phase equilibria.

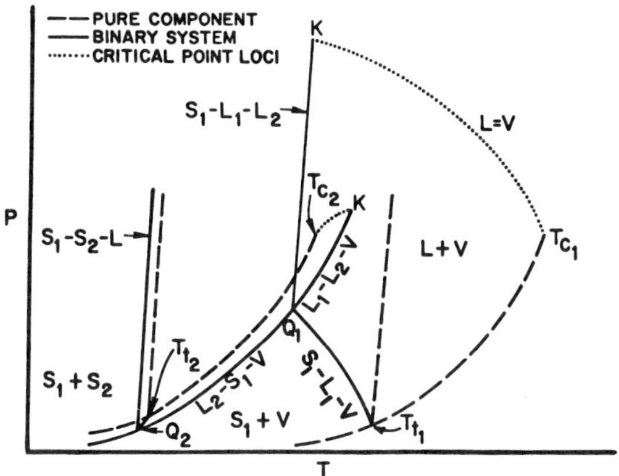

Fig. 2.   Type B binary phase equilibria.

less than that on the upper branch. The lower branch K-point is
consequently close in P-T space to the pure solvent critical point
(T_{c2}). An example of a Type A binary system is methane + n-oc-
tane.

In a Type B system, (Fig. 2), the triple point temperature of
the solute is still higher than the critical point temperature of
the solvent, but probably differs by less than 20 C^o, as in these
systems the molecular differences between solute and solvent are
smaller. The upper branch of the S-L-V locus extends from the
triple point of the solute upwards in pressure to a maximum at a
four-phase $S-L_1-L_2-V$ point (denoted Q_1). The Q_1-point is a lower
terminus of the L_1-L_2-V immiscibility locus for this system. The
lower branch of the S-L-V locus (denoted S_1-L_2-V) extends from the
Q_1 point down to a Q_2-point ($S_1-S_2-L_2-V$). The solute concentration of L_2 is often one or two orders of magnitude less than
that of L_1 at the Q_1-point. The binary system methane + n-heptane
is a Type B system.

As the molecular differences between the solute and the
solvent diminish, the phase equilibria behavior transforms into
that of Type C (Fig. 3), and Type D (Fig. 4), systems, in that
order. With respect to S-L-V behavior, Type C and D systems are
the same, with a single locus extending from the triple point of
the solute down to a Q-point (S_1-S_2-L-V). The locus has a maximum
in pressure, in the vicinity of which the steepest changes in
liquid phase solute concentration usually occur. Even in a Type D
System the liquid phase concentration of a solute can vary by
several magnitudes over the entire S-L-V locus. The Type C system
has a region of L_1-L_2-V immiscibility while the Type D system does
not.   The immiscibility phenomena illustrated in the Type B and C

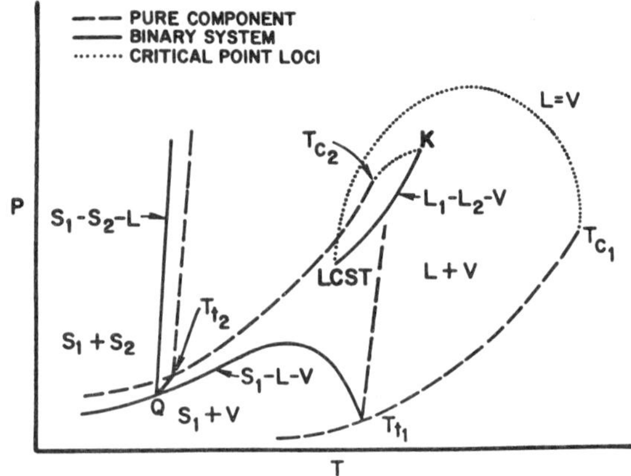

Fig. 3. Type C binary phase equilibria.

diagrams (Figs. 2 and 3) are not meant to be comprehensive.
Figures 1 to 4 are designed to illustrate the variety of S-L-V
loci. For a more complete classification of S-L-V phenomena, one
should consult van Konynenburg and Scott[1].

Figure 5 schematically illustrates how the liquid phase
solute concentration varies with reciprocal temperature along the
S-L-V loci just described. The overall drop in solute concentration along a Type A locus is usually greater than along the
others due to the molecular disparity between the solute and
solvent species. There is a gap between the K-points on this
locus; here a region of S_1-V equilibria exists. (In Type B, C,
and D systems, an S-V region is just below the S-L-V locus in P-T
space.) All solute compositions rise with increasing temperature,
an exception being near the K-point on the lower branch of the
Type A locus. Since the S-L-V system is generally very rich in
the solvent species, the liquid and vapor phases behave much like
those along the pure solvent V-L coexistence locus. Near the K-
point criticality, the liquid phase dilates, causing a decrease in
the solute composition even as temperature rises. The upper
branch of a type A locus can bend back along the temperature
coordinate with increasing pressure. This behavior is shown in
Fig. 1 but not in Fig. 5. Finally, when the solute becomes very
dilute in the liquid phase of a S-L-V system (say, less than 1%),
the locus starts to follow a straight line in ln x versus 1/T
space. This Henry's law type of dilute solution behavior is
suggested in Fig. 5 by the eventual coincidence of the S-L-V loci
(solid curves) with the dashed straight lines.

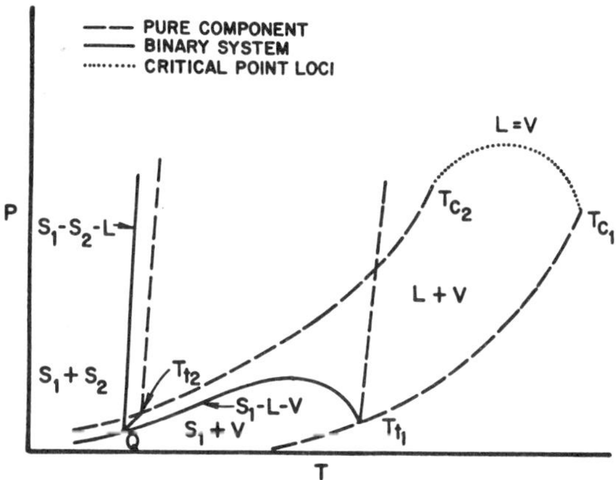

Fig. 4.   Type D binary phase equilibria.

EXPERIMENTAL METHODS

The primary experimental problem in elucidation of the phase
behavior of S-L-V loci is the determination of the composition of
the solute in the liquid phase. The solid phase is generally pure
solute, and the vapor phase is often virtually solute-free due to
the low volatility of the solute component at the temperature of
interest. An exception to the latter statement occurs with CO_2 as
a solute, since pure CO_2 has a high sublimation pressure.

The most obvious experimental method to acquire composition
information on the solute in the liquid phase would be one in
which an equilibrium device has the capability of taking liquid
samples. One version of such an apparatus is a stainless steel
bomb, usually containing an optical window, in which phase equilibrium is attained by rotating the bomb cyclically through a large
angle. The bomb is equipped with a sample line and a valve.
Another version of a sampling apparatus is a vertically mounted
equilibrium bomb containing a window, with equilibrium achieved by
recirculation of the vapor phase by means of a pump. This bomb
also has a sample line and a valve. In some cases the sample
valve is mounted in the bath and therefore is at the same temperature as the equilibrium bomb contents. In other cases the sample
valve is mounted outside the bath with a short capillary line
connecting it to the equilibrium bomb. In any case the capillary
line carrying liquid samples downstream from the sample value to
the analytical apparatus must be thoroughly heated.

A simple version of a sampling cell was developed by Kohn and
Luks[2]. It consisted of a Pyrex glass vertically mounted cell of
11 ml inside volume which contained a stainless steel capillary

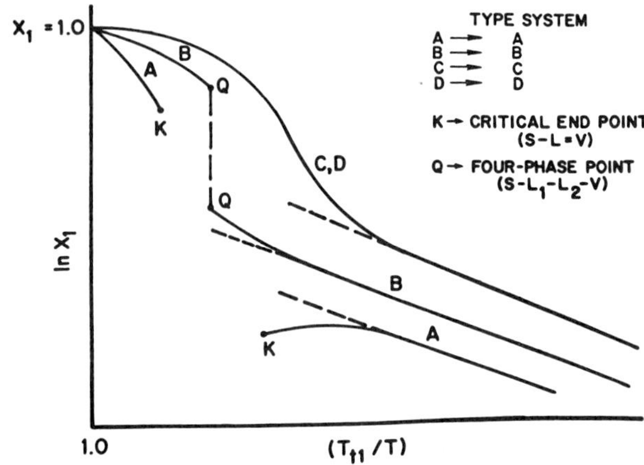

Fig. 5. Variation of liquid phase solute composition with temperature in the binary S-L-V systems of Figs. 1 to 4.

tube. The remote end of the capillary tube extended to within 6 cm of the bottom of the cell. A liquid sample could be drawn through the tube by activating a miniature valve mounted on a metal adaptor on the top of the cell. Gas phase could be added to the cell while sample was being withdrawn to keep the cell pressure constant. This sampling cell has the experimental merit of small size, permitting use of small baths and rapid temperature changes, when desired.

If any of the above sampling bomb devices are employed to sample the liquid phase along the upper branch of the Type A S-L-V locus or at temperatures above that of the pressure maximum of the other loci type (e.g. the Q_1 point in Type B systems), reliable samples may be obtained. Reliability would of course demand adequate flushing of the line from the bomb to the sample valve since the material originally in this line is not in equilibrium with the bomb contents. Also sampling must be performed slowly to maintain the sample and the bomb contents at the equilibrium pressure. Frequently this necessitates the addition of vapor phase to the equilibrium bomb while the actual liquid sample is being taken and while the bomb contents are <u>not</u> being agitated.

If liquid samples along the lower branch of the Type A locus and at temperatures <u>below</u> that of the pressure maximum of other system loci are to be taken, then severe problems will be encountered. The problem occurs whether the sample is taken with pressure downstream of the sample valve at cell pressure or at lower pressures. As the liquid sample leaves the equilibrium

locus by increasing the temperature at constant pressure or lower
pressure, it must enter the two phase S-V region. The solubility
of the solute in the vapor phase is necessarily much smaller than
it is in the liquid phase, so solid solute precipitation in the
sampling line results. A sample only fortuitously reaches the
analysis equipment intact, if at all, since plugging of the line
is highly likely.

Two approaches are reported that circumvent this analytical
problem. Kuebler and McKinley[3-5] use a single-pass continuous
flow technique with the system being at a pressure exceeding that
of the critical pressure of the solvent. In this state, the
system is a S_1-L system; however, by taking a number of different
elevated pressures, one is able to estimate roughly what the
concentration would be at a pressure where the S_1-L-V system
exists.

Kohn and Luks[2,6,7] employ a stoichimetric technique whereby
no sample is taken at all. A measured amount of solute is put
into a volume-calibrated glass cell and solvent gas is added in
known quantities. An equilibrium measurement is taken when solute
crystals are present only in a trace amount, the solute thus being
presumed to be only in the liquid phase. The amount of solvent in
the vapor phase (assumed pure) is computed by noting its volume
and state, and the composition of the liquid phase is thus determined. The largest errors in this technique lie in sighting the
"last trace" of crystals and in determining liquid phase volume,
in that order. However, crystals can be noted down to a fraction
of a milligram of solute, which allow the determination of composition down to 1000 ppm to ± 1% and down to 100 ppm to ± 5%. It
should be noted that Kuebler and McKinley, using analytical techniques, can measure compositions accurately to below 10 ppb in
their S_1-L systems at an analytical accuracy of about ± 5% independent of concentration.

The stoichiometric approach of Kohn and Luks has been successfully extended to ternary and multicomponent systems. For the
ternary case, binary V-L data on the solvent gases are used to aid
the computation of the vapor and liquid phase compositions.[8-12]
For multicomponent systems, the trace of solid phase was
identified in the presence of a liquid phase on an infinitesimal
amount of vapor phase. This technique assures the composition of
the liquid phase, but of course does not establish the composition
of the vapor phase.[13]

SURVEY OF DATA

In the liquefaction processing of natural gas, propane gas,
and ethylene, solids can precipitate from the V-L system, usually
at temperatures associated with the low concentration regions of

the S-L-V loci discussed earlier. As pointed out in the preceeding section on experimental methods, it is the accurate determination of solid solubilities in these dilute systems that can pose difficulties. Table I lists studies that have been performed on well-defined binary, ternary and higher S-L-V systems, along with the method of study employed. The "type" of the binary systems is also given.

Table I. Survey of Literature Data

System	Type	References	Experimental Method
Binary:			
CH_4 + CO_2	D	15,16,17,25	SE
+ $n-C_4H_{10}$, $n-C_5H_{12}$	D	4	SP
+ n-hexane	C	3,18	SP,CR
+ n-heptane	B	2, 3, 8	SE,SP,CR
+ n-octane	A	2, 7	SE,CR
+ cyclohexane	A	7	CR
+ benzene	A	3,18	SP,CR
+ toluene	A	3	SP
+ naphthalene	A	19	SE
C_2H_6 + CO_2	D	17,20,21	SE
+ n-octane, n-decane	D	2, 6	SE,CR
+ n-dodecane	D	6	CR
+ benzene, cyclohexane, trans-decalin	D	22	CR
+ 2-methylnaphthalene	B	23	CR
+ naphthalene	A	23	CR
C_3H_8 + CO_2	D	15,17,25	SE
+ n-decane, n-dodecane	D	23	CR
+ benzene, cyclohexane	D	24	CR
$n-C_4H_{10}$ + CO_2	D	15,17	SE
+ n-decane, n-dodecane, benzene, cyclohexane	D	24	CR
$n-C_5H_{12}$ + CO_2	D	17	SE
$i-C_4H_{10}$ + CO_2	D	25	SE
$i-C_5H_{12}$ + CO_2	D	17	SE
C_2H_4 + CO_2	D	21,25	SE
+ n-octane, n-decane, n-dodecane	D	26	CR
+ benzene, cyclohexane, trans-decalin	D	27	CR
+ 2-methylnaphthalene	B	27	CR
Ternary:			
CH_4 + C_2H_6 + n-heptane		8	CR
+ n-octane		8,10	CR
+ benzene, cyclohexane		10	CR
+ CO_2		17	SE
CH_4 + C_3H_8 + n-heptane, n-octane		9	CR
+ CO_2		17	SE
CH_4 + $n-C_4H_{10}$ + n-heptane, n-octane		9	CR
+ CO_2		17	SE
CH_4 + N_2 + n-heptane, octane		12	CR
C_2H_6 + N_2 + n-octane, n-decane, n-dodecane		11	CR
+ n-hexane + n-dodecane		2	CR
C_2H_6 + C_2H_4 + CO_2		21	SE
C_2H_6 + C_3H_8 + CO_2		17	SE
Quaternary and Higher:			
CH_4 + C_2H_6 + C_3H_8 + n-octane		13	CR
CH_4 + C_2H_6 + n-hexane + n-octane		2	CR
CH_4 + C_2H_6 + C_3H_8 + $n-C_4H_{10}$ + n-octane		13	CR
CH_4 + C_2H_6 + n-hexane + n-heptane + n-octane		2	CR
CH_4 + C_2H_6 + C_3H_8 + $n-C_4H_{10}$ + n-hexane + n-octane		13	CR
CH_4 + C_2H_6 + C_3H_8 + $n-C_4H_{10}$ + N_2 + n-octane		13	CR
CH_4 + C_2H_6 + C_3H_8 + $n-C_4H_{10}$ + N_2 + n-hexane + n-octane		13	CR

CR = cryoscopic technique
SE = sampling of liquid phase of SLV system
SP = sampling of liquid phase of SL system (i.e., an overpressured SLV system)

In constructing a model or correlation for solid solubility in S-L-V systems, one needs good binary data as a foundation. The main interest in the ternary data is in the solubility enhancement effects caused by the presence of second solvents. For example, the presence of small amounts of ethane and heavier solvent species in liquid methane induces dramatic increases in solute solubility not readily predictable through simple additivity rules. On the other hand, nitrogen as a second solvent can cause similarly dramatic reductions in solute solubility in methane and heavier solvents. The amount of solubility enhancement or reduction increases with the molecular difference between the two solvents, and is more pronounced with hydrocarbon solutes than with carbon dioxide. Some success has been realized in developing a useful multicomponent S-L-V description, using the binary and ternary data of Table I as "building blocks".[14]

REFERENCES

1. P.H. Van Konynenburg and R.L. Scott, Critical lines and phase equilibria in binary van der Waals mixtures, Phil. Trans. Roy. Soc. London, Series A, 298:1442, p. 495 (1980).
2. J.P. Kohn and K.D. Luks, Solubility of Hydrocarbons in Cryogenic LNG and NGL Mixtures, GPA Research Report RR-22 (1976).
3. G.P. Kuebler and C. McKinley, Solubility of Solid Benzene, Toluene, n-Hexane, and n-Heptane in Liquid Methane, in "Advances in Cryogenic Engineering, Vo. 19", Plenum Press, New York (1974), p. 320.
4. G.P. Kuebler and C. McKinley, Solubility of Solid n-Pentane in Liquid Methane, in "Advances in Cryogenic Engineering, Vol. 21", Plenum Press, New York (1976), p. 509.
5. G.P. Kuebler and C. McKinley, Solubility of solid methylene chloride and 1, 1, 1-trichloroethane in fluid oxygen, J. Chem. Eng. Data 23:240 (1978).
6. J.P. Kohn, K.D. Luks and P.H. Liu, Three-phase solid-liquid-vapor equilibria of binary n-alkane systems (ethane-n-octane, ethane-n-decane, ethane-n-dodecane), J. Chem. Eng. Data 21:360 (1976).
7. J.P. Kohn, et. al., Three-phase solid-liquid-vapor equilibria of the binary hydrocarbon systems methane-n-octane and methane-cyclohexane, J. Chem. Eng. Data 22:419 (1977).
8. D.L. Tiffin, K.D. Luks and J.P. Kohn, Solubility Enhancement of Solid Hydrocarbon in Liquid Methane Due to the Presence of Ethane, in "Advances in Cryogenic Engineering, Vol. 23", Plenum Press, New York (1978), p. 538.
9. C.E. Orozco, et. al., Solids fouling in LNG systems, Hydrocarbon Processing 56 (11):325 (1977).
10. D.L. Tiffin, J.P. Kohn and K.D. Luks, Solid hydrocarbon solubility in liquid methane-ethane mixtures along the three-phase solid-liquid-vapor loci, J. Chem. Eng. Data 24:306 (1979).
11. J.D. Hottovy, J.P. Kohn and K.D. Luks, Effect of Nitrogen Presence on Solid Solubility of Normal Paraffins in Liquid Ethane, in "Advances in Cryogenic Engineering, Vol. 25", Plenum Press, New York (1980), p. 609.
12. W. Chen, et. al., The effect of the presence of nitrogen of solid solubility of normal paraffins in liquid methane, J. Chem. Eng. Data 26:166 (1981).

13. K.A. Green, et. al., Solubility of hydrocarbons in LNG, NGL, Hydrocarbon Processing 58(5):251 (1979).

14. K.D. Luks and J.P. Kohn, Avoid freeze-up in LNG/LPG work, Hydrocarbon Processing, p. 135 (April, 1981).

15. H. Cheung and E.H. Zander, Solubility of carbon dioxide and hydrogen sulfide in liquid hydrocarbons at cryogenic temperatures, Chem. Eng. Prog. Symp. Ser. 64(88):34 (1968).

16. J.A. Davis, N. Rodewald and F. Kurata, Solid-liquid-vapor phase behavior of the methane-carbon dioxide system, AIChE J. 8:537 (1962)

17. F. Kurata, Solubility of Solid Carbon Dioxide in Pure Light Hydrocarbons and Mixtures of Light Hydrocarbons, GPA Research Report RR-10 (1974).

18. K.D. Luks, J.D. Hottovy and J.P. Kohn, Three-phase solid-liquid-vapor equilibria in the binary hydrocarbon systems methane-n-hexane and methane-benzene, J. Chem. Eng. Data 26:402 (1981).

19. G.A.M. Diepen and Van Hest, "The Physics and Chemistry of High Pressure", Symposium Society of Chemical Industry, London (1963), p. 10.

20. U.K. Im and F. Kurata, Phase equilibria of carbon dioxide and light paraffins in the presence of solid carbon dioxide, J. Chem. Eng. Data 16:295 (1971).

21. A.M. Clark and F. Din, Equilibria between solid, liquid and gaseous phases at low temperatures, the system carbon dioxide and ethane and ethylene, Faraday Soc. 49:202 (1953).

22. P.H. Liu, K.D. Luks and J.P. Kohn, Three-phase solid-liquid-vapor equilibria of the systems ethane-benzene, ethane-cyclohexane, and ethane-trans-decalin, J. Chem. Eng. Data 22:220 (1977).

23. D.L. Tiffin, J.P. Kohn and K.D. Luks, Three-phase solid-liquid-vapor equilibria of the binary hydrocarbon systems ethane-2-methylnaphthalene, ethane-naphthalene, propane-n-decane, and propane-n-dodecane, J. Chem. Eng. Data 24:98 (1979).

24. W. Chen, K.D. Luks and J.P. Kohn, Three-phase solid-liquid-vapor equilibria of the binary hydrocarbon systems propane-benzene, propane-cyclohexane, n-butane-benzene, n-butane-cyclohexane, n-butane-n-decane, and n-butane-n-dodecane, J. Chem. Eng. Data 26:26 (1981).

25. G. Voss and H. Knapp, Loslichkeit fester Stoffe in Flussigkeiten bei tiefen Temperaturen, Chemie-Ing.-Techn. 47(18):769 (1975).

26. D.L. Tiffin, J.P. Kohn and K.D. Luks, Three-phase solid-liquid-vapor equilibria of binary ethylene-n-alkane systems (ethylene-n-octane, ethylene-n-decane, ethylene-n-dodecane). J. Chem. Eng. Data 23:207 (1978).

27. D.L. Tiffin, J.P. Kohn and K.D. Luks, Three-phase solid-liquid-vapor equilibria of the systems ethylene-cyclohexane, ethylene-trans-decalin, ethylene-benzene, and ethylene-2-methylnaphthalene, J. Chem. Eng. Data 24:96 (1979).

DISCUSSION

Question by R. G. Scurlock, University of Southampton, England: Have you any indication from your experimental work that the initial solid precipitation is a mono-molecular surface film?

Answer by author (JPK): We can optically see a crystal in the 0.1 mg range and occasionally see films of crystals on the cell walls, but they are obviously not mono-molecular.

NEW APPROACH FOR ANALYSIS AND PREDICTION OF LIQUID-VAPOR COEXISTENCE DENSITIES INCLUDING THE CRITICAL REGION

L. J. Van Poolen* and W. M. Haynes

National Bureau of Standards
Boulder, Colorado

INTRODUCTION

Many practical engineering systems require accurate coexistence density data. Systems such as liquefiers, refrigerators, and equipment incorporating power cycles depend on saturation properties for optimization, and there is a need for accurate liquid densities in storage and custody transfer operations. Also, accurate and internally consistent liquid-vapor density data, including the critical region, are very important in the development of equations of state, especially for those that originate on the coexistence boundary.

The liquid volume fraction, i.e., the fraction of the total volume in a two-phase system that is liquid, is a new parameter that can be used to analyze and predict coexistence data. This paper describes this fraction and presents applications in analyzing internal consistency of coexistence density data for argon and propane. Another application of the liquid volume fraction is the prediction of orthobaric densities in regions in which accurate data are unavailable. The predictive capability of this approach is illustrated using ethylene data. The results from this method are compared with those using the rectilinear diameter.

An experiment to obtain, simultaneously, orthobaric liquid and vapor densities is proposed. This involves measurements of total density, temperature, and liquid volume fraction. The

---

*On leave from Calvin College, Grand Rapids, Michigan.

usefulness of these data is illustrated by analyzing the only known measurements of pure fluid liquid volume fractions, which are for ammonia.

LIQUID VOLUME FRACTION DESCRIPTION

An extensive analysis of both liquid volume and liquid mass fractions for a pure fluid at liquid-vapor coexistence was made by Van Poolen.[1] The present report focuses on the liquid volume fraction and its applications. From a simple mass balance in a two phase system, the liquid volume fraction (X_{LV}) is defined by the relation,

$$X_{LV} = (\rho_t - \rho_v)/(\rho_\ell - \rho_v) \tag{1}$$

where ρ_t is the total density, ρ_ℓ is the saturated liquid density, and ρ_v is the saturated vapor density.

Figure 1, based on methane data[2], illustrates the liquid volume fraction as a function of temperature and total density. If the ρ_t is less than the critical density (ρ_c); then, as the temperature approaches the critical point value, (T_c), X_{LV} approaches zero, i.e., the ρ_t is a saturated vapor density. Conversely, if the ρ_t is greater than ρ_c, X_{LV} will eventually go to one, i.e., the ρ_t is a saturated liquid density. From Fig. 1., it appears that the liquid volume fraction for which $\rho_t = \rho_c$, i.e., the critical liquid volume fraction ($X_{LV,c}$), approaches a value of one-half as T goes to T_c. The limit is shown to be exactly one-half by the analysis of X_{LV} that follows.

The critical density is substituted for ρ_t in Eq. 1 and the following "dome" equations taken from Green, et al.:[3]

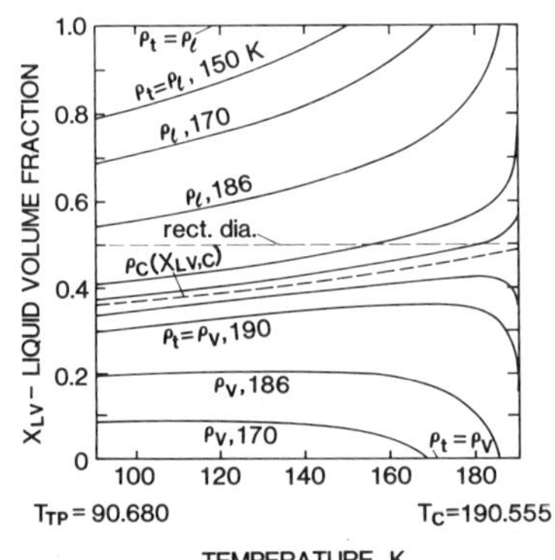

Fig. 1. X_{LV} vs T for Methane.

$$\rho_\ell - \rho_c = B_{1\ell}\,\varepsilon^{\beta_\ell} + B_{2\ell}\,\varepsilon^{\phi_\ell} + B_{3\ell}\,\varepsilon^{\psi_\ell} \tag{2}$$

$$\rho_v - \rho_c = B_{1v}\,\varepsilon^{\beta_v} + B_{2v}\,\varepsilon^{\phi_v} + B_{3v}\,\varepsilon^{\psi_v} \tag{3}$$

where: B's are constants

$\beta_\ell = \beta_v$ (equality shown analytically by Van Poolen[1]) $\cong$ .35;

and Φ_ℓ, $\Phi_v \cong 1.00$; ψ_ℓ, $\psi_v > 1.00$; $\varepsilon = T_c - T$,

are substituted into Eq. 1. The limit as $\varepsilon \to o$ is taken:

$$\lim_{T \to T_c} X_{LV,c} = (-B_{1v})/(B_{1\ell} - B_{1v}) \tag{4}$$

For first order symmetry (in the limit), $B_{1v} = -B_{1\ell}$ and then:

$$\lim_{T \to T_c} X_{LV,c} = 1/2 \ . \tag{5}$$

Note that the $X_{LV,c}$ and rectilinear diameter are different functions. The rectilinear diameter plots as a horizontal line at a constant 0.5 for all T on Fig. 1. The $X_{LV,c}$ is that curved line for which $\rho_t \ \rho_c$.

APPLICATION OF $X_{LV,c}$ TO COEXISTENCE DATA ANALYSIS

This behavior of the critical liquid volume fraction can be used to determine whether a proposed critical density is the properly correlated value. Smoothed argon data,[4] was analyzed with the results shown in Table I. The second column is the calculated $X_{LV,c}$ utilizing Eq. 1 with the published ρ_c. The values of $X_{LV,c}$ increase, go through a maximum and then decrease away from the true limit of one-half. The published $\rho_c = 13.4123$ mol/L, appears to be a saturated vapor density, i.e., this ρ_c is too low.

To obtain a properly correlated ρ_c a function:

$$X_{LV,c} = (\rho_c - \rho_v)/(\rho_\ell - \rho_v) = 0.5\,(1 + A\varepsilon^R + B\varepsilon^P + \ldots) \tag{6}$$

where A, B, R, P are unknown constants, derived from combining Eq. 1, 2, and 3, is fit to saturated density and temperature data. The ρ_c is also a parameter determined by the fit. The resulting equation for argon is shown in Table I. Only one term of the series was needed. The third column of Table I contains the values of $X_{LV,c}$ calculated using the derivied ρ_c of 13.6138 mol/L

Table I. $X_{LV,c}$ vs. T for Argon where $\rho_{c,pub} = 13.4123$ mol/L and $\rho_{c,cal} = 13.6138$ mol/L.

T (K)	$\dfrac{\rho_{c,pub} - \rho_{v,data}}{(\rho_\ell - \rho_v)_{data}}$	$\dfrac{\rho_{c,cal} - \rho_{v,data}}{(\rho_\ell - \rho_v)_{data}}$	$.5[1 - .1038 \times 10^{-2} (T_c - T)^{1.71}]$
145	.4762	.4895	.4893
146	.4777	.4920	.4923
147	.4790	.4946	.4948
148	.4795	.4971	.4969
149	.4781	.4992	.4985
150	.4693	.4988	.4996
150.86	?	–	.5000

(it is higher than the published value as predicted). These values should increase monotonically to one-half. The value at 149 K (0.4992) indicates an inconsistency in the ρ_v and ρ_ℓ as published. The fourth column shows the $X_{LV,c}$ calculated using the fitted equation. These values do increase monatonically to one-half. This equation could be used to adjust either the ρ_v or ρ_ℓ at 149 K to achieve a monatonic progression in the calculation of values from Eq. 1. (An extensive analysis, similar to that above, of handbook data was done by Van Poolen.[5])

The liquid volume fraction was used to obtain a critical density for propane that was consistent with available orthobaric density data.[6] The experimental saturated liquid and vapor densities of Thomas and Harrison,[6] covering a temperature range of 0.874 to 0.998 T_c, were fitted to Eq. 6. The critical density (5.00 mol/L) obtained was approximately one percent higher than that derived by Thomas and Harrison. Furthermore, as a check, all available orthobaric density data for propane from many independent sources were fitted to expressions for saturated liquid and vapor densities as a function of temperature (the critical density for these data sets varied over a range of almost two percent). Then these smoothed densities covering temperature ranges between 0.55 to 0.99 T_c were fitted to Eq. 6. The critical densities from Eq. 6 for these different sets of smoothed data exhibited differences of less than a few tenths percent when compared with the value of 5.00 mol/L obtained from fitting the experimental densities of Thomas and Harrison. This was true regardless of the temperature range of the fitted data.

Thus, based on the analysis of experimental and smoothed data for propane using the liquid volume fraction, a critical density of 5.00 mol/L was found to be consistent with the orthobaric liquid and vapor densities and was used in the correlation of

Goodwin and Haynes.[7] This value agreed with that selected in the correlation of Das and Eubank.[8]

Another application of the function $X_{LV,c}$ is the prediction of saturation densities where available data are minimal. To illustrate the technique, saturated liquid and vapor data taken from a preliminary set of ethylene data[9] were fit to $X_{LV,c}$ over a small temperature range (240–270 K). With one side of the saturated density curve assumed known (ρ_ℓ), this side was combined with the fitted $X_{LV,c}$ to predict the other side (ρ_v). Table II shows the results obtained.[10] When compared to the values calculated using the method of rectilinear diameters (see Table II), the results are significantly different. Using the liquid volume fraction, the fit is better, the range in temperature for reasonable prediction is larger, and the results near the critical temperature are more accurate. Thus the $X_{LV,c}$ function exhibits promise for providing consistent sets of coexistence densities where there is reliable data available only on one side of the coexistence curve and minimal data available on the other side.

The $X_{LV,c}$ fitting method is not an extrapolation technique as is that using the rectilinear diameter; rather, the $X_{LV,c}$ function is tied to a specific value (one-half) at the critical point.

Table II. Estimate of ρ_v Based pm 240–270 L Data pf Ethylene[9].

T	ρ_ℓ,data	ρ_v,data	$\rho_v$cal,rd	% Dev	$\rho_v$cal,$X_{LV,c}$	% Dev
K	mol/L	mol/L	mol/L		mol/L	
104	23.3419	.000141	− .449145	−	.054515	−
114	22.8813	.000638	− .415390	−	.030505	−
124	22.4184	.002179	− .379326	−	.013718	−
134	21.9520	.006031	− .339857	−	.005002	17.06
144	21.4804	.014185	− .295084	−	.005734	59.57
154	21.0010	.029377	− .242580	−	.017747	39.59
164	20.5110	.055036	− .179459	−	.043349	21.24
174	20.0070	.095222	− .102336	−	.085416	10.30
184	19.4850	.154620	− .007215	−	.147546	4.57
194	18.9402	.238662	.110729	53.60	.234300	1.83
204	18.3666	.353807	.257469	27.23	.351573	0.63
214	17.7566	.508084	.440636	13.27	.507166	0.18
224	17.1000	.712060	.670375	5.85	.711741	0.04
234	16.3826	.980633	.960880	2.01	.980514	0.01
*244	15.5830	1.336694	1.333582	0.23	1.336681	0.00
*254	14.6649	1.819456	1.824831	− 0.30	1.819510	− 0.00
*264	13.5534	2.508910	2.509473	− 0.02	2.508776	− 0.01
274	12.0248	3.635109	3.611187	0.66	3.637649	− 0.07
278	11.0974	4.406105	4.367912	0.87	4.416663	− 0.24
282	9.2423	6.097226	6.052212	0.74	6.139947	− 0.70

*$X_{LV,c}$ and rectilinear diameter (rd) fit in this range.

EXPERIMENTAL APPROACH TO SIMULTANEOUS DETERMINATION OF COEXISTENCE DATA

An experimental method for obtaining saturated liquid and vapor density data simultaneously is presented. Current experimental approaches obtain ρ_ℓ and ρ_v in separate experiments. Usually the ρ_ℓ is obtained directly while the ρ_v is found from the intersection of isochores with the vapor pressure curve.

The method proposed is based on the plot, derived from methane data[2], shown in Fig. 2. At a given temperature, the relationship between X_{LV} and ρ_t is linear since

$$\left. \frac{\partial X_{LV}}{\partial \rho_t} \right|_T = \frac{1}{\rho_\ell - \rho_v} \tag{7}$$

is a function only of temperature for a pure fluid.

Two pairs of liquid volume fraction-total density data at the same temperature establish a line. Simple extrapolation to $X_{LV} = 1$ for ρ_ℓ and to $X_{LV} = 0$ for ρ_v produces the saturation density values from the same set of data enhancing internal consistency.

To test the validity of using data such as illustrated in Fig. 2, X_{LV} data are needed. Unfortunately, the thermodynamic property literature contains very little of this information. However, Cragoe, et al.[11,12] in obtaining ρ_ℓ and ρ_v for ammonia, published data from which X_{LV} values could be derived. However,

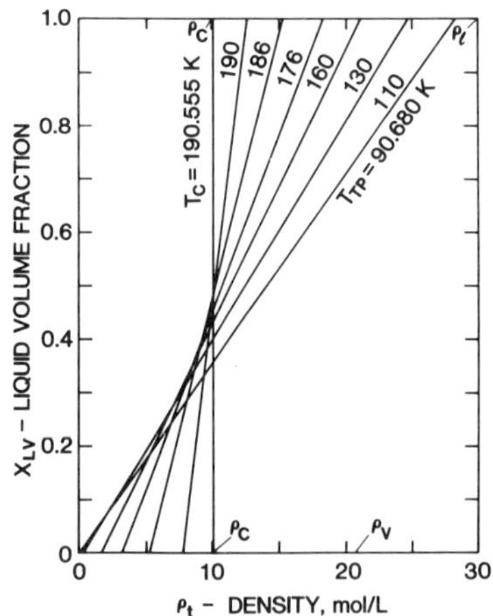

Fig. 2. X_{LV} vs ρ_t for methane.

to test the ideas presented here, saturated liquid and vapor density values are needed at the same temperature. There is only one such temperature in the data of Cragoe, et al.[11,12] Figure 3 shows X_{LV}, ρ_t values derived by Van Poolen[1] from the data. It also shows the results of the straight line extrapolations to obtain ρ_v and ρ_ℓ. Table III shows the results from the straight line method with both the Cragoe, et al.[11,12] data and with more recent values published by Haar and Gallagher.[13]

Another approach which could be taken would be to utilize accurate ρ_ℓ data ($X_{LV} = 1$) along with ($X_{LV} < 1$, ρ_t) data to obtain a linear relationship. This line could then be extrapolated to $X_{LV} = 0$ to obtain a value for ρ_v. This was done utilizing the data presented by Cragoe, et al.[11] in their report on liquid densities. The results are presented in Table IV. Good agreement is shown between the straight line method values and those that are published. The straight line values are internally consistent since they match the thermodynamic behavior of Fig. 2.

CONCLUSIONS

The usefulness of the liquid volume fraction in analyzing coexistence data including the critical region has been shown. It provides a tool to enhance internal consistency between saturated vapor/liquid densities and between these data and the critical density. Further work should be done in the development of modern measurement techniques to obtain liquid volume fraction data.

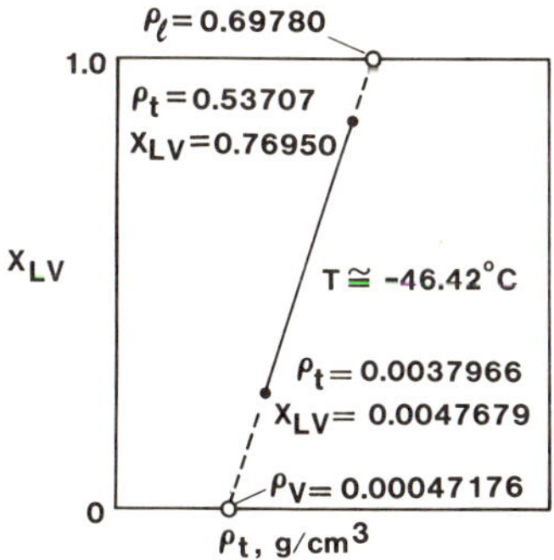

Fig. 3. X_{LV} vs ρ_t for Ammonia.

Table III. Comparison of Coexistence
Densities (g/cm^3) for Ammonia.

Van Poolen[1]	Cragoe, et al.[11,12]	Haar, et al.[13] Equation
Straight Line Method	Experimental	of State
$\rho_v = .0004717609$	$\rho_v = .0004726344$ (T = −46.42°C)	$\rho_v = .000462$ (T = −46.425°C)
$\rho_\ell = .69780$	$\rho_\ell = .69781$ (T = −46.43°C)	$\rho_\ell = .69770$ (T = −46.425°C)

Table IV. Comparison of ρ_v Values (g/cm^3) for Ammonia.

T (°C)	$\rho_{\ell,DATA}$	$\rho_{+,DATA}$	$X_{LV,DATA}$	$\rho_{v,STR} \times 10^4$ Line	$\rho_{v,CALC} \times 10^4$	$\rho_{v,PUB} \times 10^4$
Cragoe (Haar)	Based on Cragoe[11]			Van Poolen[1]	Cragoe[12]	Haar[13]
−49.96 (−50)	.702099	.576546	.821076	3.8557	3.8243	3.8100
−40.14 (−40)	.690255	.57690	.835630	6.2074	6.4057	6.440
−30.04 (−30)	.677782	.540417	.797018	10.454	10.367	10.380
−14.94 (−15)	.658592	.57627	.87464	18.977	19.706	19.670
−00.05 (0)	.638590	.553245	.865617	35.023	34.467	34.580
+10.00 (10)	.624637	.553035	.884468	48.793	48.594	48.680
+20.01 (20)	.610273	.55291	.904958	67.099	66.896	67.010
+29.98 (30)	.595295	.55275	.927413	91.772	90.258	90.050
+39.94 (40)	.579606	.53600	.923170	119.52	119.86	120.29
+49.94 (50)	.562905	.495826	.877389	158.24	157.36	157.80
+60.22 (60)	.544802	.536281	.98373	208.72	206.08	204.90
+80.58 (81)	.504322	.449203	.882783	340.94	345.88	347.44
+91.24 (91)	.479773	.44896	.929207	444.78	453.74	445.80

REFERENCES

1. L.J. Van Poolen, Analysis of Liquie Volume and Liquid Mass Fractions at Coexistence for Pure Fluids, Interagency Report NBSIR 80-1631 Nat. Bur. Stand. (U.S.) (May 1980).
2. "Methane, International Thermodynamic Tables of the Fluid State-5", IUPAC, Chemical Data Series, No. 16, Pergammon Press, Oxford (1976).
3. M.S. Green, M.J. Cooper, and J.M.H. Levelt Sengers, Extended thermodynamic scaling from a generalized parametric form, Pys. Rev. Lett. 26:492 (1971).
4. A.L. Gosman, R.D. McCarty and J.G. Hust, Thermodynamic Properties of Argon from the Triple Point to 300 K at pressures to 1000 Atmospheres, NSRDS-NBS Report No. 27 (1969).
5. L.J. Van Poolen, Analysis of property data in the critical region using the function-liquid volume fraction, ASHRAE Transactions 9:Part 11 (Dec 1977).
6. R.H.P. Thomas and R.H. Harrison, Pressure, volume, temperature relations of propane, to be published (1981).
7. R.D. Goodwin and W.M. Haynes, Thermophysical properties of propane from 85 to 700 K at pressures to 70 MPa, to be published.
8. T.R. Das and P.T. Eubank, Thermodynamic Properties of Propane, Part 1: Vapor-Liquid Coexistence Curve, in "Advances in Cryogenic Engineering, Vol. 18", Plenum Press, New York (1973), p. 208.
9. R.D. McCarty and R.T. Jacobsen, Preliminary Equations of State for Ethylene, Report to the Advisory Committee of the Joint Industry-Government Project on the Thermophysical Properties of Ethylene, Howard J. White, Jr., Program Manager, Gaithersburg, Maryland (June 6, 1979).
10. L.J. Van Poolen, Use of the Critical Volume Fraction in the Definition of coexistence Boundaries of a Pure Fluid, Research Proposal submitted to National Science Foundation, (1980).
11. C.S. Cragoe and D.R. Harper, 3rd, Specific volume of liquid ammonia, Scientific Papers of the Bureau of Standards 17:287 (1921).
12. C.S. Cragoe, E.C. McKeivey and G.F. O'Connor, Specific volume of saturated ammonia vapor, Scientific Papers of the Bureau of Standards 18:707 (March 17, 1923).
13. L. Haar and J.S. Gallagher, Thermodynamic properties of Ammonia, J. Phys. Chem. Ref. Data 7:635 (1978).

PHASE EQUILIBRIUM STUDIES FOR METHANE/SYNTHESIS GAS SEPARATION: THE HYDROGEN-CARBON MONOXIDE-METHANE SYSTEM*

J. H. Hong and R. Kobayashi

Rice University
Houston, Texas

INTRODUCTION

Vapor-liquid equilibrium (V-L-E) in the hydrogen-methane-carbon monoxide system is an important one, particularly in the synthetic fuels industry. As a primary example, in one SNG process, V-L-E in the hydrogen-carbon monoxide-methane system provides the basic data for the distillation separation of hydrogen/carbon monoxide recycle gas from the methane product.

V-L-E in the binary systems comprising the ternary system has been studied previously. Hydrogen-methane V-L-E has been most recently studied by Hong and Kobayashi[1], who also reviewed prior studies in the same binary system: Freeth and Verschoyle[2]; Steckel and Tsin[3]; Lickhter and Tikhonovich[4]; Levitskaya[5]; Benham and Katz[6]; Orientlicher and Prausnitz[7]; Kirk and Ziegler[8]; Yorizane et al[9]; Fastowsky and Gonikberg[10]; Sagara, et al[11]; and Tsang and Streett[12]. V-L-E in the hydrogen-carbon monoxide system has been studied previously by Verschoyle[13]; Ruhemann and Tsin[14]; Akers and Eubanks[15]; Yorizane et al[9]; Tsang and Street[12], and by the present authors. The methane-carbon monoxide system, while studied less extensively, has been investigated by Toyama et al.[16]; Cheung and Wang[17]; Sprow and Prausnitz[18]; and Christiansen et al[19].

---

*Supported in part by the Gas Research Institute and the Gas Processors Association.

In the present study V-L-E has been studied in the hydrogen-carbon monoxide-methane system from 108.13 K to 173.25 K at pressures to 10.34 MPa.

EXPERIMENTAL METHOD

The experimental method used in this study is essentially identical to that used in the V-L-E study of the methane-carbon dioxide system by Mraw et al[20]. A cryogenic vapor recycle V-L-E apparatus was used. The bath fluid used was a eutectic mixture of isopentane and isohexane that remained fluid to the lowest temperature of this study.

The chromatographic system used to analyze the binary compositions was modified to accommodate a dual-column and dual-detector system so that hydrogen and the other two components (CO and CH_4) could be detected in separate detectors. The original 10-port sampling valve was equipped with two sampling loops (with approximately equal volume), capable of accepting samples into each loop indiscriminately and simultaneously. The sample in each loop is swept by its carrier gas to its respective detector.

A special H_2 - He mixture was used as the carrier gas for the Co and Ch_4 analysis and argon was selected as the carrier gas for hydrogen analysis. The choice of argon was based on its high hydrogen response factor and the absence of the anomalous behavior experienced with pure helium as a carrier gas (Madison[21]).

Two 1.8 m long 3.2 mm (0.125") O.D. by 2.2 mm (0.085") I.D. stainless steel tubes packed with 120 mesh Molecular Sieve 5A were used to obtain baseline separation of H_2, CH_4 and CO. The gas-chromatographic conditions were optimized for our unit as follows: Oven temperature 40° C, the flow rate of H_2-He was 59.2 cm^3/min and of Argon was 20.6 cm^3/min, flowing through detector I and II respectively. The current through detector I was 0.240A at 67° C and that for detector II was 0.60 A at 100° C. The signal from the detectors was integrated by a digital integrator and peaks were recorded on a strip chart recorder. The calibration curves for carbon monoxide and methane were linear and could be represented by the following functions

$$A_{CO} = 448629 \; X_{CO} \tag{1}$$

$$A_{CH_4} = 391463 \; X_{CH_4} \tag{2}$$

where A is the integrated area of each peak and X is the corresponding mole fraction. On the other hand Fig. 1 shows the non-linearity of detector response for hydrogen in both binary and ternary mixtures.

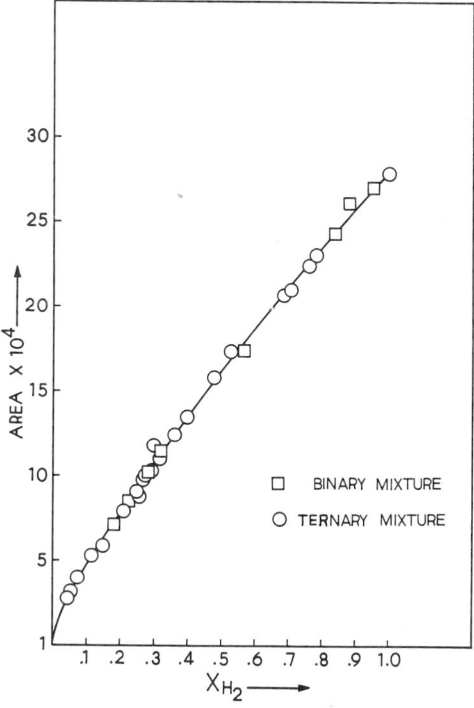

Fig. 1. Calibration curve of H_2.

MATERIALS

The pure components (hydrogen, carbon monoxide, and methane) used in the study were purchased from Matheson Gas Co., with quoted purities of 99.995, 99.99, and 99.999 mol percent, respectively. Argon with purity of 99.998 mol percent was purchased from Linde Specialty Gas. In addition, the following mixtures were purchased from the Matheson Co.: He–H_2 (91.50% – 8.5%), and H_2–CH_4 mixtures with 19.99, 39.76, 59.72 and 79.998 mole percent methane. The synthetic mixtures were prepared using a pumping method discussed by Mraw, et al.[22]

DISCUSSION

The V–L–E results from the six isotherms studied are presented in Fig. 2-7 and tabulated in Table I. Several isobars are shown in each figure. The enlargement of the two phase region with decreasing temperature is typical of mixtures containing quantum gases such as hydrogen and helium, which exhibit decreasing solubility with decreasing temperature.

Table I. Vapor-Liquid Equilibrium for the Hydrogen-Methane-Carbon Monoxide System.

P(MPa)	Liquid Mole Fraction			Vapor Mole Fraction			K-Values		
	H_2	CH_4	CO	H_2	CH_4	CO	H_2	CH_4	CO
Temperature = 173.25 K = -99.91°C									
4.137	0.0327	0.9673	-	0.2320	0.7680	-	7.095	0.7940	-
4.137	0.0286	0.9414	0.0300	0.1818	0.7588	0.0594	6.360	0.8060	1.977
4.137	0.0184	0.9030	0.0786	0.1160	0.7408	0.1432	6.314	0.8203	1.821
4.137	0.0059	0.8438	0.1503	0.0320	0.7152	0.2528	5.424	0.8476	1.682
4.137	0.0020	0.8160	0.1820	0.0099	0.7053	0.2848	4.950	0.8643	1.565
4.137	-	0.8089	0.1911	-	0.6908	0.3092	-	0.8540	1.618
5.516	0.0660	0.9340	-	0.3390	0.6610	-	5.136	0.7077	-
5.516	0.0578	0.9100	0.0322	0.2871	0.6612	0.0517	4.967	0.7266	1.606
5.516	0.0392	0.8343	0.1265	0.1750	0.6482	0.1768	4.464	0.7769	1.398
5.516	0.0165	0.7492	0.2343	0.0633	0.6499	0.2868	3.836	0.8675	1.224
5.516	critical point			0.0300	0.6800	0.2900			
6.895	0.0940	0.9060	-	0.3970	0.6030	-	4.223	0.6656	-
6.895	0.0935	0.8761	0.0304	0.3485	0.6088	0.0427	3.727	0.6949	1.405
6.895	0.0797	0.8022	0.1180	0.2510	0.6048	0.1441	3.149	0.7539	1.221
6.895	0.0756	0.7667	0.1577	0.1825	0.6314	0.1861	2.414	0.8235	1.180
6.895	0.0877	0.7373	0.1750	0.1540	0.6422	0.2038	1.756	0.8710	1.165
6.895	0.0916	0.7171	0.1913	0.1340	0.6574	0.2086	1.463	0.9167	1.090
10.342	0.1941	0.8059	-	0.4163	0.5837	-	2.145	0.7243	-
10.342	0.1976	0.7856	0.0168	0.3828	0.5984	0.0188	1.937	0.7617	1.119
10.342	critical point			0.2708	0.6831	0.0461			
Temperature = 163.17 K = -109.99°C									
2.758	0.0160	0.9840	-	0.2550	0.7450	-	15.94	0.7571	-
2.758	0.0131	0.9690	0.0179	0.2050	0.7450	0.0500	15.65	0.7688	2.793
2.758	0.0096	0.9204	0.0699	0.1063	0.7221	0.1716	11.07	0.7845	2.455
2.758	0.0089	0.9157	0.0754	0.0933	0.7221	0.1846	10.48	0.7885	2.448
2.758	0.0070	0.8943	0.0987	0.0700	0.6977	0.2323	10.00	0.7802	2.35
2.758	-	0.8570	0.1430	-	0.6717	0.3283	-	0.7838	2.296
4.137	0.0398	0.9602	-	0.4260	0.5740	-	10.70	0.5978	-
4.137	0.0350	0.9416	0.0234	0.3760	0.5792	0.0448	10.74	0.6151	1.915
4.137	0.0350	0.9416	0.0234	0.3846	0.5732	0.0422	10.99	0.6088	1.803
4.137	0.0336	0.9228	0.0436	0.3517	0.5671	0.0812	10.47	0.6145	1.861
4.137	0.0330	0.8774	0.0896	0.2871	0.5551	0.1578	8.700	0.6327	1.763
4.137	0.0322	0.8434	0.1244	0.2335	0.5491	0.2174	7.252	0.6510	1.747
4.137	0.0242	0.7971	0.1787	0.1429	0.5412	0.3159	5.905	0.6789	1.769
4.137	0.0198	0.7548	0.2254	0.1042	0.5277	0.3681	5.260	0.6991	1.633
4.137	-	0.6359	0.3641	-	0.4850	0.5150	-	0.7627	1.414
5.516	0.0650	0.9350	-	0.5080	0.4920	-	7.815	0.5262	-
5.516	0.0614	0.8874	0.0512	0.4340	0.4878	0.0782	7.068	0.5496	1.527
5.516	0.0572	0.8360	0.1068	0.3603	0.4821	0.1576	6.299	0.5767	1.476
5.516	0.0549	0.8044	0.1407	0.3125	0.4808	0.2067	5.700	0.5977	1.469
5.516	0.0458	0.7512	0.2030	0.2275	0.4864	0.2861	4.967	0.6475	1.409
5.516	0.0449	0.7067	0.2484	0.1833	0.4787	0.3380	4.082	0.6773	1.361
5.516	0.0495	0.6729	0.2775	0.1672	0.4684	0.3644	3.378	0.6961	1.313
5.516	0.0408	0.6452	0.3140	0.1403	0.4634	0.3963	3.439	0.7182	1.262
5.516	0.0405	0.6211	0.3384	0.1206	0.4617	0.4177	2.978	0.7434	1.234
5.516	0.0353	0.5931	0.3716	0.0914	0.4687	0.4399	2.589	0.7903	1.184
5.516	0.0340	0.5433	0.4227	0.0593	0.4866	0.4541	1.744	0.8956	1.074
6.895	0.0940	0.9060	-	0.5560	0.4440	-	5.915	0.4900	-
6.895	0.0919	0.8536	0.0545	0.4814	0.4464	0.0722	5.237	0.5229	1.325
6.895	0.0874	0.7214	0.1912	0.3199	0.4459	0.2342	3.660	0.6181	1.225
6.895	0.0854	0.6598	0.2548	0.2465	0.4528	0.3007	2.886	0.6862	1.180
6.895	0.0852	0.6520	0.2628	0.2398	0.4537	0.3065	2.815	0.6958	1.166
6.895	0.0904	0.6128	0.2968	0.1999	0.4633	0.3368	2.211	0.7560	1.135
6.895	0.0979	0.5769	0.3252	0.1651	0.4788	0.3561	1.686	0.8299	1.095
6.895	0.1072	0.5556	0.3372	0.1551	0.4878	0.3571	1.447	0.8780	1.059
6.895	0.1163	0.5388	0.3449	0.1421	0.5015	0.3564	1.222	0.9308	1.033
10.342	0.1650	0.8350	-	0.6000	0.4000	-	3.636	0.4790	-
10.342	0.1717	0.7594	0.0689	0.5009	0.4223	0.0768	2.917	0.5561	1.115
10.342	0.1795	0.7053	0.1152	0.4402	0.4381	0.1217	2.452	0.6211	1.056
10.342	0.1959	0.6671	0.1370	0.4078	0.4479	0.1443	2.082	0.6714	1.053
10.342	0.2146	0.6188	0.1666	0.3485	0.4784	0.1731	1.624	0.7731	1.039
10.342	critical point			0.2760	0.5410	0.1830			

Table I. (Continued)

P(MPa)	Liquid Mole Fraction			Vapor Mole Fraction			K-Values		
	H_2	CH_4	CO	H_2	CH_4	CO	H_2	CH_4	CO
Temperature = 143.08 K = -130.08°C									
2.758	0.0274	0.9726	–	0.6499	0.3501	–	23.70	0.3600	–
2.758	0.0160	0.7928	0.1912	0.3088	0.3191	0.3721	19.30	0.4025	1.946
2.758	0.0083	0.6700	0.3217	0.1609	0.2896	0.5495	19.39	0.4322	1.708
2.758	0.0078	0.6119	0.3803	0.1174	0.2757	0.6069	15.05	0.4506	1.596
2.758	–	0.4943	0.5057	–	0.2501	0.7499	–	0.5060	1.483
4.137	0.0459	0.9541	–	0.7347	0.2653	–	16.00	0.2781	–
4.137	0.0373	0.8737	0.0890	0.6118	0.2800	0.1082	16.40	0.3200	1.216
4.137	0.0380	0.8045	0.1575	0.5293	0.2700	0.2007	13.93	0.3356	1.274
4.137	0.0380	0.6800	0.2820	0.3841	0.2590	0.3569	10.11	0.3809	1.266
4.137	0.0329	0.6546	0.3125	0.3539	0.2545	0.3916	10.76	0.3887	1.253
4.137	0.0272	0.4302	0.5426	0.1711	0.2010	0.6279	6.290	0.4672	1.157
4.137	0.0196	0.3580	0.6224	0.1121	0.1879	0.7000	5.719	0.5249	1.125
4.137	0.0214	0.3360	0.6426	0.0991	0.1815	0.7194	4.631	0.5403	1.120
4.137	0.0219	0.3325	0.6456	0.0959	0.1861	0.7180	4.379	0.5597	1.112
4.137	0.0115	0.2342	0.7543	0.0323	0.1617	0.8060	2.809	0.6901	1.069
4.137	0.0394	0.7611	0.1995	0.5091	0.2681	0.2228	12.92	0.3523	1.117
5.516	0.0655	0.9345	–	0.7690	0.2310	–	11.74	0.2472	–
5.516	0.0651	0.7448	0.1901	0.5712	0.2205	0.2083	8.774	0.2961	1.096
5.516	0.0692	0.7292	0.2016	0.5610	0.2197	0.2193	8.107	0.3013	1.088
5.516	0.0630	0.7169	0.2201	0.5436	0.2185	0.2379	8.629	0.3048	1.081
5.516	0.0704	0.5034	0.4262	0.3444	0.2011	0.4545	4.892	0.3995	1.066
5.516	0.0673	0.3703	0.5624	0.2405	0.1891	0.5704	3.574	0.5107	1.014
5.516	0.0787	0.2421	0.6792	0.1366	0.1867	0.6767	1.736	0.7711	0.996
5.516	0.0941	0.2225	0.6834	0.1161	0.2001	0.6838	1.234	0.8993	1.001
6.895	0.0842	0.9158	–	0.7909	0.2091	–	9.393	0.2283	–
6.895	0.0837	0.8239	0.0924	0.6972	0.2107	0.0921	8.330	0.2557	0.9968
6.895	0.0856	0.8085	0.1059	0.6817	0.2168	0.1015	7.964	0.2681	0.9584
6.895	0.0880	0.6920	0.2200	0.5863	0.2037	0.2100	6.663	0.2944	0.9545
6.895	0.0833	0.6569	0.2598	0.5453	0.2048	0.2499	6.546	0.3118	0.9619
6.895	0.0980	0.5442	0.3578	0.4495	0.2054	0.3451	4.587	0.3774	0.9645
6.895	0.1147	0.3498	0.5355	0.2747	0.2006	0.5247	2.395	0.5735	0.9798
6.895	0.1251	0.3227	0.5522	0.2646	0.2019	0.5335	2.115	0.6257	0.9661
6.895	0.1561	0.2801	0.5638	0.2344	0.2091	0.5565	1.502	0.7465	0.9869
10.342	0.1283	0.8717	–	0.8083	0.1917	–	6.300	0.2199	–
10.342	0.1393	0.7787	0.0820	0.7364	0.1959	0.0677	5.286	0.2516	0.8256
10.342	0.1564	0.6574	0.1862	0.6394	0.2054	0.1552	4.088	0.3124	0.8335
10.342	0.1628	0.5791	0.2581	0.5701	0.2128	0.2171	3.502	0.3675	0.8411
10.342	0.2056	0.4241	0.3703	0.4336	0.2335	0.3329	2.109	0.5506	0.8990
10.342	0.2311	0.3826	0.3863	0.3991	0.2409	0.3600	1.727	0.6296	0.9319
Temperature = 123.11 K = -150.05°C									
2.758	0.0261	0.9739	–	0.8763	0.1237	–	33.60	0.1270	–
2.758	0.0233	0.9156	0.0611	0.8100	0.1210	0.0690	34.76	0.1322	1.130
2.758	0.0256	0.7739	0.2005	0.6693	0.1218	0.2089	26.14	0.1574	1.042
2.758	0.0230	0.6098	0.3672	0.5458	0.1030	0.3512	23.73	0.1689	0.9564
2.758	0.0244	0.5120	0.4636	0.4828	0.0905	0.4267	19.78	0.1766	0.9204
2.758	0.0248	0.3972	0.5780	0.3938	0.0767	0.5295	15.88	0.1931	0.9161
2.758	0.0253	0.2390	0.7357	0.2914	0.0574	0.6512	11.52	0.2402	0.8851
2.758	0.0208	0.1928	0.7864	0.2341	0.0589	0.7070	11.26	0.3055	0.8990
2.758	0.0225	–	0.9775	0.1325	–	0.8675	5.889	–	0.8875
4.137	0.0412	0.9588	–	0.9095	0.0905	–	22.08	0.0944	–
4.137	0.0397	0.9034	0.0569	0.8596	0.0936	0.0468	21.65	0.1036	0.8221
4.137	0.0423	0.8091	0.1486	0.7925	0.0895	0.1180	18.74	0.1106	0.7941
4.137	0.0495	0.6434	0.3071	0.6840	0.0825	0.2334	13.82	0.1282	0.7601
4.137	0.0496	0.6027	0.3477	0.6621	0.0812	0.2567	13.35	0.1347	0.7383
4.137	0.0503	0.5107	0.4390	0.6022	0.0734	0.3244	11.97	0.1437	0.7390
4.137	0.0538	0.3896	0.5566	0.5321	0.0722	0.3956	9.890	0.1853	0.7107
4.137	0.0539	0.3886	0.5575	0.5303	0.0685	0.4012	9.839	0.1760	0.7196
4.137	0.0667	0.1227	0.8106	0.3714	0.0334	0.5952	5.568	0.2722	0.7343
4.137	0.0750	–	0.9250	0.2720	–	0.7280	3.627	–	0.7870

Table I. (Continued)

P(MPa)	Liquid Mole Fraction			Vapor Mole Fraction			K-Values		
	H_2	CH_4	CO	H_2	CH_4	CO	H_2	CH_4	CO
5.516	0.0554	0.9446	–	0.9152	0.0848	–	16.52	0.0898	–
5.516	0.0583	0.8455	0.0962	0.8509	0.0823	0.0668	14.59	0.0974	0.6946
5.516	0.0686	0.6928	0.2386	0.7642	0.0782	0.1576	11.14	0.1129	0.6605
5.516	0.0790	0.4907	0.4303	0.6581	0.0692	0.2727	8.330	0.1410	0.6337
5.516	0.0889	0.3406	0.5705	0.5816	0.0610	0.3574	6.542	0.1791	0.6265
5.516	0.1039	0.1837	0.7124	0.4922	0.0444	0.4634	4.737	0.2417	0.6505
5.516	0.1064	0.1438	0.7498	0.4569	0.0385	0.5046	4.294	0.2678	0.6729
5.516	0.1425	–	0.8575	0.3250	–	0.6750	2.281	–	0.7872
6.895	0.0673	0.9327	–	0.9217	0.0783	–	13.70	0.0839	–
6.895	0.0664	0.8896	0.0440	0.8917	0.0787	0.0296	13.43	0.0885	0.6727
6.895	0.0681	0.8459	0.0861	0.8650	0.0785	0.0565	12.71	0.0928	0.6566
6.895	0.0784	0.7757	0.1459	0.8290	0.0763	0.0947	10.57	0.0984	0.6491
6.895	0.0771	0.7151	0.2078	0.7967	0.0790	0.1243	10.33	0.1105	0.5982
6.895	0.0873	0.6243	0.2884	0.7593	0.0729	0.1678	8.698	0.1168	0.5818
6.895	0.0990	0.5223	0.3787	0.7074	0.0700	0.2226	7.145	0.1340	0.5878
6.895	0.1217	0.3280	0.5503	0.6095	0.0602	0.3303	5.008	0.1835	0.6002
6.895	0.1470	0.1897	0.6633	0.5134	0.0539	0.4327	3.493	0.2841	0.6523
6.895	0.1734	0.1026	0.7240	0.4523	0.0398	0.5079	2.608	0.3880	0.7015
6.895	0.1708	0.0906	0.7386	0.4390	0.0366	0.5244	2.570	0.4040	0.7100
6.895	0.1879	0.0553	0.7568	0.3946	0.0266	0.5788	2.100	0.4810	0.7648
6.895	0.1840	0.0582	0.7578	0.3985	0.0296	0.5719	2.166	0.5086	0.7546
10.342	0.1015	0.8985	–	0.9236	0.0764	–	9.100	0.0850	–
10.342	0.1137	0.7876	0.0987	0.8686	0.0766	0.0548	7.639	0.0973	0.5552
10.342	0.1167	0.7268	0.1565	0.8350	0.0796	0.0854	7.155	0.1096	0.5456
10.342	0.1403	0.5658	0.2939	0.7674	0.0805	0.1521	5.470	0.1422	0.5175
10.342	0.1854	0.3730	0.4416	0.6729	0.0774	0.2497	3.629	0.2075	0.5654
10.342	0.2184	0.2528	0.5288	0.5947	0.0838	0.3215	2.723	0.3315	0.6080
10.342	0.2412	0.2115	0.5473	0.5530	0.0800	0.3670	2.293	0.3783	0.6705
10.342	0.2918	0.1425	0.5657	0.4637	0.0847	0.4516	1.589	0.5944	0.7983

Temperature = 113.14 K = -160.02°C

P(MPa)	Liquid Mole Fraction			Vapor Mole Fraction			K-Values		
	H_2	CH_4	CO	H_2	CH_4	CO	H_2	CH_4	CO
2.758	0.0250	0.9750	–	0.9335	0.0665	–	37.34	0.0682	–
2.758	0.0277	0.8845	0.0878	0.8800	0.0626	0.0574	31.77	0.0708	0.6538
2.758	0.0309	0.7770	0.1921	0.8111	0.0542	0.1347	26.25	0.0698	0.7012
2.758	0.0341	0.3058	0.6601	0.5784	0.0346	0.3870	16.96	0.1131	0.5863
2.758	0.0395	0.2483	0.7122	0.5425	0.0337	0.4238	13.73	0.1357	0.5951
2.758	0.0425	–	0.9575	0.4050	–	0.5950	9.529	–	0.6214
4.137	0.0351	0.9649	–	0.9469	0.0531	–	26.98	0.0550	–
4.137	0.0404	0.8133	0.1463	0.8725	0.0481	0.0794	21.60	0.0591	0.5427
4.137	0.0440	0.6932	0.2628	0.8225	0.0450	0.1325	18.69	0.0649	0.5042
4.137	0.0424	0.6943	0.2633	0.8221	0.0452	0.1327	19.39	0.0651	0.5040
4.137	0.0698	0.2896	0.6406	0.6696	0.0358	0.2946	9.593	0.1236	0.4599
4.137	0.0729	0.2247	0.7024	0.6264	0.0337	0.3397	8.593	0.1501	0.4836
4.137	0.0900	–	0.9100	0.515	–	0.4850	5.722	–	0.5330
5.516	0.0478	0.9522	–	0.9554	0.0446	–	19.99	0.0468	–
5.516	0.0485	0.8107	0.1408	0.8815	0.0434	0.0752	18.17	0.0535	0.5341
5.516	0.0593	0.6856	0.2551	0.8423	0.0412	0.1165	14.20	0.0600	0.4567
5.516	0.0953	0.3080	0.5967	0.7080	0.0305	0.2615	7.429	0.0990	0.4382
5.516	0.1074	0.2069	0.6857	0.6712	0.0257	0.3031	6.250	0.1242	0.4420
5.516	0.1325	–	0.8675	0.5620	–	0.4380	4.242	–	0.5049
6.895	0.0588	0.9412	–	0.9576	0.0424	–	16.29	0.0450	–
6.895	0.0666	0.8706	0.0628	0.9268	0.0444	0.0288	13.92	0.0510	0.4586
6.895	0.0819	0.7133	0.2048	0.8723	0.0407	0.0870	10.65	0.0571	0.4248
6.895	0.1259	0.3123	0.5618	0.7289	0.0320	0.2391	5.790	0.1025	0.4256
6.895	0.1467	0.1931	0.6602	0.6903	0.0255	0.2842	4.706	0.1321	0.4305
6.895	0.1850	–	0.8150	0.5700	–	0.4300	3.081	–	0.5276
10.342	0.0885	0.9115	–	0.9562	0.0438	–	10.81	0.0481	–
10.342	0.0989	0.8059	0.0952	0.9093	0.0447	0.0460	9.194	0.0555	0.4832
10.342	0.1113	0.6856	0.2031	0.8588	0.0446	0.0966	7.716	0.0651	0.4756
10.342	0.1651	0.3953	0.4396	0.7467	0.0390	0.2143	4.523	0.0987	0.4875
10.342	0.1958	0.2830	0.5212	0.7211	0.0379	0.2410	3.683	0.1340	0.4624
10.342	0.3075	0.0640	0.6285	0.5663	0.0289	0.4048	1.842	0.4516	0.6441

Table I. (Continued)

P(MPa)	Liquid Mole Fraction			Vapor Mole Fraction			K-Values		
	H_2	CH_4	CO	H_2	CH_4	CO	H_2	CH_4	CO
Temperature = 108.13 K = −165.03°C									
2.758	0.0228	0.9772	–	0.9658	0.0342	–	42.36	0.0350	–
2.758	0.0366	0.4437	0.5197	0.7457	0.0301	0.2242	20.36	0.0678	0.4314
2.758	0.0492	0.0918	0.8590	0.6444	0.0184	0.3372	13.10	0.2004	0.3925
2.758	0.0500	–	0.9500	0.5375	–	0.4625	10.75	–	0.4868
4.137	0.0333	0.9667	–	0.9649	0.0351	–	28.97	0.0363	–
4.137	0.0474	0.4616	0.4910	0.7836	0.0262	0.1902	16.53	0.0568	0.3873
4.137	0.0757	0.0907	0.8336	0.6873	0.0130	0.2997	9.079	0.1433	0.3595
4.137	0.0900	–	0.9100	0.6275	–	0.3725	6.972	–	0.4093
5.516	0.0438	0.9562	–	0.9675	0.0325	–	22.09	0.0340	–
5.516	0.0707	0.4543	0.4750	0.8129	0.0238	0.1633	11.50	0.0524	0.3438
5.516	0.0864	0.2706	0.6430	0.7520	0.0235	0.2245	8.704	0.0868	0.3491
5.516	0.1147	0.0938	0.7915	0.7161	0.0109	0.2730	6.243	0.1162	0.3449
5.516	0.1325	–	0.8675	0.6200	–	0.3800	4.679	–	0.4380
6.895	0.0544	0.9456	–	0.9685	0.0315	–	17.80	0.0333	–
6.895	0.0804	0.4698	0.4498	0.8181	0.0276	0.1543	10.18	0.0588	0.3430
6.895	0.1067	0.2758	0.6175	0.7644	0.0235	0.2121	7.164	0.0852	0.3435
6.895	0.1403	0.0921	0.7676	0.7160	0.0110	0.2730	5.103	0.1194	0.3556
6.895	0.1700	–	0.8300	0.6700	–	0.3300	3.941	–	0.3976

An equilibrium tie line on the phase envelope yields a single
point on the conjugate line, and the entire phase envelope yields
a conjugate line. Thus the conjugate line presents a sensitive
means of correlating the relationship among tie lines at a given
pressure and temperature.

For a ternary system at a constant temperature, the family of
isobaric conjugate lines describes a surface. Or, for the entire
system without constraints of constant pressure and temperature,
the conjugate line describes a volume.

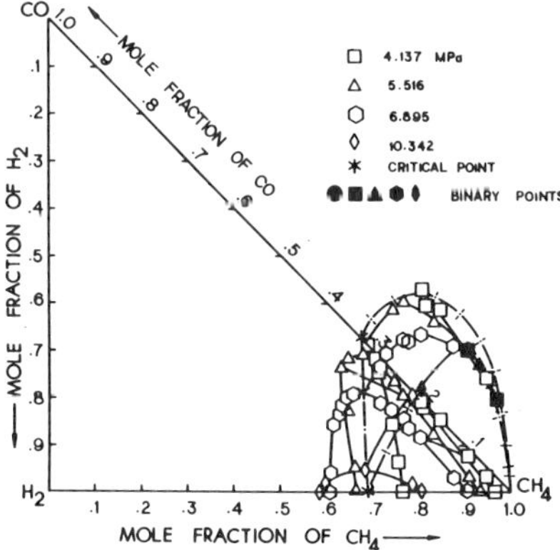

Fig. 2. V-L-E of H_2-CO-CH_4 system showing
conjugate lines: 173.25 K (−99.91° C).

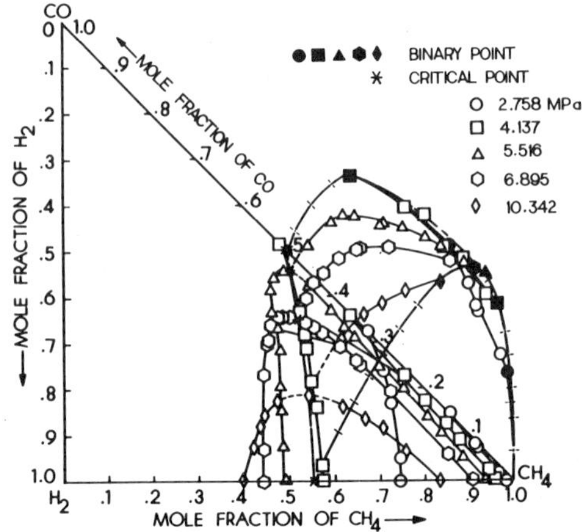

Fig. 3. V-L-E of H_2-CO-CH_4 system showing
conjugate lines: 163.17 K (-109.99° C).

   Conjugate lines relating the tie lines of each envelope are
superimposed on each figure. An attempt has been made to "gen-
eralize" the conjugate line surface. The dotted line either shows
the binary boundaries of the conjugate lines and/or the critical
point locus which terminates the conjugate lines. No three phase
conditions exist for this system in the temperature range of this
study.

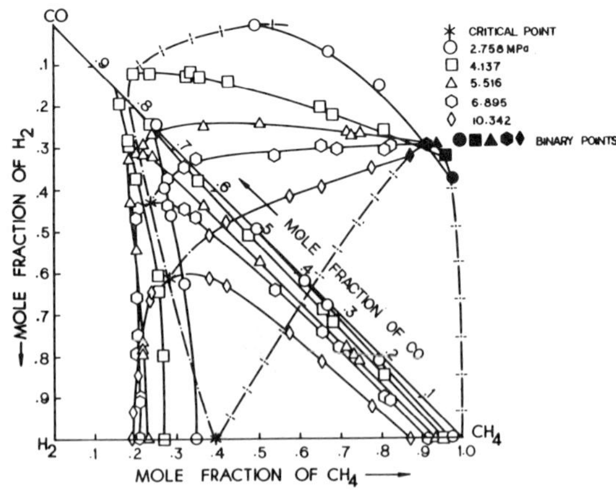

Fig. 4. V-L-E of H_2-CO-CH_4 system showing
conjugate lines: 143.08 K (-130.08° C).

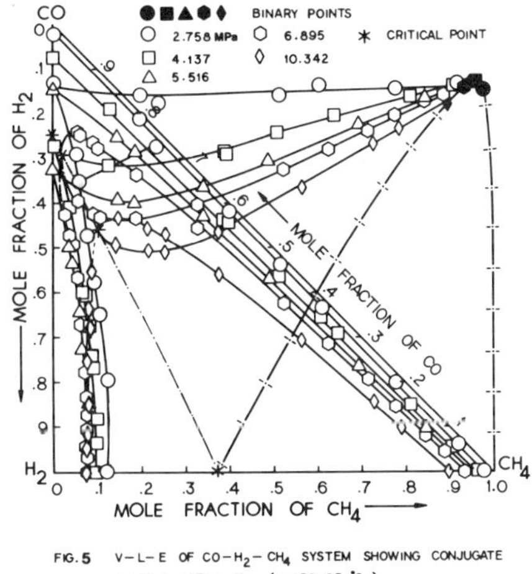

Fig. 5. V-L-E of H_2-CO-CH_4 system showing conjugate lines: 123.11 K (-150.05° C).

The conjugate lines chosen for the analysis of the experimental points were produced by the intersection of a vertical line through the liquid phase composition and a 45 degree line through the conjugate vapor phase composition. The choice of another set of equations would produce another conjugate line, however in the present case an attempt was made to project the conjugate loci away from the traffic of the V-L-E envelope. The locus described

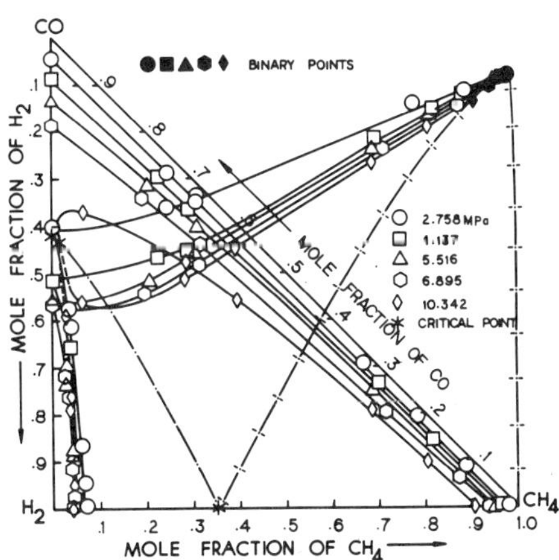

Fig. 6. V-L-E of H_2-CO-CH_4 system showing conjugate lines: 113.14 K (-160.02° C).

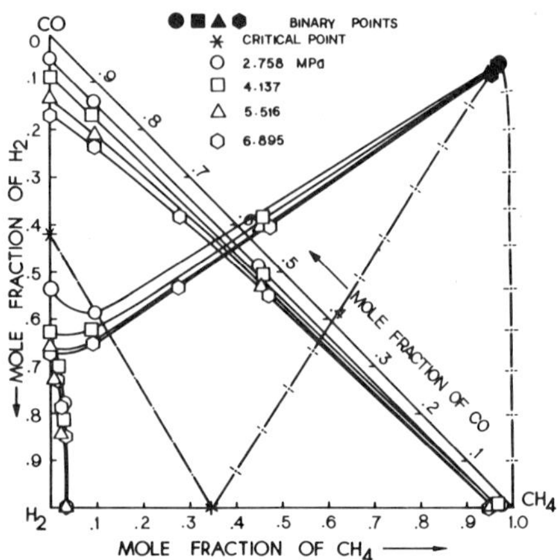

Fig. 7. V–L–E of H_2–CO–CH_4 system showing conjugate lines: 108.13 K (–165.03° C).

by –•–•–•– traces the critical points, binary or ternary, at which conjugate lines terminate if the pressures are sufficiently high.

At this writing, August 1981, correlation of the data, through an equation of state for example, has not been completed. The results will be reported in a subsequent paper.

ACKNOWLEDGEMENT

The Phillips Petroleum Company provided the Hydrocarbon bath fluid gratis. The authors acknowledge the contributions of Mr. Raymond J. Martin in the original design and construction of the apparatus and the maintenance of it thereafter.

REFERENCES

1. J.H. Hong and R. Kobayashi, J. Chem. Eng. Data 26:127 (1981).
2. F.A. Freeth and T.T.H. Verschoyle, R. Soc. London, Ser. A., 130:453 (1931).
3. F. Steckel and F. Tsin, J. Chem. Ind. (Moscow) 16:24 (1939).
4. A.I. Likhter and N.P. Tikhonovich, Zh. Tekh. 10:1201 (1940).
5. E.P. Levitskaya, J. Tech. Phys. (Leningrad) 11:197 (1941).
6. A.L. Benham and D.L. Katz, AIChE J. 3:33 (1957).
7. M. Orientlicher and J.M. Prausnitz, Chem. Eng. Sci. 19:775 (1964).
8. B.S. Kirk and W.T. Ziegler, in "Advances in Cryogenic Engineering, Vol. 10", Plenum Press, New York (1965), p. 160.

9. M. Yorizane, et. al., in "Proc. 1st Intl. Cryo. Engr.
    Conf.", Heywood Temple Industrial Publications, Ltd.,
    London (1968), p. 57.
10. M.G. Fastowsky and V.G. Gonikberg, J. Phys. Chem. (Moscow)
    14:427 (1970).
11. J. Sagara, Y. Arai and S. Saito, J. Chem. Eng. Jpn. 5:339
    (1972).
12. C.Y. Tsang and W.B. Streett, Private Communication (1980).
13. T.T.H. Verschoyle, Trans. Roy. Soc. (London) A230:189 (1931).
14. M. Ruhemann and N. Tsin, Phys. Z.d Sowjeturioni 5:455 (1937).
15. W.W. Akers and L.S. Eubanks, in "Advances in Cryogenic
    Engineering, Vol. 3," Plenum Press, New York (1960), p. 275.
16. A. Toyama, et. al., in "Advances in Cryogenic Engineering,
    Vol. 7," Plenum Press, New York (1961), p. 125.
17. H. Cheung and D.I.J. Wang, I & EC Fund. 3:355 (1964).
18. F.B. Sprow and J.M. Prausnitz, AIChE J. 12:780 (1966).
19. L.J. Christiansen, A. Fredenslund and J. Mollerup, Cryogenics
    13:405 (1973).
20. S.C. Mraw, S.C. Hwang and R. Kobayashi, J. Chem. Eng. Data
    23:135 (1978).
21. J.J. Madison, Anal. Chem. 30:1859 (1958).
22. S.C. Mraw and R. Kobayashi, J. Chromatogr. Sci. 15:191 (1977).

DETERMINATION OF MIXTURE PARAMETERS FROM ANALYTICAL-SOLUTION-OF-GROUP METHOD*

K. Tochigi and K. Kojima

*Nihon University
Tokyo, Japan*

and

W. K. Chung and B. C.-Y. Lu

*University of Ottawa
Ottawa, Canada*

INTRODUCTION

The thermodynamic methods usually used for calculating vapor-liquid equilibrium (VLE) values are based either entirely on an equation of state applicable to both liquid and vapor phases or on an expression of activity coefficient for the liquid phase together with an equation of state for the vapor phase. In both methods, mixture parameters are required in the use of the equation of state.

Recently, a cubic perturbed hard sphere equation of state was proposed and was found suitable for calculating VLE values at low temperatures.[1-3] It has the form

$$P = \frac{RT}{V} \frac{2V+b}{2V-b} - \frac{\alpha}{T^{0.5}V(V+b)} \; . \tag{1}$$

The parameters α and b for mixtures are evaluated by means of

$$\alpha = \sum_i \sum_j x_i x_j \alpha_{ij} \tag{2}$$

---

*Supported in part by the National Science and Engineering Research Council of Canada.

$$b = \sum_i x_i b_i. \tag{3}$$

Furthermore

$$\alpha_{ij} = (\alpha_{ii}\alpha_{jj})^{0.5}(1-C_{ij}). \tag{4}$$

The evaluation of the quantity C_{ij} requires binary data of all ij pairs. In this investigation, an attempt was made to evaluate the mixture parameter α from a group-contribution method so that α values for a large number of mixtures can be calculated in terms of a much smaller number of parameters which characterize the contributions of individual groups. The analytical-solution-of-group (ASOG) method was selected for this purpose because of its simplicity.

The interaction groups which were investigated were N_2, CH_4, CH_3 and CH_2. We hope to generate parameters which are suitable for VLE calculations at low temperature and high pressure conditions for systems containing nitrogen and hydrocarbons.

DEVELOPMENT OF THE PROPOSED METHOD

The relation between the excess Gibbs energy and the fugacity coefficients of a solution and its constituent pure components is given by

$$g^E = RT \left(\ln \hat{\phi} - \sum_i x_i \ln \phi_i \right). \tag{5}$$

The quantities $\ln \hat{\phi}$ and $\ln \phi_i$ can be evaluated by means of an equation of state. In this study, Eq. 1 was used for this purpose. Hence,

$$\ln \phi_i = -2\ln \frac{P(2 V_i - b_i)}{2RT} + \ln \frac{PV_i}{RT} + \frac{PV_i}{RT} -1- \frac{\alpha_{ii}}{b_i RT^{1.5}} \ln \left[\frac{(V_i+b_i)}{V_i} \right] \tag{6}$$

and

$$\ln \hat{\phi} = -2 \ln \frac{P(2V-b)}{2RT} + \ln \frac{PV}{RT} + \frac{PV}{RT} -1 - \frac{a}{bRT^{1.5}} \ln \left(\frac{V+b}{V} \right). \tag{7}$$

In the development of mixing rules relating parameters a and b of a mixture with its constituent pure component parameters, the conventional mixing rule for b as expressed by Eq. 3 was maintained; and a new quantity g^E_∞, the limiting value of g^E at infinite pressure was introduced. It can be readily shown that the quantity g^E_∞ is given by Eq. 8

$$\frac{g^E_\infty}{RT} = \ln b - \sum_i x_i \ln b_i - \frac{\ln 3}{RT^{1.5}}\left[\frac{\alpha}{b} - \sum_i x_i \frac{\alpha_{ii}}{b_i}\right] . \qquad (8)$$

Rearranging this equation yields a mixing rule for the parameters α

$$\alpha = b\left[\sum_i x_i \frac{\alpha_{ii}}{b_i} - \frac{RT^{1.5}}{\ln 3}\left(\frac{g^E_\infty}{RT} - \ln b + \sum_i x_i \ln b_i\right)\right] . \qquad (9)$$

The only quantity that needs to be known in Eq. 9 is g^E_∞, which is determined in this work by the analytical-solution-of-group (ASOG) method in terms of temperature-dependent group interaction parameters. In other words, these group interaction parameters permit the evaluation of the parameters a and b for mixtures by means of Eqs. 3 and 9. Consequently, the fugacity coefficient of component i in a mixture can be determined through the application of Eq. 10,

$$\ln \hat{\phi}_i = \frac{1}{RT} \int_\infty^V \left[\frac{RT}{V} - \left(\frac{\partial P}{\partial n_i}\right)_{T,V,n_j}\right] dV - \ln Z \qquad (10)$$

and VLE calculations can be carried out in the usual manner.

It should be mentioned that the quantity g^E_∞ was introduced to reduce the complexity of the working equations. It is not convenient to use the quantity practically because its numerical value has no useful meaning. In other words, g^E values at high pressures would serve the purpose of the proposed method.

DETERMINATION OF GROUP INTERACTION PARAMETERS

In the ASOG method, the logarithm of the activity coefficient of component i in a solution may be expressed as the sum of a size contribution, $\ln \gamma_i^S$ and a group contribution $\ln \gamma_i^G$. Hence,

$$\ln \gamma_i = \ln \gamma_i^S + \ln \gamma_i^G. \qquad (11)$$

For the sake of simplicity and following the suggestion of Ashraf and Vera,[4] the quantity $\ln \gamma_i^S$ was taken to be zero in the calculation. The group contribution is expressed by,

$$\ln \gamma_i^G = \sum_k \nu_{ki}\left(\ln \Gamma_k - \ln \Gamma_k^{(i)}\right). \qquad (12)$$

In Eq. 12, the quantity ν_{ki} refers to the total number of the k functional group of molecule i, and Γ_k and $\Gamma_k^{(i)}$ are the activity coefficients of the functional group k in the mixture and in the standard state (pure component i), respectively. The summation, $\sum_k$, extends over all functional groups of the mixture. The quantity, Γ_k, is given by

$$\ln \Gamma_k = \sum_{i=1}^{N} \sum_{j=1}^{N} \left(A_{ik} - \frac{1}{2} A_{ij} \right) X_i X_j, \tag{13}$$

where N is the number of groups, and A_{ik} and A_{ij} are the group-interaction parameters. Furthermore,

$$A_{ii} = A_{jj} = 0$$

$$A_{ij} = A_{ji}.$$

The group fraction of group ℓ, X_ℓ, in solution is given by

$$X_\ell = \frac{\sum_j x_j \, \nu_{\ell j}}{\sum_j x_j \sum_k \nu_{kj}}, \tag{14}$$

where x_j is the mole fraction of component j, and $\nu_{\ell j}$ and ν_{kj} are to the number of ℓ and k groups, respectively, in molecule j. The summation $\sum_k$ extends over all functional groups of the solution, while the summation $\sum_j$ extends over all components.

The four functional groups investigated consist of six binary interactional groups. They were all evaluated from binary VLE values. The primary information used in the evaluation is presented in Table I. In the calculation, g^E values were obtained from

$$g^E = RT(x_1 \ln \gamma_1 + x_2 \ln \gamma_2) = RT \left(x_1 \ln \gamma_1^G + x_2 \ln \gamma_2^G \right) \tag{15}$$

and used directly in the working equations without considering the pressure effect as an approximation.

With this simplification, the fugacity coefficient of component i in solution takes the form of Eq. 16,

$$\ln \hat{\phi}_i = \frac{2b_i}{2V-b} - 2 \ln \left(1 - \frac{b}{2V} \right) - \left\{ \frac{b_i}{bRT^{1.5}} \left[\sum_j x_j \frac{a_{jj}}{bj} - \frac{RT^{1.5}}{\ln 3} \left(\frac{g^E}{RT} \right. \right.$$

$$\left. \left. - \ln b + \sum_j x_j \ln b_j \right) \right] + \frac{1}{RT^{1.5}} \frac{a_{ii}}{b_i} - \frac{1}{\ln 3} \left(\ln \gamma_i - \ln b + \ln b_i \right) \right\}$$

$$\ln \frac{V+b}{V} + \frac{ab_i}{RT^{1.5} b^2} \left[\ln \left(\frac{V+b}{V} \right) - \frac{b}{V+b} \right] - \ln Z \tag{16}$$

The above equations may be applied to gas mixtures by replacing liquid mole fraction x with vapor mole fraction y.

The A_{ij} values were evaluated at given T and x by minimizing the objective function ΔP, the difference between the calculated and experimental values of system pressure. The sequence of evaluation follows the order indicated in Table I. Two of the group-interaction parameters were evaluated from VLE values of binary systems containing three groups. In these instances, the values of the third group-interaction parameter were evaluated with the assistance of the known values of the other two. The A_{ij} values obtained are plotted in Fig. 1 against temperature and are further correlated by means of Eq. 17,

$$A_{ij} = \alpha + \beta\, T. \tag{17}$$

The values obtained for the coefficients α and β are reported in Table II.

RESULTS OF CALCULATION

The VLE values of the nine binary systems used for establishing the group-interaction parameters were used to test the applicability of the correlation. The calculated P and y values using A_{ij} values obtained individually as well as those obtained from Eq. 17 are compared with the experimental values. A summary of

Table I. Primary Information Used in the Determination
of Group-Interaction Parameters A_{ij}

A_{ij}	Data	System/Group	No. of Isotherm	No. of Data Points
N_2-CH_4	VLE	Nitrogen-Methane	20	224
CH_4-CH_3	VLE	Methane-Ethane	9	89
CH_3-CH_2	VLE	Ethane-Propane	3	20
N_2-CH_3	VLE	Nitrogen-Ethane	10	115
N_2-CH_2	A_{ij}	CH_3-CH_2, N_2-CH_3		
	VLE	Nitrogen-Propane	6	50
CH_4-CH_2	A_{ij}	CH_3-CH_2, CH_4-CH_3		
	VLE	Methane-Propane	15	118
		Methane-n-Butane	11	70
		Methane-n-Pentane	7	40
		Methane-n-Hexane	3	30

Table II. Correlation of Group-Interaction Parameters
With Temperature, $A_{ij} = \alpha + \beta\ T$

Interaction Group	α	β	Temp. Range, K
N_2-CH_4	1.354	-0.00307	95.00–171.43
CH_4-CH_3	0.450	0.00105	110.93–199.92
CH_3-CH_2	0.5635		127.59–199.82
N_2-CH_3	1.720	0.00061	122.04–194.26
N_2-CH_2	6.085	-0.01630	114.10–198.15
CH_4-CH_2	-0.156	0.01237	110.93–199.88

Table III. Comparison of Calculated and Experimental
Values for Nine Binary Systems

System	n	Individual A_{ij}		Correlated A_{ij}	
		$\left\|\Delta P/P\right\|_{av.} \cdot 100\%$	$\left\|\Delta y\right\|_{av}$	$\left\|\Delta P/P\right\|_{av.} \cdot 100\%$	$\left\|\Delta y\right\|_{av.}$
Nitrogen–Methane	224	2.60	0.0131	2.91	0.0133
Nitrogen–Ethane	115	10.87	0.0283	10.91	0.0284
Nitrogen–Propane	50	2.86	0.0029	10.92	0.0028
Methane–Ethane	89	3.48	0.0088	4.30	0.0099
Methane–Propane	118	4.48	0.0027	5.92	0.0026
Methane–n–Butane	70	6.39	0.0005	7.59	0.0005
Methane–n–Pentane	40	7.17	0.0002	10.14	0.0002
Methane–n–Hexane	30	3.97	0.0005	5.06	0.0005
Ethane–Propane	20	2.35	0.0182	3.10	0.0174
Overall Average		5.35	0.0014	6.92	0.0014

Table IV. Comparison of Predicted and Experimental Values
for Two Ternary Systems[5]

System	Temp.K	n	$\left\|\Delta P/P\right\|_{av.} \cdot 100\%$	$\left\|\Delta y\right\|_{av}$
Nitrogen–Methane–Ethane	138.71	32	4.16	0.0037
	149.82	18	5.02	0.0051
Methane–Ethane–Propane	158.15	5	3.43	0.0019
	172.04	14	4.77	0.0033
	185.93	16	7.02	0.0056

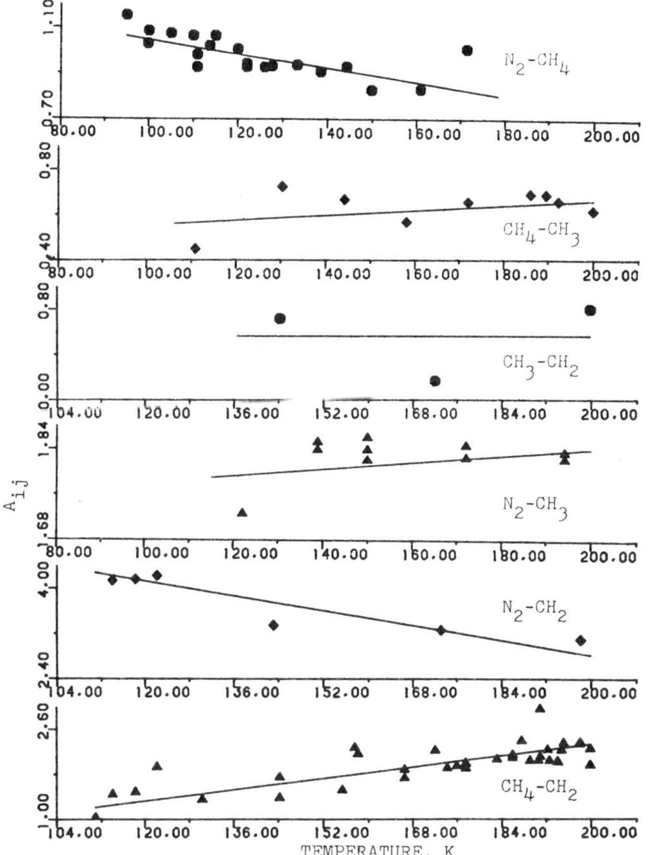

Fig. 1. Variation of A_{ij} with temperature.

the comparison is shown in Table III, indicating that the correlated values of A_{ij} are suitable for representing the data. In addition, P and y values were predicted from given T and x values for two ternary systems using the correlated A_{ij} values. A summary of the comparison between the predicted and experimental values is presented in Table IV, indicating that the results obtained are satisfactory.

DISCUSSION AND CONCLUSIONS

In the development of Eq. 8, the conditions that $b \rightarrow V/2$ and $b_i \rightarrow V_i/2$ at extremely high pressures were involved. These conditions were not strictly observed in the subsequent simplifi-

ications.   Nevertheless,   the   calculated   results   appear   to   be
satisfactory.

At present, only straight-chain aliphatics have been tested. The results obtained are excellent.   The VLE values for the
ternary system nitrogen-methane-ethane[5] have been calculated and
previously reported.[2]   A comparison of the calculated results
indicates that the values obtained in this work are as good as
those obtained directly from Eq. 1 with the combination of binary
interaction parameters C_{ij}, and are better than those obtained
from the Redlich-Kwong equation modified by Soave.[6]

It should be mentioned that the proposed approach can be
extended to other groups.   Furthermore, it is not limited to the
selected equation of state nor is it restricted to the ASOG method.

REFERENCES

1. B.C.-Y. Lu, et. al., The use of two-parameter equations of
      state for predicting vapor-liquid equilibria, Inst. Chem.
      Eng. Symposium Series No. 56, 1.2:57 (1979).
2. T. Ishikawa, W. K. Chung, and B.C. -Y. Lu, A cubic perturbed,
      hard sphere equation of state for thermodynamic properties
      and vapor-liquid equilibrium calculations, AIChE J 26:372
      (1980).
3. T. Ishikawa, W. K. Chung, and B.C. -Y. Lu, Simple and
      Generalized Equation of State for Vapor-Liquid Equilibrium
      Calculations, in "Advances in Cryogenic Engineering, Vol.
      25," Plenum Press, New York (1980), p. 671.
4. F.A. Ashraf and J. H. Vera, A simplified group method
      analysis, Fluid Phase Equilibria, 4:211 (1980).
5. R. Kobayashi and P. Chappelear, Low Temperature Data from Rice
      University for Vapor-Liquid and P-V-T Behavior, Gas
      Processes Association, Tulsa, Oklahoma, Technical
      Publication TP-4, (1974).
6. G. Soave, Equilibrium constants from a modified Redlich-Kwong
      equation of state, Chem. Eng. Sci. 37:1197 (1972).

INTERPRETATION OF ISOCHORIC P-V-T BEHAVIOR OF A NOMINAL 95-MOL%-METHANE-5-MOL%-PROPANE MIXTURE FROM NEAR-AMBIENT TO CRYOGENIC TEMPERATURES*

J. W. Magee, K. Arai, and R. Kobayashi

Rice University
Houston, Texas

INTRODUCTION

Recent advances in the experimental measurement of the volumetric behavior of fluids has led to the isochoric measurement of the P-V-T surface of methane[1], ethane[2], propane[3] and a 5.31 mol % propane–methane mixture[4]. Experiments such as these have measured volumetric behavior of a substance under the constraint of constant density, or constant average intermolecular separation. Lines of constant density or isochores are obtained which manifest simpler behavior than isotherms or isobars. Indeed, isochores have been observed to be nearly linear and the earliest equations of state assumed that they were. Any deviations from linearity carry significant messages.

For binary mixtures and mixtures of greater complexity, the comprehensive description of mixtures along isochores has seldom been conducted. Nevertheless, they will be discussed in terms of isochores of a 5.31 mol % propane–methane mixture recently reported by Arai and Kobayashi[4].

NATURE OF ISOCHORIC NONLINEARITY

A casual examination of isochores may lead an observer to see linear behavior. However, a closer examination of large-scale P-T plots plus the results of careful data processing show that iso-

---

*Supported in part by the Gas Research Institute.

chores are characterized in fact by nonlinear behavior[5,6]. For
this study the nature of the nonlinear behavior was established
for several representative isochores by analytically passing a
line tangent to the "most linear portion" of each isochore, then
calculating the deviation of the experimental data from the
line. Plots of these deviations are represented by Fig. 1 in
which the deviation from linearity along isochores is plotted
versus temperature for three isochores. It is evident that the
deviation is consistently stronger near saturation with a weaker
deviation at higher temperatures. Further, while the high temperature deviations are negative, the sign of low temperature deviations depends on density and may be zero, negative, or positive.

These observations have been expressed analytically by Magee[7]
with an equation for the pressure-temperature variation at constant density,

$$P = A + BT + f_1 + f_2 + f_3 \tag{1}$$

where, $f_1 = \exp(C_1(1-T/T_0))$, $f_2 = -\exp(C_2(T/T_0-1))$
       $f_3 = 1-\exp(D(1-T_0/T))$

In Eq. 1 the functions f_1, f_2, and f_3 give the deviation from the
linear relation,

$$P_{linear} = A + BT \tag{2}$$

and depend on temperature and density. In particular, functions
f_1 and f_2 give the deviations from linearity for densities up to
$2\rho_c$, but above $2\rho_c$ the function f_3 expresses the deviations. For
$\rho < \rho_c$ the temperature T_0 cites the condition where the thermal
pressure coefficient tends toward a minimum value; for $\rho > \rho_c$ the
thermal pressure coefficient tends toward a maximum value at T_0.
Like A, B, C_1, C_2, and D, the quantity T_0 is also a least-squares
adjustable parameter with an associated standard error. More
specifically, for densities between ρ_c and $2\rho_c$, T_0 represents the
condition where $\partial^2\rho/\partial T^2 = 0$, defining the locus of isochoric inflection points. Differentiation of Eq. 1 twice with temperature
gives, at the inflection temperature, T_0

$$\left(\frac{\partial^2 P}{\partial T^2}\right)_\rho = \left(\frac{C_1}{T_o}\right)^2 \exp(C_1(1-T_o/T_o))-\left(\frac{C_2}{T_o}\right)^2\exp(C_2(T_o/T_o-1)) \tag{3a}$$

or

$$\left(\frac{\partial^2 P}{\partial T^2}\right)_\rho = \left(\frac{C_1}{T_o}\right)^2 - \left(\frac{C_2}{T_o}\right)^2 = 0 \tag{3b}$$

when the constraint $C_1 = C_2$ is applied to Eq. 1. Hence, the non-
linear contributions to pressure diminish to zero at T_0, leaving
Eq. 2 for the pressure-temperature variation at constant density.

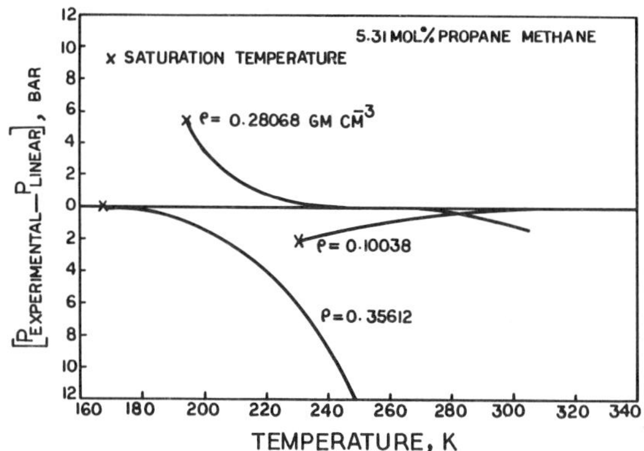

Fig. 1. Deviations from linearity for selected isochores.

RESULTS OF NON—LINEAR REGRESSION

Seventeen judiciously spaced isochores for a 5.31 mol % propane in methane mixture were regression fitted to Eq. 1. This was accomplished by a modified Gauss—Newton non—linear regression routine which minimized the sum of squared pressure errors. The results[7] of the regressions are presented in full elsewhere.

Vapor Densities

Seven vapor isochores from 0.02015 g/cm^3 to 0.12564 g/cm^3 exhibited convex behavior characterized by negative $(\partial^2 P/\partial T^2)_\rho$ at all temperatures of the study. In particular, the isochores all show evidence of significant contributions of prenucleation and the higher virial coefficients as the saturation border curve is approached. This phenomenon is most pronounced for the 0.10038 g/cm^3 (ρ_R = 0.6) isochore and less marked for higher and lower vapor densities. Equation 1 represented a total of 167 vapor phase measurements most satisfactorily, yielding an average absolute error of 0.007%.

Moderately High Densities

Eight isochores manifested sigmoidal behavior and thus each possesses an isochoric inflection point. These isochores extended from subcritical to supercritical temperatures and densities from 0.15044 g/cm^3 to 0.33015 g/cm^3 including an isochore within 0.1% of the critical density. For the 201 experimental points in this density range, the absolute error of the pressure prediction averaged 0.04%. The isochoric inflection points, T_0, where $(\partial^2 P/\partial T^2)_\rho$ = 0, yield a smooth asymmetric loop in T-ρ coordinates, Fig. 2. The graph of the locus of inflection temperatures is skewed with its maximum shifted to a relatively lower reduced density (ρ_R = 1.2).

To facilitate a direct comparison of isochoric behavior of the mixture with behavior of pure methane, sixteen methane isochores[1] were regressed with Eq. 1. Figure 3 depicts the P-T projection of the locus of inflection points for methane, 5.31 mol % propane in methane, and propane[8]. Plotted in pseudo-reduced pressure and temperature, Fig. 3 shows the propagation to higher reduced pressures and temperatures as the contribution of noncentral forces to the intermolecular interactions becomes increasingly significant.

For each sigmoidal isochore, the value of the parameter T_o in Eq. 1 is the temperature at the isochoric inflection point, where the thermal pressure coefficient $(\partial P/\partial T)_\rho$ exhibits a local maximum; hence, the isochoric curvature $(\partial^2 P/\partial T^2)_\rho$ vanishes at T_o. At temperatures below T_o isochoric curvature is positive and above T_o this derivative is negative. Implications of the sign of isochoric curvature are obtained from the relation[9]

$$\left(\frac{\partial^2 P}{\partial T^2}\right)_\rho = -\frac{\rho^2}{T}\left(\frac{\partial C_v}{\partial \rho}\right)_T \qquad (4)$$

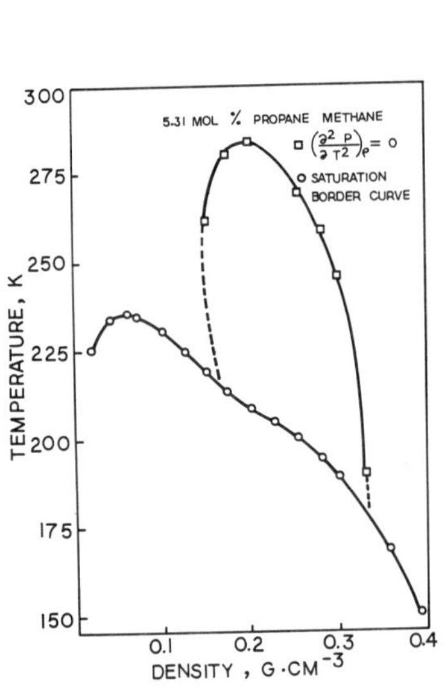

Fig. 2. Saturation & isochoric inflection loci.

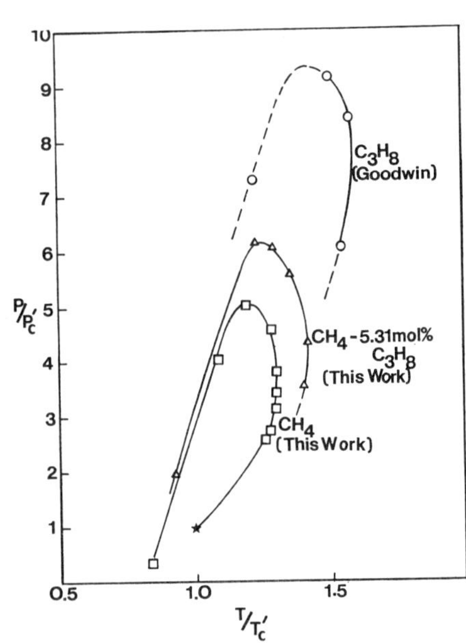

Fig. 3. Isochoric inflection loci in pseudo-reduced pressure and temperature.

Hence, a positive value of isochoric curvature implies that an isothermal density increase causes a decrease in constant volume specific heat.

Compressed Liquid Densities

Two isochores, 0.35612 g/cm^3 and 0.39081 g/cm^3, with reduced densities greater than 2.0 were regressed with Eq. 1. No inflection behavior was observed for these isochores, thus the convex behavior could be modeled solely by nonlinear function f_3. Thus the deviations from linearity depends on $\exp(1/T)$ and are negative at all temperatures for the compressed liquid isochores of this study, Fig. 1. Equation 1 yielded errors which averaged 0.08% for the 48 compressed liquid measurements.

SATURATION BORDER CURVE AND CRITICAL CONSTANTS

The intensification of measurements in the vicinity of the saturation border curve permitted accurate estimates of phase transition temperatures from spontaneous breaks in the thermal pressure coefficient. The saturation border curve is tabulated in Table I. Wichterle[10] has reported a critical temperature of

Table I. Properties of the saturation border curve for a
methane- 5.31 mol % propane mixture.

Isochore Density (g/cm^3)	Saturation Property	
	T (K)	P (bar)
0.02015	225.4	19.34
0.04055	234.1	36.53
0.05968	235.3	49.11
0.07011	234.7	54.87
0.07897	233.6	58.61
0.10038	230.6	65.69
0.12564	225.0	68.84
0.15044	218.9	67.96
0.17441	213.4	64.87
0.20001	208.4	61.04
0.22755	204.3	57.07
0.25511	199.7	51.97
0.28068	194.2	45.99
0.29999	188.8	39.61
0.35612	167.7	21.10
0.39081	149.7	9.68

213.17 K ± 0.01 K for a composition equal to that of the P–V–T
isochores. From a large scale graph of the saturation temperatures versus density a critical density of 0.1727 ± 0.0002 g/cc
(9.85 mol/ℓ) was obtained for the methane-propane mixture; the
critical density and temperature were observed to fall in the
concave portion of the T–ρ saturation locus, Fig. 2. A graphical
determination of the critical pressure from the P–V–T relations
yields 65.15 bar which is in good agreement with Wichterle's 65.09
bar (±0.1%).

CONCLUSION

Deviations from linearity along isochores have proven to be
useful in emphasizing the details of isochoric behavior for a
binary mixture of methane and propane. Three distinct regions of
isochoric behavior have been revealed – vapor, moderately dense
liquid, and compressed ($\rho_R>2$)) liquid. Further work should be
done in the development of generalized equations of state for
mixtures with improved mixing rules. Many more precise isochoric
P–V–T data on mixtures are recommended to consolidate the picture.

REFERENCES

1. A.J. Vennix, T.W. Leland and R. Kobayashi, J. Chem. Eng. Data
       15:238 (1970).
2. A.K. Pal, et al., J. Chem. Eng. Data 21:394 (1976).
3. J.F. Ely and R. Kobayashi, J. Chem. Eng. Data 23:221 (1978).
4. K. Arai and R. Kobayashi, in "Advances in Cryogenic
       Engineering, Vol. 25", Plenum Press, New York (1980), p.
       640.
5. A.J. Vennix and R. Kobayashi, AIChE J. 15:926 (1969).
6. R. Kobayashi, et. al., 176th Amer. Chem. Soc. Nat'l. Mtg.
       Miami Beach, (Sept. 10-15, 1978).
7. J.W. Magee, M.S. Thesis, Rice University, Houston, Texas,
       (May, 1981).
8. R.D. Goodwin, Provisional Thermodynamic Functions of Propane,
       From 85 to 700 K at Pressures to 700 Bar, NBSIR-77-860,
       U. S. Dept. of Commerce, (July 1977).
9. D.E. Diller, Cryogenics 11:186 (1971).
10. T. Wichterle and R. Kobayashi, J. Chem. Eng. Data 17:4 (1972).

BINARY INTERACTION COEFFICIENTS FOR VAPOR-LIQUID EQUILIBRIUM CALCULATIONS AT LOW TEMPERATURES*

C.-T. Fu, S. Yoshimura,[†] and B. C.-Y. Lu

University of Ottawa
Ottawa, Canada

INTRODUCTION

The representation and prediction of vapor–liquid equilibrium (VLE) values are frequently made by means of an equation of state. Excellent results can be obtained if the required binary interaction coefficients for the equation are available. A cubic perturbed hard sphere equation of state was recently proposed.[1] This equation was found suitable for representing and predicting VLE values at low temperatures with or without the presence of a supercritical component[1-3]. Although the two parameters of the equation for pure components have been correlated and generalized for a number of nonpolar compounds, the binary interaction coefficients of the equation have not been correlated with temperature.

The purpose of this study is threefold: (1) to determine the binary interaction coefficients from binary VLE values available in the literature for mixtures containing hydrogen, nitrogen, carbon monoxide and methane, (2) to correlate these values with temperature, and (3) to predict multicomponent VLE values using the data obtained. The compounds selected are of interest to industries concerned with natural gas, methanation process and coke oven gas.

---

*Work supported in part by Natural Sciences and Engineering Research Council of Canada.

[†]Present address: University of Hiroshima, Hiroshima, Japan.

EVALUATION OF C_{ij}

The proposed cubic equation has the form,

$$P = \frac{RT}{V} \frac{2V + b}{2V - b} - \frac{a}{T^{0.5}V(V+b)} \tag{1}$$

where the quantities a and b are related to the critical properties as follows:

$$a = \Omega_a R^2 T_c^{2.5}/P_c \tag{2}$$

and

$$b = \Omega_b RT_c/P_c \tag{3}$$

The pure-component parameters Ω_a and Ω_b are both taken to be
temperature dependent as well as substance dependent. In applying
Eq. 1 to VLE calculations, the following mixing rules are used for
gas mixtures,

$$a = \sum_i \sum_j y_i y_j a_{ij} \tag{4}$$

$$b = \sum_i y_i b_i \tag{5}$$

The same rules are used for liquid mixtures with x replacing y in
these equations. A binary interaction coefficient, C_{ij} is used to
represent the deviation of a_{ij} from the geometric mean as follows:

$$a_{ij} = (a_{ii} a_{jj})^{0.5} (1-C_{ij}) \tag{6}$$

The expressions used for calculating fugacity coefficients are
identical to those reported earlier. All the C_{ij} values are
determined from binary VLE data at isothermal conditions.

In this work, the optimum value of C_{ij} for a given system and
temperature was obtained by minimizing the difference between the
calculated and experimental values of liquid composition. The
mixtures under consideration consist of six binary systems. The
total number of 553 data points at 84 isothermal conditions for
these systems were selected for the determination of the C_{ij}
values. The C_{ij} values thus obtained together with the average
absolute deviations between the experimental and calculated equilibrium compositions are reported in Table I.

CORRELATION OF C_{ij}

The C_{ij} values obtained for the six binary systems indicate
that they are not independent of temperature. Although their
temperature dependence seems to indicate a consistent trend from

Table I.   Correlation Results of VLE Data for Six Binary Systems

System	T,K	N	C_{12}	$\left|\Delta x\right|_{av}$	$\left|\Delta y\right|_{av}$	Ref.
Hydrogen—Nitrogen	70.35	12	-0.0112	0.0223	0.0422	4
	70.55	12	-0.0237	0.0160	0.0216	4
	83.15	4	-0.0408	0.0068	0.0153	5
	83.67	5	-0.0609	0.0099	0.0593	4
	88.15	8	-0.0631	0.0114	0.0168	6
	90.03	8	-0.0272	0.0017	0.0008	7
	90.79	10	-0.0586	0.0185	0.0204	4
	95.00	3	-0.0485	0.0011	0.0010	7
	99.82	5	-0.0635	0.0076	0.0211	5
	100.00	10	-0.0641	0.0156	0.0281	4
	110.30	10	-0.0738	0.0103	0.0193	4
Hydrogen—Carbon Monoxide	77.35	6	-0.0101	0.0099	0.0105	8
	83.15	4	-0.0248	0.0040	0.0104	5
	99.82	5	-0.0625	0.0066	0.0098	5
	103.15	3	-0.0850	0.0030	0.0294	8
	122.04	2	-0.0986	0.0014	0.0192	5
	123.15	2	-0.1499	0.0226	0.0394	8
Hydrogen—Methane	90.72	9	-0.0679	0.0007	0.0007	9
	93.15	3	-0.0280	0.0087	0.0075	10
	103.07	5	-0.0925	0.0002	0.0010	9
	103.15	5	-0.0922	0.0060	0.0075	10
	103.15	5	-0.1266	0.0023	0.0083	11
	109.99	7	-0.1095	0.0004	0.0003	9
	116.48	4	-0.1656	0.0313	0.0052	12
	116.51	6	-0.1209	0.0008	0.0017	9
	123.15	6	-0.1272	0.0006	0.0210	11
	123.15	4	-0.1486	0.0038	0.0274	10
	143.05	6	-0.1653	0.0015	0.0088	11
	143.15	5	-0.0873	0.0117	0.0308	10
	144.26	5	-0.2174	0.0245	0.0147	12
	163.15	4	-0.2126	0.0171	0.0398	10
	172.04	4	-0.1119	0.0015	0.0081	12
	172.05	2	-0.2491	0.0047	0.0139	11
	173.65	9	-0.2132	0.0067	0.0172	11
Nitrogen—Carbon Monoxide	70.00	4	0.0007	0.0277	0.0279	13
	75.00	4	0.0027	0.0178	0.0176	13
	79.20	4	0.0046	0.0204	0.0277	13
	83.82	9	0.0153	0.0268	0.0355	14
	90.10	6	0.0157	0.0306	0.0411	13
	96.70	6	-0.0006	0.0421	0.0428	13
	100.00	7	0.0091	0.0182	0.0250	13
	105.10	5	0.0185	0.0230	0.0362	13

Table I. (Continued)

System	T,K	N	C_{12}	$|\Delta x|_{av}$	$|\Delta y|_{av}$	Ref.
	110.00	10	0.0071	0.0099	0.0160	13
	116.60	6	0.0074	0.0078	0.0137	13
	121.80	3	0.0077	0.0021	0.0128	13
Nitrogen–Methane	90.67	9	0.0290	0.0086	0.0034	14
	91.65	4	0.0385	0.0019	0.0420	15
	95.00	7	0.0404	0.0056	0.0008	23
	97.15	4	0.0402	0.0028	0.0402	15
	99.82	11	0.0372	0.0073	0.0122	17
	100.00	9	0.0395	0.0030	0.0009	23
	105.00	7	0.0448	0.0079	0.0005	23
	105.15	4	0.0376	0.0014	0.0156	15
	110.00	9	0.0445	0.0079	0.0009	23
	110.93	14	0.0426	0.0067	0.0037	17
	113.71	8	0.0555	0.0084	0.0020	19
	114.57	4	0.0390	0.0002	0.0115	15
	115.00	8	0.0533	0.0115	0.0028	23
	120.00	8	0.0523	0.0112	0.0027	23
	122.04	14	0.0468	0.0042	0.0069	17
	122.04	11	0.0635	0.0123	0.0033	18
	126.04	15	0.0513	0.0100	0.0067	17
	127.59	15	0.0679	0.0080	0.0120	18
	133.15	14	0.0493	0.0034	0.0095	17
	138.46	9	0.0539	0.0079	0.0095	18
	144.26	12	0.0529	0.0056	0.0105	17
	149.82	8	0.0575	0.0056	0.0059	16
	160.93	12	0.0582	0.0005	0.0122	16
	172.04	8	0.0709	0.0012	0.0056	16
Nitrogen–Methane	177.59	10	0.0867	0.0017	0.0049	16
	183.15	8	0.0995	0.0009	0.0032	16
Carbon Monoxide–Methane	90.67	9	0.0172	0.0078	0.0057	14
	91.56	5	0.0351	0.0025	0.0610	15
	97.17	4	0.0299	0.0026	0.0523	15
	105.18	5	0.0250	0.0085	0.0200	15
	113.71	1	0.0210	0.0001	0.0014	20
	114.48	5	0.0287	0.0056	0.0213	15
	122.04	2	0.0404	0.0090	0.0149	20
	123.90	3	0.0303	0.0012	0.0126	15
	130.37	4	0.0382	0.0039	0.0090	20
	138.71	1	0.0544	0.0000	0.0075	20
	144.26	5	0.0207	0.0043	0.0087	20
	152.59	1	0.0324	0.0000	0.0046	20
	158.15	4	0.0298	0.0073	0.0295	20

the same reference source, difference in numerical values between various references at similar temperatures is observed. All the C_{ij} values obtained were then plotted against temperature as shown in Fig. 1, and an attempt was made to correlate the C_{ij} values with temperature by a linear relationship

$$C_{ij} = \ell + m\, T \tag{7}$$

The results of the correlation are listed in Table II, where $|\Delta C_{ij}|_{av}$ is the average absolute deviation between the calculated and correlated C_{ij} values. In the temperature range studied, the C_{ij} values for mixtures containing hydrogen decrease with the increase of temperature. Furthermore, negative C_{ij} values are obtained for these hydrogen-containing mixtures. For the other three binary systems, the C_{ij} values increase with the increase of temperature and are positive.

APPLICABILITY OF THE CORRELATION

For testing the applicability of the proposed correlations, the calculated equilibrium compositions are compared with the experimental values for all the 553 data points of the six binary systems studied. The average absolute deviations obtained for Δx and Δy for the systems hydrogen–nitrogen, hydrogen–carbon monoxide, hydrogen–methane, nitrogen–carbon monoxide, nitrogen–methane and carbon monoxide and methane are 0.0130; 0.0218; 0.0075; 0.0163; 0.0063; 0.0117; 0.0209; 0.0274; 0.0061; 0.0075; and 0.0052, 0.0214; respectively. These deviations are somewhat higher than those reported in Table I, but are considered acceptable.

An attempt was made to compare the calculated results obtained from the proposed method with those reported by Gray[21] for the

Table II.   Correlation of C_{ij} with Temperature
By Means of Equation (7)

System	ℓ	m	$\left|\Delta C_{ij}\right|_{av}$	Temp. Range, K
Hydrogen-Nitrogen	0.0812	−0.001433	0.0091	70.35–110.30
Hydrogen-Carbon Monoxide	0.1868	−0.002549	0.0105	77.35–123.15
Hydrogen-Methane	0.0626	−0.001550	0.0281	90.72–173.65
Nitrogen-Carbon Monoxide	−0.0011	0.000096	0.0049	70.00–121.80
Nitrogen-Methane	−0.0128	0.000523	0.0061	90.67–183.15
Carbon Monoxide-Methane	0.0157	0.000126	0.0072	90.67–158.15

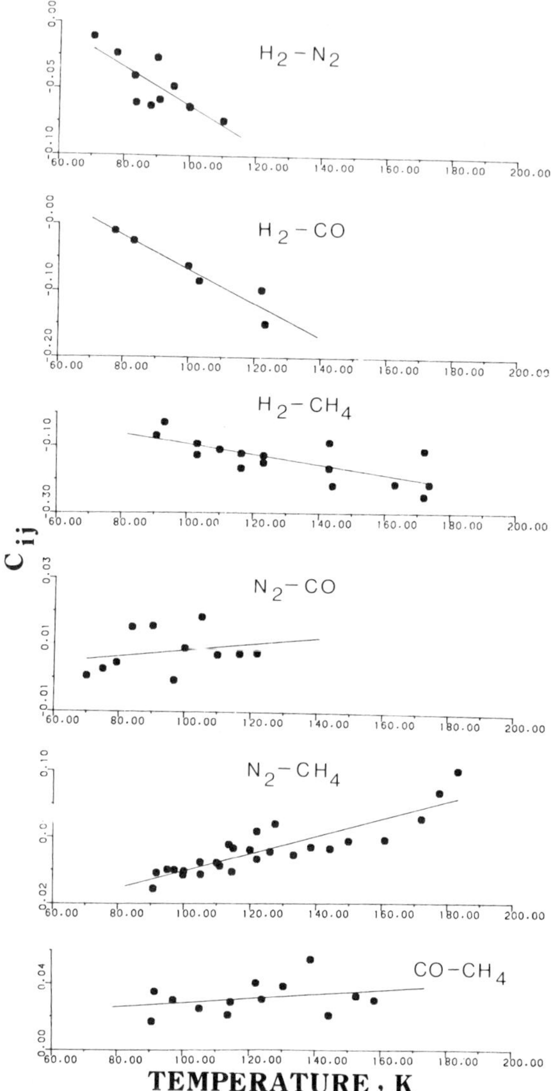

Fig. 1. Variation of C_{ij} with temperature.

hydrogen—methane and hydrogen—nitrogen binary systems using the
same literature data. The deviations obtained in the calculated K
values for hydrogen from this work are lower than those reported
by Gray approximately by a factor of two.

In addition, the C_{ij} values as represented by Eq. 7 are used
to predict equilibrium compositions for two hydrogen-containing
ternary systems at a total of four isotherms. A summary of the
satisfactory results is given in Table III.

Table III. Comparison of Predicted and Experimental Equilibrium
Compositions for Two Ternary Systems

T,K	N	Pressure Range, atm	$\left|\Delta x\right|_{av}$	$\left|\Delta y\right|_{av}$	Ref.
		Hydrogen-Nitrogen-Carbon Monoxide			
83.15	17	21.43-136.08	0.0040	0.0088	5
99.82	29	21.43- 95.26	0.0053	0.0093	5
122.04	6	34.02- 54.43	0.0039	0.0162	5
		Hydrogen-Nitrogen-Methane			
144.26	8	34.02- 68.05	0.0015	0.0182	22

REFERENCES

1. B.C.-Y. Lu, et al., The use of two-parameter equations of
     state for predicting vapor-liquid equilibria, Inst. Chem.
     Eng. Symposium Series No. 56, 1(2):57 (1979).

2. T. Ishikawa, W.K. Chung, and B.C.-Y. Lu, A cubic perturbed
     hard sphere equation of state for thermodynamic properties
     and vapor-liquid equilibrium calculations, AIChE J. 26:372
     (1980).

3. T. Ishikawa, W.K. Chung, and B.C.-Y. Lu, Simple and
     generalized equation of state for vapor-liquid equilibrium
     calculations, in "Advances in Cryogenic Engineering, Vol.
     25", Plenum Press, New York (1980), p. 671.

4. W.B. Streett and G.G. Calado, Liquid-vapor equilibrium for
     hydrogen + nitrogen at temperatures from 63 to 110 K and
     pressures to 57 MPa, J. Chem. Therm. 10:1089 (1978).

5. W.W. Akers and L.S. Eubanks, Vapor-liquid equilibria in the
     system hydrogen-nitrogen-carbon monoxide, in "Advances in
     Cryogenic Engineering, Vol. 3", Plenum Press, New York
     (1960), p. 275.

6. M. Yorizane, et al., Measurement and prediction of vapor-
     liquid equilibrium relation at low temperature and high
     pressure for the H_2-N_2 system, Kagakukugaku 35 (6):691
     (1971).

7. A. Malmoul, Liquid-vapor equilibria in the hydrogen-nitrogen
     and deuterium-nitrogen systems, AIChE J. 7:371 (1961).

8. M. Yorizane,et. al., Low temperature vapour-liquid equilibria
     of hydrogen-containing binaries, in "Proc. First Intl.
     Cryo. Engr. Conf.", Heywood Temple Industrial Publications,
     Ltd., London, (1968), p. 57.

9. B.S. Kirk and W.T. Ziegler, A phase-equilibrium apparatus for
     gas-liquid systems and the gas phase of gas-solid
     systems: application to methane-hydrogen from 66.88° to
     116.53° K and up to 125 atmospheres, in "Advances in

Cryogenic Engineering, Vol. 10," Plenum Press, New York (1965), p. 160.

10. S. Yoshimura, Measurements and prediction of vapor-liquid equilibria at low temperature and high pressure conditions, doctoral dissertation. Hiroshima, Japan (1972).

11. H. Sagara, Y. Arai, and S. Saito, Vapro-liquid equilibria of binary and ternary systems containing hydrogen and light hydrocarbons, J. Chem. Eng. (Japan) 5:339 (1972).

12. A.L. Benham and D.L. Katz, Vapor-liquid equilibria for hydrogen-light hydrocarbon systems at low temperatures, AIChE J. 3:33 (1957).

13. N.F. Yuskevich and N.S. Torocheshnikov, Investigation of the coexistence of liquid and vapor phases in solutions of nitrogen and carbon monoxide, Zh. Khim. Prom. 13:1273 (1936).

14. F.B. Sprow and J.M. Prausnitz, Vapor-liquid equilibria for five cryogenic mixtures, AIChE J. 12:780 (1966).

15. H. Cheung and D.I.-J. Wang, Solubility of volatile gases in hydrocarbon solvents at cryogenic temperatures, IEC Fundam. 3:355 (1964).

16. R. Kobayashi and P. Chappelear, Low Temperature Data from Rice University for Vapor-Liquid and P-V-T Behavior, Technical Publication TP-4, Gas Processes Association, Tulsa, Oklahoma (1974).

17. M.R. Cines, et al., Nitrogen-methane vapro-liquid equilibria, Chem. Eng. Progr. Symp. Ser. 49(6):1 (1953).

18. R. Stryjek, P.S. Chappelear, and R. Kobayashi, Low temperature vapor-liquid equilibria of the nitrogen-methane system, J. Chem. Eng. Data 19:334 (1974).

19. G.M. Wilson, Vapor-liquid equilibria of nitrogen, methane, ethane, and propane binary mixtures at lng temperatures from total pressure measurements, in "Advances in Cryogenic Engineering, Vol. 20," Plenum Press, New York (1975), p. 164.

20. A. Toyama, et al., Vapor liquid equilibria at low temperatures: The carbon monoxide-methan system, in "Advances in Cryogenic Engineering, Vol. 7," Plenum Press, New York (1962), p. 125.

21. R.D. Gray, Jr., Correlation of H_2/Hydrocarbon VLE Using Redlich-Kwong Variants, AIChE 70th Annual Meeting, Paper No. 65 c, New York (1977).

22. H.F. Cosway and D.L. Katz, Low-temperature vapor-liquid equilibria in ternary and quaternary systems containing hydrogen, nitrogen, methane and ethane, AIChE J. 5:46 (1959).

23. W.R. Parrish and M. J. Hiza, Liquid-vapor equilibria in the nitrogen-methane system between 95 and 120 K, in "Advances in Cryogenic Engineering, Vol. 19," Plenum Press, New York

DIELECTRIC CONSTANTS OF SATURATED LIQUID PROPANE, ISOBUTANE, AND NORMAL BUTANE*

W. M. Haynes and B. A. Younglove

National Bureau of Standards
Boulder, Colorado

INTRODUCTION

The dielectric constants of saturated liquid propane, isobutane, and normal butane have been measured over temperature ranges from near their triple point temperatures to approximately 300 K using a concentric cylinder capacitor. These measurements were part of a large scale program at this laboratory to determine the thermophysical properties of the major components of liquefied natural gas. Dielectric constant data have been reported previously for nitrogen[1], methane[2], and ethane[3]. Dielectric constant measurements can serve as a reliable and simple substitute for density measurements through the use of the Clausius–Mossotti (CM) function, which relates density to dielectric constant. (Typically, the CM function changes only a few percent from its ideal gas value to that for the dense liquid at the triple point.) Accurate densities of low molecular weight alkanes are needed to help provide equitable custody transfer of liquefied natural and petroleum gases.

EXPERIMENTAL EQUIPMENT AND PROCEDURES

The dielectric constant measurements were performed using a concentric cylinder capacitor of the same design as that developed by Younglove and Straty[4]. It was fabricated from copper with

---

*This work was carried out at the National Bureau of Standards under the sponsorship of Gas Research Institute.

0.05-mm-thick Kapton* film insulators. Although the dimensions of the capacitor used here were proportionally the same as that used in Ref. 4, its overall dimensions (1.8 cm outside diameter × 6.4 cm length) were significantly smaller. This resulted in a vacuum capacitance of approximately 33 pF, or about the same as that from Ref. 4. The capacitor described in Ref. 4 was used here for a few measurements on propane as a check against the results obtained with the new capacitor utilized for most of the measurements of the present study. This check was necessitated by observed large differences between experimental data for propane from this work and the results from an independent study[5], as discussed in the next section.

The apparatus described here also contains a magnetic suspension densimeter. Density data[6] have been obtained over the same temperature ranges as the dielectric constant measurements with an earlier version of the magnetic suspension densimeter. Details of the present apparatus will be presented elsewhere[7].

The dielectric constant (ε) is determined from the ratio (C/C_o) of the capacitance with liquid between the electrodes to the vacuum capacitance. The capacitance was measured to a resolution of 10^{-4} pF with a three-terminal ac bridge operated at an oscillator frequency of 5 kHz. This resulted in an estimated uncertainty of approximately 0.01% in the dielectric constant.

All of the gases were of research grade quality. The minimum purities as specified by the supplier were 99.99 mole percent for propane and 99.9 mole percent for isobutane and normal butane.

The data for each of the fluids were obtained from several fillings of the cell. Vacuum points were taken before and after each filling of the cell and were stable to better than 10^{-3} pF over the duration of this study. Short term stability was of the order of 10^{-4} pF.

RESULTS

The measured dielectric constants of saturated liquid propane, isobutane, and normal butane are presented as a function of temperature and density in Tables I–III. Densities were obtain-

---

*In order to describe materials and experimental procedures adequately, it is occasionally necessary to identify commercial products by manufacturer's or trade names. In no instance does such identifications imply endorsement by National Bureau of Standards nor does it imply that the particular product is necessarily the best available for that purpose.

Table I. Comparisons of dielectric constant data and Clausius–Mossotti values for propane from this work and Ref. 8 with Eq. 2.

Temp. K	Density mol/L	Dielectric Constant		Diff. %	C–M Function cm^3/mol		Diff. %
		expt.	calc.		expt.	calc.	
300.00	11.082	1.65110	1.65083	.02	16.092	16.087	.03
295.00	11.261	1.66348	1.66342	.00	16.083	16.082	.01
293.19	11.324	1.66789	1.66788	.00	16.080	16.080	.00
290.00	11.434	1.67559	1.67564	-.00	16.076	16.077	-.01
285.00	11.601	1.68741	1.68752	-.01	16.070	16.072	-.01
280.00	11.763	1.69893	1.69911	-.01	16.063	16.067	-.02
275.00	11.921	1.71023	1.71043	-.01	16.058	16.062	-.02
270.00	12.074	1.72124	1.72151	-.02	16.052	16.057	-.03
265.00	12.224	1.73216	1.73239	-.01	16.048	16.052	-.03
260.00	12.371	1.74281	1.74307	-.01	16.043	16.048	-.03
255.00	12.514	1.75335	1.75358	-.01	16.039	16.043	-.02
250.00	12.655	1.76375	1.76394	-.01	16.035	16.039	-.02
245.00	12.793	1.77401	1.77416	-.01	16.032	16.034	-.02
240.00	12.929	1.78443	1.78426	.01	16.033	16.030	.02
235.00	13.062	1.79456	1.79425	.02	16.031	16.026	.03
230.00	13.194	1.80454	1.80414	.02	16.028	16.022	.04
228.40	13.236	1.80752	1.80729	.01	16.024	16.020	.02
[a]228.40	13.236	1.80720	1.80729	-.01	16.019	16.020	-.01
[a]228.00	13.246	1.80785	1.80808	-.01	16.016	16.020	-.02
225.00	13.324	1.81436	1.81396	.02	16.024	16.017	.04
220.00	13.452	1.82385	1.82370	.01	16.016	16.014	.01
215.00	13.579	1.83353	1.83337	.01	16.012	16.010	.01
210.00	13.705	1.84312	1.84300	.01	16.008	16.006	.01
205.00	13.829	1.85268	1.85259	.01	16.004	16.003	.01
200.00	13.952	1.86222	1.86213	.00	16.000	15.999	.01
195.00	14.075	1.87179	1.87166	.01	15.998	15.996	.01
190.00	14.196	1.88142	1.88116	.01	15.997	15.993	.02
185.00	14.317	1.89086	1.89066	.01	15.993	15.990	.02
180.00	14.436	1.90038	1.90016	.01	15.990	15.987	.02
175.00	14.556	1.90990	1.90967	.01	15.988	15.985	.02
170.00	14.674	1.91946	1.91919	.01	15.986	15.983	.02
165.00	14.792	1.92898	1.92874	.01	15.984	15.981	.02
160.00	14.910	1.93857	1.93831	.01	15.983	15.979	.02
155.00	15.027	1.94814	1.94794	.01	15.981	15.978	.02
150.00	15.144	1.95780	1.95761	.01	15.980	15.978	.02
145.00	15.261	1.96752	1.96734	.01	15.980	15.977	.01
140.00	15.377	1.97733	1.97714	.01	15.980	15.977	.01
135.00	15.494	1.98721	1.98702	.01	15.981	15.978	.01
130.00	15.610	1.99715	1.99700	.01	15.981	15.980	.01
125.00	15.726	2.00718	2.00708	.00	15.983	15.982	.01
120.00	15.842	2.01732	2.01729	.00	15.985	15.985	.00
[a]120.00	15.842	2.01731	2.01729	.00	15.985	15.985	.00
115.00	15.958	2.02759	2.02763	-.00	15.988	15.989	-.00
[a]115.00	15.958	2.02740	2.02763	-.01	15.986	15.989	-.02
110.00	16.074	2.03804	2.03813	-.00	15.993	15.994	-.01
105.00	16.190	2.04869	2.04882	-.01	15.999	16.000	-.01
100.00	16.307	2.05952	2.05971	-.01	16.006	16.008	-.01
[a]100.00	16.307	2.05960	2.05971	-.01	16.007	16.008	-.01
95.00	16.423	2.07053	2.07084	-.02	16.014	16.017	-.02
90.00	16.540	2.08173	2.08225	-.02	16.023	16.029	-.04

[a] Data obtained using capacitor described in reference 4.

Table I. (Continued)

Temp. K	Density mol/L	Dielectric Constant		Diff. %	C-M Function cm³/mol		Diff. %
		expt.	calc.		expt.	calc.	
[b]293.19	.412	1.01995	1.01990	.01	16.050	16.006	.27
[b]313.12	.686	1.03325	1.03330	-.00	15.990	16.013	-.14
[b]323.12	.880	1.04280	1.04289	-.01	15.986	16.019	-.21
[b]343.08	1.453	1.07150	1.07157	-.01	16.024	16.040	-.10
[b]353.09	1.914	1.09510	1.09513	-.00	16.053	16.058	-.03
[b]358.10	2.245	1.11215	1.11228	-.01	16.052	16.069	-.11
[b]363.11	2.695	1.13610	1.13592	.02	16.105	16.084	.13
[b]365.60	3.022	1.15340	1.15335	.00	16.100	16.095	.03
[b]368.10	3.537	1.18150	1.18125	.02	16.130	16.109	.13
[b]369.10	3.927	1.20270	1.20271	-.00	16.118	16.119	-.00
[b]368.10	6.488	1.35240	1.35117	.09	16.202	16.151	.31
[b]365.60	7.068	1.38720	1.38661	.04	16.173	16.151	.14
[b]363.11	7.459	1.41060	1.41088	-.02	16.140	16.150	-.06
[b]358.10	8.022	1.44580	1.44638	-.04	16.128	16.146	-.11
[b]353.09	8.469	1.47470	1.47502	-.02	16.131	16.141	-.06
[b]343.08	9.157	1.51930	1.51992	-.04	16.114	16.131	-.10
[b]323.12	10.176	1.58820	1.58824	-.00	16.109	16.110	-.01
[b]313.12	10.600	1.61630	1.61731	-.06	16.078	16.100	-.14
[b]293.19	11.336	1.66790	1.66872	-.05	16.063	16.079	-.10

[b]Data from reference 8.

ed from an expression used by Haynes and Hiza[6] to represent orthobaric liquid density data for these fluids. The dielectric constant (ε) data were combined with the densities (ρ) to calculate the CM function from the following relation,

$$CM = \left(\frac{\varepsilon-1}{\varepsilon+2}\right) \frac{1}{\rho} \tag{1}$$

The CM values from this study were then combined with data of Sliwinski[8,9] at higher temperatures (and lower densities) and were fit to one of the following expressions,

$$CM = A + B\rho + C\rho^2 + D/T \tag{2}$$

or

$$CM = A + B\rho + C\rho^2 + DT \,, \tag{3}$$

where CM is in units of cm³/mol, ρ in mol/L, and T in K. Equation 2 was used for propane and isobutane while Eq. 3 was employed for normal butane.

Coefficients for each of the fluids from a least-squares analysis, along with rms deviations of fit, are given in Table IV. For isobutane and normal butane at temperatures above 300 K,

Table II. Comparisons of dielectric constant data and Clausius-Mossotti values for isobutane from this work and Ref. 9 with Eq. 2.

Temp. K	Density mol/L	Dielectric Constant expt.	Dielectric Constant calc.	Diff. %	C-M Function cm³/mol expt.	C-M Function cm³/mol calc.	Diff. %
303.15	9.360	1.73462	1.73479	-.01	21.016	21.020	-.02
300.00	9.429	1.74140	1.74149	-.01	21.017	21.019	-.01
295.00	9.536	1.75189	1.75199	-.01	21.015	21.017	-.01
293.15	9.576	1.75576	1.75583	-.00	21.015	21.016	-.01
290.00	9.642	1.76222	1.76234	-.01	21.013	21.015	-.01
285.00	9.745	1.77242	1.77255	-.01	21.011	21.014	-.01
280.00	9.847	1.78253	1.78264	-.01	21.010	21.012	-.01
275.00	9.947	1.79259	1.79263	-.00	21.010	21.010	-.00
270.00	10.046	1.80241	1.80251	-.01	21.007	21.009	-.01
265.00	10.143	1.81218	1.81230	-.01	21.005	21.008	-.01
260.00	10.238	1.82196	1.82201	-.00	21.006	21.007	-.00
255.00	10.333	1.83154	1.83166	-.01	21.003	21.006	-.01
250.00	10.426	1.84119	1.84125	-.00	21.004	21.005	-.01
245.00	10.519	1.85071	1.85078	-.00	21.003	21.004	-.01
240.00	10.610	1.86022	1.86028	-.00	21.002	21.003	-.01
235.00	10.701	1.86967	1.86973	-.00	21.002	21.003	-.01
230.00	10.791	1.87928	1.87915	.01	21.005	21.003	.01
228.40	10.819	1.88215	1.88217	-.00	21.003	21.003	-.00
225.00	10.880	1.88870	1.88856	.01	21.006	21.003	.01
220.00	10.968	1.89809	1.89796	.01	21.006	21.004	.01
215.00	11.056	1.90731	1.90734	-.00	21.004	21.004	-.00
210.00	11.143	1.91670	1.91673	-.00	21.005	21.005	-.00
205.00	11.229	1.92610	1.92611	-.00	21.006	21.006	-.00
200.00	11.315	1.93557	1.93553	.00	21.009	21.008	.00
195.00	11.401	1.94493	1.94494	-.00	21.010	21.010	-.00
190.00	11.486	1.95438	1.95439	-.00	21.013	21.013	-.00
185.00	11.571	1.96387	1.96387	-.00	21.015	21.016	-.00
180.00	11.655	1.97338	1.97340	-.00	21.019	21.019	-.00
175.00	11.739	1.98295	1.98296	-.00	21.023	21.023	-.00
170.00	11.823	1.99256	1.99259	-.00	21.027	21.028	-.00
165.00	11.906	2.00224	2.00229	-.00	21.032	21.033	-.00
160.00	11.990	2.01216	2.01206	.01	21.041	21.039	.01
155.00	12.073	2.02202	2.02191	.01	21.048	21.046	.01
150.00	12.156	2.03192	2.03187	.00	21.055	21.054	.00
145.00	12.238	2.04200	2.04192	.00	21.064	21.063	.01
140.00	12.321	2.05220	2.05210	.00	21.075	21.073	.01
135.00	12.403	2.06249	2.06241	.00	21.086	21.085	.01
130.00	12.486	2.07299	2.07288	.01	21.099	21.098	.01
125.00	12.568	2.08367	2.08353	.01	21.114	21.112	.01
120.00	12.650	2.09450	2.09436	.01	21.131	21.129	.01
115.00	12.733	2.10563	2.10542	.01	21.150	21.147	.01
[a]283.20	.101	1.00627	1.00628	-.00	20.709	20.743	-.17
[a]293.19	.137	1.00851	1.00853	-.00	20.686	20.738	-.25
[a]303.15	.181	1.01128	1.01128	.00	20.739	20.734	.02
[a]313.12	.235	1.01470	1.01468	.00	20.763	20.733	.15
[a]323.12	.303	1.01892	1.01895	-.00	20.694	20.733	-.19
[a]333.11	.385	1.02413	1.02412	.00	20.738	20.735	.02
[a]343.08	.486	1.03057	1.03055	.00	20.754	20.739	.07
[a]353.09	.612	1.03863	1.03859	.00	20.769	20.747	.11
[a]363.11	.772	1.04888	1.04883	.00	20.778	20.758	.09
[a]368.10	.866	1.05504	1.05493	.01	20.807	20.765	.20

[a] Data from reference 9, dielectric constant (ϵ) obtained from index of refraction (n) via $\epsilon = n^2$.

888   W. M. Haynes and B. A. Younglove

Table III.   Comparisons of dielectric constant data and
             Clausius–Mossotti values for normal butane
             from this work and Ref. 10 with Eq. 3.

Temp. K	Density mol/L	Dielectric Constant		Diff. %	C–M Function cm^3/mol		Diff. %
		expt.	calc.		expt.	calc.	
303.15	9.746	1.75424	1.75490	-.04	20.613	20.628	-.07
300.00	9.810	1.76031	1.76083	-.03	20.611	20.622	-.05
295.00	9.910	1.76975	1.77015	-.02	20.606	20.614	-.04
290.00	10.007	1.77898	1.77933	-.02	20.598	20.606	-.04
288.71	10.032	1.78137	1.78169	-.02	20.597	20.604	-.03
285.00	10.104	1.78823	1.78841	-.01	20.594	20.598	-.02
280.00	10.199	1.79724	1.79738	-.01	20.586	20.589	-.01
275.00	10.292	1.80626	1.80625	.00	20.581	20.581	.00
270.00	10.385	1.81504	1.81503	.00	20.573	20.573	.00
265.00	10.476	1.82385	1.82374	.01	20.566	20.564	.01
260.00	10.566	1.83253	1.83238	.01	20.559	20.556	.01
255.00	10.656	1.84117	1.84096	.01	20.551	20.548	.02
250.00	10.744	1.84967	1.84948	.01	20.543	20.539	.02
245.00	10.832	1.85818	1.85794	.01	20.535	20.531	.02
240.00	10.919	1.86671	1.86636	.02	20.529	20.522	.03
235.00	11.005	1.87511	1.87475	.02	20.521	20.514	.03
230.00	11.091	1.88354	1.88309	.02	20.514	20.506	.04
228.40	11.118	1.88640	1.88576	.03	20.515	20.503	.06
225.00	11.176	1.89200	1.89141	.03	20.508	20.497	.05
220.00	11.260	1.90021	1.89970	.03	20.498	20.489	.04
215.00	11.344	1.90840	1.90797	.02	20.488	20.481	.04
210.00	11.428	1.91661	1.91622	.02	20.479	20.472	.03
205.00	11.511	1.92482	1.92446	.02	20.470	20.464	.03
200.00	11.594	1.93297	1.93269	.01	20.460	20.456	.02
195.00	11.677	1.94109	1.94091	.01	20.450	20.447	.01
190.00	11.759	1.94928	1.94913	.01	20.442	20.439	.01
185.00	11.841	1.95741	1.95736	.00	20.432	20.431	.00
180.00	11.923	1.96556	1.96558	-.00	20.422	20.422	-.00
175.00	12.004	1.97377	1.97381	-.00	20.413	20.414	-.00
170.00	12.086	1.98187	1.98206	-.01	20.403	20.406	-.01
165.00	12.167	1.99005	1.99031	-.01	20.393	20.398	-.02
160.00	12.248	1.99821	1.99858	-.02	20.384	20.389	-.03
155.00	12.330	2.00658	2.00687	-.01	20.376	20.381	-.02
150.00	12.411	2.01483	2.01517	-.02	20.367	20.373	-.03
145.00	12.492	2.02301	2.02351	-.02	20.357	20.364	-.04
140.00	12.573	2.03139	2.03187	-.02	20.349	20.356	-.03
135.00	12.654	2.03968	2.04025	-.03	20.339	20.348	-.04
[a]283.20	.069	1.00425	1.00425	.00	20.549	20.538	-.06
[a]293.19	.092	1.00566	1.00569	-.00	20.473	20.556	-.41
[a]303.15	.123	1.00763	1.00762	.00	20.606	20.575	.15
[a]313.12	.162	1.01004	1.01001	.00	20.653	20.594	.29
[a]323.12	.210	1.01303	1.01304	-.00	20.604	20.612	-.04
[a]333.11	.269	1.01675	1.01676	-.00	20.618	20.631	-.06
[a]343.08	.342	1.02134	1.02132	.00	20.670	20.650	.10
[a]353.09	.431	1.02700	1.02698	.00	20.682	20.668	.06
[a]363.11	.543	1.03401	1.03406	-.00	20.661	20.687	-.13
[a]368.10	.606	1.03813	1.03813	-.00	20.693	20.697	-.02

[a] Data from reference 9, dielectric constant (ϵ) obtained from index of
refraction (n) via $\epsilon = n^2$.

dielectric constant data along the coexistence boundary were not available. Therefore index of refraction (n) data of Sliwinski[9] were used to calculate dielectric constants from $\varepsilon = n^2$ to establish the low density behavior of CM function. Dispersion effects should not be present in this region. Thus Eq. 2 and 3 with coefficients as given in Table IV can be used to calculate the CM function (and dielectric constant) for both gas and liquid along the coexistence boundary.

It should be noted that the experimental CM values of Sliwinski[8,9] were calculated using his density data. Therefore, for propane in the region of overlap between the data from this study and that of Sliwinski[8], the differences in the CM values result primarily from density discrepancies of approximately 0.05%.

Figures 1 and 2 show plots of the CM function as a function of density along the coexistence boundary. These figures illustrate that propane and isobutane exhibit distinctly different behavior from that of simple, nonpolar molecules. For propane and isobutane, very weakly polar fluids with relatively small dipole moments, a minimum in the CM values was observed with the CM values inversely proportional to temperature at low temperatures. Normal butane did not exhibit this type of behavior.

Also shown on Figs. 1 and 2 are comparisons with the data of independent investigators[5,8-11]. In general, the present dielectric constant data for all three fluids agree with the values from Sliwinski[8,9]; Pan, Mady, and Miller[5]; Thompson and Miller[10]; and Luo and Miller[11] within combined uncertainties with one notable exception. The dielectric constants of propane at temperatures from 91-115 K of Pan, Mady, and Miller[5] are approximately 0.8 to 1.0% larger than the results from the present

Table IV. Coefficients of Eq. 2 and 3 from least-squares analysis for propane, isobutane, and normal butane. Deviations (rms) from fit for dielectric constant and C–M function are given. Eq. 2.: CM = A + Bρ + Cρ2 + D/T. Eq. 3.: CM = A + Bρ + Cρ2 + DT. Units: CM, cm^3/mol; ρ, mol/L; T, K.

Fluid	Eq.	A	B	C	D	$\Delta\varepsilon/\varepsilon$, rms %	$\Delta CM/CM$, rms %
C_3H_8	2	15.85640	0.058615	−0.004421	37.11378	0.021	0.078
iC_4H_{10}	2	20.45394	0.121509	−0.009475	78.47544	0.005	0.069
nC_4H_{10}	3	20.00951	0.000630	0.000490	0.001865	0.018	0.091

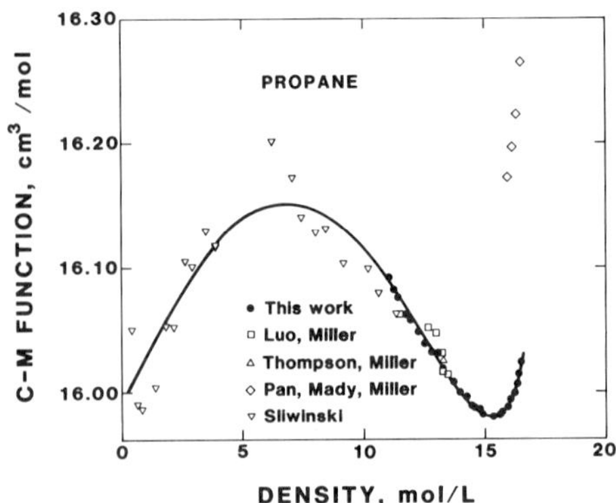

Fig. 1. Clausius Mossotti function of propane
as a function of density along coexistence boundary;
data from this work and Ref. 5,8,10,11.

work. As noted earlier, these large differences resulted in a few
measurements (see Table I) with a capacitor that had been
previously tested in this laboratory in measurements on nitrogen,
methane, and ethane.

With the same apparatus used for the measurements presented
here, dielectric constant data have recently been obtained for
compressed liquid propane, isobutane, and normal butane at
pressures to 35 MPa and at temperatures from near the triple
points up to 300 K. These data will be published in subsequent
papers.

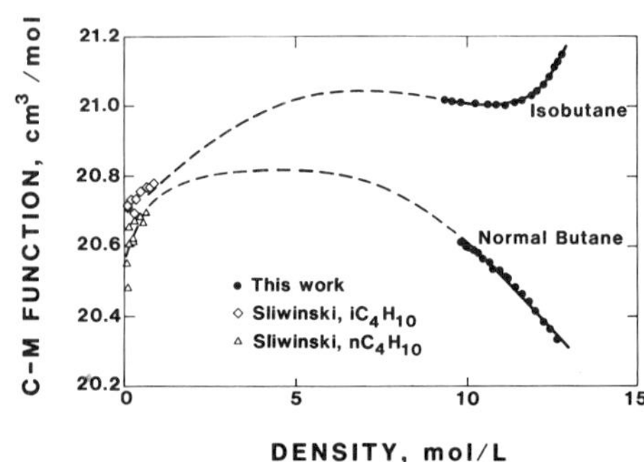

Fig. 2. Clausius Mossotti function of isobutane and
normal butane as a function of density along coexistence
boundary; data from this work and Ref. 9.

REFERENCES

1. J.F. Ely and G.C. Straty, Dielectric constants and molar polarizabilities of saturated and compressed fluid nitrogen, J. Chem. Phys. 61-1480 (1974).

2. G.C. Straty and R.D. Goodwin, Dielectric constant and polarizability of saturated and compressed fluid methane, Cryogenicis 13:712 (1973).

3. L.A. Weber, Dielectric constant data and the derived Clausius-Mossotti, function for compressed gaseous and liquid ethane, J. Chem. Phys. 65:446 (1976).

4. B.A. Younglove and G.C. Straty, A capacitor for accurate wide range dielectric constant measurements on compressed fluids, Rev. Sci. Instrum. 41:1087 (1970)

5. W.P. Pan, M.H. Mady and R.C. Miller, Dielectric constants and Clausius-Mossotti functions for simple liquid mixtures: systems containing nitrogen, argon and light hydrocarbons, A.I.Ch.E.J. 21:283 (1975).

6. W.M. Haynes and M.J. Hiza, Measurements of the orthobaric liquid densities of methane, ethane, propane, isobutane, and normal butane, J. Chem. Thermodynamics 9:179 (1977).

7. W.M. Haynes, New apparatus for density and dielectric constant measurements on cryogenic fluids, including mixtures, at pressures to 35 MPa (to be published).

8. P. Sliwinski, Die Clausius-Mossotti-Funktion fuer die gesaettiges Daempfe und Fluessigkeiten des Aethans und Propans, Z. Phys. Chem. Neue. Folge 68:91 (1969).

9. P. Sliwinski, Die Lorentz-Lorenz-Funktion von dampffoermigen und Fluessigem Aethan, Propan und Butan, Z. Phys. Chem. Neue. Folge 63:263 (1969).

10 R.T. Thompson, Jr. and R.C. Miller, Densities and Dielectric Constants of LPG Components and Mixtures at Cryogenic Storage Conditions in "Advances in Cryogenic Enginnering Vol 25," Plenum Press, New York (1980), p. 698.

11. C.C. Luo and R.C. Miller, Densities and dielectric constants for some LPG components and mixtures at cryogenic and standard temperatures, Cryogenics 21:85 (1981).

VAPOR-LIQUID EQUILIBRIUM VALUES FOR NITROGEN-ARGON-OXYGEN AT HIGH OXYGEN CONCENTRATIONS*

I. Funada

Kobe Steel, Ltd.
Kobe, Japan

S. Yoshimura, H. Masuoka, and M. Yorizane

Hiroshima University
Hiroshima, Japan

and

C.-T. Fu, and B. C.-Y. Lu

University of Ottawa
Ottawa, Canada

INTRODUCTION

Vapor-liquid equilibrium (VLE) values are established in this work for the nitrogen-argon-oxygen system at oxygen concentrations higher than 95 mole per cent in the vapor phase. These values are useful in the design of the bottom portion of the upper column used in air separation plants. The number of stages required in this region amounts to one third of all the stages in the upper column.

Although VLE values for the system have been published[1,5], very few experimental values are available at temperatures slightly below the normal boiling point of pure oxygen.

Isothermal VLE values were determined experimentally at 18 temperatures in the range of 89.97–90.16 K for a total of 60 data points. These values were successfully correlated by means of the Wilson equation using only binary parameters evaluated from the constituent binary systems.

EXPERIMENTAL

The data were determined by means of a vapor and liquid recirculation method using a visual equilibrium cell. The assembly is
comprised of the following components: equilibrium cell and
cryostat, bath pressure control system, temperature and pressure
measuring systems, vapor and liquid recirculation systems, liquid
nitrogen feeding system, charging system, sampling and vacuum
systems. A schematic flow diagram of the assembly is shown in
Fig. 1. The equilibrium cell is immersed in liquid nitrogen
contained in a glass vessel, which is placed in an air-tight
cryostat. The cryostat has a diameter of 34 cm. and a height of
100 cm. The pressure of the cryostat is controlled to within ±
0.37 torr by balancing the pressure against a reference pressure
maintained in a steel cylinder (#24 of Fig. 1) in a constant
temperature bath. By doing so, it is possible to control the
temperature of the bath fluid to within ± 0.002 K at about 90 K.
In order to keep the temperature of the equilibrium cell constant,
the cell as well as the heat exchange coils (#6 of Fig. 1) for
both the recirculating vapor and liquid are totally immersed in
liquid nitrogen in the glass vessel (#8 of Fig. 1). The liquid
nitrogen at the bottom of the cryostat is continuously pumped up
into the glass vessel by means of a nitrogen gas jet pump. As a
result, liquid nitrogen in the vessel is well stirred. The system
is shown schematically in Fig. 2. The temperature is measured
with a platinum resistor in conjunction with a potentiometer. The
equilibrium cell is equipped with two such resistors. The pressure is measured with mercury manometers. Purity of the nitrogen,
argon, and oxygen was determined as 99.99%, 99.99% and 99.95%,
respectively. The accuracies of the temperature, pressure, and
composition measurements are believed to be ± 0.01 K, ± 0.3 torr,
and ± 3% respectively. The experimentally determined VLE values
are reported in Table I. In addition, experimental values are
plotted at constant temperature and pressure conditions on a
triangular diagram for three selected isotherms in Fig. 3. The
data obtained by Wilson (4) at 88.28 K are also included in this
figure to illustrate the agreeable trend.

The relative volatilities, α, of nitrogen to oxygen and argon
to oxygen are plotted against the mole fraction of oxygen in the
liquid phase in Fig. 4. It appears that in the oxygen concentration range investigated, the relative volatility of argon to
oxygen does not vary with the concentration of oxygen, while the
relative volatility of nitrogen to oxygen gradually decreases as
the mole fraction of oxygen in the liquid phase approaches to the
pure oxygen.

Table I. Experimental VLE Values for the Ternary System
Nitrogen (1) – Argon (2) – Oxygen (3).

Temp. (K)	Press. (Torr)	Liquid Composition			Vapor Composition		
		X_1	X_2	X_3	Y_1	Y_2	Y_3
90.16	760.4	0.07	0.04	99.89	0.18	0.06	99.76
90.15	760.1	0.04	0.07	99.89	0.11	0.12	99.77
	759.6	0.05	0.11	99.84	0.14	0.19	99.67
90.14	760.1	0.20	0.16	99.64	0.73	0.28	98.99
	760.1	0.20	0.17	99.63	0.66	0.26	99.08
	760.6	0.08	1.40	98.52	0.30	2.06	97.64
90.13	760.5	0.24	0.03	99.73	0.77	0.05	99.18
	758.0	0.27	0.17	99.56	0.90	0.28	98.82
	758.6	0.19	0.14	99.67	0.72	0.19	99.09
90.12	759.7	0.19	0.13	99.68	0.65	0.21	99.14
	759.7	0.14	0.85	99.01	0.53	1.25	98.22
90.11	758.4	0.10	1.14	98.76	0.34	1.69	99.97
	758.2	0.23	0.14	99.63	0.77	0.23	99.00
	761.5	0.17	0.18	99.65	0.63	0.26	99.11
90.10	760.8	0.25	0.47	99.28	0.81	0.69	98.56
	760.2	0.09	1.04	98.87	0.35	1.56	98.09
	758.7	0.08	1.04	98.88	0.32	1.56	98.12
	759.5	0.20	0.27	99.53	0.70	0.43	98.87
90.09	758.9	0.14	0.18	99.68	0.64	0.29	99.07
	758.8	0.19	0.13	99.68	0.69	0.20	99.11
	761.5	0.17	0.74	99.09	0.61	1.11	98.28
90.08	757.7	0.35	0.86	98.79	1.22	1.28	97.50
	757.6	0.36	0.45	99.19	1.18	0.71	98.11
	757.5	0.21	0.30	99.49	0.72	0.43	98.85
	759.5	0.34	0.31	99.35	1.20	0.45	98.35
	761.6	0.10	1.79	98.11	0.37	2.73	96.90
	758.6	0.28	0.29	99.43	1.07	0.49	98.44
90.07	761.7	0.14	1.55	98.31	0.51	2.32	97.17
	760.5	0.43	0.39	99.18	1.48	0.60	97.92
90.06	757.1	0.22	0.87	98.91	0.71	1.34	97.95
	758.7	0.26	0.58	99.16	0.95	0.87	98.18
90.06	759.6	0.38	0.39	99.23	1.25	0.60	98.15
	759.0	0.33	0.48	99.19	1.20	0.72	98.08
	759.2	0.06	2.12	97.82	0.21	3.04	96.75
	759.2	0.08	2.58	97.34	0.33	3.80	95.87
90.05	760.1	0.33	0.04	99.63	1.23	0.07	98.70
	760.3	0.33	0.46	99.21	1.03	0.74	98.23
	760.4	0.10	1.99	97.91	0.36	2.91	96.73
90.04	758.5	0.37	0.56	99.07	1.18	0.83	97.99
	758.0	0.29	0.67	99.04	1.13	1.01	97.86
	759.4	0.08	2.54	97.38	0.30	3.69	96.01
	760.3	0.19	1.29	98.52	0.73	1.84	97.43
90.02	757.4	0.32	0.50	99.18	1.13	0.77	98.10
	759.6	0.27	1.06	98.67	1.00	1.42	97.58
90.01	760.6	0.34	0.40	99.26	1.33	0.59	98.08
	760.1	0.08	2.90	97.02	0.29	4.28	95.43
	760.7	0.08	2.53	97.39	0.30	3.72	95.98
90.00	759.5	0.14	2.33	97.53	0.52	3.59	95.89
	758.9	0.34	0.75	98.91	1.26	1.16	97.58
	758.6	0.43	0.71	98.86	1.32	1.00	97.68
	760.4	0.08	2.83	97.09	0.29	4.18	95.53
	759.9	0.30	1.32	98.38	1.15	1.83	97.02
	760.0	0.29	1.48	98.23	1.05	2.18	96.77
	761.0	0.13	2.30	97.57	1.05	3.37	96.11
89.98	758.4	0.34	0.46	99.20	1.23	0.71	98.06
	759.4	0.21	1.26	98.53	0.86	1.89	97.25
	760.1	0.21	1.63	98.16	0.80	2.41	96.79
89.97	759.3	0.35	1.06	98.59	1.13	1.68	97.19
	761.6	0.29	0.28	99.43	1.07	0.46	98.47
	761.9	0.26	0.87	98.87	0.97	1.27	97.76

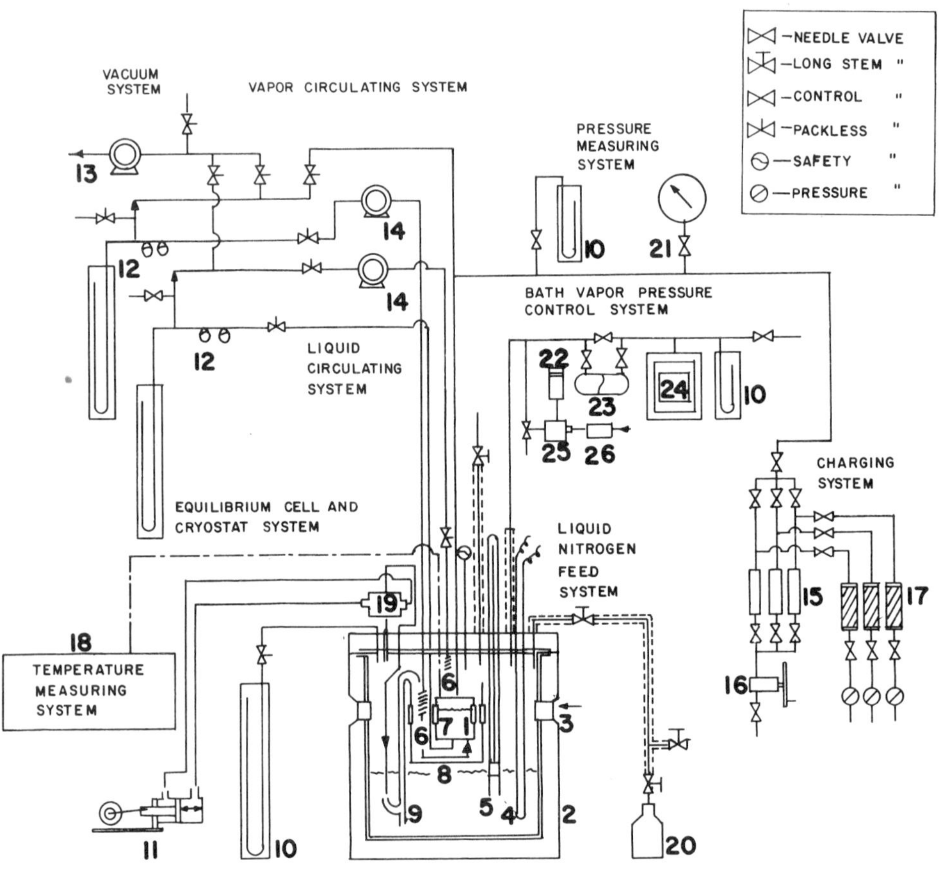

Fig. 1. Schematic flow diagram of experimental apparatus.

1. equilibrium cell	12. α type glass-sampler
2. cryostat	13. vacuum pump
3. glass window	14. diaphram pump
4. heater	15. gas reservoir
5. liquid level indicator	16. mercury pump
6. heat exchanger	17. gas dryer
7. platinum resistor	18. temperature measuring system
8. glass vessel	19. double stroke device
9. cryojet pump	20. liquid N_2 container
10. mercury manometer	21. precision pressure gauge
11. reciprocal pump	22-27. pressure control system

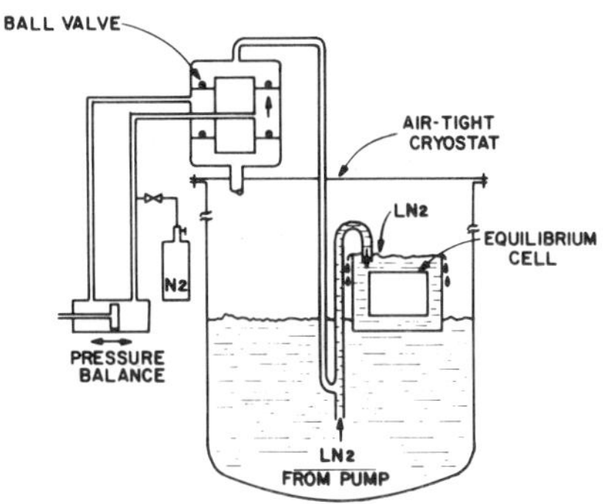

Fig. 2. Constant heat tub in air-tight cryostat with liquid nitrogen lift pump.

CORRELATION OF RESULTS

An attemp was made to represent the data obtained by means of the Wilson equation[6], using the binary parameters obtained from the data previously reported[7] (Method 1), the values reported by Wilson[4] (Method 2), and the values determined by Praunitz et al.[8] (Method 3).

The Wilson equation for a binary system is of the following form:

$$\ln\gamma_1 = \ln(x_1 + \Lambda_{12} \cdot x_2) + x_2\left[\frac{\Lambda_{12}}{x_1 + \Lambda_{12}\cdot x_2} - \frac{\Lambda_{21}}{\Lambda_{21}\cdot x_1 + x_2}\right]$$

$$\ln\gamma_2 = -\ln(\Lambda_{21}\cdot x_1 + x_2) - x_1\left[\frac{\Lambda_{12}}{x_1 + \Lambda_{12}\cdot x_2} - \frac{\Lambda_{21}}{\Lambda_{21}\cdot x_1 + x_2}\right]$$

where

$$\Lambda_{12} \equiv \frac{v_2}{v_1} \exp\left[-\frac{(\lambda_{12} - \lambda_{11})}{R\cdot T}\right]$$

$$\Lambda_{21} \equiv \frac{v_1}{v_2} \exp\left[-\frac{(\lambda_{21} - \lambda_{22})}{R\cdot T}\right]$$

The values of the binary parameters used in the calculation are listed in Table II. The calculated VLE values by means of Method I are compared with the experimental values in Fig. 5 and 6, indicating the satisfactory results obtained. The results obtained from the other two methods are also satisfactory. A comparison

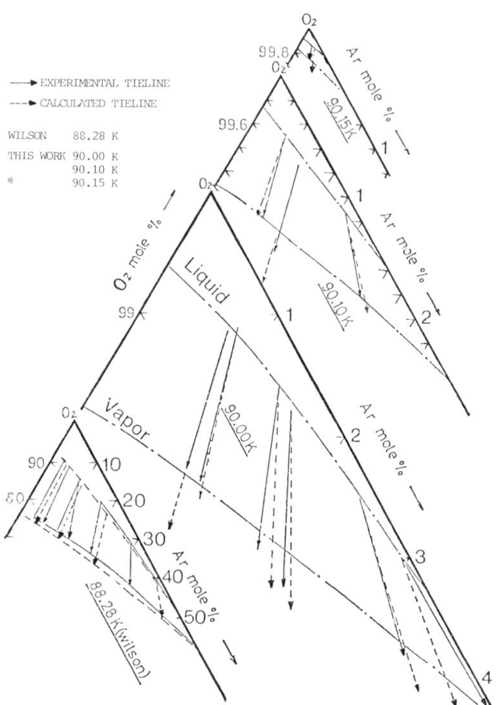

Fig. 3. Comparison of experimental and calculated V–L–E values of N_2 – AR – O_2 system by Method I at one atmospheric pressure.

of the results obtained by the three methods is shown in Table III. Some of the calculated results are also compared with the experimental values in Fig. 3. The comparison indicates that there is no significant difference between the calculated values obtained by the three sets of binary parameters. It should also be mentioned that the calculated temperatures do not always agree with the experimental values. A slight variation of the operating pressure will not, however, affect the calculated values.

Table II. Values of Wilson's Parameters

Method	N_2 – Ar		N_2 – O_2		Ar – O_2	
	$\lambda_{12}-\lambda_{11}$	$\lambda_{21}-\lambda_{22}$	$\lambda_{12}-\lambda_{11}$	$\lambda_{21}-\lambda_{22}$	$\lambda_{12}-\lambda_{11}$	$\lambda_{21}-\lambda_{22}$
I	3.14736	28.1111	2.3795	32.095	12.15778	24.02464
II	3.14736	28.1111	25.006	29.95114	13.37032	18.89195
III	−25.51	59.37	−30.98	68.66	16.03	17.57

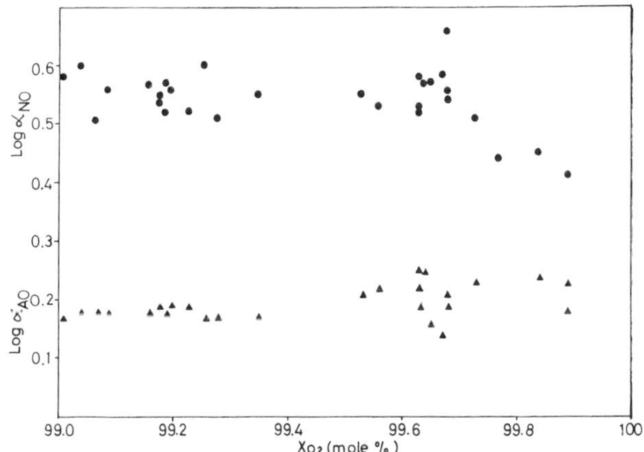

Fig. 4. Correlation of relative volatilities
to mole fraction of oxygen concentration.

Table III. Comparison of Experimental and Calculated V–L–E
    Composition of the N_2 $(1)_1$ – Ar (2) – O_2 (3) System

Temp. (K)	Calculated Error	Method 1	Method 2	Method 3
90.16	ΔY_1	0.084	0.077	0.099
~ 90.11	ΔY_2	0.041	0.035	0.047
	ΔY_3	0.115	0.098	0.113
90.10	ΔY_1	0.100	0.103	0.119
~ 90.03	ΔY_2	0.094	0.071	0.080
	ΔY_3	0.190	0.155	0.195
90.01	ΔY_1	0.079	0.070	0.099
~ 89.97	ΔY_2	0.146	0.111	0.124
	ΔY_3	0.220	0.169	0.219
90.16 ~ 89.97	ΔT	0.098	0.101	0.097

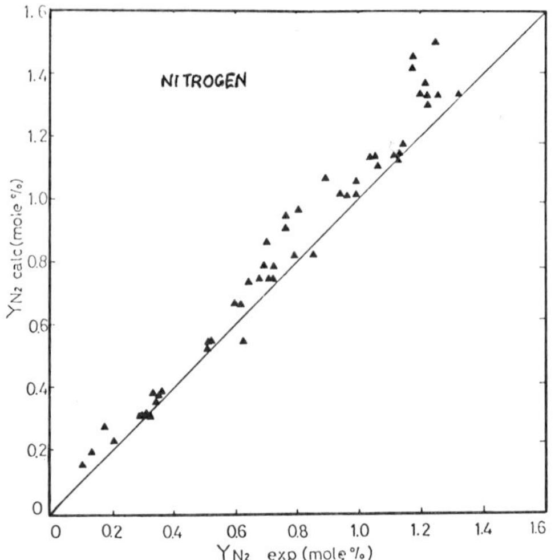

Fig. 5a. Comparison of experimental and calculated vapor composition of N_2 for the N_2-Ar-O_2 system from 89.97 to 90.16 K by means of method 1.

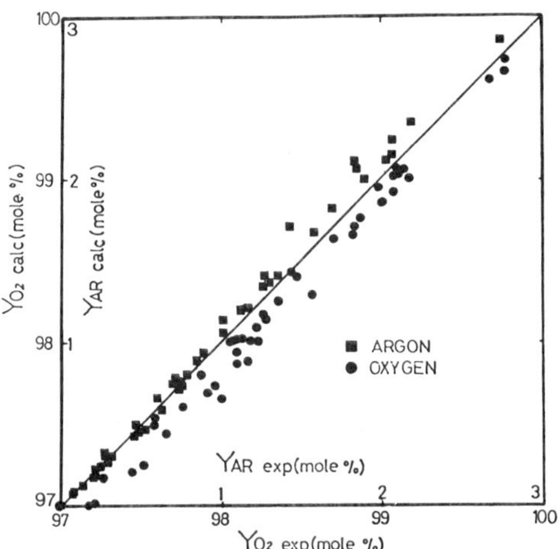

Fig. 5b. Comparison of experimental and calculated vapor compositions of Ar and O_2 for the N_2-Ar-O_2 system from 89.97 to 90.16 K by means of method 1.

NOTATION

```
R                 = gas constant
T                 = absolute temperature
v                 = molar volume of pure liquid (Wilson equation)
x                 = mole fraction in liquid phase (Wilson equation)
X                 = mole percent in liquid phase (Table 1)
Y                 = mole percent in vapor phase (Table 1)
```

Greek Symbols

```
α                 = relative volatility
γ                 = activity coefficient
Λ                 = adjustable parameter of Wilson equation
λ                 = energy parameter of Wilson equation
```

Subscripts

```
1,2,3,11,22       = component identification
12,21             = mixture identification
```

REFERENCES

1. M.L. Sagenkahn and H.L. Fink, Office of Scientific Research
    and Development Report No. 4493 (1944).
2. J. Weishaupt, Angew, Chem. B20: 1575 (1948).
3. V.G. Fastovskii and U.V. Petrovskii, Zh. Fiz. Khim. 31: 836
    (1957).
4. G.M. Wilson, P.M. Silverberg, and M.G. Zellner, in "Advances
    in Cryogenic Engineering, Vol. 10", Plenum Press, New York
    (1965), p. 192.
5. G.B. Narinskii, Zh. Fiz. Khim., 43: 408 (1969).
6. G.M. Wilson, J. Am. Chem. Soc., 86: 127 (1964).
7. M. Yorizance, et al., Chem. Engng. Science, 33: 641 (1978)
8. J.M. Prausnitz, et al., "Computer Calculations for
    Multicomponent Vapor-Liquid Equilibria", Prentice-Hall,
    Englewood Cliffs, New Jersey (1967).

THE EQUATION OF STATE AND THERMODYNAMIC PROPERTIES OF LIQUID CARBON MONOXIDE*

S. F. Barreiros, J. C. G. Calado, M. N. da Ponte, and W. B. Streett[†]

Centro de Quimica Estrutural
Lisbon, Portugal

INTRODUCTION

In view of the prominent part carbon monoxide plays in an ever increasing number of energy related projects, a comprehensive set of thermodynamic data on the liquid covering a wide range of temperature and pressure is clearly needed and, as far as we know, does not exist. On the other hand, and from the molecular standpoint, carbon monoxide is a relatively simple though slightly polar liquid, and therefore a suitable example for the testing of the modern theories of liquids. These two facts led us to perform (p,V,T) measurements on liquid carbon monoxide at temperatures from near the normal boiling point to 125 K (the critical temperature being 132.91 K) and at pressures between the vapor pressure and 140 MPa, at regular intervals of these two variables.

EXPERIMENTAL RESULTS AND DISCUSSION

The results reported here were obtained using an expansion method described by Albuquerque et al.[1] The carbon monoxide used in the experiments was supplied by Air Liquide and guaranteed to have a purity of 99.97% or better. We estimate the accuracy of the measured molar volumes to be ± 0.1%.

---

*This work was partially financed by Junta Nacional de Investigacao Cientifica e Tecnologica, under research contract 34.78.107.

† Present address: School of Chemical Engineering, Cornell University, Ithaca, New York.

"

(p,V,T) Measurements

Our experimental (p,V,T) points for liquid CO are recorded in Table I and shown in Fig. 1. These results were fitted by the Strobridge equation

$$p = RT\rho + (a_1 RT + a_2 + a_3/T + a_4/T^2 + a_5/T^4)\rho^2 + (a_6 RT + a_7)\rho^3$$

$$+ a_8 T\rho^4 + (a_9/T^2 + a_{10}/T^3 + a_{11}/T^4)\exp(a_{16}\rho^2)\rho^3 + (a_{12}/T^2$$

$$+ a_{13}/T^3 + a_{14}/T^4)\exp(a_{16}\rho^2)\rho^5 + a_{15}\rho^6 . \tag{1}$$

Table I. Experimental (p,V,T) values of liquid CO. ΔV is the difference between experimental and calculated molar volumes.

$\dfrac{P}{\text{MPa}}$	$\dfrac{V}{\text{cm}^3\text{mol}^{-1}}$	$\dfrac{\Delta V}{\text{cm}^3\text{mol}^{-1}}$	$\dfrac{P}{\text{MPa}}$	$\dfrac{V}{\text{cm}^3\text{mol}^{-1}}$	$\dfrac{\Delta V}{\text{cm}^3\text{mol}^{-1}}$	$\dfrac{P}{\text{MPa}}$	$\dfrac{V}{\text{cm}^3\text{mol}^{-1}}$	$\dfrac{\Delta V}{\text{cm}^3\text{mol}^{-1}}$	$\dfrac{P}{\text{MPa}}$	$\dfrac{V}{\text{cm}^3\text{mol}^{-1}}$	$\dfrac{\Delta V}{\text{cm}^3\text{mol}^{-1}}$
T = 81.64 K											
2.16	35.164	0.012	7.75	34.620	0.002	30.01	33.037	-0.007	52.89	31.921	-0.003
4.24	34.936	-0.008	10.99	34.352	0.010	42.83	32.382	0.007	65.57	31.433	0.009
5.37	34.835	-0.001	22.73	33.472	-0.016						
T = 90.24 K											
3.00	36.786	0.002	10.99	35.768	-0.019	51.51	32.851	0.007	94.52	31.117	-0.005
4.71	36.543	-0.004	22.29	34.719	0.004	63.78	32.273	0.005	112.44	30.584	-0.002
7.44	36.173	-0.025	33.18	33.894	-0.013	82.80	31.507	-0.012			
T = 95.21 K											
1.98	38.107	0.016	10.23	36.784	-0.007	61.98	32.829	0.010	99.76	31.336	0.007
3.22	37.886	0.021	19.53	35.736	0.022	72.88	32.327	0.000	114.78	30.859	-0.008
5.36	37.516	0.012	30.01	34.804	0.027	85.97	31.823	0.014	124.85	30.578	-0.008
7.48	37.172	-0.005	40.76	34.024	0.020	95.21	31.487	0.006			
T = 100.18 K											
3.02	39.180	0.000	10.33	37.762	-0.027	30.84	35.401	0.005	82.80	32.346	-0.010
5.41	38.662	-0.006	10.51	37.742	-0.018	51.10	33.929	0.004	111.34	31.319	-0.017
7.71	38.218	-0.015	22.43	36.209	-0.003	72.46	32.808	0.005			
T = 105.14 K											
2.60	40.791	0.012	14.30	38.182	-0.006	52.48	34.413	0.018	92.86	32.363	-0.015
5.52	39.962	0.002	19.95	37.370	0.015	61.43	33.864	0.010	108.72	31.799	0.000
7.92	39.378	-0.015	29.87	36.223	0.015	72.87	33.242	-0.010	130.36	31.115	-0.010
10.71	38.815	-0.007	40.76	35.255	0.020	81.01	32.879	0.006	138.36	30.916	0.013
T = 110.19 K											
2.86	42.469	-0.003	19.67	38.323	-0.008	71.09	33.798	-0.030	118.92	31.839	0.003
5.39	41.497	-0.023	40.48	35.929	0.013	90.39	32.872	-0.032	131.33	31.475	0.028
7.57	40.836	-0.010	64.19	34.215	0.000	108.72	32.158	-0.028	138.37	31.248	0.006
10.54	40.074	-0.002									
T = 114.88 K											
3.16	44.315	0.022	30.49	37.639	0.002	72.05	34.217	-0.011	109.27	32.492	-0.030
5.00	43.365	0.028	41.86	36.426	0.017	80.67	33.760	-0.006	120.99	32.122	0.018
11.35	41.095	0.001	52.75	35.490	0.001	91.76	33.210	-0.029	131.33	31.747	-0.020
20.84	39.042	-0.005	63.22	34.760	0.001	100.31	32.858	-0.015	139.74	31.532	0.020
T = 120.04 K											
2.96	47.284	-0.002	20.78	40.115	0.003	60.47	35.510	0.015	101.00	33.264	0.012
6.20	44.831	-0.026	29.74	38.603	0.017	72.46	34.699	-0.003	112.31	32.774	-0.016
10.54	42.855	-0.016	40.62	37.268	0.033	81.42	34.200	0.008	120.03	32.539	0.037
10.61	42.833	-0.012	51.93	36.172	0.010	92.45	33.627	-0.010	127.34	32.305	0.059
T = 125.02 K											
2.99	51.664	0.001	30.01	39.425	-0.002	71.78	35.224	-0.013	113.41	33.084	-0.029
4.86	48.720	-0.015	40.90	37.923	0.001	80.46	34.684	-0.018	118.65	32.891	-0.051
7.55	46.384	0.022	50.69	36.891	0.002	92.87	34.016	-0.017	131.06	32.476	0.023
10.81	44.542	0.030	62.54	35.886	-0.002	102.38	33.559	-0.024	138.37	32.232	0.024
20.29	41.333	-0.021									

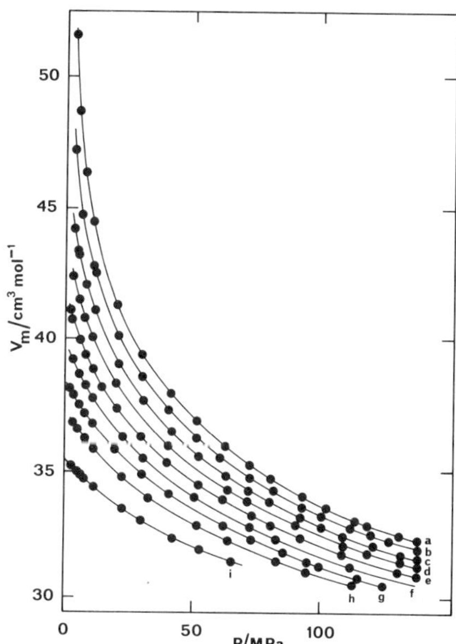

Fig. 1. Experimental (p,V,T) results for liquid CO: a,
125.02 K; b, 120.04 K; c, 114.88 K; d, 110.19 K; e,
105.14 K; f, 100.18 K; g, 95.21 K; h, 90.24 K; i, 81.64 K.

The constants, a_1 to a_{16} are listed in Table II. The
differences between experimental volumes and those calculated
from Eq. 1 are included in Table I. The curves in Fig. 1 are
those given by Eq. 1.

Vapor Pressure Correlation

We selected the vapor pressure data obtained by Clayton and
Giauque[2] and Michels et al.[3] These results were fitted by an
equation proposed by Goodwin[4]

$$\ln \Pi = \sum_{i=1}^{3} a_i x^i + a_4 x(1 - x)^{\delta}, \tag{2}$$

where $\Pi = p/p_t$ and $x = (1 - T_t/T)/(1 - T_t/T_c)$.

Both sets of data had to be corrected for the different
temperature scales used. The data of Michels et al. were obtained
on IPTS-48 and were therefore easily converted to IPTS-68. As to
the results of Clayton and Giauque, the conversion to IPTS-68 was
performed by adding 0.05 K to all their temperatures, as suggested
in Ref. 5.

Table II. Parameters for Eq. 1, where p/MPa, ρ/mol cm^{-3}, T/K. Added in proof: $A_{12} = 0.6195\ 771851 \times 10^{15}$.

i	1	2	3	4	5	6
a_i	0.4580256419×10^4	$-0.1645545494 \times 10^8$	$0.2444975899 \times 10^{10}$	$-0.1343030923 \times 10^{12}$	$0.1536137657 \times 10^{15}$	$-0.5172198660 \times 10^5$

i	7	8	9	10	11	12
a_i	0.1600949848×10^9	$-0.2759802971 \times 10^7$	$-0.4733188076 \times 10^{13}$	$0.5384479353 \times 10^{15}$	$-0.1791658222 \times 10^{17}$	$0.6196771851 \times 10^{15}$

i	13	14	15	16
a_i	$-0.9353866848 \times 10^{17}$	$0.3719857162 \times 10^{19}$	$0.1692518152 \times 10^{12}$	-0.0056

The parameters for Eq. 2 are listed in Table III. For the triple-point pressure p_t and temperature T_t we took the values obtained by Clayton and Giauque, namely 15.373 kPa and 68.14 K, the latter resulting from the addition of 0.05 K to the value quoted by these authors. The critical temperature T_c was taken as 132.91 K, resulting from the conversion to IPTS-68 of the value adopted in Ref. 6.

The largest deviation of the experimental vapor pressure values from Eq. 2, in the form of temperature differences, is 0.02 K. The standard deviation of the fit is 0.014 K. The normal boiling point calculated from Eq. 2 is 81.65 K.

Derived Quantities

Enthalpies of vaporization $\Delta_\ell^{g,\sigma} H$ were calculated at regular intervals of temperature using the Clapeyron equation, and are given in Table IV. The molar volumes of the saturated gas $V^{g,\sigma}$ were estimated with virial coefficient data. An equation of the type proposed by Guggenheim[7] was fitted to second virial coefficients reported in Ref.8, obtained by Mathot et al. and Schramm et al., and was used for interpolation and extrapolation over the temperature range under study. Since there were no third virial coefficients available for CO at those temperatures, we estimated them using a formula proposed by Chueh and Prausnitz,[9] making d = 0.68 as obtained from a correlation of d as a function of the acentric factor ω. The molar volumes of the gas so obtained, together with the values of B and C are also given in Table IV. The enthalpy of vaporization calculated at the normal boiling point of CO is (5991 $\pm$ 30) Jmol^{-1}, only slightly lower

Table III. Parameters for Eq. 2.

P_t /kPa	T_t /K	T_c /K	a_1	a_2	a_3	a_4	δ
15.373	68.14	132.91	0.4530546×10^{-1}	0.5325888×10^{1}	0.5731425×10^{-1}	0.5652296×10^{1}	0.1026520×10^{1}

than the value (6040 ± 4) Jmol^{-1} obtained through direct experimental measurement by Clayton and Giauque.[2]

The molar volumes of the saturated liquid $V^{\ell,\sigma}$, were calculated at each temperature by substituting into Eq. 1 the vapor pressure derived from Eq. 2 for that temperature, and are given in Table IV. The most recent and accurate data with which to compare these results are those of Terry et al.[10] In Fig. 2 we plot the deviations of the experimental values obtained by those authors, divided by a factor of 1.004 as explained in Ref.11, from our calculated ones, over the temperature range where the two sets of results overlap. As this figure shows, the agreement between our results and those of Terry et al. is within the combined experimental error bars ($\pm$ 0.1%) associated with both sets of data.

Table IV also includes values for the configurational internal energy of the saturated liquid $U_+^{\ell,\sigma}$, calculated from the equation

$$U_+^{\ell,\sigma} = -\Delta_\ell^{g,\sigma}H - (H^{pg} - H^{g,\sigma}) + RT - pV^{\ell,\sigma}, \tag{3}$$

where $(H^{pg} - H^{g,\sigma})$ is the enthalpy increase when the gas expands isothermally from the vapor pressure to zero pressure. This term was calculated from the equation

$$H^{pg} - H^{g,\sigma} = -RT\left[(B - TdB/dT)/V^{g,\sigma} - (2C - TdC/dT)/2(V^{g,\sigma})^2\right] \tag{4}$$

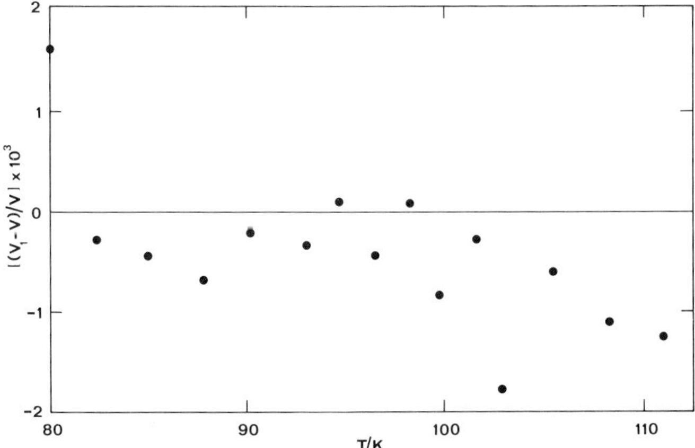

Fig. 2. Plot of the difference between the experimental orthobaric liquid volumes V_1 and for CO of Terry et al. (corrected as explained in the text), and the values V calculated from Eqs. 1 and 2.

908   **S. F. Barreiros et al.**

As can be seen in Table IV, the $(-\Delta_{\ell}^{g,\sigma}H)$ term is the most important in Eq. 3, and therefore the uncertainty in the values of $U_{+}^{\ell,\sigma}$ should be about the same as for the enthalpies of vaporization. This is estimated to be $\pm 0.5\%$ at the lower temperatures but should be larger at the higher temperatures. As the vapor pressure increases with temperature, so does the imprecision in the calculated saturated gas volumes used in the Clapeyron equation, due to the growing importance of the third and higher virial coefficients.

Equation 1 was used to derive values of the molar volumes of the liquid at regular intervals of temperature and pressure, recorded in Table V, and also to obtain the mechanical coefficients of liquid CO: the isothermal compressibility $k_T = - V^{-1} (\partial V/\partial p)_T$ and the isobaric thermal expansivity $\alpha_p = V^{-1} (\partial V/\partial T)_p$, recorded in the same table.

Configurational internal energies of the liquid U_{+}^{ℓ} are also given, as calculated at regular intervals of temperature and pressure from the equation

$$U_{+}^{\ell} = U_{+}^{\ell,\sigma} + \int_{V^{\ell,\sigma}}^{V(p)} [T(\partial p/\partial T)_V - p]dV \ . \tag{5}$$

Table IV. Thermodynamic properties of liquid and gaseous CO: vapor pressure p, second and third virial coefficients B and C, respectively, molar volumes of saturated gas $V^{g,\sigma}$ and liquid $V^{\ell,\sigma}$, molar enthalpy of vaporization to give saturated vapor $\Delta_{\ell}^{g,\sigma}H$, molar enthalpy of infinitely dilute gas H^{pg}, molar enthalpy of saturated vapor $H^{g,\sigma}_{\ell,\sigma}$ and molar configurational internal energy of saturated liquid $U_{+}^{\ell,\sigma}$.

$\dfrac{T}{K}$	$\dfrac{P}{MPa}$	$\dfrac{B}{cm^3 mol^{-1}}$	$\dfrac{C}{cm^6 mol^{-2}}$	$\dfrac{V^{g,\sigma}}{cm^3 mol^{-1}}$	$\dfrac{V^{\ell,\sigma}}{cm^3 mol^{-1}}$	$\dfrac{dP/dT}{MPaK^{-1}}\times 10^2$	$\dfrac{\Delta_{\ell}^{g,\sigma}H}{J\ mol^{-1}}$	$\dfrac{H^{pg} - H^{g,\sigma}}{J\ mol^{-1}}$	$\dfrac{U_{+}^{\ell,\sigma}}{J\ mol^{-1}}$
80	0.08359	-297.1	-19120	7645.6	35.014	0.9950	6058	76.7	-5473
81.65[a]	0.10128	-285.2	-15407	6401.5	35.373	1.1526	5991	89.3	-5405
90	0.23813	-234.9	-3704	2885.0	37.148	2.1957	5628	177.3	-5066
95	0.36809	-211.2	- 200	1908.2	38.351	3.0297	5382	253.5	-4860
100	0.54385	-191.0	1879	1307.1	39.737	4.0283	5105	352.6	-4648
105	0.77374	-173.6	3073	919.3	41.344	5.1961	4790	480.9	-4430
110	1.06636	-158.5	3716	658.6	43.232	6.5380	4426	648.6	-4206
115	1.43058	-145.3	4019	476.4	45.524	8.0627	3996	872.5	-3977
120	1.87598	-133.6	4113	343.7	48.522	9.7893	3467	1186.4	-3747
125	2.41365	-123.2	4082	240.3	53.177	11.767	2753	1682.4	-3524

[a] normal boiling temperature.

Table V. Thermodynamic properties of liquid CO as a function of temperature and pressure; molar volumes V_m, isothermal compressibilities k_T, isobaric thermal expansivities α_p and molar configurational internal energies U_+^l.

						T/K				
	p MPa	80	90	95	100	105	110	115	120	125
$\dfrac{V_m}{cm^3\ mol^{-1}}$	1	34.922	37.029	38.226	39.618	41.263				
	2	34.824	36.873	38.035	39.369	40.924	42.776	45.098	48.356	
	5	34.547	36.461	37.515	38.706	40.053	41.590	43.384	45.588	48.562
	10	34.129	35.854	36.781	37.803	38.927	40.157	41.517	43.062	44.901
	30	32.822	34.095	34.749	35.444	36.175	36.933	37.718	38.541	39.425
	50	31.867	32.896	33.421	33.927	34.543	35.126	35.718	36.323	36.953
	70	31.112	31.986	32.432	32.900	33.381	33.868	34.357	34.850	35.354
	100	30.213	30.933	31.305	31.696	32.096	32.497	32.896	33.292	33.690
	140	29.284	29.873	30.186	30.514	30.849	31.184	31.513	31.835	32.154
$\dfrac{k_T \times 10^3}{MPa^{-1}}$	1	2.874	4.202	5.182	6.584	8.707				
	2	2.802	4.050	4.950	6.207	8.041	10.901	15.930	27.427	
	5	2.609	3.659	4.373	5.314	6.581	8.344	10.924	15.085	23.078
	10	2.344	3.165	3.680	4.318	5.112	6.106	7.379	9.075	11.495
	30	1.691	2.092	2.307	2.543	2.804	3.088	3.405	3.761	4.176
	50	1.335	1.584	1.706	1.835	1.971	2.115	2.269	2.435	2.620
	70	1.110	1.282	1.363	1.446	1.532	1.622	1.715	1.816	1.926
	100	0.890	1.003	1.053	1.104	1.156	1.209	1.265	1.323	1.386
	140	0.708	0.782	0.813	0.845	0.876	0.909	0.942	0.977	1.014
$\dfrac{\alpha_p \times 10^3}{K^{-1}}$	1	6.365	6.115	6.819	7.705	8.834				
	2	6.251	5.955	6.598	7.388	8.346	9.681	11.965	17.134	
	5	5.938	5.533	6.042	6.624	7.248	8.025	9.156	11.098	14.974
	10	5.492	4.999	5.359	5.745	6.110	6.494	7.018	7.834	9.201
	30	4.328	3.768	3.936	4.081	4.168	4.216	4.300	4.453	4.744
	50	3.624	3.146	3.267	3.355	3.387	3.384	3.383	3.424	3.542
	70	3.140	2.749	2.863	2.931	2.943	2.917	2.878	2.883	2.937
	100	2.627	2.377	2.479	2.536	2.533	2.495	2.442	2.407	2.410
	140	2.162	2.056	2.162	2.213	2.207	2.157	2.091	2.033	2.001
$\dfrac{U_+^l}{J\ mol^{-1}}$	1	-5489	-5082	-4875	-4662	-4438				
	2	-5506	-5101	-4899	-4691	-4474	-4249	-4012	-3759	
	5	-5556	-5156	-4964	-4769	-4569	-4366	-4163	-3975	-3839
	10	-5630	-5236	-5058	-4879	-4697	-4515	-4341	-4198	-4143
	30	-5862	-5469	-5323	-5179	-5030	-4882	-4749	-4664	-4697
	50	-6026	-5626	-5498	-5373	-5239	-5105	-4987	-4920	-4983
	70	-6149	-5741	-5628	-5515	-5392	-5265	-5153	-5097	-5175
	100	-6282	-5866	-5771	-5674	-5561	-5440	-5334	-5284	-5374
	140	-6395	-5977	-5903	-5823	-5720	-5603	-5499	-5451	-5547

REFERENCES

1. G.M.N. Albuquerque et al, and Cryogenics, 20:601 (1980).
2. J.O. Clayton and W.F. Giauque, J. Am. Chem. Soc., 54:2610 (1932).
3. A. Michels, T. Wassenaar, and Th. N. Zwietering, Physica, 18:160 (1952).
4. R.D. Goodwin, J. Res. Nat. Bur. St. A, 73:487 (1969).
5. D. Ambrose et al., J. Chem. Thermodynamics, 11:1089 (1979).
6. A.S. Leah, Carbon Monoxide, in "Thermodynamic Functions of Gases. Vol. I", F. Din, ed. Butterworths Scientific Publications, London (1956).
7. R.D. Weir et al., Trans. Faraday Soc., 63: 1320 (1967).

8. J.H. Dymond and E.B. Smith, "The Virial Coefficients of Pure Gases and Mixtures. A Critical Compilation," 2nd ed., Clarendon Press, Oxford (1980).

9. P.L. Chueh and J. M. Prausnitz, A.I. Ch. E. J., 13:896 (1967).

10. M.J. Terry et al.,J. Chem. Thermodynamics, 1:413 (1969).

11. M. Nunes da Ponte, W. B. Streett, and L. A. K. Staveley, J. Chem. Thermodynamics, 10:151 (1978).

A THERMODYNAMIC PROPERTY FORMULATION FOR NEON FROM THE TRIPLE POINT TO 700 K WITH PRESSURE TO 700 MPa

R. T. Jacobsen, R. B. Stewart, and J. C. J. Teng

University of Idaho
Moscow, Idaho

INTRODUCTION

This work reports on the development of a new thermodynamic formulation (equation of state and associated ancillary functions) for neon . Previous correlations of thermodynamic properties of neon include the work of McCarty and Stewart,[1] for which liquid properties were estimated by application of the principle of corresponding states because no published liquid data were available, and two unpublished studies at the University of Idaho. This new formulation considers the new data published after the work of McCarty and Stewart[1] and extends the range of validity beyond that of previous formulations. This formulation is not valid for the calculation of thermodynamic properties of neon for states near the critical point. Details of this work including brief tables of calculated thermodynamic properties are reported elsewhere by Teng.[2]

THE EXPERIMENTAL DATA

The P–ρ–T and velocity of sound data used in this work are summarized in Table I. These include 1686 P–ρ–T data points at temperatures from 26 to 723 K and pressures from 0.08 to 700 MPa. The velocity of sound data considered in this work include 347 data points in the liquid ranging from 25 to 40 K with pressures from 0.03 to 15 MPa, and vapor data from 26 to 473 K with pressures from 50 to 400 MPa.

The data of Gladun[2] are only C_v data available for neon. These measurements include 210 liquid data points ranging from 26 to 44 K for densities from 36 mol/dm^3 to 61 mol/dm^3. These data

911

Table I - Selected Property Data For Neon

Source	Date	Range of Values Temp. (K)	Press. (MPa)	Phase	Number of Data Points
		P-ρ-T Data			
Crommelin[6]	1918	56-293	2.2-9.4	Vapor	151
Gibbons[7]	1969	27-70	0.2-21	Liq., Vap.	269
Gladun[8]	1967	26-43	0.5-10	Liquid	113
Holburn[9]	1926	65-673	2-10	Vapor	71
Lippold[10]	1969	28-43	9.8-58	Liquid	62
Maslennikova[11]	1976	298-423	100-700	Vapor	52
Michels[12]	1928	273-374	3.3-50	Vapor	45
Michels[13]	1960	273-423	2.5-293	Vapor	316
Onosovskii[14]	1971	44-52	0.8-25	Liq., Vap.	33
Onosovskii[15]	1970	65-273	1.0-25	Vapor	147
Rabinovich[16]	1970	298-723	12-45	Vapor	58

Source	Date	Range of Values Temp.	Press.	Phase	Number of Data Points
		Velocity of Sound Data			
Scott[17]	1967	28-70	0.1-23	Liq.,Vap.	151
Streett[18]	1971	80-130	7.4-207	Vapor	137
Sullivan[19]	1967	70-120	0.7-31	Vapor	81
Fleury[20]	1969	25-32	------	Sat. Liq.	14
Gusewell[21]	1970	25-31	------	Sat. Liq.	7
Larson[22]	1970	25-37	0.06-2.7	Liquid	144
Larson[22]	1970	25-37	--------	Sat. Liq.	31
Naugle[23]	1972	25-37	0.09-0.8	Liquid	13
Keesom[24]	1934	26-273	0.01-0.1	Vapor	16
Pashkov[25]	1978	27-39	--------	Sat. Liq.	5
Pashkov[25]	1978	27-39	0.9-15	Liquid	46
Pitaevskaya[26]	1973	298-473	50-400	Vapor	71

were used in determining the coefficients of the equation of state
of neon.

The saturated liquid density and saturated vapor density data
used in this study were calculated by the extrapolation of a
preliminary equation of state to the vapor pressure defined by the
vapor pressure equation.

THE EQUATION OF STATE

The equation of state 1 for neon was determined by a least squares fit to selected data including P-ρ-T data, velocity of sound data, isochoric heat capacity data, and saturation densities. The technique of stepwise multiple regression developed by Wagner[4] and modified for equations of state by de Reuck and Armstrong[5] was used to systematize the choice of terms for the equation of state determined by fitting to selected data for neon. The comprehensive equation of state from which the terms of the final equation were selected included 50 terms. Equation 1 includes the 28 terms selected for the representation of the data for neon:

$$P = \rho RT + \rho^2(N_1 T + N_2/T + N_3 + N_4/T + N_5/T^2 + N_6/T^3 + N_7/T^4)$$

$$+ \rho^3 (N_9 T + N_{10} + N_{11}/T + N_{12}/T^2) + \rho^4(N_{14} T + N_{15})$$

$$+ \rho^5(N_{19} T + N_{21}/T) + \rho^6 (N_{24}/T^2) + \rho^7(N_{25}/T) + \rho^9(N_{29}/T)$$

$$+ \rho^{11} (N_{31}/T) + \rho^3(N_{35}/T^4) \exp (-\gamma\rho^2)$$

$$+ \rho^9(N_{40}/T^2 + N_{42}/T^4) \exp (-\gamma\rho^2)$$

$$+ \rho^{11}(N_{44}/T^3 + N_{45}/T^4) \exp (-\gamma\rho^2)$$

$$+ \rho^{13}(N_{46}/T^2 + N_{47}/T^3 + N_{48}/T^4) \exp (-\gamma\rho^2)$$

$$+ \rho^{15}(N_{50}/T^4) \exp (-\gamma\rho^2) \tag{1}$$

Where $\gamma = 1/(\rho_c^2)$. The coefficients for Eq. 1 are given in Table II.

ANCILLARY EQUATIONS AND DATUM STATES

The isobaric heat capacity of the ideal gas, C_p^o, for neon is $(5/2)R = 20.78585$ J/mol-K. The molecular weight of neon used in this work is 20.179.[27]

The enthalpy and entropy datum states are defined for the ideal gas at $T_o = 298.15$ K and $P_o = 0.101325$ MPa.

$$H^o (298.15 \text{ K}) - H^o (0.0 \text{ K}) = 6179.0 \text{ J/mol}$$
$$\left[H^o (0.0 \text{ K}) \text{ was assigned as } 0.0 \right]$$
$$S^o (298.15 \text{ K}, 0.101325 \text{ MPa}) = 146.214 \text{ J/mol-K.}$$

The modified Simon melting curve for neon from Crawford and Daniels[27] was used in this work:

$$P = A (T + D)^C + B \qquad\qquad (2)$$

with the numerical values for A, B, c, and D given in Table II.

The data selected in fitting the vapor pressure equation for neon are from three sources.[5,29,30] The vapor pressure e-quation was selected by statistical analysis from a comprehensive function of 21 terms. The final equation includes five terms from:

$$\ln(P/P_c) = (T_c/T) \sum_{i=1}^{21} N_i \, \tau^{(i/2)} \qquad\qquad (3)$$

where P_c = 2.6638 MPa, T_c = 44.448 K, and $\tau = 1 - (T/T_c)$. The five coefficients of this vapor pressure equation determined by a least squares stepwise regression fit to the selected data are also given in Table II.

Approximating functions for the saturation densities were determined by fitting the following functions to values calculated with the equation of state at the vapor pressure given by Eq. 3. These equations were selected from initial comprehensive functions of 25 terms by statistical analysis. The saturated liquid density data are represented by the equation:

$$\rho'/\rho_c = \sum_i N_i \, \tau^{(i/3)} \qquad\qquad (4)$$

where i = 7, 8, 9, 10, 11, 12, 13, 17, 19, 23, and 24. The satur-ated vapor density equation is

$$\rho''/\rho_c = \exp \sum_i N_i \, \tau^{(i/3)} \qquad\qquad (5)$$

where i = 1, 2, 3, 4, 5, 6, 8, 9, 12, and 15,
$\qquad \rho'$ = saturated liquid density, mol/dm^3,
$\qquad \rho''$ = saturated vapor density, mol/dm^3,
$\qquad \rho_c$ = 23.93 mol/dm^3, T_c = 44.448 K and
$\qquad \tau$ = $(1 - T/T_c)$

Coefficients for Eq. 4 and 5 are given in Table II.

The triple point temperature (24.562 K) and triple point pressure (0.043379 MPa) used in this work are taken from Ancsin.[31] The critical temperature (44.448 K) and density (23.93 mol/dm^3) were selected from Peck and Fenichel[32]. The critical pressure used in this work is 2.6638 MPa.

TABLE II

Coefficients for Equations for the Calculation of
Thermodynamic Properties of Argon
(P in MPa, T in kelvins, ρ in mol/dm^3)

EQUATION OF STATE

$N_1 = 0.5653412521 \times 10^{-4}$

$N_2 = 0.3492121815 \times 10^{-2}$

$N_3 = -0.5281123734 \times 10^{-1}$

$N_4 = 0.1511353269 \times 10^{1}$

$N_5 = -0.7726939279 \times 10^{2}$

$N_6 = 0.1627781825 \times 10^{4}$

$N_7 = -0.1278521631 \times 10^{5}$

$N_9 = 0.1448956689 \times 10^{-5}$

$N_{10} = 0.1684992471 \times 10^{-3}$

$N_{11} = -0.3161566218 \times 10^{-1}$

$N_{12} = 0.1681762544 \times 10^{1}$

$N_{14} = -0.9591366803 \times 10^{-8}$

$N_{15} = 0.1130059075 \times 10^{-4}$

$N_{19} = 0.2958715965 \times 10^{-9}$

$N_{21} = -0.2914091771 \times 10^{4}$

$N_{24} = -0.2884853628 \times 10^{-5}$

$N_{25} = 0.5087713410 \times 10^{-8}$

$N_{29} = 0.6838434551 \times 10^{-12}$

$N_{31} = -0.1316369231 \times 10^{-15}$

$N_{35} = -0.8654118297 \times 10^{3}$

$N_{40} = 0.4373531010 \times 10^{-8}$

$N_{42} = -0.5948450789 \times 10^{-5}$

$N_{44} = -0.2165033503 \times 10^{-9}$

$N_{45} = 0.9601308140 \times 10^{-8}$

$N_{46} = 0.1132387373 \times 10^{-14}$

$N_{47} = 0.1010428226 \times 10^{-13}$

$N_{48} = -0.2580251681 \times 10^{-11}$

$N_{50} = 0.2140433107 \times 10^{-15}$

VAPOR PRESSURE EQUATION

$N_1 = 0.1327344571 \times 10^{-1}$

$N_2 = -0.5836125731 \times 10^{1}$

$N_3 = 0.1206896288 \times 10^{1}$

$N_{11} = -0.3635239040 \times 10^{1}$

$N_{19} = 0.2036827828 \times 10^{2}$

MELTING CURVE EQUATION

$A = 1.570774$

$B = -58.770$

$c = 1.41852$

$D = -11.685$

SATURATED LIQUID DENSITY EQ. 4

$N_7 = -0.1990179158 \times 10^{-4}$

$N_8 = 0.3274708883 \times 10^{-3}$

$N_9 = -0.1252818114 \times 10^{-2}$

$N_{11} = 0.9879028042$

$N_{12} = 0.1595198540 \times 10^{1}$

$N_{13} = 0.5945020968$

$N_{17} = 0.6741384524$

$N_{19} = -0.1264509218 \times 10^{1}$

$N_{23} = 0.4530225993 \times 10^{1}$

$N_{24} = -0.4494008449 \times 10^{1}$

SATURATED VAPOR DENSITY EQ. 5

$N_1 = 0.7530172233 \times 10^{1}$

$N_2 = -0.1291485415 \times 10^{3}$

$N_3 = 0.8380198145 \times 10^{3}$

$N_4 = -0.3127353228 \times 10^{4}$

$N_5 = 0.6680763681 \times 10^{4}$

$N_6 = -0.7044292716 \times 10^{4}$

$N_8 = 0.7771814776 \times 10^{4}$

$N_9 = -0.5942500796 \times 10^{4}$

$N_{12} = 0.1266867159 \times 10^{4}$

$N_{15} = -0.3741415254 \times 10^{3}$

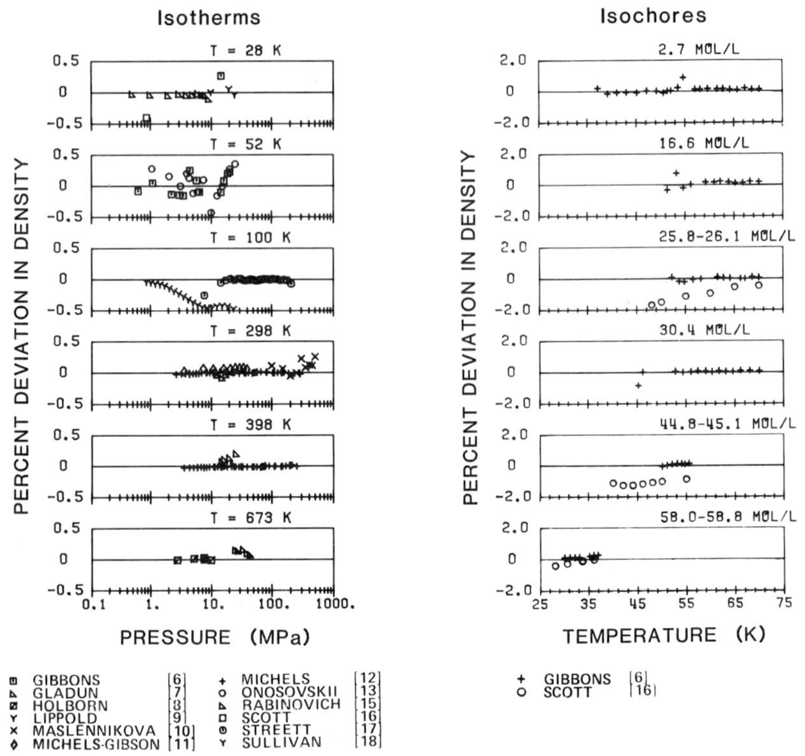

Fig. 1. Comparisons to Selected PVT Data.

SELECTED COMPARISONS TO EXPERIMENTAL DATA FOR NEON

Figures 1, 2, and 3 are abbreviated comparisons which illustrate the accuracy of the equation of state in representing
selected experimental data. In determining the equation of state,
data sets were selected based on concordance among data from

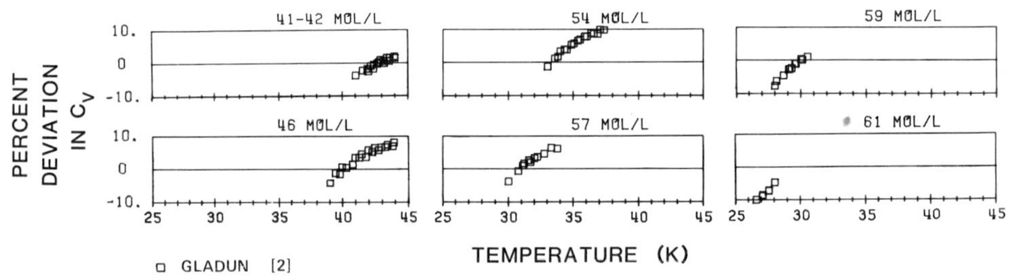

Fig. 2. Comparisons to Selected C_V Data.

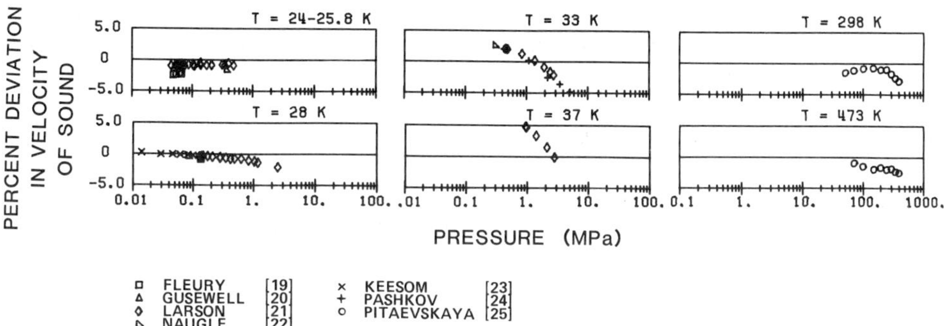

Figure 3.   Comparisons to Selected Velocity of Sound Data.

various experimenters in the same region of the P-ρ-T surface, and on relative agreement with trends established by data defining adjacent regions.   All comparisons use deviations $[(X_{data}-X_{calc})/X_{data}]$ x 100, where X is the property being compared. Further comparisons to experimental data are given by Teng[2].

The equation of state generally shows agreement with P-ρ-T data within the experimental uncertainty of the measured values, (within $\pm$ 0.25 ρ_c at temperatures within $\pm$ 0.03 T_c) except in the vicinity of the critical point.   Calculated density values are estimated to be accurate to within $\pm$ 0.1 percent except in this range.   Calculated heat capacity and velocity of sound values are estimated to have an accuracy of $\pm$ 2 percent.   This estimated accuracy for isochoric heat capacity values is not supported by the comparisons to data of Gladun[3] given in Fig. 2.

REFERENCES

1. R.D. McCarty and R.B. Stewart, in "Advances in Thermophysical Properties at Extreme Temperatures and Pressures," Serge Gratch, ed., ASME, New York (1965), p. 84.
2. J.C.J. Teng, "Thermodynamic Properties of Neon from the Triple Point to 700 K with Pressures to 700 MPa," Master's Thesis, University of Idaho (1980).
3. C. Gladun, "Spezifische Warme bei Konstantem Volumen, Temperatur, Molvolumen und Druck," Ph.D. Thesis, University of Dresden (1980).
4. W. Wagner, Eline mathematisch statistische Methode zum Aufstellen thermodynamischer Gleichungen-gezeight am Beispiel der Dampfdruck-kurve reiner fluiden Stoffe, VDI-Z, 3:39 (1974).
5. K.M. de Reuck and B. Armstrong, A Method of correlation using a search procedure, based on a step-wise least squares technique, and its application to an equation of state for propylene, Cryogenics 19 (9):505 (1979).

6. C.A. Crommelin, J.P. Martinez and H. Kammerlingh Onnes, Isotherms of monatomic substances and their binary mixtures XX. isotherms of neon from 20 C to 217 C., Commun. Phys. Lab. Univ. Leiden, No. 154a, (1918), p. 108.

7. R.M. Gibbons, The equation of state of neon between 27 and 70 K, Cryogenics 9 (4):251 (1969).

8. C. Gladun, Some thermodynamic investigations into liquid neon isotope mixtures, Cryogenics 7 (2):78 (1967).

9. L. Holborn and J. Otto, Über die Isothermen von Helium, Wasserstoff und Neon unterhalb - 200°, Z. Physik 38:259 (1926).

10. H. Lippold, Isothermal compressibility and density of liquid argon and neon up to pressure of 1,000 Kg/cm^2, Cryogenics 9 (2):112 (1969).

11. V. Ya. Maslennikova, A.N. Egorov, D.S. Tsiklis, Molar volumes and thermodynamic properties of neon at temperatures 25-125° C and pressures up to 7 Kbar, Sov. Phys. Dokl., 21:8 (1976).

12. A. Michels and R.O. Gibson, Isothermenmessungen bei hohren Drucken, Ann. Physik 87: 850 (1928).

13. A. Michels, T. Wassenaar, P. Louwerse, Isotherms of neon at temperatures between 0° C and 150° C and at densities up to 1,100 Amagat (Pressures up to 2,900 Atm), Physica 26: 539 (1960).

14. E.V. Onosovskii, Investigation of Compressibility of Neon in the Near-Critical Region, in "Thermophysical Properties of Matter and Substances, Vol. 3" edited by V.A. Rabinovich, GSSSD, Moscow (1971), p. 291. Translated for NBS and NSF, Amerind Publishing Co., New Delhi (1974).

15. E.V. Onosovskii and A.L. Moroz, Experimental P-V-T Data for Neon in the Temperature Range of 65-273 K and Pressure up to 250 Bars. in "Thermophysical Properties of Matter and Substances, Vol. 2," edited by V. A. Rabinovich, GSSSD, Moscow (1970), p. 120.

16. V.A. Rabinovich, L.A. Tokina, V.M. Bezerin, Experimental determination of the compressibility of neon and argon at 300 to 720 K for pressures up to 500 bar, High Temp. (USSR) 8 (4): 745 (1970).

17. L.R. Scott, "Density Measurements for Neon at Low Temperatures," Ph.D. Thesis, University of Michigan (1967).

18. W.B. Streett, Pressure-volume-temperature data for neon from 80-130 K and pressures to 200 atmospheres, J. Chem. Eng. Data 16 (3):289 (1971).

19. J.A. Sullivan and R.E. Sonntag, P-V-T Behavior of Neon at Temperatures from 70 to 120 K and Pressures to 300 Atmospheres in "Advances in Cryogenic Engineering, Vol. 12", Plenum Press, New York (1967), p. 706.

20. P.A. Fleury and J.P. Boon, Brillouin scattering in simple liquids: argon and neon, Phys.Review 186 (1): 244 (1969).

21. D. Gusewell, F. Schmeissner, J. Schmid, Density and sound velocity of saturated liquid neon-hydrogen and neon-deuterium mixtures between 25 and 31 K, Cryogenics 10: 150 (1970).
22. E.V. Larson, D.G. Naugle, T.W. Adair III, Ultrasonic velocity and attenuation in liquid neon, J. Chem. Phys., 54 (6): 2429 (1970).
23. D.G. Naugle, Properties of liquid neon derived from sound velocity measurements, J. Chem. Phys. 56 (11): 5730 (1972).
24. W.H. Keesom and J.A. Lammeren, Measurements on the velocity of sound in neon gas, Physica 1: 1161 (1934).
25. V.V. Pashkov and E.V. Konorodchenko, Ultrasonic velocity in liquid neon, Sov. J. Low Temp. Phys. 4 (5): 277 (1978).
26. L.L. Pitaevskaya and A.V. Bilevich, The velocity of sound in compressed neon, High Temp-High Press 5: 459 (1973).
27. IUPAC, Atomic Weight of the Elements, 1975, Pure and Appl. Chem. 47: 75 (1976).
28. R.K. Crawford and W.B. Daniels, Experimental determination of the P-ρ-T melting curve of Kr, Ne, and He, J. Chem. Phys., 55 (12):5651 (1971).
29. J. Bigeleisen and E. Roth, Vapor pressure of the neon isotopes, J.Chem. Phys. 35 (1) 68 (1961).
30. E.R. Grilly, The vapor pressure of solid and liquid neon, Cryogenics, 2 (4):226 (1962).
31. J. Ancsin, Vapor pressure and triple point of neon and the influence of impurities on these properties, Metrologia 14:1 (1978).
32. J. Peck and H. Fenichel, The critical point of neon, Phys. Letters, 49A (2):97 (1974).

A METHOD FOR CALCULATING THE DISTORTION PARAMETERS FOR A BURNETT APPARATUS*

D. Embry

Phillips Petroleum Co.
Bartlesville, Oklahoma

and

P. T. Eubank. J. C. Holste, and K. R. Hall

Texas A & M University
College Station, Texas

INTRODUCTION

Burnett data analysis[1] requires estimates for apparatus volume changes caused by pressure and temperature changes. Volume ratio correction utilizes this information. Earlier workers usually either assumed negligible distortions[2] or corrected with simplified models[3].

We have applied a more rigorous analysis to both cells and ancillary apparatus. We chose finite element analysis for pressure distortion of the cells because it has good convergence properties and handles irregular shapes. Analytical models (Love[4]) describe the ancillary apparatus. For temperature distortions, we assume isotropic materials.

The distortions are volume averages. Using available material properties, the values are accurate within 10-20%. This is an adequate level.

---

*Supported by NSF, GPA, GRI, PRF and Phillips Petroleum Co.

APPLICATION OF THE FINITE ELEMENT METHOD (FEM)

The finite element method (FEM) is used to model the volumetric distortions of the Burnett cell caused by pressure loadings applied to the cell. To simplify application, we assume that 1) welds in the material behave identically to the rest of the material, 2) holes drilled into the side of the apparatus for connection to the ancillary apparatus can be ignored, 3) the materials are not prestressed, 4) the materials are isotropic, and 5) the stress strain relationship is described by Hooke's Law. None of the assumptions are necessary for application of the FEM if sufficiently advanced element descriptions are used. The reader is referred to one of the many texts describing FEM analysis (e.g. Zienkiewicz[5], Strang and Fix[6]) for treatment of the case where these assumptions do not hold. These simplifying assumptions allow us to model our Burnett cells by axisymmetric finite elements. Linear triangular toroidal elements described by Segerlind[7] can be used, but we have found that quadratic rectangular toroidal elements (Zienkiewicz[5]) are both easier to use and generally require less computation for the same quality of results. The FEM gives the nodal displacements at a given pressure loading. These nodal displacements are converted to volumetric distortions by

$$\frac{\delta V}{V} \sim 2\left(\frac{\delta R}{R}\right) + \frac{\delta L}{L} \tag{1}$$

$$\delta R = \frac{1}{h} \int_0^L u \Big|_{r=R} dh \tag{2}$$

$$\delta L = \frac{2}{R} 2 \int_0^R \left(V\Big|_{h=L} - V\Big|_{h=o}\right) r\, dr \tag{3}$$

We have used Simpson's rule[9] to evaluate the integrals. The assumption of Hooke's Law allows the volumetric distortion coefficients, γ_v, to be determined as the change in volumetric distortion divided by the change in pressure loading, using only two pressure loadings for the entire range. It is important to check the maximum stress calculated by the FEM in each element to ensure that the assumption of Hooke's Law is valid (in some cell geometries this assumption may be violated at pressures below 1 MPa).

Mechanical properties needed for these calculations are usually obtained from the manufacturers' literature. The values determined from the literature are usually accurate and repeatable to within 10% unless the materials have undergone special treatment. In the latter case, Young's modulus and Poisson's ratio may have to be mesured. Many engineering labs have the facilities to measure these quantities.

CALCULATION OF TOTAL VOLUMETRIC DISTORTIONS

The volume weighted average of the relative distortions for each component volume is the total volumetric distortion. Love's equations for long, closed cylinders are adequate for the tubing. A length of tubing (fitting to fitting) approximates valves. Other small volume pieces of equipment are approximated either by lengths of tubing or Love's formulas.

Summing the distortions for each component volume gives

$$\delta V/V = \sum_i (V_i(\delta V/V)_i)/(\sum_i(V_i)) \tag{4}$$

where $\delta V/V$ is the total volumetric distortion. Using Hooke's Law and referencing each $\delta V/V$ to the value at zero pressure gives

$$V/V_o = 1 + \gamma P \tag{5}$$

where γ is the pressure distortion coefficient. The total temperature distortion is approximately

$$\delta V/V = \sum_i(V_i(\delta V/V)_i) = 3\alpha(\theta - \theta_{ref}) \tag{6}$$

where α is the temperature distortion coefficient and θ_{ref} is some reference temperature.

EFFECT OF BURNETT CELL GEOMETRY ON CELL DISTORTIONS

The three significant distortion coefficients are γ_r (radial), γ_a (axial), and γ_v (volume). Each is dimensionless (multiplied by Young's modulus) and a function of Poisson's ratio (taken as 0.3 in this paper) and geometry. Figure 1 presents the standard cell geometry used in this paper. We have chosen to

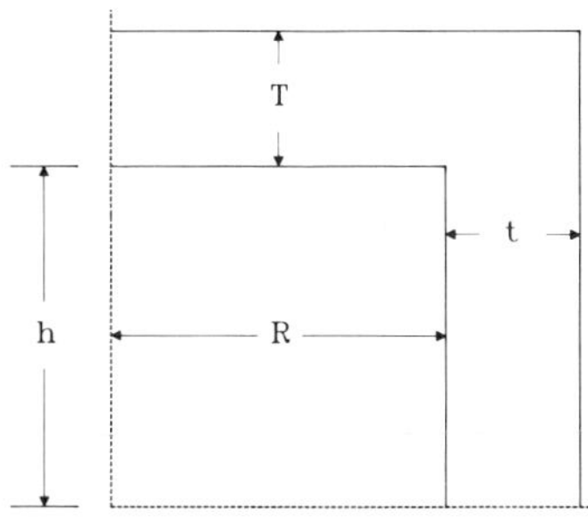

Fig. 1. Burnett Cell Standard Geometry.

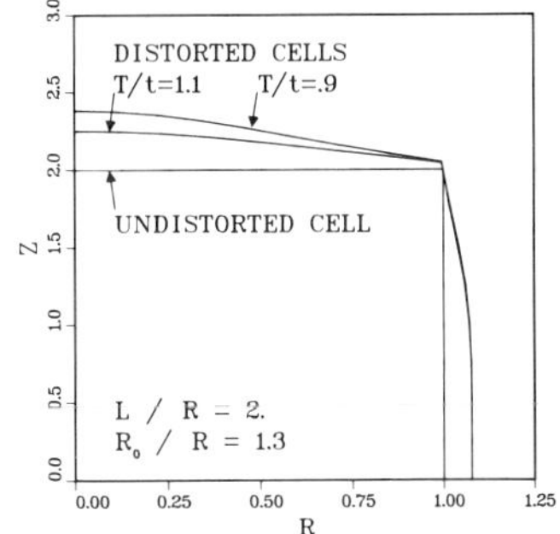

Fig. 2. Cell Pressure Distortion at two levels of T/t.

investigate the effects of varying the ratios: L/R, R_O/R, T/t. Figure 2 illustrates the effect of varying the T/t ratio.

Figure 3 shows the effect of the various ratios on the reduced axial distortion coefficient. At a fixed R_O/R, γ_a always decreases as T/t increases but at a slower rate as T/t increases. At a fixed T/t, γ_a always decreases as R_O/R increases. Doubling the wall thickness can reduce a γ_a by a factor of eight. For certain values of R_O/R, γ_a attains a minimum value at L/R of unity. The minimum becomes more pronounced as R_O/R increases and appears at lower R_O/R values for lower values of T/t.

Figure 4 shows the effects of the ratios on the reduced radial distortion coefficients. For increasing L/R, the value of γ_r tends to a constant for all T/t although the effect of T/t is more apparent at low R_O/R. Increasing T/t has little effect on γ_r.

Figures 5 and 6 present the effects of geometry γ_v. For increasing L/R, γ_v tends to a constant value. γ_v decreases as T/t increases because this change reduces bending. γ_v decreases with increasing R_O/R (although the practical limit is $R_O/R = 1.5$).

These plots indicate that a design employing the following heuristics would minimize distortions:

    a)   choose L/R less than 2
    b)   choose T/t greater than 1
    c)   choose R_O/R between 1.4 and 1.6

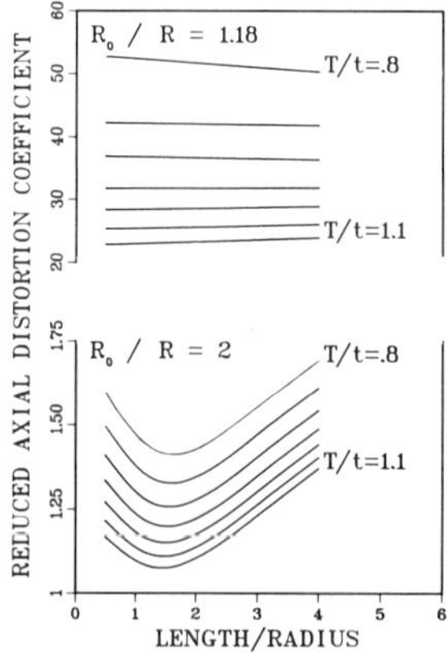

Fig. 3. L/R Effect on Axial Distortions.

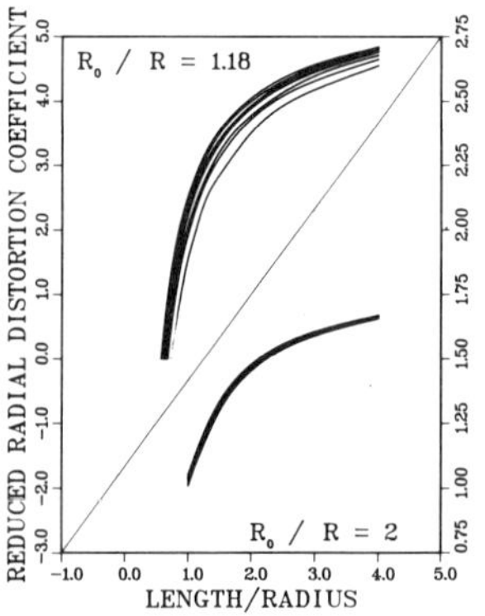

Fig. 4. L/R Effect on Radial Distortions.

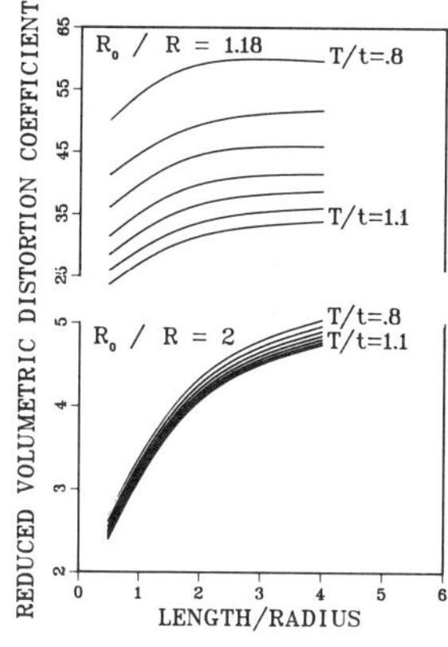

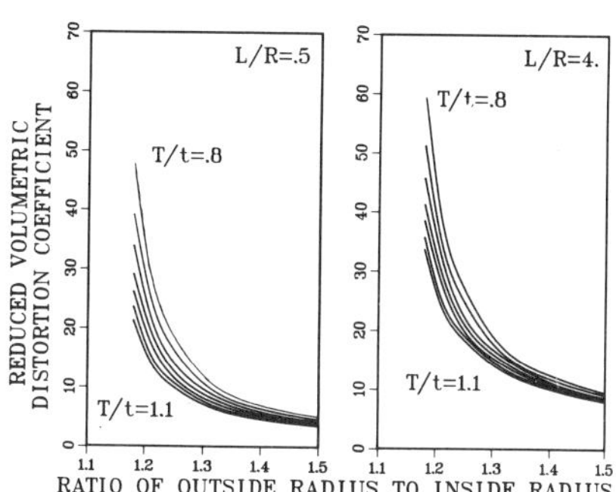

Fig. 5. L/R Effect on
Volume Distortions.

Fig. 6. R_0/R Effect on
Volume Distortions.

Using these rules will provide γ_v values which translate into
distortions of 0.5% or less at pressure of 20 MPa.

COMPARISON OF FEM RESULTS TO PREVIOUS METHODS

A comparison of the distortions calculated by our method and
those calculated by Love's equations are given in Table I. The
differences in the two methods can be substantial, significantly
more than the uncertainty due to the mechanical properties. As
demonstrated by Embry[8], this difference is a potential source of
systematic error when Love's equations are used to calculate the
distortion parameters. This comparison also shows that end effects are very important for accurate calculation of distortion
coefficients.

Table I. Comparison of Reduced Volumetric Distortion
Parameters Calculated by Love's Equation and the
Proposed Method ($\mu = 0.3$)

Geometric Quantity			Love's Equation	Proposed Method
R_0/R	L/R	T/t		
1.18	2	.8	6.1	60
1.18	2	1.1	6.1	29
1.18	4	.8	6.1	60
1.18	4	1.1	6.1	31
2.0	2	.8	3.9	4.2
2.0	2	1.1	3.9	4.0
2.0	4	.8	3.9	5.0
2.0	4	1.1	3.9	4.8

APPLICATION OF METHOD TO AN EXISTING BURNETT APPARATUS

The Burnett apparatus described by Mansoorian[10] illustrates the method. The cells, bored into a piece of 303 stainless steel bar stock, have volumes of about 767 cm^3 and 353 cm^3. Two 303 stainless steel heads close the cells using a copper gasket seal. Twelve cap screws compress the gasket. A cold-rolled steel cap holds head and screws in place. Fittings are AMINCO Super-pressure devices. The tubing is thick-walled stainless steel. A magnetic pump (described by Watson[11]) is included.

We neglect the effects of tubing entering the cell. We assume that the cap screws are a single pressure ring and that the cap, head and body are a single piece of metal at their respective interfaces.

The pressure distortion coefficients were determined for an existing Burnett apparatus by calculating cell volumes at various temperatures and pressures using the FEM. The results of these calculations can be represented by

Large cell: $10^6(\delta V/V) = 17.59 + 5.96(10^{-2})\theta + (28.3 + 4.35(10^{-3})\theta)P$

Small cell: $10^6(\delta V/V) = 16.32 + 5.53(10^{-2})\theta + (23.88 + 3.53(10^{-3})\theta)P$

where θ is in K, P is in MPa, and the numerical constants were determined by linear regression. The rest of the apparatus had a total volume of about 30.5 cm.

Combining all distortions gave

$$\frac{V_A}{V_A{}^o} = (1 + \gamma_A P)\ (1 + 3\alpha_A\ (\theta - \theta_{ref}))$$

$$\frac{V_B}{V_B{}^o} = (1 + \gamma_B P)\ (1 + 3\alpha_B\ (\theta - \theta_{ref}))$$

$$\frac{V_A + V_B}{(V_A + V_B)}{}^o = (1 + \gamma_{AB})\ (1 + 3\alpha_B\ (\theta - \theta_{ref}))$$

where A denotes the large volume, B denotes the small volume and

$$\gamma_A = (28.22 + 4.51 \times 10^{-3})\ 10^{-6}(MPa)^{-1}$$

$$\gamma_B = (23.31 + 3.76 \times 10^{-3})\ 10^{-6}(MPa)^{-1}$$

$$\gamma_{AB} = (26.35 + 4.14 \times 10^{-3})\ 10^{-6}(MPa)^{-1}$$

$$\alpha_A = (14.77 + 2.91 \times 10^{-3}(\theta - \theta_{ref}))\ 10^{-6}k^{-1}$$

$$\alpha_B = (14.78 + 2.91 \times 10^{-3} (\theta - \theta_{ref})) \; 10^{-6} K^{-1}$$

$$\alpha_{AB} = 14.77 + 2.91 \times 10^{-3} (\theta - \theta_{ref})) \; 10^{-6} K^{-1}$$

The observed and calculated (<u>a priori</u>) cell constants were

$$N(obs) = \frac{V_A^o + V_B^o}{V_A} = 1.4768$$

$$N(calc) = 1.4787$$

The agreement between N(obs) and N(calc) is an indication that our volume estimates are essentially correct.

CONCLUSIONS

We have developed an advanced method for estimating the distortion coefficients in a Burnett apparatus. The method can be applied to almost any cell geometry and requires few assumptions. It should provide results equivalent to or better than those using Love's formulas. Application of the method to an existing apparatus provides excellent results. We have also included plots which could guide design of an apparatus to minimize distortion effects.

NOTATION

E = Young's modulus (MPa)
h = axial coordinate
L = characteristic length
N = cell constant
p = pressure
r = radial coordinate
R = characteristic radius

t = wall thickness
T = end cap thickness
u = displacement in radial direction
v = displacement in axial direction
V = cell volume

α = thermal expansion coefficient
μ = Poisson's ratio

γ = pressure distortion coefficient
θ = temperature

Subscripts

i = internal value
o = external value
0 = initial value
A = larger cell
B = smaller cell

a = axial
r = radial
v = volume
AB = combined cell

REFERENCES

1. E.S. Burnett, Compressibility determinations without volume measurement, J. Appl. Mech. 3 4:A136 (1936).
2. E.S. Burnett, Application of the Burnett method of compressibility determinations to multiphase fluid mixtures, Bureau of Mines Report of Investigation 6267, (1963).
3. F.B. Canfield, Ph.D. Thesis, Rice University, Houston, Texas, (1962).
4. A.E.H. Love, "The Mathematic Theory of Elasticity", 4th ed., Cambridge Univ. Press, London (1927).
5. O.C. Zienkiewitz, "The Finite Element Method", 3rd ed., McGraw-Hill, New York (1977).
6. G. Strang and G.J. Fix, "An Analysis of the Finite Element Method", Prentice-Hall, Englewood Cliffs, New Jersey (1973).
7. L.H. Segerlind, "Applied Finite Element Analysis", Wiley, New York (1976).
8. D.L. Embry, M.S. Thesis, Texas A & M University, College Station, Texas (1978).
9. F.B. Hildebrand, "Introduction to Numerical Analysis", McGraw-Hill, New York (1974).
10. H. Mansoorian, Ph.D. Dissertation, Texas A & M University, College Station, Texas (1978).
11. M.Q. Watson, M.S. Thesis, Texas A & M University, College Station, Texas (1978).

DISCUSSION

Question by C.H. Chiu, Exxon Production Research Co.: Did you consider the interaction of the horizontal and vertical distortions i.e., are the corner points assumed to be constant?

Answer by Author: The corner points are allowed to move both vertically and horizontally as the cell distorts, but in order to calculate the distortion coefficients it was assumed that the volume contribution due to the interaction of the vertical and horizontal displacements was negligible.

LARGE-SCALE CRYOGENIC LIQUID STORAGE

J. H. Trammell and I. V. La Fave

Chicago Bridge & Iron Company
Oak Brook, Illinois

INTRODUCTION

The art and science of storing large quantities of gases as refrigerated liquids has been in a state of continuing development over the past thirty years. For the first two decades of this period the development proceeded with little or no impact from regulating bodies. This is not to imply that the development was in any way haphazard or irresponsible. Rather, it was accomplished by engineers representing owners, operators, suppliers, manufacturers and constructors working together to produce sound, practical and safe solutions to storage problems. The excellent safety record achieved by the industry during the period attests to the soundness of their work.

During the last decade considerable concern has been expressed over the safety of liquid stored and transported at cryogenic temperatures, especially liquid methane (LNG). LNG storage tanks and systems, as a group, constitute the largest cryogenic storage systems in the world. Many LNG tanks exceed 500,000 Barrels (79,500 m^3) capacity whereas large liquid oxygen (LOX) storage tanks have a capacity on the order of 17,000 Barrels (2700 m^3). A study of the regulatory and design trends for LNG can be useful in assessing what the future trends may be for large scale cryogenic storage of other useful gases.

STRUCTURE TYPES

Modern LNG storage tanks usually fall into four basic types. The oldest and most common is the free standing, double walled metal tank with an earth dike secondary containment. In its usual form, the inner tank is open topped and the outer tank doubles as the insulation jacket and as the primary pressure boundary to contain product vapor.

The second type is similar but uses a close-in, but separate, high concrete dike. This is most commonly used in congested areas to save space and/or to provide missile protection from wind or blast driven objects.

A third type, referred to as the integrated concept, combines the secondary containment with the pressure boundary (outer tank). In this concept both primary and secondary containers are on a common foundation. Through-the-roof filling and pumpout is needed which requires submerged pumps in the tank. The installation and maintenance of such pumps is a significant cost item. Since the outer tank shell is designed to contain the cryogenic liquid, this design also is referred to by some as "double integrity". However, since both tanks share a single foundation and only one pressure barrier is provided we believe "enhanced integrity" is better terminology.

The fourth type is the integrated concept either buried below grade or with earth mounded around it. Although very similar to Type 3, the surrounding earth introduces several new design considerations. Among these are the requirements for ground water control, wall and foundation heating and the weight and pressure of the surrounding earth.

Each of these concepts, represents an additional step in the search for added safety in LNG storage. Each type has advantages and disadvantages which are not always immediately apparent. In some situations what appears superficially advantageous may not be upon further study. Since each step represents an increase in complexity and cost a careful evaluation should be made. One should ask if the extra cost in time and money yields a significant safety benefit and whether the same benefit could be achieved more efficiently some other way.

Comparison of the four basic types, including relative cost factors, is given in Fig. 1. Many variations or combinations of these basic types are possible and have been used for commercial tanks. For instance, in Type 2 the height and diameter of the concrete dike can be varied. To provide added blast and missile protection, concrete roofs can be added to Types 2, 3 and 4.

The free standing tank in a dike has the advantage of simplicity and lowest construction cost. The outer tank is visible and accessible for inspection and maintenance. This visibility is an important feature since insulation breakdowns or leaks will show up as unexpected frost spots and provide a warning that a problem exists. Also, foundation settlement problems or foundation heating failure often can be detected by visual inspection. The free standing metal tank is a proven concept with an excellent safety

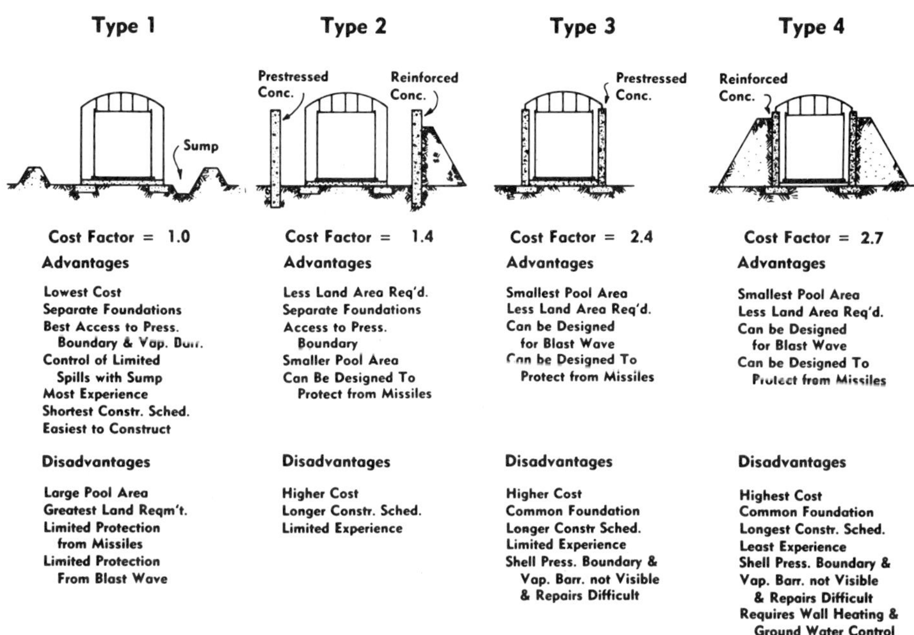

Fig. 1. Comparison of Four Basic Types
of Large Scale LNG Storage Systems.

record, and possibly should be the standard against which other
concepts are compared.

EXISTING CODES AND STANDARDS

Most cryogenic tanks built in the United States have been
built to the ASME Boiler and Pressure Vessel Code, Section VIII or
API 620, Appendix Q. Where these standards were not strictly
applicable they have been used as a guide. NFPA Standard 59A
covers LNG facilities and references API 620, Appendix Q for low
pressure metal tank design and construction.

Fracture initiation toughness is the criteria these standards
use in material selection for the primary liquid container. The
basic concept of this criteria is that at normal design stress
levels a running crack will not initiate from a flaw small enough
to avoid detection by quality control procedures and non-destructive examination. Cracks generated by fatigue or other abnormal
conditions will result in detectable leaks before catastrophic
rupture of a properly designed and constructed tank. However,
existing codes and standards are applicable to tanks which can be
visibly monitored. Other structural types should consider other
means of leak detection.

There is no known record of brittle failure of a tank designed and built to these standards. Nevertheless, a requirement for crack arrest toughness has begun to appear in some specifications and regulatory requirements for tank shells. Crack arrest toughness is that property of a material which enables it to arrest a running crack. Whether or not a crack does arrest in a given material depends on the stress level and size of flaw. Although the characteristics and units of the material property are recognized there is no universally accepted method of testing to establish a minimum acceptance criteria for a material. Crack arrest toughness has not been introduced into the codes as a requirement except for some nuclear reactor pressure vessels. Research in this area is continuing and standardization is only a matter of time.

NEW RESEARCH

Two of the materials commonly used in cryogenic storage tanks, aluminum and stainless steel, do not exhibit transition behavior hence brittle fracture is not a problem with these materials. Another suitable material, particularly for large tanks, is 9% nickel steel, which does go through a ductile to brittle transition. In order to better understand the crack arrest properties of this material, CBI funded in 1980 a research program at Materials Research Laboratory of Glenwood, Illinois. Two specimen types, and plate thicknesses of 1" and $1/2$" and plate weldments were tested at LNG temperature, -260°F (-162°C); at LOX temperature, -297°F (-183°C); and at liquid nitrogen (LN_2) temperature, -32°F (-196°C). These tests are described in detail by Bruscato[1]. Crack arrest coefficients (K_a) were obtained for plate at -297°F for 1" thick plate and at -320°F for 1" and $1/2$" thick plate but cracks could not be initiated in the Inconel weld metal or heat affected zone. A running crack could not be initiated at -260°F in either plate or weldments, and no valid K_a could be determined. From these tests it was concluded that 9% Ni in these thicknesses is on the toughness upper shelf at LNG temperature.

Other reseachers using different test specimens have had similar results for 9% Ni at LNG temperature. The material used in the CBI test was plate produced in the early 1970's which was left over from two tank projects. The material was ordered to ASTM A553 and is representative of 9% Ni delivered during that period. Today's improved production practices would generally produce plate of superior mechanical properties better (tougher) 9% Ni is available by ordering low sulfur content.

The K_a values for 9% Ni at LOX and LN_2 temperatures are high enough to provide reasonable crack arrest ability. At stresses permitted by API 620, Appendix Q, the critical crack length for $1/2$"

thick material at -320°F is about 10". This thickness would be adequate for a 70,000 Bbl (11,100 m^3) storage tank.

Recently a new requirement has been appearing in some specifications. This is the requirement to design the secondary containment to handle catastrophic rupture (sometimes referred to as "unzippering") of the primary container shell. There is concern that the surge of liquid from such a failure might overtop a low earthen dike. Designers generally have chosen a close-in concrete container or an integrated storage tank to handle this. We are not convinced that instantaneous rupture is a reasonable design criteria considering the toughness of the primary container materials being used. But if such a criteria is to be imposed, the size of the dynamic force is very important. In order to understand better what might happen in such an incident, and to get design information, CBI conducted shell rupture tests.

A 35" diameter by 17.5" high by 1/4" thick plastic container was used for the secondary container. The inner shell diameter was varied to provide different annular spaces between shells. A frangible glass vertical joint was used in the inner shell to provide fast rupture. Rupture was initiated by a blow from an air operated hammer. The hydraulic pressures were read and recorded at various points around the outer shell. Miniature pressure transducers which sampled at a rate of 2000 samples per minute were used.

The test results are summarized in Fig. 2 which shows the peak dynamic plus static pressures measured. The 2.5 times static line falls outside all but one data point and the duration of that peak was shorter than some of the lesser peaks. The highest peak pressures were recorded opposite the fracture initiation point and

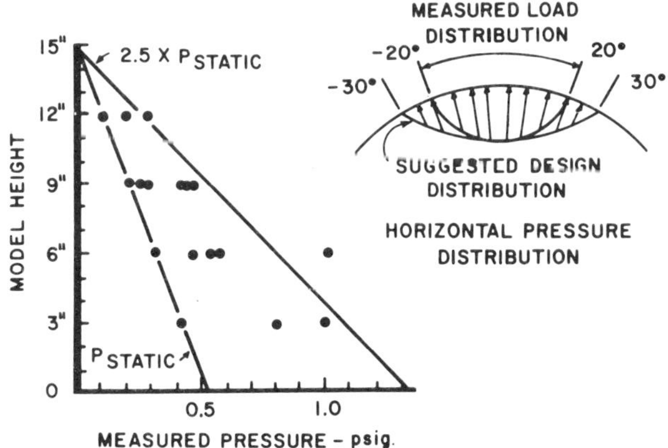

Fig. 2. CBI Rupture Test Results.

at 180° from fracture initiation. The duration was longer opposite the fracture. The peaks diminish to static pressure in the circumferential direction.

Most cryogenic tanks have loose fill perlite insulation in the annular space between shells. The effects of the perlite in an integrated design is not fully understood. When loaded dynamically over a large area perlite is quite stiff. In the event of catastrophic inner shell rupture the perlite would crush, thus absorbing energy, and would restrict the inner shell movement after rupture. The CBI tests showed that restricting or slowing the shell movement reduced the hydrodynamic forces. This was illustrated by increasing the strength of the shell-to-bottom connection which slowed the tearing away of the shell and reduced the hydrodynamic forces.

A dynamic peak load of 2.5 to 3 times static can be handled with a free standing prestressed concrete wall. Higher forces are more difficult to handle and would increase material cost and construction schedule. The CBI tests indicate a peak force of 2.5 times static would be a good upper design value for cryogenic tanks.

THE STORAGE SYSTEM

Most of the foregoing has dealt primarily with the storage tank. Since it is the largest and most visible part of a storage system, it gets a major share of the attention of owners and regulators alike. Even where extreme criteria are not required by regulations, some designers and owners have imposed them; either in an attempt at greater safety or in the hope of expediting approval of their project. However, concentration on a single item or a single failure mode in safety planning may lead to large cost increases without proportionate safety benefits.

Safety can be achieved best by a system approach. The first step should be to establish what level of safety is required. Is it safeguarding against loss of operation, against loss of a part or all of the facility, against damage to adjacent property, or against injury or loss of life to operating personnel and other people nearby? Each component of the system should be evaluated for its contribution to safety and the consequences of its failure. History has shown us that mechanical equipment failures or failures due to human error are the most common kind. Therefore particular attention should be given to selection and operation of this equipment in order to insure reliability in service. Components which should be given particular attention are filling controls, pressure control systems, level gauges and level alarms, pressure and vacuum vents, level and pressure recorders and emergency vents. Provisions for handling small leaks and spills should be considered.

SUMMARY

Careful consideration of all aspects of the storage system and environmental conditions and selection of conservative but reasonable design criteria should dictate which type of storage tank concept best suits a specific site. Unnecessary redundancy in the storage structure may not accomplish as much as hoped. An example is providing a primary liquid container with crack arresting shell material, providing blast and missile protection and then designing the secondary container for catastrophic rupture of the primary shell. This approach to the liquid container design provides no additional protection against failure due to overpressurization or overfilling.

The trend in regulations, codes and standards is toward more conservative design with regard to catastrophic shell failure even though we know of no such failure in modern cryogenic liquid storage tanks. The design engineer should guard against overemphasis of this type of failure to the exclusion of more common types of failures or problems. Insulation breakdown, small gas or liquid leaks and foundation heating system failure are examples of problems which have happened and which, if detected early, are repairable in a conventional tank. Overpressurization and foundation failure also have occurred; sometimes with disastrous results. Designs should be such that problems can be detected and repaired.

When choosing a storage concept the possibility of future maintenance or modification should be factored into the decision. The design should lend itself to taking the tank out of service and into a safe condition for repair or modification. Although this is not a frequent occurence, the authors know of several cases where major repairs or modifications have been achieved safely in tanks taken to an air atmosphere.

The conventional free standing metal tank is a proven design which should be considered where site and environmental conditions will permit its use. Where additional safety measures are required, details should be selected so that the principal advantages of the conventional above ground tank are retained.

REFERENCES

1. R.M. Bruscato, The measurement of crack arrest fracture toughness in welded 9% nickel steels in cryogenic storage tanks, The Welding Journal (July 1981).

LATEST DESIGN FEATURES AND OPERATIONAL HISTORY OF LNG INGROUND TANKS AT THE NEGISHI LNG RECEIVING TERMINAL

T. Hiro-oka

Tokyo Gas Co., Ltd.
Tokyo, Japan

STATUS OF LNG INGROUND TANK CONSTRUCTION

Tokyo Gas Co., Ltd. has developed inground storage tanks for LNG because of the advantages of safety, environmental acceptability and aethetics, and effective use of land area. Our first inground storage tank, with a capacity of 10,000 m^3, was constructed at Negishi in 1970. Since then, sixteen inground tanks, with a total capacity of over 10^6 m^3, have been constructed at Japan's two leading LNG receiving terminals, Negishi and Sodegaura, and all are in successful operation, Table I. Four large-scale tanks are now under construction; one with a capacity of 95,000 m^3 at Negishi[1], and three with capacities of 130,000 m^3 each at Sodegaura[2].

Table I. LNG Inground Tanks of Tokyo Gas

Works	Capacity K L	I.D×Liquid Height M	Quantity	Year of Completion	Contractor
Negishi	10,000	30×14.2	1	1970	I H I & Shimizu
	60,000	50×30.6	1	1972	"
	95,000	64×29.6	1	1977	I H I & Kajima
	95,000	64×29.6	1	1978	"
	95,000	64×29.6	1	1981	N K K & Shimizu
	95,000	64×29.6	1	* 1982	K H I & Shimizu
Sodegaura	60,000	60×21.3	2	1974	I H I & Shimizu
	60,000	64×18.7	5	1976	"
	60,000	64×18.7	2	1977	M H I & Kajima
	62,000	64×19.3	1	1979	I H I & Shimizu
	58,000	64×18.1	1	1979	M H I & Kajima
	130,000	64×40.5	1	* 1982	I H I & Shimizu
	130,000	64×40.5	1	* 1982	M H I & Kajima
	130,000	64×40.5	1	* 1983	N K K & Obayashi

*Under Construction

937

Our experience in the design, construction and operation of this type of tank established the basis for the general design criteria and the engineering methods to be used for the design, construction, operation and maintenance of LNG inground tanks. In 1979 the "Recommended Practice for LNG Inground Storage" was presented by the Committee on LNG Inground Storage, a part of the JGA (Japan Gas Association), which was commissioned by the MITI (Ministry of International Trade and Industry) in 1976.

LATEST DESIGN FEATURES

Soil Conditions and Concrete Structure Design

The selection of the construction methods and types of concrete structure to be used depends mainly on the surrounding soil conditions.

Soil survey. One of the most important factors in constructing an inground storage tank is the soil survey. The purpose of a soil survey is to obtain ground information necessary to confirm the suitability of the bearing layer, to give a preliminary estimate of ground settlement, and to show the stability of the surrounding ground during earthquakes. From this data we can determine the ground water treatment during and after construction, and design the storage tank structure.

Design loads for side wall and base. The following loads must be taken into consideration:

1. Dead weight
2. Gas pressure
3. Liquid pressure
4. Earth pressure on the side wall; generally, when a storage tank is constructed in rock strata, mud stone or the like, this item need not be taken into consideration.
5. Ground water pressure
6. Roof load
7. Thermal load; defined as the thermal stresses generated due to temperature differences in the side wall and base, are obtained by a thermal analysis using finite element methods (FEM).
8. Loads caused by an earthquake including; earth pressure during an earthquake, inertia forces of the side wall, base and roof and dynamic pressure of the stored liquid. The design seismic coefficients used for these calculations are determined according to the zone and ground condition of the construction site.
9. Freezing earth pressure; when the surrounding ground will become frozen, and it has been discovered from soil and frost heave tests that the ground will expand when freez-

ing, the freezing earth pressure must be taken into consideration. To analyse this pressure, we use axissymmetrical FEM.

Structural analysis. The analysis must be carried out under the load conditions specific with suitable boundary conditions and models. An FEM analysis using a cylindrical shell model is employed in most of our designs. The boundary conditions are determined from consideration of the surrounding ground and the bottom slab.

Soil Freezing and Heater Design

When the tank is put into service, the surrounding ground will freeze due to the cryogenic effect of the LNG. The strength of the soil is increased by this freezing action, which also increases its gas and water tightness, enhancing the function of the storage tank. The expansion of the surrounding ground from freezing action may cause pressure on the side wall and base or frost heaving may cause a displacement of the surrounding structures. Where it is necessary to avoid such effects, heaters must be installed.

For the 95,000 m^3 tank at Negishi, hot water circulation systems are installed as side heaters and bottom heaters, with a brine circulation system backing up the bottom water system.

Membrane Design

The membrane is a thin metal plate which transmits gas and liquid pressure to the side wall and base through the insulation; load-bearing capability is not required. However, the membrane must be designed and manufactured to ensure liquid and gas tightness. Since LNG is stored at temperatures as low as $-162^{\circ}C$, the membrane must have many corrugated parts in order to absorb thermal expansion and contraction. Many types of corrugations are now available (Fig. 1) and have been used for our inground storage tanks.

Verification of Membrane Stability

Confirmation of static load stability. It must be shown that the membrane as a whole will maintain its shape through smooth deformation or displacement of each part under static loads, such as thermal load, total pressure and mechanical load. This is done by a theoretical analysis or model test.

Unstable collapse. This is a phenomenon in which the shape of the corrugation of the membrane is not determined unconditionally under a static load; consequently the deformation process becomes

CROSS CORRUGATION TYPE

ROTATION TYPE

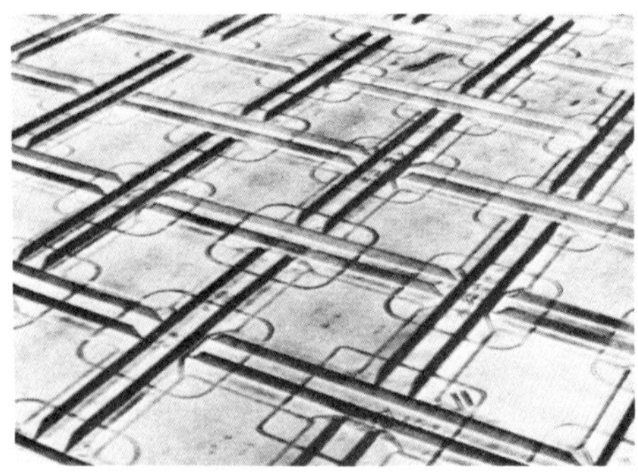

Fig. 1.   Types of Membranes

ambiguous.   It must be verified that unstable collapse cannot
occur.   This is done by theoretical elasticity analysis or a full
size model test.

Progressive deformation. This occurs when the deformations in
each portion of the membrane, especially in the corrugations,
increase progressively with cyclic loads and do not converge into
a fixed shape.   It must be shown by theoretical analysis, or a
full size model test that progressive deformation cannot occur.

Cyclic loading.   Cyclic variations are likely to occur in
liquid pressure and thermal loads, as a result of the following
operational conditions:

1. cool down to -162°C from ambient temperature
2. liquid pressure rise and fall
3. membrane cycling between liquid and gas temperature
4. warm up from liquid to ambient temperature

The number of cycles is determined by considering the design life
time and the working conditions of the storage tank. The tank must
use materials of sufficiently high fatigue strength to preclude
fatigue failure.

Earthquake Resistant Design

In the seismic design of LNG inground tanks, both the seismic-
coefficient method and the dynamic analysis method are commonly
applied to each portion of the tank. Furthermore, it is necessary
to consider the liquefaction of the ground when an earthquake
occurs and sloshing of the stored LNG during certain types of
earthquakes.

Measures against liquefaction of the ground around the tank.
In loose and fully saturated sandy soils there is a possibility of
an increase of pore water pressure and liquefaction during an
earthquake. Liquefaction will not occur unless all of the following conditions are met:

1.   the ground is fully saturated with ground water
2.   the ground water level is within 15 m of the surface
3.   the sand grain diameter is less than 2.0 mm
4.   the ground consists of sandy soil in which the fine grain
      content is less than 35%
5.   the N value determined by the standard penetration test
      for fine grain content is not higher than a certain
      value.

Methods of preventing liquefaction include increasing the
density by vibration or impact, replacing the soil by one of a
grain size less susceptible to liquefaction, and lowering the
ground water level. For the 95,000 m^3 LNG inground tank at
Negishi, the soil is silty and there is no possibility of liquefaction. However, at Sodegaura, the soil is sandy and there is a
possibility of liquefaction, so sand compaction piling is used
around the tank.

Anti-slosh design. Sloshing of the stored LNG may occur in
case of a displacement type earthquake, the period of which is
relatively long. It must be demonstrated that the stored LNG
never sloshes out in case of an earthquake. Regarding the 95,000
m^3 tank at Negishi, the suspended roof is higher than the top of
the sloshing liquid, the height of which is calculated by a dynamic analysis using several seismic waves which have occurred in
the past.

OPERATIONAL HISTORY

Special Test before Start-up

One of Tokyo Gas Company's design principles is to always verify the safety and stability of each tank before operation. To accomplish this, three special tests are executed before start up.

<u>Water test.</u> We place a water pressure loading on the base structures equivalent to the pressure when full of stored LNG. This will confirm that there is no abnormal deformation of the membrane and base, and that the membrane displacement is within the allowable design value.

<u>Pre-cool test with liquid nitrogen.</u> We then pour liquid nitrogen into the tank until the temperature of the bottom membrane reaches -162°C. This will confirm that the membrane has the designed behavior.

<u>Leak test.</u> Before the water test and after the pre-cool test, the tank is leak tested twice to check the tightness of the membrane. The ammonium sniffer method is used.

Start-up Operation

Before an LNG inground tank is ready for commercial operation, it must be confirmed that the content of oxygen, nitrogen and humidity in the tank are below allowable levels, and that the cool-down rate is within the allowable range as follows:

1. Final oxygen content: less than 0.1 vol%
2. Final humidity (dew point):   -40°C
3. Cool-down rate: below 15°C/h

Boil-off Gas Ratio

The design value of the boil-off gas ratio, one month after start-up, is 0.2% of the full capacity of the tank per day. The measured value of this ratio is 0.13% per day; far lower than the design value. The actual value will decrease with the formation of frozen soil around the tank.

Soil Freezing and Heater Operation

The formation of frozen soil varies with the type of inground tanks. At Negishi, there are two types, one is the drainage type shown in Fig. 2, the other is the temporary drainage type shown in Fig. 3. The former type of tank has both a side heater and a bottom heater, whereas the latter has only a side heater.

As shown in Fig. 2, the heater arrangements allow only a thin band of frozen soil to surround the tank, thus ensuring that frost heave will not affect the tank itself, and the frozen soil has a slight affect on the structures around the tank. This heater system requires about 232 kW, using steam as the heating medium.

The heating arrangement shown in Fig. 3 allows a mass of frozen soil to form all around the tank. This makes it necessary to anchor the surrounding structures, such as piping and pipe racks, against to the effect of frost heave and ground displacement. At Negishi, however, this effect is negligible as the soil conditions are extremely good. Thus the duration between maintenance periods is relatively long, approximately ten years. As will be obvious from the above discussion, the drainage type of tank, with both side and bottom heaters, is better at Negishi.

CONCLUSION

The Tokyo Gas Company has studied and developed LNG inground tanks since the 1960's, and has established the design criteria and engineering methods for large-capacity commercial tanks. However, pressing public issues will produce more restrictive laws and regulations regarding safety and environmental protection. Consequently, it is necessary for us to continue further technical development in the design and construction of future LNG inground installations. Cost savings and economic advantages must also be seriously considered, recognizing that the LNG inground tank is

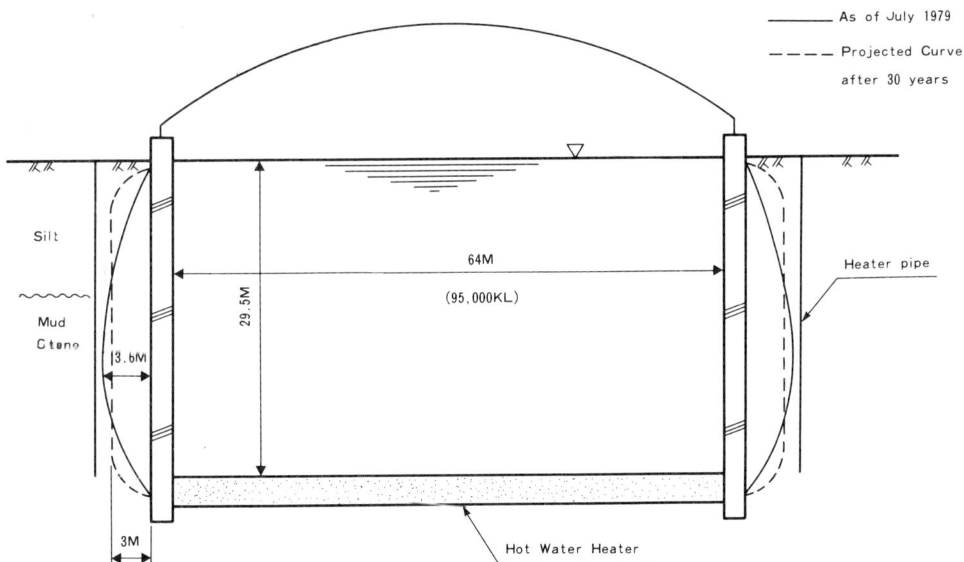

Fig. 2. LNG tank with side and bottom heaters.

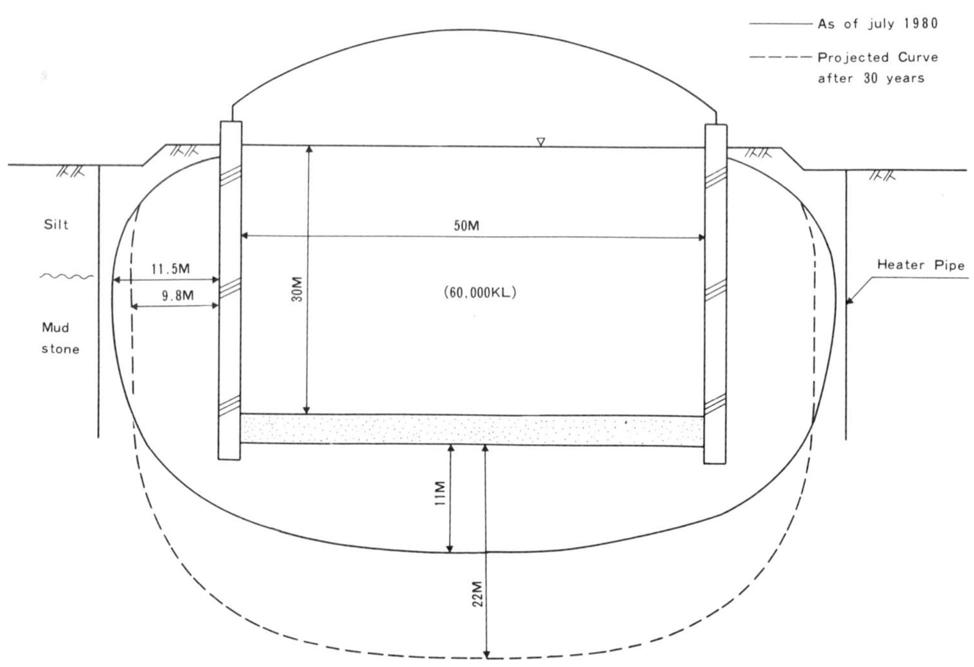

Fig. 3. LNG tank with side heater only.

the most influential part of the whole system of an LNG receiving
terminal.

REFERENCES

1. S. Takagi, Construction of the LNG inground storage tanks at
    Negishi terminal; The Foundation Engineering and Equipment
    Monthly, 7,(9):1979.
2. S. Goto, and Y. Takahashi, Construction of large LNG inground
    storage tanks at Sodegaura terminal, Soil and Foundation,
    29, (3):1981. The Japanese Society of Soil Mechanics and
    Foundation Engineering.

DISCUSSION

Question by R.C. Scurlock, University of Southampton, England:
Have you had any indication of evaporation instabilities
developing during storage periods when no mixing, filling or
decanting operations have taken place?
Answer by Author:
We have never experienced evaporation instabilities and we
think that instable evaporation does not occur during long
storage periods.

CONDENSATION IN THE ANNULUS OF A DOUBLE-WALLED CRYOGENIC STORAGE TANK

R. R. Tarakad, C. A. Durr, and D. B. Crawford

The M. W. Kellogg Company
Houston, Texas

INTRODUCTION

It has been reported by Cuperus[1] that under certain conditions condensation can occur in the annular space of a double-walled cryogenic storage tank. To our knowledge no systematic study of this phenomenon has been published. The paper by Cuperus describes the phenomenon briefly:

> Given a pure product stored under constant absolute pressure conditions, there should be no tendency of the vapor to condense against the outside of the inner tank. Furthermore, even if it would do so, the liquid would end up in the bottom of the annulus where, due to the heat inleak, the temperature is always slightly higher than that of the tank content so that it would vaporize again. In the case of a mixed product detailed investigations have shown that, due to small changes in absolute pressure, some condensation cannot be excluded. If the condensed vapor contains a component which boils at a temperature higher than the mixture in the tank this component, once condensed and in the bottom of the annulus, will only vaporize and escape very slowly, if at all...

The present study describes some of the mechanisms and identifies some situations that could lead to condensation in the annulus. The criteria for condensation to occur are explained; these same criteria can be used for devising methods to prevent condensation. The discussion emphasizes LNG tanks, but the results can be extended to LPG and NGL tanks also. The main intent of this paper is to provide an understanding of the concepts involved.

TANK FEATURES

This study is applicable to only those type of tanks where the annulus has free communication for flow of vapor to and from the inner tank. A double-walled tank with suspended deck, shown schematically in Fig. 1, is a typical case where the present study is applicable. Specifically excluded from this study are the type of double-walled tanks where the stored liquid and gas are wholly contained in the inner tank.

MECHANISMS OF CONDENSATION

An increase in the absolute pressure in the vapor space of a tank will raise the dew point of the vapor in the annulus. Because the temperature of the liquid in the tank can rise only slowly, the liquid temperature will not equilibrate instantly to the higher tank pressure. Hence the tank liquid will continue to be colder than the dew point temperature of the annulus vapor which can condense and form liquid on the outer face of the inner-tank wall.

It should be noted that for low heat leak tanks the temperature difference (ΔT) across the inner-tank wall is normally less than 0.01 degrees C. Furthermore, for pure methane at ambient pressures the dew point variation is about 1.1 degrees C per 100 mbar. Hence, for an LNG tank, an increase in pressure by 100 mbar (either operator initiated or due to rise in barometric pressure)

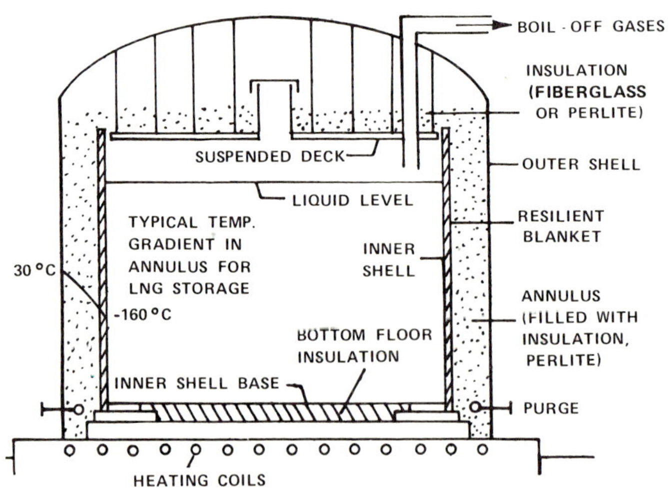

Fig. 1. Schematic Sketch of Double-Walled Cryogenic Tank with Suspended Deck.

will result in a temperature difference of 1.1 degrees C across
the inner-tank wall, thus causing a potential for condensation on
the outer face of the inner-tank wall.

Another mechanism of condensation results from variations in
liquid composition, which in turn lead to variations in bubble
point temperatures. To illustrate this mechanism suppose that the
annulus initially contains pure methane and that LNG containing
nitrogen is subsequently introduced in the tank. The presence of
one percent nitrogen lowers the LNG bubble point (at atmospheric
pressure) by about 3.5 degrees C. Thus, the LNG can be considerably colder than the methane dew point, and this will result in
condensation of methane on the outer face of the inner tank.

Consider a case where the annulus contains a vapor mixture of
methane and propane. If pure methane (LNG) is then introduced
into the inner tank the propane in the annulus will totally condense. As the propane condenses, the pressure in the annulus will
tend to drop, and methane vapor from the inner tank will flow into
the annulus. The condensed propane will trickle down the wall of
the inner tank to the bottom of the annular space where it will
vaporize because of heat inleak. The vaporized propane, because
of its higher density, will displace the methane and thus create a
propane-rich zone near the bottom of the annulus. As it rises the
propane will contact the cold inner-tank wall and recondense, and
trickle down again. Theoretically the propane could escape only
by diffusion up the warm outer wall, but the gravitational forces
would tend to cause the propane to migrate to the inner-tank wall
where it will condense over and over again. The propane would
essentially be trapped in the bottom portion of the annulus, and
it would create a continuous cycle of vaporization and recondensation. This circulatory flow can result in a strong thermal heat
transfer mode that can cause a very cold ring to form near the
bottom of the outer wall. Of course, any ethane, butane or other
heavy hydrocarbons that have condensed in the annulus will also be
subjected to cyclic vaporization and recondensation.

AMOUNT OF CONDENSATION AND FACTORS THAT LIMIT IT

Regardless of the mechanisms by which liquid condensation in
the annular space is induced, condensation cannot continue indefinitely. Some of the factors that tend to reduce the rate of
condensation or eliminate it are discussed below.

Inflow of Vapor Into Annulus

The absolute pressure in the annulus is essentially the same
as in the vapor space of the tank. Hence, any condensed vapor
will be replaced by inflow of vapor into the annulus. If this
incoming vapor has the same composition (and hence the same dew

point) as the condensing vapor, condensation may continue. But if
the incoming vapor has a dew point temperature lower than the
inner-tank-wall temperature it will not condense in the annular
space.

Consider a case where the annulus contains pure methane. If
LNG containing nitrogen is then introduced into the tank the
methane vapor in the annulus will condense because of the colder
LNG temperature. However, the boil-off vapor from the LNG will be
a mixture of nitrogen and methane, with a dew point temperature
lower than that of methane. This nitrogen-methane mixture will
replace the condensed methane vapor in the annulus, and further
condensation will stop. For this case the total amount of condensation will be limited to the amount of pure methane in the volume
of the annulus.

The condensed methane will revaporize due to heat leak.
Because of buoyancy forces the methane vapor will flow up in the
face of the outer wall and eventually escape from the annulus.
For a period of time a methane-rich zone will exist near the
bottom of the annulus and this may create a cycle of vaporization
and recondensation.

Warming of the Tank Contents

When condensation occurs in the annulus there is heat transfer from the annulus to the liquid in the inner tank. A temperature driving force (ΔT) across the inner-tank wall is therefore
essential. As the tank contents warm up, the available temperature driving force and the rate of condensation will decrease.

To illustrate this case consider an LNG tank filled with pure
liquid methane saturated at one atmosphere. The tank boil-off
vent is then closed until the tank pressure increases by 50
mbar. When this happens, the methane dew point temperature rises
by 0.6 degrees C. If the pressure rise occurs within a short
period the temperature driving force of 0.6 degrees C will be
available immediately, and condensation will begin. The condensed
methane will be replaced by more methane from the tank vapor
space, and condensation will continue so long as there is an
adequate temperature driving force across the wall. As the condensation proceeds, the liquid in the tank will gradually warm up
because the condensation process will give up heat (absorb "refrigeration") to the inner tank. In addition there will also be
normal heat leak through the tank wall, floor, and roof.

For 100,000 m^3 of pure liquid methane a subcooling of 0.6
degrees C represents enough refrigeration to condense over 165,000
kg (390 m^3) of liquid in the annulus. For a 50 m diameter tank
with a 1 m wide annulus, this condensed methane would represent a

pool of liquid about 2.5 m high. For a 50,000 m^3 pure ethane or propane storage tank of 40 m diameter the condensation corresponding to 0.6 degrees C subcooling would be 1.2 m and 1.3 m of liquid in the annulus, respectively. These figures, of course, have been derived from an oversimplified model, and it is unlikely that operators would permit such a deep pool of condensate to be formed. But it is clear that for a pure component stored in a double-walled cryogenic tank an increase in pressure can result in severe condensation in the annulus. When the liquid in the tank is a mixture the dew point of the vapor in the annulus will depend upon the vapor composition, and the severity of condensation will be less than for pure components.

Decrease in Tank Pressure

A decrease in tank pressure will cause a drop in the dew point temperature of the vapor in the annulus. The net result is a decrease in the temperature difference across the tank wall which in turn causes the condensation to either decrease or to stop. This technique is instantaneously effective for pure components.

CONDENSATION IN A STRATIFIED TANK

The case of a stratified tank is treated separately because it could lead to situations where severe condensation in the annulus is possible. Two examples will be discussed --- an LNG tank containing pure methane, and an LNG tank containing methane and nitrogen.

First consider an LNG tank partly filled with liquid methane saturated at 1100 mbar. The saturation temperature will be -160.5 degrees C. If additional liquid methane at -161.3 degrees C is bottom loaded into the tank, this additional liquid, because it is about 1.1 kg/m^3 denser, may form a stratified bottom layer.

The pressure in the annulus will be 1100 mbar, the vapor pressure of methane at - 160.5 degrees C. In the lower section of the inner tank the temperature along the outer wall will be close to -161.3 degrees C, cold enough for the methane vapor to condense. Condensation could continue until the temperatures of the top and bottom layers have equalized.

If the tank geometry, the heat leak factors, and the initial conditions of the two layers are known, it is possible to estimate the time needed for this type of stratification to be eliminated. The longer the stratification persists, the greater is the amount of condensate that can be formed.

The second example involves a tank that is partly filled with nitrogen-free LNG. A new cargo of LNG which contains 1 percent nitrogen is bottom loaded into the tank. The nitrogen-containing LNG will be about 3.5 degrees C colder and about 9 kg/m^3 heavier than the LNG already in the tank. Stratification is almost certain to occur.

There are similarities between this case and the previous case of pure methane. However, the temperature difference across the wall is much larger for this case, and the condensation rate could also be much larger. Calculations show that a condensation rate of several thousand kilograms per hour is possible for a 100,000 m^3 LNG tank.

In this example note that the methane vapor entering the annulus from the tank vapor space will be pure methane even though the LNG in the bottom layer contains nitrogen. The dew point of the vapor in the annulus will therefore remain unchanged as long as the tank pressure is not changed. Consequently, condensation can continue so long as the bottom layer is colder than the dew point temperature of methane. When stratification is finally eliminated, the vapor in the vapor space of the tank will be a nitrogen-methane mixture. This vapor will have a lower dew point than methane, and will moreover be in equilibrium with the tank liquid at the tank pressure and temperature. It will therefore not continue to condense in the annulus.

METHODS TO PREVENT CONDENSATION

Methods to prevent condensation can be selected by establishing the criteria that are necessary for condensation to occur, and then operating the facility so that condensation is avoided.

Condensation in the annulus can be prevented by ensuring that at any given tank pressure the vapor in the annular space has a dew point lower than the temperature of the tank contents. Except in the special case of a stratified tank, the temperature in the tank will nowhere be expected to be colder than the bubble point temperature at the tank operating pressure.

The dew point temperature of the annulus vapor can be controlled either by controlling the composition of the vapor or by changing the pressure in the tank (and hence in the annulus), but in practice absolute pressure control may not be a suitable method. Usually there will be other criteria that establish the operating pressure in the tank so that pressure is not normally available as a parameter for dew point control. However, as pointed out later, under some situations pressure reduction may be the only available method for lowering the dew point.

Effective control of the dew point temperature is best achieved by composition control in the annulus. The annulus is normally deadended and filled with perlite and fiberglass insulation. Hence a small continuous purge of a noncondensable gas (or gas mixture) in the annulus can ensure that a low enough dew point is always maintained.

Continuous Purge in LNG Tanks

In LNG tanks, a mixture of nitrogen and methane would be the most likely choice for purge gas. The purge rate will depend on tank size and geometry, while the composition of the gas mixture will depend on the dew point required in the annulus. Typically, a 30/70 nitrogen/methane mixture will be suitable.

The noncondensable purge can be introduced at the bottom of the annular space thus preventing condensation. The perlite and fiberglass insulation inhibit natural convection so the purge rate only needs to overcome problems due to breathing and back-diffusion in the annular space. It is also important to note that the nitrogen is more dense than the methane, and tends to stratify and minimize convective circulation in the annulus.

Special Considerations for LPG Tanks

For ethane and LPG tanks also a purge gas is needed to prevent condensation. Typically, ethane storage would need a 50/50 methane/ethane mixture; propane storage would need a 20/80 ethane/propane mixture, and butane storage would need a 20/80 propane/butane mixture. While these compositions are "needed" to prevent condensation in the annulus, they may not be "suitable" for tanks storing high purity grades of LPG and light hydrocarbons.

If the annulus of the LPG tank is a free space and not filled with perlite the problem that arises from purging the annulus with the compositions suggested above is caused by the fact that the purge gas has a lower molecular weight than the primary fluid stored in the tank. There will be a tendency for the purge gas to rise and go directly into the vapor space of the storage tank. The lighter hyrocarbon will tend to interfere with both the product purity and the operation of the vapor recovery system.

If the annular space of the LPG tank is filled with perlite a small flow of purge gas should prove to be a suitable technique to prevent condensation in the annulus. This happens because diffusional forces predominate over gravitational buoyancy forces for gas flowing through perlite.[2]

If the annulus is a free space the lighter purge gas would simply rise by natural convection, and it would not be effective in preventing condensation. Reducing the pressure may then be the only available method to lower the dew point and to prevent condensation in the annulus. An alternate solution to the problem may be to allow condensation to proceed and to continuously purge out the heavy hydrocarbons as they concentrate in the bottom zone of the annulus. In addition, the facility should be operated such that step increases in pressure are minimized, and the introduction of colder liquids that could cause stratification in the tank is avoided.

CONCLUSION

Under certain conditions condensation is expected to occur in the annular space of a double-walled cryogenic storage tank. This condensation is an undesirable phenomenon and should be prevented. A small continuous purge of a noncondensable gas mixture will, under most circumstances, prove to be an effective method to prevent condensation.

REFERENCES

1. N.J. Cuperus, Cryogenic Storage Facilities for LNG and NGL, paper presented at the 10th World Petroleum Congress, Bucharest, 1979.
2. C.C. Hanke, I. V. LaFave, and L. F. Litzinger, Purging of LNG Tanks Into and Out of Service: Considerations and Experience, paper presented at the American Gas Association Distribution Conference, May 1974.

POTENTIAL FOR CATASTROPHIC RUPTURE OF LARGE LIQUID OXYGEN STORAGE TANKS

S. M. Lainoff

Energy Incorporated
Kent, Washington

INTRODUCTION

A study of the mechanisms leading to castastrophic bursting of a liquid oxygen (LOX) tank was performed as part of a research facility safety analysis. The liquid oxygen storage and delivery systems presented one of the greatest safety concerns at the facility, and effort was made to establish a safe design and operating procedure. The study focused on generic internal causes of rupture of large aluminum dewars. The facility's tanks are 11,000 gallon dewars, operating at less than 350 psig, with 1.5 to 2 inch thick aluminum liners, carbon steel outer shells, and evacuated perlite insulation.

SIGNIFICANT HISTORICAL INCIDENTS AND EXPERIMENTAL RESULTS

While general industry experience with pressure vessel explosions comes from a number of cases, only those germane to cryogenic experience with liquid and/or gaseous oxygen were considered applicable. Clearly, very few accidents of this nature have occurred in commercial or government experience. Numerous incidents in aerospace activities involving oxygen systems were reported during the Apollo 13 investigation[1]. Of 44 incidents surveyed, 13 were due to the presence of materials incompatible with oxygen, 3 were due to equipment failures, and at least one was due to human error in failing to follow a procedure. Half of the incidents involved oxygen supply and storage systems explosions. Only a few events were due to a very high pressure (greater than 1500 psi) circumstances. A vessel explosion at a liquid oxygen production facility, described by Bayne[2], occurred in a LOX disposal tank, under conditions of very low hydrocarbon concentrations and yet ignition and detonation of the normally present contaminants occurred.

At least two tank trucks carrying LOX have been involved in sudden explosions. The first was a 1300 gallon truck which exploded in Tasmania in 1969 just before a delivery. The incident, described by Kilmartin[3], suggested the cause as the seizing of a submerged pump within the tank. The second tank truck explosion occurred in Brooklyn in 1970[4]. The incident involved a tank truck with an aluminum inner liner which had finished a delivery of LOX when the tank detonated. Several potential causes were attributed to this accident including contaminant accumulation in crevices within the tank, ignited by some mechanism causing rapid oxygen flow over the accumulation site. The reaction spread to the aluminum wall and sustained combustion of the wall led to detonation and bursting of the tank. Subsequent industry observations about the Brooklyn accident in the form of testimony by the Compressed Gas Association[5], deny that inherent design of this tank truck was the root cause of the accident. Extensive incident-free industry experience with aluminum tanks and LOX transport and storage supports this view.

The only other applicable known incident was the explosion of a 9,000 gallon LOX storage tank at the Reynolds MHD Facility in 1976[6]. Significant facts about this incident are that the tank was an aluminum dewar very similar in size to the tanks of interest, and the cause of the detonation was a reaction between LOX and the aluminum wall. As the contaminant content was within acceptable limits during LOX deliveries and no buildup was found on pieces of the tank sampled after the incident, the root cause of the explosion was not determined. However, no evidence was available to pin the cause of the accident to events other than contaminant combustion. Details of this incident are not available to public scrutiny at this time.

Numerous investigations of combustion within oxygen environments were reviewed. Some establish the ignition properties of metals including aluminum[7-10]; others survey ignition sensitivities of a broader class of contaminants[11-12]. Some experiments evaluated the susceptibility of materials to ignition due to specific mechanisms, such as impact[13]. Hust and Clark stated that it is difficult to extrapolate combustion experiments to hypothesized situations involving oxygen systems. However, certain facts pertinent to LOX storage tanks can be deduced from experimental results and provide insight into some of the mechanisms of tank burst phenomena:

1.   Aluminum has a relatively high ignition temperature compared to comtaminants likely to be in the tanks.

2.   Bulk aluminum as found in LOX storage tanks is protected by an oxide layer with a melting temperature higher than the melting temperature of the base metal. The ignition

point of the base metal is based on the transition temperature of the metal, or that temperature at which the oxide layer becomes non-protective. This value is the melting point of the oxide[9].

3. Some reviews suggest[12] that a violent reaction between LOX and aluminum can occur from the exposure of fresh aluminum by the process of crack propagation.

4. The high thermal diffusivity of aluminum allows heat to be conducted away from a source of heat very rapidly. However, once the base metal is ignited, its high heat of combustion enhances the likelihood of sustained combustion. This same property can lead to a violent reaction in a pure oxygen environment, proceeding to detonation.

5. Sensitivity of aluminum to ignition by impact is possible.

6. Sensitivity of aluminum reactions in LOX systems to adiabatic compression appears to be limited to very high pressure systems[13].

TANK RUPTURE FAULT TREE

Phenomena critical to the tank rupture event are assembled into a logical model of the rupture process by using the logic of the fault tree, shown in Figs. 1, 2, and 3. The events of the tree are combined using AND and OR logic and transfer symbols $\triangle$, $\triangle$, $\triangle$, $\triangle$. Diamonds represent basic events that may be broken down further for a more detailed representation of phenomena.

The primary approaches to the ultimate event (tank rupture) are shown in Fig. 1. Only two mechanisms have the potential to directly cause catastrophic tank rupture; 1) a pressure rise produced by a detonation within the tank and possibly including the tank, or 2) a violent reaction of the LOX contents with the aluminum wall due to crack propagation induced by stresses on a flaw in the metal. The second mechanism would be due to the exposure of fresh base metal to liquid oxygen.

The tanks in question are made of 5083 aluminum alloy as are many aluminum dewars. The alloy is chosen for its high ductility and resistance to crack propagation, especially at cryogenic temperatures[14]. These properties plus hydrotests on all vessels of this type support our view that crack propagation alone is not a significant branch of the tank rupture tree. Slow boil-off in combination with failure of redundant pressure relief devices is also considered to be an improbable root cause of tank rupture.

The primary mode of failure is then detonation within the tank. Previous accidents point out that violent detonations are rapid enough to exceed the safety relief inherent in the design. A detonation exceeding either the design strength of the vessel or

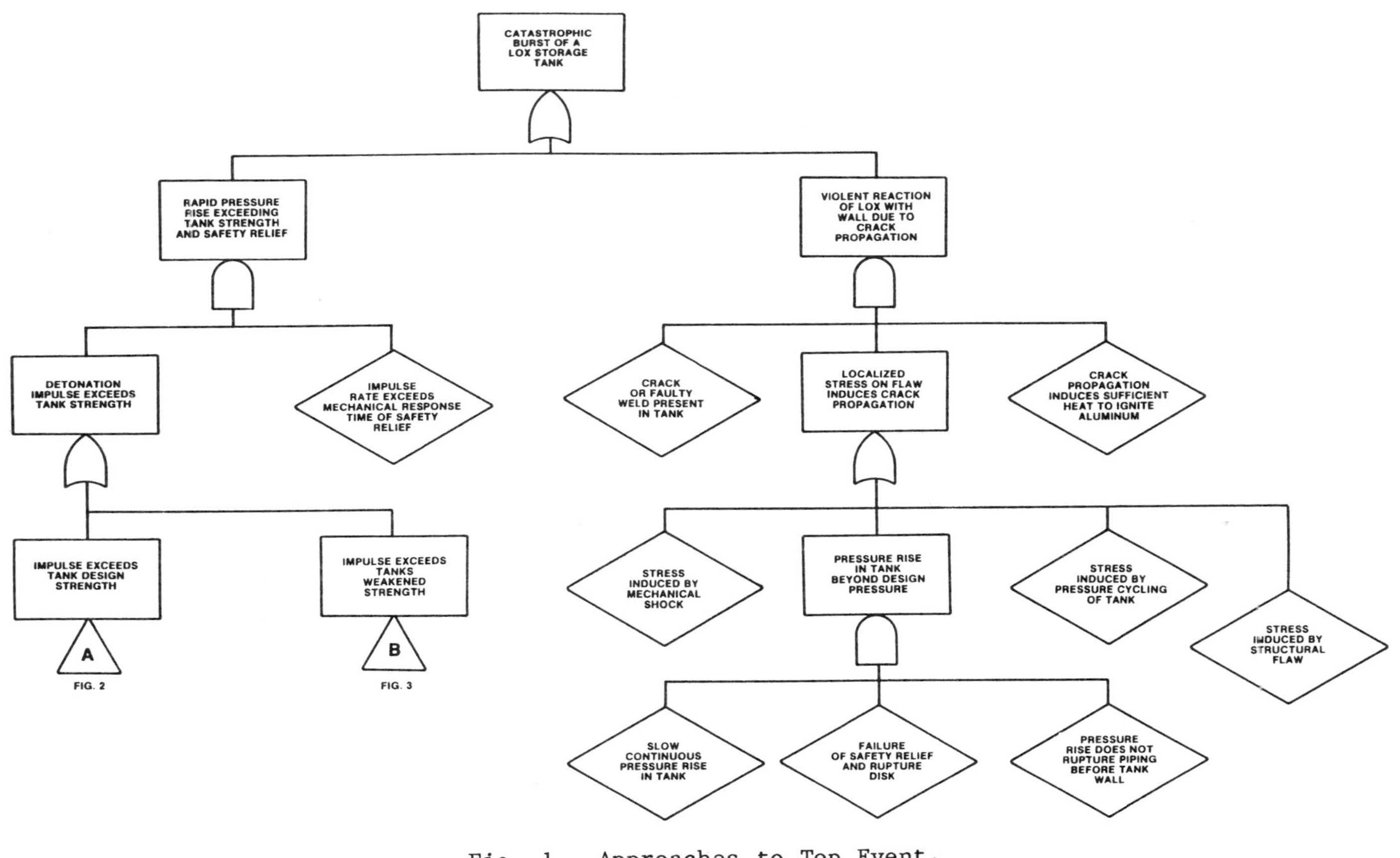

Fig. 1. Approaches to Top Event.

the vessel's strength after being weakened by an oxygen-aluminum reaction will rupture the tank when it overcomes the reliefs.

Figure 2 (Transfer A) represents detonation of a LOX and contaminant mixture. Hydrocarbons and many other compounds are known to react violently in the presence of oxygen. The reaction begins with ignition of a contaminant and proceeds to detonation of the mixture. Both a sufficient concentration of contaminant for reaction and an ignition mechanism is required. For the tanks of interest ignition by electric sparks and tank punctures were eliminated by design. Mechanical impact, a not improbable mechanism, may occur by loose parts traveling through the system.

Figure 3 (Transfer B and C) represents detonation due to a violent reaction between LOX and the aluminum wall. The wall is weakened and heated to its ignition point by localized ignition and combustion of a contaminant accumulation near the tank wall. A sustained combustion of the aluminum wall requires an intense localized heat source at the transition temperature of the base metal and a reaction site permitting continuous exposure of fresh aluminum. Reaction sites that can exist in LOX storage tanks are shown in Fig. 3. A number of ignition mechanisms have been suggested in past accident investigations. They include frictional heating due to rapid flow of GOX or LOX over the site or mechanical impact or abrasion caused by loose parts trapped near the site. Introduction of the contaminant into the tanks is a requirement for both types of a contaminant combustion in Figs. 2 and 3.

HUMAN ERROR POTENTIAL IN TANK FARM OPERATION

The most critical element linking the detonation scenarios together is the introduction of contaminants into the tanks coupled with a failure to detect critical contaminant buildup. Oxygen system design and operation was analyzed in detail to cover all fault combinations and pinpoint the more likely contributions to contaminant entry. Facility operation was segmented into on-line operation, maintenance, LOX delivery, transfer and sampling, tank cleaning, and tank cleaning verification. All physical pathways to the tanks were examined. Contaminants introduced during LOX transfer from railcar to tanks and contaminants remaining after cleaning appear to be the most likely sources of buildup.

Human error in failing to follow sampling, cleaning, or cleaning verification procedures is the most likely cause of failure to detect a buildup of contaminants. Accident history suggests that if contaminant accumulation sites exist within tanks, then buildup can occur from inadequate or infrequent cleaning. Human error overrides highly reliable dewar design as the key cause of contaminant buildup.

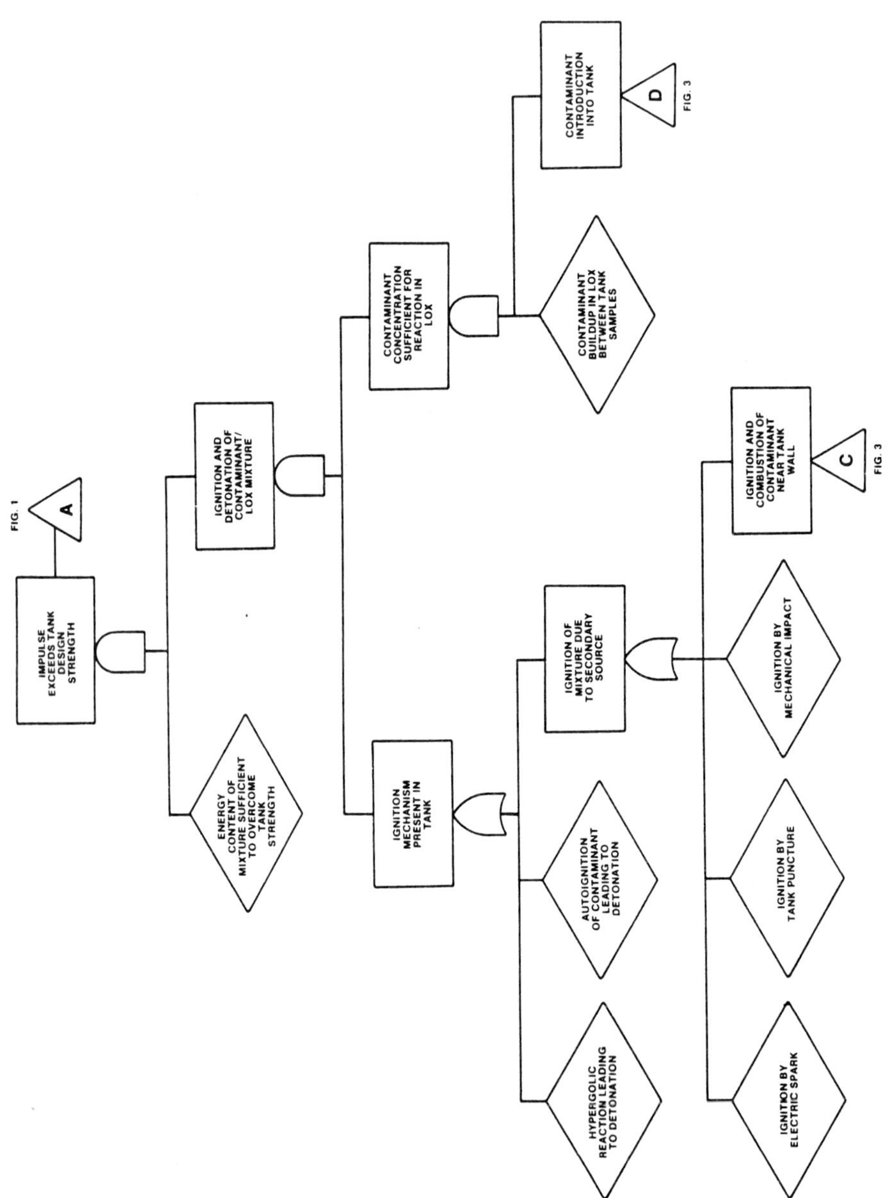

Fig. 2.   Detonation of LOX and Contaminant Mixture.

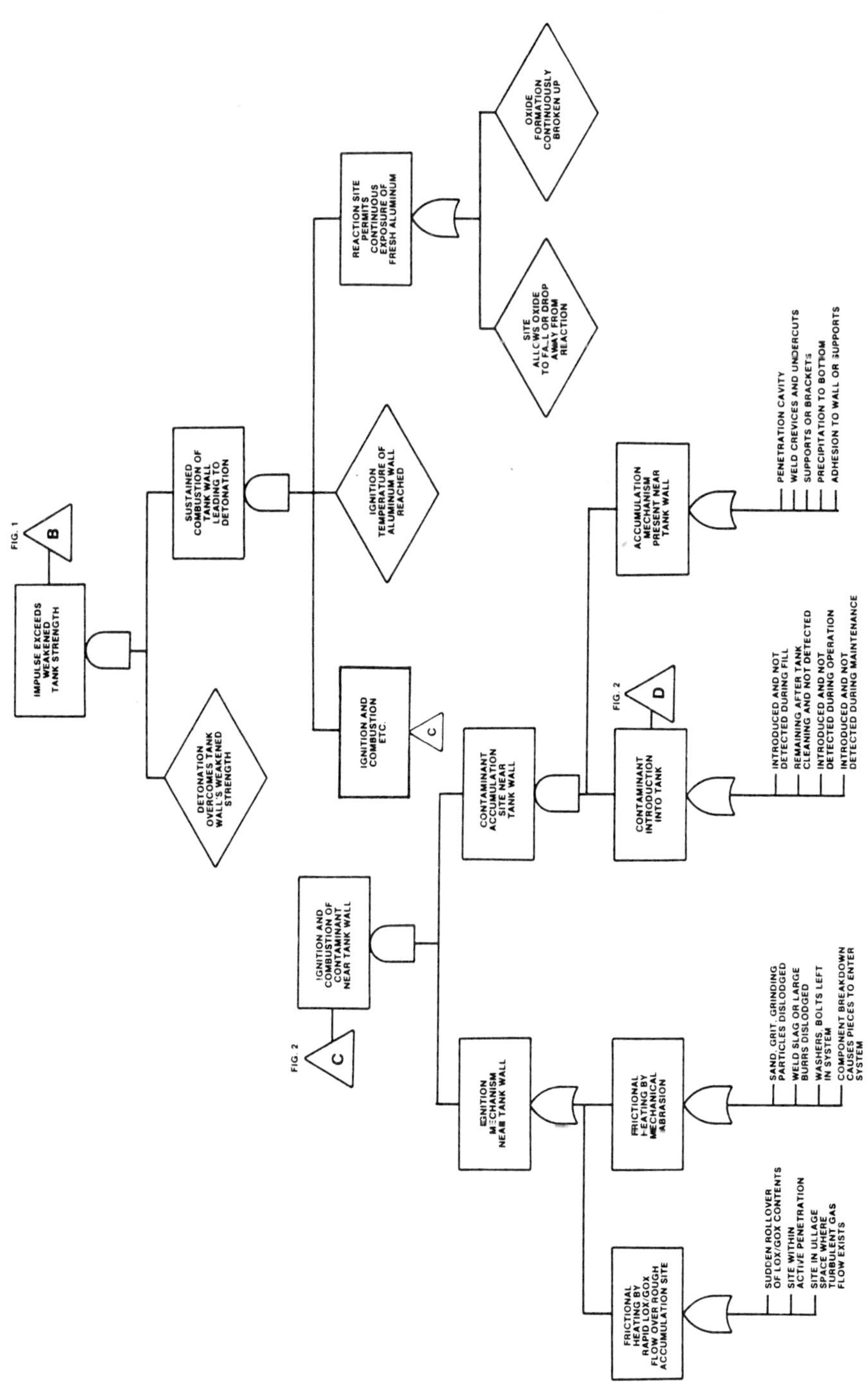

Fig. 3. Reaction of LOX and Aluminum Wall.

Elements known to reduce human errors in tank farm operation
are: 1)   the use of a written set of engineering test and in-
spections procedures, 2)   multiple cleanliness verification procedures, 3)   multiple sampling of passivating rinse after cleaning, 4)   multiple sampling of LOX tank contents periodically
during operation, 5)   use of a three-man team for all tests with
the incorporation of a checklist for the step-by-step procedure,
6)   adequate training for test personnel and, 7)   inspection of
tank cleanliness prior to initial use.

SUMMARY

The body of literature to date on oxygen system accidents and
experimental findings on oxygen-metal and oxygen-contaminant
reactions was reviewed to establish a basis for potential accident
scenarios relevant to the tanks. Using the techniques of fault
tree analysis, basic phenomena were synthesized into a logic model
representing all of the combinations of events necessary to cause
tank rupture. Methods to improve the safety of the facility by
reducing human error in tank farm operation were shown to be the
most constructive solution to the LOX tank safety question.

REFRENCES

1. NASA, MSC Apollo 13 Investigation Team, Final Report, Panel 8,
      "High Pressure Oxygen Systems Survey," May 1970, TMX 66925.
2. W.J., Bayne, Liquid oxygen disposal vessel explosion, _Safety
      in Air and Ammonia Plants (AICHe,)_ Vol. 8. 1966.
3. J. Kilmartin, Two Liquid Oxygen Explosions, _Fire Jour._ March
      1971.
4. NTSB, Highway Accident Report - Liquefied Oxygen Tank Truck
      Explosion Followed by Fires in Brooklyn, New York, May 30,
      1970, Report No. NTSB-HAR-71-6, May 12, 1971.
5. CGA, Comments and Views of the Compressed Gas Association,
      Inc., Submitted to the Dept. of Transportation in response
      to Docket No. HM-115, Notice No. 74-3 Cryogenic Liquids,
      October 25, 1974.
6. Gilbert Associates, Inc., Comments on Liquid Oxygen Storage at
      Research Facilities, Unpublished, August 17, 1978.
7. W.C. Reynolds, Investigation of Ignition Temperatures of Solid
      Metals, TN D-182, NASA, October 1959.
8. I. Glassman, et. al., A Review of Metal Ignition and Flame
      Models, AGARD Conference Proceedings 52:19 (1970).
9. I. Glassman and N.M. Laurendeau, _Ignition temperatures of
      metals in oxygen atmospheres, Comb. Sci. and Techn._, Vol.
      3, 1971.
10. C. McKinley, Experimental Ignition and Combustion of Metals,
      CGA-Oxygen Compressors and Pumps Symposium, Atlanta,
      Georgia, 1971.

11. J.G. Hust and A.F. Clark, A Survey of Compatibility of
    Materials with High Pressure Oxygen Service, NBS, 1972.
12. J.G. Hust and A.F. Clark, A Review of the Compatibility of
    Structural Materials with Oxygen, <u>AIAA Journal</u>, 12(4):
    (1974).
13. NASA, High Pressure Liquid and Gaseous Oxygen Impact
    Sensitivity Evaluation of Materials for Use at Kennedy
    Space Center, N76-17215/4G1, 1976.
14. NASA, ASRDI Oxygen Technology Survey, Vol. VII,
    Characteristics of Metals that Influence System Safety,
    Aerospace Safety Research and Data Institute, Report No SP-
    3077, 1974.

DISCUSSION

Question by R.G. Scurlock, Southampton University, England: Does
your analysis consider the case of "thermal overfill", i.e.
when $\dot{m} < Q/L$, where $\dot{m}$ is the evaporative mass flow rate, Q is the
total heat flux and L is the latent heat of vaporization?

Answer by author: The causes of thermal overfill, such as external fires, are obvious design considerations which were taken into
account for the entire oxygen storage and delivery system. It was
our task to explore the potential for LOX tank ruptures due to
more esoteric causes. Therefore we excluded thermal overfill from
our work.

COOL-DOWN FLOW-RATE LIMITS IMPOSED BY THERMAL STRESSES IN LNG PIPELINES*

J. K. Novak, F. J. Edeskuty, and J. R. Bartlit

*Los Alamos National Laboratory
University of California
Los Alamos, New Mexico*

INTRODUCTION

Warm cryogenic pipelines are usually cooled to operating temperature by a small, steady flow of the liquid cryogen. If this flow rate is too high or too low, undesirable stresses will be produced. Low flow-rate limits based on avoidance of stratified two-phase flow have been calculated earlier for pipelines cooled with liquid hydrogen or nitrogen.[1] High flow rate limits for stainless steel and aluminum pipelines cooled by liquid hydrogen or nitrogen have been determined by calculating thermal stress in thick components vs flow rate and then selecting some reasonable stress limits.[2] The present work extends these calculations to pipelines made of AISI 304 stainless steel, 6061 aluminum, or ASTM A420 9% nickel steel[3] cooled by liquid methane or a typical liquefied natural gas.

---

*Work performed under the auspices of the U. S. Department of Energy.

LOW FLOW-RATE LIMITS

The process of cooling horizontal cryogenic transfer lines to operating temperature is usually accompanied by two-phase flow. The rate of heat transfer from the metal can be different for each phase, causing unequal cooling around the circumference of the line with consequent undesired thermal stresses and/or displacements, usually upward bowing, of the line. This phenomenon has been observed, and is believed to be caused by stratification of the two-phase mixture in the line.

Baker[4] presented a correlation for predicting the type of two-phase flow in a pipe. The flow regimes are classified into several types - plug, slug, annular, wave, and stratified - the last two of which occur when the liquid flows on the bottom of the pipe while the gas flows above it. Although Baker's correlation was not devised for cryogenic fluids, the two-phase flow of hydrogen and nitrogen has been investigated[1,5] and found to follow predicted behavior. Because of the increasing importance of LNG, this work extends the prediction of stratified flow to LNG transfer.

The Baker correlation uses densities of the gaseous and liquid phases and surface tension and viscosity of the liquid phase, combined with the gas and liquid mass flow rates (G and L, respectively), to predict the flow regime. In general, stratified flows occur at low flow rates and at low ratios of L/G. Because the value of L/G must cover the range of ∞ (all liquid) to 0 (all gas) during the cooling process, the total avoidance of stratified flow is impossible.[4] However, by sufficiently raising the total flow rate (L+G) the appearance of stratified flow can be delayed until arbitrarily near completion of the vaporization process. In this work, a minimum total flow rate has been chosen to avoid stratified flow during the first 95% of the vaporization process. Figure 1 shows this minimum flow rate for liquid methane.

Making these calculations is difficult because of the variation in compositions of natural gases, and because not all the physical properties are available. Methane values were used for liquid and gas densities[6] and viscosity.[3] Values of surface tension could not be found for either LNG or methane, so this value was estimated from those of similar compounds. All properties were evaluated at 25 psia to approximate average pressures in the entire transfer line.

The effect on the calculated minimum flow rate of changing fluid properties to correspond to different LNG compositions, including the "typical" LNG (91.8% CH_4, 3.9% N_2, 4.3% C_2 compounds),[7] was evaluated. We conclude that the flow rates in Fig. 1 are valid within 10%, regardless of the LNG composition.

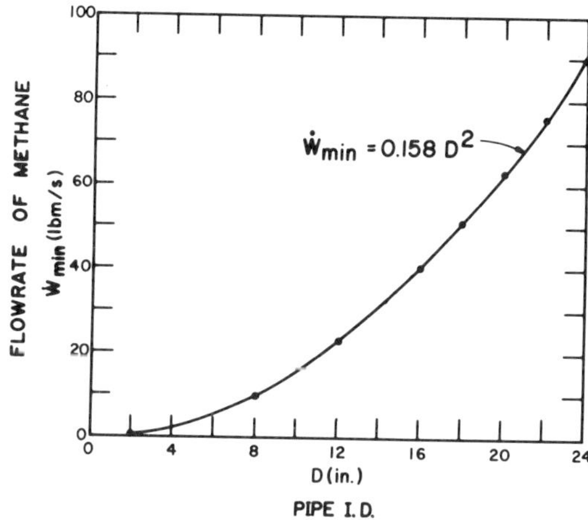

Fig. 1. Minimum flow rate of methane required to avoid stratified flow during 95% of the cool-down process.

HIGH FLOW-RATE LIMITS

At any cool-down flow rate above the minimum, the pipeline cools evenly around its circumference. Thermal stresses develop due to differences in thermal contraction associated with the radial temperature gradients. The magnitude of these stresses depends on the radial thickness of the pipeline components, the pipeline material, and the magnitude of the radial temperature gradients, which, in turn, depend on the cool-down flow rate. Thus, for a given pipeline, the greatest thermal stresses occur in the thickest components and increase as the cooldown flow rate increases.

In the previous work,[2] ASA weld neck flanges were chosen for study as being representative of thick pipeline components. Temperature distributions and thermal stresses were calculated by finite element techniques, which gave detailed results but were rather cumbersome. Several widely differing cases were investigated, and it was observed that maximum stress always occurred at the inside surface of the thickest part of the flange, and heat flow in this region was almost entirely radial.

A simple disk model with only radial heat flow was developed to permit faster calculations. The disk is divided into concentric ring-shaped elements and the calculation proceeds in short time steps. Heat is transferred by forced convection from the inside surface only; the radial and outer surfaces are perfectly insulated. The disk starts at a uniform temperature of 530 R, the cryogen flow is established instantly and held constant, the cryogen is always at the saturation temperature for 25 psia, and the Dittus-Boelter heat transfer coefficient [Nu= $0.023(Re)^{0.8}$ $(Pr)^{0.4}$] is used. At each time step the temperature of each element, the temperature of the inside surface, the yield strength of the inside surface material, and the circumferential thermal stress at the inside surface are computed. The calculation proceeds until the maximum thermal stress and the minimum difference between the thermal stress and the yield strength are found for a given flow rate.

The thermal stress in a thin disk[8] is given by

$$\sigma_{\theta a} = \psi E \left[\frac{2}{b^2 - a^2} \int_a^b r\,(\alpha T)dr - (\alpha T)_a \right] \; .$$

The matching factor ψ is used to accommodate various geometries: $\psi = 1.0$ gives stresses in a thin disk and $\psi = 1.43$ gives stresses in a long cylinder. Comparison of the thin disk results and finite element results showed that the thin disk model could be used to predict stresses in ASA flanges if $\psi = 1.1$ were used for stainless steel and $\psi = 1.43$ were used for aluminum.

Upper limits on cool-down flow rate, which will keep thermal stresses below any maximum stress value desired, can now be determined. One safe and conservative stress value is the maximum thermal stress range allowed by the American Standard Code for Pressure Piping[9]; i.e., $(1.25\,S_c + 0.25\,S_h)$, which is 28,125 psi for 304 stainless steel,[10] 14,250 psi for 6061-T4 aluminum,[11] and 47,550 psi for 9% nickel steel[3] (Table I, Code Allowable Stress). A less conservative flow-rate limit is that which causes the maximum thermal stress to just equal the local yield strength of the metal in a given disk (Table I, Stress = Yield Strength).

Limiting flow rates were calculated for every combination of the material, limiting condition, and stress matching factor (ψ) parameters listed in Table I, using liquid methane at 25 psia (212 R) as a coolant. The results can be correlated using the parameter groups $\dot{w}_{max}/D$ and t/D, and are plotted in Fig. 2 and given as fitted equations of the form $\dot{w}_{max} = D/\left[P_1(t/D)^2 + P_2(t/D) + P_3\right]$ with the coefficients P_i listed in Table I. Combinations B, J, and L do not have results shown because the thermal stress will never exceed the yield strength at any flow rate.

For calculations with liquid natural gas as a coolant, the typical LNG mixture was chosen.[7] The choice of coolant affects the heat transfer coefficient and the bulk coolant temperature; for this LNG mixture, $h_{LNG} = 0.935\,(h_{CH4})$ and $T_{bulk} = 200$ R. Calculations of limiting flow rates were made for the same combinations of parameters as for methane. The LNG results are shown compared to the methane results in Fig. 3.

Table I. Parameter combinations and coefficients of the equations giving maximum cool-down flow rates for methane.

Combination Number	Aluminum	304 Stainless	9% Nickel Steel	ψ = 1.1	ψ = 1.43	Code Allowable Stress	Stress = Yield Strength	P_1	P_2	P_3
								Coefficients of $\dot{w}_{max} = D/[P_1(t/D)^2 + P_2(t/D) + P_3]$		
A	X			X		X		-0.004126	0.185863	-0.013124
C	X				X	X		-0.012204	0.327921	-0.022615
D	X				X		X	0.000519	0.001109	-0.000113
E		X		X		X		-0.213099	5.446331	-0.374527
F		X		X			X	0.050177	1.046575	-0.093578
G		X			X	X		-0.434207	8.920028	-0.584604
H		X			X		X	-0.017061	3.075954	-0.243313
I			X	X		X		0.006577	0.016513	-0.001622
K			X		X	X		0.020195	0.273874	-0.024469
B	X		X				X	Thermal stress for these combinations will not exceed material yield strength at any flow rate.		
J		X	X				X			
L			X		X		X			

DISCUSSION

Aluminum and 9% nickel steel components can tolerate very high cool-down flow rates, based on not exceeding the material yield strength. The flow-rate limits given consider only thermal stresses and it may, in fact, be impossible to reach such high flow rates in many pipelines.

For components of certain sizes, especially thick ones in small-diameter pipelines, there exists no steady cool-down flow rate that both avoids stratified flow and keeps the maximum thermal stress within desired limits. Such pipelines may be cooled with cold gas or by intermittent cool-down schemes where small

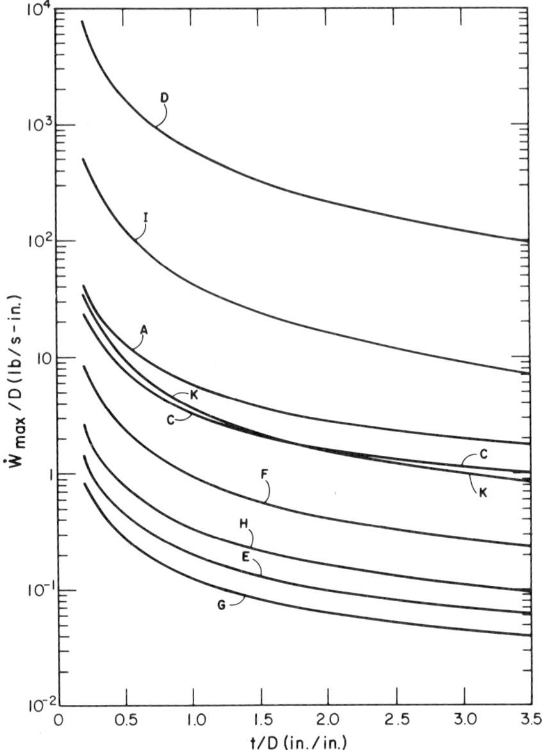

Fig. 2. Maximum flow rates of methane
for various parameter combinations.

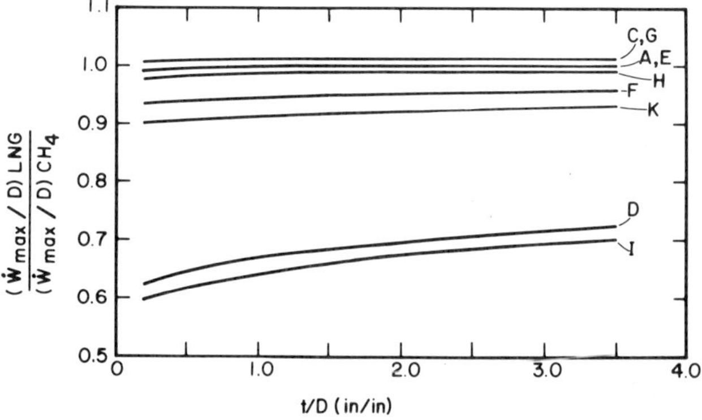

Fig. 3. The ratio of maximum LNG cool-down
flow rate to maximum CH_4 cool-down flow rate.

quantities of liquid are introduced into the pipeline and allowed to evaporate.[12]

The flow-rate limits based on yield strength are probably the least conservative that can be used comfortably for pipelines intended for long service. However, nothing catastrophic will happen immediately if these limits are exceeded slightly. The first transgression will cause the interior material of the flange to yield, and subsequent ones will lead to fatigue failure, creating small cracks that hasten deterioration by forming stress concentrations and reducing the amount of material available to carry loads. Thus, it may be feasible to exceed these flow-rate limits in piping systems intended for short lives.

NOTATION

$$
\begin{aligned}
a &= \text{inside radius of disk (in.).} \\
b &= \text{outside radius of disk (in.).} \\
C_p &= \text{specific heat of cryogen (Btu/lbm-R).} \\
D &= \text{inside diameter of flange (in.).} \\
E &= \text{elastic modulus (psi).} \\
h &= \text{convective heat transfer coefficient (Btu/s-in.}^2\text{-R).} \\
k &= \text{thermal conductivity of cryogen (Btu/s-in.-R).} \\
Nu &= \text{Nusselt number} = hD/k. \\
P_i &= \text{equation coefficients.} \\
Pr &= \text{Prandtl number} = C_p\mu/k. \\
Re &= \text{Reynolds number} = D\dot{w}/\mu \\
S_c &= \text{basic material allowable stress at cold temperature.}[9] \\
S_h &= \text{basic material allowable stress at hot temperature.}[9] \\
t &= \text{radial thickness of flange (b - a) (in.).} \\
T &= \text{temperature (R).} \\
\dot{w} &= \text{mass flow rate (lbm/s).} \\
(\alpha T) &= \text{total thermal expansion at radius r (in.).} \\
(\alpha T)_a &= \text{total thermal expansion at radius a (in.).} \\
\mu &= \text{viscosity of cryogen (lbm/in.-s).} \\
\sigma_{\theta a} &= \text{circumferential stress at radius a, inside surface} \\
&\quad\; \text{of disk (psi).} \\
\psi &= \text{matching factor.}
\end{aligned}
$$

REFERENCES

1. F.J. Edeskuty, D.H. Liebenberg, J.K. Novak, Cooldown of Cryogenic Transfer Systems, Paper No. 67-475, AIAA 3rd Propulsion Joint Specialist Conference, Washington, DC (1967).
2. J.K. Novak, Cooldown Flowrate Limits Imposed by Thermal Stresses in Liquid Hydrogen or Nitrogen Pipelines, in "Advances in Cryogenic Engineering, Vol. 15," Plenum Press, New York (1970).

3. LNG "Materials and Fluids", D. Mann, ed., Cryogenics Division,
   National Bureau of Standards, Boulder, Colorado (1977).
4. D. Baker, Design of pipelines for simultaneous flow of gas and
   oil, The Oil and Gas J. (July 1954).
5. J.C. Bronson, et. al., Problems in Cool-Down of Cryogenic
   Systems, in "Advances in Cryogenic Engineering, Vol. 7",
   Plenum Press, New York (1960).
6. J. Klosek and C. McKinley, Densities of Liquefied Natural Gas
   and Low Molecular Weight Hydrocarbons, in "Proc. 1st Intl.
   Conf. on LNG", Chicago (1968).
7. H.A. Molero, Prediction of Properties for the Estimation of
   Heat Transfer Coefficients in Two-Phase Cryogenic Heat
   Exchangers, Thesis, Illinois Institute of Technology,
   Chicago, (1974).
8. J. Goodier and S. Timoshenko, "Theory of Elasticity", 2nd Ed.,
   McGraw-Hill Book Co., New York (1951).
9. "American Standard Code for Pressure Piping", ANSI/ASME B31.1,
   par. 102.3.2, ASME, New York (1980).
10. H.P. Gibbons, R.M. McClintock, NBS Monograph 13 (1960).
11. "Cryogenic Materials Data Handbook", PB-171809, Aerojet
    General Corporation, Sacramento, CA (1966).
12. J.C. Commander, and M.H. Schwartz, Cooldown of Large-Diameter
    Liquid Hydrogen and Liquid Oxygen Lines, NASA-CR-54809,
    Aerojet General Corporation, Sacramento, (1966).

POWER GENERATION USING COLD POTENTIAL OF LNG IN MULTICOMPONENT FLUID RANKINE CYCLE

H. Shiozawa and T. Hiro-oka

Tokyo Gas Co., Ltd.
Tokyo, Japan

and

I. Aoki and T. Inoue

Chiyoda Chemical Engineering & Construction Co., Ltd.
Yokohama, Japan

INTRODUCTION

To date, almost all base load LNG imported to Japan has been regasified using the heat of seawater or warm seawater effluent from a power plant. The LNG is mainly consumed either as fuel for steam power plants or as a source of gas for local utilities. A new method of generating power efficiently from LNG cold potential in the regasification process using seawater or low level waste heat as the heat source was developed by the joint efforts of Tokyo Gas Co., Ltd. and Chiyoda Chemical Engineering & Construction Co., Ltd. This new method uses a multicomponent working fluid (MFR) extracted from the LNG in a Rankine cycle. The MFR is a mixture of methane, ethane, propane and butane.

DESCRIPTION OF PILOT PLANT

The pilot plant was constructed at the Negishi Works of Tokyo Gas Co., Ltd. located on the outskirts of Yokohama, Japan. The regasification capacity is 5 tons of LNG per hour with seawater utilized as the heating medium. A maximum of 130 kW of power was generated when the temperature of seawater was about 24°C. The process flow diagram of the plant is shown in Fig. 1.

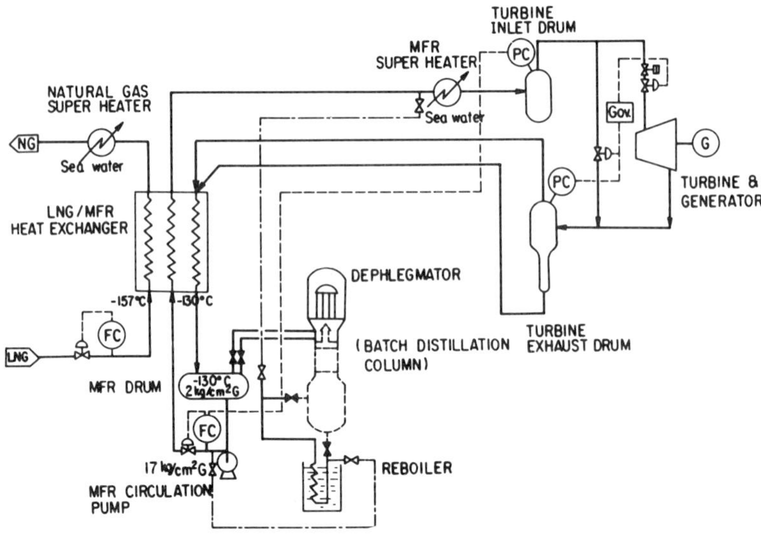

Fig. 1. Process flow diagram of MFR Process.

Features of MFR Process

The MFR process has two outstanding features:

1) The efficiency of the cycle is enhanced by adjusting the composition of the working fluid (MFR) making the condensing curve of this fluid conform to the vaporization curve of LNG so that the loss of exergy in the LNG/MFR Heat Exchanger is minimized.

2) The MFR can also be vaporized when LNG is vaporized by the condensing MFR in the multichannel LNG/MFR Heat Exchanger because the MFR is vaporized at a wide range of temperatures under the operating pressure. The loss of exergy is therefore, greatly reduced when the MFR is heated by seawater.

Process Description

The MFR is accumulated in the MFR Drum at about 0.2 MPa and $-130°C$ is pumped up to about 1.7 MPa and sent to the LNG/MFR Heat Exchanger, where LNG and MFR are vaporized simultaneously by condensing MFR. LNG which contains only a small amount of liquid at the outlet of the heat exchanger, is heated up to above $0°C$ at the Natural Gas Superheater and is transferred to the trunk line. The MFR is heated and totally vaporized by seawater in the MFR Superheater. Then the MFR drives the Turbine/Generator and is recycled to the LNG/MFR Heat Exchanger where it is condensed by the vaporizing LNG and MFR. The process can be operated in three modes.

Operation without power generation. In this mode, the process is operated only as an LNG vaporizer, with the MFR vapor totally recycled through the back pressure control valve and by-passing the turbine/generator.

Operation under speed control of turbine. In this mode, the rotating speed of the turbine is controlled by the governer actuated by the speed controller. Back pressure of the turbine is controlled by the back pressure control valve installed parallel to the turbine. The power generated is consumed using the water rheostat.

Operation under back pressure control. Here the back pressure of the turbine is controlled by the governer with the rotating speed of the turbine set to match the frequency of commercial current. The back pressure control valve installed parallel to the turbine is closed due to a biased signal from the control unit.

EXPERIMENTAL RESULTS

Make-up of the MFR fluid was started on July 16, 1980 and the plant was successfully started up on August 1, 1980. The experimental operation continued for about 6 months. The results related to process characteristics were: proof of the MFR process, start-up operation, maximum turndown ratio, and make-up of MFR. The results related to control characteristics are presented elsewhere[1].

Proof of MFR Process

The relation of power output to LNG vaporization load is shown in Fig. 2. The ratio of the turbine inlet and exhaust pressures affects the flow rate of MFR even when the LNG flow rate is kept constant. The balance of exergy is summarized in Table I. The MFR process was proved to be more efficient than the propane cycle. Power output will be doubled compared to that of a propane cycle.

Start-up Operation

The LNG and MFR Superheaters installed after the LNG/MFR Heat Exchanger are shell and tube heat exchangers made of carbon steel. Therefore it is very important to ensure that there is vapor in the channels for MFR condensation at the initial stage of start-up because LNG and MFR vaporized countercurrently must be heated to above $-45^{\circ}C$ at the outlet of the heat exchanger. A

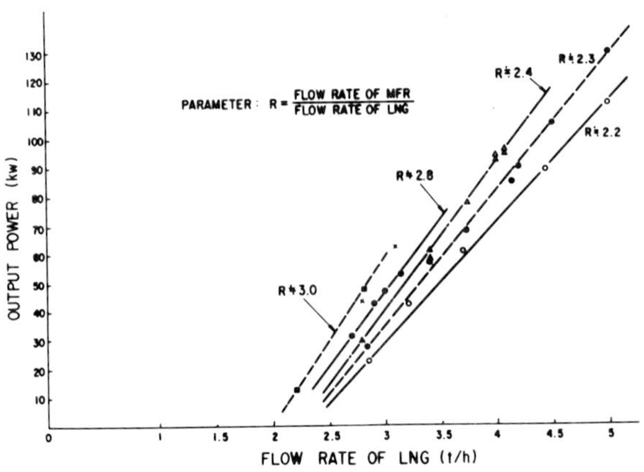

Fig. 2. Relation of power output and vaporization load of LNG

Table I. Balance of Exergy for MFR Cycle

	Design		Pilot Plant	
LNG Vaporized (ton/h)	5.0		5.0	
Vaporization Pressure of LNG (MPa)	2.25		2.28	
Temperature of Seawater (°C)	20.0		23.9	
Input of Exergy	kW	%	kW	%
LNG	1283	93.7	1258	94.0
Seawater	12	0.9	–	–
Pumps	74	5.4	81	6.0
Total	1369	100.0	1339	100.0
Output of Exergy	kW	%	kW	%
Electricity	105	7.7	130	9.7
Natural Gas[a]	585	42.7	574	42.9
Seawater	40	2.9	8	0.6
Loss of Exergy	639	46.7	627	46.8
Total	1369	100.0	1339	100.0
Breakdown of Loss of Exergy				
Turbine, Seal & Gear[b] etc.	175		122	
Heat Exchangers	371		399	
Pump, Valves & Friction	93		106	
Total	639		627	

Basis of exergy : 25°C, 1 atm.

[a]Exergy carried by natural gas is critical because it is re-gasified under the pressure of natural gas trunk line
[b]Gear loss is significant because of small turbine.

small amount of MFR liquid divided at the discharge of the MFR
Circulation Pump is vaporized at the start up heater and the vapor
is accumulated in the Turbine Inlet Drum before start-up. This
type of heat exchanger was used considering commercial plant
installation costs. Careful consideration is required for a
start-up operation, because LNG and MFR start to vaporize vertically through the channels of LNG/MFR Heat Exchanger. Therefore, a rapid increase of fluid flow is required to obtain a
stable start-up. This is shown in Fig. 3.

<u>Establishment of Start-up Procedure</u>. Some deviation of
process variables were found compared to those expected from
process design. The composition of MFR, especially the concentration of methane should be selected based on the criteria obtained from experimental results. In principal, the composition
of MFR should be based on the seawater temperature. However, some
adjustment is required to give stable start-up. The result is
shown in Fig. 4. The relation between LNG feed rate and that of
MFR liquid for stable start-up was also found to be different from

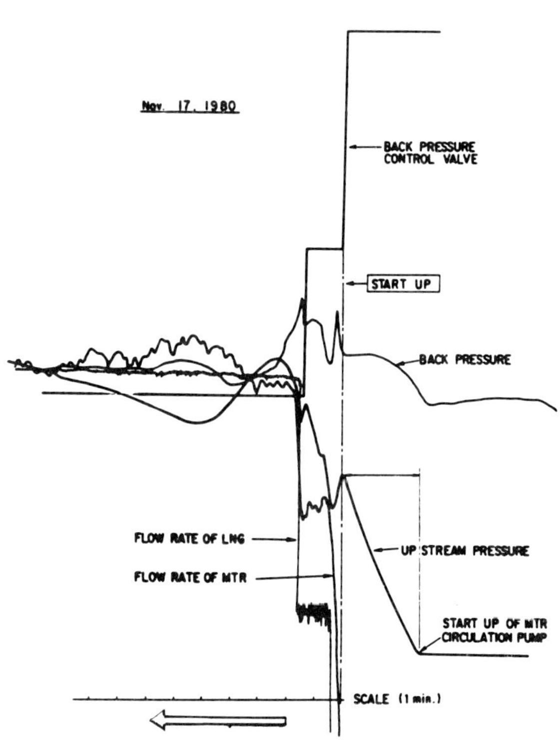

Fig. 3. History of start up operation.

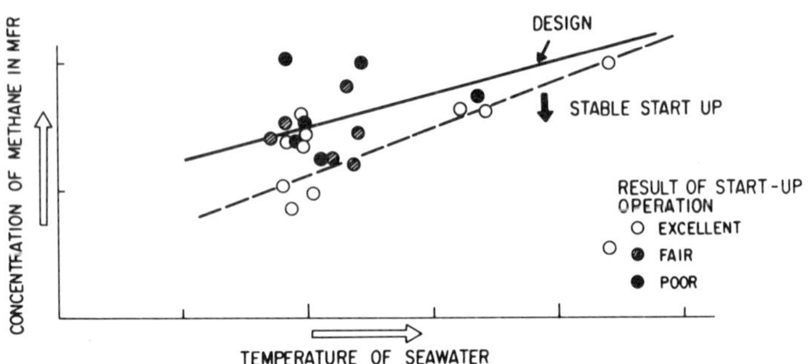

Fig. 4. Relation between concentration of
methane in MFR and seawater temperature.

when the plant was operated at a steady-state condition. This is
shown in Fig. 5.

Maximum Turndown Ratio

Since this system is to be used as an LNG vaporizer for
thermal power plants, flexibility of the process is required, and
the process must be designed to achieve stable operation at a load
of 25%. As a result of the experimental operation, unstable
performance was observed at a higher load then expected. A maximum turndown ratio was ascertained considering the minimum vaporization load required to maintain stable operation after a rapid
change in the vaporization load. Unstable vaporization of

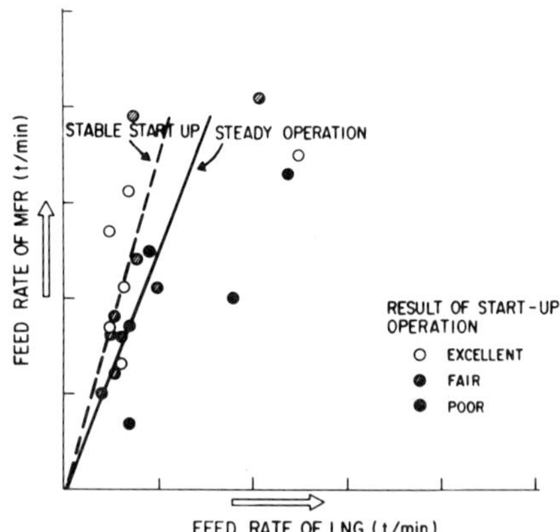

Fig. 5. Relation between feed rate of LNG
and that of MFR for stable start-up.

LNG and MFR shows sudden increases in temperature in the headers of LNG/MFR Heat Exchanger.

Make up of MFR

The MFR is the mixture of methane, ethane, propane and butane. In the United States, ethane fraction may be easily available. However, in Japan, although ethane fraction could be available in ethylene plants as a by-product, the facilities for shipping the ethane fraction as a product do not exist. Therefore, the ethane fraction has to be separated from LNG itself. The other components are easily available because a major fraction of LNG is methane and because liquefied petroleum gas is widely used as industrial or domestic fuel in Japan. Fractionation of LNG is therefore, the key operation in the preparation of MFR fluid. The conceptual process flow diagram is shown in Fig. 6. In the pilot plant, a batch distillation system is used for fractionation of LNG. After cooling down the distillation facility and MFR Drum, LNG as a refrigerant is kept in the shell side of the Dephlegmator and the boiling temperature of LNG is controlled adjusting the breeding rate of heavier fractions through the valve V-1. LNG as a feed is filled up to the maximum level of the level indicator at the bottom. A part of the reboiled

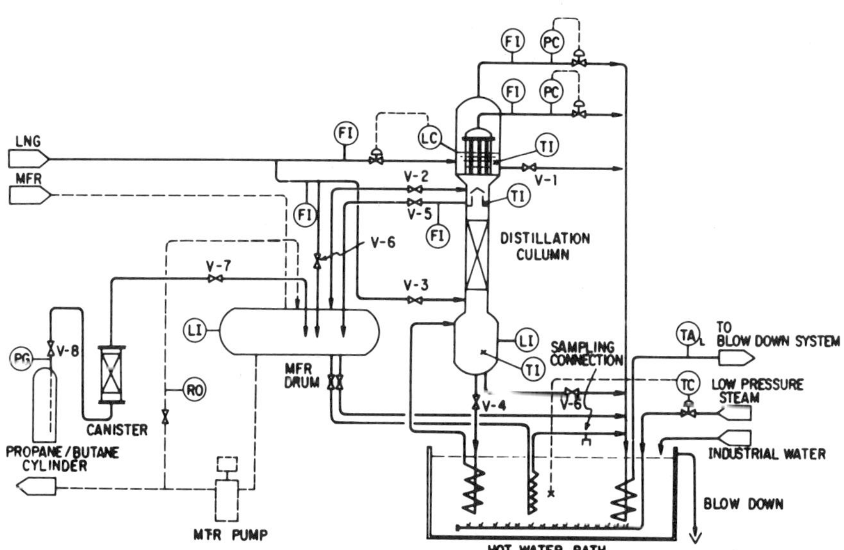

Fig. 6. Process flow diagram of batch distillation facility.

vapor which is condensed at the Dephlegmator is effective as a reflux and the residual part is vented off to the blowdown system. As the LNG is enriched, the temperature of the liquid at the bottom becomes higher and the liquid level decreases. When the temperature at the bottom increases sharply, the LNG is again fed to the maximum level at the bottom of the distillation column. After several enriching operations, a crude ethane which contains more than 50% ethane is obtained. As the overhead vapor of the column becomes rich in ethane, condensation of the vapor at the Dephlegmator is facilitated and reflux is increased. Then the residual methane is fractionated. When the composition of the overhead vapor becomes almost ethane, the temperature at the chimney shows a sharp increase. Finally, overhead vapor is totally condensed at the Dephlegmator because the composition of the vapor contains more than 98% ethane. The amount of reflux is adjusted by increasing the set pressure of the shell side of the Dephlegmator preventing flooding in the tubes of the Dephlegmator. Ethane fraction is drawn off through the valve V-5 after cool down of this piping. As drawoff proceeds, the ethane fraction gradually becomes heavier in propane. When the temperature of the bottom liquid shows a sharp increase, the drawoff is stopped. The bottom liquid mainly of propane, butane and heavier fractions is channeled to the blowdown system after heating. About 200 kg of ethane is obtained in one batch operation. The above operation was repeated until the ethane fraction necessary for the operation of pilot plant was accumulated in the MFR Drum, which took about 2 weeks. After that, each fraction, LNG as methane, propane and butane is added to the ethane fraction accumulated to make up the MFR fluid. Propane and butane are supplied from individual cylinders. The liquid is mixed by the MFR Circulation Pump operating under the minimum flow condition.

COMMERCIAL PLANT

A commercial plant which generates about 4000 kW is now being planned. The plant has a regasification capacity of 100 t/h of LNG and is planned to be operated fully automatically. The demand for LNG has been on the increase in Japan as a substitute for petroleum, with estimated annual total imports of more than 26 million tons in 1985. Provided that all of the 26 million tons of LNG is vaporized by the MFR cycle, it is estimated that more than 150 MW of electric power can be generated.

REFERENCE

1. Y. Sanga, Y. Shirasaki, and H. Shinozawa, in "Advances in Cryogenic Engineering, Vol. 27," Plenum Press, New York (1982).

CONTROL SYSTEM OF POWER GENERATING PLANT FROM LNG COLD HEAT

Y. Sanga and Y. Shirasaki

Chiyoda Chemical Engineering & Construction Co., Ltd.
Yokohama, Japan

and

H. Shiozawa and T. Hiro-oka

Tokyo Gas Co., Ltd.
Tokyo, Japan

INTRODUCTION

Since the amount of LNG consumed will increase in the future, the utilization of the LNG sensible heat, or cold heat, is becoming more important. At present, only a part of the cold heat is being utilized in air separation plants, cold warehouses, etc., because these plants are hardly able to respond to the LNG vaporization load changes which are required by power generating stations and municipal gas utilities.

A new method for generating power efficiently from LNG cold heat in the regasification process using sea water as the heat source has been developed by joint efforts of Tokyo Gas Co., Ltd. and Chiyoda Chemical Engineering & Construction Co., Ltd. This process is a Rankine cycle using a multicomponent working fluid (MFR). A power generating pilot plant for this new process was constructed with a capacity of 5000 kg/h of LNG vaporization. The plant functions as the LNG vaporizer and is able to respond faster vaporization load changes, than the usual open rack vaporizer (ORV) or submerged combustion vaporizer (SMV).

The following results were achieved during the experimental operation of this pilot plant:

(1) A quick response to the LNG vaporization load change at the

rate of 15% per minute which is more severe than that required by power stations.

(2) Keeping a steady LNG vaporization rate following a turbine or generator trip.

(3) The automatic start-up of the turbine operation.

An optimum control strategy for the maximum recovery of LNG cold heat was also developed during the experimental operation. This paper summarizes the optimal control strategy, the advanced control system implemented in this plant to achieve the above functions, and the results obtained from such control. The control system is a distributed microprocessor-based system whose software is systemized to permit easy control configuration with the modular software packages.

OVERALL CONTROL FLOW

The overall control flow diagram is shown in Fig. 1. The overall control system is divided into two feedback control loops i.e., those for the LNG stream and for the multicomponent working fluid (MFR) stream.

Control of the LNG Stream

The LNG at approximately $-160^{\circ}C$, 33 ata is fed to the cold box where most of it is vaporized by the heat extracted from the MFR stream recycled from the turbine exhaust. The LNG is further heated up by sea water in the Natural Gas Superheater to approximately $0^{\circ}C$, 25 ata and enters the main gas header.

The control of this stream is as follows: The master pressure controller, PC-01, in the main gas header changes the set points of the LNG flow controllers, FC-01, etc, in a number of vaporizers in accordance with vaporization load change demanded by the municipal utility.

Control of the MFR Stream

The MFR in the accumulator at approximately $-130^{\circ}C$, 4 ata is compressed by the MFR feed pump to approximately 18 ata and enters the cold box where it is warmed to approximately $-35^{\circ}C$ and is partially vaporized by condensing the exhaust fluid from the turbine. From the cold box, the MFR passes into the MFR Superheater where it is completely vaporized by sea water to gas at approximately $10^{\circ}C$, 12 ata and is then routed to the turbine through the turbine control valve. After the power recovery by expansion in the turbine, the gas is exhausted at approximately $-22^{\circ}C$, 4 ata and is recycled into the cold box where it is cooled and condensed by heat exchange with both the LNG and the MFR. The

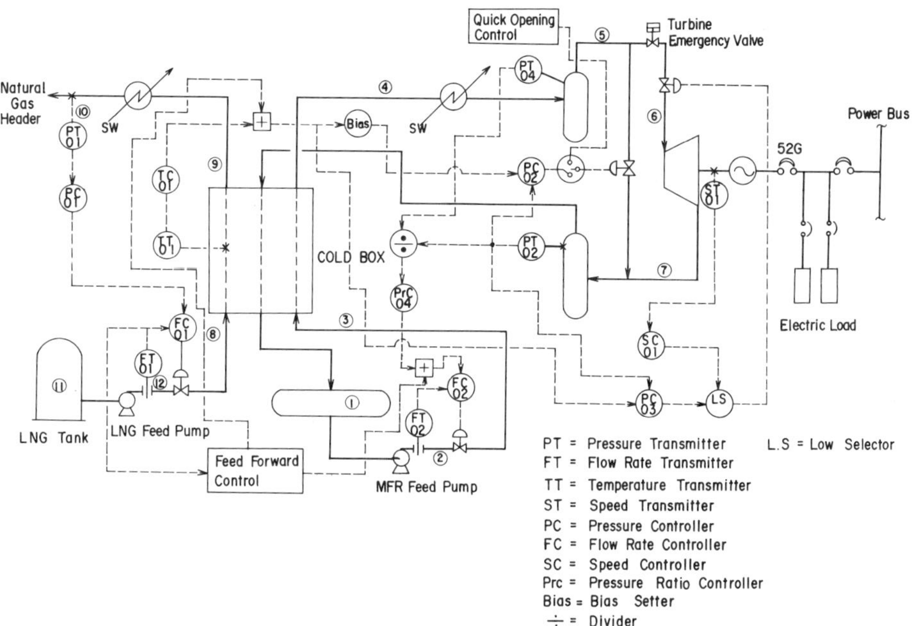

Fig. 1. Overall control flow diagram.

condensed MFR leaves the cold box at approximately -130°C, 4 ata and is returned to the MFR accumulator.

The control of this stream is as follows: The turbine outlet pressure is controlled by either the turbine outlet pressure controllers, PC-02 or PC-03, which manipulate the turbine bypass valve or the turbine control valve respectively. In the case of turbine bypass operation and turbine speed control operation, PC-02 is used to control the turbine outlet pressure. In the case of the turbine pressure control operation, PC-03 is used.

The heat imbalance in the cold box caused by the MFR flow rate not matching a current LNG flow rate is reflected first as the turbine outlet pressure change which causes the deviation of the pressure from the setpoint of PC-2 or PC-3. This actuates the turbine bypass valve or the turbine control valve to correct the deviation, and this subsequently causes the deviation of the turbine pressure ratio from the setpoint of the turbine pressure ratio controller, PrC-04. From the PrC-04, the MFR flow rate is eventually adjusted to a proper flow rate matching the LNG flow rate through the MFR flow rate controller, FC-02. The temperature controller, TC-01, at a mid point of the LNG stream in the cold box has the function to trim the heat imbalance in the cold box caused by other disturbances, e.g. a composition change in the multicomponent working fluid, by means of regulating the turbine outlet pressure level.

ADVANCED CONTROL

Since conventional feedback control was considered to be unable to achieve the desired plant performance, the advanced control system illustrated briefly in Fig. 1 was implemented in this plant.

Feed Forward Control

This control was implemented to respond quickly to rapid LNG vaporization load changes. The algorithm is as follows:

(1)   The LNG flow change, Δ Fl, is calculated from the LNG flow rate, Fl (t),

$$\Delta Fl = Fl\ (t_n) - Fl\ (t_{n-k})\quad \text{for } |Fl\ (t_n) - Fl\ (t_{n-k})\ | > \varepsilon \quad (1)$$

$$= 0 \quad\quad\quad \text{for } |Fl\ (t_n) - Fl\ (t_{n-k})\ | \leq \varepsilon \quad (2)$$

where ε = a low limit constant to avoid flow noise.

Fl (t_n) = the sampled data at n-th sampling instant

(2)   The feed forward calculation values, ΔF ΔP, are added to the setpoints of the MFR flow rate controller and the turbine outlet pressure controllers respectively:

$$\Delta F = Kl\ \Delta Fl \quad\quad\quad\quad\quad (3)$$

$$\Delta P = K2\ \Delta Fl \quad\quad\quad\quad\quad (4)$$

where Kl = the ratio constant of MFR flow rate to LNG flow rate

        K2 = the coefficient of the turbine outlet pressure change per unit change of the LNG flow.

(3)   The algorithm is repeated after some sampling interval.

Quick Opening Control

This plant is required to continue to operate as an LNG vaporizer, without generating power, following a turbine or generator trip, because the primary function of this plant is to regasify LNG. In order to meet this requirement, the quick opening

control is implemented by means of instantaneously opening the
turbine bypass valve to the desired position following a turbine
or generator trip. The algorithm for this is shown in Fig. 2 and
3, where the process variables measured immediately before the
trip are used to calculate the desired positions.

OPTIMAL CONTROL FOR MAXIMUM RECOVERY OF LNG COLD HEAT

The optimal control strategy of the plant is to extract
maximum power from the cold heat contained in the LNG flow. The
power generated by the turbine is:

$$P = \eta G \, \Delta h \, /860 \tag{5}$$

where P = power generated
         η = turbine efficiency
         G = mass flow passing through turbine
         Δh = adiabatic specific enthalpy drop across turbine

From this, the optimal control strategy may be expressed by
the way both the flow and the adiabatic specific enthalpy drop are
maximized under constant conditions of the LNG flow rate and
seawater temperature.

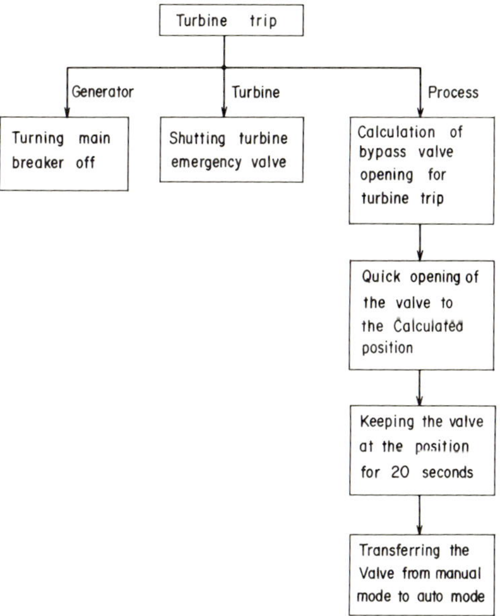

Fig. 2. Quick opening control block diagram for turbine trip.

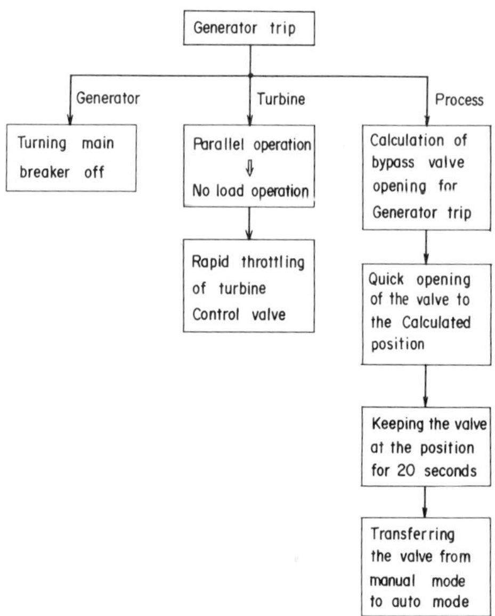

Fig. 3. Quick opening control block diagram for generator trip.

It was observed in the experimental operation that the MFR mass flow is increased as the turbine pressure ratio is decreased under these constant conditions, until the turbine control valve is opened to the maximum valve position corresponding to the LNG flow rate. In this paper the turbine pressure ratio is the ratio of the absolute pressure throttled by the turbine control valve to the absolute pressure in the turbine outlet. The turbine inlet pressure is the pressure throttled by the turbine control valve.

In this plant, the increase of the MFR mass flow results in the increase of Δh, because the turbine control valve is the throttling type and the turbine outlet pressure is usually maintained constant.

Figure 4 is the pressure–enthalpy diagram with the operational data under two steady-state conditions with different turbine pressure ratios depicted schematically. It is noted that the temperature of the vaporizing MFR at the outlet of the cold box does not vary with the turbine pressure ratio. This means that the heat flow exchanged between the condensing MFR and the vaporizing MFR increases, because the isothermal curve in the mixture of MFR liquid and vapor has a negative slope.

On the other hand, the heat balance in the cold box is represented by:

$$F \, \Delta h_L + G \, \Delta h_V = G \, \Delta h_c \qquad (6)$$

where F is the LNG mass flow, G is the MFR mass flow and Δh_L, Δh_V, Δh_c is specific enthalpy difference across the cold box of the LNG, the vaporizing MFR and the condensing MFR respectively. The quantity G is derived from Eq 6:

$$G = F \, \Delta h_L / (\Delta h_c - \Delta h_V) \qquad (7)$$

As shown in Fig. 4, Δh_c does not vary with the turbine pressure ratio and also Δh_L is proven in the experimental operation to hardly vary. Therefore it is apparent from the Eq. 7 that the increase of the MFR mass flow results from the increase of Δh_V with a decrease in the turbine pressure ratio.

As to the reason for the increase of Δh_V with the decrease in the pressure ratio, it is assumed that the surface area for heat transfer is sufficient for partial plant load to permit the increase of Δh_V in spite of the smaller temperature difference for heat transfer in the cold box with the decrease of the ratio.

As understood above, the optimal control strategy is found to be realized in the thermodynamic theory by means of continuously decreasing the turbine pressure ratio so long as the temperature of the vaporizing MFR leaving the cold box does not tend to fall.

In practice, however, it is preferable that the optimal control strategy be implemented by means of regulating the turbine pressure ratio so that the opening of the turbine control valve is

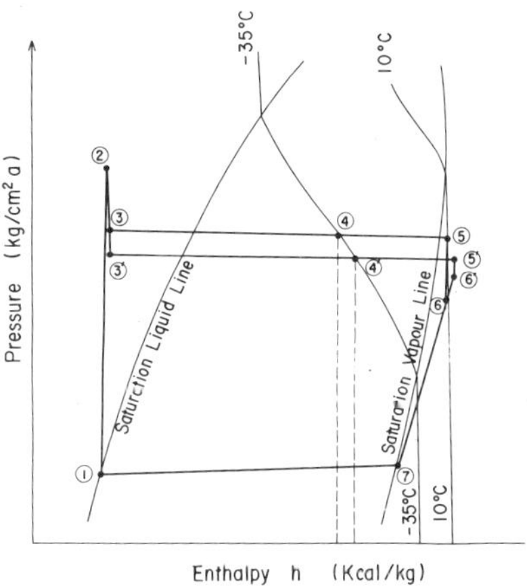

Fig. 4.   MFR pressure—enthalpy diagram.

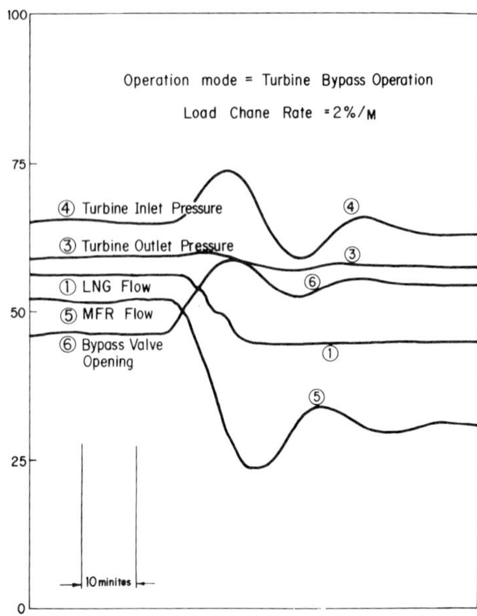

Fig. 5. Response of feed back control to LNG load change.

within the maximum domain corresponding to the LNG vaporization load. This is due to the temperature of the vaporizing MFR falling too quickly to be restored.

EXPERIMENTAL RESULTS

It was confirmed in the experimental operation that this advanced control system functions successfully.

LNG Vaporization Load Change

Figure 5 shows the process behavior under the LNG vaporization load change at a rate of 2% per minute from 53% to 67% of the full load in the turbine bypass operation using only the feed back control. It shows that the large pressure rise of 1.8 kg/cm^2 occurs with hunting at the turbine inlet under the load change rate of only 2% per minute and so it is predicted that an LNG load change of more than 2% per minute will cause this plant to go to the upset condition.

On the other hand, Fig. 6 and 7 show the process behavior under the LNG vaporization load change at a rate of 15% per minute from 44% to 100% of the full load in the turbine bypass operation and from 57% to 81% of the full load in the turbine pressure control operation using both feed back and feed forward control.

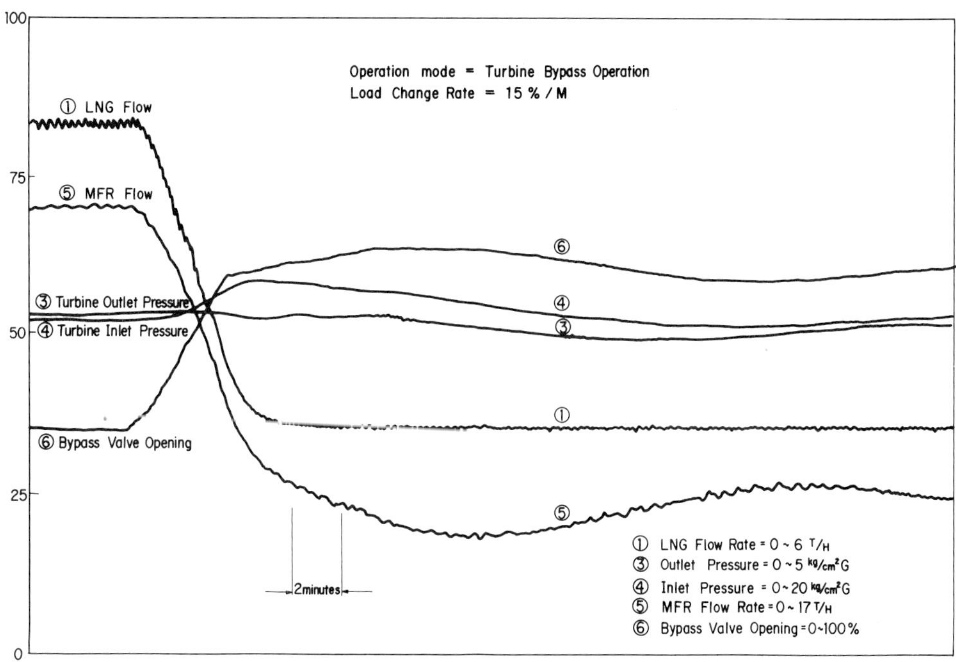

Fig. 6.   LNG load change test.

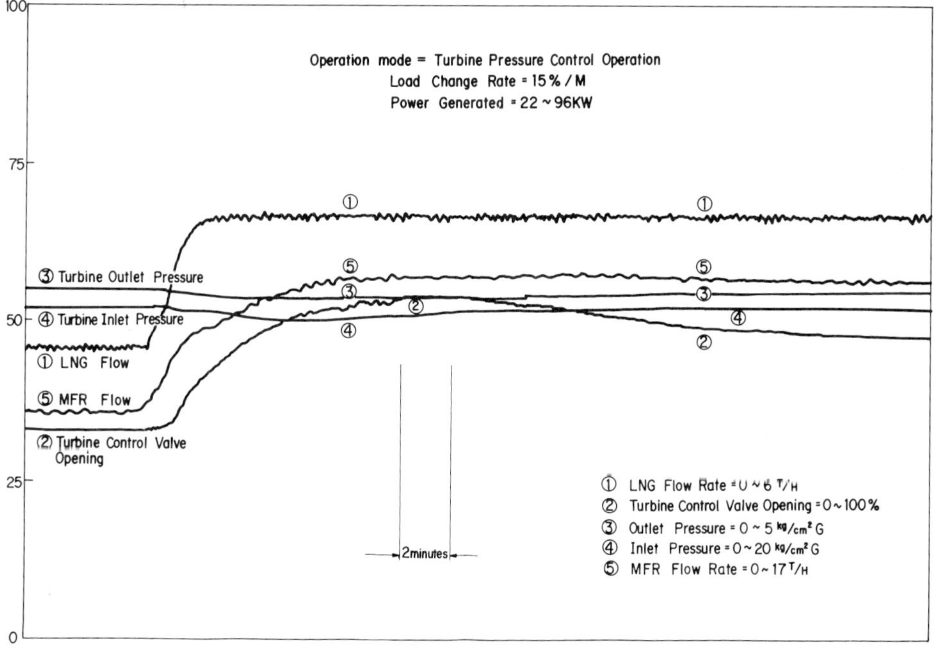

Fig. 7.   LNG load change test.

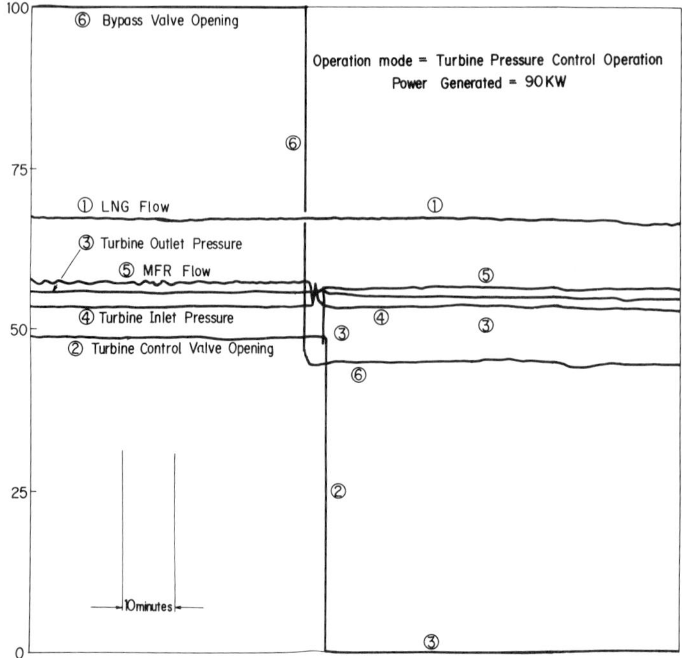

Fig. 8. Turbine trip test.

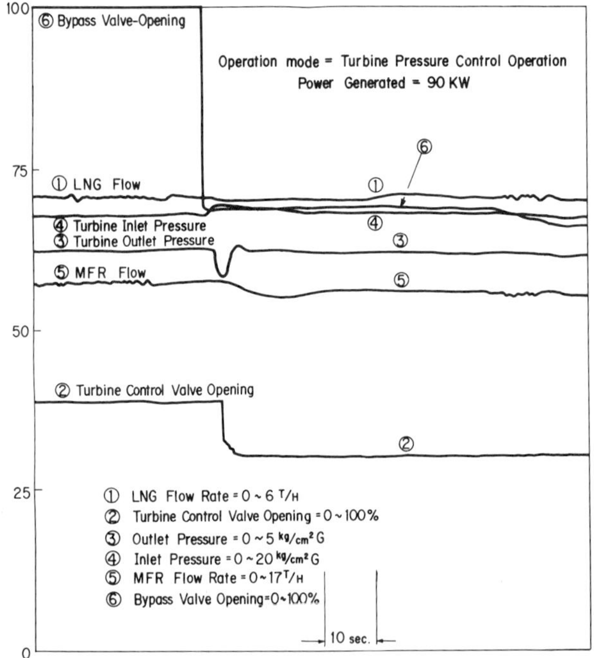

Fig. 9. Electric load damp test.

It is seen in Fig. 6 that the maximum pressure rise of 1.3 kg/cm^2 takes place at the turbine inlet, but this is confirmed to have little effect on stable operation.

Turbine or Generator Trip

Figures 8 and 9 show the responses of this plant to the trip of the turbine and generator respectively. It can be seen that a relatively larger variation of turbine outlet pressure occurs than of turbine inlet pressure and also that this variation diminishes within approximately 10 seconds. It was also confirmed in the experimental operation that the above pressure variations do not prevent stable operation of LNG vaporization because they are diminished in a short period.

Automatic Operation of Plant Start-up

The automatic start-up from the turbine bypass operation to the turbine pressure control operation, which is the normal operation in this plant, was conducted using sequencing control, while the start-up for the turbine bypass operation was operated manually. The automatic start-up for the turbine bypass operation is planned for the commercial plant.

CONCLUSION

This power generating plant with the advanced control system has nearly the same capability and reliability as ORV and SMV. The recovery efficiency of LNG cold heat is enhanced by applying the optimal control strategy developed in this plant. It is expected from the standpoint of energy conservation that the active utilization of LNG cold heat will be economical for commercial plants with capacities of $\sim 10^5$ kg/h of LNG vaporization and electric power ratings of $\sim$ 4000 kWh.

NITROGEN REJECTION FACILITIES FOR INERT-GAS ENHANCED RECOVERY PROJECTS

H. L. Vines, J. B. Peeples, and C. L. Newton

Air Products and Chemicals, Inc.
Allentown, Pennsylvania

INTRODUCTION

A number of oil and gas condensate enchanced recovery processes employ injection of an inert gas into the reservoir for displacement. High purity nitrogen has been used for some of these processes. Natural gas subsequently produced from these reservoirs will gradually become diluted with the injected gas; the nitrogen diluent must be removed from the produced gas so that sales gas can meet pipeline specifications for heating value.

Nitrogen rejection facilities for enhanced recovery projects must be able to accept the gradual increase in nitrogen concentration of the feed gas which occurs as the injected gas dilutes the production wells. A typical design requirement for such a project may cover a range from 5% to 80% nitrogen in the feed. Major factors in the design of such facilities may be: the life of the project, plant size, natural gas liquids (NGL) recovery, feed impurities, etc. There are also uncertainties in the behavior of the gas within the reservoir in terms of the resultant compositions which will be obtained from the well over time. This paper focuses on the effects of these variable feed condition requirements on facility design and economics.

The relationship of the nitrogen rejection facility to other gas handling units can have significant effects on the design and economics of the enhanced recovery project. Some of these factors are: (1) efficient use of preexisting gas handling facilities, including pretreatment facilities for impurities removed and/or existing NGL recovery plants, (2) the possibility of using rejected nitrogen for reinjection into the reservoir, and (3) reliability of the rejection facility.

DESIGN CONSIDERATIONS

Boundary Conditions

Boundary conditions represent constraints on the design of the plant which are not directly connected with economic trade-offs.

Reliability. In many cases, 100% reliability of the nitrogen rejection facility is essential for the operation of the oil field, since the facility may be required to produce fuel gas for the field throughout the life of the project.

Prepurification. The feed gas may contain trace impurities which must be removed for the efficient operation of the facility. These impurities must be dealt with either by pretreatment of the gas prior to the cryogenic plant, or by design of the plant to handle the impurities without solidification in the coldest sections of the plant.

Onsite facilities. Existing equipment may require refitting or bypassing, and some utilities may not be available.

Environmental. Some economic options may not be acceptable from an environmental view.

Economic Factors

In examining the economic factors for nitrogen rejection facilities, it is necessary to examine not only design for a particular gas composition and condition, but to consider the life of the injection project. Several items are of importance in this regard.

BTU value of fuel. The value of the BTU content of the natural gas stream is the key to determining the optimum level of recovery of methane and other contained hydrocarbons.

Liquids recovery. This is important if there is a substantial incremental value for natural gas liquids over their heating value.

Utilities. This factor is important to consider, especially for the cost of utilities vs methane recovery.

Nitrogen. A difficult factor to evaluate in many cases is the value of nitrogen for reinjection or export. Nitrogen can often be recovered at high pressure and its value at pressure may determine the decision to choose one cycle over another.

Economic factors and boundary conditions will always interact to determine a particular choice of cycle or equipment design, and in some cases an entire plant may be dictated by a single customer constraint. For example, in a particular facility, when the cost of recovering the contained hydrocarbon in the gas stream becomes high due to high nitrogen content, it may no longer be economical to attempt to recover the hydrocarbon except as plant fuel. In such cases it may be desirable to reinject all of the feed except the portion extracted for fuel.

Range of Nitrogen Composition

The difficulty in designing of rejection facilities is compounded by the requirement that both cycle and equipment handle the wide ranges of natural gas flow and composition. Because of this, the process design efforts must concentrate on the controlling conditions for the plant design. At a minimum, it is necessary to develop a design suitable for initial low nitrogen content, and also for the highest expected nitrogen content. Some mid-range cases should also be examined to predict the behavior of the plant over the complete range of expected operating conditions.

Figure 1 shows a nitrogen breakthrough curve which could be experienced over the life of an injection project. The lag between start-of-injection and start-of-breakthrough is dependent on specific reservoir conditions, as is the speed with which the nitrogen concentration increases. In some cases nitrogen may increase by 10% per year or greater, while other situations have increases of only 1 to 2% per year. Reservoir models can be used to try to predict this behavior, but uncertainties implicit in the

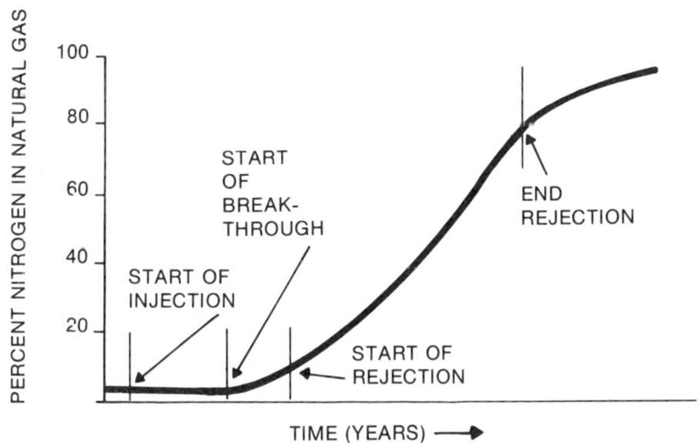

Fig. 1. Typical nitrogen breakthrough pattern.

models require that the nitrogen rejection facility be capable of meeting faster breakthrough than predicted.

POSSIBLE PLANT CONFIGURATIONS

If liquids recovery is desired in the course of natural gas processing, there are several possible arrangements for integrating the Nitrogen Rejection Unit (NRU) with the liquids extraction.

Tail-end NRU

In the tail-end NRU (Fig. 2), the natural gas residue from a NGL plant is processed in the NRU after rewarming through the gas-gas exchangers. The NGL plant must provide adequate pressure to the NRU in order to process the gas for nitrogen rejection. One problem associated with this configuration is that the NGL plant must be onstream for as long as the facility is in operation. This limits the option of shutting the NGL plant down when hydrocarbon content of the feed gas may no longer justify NGL recovery.

Front-end NRU

In this cycle configuration (Fig. 3), the NRU preprocesses the feed to the NGL plant and sends all (C_1^+) or a part (C_2^+) of the contained hydrocarbon content to the natural gas liquids plant. If the NGL plant must shut down at some point during the course of the project, it will not affect the operation of the nitrogen rejection plant. However, some modifications may be required to an existing NGL plant for this arrangement.

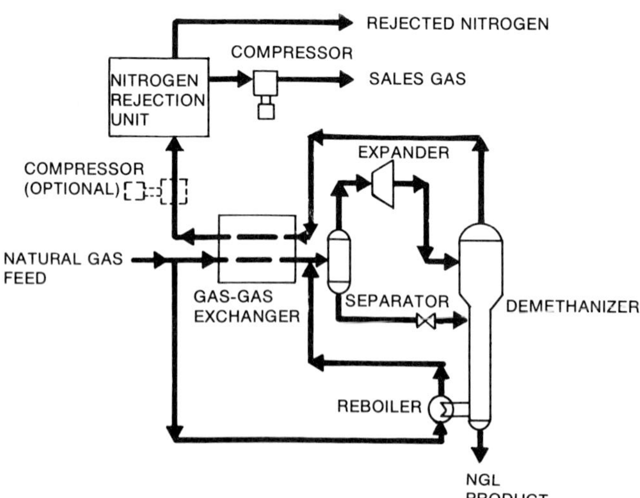

Fig. 2. Tail-end nitrogen rejection unit.

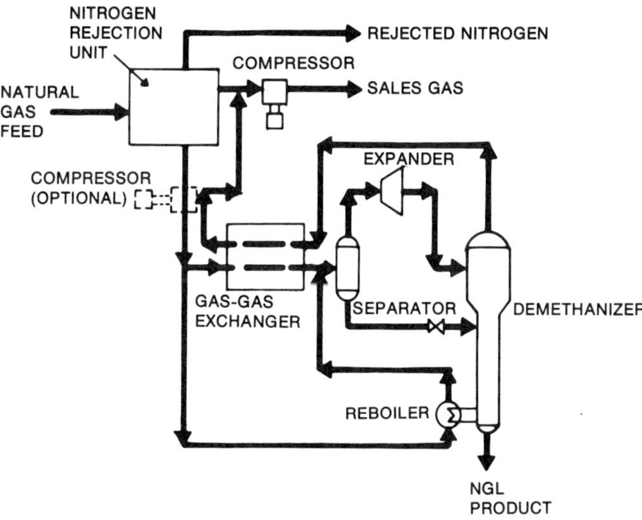

Fig. 3. Front-end nitrogen rejection unit.

Cold-end NRU

This cycle configuration can also be called an integrated plant (Fig. 4). In this plant, the cold demethanizer overhead from the NGL plant is used to feed the NRU. This saves on heat exchange and is the most thermodynamically efficient manner in

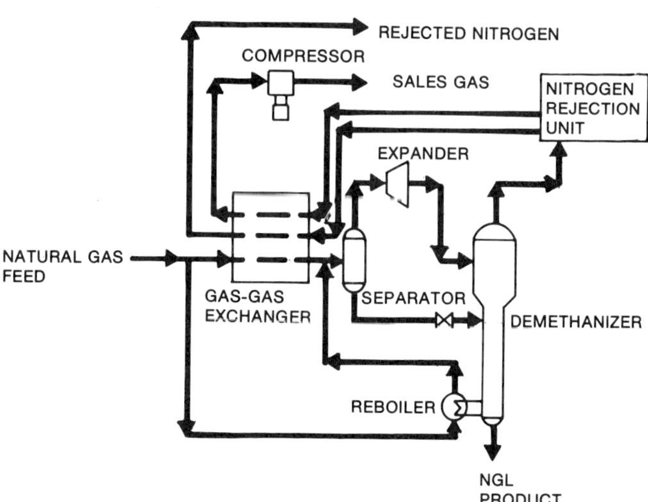

Fig. 4. Cold-end nitrogen rejection unit (integrated cycle).

which the two plants can be integrated. It may also be possible
to use multiple feeds from the natural gas liquids plant to improve on the operation of the nitrogen rejection plant.

PROPOSED CYCLES FOR NITROGEN REJECTION

For variable-content nitrogen rejection plants, an important
aspect of design is the "trajectory" of various design parameters
over the life of the project. It is essential to understand these
parameters well enough so that the selected conditions for determining equipment sizes and energy requirements cover the complete
range of expected operating conditions. In addition, if a
"switchover" mode occurs, e.g., a plant requiring an added source
of refrigeration, the plant operation must be simulated before and
after switchover since there are no sharp transitions in feed
compositions

Heat Pumped Cycle

The heat-pumped cycle is shown schematically in Fig. 5. Feed
to the distillation column is precooled in exchanger E-1 and
flashed to column pressure. Condensing and reboiling duties are
provided by the closed loop methane circulation system (heat
pump). Details of heat interchange to reduce power requirements
are not shown. In order to balance the condensing and reboiling
duties, the vapor content of the feed into the column must be

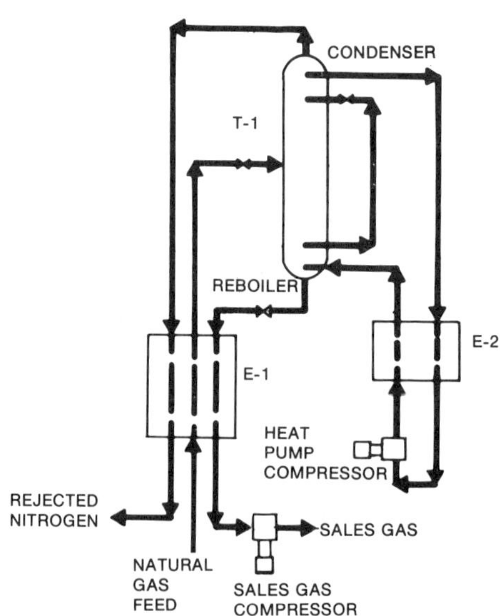

Fig. 5. Single column heat-pumped cycle.

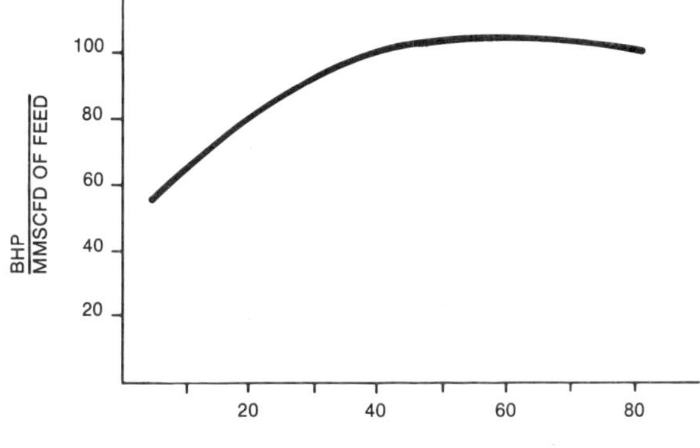

Fig. 6. Typical power consumption for
single column cycle.

approximately the same as the gaseous overhead product (nitrogen). Figure 6 shows approximate power requirement per million
standard cubic feet per day (units: BHP/MMSCFD) over a typical
range of compositions. Power requirement is low at first but
increases at higher nitrogen concentrations in the feed.

This cycle has several advantages. All of the contained
nitrogen is available at high pressure for reinjection or for
export, over the life of the project. The cycle is also capable
of providing good methane recovery over all ranges of feed nitrogen content by increasing the heat pump flow rate to supply more
reflux to the column. Higher levels of carbon dioxide content may
be handled (up to several hundred parts per million) over the life
of the project. By keeping the streams containing CO_2 at sufficiently high pressure and warmer temperatures, solid formation is
avoided.

However, if there is no immediate use for the high pressure
nitrogen, this cycle may not be economical. Much of the work of
separation provided by the heat pump is needed to separate and
reject the nitrogen at high pressure. For short-lived projects
where nitrogen content of the feed rises quickly, the installation
of a large amount of horsepower to recover high-pressure nitrogen
may not be economical.

Double Column Cycle (Cascade System)

The double column cycle shown in Fig. 7 is familiar to those
who have worked with air separation cycles. The feed gas is
cooled and partially condensed in exchanger E-1, then flashed to
the pressure of the high pressure column, T-1. The high pressure

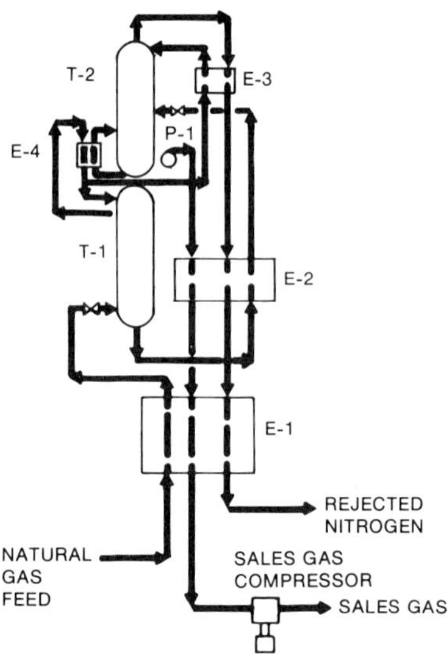

Fig. 7. Double column cycle (cascade system).

column performs a partial separation of the nitrogen–methane
mixture, generating a stream of nearly pure liquid nitrogen as an
overhead product. The remaining nitrogen and methane is withdrawn
as a bottom product. Both streams are flashed to the pressure of
the low-pressure column, T-2, where the final separation takes
place. The high purity liquid nitrogen from the high pressure
column provides a reflux to the low pressure column while reboiling duty for the low pressure column is provided by the reboiler-
condenser, E-4. (Note: The concept of a double column, which is
old in the air separation industry, is now being considered in
multiple product distillation processes, where energy conservation
is becoming increasingly important.) Product methane is raised in
pressure by the LNG pump P-1, and rewarmed to ambient temperature. Subcoolers E-2 and E-3 increase the liquid-to-vapor ratios
in the low-pressure column and thereby improve the degree of
separation of the mixture. The nitrogen from the low pressure
column is rewarmed to ambient temperature.

   ·   The refrigeration requirement for operating the plant is
provided in part by the efficient expansion of the nitrogen
contained in the feed. Thus, as the nitrogen content increases,
the net power required for compression of product methane
decreases. Figure 8 shows specific power consumption vs feed
nitrogen composition for a typical application.

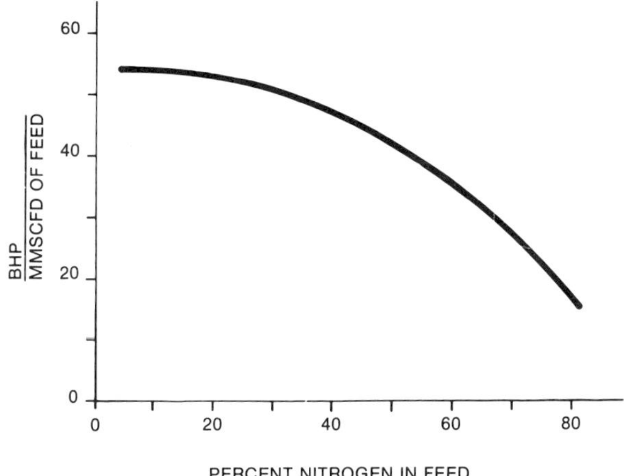

Fig. 8. Typical power consumption for double column cycle.

One principal advantage of the double column cycle is that good separation is performed relatively efficiently. At high nitrogen contents, increased methane recovery can be obtained with decreased power consumption, as may be inferred from Fig. 8. Nitrogen is recovered at a low pressure; the decision to compress and reinject this low pressure nitrogen can be made at any time, based on the prevailing economics.

There are two main disadvantages to this cycle. Because the double column performs the bulk of the separation at lower pressures, and therefore, lower temperatures, the system can only accommodate feeds with carbon dioxide concentrations of about 20 ppm or less, without solid formation. Pretreatment to these levels is required for operation. Additionally, the double column system would require additional power and complexity to obtain high methane recoveries at feed compositions of less than 20% nitrogen. For the process drawn in Fig. 7, reflux to the low pressure column is obtained entirely from the feed. At low levels of contained nitrogen, the reflux generated in the process is insufficient to reduce the methane concentration in the nitrogen product to levels consistent with high recoveries.

Triple Column Cycle

Triple-column cycles combine the principles of the heat-pumped and cascade cycles, and may be suited to particular customer requirements. One combination in actual operation for a fixed nitrogen composition uses a front-end high pressure column to pre-separate the feed gas mixture and to remove the carbon dioxide to an acceptably low concentration for further processing at the colder temperatures. Either direct or indirect refluxing of the front-end column is feasible. The cycle may be especially suited for cases in which carbon dioxide handling is required but high pressure nitrogen product cannot be conveniently used.

CONCLUSIONS

Many combinations of the heat pumped and cascade cycles which have been described are possible for dealing with specific inlet conditions and performance requirements. The design of plants for processing natural gas with variable nitrogen content is complicated by the need to consider operation over a continuous range of conditions. Selection of the proper cycle must also take into consideration the boundary conditions, economic criteria, and inherent uncertainties in predictive models of reservoir behavior.

DISCUSSION

Question by R. Tarakad, M. W. Kellogg Co.: What is the most economical concentration of methane in the rejected nitrogen at the top of the N_2-C_1 column?

Answer by author: This is a strong function of the value attributed to methane as a fuel gas. In general, high values for fuel gas will make methane recoveries in excess of 97 percent economical.

OPERATION OF AN AIRCRAFT ENGINE USING
LIQUEFIED METHANE FUEL

J. A. Raymer

Beech Aircraft Corporation
Boulder, Colorado

INTRODUCTION

In recent years the world has experienced liquid fuel short-ages and ever increasing fuel prices. The fuels derived from crude oil have in the past been considered inexhaustable and very little effort expended in investigating other fuels for world transportation needs. Interest in alternatives to gasoline has increased in the last few years. One of these alternatives is methane, currently obtained from natural gas which is about 85 percent methane. Other sources of methane not yet being used are coal mine gas, land fill decomposition gas and sewage treatment gas.

Methane has several advantages over gasoline for use in reciprocating engines. It is cleaner burning, it produces less exhaust pollution, and can extend engine life. It also has better cold weather starting characteristics and an equivalent octane rating of about 130.

The goal of this project was to demonstrate the operation of a reciprocating aircraft engine on methane fuel. Since storage of the methane fuel in the gaseous state would be impractical for a flight fuel system, a liquid storage system was used. System valving was configured to deliver only liquid methane to the engine supply line.

EQUIPMENT DESCRIPTION

The engine selected for this program was a Lycoming 0-360 four cylinder 180 horsepower engine similar to those installed in several Beech Aircraft single and twin engine light aircraft. For the test program the basic engine was not modified. A four bladed test club propeller was used to provide engine loading, and a cooling shroud was installed to provide adequate cooling air during test stand operation of the engine.

The conversion of the fuel system to operate on liquid methane was straightforward. Figure 1 schematically shows the system. The required components included: the liquid methane storage dewar, shut-off valve, compact heat exchanger/vaporizer, pressure regulator and air-fuel mixer.

The liquid methane storage dewar (shown in Fig. 2) was designed by Beech Aircraft for automotive applications. Normal operating pressure was 20 to 30 psig with a relief valve set at 60 psig. The dewar had a usable volume of 18 gallons and about ten percent ullage space. Fluid temperature at normal operating conditions was about 221° R. Normal standby time for a full dewar to reach relief pressure without usage was greater than seven days.

The vaporizer was a compact plate-fin type cross flow heat exchanger. The energy source was air heated by exhaust gases within a shroud surrounding the muffler. Approximately 30,000 BTU/h were needed at maximum flow rate to vaporize the fuel and raise its temperature to near ambient. The physical size of the heat exchange was five inches by ten inches by seven inches and is shown in Fig. 3.

Pressure control of the system was achieved by a pair of off-the-shelf liquefied petroleum gas regulators. The regulators are operated in parallel to assure adequate flow at high power conditions. The fuel pressure was controlled to a positive five inches water column, which was the design requirement of the air-fuel-mixer. A single pressure regulator is shown in Fig. 4.

The air-fuel mixer (Fig. 5) was also an off-the shelf liquefied petroleum gas component. This unit was capable of providing 21,600 cubic feet of air per hour to the engine. The mixer

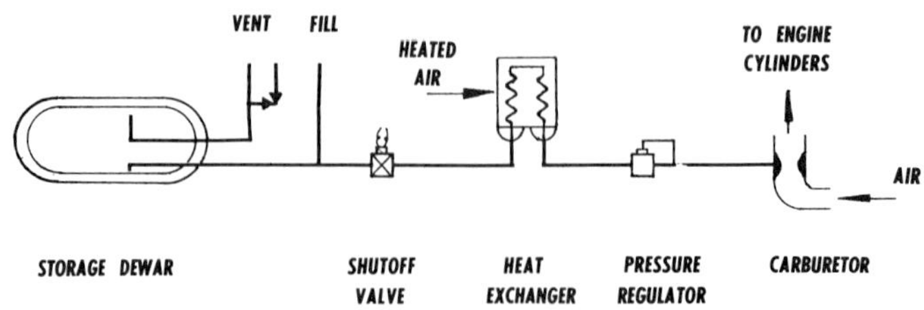

Fig. 1. Liquid methane fuel system.

Fig. 2. Liquid methane storage dewar.

controls fuel flow to the engine based on the total air flow. It
had tuning adjustments to vary the mixture ratio at both idle and
maximum power for best performance.

TEST PROGRAM

The test program was developed to provide a comparison between gasoline and methane operation of the same engine. All
testing was accomplished with the engine mounted in a static test
fixture. In order to record engine output power, a torque meter
was installed between the engine and propeller.

A series of tests were performed prior to conversion of the
fuel system to liquid methane.

With the engine operating at the maximum speed of 2700 RPM,
out-put torque, fuel flow rate, engine temperature and exhaust gas
temperature were recorded. Three-five minute runs were made to
obtain a series of data points which could be averaged. Following
the maximum power tests, two cruise power conditions were tested,
75 percent power or performance cruise and 65 percent power or
economy cruise. In both situations the same data items were
recorded as in the first test. With completion of the gasoline
tests, the gasoline fuel system was removed and the liquid methane
components were installed.

The engine was initially operated on liquid methane to retune
the system to achieve the best power output. The ignition timing
was advanced from 25 degrees to 35 degrees before top dead center

Fig. 3. Methane heat exchanger.

to compensate for the slower burning methane. This change also reduced the exhaust gas temperatures from the initial run. Once the engine was retuned for operation with methane, the test sequences as used with gasoline were repeated, duplicating the engine RPM as near as possible.

RESULTS

Data comparison showed that the engine performed very well when operated on the methane fuel. Cruise power conditions were easily attainable and specific fuel consumption was improved. At

Fig. 4. Methane pressure regulator.

Fig. 5. Methane air-fuel mixer.

full throttle, the engine did not develop the same power as produced on gasoline. This reduction in maximum power was due to using a vaporized fuel since, based on the air-fuel ratio, the methane fuel displaced about ten percent of the normal air flow into the engine. This fact translated directly to the observed power reduction of ten percent.

Specific fuel consumption showed significant improvement for all tests performed. Table I lists the pertinent data for each of the tests. The economy cruise power conditions would show a

TABLE I. Engine Test Results

Test Conditions	Gasoline			Methane		
RPM	2700	2450	2350	2650	2450	2350
Horsepower	136.4	101.5	88.7	125.1	99.8	90.6
Fuel Flow (1b/hr)	76.0	58.3	53.9	54.2	42.0	32.8
SFC (1b/Hp-hr)	0.557	0.58	0.608	0.433	0.421	0.362
Specific Energy Consumption (Btu/Hp-hr)	10583	11020	11552	9310	9052	7783
Standard Conditions Horsepower	179.9	105.6	92.3	161.7	103.8	92.6

*Based on average lower heating values of iso-octane and methane.

significant reduction of fuel costs per mile based on current costs of aviation gasoline and liquid methane.

CONCLUSIONS

The conversion of an aircraft reciprocating engine to operate on liquified methane fuel is possible with very satisfactory results. A reduction of maximum power output did occur. However, for most operational conditions this would be acceptable since full power could be obtained by boosting the air intake pressure slightly. Engine operation appears to be smoother and hot starting is easily accomplished. The significant improvement in the specific fuel consumption would result in reduced fuel costs at current prices. Since methane is obtainable from a variety of renewable sources, reciprocating engine powered aircraft operation using essentially current designs is not solely dependent on petroleum supplies.

DISCUSSION

Question by J. L. Smith, Jr. Massachusetts Institute of Technology: What detonation problems were found with the methane fuel?

Answer by author: None, since the methane fuel has an equivalent octane rating of about 130.

Question by R. C. Longsworth, Air Products and Chemicals, Inc: What is the specific fuel consumption of methane vs gasoline when the container weight is included?

Answer by author: This point has not been evaluated. The tanks used in this demonstration are of steel construction and do not represent a design that would be applicable to aircraft installation.

Question by J. Lester, Ball Aerospace Systems Div.: Do you plan to do studies to determine if the weight advantage holds up when packaging of the tanks into an airplane is considered?

Answer by author: At this time there are not any plans to evaluate this concept for an all out production program. The planned flight test program of this system will only evaluate the aircraft/engine performance characteristics.

Note added in proof: The first flight of the Beech Aircraft Sundowner powered by the liquid methane system was on September 15, 1981.

HISTORY, STATUS AND FUTURE APPLICATIONS
OF SPACEBORNE CRYOGENIC SYSTEMS

Allan Sherman

NASA Goddard Space Flight Center
Greenbelt, Maryland

INTRODUCTION

There are increasing numbers of space instruments that require cryogenic cooling to accomplish their objectives. These
include instruments for Earth observation, atmospheric science,
gammaray and X-ray astronomy, infrared astronomy, space surveillance, and basic research. Potential future space applications
for cryogenic cooling include instruments for high energy astronomy, radioastronomy, relativity measurements, and a variety of
instruments and systems employing superconducting devices.

The elements of the instruments that may require cooling are
the radiation detectors, optical components, baffles, or, in some
cases, the whole instrument. Cryogenic cooling is necessary to
provide the required detector response, reduce pre-amplifier
noise, and/or reduce background radiation.

Table I shows an approximate breakdown of the cooling requirements for the various applications mentioned. To provide for
these requirements, there are a variety of techniques available or
under development. Table II lists the cooling techniques with
approximate ranges of practical temperatures and refrigeration
load.

The development of spaceborne cryogenic systems presents a
considerable engineering challenge. The objective of this
discussion is to describe the systems' history, status, anticipated technology development and future applications.

1007

Table I. General Mission Categories Requiring Cryogenic Cooling
in Space (Values Approximate)

Discipline	Temp. Range, K	Refrigeration Load
Applications Missions (weather, atmospheric science, Earth resources pollution monitoring etc.)	10-100	milliwatts to 10 W
Gamma-ray and X-ray Astronomy	80-125	milliwatts to watts
Cosmic Ray Measurements	~4	milliwatts
Space Surveillance (as published)	12-100	wide range
IR Astonomy	0.3-5	microwatts to 100 mW
Relativity Measurements	~2	microwatts to milliwatts
Superconducting Devices	1-10	wide range
Basic Research Experiments	1-10	100 mW

Table II. Spacecraft Cryogenic Cooling Techniques

Cooling Technique	Temp. Range,[a] K	Usable Refrigeration Load for One Year Mission[b]
Radiant Coolers	70-100	0-90 mW to date: higher capacity possible in future
Stored Solid Cryogen Coolers	10-125	0-800 mW[b] plus vent gas cooling
Stored Liquid-Helium Coolers	1.5-5.2	0-100 mW[b]
Mechanical Coolers	4-100	0-300 W
He^3 Coolers	0.3	0-100 μW
Adiabatic Demagnetization Refrigeration	0.001-0.3	0-100 μW
Higher Temp. Adsorption and magnetic Systems	not yet defined; possibly wide range.	

[a]These values are not theoretical limits, but are estimates of temperatures and loads based on the designs as they appear to be evolving.
[b]For missions of shorter duration (e.g., 7-30 day Shuttle sortie), higher cooling loads could be accommodated.

RADIANT COOLERS

To date, the most widely used system for obtaining cryogenic temperatures aboard spacecraft have been radiant coolers. These devices were developed for cooling IR (1-25 µm) detectors for NASA Earth-observation instruments and as part of the Defense Meteorological Satellite Program.[1]

As shown in Fig. 1, a radiant cooler is usually comprised of two stages. The first consists of a cone with a highly reflective, specular inner surface and diffusely coated external radiator. The cone shields the patch (the detector stage) from the spacecraft and Earth or reflects shallow-angle sunlight out of the cooler before it reaches the patch. Mounting the patch to the first stage with low conductive supports further isolates it from the spacecraft. Without a need for shielding from the spacecraft, and with no sun input to the cooler, the first-stage cone can be eliminated (Fig. 1 bottom). A deployable (or fixed) door attached to the mouth of the first stage can then block the patch view of the Earth. A deployable door can also be used with a cone-type cooler to limit Earth input or to warm the cooler (door closed) periodically while in orbit. The latter mode of operation would be employed for decontamination.

Typical operating temperatures for present day radiant coolers run about 90-100 K for the patch, and 160 K for the first stage, although a patch temperature of 71 K was attained by a Santa Barbara Research Corporation cooler aboard a meteorological satellite (SMS-1) in a geosynchronous orbit. The patches of radiant coolers have typical radiative capacities on the order of 30 mW, with the sensor heat dissipation but a fraction of the total. An exception is the Thematic Mapper cooler (Hughes Aircraft) that will fly on Landsat-D which has a sensor load of 85 mW. Typical patch and cone mouth-area values are 60 cm^2 and 1000 cm^2, respectively.

Radiant coolers offer a relatively simple, passive, low-weight technique for cooling sensors aboard spacecraft. Unfortunately, thus far they have been severely constrained with respect to temperature (>70 K), cooling load (milliwatts), placement on spacecraft, and spacecraft orbit.

The Air Force has funded a technology program to improve structural materials and assembly techniques with the objective of developing much larger radiant coolers with higher capacity.[2] As part of this program, Rockwell International has fabricated and tested a 9 m^2, two stage radiant cooler (Fig. 2). The cooler employs oxygen heat pipes to transport the energy from the source to the radiator and to isothermalize the second stage radiator

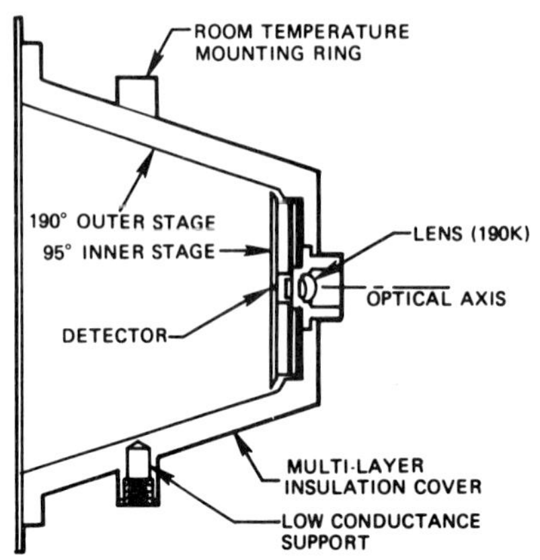

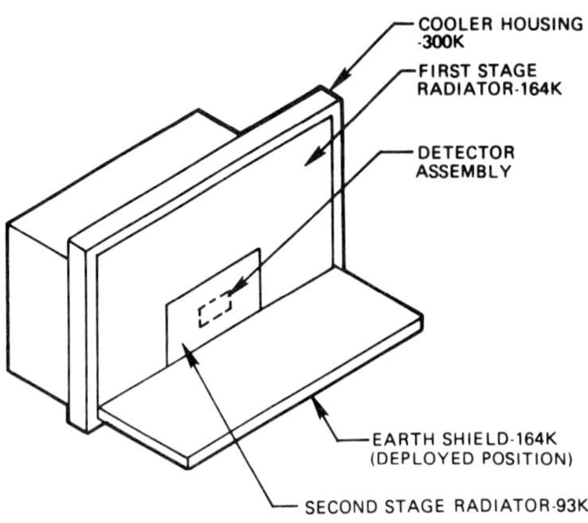

Fig. 1.   Radiant Coolers: A. D. Little (top), ITT (bottom).

plate.   Later, a third stage was added to the system, and the
cooler was tested again.

A large amount of the data for the advanced radiator has been
obtained. As an example, for the two-stage cooler, a second stage
temperature of 85 K and a first stage temperature of 148 K were
obtained with cooling loads of 10 W and 20 W, respectively. For
the three-stage radiator, with zero first stage load, third and
second stage temperatures of 41 K and 84 K were obtained for loads
of 0.2 W and 2 W, respectively. These results, however, must be
interpreted very carefully. During the testing, there was no
simulated Earth, sun or spacecraft loads, while the potential
input from these sources are much greater than the measured heat
rejection.   Thus, to apply the results of this testing to a real

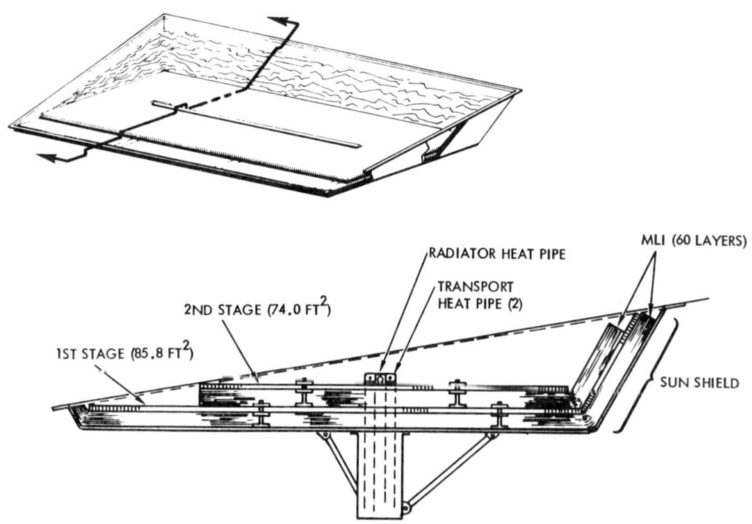

Fig. 2. Advanced Radiator Design: Rockwell International.

situation, one must be sure to include all of the heat loads from
the spacecraft and orbital conditions.

There is little doubt that radiant coolers will continue to
be widely utilized in the future. However, regardless of the size
of the system, the necessity of shielding and/or constraints on
spacecraft orbit and orientation, in order to limit heat input,
will continue to be a problem. From a practical viewpoint, their
use will probably be limited to temperatures above 70 K with
relatively low heat loads except, perhaps, in rather unique cases.

SOLID CRYOGEN COOLERS

Solid cryogen cooler systems utilize the heat of the sublimation of a cryogen to absorb the cooling load at constant temperature. As shown in Fig. 3, the operating point of the cooler must
be below the triple point. The lower operating temperature of the
cooler will be limited by the pressure that can be maintained over
the solid cryogen while still allowing enough of a potential for
venting the effluents. This pressure, around 0.1 torr, is a
function of vent lines and effluent mass flow.

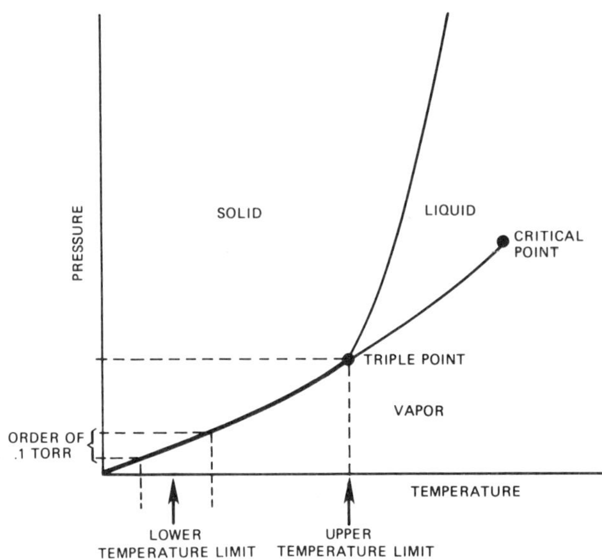

Fig. 3. Solid Cryogen Cooler Operating Range.

For space applications, stored solid cryogen systems offer
distinct advantages over stored liquid systems. Cryogens for this
temperature range have heats of sublimation about 10-15 percent
higher than their heats of vaporization. Also, the solid cryogen
has a density 10-15 percent higher than the liquid, permitting a
higher capacity cooler per unit volume of cryogen. By proper
design of fins or other heat exchange surfaces that penetrate and
contain the solid cryogen, the phase separation in zero gravity is
not a problem. A solid cryogen also, eliminates possible sloshing
problems that could affect the performance of a scanning or fine
pointing spacecraft. Advantages of solid cryogen coolers over
mechanical refrigerators are that they are vibration-free and
require no spacecraft power. The advantages of solid cryogen
systems over radiant coolers are that they are generally not
limited by orbit, spacecraft orientation, or locations within the
spacecraft.

Table III lists possible cryogens which could be used in
solid cryogen coolers. Listed are operating temperature ranges
defined by the triple point at the high end and the somewhat
arbitrary 0.1 torr venting limit at the low end. For comparison
purposes, the heat of vaporization of helium is also shown. The
table shows the advantage of utilizing a two-stage cooler in which

Table III. Candidate Cryogens for Solid Cryogen Coolers

Cryogen	Temperature Range* (K)	Approximate Volumetric Heat of Sublimation** cal/cm^3	Approximate Weight Heat of Sublimation** cal/g
Hydrogen	13.8-8.3	9.7	108
Neon	24.5-13.5	36.4	25.1
Nitrogen	63.1-43.4	59	60.2
Carbon Monoxide	68.1-45.5	75.1	72.2
Argon	83.9-47.8	78.3	48.3
Methane	90.7-59.8	73.5	147
Nitrogen Oxide	110-78.1	174.3	129.1
Ethylene	104-95	123.5	169.2
Carbon Dioxide	217.5-125	246.7	145.1
Ammonia	195.4-150	315.9	432.7
Liquid Helium (For Comparison)	1.5-5.2	0.8	5.5

*Lower temperature corresponds to 0.1 torr
**Values at lower temperature point

a higher temperature cryogen, such as ammonia, cools an intermediate thermal shield guarding a lower temperature cryogen. The advantages of using solid hydrogen as compared to liquid helium for 10 K cooling is also apparent.

An example of a flight cooler design is shown in Fig. 4. This two stage methane/ammonia cooler successfully flew on the High Energy Astronomical Observatory (HEAO-B and HEAO-C) spacecraft. It was developed by the Ball Aerospace Systems Division, had a loaded weight of 75 kg and measured 76 cm long by 56 cm in diameter. The cooler has tubes attached to the cryogen tank walls to flow LN_2 (77 K) during the cryogen fill operation. After the final servicing before flight, the cryogens gradually warm up

Table IV.   Flight Solid Cryogen Coolers

Cooler Identification	Cryogens	Total Loaded Weight (kg)	Length (cm)	Diameter (cm)	Low Temp. (K)	High Temp. (K)	Low Temp. Stage Inst. Load (mW)	High Temp. Stage Inst. Load (mW)	Orbit Lifetime
DOD/Lockheed SESP 72-2 (Two launched on same spacecraft in 1972)	carbon dioxide	21.9	35.6	40.6	126	Single Stage Cooler	230	Single Stage Cooler	8 mo. (1 Unit) 7 mo. (second unit)
NASA/Lockheed LRIR Inst. on NIMBUS-F; Launched in 1975	methane/ ammonia	24.1	62.2	33	65	152	52	91	7 mo.
NASA/Lockheed LIMS Inst. on NIMBUS-G; Same cooler as LRIR; Launched in 1978	methane/ ammonia	24.1	62.2	33	65	152	52	91	7 mo.
NASA/Ball HEAO-B launched in 1978	methane/ ammonia	75	81	56	80	150	200*	200	11 mo.
NASA/Ball HEAO-C launched in 1979	methane/ ammonia	75	81	56	80	150	359	367	8 mo.
DOD/Lockheed Teal Ruby to be launched in 1983; performance numbers are estimates	neon/ methane	159	123	80	<16.8	75	38	252	17 mo.**

*Does not include heater cycling for decontamination
**Projected with radiant-cooled 150 K outer shell

ntil the vent seal is opened in orbit. The cryogens then vent to space while drifting down to steady-state temperatures. Heat exchange surfaces are located in both the primary and secondary tanks to promote even heat absorption by the cryogen.

To date, six solid cryogen coolers have been placed in orbit aboard spacecraft, and another will be launched in the near future. A summary of the basic design parameters and performance of the coolers is in Table IV.

Solid cryogen coolers do not have the inherent boundary constraints of radiant coolers, although it is always advantageous to have the outer shell operate as low in temperature as possible. Compared to a mechanical cooler, the solid cryogen system has the advantage of eliminating possible vibration and microphonics problems. The prime disadvantage of a solid cooler is that its weight and volume increase rapidly with cooling load and lifetime requirements. Other disadvantages include the need for bulky ground support equipment and a relatively high cost.

Thus far, the coolers have not performed as well as originally anticipated. Areas where technology studies could improve performance are in the design of the multilayer insulation and perhaps, in controlling internal contamination. Although little technology development is underway in these areas, an important study on the characteristics of a solid hydrogen cooler has been completed.[3] In this study, the fill and vent of a solid hydrogen cooler was demonstrated and a detailed safety analysis was performed. As an add-on to this effort, a para-to-ortho hydrogen catalytic converter to greatly improve effluent gas cooling is being developed. Solid hydrogen has now been proposed as the cooling system for two instruments for the Upper Atmospheric Research Satellite (UARS).

In the future, there is no doubt that solid cryogen coolers will continue to find widespread use. Efforts to reduce their cost for short-term Shuttle flights are already underway. For free-flyer satellites, the established coolers, as well as solid hydrogen, will be employed. However, for long-duration and/or high cooling load flights, it is clear that a closed cycle cooling system will be required.

LIQUID HELIUM SYSTEMS

A new adventure in space cryogenic cooling systems is the development of liquid helium systems. These systems, which cover the 1.6 K – 4.2 K temperature range, will find wide application. Thus far, the United States has not orbited a liquid helium system. Pravda has reported[4] successful flights of 1.8 liter and 10 liter normal helium cryostats aboard the Soviet series of

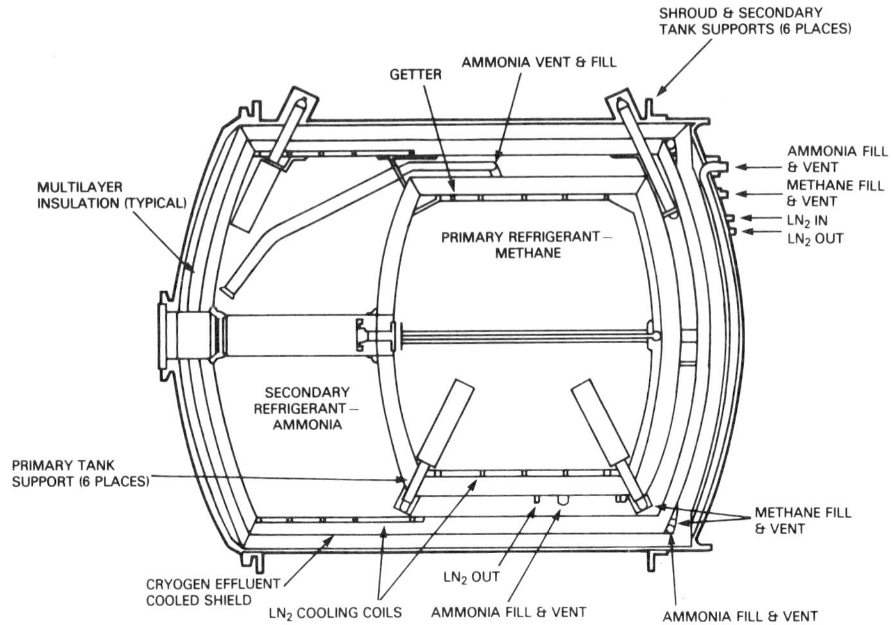

Fig. 4. HEAO B and C Solid Cryogen Cooler: Ball
Aerospace Systems Division.

spacecraft.   The Soviet Union also reports of the successful
flight of a helium liquefier aboard Salyut-6.[5]

There are numerous NASA liquid helium space cooling systems
under development for flights in the near future.   These include
long   duration,   superfluid   helium   dewars   for   the   Infrared
Astronomical Satellite (IRAS)* and the Cosmic Background Explorer
(Cobe) missions, and shorter duration systems for the Small
Helium-Cooled Infrared Telescope and the Superfluid Properties
Experiment.

A schematic of the IRAS cryogenic system (described in detail
in Ref. 6) is shown in Fig. 5.   The 1.8 K superfluid helium is
contained in a toroidal shaped, 5083 aluminum alloy main cryogen
tank.   The infrared telescope system fits into the center of the
toroid and is mounted to the telescope mounting ring.   Structural
support is achieved through nine fiberglass/epoxy bands which
terminate at the main shell.   Additional thermal protection is
provided by three vapor-cooled shields and multilayer insulation

---

*Joint U.S., Netherlands, U.K. Project.

in the vacuum space. The main shell will operate at 170 K via radiant cooling in orbit to further reduce heat leaks to the main cryogen tank. Liquid-to-vapor phase separation in zero gravity is accomplished by a porous plug located at the main cryogen tank/helium vent line junction. A supercritical helium tank is employed to cool the main dewar cover, which is jettisoned during the initial stages after achieving orbit. From ground test data and mathematical modeling, the operational orbital lifetime for the IRAS dewar is predicted to be 10 months. The IRAS cryogenic system has completed performance and space qualification testing at the Ball Aerospace Systems Division. It is presently (August, 1981) in the Netherlands where it is being integrated with the spacecraft, after which it will be shipped back to the United States. The planned launch date is August, 1982.

The Cosmic Background Explorer mission will use the basic IRAS dewar, with some modifications. These include deletion of the aperture cover tank, additional support straps for Shuttle launch (the IRAS is launched via a Delta rocket), and improved valves. The contract for the dewar should be issued to the Ball Aerospace Systems Division in 1982. The launch date for the satellite is in 1987.

A schematic of the Small Infrared Telescope system, which will fly on a 7 day Shuttle mission, is shown in Fig. 6. Details of this complex system are in Ref. 6. The system is comprised of a 250 liter dewar, a helium transfer assembly, and a cryostat surrounding the telescope. By utilizing a vacuum sealed rotary

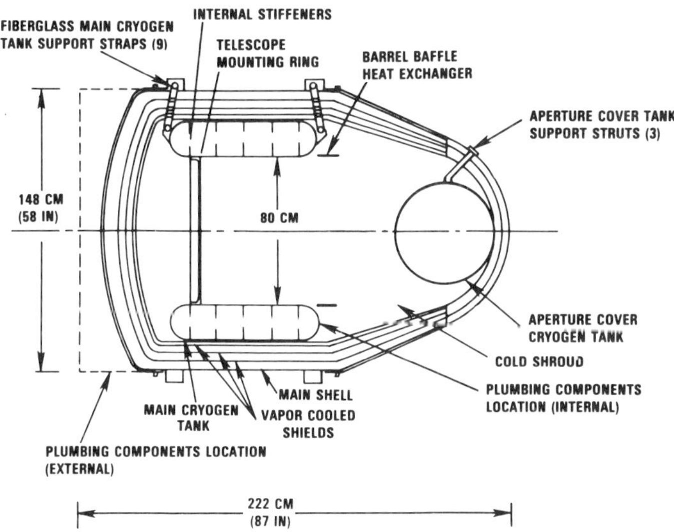

Fig. 5. IRAS Dewar System: Ball Aerospace Systems Division.

joint, the telescope can scan in one axis while the dewar remains
stationary.   The porous plug phase separator is located in the
transfer assembly and, thus, only gas is delivered to the cryostat
assembly.   The helium in the dewar is maintained at 1.6 K while
temperature requirements for the cryostat vary from 2.5 K at the
detectors to 60 K at the upper telescope sections.   During ground
hold, a pump located on the Shuttle Pallet will maintain the
helium in the superfluid state.   The pump will be shut off prior
to launch, but will fly with the Shuttle.   The dewar and cryostat
for the small Infrared Telescope system have been manufactured by
Cryogenics Associates, while NASA/Marshall Space Flight Center and
the University of Alabama are responsible for the helium transfer
assembly and subsequent integration.   At the present time, the
cryogenic system is being assembled at the Marshall Space Flight
Center.   Completion of the assembly is expected early in 1982.
The Spacelab 2 flight is scheduled for November, 1984.

A schematic of the Superfluid Helium Properties Experiment
mounted on the Shuttle Pallet is shown in Fig. 7.   Details of the
experiment are described in Ref. 9.   The objectives of the experiment are (1) to measure fluid motion and temperature fluctuations
in superfluid helium, and (2) to contribute to the basic under-

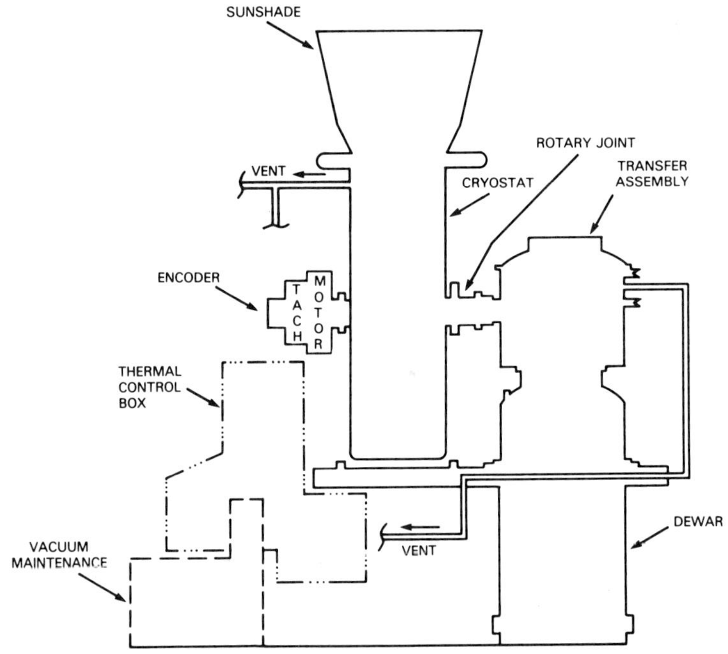

Fig. 6.   Small Helium-Cooled Infrared Telescope.

standing of the quantum nature of He II and the interactions between the normal and superfluid components. The dewar, built by Ball Aerospace Systems Division, contains the necessary flight instrumentation. Vapor-cooled shields and multilayer insulation in a guard vacuum are again employed to reduce boil-off, and a porous plug is utilized for phase separation. Testing of the dewar for the superfluid Helium Properties Experiment is being completed at Ball Aerospace Systems Division and it will be delivered to the Jet Propulsion Laboratory in September, 1981, at which time the integration period for the instruments will begin. The experiment will fly with the 7 day Spacelab 2 mission scheduled for launch in November, 1984.

What does the future hold for liquid helium systems? It is clear that liquid helium coolers will be utilized a great deal in the future. Two missions that are in the planning stages[10] include the Shuttle Infrared Telescope Facility (SIRTF) and the Gyro Relativity Satellite. The SIRTF would use a combination of super-critical and superfluid helium cryostats for short duration Shuttle missions, while the Gyro Relativity Satellite would use a dewar similar to IRAS. In addition, in the future, superconducting devices may have wide ranges of applications in space. Technical improvements in liquid helium systems will occur as future designs evolve. Under a recent contract awarded to the Lockheed Palo Alto Research Laboratories from the NASA/Ames Research Cen-

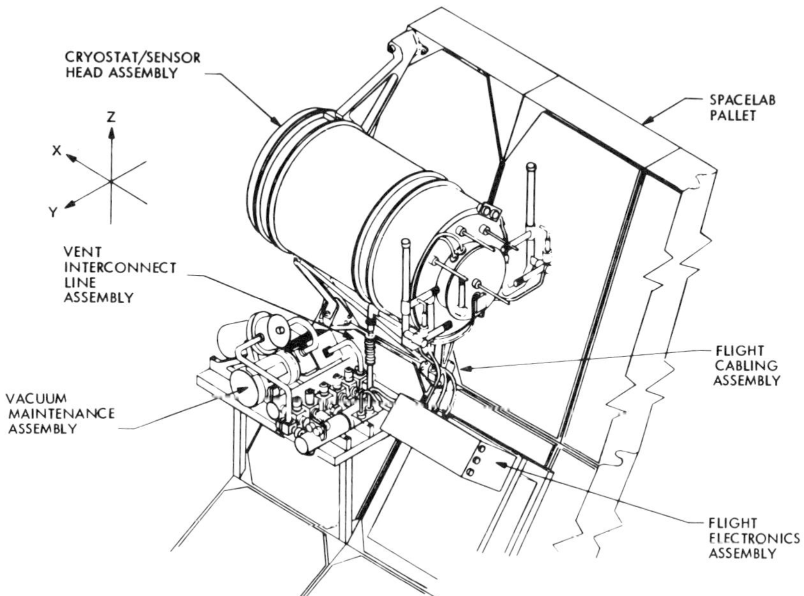

Fig. 7. Superfluid helium experiment for Spacelab 2.

ter, technical improvements and alternatives relative to lifetime
extension will be studied.

MECHANICAL COOLERS

As payload refrigeration loads and lifetime requirements are
increasing, the need for a closed-cycle mechanical cooler system
is becoming more critical. A great deal of effort has been
expended in an attempt to develop a long lifetime (3-5 years)
mechanical cooler with stable performance. While some progress is
being made, much remains to be done. The following is a brief
status report on various mechanical cooler development programs
and significant recent accomplishments.

Stirling Cycle Cooler Aboard P-78-1 Spacecraft

One of the most significant events in recent years in mechanical cooler technology is the orbiting of four Stirling cycle
coolers aboard the DOD P-78-1 satellite.[11] Launched on February
24, 1979, these units have successfully provided cooling for two
gamma ray spectrometer detectors.

The coolers were developed by the Philips Laboratories with
the Johns Hopkins Applied Physics Laboratory providing the electronics controls and space qualification program. A schematic of
the two-stage system is shown in Fig. 8. The first stage runs at
140 K with a 1.5 W load, while the second stage runs at <75 K with
a 0.3 W load. Power consumption at this condition is 30 W. The
weight of each cooler is 7.2 kg, and overall dimensions are 15.4
cm x 18 cm x 30.7 cm. The helium charge pressure is 0.48 MPa.
Two permanent-magnet, brushless dc, counter rotating motors drive
the machine at about 1150 rpm. Krytox grease is used in the
crankcase for lubrication, while piston and displacer basic seal
materials include glass-filled Teflon and Roulon. The machine
employs the Philips rhombic drive mechanism, which provides the
correct phase relationship between the piston and displacer. The
rhombic drive also provides dynamic balance by utilizing counter
rotating weights, although the results are not perfect. Ground
test data showed some degradation in detector performance caused
by machine microphonics, but it was an acceptable amount.

As of July, 1980, when the spacecraft had been in orbit for
500 days, the cumulative operating time of each of the four coolers was 2657,5134,9542 and 6518 hours. The latest unpublished
data shows that one of the machines had surpassed 15 000 hours of
run time. Although the total run time of the refrigerators is
impressive, significant degradation in performance occurred. The
reason for this degradation cannot fully be discerned from the

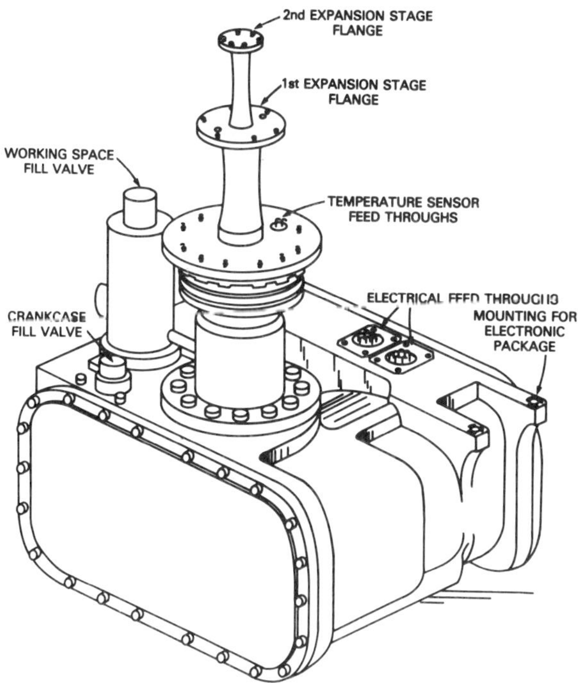

Fig. 8.   Philips Rhombic Drive Cooler.

data, but it is clear that a significant factor is helium gas leakage. Other factors may be seal wear and internal contamination. Consequently, after 130 days into the flight, it was necessary to operate two refrigerators per instrument to provide the required cooling, whereas it was planned to operate only one per instrument. Further details on the orbital performance of the machines is in Ref. 11.

Notwithstanding the performance degradation of the coolers, the utilization of the Stirling machines to provide instrument cooling over a long period of time in space is indeed a significant accomplishment. Heretofore, no mechanical cooler had operated in space for any significant duration.*

---

*Short-duration operations of mechanical coolers in space include a Stirling cooler on Skylab and a Vuilleumier cooler operated in space in 1973.

3-5 Year Magnetic Bearing Stirling Cycle Cooler

An innovative approach to the long lifetime mechanical cooler problem is the magnetic bearing Stirling cycle cooler that has been under development for the past 3 years. This cooler is being developed by the North American Philips Corporation under the management of NASA/Goddard Space Flight Center. The basic design philosophy of the cooler is (1) to have no rubbing surfaces in the machine, (2) to electronically control axial position and piston/displacer phase angles, (3) to utilize a linear drive system, and thus, eliminate mechanical linkages, (4) to reduce the potential for internal contamination by eliminating all organics inside the working gas, and (5) to provide an essentially dynamically balanced machine. A schematic of the machine to accomplish these objectives is shown in Fig. 9.

The baseline cooler is a single-stage device which will maintain a cold temperature of 65 K with a 5 W cooling load. Input power under these conditions will be about 180 W, as the machine operates a 1700 strokes/minute. The helium charge pressure is 1.6 MPa. The design lifetime is 3-5years with either continuous operation or 1 000 stop/start cycles. Noncontact between moving and static components can be achieved for startup, steady operation, and shutdown because the bearings can be actuated independent of the linear motor. Piston and displacer seals merely consist of sections with 0.025 mm clearances.

The engineering model cooler is due to be delivered to Goddard Space Flight Center in October, 1981, although it is hoped that initial system testing will begin earlier at Philips. This model will be quite sophisticated and, except for the electronics, is required to meet the 3-5 year lifetime. A follow-on effort to produce an upgraded prototype model will then follow. Thus far, the piston motor with associated electronics has been thoroughly tested. Static and dynamic testing of the bearings has also been completed, as have displacer motor static tests. All tests have been highly successful. On the negative side, a series of manufacturing problems, coupled with the need to do more component level tests, has caused schedule slips. A detailed update of the status of the development is in Ref. 12. If the systems level testing of the magnetic bearing machine are successful, a great step forward in the development of reliable mechanical coolers will have been made. Extension of the technology to machines with different load and temperature requirements can then follow.

Vuilleumier (VM) Coolers

A noteworthy early program in Vuilleumier cooler technology was the NASA/Goddard Space Flight Center effort in 1969 and the early seventies. The program resulted in a 5 W, 65 K machine

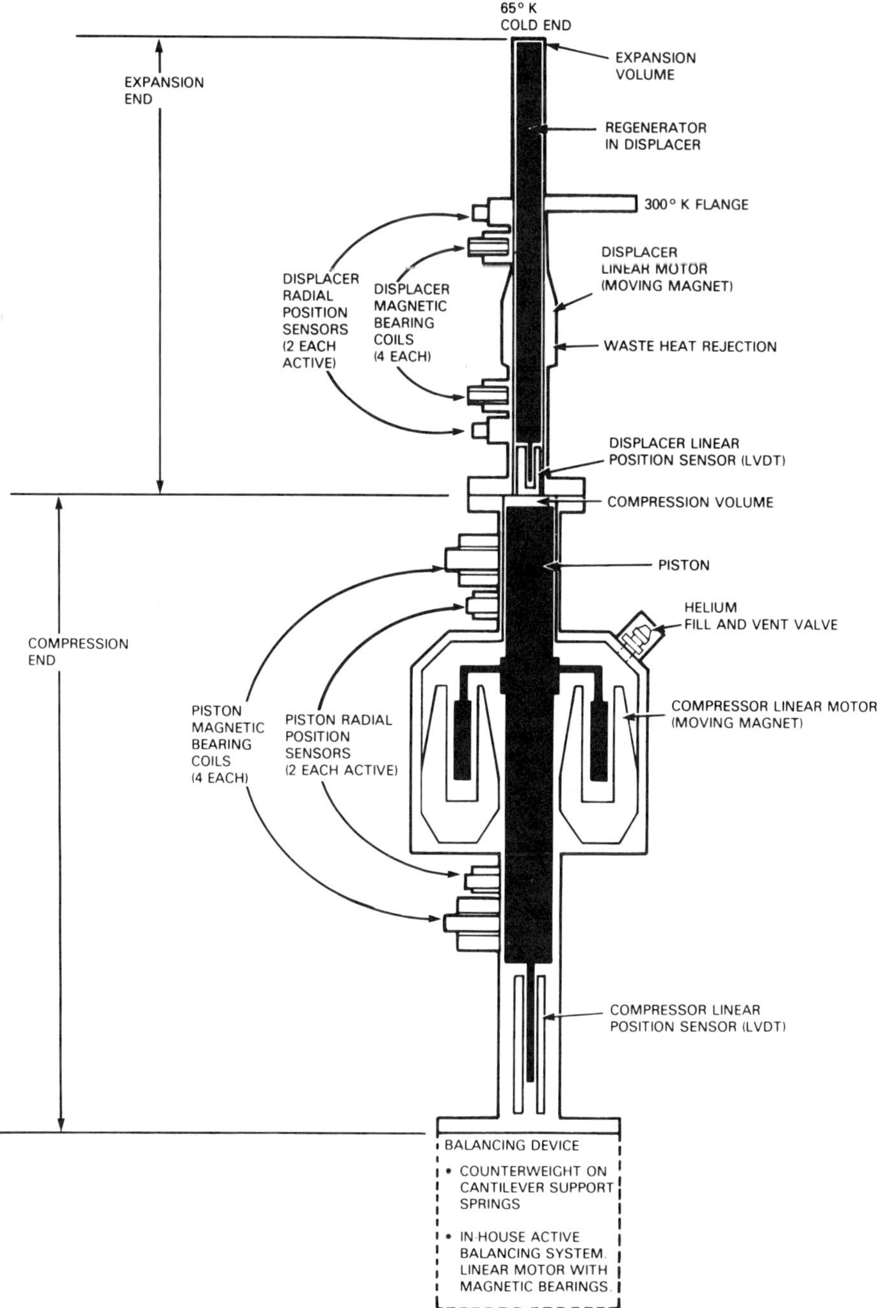

Fig. 9. Single Expansion Cryogenic Cooler with Linear Magnetic Suspension: North American Philips Corp.

(built by Garrett Airesearch) that had many interesting features, such as solid bearings composed of Boeing 6-84-1 running against flame-sprayed Inconel 718. The machine was tested for about 6 000 hours and met all thermodynamic goals, but exhibited some mechanical clearance problems.

The Air Force has had an extensive Vuilleumier cooler development program over the past decade. At the present time, the major effort is in the development of the Hughes Aircraft Corporation Hi Cap machine (Fig. 10). Design requirements for the three-stage cooler are 12 W, 10 W, and 0.3 W cooling loads at 75 K, 33 K, and 11.5 K, respectively. Long lifetime operation is intended (~20 000 hours) with a maximum input power of 2 700 W. With the Vuilleumier cycle, most of this input power is in the form of heat.

The Hughes approach uses dry lubricated ball bearings (MoS_2) with a bearing retainer made of Roulon A (filled Teflon) with five percent MoS_2. The machine employs rubbing seals on both the hot and cold displacers and flexure pivots at the displacer drive-rod interfaces. Past problems included metal fatigue, internal contamination and seal wear.

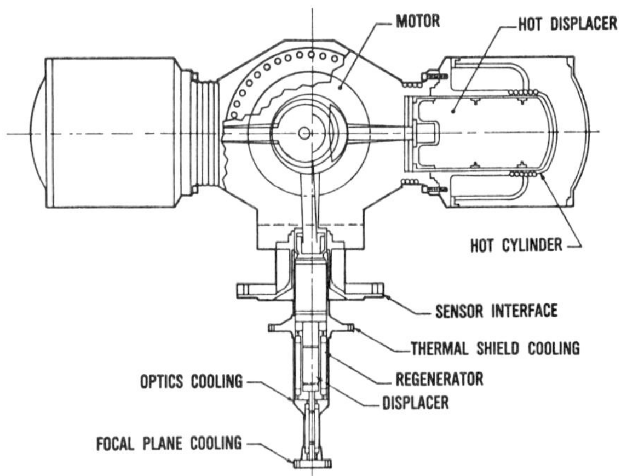

Fig. 10. AF Vuilleumier Cooler: Hughes Aircraft Corp.

A major renewed effort to improve the reliability of the Hi Cap machine is now underway.[13] This effort consists of extensive testing of the three existing coolers to evaluate seals, regenertors, and other components. Several internal reports have been written, which give the results of the exhaustive seal materials test program. In addition, the Hi Cap program includes the fabrication of three improved flight coolers. It is expected that the flight coolers will be part of the SIRE (Space Infrared Sensor) experiment, which will fly on short-duration Shuttle flights.

It is difficult, at this point in time, to predict the ultimate success of the Air Force Vuilleumier cooler program. However, much valuable data is presently being collected.

Rotary Reciprocating Coolers

The Air Force also has an extensive program in rotary reciprocating cooler development. This effort began in the 1960's. The system utilizes self-actuating, rotary gas bearings in both the compressor and expander sections, and linear motion to achieve the refrigeration. Thus, each moving member is both rotated and reciprocated. The machine also utilizes clearance seals; thus, after startup, there are no rubbing surfaces in the machine. The thermodynamic cycle employed in the current rotary reciprocating cooler is the reverse-Brayton.

To date, all testing has been done on a two-stage cooler with cooling capacities of 1.5 W at 12 K and 40 W at 60 K. The effort has been rather long and arduous. At the present time, the most pressing problem is bearing seizure in the first-stage expansion engine.[13] Modifications to the design are planned which will, hopefully, alleviate the problem.

Turborefrigerators

Turbomachinery employing rotary gas bearings offers a high potential for long lifetime operation in space. Consequently, a reverse Brayton turborefrigerator for space was designed under DOD sponsorship in the early seventies. The system goals were cooling loads of 1.5 W and 30 W at 12 K and 60 K, respectively, 30 000 hour lifetime, and a maximum power consumption of 4 KW. The program resulted in a design of a four-stage turborefrigerator with gas bearings and included some component fabrication and test. The contractor for the work was the General Electric Corporation.

The Defense Advanced Research Project Agency (DARPA) initiated a turborefrigerator program in 1978 with a contract to the Garrett Airesearch Corporation.[13] A system was designed, fabricated and performance tested. The tests exhibited satisfactory operation over a range of conditions and over a time period of hundreds of hours. Full performance testing will continue in 1981 with endurance testing scheduled for 1982. The results of this program to-date are indeed encouraging.

The main problem with turborefrigerators is that their efficiency drops off drastically with gas flow. As a result, they are simply neither competitive nor applicable to many of the mission requirements. Reciprocating, or positive displacement, machines are needed in the low to moderate cooling load regimes.

Other Mechanical Cooler Efforts

Although not specifically applicable to space, the Office of Naval Research has recently initiated a cryogenic cooler program that may have some spin-off. Also, there is a recent effort at Oxford University to develop a 0.5 W, 2 year lifetime, Stirling cycle cooler. This machine employs a linear drive and control circuitry similar to the NASA/Goddard effort, but utilizes a spring support system in lieu of magnetic bearings.

ADSORPTION AND MAGNETIC SYSTEMS

Several investigators have developed, or are working in, 0.3 K, ^{3}He adsorption systems for long wavelength infrared bolometers. Balloon experiments utilizing such systems have already flown,[14] and other flights are planned by NASA. The systems typically use charcoal or zeolite to provide pumping on a container of ^{3}He. A system of this nature, designed and fabricated at the Goddard Space Flight Center for the ground testing of bolometers,[15] is shown in Fig. 11. Note the heat switch to permit recycling of the helium.

NASA/Ames Research Center is working on a ^{3}He system[16] with a copper sponge type of fluid maintenance device for zero-g operations. Testing of the device with an adverse gravity orientation is planned, but has not yet been accomplished.

The Jet Propulsion Laboratory is working on adsorption, Joule-Thomson expansion, refrigerators[17] for possible use on planetary missions. Such systems are generally inefficient. However, the power input is in the form of heat, which could be obtained from an isotope source or a solar collector. To date analytical studies of system designs have been conducted and potential compressor designs are being tested.

Efforts to develop an adiabatic demagnetization refrigerator for space application have been proceeding at the Ames Research Center and Goddard Space Flight Center. The system offers the possibility of cooling infrared bolometers to the 100 mK range

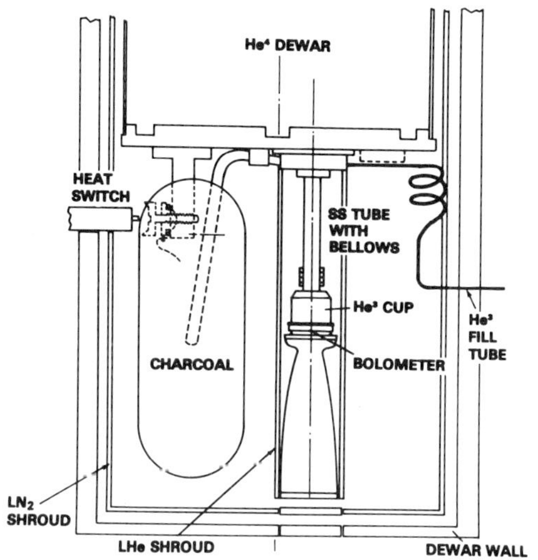

Fig. 11. Adsorption ^{3}He Cooler.

without any unusual gravity effects. The basic approach differs little from laboratory systems except that the structural design must survive launch loads. Considerable effort has been expended in this area at Goddard and, hopefully, a sound structure has been designed. However, testing of the refrigerator is not yet underway. At Ames, performance demonstration tests of a laboratory system have been completed.

Los Alamos Scientific Laboratory has been exploring the development of magnetic cooling for higher temperature/higher cooling load systems,[18,19] as compared to the described adiabatic demagnetization systems. Under Jet Propulsion Laboratory finding, Los Alamos has examined their systems for space application. These systems still require a great deal of development in material technology and in the overall design scheme. Another problem is that superconducting magnets are required; thus, the cooling system itself requires very low temperatures with all of the inherent problems for space application.

CONCLUSION

A variety of systems will be used to provide for the increasing demand for spaceborne instrument cryogenic cooling.

Radiant coolers will continue to find widespread application for low cooling-load/high-temperature situations. Advanced radiant coolers will provide larger heat rejection, but will still suffer from spacecraft and orbit constraints.

Solid cryogenic coolers are less sensitive to their environment but, as an open system, are lifetime limited. Future utilization of 10 K solid hydrogen is envisioned.

For very low temperature requirements, liquid helium systems are being developed. The most notable effort is the design, fabrication and test of the nominal one-year lifetime superfluid helium dewar for the Infrared Astronomical Satellite. The same basic dewar will also be used for the Cosmic Background Explorer Satellite.

A long-lifetime closed-cycle, mechanical cooler is one of the most critical space technological needs. Numerous programs are underway to develop these systems.

Other advanced cooler efforts include development work on adsorption and magnetic systems.

The next five years should indeed be very exciting, as many new cryogenic systems will fly and others may exhibit technological breakthroughs. Studies on optimizing system performance by using combinations of the basic cooling systems will also be performed.

REFERENCES

1. M. Donohoe, A. Sherman, and D. Hickman, Radiant coolers-theory, flight histories, design comparisons, and future applications, AIAA Preprint 75-184, AIAA 13th Aerospace Sciences Meeting, Pasadena, Calif., (Jan. 1975).
2. J.P. Wright, Development of a 5-watt 70° K passive radiator, AIAA paper 80-1512, presented at the AIAA Thermophysics Conference, July, 1980.
3. T. Nast, Study of solid hydrogen cooler for spacecraft instruments and sensors, LMSC-D 766177, Final Report for NAS5-25792, (August, 1980).
4. A. Fradkov and Troitskii, Helium cryostats for physical studies in space, Cryogenics, (August, 1975).
5. PRAVDA, February 21 and 26, 1978.
6. A.R. Urbach, P.V. Mason, and J.W. Vorreiter in "Proceedings of the Seventh Int. Cryo. Engr. Conf." IPC Science and Technology Press, London (1978).
7. W.F. Brooks and R. Hopkins, Orbital Performance of a One Year Lifetime Superfluid Helium Dewar Based on Observed Ground Performance and Computer Modeling Techniques, in "Advances in Cryogenic Engineering, Vol. 27," Plenum Press, New York (1982).

8. E.W. Urban, et al, A cryogenic system for the small infrared telescope for Spacelab 2, *Refrigeration for Cryogenic Sensors and Electronics*, NBS Special Publication 607, (May 1981).

9. P. Mason, Superfluid Helium Experiment for Spacelab 2 in "Advances in Cryogenic Engineering, Vol. 25," Plenum Press, New York (1980), p.80.

10. J. Vorreiter, Cryogenic systems for spacecraft, *Contemporary Physics*, 21(3):201.

11. L. Naes and T. Nast, Long-life orbit operation of stirling cycle mechanical refrigerators, *Proc. of the Society of Photo-optical Inst. Engr.* 245(23):126 (1980).

12. A. Daniels, M. Gasser, and A. Sherman, Magnetically Suspended Stirling Cryogenic Space Refrigerator Static Report, in "Advances in Cryogenic Engineering, Vol. 27," Plenum Press, New York (1982).

13. R. Harris, J. Chenoweth and R. White, Cryo-cooler development for space flight applications, *SPIE Technical Symposium Paper 280-10*, Washington, D.C. (April, 1981).

14. I. Radostitz, et al, Portable ^{3}He vector cryostat for the far infrared, *Rev. Sci. Inst.*, 49:1 (1978).

15. A. Sherman and O. Figueroa, A Portable ^{3}He Cryostat for Studies in Astrophysics, in "Advances in Cryogenic Engineering, Vol. 25," Plenum Press, New York (1980), p.775.

16. P. Kittel and W.F. Brooks, Demountable Self-Contained ^{3}He Refrigerator, in "Advances in Cryogenic Engineering, Vol. 27," Plenum Press, New York (1982).

17. C. Chan, E. Tward and D. Ellemann, Miniature JT Refrigerators using Adsorption Compressors, in "Advances in Cryogenic Engineering, Vol. 27," Plenum Press, New York (1982).

18. W. Steyert, Rotating current-cycle magnetic refrigerators for use near 2 K, *J. Appl. Phys.* 49:3 (1978).

19. W.A. Steyert, Stirling-cycle rotating magnetic refrigerators and heat engines for use near room temperature, *J. Appl. Phys.* 49:3 (1978).

CRYOGENIC PARTICLE COLLECTION ON A COMETARY MISSION*

R. J. Szara, T. E. Economou, E. Blume, and A. L. Turkevich

The University of Chicago
Chicago, Ilinois

INTRODUCTION

Comets have always been a fascination to man, but modern
scientists have more than just a curious interest in these
celestial bodies. Knowledge about comets may very well contribute
to the overall knowledge about the origin of the universe. Comets
seem to have an origin outside our solar system, and indeed, many
have an orbit that takes them far beyond the orbit of Pluto before
traversing an ellipse around the sun. Their composition may very
likely be primordial matter. Although they appear as fireballs in
the sky, comets are cryogenic in nature. They are really spheres
of solidified gases in which are mixed liberal amounts of small
solid particles. Their tails are the result of surface sublimation as a direct consequence of solar thermal radiation when
their orbit takes them to the proximity of the sun. Solar wind is
responsible for the geometry of the tail and thus the tail always
faces away from the sun.

The purpose of this research is to devise a method to capture
effectively particulate matter as well as volatiles from the tail
of a comet for the purpose of chemical analysis. This would, of
course, be done on a space mission which would have a means of
propulsion so that it could be maneuvered into the coma. Since it
is expected that a variety of other experiments will be aboard the
spacecraft, some of which will be adversely affected by dust (the
particulate matter) it will be necessary for most of the trajectory to remain in areas of low particle flux. This neces-

---

*This project was supported by NASA Grant NSG001-135 and NSG-75-
32.

sitates a large surface area for gathering enough particles and a
method of concentration so an analysis can be performed. The
collecting medium has to be chemically inert and easily dis-
tinguishable from the particles being collected. Since the light
chemical elements (C,N,O) are of particular interest, these elements must be specifically avoided.

We are proposing a film of solid xenon at at temperature of
65 K as the collecting medium. Sublimation of this film and the
resultant pressure will be used to concentrate the particles on a
small area suitable for analysis. This is accomplished by closing
off the xenon film before heating and directing the stream of gas
bearing the captured particles through a small orifice. The
stream of gas with its particles would be aimed at a small plate
to which particles would adhere. This would permit a concentration of $\sim 2 \times 10^4$ to 1.

LABORATORY APPARATUS

A vacuum chamber capable of 10^{-7} torr was converted into a
space environment simulator (Fig. 1). An Xe cell with flexible
lines for cryogenic cooling was built into the system. The Xe
cell could be traversed through a 90° arc providing two orthogonal
operating positions. In one position the Xe film could be formed
and later studied with a long focus microscope through a quartz
window at the top of the chamber. In the other position the Xe

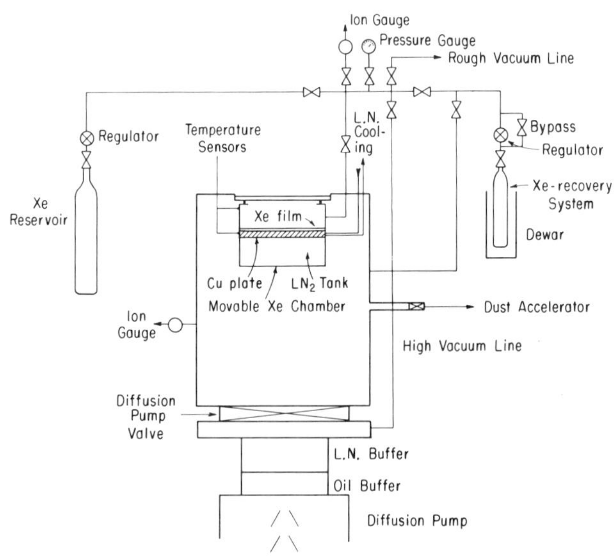

Fig. 1. Schematic of laboratory experiment.

cell was in line with an accelerated stream of particles generated by a particle accelerator. The cell is made of copper and consists of a 4" diameter plate on which Xe is deposited. It is cooled by a 1 L tank filled with liquid nitrogen. Within the tank there are copper fins in thermal contact with the plate so there is no interruption in cooling when the tank is traversed from horizontal to vertical position. Flexible metal lines are used for the transfer of liquid nitrogen and for pumping on the liquid to reduce the temperature. Within the LN_2 reservoir are two carbon resistors which detect the liquid level. In operation the liquid level is kept between the two resistors which are 2.5 cm apart. A gravity feed system is used with a micrometer valve for accurate control. Filling is done at the boiling point as well as at 65 K when the liquid nitrogen is being pumped. In order to prevent excessive splashing when liquid is introduced to the pumped reservoir, the fresh liquid is admitted through a coil immersed in the cooler liquid already in the reservoir. This affects an even distribution of the added heat load of the new and warmer liquid.

Temperature is measured to within a 0.1 K by a copper-constantan thermocouple imbedded in the copper plate. Temperature is controlled by regulating the pumping speed with a valve.

The particle accelerator[1] consists of a gas and particle insertion chamber, a 1.5 m long 1mm diameter capillary tube and 3 skimmers, each having a 600 L/min vacuum pump. The particle accelerator is connected to the vacuum chamber through a ball valve. When the valve is open the stream of particles, (with the propulsive gas skimmed off), is directed at the Xe film target. The velocity of the particles is a function of the mass of the propelling gas. The length of the capillary tube must be long enough so that the gas emerges at sonic velocity. Once the length of the capillary tube was determined, velocity was varied by the choice of gas used. Helium gives a velocity of 1000 m/s, argon gives a velocity of 300 m/s. Control of the pumping speed in the first skimmer chamber was also used to control the velocity of particles. This allowed a velocity of particles within a 50 to 1000 m/s range.[2]

Since the cost of Xe for experiments was a significant portion of the development project cost a recovery system was devised. At the conclusion of an experiment the sublimed Xe was directed into a stainless steel pressure cylinder. The bottom portion of this cylinder was immersed in liquid nitrogen. At 77 K all of the Xe was transferred to the cylinder. The Xe can be used in future experiments.

EXPERIMENTAL WORK

The sequence of a laboratory experiment is as follows:
1. Pump down the vacuum chamber to ~ 10^{-6} torr.
2. Cool down Xe cell to 77 K by filling tank with liquid nitrogen
3. Seal the cell into the horizontal position and admit the prescribed amount of Xe gas to form a film of desired thickness.
4. Start pumping on the LN_2 to reach ~ 65 K cell temperature.
5. Lower cell into vertical position in line with the particle beam.
6. Start the particle accelerator system and open valve between systems.
7. Release particles into the gas stream. Particles are of phosphorescent material, such as ZnS or fluorescent chalk in the micron size range.
8. After collection period, close valve and bring Xe cell to sealed upright position for observation.
9. Particle patterns and numbers are observed with a long focus microscope using UV light for excitation. Photo micrographs are taken for permanent records.
10. After all data is taken, valves are opened to Xe recovery system and the cell is allowed to warm up. When the Xe is evaporated valves are closed and recovery system is warmed up.

Control runs are also performed using oil coated metal plates. Comparisons are made between these and actual Xe runs to determine collection efficiency of the Xe target.

EXPERIMENTAL RESULTS

Two forms of solid Xe were deposited reproducibly on the Cu plate of the Xe cell. Glassy solid was made by reaching the temperature of 160.5 to 165.3 K (the liquid range of Xe) and condensing all of the gas on the horizontal surface. This temperature was achieved by circulating N_2 gas through a cooling coil in thermal contact with the Cu plate. The N_2 gas pre-cooled with LN_2 in an external heat exchanger. After condensation was complete, temperature was lowered gradually until operating temperature (~ 65 K) was reached.

Microcrystalline solid (Fig. 2) was formed by cooling the condensing surface to 77 K and then admitting all of the Xe gas. This type surface was superior in trapping micron-size particles. The increase in surface area resulting from microcrystals instead of a smooth glassy surface proved not to be a serious additional sublimation problem in the 65 to 68 K temperature region. This was a matter of concern, however, because of

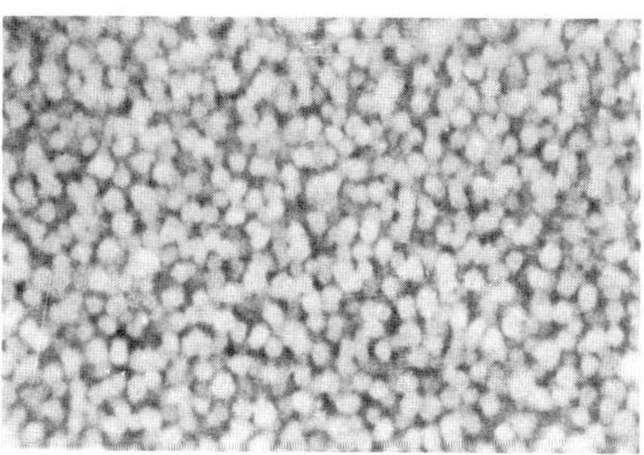

Fig. 2. Photomicrograph of microcrystalline solid xenon.

the reported low thermal conductivity of Xe.[3] Runs of 3 hours or more exposure of the Xe surface to high vacuum were done repeatedly. Measured loss of Xe was ~ 10% per run. The Xe surface was, of course, exposed to ambient temperature radiation (~ 300 K). In an actual comet mission the Xe surface would be exposed to black space (3 K) with only slight amounts of radiation from the comet nucleus so no significant losses are anticipated.

Numerous runs were made actually collecting particles generated by the particle accelerator. Figure 3 shows examples of collected particles as seen under microscope using UV light for excitation. Runs were made using an oiled aluminum plate as a collecting medium (Fig. 4). Comparisons were made by studying the photomicrographs taken in comparable runs. These runs were repeated using particles of varying velocities. The results indicated that a microcrystalline Xe surface was as good a collector as an oiled surface. The glassy Xe surface was inferior in the velocity range tested (50 to 1000 m/s). Future tests using

Fig. 3. Collected particles under UV light.

Fig. 4. Particles collected on oiled aluminum plate.

higher velocity particles are planned.   These, however, will require a complete redesigning of the particle accelerator.

Experiments have also been performed to concentrate micron size particles on a small surface. This was accomplished by accelerating particles suspended in a gas through a small orifice forcing a collision with a small plate.[4] A number of experiments was done to determine the best plate position and orifice size. Particles were concentrated on a diameter between 1 or 2 mm (Fig. 5). This concentration is sufficient for a detailed chemical analysis.

CRYOGENIC REQUIREMENTS FOR A SPACE MISSION

In an actual space mission it will be necessary to take particle samples in a zone of low particle flux to protect the spacecraft and other payload from being damaged. This necessitates collection on a rather large surface area and long exposure times. A realistic requirement for chemical analysis is a sample of $\simeq$ 100 mg. This represents 10^8 particles of $1\mu m$ diameter with a density of 2 gm/cm^3.   A 0.2 m^2 collecting area and an exposure of 24 h allows collecting in a flux low enough to prevent any damage to other experiments on the spacecraft. A solid Xe thickness of 0.1 mm is adequate to entrap particles of the velocities expected (50–1000 m/s).   This amounts to a Xe requirement of 300 g per sample. The best choice for the material of the collecting plate was Mg, chosen because of its light weight, structural strength, and thermal conductivity. The total weight of the entire experimental package was estimated to be 28 kg of which 4 kg required cryogenic refrigeration.

The chief components of the experimental package include the Xe storage, the Mg collection plate, the cover plate, a mechanism for operating the cover plate, the cryogenic generator, the

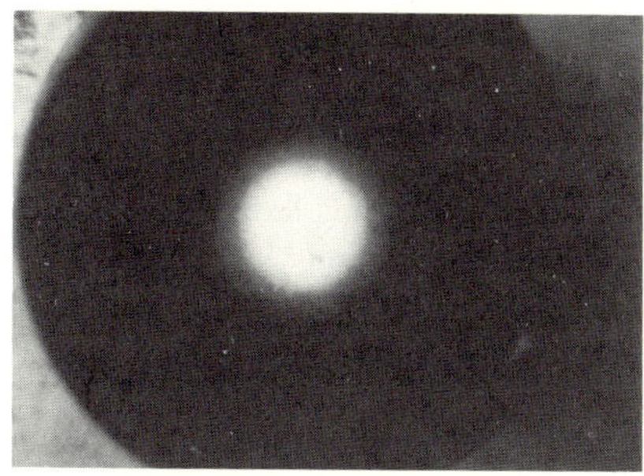

Fig. 5. Concentration of particles.

concentrator, the chemical analysis instrument, and the associated electronics and control circuits. One hundred layers of 6.4 µm crinkled aluminized M was chosen to insulate the collection cell from the anticipated 300 K temperature of the spacecraft.

The proposed nominal sequence of the cometary experiment is:

T - 50 h   - Passive precooling of collector to 110 K
T = 0   - Start Xe film preparation; cool to 65 K
T + 10   - End film preparation; open cover and
        begin collecting
T + 34   - End collection; close cover; begin
        heating collector
T + 40   - Transfer particles to concentrator for
        subsequent chemical analysis

The total amount of refrigeration needed for each complete cycle is 22 Wh. We expect to take advantage of passive radiative cooling into black space to the extent that the Mg substrate, insulation and supports will be at 110 K before active refrigeration is begun. The breakdown of refrigeration requirements is as follows:

Cooling of Xe gas, liquid and solid from 300 to 65 K   2.8 Wh
Cooling of the Mg substrate from 110 to 65 K   8.4 Wh
Cooling of the super insulation from 110 to 65 K   1.2 Wh
Cooling of the structural supports from 110 to 65 K   2.8 Wh
Heatload from spacecraft and comet nucleus during
collection period   7.0 Wh

Three basic methods are being considered for providing the required refrigeration. As in all space ventures, limitations are set on weight, but the usual power limitations will not be imposed on this experiment. To follow a comet requires solar electric propulsion which, in turn, necessitates a solar power supply

capable of 25 kW. This power would not be used for propulsion at
the time of data taking and, is therefore, available for the
experimental apparatus. The first method considered, was an
electrically operated mechanical refrigerator. State of the art
refrigeration equipment is available to fulfill experimental
requirements. The possibility exists that with some development
work a mechanical refrigerator fulfilling the requirement, but of
lesser weight, can be constructed.

The second method considered was a hybrid cryogenic cooler
which required only compressed gas storage, with cooling achieved
by a Joule-Thomson expansion of the compressed gas.

The final method involved only passive cooling by radiation
into black space. The cooling requirement can be achieved by any
of these methods. The problem is to optimize the dissipation of
heat with the smallest amount of mass within the time constraints
of the experiment.

CONCLUSIONS

A microcrystalline solid Xe surface has been determined to be
an excellent noncontaminating receiver for micron-size solid
particles striking the surface at velocities in the range of 50 to
1000 m/s. Determination has also been made that a solid Xe surface at 65 K exposed to vacuum and black space can be kept for
long periods of time (24 hours or more) without significant losses
by sublimation., It is, therefore, deemed as an ideal medium for
the collection of cometary particles on a space mission. The
evaporated Xe gas with entrained particles is also useful in the
concentration of particles for subsequent analysis. State of the
art cryogenic refrigeration is available for repeatedly generating
solid Xe films on a space mission. We, therefore, conclude that
the collection and concentration of cometary particles on a space
mission with a Xe target is feasible.

REFERENCES

1. A. L. Turkevich, et. al., Cryogenic collection of particles by
    solid xenon surfaces, Lunar and Planetary Science, XI:1172.
    (1980).
2. H. Dautet, et. al., A gas jet transport system for the radio-
    active products of fast-neutron induced fission, Nucl.
    Instr. and Meth. 107:49 (1973).
3. I.N. Krupskii, V.G. Manzhelii, Multiphonon interactions and
    the thermal conductivity of crystalline argon, krypton, and
    xenon, Soviet Physics JEPT, 26 (6): 1097 (1969).
4. A.R. MacFarland and H.W. Zeller, A study of a large volume
    impactor for high altitude aerosol collection, U.S. Atomic
    Energy Commission Document TID - 8624 (April 1,1963).

PROGRESS REPORT ON THE INFRARED ASTRONOMICAL SATELLITE CRYOGENIC SYSTEM

A. R. Urbach

*Ball Aerospace Systems Division
Boulder, Colorado*

P. V. Mason

*Jet Propulsion Laboratory
California Institute of Technology
Pasadena, California*

and

W. F. Brooks

*NASA Ames Research Center
Moffett Field, California*

INTRODUCTION

The goal of the Infrared Astronomical Satellite (IRAS) is to conduct an all-sky survey in the infrared bands from 8 to 120 μm, and auxilliary observations from 8 to 300 μm. The satellite consists of a spacecraft and solar panels provided by the Netherlands, and the cryogenic system and telescope supplied by Ball Aerospace Systems Division (BASD) under contract to NASA/Ames Research Center (ARC). Jet Propulsion Laboratory (JPL) has US project management responsibility as well as supplying the detector focal plane assembly for the telescope.

The cryogenic system is required to maintain the telescope optics below 10 K, and the focal plane below 3.5 K. The design lifetime is one year in a 900 km sun-synchronous orbit. Two cryogenic systems are used: (1) a dewar to surround the telescope with 70 kg of superfluid helium, and (2) a cover containing 6 kg of supercritical helium to seal the aperture of the dewar. The cover is ejected in space once the satellite outgassing rate has reached an acceptable level.

The IRAS dewar will be the first large cryogenic system using superfluid helium in zero gravity. Important features described include: the multilayer insulation system; a porous plug to retain liquid in the cryostat; motor valves operating at liquid helium temperatures; and gold foil thermal joints.

The preliminary design of the cryogenic system was described at the 7th International Cryogenic Engineering Conference in 1978. Since that time, the design has been completed, the dewar fabricated and partially tested. The following paragraphs describe the final system and its performance.

CRYOGENIC SYSTEM DESCRIPTION

Figure 1 presents an isometric of the IRAS satellite. The cutaway exposes elements of the cryogenic system as described in the following sections.

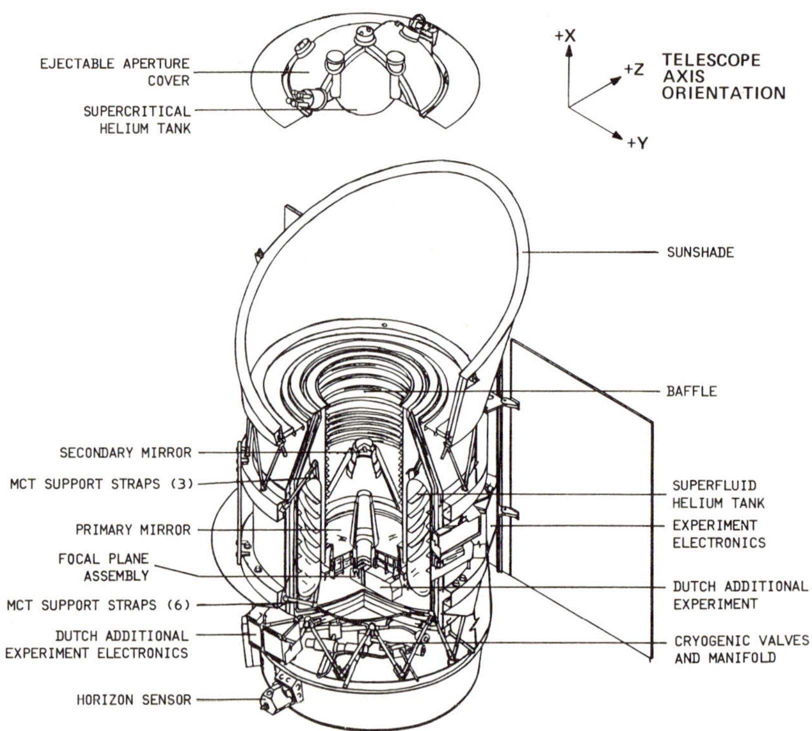

Fig. 1. IRAS telescope subsystem.

Main Cryogen Tank

The main cryogen tank (MCT) is sized to contain 70 kg of superfluid helium with 12% ullage at a temperature of 1.8 K. ·It is annular in shape, constructed of 5083 aluminum alloy, and contains internal stiffening rings. The telescope mounting ring is welded to the surface of the 75 cm cavity and the entire cavity is surrounded by a thin aluminum thermal shroud. The shroud provides a 2 K environment for the experiments and also prevents stray thermal radiation from impinging on the detectors. The main cryogen tank and insulation system are supported at the top by three supports and at the bottom by six supports. Anchored to one of the upper support brackets is a getter cup containing 50 g of charcoal.

Insulation System

The insulation system surrounds the main cryogen tank with four blankets of multilayer insulation (MLI) spaced by three vapor cooled shields (VCS). The multilayer insulation is 6.4 μm double-aluminized Mylar separated by polyester net. A black ring is thermally attached to the outer vapor cooled shield which during normal operation in orbit will radiate approximately 430 mW of heat to space.

Main Shell

The primary structural integrity of the system is provided by the main shell. It serves as a vacuum vessel for the insulation system and also, through the use of two box like girth rings, provides rigidity for the support system. On orbit the main shell is cooled to 170 K by using insulation blankets on one side and second surface paint on the opposite side. The insulation blankets reject the heat on the side facing the earth and the sun while the paint radiates to space thereby reducing the temperature of the outer shell.

Aperture Cover

The aperture cover subassembly, Fig. 2, is the vacuum seal for the main shell during ground operation. It is also a gas condensation trap prior to and during cooldown of the main cryogen tank and it minimizes heat leak to the main cryogenic tank during launch hold. It contains 6 kg of supercritical helium, which allows fourteen days on orbit with a 48 h launch pad hold. After a minimum of four days the cover will be ejected into space and the all-sky survey started. The aperture cover subassembly consists of a spherical tank, two multilayer insulation blankets, three fiberglass support struts, one vapor cooled-shield, two valves, one back pressure regulator, and instrumentation.

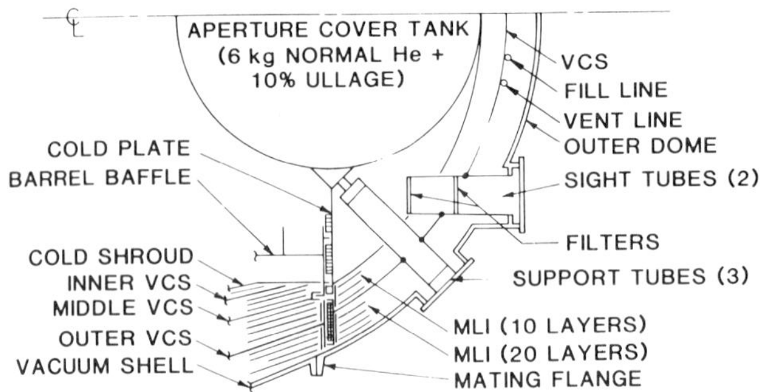

Fig. 2. Aperture cover configuration.

Fluid Management System

The fluid management system consists of the components schematically shown in Fig. 3. The fluid management system includes
three internal valves: V2 is an internal fill valve, V3 is a
cross over valve to permit venting through either the fill or vent
line, and V4 is a porous plug bypass valve used during tank fill
operations. Valves V2 and V4 prevent superfluid helium from
creeping into the plumbing thereby eliminating thermal acoustic
oscillations. A sintered stainless steel porous plug is used to
retain superfluid helium in the tank and two external valves are
used to close off the fill and vent lines. Burst disks protect
the tank and plumbing against rupture and low thrust vents eliminate a disturbing torque during venting. The instrumentation
monitors fluid condition and level. Normal operating pressure of

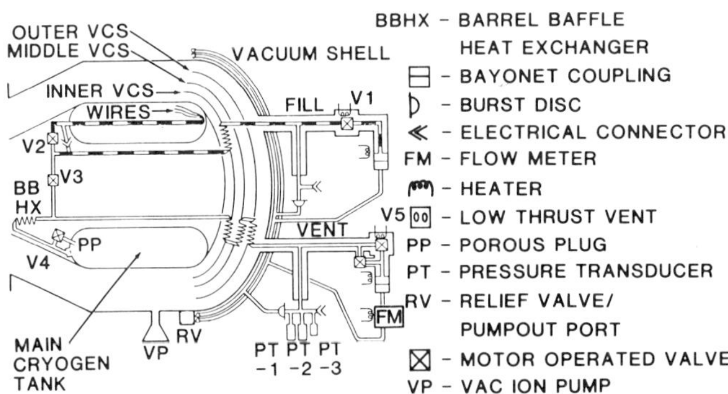

Fig. 3. Main dewar fluid management and component schematic.

the main cryogen tank is < 24 torr, the burst disk pressure differential is 75.8 kPa (11 psi).

KEY COMPONENT DEVELOPMENT

The three most important components used in the IRAS dewar are
the porous plug, the motor operated valves, and the thermal contact joints. These proved to be the most difficult to develop.

Porous Plug

The initial plug selection had to be revised and the original
plug replaced. Filter material specified as 5 μm pore size was in
fact between 25 to 30 μm in diameter. After a series of tests the
plug was resized to one with a pore size of 3.9 μm, a surface area
of 3.5 cm^2, and a thickness of 0.64 cm. Heat load capability is
from 22 mW to 350 mW. This size is satisfactory for all modes of
IRAS operation.

Motor Operated Valves

The motor operated valves development was key to the success
of the IRAS program. Failure of the vent valve V5 (Fig. 3) could
cause failure of the mission or at least a significant reduction
in lifetime. To increase reliability of the total system, a
pyrotechnic valve has been placed in parallel with V5.

The motor operated valves consist of a ball driven by a permanent magnet electric motor/drive train made by Globemotors. The
ball is supported on each side by a bearing and has a cam attached
at the upper end of the drive shaft which contacts position indication switches. Sealing of the ball is accomplished by a KEL-F
ring seal in combination with Teflon on the ball. At room temperature, the seal leakage is less than 1×10^{-6} sccs, at 2 K, the
leakage is less than 1×10^{-1} sccs. Initial design of the valve
was by Flodyne Controls Inc., with the final configuration developed by BASD.

Thermal Interface Contact Joints

The focal plane assembly (FPA) must be maintained at a temperature of 3.5 K or less. The joint designs developed to accomplish
this consist of a copper strap with a 2.54 cm square block on each
end. Annealed gold foil 0.127 mm (0.005 inch) thick provides the
thermal contact on one end of the strap, while 0.076 mm (0.003
inch) indium is used on the other end. Electrical isolation is
accomplished by coating one block with Paraline and bonding the
gold foil to it. Heat conducted across this strap was 50 mW
during development tests. A second series of metal to metal
thermal joints are made between the optics/FPA assembly and the

dewar mounting ring. These joints are twelve 2.54 cm square
pieces of 0.127 mm (0.005 inch) annealed gold foil mounted between
the two aluminum rings. Loading is applied by torquing a bolt at
each joint to approximately 0.7 MPa (100 psi).

PERFORMANCE

To date (August, 1981) IRAS dewar has been cooled to helium
temperature twice. During the first cold cycle a mass model was
substituted for the experiment and optical subassembly. The dewar
was subjected to a series of vibration tests along three axes at
superfluid temperature with a full 70 kg cryogen load in the
dewar. The dewar and cover thermal performance was not affected
by the shake test. The second cold cycle is still underway with
the IRAS Satellite in the Netherlands.

Vacuum Acquisition

The initial vacuum acquisition lasted 18 d. The system was
backfilled with dry nitrogren and roughed out with a large capacity zeolite pump. The second state pump was mechanically
refrigerated, activated charcoal sorption pump with a rated speed
of 1000 L/S at 10^{-6} torr. These pumps were utilized to avoid
exposing the optics to hydrocarbon oils. The pumpdown for the
second cooldown lasted 14 d, the system, however, contained two
new components which increased the outgassing. The first was the
50g charcoal trap and the second was the black anodized barrel
baffle. Figure 4 is a plot of the pressure vs. time during both
pumpdowns. The first pumpdown which was baked at $40^{\circ}C$ reached a
pressure almost four times less than the second pumpdown. The
spikes on each curve represents the static pressure after 12 min
with the vacuum station valved off. During the first cold cycle
the lowest pressure was approximately 2×10^{-7} torr. During the
second cold cycle with the activated charcoal, a pressure of ~1 x
10^{-8} torr has been achieved.

Cryogenic Fill

Cooldown and fill of the IRAS cryogenic system requires 48
h. The dewar can be cooled and filled in approximately 10 h but
the telescope and detector system have a longer cooldown due to
joint conductance limitations. The gold foil thermal joints
between the telescope interface ring and the dewar mounting ring
have a conductance of approximately 8.0 W/K at 150-300 K. This
conductance drops to approximately 1.5 W/K at 20 K and below.
Cooldown flow rate is approximately 50 L/h with the actual fill
being approximately at 200 L/h. After the dewar is filled with
normal LHe at 4.2 K, it is pumped to superfluid at 15 torr pressure and 1.8 K. Mass loss is approximately 30%. At the same time

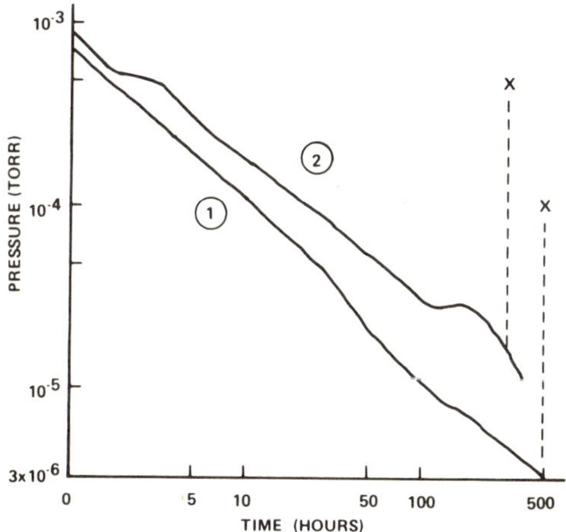

Fig. 4. Dewar annulus pressure.

the 1000 L supply dewar is pumped to a pressure of 35 torr or approximately 2.1 K.

The internal fill and vent valves V2 and V4 (Fig. 3) are closed and V3 opened. This permits cooldown of the transfer lines and fill plumbing without introducing this warm gas into the stored liquid helium. Cooldown of the lines and plumbing takes approximately 30 min at which time topoff is initiated. Topoff to 100% of volume is easily achieved in 40 min using a transfer pressure of 150 torr on the subcooled supply dewar. Pumpdown to 15 torr takes nine hours and results in 73 kg of stored superfluid helium; the launch requirement is 70 kg.

Dewar Thermal Performance

During the first cold cycle, it was noted that the porous plug sensors at the top of the main cryogen tank were wetted with superfluid until the bulk liquid fell to the 60% level. During the second cold cycle, after the cover had been modified and the telescope installed, the porous plug remained wetted until the bulk reached the 37% level. The critical heat flow of unsaturated films follows the relationshp[2]:

$$Q_c = k(L_v + TS) \rho_s \rho \, PV_c / H^{0.435} \qquad (1)$$

where
L_v = latent heat; T = temperature; S = entropy; ρ_s = density of superfluid component
P = perimeter of tank; V_c = critical velocity; k = constant
H = height of film; ρ = total helium density

The difference in heights implied a 20% reduction in tank top heat during the second run. Using the appropriate values in Eq. (1) yields an estimated tank top heat of 35 mW.

During the second cold cycle the dewar was undisturbed for several days with the experiment power shut off. The equilibrium temperatures are shown in Fig. 5. The calibrated wet test meter reading of 6 mg/sec with the power off to the FPA is within 5% of the 0.7%/day indicated liquid level loss. With the tank topped to 88% or 70 kg we can expect 125 d ground hold for the main cryogen tank. Extrapolation of the observed ground behavior to flight conditions yields an anticipated lifetime of 320 d.[3] The individual components of the parasitic heat loads are also given in Fig. 5.

The focal plane power dissipation was 13.5 mW. The observed increase in flow with the power on was approximately 0.6 mg/sec. With power on the experiment the focal plane sensor indicated 3.05 K with a bath temperature of 1.61 K. This implies a conductance from the focal plane modules to the tank of 9.5 mW/K. This is significantly below the 50 mW/K measured for two copper straps in

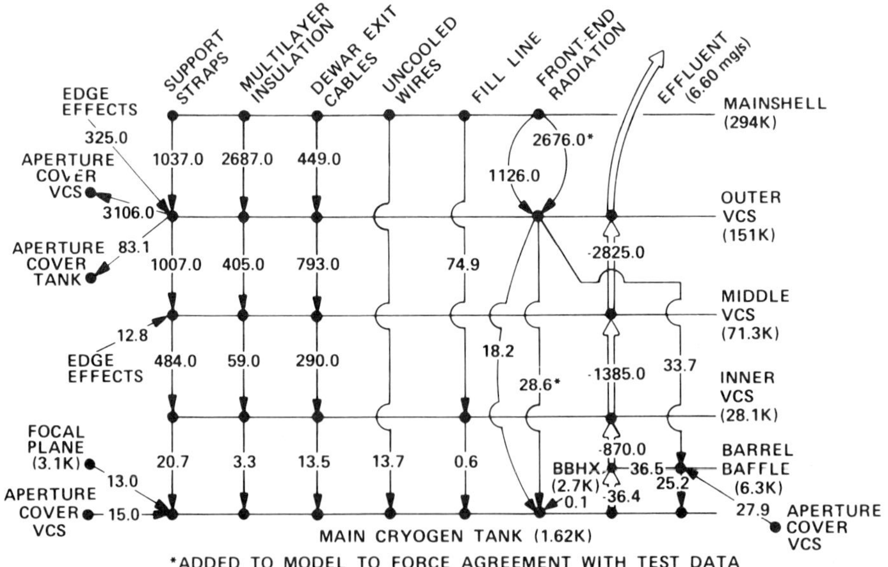

Fig. 5. Equilibrium temperatures and predicted heat flows (mW) during ground testing.

an offline test.   A development test of the new focal plane indictes a conductance of 26 mW/K.

Aperture Cover Thermal Performance

On the first cold cycle the thermal performance of the aperture cover fell short of the predicted five day lifetime.   The cover was modified and retested.   As a result of the cover modifications the differential temperature across the cold plate is 2 K.   Thermal acoustic oscillations in the fill line have been eliminated, and the ground lifetime extended from one day to four days.   On orbit lifetime is expected to be 14 d.

IRAS CRYOGENIC SYSTEM STATUS

The IRAS telescope was shipped cold to the Netherlands in May, 1981 for integration with the spacecraft and additional performance tests.   After several weeks there the vacuum annulus pressure level had dropped to $\sim 1 \times 10^{-8}$ torr.   Concurrently, the measured gas flow is now lower, approximately 4.5 mg/sec.   This indicates that the MLI had a larger gas conduction term than was used in the thermal model.   With this lower value the lifetime on the ground is extended to approximately 150 d.   In September, 1981 the satellite will be returned to JPL for solar thermal vacuum and structural tests.   Launch is scheduled for August 1982 from Vandenburg, California.

REFERENCES

1. A.R. Urbach, J. Vorreiter and P. Mason, Design of a Super-
        fluid Helium Dewar for the IRAS Telescope in " Proc. 7th
        Int'l. Cryo. Engr. Conf."   IPC Science and Technology Press
        Limited, (1978).
2. J. Wilks, "The Properties of Liquid and Solid Helium",
        Clarendon Press, (1967).
3. R.A. Hopkins and W.F. Brooks, Orbital Performance of a One-
        Year Lifetime Superfluid Helium Dewar Based on Ground
        Testing and Computer Modelling, in "Advances in Cryogenic
        Engineering, Vol. 27", Plenum Press (1982).

PERFORMANCE OF A SUPERFLUID HELIUM FACILITY
FOR SPACELAB PAYLOADS

J. Lizon-Tati

European Space Agency
Noordwijk, The Netherlands

M. Girard and A. Tardivo

S.N.I. Aerospatiale
Cannes, France

and

P. Delacour and M. Till

L'Air Liquide
Sassenage, France

INTRODUCTION

Open-cycle helium cryostats, a well proven technology for terrestrial applications, are the most suitable systems for cooling to the 2 Kelvin region space experiment payloads with low-heat dissipation and short mission duration. This is applicable to most of the infra-red astronomy, atmospheric physics and fluid physics experiments proposed as future missions on Spacelab flights. It is with this application in view that the development of a superfluid helium cryostat has been undertaken by the European Space Agency.

Similarities between the proposed scientific experiments, in terms of geometrical, mechanical and thermal requirements have resulted in the definition of a standard cryostat to be proposed as a Spacelab facility. The following basic requirements have been defined: 1) provision to users, of an operating temperature, in orbit, lower than 2 K for a mission up to 30 days in length and for an average heat load of 40 mW, 2) ability to accommodate a medium-sized experiment, for example, an internally mounted telescope of 20 cm diameter and 55 cm in length, while maintaining the flexibility for adaptation to possible other passenger payloads,

(3) compliance with the Spacelab/Shuttle launch environment and the reusability (10 flights) constraints.

PROGRAM PLAN

The CRHESUS (Cryostat à Hélium Superfluide pour Utilisation Spatiale) program was started in 1978. Two development models, a Structural Model (SM) and a Thermal/Engineering Model (TM), have been built and tested. The last phase of the program, dedicated to the manufacture and testing of the protoflight unit, which will be delivered to ESA mid-1982 is currently in progress. A qualification flight of this unit on Spacelab, will provide the opportunity to incorporate experimental payload for a scientific mission.

PROTOFLIGHT SYSTEM

A schematic of the Protoflight cryostat is given in Fig. 1. The overall dimensions are 75 cm diameter and 123 cm length. The total mass of the facility is about 118 kg at launch including 8.5 kg of helium but excluding the mass of the passenger experiment.

The external vacuum-tight vessel built of 5056 aluminum alloy, is composed of a cylindrical central shell joined to two identical domes by flanges.

The protoflight unit is configured as a closed cryostat with flanges on the top of the upper dome for mounting an optical window, lid or viewing port system for experiments located inside the cryostat and viewing outwards.

The helium management system, radiation shields and insulation, are suspended on the central part of the outer vessel, enabling the substitution of the present upper dome with any other cover specially designed for the passenger experiment. This configuration provides maximum flexibility with respect to experimental accommodations.

The helium tank assembly is composed of two toroidal stainless steel reservoirs, containing 58 and 23 litres of helium. The tank wall provides the 2 K environment to the experimental cavity within the torus and on the top of the main reservoir (MR). The cryogenic components are located below the auxiliary reservoir (AR).

The thermal protection surrounding the tank and the experimental cavity is composed of two vapour cooled shields, constructed of high conductivity aluminium alloy. Each shield is wrapped in a multilayer insulation blanket (MLI) consisting of aluminized Mylar and "bolloré", a cellulose acetate material.

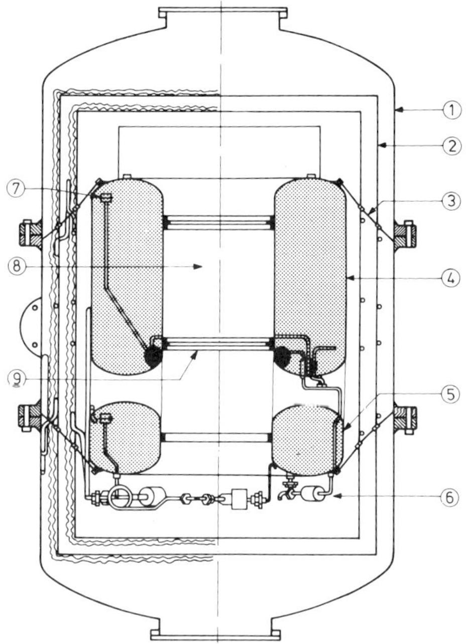

Fig. 1. CRHESUS protoflight configuration, 1) external
vacuum vessel, 2) vapour cooled shields, 3) support struts,
4) main reservoir (MR), 5) auxiliary reservoir (AR), 6)
valves and cryogenic components, 7) porous plugs, 8) experiment dedicated cavity, 9) experiment mounting flanges.

The helium tank assembly and the thermal protection system
are supported by a network of 24 fiberglass struts (Fig. 2). Each
strut is composed of three elements connected together via an
aluminum flange included in each shield. The tank assembly is
clamped between two flanges to which the strut tension devices are
connected.

HELIUM MANAGEMENT SYSTEM AND OPERATIONS

The driving aspects in designing the helium management system
were the in-orbit phase separation and the ground operations
constraints of the Space Shuttle.

As opposed to normal helium (HeI) and other cryogenic fluids
for which phase separation is achieved by sophisticated devices[1],
a specific property of superfluid helium (HeII), the fountain
effect, which takes place in porous media or narrow channels, can

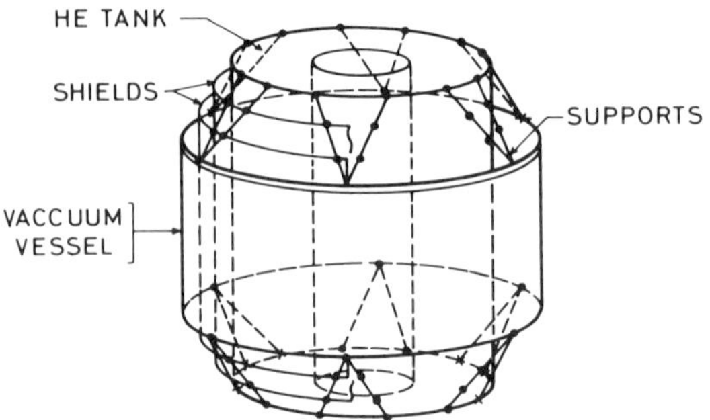

Fig. 2.   Helium tank support schematic.

be used to solve the problem of phase separation in a zero-g
environment[2,3] for this fluid.

It was considered desirable to launch the cryostat with the
helium already in a superfluid state since this resulted in reduced complexity, mass savings, and also prevented any liquid escap-
ing before HeI/HeII conversion. The adverse aspect of this choice
is that maintaining a helium bath in the superfluid state i.e.
below 2.17 K (λ point), implies keeping, by pumping, the pressure
above the helium bath below 50 mbar (Fig. 3). This requirement is
very difficult to reconcile with Space Shuttle ground operations,
as the latest possible access to the cryostat for filling and top-
off occurs 10 days before lift-off. The access at that time is
very limited and makes proper utilization of the ground support
equipment for pumping very difficult. In addition the electric
power supply of the Orbiter is not available continuously to
operate an on-board vacuum pump (which could have been included in
the flight package). Nevertheless this last approach is considered as the baseline for cryogenic experiments on the Spacelab II
mission.

The concept of a "Cryogenic Facility" selected for CRHESUS,
which implies a simplification of the cryostat operation and
requires a system as automatic as possible, has led to a different
approach with a strong impact on the helium management system.

The selected solution is based on the Roubeau–Claudet double
bath principle[4] and consists of two helium reservoirs: an auxiliary reservoir filled with normal helium and kept at normal
atmospheric pressure on the ground (T $\simeq$ 4.2 K) and a main reservoir, filled with superfluid helium kept at a temperature below
the λ point but also at normal atmospheric pressure.

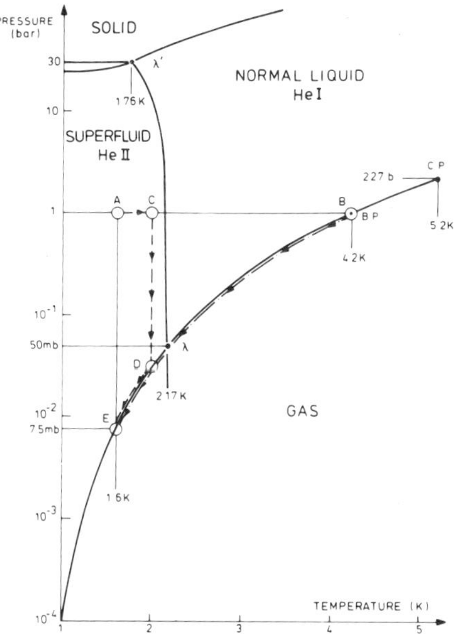

Fig. 3. Cryostat set point on helium phase diagram. Main
reservoir set point: A, end of filling – A to C, on-ground
standby phase – C to E, launch phase. Auxiliary reservoir
set point: B, end of filling and on ground phase – B to E,
launch phase.

The helium management system of the protoflight cryostat,
illustrated in Fig. 4, includes modifications resulting from the
tests carried out with the Thermal/Engineering Model. The auxiliary reservoir, AR, and the main reservoir, MR, are filled
initially with HeI through fill valve V3, communication valve V2
and the experiment cooling line. Then the MR bath subcooling and
HeI/HeII conversion is started by pumping the AR bath via the
Joule-Thomson expansion valve V1 and the evaporator included in
MR. Topping of AR and HeII conversion in MR are simultaneous.
During the standby phase, the vapour from AR (HeI bath at 4.2 K
and 1 Atm) is used for cooling of the radiation shields and for
limiting the MR (HeII initially at 1.6 K and 1 Atm) to a temperature lower than the λ point (< 2.17 K).

Figure 3 gives the set-point of each reservoir on the helium
phase diagram during the various sequences. While in orbit both
reservoirs are connected together via V2. Vent lines on each
reservoir are fitted with porous plugs for a proper phase separation in orbit.

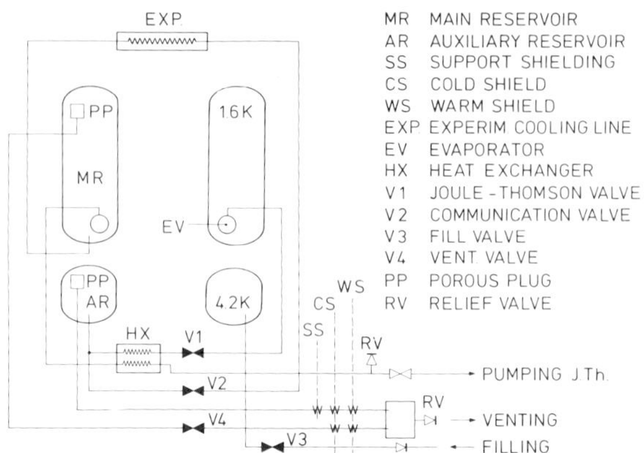

Fig. 4. Helium management system schematic.

All the cold valves are operated by helium gas pressure during filling and subcooling and the whole system is self-contained in orbit.

TESTS ON CRYOSTAT DEVELOPMENT MODELS

Mechanical and thermal tests have been performed on two full-scale models: a Structural Model (SM) has been used for mechanical testing at room temperature and a Thermal/Engineering model (TM), illustrated in Fig. 5, for functional, thermal and vibration tests at liquid helium temperature.

Fig. 5. View of thermal engineering model.

Results of the SM test program which included static load, modal survey and vibration tests have been described in a previous paper[5], and have been considered as satisfactory.

The tests on this Thermal Engineering model including functional and thermal performance tests, were the main activities of the second phase of the program.

The functional aspect was related to the verification of the proper operation of the helium management system during the simulation of each phase of cryostat separation: evacuation, filling, sub-cooling standby, Orbiter erection, launch including vibrations, orbital phase and return to Earth. The thermal aspect was limited to the verification of the heat balance during standby and on-orbit phases, correlation with thermal analysis and derivation of the ground and orbital lifetime.

Most of these objectives have now been achieved after a period of tests covering about six months and involving a continuous adaptation of the test procedure, instrumentation and even the test specimen itself. The various improvements implemented following each test resulted in satisfactory results for both the functional and thermal performance aspects. The main outcome of these tests is summarized below :

- the pre-cooling and helium transfer is achieved very easily and within 8 hours. The helium consumption is less than 150 L during this procedure.

- the minimum temperature obtained after helium conversion using the Joule-Thomson operation is 1.62 K. The duration of this phase is less than 36 hours.

- the maximum duration of the on-ground hold time (standby phase) is 8 days with the present TM configuration and tank volume, instead of 10 days as originally specified.

- however, with the same TM configuration, the lifetime during the orbital phase at a temperature below 1.7 K is 24 days, instead of 20 days requested.

- the structural behavior of the cryostat during vibration tests at liquid helium temperature (Fig. 6) is satisfactory and no degradation in thermal performance has been observed.

CONCLUSIONS AND FURTHER ACTIVITIES

The series of cryogenic tests performed on the thermal model has fully confirmed the viability of the proposed design. Most of the major critical aspects encountered during these tests are now

Fig. 6 Vibrations tests at low temperature
on thermal engineering model.

under control and the protoflight configuration reflects the
experience gained. Potential solutions to further improve the
design have been identified. Compliance with the required standby
phase has been tackled by two different approaches: on one hand
increasing the thermal inertia of the main reservoir by a 12%
increase of its volume and on the other hand by reduction of the
heat leak into this same reservoir. This second aspect, which is
the most critical, has been implemented by modification of the
internal layout in order to improve the multilayer insulation
efficiency, by conductively coupling the tank supports to the vent
pipes and also by improvement of the valve design.

Performance improvement brought by these modifications will
be assessed during a series of tests to be performed on the protoflight unit. These results are expected to be available by mid-
1982.

REFERENCES

1. R. Mitchel, et al., Study of zero gravity vapour/liquid
    separation, NASA – CR 71624 (1966).
2. D. Petrac and P. Mason, Evaluation of Porous Plug Liquid-
    Vapour Separators for Space HeII Systems, in "Proc. 7th
    Intl Cryo. Engr. Conf.," IPC Science and Technology Press,
    Guildford (1978), p. 120.
3. H. Denner, et al., Performance of an Active Phase Separator
    for Helium, in "Proc. 8th Intl. Cryo. Engr. Conf," IPC
    Science and Technology Press, Guildford (1980), p. 32.

4. G. Bon Mardion, et al., Helium II in Low Temperature and
    Superconductive Magnet Engineering, in "Advances in
    Cryogenic Engineering, Vol 23," Plenum Press, New York
    (1977).
5. J. Lizon-Tati, Design and Development Test of CRHESUS, in
    "Proc. 8th Intl Cryo. Engr. Conf.," IPC Science and
    Technology Press, Guildford, (1980).

INTEGRATING AND TESTING THE THERMAL MODEL
OF THE GERMAN INFRARED LABORATORY (GIRL)

A. Seidel, M. Mecker, J. Carl, and H. Lieb

Messerschmitt-Boelkow-Blohm GmbH
Ottobrunn, Federal Republic of Germany

SCOPE OF THE THERMAL MODEL

Since the cooling and thermal stability requirements for the scientific experiments and the telescope are very stringent, a thermal model (TM) of GIRL was built and tested prior to design of the flight model. The TM simulates the active part of GIRL i.e., all the He operations. The goals of the TM are:

o To give evidence that the concept basically meets requirements,
o To show that experiments and telescope can be adequately cooled,
o To demonstrate operational feasibility,
o To provide data for updating the mathematical thermal model
o To provide a solid basis for design of the qualification/flight model.

The scientific goal and basic design of GIRL have been described earlier.

DESCRIPTION OF THE THERMAL MODEL

He System and Vapor Cooled Shields

Like the flight model, the TM contains a toroidal cylinder 300 L HeII tank made of AlMg4.5Mn-alloy, equipped with a massive flange for carrying the instrument platform (IPFM) on one side and the inner stray light baffle on the upper side. All the valves and tubing for filling and venting the tank as well as a spring loaded safety valve and two burst discs are mounted to the upper tank side. The valves are flat seat types spring loaded closed and opened from outside the TM, avoiding a cold actuator. The

HeII-tank is also equipped with a laboratory model of the "Active Phase Separator (APS)" developed by G. Klipping et al.[3]

Besides these active He components the TM is equipped with 3 vapor cooled shields, 2 of which are superinsulated, and the inner baffle cooler. Due to the length of the inner baffle only its cooler, equipped with a heater simulating the baffle heat loads was installed. It is located on a glass-fiber composite (GFC) cylinder which insulates it from the tank. For simulation of the baffle radiation shields, not present in the TM, the vapor cooled shields are heated at the interfaces with the baffle shields. For optimum leak tightness, all connections of the piping between tank and exit of the TM are welded with one exception at the APS interface. The shields are cooled by finned aluminum tubes of 15 mm i.d. and 1 mm wall thickness which are spot welded to the shields. The tubing between the shields is stainless steel of the same i.d. but with a wall thickness of 0.5 mm to minimize conduction. Diffusion bonded metal transition joints are used between the aluminum and steel sections. Figure 1 shows the upper tank side with part of the tubing and shields after installation of the cylindrical shields and the superinsulation.

Instrument Platform with Primary Mirror and Experiment Models

The TM is equipped with an IPFM made of AlMgSi to carry thermal models of four scientific experiments on the underside and the primary telescope mirror on the upper side. Thermal models of two "optical beam guides" (OBG) which direct the IR-beam to the experiments are also mounted on the IPFM. For maximum thermal contact with the tank, the IPFM is attached to the tank flange by 40 Cu/Ni screws of 8 mm diameter.

Fig. 1. Upper He tank side after shield integration.

The thermal models of the experiments and the OBGs are equipped with heaters which simulate the waste heat of active elements (e.g. filter wheel drives, choppers, laser diodes, motors) within the experiments. Since the secondary telescope mirror and its chopper are not present in the TM their heat load to the bath is also electrically simulated. Critical elements of the experiments, e.g. locations of the IR-detectors, are thermally insulated inside the experiments and connected to the tank by high purity copper straps. In order to cool the "zerodur" primary mirror, 42 points on its backside are connected by copper braid and straps with the rim of the IPFM. The braids are partly brazed or glued to the mirror. All elements are mounted and aligned to the IPFM prior to its integration into the TM. After attachment of the IPFM to the tank flange the copper straps are mounted to the tank wall. Figure 2 shows the integration of the fully equipped IPFM into the TM; the primary mirror and the thermal models of the experiments can be clearly identified. Figure 3 shows the inner space of the HeII tank after IPFM integration; the Cu straps attached to the tank can be seen.

Electrical Cables

Many (927) cables connecting the experiments with the cryostat outside are required. Many of the cables must be coax types and some are power cables for up to 1 A. Cables for temperature

Fig. 2. Integration of fully equipped
instrument platform into the TM.

measurement of the insulation system, the IPFM, the telescope and
the experiments were added. A total of 1145 cables run from the
outer vessel to the HeII tank. In order to minimize the heat load
by conduction through the cables, a compromise between cable
length, cross section, electrical and thermal conductivity had to
be found. Except for the power cables, all the cables were made
of Manganin with an average wire diameter of 0.1 mm.

The cables mated with connectors at the tank interface and
with vacuum feed-throughs at the outer vessel. To provide rigidity the cables were bundled and fixed to a GFC structure. The
cable trees were manufactured outside the TM and then integrated
into it. In order to intercept the incoming heat, the cable trees
are thermally connected for a short distance to the outer and the
inner vapor-cooled shields. The cable trees are spirally wound
from the outer vessel to a ring carrying the connectors at the
HeII tank (see Fig. 3), resulting in an average length of 7 m.

TESTING THE THERMAL MODEL

Helium and Insulation Systems

After closing the outer vessel of the TM the insulating vacuum
was established with two turbo-molecular pumps. At ambient temperature a vacuum of 3×10^{-5} mbar was achieved after 40 h. With
this vacuum a final leak check of the He-system was conducted
showing a total He leak rate of $< 2 \times 10^{-7}$ mbar L/sec. In order to
prove that no unacceptable outer leak was present, the two pumps
were shut off and the pressure increase was monitored over 10
days, reaching an equilibrium value of 1×10^{-2} mbar. This
pressure could then be pumped down to 5×10^{-6} mbar in only 20
h. After connection of the TM to the He-supply and the
measurement systems, cryogenic testing of the TM was initiated.

To prevent excess thermal stresses in the "zerodur" primary
mirror a minimum cooling time of 25 h was required. A maximum
temperature difference between two extreme points of the mirror
should not exceed 15 K during cooling. During the first cool down
this temperature difference was not reached and the tank and the
IPFM were cooled down much quicker than the mirror. Figure 4
gives the real temperature curves for the tank interior, the IPFM
and the primary mirror. This showed that filling with normal He
could be initiated after 10 h without affecting mirror cooling.
The final mirror temperature is reached after about 65 h. Due to
the remaining temperature difference between the tank and the
mirror, the tank must be refilled during this period. Total LHe
consumption for cooling down and filling the tank with HeI was
about 1750 L. The insulating vacuum was then 2 to 3×10^{-8} mbar.

Fig. 3. View to lower tank side, experiments and cable tree.

After cooling and filling with HeI, the production of HeII was started by reducing the tank pressure to 7 mbar with a vacuum Roots pump. Starting from 95 % full tank, the required HeII temperature of 1.6 K was reached with a 55 % residual. Refilling

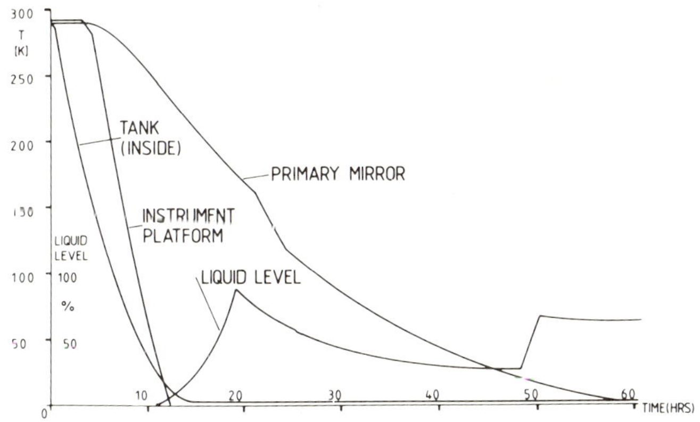

Fig. 4. Cool down of tank, IPFM and telescop mirror.

of the tank with HeII was then repeatedly achieved by throttling HeI from the LHe storage tank by a valve within the transfer line. This throttling cannot be done to the final pressure and therefore the pressure and temperature in the tank increases when refilling starts. Both, are then gradually reduced again by a process of pumping down from 100 % to about 95 % filling, followed by a new refilling to 100 %. By that means a final filling of 95 % can be achieved at 1.6 K in 10 hours.

After production of HeII in the tank the cryostat was tested for different heat load cases. Since theoretical calculation of the heat loads on the radiation shields may be uncorrect, not only nominal values were tested but also ± 50 % levels.

Based on the results of tests without shield heating a preliminary adjustment of the mathematical model was obtained for comparison of calculated and measured temperatures and assessment of the contribution of different elements to the tank heat load. As Table I shows, the heat load by the cables is by far the highest contribution.

Active Phase Separator (APS)

Prior to installation into the TM the APS was tested under laboratory conditions at the Free University of Berlin. Testing under real conditions in the TM proved that the APS was able to

Table I.   Contribution of components to tank heat load; temperatures and He flow rate.

	Cable Conduct.	GFC straps	Valve Actuat.	Baffle	Radiation	Tubes & filling	Phase Separ.
Calculated tank heat load (mW)	172.3	78.9	6.6	8.9	100.0	37.0	10.0
Contrib. (%)	41.7	19.1	1.6	2.1	24.2	8.9	2.4

Mean temp.:	calculated	measured
inner shield	66.4 K	62 K
middle shield	160.2 K	123 K
outer shield	234.1 K	209 K
outer vessel	290 K	294 K
He-flow rate $\left[\frac{mg}{s}\right]$	18.3	17.5

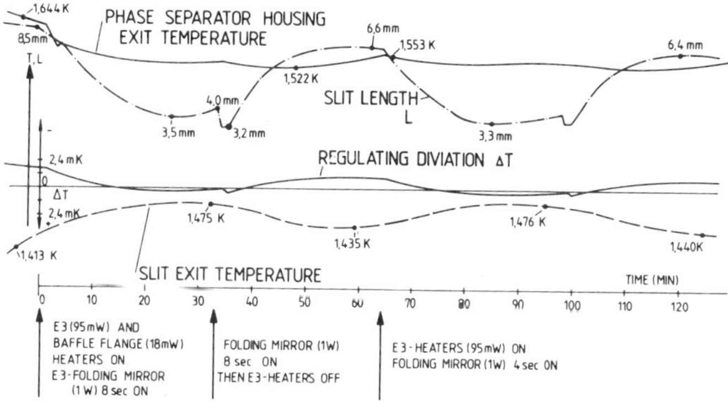

Fig. 5. Active phase separator regulating diagram (typical).

stabilize the HeII bath temperature to 1.6 K or slightly below within ± 3 mK or less for all simulated experiment operations and test runs for astronomical mission cases. Figure 5 shows the regulating diagram of such an experiment simulation with the Ebert-Fastie-Spectrometer (E3). Since the ordinate is not linear, only extreme values are given. For operation of E3 in orbit, the baffle needs much more He for adequate cooling. Therefore the HeII bath must be first heated to about 1.9 K to reach a new range where the APS can again stabilize the bath temperature.

Simulated Scientific Experiments Operation

For testing of the thermal behavior of the models of the scientific experiments, the following types of simulated operations were conducted:

o Steady state runs at average single experiment heat loads,
o Simplified time dependent experiment operations with varying or pulsed heat loads on single experiments,
o Simulated experiment operation during two orbits with all experiments.

During this test series the influence of the two OBGs and the secondary telescope mirror (S2) chopper was also evaluated. The goal of these tests was the determination of equilibrium or transient maximum temperatures at different critical points in or at the experiments and the times required for achieving equilibrium. Although the experiments proved to have fairly short periods of unacceptably high temperature gradients, the insulation of critical areas within the experiments and their coupling to the HeII tank wall must be improved in some cases.

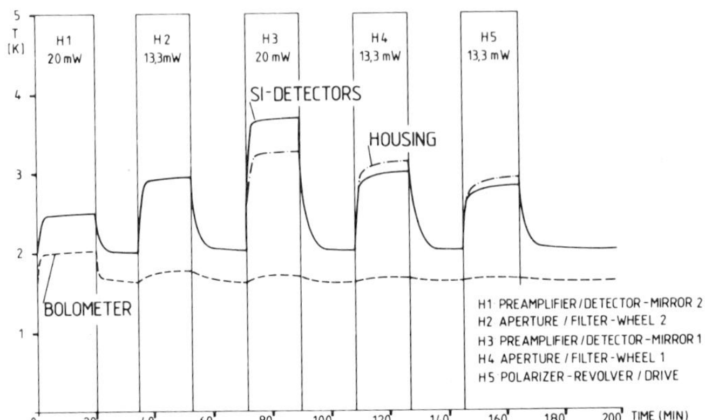

Fig. 6. Photometer/polarimeter temperature-history.

Typical test results

Figure 6 shows the temperature history of the Photometer/ Polarimeter (E2)-thermal model during interrupted operation of the different aperture/filter wheel drives until new equilibrium is reached. The critical bolometer is only strongly affected when the preamplifier/detector-mirror in its direct neighborhood is operated.

The temperature history of the Ebert-Fastie-Spectrometer (E3)-thermal model during a simulated astronomical measurement cycle is shown in Fig. 7. Equilibrium temperature at the detectors is reached in about 3 min. The temperature of the transducers measuring the diffraction gate movement need considerable time for equilibrium but does not influence the detectors.

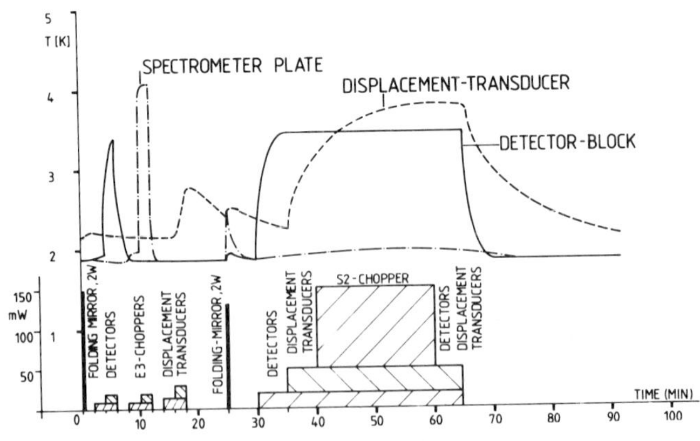

Fig. 7. Simulated Ebert-Fastie-spectrometer operation.

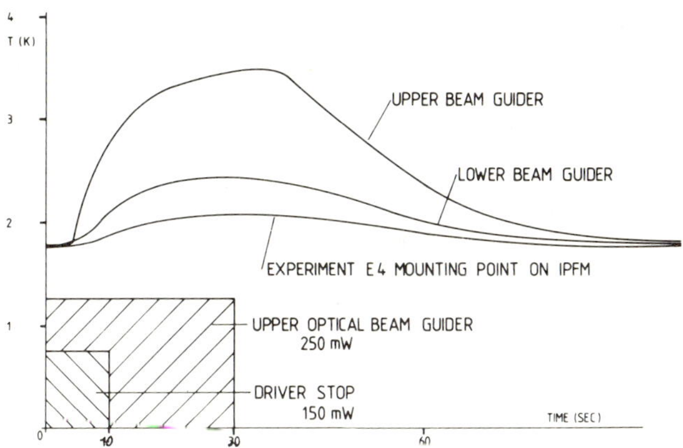

Fig. 8. Optical beam guider influence on instrument platform.

Figure 8 indicates the influence of an OBG on the thermal behavior of the IPFM and the experiments. The OBGs are driven by tumbling disc motors which are active for a maximum of 1 min each, producing losses of about 250 mW. It can be seen that experiments are affected for only 2 min.

Figure 9 shows the total HeII bath heat load caused by a simulated measurement program of all four experiments and the operation of the S2-chopper and the OBGs during two orbits. The He flow rate controlled by the APS follows the heat load curve with some delay due to the thermal inertia of the system.

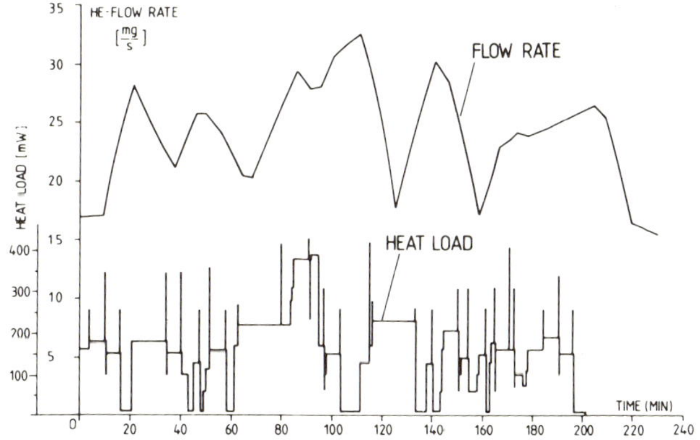

Fig. 9. Simulated measurement program over two orbits.

SUMMARY

Cooling of the thermal model to He temperatures, production of HeII, refilling of HeII and simulated scientific experiments operation could be done without major problems. Although thermal equilibrium of the experiments can be achieved in reasonable time, local cooling has still to be improved. Tests show that an on-orbit lifetime of 13 days can be expected without improvement of insulaton. Evaluation of the tests indicate that 42 % of the HeII heat load is caused by the electrical cables. The active phase separator proved to stabilize the bath temperature within a few milli-Kelvin at varying heat loads.

REFERENCES

1. D. Lemke et al., The German Infrared Laboratory GIRL, in "Proc. Conf. on Optical and Infrared Telescopes for the 1990's", A. Hewitt, editor, Kitt Peak National Observatory, Tucson, USA (1980).
2. A. Seidel, The German Infrared Laboratory (GIRL), It's Overall Concept and Cooling System in "Proc. 8th Int'l. Cryo. Engr. Conf.", IPC Science & Technology Press, Guildford (1980).
3. G. Klipping et al., Performance of an Active Phase Separator for Helium, ibid.

SUPERFLUID-SUPERCRITICAL HELIUM TRADEOFF ANALYSIS FOR THE SHUTTLE INFRARED TELESCOPE FACILITY (SIRTF)*

H. L. Gier

Beech Aircraft Corp.
Boulder, Colorado

R. Stoll

Perkin-Elmer Corp.
Danbury, Connecticut

and

W. F. Brooks

NASA Ames Research Center
Moffett Field, California

INTRODUCTION

The NASA Shuttle Infrared Telescope Facility (SIRTF) will fly missions of two weeks or longer starting in 1987. Each mission will provide experimenters with as many as six separate experimental arrangements. A feasibility study[1] and a design optimization study[2] have been performed to define the capabilities of the SIRTF. This paper is a partial report of work done during the design optimization study.

There are many advantages of basing infrared telescopes in orbit above the earth's atmosphere, including, reduction in sky brightness to the zodiacal level. However, to reduce the thermal noise to a similar level it is necessary to cool the entire

---

*Work done under NASA Contract No. NAS2-10066, Ames Research Center.

telescope. For wavelengths greater than 50 m the telescope must
be cooled to below 20 K.

The SIRTF telescope (Fig. 1) will be mounted on a two-pallet
train in the Space Shuttle and is designed to be a general purpose
infrared telescope. It will be fabricated primarily of aluminum
with the outer vacuum housing as the principal structural member.

Design of the aperture shade is determined by the requirement
that viewing within 45° of sun, moon, or earth limb be possible
without specularly reflecting energy into the telescope aperture. The energy into the aperture is thus diffuse or infrared
radiation from the inner shade surface. Pre-launch particle
contamination which might degrade the shade performance is
prevented by the dustcover.

The SIRTF cryogenic system exists to provide the operating
conditions for the scientific instruments in the Multiple
Instrument Chamber (MIC). These operating conditions have been
defined as: (1) a temperature base of approximately 2 K shall be
available in the MIC; (2) the MIC case shall be uniformly
maintained at near reservoir temperature, but no higher than 10 K,
despite the presence of heat sources (the Charge Coupled Device
star tracker) as high as 100 K; (3) the mirror and associated
focal stops shall be maintained at approximately 10 K; and (4) the
baffles shall not be at a sufficiently high temperature to create
noise or off-axis rejection problems. All other cryogenic factors

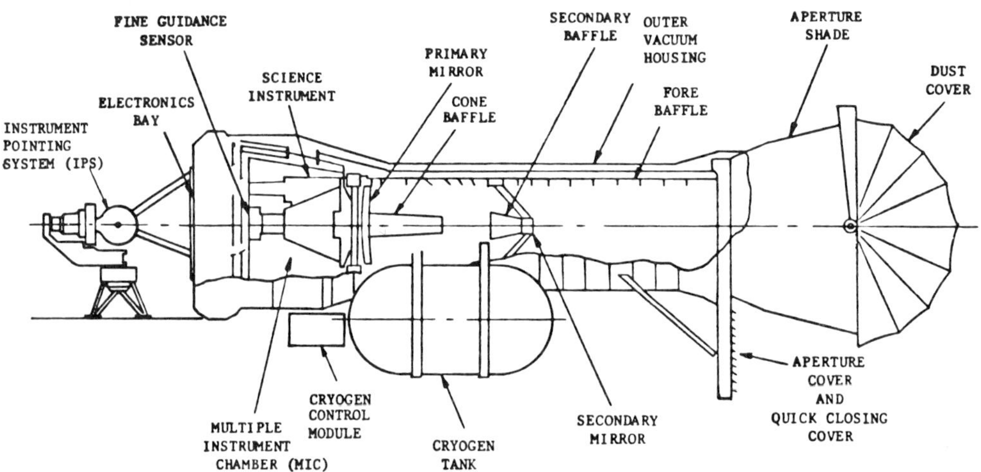

Fig. 1. SIRTF Components.

are dependent on maintaining these four operating conditions in the Shuttle environment.

This paper compares the methods to obtain the required cooling of the SIRTF and recommends one method for implementation. Three different systems were studied: (1) a supercritical helium (ScHe) system in which the 2 K temperatures are obtained by a Joule-Thomson Expander (JTX); (2) a superfluid (He II) helium (SfHe) system, and (3) a hybrid system in which supercritical helium provides the major cooling and small He II reservoirs supply specific detector cooling.

The flow routing diagram for the hybrid system is shown in Fig. 2. The routing of the flow, such that the total cryogen is available for cooling all critical components, makes the flow rate required only weakly dependent on either cryogen tank Vapor-Cooled Shield (VCS) flow rate or mirror purge gas flow rate. Small reservoirs of superfluid helium (He II) are located in the MIC to provide 2 K cooling. The small reservoirs provide low temperature cooling only up to 50 mW continuous. The vent gases are removed from the telescope immediately, cooling only the small reservoir fill and vent lines.

Figure 3 compares the flow routing for the ScHe and SfHe systems. These three systems were evaluated for thermal performance, technical feasibility, and overall cost during the study.

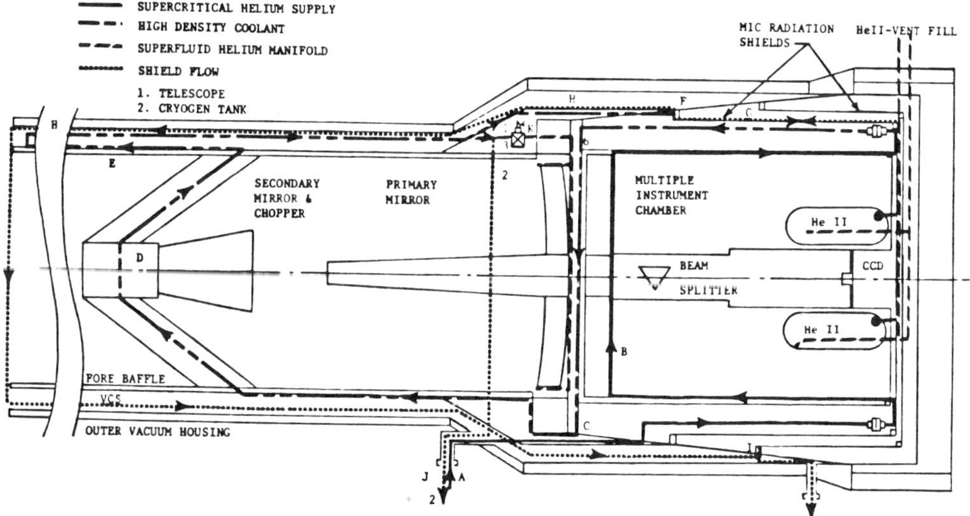

Fig. 2. SIRTF Cryogenic Routing – Hybrid System

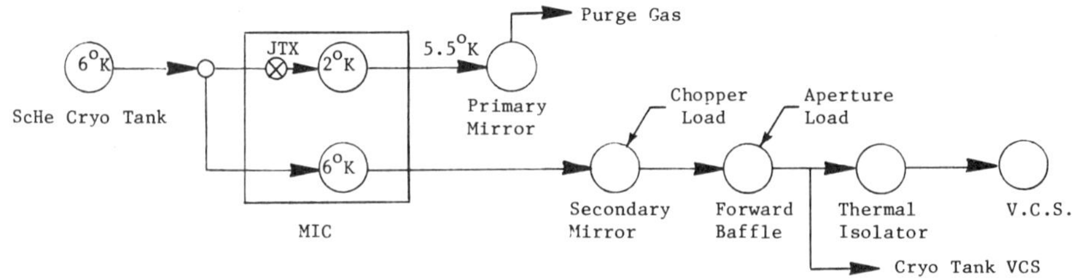

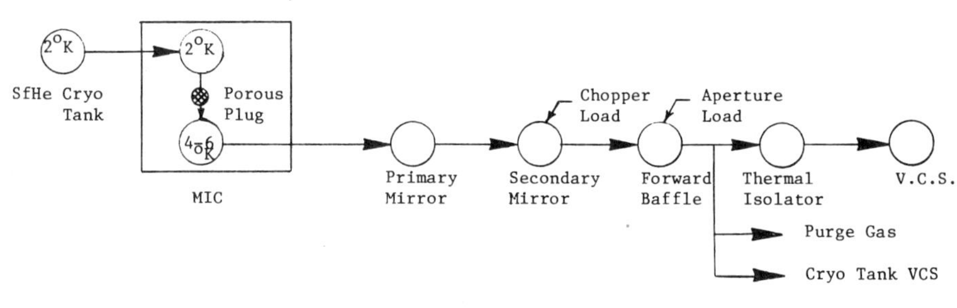

Fig. 3. Comparative cryogen routing.

THERMAL PERFORMANCE

For purposes of thermal modeling, the SIRTF telescope system was divided into five major subsystems: The Cassegrain telescope, the MIC, the outer shell, the radiation shields and the thermal isolator. The telescope was further divided into primary support structure, telescope barrel and optics. The MIC consists of the housing, an instrument support structure and the instruments, but for purposes of this analysis only the MIC housing was modeled. Both passive multilayer insulation (MLI) and active VCS radiation shields were used.

The effect of cryogen mass flow rate on the secondary mirror temperatures, for different aperture loads, is demonstrated by Fig. 4. Notice that as the total mass flow rate decreases below 0.07 g/s, an abrupt change in the slope of the curves occurs and the telescope temperatures rise dramatically. This is a result of the purge gas flow (.038 g/s) and cryogen tank VCS flow (.013 g/s) which are constant being diverted from the flow through the telescope.

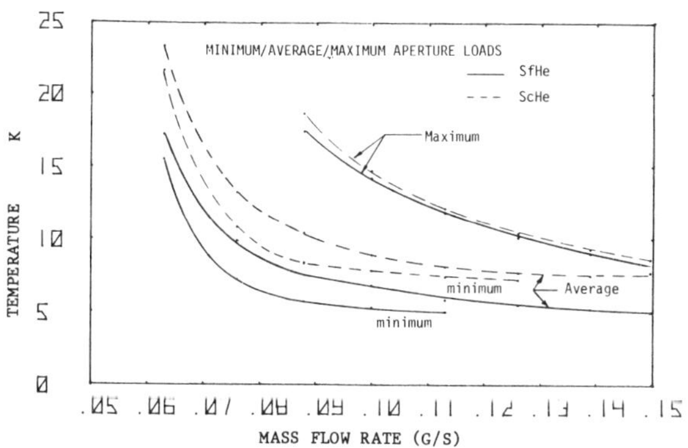

Fig. 4. Secondary mirror temperatures.

The average cryogen mass flow rate is based upon the nominal orbit mission parameters of aperture loads and nominal MIC, chopper and CCD power levels of 1.0, 0.2 and 1.0 watts respectively. The telescope temperature requirements used to determine the cryogen mass flow rates are a maximum secondary optics temperatures of 10 K and a maximum forward baffle temperature of 20 K. Analysis of cryogen requirements for both the 14 and 30 day missions reveals that less fluid is required in the superfluid system than in either supercritical hybrid systems.

The dewar sizes and weights are shown in Table I for a 14-day mission. A 30-day mission would use a second dewar of the same size. To meet ground conditions, the superfluid tank must be

Table I. Dewar Requirements

	ScHe/SfHe Hybrid	ScHe	SfHe
Density, g/cm^3	0.119 (at 20 Torr)	0.119 (at 20 Torr)	0.146 (at 3 atm)
Percent Fill Achieved	100	100	85
Tank Volume, L	1332	1400	1050
Tank Weight, kg	172	181	155
Insulation System:			
VCS Flow Rate Tank, g/s	0.013	0.013	0.013
Insulation	VCS-33MLI	VCS-33MLI	VCS-44MLI
Tank Heat Leak, W	0.47	0.47	0.38
Tank Dimensions:			
Pressure Vessel, cm	99.6 D × 208.1 L	99.6 D × 213.4 L	98 D × 167.6 L

better insulated with multilayer insulation (MLI) than the
supercritical tank, however it is still a lighter and smaller
dewar.

After sizing the tanks it was possible to simulate a mission
profile for the cryogen tanks, where launch occurs at approximately 80 hours. (Fig. 5 and 6). Pressure control is done using
an on-off control with a deadband on the dewar vapor cooled
shield. This results in the pressure and temperature fluctuations
shown. The 14-day mission is complete before the systems enter
their final decay mode at 450 hours. The steady temperature rise
of the ScHe system is inherent in the single phase dewar.

OPERATIONS

A review of the status of helium technology was made during the
study in 1979 (Fig. 7).

No major technical problems were seen in the operation of the
hybrid system as an extrapolation of current development. Problems were foreseen, however, for both the ScHe-JTX and SfHe,
unless additional development programs were undertaken.

In-flight operations offer little choice between the
systems. The ground operations provide a major obstacle to the
SfHe system because of the large ground-based vacuum system which
must be available at intervals during the prelaunch phase.

COST

A costing analysis was run on the SIRTF as part of this
study. The cryogen subsystem is a fraction of the total cost of
the program and differences between the refrigeration modes are

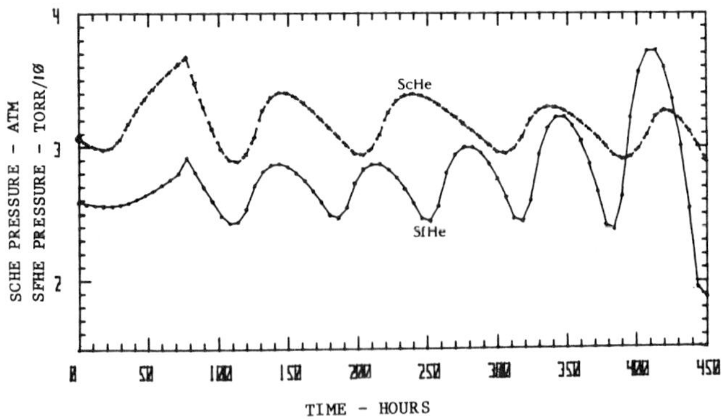

Fig. 5. Dewar mission pressure profile.

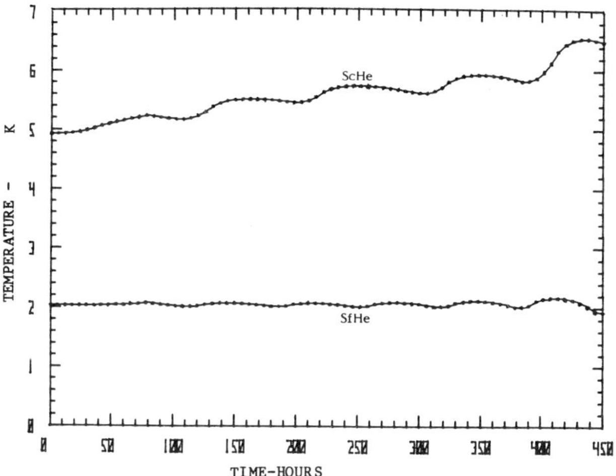

Fig. 6. Dewar mission temperature profile.

only a fraction of the subsystem cost. Additional development costs must be applied to the ScHe-JTX and SfHe systems. Ground support equipment for both supercritical helium systems can be modified from previous programs such as the Lunar Module (LM) and Infrared Astronomical Satellite (IRAS) whereas the SfHe system will require considerable new equipment.

Item \ Condition	SUPERCRITICAL		SUPERFLUID
	HYBRID	JTX	
PREVIOUS PROGRAMS	GEMINI, APOLLO, ETC.	NONE	SUBORBITAL MSFC, JPL
FLIGHTS	MULTIPLE		2
PLANNED FLIGHTS	CLIR	NONE	SUBORBITAL – FIRSSE
			IRAS; SPACELAB II IRT, JPL; COBE.
LABORATORY EXPERIENCE	EXTENSIVE	SLIGHT	MODERATE
COMPONENTS			
VALVES	GOOD	GOOD	FAIR, NEEDS WORK ON SEAT SEALING
COLD VALVES	–	–	FAIR, NEED DEVELOPMENT
JOULE THOMPSON EXPANDER	–	UNPERFECTED	—
POROUS PLUG	–	–	LAB EXPERIENCE, GOOD
			LARGE SIZE UNUSUAL
INSTRUMENTATION			
QUANTITY GAGING	FAIR	FAIR	GOOD, 1G ENVIRONMENT; POOR, 0 G
PRESSURE	EXCELLENT	VERY GOOD	GOOD
TEMPERATURE	EXCELLENT	VERY GOOD	GOOD
DESIGN DATA	MATERIALS, FAIR	MATERIALS, FAIR	FLUID PROPERTIES, GOOD
	FLUIDS, VERY GOOD	FLUIDS, GOOD	HEAT TRANSFER DATA, FAIR
	NO SIGNIFICANT VOIDS	JTX PERFORMANCE, POOR	POROUS PLUG DATA, FAIR
			MATERIALS PROPERTIES, POOR
			SLOSHING DATA, POOR
PROBLEM AREAS		JTX PERFORMANCE	COLD VALVE; POROUS PLUG SIZE

Fig. 7. Current helium requirements.

SUMMARY AND CONCLUSIONS

Table II shows the final ratings for each of the systems for performance operations and cost.

The ratings went from the best (1) to worst (3) with ties allowed but no fractional ratings. The SfHe system offers superior performance and would be the system to use if the funding were available to support it. Study estimates considered the categories of performance, operations, and cost, to be of equal importance, and the hybrid system was selected as the best compromise to obtain an operational SIRTF.

Table II. Trade-off Summary

	Supercritical He		Superfluid He
	Hybrid	JTX	
Performance			
Cooling Capability	2	3	1
Temperature Stability	2	3	1
System Weight	2	3	1
TOTAL PERFORMANCE RATING	6	9	3
Operations			
Technology	1	3	2
Complexity	1	2	3
Reliability	1	3	2
TOTAL OPERATIONS RATING	3	8	7
Cost			
Subsystem Cost (Ground Support Equipment)	1	2	3
GSE Usability	1	2	3
Development Cost	1	2	2
TOTAL COST RATING	3	6	8
Summary			
Performance	2	3	1
Operations	1	3	2
Cost	1	2	3
TOTAL	4	8	6
OVERALL RATING	FIRST	THIRD	SECOND

ACKNOWLEDGMENT

The assistance of Lou S. Young, project manager has been appreciated.

REFERENCES

1. Anon., Shuttle Infrared Telescope Facility (SIRTF) Preliminary Design Study Final Report, Hughes Aircraft Corporation, Data Item I.D.-9, Culver City, Calif. (August 1976).
2. W.C. Marlow, et al., SIRTF System Design Summary Document, Report No. ER-399A, Perkin-Elmer Corporation (Ridgefield, Conn.) (August 1979).

IMPROVED ACTIVE PHASE SEPARATOR FOR He II
SPACE COOLING SYSTEMS*

H. D. Denner, G. Klipping, I. Klipping, K. Lüders, T. Oestereich,
U. Ruppert, U. Schmidtchen, Z. Szücs, and H. Walter

Freie Universität Berlin
Berlin, Federal Republic of Germany

INTRODUCTION

The active phase separator (APS) for the He II cooling system
of the German Infrared Laboratory (GIRL) covers a wide range of
operating conditions with regards to heat load and temperature and
it provides a high degree of temperature stability. For confinement of the bulk liquid, use is made of the thermomechanical
effect of a narrow annular gap of adjustable length formed by a
bushing and a moveable pin driven by an electrodynamic system.
The APS is provided with a heat exchanger arranged directly behind
the outlet of the annular flow channel.[1] For GIRL, temperature
control of the bulk liquid has been chosen in order to achieve
maximum versatility with regard to the instruments to be cooled
and the range of application of the telescope.[2] Therefore, the
APS heat exchanger is in contact with the bulk liquid. The heat
exchanger allows operation of the APS also in the dissipative flow
regime where controlled amounts of liquid pass through the flow
gap and evaporate behind it. In this mode of operation the range
of mass flow available from thermomechanical phase separators is
considerably increased.[3-5]

This type of APS is capable of keeping the temperature
constant for considerable change in the heat load. The temperature stability required for GIRL, with He II operation as well as
with venting of He gas only is, $\Delta T/T < 10^{-4}$ at stationary

---

*This work was supported by the Bundesministerium fur Forschung
und Technologie under contract DFVLR–BPT Ø 1QV748–AA/WF/WRKO
275/4/13.7.

conditions and $\Delta T/T \leqslant 10^{-3}$ during change of heat load.[1] The stability, however is poorer whenever the pressure drop along the flow channel falls below a certain value.

Stationary conditions can possibly be expected during pre-launch and launch when for some time the cryostat cannot be pumped and the actual increase of the cryostat pressure cannot be predicted. The heat load will change after filling the cryostat to the maximum possible level, that is when the heat exchanger of the APS is completely flooded and in temperature equilibrium with the surrounding bulk liquid. This can, also occur during the mission if the downstream (upstream) temperature accidentally increases to values above (considerably above) $T\lambda$. Tests have shown that under these conditions evacuation of the liquid inside the heat exchanger and establishment or reestablishment of the pressure drop necessary for operation is problematic.

Improvement of the APS with the aim of enlarging its control range as well as increasing its response speed was, therefore, felt to be desirable not only in view of optimizing its performance in the GIRL cryostat, but also in order to increase its versatility in general. The APS should be suitable for various functions within the valve system of a He II cryostat and should as well be applicable for cryogenic fluids other than superfluid helium, for example He I, H_2 or others which are of interest for hybrid cooling systems.

MODIFIED ACTIVE PHASE SEPARATOR

Figure 1a is a schematic of the modified APS (Laboratory Model III). As in the previous version[1] the APS is inserted into the LHe tank (4) by means of flanges (1,2,3). The annular flow gap of variable length is formed by the bushing (5) and the moveable pin (6). Again the insert tube (7) serves for heat exchange with the bulk liquid in the tank.

However, the flow path of the exhaust gas inside the heat exchanger has been changed. The inner heat exchanger tube (8) extends only over about two thirds of the length of the insert tube and its entrance is surrounded by a cap (9) which is fixed to the suspension (10) of the moveable pin (6). Moreover, the upper end of the annular space between the heat exchanger tubes (7,8) is now closed which means, the exhaust gas must flow through the interior of the tube (8) into the vent gas line. This design ensures that independent of the orientation of the APS or the place where liquid collects inside the heat exchanger, no liquid will pass into the vent line as long as the system remains within the proper control range.

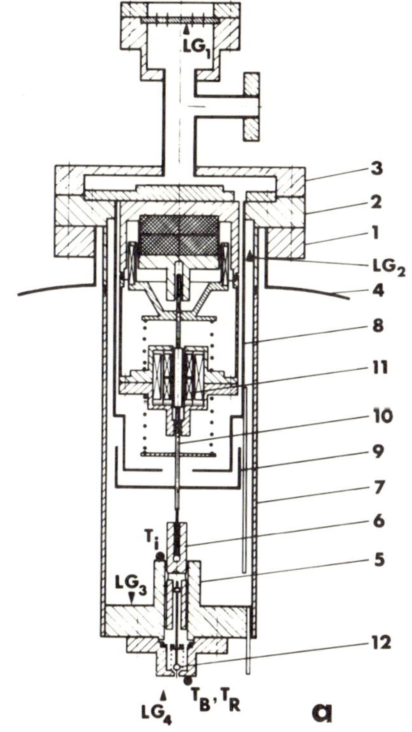

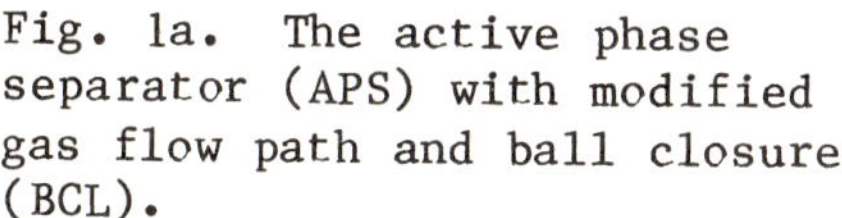

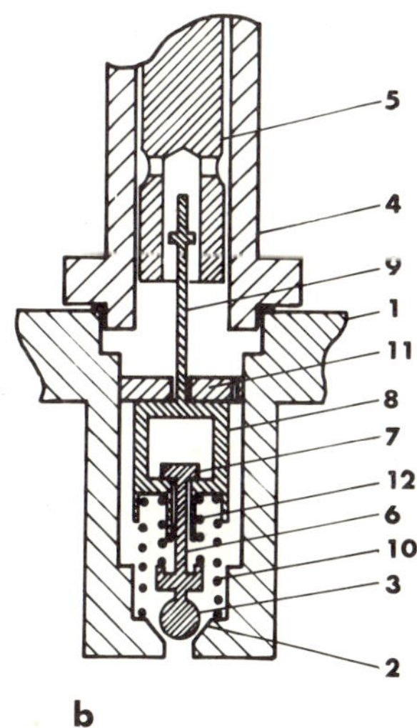

Fig. 1a. The active phase separator (APS) with modified gas flow path and ball closure (BCL).

Fig. 1b. Details of the ball closure.

In order to make the APS more compact, the electrodynamical driving system has also been redesigned and optimized, and the displacement transducer (11) has been incorporated into it. Another new feature is a ball closure (BCL; 12) situated before the entrance of the annular flow channel. Its design details are shown in Fig. 1b. A housing (1) with a seat (2) for the ball (3) is mounted to the base plate of the APS (see Fig. 1a) carrying the bushing (4), which together with the pin (5) forms the flow channel for mass flow rate control. The ball (3) is connected to a stem (6) which ends in a disk (7). The latter is enclosed in a cage (8) which in turn is connected to a rod (9), the free end of which extends into the central bore of the pin (5).

As long as the pin (5) during operation keeps positions other than the maximum possible intrusion into the bushing (4), a spring (10) pressing the cage (8) against a guidance and stopping plate

(11) within the housing (1) keeps the BCL (2,3) open. Only when the pin (5) moves deep enough into the bushing (4) to make the free end of the rod (9) touch the upper end of the bore within the pin, i.e., when the mass flow rates required approach the minimum value (maximum gap length), is the ball (3) pressed against seat (2). A second spring (12) serves for making the connection of the BCL to the electromagnetic actuating mechanism elastic in order to avoid damage by too hard closing.

The BCL is by no means a lambda-tight valve as has been described in the literature.[6] It is just another, "passive" kind of flow restriction which by means of the thermomechanical effect reduces the mass flow rate of He II by about two orders of magnitude compared to the values required, for example, for temperature control within the normal range of operation of GIRL. At the same time, the BCL acts as closing means for He I with negligible leakage even at pressure drop values as high as 1000 mbar (Table I).

Since the BCL allows one to reduce the mass flow rate to extremely small values, it is no longer a problem to (re)establish operating conditions after decrease of the pressure drop and flooding of the heat exchanger. This means increased safety for operation during the mission, and gives the possibility for

Table I. Mass flow rates obtained with the
BCL closed (1 kp controlling force)

Fluid	Pressure at entrance p_B mbar	Pressure drop Δp mbar	Mass flow rate $\dot{M}$ mg/s
He gas at 4.2 K	1000	80	0.1
	1000	160	0.3
	1000	1000	0.3
	1000	50	0.3
	100	100	0.3
	1000	180	0.8
He I	1050	300	1.1
	1030	530	1.3
	1020	1015	1.4
He II	35	35	0.02
	50	50	0.04

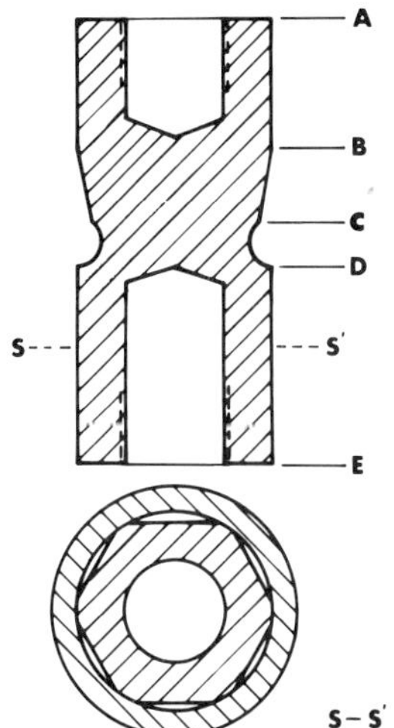

Fig. 2. Shape of the pin with cylindrical and tapered
control section.

operating the APS in any position in ground experiments to
simulate flight conditions.

ACTIVE PHASE SEPARATOR WITH EXTENDED CONTROL RANGE

A "closing" function of the APS, at least for He II
operation, can also be obtained by changing the shape of the
moveable pin (item 6 in Fig. 1) such that various control ranges
are obtained. In the example shown in Fig. 2 the pin consists of
a cylindrical upper section AB followed by a tapered section BC,
the circumferential liquid supply groove CD and the lower guidance
section DE. Bypassing of the guidance section, where the gap
width between pin and bushing is 5 µm, is here achieved by means
of a hexagonal shape of the pin with only small ridges left of the
cylindrical circumference yielding guidance (see cross section
SS'). The gap width in the upper cylindrical section AB is also 5
µm while in the tapered section BC the gap width increases
continuously from 5 µm at B to 15 µm at C. Here, a simple conical
shape has been chosen for the tapered section BC so as to yield a
conical flow passage. However, any other, more complicated shape
with decreasing pin diameter can also be used if it is advantageous with regard to the control properties of the APS in a
special application.

When the pin is moved into the bushing to such an extent that
the 5 µm section AB of the flow passage becomes effective for mass
flow rate control, values in the range of some tenth of a mg/s to

a few mg/s are obtained. However, even within this region the
mass flow rate is "actively" controlled, that is, it remains
adjustable by the control system. The extension of the control
range towards low mass flow values results in a quicker response
of the control.

While the upper cylindrical section AB of the pin serves for
extension of the control range far into the ideal flow region, the
tapered section BC takes control within the range of comparatively
large mass flow rates required for the operation of GIRL (at T_B =
1.6 K: 10 mg/s $\leqslant$ $\dot{M}$ $\leqslant$ 25 mg/s; at T_B = 2.0 K: 55 mg/s $\leqslant$ $\dot{M}$ $\leqslant$ 85
mg/s), which fall into the dissipative flow region[4,7]. A conical
shape of the flow passage means that any change in the gap length
is accompanied by a change in the effective gap width. A change
in the mass flow rate is, therefore, achieved by smaller
displacements of the pin than in case of a straight annular flow
passage. The wide end of the conical gap is intended for controlling the mass flow rates when operating with gaseous helium.
This operation mode will occur during ground tests as well as
during the mission when, for instance, the amount of LHe is
decreasing.

Experimental results obtained with a flow passage of this
design (gap width as mentioned above, gap lengths 1_{AB} = 13 mm and
1_{BC} = 7 mm, pin diameter D = 20 mm) are shown in Fig. 3. The mass
flow rate $\dot{M}$ was measured as a function of the gap length 1 with
manual setting of 1 and without heat exchanger behind the flow
passage (for experimental set-up and procedure see Ref. 8). For
each curve the bath temperature T_B and the pressure drop along the
flow passage Δp were kept constant as indicated in the figure.
The arrow at the abscissa marks the transition between the
straight annular and the conical flow passage.

It can be seen that as long as the mass flow rate is
controlled by the 5 μm straight annular flow passage it is very
small. For a given value of Δp, the flow rate is less sensitive
to changes in the gap length 1 the lower the bath temperature.
Even with the pressure drop Δp = $p_{upstream} - p_{downstream}$ + ρgh
increased up to 9 mbar at T = 1.6 K similar results were ob-
tained. Under these conditions the flow thus remains mostly
within the ideal region, which is in accordance with our previous
results[3] as well as with the theoretical interpretation[5,7].

The mass flow rate can be increased within the control range
of the straight 5 μm passage only at higher bath temperatures or
by using the conical flow passage for control. Within the control
range of the latter the mass flow rate is much more sensitive to
variations of the gap length, and the shape of the curves is no
longer hyperbolic as for a straight annular flow passage.
Therefore, length variation within a much shorter region is

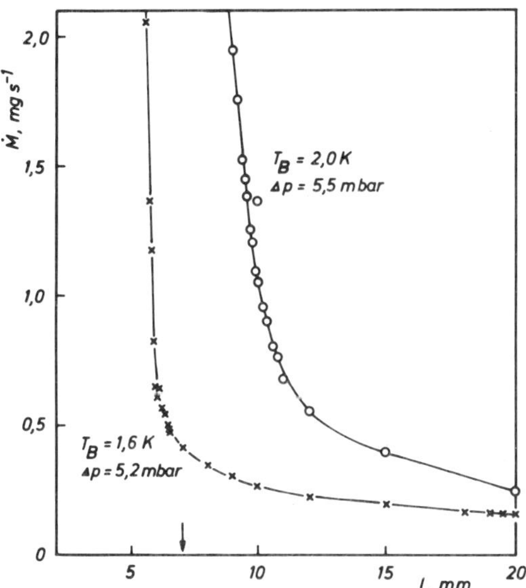

Fig. 3. Mass flow rate as a function of the gap
length at different bath temperatures.

sufficient for covering, for example, the mass flow rates required
for GIRL.

This can also be seen from Fig. 4, where the results obtained
with the combined straight/conical flow passage of 5/5-15 μm width
(pin diameter D = 20 mm) are compared with previous results[1] for a
straight flow passage of 14 μm width (pin diameter D = 10 mm).
Here again the arrow at the abscissa indicates the transition
between straight and conical shape of the flow passage. The
considerable extension of the control range by combining flow
passages of different shape is clearly demonstrated. As mentioned
before, the increased steepness of the mass flow curves means that
smaller changes in the gap length are sufficient for controlling
the mass flow within a given region. This can be of special
advantage, for example, with regard to the actuating system.
Whether it is advantageous or disadvantageous for temperature
control in case of the GIRL cryostat can only be answered by
performance tests which yet are to be made. In any case, these
results show that by proper choice of the pin shape the APS can be
tailored to special requirements regarding cooling power, control
range and the stabililty of the control.

The temperature control performance of this APS will be
investigated using a large cryostat suitable for simulating GIRL
operation conditions. Moreover, it will in the near future be
tested under zero-g conditions in the KC-135 flight test facility.

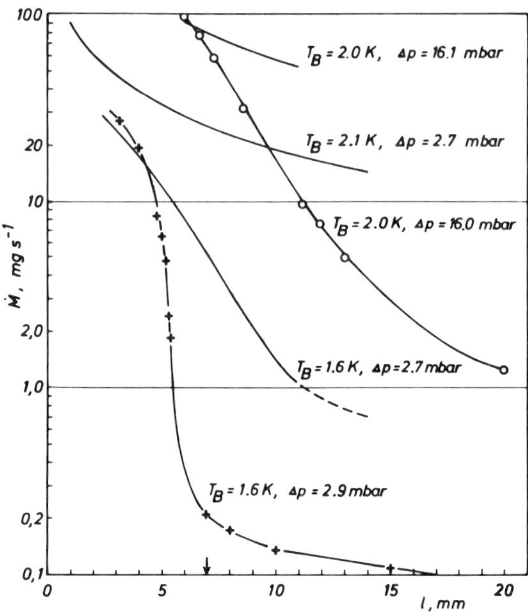

Fig. 4. Comparison of the mass flow rates $\dot{M}$ = f(1) at different bath temperatures obtained with the APS.

-       with pin diameter D = 20 mm and a combination of
        straight 5 μm and conical 5 to 15 μm flow
        passage (curves + and o), and
-       with pin diameter D = 10 mm and a straight 14 μm flow
        passage (full lines)[1].

REFERENCES

1. H.D. Denner, et al., in "Proc. 8th Intl. Cryo. Engr. Conf.,"
        IPC Science and Technology Press, Guildford (1980), p. 32.
2. D. Lemke, et al., Space Optics, 183:31 (1979).
3. H.D. Denner, et al, in "Advances in Cryogenic Engineering, Vol
        25," Plenum Press, New York (1980), p. 783.
4. U. Schotte and H.D. Denner, in "Proc. 8th Intl. Cryo. Engr.
        Conf.," IPC Science and Technology Press, Guildford (1980),
        p. 27.
5. U. Schotte and H.D. Denner, Z Phys, B41:139 (1981).
6. B.S. Blaisse, J.M. Goldschwartz and W.P. van de Merewe, in
        "Proc. 6th Intl. Cryo. Engr. Conf," IPC Science and
        Technology Press, Guildford (1976), p. 172.
7. U. Schotte, in "Advances in Cryogenic Engineering, Vol 27,"
        Plenum Press, New York (1982).
8. H.D. Denner, et. al., Cryogenics 18:166 (1978).

ORBITAL PERFORMANCE OF A ONE-YEAR LIFETIME SUPERFLUID HELIUM DEWAR BASED ON GROUND TESTING AND COMPUTER MODELLING

R. A. Hopkins

Ball Aerospace Systems Division
Boulder, Colorado

and

W. F. Brooks

NASA Ames Research Center
Moffett Field, California

INTRODUCTION

Fabrication of the superfluid helium dewar for the Infrared Astronomical Satellite (IRAS) has been completed, and the dewar has been extensively ground testing with the vacuum shell at 300 K. The temperature of the vacuum shell during orbital operation is predicted at 170 K. Consequently from ground testing to orbital operation there is a shift in the relative magnitudes of the major heat fluxes into the dewar. On orbit, radiative heating is greatly diminished, and conduction through the internal support system and electrical wiring becomes dominant. Orbital lifetime of the dewar has been predicted at 11 months through careful evaluation of test data and extrapolation of test performance to the orbital boundary conditions using a thermal network computer model.

A schematic of the dewar is shown in Fig. 1.[1-3] With a reservoir capacity of 550 L and four vapor cooled shields (VCS), the system was designed to maintain the internal instrument assembly at ~3 K for a period of one year on orbit. Heat conduction into the dewar is effectively minimized by using fiberglass/epoxy tension bands to support the helium tank and internal instrument assembly. The aperture heat load into the telescope is reduced to about 8 mW by a sunshade using three stages of radiative cooling. The orbital vacuum shell temperature is reduced using

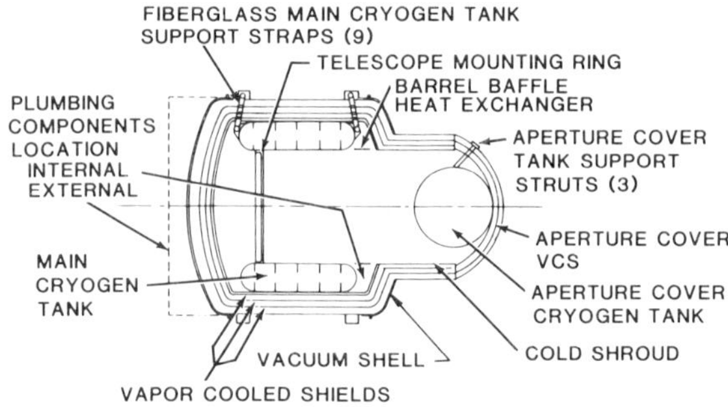

Fig. 1. Schematic of the cryogenic system.

multilayer insulation blankets, appropriate surface finishes, and
an earth radiation shield.

The dewar has an ejectable aperture cover which includes a 54
L supercritical helium tank. This cover closes the dewar vacuum,
provides a low radiation background (~7 K) for instrument checkouts on the ground and on orbit, and protects the telescope from
contamination during the first two weeks on orbit.

DESCRIPTION OF THE THERMAL MODEL

Performance of the dewar is predicted with a lumped-parameter
thermal network computer model which is input to the Ball
Cryogenic/Thermal Analyzer Program (BCAP). This program is a
general purpose thermal analysis procedure which uses a block
relaxation technique for steady-state solutions and forward finite
differencing for transient solutions. Specific capabilities have
been added to account for vapor cooling and the thermodynamics of
stored cryogens. All material properties can be considered temperature dependent.

Development of the Thermal Network Model

During the several years over which cryogenic analysis of the
IRAS dewar was conducted, many slightly different models were used
to investigate various cases and conditions of interest. The
following important assumptions and approaches were common to all
models:

- the vacuum shell, the VCS, and the helium tanks are each
  isothermal
- the temperature of effluent in the vent line reaches that
  of a given VCS before the vent line leaves that shield
- there is no gaseous conduction

- the support straps are perfectly attached (thermally) to the VCS
- the thermal conductivity of the multilayer insulation is given by

$$K = 19.9 \times 10^{-12} (T_1^2 + T_2^2)(T_1 + T_2) + 1.5 \times 10^{-6} \text{ mW/cm-K}.$$

The thermal network model is shown schematically in Fig. 2. The major paths of heat flow into the dewar are through the VCS. Conduction through the center conductors of the 400 stainless steel coaxial cables is assumed to pass straightway into the helium tank (without vapor cooling). Wires inside the fill lines are also not vapor-cooled. The rest of the electrical cabling is vapor cooled by spot bonding to the support straps at the locations where they are attached to the VCS.

FIRST DEWAR TEST

Initial Observations

During initial dewar testing in August 1980, thermal/acoustic oscillations were observed in the aperture cover. After they were subdued, flowrates from the main tank and aperture cover tank were still 60% and 600% higher than predicted, respectively. The temperature of the inner VCS was about half that predicted by the original computer analysis, in which almost all the heat load to the main tank came from the inner VCS, indicating that this was not the case. Possible explanations for these discrepancies were: ineffective heat sinking of the support straps and electrical cabling to the vapor-cooled shields, gaseous conduction in the guard vacuum, underestimate of radiation input at the front-end of the dewar and through the aperture cover viewports, thermal shorts, and poor heat sinking of vent lines to the VCS.

Since vapor was exiting the aperture cover system at a temperature considerably below that of the VCS, it was apparent that serious aperture cover repairs were needed. The following series of tests were then conducted to gain additional understanding of the main dewar performance.

Flow Swap Test. In this test the main tank was vented through the fill line, which is attached only to the inner VCS. Shield temperatures after four days are shown in Table I. As expected, they increased significantly above the values during normal operation.

Table II shows heat loads to the main tank based on measured flowrates for the normal and flow swap configurations and also the heat loads predicted by the model with shield temperatures of the observed values. By subtracting the latter from the former, we

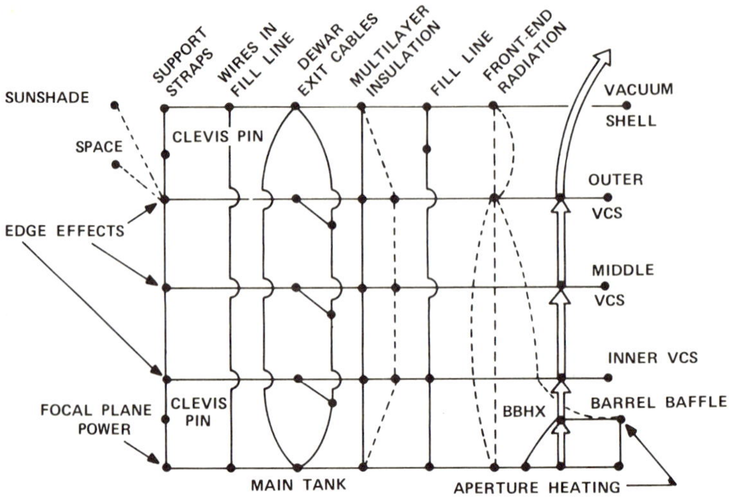

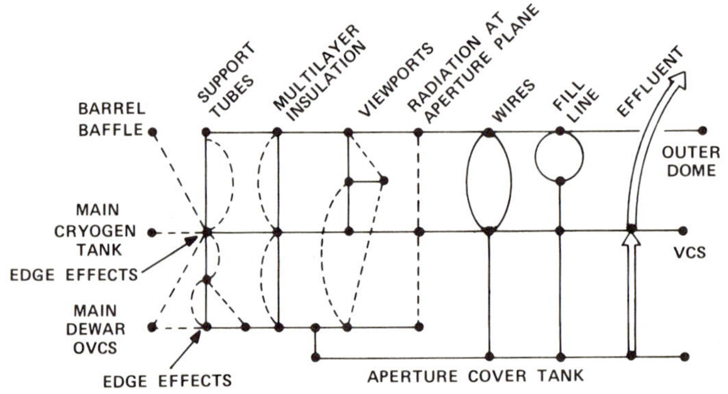

Fig. 2. Dewar thermal network model.

arrive at the heat load not accounted for by the model. The ratio
of the heat load not accounted for in going from the normal to the
flow swap configuration is 2.02, which matches well with the ratio
of outer VCS temperatures to the fourth power. This supports the
idea that the heat load not accounted for by the model is coming
from the interface area between the main dewar and the cover,
where the line of sight into the dewar is blocked by a black
labyrinth formed between the outer VCS and the aperture cover VCS.

Cold Flow Test. LHe was circulated continuously through the
aperture cover plumbing in an attempt to reduce the VCS temperature and thereby the temperature of the main dewar outer VCS to
which it is strongly coupled by radiation. This would further

Table I. Temperatures Measured During the Flow Swap Test

	Normal (N)	Flow Swap (F)	$(T^F/T_N)^4$
Outer VCS	155 K	190 K	2.25
Middle VCS	65	135	18.0
Inner VCS	25	50	16.0

verify the conclusions of the flow swap test. Unfortunately, this test only further verified that the aperture cover vent line was very poorly attached to the VCS; the out VCS temperature was not significantly perturbed.

Power Off Test. All power was turned off and cables disconnected. The dewar flowrate remained unchanged.

Hot/Cold Viewport Test. The temperature of the viewport outer windows was raised and lowered 20 K, and no significant change in main tank flowrate was seen. If radiation through the viewports was a major source of the unaccounted heat load, a noticeable perturbation in main tank flowrate would have taken place.

Thermal efficiency of straps. Temperature sensors on one cable bundle indicated a temperature drop of 10 K where the cable is attached to the inner VCS.

Adjustments to model. Considering the test results, through inspection of the model inputs, and through review of the dewar design, it was evident that the model was underestimating the radiative heat loads at the front-end of the dewar (i.e., at the interface between the main dewar and the cover). Accurate determination of radiation exchange factors in that area were difficult to make because of the complexity of the area and the uncertainty regarding the size of the gaps between the edges of the insulation layers and the surfaces they butt up against. Consider for example, that a 1-cm gap would result in about the same magnitude of radiation from the vacuum shell to the outer VCS as is predicted through the 35-layer multilayer insulation blanket.

Table II. Comparison of Heat Flows from the Flow Swap Test

	Normal	Flow Swap
Observed heat load	143 mW	288 mW
Heat load predicted by model	50	100
Excess not accounted for by model	93	188

After the first phase of testing, the area at the interface
between the main dewar and cover was carefully inspected, and a
computer model of that area was then constructed to make a more
rigorous evaluation of radiation exchange factors for input to the
model. Also, the conductances representing the support strap end
fittings were adjusted on the basis of temperatures measured
during dewar testing. With values based on the observed temperature drop added to the model between the cabling and each vapor-
cooled shield, there was only a slight degradation in predicted
flowrate.

MODIFICATIONS AND SECOND TEST

Modifications

After four months of testing, the dewar was warmed up and the
aperture cover removed. Inspection revealed that the adhesive
providing the thermal contact between the vent line and aperture
cover VCS had almost entirely popped loose. The VCS and line
attachment were completely redesigned and the aperture cover
rebuilt. After adding several radiation barriers at the interface
between the main dewar and the cover, the cover was again installed and the system evacuated.

The dewar was filled with helium for the second time in March
1981. The dewar now contained the optical subassembly, whereas for
the first cooldown a mass model had been installed. The optical
subsystem is instrumented with more thermometers than was the mass
model. Also, thermometers had been added at several locations on
the cold end of one of the dewar electrical cables to evaluate the
efficiency of the thermal attachment of the cabling to the support
strap at the inner VCS.

Observations

During the second phase of testing, electrical subsystem
checkouts were performed, and the dewar was vented only through
the vent line. Ground life of the aperture cover tank had improved about a factor of five to a value of 4-5 days. The main
tank flowrate had improved about 10 percent to a value of 6.6 mg/s
with the focal plane turned on; this improvement was probably due
mostly to the rebuild of the aperture cover. Also, the top of the
tank remained wetted by the superfluid film at a liquid level 50
percent lower than before. This indicated a reduction in heat
load to the top end of the tank.

The upper end of the telescope barrel baffle forms a labyrinth
along with the aperture cover to block radiation into the instrument cavity. Heat is conducted from the barrel baffle to the main
tank via titanium flexures which support the barrel baffle and

also to the barrel baffle heat exchanger (where it is intercepted
by vapor cooling) via a copper thermal strap. About half of the
heat load to the barrel baffle leaves through each path. Based on
the computed conductance of the flexures and temperatures measured
on the barrel baffle, about 30 mW was flowing through the flexures. This indicates a total heat load of about 60 mW to the
barrel baffle, which is much bigger than predicted with the original computer model.

Comparison with Model Calculations

The superfluid flowrate measured was 20 percent greater than
predicted with the updated model, and shield temperatures agreed
fairly well. Predicted aperture cover performance closely matched
the test data. In order to force the model predictions to agree
with the test data, two radiation paths were added to the model:
one from the vacuum shell to the outer vapor-cooled shield and one
from the outer vapor-cooled shield to the instrument cavity shroud
(i.e., the helium tank). These two radiation paths were adjusted
so that predicted shield temperatures and boil-off rate exactly
matched the test data. It was then verified that the model reasonably matched the previous flow swap test results.

FINAL THERMAL ANALYSIS

A detailed breakdown of steady-state heat flows and component
temperatures predicted by the model for the ground test boundary
conditions is given in Fig. 3. Flowrates and temperatures shown
match exactly with the test data. The relative heat fluxes
through the two radiation paths which were added to the model to
force agreement with the test data are not suprising considering
the uncertainties involved in modelling radiation transfer in that
region. The total radiation load to the upper end of the main
tank assembly is 61.8 mW and to the upper end of the barrel baffle
is 61.6 mW. These radiation inputs are much larger than predicted
with the original model (before updating the radiation exchange
factors in that area).

Heat flows and temperatures using the network model are shown
in Fig. 4 for timelined boundary conditions corresponding to the
orbital environment. The diminished magnitudes of radiation
transfer resulting from the lower vacuum shell temperature are
apparent. The dominant sources of heat input to the helium tank
are focal plane power dissipation and conduction through support
straps and electrical wiring. Notice the importance of the heat
rejection to space from the black painted radiation disc which is
attached to the exposed end of the outer VCS.

Helium loss from the start of ground hold (12 h before launch)
until the cover is ejected was determined with a transient

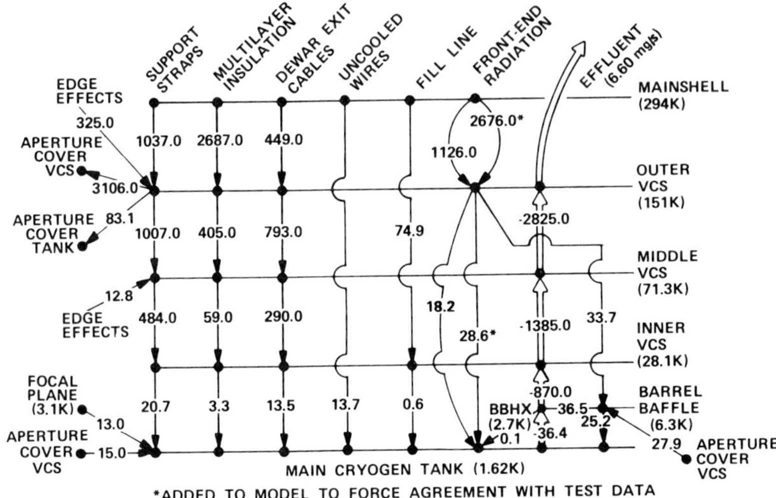

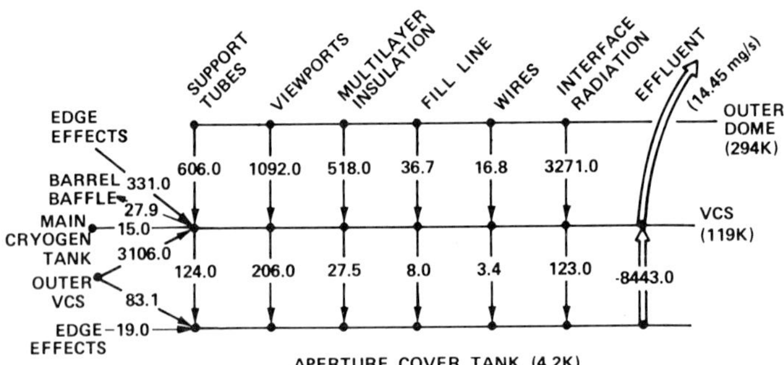

Fig. 3. Predicted heat flows (mW) during ground testing.

computer model using the same thermal network already described. At this time the dewar contains about 73 kg of superfluid helium at a temperature of 1.7 K, and the aperture cover tank has just been topped off. The main tank vent valve is opened about one hour after launch. The back pressure regulator, through which the aperture cover tank is vented, opens when pressure in the tank reaches 2.6 atm (about 12 hours after top-off).

Figure 5 shows profiles of flowrates, helium loss, and aperture cover tank temperature during the period before cover ejection. The cover is ejected approximately two weeks after launch when the tank temperature reaches 18 K. Helium mass loss from the main tank at this time is 2.6 kg. Based on this, it is considered that 70 kg remains for telescope cooling during the survey.

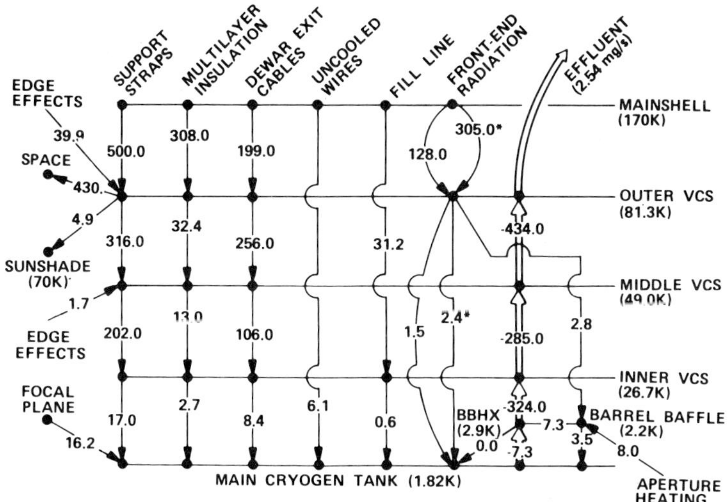

Fig. 4. Predicted Heat Flows (mW) and Temperatures During Survey Operation.

With the predicted helium loss rate of 2.54 mg/s, a survey lifetime of 319 days is predicted.

The driving boundary conditions during the orbital operation are vacuum shell temperature, aperture heating, and focal plane power dissipation. Nominal timelined values of these parameters are 170 K, 8.0 mW, and 16.2 mW. The values of vacuum shell temperature and aperture heating are the most uncertain because of analytic inaccuracy and potential degradation of external thermal control surfaces due to contamination buildup. The effects of variations in these parameters are shown in Fig. 6. Variations of ±10 K in vacuum shell temperature and ± 200 percent in aperture heating result in a lifetime perturbation of about 20 days. These variations are within the estimated margin of uncertainty involved in evaluating these boundary conditions.

FUTURE EFFORTS

The IRAS is currently (August, 1981) in Holland for spacecraft integration and vibration testing and is scheduled to undergo thermal/vacuum testing in early 1982 and launch in August of 1982. Dewar performance throughout these final phases of testing will be used to further substantiate or perhaps to modify the computer model of the cryogenic system. The experience gained during the IRAS program will significantly aid the design and

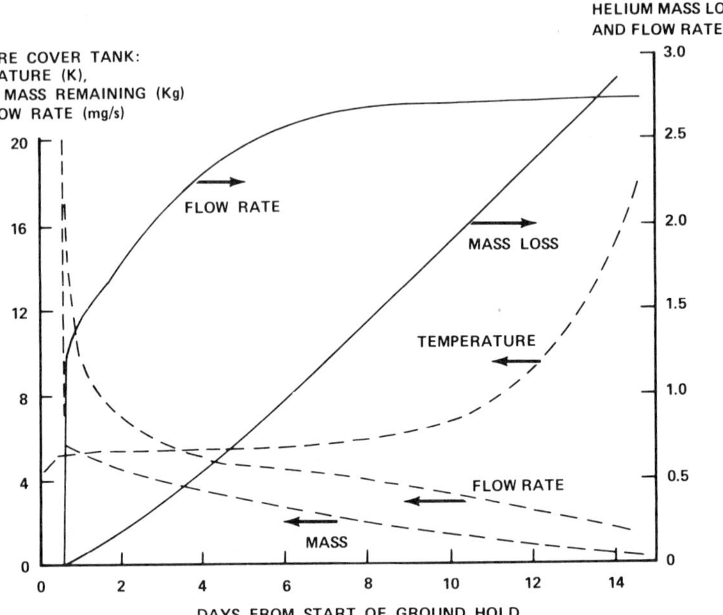

Fig. 5. Dewar operation from ground hold through cover ejection.

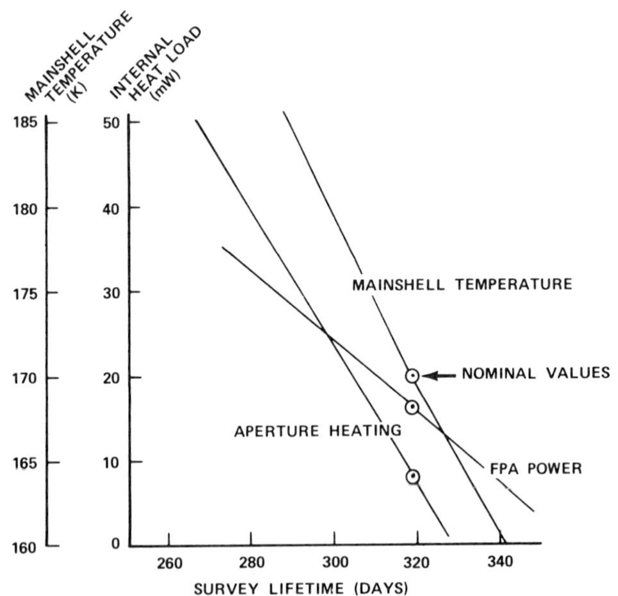

Fig. 6. Impacts of critical boundary conditions on survey lifetime.

performance predictions of future long-lifetime, spaceborne cryogenic systems.

REFERENCES

1. R.A. Hopkins, Design of a one-year lifetime spaceborn
       superfluid helium dewar, <u>ASME Publicaton</u> 79-ENAs-23 (1979).
2. R.R. Anderson and L.J. Hilton, Thermal Design for the IRAS
       Telescope System, <u>ASME Publication</u> 79-ENAs-38 (1979).
3. A.R. Urbach, P.V. Mason, and W.F. Brooks, Progress Report on
       the IRAS Cryogenic System, "Advances in Cryogenic
       Engineering, Vol. 27", Plenum Press, New York (1982)

TOP-OFF PROCEDURE FOR SPACE-BOUND SUPERFLUID HELIUM CRYOSTATS*

D. Petrac

*Jet Propulsion Laboratory
California Institute of Technology
Pasadena, California*

INTRODUCTION

The lifetime of a superfluid helium cryostate in space is in part determined by the total mass and temperature of the coolant at launch. This is particularly valid for cryostat in satellites with no top-off capabilities. The same constraints also hold for the cryostats to be flown on the Space Shuttle, for which retanking capabilities will be developed some time in the future if the needs require it.

In the laboratory, top-off procedures for cryostats using normal helium are well developed, although each specific case may be different. The complexities arise when top-off of superfluid helium[1,2] is required. The transfer of superfluid helium from the storage dewar to the experimental dewar when both dewars are at temperatures below the lambda temperature, $T\lambda$, is possible for relatively short vacuum jacketed transfer lines and for specific temperature and presure differences.[3] A specific Δp is necessary for the transfer; otherwise only vapor is transferred due to boiling in the transfer line. For long transfer lines without any special arrangement for cooling the jacket separately or without vapor counterflow, the needed presure differences for liquid transfer to top the dewar exceed the saturated vapor pressure at $T\lambda$ (~ 38 torr). Therefore, superfluid cannot be transferred under saturated vapor pressure conditions except with a pump, which is impractical at this temperature. We conducted tests on the

---

*Work supported by the National Aeronautics and Space Administration.

transfer of pressurized liquid helium, slightly above Tλ, in order to determine the optimum transfer parameters for a ground-based top-off just prior to launch.

DESCRIPTION OF THE TESTS

Pressurization of Superfluid Helium Test

The tests were performed in a 6-liter glass cryostat. The temperature of the liquid was measured with five submerged calibrated germanium thermometers attached along a vertically adjustable rod. The initial temperature was achieved by pumping on the helium bath. Pressurization with warm helium gas was performed at several initial temperatures and was controlled by pressure-regulating valves. The pressure in the vapor phase was recorded as a function of time. The rate of rise of temperature in the superfluid helium bath was determined by the overpressure above the liquid and was found to be quite rapid. The rise time was intentionally chosen to be slow by controlled flow of helium gas. When the temperature of the liquid reached the λ point, the pressure was increased to a pressure level regulated by a manostat.

Test Outcome

The pressure history of the gaseous phase and the temperature changes of the liquid phase when helium was introduced into a volume containing superfluid helium were determined. When warm helium gas was added to the ullage volume of a cryostat containing superfluid helium, the temperature of the liquid rose rapidly to the λ temperature, 2.2 K, and then increased very slowly. As long as the liquid was superfluid, the pressure of the gas was no more than a few torr above the saturated vapor pressure (SVP) of the superfluid. Steady-state overpressurization of the superfluid was not possible. When the liquid was slightly above the λ point, the ullage pressure could be increased above the SVP of the liquid bath. We pressurized it up to 760 torr with the liquid remaining slightly above 2.2 K, corresponding to a SVP of 30 to 50 torr.

INTERPRETATION OF THE PRESSURIZATION TEST RESULTS

The initial temperature was 1.5 to 2 K. Figure 1 shows a typical plot of the temperature of the helium bath as a function of time. The dashed curve is for an initial temperature of 1.7 K. The λ temperature could have been reached rapidly; however, we chose to increase the temperature at a relatively slow rate, achieving T_λ in about 5 minutes. As shown in Fig. 2, the pressure difference for $T < T_\lambda$ was very small. It depended on the rate of introduction of the gas. For very slow rates the pressure difference would be zero. Pressure in the gas when $T < T_\lambda$ can be much

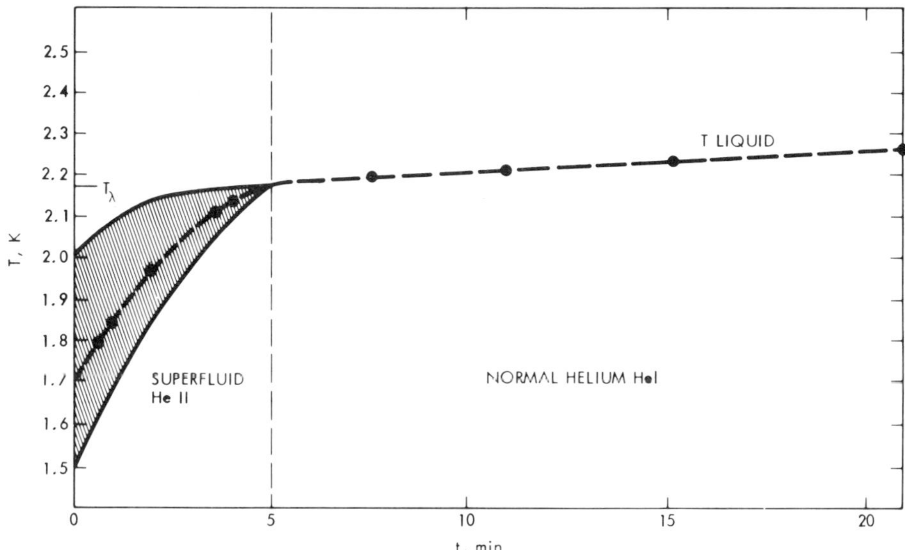

Fig. 1.   Temperature of pressurized liquid helium.

higher than the SVP (shown in Fig. 2). This different response
for $T < T_\lambda$ and $T > T'_\lambda$ is explainable by considering the thermal
properties of helium.

Below the λ point the gas and the entire volume of liquid are
in good thermal contact. Helium II has an extremely high thermal
conductivity so that the whole volume of liquid is at the same
temperature and no stratification can occur. Thus the warm gas is
condensed to liquid and the heat of condensation is taken from the
whole liquid and not just from the top layers.

Above the λ point the helium liquid has low thermal conductivity, and a warmer, insulating layer is formed on the liquid
surface (an example of this is shown in Fig. 3). This is possible
because the density of liquid decreases with temperature (Fig.
4). As a result the overpressure can be large with respect to the
saturated vapor pressure, corresponding to the temperature of the
bulk liquid. Figure 2 is for an overpressure of about 100 torr.
Only small increases in bulk temperature occur due to heat transfer from the warmer gas. The temperature gradient, in the absence
of mechanical stirring, is limited to a thin top layer. The
pressure above the liquid is very close to the saturated vapor
pressure of the liquid in this layer. Conduction of heat in the
normal helium is many many times smaller than in the superfluid
and the warming of the main mass of the liquid is slow.

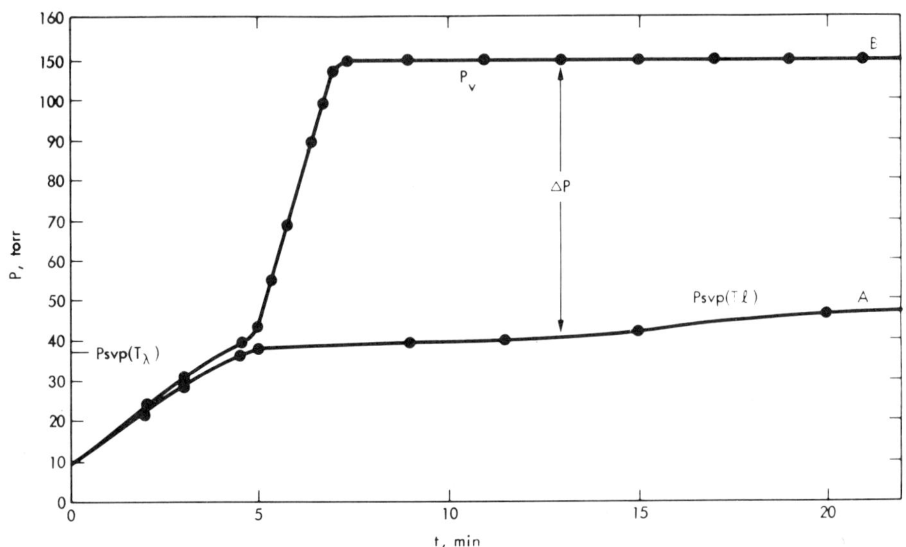

Fig. 2. Overpressure vs time.

In summary, below the λ point, liquid helium cannot be
pressurized above its saturated vapor pressure due to the high
thermal conductivity of superfluid helium as the condensing vapor
warms the whole liquid volume.   Above the λ point the overpressurization with gas is possible because of stratification, and an
insulating layer is formed on the liquid surface.   This overpressure can be maintained for an extended time.

APPLICATION TO THE TRANSFER FOR TOP-OFF

Overpressurized liquid helium can be transferred at a temperature much lower than if the liquid were at the saturated vapor
pressure.   The transfer line inlet must be positioned near the
bottom of the supply dewar in order to transfer the coldest
liquid.   The main steps in the top-off procedure in order to
achieve high fills are:

(1)   The cryostat is filled with liquid helium by transfer of
      normal helium and subsequently pumped down to an equilib-
      rium temperature below T$_\lambda$.

(2)   A sufficiently large supply dewar is pumped down to
      temperatures below T$_\lambda$ also in order to ensure constant
      temperature within the whole liquid volume.

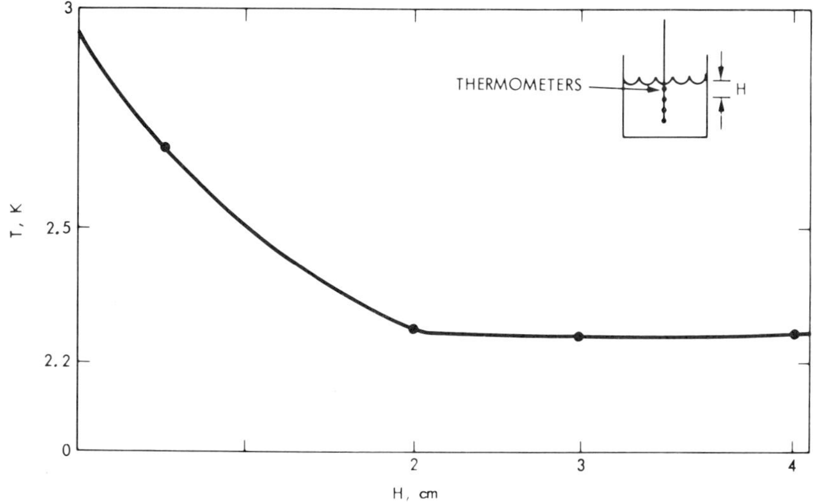

Fig. 3. Temperature vs depth (H).

(3) With the transfer line connecting the supply (storage) dewar and the cryostat, pumping on the cryostat is continued to maintain temperature below T_λ.

(4) Helium gas is introduced into the storage dewar to raise the temperature in the storage slightly above T_λ. The overpressure can then be raised to the value needed for transfer. The temperature of liquid in the storage dewar entering the transfer line will be about 2.2 - 2.4 K. It will be warmed up in the transfer line and added to the colder liquid of the cryostat. The temperature in the cryostat will therefore rise but the increase will be slowed due to the continuous pumping on the cryostat.

The temperature in the cryostat after the top-off is estimated to be around 2.2 K. The exact value will depend on the details of the plumbing, the precooling process, the rate of introduction of the gas and liquid, and the total amount of liquid transferred. After the transfer, the helium in the cryostat will be conditioned to the lowest possible temperature before the closure of the vent line in the final preparation for launch.

These top-off procedures have been tried in the laboratory for two different sized cryostats. One cryostat used in low-g experiments,[4] was small, with 250 mL volume. The final fill at 1.8 K was to about 85% of capacity. In 7-liter cryostats the top-off can reach 100%. On a larger scale, these procedures were used

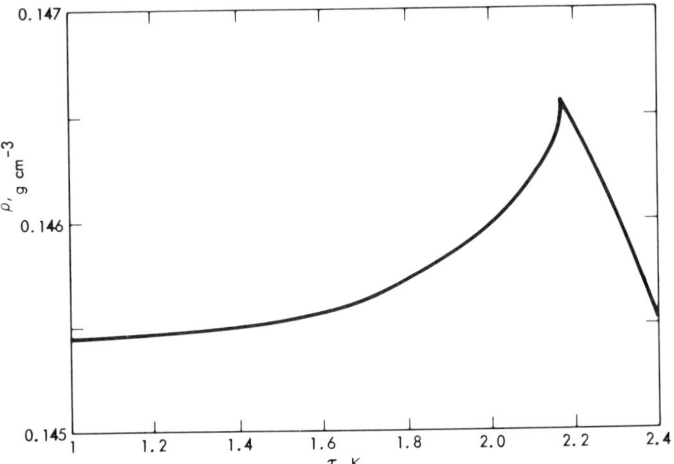

Fig. 4. Density of liquid helium at SVP.

with a dewar-to-dewar transfer at Ball Aerospace System Division, Boulder, Colorado. At least two top-offs with over 90% fill at below T were also done on the IRAS (Infrared Astronomy Satellite) main cryogen Tank.[5]

CONCLUSION

The maximum mass fill of a spaceborne cryostat can be accomplished with the low pressure top-off after initial fill with normal helium. The realistic maximum mass fill at temperatures below T_λ can be expected to be at least 90%, which results in about 40 to 50% more mass at launch than without the top-off. Each case will require specific ground support equipment to satisfy the peculiar individual cryostat requirement.

The transfer procedure must be worked out in detail and extreme care is necessary, particularly during the transient periods after initiation of the transfer.

ACKNOWLEDGEMENTS

Thanks are due P. Ludtke, P. Mason and R. Urbach for useful discussions and J. Gatewood for excellent technical support.

REFERENCES

1. J. Lizon-Tati et. al., Design and Development Tests of a
    Superfluid Helium Space Cryostat (CRHESUS), in "Proc. 8th
    Intl. Cryo. Engr. Conf.," IPC Science and Technology Press,
    Guildford (1980), p. 189.
2. G. Bon Mardion, et. al., Helium II in Low-Temperature and
    Superconductive Magnet Engineering, in "Advances in
    Cryogenic Engineering, Vol. 23," Plenum Press, New York
    (1978), p. 358.
3. L.C. Yang, et. al., Characterization of superfluid helium
    transfer, Cryogenics 21:207 (1981).
4. P. Mason, et. al., The Behavior of Superfluid Helium in Zero
    Gravity, in "Proc. 7th Intl. Cryo. Engr. Conf.," IPC
    Science and Technology Press, Guildford (1978), p. 99.
5. R. Urbach, Ball Aerospace Systems Division, private
    communication.

AN EXPERIMENT TO EVALUATE LIQUID HYDROGEN STORAGE IN SPACE

R. N. Eberhardt, D. A. Fester, W. A. Johns,
and J. S. Marino

Martin Marietta Denver Aerospace
Denver, Colorado

INTRODUCTION AND BACKGROUND

Many applications have been identified for cryogenic liquid storage and supply in space, including life support, power generation, space-based orbital transfer vehicle propulsion, laser feed systems and orbital resupply. The low-g behavior of the liquid and ullage within the storage container influences the selection of specific elements which must be included in the tank to meet the system fluid management and thermal control requirements. Fluid management requirements generally include single-phase liquid feed to the user subsystem over a broad range of flowrates under fast response (possibly less than one second) conditions. Controlling residual forces exerted on the storage tank and other subsystems by fluid slosh may also be required. Thermal control involves efficient (gas-only) venting to provide tank pressure control in response to the input heat leak.

Fluid behavior in space is controlled by surface tension forces which are generally insignificant in a gravity dominated environment. Basic phenomena such as buoyancy, settling, convection, mixing and diffusion are considerably different when gravity is not present. The shape of a gas/liquid interface in low-gravity is determined solely by capillary forces. The Young-Laplace equation relates liquid surface tension and the curvature of the interface to the pressure differential between the gas and liquid,

$$P_G - P_L = \sigma \left(\frac{1}{R_1} + \frac{1}{R_2} \right)$$

where the subscripts G and L refer to the gas and liquid respectively, σ is the surface tension, and R_1 and R_2 are the radii

of curvature of the interface, as shown in Fig. 1. For a circular
pore of radius r, the pressure differential across the interface
becomes 2σ/r.

Liquid hydrogen has the lowest surface tension (1.8 x 10^{-3}
N/m at 138 kPa) of any of the fluids of interest (excluding liquid
helium).   To take advantage of the forces exerted at the
liquid/vapor interface, a liquid acquisition device (LAD) is
configured with an extremely small pore diameter, on the order of
microns. Fine mesh screen covered channels similar to those shown
in Fig. 2 are used in close proximity to the tank wall to acquire
and feed single phase liquid. The pressure difference across the
interface must be greater than the sum of hydrostatic and flow
losses to prevent premature gas ingestion into the channel, and
thence to the outlet.

One concept for thermal control is to supply single-phase
liquid from the LAD to a thermodynamic vent system (TVS) and a
vapor-cooled shield (VCS) surrounded by multilayer insulation.
The liquid undergoes an isenthalpic expansion of saturated or
subcooled liquid, resulting in two-phase flow in heat exchangers
mounted on the VCS. The heat exchanger configurations are shown
schematically in Fig. 3. Heat exchanger 1 (HX1) with a bottom-to-
top flow direction, intercepts 60-80 percent of the total input
heat leak. Heat exchanger 2 (HX2), with top-to-bottom flow has
the capability to intercept up to 200 percent of the heat leak.
Surface tension forces acting on the two-phase fluid in the heat
exchangers can be significant when compared to viscous, inertial
and gravitational forces. The heat transfer coefficient of

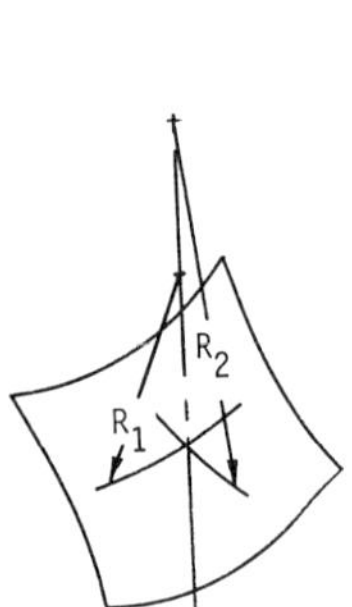

Fig. 1. Liquid/Vapor
interface.

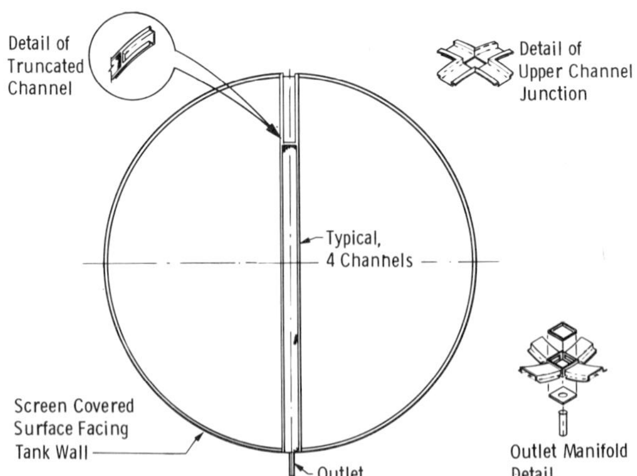

Fig. 2. Liquid acquisition
device (LAD).

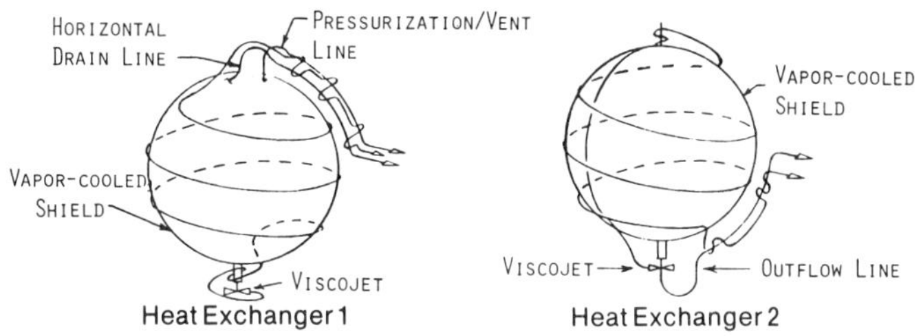

Fig. 3. Thermodynamic vent system (TVS).

Rohsenow and Dougall[1], as further developed by Tong[2], was selected
for qualities greater than 0.1:

$$\bar{h} = \frac{0.023k_v}{D} \left[Re\right]^{0.8}\left[Pr_v\right]^{0.4}$$

$$\text{where} \quad Re = \frac{\rho_v D(Q_1 + Q_v)}{\mu_v A}$$

EXPERIMENT APPROACH

There are various approaches for conducting cryogenic testing
to verify design and performance parameters. They include one-g
tests, and low-g tests, which can range from drop tower testing,
to experiments in the KC-135 aircraft facility flying Keplerian
trajectories, to experiments conducted in the Earth-orbital space
environment. Ground tests do not produce a representative highly-
curved low-g liquid/vapor interface shape with the liquid preferentially wetting the storage tank walls. Low-g testing, therefore, is preferred.

Drop tower tests are characteristically conducted at 1/10 to
1/40 scale while KC-135 tests can range between 1/10 and full
scale, depending on system size. Both of these test methods,
however, provide test times of only two to 30 seconds which are
not nearly adequate for thermal stabilization or quality/density
measurements. Therefore, an orbital experiment becomes the only
viable approach for providing thorough resolution of the fluid
management and thermal control issues and verification of designs. Martin Marietta pursued this approach for a liquid oxygen
system under NASA contract in the 1970's.[3] With the Space Shuttle
now approaching operational status, the selected approach was an
experiment mounted on a Spacelab (or other) pallet and having a
duration of a nominal seven-day Shuttle mission.

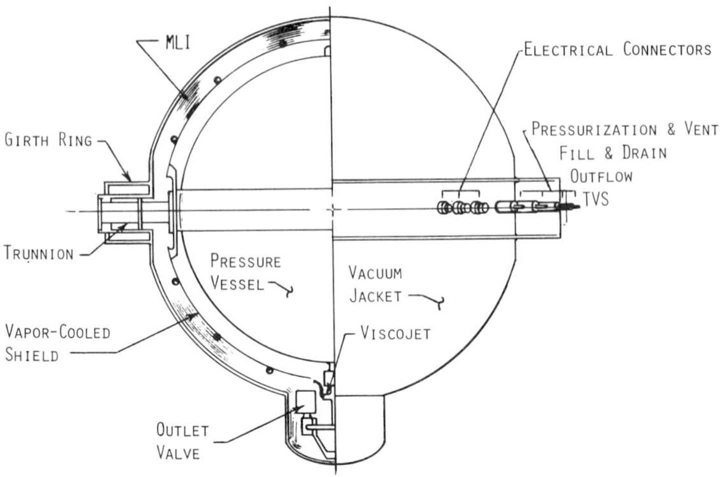

Fig. 4. Liquid hydrogen storage and supply tank assembly.

ORBITAL EXPERIMENT DESIGN

The orbital experiment was designed to obtain engineering data and establish design criteria for storage and supply tankage.[4] The Cryogenic Fluid Management Experiment (CFME) tank, shown in Fig. 4, will provide LAD and TVS performance data over a heat flux range of 0.8 to 63 W/m^2. A simplified experiment schematic is presented in Fig. 5 and an artist's concept of the experiment package mounted on a Spacelab pallet is shown in Fig. 6. A microprocessor controls the TVS, and collects and stores the data on a self-contained tape recorder. This low-g test data will be correlated with results from analytical models and used for the analysis and design of future storage and supply tankage.

EXPERIMENT OPERATION

A typical time sequence of the CFME during a seven day orbital mission is presented in Fig. 7. Tank pressure is shown as a function of time, and the mass of helium injected to maintain the pressure at its prescribed level is indicated with time during periods of liquid hydrogen outflow. The baseline CFME operational sequence is as follows. The experiment is activated at lift-off (time zero) with 39 kg of liquid hydrogen in the storage vessel at a pressure of 103 kPa. A four-day storage period is provided so that thermal conditions within the storage vessel can stabilize and data can be gathered on the TVS performance.

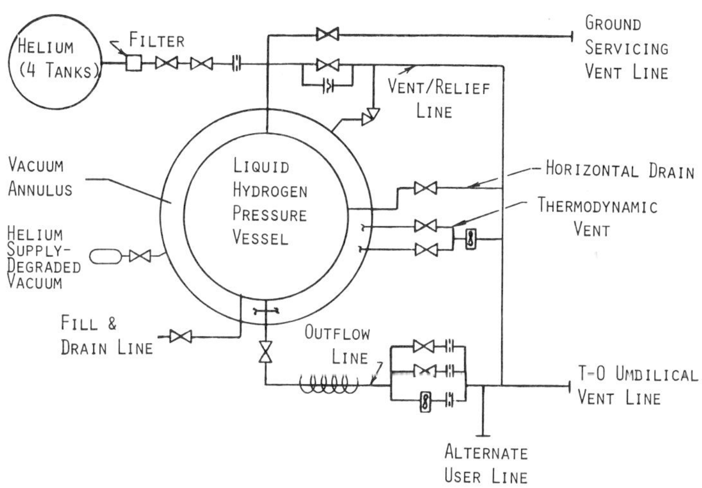

Fig. 5. Experiment schematic.

At 96 hours into the mission, helium pressurization of the
tank is performed to raise the tank pressure to 317 kPa and liquid
is expelled under constant pressure conditions. The tank pressure
is again raised to a pressure of 380 kPa at 109 hours, followed by
a storage period until 140 hours. At this time outflow is initiated without pressurant being added to the tank (blowdown ex-
pulsion), and the pressure is allowed to decrease to a level
closely corresponding to the saturation pressure level of the bulk
liquid within the tank. Unpressurized outflow is terminated,
followed shortly thereafter by helium pressurized outflow to
depletion.

Two additional pressure scenarios are shown in Fig. 7, the
first in which HX1 only is used throughout the entire operational
sequence, and the second in which both HX1 and HX2 are on continuously. The flowrate for HX1 is 0.023 kg/h and for HX2 is 0.08
kg/h. Analytical models correlated only with the limited ground
test data provide a basis for determining these TVS flowrates.

An indication of the low-g thermal transients within the tank
thermal control system is indicated in Fig. 8, where the simulated
temperature response of the VCS is plotted for a typical 7-day run
with various heat exchanger operations. After 80 hours, the
temperatures at the top and middle of the VCS are still decreasing. When HX2 is activated, all segments of the VCS quickly drop
to 18 K. This is not an optimum operating condition for the VCS
since the fluid exiting the heat exchangers is at a colder temper-

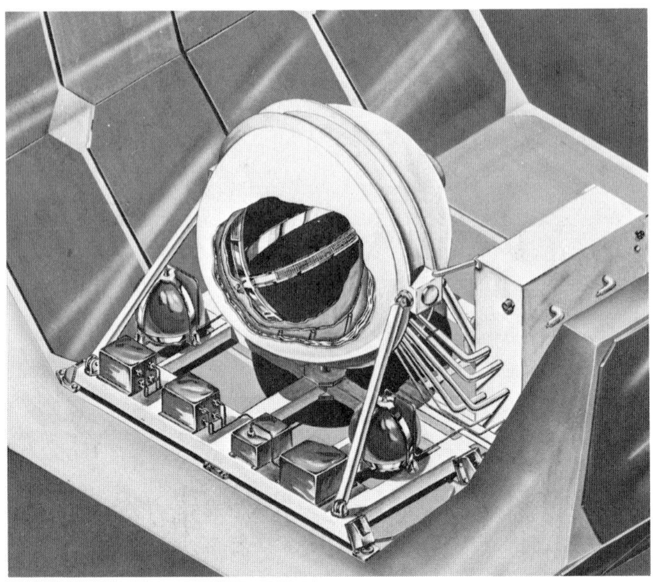

Fig. 6.   Cryogenic fluid management experiment (CFME).

ature than the surrounding surfaces. The capability to quickly
reduce heat input and minimize tank pressure increases is
provided, however.

GROUND TEST VERIFICATION

To gain some insight into experiment operation, a subscale
ground test program is being conducted with liquid nitrogen and
liquid hydrogen to assess the influence of gravity on two-phase
flow heat transfer coefficients and to investigate integrated
storage tank thermal performance. The heat exchanger tube, shown
in Fig. 9, can be tested in an up, down or horizontal attitude. A
1/3-scale CFME-type storage tank, complete with LAD and a TVS
composed of two heat exchangers attached to a vapor-cooled shield,
is being tested. This system is shown in Fig. 10 prior to installation of the MLI and vacuum jacket. The one-g test program is in
progress; results will be published later.

CONCLUDING REMARKS

Design of the CFME storage and supply tank has been completed.
The CFME will be the supply tank for the Shuttle Cryogenic Test
Facility. In addition to providing the previously discussed CFME
data, the facility will be used to investigate: 1) Transfer line
and receiver tank chilldown, 2) receiver tank fill in low-g, 3)

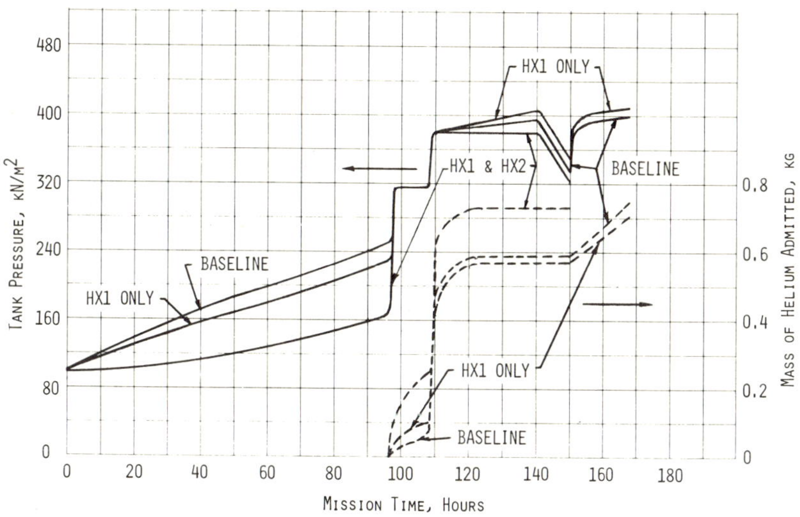

Fig. 7. Typical experiment pressure response.

LAD fill in low-g and, 4) mass gauging and fluid density measurements. In-house, one-g testing is presently underway to substantiate as completely as possible our analytical models. The next
step for CFME is fabrication, followed by CFME ground and flight
test data correlation with the analytical models.

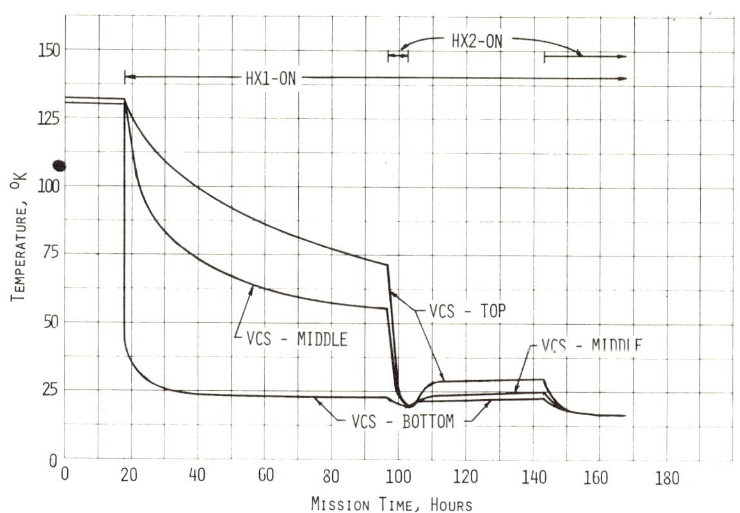

Fig. 8. Vapor-cooled shield (VCS) temperature response.

Fig. 9. Heat exchanger test article.

Fig. 10. Subscale storage tank test article.

NOTATION

$\bar{h}$ — Mean effective heat transfer coefficient, $W/m^2 \cdot K$

A — Cross-sectional area of tube, m^2

D — Diameter of tube, m

k_v — Thermal conductivity of saturated vapor, $W/m \cdot K$

Pr_v — Prandtl Number of saturated vapor

Q_l — Volumetric flowrate of Liquid, m^3/sec

Q_v — Volumetric flowrate of vapor, m^3/sec

ρ_v — Density of saturated vapor

μ_v — Viscosity of saturated vapor, $kg/m \cdot sec$

ACKNOWLEDGEMENT

The authors wish to thank Mr. John C. Aydelott and Mr. Eugene P. Symons of the NASA-Lewis Research Center for their guidance on the CFME Program, Contract NAS3-21591, and Mr. William J. Bailey, Principal Investigator of the Martin Marietta Research Task on Cryogenic Storage Technology.

REFERENCES

1. W.M. Rohsenow and R.S. Dougall, Film boiling on the inside of vertical tubes with vertical flow of the fluid at low qualities, MIT Report 9079-26 (1963).
2. L.S. Tong, "Boiling Heat Transfer and Two-Phase Flow," John Wiley & Sons, Inc., New York (1965).
3. R.C. Tegtmeyer, Acquisition/expulsion system for earth orbital propulsion system, Volume IV – flight test article, Martin Marietta Report, MCR-73-97, Denver, Colorado (October 1973).
4. R.N. Eberhardt, D.A. Fester and J.C. Aydelott, A liquid hydrogen experiment as a shuttle payload, AIAA Paper No. 80-1096, Presented at AIAA/SAE/ASME 16th Joint Propulsion Conference, Hartford, Conn. (June 30-July 2, 1980).

DEVELOPMENT OF GAS GAP CRYOGENIC
THERMAL SWITCH

T. Nast, G. Bell, and C. Barnes

Lockheed Palo Alto Research Laboratories
Palo Alto, California

INTRODUCTION

Cryogenically cooled instruments, including infrared and gamma
ray dectectors, are used for various missons in space. When data
collection is not required, it may be desirable to decouple the
sensor from the cryogenic cooler, particulary in the case of an
open cycle system which uses solid or liquid cryogens as the
coolant. The advantages of this decoupling are: (i) cryogen
usage will be reduced during "off" periods allowing an extended
lifetime; and (ii) contaminants which have collected on the sensor
during cryogenic operation may be removed by the resulting warm
up.

Various types of cryogenic switches to perform this decoupling
have been developed in the past, some operating on the same helium
gas-gap principle as the unit described in this paper. The thermal switch development described here is more comprehensive than
that previously reported and includes cycle testing.

PRIOR SWITCH DEVELOPMENT

Various types of heat switches have been developed with varying degrees of success. Most of these switches are of the "demand" type; i.e., they are switched on and off by some action of a
solenoid valve, magnetic field, or temperature change. Earlier
switches include: (i) a solenoid-activated metal-to-metal contact
type[1] which was used on the Surveyor and Viking spacecraft with
limited success; (ii) a single gallium crystal which was operated
at helium temperature[2] and switched by a magnetic field; (iii) a
heat pipe which used two separate wicks and transferred a fixed
amount of energy until the nitrogen working fluid was transferred
from one end of the pipe to another[3]; and (iv) a cryogenic switch

utilizing dissimilar materials which was developed[4] for use on redundant mechanical refrigerators where it was desirable to reduce the parasitic heat load from the "off" refrigerator during operation of a second unit. This switch had a switching ratio of 300 and was activated by a temperature change resulting from turning a refrigerator on or off.

The gas-gap type of switch utilizes concentric cylinders separated by a small gap which is filled or emptied with a conductive gas to achieve the switching action. W. E. Gifford describes[5] a "thermal valve" utilizing this principle. Another gas-gap switch development similar to the one described here is described by Bywaters et al.,[6] in which they achieved a switching ratio of 512. The use of a getter to absorb and release gas to the switch, which potentially avoids the gas supply bottle and valves, is described in Ref. 7.

GAS–GAP SWITCH DESIGN

The primary design requirements of the switch are summarized in Table I.

The basic approach to the switch design is shown in Fig. 1. In this design the "cold" end of the switch is thermally grounded to the cooling system (a solid cryogen cooler for this study). The "hot" end of the switch is thermally grounded to the sensor. Thermal separation between the switch ends is achieved by a thin wall epoxy fiberglass tube which connects the switch halves, provides the alignment, and retains the helium exchange gas. The two cylindrical halves of the switch are separated by a small gap across which the exchange gas effects heat transfer during operation. The width of this gap is a key parameter in the switching ratio. The copper surfaces as well as the inside of the fiberglass tube are gold plated to minimize radiative coupling in

Table I. Primary Design Requirements

o	Working temperature	60 K
o	"On" Resistance at 80 K	1 K/W maximum
o	"Off" Resistance at 230 K	1500 K/W minimum
o	Switching Ratio (R off/R on):	1500 or greater
o	Fill Response:	"On" resistance attained within 1 minute
o	Life-Cycle Capacities:	2500 temperature and pressure cycles
o	Weight	low – for space flight

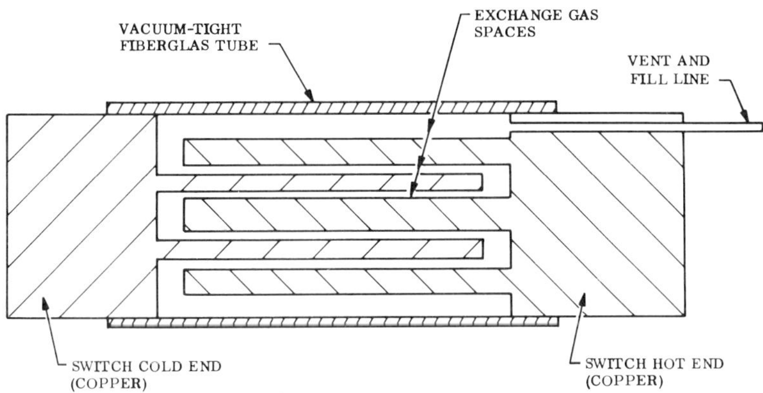

Fig. 1. Thermal switch concept.

the switch. The gold plating on the fiberglass tube also reduces the permeation rate of the helium gas during operation.

A gas supply bottle and a valve system are required to operate the switch. In considering the relative merits of alternative exchange gas handling schemes, one of the key requirements was to attain the high reliability required for a long-term space mission.

Three schemes were studied: (1) direct gas flow from a high pressure helium bottle to the switch element, requiring an excessive amount of gas storage and resulting in full bottle pressures on the switch, and stresses on the fiberglass tube than is other schemes; (2) using a pressure regulator to reduce gas usage and pressure, but having inadequate regulator reliability; (3) using redundant metered gas volumes to reduce pressure and usage. The latter was selected since it avoided the undesirable features of the other schemes. Figure 2 shows the redundant arrangement which was selected for the final design and which had adequate reliability.

To attain an optimum switch element design for a low "on" resistance, a computer model was set up to optimize the design variables, e.g., gas-gap thickness, over-all length and diameter. The study shows that near the optimum design point the choice of material is important. The copper leads to an "on" resistance of one-fourth that of aluminum.

The final design was based on computer studies and on considerations of manufacturing and assembly. The principal design values are summarized in Table II.

Table II   Principal Design Values

Fiberglass Tube Length	3.0 in.
Overall Length	3.7 in.
Heat Transfer Length	2.2 in.
Outer diameter	1.14 in.
Fiberglass Tube Length	2.2 in.
Annular Gap	0.003 in.
Weight	1.5 lb.

FABRICATION OF SWITCH

The two major considerations in the construction of the switch element were to use assembly techniques which allow the annular gap of 0.003 in. to be maintained, and to use a conservative joint and fiberglass tube design to prevent leakage of helium gas over the required 2500 life cycles. These considerations were met by the selection of filament-wound tube, vacuum tight to a level of 1 x 10^{-8} cc/s, gold plated to reduce the radiant heat loads and gas diffusion. The switch halves were assembled using epoxy* bonding techniques.

TEST RESULTS

The following operational and life characteristics of the thermal switch were measured: switch leak integrity at cryogenic

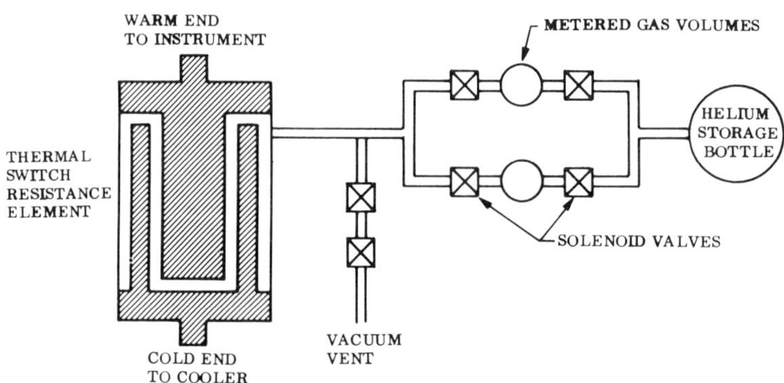

Fig. 2.   Thermal switch design concept.

*Furane Plastics Co., Los Angeles; resin: Epibond 123, hardener: Epibond 9615-10.

temperatures, switch "on" resistance, switch "off" resistance, switch fill tests, and switch life-cycle tests.

Test Setup

A schematic drawing of the test setup is shown in Fig. 3. The switch to be tested is threaded into a thermal link assembly, which is surrounded by a liquid nitrogen cooled shield to prevent stray radiation loads from impinging on the assembly and negating the thermal link calibration. To approximate the radiation boundary conditions that the switch will encounter in operation, the switch is surrounded by a shield temperature controlled to ~ 150 K. As shown in Fig. 3, the switch cryogen tank assembly is mounted inside a liquid nitrogen guarded dewar, the interior of which was evacuated using a liquid nitrogren trapped diffusion pump. During life testing, the pressure and temperature cycling control is accomplished using a 4-cam recycling timer coupled to two NRC solenoid-actuated valves and a Research Incorporated "Thermac" controller. A "Lab Ac" power supply provides controlled heat-up of the switch hot end (Fig. 4).

Experimental Results: Discussion

The experimental results are compared to the requirements and predictions in Table III.

Both the "on" resistance, measured at ~100 K, and the "off" resistance, measured at a hot end temperature of ~154 K, compared favorably to the predicted values. To test for the switch fill

Table III. Switch Performance Versus Requirement

	Requirement	Prediction from Computer Model	Measured
"on" Resistance	1 K/W Maximum	0.75 K/W	0.82 K/W
"off" Resistance	1500 K/W Minimum	2110 K/W	1880 K/W
Switching Ratio	1500	2880	2293
Fill Time	1 min	1 min	1 min
Vent Time	No Requirement	10 min	Not Determined
Life Cycle	2500 min	--	525*

*Leak developed at cycle 525; area of leak was redesigned to eliminate dissimilar material joint and cycle additional 60 cycles before termination of tests.

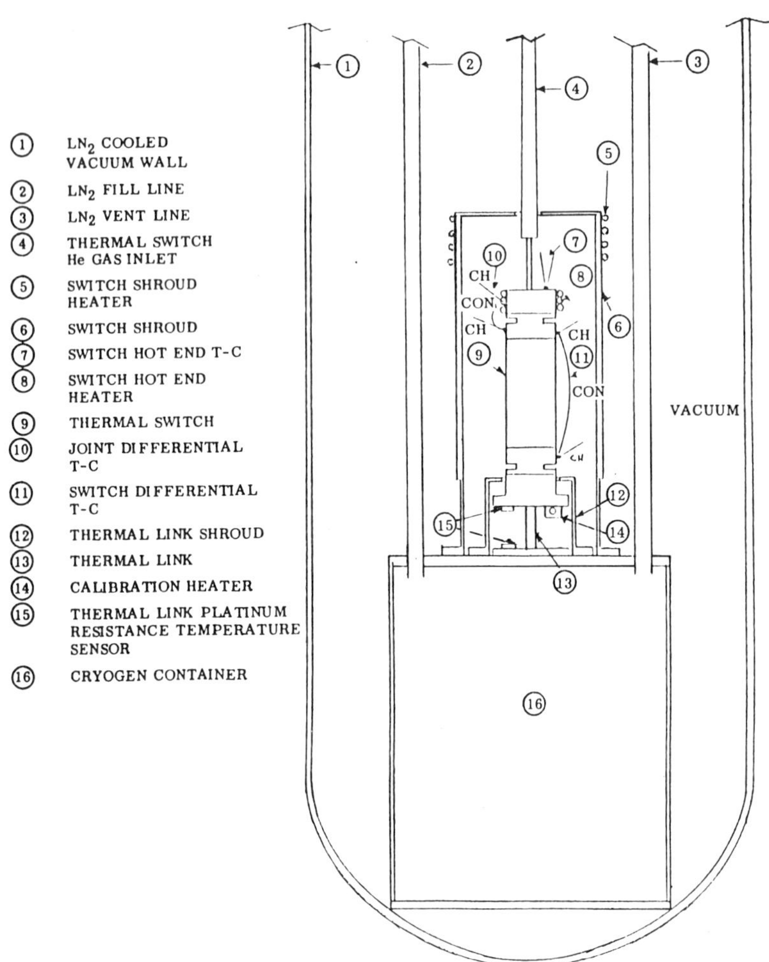

Fig. 3. Thermal switch test setup.

time, the "on" resistance measurement was repeated after filling the evacuated switch for one minute with a regulator pressure of 17.0 psia using helium gas. The resulting value was 0.88 K/W at a temperature of ~100 K. A test for measuring the switch "off" resistance after evacuation from the "on" condition was planned. However, no results were obtained, as internal outgassing from the warmer portion of the switch fill line required active pumping to maintain the switch in the full "off" position.

Life cycle tests were run using the setup shown schematically in Fig. 4; the pressure to the switch was cycled from 0 to 17 psia, while the temperature was cycled between 80 K and 230 K

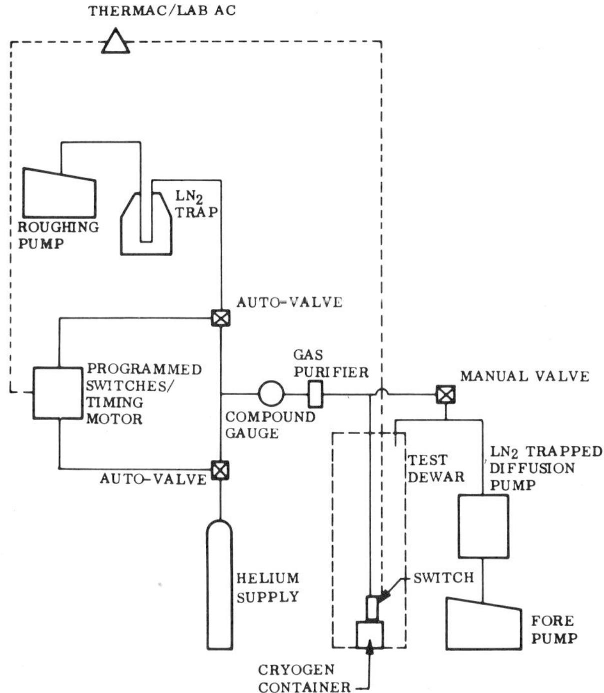

Fig. 4. Thermal switch life-cycling apparatus.

during a 30-minute interval. After 10 cycles, the switch was
checked for helium leakage from the interior. No leaks were
found. At some point between cycle No. 480 and cycle No. 525, the
switch began leaking, exhibiting itself as an increase in pressure
in the vacuum space surrounding the switch during the periods when
the switch interior was pressurized at ~17.0 psia. To eliminate
the leak, found to be at the joint between the stainless steel
switch fill line and the upper hot end, the dissimilar metal joint
was eliminated. Testing continued up to 585 cycles without incident, and was then terminated.

A portion of the actual life-cycle temperature data as recorded appears in Fig. 5. Note the discontinuity in the curve during
the warmup from ~80 K to 220 K. One possible explanation is a
temporary shorting due to the very rapid warming of the switch hot
end, necessary to reduce the total test time. In actual use, the
warmup period would be much slower than the 15 minutes on these
tests. At no time during the switch testing were other positive
indications of shorting under steady-state-boundary conditions
present.

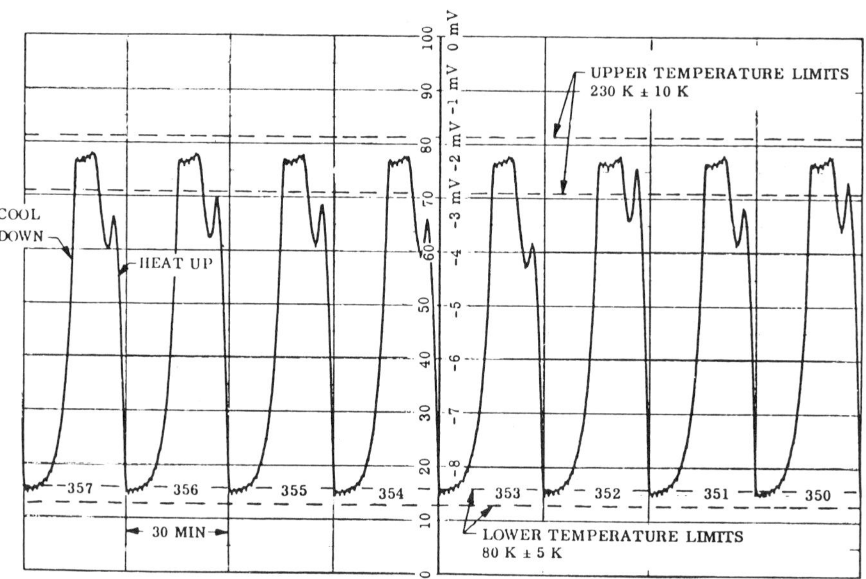

Fig. 5. Thermal switch temperature response during life-cycling tests.

CONCLUSIONS

It is concluded that a thermal switch of comparatively small size and weight can be made to operate reliably at cryogenic temperatures. A thermal switching ratio of 2300 was attained and 525 on-off cycles were demonstrated. This type of switch is feasible for long-life orbital operation; for some missions the lower duty cycles can reduce cryogen consumption and allow outgassing of sensor systems.

REFERENCES

1. "Surveyor VII Mission Report," NASA TR 32-1264 (JPL) (1969), 57.
2. J.M.L. Engels et al., Cryogenics 12:141 (1972).
3. A. Basiulis, ASME Paper 71-AV-29 (1971).
4. L.G. Naes and T.C. Nast, Long Life Orbital Operation of Stirling Cycle Mechanical Refrigerators, presented at SPIE 24 Annual Technical Symposium, 28 July -- 1 Aug 1980, San Diego, Calif.
5. W.E. Gifford, et al., in "Advances in Cryogenic Engineering, Vol. 15", Plenum Press, New York (1970), p. 424.
6. R.P. Bywaters and R.A. Griffin, A gas-gap thermal switch for cryogenic applications, Cryogenics 13:344 (1973).
7. S.H. Castles, Heat switch has no moving parts, NASA Technical Briefs (Fall, 1980).

CRYOPUMPING OF HYDROGEN AND HELIUM

H. J. Halama, H. C. Hseuh, and T. S. Chou

*Brookhaven National Laboratory**
Upton, New York

INTRODUCTION

Cryopumping is of interest to designers of both fusion reactors and particle accelerators since it can provide the following features: (1) a very high pumping speed over a large pressure range, (2) very low equilibrium pressure, (3) very large capacity, (4) cleanliness, (5) the ability to store the pumped species on the cryosurface for later recovery and (6) neither high voltage nor a magnetic field is required, which makes a cryopump ideal for operation in strong magnetic fields.

In Tokamak fusion reactors, the major gas species to be pumped are hydrogen isotopes from neutral beam injectors, and a mixture of hydrogen isotopes and helium from the torus. Multiampere neutral beam injectors will require vacuum systems capable of pumping large fluxes (>10 torr L/s) of deuterium and tritium while maintaining a pressure in the 10^{-5} torr range during neutral beam pulses. The gas flux from the torus will be lower than that from the neutral beam injectors. However, the necessity to maintain an equilibrium pressure of 10^{-8} torr between the neutral beam pulses requires a more complex vacuum system, such as a compound cryopump using cryocondensation and cryosorption.

In particle accelerators and storage rings, a low equilibrium pressure is more critical than the pumping speed and capacity. The residual gas inside the beam vacuum system is mainly H_2 or H_2 and He if superconducting magnets are used. A vacuum system with a cryosorption pump will pump these gases down to $\sim 10^{-11}$ torr.

---

*Work performed under the auspices of the U.S. Department of Energy.

We will first summarize our results obtained in numerous tests on small (~1000 L/s) and large (~10^5 L/s) cryopumps cooled to 4.2 by LHe or to between 10 to 20 K by closed-cycle helium refrigerators. We will also present our measurements of the compound cryopump, designed to work on Tokamak to handle the exhaust gases from the torus. A more detailed description of each cryopump and measurements can be found elsewhere.[1]

CRYOCONDENSATION PUMPING OF H_2, D_2 AND T_2[2-6]

Cryopumps

Several cryopumps having cooled surfaces from ~10^2 cm^2 to ~1m^2 and pumping speed for H_2 from ~1000 L/S to ~10^5 L/S have been constructed and operated. These pumps all contain the basic structure of a LHe cooled polished metal surface serving as a cryocondensation panel for hydrogen isotopes and a LN_2 cooled radiation shielding and chevron baffle surrounding the LHe reservoir and cryopanel.

Measurements

Details of each measurement were given in Ref. 1-5. Basically, known amounts of gas or mixture of gases were fed into the vacuum chamber and pumped by the cryopump. The partial pressure of each gas species at the cryopump side was measured to determine the pumping speed, surface coverage and equilibrium pressure as well as other parameters.

Results

Pumping speed. The measured pumping speeds of the LHe cooled surface for H_2, D_2 and T_2 are about 11, 8 and 6.5 L/S cm^2 respectively, for typical chevrons with a molecular transmission coefficient of ~ 25%. These pumping speeds correspond to an effective sticking coefficient of almost unity. Within 10% accuracy the pumping speeds for H_2 and D_2 are independent of surface coverage up to at least 45 torr L/cm^2 (1.5 x 10^{21} molecules/cm^2 or 2 x 10^6 monolayers or a condensed film 0.56 mm thick). The pumping speed is also independent of the input gas flux up to 3 x 10^{-3} torr L/s cm^2.

Equilibrium pressure. Pressures of approximately 1 x 10^{-6} and 5 x 10^{-8} torr (N_2 equivalent) are typical in our vacuum chambers when H_2 and D_2, respectively, are pumped and are independent of surface coverage up to 1.5 x 10^{21} molecules/cm^2. After correcting for gauge sensitivity and thermal transpiration, these pressures correspond to a real pressure of 5 x 10^{-7} torr for H_2 and 2 x 10^{-8}

torr for D_2 on the 4 K surface. The vapor pressures of H_2, D_2 and T_2 at 4.2 K are 5×10^{-7}, 5×10^{-11} and $< 1 \times 10^{-11}$ torr, respectively. This indicates that the saturated vapor pressure is reached above the condensed film of hydrogen. The contamination of hydrogen and helium gives the higher equilibrium pressure over the deuterium condensate. The pressure during tritium pumping is dominated by the presence of ^{3}He which originates from the beta decay of tritium.

Radiation effect and thermal conductivity. Both the equilibrium pressure and LHe boil-off rate show little change with increasing film thickness up to ~ 0.5 mm which equals to ~ 12 wavelengths of the most intense infrared radiation from the 77 K chevron and shield. Therefore the frozen H_2 and D_2 are transparent to the infrared radiation and will neither adversely effect the emissivity of the cryopanel nor raise LHe consumption. The total thermal radiation adsorption of 0.5 mm thick H_2 or D_2 film is estimated to be $< 5 \times 10^{-6}$ W/cm^2. The neutron flux estimated for a Tokamak will not exceed 10^{10} neutrons/s cm^2. A thick deuterium film condensed on the 4 K cryopanel was bombarded with either slow ($\sim$ eV) neutrons or fast ($\sim$ MeV) neutrons with fluxes up to 5×10^9 neutrons/s cm^2 for up to 1 h.[5] No increase in deuterium partial pressure (we could down to $\sim 10^{-12}$ torr) was observed, which indicates the desorption caused by neutron or neutron induced radiation is negligible.

Trapping of He by H_2 or D_2 and of H_2 by D_2. We have measured the pumping behavior of mixtures of D_2, H_2, and He at 4 K. The results are expressed as a trapping ratio Q_x/Q_y, where Q_x is the minimum flux of trapping gas X (the lower vapor pressure one), for which no trapped gas, Y (the higher vapor pressure one), diffuses from the composite film after the gas flow has been stopped. A value of $Q_{D2}/Q_{H2} \sim 9$ was obtained, indicating that an equilibrium pressure lower than 1×10^{-6} torr is reached when pumping a mixture of $> 90\%$ D_2 and $< 10\%$ H_2. The trapping ratios Q_{D2}/Q_{He} and Q_{H2}/Q_{He} of $\sim 5 \times 10^4$ indicates that one cannot use hydrogen isotopes to trap He.

The behavior of tritium. Compared with D_2 and H_2, tritium behaves very differently chemically in our stainless steel system. Tritium readily undergoes isotope exchange reaction with the hydrogen of the physisorbed H_2O molecules on the stainless steel wall and forms HTO, which explains the disappearance of T_2 at room temperature and the presence of mass 20 (presumably HTO) in the vacuum system. At T_2 pressure of 10^{-3} torr the partial pressure ratio HTO/H_2O is ~ 0.5. Our results suggest that the future cryopumps for fusion reactors must take into consideration the problem of surface treatment of the pump.

CRYOSORPTION PUMPING[7-9]

Cryopumps

Several cryosorption pumps have been constructed and operated to study the pumping behavior of He at 4 K and between 10 and 20 K. They are shown schematically in Figs. 1 and 2. The 4 K pumps consist of two chevrons and radiation shields, the outer ones LN_2 cooled to about 80 K, while the inner ones are maintained at about 20 K by the boil off from LHe reservoir. Both activated charcoal and molecular sieve (Linde 5A type) have been used as adsorbent.

A closed-cycle He refrigerator is used to cool down the cryosorption panel to 10–20 K. Only charcoal has been tested as the adsorbent at this temperature range. Although most measurements are done with epoxy-bonded charcoal, a 3.5% Ag–Sn alloy has been successfully used as a bond. In this technique charcoal is mechanically embedded in melted alloy and formed into a 3.2 mm thick plate. Metalized bonding has better thermal conductivity, higher resistance to radiation damage, smaller thermal expansion and does not block the porous paths on the surface of charcoal granules.

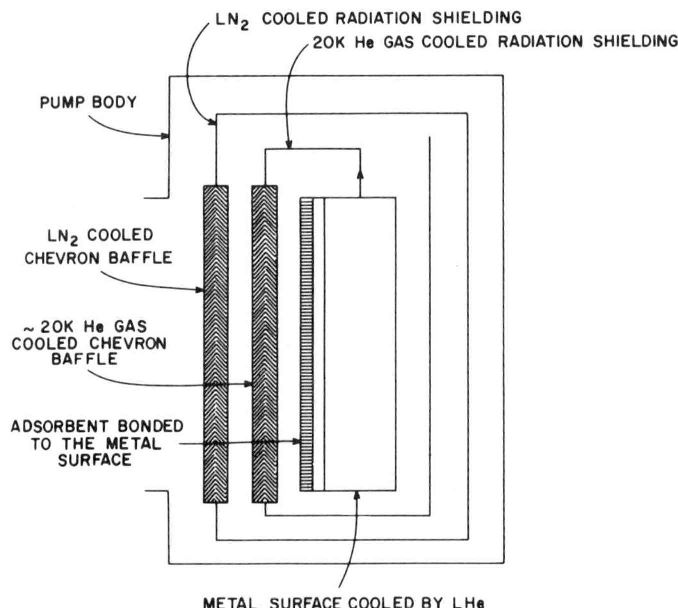

Fig. 1. Schematics of the cryosorption pumps cooled by LHe.

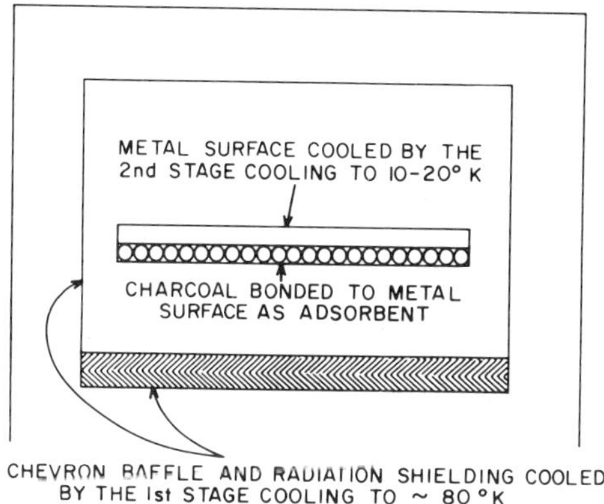

Fig. 2. Schematics of the cryosorption pumps
cooled by closed-cycle He refrigerator to 10-20 K.

Measurements

The pumping speed and the capacity of the adsorbent are very
much dependent on the number of available active sites on the
surface. A thorough reactivation to remove adsorbed gases, not-
ably water vapor, from these active sites is necessary to ensure
the proper operation of the cryosorption panel. To activate
molecular sieve, a vacuum bake at 300°C for at least 24 hours is
necessary. Charcoal can be fully activated with vacuum bake at
100°C for several hours. The procedures used in studying the

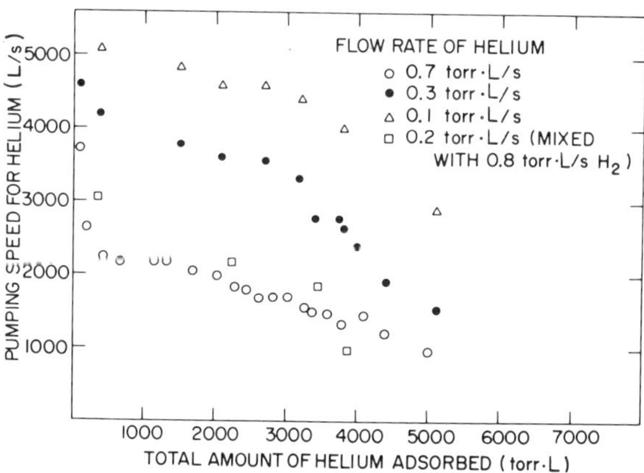

Fig. 3. Helium Pumping Speed of the Compound
Cryopump at Various Flow Rate and Surface Coverage.

pumping behavior of He are similar to those described before and will not be repeated.

Results

Pumping behavior of He by molecular sieve at 4 K. The measured pumping speed of ~ 4.5 L/s is virtually independent of surface coverage up to ~ 7 torr L/cm^2 and flow rate up to $\sim 2 \times 10^{-7}$ torr L/s cm^2. Beyond this, the pumping speed drops slowly with increasing surface coverage and reaches ~ 1 L/s at ~ 15 torr L/cm^2. At a flow rate higher than 1×10^{-5} torr L/cm^2, the pressure increases slowly until thermal instability occurs. The equilibrium pressure over the surface of molecular sieve follows the adsorption isotherm predicted by the theory of Dubinin and Radushkevich[10] down to $\sim 1 \times 10^{-10}$ torr and torr ~ 1 torr L/s cm^2. We have also studied the adsorption of H_2, equilibrium pressure of $\sim 3 \times 10^{-12}$ torr has been obtained at surface coverage of ~ 2 torr L/cm^2.

Pumping behavior of He by activited charcoal at 4 K and 10-20 K. An initial pumping speed of ~ 4 L/s cm^2 for He is achieved at temperature up to 17 K and at flow rate up to $\sim 7 \times 10^{-4}$ torr L/s cm^2. At 4 K the pumping speed is independent of surface coverage up to ~ 8 torr L/cm^2, and at higher temperature it drops off more rapidly with increasing surface coverage. The adsorption capacity decreases with increasing temperature from ~ 10 torr L/cm^2 at 4 K to ~ 1 torr L/cm^2 at 17 K. The difference in pumping behavior at different temperatures can be explained by the number of active sites available for the adsorption of He, which decrease with increasing temperature. Less than 30% decrease in pumping speed was observed at 4 K when flow rate was increased by a factor of 35 from $\sim 2 \times 10^{-5}$ to $\sim 7 \times 10^{-4}$ torr L/s cm^2, indicating that the rate of diffusion of helium from the surface of charcoal into the pores is reasonably fast compared with the sticking probability. The equilibrium pressures of He at 4 K increase from $\sim 1 \times 10^{-9}$ torr to $\sim 1 \times 10^{-8}$ torr for surface coverage from 0 to ~ 10 torr L/cm^2. The heat of adsorption derived from adsorption isosters decreases slowly with increasing surface coverage from 940 cal/mole at 1.7×10^{-2} torr L/s cm^2 to 760 cal/mole at 0.5 torr L/cm^2, which indicates that some adsorption sites have higher heat of adsorption and are occupied first. The low values of heat also indicate that the heat load to cryopanel from heat of adsorption is negligible compared to other sources like radiation and conduction.

The pumping behavior of the mixture of H_2 and He was also studied at 4 K and 13.5 K. No significant change in He pumping speed is observed when the fraction of H_2 in the gas mixture was small (< 5%). At higher percentage of H_2, the pumping speed of He dropped accordingly and the percentage of decrease (compared with pumping pure He) is more severe at higher temperature. The combined adsorption capacity of He and H_2 is very close to the capacity of pure He. This suggests that H_2 is competing with He for active sites on charcoal surface but not inhibiting the adsorption of He.

Comparison of charcoal and molecular sieve as adsorbent for He. The pumping speed and capacity of He at 4 K are comparable for both adsorbents, which is not surprising since both of them have comparable total surface area ($\sim$ 600 m^2/gm for Molecular Sieve 5A and $\sim$ 900 m^2/gm for charcoal). However, charcoal can tolerate a flow rate $\sim$ 1000 times higher than molecular sieve, and will make a much better crysorption pump for fusion reactors which generate a large flux of He between neutral beam pulses. Molecular Sieve can achieve a much lower equilibrium pressure ($\sim 10^{-11}$ torr) than charcoal, which makes it a good candidate for particle accelerators or storage rings where ultra high vacuum is required but the gas load is low. The presence of H_2 or other impurity gases has smaller effect on charcoal than on molecular sieve. In molecular sieve, the uniform distribution of small size pores (5 A in diameter) prevents the bigger impurity molecules like H_2O from diffusing into the bulk. They condense on the sieve surface and impede the diffusion of H_2 and He onto the adsorption sites. Therefore a thorough reactivation is necessary to remove all these impurity gases. Charcoal has nonuniform pores with size up to $\sim$ 20 Å , and the heat of adsorption is also less ($< 10^3$ cal/mole), which give higher mobility to adsorbed gases.

COMPOUND CRYOPUMP FOR EVACUATION OF FUSION TORUS

In order to pump plasma exhaust gases in fusion reactors a special compound cryopump was developed and tested. The construction details have been presented elsewhere.[1,9]

Cryosorption Pumping of He

The pumping speed of He as a function of flow rate and surface coverage is shown in Fig. 3. Values of $\sim$ 2000 L/s (L/s cm^2) are obtained with flow rate $\sim$ 0.7 torr L/s ($\sim$ x 10^{-3} torr L/s cm^2), and surface coverage up to $\sim$5000 torr ($\sim$ 8 torr L/cm^2), which are consistent with our results on the study of a smaller cryosorption pump. However, the pumping speed at smaller flow rate is higher for the compound cryopump (i.e. $\sim$ 5000 L/s at 0.1 torr L/s) and may be due to the higher thermal conductivity of the metalized charcoal, which enhances the diffusion rate of He from the surface of the charcoal granules to inner bulk.

Cryopumping of a mixture of He and H_2

Test runs were made on a mixture of 80% H_2 and 20% He. The initial pumping speed for H_2 was ~ 16000 L/s, as several monolayers of H_2 were condensed, the partial pressure of H_2 increased to ~ 4×10^{-6} torr and made the pumping speed measurements rather difficult. The pumping speeds for He as shown in Fig. 3 were lower than those of pure He (~ 30% less), which might be caused by the large uncertainty in measuring the He partial pressure when H_2 partial pressure reached ~ 4×10^{-6} torr, or small amount of H_2 condensate on the surface of charcoal granules.

The idea of using a compound cryopump as the heart of the torus vacuum system has been successfully demonstrated. The needs to separate and recover the unburned tritium from helium ashes make the combination of cryocondensation and cryosorption panels necessary. These could be achieved by using separate LHe reservoirs, transfer lines and temperature control devices. The activated charcoal has been proven to be a superior adsorbent for cryosorption pumping of He in a complex environment. Unlike Molecular Sieve, no high-temperature bake will be required to recondition the charcoal, and the presence of hydrogen and other impurities has little effect on the pumping speed and adsorption capacity of charcoal.

REFERENCES

1. H.C. Hseuh, H.A. Worwetz, J. Vac. Sci. Technol., 18:1131 (1981).
2. H.J. Halama, J.A. Bamberger, in "Proc. 6th Symp. Eng. Problems of Fusion Research", IEEE, New York (1975) p. 202.
3. H.J. Halama, C.K. Lam, J.A. Bamberger, J. Vac. Sci. Technol., 14:1201 (1977).
4. T.S. Chou, H.J. Halama, in "Proc. 7th Symp. Eng. Problems of Fusion Research", IEEE, New York (1977), p. 1790.
5. T.S. Chou, H.J. Halama, in "Proc. 7th Intl. Vac. Congr.", Vienna, (1977), p. 65.
6. T.S. Chou, H.J. Halama, J. Vac. Sci. Technol., 16:81 (1979).
7. H.J. Halama, J.R. Aggus, J. Vac. Sci. Technol., 11:333 (1974).
8. H.J. Halama, J.R. Aggus, J. Vac. Sci. Technol., 12:532 (1975).
9. H.C. Hseuh, et. al., in "Proc. 8th Symp. Eng. Problems of Fusion Research", IEEE, New York (1979), p. 1568.
10. P.A. Redhead et al., "The Physical Basis of Ultra High Vacuum", Chapman and Hall, London, (1968) p. 45.

DISCUSSION

Question by J.R. Coupland, Culham Laboratory, England: What gauge factor did you use in the calculation of pumping speed for H_2?

Answer by Author: Based on an anode to cathode potential of 150 V (Bayard-Alpert gauge), a gauge correction factor of 2.2 for H_2 relative to N_2 was used in our calculation of pumping speeds and flow rates.

Question by J.R. Coupland: What is the sticking coefficient for your surface?

Answer by Author: The sticking coefficients will depend on the surface temperature and coverage. At 4.2 K, $s(H_2, D_2, T_2) > 0.9$ for more than 3 monolayers on a metal surface. In helium at 4.2 K $s(He) \sim 0.8$ at $\theta \ll 0.1$ and $s(He) < 0.5$ at $\theta \sim 0.2$.

Question by B. A. Hands, University of Oxford, England: How was the Molecular Sieve bonded to the panel?

Answer by Author: The Molecular Sieve 5 A in our Excaliber CVR-1008 pump was engrooved into at 1.6 mm wide channel on a stainless steel plate.

CHARACTERISTICS OF A 550-mm ID LOW PROFILE REFRIGERATED CRYOPUMP

R. C. Longsworth

Air Products and Chemicals, Inc.
Allentown, Pennsylvania

INTRODUCTION

Small cryopumps cooled by closed-cycle cryogenic refrigerators such as described in 1977[1] have been widely accepted in the semi-conductor industry[2,3] and research applications requiring small clean high vacuum pumps.[4] The most common size pumps at present have inlet ports that are 193 mm and 295 mm. As a result of the increased use of cryopumps, there has been a growing interest for larger size units for larger research chambers and for some indus-trial applications such as metallizing chambers.

In 1978 we at Air Products developed a 550 mm ID cryopump which has a geometry similar to the smaller ones which can be described as a cup inside a cup.[1,5] A number of these pumps have been installed on high vacuum chambers where they are performing well. It was found while designing this unit that while the heat load increases with D^2, D being the inside diameter of the port, the mass of the panels increases approximately with D^3. This results in longer cooldown times for large cryopumps compared with small cryopumps of this geometry, e.g., 5.5 hours for the 550 mm ID unit versus 1.5 hours for the 193 mm ID unit.

This paper describes the results of work that has been done during the past year to increase the capacity of the expander, improve the geometry of the 550 mm ID cryopump, reduce its cool-down time, and measure cryopump performance characteristics under conditions of high heat loads, and mixtures of gases.

REFRIGERATOR

The 550 mm ID cryopump is cooled by the Model CS-208R Displex refrigerator which operates on a modified Solvay cycle. The compressor is a four cylinder two stage oil lubricated water or aircooled unit which draws 6 kW of power. Recent work on optimizing the timing and gas dynamics within the expander has resulted in increasing the ratings from 65 W at 77 K plus 6 W at 20 K to 80 W at 77 K plus 7.5 W at 20 K.

Figure 1 shows the measured refrigeration capacity of the unit used for the test program operating on 60 Hz power. This unit has no load temperatures of 30 K and 8.5 K and produces about 90 W and 7.8 W at 77 K and 20 K on the first and second stages respectively. The first stage intercepts most of the thermal radiation from room temperature 294 K which can be 100 W if both the emitting and absorbing surfaces are effectively black. Testing was thus extended to higher loads on the first stage so that the capacity curves could be used as a calorimeter for determining cryopanel heat loads over the range that was anticipated.

CRYOPUMP CONFIGURATION

Figure 2 shows the configuration of the cryopump. The feature which permits the cooldown time and size to be reduced, compared with the previous design, is the orientation of the expander parallel to the inlet louver.[6] Having the first stage heat station near the center of the louver permits a fairly direct radial heat

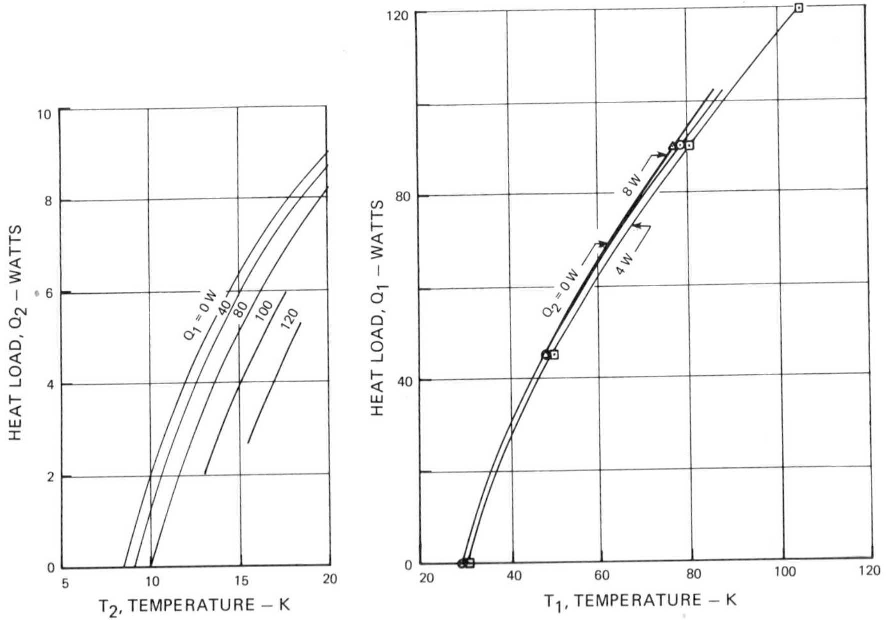

Fig. 1. Measured Refrigeration Capacity of the Model CS-208R Refrigerator Used for Test Program. Q_1, T_1, Q_2, T_2 are the refrigeration and temperature of the first and second stages.

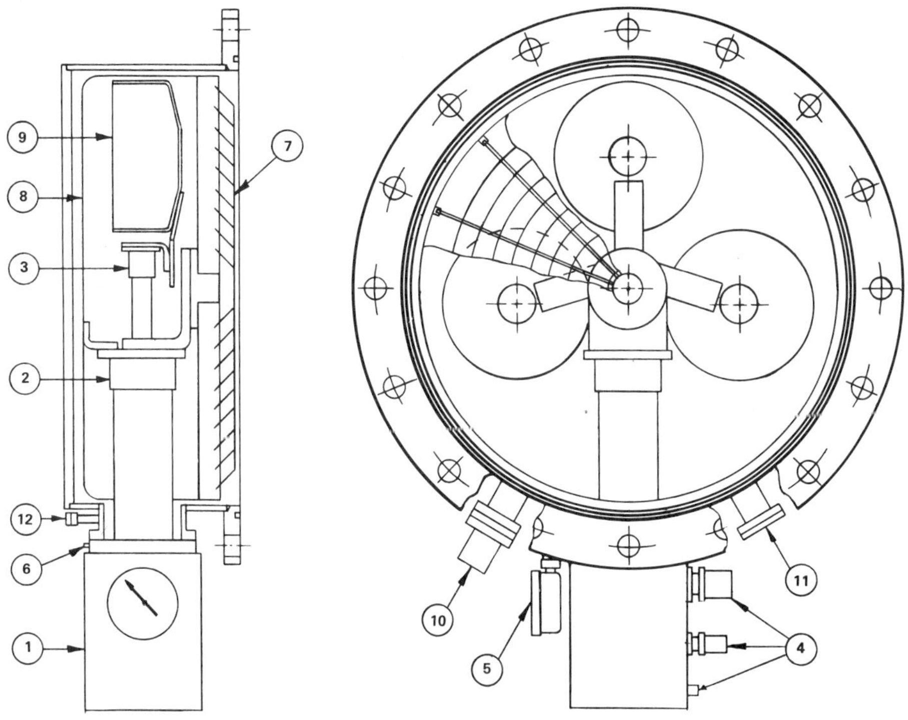

Fig. 2.   Model LP208-20C 550 mm ID Cryopump Configuration.

transfer path into the refrigerator. Weight of the inlet louver is
established by designing for a maximum temperature difference
between the edge of the louver and the refrigerator at 30 K at
rated heat load. The intent is to stay well below the temperature
of 145 K required to have an equilibrium pressure for water of
less than 10^{-8} Torr. The cold panel consists of three cups which
face away from the louver and are lined on the inside with charcoal. The cold panel is surrounded by the warm panel which is
attached at the center to the first heat station and at the edge
to the louver. Its inside bottom is painted black to shield the
charcoal from room temperature radiation.[7]

In operation it is typical to have water condense on the
louver, nitrogen, oxygen, and argon condense on the outside of the
cold panel, and hydrogen and helium be adsorbed on the charcoal.

Having the expander parallel to the louver permits the housing to be only 230 mm deep compared with about 1,500 mm for a comparable diffusion pump. Because of its compact geometry it is referred to as a low profile cryopump.

INSTRUMENTATION

The test dome, pressure gauges, and gas injection system are designed in accordance with the American Vacuum Society standard for testing diffusion pumps.[8]

Batch calibrated iron doped gold vs. chromel thermocouples are mounted on the refrigerator heat stations and at several points on the louver and cold panel. A hydrogen vapor bulb thermometer is also mounted on the cold heat station.

It is good vacuum practice to electropolish the inside of the stainless steel test dome in order to have a "clean" system, minimize heat loads, and achieve the maximum performance possible from a cryopump. Pumping speeds, gas capacity, maximum throughputs, cooldown time, etc., which are published as ratings are usually measured in a "clean" test dome and also with only one gas species at a time in the cryopump. In order to determine the effect of operating in a "dirty" chamber, the inside of the test dome was painted dull black.

TEST RESULTS

Cooldown

Tests were run to determine the effect of initial pressure in the test chamber on cooldown time with water being the principal residual gas in the chamber. Cooldown time to 20 K repeated within a few minutes for initial pressures from 1.0×10^{-2} to 2.0×10^{-1} Torr. Cooldown time is measured to be 210 minutes in the black test dome compared with a predicted time of 180 minutes in an electropolished test dome.

Heat Loads

Several series of tests were run under different conditions to measure the heat load on the refrigerator and the temperature differences across the cryopanels. Table 1 summarizes data from the following series:
1. Radiant heat load versus amount of water pumped.
2. Heat load versus argon flow rate after coating panels with water.
3. Heat load versus argon flow rate after loading pump with argon.

All of these tests were run at pressure below 1×10^{-3} Torr so
that conduction heating was negligible. Run 1a shows that the
heat load on the first stage of the refrigerator before introducing any gases into the test chamber is about half of black body
radiation from 294 K. The inlet louver is polished so the black
paint behind the louver is the primary absorber of the heat.
Adding 45 g of water to the pump converts the inlet louver to an
essentially "black" surface,[9] run 1b, which is not changed by
adding more water, run 1c. The absorption of radiant heat from
the test dome results in its temperature being an average of about
10°C colder than the room ambient which averaged 24°C. Water was
introduced at a rate of about 10 g/h which results in a chamber
pressure of 1×10^{-4} Torr and a heat load on the first stage of
approximately 10 W.

Runs 2a and 2b show the same high heat loads on the first
stage due to absorption of radiation by water on the louver plus
additional heat load on the second stage due to the cryocondensing
of argon. An argon flow rate of 1 SL/min imposes a heat load of
9.5 W on a cryopump of which about 7.3 W is carried by the second
stage.

Run 3a shows that a large cryodeposit of argon on the cold
panel does not result in appreciably larger heat loads than the
bare pump of run 1a. It was found that pumping on the residual
water vapor in the test dome over a period of several days would
change the emissivity of the louver enough to account for the
difference in heat load on the first stage that is reported between runs 1a and 3a. Run 3b shows the results of flowing argon
at 0.4 SL/min while the pump is loaded with 1,000 SL of argon.

Pumping Speed and Capacity

Studies by Bentley and Hands[10] show that argon and nitrogen
have capture coeffecients of 1.0 when they impinge on a surface
that is below the saturation temperature corresponding to the
partial pressure of the incident gas. This means that over a wide
range of pressures, cold panel temperatures, and cryodeposit
thickness, the pumping speed for these gases should be constant.
Speed for argon was measured to have an average value of 7,300 L/s
with a mean deviation of 600 L/s. The average speed is slightly
below the predicted speed of 8,000 L/s and the spread in data was
greater than has been seen during previous testing. Based on
argon speed, the speed for nitrogen should be 9,000 L/s. Water
speed is calculated to be 31,000 L/s. Helium speed was measured
to be 5,300 L/s initially dropping to 75% of this value with 0.22
SL adsorbed.

The ability to pump large quantities of argon by itself was
demonstrated by setting a flow rate of 0.45 SL/min. and allowing

Table I. Temperature Data and Heat Loads For Selected Operating Conditions

			Run No.				
	1a	1b	1c	2a	2b	3a	3b
Temperatures							
T_1 - K	52	84	86	83	89	60	62
ΔT_1 - K	17	41	42	38	41	14	16
T_2 - K	11.0	14.4	14.4	15.1	18.7	12.8	16.5
ΔT_2 - K	1	2	2	2	2	2	3
Gas Load							
H_2O - g	Trace	45	150	100	100	Trace	Trace
Ar - SL	0	0	0	.25	.24	1,000	1,000
H_2 - SL	0	0	0	.3	.2	0	0
Gas Flow							
Ar - SL/min.	0	0	0	.20	.40	0	.40
Heat Loads							
q1 - W	52	95	98	94	102	63	66
q2 - W	2.5	4.0	4.0	4.5	6.5	3.5	6.5

2,800 SL of argon to accumulate before shutting it down. Even with gas continuing to flow into the pump it took 112 minutes to warm up and vent. The argon vented in less than 1 minute and reached a peak pressure of 0.87 kPa (12.7 psig).

The effect of pumping water, argon, and hydrogen together was explored by first introducing 100 g of water then introducing argon and interrupting the flow of argon to measure hydrogen speed. The water subjects the pump to maximum radiant heat load and the argon reduces the speed and capacity for hydrogen by getting into the charcoal. Figure 3 shows the relation between hydrogen speed and the amount of argon that has been pumped. The standard cryo pump as shown in Fig. 2 has an initial speed of 12,000 L/s which drops to 8,700 L/s with 100 SL of argon. It was found while running these tests that the argon flow rate had to be reduced from 0.4 SL/min. with a small quantity of H_2 in the pump to 0.2 SL/min. towards the end of the run. The effect of excess argon flow rate is shown by the discontinuity in the curve when the pump warmed briefly to 32 K and allowed more argon to transfer to the charcoal.

A baffle was added to the cold panel to reduce the probability of argon getting into the charcoal. This reduced the initial speed of hydrogen to 9,500 L/s but enabled higher speeds to be maintained with more argon. Each of these runs was terminated by the pump warming up with argon flow rates of about 0.2 SL/min.

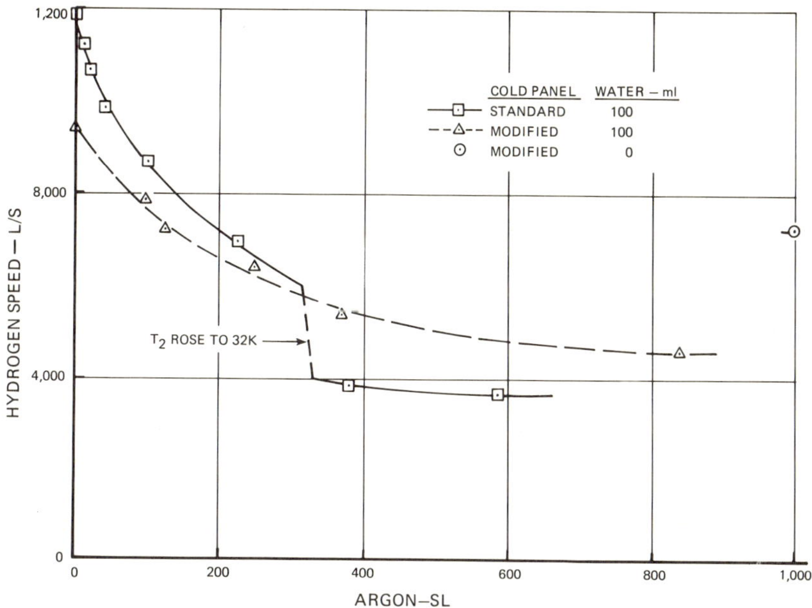

Fig. 3. Hydrogen Speed vs. Quantity of Argon Pumped.

Figure 3 shows one point for hydrogen speed with 1,000 SL of argon but no water in the pump. It is not known why the speed is appreciably higher than when water has been pumped.

Figure 4 shows a comparison of hydrogen speed vs. quantity of hydrogen absorbed as a continuation of the previous test with 1,000 SL of argon in the pump and compares it with the standard pump with no other gases present. With no other gases present, the charcoal adsorbed 14.4 SL of hydrogen before the speed decreased by 25%. With 1,000 SL of argon, the charcoal adsorbed 10.8 SL of hydrogen before the speed decreased 25%. This indicates that the argon has loaded about 25% of the charcoal. Two other points are noted in comparing these test results: first, it is possible to supply hydrogen at a much higher flow rate when it is the only gas being pumped and, second, the base pressure is lower.

SUMMARY

The 550 mm ID low profile cryopump was tested under conditions of high radiant heat loads, high gas loads, and pumping mixtures of gases. It takes only a small amount of water to convert the intially reflective cryopump louver to a condition where it ad-

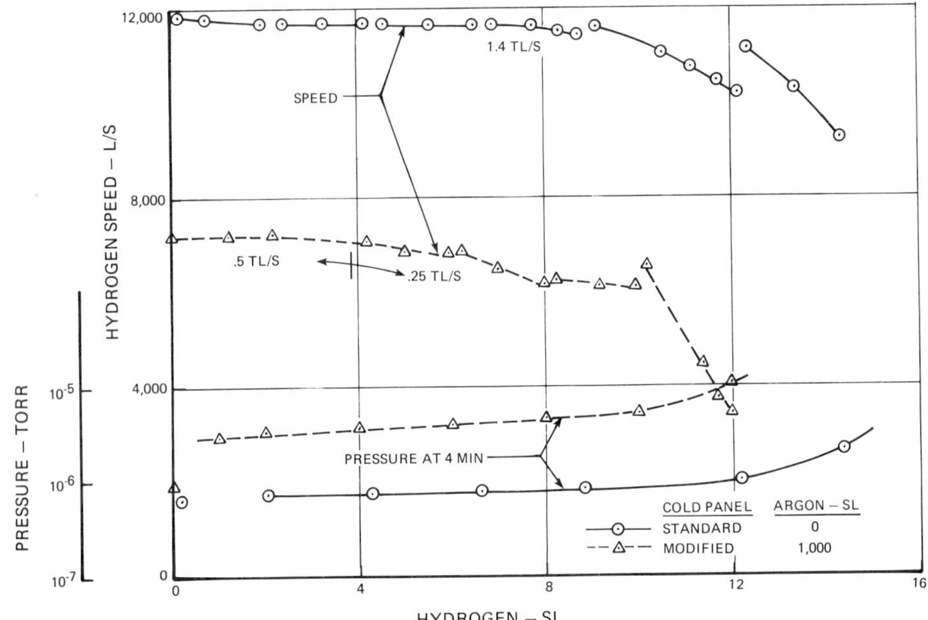

Fig. 4. Hydrogen Speed and Pressure Four Minutes
After Stopping Flow vs. Quantity of Hydrogen Adsorbed.

sorbs most of the incident thermal radiation. This reduces the
refrigeration available for pumping other gases and limits
throughputs to lower values than when tested individually. Argon
has a tendency to get into the charcoal and thus reduce the speed
and capacity of the pump for pumping hydrogen. The hydrogen and
water in turn reduce the total amount of argon that can be pumped.

Pumping speeds for gases that are cryocondensed are not appreciably effected by high heat loads, high gas loads, and gas mixtures but speeds for hydrogen and helium, maximum throughputs, and
total gas accumulations are significantly reduced. It is recommended that whenever possible the cryopump be located such that is
facing a clean reflective surface which will minimize the radiant
heat load.

REFERENCES

1. R.C. Longsworth, Performance of a Cryopump Cooled by a
    Small Closed-Cycle 10 K Refrigerator, in "Advances in
    Cryogenic Engineering, Vol. 23," Plenum Press, New York
    (1978) p. 658.

2.   J.E. deRijke, Performance of a cryopump-ion pump system, J. Vac. Sci. Technol., 15:765 (1978).
3.   R.W. Dennison and G.R. Gray, Cryogenic Versus turbomolecular pumping in a sputtering application, J. Vac. Sci. Technol., 16:728 (1979).
4.   M. Michaud and L. Sanche, Characteristics of a bakeable ion-cryopumped UHV system, J. Vac. Sci. Technol., 17:214 (1980).
5.   R.C. Longsworth, Cryopumping Method and Apparatus, U.S. Patent 4,150,549 (1979).
6.   R.C. Longsworth, Cryopumping Apparatus, U.S. Patent 4,277,951 (1981).
7.   R.C. Longsworth, Method for Coating Cryopumping Apparatus, U.S. Patent 4,219,588 (1980).
8.   American Vacuum Society Standard 4.8.
9.   R.P. Caren, A.S. Gilcrest and C.A. Zierman, Thermal Adsorptances of Cryodeposits for Solar and 290 K Black Body Sauces, in "Advances in Cryogenic Engineering, Vol. 9", Plenum Press, New York (1964) p. 457.
10. P.D. Bentley and B.A. Hands, The condensation of atmospheric gases on cold surfaces, Proc. R. Soc. Lond., A359:319 (1978).

CRYOPUMP OPERATIONS WITH THE TOKAMAK NEUTRAL BEAM INJECTOR PROTOTYPE*

R. A. Byrns and G. A. Newell

*Lawrence Berkeley Laboratory
University of California
Berkeley, California*

INTRODUCTION

In January 1977 the Lawrence Livermore and Berkeley Laboratories (LLNL/LBL) were commissioned to develop the prototype neutral beam line for the Tokamak Fusion Test Reactor (TFTR), at the Princeton Plasma Physics Lab (PPPL). Construction and installation proceeded during 1978-79 at Berkeley and the first cooldown of the cryosystem was in July 1979. The design of the prototype has been described earlier.[1-4] Each ion source is rated at 120 kV, 65 A and 0.5 s pulse length with deuterium (D_2) ($\sim$ 23 MW for 3 sources, $\sim$ 5.5 MW at target).

Starting from the right side of the beam line (Fig. 1) a brief description of the process follows. For the three sources, total deuterium gas flow is 90 Torr-L/s plus 10% from the ion dump. Electrons are stripped in the source, and the D^+ is accelerated to 120 kV. The energetic beam travels through the dense gas gradient created by excess D_2 source gas in the neutralizer and, through charge exchange, becomes neutral (D^o). The magnet permits the D^o neutrals to pass through and deflects the remaining D^+ ions into the ion dump. The beam then passes through the photo diode array into the calorimeter, both of which are used for diagnostics. With the calorimeter retracted, the beam can pass

---

*This work was supported by the Director, Office of Energy Research, Office of Fusion Energy, Development and Technology Division, of the U.S. Department of Energy under Contract No. W-7405-Eng-48.

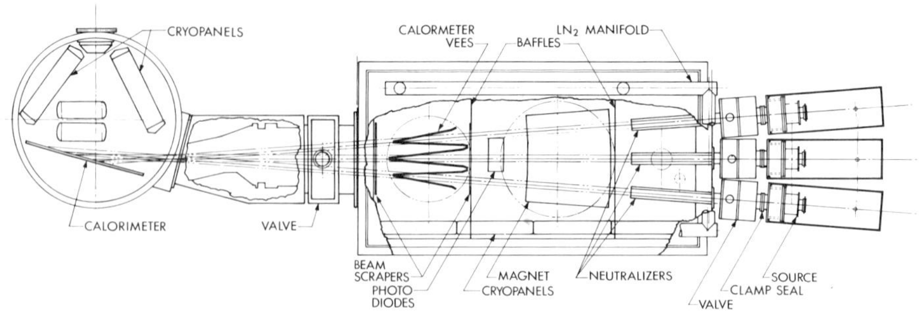

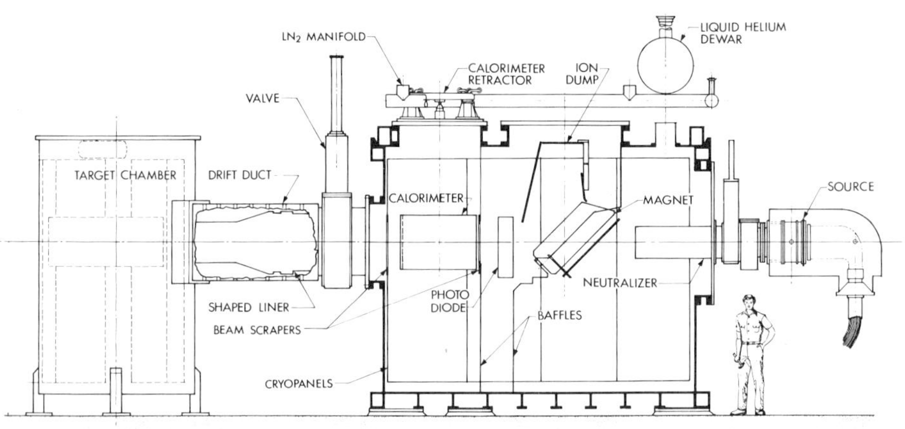

Fig. 1. Neutral beam injector with target tank.

through the duct, which models the Tokamak injector. Linear passage of a pure neutral D^0 beam, through the toroidal magnetic field and entry into the plasma is possible since any D^+ ions would be deflected. The beam provides fuel and heating for the fusion reaction plasma. At LBL the beam terminates on a water-cooled beam dump in the target tank.

The high gas flow required for optimum ion source performance, coupled with the proper neutralizer gradient and the low gas density downstream to prevent beam loss present a unique design challenge. This can be solved with cryopumping and conductance baffling.

CRYOSYSTEM COMPONENTS   (listed in Table I)

Table I. Neutral Beam Injector (NBI) Cryopump Equipment List

1. Injector Vacuum Vessel, 5.5 m long x 4.0 m high x 3.0 m wide, Type 316 stainless
   steel. Gas volume (less enclosed apparatus) 50,000 liters. Vessel weight,
   30,000 kg (66,000 lb). Lid weight, 17,300 kg (38,000 lb).

2. Mechanical Vacuum Pumps.

   a) One Stokes Model 1722 two-stage consisting of a 2313 m^3/h (1300 CFM) rotary
      dry lobe high vacuum booster and a Model 412 510 m^3/h (300 CFM) rotary oil-
      sealed pump.
   b) One Stokes Model 412 510 m^3/h (300 CFM) rotary oil-sealed pump.
   c) One Leybold-Heraeus turbomolecular pump. 3500 liter/sec (air) backed by a
      Leybold 26 liter/sec rotary oil-sealed pump.

3. Eight cryopanels, each module 1.2 m (48 in.) x 3.6 m (144 in.); total active pump
   area 30 m^2, design pump speed on D_2 ~ 2.5 x 10^6 liter/sec at 4.5 K. Mass of
   one LHe quilt ~ 91 kg, volume of one LHe quilt ~ 15 liters, weight of one module
   ~590 kg (two quilts 91 kg each, plus 409 kg frame with chevrons). Mass to be
   cooled to LN_2 ≃ 5900 kg, includes 1070 kg to LHe and requires minimum of
   3500 ℓ LN_2; 210 ℓ LHe.[1]

Heat Load to Panels	LBL (meas.)	PPL (meas.)[5]	LLNL Design[1]
Eight panels – LHe	30 liter/hr (21 W)	32 liter/hr (23 W)	45 W
Eight panels – LN	75 liter/hr	80 liter/hr (est)	104 liter/hr (4700 W)

4. One 750 liter LHe dewar, with LN shield, 40 layers of superinsulation to 80 K,
   10 to 4.5 K, vacuum insulated. Top and bottom co-axial bayonets.

5. One diverter valve contained within the 750-liter dewar.

6. One LN dewar, 20 cm diam. tube, 12 m long (~380 liters), vacuum insulated.

7. One Sullair oil lubricated helical screw compressor, 1500 m^3/h (880 CFM)
   displacement; delivery 55 g/s, 18 atm discharge, 3600 RPM, 300 kW, 2 pole
   induction motor.

8. One coldbox CCI (Cryogenic Consultants, Inc.) LN precooled. 400 W at 4.5 K with one
   dry (20 K) reciprocating expander engine and one wet (6 K) reciprocating engine.
   250 W with only the dry engine.

9. External LHe distribution system,[8] 72 m of separate supply and return 2.5 cm diam.
   vacuum insulated transfer line with 40 layers of superinsulation (18 W), 20 bayonet
   connections (25 W) and 4 broken stem valves (3 W) for flow control to the NBI and
   target tank. A 9 kW heater tied to GHe storage provides for warm-up/cooldown
   transients.

Cryopanels and Heat Loads

Eight modular cryopanels line both sides of the injector
vessel, four panels to a side. In addition, two cryopanel modules
are installed in the target tank. Panel construction (Fig. 2)

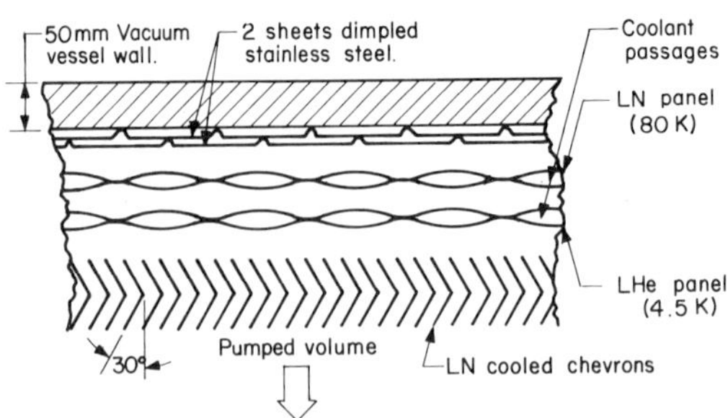

Fig. 2. Cryopanel assembly section.

consists of two stainless-steel quilts and a copper chevron assembly. The quilts are fabricated by joining two sheets of stainless steel with spot and edge seal welds and then hydroformed at 2.8 MPa (400 psi). For heat load reduction, the chevrons are painted black and the vacuum tank wall is shielded with two dimpled stainless steel sheets. The cryopanel heat loads as measured at LBL are 30.3 L/h (21 W) LHe by a 5-hour level decay in the LHe dewar. LN losses were measured by level decay at 75 L/h. These values correlate well with independent measurements by Garzotto[5] at PPPL of boil-off gas for a single panel, 4.5 L/h LHe and 11 L/h LN, and 32 L/h LHe for eight panels and the LHe dewar. Considering 60 m^2 of 4.5 K surface looking at 80 K, the 22 W/60 m^2 or 0.36 W/m^2 indicates a realistic value for emissivity ($\varepsilon < 0.2$).

Diverter Valve[6,7]

The diverter valve is a single package removable from the dewar as a unit. The valve has all-metal seats with varied spring loads and requires careful seat lapping, alignment and flow/leak calibration prior to installation. It combines four valve functions, and includes a phase separator for return flow from the panels. The diverter valve has three operating modes: 1) cooldown from 300 K where cold helium gas flow is forced through the bottom of the cryopanel 2) circulation, where phase separation occurs in the 750-liter dewar and the cryopanels are gravity fed, and 3) regeneration where the panels are emptied of LHe, warmed by gas conduction in the vessel, then the frozen deuterium is sublimed and pumped out, and LHe is transferred back to the panels for rapid cooldown. Figure 3 schematically shows the diverter valve (V-1, 2, 3, 4) located in the LHe dewar. It can be remotely operated by electric motors from the control room console.

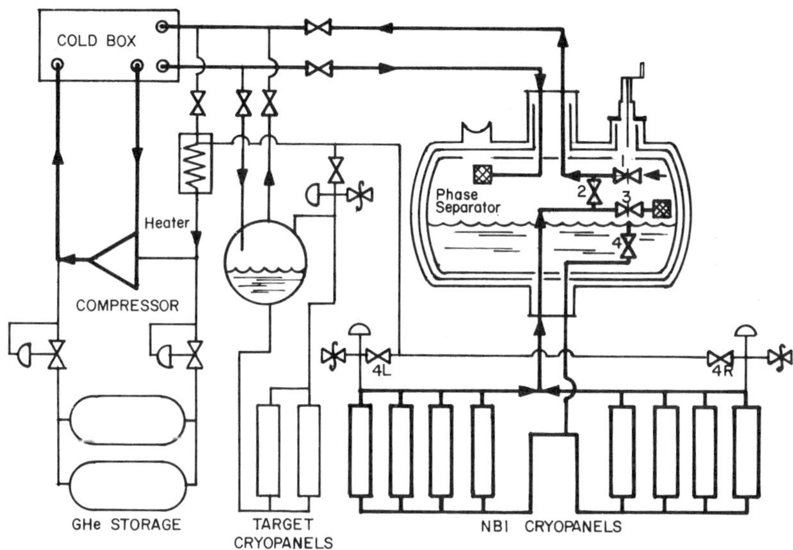

Fig. 3. Helium system schematic. For cooldown V-1 is closed, V- 2 and V-4 open, V-3 closed acts as 4 psi relief. For circulate V-1 is open, V-2 as 4 psi relief, V-3 and V-4 open. For regenerate V-1 is open, V-2 as a 10 psi relief, V-3 closed, and V-4 closed acts as a 5 psi relief.

REGENERATION

For safety reasons it is necessary to de-rime or regenerate the panels. The administrative limit[4] for frozen H_2 or D_2 has been set at 12 Torr (at 300 K in the 50 000-liter volume this is 600 000 Torr-L). For regeneration, valve sealing is critical to prevent leakage and provide pressure to transfer LHe from the panels into the dewar. Heat must be added to the panels to boil enough LHe to make the transfer and empty the panels, raise the temperature, and sublime the frozen D_2. The first "REGEN" test with frozen H_2 required 15 minutes after valve closing to initiate the cycle, the next test on D_2 required almost an hour. Since then we have done many "REGEN" tests with H_2, D_2 and He for initiation of panel de-rime.[6]

The preferred technique (Fig. 4) for handling the administrative limit is to admit 12 Torr-L of GHe to the vacuum. Response is within 10 minutes. The temperature goes to 66 K, the pressure to 4 Torr, pumps are valved in, valve V-4 is opened and the pressure drops to 10^{-7} Torr within 35 minutes. The pressure

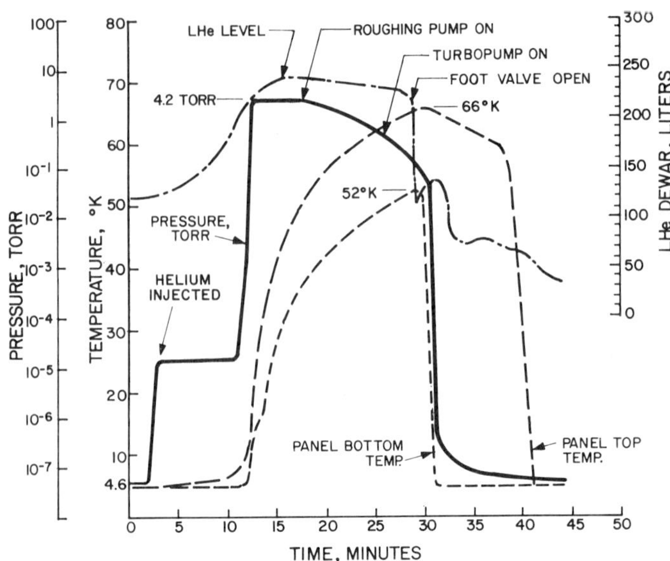

Fig. 4. Cryopanel regeneration with diverter valve.
12 Torr-L GHe injected to remove 13 Torr @ 273 K
frozen D_2

rise to 4 Torr shows the gas is very cold, 100 K, which is also
significant for beam line gas density. About 90% of the frozen D_2
is removed by the mechanical pumps before the foot-valve, V-4 is
opened.

VACUUM PUMPING SYSTEM

Design values for the 30 m^2 of cryopanel were 2.5 x 10^6 L/s
pump speed using a specific speed of 8.3 L/s-cm^2 for deuterium.[1]
For a gas load of 100 Torr-L/s pressure would be 4 x 10^{-5} Torr.
To minimize pressure at the entrance to the Tokamak, the beam line
vessel is divided into three differentially pumped sections. The
baffles, shown in Fig. 1 block each section except for beam
passage apertures at the magnet and the end section. Linear
density of gas along the beam path to minimize beam loss is of
more importance than pumping speed per se. Thus, location of the
baffles is critical and can be optimized.

During gas loading of the cryopanels for the de-riming tests,
hydrogen gas pulses up to 40 Torr-L/s for periods up to 100 s were
injected. Pressures at the exit end remained constant at 2 x 10^{-5}
Torr. Similar pulses for D_2 of 30 Torr-L/s for 10 s gave 2 x 10^{-6}
Torr (from 1.5 x 10^{-7} Torr) at the exit end, with a rise to 5 x
10^{-5} Torr and return to 2 x 10^{-7} Torr at the neutralizer. These

values seem consistent with H_2/D_2 vapor pressures, and show the cryosystem meets acceptable criteria for beamline pressure gradient. A more detailed report of beam line performance will be published by J. Feist, Max Planck Institute, Garching.[8]

REFRIGERATOR AND OPERATIONS

Initial refrigerator capacity selection was based on 1) an estimated 100 W for the NBI plus external distribution system, 2) simultaneous test operation of the General Atomic Doublet III injector, and 3) cryopanel loads for a target tank on the end of the NBI. The first specification called for 400 W or 80 L/h lowered to 300 W or 80 L/h to permit all turbo system bidders to quote. Four firms quoted in 1977 and Cryogenic Consultants, Inc. of Allentown, Pa. was low at $300,000. The refrigerator package included a cold box, an 80 K adsorber, two expander engines and an oil lubricated screw compressor with oil clean-up and controls.

Both LBL and PPPL[5] measured the heat load at 22 W for eight cryopanels and the LHe dewar. External heat loads were calculated[9] at 46 W, and the target tank cryopanel heat load is estimated at 8 W when on-line. Total heat load is then estimated at 68-76 W. The LN pre-cooled refrigerator, with 47 g/s mass flow at 18 atm, has delivered 85 L/h while holding the 68 W static heat load. On this basis, performance should be in excess of 475 W or 95 L/h and easily meets the original specification of 400 W or 80 L/h.

The refrigerator is run at maximum capacity mass flow for cooldown; LHe filling of eight panels with total volume of 120 L plus 200 L in the dewar matches the gas storage capacity and requires about 4 hours. Then the refrigerator is backed down to a flow of 21 g/s at 11 atm where the compressor draws about 150 kW. The helium system as of October, 1981 has about 4300 clock hours and only requires occasional operator time. The simplified control system[10] using a standard industrial controller (Barber Colman) and a suction pressure transducer (Gulton Corp.) permits long periods of unattended night and week-end operation.

The excess refrigerator capacity has been a great blessing, providing rapid cooldown and liquefaction, great operating time economy and enhanced system reliability due to a wide margin for error. The screw compressor with its slide valve and pressure controls together with the cold box reciprocating expanders permit an infinite number of pressure and mass flow combinations, minimizing power draw for various operating modes.

REFERENCES

1. L. E. Valby, Cryopumping System for the TFTR Neutral Beam
    Injectors, in "Proc. 7th Symposium on Engr. Problems of
    Fusion Research", IEEE, New York (1977).
2. L. C. Pittenger, et. al., Final Design of the Neutral Beam
    Lines for TFTR, in "Proc. 8th Symposium on Engr. Problems
    of Fusion Research", IEEE, New York (1979).
3. K. Lou, et. al., "TFTR Neutral Beam Injector Prototype
    Hardware," in "Proc. 8th Symposium on Engr. Problems of
    Fusion Research", IEEE, New York (1979).
4. L. C. Pittenger Large Scale Cryopumping for Controlled Fusion,
    in "Advances in Cryogenic Engr., Vol 23", Plenum Press, New
    York (1977).
5. V. Garzotto (PPPL), private communication, February 1981.
6. R. Byrns, G. Newell, Diverter valve operation, LBL Eng. Note
    M5569, LBID 270, (March, 1981).
7. G. Newell, W. Oglesby, Diverter valve rework, LBL Eng. Note
    M5610, (October, 1980).
8. J. Feist, in "Proc. 9th Symposium on Engr. Problems of Fusion
    Research", IEEE, New York (1982).
9. J. Carrieri, He transfer lines, LBL Eng. Note M5356, (April
    1979).
10. R. Curtis, Automation of the refrigeration system, LBL Eng.
    Note M5700, LBID 376, (March, 1981).

DISCUSSION

Question by W. J. Schneider, Brookhaven National Laboratory: What
was the performance of the transfer lines?

Answer by author: We have measured a heat leak of 0.25 W/m on
test sections of 50 mm LHe lines and expect the 25 mm lines for
the NBI to have a heat leak of 0.125 W/m.

Question by C. B. Hood, CVI Corp: Are you able to maintain closed
loop operation when you recool after regeneration?

Answer by author: Yes, helium is conserved and recovery is only
slightly dependent in the opening speed of valve V4.

CRYOGENICS FOR NEUTRAL BEAM TESTING*

G. W. Tasoulias

*Grumman Aerospace Corporation
Bethpage, New York*

and

K. E. Wright

*Princeton University
Princeton, New Jersey*

INTRODUCTION

Princeton Plasma Physics Laboratory (PPPL) will utilize cryocondensing techniques to attain high vacua at extremely fast pumping speeds for the four Neutral Beam Lines, a plasma heating device for the Tokamak Fusion Test Reactor (TFTR). Each Neutral Beam Line (NBL) contains eight cryopanel arrays of which half are located on opposite sides. Each array consists of a liquid helium quilted panel protected with nitrogen cooled chevrons on the inside and a quilted panel on the outside surface. The NBL helium panel is operated at 4.5 K or 3.8 K for pumping either deuterium or hydrogen respectively, reflecting the differences in vapor pressures. The pumping speed of 2.5×10^6 L/s for each NBL is required to maintain 10^{-5} to 10^{-6} torr pressure levels within the vacuum enclosure while injecting 45 to 90 L/s during the 1.5 s NBL pulse. To reach reaction temperature, the TFTR will employ four NBLs for injection of neutral deuterium or hydrogen particles through a connecting duct into the plasma contained in a toroidal chamber. The toroidal chamber has a volume of 71,000 L approximately the same volume as that of each NBL. The NBL is evacuated by turbo-molecular and mechanical pumps to obtain 10^{-4} torr for thermal insulation prior to flow of cryogens.

---

*This work was supported by the U.S. Department of Energy under Contract DE-AC02-76-CHO-3073 with Princeton University, Princeton, New Jersey.

The four NBLs[1] are connected to the toroidal chamber in the experimental test cell. The Neutral Beam Test Cell (NBTC), located in an adjoining area is used for the testing of the complete (three ion source) NBL connected to a test target chamber through a duct as shown in Fig. 1. The duct contains a vacuum valve to isolate the NBL from the test target chamber and simulate the connecting duct for TFTR. The test chamber contains two cryo arrays, a water cooled beam dump and beam diagnostic apparatus. This complete assembly is housed within a concrete enclosure for radiation shielding. An adjacent area is provided for the check-out and testing cryogenic components of the NBL.

The objectives for the NBTC are to simultaneously operate three ion sources within a single beam line, to study beam transmission through the confining apertures of the duct and to gain operational experience.

The TFTR facility contains a common liquid nitrogen system, and two independent helium refrigerator systems[2,3]. One helium system is for the experimental test cell while the other one for the testing of the 3 ion source NBL in the NBTC. The present paper describes the cryogenic equipment and design for the NBTC.

HELIUM REFRIGERATOR DESIGN FEATURES

The nominal 225 W helium refrigerator unit provides liquefaction/refrigeration at 3.8 K or 4.5 K. The cold box contains two delivery lines with individual J-T expansion valves, one to supply

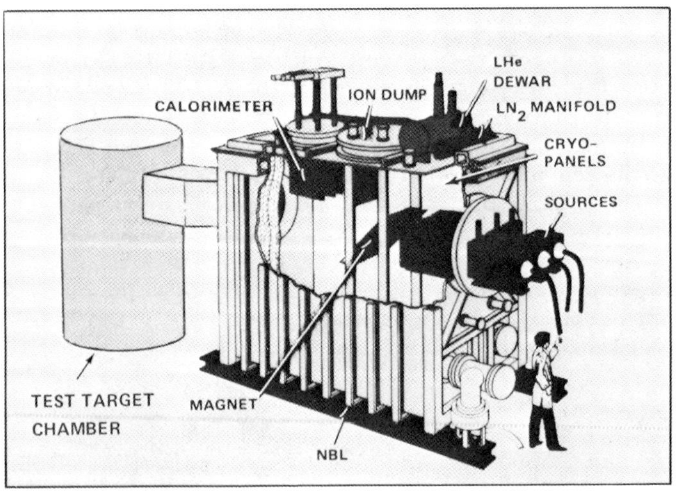

Fig. 1. NBL connected to test target chamber.

the NBTC and the other for additional tests. The primary delivery line is connected to a storage dewar through a coaxial transfer tube, while the second port will be connected using a conventional transfer line. The liquid helium supplied to the dewar normally operates slightly above atmospheric pressure, except when operating at 3.8 K. Subatmospheric operation is accomplished by lowering the compressor suction which also decreases dewar pressure. This normally would have a detrimental effect on refrigerator capacity, but an ejector circuit replaces the J-T valve to increase capacity. The ejector greatly increases the compressor suction pressure, in comparison to a J-T system, while maintaining lower pressure and temperature in the NBL and test target chamber. In both operations at 4.5 K and at 3.8 K with the ejector, the helium is transferred via the coaxial line to the storage dewar where the liquid and gas mixture is separated. The gas returns through the coaxial line and the liquid is delivered to the NBL through an expansion valve. The reduced pressure gas from the NBL is returned to the suction side of the ejector where it is compressed and added to the helium stream exiting the ejector nozzle for delivery to the storage dewar.

Two series, centrifugal gas-bearing turbines are employed, with cooled brake circuits for work extraction. Besides isolation valves on both sides of each turbine, cross-over valves are provided to interchange turbines in the event of failure. This condition is most advantageous when both turbines are identical. Currently, at 16 mm. first-stage turbine and 14 mm second-stage are employed to increase capacity. Interchangeability of spare cartridges permits the installation of a 14 mm turbine into the 16 mm housing.

In the event the 16 mm turbine fails, refrigeration will cease, while a 14 mm turbine failure only reduces capacity. The ability to replace spare cartridges in a few hours minimizes shutdown time or reduced load operations.

REFRIGERATION

The refrigerator (Table I) supplies two-phase helium through the coaxial transfer line at above atmospheric pressures to the storage dewar for 4.5 K or 3.8 K operation. The 1000 liter storage dewar is vacuum jacketed with multi-layer insulation. Helium is delivered from the dewar at 2.99 g/s to the NBL and 2.69 g/s to the test target chamber with maximum 50% quality. The refrigeration capacity is able to provide for line heat leak plus steady-state load requirements. Return helium from the load and storage dewar is returned through the cold box to the compressor suction. The turbine expander process stream is also returned to the compressor suction. During 3.8 K ejector operation the load return is to the ejector throat and not to the compressor suction

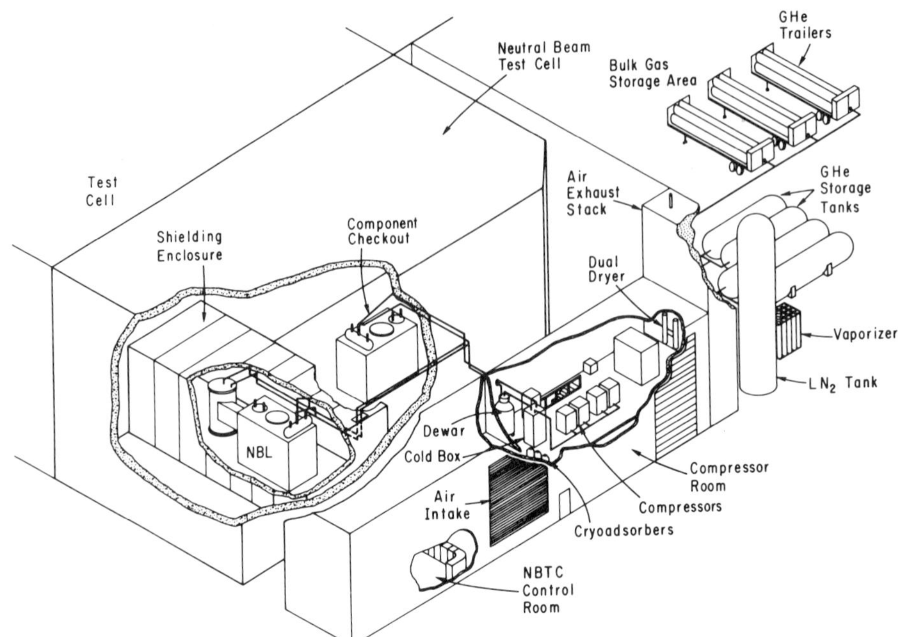

Fig. 2. Schematic of north east section of the TFTR facility.

In this configuration the boil off from the load is returned at 3.88 K and 0.618 atm.

Helium gas is supplied from two 81 m^3 external storage tanks connected to storage trailers for make-up gas. Although the refrigerator is capable of liquefaction at 3.8 K and 4.5 K, the latter mode will yield greater capacity. Liquefaction will be performed at 4.5 K and once the dewar is full, the machine is adjusted to either the 4.5 K or the 3.8 K refrigeration mode as needed. The piping configuration does not warrant a separate warm return line from the independent loads. Primary operation will be with both loads simultaneously cooled down, the return helium introduced for effective cooling according to its temperature into the appropriate cold box heat exchanger. In the event an independent load is cooled while the others are operating, the return gas from both loads will be mixed and introduced to the proper temperature heat exchanger. During this time the refrigeration will be in a hybrid mode with decreased capacity until the cooldown is completed.

Table I. Refrigeration Description

Liquid Nitrogen:		
Operation Pressure, psia		35 - 70
NBTC Consumption, GPH		90
Test Cell Consumption, GPH		150
Storage Capacity, Gallons		11,000
Liquid Helium:		
Liquefaction:	4.5K, l/h	60
Refrigeration:	3.8K, g/s	9.66
	4.5K, g/s	12.12
Compressor Power:		
Liquefaction:	4.5K, kW	305
Refrigeration:	3.8K, kW	193
	4.5K, kW	305
Refrigeration Minimum Turndown, %		30
Compressor Performance:		
3 Compressors with Ejector 3.8K:		.652
Ejector Suction, atm		1
Suction, atm		1
Discharge, atm		17
Flow, g/s		40
3 Compressors. 4.5K:		
Suction, atm		1
Discharge, atm		18
Flow, g/s		45
Utilities:		
Liquid Nitrogen:		
Liquefaction, l/h		60
Refrigeration, l/h		20
Cooling Water;		
Turbines, GPM		5
Compressors, GPM		80

Slide valves on the semi-hermetically sealed oil-flooded rotary compressors provide adjustable control of suction flow. Turn-down capacity of the compressors permit balancing the load to refrigerator capacity. With four 150 HP compressors connected in parallel, variable capacity is obtainable by slide valve control and by shutting down individual compressors. For 4.5 K, only three compressors are required, allowing the fourth compressor to be an in-situ spare. Operation at 3.8 K may require using 4 compressors. In the event of a compressor failure, the failed compressor is isolated and repaired without affecting operations. An overhead monorail is provided for equipment removal.

Liquid nitrogen is provided from 50 m^3 storage dewar for pressurized transfer to the loads. The LN_2 piping system is shared between the test cell and NBTC, supplying cryogen to NBLs, test target chamber and refrigerators. In the NBTC, 100 L/h is supplied to the NBL, 25 L/h to the test target chamber and 100 L/h for component checkout. An additional 90 L/h is supplied to the refrigerator cold box and cryoadsorbers in the compressor room. Heat leak of supply piping has no appreciable consumption effect. The system load parameters are shown in Table II.

Interlocks are provided in the system to prevent equipment damage and to safeguard against personnel hazards. The piping, test target chamber and NBL is protected with these interlocks, while the refrigerator is provided with independent controls for self-protection. Refrigerator operation is accomplished through local manual controls from the refrigerator control panel, while the piping is controlled from the NBTC control room, either manually or computer-assisted.

PIPING TRANSFER SYSTEM

The implementation of a direct pipe line configuration was not possible due to constraints imposed by the facility, equipment interferences, radiation enclosure and the location of the refrigerator. The main helium piping headers cross into the Neutral Beam Test Cell from the compressor room, where the pipes run along the east wall beneath the overhead handling equipment. The pipes drop vertically through one penetration in the shielding enclosure ceiling and run into the area for manifolding to the usage and

Table II.   Primary Load Parameters

Liquid Nitrogen Load	
NBL, l/h	100
Test Target Chamber, l/h	25
Component Checkout, l/h	100
Refrigerator, l/h	60
Cryoadsorbers, l/h	6
Total Nitrogen, l/h	291
Helium Supply Load	
NBL, W	40
Test Target Chamber, W	30
Supply Piping, W	37
Supply Load	107
Helium Return Superheat Load	
NBL, W	5
Return Piping, W	55
Return Load	60
Total Supply and Return Load, W	167
3.8K Refrigeration Capacity, W	200
4.5K Refrigeration Capacity, W	225

points. Supply and return jumpers connect the piping to the NBL
and test target chamber. The jumpers are inverted U shape, consisting of hard and flexible sections to allow for expansion/con-
traction and misalignment. The bayonets permit the removal of the
jumpers for servicing of the NBL.

The shielding ceiling consists of nearly identical rectangular blocks, which are removable for access to the equipment.
Removal of any ceiling or side wall blocks does not require dis-
connecting cryogenic lines. The height of lines is dictated by
the need to remove the NBL over the piping and shielding enclosure
walls. The piping is fabricated in spool sections to enable
transportation and installation within the facility. Transition
joints have been located in designated sections of the piping for
removal, repair or future reuse of the piping in a different
configuration.

The nitrogen piping enters the NBTC from the south wall and
follows a similar path as the helium piping into the shielding
enclosure. A supply of nitrogen and helium for the component
checkout is provided from the main headers. Longer jumpers are
required to supply and return helium and nitrogen to the checkout
area. It was impractical to design the jumpers to be interchangeable. Each jumper is designed for a specific application. The
helium connection at the NBL is a coaxial type bayonet with the
supply jumper connected to the top and return jumper to the
side. Removal of this bayonet requires simultaneous removal of
two jumpers. The nitrogen supply and return jumpers are individually connected. Each of the pipe spool sections is manufactured
of 304 stainless steel and contains an individual vacuum enclosure
with a vacuum pump-out port and thermocouple gage. Although
vacuum integrity is generally insured, an occasional leak over a
period of time into the vacuum enclosure may occur. A maintenance
program with periodic readout of the vacuum levels, assures continuous operation with minimum loss of operational time. The
cryogenic nitrogen and helium valves are vacuum jacketed. Each
load contains an inlet and outlet isolation valve. A bellows-
sealed type for the helium lines. Bayonets are limited to
jumpers, where removal of load is required. Downstream of
isolation valves, purge ports are provided to allow evacuation and
back filling of the load to remove gaseous contaminants prior to
cool down.

On the return nitrogen line, a multi-element array heater
with spare elements heats the nitrogen gas to ambient temperature
prior to venting through the 18 m exhaust stack to prevent cold
nitrogen accumulation at ground level.

PURIFICATION SYSTEMS

Redundant 80 K and 20 K purifiers are located in the cold box. Both purifiers are provided with isolation valves to permit regeneration of one at each temperature level without affecting operation. The purifiers contain activated charcoal and can be regenerated in approximately 4 to 6 hours. External purification units consisting of dryers and 80 K adsorbers are installed on the helium inlet supply and compressor discharge to the cold box. This additional equipment provides protection against contaminated gas entering from the storage trailer and storage tanks into the refrigerator system. Precautions have been taken in the selection of components to prevent the occurrence of air in-leaks. Moisture sensors and an air purity monitor are provided in some flow streams for monitoring the helium gas purity. The helium inlet purifier has sufficient capacity to clean initial charges and the makeup gas. The compressor discharge purifier will take only a partial stream due to its limited capacity.

HELIUM STORAGE

The total system helium inventory can be stored in the 1000 liter storage dewar as liquid. The NBL can contain 100 liters in the cryopanels and up to 750 liters in its dewar. The test chamber cryopanels contain 25 liters with a smaller dewar of 35 liters. Helium piping will account for about 75 liters. During operation the majority of the liquid helium inventory will be in the NBL and test target chamber. During short duration shut downs, the NBL and test target chamber are allowed to warm up and the returned helium is reliquefied and stored in the 1000 liter dewar. The remaining gas is returned to the gaseous helium storage tanks. Dewar helium loss rate due to boil-off is one percent per day. This boil-off can be vented to atmosphere or reliquefied. For extended maintenance periods the NBLs and test target chamber are allowed to warm up and the gaseous helium compressed and stored in the external helium tanks. The gaseous helium storage tanks with 2500 m^3 volume, operating at 1.8 MPa maximum have the capability to store the total system helium inventory. In the event of power failure, emergency power is available for one compressor to re-compress the return boil-off gas to the storage tanks. During operation, each of the three trailers store 1250 m^3 of gas at 16.5 MPa to make up losses. This inventory is sufficient for initial startup of either the large refrigerator for the experimental test cell or for the NBTC refrigerator. The storage trailers are road transportable and will be resupplied as the need arises. Commercial suppliers can also refill the trailers locally through a manifold.

The gaseous helium storage tanks are provided with ports and isolation valves for the future addition of an external purifica-

tion system. This system will provide clean-up of individually
contaminated tanks without affecting refrigeration operation.

SUMMARY

The refrigeration system of the Neutral Beam Test Cell is
expected to be versatile, providing independent operation from the
larger refrigerator, 3.8 K and 4.5 K temperature levels and flexibility for component testing. The use of the ejector circuitry at
the 3.8 K greatly increases the capacity. This system is to be
locally operated with design capability built in for hookup to the
central computer for automated control. Initial testing of the
test target chamber with 3 ion source NBL is planned to begin in
the spring of 1982. Considerable safety is designed into the
system and redundancy provides a highly reliable operation to
prevent costly down time periods.

REFERENCES

1. L.E. Valby, in "Proc of the Seventh Symp. on Engr. Problems of
      Fusion Research, Vol. II.", IEEE Pub. No. 77CH1267-4-NPS,
      New York, (1977), p. 1040.
2. G.R. Pinter and G.W. Tasoulias, in "Proc. of the Eighth Symp
      on Engr. Problems of Fusion Research, Vol. IV.", IEEE Pub.
      No. 79CH1441-5NPS, New York (1979), p. 2147.
3. G.R. Pinter, in "Proc. of the Seventh Symp. on Engr. Problems
      of Fusion Research, Vol. II.", IEEE Pub. No. 77CH1267-4-
      NPS, New York (1977), p. 1035.

PRODUCTION AND USE OF HIGH GRADE SILICON
DIODE TEMPERATURE SENSORS

J. Sondericker

*Brookhaven National Laboratory**
Upton, New York

INTRODUCTION

Because of tight budgets and increasing costs, helium refrigeration system engineers must design for increased cycle efficiency and shave excess refrigerator capacity to reduce capital costs and operating expenses. For a project such as ISABELLE this task is noteworthy since the load is not one easily calculable dewar but 1100 magnets supplied by long transfer lines, interconnecting links and multitudes of lead pots. Realizing the consequences, the ISABELLE cryogenics group has been measuring heat loads of magnets and associated equipment.

Since temperature measurement is a kind of common denominator in instrumenting cryogenic systems, work was started in the mid 1970's by the Brookhaven 80" Bubble Chamber group, to find a sensor that had all the features of an "optimal" sensor, i.e., one which:

a) Covers the range from above 300 K to below 4.2 K
b) Has high output (dV/dT)
c) Is completely interchangeable
d) Uses two wire readout
e) Has stability at helium temperature > $\left| 0.05 \text{ K} \right|$
f) Is unaffected by magnetic fields
g) Is readily available and inexpensive

---

*Operated by Associated Universities, Inc. under contract with the U.S. Department of Energy.

Finding an "optimal" sensor is admittedly an impossible task since there is no known single thermometer with all these character- istics. However, if requirements c) and f) are waived the silicon diode comes very close to optimal.

Early work concentrated on finding an easy method to make the silicon diode sensors interchangeable and to enchance their sta- bility. Work had barely begun when it was realized that with todays "smart" electronic circuits, the readout system could shoulder the burden of interchangeability rather than the sensor. The result of this decision is described elsewhere[1]. It was also realized that the influence of magnetic fields on the sensor could be minimized by judicious placement.

SILICON DIODE STABILITY

Work then focused on the problem of diode stability. This can best be demonstrated by referring to diode junction voltage drop data derived from a series of twenty dip cycles carried out on commercially calibrated silicon diode temperature sensors. Voltage drop variations of fourteen sensors ranged from + 13 mV to − 3 mV., with one standard deviation of 4.5 mV, an equivalent temperature error of ~ 0.09 K. This absolute error was worrysome for high accuracy work but far worse was the continual changes throughout thermal cycling.

Groups of commercial signal diodes and transistors were checked for stability by similar twenty dip tests. A total var- iation in the forward voltage drop, V_j of 2 mV was considered acceptable including all variations from the first dip through the twentieth. This corresponds to a ΔV_j equivalent to 0.05 K. This was enlightening for it was found that:

a)   Many commercial semiconducting devices will function as temperature sensors.
b)   Similar type devices supplied by different manufacturers exhibited different characteristics at low temperatures.
c)   Sensor yield (the fraction satisfying the 0.05 K criterion) varied from batch to batch supplied by a single manufacturer.
d) ΔV_j from the first dip to the last appeared as either a positive or negative function with successive dip data oscillating around the general trend line.

One of the better devices tested was the Motorola 2N3906 transistor. A standard stock-room item at Brookhaven, it comes housed in a TO-92 plastic case at a cost of eleven cents each in quantity. A typical collector-base voltage-temperature character- istic is shown in Fig. 1. The first derivative of V_j is shown in Fig. 2. The yield of sensor grade diodes after the twenty-dip

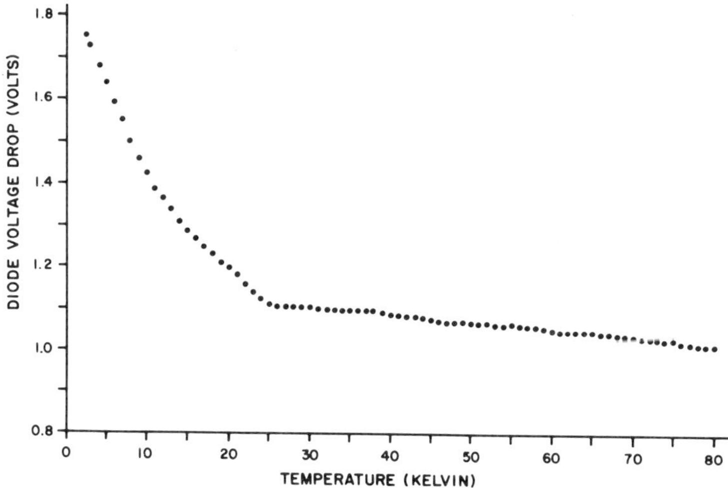

Fig. 1. Si diode forward voltage drop at 10μ A
vs temperature for Motorola 2N3906 C-B.

seasoning runs between 15 and 40% for the Motorola 2N3906 whereas
2N3906 transistors from other manufacturers have yields of less
than 10%.

In order to investigate long term stability, twenty three
previously seasoned and selected sensors were shelved for a period
of months and then cycled through a second series of twenty
dips. A factor of two improvement in stability was obtained as a
result of the additional seasoning.

Results of the seasoning tests demonstrated that fully
documented, stable silicon diodes can be selected but only after a
considerable expenditure of time and effort. To make the task
less labor intensive a computer based data acquisition system was
designed and installed. The system enables an operator to season
30 devices at a time while checking liquid bath temperature and
electonics calibration. Dip cycle times were shortened by rapidly
warming the diodes to 50°C in hot water as opposed to letting them
drift up in the air. The harsher treatment reduced the yield a
few percent but it is thought to be beneficial since high thermal
stresses tend to be effective in weeding out marginal devices.

CALIBRATION CRYOSTAT

A calibration cryostat (shown in Fig. 3) was constructed to
complete the system. It consists of an evacuated pot suspended in
a liquid helium filled dewar. Two copper blocks inside the pot

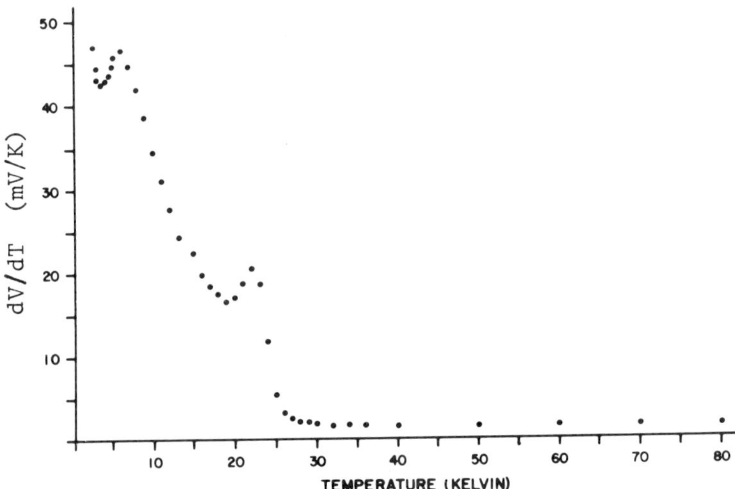

Fig. 2. Si diode sensitivity dv/dt vs temperature
at 10 μA for Motorola 2N3906 C-B.

were hung from the top flange thermally series connected to the
helium reservoir. The bottom mass holds all reference sensors and
provides space for twenty four devices to be simultaneously calibrated. Superinsulation totally surrounds both blocks to minimize
thermal radiation effects. All sensor wiring is thermally heat
sunk to the bottom block, threaded up through connecting copper
tubing and heat sunk again at the top block. Wiring then exits
the pot via a multipin hermetic connector and is brought out
through the top flange of the dewar to electronic apparatus.
Complete thermal isolation of the bottom block is accomplished by
an electronic control loop designed to heat the upper copper mass
to be in thermal equilibrium with the lower block. Sensor temperature is changed by the addition of energy to the bottom block by
means of a finely controlled electrical heater.

A typical calibration run of twenty four sensors is carried
out over a two day period. Sensors are installed, checked electrically and measuring equipment calibrated. The pot is usually
precooled to N_2 temperature overnight, helium filled and subcooled
to 2.4 K the next morning. Energy is then added to the lower
block to drive the mass up in discrete temperature steps. At each
step sensor data is stored in microprocessor memory and echoed via
character printed output. A germanium sensor is used as a reference to 50 K and a platinum resistance temperature standard for
higher temperatures. All stored data is transferred to the computer data acquisition system which generates a calibration sheet
for individual devices (Fig. 4) and produces floppy disc data

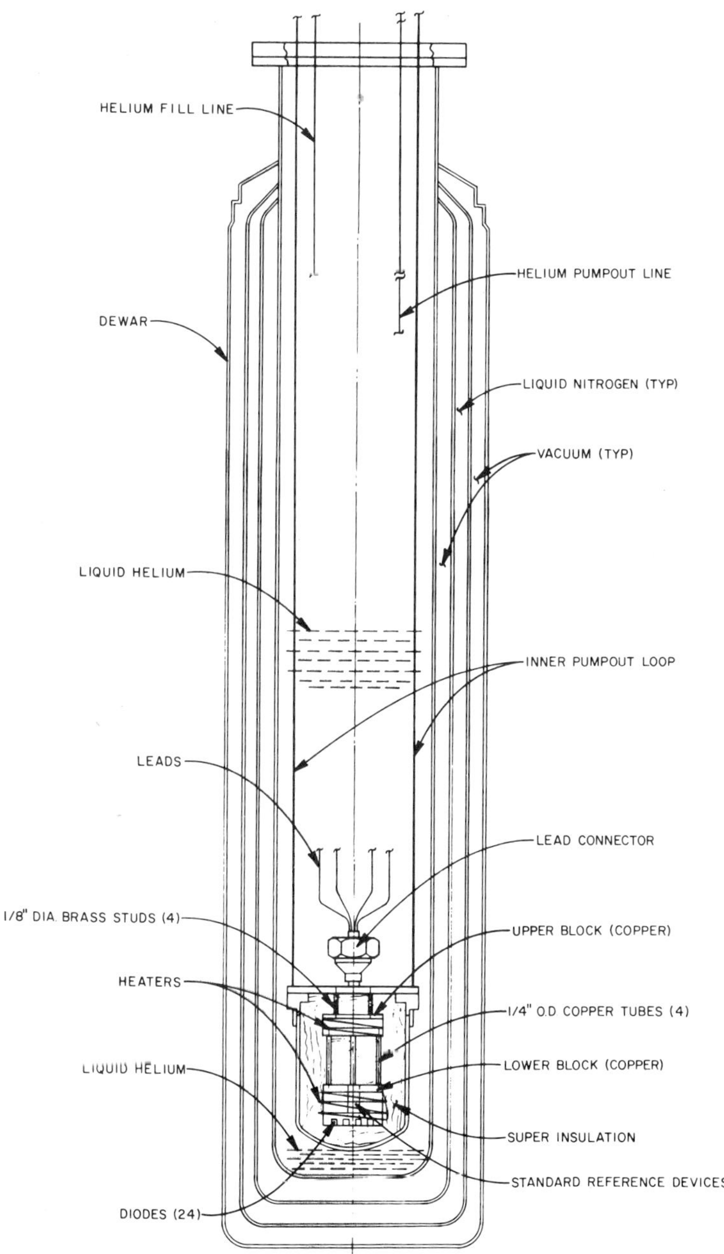

Fig. 3. Temperature sensor calibration cryostat.

1168 J. Sondericker

```
DATA FROM SENSOR MODEL 3906
SERIAL NUMBER 422
TAKEN FROM DX1:A00422.DIO
```

TEST TEMP	VOLTAGE (@10UA)	DV/DT (MV/DEGREE)	FREQ (KHZ)	CALCULATED TEMPERATURE	ERROR	BREAKPOINT FREQUENCY(KHZ)		TERMS AND COEFFICIENTS
2.4	1.728100		38.656	2.400036	0.000036	******** 3.8655998E+01		
2.6	1.723100	-25.75	38.906	2.600916	0.000916			
2.8	1.717800	-27.25	39.171	2.800270	0.000270			
3.0	1.712200	-30.00	39.451	2.998055	-0.001945			
3.4	1.699800	-32.75	40.070	3.398710	-0.001290			
3.8	1.686000	-35.50	40.759	3.802773	0.002773		0	2.4000361E+00
4.2	1.671400	-37.50	41.489	4.200661	0.000661		1	8.3115423E-01
4.6	1.656000	-39.12	42.258	4.599547	-0.000453		2	-1.1627522E-01
5.0	1.640100	-40.79	43.052	4.998335	-0.001665		3	2.3682185E-02
6.0	1.598900	-41.40	45.110	6.000673	0.000673		4	-3.0989363E-03
7.0	1.557300	-41.05	47.188	6.999846	-0.000154		5	2.5311470E-04
8.0	1.516800	-39.00	49.211	8.001766	0.001766		6	-1.2229141E-05
9.0	1.479300	-36.10	51.084	8.996255	-0.003745		7	3.2171664E-07
10.0	1.444600	-32.90	52.817	10.003500	0.003500		8	-3.5508954E-09
11.0	1.413500	-29.85	54.371	10.997510	-0.002490			
12.0	1.384900	-27.15	55.799	12.003123	0.003123			
13.0	1.359200	-24.80	57.083	12.996351	-0.003649			
14.0	1.335300	-23.05	58.277	14.002087	0.002087			
15.0	1.313100	-21.50	59.386	15.000010	0.000010	******** 5.9386002E+01		
16.0	1.292300	-20.15	60.425	15.999886	-0.000114		0	1.5000010E+01
17.0	1.272800	-19.05	61.399	17.000463	0.000463		1	8.6008513E-01
18.0	1.254200	-18.25	62.328	17.999065	-0.000935		2	1.8682237E-01
19.0	1.236300	-17.80	63.222	19.000896	0.000896		3	-1.2946531E-01
20.0	1.218600	-18.00	64.106	19.999851	-0.000149		4	5.4102384E-02
21.0	1.200300	-19.06	65.020	20.999496	-0.000504		5	-1.2522629E-02
22.0	1.180480	-21.28	66.010	22.000528	0.000528		6	1.6191467E-03
23.0	1.157730	-23.08	67.146	22.999741	-0.000259		7	-1.1140046E-04
24.0	1.134310	-18.69	68.316	24.000038	0.000038		8	3.1920840E-06
25.0	1.120340	-9.70	69.014	25.000027	0.000027	******** 6.9014000E+01		
26.0	1.114910	-4.41	69.285	25.999517	-0.000483			
27.0	1.111530	-3.04	69.454	27.001694	0.001694		0	2.5000027E+01
28.0	1.108830	-2.53	69.589	27.999111	-0.000889		1	2.3232248E+00
29.0	1.106460	-2.27	69.707	28.996387	-0.003613		2	4.9521379E+00
30.0	1.104280	-2.04	69.816	30.004654	0.004654		3	1.2924147E-01
32.0	1.100330	-1.92	70.014	31.998419	-0.001581		4	1.6316586E+00
34.0	1.096600	-1.83	70.200	34.000298	0.000298		5	-4.3406897E+00
36.0	1.093020	-1.77	70.379	35.999512	-0.000488		6	3.1937273E+00
38.0	1.089510	-1.74	70.554	38.000702	0.000702		7	-1.0105220E+00
40.0	1.086070	-1.72	70.726	39.999657	-0.000343		8	1.1961359E-01
45.0	1.077450	-1.72	71.156	45.000023	0.000023			
50.0	1.068820	-1.74	71.587	50.014240	0.014240	******** 7.1586998E+01		
55.0	1.060010	-1.78	72.028	54.991825	-0.008175			
60.0	1.051000	-1.82	72.478	59.984299	-0.015701			
65.0	1.041820	-1.85	72.936	64.983406	-0.016594			
70.0	1.032470	-1.88	73.403	70.002525	0.002525			
75.0	1.023000	-1.90	73.876	75.011948	0.011948			
77.5	1.018220	-1.92	74.115	77.516586	0.016586			
80.0	1.013410	-1.95	74.355	80.014763	0.014763			
90.0	0.993890	-1.97	75.330	90.000557	0.000557			
100.0	0.973920	-2.02	76.328	99.979088	-0.020912			
110.0	0.953440	-2.07	77.351	109.986168	-0.013832		0	5.0014240E+01
120.0	0.932510	-2.11	78.396	120.007729	0.007729		1	1.1388309E+01
130.0	0.911280	-2.14	79.457	130.003143	0.003143		2	-2.3700513E-01
140.0	0.889680	-2.18	80.535	139.999756	-0.000244		3	1.7073220E-02
150.0	0.867770	-2.20	81.630	150.013824	0.013824		4	-1.4151566E-03
160.0	0.845640	-2.23	82.735	159.997574	-0.002426		5	1.0131003E-04
170.0	0.823250	-2.25	83.854	170.001083	0.001083		6	-4.6686232E-06
180.0	0.800660	-2.27	84.982	179.991180	-0.008820		7	1.1576393E-07
190.0	0.777860	-2.29	86.121	189.994522	-0.005478		8	-1.1551893E-09
200.0	0.754880	-2.30	87.269	200.000931	0.000931			
210.0	0.731760	-2.32	88.424	210.000763	0.000763			
220.0	0.708480	-2.33	89.586	220.003876	0.003876			
230.0	0.685130	-2.34	90.753	230.008179	0.008179			
240.0	0.661760	-2.34	91.920	239.993042	-0.006958			
250.0	0.638380	-2.33	93.088	249.997696	-0.002304			
260.0	0.615120	-2.32	94.250	259.999451	-0.000549			
270.0	0.592060	-2.29	95.402	270.002777	0.002777			
280.0	0.569280	-2.26	96.539	279.996613	-0.003387			
290.0	0.546810	-2.23	97.662	290.003967	0.003967			
300.0	0.524620		98.770	299.998840	-0.001160	******** 9.8769997E+01		

Fig. 4. Typical calibration sheet.

files for recordkeeping purposes. Each sensor is then removed from the pot and dipped in helium once more as a final check.

A computer program was written to plot point to point slopes for each diode calibrated. It was found that the sensors seem to exhibit three distinct signatures (Fig. 5) through their lower temperature range. Data is being accumulated to determine if there is any statistical correlation between signature and stability or failure.

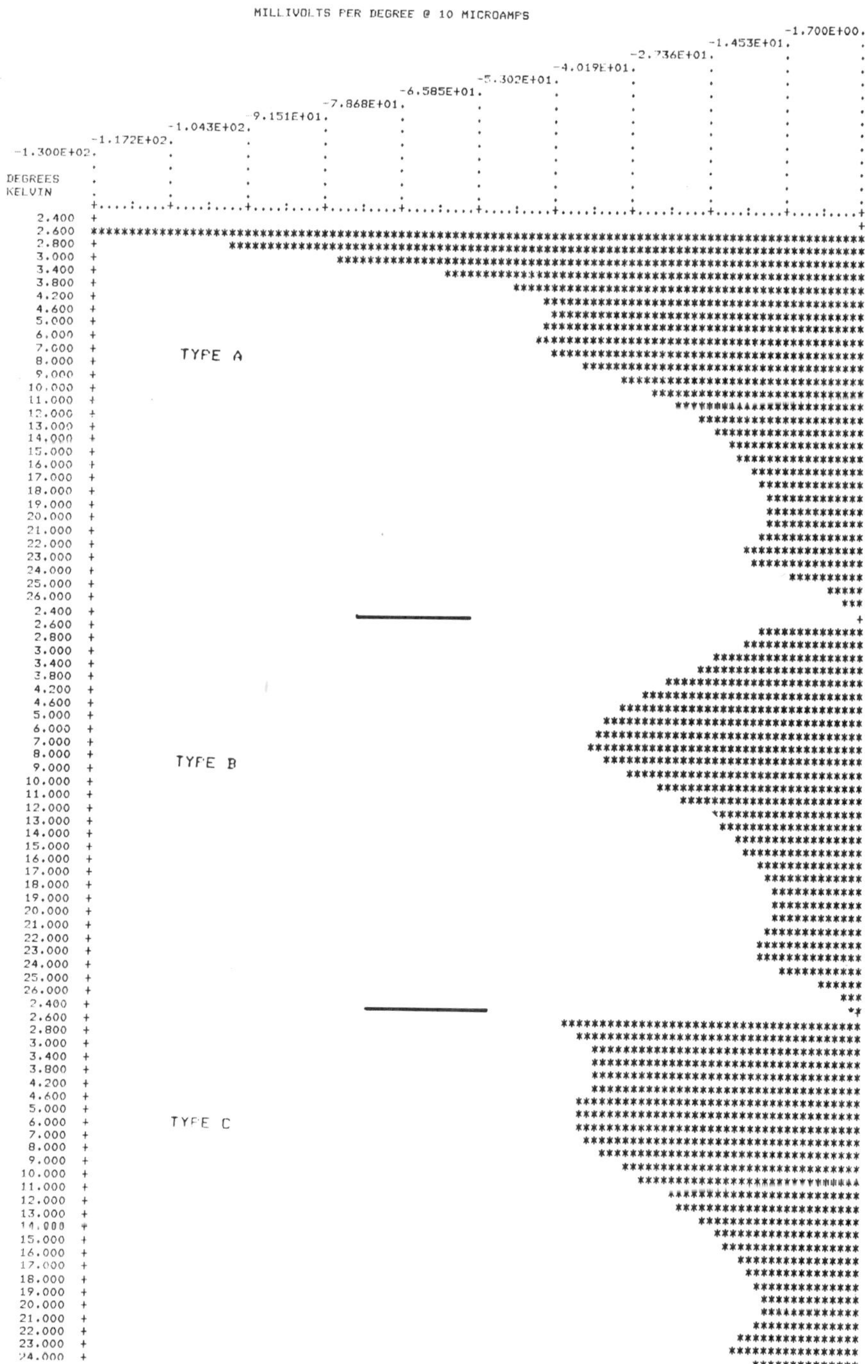

Fig. 5. Signatures of silicon diodes at low temperatures.

The cryostat system has been used to calibrate hundreds of various types of sensors and aside from occasional vacuum leaks has performed satisfactorily.

INDUSTRIAL EXPERIENCE

Testing and calibration of sensors under laboratory conditions is necessary and enlightening but ones education is not complete until the devices are put to work in an industrial environment. About six hundred sensors of both commerical and inhouse types have been used in ISABELLE projects at Brookhaven.

The accumulated failure rate of all devices in service is less than one percent of the total. However, some of the problems associated with cryogenic thermometry can be more subtle than outright catastrophic failure. These problems are discussed below.

a) Sensors mounted in welded transistor cans have higher failure rates on the average, increasing to 10 to 20% of those in service. The problem is caused when a small seal leak permits the volume inside the can to equalize with process conditions. A sudden warmup or even slow rise, depending upon the magnitude of the leak, over pressurizes the case and causes the cap to blow off, destroying the sensor.

b) Typically all sensors at ISABELLE are installed in the process stream at some intermediate time in the construction phase. Assembly, welding, and leak checking then continues. Final check out revealed that groups of well seasoned, stable silicon diodes would sometimes be found to have failed catastrophically, have finite front to back resistance ratios or were obviously out of calibration upon cooldown. The source of damage was traced to the rf start of the heliarc welding that followed installation. The problem should be more fully investigated but it appears that enough energy is coupled into the diode to damage the thermometer if its leads are short circuited or terminated in a low impedance. Damage is caused by either high circulating forward currents or by exceeding the reverse breakdown voltage rating of the semiconductor junction. Open circuit leads eliminate the problem but to be absolutely safe an rf bypass should be installed as close to the diode as possible.

c) Cryogenic thermometers at helium temperature are usually high impedance devices. By observing accepted wiring and grounding practices hum, pick-up, and noise can be minimized but in some areas because of poor grounds or local high power equipment, the minimized hum may still be appreciable. It is therefore good practice to always use voltage to frequency or dual-slope analog to digital converters to take advantage of their ability to reject ripple.

d) Analog to digital converter ripple rejection is perfect only when positive and negative ripple envelopes are equal and average to zero. When silicon diode sensors are utilized and hum is present on the normal 10μ A constant current drive a distorted hum voltage envelope will appear across the diode junction producing a negative dc error voltage. The negative offset is added to the positive junction voltage drop which produces an output signal that favors higher temperature readings. The effect is inherent in all junction devices because of their non-linear dynamic impedance around the dc operating point. In bench tests millivolt level offsets were generated which can correspond to errors of many tenths of degrees depending upon sensor dv/dT. The addition of a 0.01 μF high quality capacitor across the diode completely eliminates the ripple and is standard on all BNL installations.

ACKNOWLEDGEMENTS

The author wishes to thank Dieter Zantopp for his dedication and quality work in cryostat construction and sensor calibration, Bill Kollmer for his continuing help since 80" Bubble Chamber days, Ray Atkins for diode seasoning and Jim Osterlund for his computer hardware and software development.

REFERENCES

1. J. H. Sondericker, A very high accuracy cryogenic digital thermometer, _Internal Technical Note_, Brookhaven National Laboratory.

THE CHOICE OF STRAIN GAUGE FOR USE IN A LARGE SUPERCONDUCTING ALTERNATOR

C. Ferrero, C. Marinari, and S. Desogus

Istituto di Metrologia—"G. Colonnetti"
Torino, Italy

INTRODUCTION

As part of the Italian Finalized Project on Superconducting Materials and Alternators, electrical strain gages were investigated from ambient to liquid-helium temperatures. Special care was taken in the experimental determination of the curves of apparent strain (AS) vs. temperature (T), because of the important role of thermal and mechanical stresses in a superconducting rotor in the cooling (from room to LHe temperature) and operational phases.

The investigation was carried out with commercially available Karma and modified-Karma alloy foil strain gages made by different manufacturers (Micro Measurements WK; BLH Electronics FSM; Kyowa KFL) which were either applied on the surface of supports of Cu, Al, Invar, AISI 304L, Araldite, and Nb, or embedded inside the specimen.

OBJECT OF THE PRESENT WORK

High scattering in the apparent-strain data is known to be found between room and LHe temperature, even with strain gages from the same lot, and the uncertainty in the determination of actual strain is consequently high. AS vs T curves of Karma alloy foil gages show near room temperature an initial slope $\partial\varepsilon/\partial T$, the absolute value and the sign of which depend on the mismatch between the temperature resistance coefficient (TRC) of the gage and the thermal expansion coefficient (TEC) of the material the gage is applied on. At cryogenic temperature these curves exhibit a slope reversal, i.e., strain increase with temperature decrease, which was thoroughly investigated by Telinde[1]. As Walstrom[2]

pointed out the uncertainty depends mainly on the uniformity of
the heat treatment for temperature compensation, on induced stresses during gage application, and on the uniformity of bonding
conditions. This lack of reproducibility near 4.2 K, may lead to
sizable errors, unless the gage temperature is measured with
sufficient accuracy.

The object of the present work is to show how experimental
results can fit the theoretical model concerning the reversal
effect, based on the metallurgical composition of the alloys used
in gage construction, and how the scattering is reduced in the
range between the reversal temperature and LHe temperature.

REVERSAL EFFECT: A BRIEF SURVEY

At the end of the 1950's and the beginning of the 1960's
considerable stimulus was given to research work on alloys for
cryogenic strain gages, in order to obtain minimal AS and small
temperature variations of the k calibration factor.

The alloys investigated were those used for room temperature
in strain gages slightly modified as to composition. With Karma
and modified-Karma alloys the slope reversal effect occurs near
the temperature of liquid hydrogen. In recent studies[2,3] closer
attention is given to the shape of the curve where slope reversal
takes place, giving rise to interesting speculations.

The fields of solid-state physics and metallurgy working, in
parallel, on electrical and thermal conductivity, susceptibility,
etc. for the characterization of pure metals or alloys, starting
in 1933 with the studies of De Haas and Van den Berg in Leiden,
have given increasing attention to the phenomenon of "minimum
resistance" at low temperatures[4]. The existence of a minimum
resistance was later established on numerous materials containing
transitional atoms as dilute impurities (Fe, Mg, Cr). It proved
to be particularly interesting both for theoretical physicists, as
the very existence of the "minimum" was not amenable to previous
theories, and for metallurgists, owing to the new possibilities
opened up in the realization of alloys having a temperature resistance coefficient (TRC) approaching zero at cryogenic temperatures and for a limited ΔT.[5]

Finally, in 1964 J. Kondo[6] formulated a theoretical model for
the behavior of those alloys with a logarithmic increase of
resistivity with decrease in temperature. He demonstrated theoretically that the existence of a minimum was connected to the
presence of a magnetic moment in an impurity atom, and that "the
scattering of an electron from a magnetic impurity cannot be
considered as a one-electron problem. A singular term proportional to lnT has been obtained as a result of the Fermi sta-

tistics, which takes account of the many-body effect". The spin-dependent contribution to resistivity (ρ) is proportional to the concentration of 'dilute' magnetic impurities (c) and, if resistivities are additive, the total resistivity is

$$\rho = AT^5 + c\,\rho_o - Bc\,(\ln T)$$

where AT^5 represents the phonon contribution, $c\rho_o$ the impurity contribution, and $Bc\,(\ln T)$ the spin-dependent contribution, with a minimum at $Tmin = (cB/5A)^{1/5} \propto c^{1/5}$.

APPARATUS AND PROCEDURES

Most of the tests made to obtain the AS vs T curves for different combinations of strain gages support materials were made by using rectangular-shape specimens (5x20x50) mm, whereas other tests for lot characterization (8 to 10 strain gages) were carried out at different stress levels with AISI 304L specimens. For these tests, a calibration system of double-actuator uniform-strength bar-type, ensuring suitable measurement reproducibility, was designed and checked[7]. On the back of each specimen was mounted a temperature sensor (Micro-Measurement, CLTS).

The whole measurement process was controlled by a scanning-conditioning system[8] allowing up to 100 channels to be monitored, with an error less than 0.15 μ E. The system was connected to an intelligent minicomputer via a IEEE-488 bus. The whole system can operate without interruption, to optimize its use and minimize data acquisition time. Tests were carried out both with contin-uous and pulsed powers at different voltage values, V_s at which the dissipated power per unit area of the gage grid was made to vary between 10 and 425 mW/cm^2.

EXPERIMENTAL RESULTS

Figure 1 gives AS vs T for the three types of strain gages tested, indicated by A, B and C, which belong to different lots and were applied to different support materials. The curves to permit comparison in a single figure, were referenced to 4.2 K instead of to room temperature, namely, AS was made equal to zero at LHe temperature. Figure 2 collects results obtained with Type A strain gages applied on Cu, Al, AISI 304L, Nb, Invar, Araldite. In this figure values of the measured strain are also referenced to 4.2 K. It is to be kept in mind that variation of the total AS lies between + 1500 $\mu\varepsilon$ and - 12000 $\mu\varepsilon$ from room to LHe temperature, for the various combinations of strain gages and support materials.

Table I compares scattering of total AS vs T curves with respect to their scattering as referenced to LHe temperature for

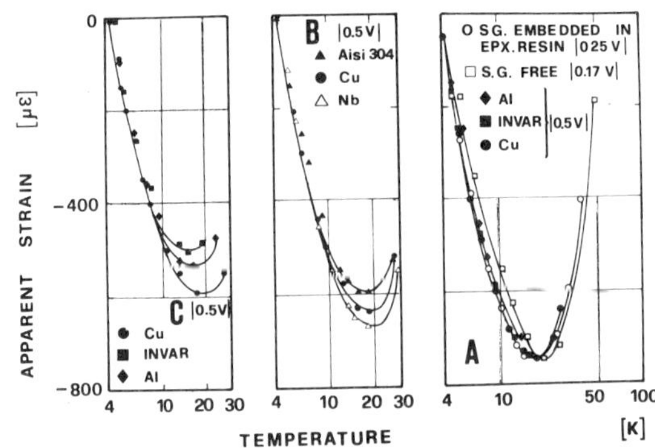

Fig. 1. AS vs T for different strain gages and materials.

one lot of type A strain gages applied on AISI 304L. Figure 3
gives differences of AS values vs T obtained when the supply
voltage is changed from V_o = 0.5 V to V_1 = 5V (and therefore
for two levels of power dissipated in gage grid, with ratio P_1/P_o
= 100).

ANALYSIS OF RESULTS

4.2 TO 7.2 K Range

In the 4.2 K to 7.2 K range AS vs. T curves appear to be
linear on a semilogarithmic scale; they can be interpolated by
equations of the type: y = apparent strain = b_o +b_1 lnT.
Coefficients b_o, b_1, and the values of R-square and maximum

TABLE I. AS vs T for one lot of A-type gages
(row a, b) and reversal effect (row c,d) for
two different levels of applied voltage

Apparent strain	$\bar{x}$ ($\mu\varepsilon$)	max deviat. ($\mu\varepsilon$)	S ($\mu\varepsilon$)	$S/\bar{x}$
AS^{5V} (a)	−2489	233	89	3.6×10^{-2}
$AS^{.5V}$ (b)	−2267	241	93	4.1×10^{-2}
Δ^{5V} (c)	−500	10	3.8	7.6×10^{-3}
$\Delta^{.5V}$ (d)	−711	15	5	7×10^{-3}
b+d	−2984	246		
a+c	−2989	239		

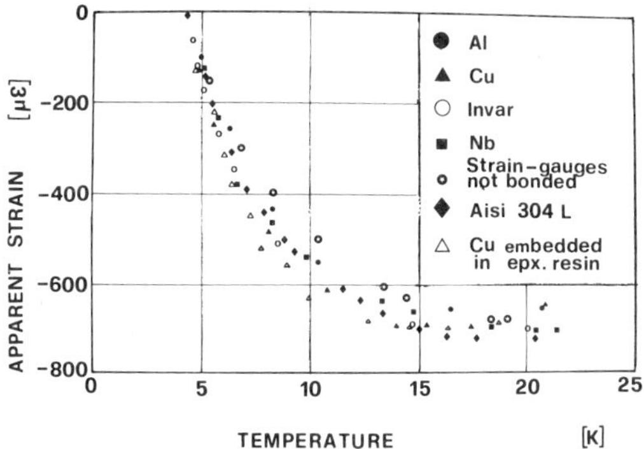

Fig. 2. AS vs T for type A gages applied on several materials.

deviation between experimental points and fitting curves for type A strain gages and different support materials are in given in Table II.

All the other data shows that complete homogeneity does not exist for the three series of the strain gages investigated. For strain gages A, B, C, applied on a Cu specimen, at temperature values lower than 5.8 K, the joint confidence ellipse for b_o and b_1 is calculated by means of the following equation:

$$\sum_{i=1}^{n} x_i^2 (\beta_1 - b_1)^2 + 2 \sum_{1}^{n} x_i (\beta_o - b_o)(\beta_1 - b_1) + n(\beta_o - b_o)^2 = 2 \ F_{2,n-2} \cdot \alpha S_T^2$$

One can see in Fig. 4 which gives the 95% confidence-level ellipsis, that there is no inclusion of representative points of

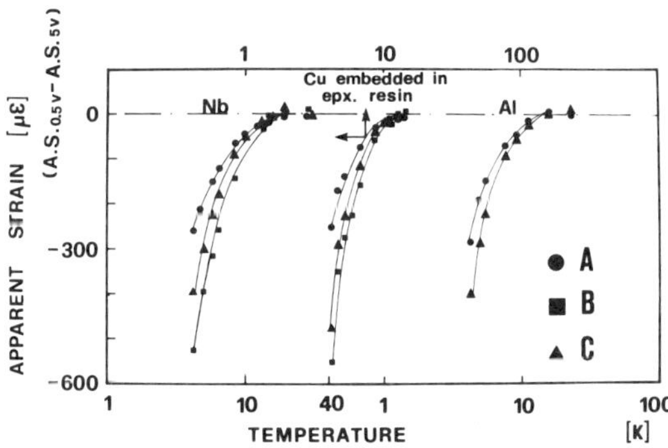

Fig. 3. AS vs T variation with applied voltage of 0.5 V and 5 V.

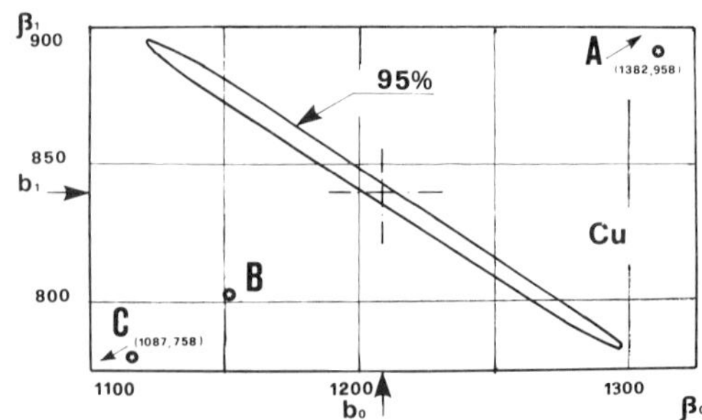

Fig. 4. Joint confidence region.

the various single regressions in the confidence region of the
overall regression, as a confirmation of the lack of homogeneity.

4.2 to 30 K Range

The wider range between 4.2 to 30 K, results can be fitted by
4th degree polynomial curves of the type:

$$y = a_o + a_1 T + a_2 T^2 + a_3 T^3 + a_4 T^4$$

Table III supplies the coefficients of this equation for type A
strain gages on different materials. The values of $S/\bar{x}$ calculated
on overall curves are also given (from Table I). They show, on
the curve portion concerning the reversal zone, a variation be-
tween 6×10^{-2} and 7×10^{-3} or on absolute strain variation between $\pm$
100 and $\pm$ 10 $\mu\varepsilon$.

Table II. Coefficients of logarithmic equation
$y = b_o\ b_1\ \ln T$, R-square and max deviat.

Specimen materials	Strain gages type A			
	b_o	b_1	R-square	max deviat. ($\mu\varepsilon$)
Cu	1382	−957.6	0.9927	20
Al	1327.4	−912.64	0.9973	12
Nb	1304.1	−896.68	0.998	10
AISI304L	1290.3	−888.64	0.9973	10

Table III. Coeffficients of polynomial equation for
A type gages on different support materials, and max
deviation

| Specimen | Coefficients | | | | | max |
materials	a_o	a_1	a_2	a_3	a_4	deviat.($\mu\varepsilon$)
Al	1304.66	−461.02	42.80	−1.83	0.03	+12
Cu	1556.22	−571.42	50.17	−2.74	0.05	−14
Invar	1352.99	−478.60	43.75	−1.85	0.03	+10
Nb	1296.96	−457.95	42.62	−1.84	0.03	−6
Inox	1185.04	−413.35	36.95	−1.57	0.02	+15

Reversal temperature

The reversal temperature (T_{REV}) results are constant and
repeatable in agreement with Kondo's theory: T_{REV} does not depend
on the type of strain gage (variable as to manufacturer, TRC, and
grid area), nor does it depend, in a first approximation, on the
TEC of support materials, and consequently on strain gage substrate mismatch. The observed variation for different specimens
(T_{REV} = 15± 1.5 K) seems to depend rather on the different
thermal conductivity of support materials and therefore on the
real gage grid temperature, which is not detectable with the
temperature transducers used. This is confirmed by Fig. 3.
Variation of the power dissipated in the gage grid involves, in
fact, variation in its temperature with respect to that of the
support. The curves get nulled at the minimum point of the AS vs.
T characteristic, so that reversal temperature can be defined with
greater accuracy.

Behavior with strain

The behavior of AS vs T is not affected by strain level;
tests made on two lots of type-A and type-C gages applied on a
AISI 304L specimen at three strain levels (0, 150, 1000$\mu\varepsilon$), did
not show any variations within the limits of measurement repeatability.

Power dissipation Effects

With dissipated-power values lower than 10 mW/cm^2, AS curves
are independent of the dissipated power and the support material,
and consequently, of thermal exchange conditions. At higher
values (>10 mW/cm^2) they are dependent on the power dissipated in
gage grids because the same temperature of both grid and support
is not ensured[9].

CONCLUSIONS

Below reversal temperature the effect of TRC on TEC predomin-
ates; consequently, the behavior of AS vs T curves mainly depends
on the concentration of magnetic impurities in the gage-alloy
material, and can be described analytically. The value of the
temperature at which reversal occurs of the characteristic (T_{REV}),
is stable and repeatable according to Kondo's theory.

In the 4.2 K - 7.2 K interval, variation of AS vs T can be
expressed by a logarithmic-type relation, such as $y = b_o + b_1 \ln T$,
whereas over the 4.2 K - 30 K temperature interval it is advisable
to use a polynomial equation of the type $y = a_o + a_1 T + a_2 T + a_3 T^3
+ a_4 T^4$.

In applying experimental results to a wide range of technological applications (superconducting alternators, high-field
magnets) which require stress measurement and control in a narrow
temperature range near 4.2 K, it is advisable to use AS analytical
curves referred to 4.2 K instead of overall AS curves with a
correction due to temperature. In this way, in fact, a factor 10,
or greater, reduction is obtained in the scattering between the
various AS curves concerning the same lot of strain gages, and
consequently in the uncertainty of results.

REFERENCE

1. J.C. Telinde, Strain gages in cryogenic environment, _Exp._
      Mechan., 10:394 (1970).
2. P.L. Walstrom, Strain gauges for superconducting magnet
      testing, _Cryogenics_, 9:509 (1980).
3. R.S. Freynik, et. al., Evaluation of metal-foil strain gages
      for cryogenic application in magnetic fields, in "Advances
      in Cryogenic Engineering, Vol. 23" Plenum Press, New York
      (1978) p. 23.
4. G.K. While, S. B. Woods, Electrical and thermal resistivity of
      the transition elements at low temperatures, _Trans Royal_
      Soc London,(1959).
5. C. Rizzuto, Formation of localized moments in metals:
      experimental bulk properties, _Rep. Prop. Phys._, 37:147
      (1974).
6. J. Kondo, Theory of dilute magnetic alloys, in: "Solid State
      Physics", F.Seitz et al., ed., Academic Press, New York and
      London (1969).
7. S. Desogus, C. Ferrero, C. Marinari, Analisi comparata di
      sistemi per la caratterizzazione di estensimetri elettrici
      a reistenza, in "Proc. VIII Conv. Naz. AIAS," Firenze,
      Italy (1980).

8. F. Pavese, et. al., Automatic test facility for cryogenic
    transducers in the range 2 K to 300 K and up to 7 T, in
    "Proc. 8th Int'l. Cryogenic Engr. Conf.," IPC Science and
    Technology Press, Guildford (1980) p. 812.
9. C. Ferrero, C. Marinari, S. Desogus, An apparatus for
    characterizing strain gauges: optimal dissipated power to
    limit self-heating apparent strain at cryogenic
    temperature, in "Proc. I Congr. A. Cr. I.," Bressanone,
    Italy (1981).

STATE OF THE ART MICROPROCESSOR TECHNOLOGY APPLIED TO CRYOGENIC TEMPERATURE READOUT AND CONTROL

D. Sheats and G. Halverson

Scientific Instruments, Inc.
West Palm Beach, Florida

INTRODUCTION

The requirements for measurement, indication, and control of temperature in a cryogenic environment continue to increase in complexity with the growth of the entire cryogenic field. For many years, analog systems have met the needs of the industry quite adequately, but with the increasing demand for fully automated systems, standard analog techniques are no longer sufficient.

Most analog systems require considerable adjustment to reach the precise temperature desired, and to maintain the proper feedback gain and time response to yield optimum control. Any change in temperature affects the inherent physical parameters and often requires a new set of adjustments to achieve proper control. True remote control of these parameters through analog techniques, can become quite cumbersome, and complete automation, including programmability, is difficult to achieve.

Applying digital techniques to these applications has met with varying degrees of success in the past. There are several reasons for this. Probably the principal difficulty in achieving the same performance with digital techniques as with a properly adjusted analog controller has been insufficient resolution and slow response in the feedback loop. In addition, the first microprocessors available were 4-bit devices which were difficult to program when mathematical operations were required. However, 16-bit microprocessors are now available which are much easier to program. In addition, low cost 16-bit A/D converters are now available which provide the needed resolution of the sensor input.

DESIGN CONSIDERATIONS

The initial design objectives of a microprocessor based cryogenic temperature controller were as follows:

1. Easy remote operation
2. True programmability
3. Ramp control of temperature
4. Minimimum adjustment to reach exact temperature
5. Automatic adjustment to varying control environments
6. Controllability equal to accuracy of display
7. Repeatability equal to accuracy of display
8. Absolute accuracy equal to accuracy of sensor calibration
9. Multiple program storage
10. Multiple sensor readout
11. Mixed sensor type

In summary, the design objective was to duplicate the capabilities of a good quality analog controller while eliminating operator adjustments as much as possible. The intent was to design a system which would limit operator involvement to in putting the temperature set point and the time/temperature profile, with no need to consider offset, bandwidth, rate, etc. The design should allow the operator to specify a time/temperature profile and then execute that profile as many times as desired with repeatable results.

PRODUCT DESCRIPTION

The design result (designated the Model 3100 and shown in Fig. 1.) was implemented using a Texas Instruments microprocessor (TMS9900) as the CPU. This device is quite versatile, and lends itself to real time processing with no difficulty. A 16-bit A/D converter is used to read the temperature from a silicon diode temperature sensor providing an internal resolution of 1 m K from 2 to 35 K. A 2-line × 40 character alpha/numeric display coupled with two small membrane keypads for function control and data input were chosen for the operator interface. The functions available were divided into two categories (controller modes and edit functions) and were implemented on a time-sharing basis to allow the operator to run a time/temperature profile and at the same time to edit or create a new program. Other features include an RS232/ASCII link for remote control with a terminal or host computer and printout of pertinent data on request or at one-minute intervals. Those parameters which are directly available on the display or remotely transmitted include:

1. the current program point
2. the two end points of each segment of the time/temperature profile
3. the total time of each segment
4. the time remaining on the current program segment
5. the current set point
6. the actual temperature
7. the difference between the current set point and the actual temperature
8. the percent heater power
9. the time of day
10. the modes of operation.

Options provide non-volatile memory for storage of 20 separate profiles of up to 30 points each, interface with other instrumentation via the IEEE-488 GPIB, and dual sensor readout. Although the unit was initially designed using a silicon diode sensor with a temperature range of 2 - 300 K and a display resolution of 0.1 K, any of the standard cryogenic sensors such as a germanium RTD can be used depending on the temperature range and resolution required.

The physics and ideal mathematical control model of a cryogenic controller has been shown by Forgan[1] to include proportional, derivative, and integral control. This design philosophy has been followed in the Model 3100, and a functional block diagram of the hardware used is shown in Fig. 2. The actual control function is implemented in a three term equation in software by converting the set point temperature into the desired A/D reading

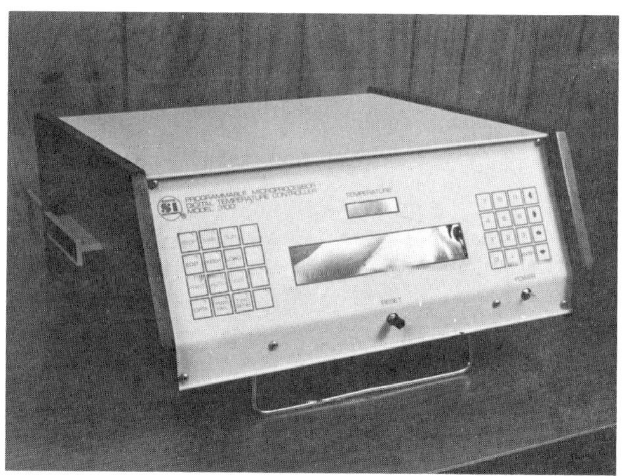

Fig. 1. Model 3100 programmable microprocessor digital Temperature Controller.

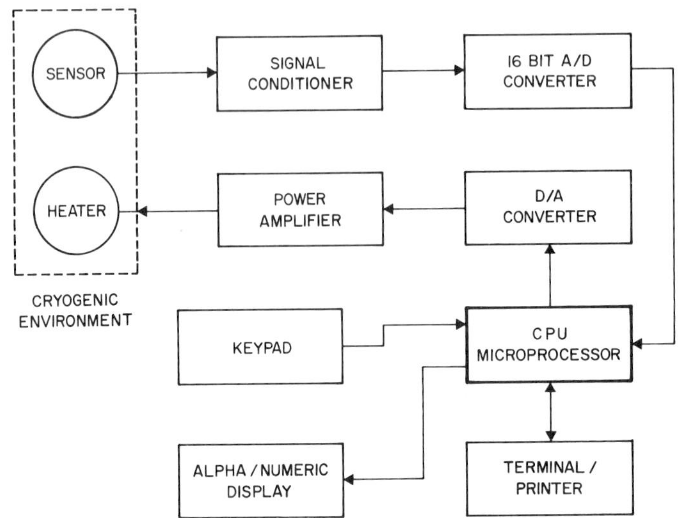

Fig. 2.   Functional block diagram of the model 3100.

and then controlling on the difference in order to achieve maximum control from the sensor in use. The ramp function was implemented by varying the set point according to the time/temperature node points input by the operator, and then allowing the controller to control at the set point. This set point is updated approximately 4 times a second. By limiting the maximum rate to a reasonable rate (7 K/min), excessive overshoot is minimized. Further tests are being conducted on varying cryogenic environments with the goal in mind of automatically adjusting equation parameters to optimize the controllability for any given configuration. By utilizing the capability of a 16-bit microprocessor, system adjustments can be accomplished automatically that would normally be done by the operator through trial and error. The end result is truly a "hands off" instrument that will reach and maintain a given temperature in a variety of environments.

EVALUATION RESULTS

Evaluation of the Model 3100 design concept was first performed using a small cryogenic refrigerator manufactured by Air Products and Chemicals, Inc. (Model 202 Displex). Controllability (both at a fixed point and during a ramp function), repeatability, absolute accuracy, and setability were of primary concern. The results of this evaluation are shown in Table I. The results are divided into two due to the silicon diode sensor used. The slope of the temperature/voltage curve is a factor of 10 or better below 35 K, thus giving much greater resolution and control in that temperature range.

Table I. Test Results on a Small Cryogenic Refrigerator

Temperature Range	12 − 35 K	35 − 300 K
Resolution	0.1 K	0.1 K
Controllability	±0.01 K	± 0.1 K
Repeatability	± 0.1 K	± 0.1 K
Accuracy	±0.02 K	±0.05 K
Setability	± 0.1 K	± 0.1 K

Initial results indicate that the design objectives were reached. Of particular interest was the ramp function, since there is no known cryogenic controller which attempts to accomplish this function.

Figure 3 shows a plot of two temperature regions where a time/temperature profile was run below 35 K. While operating at rates less than 4 K/min., the maximum error observed was 2 K. It stands to reason that the faster one attempts to reach a given temperature, the greater will be the observed error, primarily at the transition point from one rate to another. There are many factors which affect this error such as temperature, rate, and the amount of rate change. Once the unit settled on a particular rate, the error observed was no more than if the set point were fixed. In the lower region (below 35 K), the temperature can be changed at a rate as slow as one m K/minute with very good accuracy.

POSSIBLE APPLICATIONS

This instrument will find application in any situation where a certain time/temperature profile must be followed exactly or where such a profile must be repeated numerous times. By virtue of the microprocessor-based controller, the Model 3100 provides automatic control, data collection, and recording in a simplified manner. In addition, the operator is relieved from concentrating on temperature control, and is free instead to concentrate on the particular experiment or process being observed.

A few specific applications that would lend themselves to this type of controller would be a study of the release of gases as temperature increases, a calibration process, or a manufacturing process that requires a certain temperature profile. A further application using basically the same instrument with a few modifications would be a differential controller, utilizing two sensors and two heaters. The first stage would achieve a set point temperature of an outside environment, while the second stage would hold a temperature of the inner environment a few

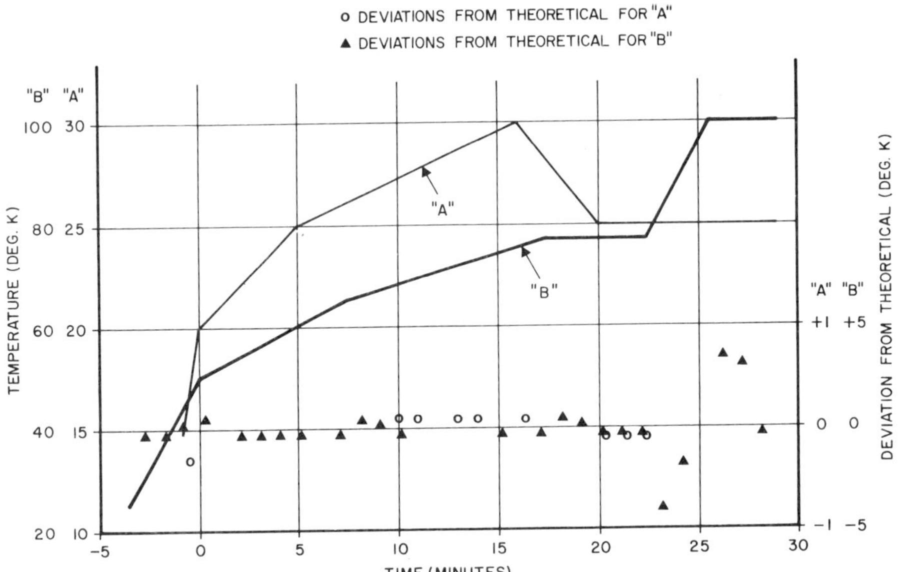

Fig. 3. Time/Temperature profile results for two
temperature ranges A, 15–30 K and B 25–100 K.

degrees above the outside, thus allowing for more even temperature
distribution in the area of concern.

SUMMARY

The tests thus far conducted indicate that the design objectives have been met. The present state-of-the-art microprocessor technology has made it possible to duplicate the control
capability of an analog controller with digital techniques. In so
doing, all of the advantages available in digital systems can be
easily applied to provide system capabilities which before were
not possible. It is anticipated that such a controller will be
able to fill many needs in the cryogenic engineering field.

ACKNOWLEDGEMENTS

The authors wish to thank Dr. Ralph Longsworth of Air Products and Chemicals, Inc., and Mr. Tom Tucker, Mr. Joe Rawlins,
and the staff of Scientific Instruments, Inc. for their contributions to this project.

REFERENCES

1. E.M. Forgan, On the use of temperature controllers in
    cryogenics, Cryogenics, 14:207 (1974).

MICROPROCESSOR CONTROL OF CRYOGEN PRESSURE*

P. L. DePerro and H. L. Gier

*Beech Aircraft Corporation
Boulder, Colorado*

INTRODUCTION

The application of microprocessors to the control of cryogenic parameters offers the advantages of increased design flexibility, computer compatibility and autonomous experiment control. Beech Aircraft Corporation has proposed the use of microprocessor based controllers for the Shuttle Infrared Telescope Facility (SIRTF) and the Cryogenic Fluid Management Experiment (CFME). This paper reports the results of the use of a microprocessor based controller (MBC) to control the pressure of cryogenic helium. The work was performed as a Beech funded Industrial Research and Development Program with a supercritical helium (ScHe) dewar (Fig. 1) on loan to Beech from NASA-Ames. The dewar was designed and built by Beech under contract to Grumman Aerospace Corporation to provide a supply of ScHe to cool an experimental heat load while maintaining the pressure of the He with a controlled heat leak. This method of pressurization was chosen because of its high reliability and because it would be capable of providing pressurization at the required He flow rate of a few grams per second.

The MBC (Fig. 2) maintained the pressure of the 1270 ·L (45 cu ft) dewar at 3.20 atm while supply flows of He were withdrawn at rates of 0, 0.02, 0.05 and 0.11 g/s. The microprocessor is a 6502[+] and is incorporated in a KIM I[+] minicomputer.

---

*Supported in part by the National Aeronautics and Space Administration.

+Product of Commodore Semiconductor Group.

Fig. 1. Microprocessor based controller.

EXPERIMENT SET UP

The MBC was tested in the experimental set up shown schematically in Fig. 3. The fill and vent lines of the dewar were used on ventionally to fill and empty the dewar and to provide excess pressure relief. The supply line was used to simulate the flow of helium from the dewar to a device which required cryogenic cooling. The pressure of the He was controlled by varying the heat leak from room temperature to the helium pressure vessel. The heat leak was determined by the temperature of a vapor cooled shield (VCS), the temperature distribution through the multilayer insulation and conduction along the fill, vent and supply lines and the dewar supports. The heat leak was varied within the allowable range by adjusting the flow of He gas through the VCS. The maximum heat leak is obtained with no flow after the VCS and multilayer insulation have reached equilibrium temperature and the minimum heat leak is obtained with maximum flow. The MBC was used to modulate the flow of He through the VCS after the control pressure had been reached.

MICROPROCESSOR BASED CONTROLLER

The MBC consists of an analog pressure to digital signal circuit, the KIM-I, a digital to pneumatic valve control circuit and a cassette tape recorder (Fig. 4). The cassette tape recorder was used to record and load the control program and would be

Fig. 2. SC He dewar.

replaced by a read only memory (ROM) in an application requiring
unattended control. Pressure control is initiated by a blank and
convert (B/C) signal from the KIM-I1 (Fig. 5) to the analog to
digital (A/D) converter which clears the last signal from the A/D
and obtains a new reading from the pressure transducer. Upon
completion of the conversion cycle the KIM-I[1] reads the A/D output
and compares it to preprogrammed limits. The KIM-I[1] then generates a stepwise linear digital signal proportional to the difference between the fluid pressure and the minimum pressure
limit. The linear algorithm provided the necessary control in
this application but another algorithm could have been substituted
in the progam which would in effect change the transfer function
of the valve without changing the valve or the valve stem. The
signal is output with the least significant bit (LSB) set high to
blank the A/D converter and then with the LSB set low to initiate
the next A/D conversion cycle. Using the LSB of the valve signal
for the B/C signal does not change the valve response because the
valve response time is so much slower than the run time of the
program that the signal is repeated many times prior to the valve
completing its response and the increments in valve movement are
large compared to the LSB. The valve control signal is then
converted to a pneumatic signal through a digital to analog converter (D/A), a voltage to current (E/I) converter and a current
to pneumatic (I/P) converter. The pneumatic signal controls the
position of the VCS flow control valve.

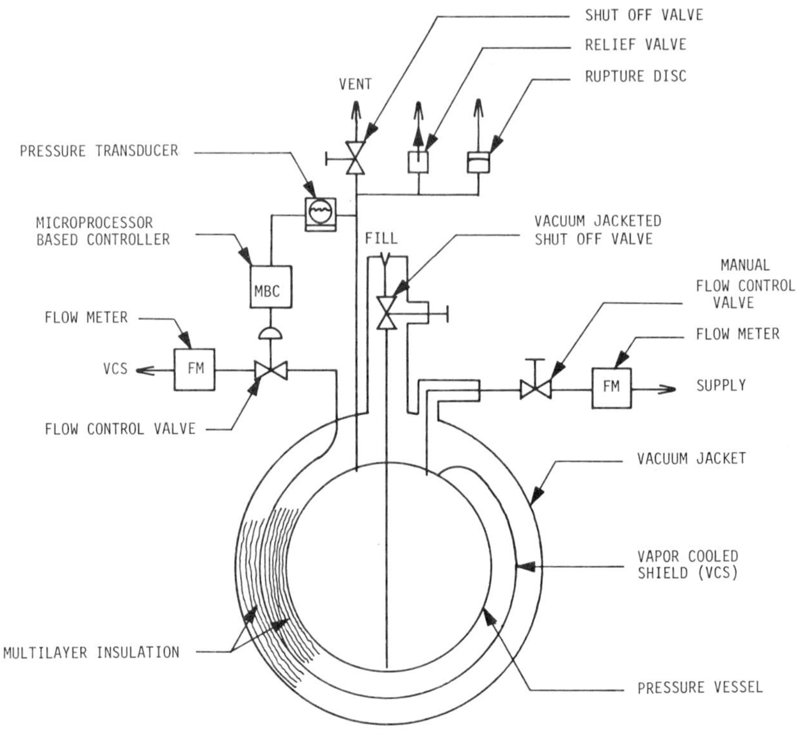

Fig. 3. MBC experiment.

EXPERIMENT RESULTS

The MBC performed so well that no error analysis of the amplifier and the four converters was necessary. The actual performance of the MBC is compared to the predicted results in Fig.
6. There is reasonably good agreement with the pressure rise
after lock-up except for a difference in the starting pressure and
the appearance of an underdamped oscillitory response in the
predicted results when pressure control was initiated. The analytical results predicted a much faster response time of the
pressure to changes in the supply flow rate than was actually
observed. It is believed that this is due to the fact that the
analysis assumed a uniform thermodynamic equilibrium in the
dewar. The analytical prediction also shows a tailing off of
pressure at the end of the 12 day experiment. This pressure drop
would have occurred except for an anomously high heat leak which
resulted from a vacuum leak in the supply line vacuum jacket. The
important factor is that the MBC was able to maintain control
after the vacuum leak occurred.

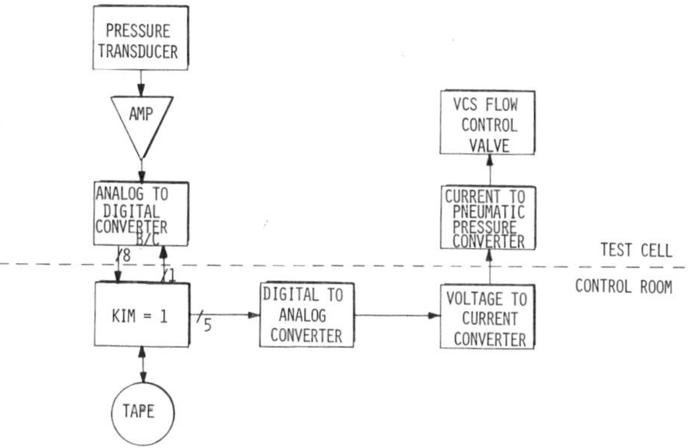

Fig. 4. MBC block diagram.

A magnified view of the transient response of the dewar to
changes in supply flow rate is shown in Fig. 7. The initial
supply flow was the boil off of He from the dewar prior to
lockup. Once the tank had pressurized, a short test of the response was run by initiating a supply flow of 0.11 g/s and the
control responded as anticipated. The dewar was then allowed to
come to equilibrium under MBC control before the first simulated

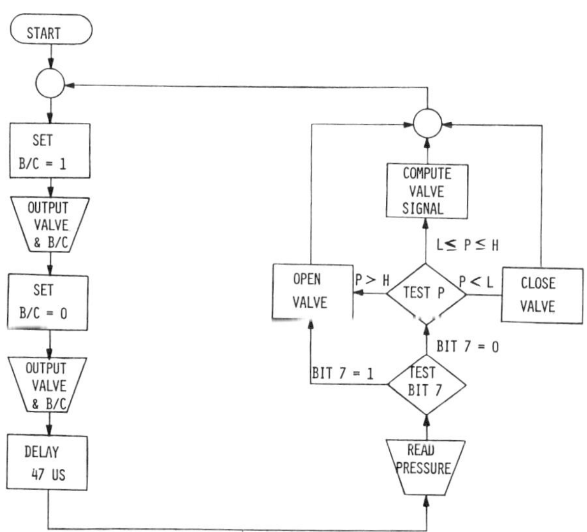

Fig. 5. Control program flow chart.

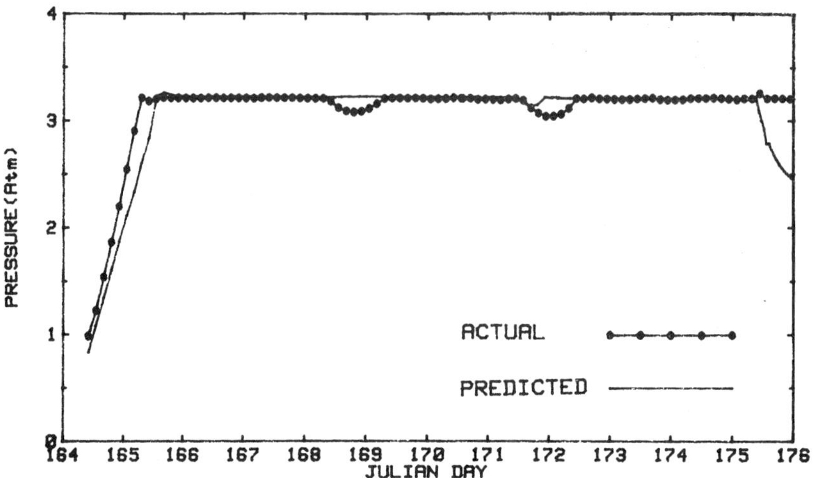

Fig. 6.   Predicted performance.

supply flow was initiated on the fifth day (Julian Date 168). The
supply flow valve was opened manually and supply flow fluctuated
slightly with pressure while the VCS warmed and the dewar repres-
surized. The dewar was again allowed to stabilize until Day 171
and the supply flow as increased again. The response time of both
transients was measured as the width of the two pressure troughs

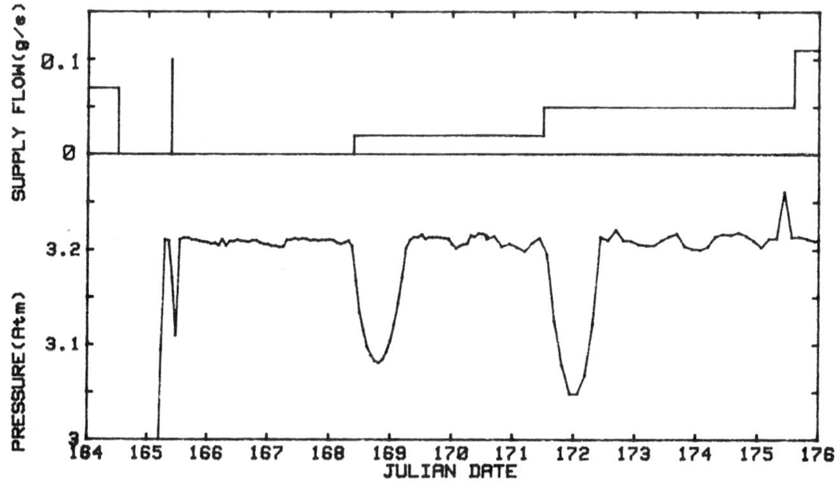

Fig. 7.   Transient response.

at half the trough depth and they were 15.24 h and 15.42 h respectively. The pressure peak on the twelfth day of the experiment (Day 175) was due to inadvertently turning the MBC off; control was resumed as soon as it was turned on again.

CONCLUSIONS

The control program which was used consisted of only 80 eight bit bytes and the maximum run time of the program was 217 μs. demonstrating that microprocessor based control can be implemented at very low memory and cycle time costs. This is especially important when other aspects of the experiment can use the microprocessor data acquisition. An MBC can also provide greater design flexibility because the response of the components can be varied with changes in the MBC software. Similar and even more precise control of other system parameters can be achieved with MBC's. For example, by adding an experiment clock to the programs the supply flow changes can be anticipated, thus reducing the small pressure changes which did occur and then using this steadier pressure to limit temperature excursions in a cryogenically cooled device to a few millikelvin by controlling the supply flow based on the device temperature.

Applications like this will be of great importance to the success of programs like SIRTF and CFME.

ACKNOWLEDGMENTS

We would particularly like to thank Arthur L. Kay and Stephen P. Reimer for their help with the microprocessor based controller.

DISCUSSION

Question by T. Nast, Lockheed Palo Alto Research Laboratory:

What was the degree of temperature stratification in the tank?

Answer by the Author:

The degree of stratification could not be measured because the tank was designed for flight conditions and so had only two temperature sensors, neither of which was operational for this experiment. From prior heat flux measurements and the helium outflow rate, it was deduced that the helium flowing from the top of the pressure vessel had an enthalpy approximately four times the enthalpy of the bulk fluid.

INSTRUMENTATION, DATA ACQUISITION AND REDUCTION FOR A LARGE SPACEBORNE HELIUM DEWAR

W. F. Brooks

NASA Ames Research Center
Moffett Field, California

C. Banda

Informatics
Palo Alto, California

and

C. Robertson

Ball Aerospace Systems Division
Boulder, Colorado

INTRODUCTION

For the Infrared Astronomical Satellite (IRAS), [1,2] being constructed jointly by the U.S., U.K., and the Netherlands, system safety, personnel protection, and performance assessment are important project requirements. Since the cooled infrared (IR) telescope is a high-cost, one-of-a-kind Dewar. The telescope uses a main cryogen tank with 500 L of superfluid helium and a deployable aperture cover whose tank contains 54 L of supercritical helium. The main tank is designed to last for one year in space while the aperture tank will be in place for approximately one week after launch, during cleanup and checkout.

The observational strategy for the mission is dependent on a best estimate of the Dewar orbital lifetime. This estimate is made based on a thermal model[3] which has been correlated with ground performance.

Dewars which store liquid helium are potentially hazardous due to the large amount of energy stored in the fluid. If the Dewar vacuum is degraded or a structural component fails, the high heat leak result in over pressures which can damage the apparatus and the Dewar or, in extreme cases, pose a hazard to personnel.

The Dewar was equipped with temperature, pressure, liquid level, and flow rate transducers to record and monitor the condition of the system. The ground support data acquisition problem is to stimulate and read a subset of these sensors, convert the raw voltages to engineering values using the appropriate calibrations, display the data in real time, and finally, store the raw data on a magnetic medium for further off-line processing. System requirements imposed for the ground acquisition and processing of IRAS thermal data included real-time acquisition display and storage of a set of operator designated transducers. Short-term storage on hardcopy printout and long-term storage on magnetic tape were required. At least ten channels were required to have programmable limits with warning indicators on the display screen. The off-line processing requirements included full listing of converted or raw data and plotting of converted or raw data at user specified intervals ranging from minutes to months for detailed or overview analysis of the Dewar behavior.

DATA ACQUISITION SYSTEM

Hardware

A single electronics rack called the cryogenic monitor cabinet (CMC) contains the equipment required for data acquisition and valve control of the IRAS Dewar. A block diagram of the CMC is shown in Fig. 1. An HP-9825A desktop computer[4] is the system controller. It continuously executes a program which commands the IEEE-488 bus linked electronics to generate and measure voltage drops across the transducers in the telescope. The desktop computer includes a magnetic cassette tape drive and small thermal printer.

The acquisition system monitors five types of transducers: (1) platinum and germanium resistance thermometers, (2) superconducting liquid level indicators for both cryogen tanks, (3) pressure transducers for both tanks, (4) strain gauges at critical structural locations, and (5) flowmeters for monitoring cryogen tank boiloff which are linear voltage output transducers (not as yet been incorporated into the system). All transducers produce a voltage proportional to the parameter they measure which is routed to a single 6-1/2 digit voltmeter via an 80-channel multiplexer (MUX). To stimulate the germanium (GRT) and platinum (PRT) sensors, a programmable current source was constructed using a programmable D/A to drive an operational amplifier. The output voltage of the D/A is applied to the positive input of an LM108 to produce 10 and 100μ A, for the GRTs and PRTs respectively. The thermometers were stimulated only while being read; this was

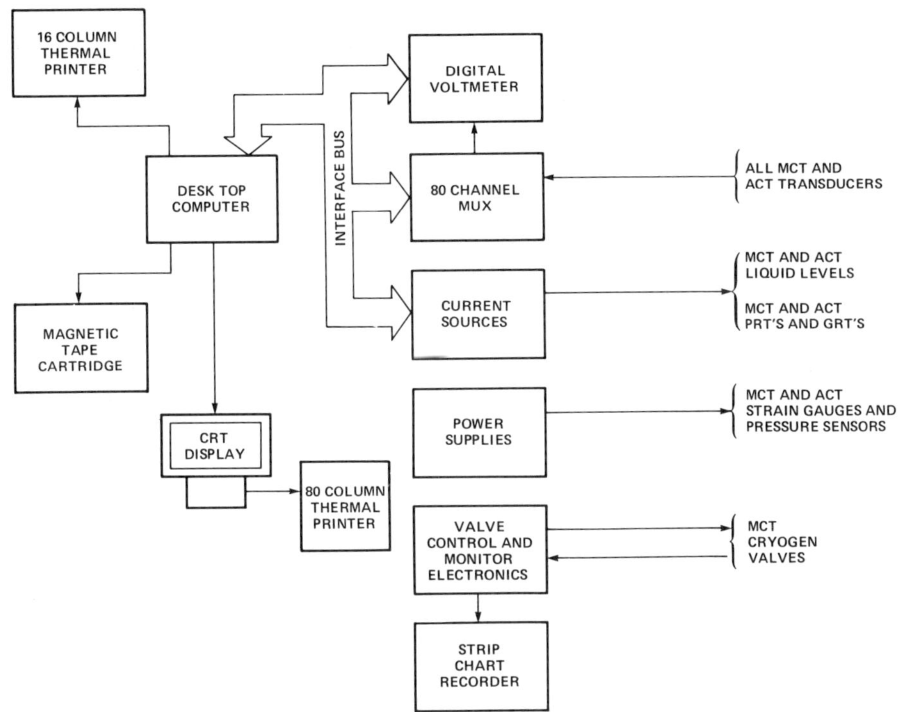

Fig. 1. Block diagram of data acquisition hardware. MCT and ACT refer to main cryogen tank and aperture cover tank, respectively.

accomplished by trying one current lead from all thermometers to the high side of the current source of the CMC and using the three relay per channel capability of the HP-3495A scanner to switch V^+, V^-, and I^-.

Four-wire read out of the sensors eliminates the lead wire resistance, which is sometimes significant. Four wires per channel is a heavy burden to bear when we have over 50 sensors. Therefore, this approach was only used in critical instrument locations. The wiring overhead was reduced significantly by incorporating reference loops for groups of two-wire sensors. The round-trip length and location of the reference loop are close to that of the two-wire sensors with which the loop is to be used. The resistance value of this loop is subtracted from the raw sensor value.

The liquid level (LL) sensors are the superconducting wire type whose resistance is proportional to the length submerged in liquid with a maximum resistance when the Dewar is empty. The LL sensors require ~ 70 MA. This is inconsistent with the LM108 output, and a separate power supply was used for the LL sensors. The liquid level current is also too high for the reed relays in the MUX. Therefore the scanner is used to control the base of a PNP transistor which activates a switch relay for the 70 MA supply.

Valve Control

The valve control and monitoring electronics operate the five motorized cryogenic valves in the Dewar. This subsystem allows each of the five valves to be opened and closed with a display of their status. Because the system is manned 24 hours per day whenever it is at cryogenic temperatures, the valve control circuits were not automated. Although this is a straightforward electrical change, it would put a considerable load on the controller memory since the decision tree for all five flow valves is involved.

Data Acquisition Software

The CMC II data acquisition software is a 315-line program written in HPL for the 9825A controller. The program is designed to run continuously whenever the Dewar is cold and electrical power is available during ground operations. The program has six primary functions: (1) Definition and modification of the channels to be stimulated (a specified subset of the available channels constitutes a "test"); (2) sensor stimulation and measurement; (3) raw data storage; (4) raw data conversion; (5) real-time formatting and output of raw or converted data; and (6) editing.

The heart of the program is a driver section which represents about 20% of the code. The driver mode sequences and controls the six primary functions. The driver is completely table driven. It utilizes data structures to determine program flow and use of channels and peripherals. A real-time clock determines when each of the four display cycles is to be invoked by interrupting the driver program. At this time, August 1981, formatted converted or raw data is routed to any combination of the output devices: (1) CRT; (2) hardcopy printer; (3) 9825A thermal printer; and (4) magnetic tape. This table-driven design differs significantly from the more common data acquisition approach in which program flow and channel assignments are hardwired into the program logic. This flexibility is extremely important when software is being developed in parallel with the proto-flight hardware in which wiring and transducer changes are made during construction or refurbishment.

A second "edit" mode of the acquisition program is invoked when the acquisition program is initiated or later on by operator

intervention. In this mode the program ignores clock interrupt
but recognizes 32 special function keys which allow the operator
to: (1) Construct a new test by specifying channels or modify the
current test; (2) change the time interval for all output devices;
(3) list the test; (4) request immediate output on any device; (5)
inhibit or enable the magnetic tape; (6) modify limit channels;
(7) invoke debug functions; and (8) return to "driver." Once back
in the driver, a master activity list, which is unique for each
"test," determines how each of the 80 channels (transducers) will
be stimulated and processed.

The magnetic tape is formatted so that each data collection is
an individual file. This makes for a reliable system since power
outages and equipment failures will result in the loss of only the
most recent scan of the transducers. The unwieldiness of having
thousands of data files is overcome during off-line data processing.

The real-time displays include a test identifier, time and
date, the eight-character mnemonic for each displayed channel, the
raw or converted transducer reading, and the appropriate engineering unit. Any or all active channels can be selected independently for each output device.

DATA REDUCTION SYSTEM

Hardware

Since IRAS is a flight project, many of the systems have
backup units, and this is true of the CMC, since a failure in this
unit would make it impossible to monitor the Dewar status. The
availability of a second unit was capitalized on for data reduction. The first CMC can continuously record raw data while the
second unit processes the data off-line.

From the second unit only the controller, HP IB Bus, and line
printer are required. In addition, a floppy disk and plotter were
added to assist in the processing. The hardware configuration is
shown in Fig. 2. The processing hardware is modular and, set up,
requires that some HP IB connections and output ports be interchanged.

Off-Line Reduction Software

The task of the off-line data reduction software is to take
packed raw voltage files from a cassette, convert them to engineering values, and produce concatenated raw and converted disk
files, listings, and plots. The disk files are used both to

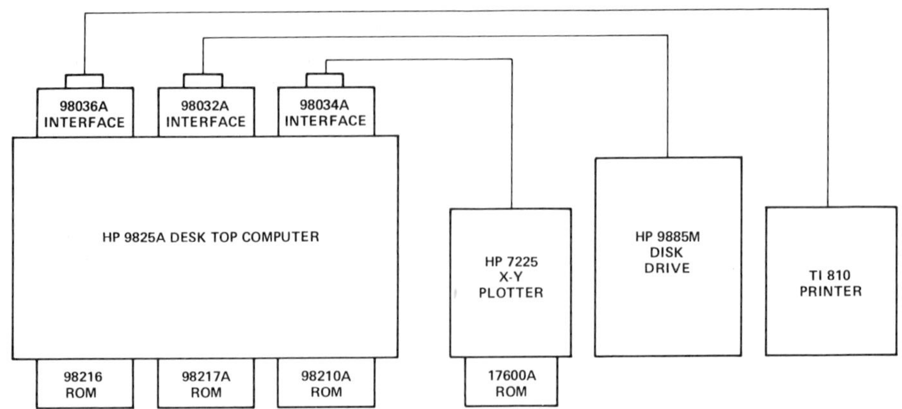

Fig. 2. Block diagram of data reduction hardware.

archive the data and to provide a rapid access medium for listing and plotting the data. All data tape files are pre-formatted to a standard length for simplicity in data reduction; each file will hold up to 50 channels plus five reference voltage channels. Inactive channels are preset in the acquisition software to a special value. This approach requires more tape space than a variable length file approach where only active channels are stored, but is considerably simpler for data reduction programs to analyze and process. Fixed length files also eliminate the need for "header files" which would be necessary to signal a change in the number of active channels in a variable-length files.

Due to the memory size limitations of the HP-9825 with Option I (13.5 Kbytes are available with the necessary ROMs installed), it was decided to divide the processing into four functions to be performed by four separate programs which would be chained into memory. Although forced upon us by memory limitations, this approach afforded easy break points in the processing which proved to be a useful feature.

The first program is the Data Tape Analyzer which scans the entire data tape, prints header information (including conversion factors) and a summary in the form of bit maps of the active channels. Since each file on a data tape may contain data from different sets of active channels, every time the set of active channels is changed, a new bit map is printed. This feature is important to the off-line operator, because without these bit maps as guides to channel use, plotting can be frustrating.

The second program is the Disk File Builder. This program transfers raw data from a tape or a tape segment to a disk file.

For each file, the user selects the set of active channels he wishes to transfer to disk. A sequential file organization was chosen rather than direct access because of the nature of the data to be stored: data collections are sequential in time. Also, the physical length of records for direct access files on the HP-9825 system is fixed at 256 bytes each, which could result in wasted disk space. The disk file builder concatenates one or more tapes containing 400 files into a single file which facilitates plotting and listing.

The third program is a List and Convert program. This program directs the user to choose a listing format which specifies the channels to appear together on a page. The program then makes two passes through the disk file of raw data. During the first pass, raw data is listed; during the second pass, the data is converted, listed, and saved in a concatenated data file to be input to the plotter. Listings are made according to user direction, with up to 55 consecutive collections per page. This variable format concept was selected to enable the user to easily follow trends between groups of channels.

The fourth program is the Plotter, which enables the user to create plots of selected data over a designated time interval. This allows a quick look at trends or detailed look over a short time interval. The user specifies a channel selection, a plot title, axes labels, and a time interval for each plot. Scaling for the Y-axis is semi-automatic: the plotter scans the data and prints minimum and maximum values, then prompts the user to choose a convenient scale.

After exercising the programs, the following features were added so that:

1. The user can read in additional channels and display them on the same plot; this allows an unlimited number of channels per plot.

2. The user can concatenate data from subsequent tapes to graph a long time history of selected channels. (Note: This is mutually exclusive on a given plot.)

3. The user can make new plots without re-entering all the plot parameters.

4. The plot title and axes labels are automatically centered.

5. Many defaults are available for user convenience, some of which are displayed.

CONCLUSION

For the relatively low-speed data acquisition associated with a large Dewar system, a desktop computer with an IEEE-488 bus is an excellent choice. The ease with which components can be interfaced or interchanged is a valuable feature. High-level languages, especially a structured language, also appear to be the choice when dealing with developing hardware and software needs. In choosing a system memory expansion should be an important consideration. During the time we worked with this system 32 Kbytes was the maximum available memory. We often ran into serious limitations with our file driven approach because of limited memory. Memory is relatively inexpensive, and a generous overhead should be allotted to allow for the use of structured techniques. Many desktop computers now offer 320 or more Kbytes of storage.

Another essential feature of the computer language is the ability to comment. Our inability to comment the programs caused serious problems with software maintenance. Other factors to consider are processor speed, universal rather than specialized high-level language, lexicographical level of statements, type and capacity of available disk, and display capability (CRT, printer, LED).

Although data processing software is often developed after the acquisition software, the parallel development of these programs will usually result in faster and more accessible data storage and retrieval. Finally, in any system where calibration and conversion of the acquired data is required, the specification of conversion accuracies should not be done without consideration of the availability of accurate algorithms or the penalty in memory which is incurred by overspecifying temperature resolution and accuracy requirements.

REFERENCES

1. A.R. Urbach and R. N. Herring, A Long Term Helium Dewar for Space Experiments, in: "Proc. 6th Int'l. Cryo. Engr. Conf." IPC Science and Technology Press, Guildford, (1976).
2. R.A. Hopkins, Design of a one year lifetime, spaceborne superfluid helium dewar, ASME Publication, 79-ENAs23 (1979).
3. R.A. Hopkins and W. F. Brooks, Orbital performance of a One Year Lifetime Superfluid Dewar Based on Ground Test and Computer Modeling, in "Advances in Cryogenic Engineering, Vol. 27", Plenum Press, New York (1982).
4. V.L. Laing, Instrument system provides precision measurement and control capabilities, Hewlett Packard Journal (July 1981).

AUTOMATIC MEASUREMENTS OF HEAT LOAD

M. Kuchnir

Fermi National Accelerator Laboratory *
Batavia, Illinois

INTRODUCTION

The components of a large system of superconducting magnets, such as the Fermilab Energy Saver are usually refrigerated by flowing helium. The heat load into the helium temperature region of a component can be obtained in a straightforward way by measuring the pressure, the mass flow rate and the increase in the temperature of the cold helium gas flowing through it. The heat load is then the product of the corresponding increase in enthalpy by the mass flow rate. This method[1] is used to determine heat loads of individual Energy Saver components, dipole magnets, quadrupole magnets and "spool pieces" which are cryostats containing the higher order correction magnets and most of the accessories of a cryogenic nature.

PROCEDURE

The preparation for the tests involves connecting together one to six components with special inserts between them, a turn-around-box at one extremity and a refrigeration system at the other. This forms a cryogenic circuit similar to an Energy Saver cryoloop. The special inserts introduce two thermometers in the single phase helium flow stream, one before and one after the junction formed by consecutive components. After leak checking, the train of components is cooled by starting the flow of liquid nitrogen through the shields and cold helium gas through the single phase helium path.

---

*Operated by Universities Research Association Inc. under contract with the U.S. Department of Energy.

The two-phase (boiling) helium path used in the operating mode[2] is kept valved-off. The gas exiting from the turn-around-box is warmed to room temperature and returned through a positive displacement-type gas meter to the refrigerator compressors. Cooling proceeds until the helium gas enters the train at ~ 5.5 K, at which time stability is achieved by electronically controlling the inlet temperature, manually adjusting the refrigerator for a steady mass flow rate and waiting for equilibrium. Figure 1 presents a diagram of a typical train of six spool pieces with the sensor location indicated. For heat load measurements, dipole and quadrupole magnets can replace spool pieces in the train.

DATA ACQUISITION SYSTEM

The data acquisition system, with a digital multimeter[a] and a scanner[b] uses components available in the microprocessor market. The microcomputer[c] has an S-100 bus structure with a 64 K memory, a TU-ART interface, a 100 000-day clock[d], two 5-inch disk drivers and a pulse counter. A graphics terminal[e], a printer[f], plus a few homemade electronic modules complete the system. Data transmission follows the RS232 standard.

The pressure transducers used are based on integrated circuit technology[g]. Two precautions make these inexpensive units quite acceptable: the use of those which measure gauge pressure (helium diffusion spoils the reference pressure in absolute-pressure gauges); and the design of special packaging. We have a transducer installed in the insulating vacuum of the turn-around box which provides the barometric pressure for the calculation of absolute pressure. A Baratron[h] has been used to check transducer

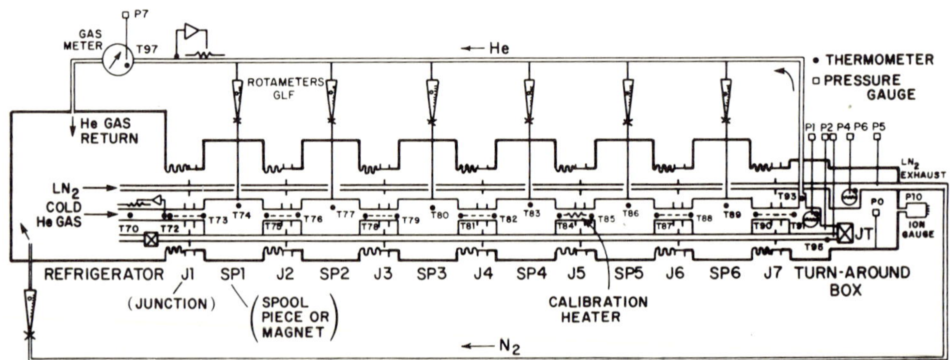

Fig. 1. Schematic for heat load measurements of testing facility, showing train with six spool pieces (SP1-SP6).

calibration. The insulating vacuum is monitored with an ion gauge,[i] whose logarithmic output is recorded at every interval.

Most of the thermometers are 100 Ω 1/8W carbon resistors[j] specially mounted and calibrated against a carbon-glass[k] unit. The resistors are connected in series with a 10μ A current source whose polarity is computer controlled. The voltage leads are connected to low signal relay cards[m] in the scanner.

A gas meter[n] actuates a switch for every 0.05 ft^3 of gas. This action is converted into an electronic pulse and these pulses are counted by an S-100 scaler based on an integrated circuit[o]. The application program uses this scaler to calculate the average flow rate between readings. A thermistor[p] and a pressure transducer at the inlet of the gas meter provide the remaining information for the calculation of the total mass flow rate. Part of this flow comes through rotameters from the vapor-cooled leads. This flow is manually set and the reading entered manually into the computer. A heater and a control system ensure that the temperature of the gas entering the gas meter is close to room temperature.

The application program, written in BASIC[q], is a series of specialized subroutines and several driver programs. For special monitoring of the system, driver programs can be entered quickly. Since the program is executed through a BASIC interpreter, there is no compiling after revisions or additions, making their implementation very fast and easy. This method of execution is negligibly slower because the revelant time constants of the system are long. Comments can be easily entered into the recorded data through the terminal. Day and time are automatically included in the recorded comment. At the start of a run, prompting is made for comments identifying the components of the train.

In order to have a simple starting procedure, the master diskette, on the first disk driver, holds the disk operating system, the console processor, the BASIC interpreter and the application program. The second disk driver has the data diskette containing two files, one storing the data and another the latest values of the application program parameters. This permits full recovery from application program modifications, power failures, etc. The full calibration tables for the 30 carbon thermometers are also stored in the master diskette, however, the program stores for each thermometer the interpolation parameters of the last temperature interval used in order to reduce the number of disk accesses. Special programs, stored in a utilities diskette, can be merged in prior to a run, in order to facilitate changing the calibration tables, since each new run involves new thermometers and old ones in new locations.

A conspicuously available listing of the application program, well documented with remarks, an understanding of BASIC by the technicians operating the facility, and a number of lines reserved for their own driver programs are important factors contributing to the utility of this data acquisition system, not only for heat load measurements, but for the general operation of the facility.

HEAT LOAD MEASUREMENTS

For heat load measurements the data acquisition system described above reads the thermometers, pressure gauges and the gas meter at regular time intervals. It processes these data and computes the heat loads between pairs of thermometers for each time interval. A typical scan includes day and time, mass flow rate, 24 temperatures, 13 heat loads, and takes ~ 5 minutes.

The data are valid when the system of components plus refrigerator reaches steady-state. Several steady-states, differing in mass flow rate, power in a calibration heater or lead flow rate (the higher order correction magnets have vapor-cooled leads), are used to verify the consistency of the data.

For the calculation of the heat loads, part of the helium enthalpy table[4] stored in the program is used to interpolate for pressure and temperature. A further correction might be needed when the heat load is largely due to conduction from the N_2 shield, since the value just determined corresponds to a temperature difference from 78 K to the average component temperature during the measurement, which is different from its average temperature in actual operating conditions. Since this conduction in the case of Fermilab superconducting dipoles is mostly through epoxy-fiberglass composites, the expression for the integrated thermal conductivity

$$\frac{\int_{4.6\ K}^{78\ K} \kappa\ dT}{\int_{T_{average}}^{78\ K} \kappa\ dT} = \frac{679.01}{688.88 - T_{average}^{1.5}}$$

based on the data of Kasen[4] is used as a correction factor. This factor usually amounts to less than a 10% correction to the heat load.

The results for one Energy Saver dipole magnet (Serial No. TC-0361) and one Energy Saver quadrupole magnet (Serial No. TQ-D-86) are 9.4 ± 3W and 3.0 ± 3W respectively. About two dozen spool pieces have been measured to date (July 1981) in eight runs.

Their heat load, a function of the lead flow rate, varies from 10 to 18W. The lower value corresponds to the design lead flow rate of 30 scfh.

This capability of taking heat load readings in real-time so to speak has simplified considerably the testing of components for the Fermilab superconducting accelerator.

VENDORS

a. John Fluke Mfg Co., Inc., P.O. Box 43210, Mountlake Terrace, Washington 98043, Model 8502 A.
b. John Fluke Mfg. Co., Inc., Model 2204 A.
c. Cromenco, Inc., 280 Bernardo Ave., Mountain View, California 94043, Model Z2D.
d. Mountain Computer, Inc., 300 Harvey W. Blvd., Santa Cruz, California 95060.
e. Lear Siegler, Inc., Annaheim, California 92804, Model ADM-3.
f. Integral Data Systems, Inc., 14 Tech Circle, Natick, Massachusetts 01760, Model 440.
g. National Semiconductor Co., 2900 Semiconductor Dr., Santa Clara, California 95051, Model L x 16 xx 6.
h. MKS Instruments, Inc., 22-24 Third Ave., Burlington, Massachusetts 01803, Model 310 BHS-10 K.
i. Granville-Phillips Co., Boulder, Colorado 80303.
j. Allen Bradley Co., 1200 S. 2nd St., Milwaukee, Wisconsin 53204.
k. Lake Shore Cryotronics, Inc., 64 E. Walnut St., Westerville, Ohio 43081, Model CGR-1-2000.
m. John Fluke Mfg. Co., Inc. Model 2200A-06.
n. American Meter Co. (Singer), 300 N. Gilbert St., Fullerton, California 92633, Model AL-800.
o. Intel Co., 3065 Bowers Ave., Santa Clara, California 95051, Model 8253.
p. Yellow Springs Instrument Co., Yellow Springs, Ohio 45387, Model YSI-44001
q. Microsoft, 10800 NE Eighth St., Bellevue, Washington 98004.

REFERENCES

1. M. Kuchnir, et al., IEEE Trans. on Magnetics, NS-28:3325 (1981).
2. W.B. Fowler and P.C. VanderArend, U.S. Patent #4 048 437 (1977).
3. R.D. McCarty, Thermophysical Properties of Helium-4 from 2 to 1500 K With Pressures to 1000 Atmospheres, NBS Technical Note 631, National Bureau of Standards (1972).
4. M.B. Kasen, et al., in "Advances in Cryogenic Engineering, Vol 26," Plenum Press, New York (1980), p. 235.

SUBJECT INDEX